Intellectual Property
in the
New Technological Age

Intellectual Property in the New Technological Age

Second Edition

Robert P. Merges
Wilson, Sonsini, Goodrich, and Rosati
Professor of Law and Technology
University of California at Berkeley

Peter S. Menell
Professor of Law
University of California at Berkeley

Mark A. Lemley
Professor of Law
University of California at Berkeley

ASPEN LAW & BUSINESS
A Division of Aspen Publishers, Inc.
Gaithersburg   New York

Printed in the United States of America

Library of Congress Cataloging-in Publication Data

Merges, Robert P.
Intellectual property in the new technological age / Robert P. Merges, Peter S. Menell, Mark A. Lemley.—2nd ed.
    p. cm.
    Rev. ed. of: Intellectual property in the new technological age / Robert P. Merges . . . [et al.]. c1997.
    Includes index.
    ISBN 0-7355-1226-4 (casebound)
    1. Intellectual property — United States. 2. Technological innovation — Law and legislation — United States. I. Menell, Peter Seth. II. Lemley, Mark A., 1966– III. Title.
KF2979 I432 2000
346.7304'8—dc21   99-089259

1   2   3   4   5   6   7   8   9   0

About Aspen Law & Business
Legal Education Division

With a dedication to preserving and strengthening the long-standing tradition of publishing excellence in legal education, Aspen Law & Business continues to provide the highest quality teaching and learning resources for today's law school community. Careful development, meticulous editing, and an unmatched responsiveness to the evolving needs to today's discerning educators combine in the creation of our outstanding casebooks, coursebooks, textbooks, and study aids.

ASPEN LAW & BUSINESS
A Division of Aspen Publishers, Inc.
A Wolters Kluwer Company
www.aspenpublishers.com

Summary of Contents

Contents

4

Copyright Law   **345**

5   *Trademarks and Trade Dress*   *557*

6

State Intellectual Property Law and Federal Preemption   *795*

7   *Protection of Computer Software*   *893*

⊗8

Intellectual Property and Competition Policy   *1101*

Preface to the
Second Edition

The impetus for writing Intellectual Property in the New Technological Age was our desire to have a book that integrated the different modes of intellectual property protection while emphasizing the major technological advances driving the U.S. economy and changes in intellectual property law. The rapid diffusion of new technology has continued since the first edition of this book was released in 1997. Courts, Congress, and international bodies have been busy responding to these developments. The pace of change necessitates the publication of this Second Edition.

Among the notable changes in this edition are the following:

Chapter 2. Trade Secret Protection: We have expanded coverage of the Economic Espionage Act of 1996 and revised the section on remedies.

Chapter 3. Patent Law: This edition thoroughly updates the case law, incorporating recent Supreme Court cases on patent validity (on-sale bar), patent claim interpretation, and infringement (doctrine of equivalents). We have added major new sections on both the written description requirement and claim interpretation.

Chapter 4. Copyright Law: We have incorporated material on new legislation extending the term of copyright protection, affording protection for the design of boat hulls, expanding the exemption from the public performance right for small businesses, and creating a new exclusive right to perform sound recordings through digital audio transmissions. We have expanded coverage of the Audio Home Recording Rights Act of 1992 and substantially revised the cases and materials on copyright infringement.

Chapter 5. Trademarks and Trade Dress: We have substantially reorganized the materials on distinctiveness and priority and defenses for easier teachability. In addition, we have included important new cases and material relating to the Federal Antidilution Act of 1996, trademark disparagement (the Washington Redskins case), protection of product configurations, and the application of trademark law to dispute over Internet domain names.

Chapter 6. State Intellectual Property Law and Federal Preemption: This edition contains expanded and updated treatment of the right of publicity.

Chapter 7. Protection of Computer Software: We have incorporated new material on software and Internet "business method" patents, summarized the Digital Millennium Copyright Act, and provided expanded treatment of copyright licensing and derivative works. We have also greatly expanded the discussion of prior art and enablement issues in software patents and the protection of Internet domain names. We have moved material describing computers and the development of computer software to the Statutory Supplement/Technology Primer.

Chapter 8. Intellectual Property and Competition Policy: This edition provides coverage of the antitrust litigation between the federal and state governments and Microsoft Corporation and the cases involving Intel Corporation.

We expect to be making periodic updates of the material on our website ⟨provide URL⟩. We also recommend that professors and students subscribe to the Annual Review of Law and Technology, published by the Berkeley Technology Law Journal ⟨http://www.law.berkeley.edu/journals/btlj/⟩.

Robert P. Merges
Peter S. Menell
Berkeley, California   *Mark A. Lemley*
January 2000

Preface to the First Edition

New technologies have dramatically altered the shape of intellectual property practice. Not too long ago this area of practice was neatly divided along the lines of traditional casebooks — copyright practice focusing upon the protection of literary, musical, and artistic works (representing publishing, recording, and motion picture interests); patent and trade secret practice focusing upon the protection of technological works; and trademark and antitrust practice focusing upon separate aspects of business activities. These orderly divisions have been profoundly and permanently changed by new technologies, in particular computers.

The blurring of traditional doctrinal lines can be seen throughout the practice of intellectual property law. The major transactions and cases today affecting high technology firms typically involve complex technologies, multiple areas of intellectual property law, and antitrust issues. Computer technology matters frequently require lawyers to simultaneously address trade secret, copyright, patent, and antitrust issues. Biotechnology raises a host of challenging patent and trade secret issues. Both technologies strain the existing intellectual property system and raise novel issues of "technology policy." Other new technologies, such as the Internet and interactive multi-media products, also press the existing doctrinal frameworks to and perhaps beyond their limits. In order to appreciate these issues and obtain an intuitive grasp of how the law accommodates new technologies, students must understand the main features of the technologies. They must also possess an integrated command of intellectual property law, antitrust, and the role of government. In addition, the increasing globalization of markets and the borderless realm of the Internet increasingly require intellectual property lawyers to understand the international dimensions of the field.

This casebook is designed to prepare students to meet the challenges of law practice and public policymaking into the next century. The book provides a basic survey of the modes of protecting intellectual property while integrating and supplementing these areas of law and policy with their technological background, philosophical underpinnings, and some of the analytical insights of economics. It emphasizes the complementarity and interplay of the intellectual property modes and pushes students to use these doctrines to solve complex, real-world problems. At the same time, the scholarly orientation of the materials prepares future lawyers to address emerging issues as well as contribute to the broader public policy questions generated by new technologies.

Overview

The first chapter of the book provides an introduction to the key philosophical perspectives justifying intellectual property protection. The second chapter covers trade secret protection, emphasizing the role of state statutory and common law and employment practices in protecting innovation. The third chapter provides a thorough survey of patent law. The main sections of the chapter emphasize traditional doctrines, integrating academic perspectives and problems. The statutory supplement accompanying this casebook contains an introduction to biotechnology designed to be read in conjunction with this chapter. The fourth chapter covers copyright law, emphasizing the post-1976 copyright regime and the challenges of copyright scope in the new technological era. The fifth chapter examines trademark, trade dress, and related areas of unfair competition law, which covers both traditional doctrines and the emerging trademark law of the Internet. The sixth chapter surveys the burgeoning field of state intellectual property law and the interplay between federal and state systems of protection.

As a means of integrating the various fields of intellectual property law and coherently teaching one of the most challenging subject matter areas, Chapter 7 focuses on legal protection for computer software. More than any other technology, computer software blurs the traditional lines of intellectual property protection. This chapter begins with an overview of technical and economic aspects of computer programming. It then reviews the key cases arising in each of the various modes of legal protection. It concludes with an analysis of proposed sui generis systems of protection. Chapter 8 focuses on the intersection between intellectual property and antitrust law, providing both an overview of the general landscape of antitrust law and an in-depth examination of the doctrines most relevant to advising firms and people working in technological fields.

Analytical Approach

Recent advances in economics and public policy analysis have shed new light on the way many of the legal and policy issues raised by new technologies are addressed in courts and legislative fora. Microeconomic theorists and economic historians have refined the analysis of incentive systems for fostering innovation. Recent theory, for example, highlights the complex incentives facing inventors who follow in the wake of a pioneering breakthrough. Industrial organization economists have developed models for understanding the behavior of markets featuring network externalities, i.e., those in which standardization of products is valued by consumers. This work has important implications for intellectual property protection in a number of industries, as well as for antitrust analysis. "Chicago School" economists have had a profound impact on the shape of antitrust law. The "New Institutional Economics" has developed insight into the innovative process that has important implications for antitrust law and policy, particularly in the area of joint ventures. This book integrates these concepts and frameworks into the study of intellectual property law. At the same time, it explores noneconomic explanations for various intellectual property regimes. We relate these analytical approaches to legal doctrines, asking what theories underlie various legal rules.

Pedagogical Approach

We have also sought to bring a functional approach to the presentation of material. The practice of intellectual property law is heavily problem-oriented. Throughout the book, we use sophisticated problems that reflect both the challenges of practicing in intellectual property fields and the legal and policy questions on the cutting edge of legal doctrine and legislative reform efforts. At the same time, the problems are designed to supplement rather than direct the course of the class, so that the reader can make as much or as little use of them as desired.

Flexibility

Another distinguishing feature of the book is its adaptability to a variety of courses on technology and the law. The entire book can be taught as a comprehensive four-credit introductory course on intellectual property. Through selective emphasis, professors can offer a variety of three-credit courses: a broad survey of intellectual property law with emphasis on technology generally; a more focused survey of intellectual property with in-depth emphasis on computer technology or biotechnology; a course on the protection of new technologies; or a more traditional copyright, trademark, and state law course. In addition, the book can be used for specialized two-credit classes on law and technology, advanced antitrust focusing on new technology, or trade regulation law.

> *Robert Patrick Merges*
> *Peter S. Menell*
> *Mark Alan Lemley*
> *Thomas M. Jorde*

Acknowledgments

We are indebted to a great many people who have helped us since this project began in 1991. We would like to thank our many colleagues who reviewed earlier drafts of the book and provided helpful guidance. While many of these reviews were anonymous, we have also benefitted from the advice of Lynn Baker, Paul Heald, and Pam Samuelson, each of whom read several different drafts of the book as it made its way through the editorial process. We gratefully acknowledge the research assistance of Evelyn Findeis, Edwin Flores, Ryan Garcia, Shari Heino, Toni Moore Knudson, Christopher Leslie, and Barbara Paris. We would also like to thank Michele Co for exceptional secretarial and administrative assistance in completing this text.

We are grateful to Ines Gonzalez, Pilar Osserio, Laurence Trask, John Sasson, and Ryan Garcia for their assistance in completing this volume. We are also grateful to Julie Cohen, Ken Dam, Terry Fisher, Paul Heald, Marshall Leaffer, Glynn Lunney, Ron Mann, Ruth Okediji, Malla Pollack, Peggy Radin, Jerry Reichman, and a large number of anonymous reviewers for their comments and suggestions in preparing the second edition.

Finally, we acknowledge the authors of the following images and excerpts used in this volume with their permission:

Bowman, Ward S., Patent and Antitrust Law: A Legal and Economic Appraisal (U. Chicago Press 1973). Reprinted by permission of The University of Chicago Press.

Brown, Vance, three figures showing source code, object code, and assembly code, from The Incompatibility of a Copyright and Computer Software: An Economic Evaluation and Proposal for a Marketplace Solution, 66

N.C.L. Rev. 977, 1018–21, reprinted by permission of the North Carolina Law Review.

Gasaway, Laura N., table on duration of copyright courtesy of Laura N. Gasaway.

Goldstein, Paul, Derivative Rights and Derivative Works in Copyright, 30 J. Copyrt. Society 209 (1983), copyright © 1982 Paul Goldstein. Reprinted courtesy of Professor Goldstein.

Kellner, Lauren Fisher, Trade Dress Protection for Computer User Interface "Look and Feel," 61 U. Chi. L. Rev 1011 (1994). Reprinted by permission of the University of Chicago Law Review.

Menell, Peter S., The Challenges of Reforming Intellectual Property Protection for Computer Software, 94 Colum. L. Rev. 2644, 2651–54 (1994), reprinted by permission of the Columbia Law Review.

Menell, Peter S., Tailoring Legal Protection for Computer Software, 39 Stan. L. Rev. 1329 (1987), © 1987 by the Board of Trustees of the Leland Stanford Junior University.

Radin, Margaret Jane, Property and Personhood (1982), as revised in Margaret Jane Radin, Reinterpreting Property (University of Chicago Press, 1993). Reprinted courtesy of Professor Radin.

Samuelson, Pamela, Randall Davis, Mitchell Kapor, and J. H. Reichman, A Manifesto Concerning the Legal Protection of Computer Programs. This article originally appeared at 94 Colum. L. Rev. 2308 (1994). Reprinted by permission of Pamela Samuelson and the Columbia Law Review.

Steinberg, Saul, New Yorker's View of the World. Artwork copyright © 1976 Saul Steinberg. Cover reprinted by special permission of Saul Steinberg, Artists' Rights Society, and The New Yorker. All Rights Reserved.

Xerox advertisement courtesy of Xerox Corporation.

Note: We have selectively omitted citations and footnotes from cases without the uses of ellipses or other indications. All footnotes are numbered consecutively within each chapter, except that footnotes in cases and other excerpts correspond to the actual footnote numbers in the published reports.

Many of the problems in this text are taken from actual cases. However, in many instances we have altered the facts of the case. In most cases we have also altered the names of the parties involved. In a few cases, however, particularly in the trademark and antitrust chapters, we felt that it was important to the problem to use the name of a product or company with which the reader would be familiar. Readers should understand that the problems are hypothetical in nature and that we do not intend them to represent the actual facts of any case or situation.

*Intellectual Property
in the
New Technological Age*

1
Introduction

The concept of property is well understood in Western society. It is among the oldest institutions of human civilization. It is widely recognized that people may own real property and tangible objects. The common law and the criminal law protect private property from interference by others. The Fifth Amendment to the U.S. Constitution protects private property against takings by the government without just compensation. The philosophical bases for protection of private property are well entrenched in our culture. Private property has been viewed as resulting when labor is applied to nature, as an incentive for discovery, as an essential part of personhood, and as a foundation for an ordered economic system.

Ideas, by definition, are less tangible. They exist in the mind and work of humans. Legal protection for intellectual work evolved much later in the development of human society than did protection for tangible property. The protection of such "intellectual property" raises complex philosophical questions. Should the first person to discover a way of performing an important task — for example, a procedure for closing a wound — be entitled to prevent others from using this procedure? Should the first person to pen a phrase or hum a melody be entitled to prevent others from copying such words or singing the song? Should such "intellectual property rights" be more limited than traditional property rights (i.e., the fee simple)? This book explores the legal institutions and rules that have developed to protect intellectual property.

This chapter has two principal purposes. It first explores the principal philosophical foundations for the protection of intellectual property. Understanding the reasons why we protect intellectual property — and how those reasons differ from the justifications for real property — will help the reader grasp the details of the many legal rules that will follow in this book. The second section provides a comparative overview of the principal modes of

1

intellectual property protection: patent, copyright, trademark/trade dress, and trade secret. Understanding the intellectual property landscape requires thinking about each form of intellectual property not just in isolation but as it interacts with all the others. The remainder of this book will explore these areas in detail, highlighting their logic and interplay in promoting progress in technology and the arts.

A. PHILOSOPHICAL PERSPECTIVES

All justifications for intellectual property protection, whether based in economics or morality, must contend with a fundamental difference between ideas and tangible property. Tangible property, whether land or chattels, is composed of atoms, physical things that can occupy only one place at any given time. This means that possession of a physical thing is necessarily "exclusive" — if I have it, you don't. Indeed, the core of the Western concept of property lies in the right granted to the "owner" of a thing or a piece of land to exclude others from certain uses of it. Settled ownership rights in land and goods are thought to prevent both disputes over who can use the property for what purpose, and the overuse of property that would result if everyone had common access to it.

Ideas, though, do not have this characteristic of excludability. If I know a particular piece of information, and I tell it to you, you have not deprived me of it. Rather, we both possess it. The fact that the possession and use of ideas is largely "nonrivalrous" is critical to intellectual property theory, because it means that the traditional economic justification for tangible property does not fit intellectual property. In the state of nature, there is no danger of overusing or overdistributing an idea, and no danger of fighting over who gets to use it. Everyone can use the idea without diminishing its value. See generally Peter S. Menell, Intellectual Property: General Theories, Encyclopedia of Law and Economics (B. Bouckaert & G. De Geest, eds. 2000).

Theorists have therefore turned elsewhere to justify exclusive rights in ideas. Over the course of human history, numerous theories have been put forth to explain intellectual property protection. The principal basis for such protection in the United States is the utilitarian or economic incentive framework. Nonetheless, other theories — most notably the natural rights and personhood justifications — have been important in understanding the development and scope of intellectual property law, both here and abroad.

The two excerpts that follow are written as justifications for tangible property. Consider how well they apply to ideas.

1. The Natural Rights Perspective

John Locke, Two Treatises on Government
Third Edition, 1698

Though the earth and all inferior creatures be common to all men, yet every man has a "property" in his own "person." This nobody has any right to but himself. The "labour" of his body and the "work" of his hands, we may say, are properly his. Whatsoever, then, he removes out of the state that Nature hath provided and left it in, he hath mixed his labor with it, and joined to it something that is his own, and thereby makes it his property. It being by him removed from the common state Nature placed it in, it hath by this labour something annexed to it that excludes the common right of other men. For this "labour" being the unquestionable property of the labourer, no man but he can have a right to what that is once joined to, at least where there is enough, and as good left in common for others.

He that is nourished by the acorns he picked up under an oak, or the apples he gathered from the trees in the wood, has certainly appropriated them to himself. . . . That labour put a distinction between them and common . . . . And will any one say he had no right to those acorns or apples he thus appropriated because he had not the consent of all mankind to make them his? Was it a robbery thus to assume to himself what belonged to all in common? If such a consent as that was necessary, man had starved, notwithstanding the plenty God had given him. We see in commons, which remain so by compact, that it is the taking any part of what is common, and removing it out of the state Nature leaves it in, which begins the property, without which the common is of no use. And the taking of this or that part does not depend on the express consent of all the commoners. . . .

It will, perhaps, be objected to this, that if gathering the acorns or other fruits, of the earth, etc., makes a right to them, then any one may engross as much as he will. To which I answer, Not so. The same law of Nature that does by this means give us property, does also bound that property too. . . . As much as any one can make use of to any advantage of life before it spoils, so much he may by his labor fix his property in. Whatever is beyond this is more than his share, and belongs to others. . . .

As much land as a man tills, plants, improves, cultivates, and can use the product of, so much is his property. He by his labor does, as it were, enclose it from the common. . . .

Nor was this appropriation of any parcel of land, by improving it, any prejudice to any other man, since there was still enough and as good left, and more than the yet unprovided could use. So that, in effect, there was never the less left for others because of his enclosure for himself. For he that leaves as much as another can make use of does as good as take nothing at all.

COMMENTS AND QUESTIONS

1. How do Locke's theories of real property apply to intellectual property? Should we treat the two as the same for ownership purposes? What would Locke say about the exclusive rights granted by the patent laws, to prevent others from using the claimed invention for up to 20 years, whether or not they discovered the invention on their own? Surely if Locke considered the working of land and raw materials to be "labor" that justified ownership of the resulting product, he would have considered labor toward the creation of a new *idea* — the "sweat of the brow" — to be equally deserving of protection. Or do the differences between real and intellectual property mean that Locke's arguments shouldn't apply to intellectual property?

For illumination on these and related points, see Edwin C. Hettinger, Justifying Intellectual Property, 18 Phil. & Pub. Aff. 31 (1989). Hettinger critiques the major theories of intellectual property rights. As to Lockean labor theory, Hettinger makes these useful observations:

- [A]ssuming that labor's fruits are valuable, and that laboring gives the laborer a property right in this value, this would entitle the laborer only to the value she added, and not to the *total* value of the resulting product. Though exceedingly difficult to measure, these two components of value (that attributable to the object labored on and that attributable to the labor) need to be distinguished. [p. 37]
- Property rights in the thing produced are . . . not a fitting reward if the value of these rights is disproportional to the effort expended by the laborer. "Effort" includes (1) how hard someone tries to achieve a result, (2) the amount of risk voluntarily incurred in seeking this result, and (3) the degree to which moral considerations played a role in choosing the result intended. The harder one tries, the more one is willing to sacrifice, and the worthier the goal, the greater are one's deserts. [pp. 41-42]

The philosopher Robert Nozick made the first point by means of *reductio ad absurdum:* he asks whether the owner of a can of tomato juice who dumps it into the ocean can thereafter claim ownership of all the high seas. See Robert Nozick, Anarchy, State and Utopia 175 (1984).

2. An application of the Lockean approach with an especially detailed consideration of the Lockean "proviso" (i.e., that "as much and as good" be left for others after appropriation) can be found in Wendy J. Gordon, A Property Right in Self-Expression: Equality and Individualism in the Natural Law of Intellectual Property, 102 Yale L.J. 1533 (1993). The proviso poses a problem we will see throughout the course: how to delimit the rights of a creator in the face of claims by consumers and other members of the public at large.

More particularly, Gordon challenges the extreme view taken by some commentators that a creator's rights should be absolute. The absolutist view proceeds from the idea that since the creator is solely responsible for the creation, no one is harmed if the creation is withheld from the public entirely. And since the creator can withhold it entirely, he or she can naturally restrict its availability in any manner, including a high price or conditions on its purchase. See, e.g., John Stuart Mill, Principles of Political Economy 142 (1872); Steven N. S. Cheung, Property Rights and Invention, in 8 Research in Law and Economics: The Economics of Patents and Copyrights 5, 6 (John Palmer & Richard O. Zerbe, Jr., eds., 1986). Gordon's example — which she says illustrates "a reliance argument" — shows that, sometimes at least, the public can be worse off if a creation is offered and then removed than it would have been had the creation never been made.

Should it matter whether Locke's hypothetical creator was the only one likely to come up with his particular invention or discovery? If others would have discovered the same phenomenon a few years (or a few weeks) later, does Locke's argument for property rights lose its force?

For detailed discussions of the "Lockean proviso," see Jeremy Waldron, Enough and as Good Left for Others, 29 Phil. Q. 319 (1979); Robert Nozick, Anarchy, State and Utopia 175-182 (1984); David Gauthier, Morals by Agreement 190-232 (1986). For more detailed treatment of another problem addressed by Gordon — the follow-on creator who builds on a preexisting work — see Lawrence C. Becker, Deserving to Own Intellectual Property, 68 Chicago-Kent L. Rev. 609 (1993).

3. Nozick's book (cited above) offers a different philosophical perspective on intellectual property, one rooted in the libertarian tradition. It is not clear how libertarians should think of intellectual property rights. On the one hand, ownership of property seems necessarily to underlie the market exchange that is at the heart of the libertarian model of society. On the other hand, one might view the free flow of information unfettered by property rights as the norm, and view government-enforced intellectual property rights as an unnecessary aberration. For a libertarian approach that is decidedly hostile to intellectual property, see John Perry Barlow, The Economy of Ideas, 2.03 Wired 84 (Mar. 1994).

4. Natural rights are strongly emphasized in the continental European justifications for intellectual property. Those justifications to some extent parallel Locke's arguments, but there are important differences. Continental scholars emphasize the importance of reputation and noneconomic aspects of intellectual property, factors that lead them to support moral rights in copyright law. Professor Alfred Yen presents a thorough account of the role of natural rights in American copyright law. See Alfred Yen, Restoring the Natural Law: Copyright as Labor and Possession, 51 Ohio St. L.J. 517 (1990).

PROBLEM

Problem 1-1.   How does Locke's "labor theory" of property apply to the following scenario: You are a botanist exploring a remote region of a small tropical country when you stumble across a field of beautiful flowers, the likes of which you have never seen before. You pluck one of the flowers and, when you return to your encampment that night, you show it to a fellow explorer, Dr. *X*, who is an expert in the bio-chemistry of plants. After smelling and tasting the flower, she says it smells faintly like Substance *P*, a medicine widely used to treat a variety of serious diseases. She remarks that Substance *P* is easy to detect; it turns a bright yellow when exposed to intense heat. You put the flower over the campfire that night, and sure enough it turns bright yellow.

Upon your return home, you work for months to isolate the active ingredient in the flower, a chemical different from Substance *P* but a close structural analog. In various lab experiments you discover that it shows an amazing degree of activity in fighting many of the same diseases that Substance *P* is used to treat.

a) What rights should you have in the new chemical? Should your rights prevent anyone else from going back to the tropical country, finding the flower, and isolating the active chemical you have discovered?

b) What if a highly skilled chemist could have isolated the active ingredient in less than a day, at essentially no cost, instead of the months it took you? Should this affect the existence or scope of your rights in the substance?

c) Assume native tribespeople in the tropical country had been using this same flower to treat various diseases for centuries. Should they have the right to receive a portion of your profits from the sale of the new chemical? Should you be able to charge these tribespeople a royalty for using their traditional medicines because you have now isolated the active chemical?

d) Assume that some members of the public who come to rely on the new chemical as a treatment for their diseases become addicted to it. After using it for a short period of time, they cannot go back to preexisting treatments without serious risk of severe medical complications, even death. Do these factors affect your ability to remove the chemical from the market if you wish, or to charge whatever price you want for it?

2. The Personhood Perspective

≡ *Margaret Jane Radin, Property and Personhood*
34 Stan. L. Rev. 957 (1982)

This article explores the relationship between property and personhood, a relationship that has commonly been both ignored and taken for granted in legal thought. The premise underlying the personhood perspective is that to achieve proper self-development — to be *a person* — an individual needs some control over resources in the external environment. The necessary assurances of control take the form of property rights. Although explicit elaboration of this perspective is wanting in modern writing on property, the personhood perspective is often implicit in the connections that courts and commentators find between property and privacy or between property and liberty. In addition to its power to explain certain aspects of existing schemes of property entitlement, the personhood perspective can also serve as an explicit source of values for making moral distinctions in property disputes and hence for either justifying or criticizing current law. . . .

In what follows I shall discuss the personhood perspective as Hegel developed it in *Philosophy of Right*, trace some of its later permutations and entanglements with other perspectives on property, and try to develop a contemporary view useful in the context of the American legal system. . . .

I. Property for Personhood: An Intuitive View

Most people possess certain objects they feel are almost part of themselves. These objects are closely bound up with personhood because they are part of the way we constitute ourselves as continuing personal entities in the world. They may be as different as people are different, but some common examples might be a wedding ring, a portrait, an heirloom, or a house.

One may gauge the strength or significance of someone's relationship with an object by the kind of pain that would be occasioned by its loss. On this view, an object is closely related to one's personhood if its loss causes pain that cannot be relieved by the object's replacement. If so, that particular object is bound up with the holder. For instance, if a wedding ring is stolen from a jeweler, insurance proceeds can reimburse the jeweler, but if a wedding ring is stolen from a loving wearer, the price of a replacement will not restore the status quo — perhaps no amount of money can do so.

The opposite of holding an object that has become a part of oneself is holding an object that is perfectly replaceable with other goods of equal market value. One holds such an object for purely instrumental reasons. The archetype of such a good is, of course, money, which is almost always held only to buy other things. A dollar is worth no more than what one chooses to buy with it, and one dollar bill is as good as another. Other examples are the wedding ring in the hands of the jeweler, the automobile in the hands of the dealer, the land in the hands of the developer, or the apartment in the hands of the commercial landlord. I shall call these theoretical opposites — property that is bound up with a person and property that is held purely instrumentally — personal property and fungible property respectively. . . .

III. Hegel, Property, and Personhood

A. Hegel's Philosophy of Right . . .

Because the person in Hegel's conception is merely an abstract unit of free will or autonomy, it has no concrete existence until that will acts on the external world. . . .

Hegel concludes that the person becomes a real self only by engaging in a property relationship with something external. Such a relationship is the goal of the person. In perhaps the best-known passage from this book, Hegel says:

> The person has for its substantive end the right of placing its will in any and every thing, which thing is thereby mine; [and] because that thing has no such end in itself, its destiny and soul take on my will. [This constitutes] mankind's absolute right of appropriation over all things.

Hence, "property is the first embodiment of freedom and so is in itself a substantive end." . . .

Hegel seems to make property "private" on the same level as the unit of autonomy that is embodying its will by holding it. He argues that property is private to individuals when discussing it in the context of the autonomous individual will and that it is essentially common within a family, when discussing it in the context of the autonomous family unit. He does not make the leap to state property, however, even though his theory of the state might suggest it. For Hegel, the properly developed state (in contrast to civil society) is an organic moral entity . . . and individuals within the state are subsumed into its community morality. . . .

B. Hegel and Property For Personhood

[A] theory of personal property can build upon some of Hegel's insights. First, the notion that the will is embodied in things suggests that the entity we know as a person cannot come to exist without both differentiating itself from the physical environment and yet maintaining relationships with portions of that environment. The idea of embodied will, cut loose from Hegel's grand scheme of absolute mind, reminds us that people and things have on-going relationships which have their own ebb and flow, and that these relationships can be very close to a person's center and sanity. If these relationships justify ownership, or at least contribute to its justification, Hegel's notion that ownership requires continuous embodiment of the will is appealing.

Second, Hegel's incompletely developed notion that property is held by the unit to which one attributes autonomy has powerful implications for the concept of group development and group rights. Hegel thought that freedom (rational self-determination) was only possible in the context of a group (the properly organized and fully developed state). Without accepting this role for the state, one may still conclude that in a given social context certain groups are likely to be constitutive of their members in the sense that the members find self-determination only within the groups. This might have political consequences for claims of the group on certain resources of the external world (i.e., property).

Third, there may be an echo of Hegel's notion of an objective community morality in the intuition that certain kinds of property relationships can be presumed to bear close bonds to personhood. If property in one's body is not too close to personhood to be considered property at all, then it is the clearest case of property for personhood. The property/privacy nexus of the home is also a relatively clear case in our particular history and culture. . . .

[T]he personhood theory helps us understand the nature of the right dictating that discrete units [i.e., an undivided, individual asset] ought to be protected.

An argument that discrete units are more important than total assets takes the following form. A person cannot be fully a person without a sense of continuity of self over time. To maintain that sense of continuity over time and to exercise one's liberty or autonomy, one must have an ongoing relationship with the external environment, consisting of both "things" and other people. One perceives the ongoing relationship to the environment as a set of individual relationships, corresponding to the way our perception separates the world into distinct "things." Some things must remain stationary if anything is to move; some points of reference must be constant or thought and action is not possible. In order to lead a normal life, there must be some continuity in relating to "things." One's expectations crystallize around certain "things," the loss of which causes more disruption and disorientation than does a simple decrease in aggregate wealth. For example, if someone returns home to find her sofa has disappeared, that is more disorienting than to discover that her house has decreased in market value by 5%. If, by magic, her white sofa were instantly replaced by a blue one of equal market value, it would cause no loss in net worth but would still cause some disruption in her life.

This argument assumes that all discrete units one owns and perceives as part of her continuing environment are to some degree personal. If the white sofa were totally fungible, then magically replacing it with a blue one would cause no disruption. In fact, neither would replacing it with money. . . .

But the theory of personal property suggests that not all object-loss is equally important. Some objects may approach the fungible end of the continuum so that the justification for protecting them as specially related to persons disappears. They might just as well be treated by whatever general moral rules govern wealth-loss at the hands of the government. If the moral rules governing wealth-loss correspond to Michelman's utilitarian suggestion — government may take whatever wealth is necessary to generate higher welfare in which the individual can confidently expect to share — then the government could take some fungible items without compensation. In general, the moral inquiry for whether fungible property could be taken would be the same as the moral inquiry for whether it is fair to impose a tax on this particular person.

On the other hand, a few objects may be so close to the personal end of the continuum that no compensation could be "just." That is, hypothetically, if some object were so bound up with me that I would cease to be "myself" if it were taken, then a government that must respect persons ought not to take it. If my kidney may be called my property, it is not property subject to condemnation for the general public welfare. Hence, in the context of a legal system, one might expect to find the characteristic use of standards of review and burdens of proof designed to shift risk of error away from protected interests in personal property. For instance, if there were reason to suspect that some object were close to the personal end of the continuum, there might be a prima facie case against taking it. That prima facie case might be rebutted if the government could show that the object is not personal, or perhaps that the object is not "too" personal compared with the importance to the government of acquiring that particular object for social purposes.

COMMENTS AND QUESTIONS

1. How well does Professor Radin's theory of real property apply to intellectual property? Can an individual be so "bound up" in her own inventions or works of authorship that their loss would occasion more than economic damage? Does it affect your answer that intellectual property can be used simultaneously by many people without depleting its functional value to any one — so that an author's "loss" is not the physical deprivation of stolen chattels, but the less personal fact that someone else has copied her work?

It may be that the investment of "personhood" in intellectual property varies greatly, both with the type of intellectual property at issue and with the time and effort the owner put into developing it. For example, a novel on which one has worked for several years may have more personal value than a company's customer list. Should the law take account of these differences, giving greater protection to more personal works?

For applications of Radin's "personhood" ideas in the context of intellectual property, see Neil W. Netanel, Copyright Alienability Restrictions and the Enhancement of Author Autonomy: A Normative Evaluation, 24 Rutgers L.J. 347 (1993); Steven Cherensky, Comment, A Penny for Their Thoughts: Employee-Inventors, Preinvention Assignment Agreements, Property, and Personhood, 81 Cal. L. Rev. 595, 641 (1993). Cherensky argues, using personhood theory and the related idea of market-inalienability, that employee-inventors should retain greater property interests in their inventions than they typically do under conventional employee assignment contracts. On market-inalienability, i.e., things which should not be subject to market exchange at all, see Margaret Jane Radin, Market-Inalienability, 100 Harv. L. Rev. 1849 (1987). Cf. Justin Hughes, The Philosophy of Intellectual Property, 77 Georgetown L.J. 287, 350-53 (1988) (suggesting various strains of the personhood justification in American copyright law).

2. For a critique of Radin's overall theory, see Stephen J. Schnably, Property and Pragmatism: A Critique of Radin's Theory of Property and Personhood, 45 Stan. L. Rev. 347 (1993) (challenging Radin's appeal to consensus and arguing that this focus obscures issues of power and the like). See also Margaret Jane Radin, Reinterpreting Property (1993) (a collection of related essays); A. John Simmons, The Lockean Theory of Rights (1992); Jeremy Waldron, The Right to Private Property (1991) (comparing Lockean and Hegelian property rights theories); Symposium, Property Rights, 11 Soc. Phil. & Policy 1-286 (1994).

3. A radical critique of some bedrock notions implicit in intellectual property — most notably, the concept of "*the* author" herself — has grown up in recent years, fueled by a general deconstructive trend in literary criticism. While not aimed directly at Radin's personhood approach to certain forms of property, these ideas do pose a challenge to the application of Radin's approach in the intellectual property context. They suggest that the concept of authorship is so malleable, contingent, and "socially constructed" that we should be wary about identifying a creative work too closely with a particular author, let alone her personality. In this view, all creations are largely a product of communal forces. Dividing the stream of intellectual discourse into discrete units — each owned by and closely associated with a particular author — is therefore a logically incoherent exercise subject more to the political force of asserted authors' groups than to recognition of inherent claims of "personhood." See, e.g., Martha Woodmansee & Peter Jaszi, eds., The Construction of Authorship: Textual Appropriation in Law and Literature (1994); Peter Jaszi, Toward a Theory of Copyright: The Metamorphoses of "Authorship," 1991 Duke L.J. 455; James Boyle, A Theory of Law and Information: Copyright, Spleens, Blackmail and Insider Trading, 80 Cal. L. Rev. 1413 (1992).

If authorship is an inchoate concept, is there any role at all for copyright law? How can one protect the rights (natural, moral, or economic) of the author, if in fact there is no author? Does the literary critique answer the charge that authors will not create in the absence of economic reward? Or is it directed only at personality-based theories of intellectual property?

4. Consider the observations of philosopher Lawrence Becker:

> So if property-as-personality [à la Hegel] again turns out to be a dead end, perhaps we should dispense with the search for a deep justification for property rights (from metaphysics, moral psychology, sociobiology, or whatever) and focus on the behavioral surface: the observed, persistent, robust behavioral connections between various property arrangements and human well-being, broadly conceived. This may provide a foundation for egalitarian arguments that is more secure than speculative metaphysics, and a foundation for private property that is more stable than a pluralistic account of the standard array of bedrock justifications for it.

Lawrence Becker, Too Much Property, 21 Phil. & Pub. Aff. 196, 206 (1992).

3. The Utilitarian/Economic Incentive Perspective

Intellectual property in the United States is fundamentally about incentives to invent and create. Although we have considered a number of non-economic theories offered to explain copyright and patent law, both the United States Constitution and judicial decisions seem to acknowledge the primacy of incentive theory in justifying intellectual property. The Constitution expressly conditions the grant of power in the patent and copyright clause on a particular end, namely "to Promote the Progress of Science and useful Arts." U.S. Const., art. I, cl. 8. As the Supreme Court explained in the landmark case of Mazer v. Stein, 347 U.S. 201 (1954):

> "The copyright law, like the patent statutes, makes reward to the owner a secondary consideration." United States v. Paramount Pictures, 334 U.S. 131, 158. However, it is "intended definitely to grant valuable, enforceable rights to authors, publishers, etc., without burdensome requirements: 'to afford greater encouragement to the production of literary [or artistic] works of lasting benefit to the world.' " Washington Pub. Co. v. Pearson, 306 U.S. 30.
>
> The economic philosophy behind the clause empowering Congress to grant patents and copyrights is the conviction that it is the best way to advance public welfare through the talents of authors and inventors in "Science and useful Arts." Sacrifical days devoted to such creative activities deserve rewards commensurate with the services rendered.

To understand why the Framers thought exclusive rights in inventions and creations would promote the public welfare, consider what would happen absent any sort of intellectual property protection. Invention and creation require the investment of resources — the time of an author or inventor, and often expenditures on facilities, prototypes, supplies, etc. In a private market economy, individuals will not invest in invention or creation unless the expected return from doing so exceeds the cost of doing so — that is, unless they can reasonably expect to make a profit from the endeavor. To profit from a new idea or work of authorship, the creator must be able either to sell it to others for a price or to put it to some use that provides her with a comparative advantage in a market.[1]

But ideas (and writings, for that matter) are notoriously hard to control. Even if the idea is one that the creator can use herself, for example to boost productivity in her business, she will reap a reward from that idea only to the extent that her competitors don't find out about it. A creator who depends on secrecy for value therefore lives in constant peril of discovery and disclosure. Competitors may steal the idea or learn of it from an ex-employee. They may be able to figure it out by watching the creator's production process or by examining the products she sells. Finally, they may come upon the idea on their own or discover it in the published literature. In all of these cases, the secrecy value of the idea will be irretrievably lost.

The creator who wants to sell her idea is in an even more difficult position. Selling information requires disclosing it to others. Once the information has been disclosed outside a small group, however, it is extremely difficult to control. Information has the characteristics of what economists call a "public good" — it may be "consumed" by many people without depletion, and it is difficult to identify those who will not pay and prevent them from using the information. See Kenneth J. Arrow, Economic Welfare and the Allocation of Resources for Invention, in The Rate and Direction of Inventive Activity: Economic and Social Factors 609, 614-616 (Natl. Bureau of Economic Research ed., 1962). Once the idea of the intermittent windshield wiper is disclosed, others can imitate its design relatively easily. Once a book is published, others can copy it at low cost. It is difficult to exclude nonpurchasers. Ideas and information can also be used by many without depleting the enjoyment of others. Unlike an ice cream cone, a good story or the concept of intermittent windshield wipers can be enjoyed by many without diminishing the enjoyment of these creations by others.[2]

If we assume that it is nearly costless to distribute information to others — an assumption that was once unrealistic but now much more reasonable with the development of the Internet — it will prove virtually impossible to charge for information over the medium run. If the author of a book charges more than the cost of distribution, hoping to recover some of her expenditures in writing the work, competitors will quickly jump in to offer the book at a lower price. Competition will drive the price of the book toward its marginal cost — in this case the cost of producing and distributing one

---

1. The latter may occur, for example, where an idea for a more efficient machine is used to reduce the cost of producing goods, allowing the owner of the idea to compete more effectively in selling those goods.

2. To some extent this statement oversimplifies the problem by ignoring possible second-order distorting effects. In practice, if you taught several hundred million people to fish, the result might be depletion of a physical resource (fish) that would otherwise not have occurred. Similarly, wide dissemination of information may have particular effects on secondary markets, depending on what the information is.

additional copy. In such a competitive market, the author will be unable to recoup the fixed cost of writing the book. More to the point, if this holds true generally, authors may be expected to leave the profession in droves, since they cannot make any money at it. The result, according to economic theory, is an underproduction of books and of other works of invention and creation with similar public goods characteristics.[3]

Information is not the only example of a public good. Economists generally offer lighthouses and national defense as examples of public goods, since it is virtually impossible to provide the benefits of either one only to paying clients. It is impossible, for example, to exclude some ships and not others from the benefits of a lighthouse. Furthermore, the use of the lighthouse by one ship does not deplete the value of its hazard warning to others. As a result, it would be inefficient to exclude nonpayers from using the lighthouse's warning system even if we could, since consumption of this good is "nonrivalrous" (meaning that everyone can benefit from it once it is produced). For these reasons, the market will in theory undersupply such goods because producers cannot reap the marginal (incremental) value of their investment in providing them.[4]

Can you see why broadcast television signals, beautiful gardens on a public street, and national defense are also public goods?

By contrast, markets for pure private goods, such as ice cream cones, feature exclusivity and rivalrous competition — the ice cream vendor provides the good only to those who pay the price, and the consumer certainly depletes the amount of the good available to others. Thus the market system provides adequate incentives for the creation of ice cream cones: sellers can exact their cost of production, and the value of the product is fully enjoyed by the purchaser.

In the case of national defense (and most lighthouses), we avoid the underproduction that would result from leaving it to the market by having the government step in and pay for the public good. For a variety of good reasons, discussed in note 3 following this section, we have not gone that route with information. Instead, government has created intellectual property rights in an effort to give authors and inventors control over the use and distribution of their ideas, thereby encouraging them to invest in the production of new ideas and works of authorship. Thus the economic justification for intellectual property lies not in rewarding creators for their labor but in assuring that they (and other creators) have appropriate incentives to engage in creative activities.

3. See, e.g., F. M. Scherer, Industrial Market Structure and Economic Performance 444 (2d ed. 1980) ("If pure and perfect competition in the strictest sense prevailed continuously . . . incentives for invention and innovation would be fatally defective without a patent system or some equivalent substitute."). Professor Scherer goes on to note, however, that natural market imperfections may give advantages to first movers, reducing the need for intellectual property protection. Id. at 444-445.

4. Ronald Coase has offered evidence casting doubt on the economic assumption that lighthouses must be publicly provided. See Ronald H. Coase, The Lighthouse in Economics, in The Firm, the Market, and the Law (1988).

Unfortunately, this approach comes at a cost. Granting authors and inventors the right to exclude others from using their ideas necessarily limits the diffusion of those ideas and so prevents many people from benefiting from them. In economic terms, intellectual property rights prevent competition in the sale of the particular work or invention covered by those rights, and therefore may allow the intellectual property owner to raise the price of that work above the marginal cost of reproducing it. This means that in many cases fewer people will buy the work than if it were distributed on a competitive basis, and they will pay more for the privilege. A fundamental principle of our economic system is the proposition that free market competition will ensure an efficient allocation of resources, absent market failures. See generally Paul Samuelson & William D. Nordhaus, Economics 678 (12th ed. 1985). In fact, the principal thrust of the antitrust laws serves precisely this goal. In this limited sense, then, intellectual property rights appear to run counter to free market competition: they limit the ability of competitors to copy or otherwise imitate the intellectual efforts of the first person to develop an idea. These rights enable those possessing intellectual property rights to charge monopoly prices or to otherwise limit competition, such as by controlling the use of the intellectual work in subsequent products.

Because intellectual property rights impose social costs on the public, the intellectual property laws can be justified by the public goods argument only to the extent that they do on balance encourage enough creation and dissemination of new works to offset those costs. One of the reasons that intellectual property rights are limited in scope, duration, and effect is precisely in order to balance these costs and benefits. For example, the limited term of intellectual property rights ensures that inventions will be freely available after that fixed term. The key to economic efficiency lies in balancing the social benefit of providing economic incentives for creation and the social costs of limiting the diffusion of knowledge. We will encounter this critical trade-off throughout our study of intellectual property. The two examples below highlight some of the major issues.

Economic Incentive Benefit. Intellectual property protection is necessary to encourage inventors, authors, and artists to invest in the process of creation. Without such protection, others could copy or otherwise imitate the intellectual work without incurring the costs and effort of creation, thereby inhibiting the original creators from reaping a reasonable return on their investment. Consider the following example:

> After years of effort and substantial expense, Earnest Inventor develops the Mousomatic, a significantly better mousetrap. Not only does it catch mice better than the competition's trap, but it also neatly packages the dead mice in sanitary disposable bags. Consumers are willing to pay substantially more for this product than for the competition's. The Mousomatic catches the attention of Gizmo Gadget Incorporated. Gizmo copies the basic design of the Mousomatic and offers its version of the Mousomatic at a substantial discount. (Gizmo can still

earn a profit at this lower price because it had minimal research and development expense.) In order to stay in business, Earnest is forced to lower his price. Market competition pushes the price down to the cost of production and distribution. In the end, Earnest is unable to recover his cost of research and development and suffers a loss. Although he has numerous other interesting ideas, he decides that they are not worth pursuing because Gizmo, or some other company, will simply copy them if they turn out well.

The existence of intellectual property rights encourages Earnest and other inventors to pursue their creative efforts. If Earnest can obtain the right to prevent others from copying his inventions, then he stands a much better chance to reap a profit. Hence he will be much more inclined to make the initial investment in research and development. In the end, not only will Earnest be wealthier, but the public will be enriched by the new and better products brought forth by intellectual property protection.

Costs of Limiting Diffusion. Legal protection for ideas and their expression prevents others from using those works to develop similar works that build upon them. Knowledge in society is cumulative. Hence society at large can be harmed by intellectual property protection to the extent that it raises the cost of acquiring a product (through monopoly pricing by the right holder) and limits others from making further advances. Consider the following scenario:

Professor Lee conducts research on drug treatments at University College. Her laboratory is generously funded by grants from the federal government. For the past decade, Professor Lee has competed with colleagues at other laboratories to discover the cure for a prevalent form of cancer. It is likely that the first person to discover the cure will win a Nobel prize, as well as numerous other financial and professional rewards. In early 1995, Professor Lee hits upon the Alpha drug, which cures the disease. She files for and receives a patent. Professor Hu, a researcher at another research institution, independently discovers the identical cure a few months later. With patent in hand, Professor Lee starts a company to sell Alpha and charges a price 100 times the cost of production. Because Alpha is a life-saving cure, those stricken with the disease who can afford the treatment are more than willing to pay the price. Then, to relieve the suffering of millions, Professor Hu begins selling Alpha at the cost of production. Moreover, she has developed an improvement on the Alpha drug, Alpha +, which reduces the side effects of the treatment. Professor Lee quickly obtains an injunction preventing Hu from selling either version of Alpha for the life of the patent.

This example raises serious questions about whether intellectual property protection is desirable, at least for this class of invention. Professor Lee does not bear significant risk in pursuing the invention because her research is generously supported by the government and the university. Furthermore, the potential for a Nobel prize, expanded research funding, and professional recognition provide substantial encouragement for Professor Lee to pursue a cure whether or not she gains financially from sales of Alpha. Moreover, other researchers were poised to make the same discovery at about the time that Professor Lee made her discovery. Yet she has the right not only to block sales of Alpha by competitors, but also to block sales of improvements such as Alpha +. Does such a system benefit society? One must also consider that without the financial incentive of a patent, there would perhaps have been less competition to discover and market any cure for cancer. Problems such as this one have led some scholars to question the economic efficiency of the patent system in particular circumstances. See F. M. Scherer, Industrial Market Structure and Economic Performance 445-455 (2d ed. 1980).

In applying a utilitarian framework, the economic incentive benefits of intellectual property rights must be balanced against the costs of limiting diffusion of knowledge. A critical issue in assessing the need for intellectual property protection is whether innovators have sufficient means to appropriate an adequate return on investment in research and development. In this regard, the market itself often provides means by which inventors can realize sufficient rewards to pursue innovation without formal intellectual property rights beyond contract law. The first to introduce a product can in many contexts earn substantial and long-lived advantages in the market. In many markets, the costs or time required to imitate a product (for example, to reverse engineer a complex machine) are so great that the first to market a product has substantial opportunity for profit. Moreover, as we will see in Chapter 2, inventors can often prevent imitation through contractual means, such as trade secrecy and licensing agreements with customers. Where the invention relates to a manufacturing process, the innovator may be able to maintain protection through secrecy even after the product is on the market. Alternatively, a producer may be able to bundle products with essential services and contracts for updates of the product. In addition, the producer may be able to spread the costs of research and development among a group of firms through research joint ventures.

In those areas in which economic incentives for innovation are inadequate and the creation of intellectual property rights is the most efficacious way of encouraging progress, society must determine the appropriate requirements for, duration and scope of, and set of rights afforded intellectual property. Over the past several decades, economists have developed and refined models to assess the appropriate trade-off between the social benefits of providing economic incentives for innovation through intellectual property rights and the social costs of limiting diffusion of knowledge. Professor William Nordhaus developed the first formal model analyzing the optimal duration of intellectual property. His classic model of the innovative process assumed that investments in research produced a single independent innovation. W. Nordhaus, Invention, Growth, and Welfare: A Theoretical Treatment of Technological Change 3-7 (1969). The principal policy implication of this model is that the term of intellectual property protection should be calibrated to balance the incentive benefits of protection against the dead-weight loss of monopoly pricing and the resulting limitations on dissemination.

Since Nordhaus's important early work, economic historians and economic theorists have greatly enriched our understanding of the innovative process and the implications for public policy. Historical and industry studies of the innovation process find that inventions are highly interdependent: "Technologies . . . undergo a gradual, evolutionary development which is intimately bound up with the course of their diffusion."[5] In fact, "secondary inventions" — including essential design improvements, refinements, and adaptations to a variety of uses — are often as crucial to the generation of social benefits as the initial discovery.[6] Economic theorists have more recently developed models of the innovative process incorporating concepts of rivalrous and cumulative innovation.[7] These models have uncovered a range of important effects beyond the reach of the Nordhaus framework. These more complex accounts of the innovative process alter the basic policy conclusions of the simpler framework. Excessive protection for first-generation innovation can impede later stages, thereby undermining some of the salutary effects of strong intellectual property protection.[8] More generally, these models cast doubt on the notion that society can perfectly calibrate intellectual property rewards for each innovation.

As is increasingly evident, the range of innovative activity and creative expression in our society is vast and ever changing. As the materials in this book will highlight, the intellectual property institutions and rules that have evolved to promote technology and the arts are intricate. It will be the challenge of future generations of policymakers, judges, and lawyers to refine the ability of the intellectual property system to enhance the public welfare.

COMMENTS AND QUESTIONS

1. One significant difference between the natural rights perspective and the utilitarian perspective relates to who is entitled to the fruits of productive labor. In the natural rights framework, the inventor or author is entitled to

---

5. Paul David, New Technology, Diffusion, Public Policy, and Industrial Competitiveness 20 (Center for Economic Policy Research, Publication No. 46, Apr. 1985). See generally, Richard Nelson & Steven Winter, An Evolutionary Theory of Economic Change (1982); Nathan Rosenberg, Factors Affecting the Diffusion of Technology, 10 Explorations Econ. Hist. 3 (1972).

6. See, e.g., Enos, A Measure of the Rate of Technological Progress in the Petroleum Refining Industry, 6 J. Indus. Econ. 180, 189 (1958); Mak & Walton, Steamboats and the Great Productivity Surge in River Transportation, 32 J. Econ. Hist. 619, 625 (1972).

7. See Suzanne Scotchmer, Standing on the Shoulders of Giants: Cumulative Research and the Patent Law, 5 J. Econ. Perspectives 29 (1991); Robert P. Merges & Richard Nelson, On the Complex Economics of Patent Scope, 90 Colum. L. Rev. 839, 843, 868-879 (1990); M. Kamien & N. Schwartz, Market Structure and Innovation (1982).

8. See Scotchmer, supra, at 38.

all the social benefits produced by his or her efforts. In the utilitarian framework, reward to the inventor or author is a secondary consideration; the principal objective is to enrich the public at large. Which view is more compelling? Consider in this regard the optimal division of benefits from the invention of Alpha among Professor Lee, Professor Hu, and the public at large. Is Professor Lee entitled to all or even a lion's share of the benefits?

2. In 1966, the Report of the President's Commission on the Patent System identified four major economic justifications for the patent laws.

> First, a patent system provides an incentive to invent by offering the possibility of reward to the inventor and to those who support him. This prospect encourages the expenditure of time and private risk capital in research and development efforts.
>
> Second, and complementary to the first, a patent system stimulates the investment of additional capital needed for the further development and marketing of the invention. In return, the patent owner is given the right, for a limited period, to exclude others from making, using, or selling the invented product or process.
>
> Third, by affording protection, a patent system encourages early public disclosure of technological information, some of which might otherwise be kept secret. Early disclosure reduces the likelihood of duplication of effort by others and provides a basis for further advances in the technology involved.
>
> Fourth, a patent system promotes the beneficial exchange of products, services, and technological information across national boundaries by providing protection for industrial property of foreign nationals.

While directed specifically at the patent system, many of these arguments have application to all forms of intellectual property.

Are these incentives necesary to invention and creation? Using cost and other data from publishing companies, Professor (now Justice) Stephen Breyer contended that lead time advantages and the threat of retaliation reduce the cost advantages of copiers, thus obviating if not eliminating the need for copyright protection for books. See Stephen Breyer, The Uneasy Case for Copyright: A Study in Copyright of Books, Photocopies and Computer Programs, 84 Harv. L. Rev. 281 (1970). Cf. Tyerman, The Economic Rationale for Copyright Protection for Published Books: A Reply to Professor Breyer, 18 UCLA L. Rev. 1100 (1971); Stephen Breyer, Copyright: A Rejoinder, 20 UCLA L. Rev. 75 (1972). This debate took place more than a quarter-century ago. Have advances in technology strengthened or weakened Breyer's argument?

3. Professor George Priest argues that economic analysis of patent law is "one of the least productive lines of inquiry in all of economic thought" because the lack of an adequate empirical basis for assessing the theoretical models of innovation severely limits what lawyers and policymakers can learn from economic analysis. See George Priest, What Economists Can Tell Lawyers About Intellectual Property, in 8 Res. Law & Econ. 19 (J. Palmer & R. Zerbe, eds. 1986). Does this suggest that society should give up on a utili-

tarian perspective? Or is it still better than anything else? What alternatives would you suggest?

Priest's observation is bolstered by the fact that numerous institutional mechanisms exist for addressing the public goods problem inherent in the production of ideas and information — direct government funding of research, government research subsidies, promotion of joint ventures, and prizes. The case for intellectual property rights ideally compares all of these options. Nonetheless, intellectual property rights have the advantage of limiting the government's role in allocating resources to a finite set of decentralized decisions: whether particular inventions are worthy of a fixed period of protection. The market then serves as the principal engine of progress. Decentralized consumers generate demand for products and competing decentralized sellers produce them. By contrast, most other incentive systems, especially large-scale research funding, require central planning on a mass scale. Most economists place more confidence in the former means of allocating resources. The case for intellectual property rights, then, is based more on a generalized perception of institutional choice than on strong direct evidence of the superiority of intellectual property rights relative to the alternatives.

4. Recently, some observers have posed a radical critique of the entire concept of owning ideas in an information economy. This body of thought is not fully formed, but it has a number of articulate adherents in the computer community. This approach, essentially libertarian in nature, is often summed up in the computer hacker credo "information wants to be free." See John Perry Barlow, The Economy of Ideas, 2.03 Wired 84 (Mar. 1994).

What is likely to happen if we abolish the concept of ownership of information? Will people stop producing ideas, as the incentive theory of intellectual property would suggest? Or will other types of incentives continue to encourage invention and creation?

PROBLEM

Problem 1-2. Building on your lifelong interest in music and computers, you have just spent ten years creating a computer-based Encyclopedia of Music. The encyclopedia, based on a CD-ROM, includes text, snippets of recorded music (no longer than seven seconds each), pictures, and other graphics, all of which can be called up by the user. It is thus an instance of the new generation of "multimedia" computer products.

Like many other multimedia products, your encyclopedia allows the user to select a topic of interest and see (or hear) more about it. For example, from the opening menu a user can select classical composers, then select Mozart, and then choose from a biographical sketch (with text, pictures, and samples of sheet music) or musical samples that actually play on the computer's speakers.

To make it easier for computer users to access all the information in your encyclopedia, you have used an "access interface" similar to the widely-used "MediaMate" interface written and made popular by multigazillionaire computer impresario Gil Bates. (Bates adapted the "MediaMate" interface from a publicly available interface designed by the U.S. Army to teach electronics and other technical subjects to new recruits.) Although MediaMate has now become the standard interface program, Bates, because of his awesome programming talent, wrote the program in his spare time over several weeks while watching soap operas. According to his autobiography, he "never broke a sweat" while writing it; "in fact, it was a breeze."

Since Bates's company, MacroLoft, also owns many of the legal rights needed to play music from such musicians as Jimi Hendrix and Aretha Franklin, you have decided it is time to negotiate directly with Bates prior to placing your encyclopedia on the market. Note that although Bates has received valid licenses in the relevant copyrights for these songs, it is widely known that some of the musicians who created the songs are not pleased with the use of short snippets on inadequate computer sound systems, some of which sound "tinny."

You are aware that negotiating with Bates is far different from standard business practice. He is known to do all negotiating standing up in an unheated room in a mountain cabin, with no lawyers or assistants of any kind. Most unusually, he is not interested at all in formal legal rights. Reflecting his training as a serious student of philosophy, he insists on negotiating from what he calls "first principles." This usually involves long discussions of who should have which rights in which situations, and why.

You would like to convince Bates to let you market your encyclopedia without paying any royalties to him for songs, the access interface, or anything else. Failing this, you would like to keep these royalties to a minimum. Draw up a negotiating strategy that will convince him.

B. OVERVIEW OF INTELLECTUAL PROPERTY

Intellectual property law has traditionally been taught along doctrinal lines. Separate courses have covered patent, copyright, and trademark law, with trade secrets often lost between the gaps. Yet the practice of intellectual property law increasingly cuts across these lines. Computer technology matters, for example, frequently require lawyers to address trade secret, copyright, patent, trademark, and antitrust issues simultaneously. Moreover, from a purely practical standpoint, clients are ultimately interested in appropriating a return from their investments, not in how many patents, copyrights, trademarks, or trade secrets their lawyers can obtain. Thus intellectual property lawyers must possess an integrated understanding of these various fields in order to provide sound advice.

With this objective in mind, our book integrates the various modes of intellectual property in a functional manner. Nonetheless, it is still necessary to devote significant time to mastering each of the distinct fields. Therefore, the next five chapters survey the principal modes of intellectual property — trade secret, patent, copyright, trademark/trade dress, and related state law doctrines — while emphasizing the overlaps and interactions of the various bodies of law. Chapter 7 integrates these fields directly in exploring legal protection for computer technology. Chapter 8 surveys the interplay between intellectual property law and antitrust law.

Before we begin this more detailed study, however, a brief survey of the overall system of intellectual property is in order. The following section sketches the elements of each of the principal modes of intellectual property protection in a comparative framework. These elements are usefully captured in Table 1-1. As an initial exploration, we conclude the chapter with a problem highlighting the integrated nature of intellectual property law and the challenges of applying its many branches to a real-world problem.

1. Trade Secret

Trade secret laws are state law doctrines that protect against the misappropriation of certain confidential information. As such, they are more akin to traditional tort and contract law than to patent or copyright law. While protection for trade secrets has long been a part of the common law, most states today protect trade secrets by statute. The basic purpose behind protecting trade secrets is to prevent "theft" of information by unfair or commercially unreasonable means. In essence, trade secret law is a form of *private* intellectual property law under which creators establish contractual limitations or build legal "fences" that afford protection from misappropriation.

The definition of subject matter eligible for protection is quite broad: business or technical information of any sort. To benefit from trade secret protection, the information must be a secret. However, only relative and not absolute secrecy is required. In addition, the owner of a trade secret must take reasonable steps to maintain its secrecy. Trade secrets have no definite term of protection but may be protected only as long as they are secret. Once a trade secret is disclosed, protection is lost.

There is no state agency in charge of "issuing" (or even registering) secrets. Rather, any information that meets the above criteria can be protected. Courts will find misappropriation of trade secrets in two circumstances: where the secrets were obtained by theft or other improper means, or where they were used or disclosed by the defendant in violation of a confidential relationship. However, trade secret laws do not protect against independent discovery or invention. Nor do they prevent competitors from "reverse engineering" a legally obtained product in order to determine the secrets contained inside. Violations of trade secret law entitle the owner to damages and in some cases injunctions against use or further disclosure.

2. Patent

Patent law is the classic example of an intellectual property regime modeled on the utilitarian framework. Following the constitutional authorization, patent law creates a limited monopoly to encourage the production of inventions — processes, machines, and compositions of matter. The public benefits directly through the spur to innovation and disclosure of the patented invention. After the term of the patent expires, the innovation becomes part of the public domain, freely available to all.

To obtain a utility patent, an inventor must submit an application to the Patent and Trademark Office (PTO) that meets four requirements: patentable subject matter, novelty, non-obviousness, and usefulness. The patentee must also disclose the innovation to the public in a way that would enable others to make and use the invention. While the threshold for usefulness is low, the novelty and non-obviousness standards are exacting, and the PTO conducts an independent review of the application to ensure that it meets these requirements. If the PTO grants the patent, the inventor obtains exclusive rights to make, use, and sell the innovation for a term of up to 20 years. The patent grant is nearly absolute, barring even those who independently develop the invention from practicing its art.

The PTO also issues plant patents for distinctive plants and design patents for ornamental designs for articles of manufacture. Design patents have a term of 14 years.

3. Copyright

Although the copyright and patent laws flow from the same constitutional basis and share the same general approach — statutorily created monopolies to foster progress — they feature different elements and rights, reflecting the very different fields of creativity that they seek to encourage. In general, copyrights are easier to secure and last substantially longer than patents, although the scope of protection afforded copyrights is narrower and less absolute than that given to patents.

Copyright law covers the broad range of literary and artistic expression — including books, poetry, song, dance, dramatic works, computer programs,

TABLE 1-1
Principal Modes of Legal Protection for Intellectual Work

	Trade secret	Patent	Copyright	Trademark/dress
Underlying Theory	Freedom of contract; protection against unfair means of competition	Limited monopoly to encourage production of utilitarian works in exchange for immediate disclosure and ultimate enrichment of the public domain	Limited (although relatively long-lived) monopoly to encourage the authorship of expressive works; developed initially as a means of promoting publishing	Perpetual protection for distinctive nonfunctional names and dress in order to improve the quality of information in the marketplace
Source of Law	State statute (e.g., Uniform Trade Secrets Act); common law	Patent Act (federal)	Copyright Act (federal); common law (limited)	Lanham Act (federal); common law (unfair competition)
Subject Matter	Formula, pattern, compilation, program, device, method, technique, process	Process, machine, manufacture, or composition of matter; plants (asexually reproducing); designs—*excluding*: laws of nature, natural substances, business methods, printed matter (forms), mental steps	Literary, musical, choreographic, dramatic and artistic works *limited by* idea/ expression dichotomy (no protection for ideas, systems, methods, procedures); no protection for facts/research	Trademarks; service marks; certification marks (e.g., Good Housekeeping); collective marks (e.g., Toy Manufacturers of America); trade dress (§43(a)); *no protection for* functional features, descriptive terms, geographic names, misleading aspects, or "generic" names (e.g., thermos)
Standard for Protection	Information not generally known or available; reasonable efforts to maintain secrecy; commercial value	Novelty; non-obviousness; and utility (distinctiveness for plant patents; ornamentality for design patents)	Originality; authorship; fixation in a tangible medium	Distinctiveness; secondary meaning (for descriptive and geographic marks); use in commerce (minimal); famous mark (for dilution cases)

	Trade Secret	Patent	Copyright	Trademark
Scope of Protection	Protection against misappropriation—acquisition by improper means or authorized disclosure	Exclusive rights to make, use, sell innovation as limited by contribution to art; extends to "equivalents"	Rights of performance, display, reproduction, derivative works	Exclusive rights in U.S.; likelihood of confusion; false designation of origin (§43(a)); dilution (for famous marks)
Period of Protection	Until becomes public knowledge	20 years from filing (utility); extensions up to 5 years for drugs, medical devices and additives; 14 years (design)	Life of author + 70 years; "works for hire": minimum of 95 years after publication or 120 years after creation	Perpetual, subject to abandonment
Disclosure	Loss of protection (unless sub rosa)	Right to patent lost if inventor delays too long after publishing before filing application; full disclosure is required as part of application; notice of patent required for damages	© notice and publication no longer required, but confer certain benefits	® notice optional; establishes prima facie evidence of validity, constructive knowledge of registration, confers federal jurisdiction, becomes incontestable after 5 years of continuous use, authorizes treble damages and attorney fees, and right to bar imports bearing infringing mark
Rights of Others	Independent discovery; reverse engineering	Only if licensed; can request reexamination of patent by Patent and Trademark Office	Fair use; compulsory licensing for musical compositions, cable TV, et al.; independent creation	Truthful reflection of source of product; fair and collateral use (e.g., comment)
Costs of Protection	Security expenses; personnel dissatisfaction; litigation costs	Filing, issue, and maintenance fees; litigation costs	None (protection attaches at fixation); publication requires notice; suit requires registration; litigation costs	Registration search; marking product (optional—see above); litigation costs

TABLE 1-1
(*Continued*)

	Trade secret	*Patent*	*Copyright*	*Trademark/dress*
Licensing and Assignment	Discouraged by inherent nature of bargaining (seller wants guarantee before disclosure; buyer wants to know what is offered)	Encouraged by completeness of property rights, subject to antitrust constraints	Assignor has termination right between 36th and 41st years (of notice given)	No naked licenses (owner must monitor licensee); no sales of trademark "in gross"
Remedies	Civil suit for misappropriation; conversion, unjust enrichment, breach of contract; damages (potentially treble) and injunctive relief; criminal prosecution for theft	Injunctive relief and damages (potentially treble); attorney fees (in exceptional cases)	Injunction against further infringement; destruction of infringing articles; damages (actual or profits); statutory ($200–$100,000 damages within court's discretion); attorney fees (within court's discretion); criminal prosecution	Injunction; accounting for profits; damages (potentially treble); attorney fees (in exceptional cases); seizure and destruction of infringing goods; criminal prosecution for trafficking in counterfeit goods or services

movies, sculpture, and paintings. Ideas themselves are not copyrightable, but the author's particular expression of an idea is protectable. A work must exhibit a modicum of originality and be fixed in a "tangible medium of expression" to receive protection. Copyright protection attaches as soon as a work is fixed. There is no examination by a governmental authority, although the Copyright Office registers copyrightable works. Such registration is no longer required for validity, but U.S. authors must register their works prior to filing an infringement suit. A copyright lasts for the life of the author plus 70 years, or a total of 95 years in the case of entity authors.

The breadth and ease of acquisition of copyright protection are balanced by the more limited rights that copyright law confers. Ownership of a valid copyright protects a copyright holder from unauthorized copying, public performance, and display, and it entitles the holder to make derivative works and to control sale and distribution of the work. These rights, however, are limited in a number of ways. Others may make "fair use" of the material in certain circumstances. The Copyright Act also establishes compulsory licensing for musical compositions and cable television.

Copyright law protects only against *copying* of protected expression. Independent creation of a copyrighted work does not violate the Copyright Act, nor does copying the unprotected elements of a work. Therefore, copyright law must have some mechanism for determining when a work has been copied illegally. While in rare cases direct proof of copying may be available, usually it is not. In its place, courts infer copying from proof that the defendant has had *access* to the plaintiff's work combined with evidence that the two works are *similar*. Even if copying is established, it must be further shown that the defendant's work is *substantially similar* to protected elements (e.g., excluding ideas) of the plaintiff's work.

4. Trademark/Trade Dress

Trademarks are also protected by federal statute, although the source of constitutional authority is different from that of the Patent and Copyright Acts. Rather than deriving from a specific grant of constitutional power, federal power to regulate trademarks and unfair competition stems from the Commerce Clause of the Constitution, which authorizes Congress to regulate interstate commerce. Unlike patent and copyright protection, trademark law did not evolve from a desire to stimulate particular types of economic activity. Rather, its original purpose was to protect consumers in a world of mass merchandising from unscrupulous sellers attempting to fly under the banner of someone else's well-known logo or identifying symbol. Only in recent years has trademark law begun to embrace the incentive, personhood, and natural rights rationales.

The Lanham Act (the federal trademark statute) protects words, symbols, and other attributes that serve to identify the nature and source of goods or services. Examples of marks protectable under the Lanham Act include corporate and product names, symbols, logos, slogans, pictures and designs, product configurations, colors, and even smells. Not all such marks are protectable, however. To receive trademark protection, an identifying mark need not be new or previously unused, but it must represent to consumers the source of the good or service identified. It cannot be merely a description of the good itself or a generic term for the class of goods or services offered. Further, the identifying mark may not be a functional element of the product itself but must serve a purely identifying purpose. Since 1996, famous marks also receive federal protection against "dilution" by blurring or tarnishment. Finally, trademark protection is directly tied to the use of the mark to identify goods in commerce. Trademarks do not expire on any particular date but continue in force until they are "abandoned" by their owner or become unprotectable.

The PTO examines trademark applications and issues trademark registrations that confer significant benefits upon the registrants, including: prima facie evidence of validity; constructive notice to others of the claim of ownership; federal subject matter jurisdiction; incontestability after five years, which confers exclusive right to use the mark; authorization to seek treble damages and attorney fees; and the right to bar importation of goods bearing the infringing mark. Federal trademark registration, however, is not necessary to obtain trademark protection. A trademark owner who believes that another is using the same or a similar mark to identify competing goods can bring suit for trademark infringement. Unlike patent and copyright law, the outcome does not turn on the similarity between the marks or on whether the defendant copied the mark from the trademark owner. Rather, infringement turns on whether consumers are likely to be confused as to the origin of the goods or services. If so, the trademark owner is entitled to an injunction against the confusing use, damages for past infringement, and in some cases the seizure and destruction of infringing goods.

PROBLEM

Problem 1-3.

MEMORANDUM

To: Associate
From: Senior Partner
Re: HEALTHWARE Inc.

Janet Peterson called me yesterday about a new venture that she plans to try to get off the ground. As you may know, Janet is a computer programmer and a registered nurse. She has an interesting idea for a new venture and would like our advice on how she might structure the business so as to have the best potential for success.

The proposed venture will be called Healthware. Janet believes that she can tap into the current diet/health/environmental/personal computer craze by developing a user-friendly computer program that would monitor the user's diet and fitness activity. The user would input information on his or her health (e.g., age, weight, medical history, dietary restrictions). Each day, the user would input information on diet and physical exercise. The program would have simple "pull down" menus for making this quick and easy; nutritional information on all foods would be stored in the program. The computer would periodically provide an analysis of the user's health, as well as suggestions for achieving the user's goals, whether weight reduction, better fitness, or general health. In addition, the program would compile a record of the user's activities which could be brought to annual physicals. Other subroutines would be available for pregnant and lactating women, children, the elderly, diabetics, vegetarians, triatheletes, etc.

Janet thinks that she could put together the diverse people necessary to pull this project off: programmers, a nutritionist, a physician, a fitness consultant. She is concerned, however, that any one of these people could, after they are familiar with the product, develop a competing program.

What are the options for structuring Healthware? What problems do you foresee in structuring this venture? Assuming that the product is popular, what are the major risks to Healthware's success? How can we structure Healthware so as to overcome these problems?

2

Trade Secret Protection

A. INTRODUCTION

1. History

The idea that information should be protected against "theft" (which may include the physical taking of tangible goods containing information or simply the copying or memorization of data) is a venerable one in the law. One scholar traces the earliest legal protection against "misappropriation of trade secrets" to the Roman empire.[1] The Roman courts created a cause of action called "actio servi corrupti" — literally, an action for corrupting a slave.[2] According to Schiller, the actio servi corrupti was used to protect slave owners from third parties who would "corrupt" slaves (by bribery or intim-

---

1. Arguably, trade secrets existed before this time, albeit in unusual forms. Consider Mark C. Suchman, Invention and Ritual: Notes on the Interrelation of Magic and Intellectual Property in Preliterate Societies, 89 Colum. L. Rev. 1264, 1274 (1989):

> [L]et us imagine a hypothetical preliterate inventor who, through diligence or good fortune, discovers that her maize crop is larger when she plants a small fish next to each kernel of corn. . . . Clearly, this technique has economic value and could garner its creator material and social rewards if she could monopolize it and license it for a fee. Unfortunately, the odds of keeping such a discovery secret are slight. . . . [T]he procedure is so simple that even a casual observer could replicate the process. . . . Magic, however, provides a way out of the dilemma. By claiming, for example, that the power of the fish is activated by a talisman that she alone possesses, the inventor can remove her idea from the public domain. . . . [T]he magicked process is far easier to monopolize than the simple technology alone.

2. For a discussion of the Roman law, see A. Arthur Schiller, Trade Secrets and the Roman Law: The Actio Servi Corrupti, 30 Colum. L. Rev. 837 (1930). For a dissenting view on the role of Roman law in protecting trade secrets, see Alan Watson, Trade Secrets and Roman Law: The Myth Exploded, 11 Tul. Eur. & Civ. L.F. 19 (1996).

idation) into disclosing their owners' confidential business information. The law made such third parties liable to the slave owner for twice the damages he suffered as a result of the disclosure.

While recent scholarship has cast some doubt on the enforcement of trade secret protection in the Roman empire, the concept that so-called business or "trade secrets" were entitled to legal protection spread rapidly throughout the world. As early as the Renaissance, most European nation-states had laws that protected businesses (notably, the Guild cartels) from those who used their secret processes and ideas without permission. These early laws were translated during the Industrial Revolution into statutes that protected "industrial secrets." Many of these statutes are still in force today, albeit in modified form.

The roots of trade secrecy in slavery law were further evident in the treatment of employees in the centuries before the Industrial Revolution. Both commerce and foreign policy included a strong dose of "mercantilism." Governments and private guilds attempted to keep "their" intellectual property within their grasp, using a combination of rewards to inventors and rules that reduced employee mobility. These developments are well traced in Carlo M. Cipolla, Before the Industrial Revolution: European Economy and Society 1000-1700 (2d ed. 1980); David J. Jeremy, Transatlantic Industrial Revolution: The Diffusion of Textile Technologies Between Britain and America, 1790-1830's, at 185-189 (MIT Press 1981). The authors provide such examples as restrictive British secrecy laws, city rewards to woollen craftsmen in thirteenth-century Bologna and, on the other side, the kidnapping of skilled Swedish ironworkers by France in 1660.

This patchwork of public and private regulation arguably served as a substitute for strong intellectual property protection by protecting a country's or a corporation's "human capital" and so encouraging investment in employee skills and training. The guild system of apprenticeship and mandatory periods of servitude operated to prevent excessive "spillovers" of guild knowledge to competitors and hence may have encouraged innovation. On this point, compare Paul A. David, Intellectual Property and the Panda's Thumb: Patents, Copyrights, and Trade Secrets in Economic Theory and History, in Global Dimensions of Intellectual Property Rights in Science and Technology 19, 45 (Natl. Academy Press 1993) (guild system encouraged investment in innovation) with Cipolla, supra, at 261 (guilds restricted competition and therefore restrained business and technological progress). In effect, guilds and countries solved the public goods problem by doing their best to privatize the good — keeping information secret from prying eyes.

Unfortunately, this form of private protection comes at a price. Companies (or countries) that must rely on current and former employees to keep secrets run into a host of problems. Companies may not disclose to employees all the information they need to know. They may take inefficient physical security precautions. They may hire employees based on their loyalty rather than their productivity. And finally, they may simply choose to prey on other companies rather than investing in innovation themselves. Robert Sherwood makes the case that many of these problems occurred in the recent past in Mexico and Brazil as a result of inadequate trade secret protection. See Robert M. Sherwood, Intellectual Property and Economic Development 113 (1990).

Obviously, it would be preferable to have a legal system that could protect an employer's secrets while allowing employee mobility and the dissemination of information through licensing. And such rules have generally developed as legal systems and norms became stable enough to support them. European countries, for example, developed trade secret protection quite early.[3]

By contrast, Anglo-American jurisprudence was a relative latecomer to the protection of trade secrets. English and American courts first recognized a cause of action for damages for misappropriation of trade secrets in 1817 and 1837, respectively;[4] injunctive relief against actual or threatened misappropriation came still later.[5] These early decisions concerned issues that are still debated in trade secret cases today: the circumstances in which an employee may continue her business after departing her employer, the circumstances in which a competitor may copy another's publicly sold product, and whether courts will enforce a contract requiring that business information be kept confidential.

Although certain fact patterns have not changed much since the nineteenth century, the overall incidence of trade secret litigation has. A recent study of 530 manufacturing firms in Massachusetts, for example, found that 43 percent of the intellectual property disputes involving these firms had a trade secret component. Josh Lerner, The Importance of Trade Secrecy: Evidence from Civil Litigation, Harv. Bus. School working paper #95-043 (Dec. 1994). Indeed, practicing lawyers often note that trade secret claims crop up more frequently in litigation than any other form of intellectual property.

Trade secret litigation is particularly important to small companies. The study by Lerner found that small firms spent a greater proportion of their litigation resources on trade secret cases than did large firms, perhaps because of the enormous cost associated with patent litigation. Lerner's data indicate that trade secrets, though important to all firms, are absolutely crucial for the small companies that drive innovation in many developing fields.

---

3. Interestingly, Japan has only recently enacted a trade secret protection statute. See Hideo Nakoshi, New Japanese Trade Secret Act, 75 J. Pat. & Trademark Off. Society 631 (1993). Nakoshi provides some interesting thoughts on why Japan considered such a trade secret statute unnecessary for so long. He argues that cultural norms of assumed trust made it socially uncomfortable to insist on confidentiality, and that long-term or lifetime employment eliminated many trade secret issues associated with employee mobility.

4. See Newberry v. James, 35 Eng. Rep. 1011, 1013 (Ct. Ch. 1817); Vickey v. Welch, 36 Mass. 523, 527 (1837).

5. See Yovett v. Winyard, 37 Eng. Rep. 525 (Ct. Ch. 1820); Taylor v. Blanchard, 95 Mass. 370 (1866).

2. Overview of Trade Secret Protection

Today, every one of the United States protects trade secrets in some form or another. Improper use or disclosure of a trade secret is generally a common law tort. In this century, the principal document setting forth the "law" of trade secrets was the Restatement of Torts, published in 1939. Sections 757 and 758 of the Restatement set forth basic principles of trade secret law that were adopted by courts in most states in the country. National protection for trade secrets is also compelled by United States adherence to the intellectual property agreement in the Uruguay Round of the General Agreement on Tariffs and Trade (GATT) Trade-Related Aspects of Intellectual Property Rights (TRIPs) in 1994. GATT TRIPs article 39 requires member nations to protect trade secrets.

The Restatement protected as a trade secret any information "used in one's business" that gives its owner "an opportunity to obtain an advantage over competitors who do not know or use it," so long as the information was in fact a secret. Restatement §757, Comment *b*.

When the Restatement (Second) of Torts was published in 1979, the authors decided not to include sections 757 and 758, on the grounds that the law of trade secrets had developed into an independent body of law that no longer relied on general principles of tort law. The influence of the original Restatement remained, however, because it had been adopted by so many courts.[6]

Beginning in 1979, the National Conference of Commissioners on Uniform State Laws promulgated a model state statute, the Uniform Trade Secrets Act, which differed in some respects from the common law. The Uniform Trade Secrets Act (UTSA) has now been enacted (in one form or another) by 40 states and the District of Columbia. See Restatement (Third) of Unfair Competition §39, at 437-438 (listing state statutes). Because of the Uniform Act's importance, we reproduce its primary provisions here.

Uniform Trade Secrets Act, with 1985 Amendments

§1. Definitions
     As used in this [Act], unless the context requires otherwise:
          (1) "Improper means" includes theft, bribery, misrepresentation, breach or inducement of a breach of a duty to maintain secrecy, or espionage through electronic or other means;
          (2) "Misappropriation" means:
               (i) acquisition of a trade secret of another by a person who knows or has reason to know that the trade secret was acquired by improper means; or

---

6. The authors of the Restatement (Third) of Unfair Competition, published in 1994, once again included a section on the law of trade secrets. The section is organized to follow the Uniform Act, and many of its provisions are consistent with the act. Nonetheless, there are some differences between the act and the Restatement. They are discussed below.

(ii) disclosure or use of a trade secret of another without express or implied consent by a person who

(A) used improper means to acquire knowledge of the trade secret; or

(B) at the time of disclosure or use, knew or had reason to know that his knowledge of the trade secret was

(I) derived from or through a person who had utilized improper means to acquire it;

(II) acquired under circumstances giving rise to a duty to maintain its secrecy or limit its use; or

(III) derived from or through a person who owed a duty to the person seeking relief to maintain its secrecy or limit its use; or

(C) before a material change of his [or her] position, knew or had reason to know that it was a trade secret and that knowledge of it had been acquired by accident or mistake. . . .

(4) "Trade secret" means information, including a formula, pattern, compilation, program, device, method, technique, or process, that:

(i) derives independent economic value, actual or potential, from not being generally known to, and not being readily ascertainable by proper means by,[7] other persons who can obtain economic value from its disclosure or use, and

(ii) is the subject of efforts that are reasonable under the circumstances to maintain its secrecy.

A trade secret claim can be broken down into three essential elements. First, the subject matter involved must qualify for trade secret protection; it must be the type of knowledge or information that trade secret law was meant to protect, and it must not be generally known to all. On eligible subject matter, the current trend, exemplified once again by the UTSA, is to protect as a trade secret *any* valuable information. So long as the information is capable of adding economic value to the plaintiff, it can be protected by trade secret law. The requirement that the information not be generally known follows from the label *secret*. The requirement is meant to insure that no one claims intellectual property protection for information commonly known in a trade or industry.

Second, a trade secret plaintiff must also prove that the defendant acquired the information wrongfully — in a word, that the defendant *misappropriated* the trade secret. Just because a person's information is valuable does not make it wrong for another to use it or disclose it. But use or disclo-

---

7. Some states, including California, do not include the "readily ascertainable" language in the definition of a trade secret. See Cal. Civil Code §3426.1. — EDS.

sure is wrong, in the eyes of trade secret law, when the information is acquired through deception, skulduggery, or outright theft. As we will see in the cases that follow, close cases abound in this area because of the creativity of competitors in rooting out information about their rivals' businesses and products.

In many cases a defendant's use or disclosure is wrongful because of a preexisting obligation to the plaintiff not to disclose or appropriate the trade secret. Such an obligation can arise in either of two ways: explicitly, by contract; and implicitly, because of an *implied duty*. A classic example of an implied duty is the case of an employee. Even in the absence of an explicit contract, most employees are held to have a duty to protect their employers' interests in the employers' secret practices, information, and the like. Even where the duty arises by explicit contract, however, public policy limitations on the scope and duration of the agreement will often come into play, in some cases resulting in substantial judicial modification of the explicit obligations laid out in the contract.

The third element to be established by the plaintiff in a trade secret case is that the plaintiff, holder of the trade secret, took *reasonable precautions* under the circumstances to prevent its disclosure. Courts have shown some confusion over the rationale for this requirement. Some see in it evidence that the trade secret is valuable enough to bother litigating; others argue that where reasonable precautions are taken, chances are that a defendant acquired the trade secret wrongfully. Whatever the justification, it is clear that no one may let information about products and operations flow freely to competitors at one time and then later claim that competitors have wrongfully acquired valuable trade secrets. To establish the right to sue later, one must be consistently diligent in protecting information. As always, however, the presence of the term "reasonable" assures close cases and difficult line-drawing for courts, a theme reflected in several of the cases that follow.

3. Theory of Trade Secrets

Legal protection for trade secrets is premised primarily on two theories that are only partly complementary. The first is utilitarian. Under this view, protecting against the theft of proprietary information encourages investment in such information. This idea is sometimes associated with the view that trade secrets are a form of property. The second theory emphasizes deterrence of wrongful acts and is therefore sometimes described as a tort theory. Here the aim of trade secret law is to punish and prevent illicit behavior, and even to uphold reasonable standards of commercial behavior. Cf. Kim Lane Scheppele, Legal Secrets: Equality and Efficiency in the Common Law (1988) (arguing that cases involving legal secrets — including trade secrets cases — are better explained in terms of principles all would be willing to agree to rather than in the efficiency terms of law and economics). Although under the tort theory trade secret protection is not explicitly about encouraging investments, it is plain that one consequence of deterring wrongful behavior would be to encourage investment in trade secrets. Hence, despite their conceptual differences, the tort and property/incentive approaches to trade secrets may well push in the same direction.

Property Rights. Some jurists have conceptualized "intellectual property" as a species of the broader concept of "property." The Supreme Court adopted this view of trade secret law in Ruckelshaus v. Monsanto Co., 467 U.S. 986, 1001-1004 (1984). There the Court faced the question (considered in more detail below) of whether a federal law that required Monsanto to publicly disclose its trade secrets was a "taking of private property" for which the Fifth Amendment required compensation. The Court, in finding that trade secrets were "property," reasoned in part that "[t]rade secrets have many of the characteristics of more tangible forms of property. A trade secret is assignable. A trade secret can form the res of a trust, and it passes to a trustee in bankruptcy." Id. at 1002-1004. Treatment of trade secrets as property rights vested in the trade secret "owner" is consistent with a view of trade secrets law as providing an additional incentive to innovate, beyond those provided in patent law. The Supreme Court has offered some support for this view as well, in cases such as Kewanee Oil Co. v. Bicron Corp., 416 U.S. 470, 481-485 (1974).[8]

Tort Law. A powerful alternate explanation for much of trade secrets law is what might be described as a "duty-based" theory, or what Melvin Jager calls "the maintenance of commercial morality." 1 Melvin Jager, Trade Secrets Law §1.03, at 1-4. The Supreme Court adopted this view in a famous early decision in which (unlike *Monsanto*) the Court was actually called upon to construe the trade secret laws:

> The word "property" as applied to trademarks and trade secrets is an unanalyzed expression of certain secondary consequences of the primary fact that the law makes some rudimentary requirements of good faith. Whether the plaintiffs have any valuable secret or not, the defendant knows the facts, whatever they are, through a special confidence that he accepted. The property may be denied, but the confidence cannot be. Therefore the starting point for the present matter is not property or due process of law, but that the defendant stood in confidential relations with the plaintiffs. . . .

---

8. In *Kewanee*, the Court held that:

Certainly the patent policy of encouraging invention is not disturbed by the existence of another form of incentive to invention. In this respect the two systems are not and never would be in conflict. . . .

Trade secret law will encourage invention in areas where patent law does not reach, and will prompt the independent innovator to proceed with the discovery and exploitation of his invention. Competition is fostered and the public is not deprived of the use of valuable, if not quite patentable, invention.

E.I. du Pont & Co. v. Masland, 244 U.S. 100, 102 (1917).[9] Closely related to *Masland*'s theory of "breach of confidence" is the contract basis for trade secret law. While not always applicable, many trade secret cases arise out of a "duty" explicitly stated in a contract, such as a technology license or an employment agreement. The tort-based theory of breach of duty merges in those cases with a standard common law action for breach of contract.

For an argument that there is no theoretical justification for trade secret law, and that other contract and tort doctrines can do just fine in protecting confidential information, see Robert G. Bone, A New Look at Trade Secret Law: Doctrine in Search of Justification, 86 Calif. L. Rev. 241 (1998).

———————

The different theories of trade secret law reflect different views of the nature of intellectual property law as a whole, discussed in Chapter 1. These debates recur throughout this book, because the reason that we protect intellectual property turns out to matter a great deal in deciding whether and to what extent we protect it.

You need not decide at the outset how or why (or even whether) trade secrets should be protected. Rather, keep in mind the different rationales for trade secret protection as you read the material that follows. Consider the extent to which each of these theories can explain the law of trade secrets.

B. SUBJECT MATTER

1. Defining Trade Secrets

≡ *Metallurgical Industries Inc. v. Fourtek, Inc.*
United States Court of Appeals for the Fifth Circuit
790 F.2d 1195 (5th Cir. 1986)

GEE, Circuit Judge:
Today's case requires us to review Texas law on the misappropriation of trade secrets. Having done so, we conclude that the district court misconceived the nature and elements of this cause of action, a misconception that led it to direct a verdict erroneously in favor of appellee Bielefeldt. We also

———————

9. The *Monsanto* Court attempted to distinguish *Masland* in a footnote, claiming that "Justice Holmes did not deny the existence of a property interest; he simply deemed determination of the existence of that interest irrelevant to the resolution of the case." *Monsanto*, 467 U.S. 1004 n.9.

conclude that the court abused its discretion in excluding certain evidence. Accordingly, we affirm in part, reverse in part, and remand the case for a new trial.

I. Facts of the Case

We commence with a brief description of the scientific process concerned. Tungsten carbide is a metallic compound of great value in certain industrial processes. Combined with the metal cobalt, it forms an extremely hard alloy known as "cemented tungsten carbide" used in oil drills, tools for manufacturing metals, and wear-resistent coatings. Because of its great value, reclamation of carbide from scrap metals is feasible. For a long time, however, the alloy's extreme resistance to machining made reclamation difficult. In the late 1960's and early 1970's, a new solution — known as the zinc recovery process — was devised, a solution based on carbide's reaction with zinc at high temperatures. In the crucibles of a furnace, molten zinc will react with the cobalt in the carbide to cause swelling and cracking of the scrap metal. After this has occurred, the zinc is distilled from the crucible, leaving the scrap in a more brittle state. The carbide is then ground into a powder, usable in new products as an alternative to virgin carbide. This process is the generally recognized modern method of carbide reclamation.

Metallurgical Industries has been in the business of reclaiming carbide since 1976, using the more primitive "cold-stream process." In the mid-1970's, Metallurgical began to consider using the zinc recovery process. In that connection, it came to know appellee Irvin Bielefeldt, a representative of Therm-O-Vac Engineering & Manufacturing Company (Therm-O-Vac). Negotiations led to a contract authorizing Therm-O-Vac to design and construct two zinc recovery furnaces, the purchase order for the first being executed in July 1976.

The furnace arrived in April 1977. Dissatisfied with its performance, Metallurgical modified it extensively. First, it inserted chill plates in one part of the furnace to create a better temperature differential for distilling the zinc. Second, Metallurgical replaced the one large crucible then in place with several smaller crucibles to prevent the zinc from dispersing in the furnace. Third, it replaced segmented heating elements which had caused electric arcing with unitary graphite heating elements. Last, it installed a filter in the furnace's vacuum-pumps, which zinc particles had continually clogged. These efforts proved successful and the modified furnace soon began commercial operation.

. . . [M]etallurgical returned to Therm-O-Vac for its second furnace. A purchase order was signed in January 1979, and the furnace arrived that July. Further modifications again had to be made, but commercial production was allegedly achieved in January 1980.

In 1980, after Therm-O-Vac went bankrupt, Bielefeldt and three other former Therm-O-Vac employees — Norman Montesino, Gary Boehm, and Michael Sarvadi — formed Fourtek, Incorporated. Soon thereafter, Fourtek agreed to build a zinc recovery furnace for appellee Smith International, Incorporated (Smith). . . .

Metallurgical . . . brought a diversity action against Smith, Bielefeldt, Montesino, Boehm, and Sarvadi in November 1981. In its complaint, Metallurgical charged the defendants with misappropriating its trade secrets. . . . [Trial] testimony indicated Metallurgical's frequent notices to Bielefeldt that the process was a secret and that the disclosures to him were made in confidence. Another witness recounted meetings in which the modifications were agreed to. Bielefeldt was allegedly unconvinced about the efficacy of these changes and contributed little to the discussion. Metallurgical also presented evidence that it had expended considerable time, effort, and money to modify the furnaces.

[The district court nonetheless granted defendants' motions for directed verdicts.] The principal reason advanced was the court's conclusion that no trade secret is involved. At trial, Metallurgical acknowledged that the individual changes, by themselves, are not secrets; chill plates and pump filters, for example, are well-known. Metallurgical's position instead was that the process, taken as a whole, is a trade secret in the carbide business. The court, however, refused to recognize any protection Texas law provides to a modification process. It also concluded that the information Bielefeldt obtained from working with Metallurgical is too general to be legally protected. Finally, it ruled that "negative know-how" — the knowledge of what not to do — is unprotected. . . .

III. Defining a "Trade Secret"

We begin by reviewing the legal definition of a trade secret. Of course, to qualify as one, the subject matter involved must, in fact, be a secret; "matters of general knowledge in an industry cannot be appropriated by one as his secret." Wissman v. Boucher, 150 Tex. 326, 240 S.W.2d 278, 280 (Tex. 1951); see also Zoecon Industries v. American Stockman Tag Co., 713 F.2d 1174, 1179 (5th Cir. 1983) ("a customer list of readily ascertainable names and addresses will not be protected as a trade secret"). Smith emphasizes the absence of any secret because the basic zinc recovery process has been publicized in the trade. Acknowledging the publicity of the zinc recovery process, however, we nevertheless conclude that Metallurgical's particular modification efforts can be as yet unknown to the industry. A general description of the zinc recovery process reveals nothing about the benefits unitary heating elements and vacuum pump filters can provide to that procedure. That the scientific principles involved are generally known does not necessarily refute Metallurgical's claim of trade secrets.

Metallurgical, furthermore, presented evidence to back up its claim. One of its main witnesses was Arnold Blum, a consultant very influential in the decisions to modify the furnaces. Blum testified as to his belief that Metallurgical's changes were unknown in the carbide reclamation industry. The evidence also shows Metallurgical's efforts to keep secret its modifications. Blum testified that he noted security measures taken to conceal the furnaces from all but authorized personnel. The furnaces were in areas hidden from public view, while signs warned all about restricted access. Company policy, moreover, required everyone authorized to see the furnace to sign a non-disclosure agreement. These measures constitute evidence probative of the existence of secrets. One's subjective belief of a secret's existence suggests that the secret exists. Security measures, after all, cost money; a manufacturer therefore presumably would not incur these costs if it believed its competitors already knew about the information involved. In University Computing Co. v. Lykes-Youngstown Corp., 504 F.2d 518, 535 (5th Cir. 1974), we regarded subjective belief as a factor to consider in determining whether secrecy exists. Because evidence of security measures is relevant, that shown here helps us conclude that a reasonable jury could have found the existence of the requisite secrecy.

Smith argues, however, that Metallurgical's disclosure to other parties vitiated the secrecy required to obtain legal protection. As mentioned before, Metallurgical revealed its information to Consarc Corporation in 1978; it also disclosed information in 1980 to La Floridienne, its European licensee of carbide reclamation technology. Because both these disclosures occurred before Bielefeldt allegedly misappropriated the knowledge of modifications, others knew of the information when the Smith furnace was built. This being so, Smith argues, no trade secret in fact existed.

Although the law requires secrecy, it need not be absolute. Public revelation would, of course, dispel all secrecy, but the holder of a secret need not remain totally silent:

> He may, without losing his protection, communicate to employees involved in its use. He may likewise communicate it to others pledged to secrecy. . . . Nevertheless, a substantial element of secrecy must exist, so that except by the use of improper means, there would be difficulty in acquiring the information.

Restatement of Torts, §757 Comment b (1939). We conclude that a holder may divulge his information to a limited extent without destroying its status as a trade secret. To hold otherwise would greatly limit the holder's ability to profit from his secret. If disclosure to others is made to further the holder's economic interests, it should, in appropriate circumstances, be considered a limited disclosure that does not destroy the requisite secrecy. The only question is whether we are dealing with a limited disclosure here. . . .

Looking . . . to the policy considerations involved, we glean two reasons why Metallurgical's disclosures to others are limited and therefore insufficient to extinguish the secrecy Metallurgical's other evidence has suggested. First, the disclosures were not public announcements; rather, Metallurgical divulged its information to only two businesses with whom it was dealing. This case thus differs from Luccous v. J.C. Kinley Co., 376 S.W.2d 336 (Tex. 1964), in which the court concluded that the design of a device could not be a trade secret because it had been patented — and thus revealed to all the world — before any dealing between the parties. Second, the disclosures were made to further Metallurgical's economic interests. Disclosure to Consarc was made with the hope that Consarc could build the second furnace. A long-standing agreement gave La Floridienne the right, as a licensee, to the information in exchange for royalty payments. Metallurgical therefore revealed its discoveries as part of business transactions by which it expected to profit.

Metallurgical's case would have been stronger had it also presented evidence of confidential relationships with these two companies, but we are unwilling to regard this failure as conclusively disproving the limited nature of the disclosures. Smith correctly points out that Metallurgical bears the burden of showing the existence of confidential relationships. Contrary to Smith's assertion, however, confidentiality is not a requisite; it is only a factor to consider. Whether a disclosure is limited is an issue the resolution of which depends on weighing many facts. The inference from those facts, construed favorably to Metallurgical, is that it wished only to profit from [disclosure of] its secrets to the public. We therefore are unpersuaded by Smith's argument.

Existing law, however, emphasizes other requisites for legal recognition of a trade secret. In Huffines, 314 S.W.2d 763, a seminal case of trade secret law, Texas adopted the widely-recognized pronouncements of the American Law Institute's Restatement of the Law. The Texas Supreme Court quoted the Restatement's definition of a trade secret:

> A trade secret may consist of any formula, pattern, device, or compilation of information which is used in one's business, and which gives him an opportunity to obtain an advantage over competitors who do not know or use it. It may be a chemical compound, a process of manufacturing, treating or preserving materials, a pattern for a machine or other device or a list of customers.

Id. at 776, quoting Restatement of Torts, §757 Comment b (1939). From this the criterion of value to the holder of the alleged secret arises. . . .

Metallurgical met the burden of showing the value of its modifications. Lawrence Lorman, the company's vice president, testified that the zinc recovery process gave Metallurgical an advantage over its two competitors by aiding in the production of the highest quality reclaimed carbide powder. The quality of the powder, in fact, makes it an alternative to the more costly virgin carbide. Lorman testified that customers regarded Metallurgical's zinc reclaimed powder as a better product than that reclaimed by the coldstream process used by others. This evidence clearly indicates that the modifications that led to the commercial operation of the zinc recovery furnace provided a clear advantage over the competition.

Another requisite is the cost of developing the secret device or process. In Huffines' companion case, K&G Oil, Tool & Service Co. v. G&G Fishing Tool Service, 158 Tex. 594, 314 S.W.2d 782, 790, 117 U.S.P.Q. (BNA) 471 (Tex. 1958), the court recognized the cost involved in developing the device in question; "the record shows . . . that much work and ingenuity have been applied to the development of a practical and successful device." See also Zoecon Industries, 713 F.2d at 1179 ("even if the names and addresses were readily ascertainable through trade journals as the defendants allege, the other information could be compiled only at considerable expense"). No question exists that Metallurgical expended much time, effort, and money to make the necessary changes. It clearly has met the burden of demonstrating the effort involved in making a complex manufacturing process work.

That the cost of devising the secret and the value the secret provides are criteria in the legal formulation of a trade secret shows the equitable underpinnings of this area of the law. It seems only fair that one should be able to keep and enjoy the fruits of his labor. If a businessman has worked hard, has used his imagination, and has taken bold steps to gain an advantage over his competitors, he should be able to profit from his efforts. Because a commercial advantage can vanish once the competition learns of it, the law should protect the businessman's efforts to keep his achievements secret. As is discussed below, this is an area of law in which simple fairness still plays a large role.

We do not say, however, that all these factors need exist in every case. Because each case must turn on its own facts, no standard formula for weighing the factors can be devised. Secrecy is always required, of course, but beyond that there are no universal requirements. In a future case, for example, should the defendant's breach of confidence be particularly egregious, the injured party might still seek redress in court despite the possibility that the subject matter was discovered at little or no cost or that the object of secrecy is not of great value to him. The definition of "trade secret" will therefore be determined by weighing all equitable considerations. It is easy to recognize the possibility of a trade secret here, however, because Metallurgical presented evidence of all three factors discussed above. . . .

COMMENTS AND QUESTIONS

1. Sales of goods in mass retail markets are made to further the seller's economic interests. Under the rationale in *Fourtek*, why do such sales destroy the value of the secret?

2. If the defendants had acquired information about Metallurgical's process from Consarc or La Floridienne — the Metallurgical licensees — would the case have come out differently? Should it matter whether they acquired the information with knowledge that Metallurgical still considered it a trade secret?

3. The categories of information *eligible* for protection as trade secrets are quite expansive. As *Fourtek* makes clear, they include secret combinations of items which by themselves are publicly known. They also include both scientific and technical information and business information, such as customer lists and business plans. Can you think of any type of information that should be ineligible for trade secret protection?

Note that eligibility for protection is only the first of many hurdles required to establish the existence of a trade secret. Many alleged trade secrets that are *eligible* for protection do not in fact receive protection because they do not meet one of the other standards described in this section. Courts have frequently held certain basic ideas or concepts incapable of protection as secrets on the grounds that they were too well known to derive value from secrecy. For example, in Buffets, Inc. v. Klinke, 73 F.3d 765 (9th Cir. 1996), the court held that the plaintiff could not claim its relatively straightforward recipes for barbequed chicken and macaroni and cheese as trade secrets. On the other hand, in Camp Creek Hospitality Inns v. Sheraton Franchise Corp., 139 F.3d 1396 (11th Cir. 1997), the court held that a hotel could protect information about its prices, discounts, and occupancy levels as a trade secret where it was closely guarded information in the industry.

4. Courts have made it clear that strict novelty is not required for trade secret protection. The idea may have occurred to someone before; it may even be in use by another. But if it is not generally known or readily ascertainable to the competitors in an industry, it may still qualify for trade secret protection. One widely cited decision described the standard for protectable ideas as follows:

> [U]niqueness in the patent law sense is not an essential element of a trade secret, for the patent laws are designed to encourage invention, whereas trade secret law is designed to protect against a breach of faith. However, the trade secret must "possess at least that modicum of originality which will separate it from everyday knowledge." Cataphote Corporation v. Hudson, 444 F.2d 1313, 1315 (5th Cir. 1971). As stated in an authoritative treatise on this subject:
>
>> As distinguished from a patent, a trade secret need not be essentially new, novel or unique; therefore, prior art is a less effective defense in a trade secret case than it is in a patent infringement case. The idea need not be complicated; it may be intrinsically simple and nevertheless qualify as a secret, unless it is in common knowledge and, therefore, within the public domain.
>
> 2 Callman, Unfair Competition, Trademarks and Monopolies §52.1 (3d ed. 1968).

Forest Laboratories v. The Pillsbury Co., 425 F.2d 621, 624 (7th Cir. 1971). Some courts have gone even further, suggesting that "[a] trade secret may be no more than 'merely a mechanical improvement that a good mechanic can make.' " SI Handling Systems, Inc. v. Heisley, 753 F.2d 1244, 1256 (3d Cir. 1985).

Why not require novelty in order to protect a trade secret? That is, why should the law protect the "secrecy" of a piece of information if others have already discovered it?

5. The Restatement of Torts offers further guidance in determining whether information constitutes a trade secret. It lists six factors to be considered:

- The extent to which the information is known outside the claimant's business.
- The extent to which it is known by employees and others involved in the business.
- The extent of measures taken by the claimant to guard the secrecy of the information.
- The value of the information to the business and its competitors.
- The amount of effort or money expended by the business in developing the information.
- The ease or difficulty with which the information could be properly acquired or duplicated by others.

Each of these factors is open-ended. Together, they provide the basis for an inquiry into how *secret* the information really is. This is also the point of *Forest Laboratories'* truncated inquiry into novelty. To be protectable, information must not be "generally known" or "readily ascertainable" by competitors in an industry.

6. Why should we bother to protect secrets that were stumbled upon with little or no investment in research but that happen to have "value"? Does the economic rationale for intellectual property suggest that such secrets will be underproduced absent protection?

7. Why is secrecy required at all? Trade secrets are not misappropriated unless information is taken by improper means or from a confidential relationship. Why aren't those tortious elements enough? It is certainly possible to envision a "misappropriation" tort that punishes diversion of information, whether or not it is secret.[10] Indeed, a number of controversial cases discussed in Chapter 6 have created just such a common law tort. In addition to the cases cited there, see United States Sporting Products v. Johnny Stewart Game Calls, 865 S.W.2d 214 (Tex. Ct. App. 1993) (publicly sold uncopyrighted recordings protectable under a "labor theory"); Note, The "Genetic Message" from the Cornfields of Iowa: Expanding the Law of Trade Secrets, 38 Drake L. Rev. 631 (1989) (similar case involving publicly sold grain).

One possible objection to such a scheme is that it may chill the legitimate acquisition of information from competitors. But the only information protected by a misappropriation tort that is not also protected by trade secret law is public information. Since it is public, the need for competitors to acquire it directly from another company or through dubious means is presumably low.[11]

---

10. For suggestions along these lines, see Dennis Karjala, Misappropriation as a Third Intellectual Property Paradigm, 94 Colum. L. Rev. 2594 (1994); J. H. Reichman, Legal Hybrids Between the Patent and Copyright Paradigms, 94 Colum. L. Rev. 2432 (1994).

11. Another reason for limiting common law protection to secret information may be concern over preemption of state common law by the federal intellectual property laws. While the course of the law is not completely clear, federal courts have generally held that state laws that create property rights in public information are preempted by the patent and copyright laws. The

But there may be a wide gulf between "secret" information and "public" information. If three large companies all use the same process but guard it closely, is it a secret? If a company guards a process closely as a secret, but an account of the process is available in an obscure published source, does the company have a protectable trade secret? Does it matter how obscure the published source is, if the defendant in fact steals the information from the plaintiff rather than going to the public source? What theory of trade secrets would support a finding of liability in such a case?

In Rohm & Haas Co. v. Adco Chemical Co., 689 F.2d 424 (3d Cir. 1982), defendant Harvey was a former Rohm & Haas employee who was hired by Adco to duplicate a process for producing "paint delivery vehicles," the chemicals added to paint that allow it to be applied to surfaces easily. There seems no question in the case that Harvey did in fact memorize the plaintiff's formula and take it to Adco. In their defense, the defendants offered evidence that Rohm & Haas's "secret" process was in fact disclosed in a series of prior publications. The court nonetheless concluded that it was a protectable trade secret, in part because the defendants *did not in fact* obtain the information from those publications. Why should this matter? Certainly the defendants' conduct proves that they took the information from the plaintiff, but that should be irrelevant if the plaintiff does not have a protectable secret. Has the *Rohm & Haas* court in effect abolished the requirement of secrecy by requiring only that the defendants in fact obtain the information from the plaintiff?

Rohm & Haas is at the center of a critical debate in trade secret law. At issue is whether information must actually be "known" to competitors or merely be "knowable" for the court to conclude that it is not secret. Many courts following the Restatement have taken the former view, as *Rohm & Haas* does. This view gives broad scope to trade secret law, because it protects information that could have been acquired properly but in fact was not. It also underscores the unfair competition rationale for trade secret protection — the problem is not that the defendant acquired the information at all, but the way in which it was acquired.

In a significant break with the old Restatement rule, the Uniform Trade Secrets Act provides that information is not a trade secret if it is "generally known" *or* "readily ascertainable by proper means." This is a narrower view of what constitutes a trade secret.[12] Under this view, once a secret is readily

---

policy being served by preemption is to protect the balance struck by federal intellectual property laws. See, e.g., Synercom Technology v. University Computing Co., 474 F. Supp. 37 (N.D. Tex. 1979) (state unfair competition laws cannot prevent copying of public information).

There are two rationales for such preemption. First, it may be that the federal laws reflect a judgment that unpatentable inventions ought to belong to the public. On this rationale, see Bonito Boats v. Thunder Craft Boats, 489 U.S. 141 (1989). Second, preemption may serve to erect barriers between different types of intellectual property protection, "channeling" inventions. into one or another form of protection. We discuss federal preemption in more detail in Chapter 6.

12. Some states, however, refused to accede to this change and have modified the language of the act to remove the reference to "readily ascertainable" information. California is one prom-

available through public sources, it loses all trade secret protection. At this point, the defendant is free to obtain the information from the public source *or from the plaintiff herself.* See Restatement (Third) of Unfair Competition §39 Comment *f,* at 433 ("When the information is readily ascertainable from such [public] sources, however, actual resort to the public domain is a formality that should not determine liability.").

Even jurisdictions that follow the Restatement of Torts view place some limit on what can qualify for trade secret protection. If information is generally known to the public, or even within a specialized industry, it does not qualify for protection. No company can claim that "$E=mc^2$" is a trade secret, for example, even if it keeps the formula under lock and key, and even if the defendant steals it from the company rather than obtaining it elsewhere. We consider the rather different issue of whether two parties could *agree* to treat the formula as a secret in section D, below.

8. What relevance do the competing theories of trade secret protection have for the known vs. knowable debate? If the purpose of trade secret law is to promote innovation, why should a company be entitled to protect an idea that already exists in the literature and can readily be found there? On the other hand, doesn't the property theory compel the conclusion that a company should be able to protect an idea against being stolen *from it,* even if the idea is commonly known elsewhere?

9. The Restatement factors used in *Forest Labs* seem to focus on how widely the information is known, suggesting that information used by many people cannot be a secret. The secrecy requirement, therefore, may reflect a policy judgment in favor of the distribution of information once it has reached a certain "critical mass." *Metallurgical Industries,* by contrast, focuses on attempts to prevent disclosure of the information. This approach suggests that even widely known information (or published information, as that in *Metallurgical Industries* was) may receive trade secret protection.

Compare the definition of a trade secret offered in the new Restatement (Third) of Unfair Competition:

> A trade secret is any information that can be used in the operation of a business or other enterprise and that is sufficiently valuable and secret to afford an actual or potential economic advantage over others.

Restatement (Third) of Unfair Competition §39. Under the new Restatement, many different companies can possess the same information and each protect it as a secret. Id. at illustration 1.[13] In a comment interpreting this

---

inent example. See Cal. Civ. Code §3426.1(d)(1); ABBA Rubber Co. v. Seaquist, 286 Cal. Rptr. 518 (Ct. App. 1991) (customer list qualified for trade secret protection even though information was available in trade directories). California law does permit proof of ready ascertainability as a defense, however.

13. The Restatement does note, however, that "[w]hen information is no longer sufficiently secret to qualify for protection as a trade secret, its use should not serve as a basis for the imposition of liability," id. at 433 Comment *f,* suggesting that at some point information is not secret even though companies protect it as if it were.

provision, the drafters note that "[t]he concept of a trade secret as defined in this Section is intended to be consistent with the definition of "trade secret" in §1(4) of the [Uniform Trade Secrets] Act." Is it?

PROBLEMS

Problem 2-1. Company *X* possesses a valuable piece of information about the process for making its product. That information is not known at all outside company *X*. Suppose *X* discloses the information to two companies, *A* and *B*. *A* receives the information in confidence, and subject to a written agreement that it will not use or disclose the information outside the bounds of the relationship. *B* receives the information without any restrictions whatever on its use. Does *X* have a protectable trade secret that it can assert against *A*? Against *B*? Against *C*, an individual who steals the information from *X*'s computer network?

Does your answer change if *X, A, B,* and *C* are the only companies in the industry?

Problem 2-2. StartUp, Inc., is the only participant in a new market. The market is based on a product for which StartUp has a nonexclusive license from the inventor; that is, StartUp cannot prevent others from obtaining a similar license. Nonetheless, StartUp has exhaustively researched the demand for the product, has concluded that a market exists, and has worked to stimulate demand. As a result, it has both "made" a market for the product and developed a comprehensive list of customers.

Thaddeus, a sales representative for StartUp, leaves to found his own company. He gets a license to make the product from its inventor. He takes with him from StartUp the customer list he worked with as an employee, his personal knowledge of and contacts with specific customers, and StartUp's knowledge of the market. StartUp sues Thaddeus for misappropriation of trade secrets. What result?

Problem 2-3. Research Co. is a major pharmaceutical company working on a cure for certain types of cancer. Derek is a molecular biologist employed by Research. After several years on the job, Derek leaves Research for Conglomerate, Inc., another pharmaceutical company, which has decided to work on the same cancer cure. At the time Derek leaves, Research has not been successful in finding a cancer cure. However, as a result of his work at Research, Derek is able to help Conglomerate avoid several unproductive avenues of research. Aided in part by this knowledge, Conglomerate (using scientists other than Derek) develops a cancer cure before Research. Research sues Conglomerate, alleging misappropriation of trade secrets. Does Research have a case?

Problem 2-4. The Church of True Belief is a religious group founded around a set of closely guarded scriptural materials supposedly handed down to Church's elders from Church's deity. After a bitter theological dispute, a group of adherents leave the church to form the House of Absolute Belief. They take with them a copy of Church's confidential scriptures, which they rely on in gaining adherents and founding the new House. Church sues House for misappropriation of trade secrets.

At trial, the issue is whether the scriptures qualify as a trade secret. The evidence indicates that the scriptures had never before been removed from the confines of Church, that both Church and House are tax-exempt nonprofit organizations which rely on donations for their funding, and that Church (but not House) rations access to the scriptures in proportion to the size of an adherent's donation.

Can the scriptures qualify as a trade secret? Does your answer depend on whether the governing law is the UTSA, the Restatement of Torts, or the Restatement (Third) of Unfair Competition?

2. Reasonable Efforts to Maintain Secrecy

Besides the existence of a trade secret, plaintiffs must show under the Uniform Act that they have taken "reasonable measures" to protect the secrecy of their idea. Certainly, a plaintiff cannot publicly disclose the secret and still expect to protect it. But precautions must go further than that. Generally, they must include certain efforts to prevent theft or use of the idea by former employees.

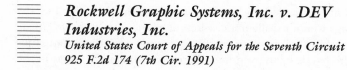

Rockwell Graphic Systems, Inc. v. DEV Industries, Inc.
United States Court of Appeals for the Seventh Circuit
925 F.2d 174 (7th Cir. 1991)

POSNER, Circuit Judge

This is a suit for misappropriation of trade secrets. Rockwell Graphic Systems, a manufacturer of printing presses used by newspapers, and of parts for those presses, brought the suit against DEV Industries, a competing manufacturer, and against the president of DEV, who used to be employed by Rockwell. . . .

When we said that Rockwell manufactures both printing presses and replacement parts for its presses — "wear parts" or "piece parts," they are called — we were speaking approximately. Rockwell does not always manufacture the parts itself. Sometimes when an owner of one of Rockwell's presses needs a particular part, or when Rockwell anticipates demand for the part, it will subcontract the manufacture of it to an independent machine shop, called a "vendor" by the parties. When it does this it must give the vendor a "piece part drawing" indicating materials, dimensions, tolerances, and methods of manufacture. Without that information the vendor could not manufacture the part. Rockwell has not tried to patent the piece parts. It believes that the purchaser cannot, either by inspection or by "reverse engineering" (taking something apart in an effort to figure out how it was made), discover how to manufacture the part; to do that you need the piece part drawing, which contains much information concerning methods of manufacture, alloys, tolerances, etc. that cannot be gleaned from the part itself. So Rockwell tries — whether hard enough is the central issue in the case — to keep the piece part drawings secret, though not of course from the vendors; they could not manufacture the parts for Rockwell without the drawings. DEV points out that some of the parts are for presses that Rockwell no longer manufactures. But as long as the presses are in service — which can be a very long time — there is a demand for replacement parts.

Rockwell employed Fleck and Peloso in responsible positions that gave them access to piece part drawings. Fleck left Rockwell in 1975 and three years later joined DEV as its president. Peloso joined DEV the following year after being fired by Rockwell when a security guard caught him removing piece part drawings from Rockwell's plant. This suit was brought in 1984, and pretrial discovery by Rockwell turned up 600 piece part drawings in DEV's possession, of which 100 were Rockwell's. DEV claimed to have obtained them lawfully, either from customers of Rockwell or from Rockwell vendors, contrary to Rockwell's claim that either Fleck and Peloso stole them when they were employed by it or DEV obtained them in some other unlawful manner, perhaps from a vendor who violated his confidentiality agreement with Rockwell. Thus far in the litigation DEV has not been able to show which customers or vendors lawfully supplied it with Rockwell's piece part drawings.

The defendants persuaded the magistrate and the district judge that the piece part drawings weren't really trade secrets at all, because Rockwell made only perfunctory efforts to keep them secret. Not only were there thousands of drawings in the hands of the vendors; there were thousands more in the hands of owners of Rockwell presses, the customers for piece parts. The drawings held by customers, however, are not relevant. They are not piece part drawings, but assembly drawings. . . . An assembly drawing shows how the parts of a printing press fit together for installation and also how to integrate the press with the printer's other equipment. Whenever Rockwell sells a printing press it gives the buyer assembly drawings as well. These are the equivalent of instructions for assembling a piece of furniture. Rockwell does not claim that they contain trade secrets. It admits having supplied a few piece part drawings to customers, but they were piece part drawings of obsolete parts that Rockwell has no interest in manufacturing and of a safety device that was not part of the press as originally delivered but that its customers were clamoring for; more to the point, none of these drawings is among those that Rockwell claims DEV misappropriated.

. . . DEV's main argument is that Rockwell was impermissibly sloppy in its efforts to keep the piece part drawings secret.

On this, the critical, issue, the record shows the following. (Because summary judgment was granted to DEV, we must construe the facts as favorably to Rockwell as is reasonable to do.) Rockwell keeps all its engineering drawings, including both piece part and assembly drawings, in a vault. Access not only to the vault, but also to the building in which it is located, is limited to authorized employees who display identification. These are mainly engineers, of whom Rockwell employs 200. They are required to sign agreements not to disseminate the drawings, or disclose their contents, other than as authorized by the company. An authorized employee who needs a drawing must sign it out from the vault and return it when he has finished with it. But he is permitted to make copies, which he is to destroy when he no longer needs them in his work. The only outsiders allowed to see piece part drawings are the vendors (who are given copies, not originals). They too are required to sign confidentiality agreements, and in addition each drawing is stamped with a legend stating that it contains proprietary material. Vendors, like Rockwell's own engineers, are allowed to make copies for internal working purposes, and although the confidentiality agreement that they sign requires the vendor to return the drawing when the order has been filled, Rockwell does not enforce this requirement. The rationale for not enforcing it is that the vendor will need the drawing if Rockwell reorders the part. Rockwell even permits unsuccessful bidders for a piece part contract to keep the drawings, on the theory that the high bidder this round may be the low bidder the next. But it does consider the ethical standards of a machine shop before making it a vendor, and so far as appears no shop has ever abused the confidence reposed in it.

The mere fact that Rockwell gave piece part drawings to vendors — that is, disclosed its trade secrets to "a limited number of outsiders for a particular purpose" — did not forfeit trade secret protection. On the contrary, such disclosure, which is often necessary to the efficient exploitation of a trade secret, imposes a duty of confidentiality on the part of the person to whom the disclosure is made. But with 200 engineers checking out piece part drawings and making copies of them to work from, and numerous vendors receiving copies of piece part drawings and copying them, tens of thousands of copies of these drawings are floating around outside Rockwell's vault, and many of these outside the company altogether. Although the magistrate and the district judge based their conclusion that Rockwell had not made adequate efforts to maintain secrecy in part at least on the irrelevant fact that it took no efforts at all to keep its assembly drawings secret, DEV in defending the judgment that it obtained in the district court argues that Rockwell failed to take adequate measures to keep even the piece part drawings secret. Not only did Rockwell not limit copying of those drawings or insist that copies be returned; it did not segregate the piece part drawings from the assembly drawings and institute more secure procedures for the former. So Rockwell could have done more to maintain the confidentiality of its piece part drawings than it did, and we must decide whether its failure to do more was so plain a breach of the obligation of a trade secret owner to make reasonable efforts to maintain secrecy as to justify the entry of summary judgment for the defendants.

The requirement of reasonable efforts has both evidentiary and remedial significance, and this regardless of which of the two different conceptions of trade secret protection prevails. . . . [T]he two different conceptions of trade secret protection are better described as different emphases. The first emphasizes the desirability of deterring efforts that have as their sole purpose and effect the redistribution of wealth from one firm to another. The second emphasizes the desirability of encouraging inventive activity by protecting its fruits from efforts at appropriation that are, indeed, sterile wealth-redistributive — not productive — activities. The approaches differ, if at all, only in that the second does not limit the class of improper means to those that fit a preexisting pigeonhole in the law of tort or contract or fiduciary duty — and it is by no means clear that the first approach assumes a closed class of wrongful acts, either.

Under the first approach, at least if narrowly interpreted so that it does not merge with the second, the plaintiff must prove that the defendant obtained the plaintiff's trade secret by a wrongful act, illustrated here by the alleged acts of Fleck and Peloso in removing piece part drawings from Rockwell's premises without authorization, in violation of their employment contracts and confidentiality agreements, and using them in competition with Rockwell. Rockwell is unable to prove directly that the 100 piece part drawings it got from DEV in discovery were stolen by Fleck and Peloso or obtained by other improper means. But if it can show that the probability that DEV could have obtained them otherwise — that is, without engaging in wrongdoing — is slight, then it will have taken a giant step toward proving what it must prove in order to recover under the first theory of trade secret protection. The greater the precautions that Rockwell took to maintain the secrecy of the piece part drawings, the lower the probability that DEV obtained them properly and the higher the probability that it obtained them through a wrongful act; the owner had taken pains to prevent them from being obtained otherwise.

Under the second theory of trade secret protection, the owner's precautions still have evidentiary significance, but now primarily as evidence that the secret has real value. For the precise means by which the defendant acquired it is less important under the second theory, though not completely unimportant; remember that even the second theory allows the unmasking of a trade secret by some means, such as reverse engineering. If Rockwell expended only paltry resources on preventing its piece part drawings from falling into the hands of competitors such as DEV, why should the law, whose machinery is far from costless, bother to provide Rockwell with a remedy? The information contained in the drawings cannot have been worth much if Rockwell did not think it worthwhile to make serious efforts to keep the information secret.

The remedial significance of such efforts lies in the fact that if the plaintiff has allowed his trade secret to fall into the public domain, he would enjoy a windfall if permitted to recover damages merely because the defendant took the secret from him, rather than from the public domain as it could have done with impunity. Brunswick Corp. v. Outboard Marine Corp., supra, 79 Ill. 2d at 479, 404 N.E.2d at 207; Van Products Co. v. General Welding & Fabricating Co., 419 Pa. 248, 267-68, 213 A.2d 769, 779-80 (1965) (repudiating the interpretation of Pennsylvania law that this court had adopted in Smith v. Dravo Corp., 203 F.2d 369, 374-75 (7th Cir. 1953)). It would be like punishing a person for stealing property that he believes is owned by another but that actually is abandoned property. If it were true, as apparently it is not, that Rockwell had given the piece part drawings at issue to customers, and it had done so without requiring the customers to hold them in confidence, DEV could have obtained the drawings from the customers without committing any wrong. The harm to Rockwell would have been the same as if DEV had stolen the drawings from it, but it would have had no remedy, having parted with its rights to the trade secret. This is true whether the trade secret is regarded as property protected only against wrongdoers or (the logical extreme of the second conception, although no case — not even *Christopher* — has yet embraced it and the patent statute may preempt it) as property protected against the world. In the first case, a defendant is perfectly entitled to obtain the property by lawful conduct if he can, and he can if the property is in the hands of persons who themselves committed no wrong to get it. In the second case the defendant is perfectly entitled to obtain the property if the plaintiff has abandoned it by giving it away without restrictions.

It is easy to understand therefore why the law of trade secrets requires a plaintiff to show that he took reasonable precautions to keep the secret a secret. If analogies are needed, one that springs to mind is the duty of the holder of a trademark to take reasonable efforts to police infringements of his mark, failing which the mark is likely to be deemed abandoned, or to become generic or descriptive (and in either event be unprotectable). The trademark owner who fails to police his mark both shows that he doesn't really value it very much and creates a situation in which an infringer may have been unaware that he was using a proprietary mark because the mark had drifted into the public domain, much as DEV contends Rockwell's piece part drawings have done.

But only in an extreme case can what is a "reasonable" precaution be determined on a motion for summary judgment, because the answer depends on a balancing of costs and benefits that will vary from case to case and so require estimation and measurement by persons knowledgeable in the particular field of endeavor involved. On the one hand, the more the owner of the trade secret spends on preventing the secret from leaking out, the more he demonstrates that the secret has real value deserving of legal protection, that he really was hurt as a result of the misappropriation of it, and that there really was misappropriation. On the other hand, the more he spends, the higher his costs. The costs can be indirect as well as direct. The more Rockwell restricts access to its drawings, either by its engineers or by the vendors, the harder it will be for either group to do the work expected of it. Suppose Rockwell forbids any copying of its drawings. Then a team of engineers would have to share a single drawing, perhaps by passing it around or by working in the same room, huddled over the drawing. And how would a vendor be able to make a piece part — would Rockwell have to bring all that work in house? Such reconfigurations of patterns of work and production are far from costless; and therefore perfect security is not optimum security.

There are contested factual issues here, bearing in mind that what is reasonable is itself a fact for purposes of Rule 56 of the civil rules. Obviously Rockwell took some precautions, both physical (the vault security, the security guards — one of whom apprehended Peloso in flagrante delicto) and contractual, to maintain the confidentiality of its piece part drawings. Obviously it could have taken more precautions. But at a cost, and the question is whether the additional benefit in security would have exceeded that cost. We do not suggest that the question can be answered with the same precision with which it can be posed, but neither can we say that no reasonable jury could find that Rockwell had done enough and could then go on to infer misappropriation from a combination of the precautions Rockwell took and DEV's inability to establish the existence of a lawful source of the Rockwell piece part drawings in its possession.

This is an important case because trade secret protection is an important part of intellectual property, a form of property that is of growing importance to the competitiveness of American industry. Patent protection is at once costly and temporary, and therefore cannot be regarded as a perfect substitute. If trade secrets are protected only if their owners take extravagant, productivity-impairing measures to maintain their secrecy, the incentive to invest resources in discovering more efficient methods of production will be reduced, and with it the amount of invention. . . .

Reversed and remanded.

COMMENTS AND QUESTIONS

1. There is obviously an intuitive relationship between the existence of a secret and reasonable efforts to protect a secret. After all, if something is not a secret, there would not seem to be any point to protecting it. And the fact that an idea is well protected may be evidence that it is in fact a secret. Nonetheless, the requirements are conceptually distinct. Information in the public domain cannot be turned into a secret merely by treating it like a secret, a point that lawyers and even courts sometimes forget. This distinction is

made clear in the Uniform Trade Secrets Act, which defines a trade secret as information that is both "not generally known" *and* the subject of reasonable efforts to maintain secrecy. UTSA §1(4).

Consider whether the opinion in *Rockwell* conflates these two into a single requirement. The court seems to emphasize the evidentiary significance of the precautions Rockwell took in proving misappropriation. Since it was clear (to the court, anyway) that the DEV employees did in fact take the information from Rockwell, the court did not consider the precautions to be that important. This approach was apparently followed in the Restatement (Third) of Unfair Competition. Unlike the Uniform Act, the new Restatement does not contain a requirement that plaintiffs take reasonable precautions to protect their secrets. The comment to section 39 takes the position that, while "[p]recautions taken to maintain the secrecy of information are relevant in determining whether the information qualifies for protection as a trade secret," "if the value and secrecy of the information are clear, evidence of specific precautions taken by the trade secret owner may be unnecessary." Restatement (Third) of Unfair Competition §39, Comment *g*, at 435-436.

Contrast *Rockwell* with the Minnesota Supreme Court's decision in Electro-Craft Corp. v. Controlled Motion, Inc., 332 N.W.2d 890 (Minn. 1983), a case that also involved information taken by former employees and used in starting a competing company. The court found that the information the employees took was not generally known or readily ascertainable in the industry. However, it found that the information did not constitute a trade secret:

> (c) Reasonable efforts to maintain secrecy. It is this element upon which [plaintiff Electro-Craft Corp., or "ECC"]'s claim founders. The district court found that, even though ECC had no "meaningful security provisions," ECC showed an intention to keep its data and processes secret. This finding does not bear upon the statutory requirement that ECC use "efforts that are reasonable under the circumstances to maintain . . . secrecy." Minn. Stat. §325C.01, subd. 5(ii). . . . [E]ven under the common law, more than an "intention" was required — the plaintiff was required to show that it had manifested that intention by making some effort to keep the information secret.
>
> This element of trade secret law does not require maintenance of absolute secrecy; only partial or qualified secrecy has been required under the common law. What is actually required is conduct which will allow a court acting in equity to enforce plaintiff's rights. . . .
>
> In the present case, even viewing the evidence most favorably to the findings below, we hold that ECC did not meet its burden of proving that it used reasonable efforts to maintain secrecy as to [the subject matter of the suit, a product called the ECC 1125]. We acknowledge that ECC took minimal precautions in screening its Handbook and publications for confidential information and by requiring some of its employees to sign a confidentiality agreement, but these were not enough.
>
> First, ECC's physical security measures did not demonstrate any effort to maintain secrecy. By "security" we mean the protection of information from discovery by outsiders. Security was lax in this case. For example, the main plant

had a few guarded entrances, but seven unlocked entrances existed without signs warning of limited access. Employees were at one time required to wear badges, but that system was abandoned by the time of the events giving rise to this case. The same was generally true of the Amery, Wisconsin plant where ECC 1125 and brushless motors were manufactured. One sign was posted at each plant, however, marking the research and development lab at Hopkins and the machine shop at Amery as restricted to "authorized personnel." Discarded drawings and plans for motors were simply thrown away, not destroyed. Documents such as motor drawings were not kept in a central or locked location, although some design notebooks were kept locked.

The relaxed security by itself, however, does not preclude a finding of reasonable efforts by ECC to maintain secrecy. Other evidence did not indicate that industrial espionage is a major problem in the servo motor industry. Therefore, "security" measures may not have been needed, and the trial court could have found trade secrets if ECC had taken other reasonable measures to preserve secrecy.

However, ECC's "confidentiality" procedures were also fatally lax, and the district court was clearly in error in finding ECC's efforts to be reasonable. By "confidentiality" in this case we mean the procedures by which the employer signals to its employees and to others that certain information is secret and should not be disclosed. Confidentiality was important in this case, for testimony demonstrated that employees in the servo motor business frequently leave their employers in order to produce similar or identical devices for new employers. ECC has hired many employees from other corporations manufacturing similar products.[16] If ECC wanted to prevent its employees from doing the same thing, it had an obligation to inform its employees that certain information was secret.

ECC's efforts were especially inadequate because of the non-intuitive nature of ECC's claimed secrets here. The dimensions, etc., of ECC's motors are not trade secrets in as obvious a way as a "secret formula" might be. ECC should have let its employees know in no uncertain terms that those features were secret.

Instead, ECC treated its information as if it were not secret. None of its technical documents were marked "Confidential," and drawings, dimensions and parts were sent to customers and vendors without special marking. Employee access to documents was not restricted. ECC never issued a policy statement outlining what it considered to be secret. Many informal tours were given to vendors and customers without warnings as to confidential information. Further, two plants each had an "open house" at which the public was invited to observe manufacturing processes. . . .

In summary, ECC has not met its burden of proof in establishing the existence of any trade secrets. The evidence does not show that ECC was ever consistent in treating the information here as secret.

In a part of the opinion not reprinted here, the court noted that the information in question was not in fact known at all outside ECC. Given that,

---

16. One ECC employee actually prided himself on the information he had brought with him from his former employer. One day, just before that employee left ECC to join another company, the president of ECC found him copying documents after hours. ECC never questioned the employee or warned him or his new employer that certain information was confidential.

why shouldn't the company be able to prevent its employees from using the information they acquired there? Should the laxity of ECC's precautions matter if no one in fact took advantage of it?

2. In *Rockwell*, the court apparently assumed that the manufacturer was in a confidential relationship with the subcontractors that it sent drawings to. Is it reasonable to assume that there was an implied relationship of confidentiality in the absence of an express agreement? Two decisions by the Ninth Circuit set the parameters of this debate. In Entertainment Research Group v. Genesis Creative Group, 122 F.3d 1211 (9th Cir. 1997), the court found no confidential relationship between a manufacturer of inflatable costumes and the marketing firm hired to distribute them. By contrast, in IMAX Corp. v. Cinema Technologies, 152 F.3d 1161 (9th Cir. 1998), the court concluded that the proprietor of an IMAX movie theatre had a duty to IMAX to protect the specifications of the projector as trade secrets from a competitor who was attempting to reverse engineer the IMAX technology.

3. How much effort should be required of trade secret owners? Obviously, the best way to protect a secret is not to tell anyone at all. In the modern commercial world, however, this is normally impractical. Companies with trade secrets must tell the secret to their employees, their business partners, and often their distributors and customers as well. But the risk of *inadvertent* use or disclosure can be reduced in a number of ways: for example, by requiring employees, licensees, and even customers to sign confidentiality agreements; by investing in physical security measures against theft, such as fences, safes, and guards; and by designing products themselves so that they do not reveal their secrets upon casual (or even detailed) inspection.

4. Will reasonable precautions *always* be a question of fact? Or are certain activities (publishing a secret formula, for example) so inconsistent with trade secret protection that they automatically preclude a successful trade secret suit?

One issue that arises frequently is whether companies which sell products on the open market that embody their trade secret have disclosed the secret. As we shall see, customers who buy a product on the open market are entitled to break it apart to see how it works. This process is called "reverse engineering" the product. Trade secret law does not protect owners against legitimate purchasers who discover the secret through reverse engineering, absent a valid nondisclosure agreement. But does the possibility that a product might be reverse engineered foreclose *any* trade secret protection, even against people who have not actually reverse-engineered the product? At least one court has said no. In Data General Corp. v. Grumman Systems Support Corp., 825 F. Supp. 340, 359 (D. Mass. 1993), the court upheld a jury's verdict that Grumman had misappropriated trade secrets contained in object code form in Data General's computer program, despite the fact that many copies of the program had been sold on the open market. The court reasoned: "With the exception of those who lawfully licensed or unlawfully misappropriated MV/ADEX, Data General enjoyed the exclusive use of MV/ADEX.

Even those who obtained MV/ADEX and were able to *use* MV/ADEX were unable to discover its trade secrets because MV/ADEX was distributed only in its object code form, which is essentially unintelligible to humans."

Data General suggests that reasonable efforts to protect the secrecy of an idea contained in a commercial product — such as locks, black boxes, or the use of unreadable code — may suffice to maintain trade secret protection even after the product itself is widely circulated. Does this result make sense?

On the basis of the reasoning in *Data General* — that the source code, which contained the secrets, was not widely disclosed — would the distribution of the Data General software in *object* code form to the general public constitute misappropriation of those trade secrets?

PROBLEM

Problem 2-5. Smith, a bar owner in rural Alabama, develops by accident one night the relatively simple formula for a new alcoholic beverage. The drink is simply a mixture of three common ingredients. Smith begins selling the drink, which he calls "Mobile Mud," in his bar. However, he instructs his bartenders not to reveal the formula to anyone and has them premix "Mud" in the back of the bar, out of sight of customers. Smith is outraged when he learns that Jimmy Dean, an international distributor of alcoholic beverages, has copied his formula and is marketing it under a different name. At trial, Dean employees and independent experts unanimously testify that it is possible for someone with experience in the beverage industry to determine the formula for Mud by looking at, smelling and tasting the drink.

Has Smith taken reasonable precautions? What more could he have done to protect the "secret formula" of Mobile Mud? Is the secret so obvious to consumers that selling the product on the open market destroys protection? Does your opinion of the case change if you learn that Dean's representative went to Smith's bar and bribed a bartender to disclose the formula?

Note on Fencing Costs and Trade Secrets

One way to make sense of the "reasonable precautions against disclosure" standard is to imagine a physical analogy: erecting a fence on a piece of land. Imagine that after the expenditure of $100 on research you have developed a piece of information important for your industry. You estimate its ultimate value at $200. Imagine also that the cost of building a fence to keep this information secret is $150. A good way to envision this is to assume that competitors are located nearby, and they pass by your research facility periodically.

Notice that with these figures, you are in a pickle. While you have in hand technology with a net positive value (it is worth $200 but cost only $100), putting it into practice without first constructing a fence would reveal it first to one competitor and eventually to them all. This disclosure may not be a problem if it does not reduce your private payoff from the information too much. In our example, as long as *you* add more than $100 to your revenues, you will at least break even, since the cost of the R&D was $100. (For example, if you make an extra $120 from the information before your competitors catch on and copy it, you will still be $20 ahead.) But if revealing the information erodes all your net profit — i.e., if the lack of a fence keeps your added revenues below $100 — stockholders will be disappointed. Your hardfought R&D results would bring you only a loss. And of course, you and your competitors as well will probably learn from your failure to make a profit; not many R&D projects will be started in the future.

Now add the element of trade secret law to the story. If legal protection is available as a *supplement* to physical fencing, your economic situation might look different. Specifically, if the legal rule specified that competitors would be enjoined from using the information if you had spent somewhere close to $100 to protect it, then you would (1) spend close to that amount, and (2) police your competitors to see whether they had in fact taken your information.

Adding a few more numbers to the discussion should clarify the point. If you spend, say, $70 constructing a fence, your total outlay will be $170; this leaves a net profit of $30. As we saw, however, you cannot get enough fence for $70 to make the invention secure. (Picture a fence six feet high, which deters most but not all competitors from peering in.) But trade secret law comes to your assistance here. If a competitor takes the information you created — i.e., looks over the six-foot fence — you will have a cause of action against him or her. The legal rule, in effect, displaces the need to construct a totally effective fence. Legal protection is essentially equivalent to having some benevolent third party — e.g., the government — adding on to your fence to make it effective.

On this view, an optimal standard would encourage you to spend money on a fence up to the point where your expected profit is maximized. (Under some circumstances, this would be the break-even point for the entire project). Past that point, you are spending too much; before that point, too little. For an elaboration of this concept, see Note, Trade Secret Misappropriation: A Cost-Benefit Response to the Fourth Amendment Analogy, 106 Harv. L. Rev. 461 (1992).

COMMENTS AND QUESTIONS

1. This example ignores the costs of conducting a trade secret infringement suit. These legal costs will be far from negligible. Assuming you elected to build the $70 fence, would it matter if the legal costs of enforcing your secret exceed $30? That is, should a wise policymaker care if the total enforcement cost is higher than the total value to you of the trade secret?

Probably not, for several reasons. First, the private value to you is not the same as the total social value of the trade secret. For example, the information may have applications beyond those your firm is capable of discovering or pursuing. Permitting expenditures beyond your current private valuation might make sense in light of the eventual total social value of the secret. (It might eventually leak out, for example, to others who are in a better position to develop applications that you had to ignore.)

Second, and more important, protecting *your* secret might serve as an incentive for others to invest in the creation of other new trade secret technology or information. Litigation often produces this sort of positive externality; this makes it worth tolerating in some cases where the cost of resolving the particular dispute at issue exceeds the value of the asset or right being fought over.

On the other hand, what if the cost of the lawsuit was higher than the total value of the trade secret?

2. The question of incentives raises an interesting point that was overlooked in the preceding discussion. If the idea is to encourage investment in trade secrets, why require any degree of "reasonable precautions" at all? One sometimes hears in this regard that *all* "fencing" expenditures are inefficient. Cf. Edmund W. Kitch, The Law and Economics of Rights in Valuable Information, 9 J. Legal Stud. 683 (1980) (arguing that reasonable precautions make sense only as evidence of the existence of a trade secret). Why not simply require explicit notice — large neon signs, stamps on all documents, or publication of a secrecy policy — in place of physical precautions? For a suggestion that proceeds along these lines to some extent, see J. H. Reichman, Legal Hybrids Between the Patent and Copyright Paradigms, 94 Colum. L. Rev. 2432 (1994).

Is it appropriate to punish the creator of a valuable secret who has not invested in "reasonable precautions?" What value is there in a legal rule that requires, as does the Harvard Law Review Note cited above, investment in precautions up to the level that would be rational *in the absence of* the legal rule? Are prospective trade secret thieves actually encouraged by the reasonable precautions argument to steal ideas when they observe a lapse in security, and does this rule give them an incentive to search out such lapses? Professor Kitch asks the related question of why these expenses should be required in addition to the expense of bringing a trade secret lawsuit.

Note by the way that we could take legal enforcement costs into account in other ways. We could, for example, refuse to enforce any trade secret, regardless of how much you spend on fencing and the like, if its value were less than the average cost of a trade secret infringement suit. But recall in this connection our discussion of the occasions when trade secret litigation, though not efficient in light of the short-term value of the secret and the litigation costs, ought to be tolerated for the sake of the signal it sends to future trade secret inventors.

3. David Friedman, William Landes & Richard Posner, Some Economics of Trade Secret Law, 5 J. Econ. Perspectives 61, 67 (1991), makes the related point that a trade secret cause of action which yields a legal remedy ought to be available when it is cheaper than the physical precautions that would be necessary to protect a piece of information. They also state:

> Where on the contrary the social costs of enforcing secrecy through the legal system would be high, the benefits of shared information are likely to exceed the net benefits of legal protection.

Should it be a defense to a trade secret action that the plaintiff could more easily have protected the secret through physical precautions?

Kitch notes by way of analogy that we do not prohibit criminal complaints for larceny just because a property owner was careless. (On the other hand, many states do reduce recovery in tort suits on the basis of "comparative negligence.") He also suggests that reasonable precautions are required only to put prospective infringers on notice about the existence of a right and to serve as evidence of the fact that the secret is worth protecting legally. The fencing thus serves a notice function, akin to "marking" products with patent numbers, copyright symbols, or trademark symbols. Kitch, supra, at 698.

The Court in duPont & Co. v. Christopher, 431 F.2d 1012 (5th Cir. 1970), *cert. denied,* 400 U.S. 1024 (1971), pays significant attention to the role of fencing costs in trade secrets suits. We will return to that case when we consider misappropriation of trade secrets in section C.

4. One might imagine other rationales for requiring reasonable protection. For example, one might treat the requirement of reasonable precautions as serving a gatekeeper function that helps to weed out frivolous trade secret claims by requiring evidence of investment by the plaintiff in protecting the secret.

3. Disclosure of Trade Secrets

It is axiomatic that public disclosure of a trade secret destroys the "secret," and therefore ends protection forever. The corollary to this rule is that as long as a trade secret remains secret, it is protectable. Thus, trade secrets do not last for a specific term of years but continue indefinitely until the occurence of a particular event — the public disclosure of the secret.

Disclosure of a once-protected trade secret can occur in a number of ways:

- First, a trade secret owner may publish the secret, whether in an academic journal or any other forum. In that case, secrecy is lost. This loss might reasonably be considered a substantial disincentive to publication of scientific or technical advances. But publication of secret information regularly occurs, either because the inventors have not thought through the consequences of their actions or because the value or prestige of first publication is deemed to outweigh the potential loss of commercial trade secret protection.

  One common form of disclosure is the publication of an issued patent. Because (as we shall see) patent law requires the public disclosure of an invention with sufficient specificity to enable one of ordinary skill in the art to make it, obtaining a patent on an invention destroys trade secret protection. See Ferroline Corp. v. General Aniline & Film Corp., 207 F.2d 912 (7th Cir. 1953), *cert. denied,* 347 U.S. 953 (1954). Thus an inventor must "elect" either patent or trade secret protection, for the two cannot protect the same invention simultaneously.

- Second, a trade secret owner may in some cases disclose the secret by selling a commercial product that embodies the secret. As one court explained, both the Restatement of Torts and the Uniform Trade Secrets Act "necessarily compel the conclusion that a trade secret is protectable only so long as it is kept secret by the party creating it. If a so-called trade secret is fully disclosed by the products produced by use of the secret then the right to protection is lost." Vacco Indus. v. Van den Berg, 6 Cal. Rptr. 2d 602, 611 (Ct. App. 1992) (citations omitted). Further, disclosure may occur even without sale of the product itself, if the secret is disclosed freely and without restriction during the manufacturing or development processes.

  However, sales of a product to the public do not necessarily disclose a trade secret simply because the product embodies the trade secret. Rather, the question is whether the secret is apparent from the product itself. Secrets that are apparent to the buyers of a product are considered disclosed by the product, but secrets contained in undecipherable form within the product (such as object code in a computer program) are considered secret even when the product is sold.

  Of what relevance is the motivation behind the disclosure of a secret? Recall that in *Metallurgical Industries,* the court found the fact that Metallurgical had disclosed its secrets only for profit to weigh in favor of trade secret status. Why should this be the case? On the one hand, licensing is evidence that a secret has value and is worth protecting. On the other hand, one could argue that the fact that a secret holder has sold its information for profit suggests that it is not trying to keep this information secret at all but rather is attempting to profit from its disclosure. Which of these arguments you find persuasive may depend on the view you take of the reasons for trade secret protection.

- Third, trade secrets may be publicly disclosed (through publication or the sale of a product) by someone other than the trade secret owner. Commonly, this occurs when someone other than the trade secret owner has independently developed or discovered the secret. Call the first trade secret "owner" *A,* and the independent developer *B. A* has no control over what *B* does with her independent discovery; if she chooses to publish the secret, she defeats not only her rights to trade secret protection, but *A*'s rights as well.

  Suppose *B* did not develop the secret independently of *A* but in fact stole it from *A.* What happens if *B* publishes the secret? Can *A* still protect it? If so, what happens to *C,* who began using the secret after reading *B*'s publication? This issue was addressed in Religious Technology Center v. Lerma, 908 F. Supp. 1362 (E.D. Va. 1995). In that case, the Church of Scientology sued (among others) the Washington Post, which had quoted from part of its confidential "scriptures." The court concluded that the fact that the scriptures were posted on a Usenet newsgroup for ten days defeated any claim of trade secrecy:

  > [For ten days, the documents] remained potentially available to the millions of Internet users around the world.
  >
  > As other courts who have dealt with similar issues have observed, "posting works to the Internet makes them generally known" at least to the relevant people interested in the news group. Once a trade secret is posted on the Internet, it is effectively part of the public domain, impossible to retrieve. Although the person who originally posted a trade secret on the Internet may be liable for trade secret misappropriation, the party who merely down loads Internet information cannot be liable for misappropriation because there is no misconduct involved in interacting with the Internet.

  908 F. Supp. at 1368; accord American Red Cross v. Palm Beach Blood Bank Inc., 143 F.3d 1407 (11th Cir. 1998) (Red Cross donor list lost trade secret status because it was posted on a publicly accessible computer bulletin board).

- Fourth, trade secrets may be disclosed inadvertently (for example, by being left on a train or elsewhere in public view). While the case law on this issue is sparse, it seems reasonable to argue that a truly accidental disclosure should not defeat trade secret protection if reasonable precautions have been taken. On the other hand, if the inadvertent disclosure is widespread, it would seem unfair (as well as impracticable) to require the public as a whole to "give back" the secret. Note that section 1(2)(ii)(C) of the Uniform Trade Secrets Act provides that it is misappropriation for someone to disclose a secret that they have reason to know has been acquired "by accident or mistake." The Restatement (Third) of Unfair Competition §40(b)(4) takes the same position, "unless the [accidental] acquisition was the result of the [trade secret owner]'s failure to take reasonable precautions to maintain the secrecy of the information." See also Williams v. Curtis-Wright Corp., 681 F.2d 161 (3d Cir. 1982) (user of secrets disclosed by mistake was liable for misappropriation because he had constructive notice of the secrecy of the information).

- Finally, government agencies sometimes require the disclosure of trade secrets by private parties in order to serve some other social purpose. See Corn Products Refining Co. v. Eddy, 249 U.S. 427 (1919) (requiring a food manufacturer to label its product with an accurate list of ingredients). Health and environmental concerns are a very common reason for the government to require disclosure of product contents. For example, the Federal Insecticide, Fungicide and Rodenticide Act (FIFRA), 7 U.S.C. §136 et seq., requires disclosure of the contents of pesticides as well as a great deal of other information. FIFRA makes two concessions to trade secret protection, however. First, it limits public disclosure of information concerning manufacturing processes and inert (as opposed to active) contents. Second, it provides for compensation to be paid to the inventors of trade secrets which the government appropriates by public disclosure. See also Ruckelshaus v. Monsanto Co., 467 U.S. 986 (1984) (federal requirement that private parties disclose trade secrets may constitute a taking under the Fifth Amendment).

COMMENTS AND QUESTIONS

1. Under what circumstances does the sale of a commercial product embodying a trade secret destroy the secret? Is the answer different for a commercially available product produced by a secret manufacturing process?

2. Patent applications are kept secret by the U.S. Patent Office until the patent issues. See 35 U.S.C. §122. If a patent application is not actively prosecuted, or if the patent does not issue, it is declared abandoned by the Patent Office. See 37 CFR §114 (1995). Note that abandoned applications are *not* available to the public. In fact, the application itself is destroyed after 20 years. Id.

In Europe, Japan, and elsewhere, patent applications are published 18 months after they are filed, regardless of whether the patent has yet issued. Normally, however, an applicant is notified before the application is published, allowing him or her to "elect" whether or not to continue to pursue the patent (and have the application disclosed to the world) or to forego prosecuting the patent (and retain the secrecy of the application). In the United States, Congress has considered a number of bills in recent years that would require publication of patent applications after 18 months. Can you see advantages to this system? Disadvantages? Does your answer depend on how fast the Patent Office works?

The notion of an "election" between trade secret and patent protection assumes that the patent application actually describes all the details of an invention. For more on this issue — known as the "enablement" requirement in patent law — see 35 U.S.C. §112, and Chapter 3 in this book. Certainly there is evidence that firms sometimes make precisely such an election. For an interesting historical example, see Henry Petroski, The Pencil 114-115 (1990) (describing how the family of Henry David Thoreau kept its pencil-making technology secret rather than disclose it by obtaining a patent).

Most patent licenses also allow the licensee to use trade secrets and know-how related to or associated with the patent. If both a patent and its associated secrets are licensed, should there be any limits on the duration of the license? Do these licenses suggest that an inventor can avoid the "election" doctrine and "have it both ways" in some respects? See United States v. Pilkington plc, No. 94-345-TUC-WDB (D. Ariz. May 25, 1994) (government antitrust action against a company that sought to enforce trade secrets against competitors after a related patent had expired).

3. To what extent does the incorporation of a secret in a public governmental record preclude trade secret protection? See Frazee v. U.S. Forest Service, 97 F.3d 367 (9th Cir. 1996) (information was not a trade secret because it could be obtained from the government under the Freedom of Information Act); Weygand v. CBS, Inc., 43 U.S.P.Q.2d 1120 (C.D. Cal. 1997) (depositing a work with the U.S. Copyright Office destroys trade secrecy).

C. MISAPPROPRIATION OF TRADE SECRETS

Not all uses of another's trade secrets constitute misappropriation. Acquisition or use of a trade secret is illegal only in two basic situations: where it is done through improper means, or where it involves a breach of confidence. Uniform Trade Secrets Act §1; Restatement (Third) of Unfair Competition §40.

1. Improper Means

E. I. duPont deNemours & Co. v. Rolfe Christopher et al.
United States Court of Appeals for the Fifth Circuit
431 F.2d 1012 (5th Cir. 1970), cert. denied, 400 U.S. 1024
(1971)

GOLDBERG, Circuit Judge:

This is a case of industrial espionage in which an airplane is the cloak and a camera the dagger. The defendants-appellants, Rolfe and Gary Christopher, are photographers in Beaumont, Texas. The Christophers were hired by an unknown third party to take aerial photographs of new construction at the Beaumont plant of E. I. duPont deNemours & Company, Inc. Sixteen photographs of the DuPont facility were taken from the air on March 19, 1969, and these photographs were later developed and delivered to the third party.

DuPont employees apparently noticed the airplane on March 19 and immediately began an investigation to determine why the craft was circling over the plant. By that afternoon the investigation had disclosed that the craft was involved in a photographic expedition and that the Christophers were the photographers. DuPont contacted the Christophers that same afternoon and asked them to reveal the name of the person or corporation requesting the photographs. The Christophers refused to disclose this information, giving as their reason the client's desire to remain anonymous.

Having reached a dead end in the investigation, DuPont subsequently filed suit against the Christophers, alleging that the Christophers had wrongfully obtained photographs revealing DuPont's trade secrets which they then sold to the undisclosed third party. DuPont contended that it had developed a highly secret but unpatented process for producing methanol, a process which gave DuPont a competitive advantage over other producers. This process, DuPont alleged, was a trade secret developed after much expensive and time-consuming research, and a secret which the company had taken special precautions to safeguard. The area photographed by the Christophers was the plant designed to produce methanol by this secret process, and because the plant was still under construction parts of the process were exposed to view from directly above the construction area. Photographs of that area, DuPont alleged, would enable a skilled person to deduce the secret process for making methanol. DuPont thus contended that the Christophers had wrongfully appropriated DuPont trade secrets by taking the photographs and delivering them to the undisclosed third party. In its suit DuPont asked for damages to cover the loss it had already sustained as a result of the wrongful disclosure of the trade secret and sought temporary and permanent injunctions prohibiting any further circulation of the photographs already taken and prohibiting any additional photographing of the methanol plant. . . .

. . . [T]he Christophers argue that for an appropriation of trade secrets to be wrongful there must be a trespass, other illegal conduct, or breach of a confidential relationship. We disagree.

It is true, as the Christophers assert, that the previous trade secret cases have contained one or more of these elements. However, we do not think that the Texas courts would limit the trade secret protection exclusively to these elements. On the contrary, in Hyde Corporation v. Huffines, 1958, 158 Tex. 566, 314 S.W.2d 763, the Texas Supreme Court specifically adopted the rule found in the Restatement of Torts which provides:

> One who discloses or uses another's trade secret, without a privilege to do so, is liable to the other if
> (a) he discovered the secret by improper means, or
> (b) his disclosure or use constitutes a breach of confidence reposed in him by the other in disclosing the secret to him. . . .

Restatement of Torts §757 (1939). Thus, although the previous cases have dealt with a breach of a confidential relationship, a trespass, or other illegal conduct, the rule is much broader than the cases heretofore encountered. Not limiting itself to specific wrongs, Texas adopted subsection (a) of the Restatement which recognizes a cause of action for the discovery of a trade secret by any "improper" means. . . .

The question remaining, therefore, is whether aerial photography of plant construction is an improper means of obtaining another's trade secret. We conclude that it is and that the Texas courts would so hold. The Supreme Court of that state has declared that "the undoubted tendency of the law has been to recognize and enforce higher standards of commercial morality in the business world." Hyde Corporation v. Huffines, supra 314 S.W.2d at 773. That court has quoted with approval articles indicating that the proper means of gaining possession of a competitor's secret process is "through inspection and analysis" of the product in order to create a duplicate. K & G Tool & Service Co. v. G & G Fishing Tool Service, 1958, 158 Tex. 594, 314 S.W.2d 782, 783, 788. Later another Texas court explained:

> The means by which the discovery is made may be obvious, and the experimentation leading from known factors to presently unknown results may be simple and lying in the public domain. But these facts do not destroy the value of the discovery and will not advantage a competitor who by unfair means obtains the knowledge without paying the price expended by the discoverer.

Brown v. Fowler, Tex. Civ. App. 1958, 316 S.W.2d 111, 114, *writ ref'd n.r.e.*.

We think, therefore, that the Texas rule is clear. One may use his competitor's secret process if he discovers the process by reverse engineering applied to the finished product; one may use a competitor's process if he discovers it by his own independent research; but one may not avoid these labors by taking the process from the discoverer without his permission at a time when he is taking reasonable precautions to maintain its secrecy. To obtain knowledge of a process without spending the time and money to discover it independently is improper unless the holder voluntarily discloses it or fails to take reasonable precautions to ensure its secrecy.

In the instant case the Christophers deliberately flew over the DuPont plant to get pictures of a process which DuPont had attempted to keep secret. The Christophers delivered their pictures to a third party who was certainly aware of the means by which they had been acquired and who may be planning to use the information contained therein to manufacture methanol by the DuPont process. The third party has a right to use this process only if he obtains this knowledge through his own research efforts, but thus far all information indicates that the third party has gained this knowledge solely by taking it from DuPont at a time when DuPont was making reasonable efforts to preserve its secrecy. In such a situation DuPont has a valid cause of action to prohibit the Christophers from improperly discovering its trade secret and to prohibit the undisclosed third party from using the improperly obtained information.

We note that this view is in perfect accord with the position taken by the authors of the Restatement. In commenting on improper means of discovery the savants of the Restatement said:

> f. Improper means of discovery. The discovery of another's trade secret by improper means subjects the actor to liability independently of the harm to the interest in the secret. Thus, if one uses physical force to take a secret formula from another's pocket, or breaks into another's office to steal the formula, his conduct is wrongful and subjects him to liability apart from the rule stated in this Section. Such conduct is also an improper means of procuring the secret under this rule. But means may be improper under this rule even though they do not cause any other harm than that to the interest in the trade secret. Examples of such means are fraudulent misrepresentations to induce disclosure, tapping of telephone wires, eavesdropping or other espionage. A complete catalogue of improper means is not possible. In general they are means which fall below the generally accepted standards of commercial morality and reasonable conduct.

Restatement of Torts §757, Comment *f* at 10 (1939).

In taking this position we realize that industrial espionage of the sort here perpetrated has become a popular sport in some segments of our industrial community. However, our devotion to free-wheeling industrial competition must not force us into accepting the law of the jungle as the standard of morality expected in our commercial relations. Our tolerance of the espionage game must cease when the protections required to prevent another's spying cost so much that the spirit of inventiveness is dampened. Commercial privacy must be protected from espionage which could not have been reasonably anticipated or prevented. We do not mean to imply, however, that everything not in plain view is within the protected vale, nor that all information obtained through every extra optical extension is forbidden. Indeed, for our industrial competition to remain healthy there must be breathing room for observing a competing industrialist. A competitor can and must shop his competition for pricing and examine his products for quality, components, and methods of manufacture. Perhaps ordinary fences and roofs must be built to shut out incursive eyes, but we need not require the discoverer of a trade secret to guard against the unanticipated, the undetectable, or the unpreventable methods of espionage now available.

In the instant case DuPont was in the midst of constructing a plant. Although after construction the finished plant would have protected much of the process from view, during the period of construction the trade secret was exposed to view from the air. To require DuPont to put a roof over the unfinished plant to guard its secret would impose an enormous expense to prevent nothing more than a school boy's trick. We introduce here no new or radical ethic since our ethos has never given moral sanction to piracy. The market place must not deviate far from our mores. We should not require a person or corporation to take unreasonable precautions to prevent another from doing that which he ought not do in the first place. Reasonable precautions against predatory eyes we may require, but an impenetrable fortress is an unreasonable requirement, and we are not disposed to burden industrial inventors with such a duty in order to protect the fruits of their efforts. "Improper" will always be a word of many nuances, determined by time, place, and circumstances. We therefore need not proclaim a catalogue of commercial improprieties. Clearly, however, one of its commandments does say "thou shall not appropriate a trade secret through deviousness under circumstances in which countervailing defenses are not reasonably available." . . .

COMMENTS AND QUESTIONS

1. Improper means has a substantial overlap with basic common law protections. In this case, the overlap is with other torts. See Comment *f* to the Restatement, cited in *duPont*, and the Restatement (Third) of Unfair Competition §43 (defining improper means as including "theft, fraud, unauthorized interception of communications, inducement of or knowing participation in a breach of confidence, and other means either wrongful in themselves or wrongful under the circumstances of the case.").

DuPont itself provides an example of conduct prohibited by trade secret law that is probably not otherwise tortious. Should otherwise legitimate conduct be prohibited because it will disclose a trade secret? Most people would probably say yes in the context of the *duPont* case. As Judge Posner noted in *Rockwell,* misappropriation of trade secrets is largely redundant if it does not reach any further than other torts. Nonetheless, the reach of the case is troubling. Consider the last sentence of the *duPont* opinion. What is wrong with "deviousness"? Is reverse engineering (which is legal under both the Restatement and the Uniform Act) any less devious than aerial photography? And why should "countervailing defenses" enter the picture?

2. There is general agreement that "reverse engineering" — that is, buying a product and taking it apart to see how it works — is not a misappropriation of a trade secret.[14] But while it is easy to see how to reverse engineer a product, how does one reverse engineer a process? Are there realistic alternatives to the sort of espionage condemned in *duPont* for those who seek to reproduce a process? Should there be some legally protected way of discovering a competitor's process?

3. Note how the decision in *duPont* dovetails with the "fencing costs" rationale presented in section B.2 above. DuPont could have protected itself against aerial photography by building a roof over its plant area before beginning internal construction. The Fifth Circuit rejected this alternative because it would "impose an enormous expense to prevent nothing more than a school boy's trick." On the other hand, the courts are clearly willing to put duPont to *some* expense to protect its secrets from prying eyes. If duPont had allowed its engineers to leave copies of the plant blueprints on subways, the result of the case might be very different.

How much expense is duPont required to incur to protect itself? The uniform answer of the courts is that only "reasonable" precautions must be taken. While this is not a terribly helpful answer, it may be a fairly practical one in any given industry, where companies can protect themselves by taking those precautions that are customary.

On the other hand, consider the Second Circuit's statement in Franke v. Wiltschek, 209 F.2d 493, 495 (2d Cir. 1953):

> It matters not that the defendants could have gained their knowledge from a study of the expired patent and plaintiff's publicly marketed product. The fact is they did not. Instead they gained it from plaintiffs via their confidential relationship, and in doing so incurred a duty not to use it to plaintiff's detriment. This duty they have breached.

Is this an accurate statement of the law?

4. An interesting case involving only circumstantial evidence of misappropriation is Pioneer Hi-Bred International v. Holden Foundation Seeds, Inc., 35 F.3d 1226, 31 U.S.P.Q.2d 1385 (8th Cir. 1994). Plaintiff Pioneer could not establish a specific act of misappropriation, but it did show (through the use of three sophisticated, scientific tests) that it was highly unlikely that defendant's hybrid seeds had been developed independently of plaintiff's seeds. Instead, due to genetic similarities, Pioneer showed that it was much more likely that those seeds were "derived from" a popular Pioneer hybrid. Pioneer was found to have maintained reasonable measures to guard against

---

14. As we shall see, however, reverse engineering of a patented invention is generally not permissible.

disclosure, including putting experimental seeds in bags marked with a secret code, allowing seeds to be grown only under strict nondisclosure arrangements, and leaving unmarked the fields in which the seeds had been planted. Pioneer was awarded $46.7 million in damages.

PROBLEM

Problem 2-6. Suppose that, at trial in the *duPont* case, the Christophers proved that they could have discovered the secrets contained in the layout of the duPont plant from another source (such as a copy of the blueprints on file with the Environmental Protection Agency and available under the Freedom of Information Act) that placed no restrictions on their use of the information. Are the Christophers then relieved from liability for misappropriation of trade secrets? Would your answer change if they had obtained the photos by sneaking onto duPont property on the ground?

---

2. Confidential Relationship

The previous section concerned misappropriation of trade secrets by improperly obtaining them from their owner. But secrets that have been properly obtained may still be misappropriated if they are improperly used or disclosed. Most often, this occurs when the secrets are used or disclosed in violation of a confidential relationship.

What is a confidential relationship? Obviously, the easiest way to create a confidential relationship is to sign a contract to that effect. Agreements to keep certain information confidential are generally enforceable, at least if the information meets the definition of a trade secret. See Restatement (Third) of Unfair Competition §39 Comment *d*, at 430. But confidential relationships may also arise without any express agreement. Consider the following case.

≡
≡ **Smith v. Dravo Corp.**
≡ *United States Court of Appeals for the Seventh Circuit*
≡ *203 F.2d 369 (7th Cir. 1953)*

LINDLEY, J.

Plaintiffs appeal from a judgment for defendant entered at the close of a trial by the court without a jury. The complaint is in four counts: 1 and 2 charge an unlawful appropriation by defendant of plaintiffs' trade secrets relating to the design and construction and selling and leasing of freight containers; 3 and 4 aver infringement of plaintiffs' patents Nos. 2,457,841 and 2,457,842. . . .

In the early 1940s Leathem D. Smith, now deceased, began toying with an idea which, he believed, would greatly facilitate the ship and shore handling and transportation of cargoes. As he was primarily engaged in the shipbuilding business, it was quite natural that his thinking was chiefly concerned with water transportation and dock handling. Nevertheless his overall plan encompassed rail shipping as well. He envisioned construction of ships especially designed to carry their cargo in uniformly sized steel freight containers. These devices (which, it appears, were the crux of his idea) were: equipped with high doors at one end; large enough for a man to enter easily; weather and pilfer proof; and bore collapsible legs, which (1) served to lock them (a) to the deck of the ship by fitting into recesses in the deck, or (b) to each other, when stacked, by reason of receiving sockets located in the upper four corners of each container, and (2) allowed sufficient clearance between deck and container or container and container for the facile insertion of a fork of a lift tractor, and (3) were equipped with lifting eyelets, which, together with a specially designed hoist, made possible placement of the containers upon or removal from a ship, railroad car or truck, while filled with cargo. The outer dimensions of the devices were such that they would fit compactly in standard gauge North American railroad cars, or narrow gauge South American trains, and in the holds of most water vessels.

[At the end of World War II, Smith's company — Safeway Containers — had some success building and selling such containers.]

On June 23, 1946, Smith died in a sailing accident. The need for cash for inheritance tax purposes prompted his estate to survey his holdings for disposable assets. It was decided that the container business should be sold. Devices in process were completed but no work on new ones was started.

Defendant was interested in the Safeway container, primarily, it appears, for use by its subsidiary, the Union Barge Lines. In October 1946 it contacted Agwilines [one of Smith's customers] seeking information. It watched a loading operation in which Agwilines used the box. At approximately the same time, defendant approached the shipbuilding company and inquired as to the possibility of purchase of a number of the containers. It was told to communicate with Cowan, plaintiffs' eastern representative. This it did, and, on October 29, 1946, in Pittsburgh, Cowan met with defendant's officials to discuss the proposed sale of [containers]. But, as negotiations progressed, defendant demonstrated an interest in the entire container development. Thus, what started as a meeting to discuss the purchase of individual containers ended in the possible foundation for a sale of the entire business.

Based upon this display of interest, Cowan sent detailed information to defendant concerning the business. This included: (1) patent applications for both the 'knock-down' and 'rigid' crates; (2) blue prints of both designs; (3) a miniature Safeway container; (4) letters of inquiry from possible users; (5) further correspondence with prospective users. In addition, defendant's representatives journeyed to Sturgeon Bay, Wisconsin, the home of the ship-building company, and viewed the physical plant, inventory and manufacturing operation.

Plaintiffs quoted a price of $150,000 plus a royalty of $10 per unit. This was rejected. Subsequent offers of $100,000 and $75,000 with royalties of $10 per container were also rejected. Negotiations continued until January 30, 1947, at which time defendant finally rejected plaintiffs' offer.

On January 31, 1947 defendant announced to Agwilines that it "intended to design and produce a shipping container of the widest possible utility" for "coastal steamship application . . . [and] use . . . on the inland rivers and . . . connecting highway and rail carriers." Development of the project moved rapidly, so that by February 5, 1947 defendant had set up a stock order for a freight container which was designed, by use of plaintiffs' patent applications, so as to avoid any claim of infringement. One differing feature was the use of skids and recesses running the length of the container, rather than legs and sockets as employed in plaintiffs' design. However, Agwilines rejected this design, insisting on an adaptation of plaintiffs' idea. In short defendant's final product incorporated many, if not all, of the features of plaintiffs' design. So conceived, it was accepted by the trade to the extent that, by March 1948, defendant had sold some 500 containers. Many of these sales were made to firms who had shown considerable prior interest in plaintiffs' design and had been included in the prospective users disclosed to defendant.

One particular feature of defendant's container differed from plaintiffs: its width was four inches less. As a result plaintiffs' product became obsolete. Their container could not be used interchangeably with defendant's; they ceased production. Consequently the prospects of disposing of the entire operation vanished.

The foregoing is the essence of plaintiffs' cause of action. Stripped of surplusage, the averment is that defendant obtained, through a confidential relationship, knowledge of plaintiffs' secret designs, plans and prospective customers, and then wrongfully breached that confidence by using the information to its own advantage and plaintiffs' detriment.

[The court found that, notwithstanding certain disclosures of information during the operation of Safeway, plaintiffs' information about how to design its containers remained a trade secret.]

(3) Was Defendant in a Position of Trust and Confidence at the Time of the Disclosure?

Mr. Justice Holmes once said that the existence of the confidential relationship is the "starting point" in a cause of action such as this. E. I. DuPont de Nemours Powder Co. v. Masland, 244 U.S. 100, 102, 37 S. Ct. 575, 61 L. Ed. 1016. While we take a slightly different tack, there is no doubt as to the importance of this element of plaintiffs' case.

Certain it is that a non-confidential disclosure will not supply the basis for a law suit. Plaintiffs' information is afforded protection because it is secret. Promiscuous disclosures quite naturally destroy the secrecy and the corresponding protection. But this is not true where a confidence has been reposed carrying with it communication of an idea.

It is clear that no express promise of trust was exacted from defendant. There is, however, the further question of whether one was implied from the relationship of the parties. Pennsylvania has not provided us with a decision precisely in point but Pressed Steel Car Co. v. Standard Car Co., 210 Pa. 464, 60 A. 4, furnishes abundant guideposts. There plaintiff delivered its blue prints to customers in order that they might acquaint themselves more thoroughly with the railroad cars they were purchasing; from these customers, defendant obtained the drawings. In holding that the customers held the plans as a result of a confidence reposed in them by plaintiff, and that the confidence was breached by delivery of the blue prints to defendant, the court said, 60 A. at page 10: "While there was no expressed restriction placed on the ownership of the prints, or any expressed limitation as to the use to which they were to be put, it is clear . . . that the purpose for which they were delivered by the plaintiff was understood by all parties. . . ."

The quoted language is applicable and determinative. Here plaintiffs disclosed their design for one purpose, to enable defendant to appraise it with a view in mind of purchasing the business. There can be no question that defendant knew and understood this limited purpose. Trust was reposed in it by plaintiffs that the information thus transmitted would be accepted subject to that limitation. "[T]he first thing to be made sure of is that the defendant shall not fraudulently abuse the trust reposed in him. It is the usual incident of confidential relations. If there is any disadvantage in the fact that he knew the plaintiffs' secrets, he must take the burden with the good." E. I. DuPont de Nemours Powder Co. v. Masland, 244 U.S. 100 at page 102, 37 S. Ct. 575, at page 576, 61 L. Ed. 1016.

Nor is it an adequate answer for defendant to say that the transactions with plaintiffs were at arms length. So, too, were the overall dealings between plaintiffs and defendants in Booth v. Stutz Motor Car Co., 7 Cir., 56 F.2d 962; Allen-Qualley Co. v. Shellmar Products Co., D.C., 31 F.2d 293, affirmed, 7 Cir., 36 F.2d 623 and Schavoir v. American Rebonded Leather Co., 104 Conn. 472, 133 A. 582. That fact does not detract from the conclusion that but for those very transactions defendant would not have learned, from plaintiffs, of the container design. The implied limitation on the use to be made of the information had its roots in the "arms-length" transaction.

(4) The Improper Use by Defendant of the Secret Information

Defendant's own evidence discloses that it did not begin to design its container until after it had access to plaintiffs' plans. Defendant's engineers admittedly referred to plaintiffs' patent applications, as they said, to avoid infringement. It is not disputed that, at the urging of Agwilines, defendant revised its proposed design to incorporate the folding leg and socket principles of plaintiffs' containers. These evidentiary facts, together with the striking similarity between defendant's and plaintiffs' finished product, were more than enough to convict defendant of the improper use of the structural information obtained from plaintiffs.

COMMENTS AND QUESTIONS

1. Since the shipping containers were available on the open market, couldn't Dravo have argued that they were not a trade secret? Did Smith's agent, Cowan, disclose anything that was *not* readily apparent from inspection of the containers?

2. In Van Prod. Co. v. General Welding & Fabricating Co., 213 A.2d 769, 779-780 (Pa. 1965), the Pennsylvania Supreme Court criticized *Smith* for focusing on the existence of a confidential relationship to the exclusion of whether there was a trade secret at all. The court said: "The starting point in every case of this sort is not whether there was a confidential relationship, but whether, in fact, there was a trade secret to be misappropriated."

3. Compare Smith v. Dravo with Omnitech Intl. v. Clorox Co., 29 U.S.P.Q.2d 1665 (5th Cir. 1994), *cert. denied,* 115 S. Ct. 71 (1994), where the court held that it was not an actionable "use" of a trade secret for the defendant to *evaluate* it in the course of trying to decide whether to (a) acquire the company or (b) take a license to use the trade secret. A finding of no liability here makes sense, because it allows the potential licensee to make an informed judgment and therefore promotes efficient licensing. But how far does it extend? Is the potential licensee entitled to replicate the research or build models in an effort to evaluate it? Are there special limits that should be placed on companies engaged in a "make or buy" decision — that is, who are considering *either* licensing the plaintiff's technology *or* entering the market themselves?

Smith and *Omnitech* both involved disclosures in the course of licensing negotiations. Cases such as these form part of the amorphous law of "precontractual liability." See E. Allen Farnsworth, Precontractual Liability and Preliminary Agreements: Fair Dealing and Failed Negotiations, 87 Colum. L. Rev. 217 (1987). Both the *Smith* and *Omnitech* cases present the problem of Arrow's Information Paradox. What the plaintiff has to sell to the defendant is information that is valuable only because it is secret. If there is no legal protection and if the plaintiff discloses the secret to the defendant, its value will be lost. But the defendant cannot be expected to pay for information unless it can see the information to determine its value. Thus, absent some form of legal protection for confidential disclosures, sale or licensing of trade secrets may not occur. Note that this problem does not arise in other areas of intellectual property, such as patent law, since the inventions being licensed are already publicly disclosed.

4. The Restatement (Third) of Unfair Competition §41 holds that a confidential relationship is established in the following circumstances:

> (a) the person made an express promise of confidentiality prior to the disclosure of the trade secret; or
> (b) the trade secret was disclosed to the person under circumstances in which the relationship between the parties to the disclosure or the other facts surrounding the disclosure justify the conclusions that, at the time of the disclosure,
>> (1) the person knew or had reason to know that the disclosure was intended to be in confidence, and
>> (2) the other party to the disclosure was reasonable in inferring that the person consented to an obligation of confidentiality.

On the issue of knowledge, compare this "had reason to know" standard with the Eleventh Circuit's decision in Bateman v. Mnemonics, Inc., 79 F.3d 1532 (11th Cir. 1996). In that case, the court indicated that it was "wary" of trade secret claims based on implied confidential relationships because they were subject to abuse. The court rejected Bateman's allegation that such a relationship existed because Bateman had not "made it clear to the parties involved that there was an expectation and obligation of confidentiality." Thus the court seemed to create a standard of actual knowledge on the part of the recipient of confidential information that the discloser of such information intended the disclosure to be confidential. At the other extreme, the Fifth Circuit in Phillips v. Frey, 20 F.3d 623, 631-632 (5th Cir. 1994) found an implied confidential relationship to exist in the course of negotiations over the sale of a business, despite the fact that the disclosing party never requested that the information remain confidential.

Which approach makes more sense? Who is in the best position to clarify the question, the discloser or the recipient?

5. Is there any way for a potential licensee to prevent the formation of a confidential relationship in such a situation?

6. The defendant in *Smith* actually sold a device that was not identical to the plaintiff's. Under what circumstances can a defendant be liable for misappropriation without literally copying the trade secret? In Mangren Res. & Dev. Corp. v. National Chem. Co., 87 F.3d 1339 (7th Cir. 1996), the court defined improper "use" broadly, stating that "the user of another's trade secret is liable even if he uses it with modifications or improvements upon it effected by his own efforts, so long as the substance of the process used by the actor is derived from the other's secret. . . . If trade secret law were not flexible enough to encompass modified or even new products that are substantially derived from the trade secrets of another, the protections that the law provides would be hollow indeed."

To similar effect as *Mangren* is Texas Tanks Inc. v. Owens-Corning Fiberglas Corp., 99 F.3d 740 (5th Cir. 1997). In *Texas Tanks*, the Fifth Circuit held that any improper "exercise of control and domination" over a secret constituted a commercial use of that secret. It rejected the defendant's argument that it could not be liable for taking a secret unless the secret was actually incorporated in a commercial product. The court noted that Owens Corning's awareness of the secret would likely influence the development of its own competing product, and that this was enough to demonstrate improper "use" of the secret.

PROBLEMS

Problem 2-7. Solomon, a regular customer of ToolCo's products, comes up with an idea for a new tool. He sends the idea to ToolCo, suggesting that they manufacture it. ToolCo does in fact produce and sell the new tool. Is Solomon entitled to compensation for ToolCo's use of his idea? Does it make any difference if (1) ToolCo actively solicited the idea from Solomon? (2) Solomon sent the suggestion to ToolCo along with a letter saying he wanted to open negotiations over a possible licensing arrangement to use the idea? (3) ToolCo has in the past had an informal, unwritten policy of compensating inventors who submit good ideas?

Problem 2-8. VenCo, a venture capitalist in the business of financing start-up companies, investigates TechCo in an effort to decide whether or not to finance it. To aid in its investigation, VenCo asks for and receives confidential information about TechCo's products and market position. VenCo eventually decides not to finance TechCo because of concerns about its management, but it does finance a start-up competitor of TechCo in the same field. Are VenCo or the start-up liable to TechCo? What obligations, if any, does VenCo undertake as a result of its exposure to TechCo's secrets?

---

3. Reverse Engineering

Not all use or disclosure of someone's secret is actionable misappropriation. Rather, as the commissioners who drafted the Uniform Trade Secrets Act noted, there are several categories of "proper means" of obtaining a trade secret. We classify these "proper means" as defenses, because they do not directly deny the existence of a trade secret or the defendant's use of that secret. Rather, they are *legitimate* uses of trade secrets by a competitor.

The commissioners' Comment to section 1 of the act had this to say on the subject of defenses:

One of the broadly stated policies behind trade secret law is "the maintenance of standards of commercial ethics." The Restatement of Torts, Section
757, Comment (f), notes: "A complete catalogue of improper means is not
possible," but Section 1(1) includes a partial listing.

Proper means include:

1. Discovery by independent invention;

2. Discovery by "reverse engineering," that is, by starting with the known
product and working backward to find the method by which it was developed.
The acquisition of the known product must of course, also be by a fair and
honest means, such as purchase of the item on the open market for reverse
engineering to be lawful; . . .

4. Observation of the item in public use or on public display;

5. Obtaining the trade secret from published literature.

See also Restatement (Third) of Unfair Competition §43 ("Independent
discovery and analysis of publicly available products or information are not
improper means of acquisition.").

≡≡≡ *Chicago Lock Co. v. Fanberg*
≡≡≡ *United States Court of Appeals for the Ninth Circuit*
≡≡≡ *676 F.2d 400 (9th Cir. 1982)*

ELY, Circuit Judge:

The Chicago Lock Company ("the Company"), a manufacturer of "tubular" locks, brought suit against Morris and Victor Fanberg, locksmiths and
publishers of specialized trade books, to enjoin the unauthorized dissemination of key codes for the Company's "Ace" line of tubular locks. The
District Court granted summary judgment in favor of the Fanbergs as to the
Company's federal claims of trademark infringement and unfair competition,
but held trial on the common law claim of unfair competition under former
Cal. Civ. Code §3369.

The court concluded that the key codes for the Company's tubular locks
were improperly acquired trade secrets and enjoined distribution of the Fanbergs' compilation of those codes. For the reasons set forth in this opinion,
we reverse the District Court and order that judgment be entered in favor of
the Fanbergs.

The Facts

Since 1933 the Chicago Lock Company, a manufacturer of various types
of locks, has sold a tubular lock, marketed under the registered trademark
"Ace," which provides greater security than other lock designs. Tubular Ace
locks, millions of which have been sold, are frequently used on vending and
bill changing machines and in other maximum security uses, such as burglar

alarms. The distinctive feature of Ace locks (and the feature that apparently makes the locks attractive to institutional and large-scale commercial pur-chasers) is the secrecy and difficulty of reproduction associated with their keys.

The District Court found that the Company "has a fixed policy that it will only sell a duplicate key for the registered series 'Ace' lock to the owner of record of the lock and on request of a bona fide purchase order, letterhead or some other identifying means of the actual recorded lock owner." In ad-dition, the serial number-key code correlations are maintained by the Com-pany indefinitely and in secrecy. The Company does not sell tubular key "blanks" to locksmiths or others, and keys to Ace locks are stamped "Do Not Duplicate."

If the owner of an Ace lock loses his key, he may obtain a duplicate from the Company. Alternatively, he may have a proficient locksmith "pick" the lock, decipher the tumbler configuration, and grind a duplicate tubular key. The latter procedure is quicker than the former, though more costly. The locksmith will, to avoid the need to "pick" the lock each time a key is lost, record the key code (i.e., the tumbler configuration) along with the serial number of the customer's lock. Enough duplicate keys have been made by locksmiths that substantial key code data have been compiled, albeit noncom-mercially and on an ad hoc basis.

Appellant Victor Fanberg, the son of locksmith Morris Fanberg and a locksmith in his own right, has published a number of locksmith manuals for conventional locks. Realizing that no compilation had been made of tubular lock key codes, in 1975 Fanberg advertised in a locksmith journal, Locksmith Ledger, requesting that individual locksmiths transmit to him serial number-key code correlations in their possession in exchange for a copy of a complete compilation when finished. A number of locksmiths complied, and in late 1976 Fanberg and his father began to sell a two-volume publication of tu-bular lock codes, including those of Ace locks, entitled "A-Advanced Lock-smith's Tubular Lock Codes." In 1976 and 1977 Fanberg advertised the manuals in the Locksmith Ledger for $49.95 and indicated that it would be supplemented as new correlations became known. About 350 manuals had been sold at the time of trial. The District Court found that Fanberg "had lost or surrendered control over persons who could purchase the books," meaning that nonlocksmiths could acquire the code manuals.

The books contain correlations which would allow a person equipped with a tubular key grinding machine to make duplicate keys for any listed Ace lock if the serial number of the lock was known. On some models, the serial numbers appear on the exterior of the lock face. Thus, Fanberg's manuals would make it considerably easier (and less expensive) for a person to obtain (legitimately or illegitimately) duplicate keys to Ace locks without going through the Company's screening process. This is what caused consternation to the Company and some of its customers. At no time did Fanberg seek, or the Company grant, permission to compile and sell the key codes. Nor did the individual locksmiths seek authorization from the Company or their cus-tomers before transmitting their key code data to Fanberg. . . .

The Trade Secrets Claim

Appellants argue that the District Court erroneously concluded that they are liable under Section 3369 for acquiring appellee's trade secret through improper means. We agree, and on this basis we reverse the District Court.

Although the District Court's Findings of Fact and Conclusions of Law are lengthy, the thrust of its holding may be fairly summarized as follows: appellants' acquisition of appellee's serial number-key code correlations through improper means, and the subsequent publication thereof, constituted an "unfair business practice" within the meaning of Section 3369. Even though the court did not make an explicit finding that appellee's serial number-key code correlations were protectable trade secrets, both appellants and appellee premise their appeal on such an "implicit" finding. We think it clear that the District Court based its decision on a theory of improper acquisition of trade secrets, and in the following discussion we assume arguendo that appellee's listing of serial number-key code correlations constituted a trade secret. . . .

Trade secrets are protected . . . in a manner akin to private property, but only when they are disclosed or used through improper means. Trade secrets do not enjoy the absolute monopoly protection afforded patented processes, for example, and trade secrets will lose their character as private property when the owner divulges them or when they are discovered through proper means. "It is well recognized that a trade secret does not offer protection against discovery by fair and honest means such as by independent invention, accidental disclosure or by so-called reverse engineering, that is, starting with the known product and working backward to divine the process." Sinclair, 42 Cal. App. 3d at 226, 116 Cal. Rptr. at 661.

Thus, it is the employment of improper means to procure the trade secret, rather than mere copying or use, which is the basis of liability. Restatement (First) of Torts, sec. 757, comment a (1939). The Company concedes, as it must, that had the Fanbergs bought and examined a number of locks on their own, their reverse engineering (or deciphering) of the key codes and publication thereof would not have been use of "improper means." Similarly, the Fanbergs' claimed use of computer programs in generating a portion of the key code-serial number correlations here at issue must also be characterized as proper reverse engineering. The trial court found that appellants obtained the serial number-key code correlations from a "comparatively small" number of locksmiths, who themselves had reverse-engineered the locks of their customers. The narrow legal issue presented here, therefore, is whether the Fanbergs' procurement of these individual locksmiths' reverse engineering data is an "improper means" with respect to appellee Chicago Lock Company.

The concept of "improper means," as embodied in the Restatement, and as expressed by the Supreme Court, connotes the existence of a duty to the trade secret owner not to disclose the secret to others. See Restatement (First) of Torts, sec. 757, comment h (1939). "The protection accorded the trade secret holder (i.e., in this case the Company) is against the disclosure or unauthorized use of the trade secret by those to whom the secret has been confided under the express or implied restriction of disclosure or nonuse." Kewanee Oil Co. v. Bicron Corp., 416 U.S. 470, 475, 94 S. Ct. 1879, 1883, 40 L. Ed. 2d 315 (1974) (emphasis added).

Thus, under Restatement sec. 757(c), appellants may be held liable if they intentionally procured the locksmiths to disclose the trade secrets in breach of the locksmiths' duty to the Company of nondisclosure. See Restatement (First) of Torts, sec. 757, comment h (1939). Critical to the District Court's holding, therefore, was its conclusion that the individual locksmiths, from whom the Fanbergs acquired the serial number-key code correlations, owed an implied duty to the Company not to make the disclosures.

We find untenable the basis upon which the District Court concluded that the individual locksmiths owe a duty of nondisclosure to the Company. The court predicated this implied duty upon a "chain" of duties: first, that the locksmiths are in such a fiduciary relationship with their customers as to give rise to a duty not to disclose their customers' key codes without permission; and second, that the lock owners are in turn under an "implied obligation (to the Company) to maintain inviolate" the serial number-key code correlations for their own locks.

The court's former conclusion is sound enough: in their fiduciary relationship with lock owners, individual locksmiths are reposed with a confidence and trust by their customers, of which disclosure of the customers' key codes would certainly be a breach. This duty, however, could give rise only to an action by "injured" lock owners against the individual locksmiths, not by the Company against the locksmiths or against the Fanbergs.[3]

The court's latter conclusion, that lock owners owe a duty to the Company, is contrary to law and to the Company's own admissions. A lock purchaser's own reverse-engineering of his own lock, and subsequent publication of the serial number-key code correlation, is an example of the independent invention and reverse engineering expressly allowed by trade secret doctrine.[4] Imposing an obligation of nondisclosure on lock owners here would frustrate the intent of California courts to disallow protection to trade secrets discovered through "fair and honest means." See id. Further, such an implied obligation upon the lock owners in this case would, in effect, convert the

---

3. The Company premised its complaint on a violation of its own rights under trade secret law; the Company did not allege violations of the lock owners' rights. Cf. Barquis v. Merchants Collection Assn., 7 Cal. 3d 94, 110, 101 Cal. Rptr. 745, 756, 496 P.2d 817 (1972) (Section 3369 permits members of the public to sue on behalf of the public generally); Hernandez v. Atlantic Finance Co. of Los Angeles, 105 Cal. App. 3d 65, 72-73, 164 Cal. Rptr. 279, 284 (1980) (statute authorizes suits for injunction by individuals on behalf of the general public).

4. If a group of lock owners, for their own convenience, together published a listing of their own key codes for use by locksmiths, the owners would not have breached any duty owed to the Company. Indeed, the Company concedes that a lock owner's reverse engineering of his own lock is not "improper means."

Company's trade secret into a state-conferred monopoly akin to the absolute protection that a federal patent affords. Such an extension of California trade secrets law would certainly be preempted by the federal scheme of patent regulation.

Appellants, therefore, cannot be said to have procured the individual locksmiths to breach a duty of nondisclosure they owed to the Company, for the locksmiths owed no such duty. The Company's serial number-key code correlations are not subject to protection under Restatement sec. 757, as adopted by the California courts, because the Company has not shown a breach of any confidence reposed by it in the Fanbergs, the locksmiths, or the lock purchasers — i.e., it has failed to show the use of "improper means" by the Fanbergs required by the Restatement.

COMMENTS AND QUESTIONS

1. Why is reverse engineering lawful? If, as the Commissioners suggested, one purpose of trade secret law is to promote standards of commercial ethics, doesn't there seem to be something wrong with taking apart a competitor's product in order to figure out how to copy it (or, as in *Chicago Lock*, how to render it useless)? Does reverse engineering benefit only those competitors who are not smart enough to develop ideas or products for themselves?

Does reverse engineering by one or a few firms mean that the plaintiff's information is no longer secret? See Barr-Mullin, Inc. v. Browning, 442 S.E.2d 226 (Ct. App. N.C. 1993). In Reingold v. Swiftships, 126 F.3d 645 (5th Cir. 1997), the court rejected a claim that a boat hull mold could not be a trade secret because it could readily be reverse engineered. While reverse engineering may protect one who engages in it, the court held, it does not protect those who actually acquire the secret by improper means. Should it matter whether the reverse engineering is very difficult or easy enough that any competitor could do it? Reverse engineering may be explained as a legal rule designed to weaken trade secret protection relative to patent protection. Can you think of reasons why we would want to weaken trade secret protection? See Kewanee Oil Co. v. Bicron Corp., 416 U.S. 470 (1974) (patent law does not preempt trade secret law, in part because the reverse-engineering rule "weakens" trade secret law).

2. Consider the recurring question of whether the parties can agree to override trade secret law. Suppose that the owner of a trade secret includes in a license or sale contract a provision prohibiting the buyer from reverse engineering the product. Is that contractual provision enforceable? The courts are split. Compare K & G Oil & Tool Service Co. v. G & G Fishing Tool Service, 158 Tex. 94, 314 S.W.2d 782, 785-86 (1958) (contract preventing disassembly of tools to protect trade secrets was enforceable) with Vault Corp. v. Quaid Software Ltd., 847 F.2d 255, 265-267 (5th Cir. 1988)

(contract provision prohibiting reverse engineering of software void as against public policy).

3. If a company is looking for a piece of information — say, the solution to a given technical problem — society is very much concerned that the company be allowed free access to information in the public domain. The reason is that minimizing the search costs of the company is thought to be a social good, consistent with some level of protection for the prior investment of others. In a similar vein, suppose the costs of obtaining that information from the public domain are very high, but the costs (to the company) of stealing it from a competitor are very low. Should society punish — and therefore deter — the theft? Wouldn't that simply create an inefficiency, since the company could get the information but would have to incur greater search costs? Cf. Friedman, Landes & Posner, Some Economics of Trade Secret Law, 5 J. Econ. Perspectives 61, 62 (1991) (arguing that trade secret law prohibits only costly means of obtaining competitors' information, while encouraging cheaper forms of obtaining information such as reverse engineering). Are there other considerations that militate in favor of requiring such a search?

PROBLEMS

Problem 2-9. Atech and Alpha both manufacture complex medical devices used in diagnosing a variety of ailments. These devices are sold almost exclusively to hospitals, since they cost in excess of $100,000 each. There are several hundred Atech devices currently in use in hospitals throughout the country. Atech, which claims a trade secret in the internal workings of its device, carefully monitors the purchasers of its device. Alpha pays a third party to buy a device from Atech without disclosing that it will be given to Alpha. Once it has obtained the device, Alpha disassembles it and studies it in order to compare it to Alpha's own device. In the course of opening it up, Alpha's engineers pick two internal padlocks on the Atech device. When Atech discovers that Alpha has obtained the device, it demands the unit's return, offering to refund the purchase price. Alpha refuses, and Atech sues for misappropriation of trade secrets.

a) Assume that Atech's trade secrets were worth $5 million. Assume further that the padlocks cost $5 each, and that it costs $100 per lock to pick these padlocks. Has Atech taken reasonable precautions to preserve its secrets? What other security measures must Atech take — and at what cost — both to deter Alpha (and others) from reverse engineering and to preserve its secrecy? Does Atech's sale of the products on the open market automatically preclude a finding of secrecy?

b) Assume Atech has presented the buyer of the machine with a contract that licenses (rather than sells) the machine, subject to the following restrictions: (a) the buyer is prohibited from disclosing anything it learns during the course of using the machine; (b) the buyer is prohibited from reselling the machine; and (c) the buyer is liable for Atech's damages in the event that any third party learns of Atech's secrets from the buyer or the buyer's machine. Is such an agreement enforceable? Would you advise a client thinking of licensing an Atech machine to sign this contract? How would you redraft the agreement to protect the buyer?

c) Assume that after Alpha's engineers picked the padlocks, they gained access to the inner workings of the machine. Assume further that they discovered numerous flaws in the imaging mechanism that caused potentially serious defects in the images (and hence the diagnoses) stemming from the Atech machine. Finally, assume that Alpha's engineers not only fixed these problems but significantly improved on Atech's design and hence the reliability of the machine. Should these facts affect Alpha's liability? Atech's remedy?

Problem 2-10. Bonnie Bluenote, a world-famous blues guitarist, is noted for her distinctive sound, which she gets by tuning her guitar specially every time before she plays. Although the guitar itself is a standard professional model, Bonnie's adjustments of settings on the guitar, amplifier, and sound system combine to produce a distinctive sound. Because the sound is so important to her image, Bonnie guards it carefully. While she is tuning her guitar, only band members and close associates are allowed in the room. When she records in a new studio, she has the sound engineers sign nondisclosure agreements.

One day Freddie Fender-Rhodes, a big fan of Bonnie's and a budding bluesman himself, is hanging around outside the studio where Bonnie is recording her newest album. He happens to see an ID tag, worn by all guests in the studio, in the wastebasket. He fills in his name and walks into the studio. The band members and recording engineers, seeing the tag, let Freddie stay. He observes how Bonnie tunes her guitar, sees the sound board settings, and makes extensive mental notes.

Five months later, Bonnie is shocked to see in a record store a CD by Freddie titled "The Bonnie Bluenote Sound." Then she discovers that Freddie is planning to publish an article in Blues Guitar magazine revealing the secrets to Bonnie's sound. Does she have any recourse against Freddie? What additional precautions should she have taken?

4. The Special Case of Departing Employees

Many of the thorniest issues in trade secret (and contract) law arise when employees leave a company in order either to start their own business or to take a job elsewhere. Such cases present a fundamental clash of rights between an employee and an employer, as Judge Adams suggests in the following excerpt from SI Handling Systems v. Heisley, 753 F.2d 1244, 1266-1269 (3d Cir. 1985) (Adams, J., concurring):

When deciding the equitable issues surrounding the request for a trade secret injunction, it would seem that a court cannot act as a pure engineer or scientist, assessing the technical import of the information in question. Rather, the court must also consider economic factors since the very definition of "trade secret" requires an assessment of the competitive advantage a particular item of information affords to a business. Similarly, among the elements to be weighed in determining trade secret status are the value of the information to its owner and to competitors, and the ease or difficulty with which the information may be properly acquired or duplicated.

While the majority may be correct in suggesting that the trial court need not always "engage in extended analysis of the public interest," the court on occasion must apply the elements of sociology. This is so since trade secret cases frequently implicate the important countervailing policies served on one hand by protecting a business person from unfair competition stemming from the usurpation of trade secrets, and on the other by permitting an individual to pursue unhampered the occupation for which he or she is best suited. "Trade secrets are not . . . so important to society that the interests of employees, competitors and competition should automatically be relegated to a lower position whenever trade secrets are proved to exist." Robison, The Confidence Game: An Approach to the Law About Trade Secrets, 25 Ariz. L. Rev. 347, 382 (1983).

These observations take on more force, I believe, when a case such as the present one involves the concept of "know-how." Under Pennsylvania law an employee's general knowledge, skill, and experience are not trade secrets. Thus in theory an employer generally may not inhibit the manner in which an employee uses his or her knowledge, skill, and experience — even if these were acquired during employment. When these attributes of the employee are inextricably related to the information or process that constitutes an employer's competitive advantage — as increasingly seems to be the case in newer, high-technology industries — the legal questions confronting the court necessarily become bound up with competing public policies.

It is noteworthy that in such cases the balance struck by the Pennsylvania courts apparently has favored greater freedom for employees to pursue a chosen profession. The courts have recognized that someone who has worked in a particular field cannot be expected to forego the accumulated skills, knowledge, and experience gained before the employee changes jobs. Such qualifications are obviously very valuable to an employee seeking to sell his services in the marketplace. A person leaving one employer and going into the marketplace will seek to compete in the area of his or her greatest aptitude. In light of the highly mobile nature of our society, and as the economy becomes increasingly comprised of highly skilled or high-tech jobs, the individual's economic interests will more and more be buffeted by competitive advantage. Courts must be cautious not to strike a balance that unduly disadvantages the individual worker. . . .

In my view a proper injunction necessarily would impose the minimum restraint upon the free utilization of employee skill consistent with denying unfaithful employees an advantage from misappropriation of information. Thus, as I see it, the district court, on remand, should fashion an injunction that extends only so long as is essential to negate any unfair advantage that may have been gained by the appellants.

The majority opinion in *SI Handling* partially upheld a finding that two former employees had misappropriated trade secrets, but it vacated an injunction against them in order for the district court to reconsider its scope.

Most companies require their employees to sign some sort of employment agreement, either when they are hired or at some point during their tenure. Employment agreements generally fall into one or more of three categories: confidentiality agreements, invention assignments, and noncompetition agreements. Confidentiality agreements, with which we are concerned here, generally recite that the employee will receive confidential information during her employment, and that she undertakes to keep it secret and not to use it for anyone other than the employer. Invention assignments give the employer the right to intellectual property created by the employee while she was employed. Noncompetition agreements limit the circumstances in which former employees can compete for customers with their former employers.

Presumably, employers may require their employees to agree not to use or disclose trade secrets to which they have access during employment. The problem comes when employers attempt to limit an employee's use of information that does not constitute a trade secret under the legal standards described above. Can employers and employees "agree" that the employee will not use *any* information obtained during employment, whether it is public or not? This question parallels the problem of restrictive license provisions — are employers limited by the intellectual property laws to protecting only trade secrets that meet the statutory requirements, or are they free to impose additional restrictions on their employees so long as they do so by agreement? Courts struggle with this issue, with the majority concluding that "reasonable" contract restrictions on employees are enforceable even in the absence of a protectable trade secret. See 12 Roger Milgrim, Milgrim on Trade Secrets §3.05[1][a], at 3-209 to 3-210. State statutes that address the issue have generally been interpreted to allow such "reasonable" employee agreements, regardless of how the statutes themselves are worded.[15] Of course, what agreements are "reasonable" is far from clear in this context, and is the subject of considerable litigation.

The issues regarding the rights of departing employees are here divided into three categories: the right to "raid" a former employer by hiring away

---

15. An example of such a statute is Tex. Bus. & Com. Code §15.50 (employee agreements are enforceable if they "are reasonable and do not impose a greater restraint than is necessary to protect the goodwill or other business interest of the promisee."). In Light v. Centel Cellular Co. of Texas, 883 S.W.2d 642 (Tex. 1994), the Texas Supreme Court held that to be enforceable, such agreements must be ancillary to an otherwise valid agreement with consideration on both sides. Thus the court held that noncompetition agreements could not be enforced in an employment-at-will contract because the employer had no continuing obligation.

its employees; the right to compete with a former employer, including the right to use information obtained from the former employer; and the ownership of an employee's own inventions made during employment.

a. Can Employees "Raid" Their Company?

The practice of "raiding" — hiring away some or all of the employees in a company to start a directly competing company — is a subset of the question whether departing employees can take their employer's ideas. Stated this way, the answer to the question seems simple: no. General principles of trade secret law state that you cannot use or disclose the trade secrets of another when you obtained those secrets in a confidential relationship (such as an employer-employee relationship). These principles go a long way toward resolving the issues that arise when a departing employee attempts to take an employer's information with her.

On the other hand, ex-employees are free to take with them the "general skill, knowledge, training and experience" that they obtained from working in the industry. See, e.g., Restatement (Third) of Unfair Competition §42 Comment *c*, at 481. Balancing these competing considerations is not always easy. In Stampede Tool Warehouse Inc. v. May, 651 N.E.2d 209 (Ill. Ct. App. 1995), for example, the plaintiff alleged that former employees who started a competing business had misappropriated its trade secrets by memorizing a list of plaintiff's customers and soliciting those customers. The court found the defendants liable, reasoning that "memorization is one method of misappropriation." But if the case involved merely a former employee contacting those she knew from experience to be potential customers, rather than explicitly attempting to memorize a list of customers, it is hard to fault the employee's conduct.

If the departing employees can avoid using the employer's trade secrets when they leave, the question becomes a rather different one — is it acceptable to hire away a group of employees? This question has a very different answer. In Diodes, Inc. v. Franzen, 260 Cal. App. 2d 244, 67 Cal. Rptr. 19 (1968), the president and vice-president of Diodes left to form a competing company, called Semtech. Before they left, the officers solicited a number of Diodes employees to join them. Diodes sued the departing employees, alleging a number of claims centering on unfair competition and breach of fiduciary duty. The court dismissed the complaint, stating:

> As a general principle, one who unjustifiably interferes with an advantageous business relationship to another's damage may be held liable therefor. The product is bottled under a variety of labels, including unfair competition, interference with advantageous relations, contract interference, and inducing breach of contract.
>
> Even though the relationship between an employer and his employee is an advantageous one, no actionable wrong is committed by a competitor who so-

licits his competitor's employees or who hires away one or more of his competitor's employees who are not under contract, so long as the inducement to leave is not accompanied by unlawful action. In the employee situation the courts are concerned not solely with the interests of the competing employers, but also with the employee's interest. The interests of the employee in his own mobility and betterment are deemed paramount to the competitive business interests of the employers, where neither the employee nor his new employer has committed any illegal act accompanying the employment change.

67 Cal. Rptr. at 25-26 (citations omitted).

Can employers change this result by forcing their employees to sign "nonsolicitation agreements" that prevent a departing employee from soliciting other employees to join him? What if such agreements also prohibit departing employees from soliciting their former customers? In a case that has been heavily criticized on other grounds, the Ninth Circuit has upheld (albeit without discussion) an injunction against solicitation of employees or customers based on such a nonsolicitation agreement. See MAI Systems Corp. v. Peak Computing, Inc., 991 F.2d 511, 523 (9th Cir. 1993). Other courts and commentators take a more guarded approach, prohibiting solicitation only where there is active inducement or where departing employees have taken a customer list which is itself a trade secret. Cf. James H. A. Pooley, Restrictive Employee Covenants in California, 4 Santa Clara Computer & High Tech. L.J. 251, 259 (1988).

Should former employees be allowed to compete for the business of their old customers? Should they be allowed to hire away their coworkers? Does your answer depend on whether the employees are using trade secrets in the solicitation? On whether the employee has signed a nonsolicitation agreement?

Note that while there is a general legal policy in favor of employee mobility, some economists have argued that the United States encourages too many start-up companies and thereby dissipates technological concentration, integrative synergy, and economic advantage. See Richard L. Florida & Martin Kenney, The Breakthrough Illusion: Corporate America's Failure to Move from Innovation to Mass Production (1990); F. M. Scherer, Entrepreneurship, Technological Innovation, and Economic Growth: Studies in the Schumpeterian Tradition (1992).

b. The Right to Compete

Employees inevitably learn a great deal about a business on the job. They may learn the trade secrets of their employer, but they may also learn such basic information as how many people it takes to make a product, who is likely to buy the product, where to advertise, etc. Further, employees develop personal relationships with vendors, customers, and others that are immensely useful in business. For this reason, many employers would like to prevent their employees from competing against them at all. Such employers often ask their employees to sign "noncompetition agreements," which prevent the employee from competing with his former employer for customers for a set period of time. Should such agreements be enforceable?

In Comprehensive Technologies Intl. v. Software Artisans, Inc., 3 F.3d 730 (4th Cir. 1993), CTI brought suit for copyright infringement and misappropriation of trade secrets against a group of former employees who left the company to form a competing company which shortly thereafter came out with a new product. The court concluded that the departing employees had not infringed CTI's copyrights or misappropriated any CTI trade secrets. Nonetheless, the court enforced an agreement signed by one of the employees, Dean Hawkes. The agreement provided that for a period of twelve months after he left CTI, Hawkes would not

> engage directly or indirectly in any business within the United States (financially as an investor or lender or as an employee, director, officer, partner, independent contractor, consultant or owner or in any other capacity calling for the rendition of personal services or acts of management, operation or control) which is in competition with the business of CTI. For purposes of this Agreement, the "business of CTI" shall be defined as the design, development, marketing, and sales of CLAIMS EXPRESS) and EDI LINK) type PC-based software with the same functionality and methodology. . . .

The court stated the general legal standard governing covenants not to compete:

> Virginia has established a three-part test for assessing the reasonableness of restrictive employment covenants. Under the test, the court must ask the following questions:
>
>> (1) Is the restraint, from the standpoint of the employer, reasonable in the sense that it is no greater than is necessary to protect the employer in some legitimate business interest?
>> (2) From the standpoint of the employee, is the restraint reasonable in the sense that it is not unduly harsh and oppressive in curtailing his legitimate efforts to earn a livelihood?
>> (3) Is the restraint reasonable from the standpoint of a sound public policy?
>
> Blue Ridge Anesthesia & Critical Care, Inc. v. Gidick, 239 Va. 369, 389 S.E.2d 467, 469 (Va. 1990). If a covenant not to compete meets each of these standards of reasonableness, it must be enforced. As a general rule, however, the Virginia courts do not look favorably upon covenants not to compete, and will strictly construe them against the employer. The employer bears the burden of demonstrating that the restraint is reasonable.

The court found that Hawkes's agreement, which prevented him from competing with CTI anywhere in the United States, was reasonable because CTI had offices, clients, or prospects in many (though not all) states throughout the country. Further, the court noted:

. . . As the individual primarily responsible for the design, development, marketing and sale of CTI's software, Hawkes became intimately familiar with every aspect of CTI's operation, and necessarily acquired information that he could use to compete with CTI in the marketplace. When an employee has access to confidential and trade secret information crucial to the success of the employer's business, the employer has a strong interest in enforcing a covenant not to compete because other legal remedies often prove inadequate. It will often be difficult, if not impossible, to prove that a competing employee has misappropriated trade secret information belonging to his former employer. On the facts of this case, we conclude that the scope of the employment restrictions is no broader than necessary to protect CTI's legitimate business interests.

Most courts similarly apply an overarching requirement of "reasonableness" to covenants not to compete. They may disagree, however, on what restrictions are reasonable. In Gateway 2000 Inc. v. Kelley, 9 F. Supp. 2d 790 (E.D. Mich. 1998), the court invalidated an agreement that was similar to the one upheld in CTI. The court relied in part on the fact that the company had later adopted a less restrictive noncompetition provision, suggesting that the older, broader provision was not necessary to protect its interests.

COMMENTS AND QUESTIONS

1. There seems to be no question in the court's mind that none of the defendants misappropriated any CTI trade secrets, infringed any copyrights, or otherwise "took" anything belonging to CTI in starting Software Artisans. Why doesn't that dispose of the case? What social purpose is served by enjoining former employees from pursuing their livelihood? Shouldn't the mobility and liberty of individuals be the paramount consideration, as California courts have suggested? See Diodes, Inc. v. Franzen, 260 Cal. App. 2d 244 (1968). One author has suggested that there is a more practical economic motivation for precluding such noncompetition agreements. She argues that the relative success of California's Silicon Valley compared to Boston's Route 128 is directly attributable to the prevalence of noncompetition agreements in Route 128 companies, which prevented the free movement of employees and therefore discouraged start-up companies. See Annalee Saxenian, Regional Advantage: Culture and Competition in Silicon Valley and Route 128 (1994).

Some courts have limited the enforcement of noncompetition agreements to situations where trade secrets are likely to be used or disclosed if an employee is allowed to compete. The New York Court of Appeals, for example, took the following view:

> Undoubtedly judicial disfavor of these covenants is provoked by "powerful considerations of public policy which militate against sanctioning the loss of a man's livelihood" (Purchasing Assoc. v. Weitz. . . .) Indeed, our economy is premised on the competition engendered by the uninhibited flow of services, talent and ideas. Therefore, no restrictions should fetter an employee's right to apply to

his own best advantage the skills and knowledge acquired by the overall experience of his previous employment. This includes those techniques which are but "skillful variations of general processes known to the particular trade" (Restatement, Agency 2d, §396 Comment *b*.

Of course, the courts must also recognize the legitimate interest an employer has in safeguarding that which has made his business successful and to protect himself against deliberate surreptitious commercial piracy. Thus restrictive covenants will be enforceable to the extent necessary to prevent the disclosure or use of trade secrets or confidential customer information. In addition injunctive relief may be available where an employee's services are unique or extraordinary and the covenant is reasonable. This latter principle has been interpreted to reach agreements between members of the learned professions.

Reed Roberts Assoc. v. Strauman, 40 N.Y.2d 303, 353 N.E.2d 590 (Ct. App. 1976). Does the *Reed Roberts* approach in essence hold noncompetition agreements unenforceable, since it allows them to operate only when trade secret laws also provide relief? Are there sound reasons to enforce an employer-employee agreement that prevents the employee from competing after termination? Many courts have done so. See, e.g., Picker, Intl. v. Blanton, 756 F. Supp. 971 (N.D. Tex. 1990); but see Light v. Centel Cellular Co. of Texas, 883 S.W.2d 642 (Tex. 1994) (covenants not to compete are enforceable only in very limited circumstances).

2. Other states, notably California, have an even more restrictive rule. For example, Cal. Bus. & Prof. Code §16600 provides that "every contract by which anyone is restrained from engaging in a lawful profession . . . is to that extent void." California courts have interpreted this statute to bar noncompetition agreements altogether in employee contracts but to permit such agreements if they are ancillary to the sale of a business, so long as the terms of the agreement are "reasonable." See Monogram Indus., Inc. v. SAR Indus., Inc., 64 Cal. App. 3d 692, 134 Cal. Rptr. 714, 718 (1976). Further, while California courts will not enforce a noncompetition agreement, they will prevent departing employees from using or disclosing their former employer's trade secrets. See State Farm Mutual Automobile Ins. Co. v. Dempster, 344 P.2d 821 (Cal. App. 1959); Gordon v. Landau, 49 Cal. 2d 690, 321 P.2d 456 (Cal. 1958).

The strength of California's commitment to the free movement of employees was demonstrated in The Application Group, Inc. v. The Hunter Group, Inc., 72 Cal. Rptr. 2d 73 (Ct. App. 1998). There, the California Court of Appeals held that § 16600 precluded the enforcement of a noncompetition agreement entered into in Maryland between a Maryland employer and employee, where the employee subsequently left to take a job telecommuting from Maryland for a California company. Despite the fact that Maryland courts would enforce the agreement, the California court concluded that California's interests were "materially stronger" than Maryland's in this case.

3. What are the competing policy interests at stake in noncompetition clauses? On the one hand, such restrictions seem onerous burdens to impose on employees. Imagine how you would feel as an attorney if you left a firm

only to find that you were prevented from practicing law in the same field or geographic region for the next two years. (In this regard, it is significant that even *Reed Roberts* expressed the view that the "learned professions" were properly subject to noncompetition agreements.) In addition, it is not completely clear that such provisions benefit companies in the long run. Strauman, the defendant in *Reed Roberts,* came to Reed Roberts after having worked for a competitor for four years. He was hired in part because of his valuable experience in the industry. What if Strauman's former employer had required him to sign an enforceable noncompete agreement?

On the other hand, it seems unfair to employers to simply allow their employees to do whatever they want upon leaving. Particularly where the employees were in positions of importance, their knowledge of the employer's trade secrets may leave the former employer at a competitive disadvantage. In a competitive industry, preventing the disclosure of trade secrets is far preferable to suing for misappropriation after they have already been disclosed. A noncompetition agreement may be a reasonable way for an employer to prevent a problem — and a lawsuit — before it starts.

At least one commentator has suggested that the difficulty in these cases is precisely the uncertainty of trade secret litigation. She recommends that employer-employee disputes be treated in a manner analogous to the dissolution of concurrent real property interests. Suellen Lowry, Note, Inevitable Disclosure Trade Secret Disputes: Dissolutions of Concurrent Property Interests, 40 Stan. L. Rev. 519 (1988).

4. When a firm requires a new employee to sign an employment agreement containing a covenant not to compete, the employee is giving up something substantial. What is the employer giving up? Some cases have raised the issue of consideration (in the contract law sense) in such an agreement on the part of the employer; they generally conclude that there is consideration, on one theory or another. See, e.g., Central Adjustment Bureau v. Ingram, 678 S.W.2d 28 (Tenn. 1984) (consideration in the form of continuous employment over a long period of time).

Why not require, out of fairness, that an employer who insists on such a covenant must pay the employee's salary during the term of the noncompete provision?[16] How would this change the behavior of an ex-employer vis-à-vis the newly departed employee? Would you expect the rate of employee mobility to be higher under such a system?

---

16. In the United States, some companies have adopted such a rule voluntarily. See Marcam Corp. v. Orchard, 885 F. Supp. 294 (D. Mass. 1995) (enforcing a contractual provision preventing Orchard from working for any competitor in the country for one year, provided that Marcam paid 110 percent of the salary offered by the competitor).

Note on the "Inevitable Disclosure" of Trade Secrets

Are there circumstances in which an employee's use or disclosure of trade secrets is "inevitable"? Consider the position of a former head of research and development at a computer company. Is there any way that that employee can "keep separate" the ideas and projects he was working on for his old employer from the ideas and projects he will be asked to develop for his new employer? If not, should the employer be entitled to enforce a noncompetition agreement even if it cannot show that the employee *intends* to use its trade secrets? IBM made just this argument when one of its disk drive specialists, Peter Bonyhard, left IBM to work for Seagate Technology in the same field. A preliminary injunction granted by the district court against Bonyhard's employment was reversed by the Eighth Circuit in an unpublished decision. IBM v. Bonyhard, 962 F.2d 12 (8th Cir. 1992). Several other decisions have also rejected inevitable disclosure as a basis for an injunction against employment. Campbell Soup Co. v. Giles, 47 F.3d 467 (1st Cir. 1995); FMC Corp. v. Cyprus Foote Mineral Co., 899 F. Supp. 1477 (W.D.N.C. 1995). In *Campbell*, the court grounded its rejection of inevitable disclosure theory in the public policy that favors employee mobility. It quoted the district court opinion in the case, holding that public policy "counsels against unilateral conversion of non-disclosure agreements into non-competitive agreements. If Campbell wanted to protect itself against the competition of former employees, it should have done so by contract. This court will not afford such protection after the fact." See also Carolina Chem. Equip. Co. v. Muckenfuss, 471 S.E.2d 721 (S.C. App. 1996) (agreement entitled "Covenant Not to Divulge Trade Secrets" was actually an overly broad and unenforceable covenant not to compete, because the definition of trade secrets in the agreement was so broad as to effectively prevent any competition).

On the other hand, the Seventh Circuit recently ordered the issuance of a preliminary injunction preventing Quaker Oats Co. from employing Redmond, a former general manager for PepsiCo North America. Redmond was general manager of PepsiCo's California business unit for ten years, until in 1994 he accepted Quaker's offer to become the chief operating officer of its Gatorade and Snapple Co. divisions. The court held that Redmond would inevitably be forced to use PepsiCo trade secrets for his new employer:

> PepsiCo asserts that Redmond cannot help but rely on PCNA [PepsiCo North America] trade secrets as he helps plot Gatorade and Snapple's new course, and that these secrets will enable Quaker to achieve a substantial advantage by knowing exactly how PCNA will price, distribute, and market its sports drinks and new age drinks and being able to respond strategically. This type of trade secret problem may arise less often, but it nevertheless falls within the realm of trade secret protection under the present circumstance.
>
> Quaker and Redmond assert that they have not and do not intend to use whatever confidential information Redmond has by virtue of his former em-

ployment. They point out that Redmond has already signed an agreement with Quaker not to disclose any trade secrets or confidential information gleaned from his earlier employment. They also note with regard to distribution systems that even if Quaker wanted to steal information about PCNA's distribution plans, they would be completely useless in attempting to integrate the Gatorade and Snapple beverage lines.

The defendants' arguments fall somewhat short of the mark. Again, the danger of misappropriation in the present case is not that Quaker threatens to use PCNA's secrets to create distribution systems or coopt PCNA's advertising and marketing ideas. Rather, PepsiCo believes that Quaker, unfairly armed with knowledge of PCNA's plans, will be able to anticipate its distribution, packaging, pricing, and marketing moves. Redmond and Quaker even concede that Redmond might be faced with a decision that could be influenced by certain confidential information that he obtained while at PepsiCo. In other words, PepsiCo finds itself in the position of a coach, one of whose players has left, playbook in hand, to join the opposing team before the big game. Quaker and Redmond's protestations that their distribution systems and plans are entirely different from PCNA's are thus not really responsive. . . .

Quaker and Redmond do not assert that the confidentiality agreement is invalid; such agreements are enforceable when supported by adequate consideration.[10] Rather, they argue that "inevitable" breaches of these contracts may not be enjoined. The case on which they rely, however, R. R. Donnelley & Sons Co. v. Fagan, 767 F. Supp. 1259 (S.D.N.Y. 1991) (applying Illinois law), says nothing of the sort. The R. R. Donnelley court merely found that the plaintiffs had failed to prove the existence of any confidential information or any indication that the defendant would ever use it. Id. at 1267. The threat of misappropriation that drives our holding with regard to trade secrets dictates the same result here.

PepsiCo, Inc. v. Redmond, 54 F.3d 1262, 35 U.S.P.Q.2d 1010 (7th Cir. 1995). The court distinguished two prior cases refusing to find that an ex-employee's disclosure of trade secrets was inevitable:

In *Teradyne,* Teradyne alleged that a competitor, Clear Communications, had lured employees away from Teradyne and intended to employ them in the same field. In an insightful opinion, Judge Zagel observed that "threatened misappropriation can be enjoined under Illinois law" where there is a "high degree of probability of inevitable and immediate . . . use of . . . trade secrets." *Teradyne,* 707 F. Supp. at 356. Judge Zagel held, however, that Teradyne's complaint failed to state a claim because Teradyne did not allege "that defendants have in fact threatened to use Teradyne's secrets or that they will inevitably do so." [The *Teradyne* court held]:

10. The confidentiality agreement is also not invalid for want of a time limitation. See 765 ILCS 1065/8(b)(1) ("[A] contractual or other duty to maintain secrecy or limit use of a trade secret shall not be deemed to be void or unenforceable solely for lack of durational or geographic limitation on the duty."). Nor is there any question that the confidentiality agreement covers much of the information PepsiCo fears Redmond will necessarily use in his new employment with Quaker.

the defendants' claimed acts, working for Teradyne, knowing its business, leaving its business, hiring employees from Teradyne and entering the same field (though in a market not yet serviced by Teradyne) do not state a claim of threatened misappropriation. All that is alleged, at bottom, is that defendants could misuse plaintiff's secrets, and plaintiffs fear they will. This is not enough. It may be that little more is needed, but falling a little short is still falling short.

Id. at 357.

In *AMP,* we affirmed the denial of a preliminary injunction on the grounds that the plaintiff AMP had failed to show either the existence of any trade secrets or the likelihood that defendant Fleischhacker, a former AMP employee, would compromise those secrets or any other confidential business information. AMP, which produced electrical and electronic connection devices, argued that Fleishhacker's new position at AMP's competitor would inevitably lead him to compromise AMP's trade secrets regarding the manufacture of connectors. *AMP,* 823 F.2d at 1207. In rejecting that argument, we emphasized that the mere fact that a person assumed a similar position at a competitor does not, without more, make it "inevitable that he will use or disclose . . . trade secret information" so as to "demonstrate irreparable injury." Id.

The ITSA, *Teradyne,* and *AMP* lead to the same conclusion: a plaintiff may prove a claim of trade secret misappropriation by demonstrating that defendant's new employment will inevitably lead him to rely on the plaintiff's trade secrets. See also 1 Jager, . . . §7.02[2][a] at 7-20 (noting claims where "the allegation is based on the fact that the disclosure of trade secrets in the new employment is inevitable, whether or not the former employee acts consciously or unconsciously"). . . .

PepsiCo presented substantial evidence at the preliminary injunction hearing that Redmond possessed extensive and intimate knowledge about PCNA's strategic goals for 1995 in sports drinks and new age drinks. The district court concluded on the basis of that presentation that unless Redmond possessed an uncanny ability to compartmentalize information, he would necessarily be making decisions about Gatorade and Snapple by relying on his knowledge of PCNA trade secrets. It is not the "general skills and knowledge acquired during his tenure with" PepsiCo that PepsiCo seeks to keep from falling into Quaker's hands, but rather "the particularized plans or processes developed by [PCNA] and disclosed to him while the employer-employee relationship existed, which are unknown to others in the industry and which give the employer an advantage over his competitors." *AMP,* 823 F.2d at 1202. The Teradyne and AMP plaintiffs could do nothing more than assert that skilled employees were taking their skills elsewhere; PepsiCo has done much more.

Id.

COMMENTS AND QUESTIONS

1. What Pepsi trade secrets are threatened by Redmond's "defection" to Quaker? The information Redmond possesses includes (1) new flavor and product packaging information; (2) pricing strategies; (3) Pepsi's "attack plans" for specific markets; and (4) Pepsi's new distribution plan, being pilot

tested in California. What advantages would Quaker obtain by knowing this information? How long would it take Quaker to find out about each in the absence of inside knowledge from Redmond? If all these items would become readily apparent the moment Pepsi's plans were implemented, does this circumstance suggest a limit on the appropriate remedy?

Assuming that Quaker will learn of Pepsi's strategies from Redmond unless enjoined, how expensive would it be for Pepsi to develop a new marketing strategy? Could the new strategy take advantage of the fact that Quaker *thinks* it knows what Pepsi will do? How would this possibility affect Quaker's use of the information?

2. Assume that the market for sports and new age drinks is increasingly concentrated in the hands of two companies, Pepsi and Quaker. Should this affect the outcome of the case? Given the court's decision, what can one predict about future salary and benefits negotiations in this industry for employees like Redmond?

3. Is it fair to preclude former employees from doing any work for a competitor simply because they would be incapable of not using the information they obtained from their former employer? Note that the Seventh Circuit upheld an injunction against Redmond with no time limit, apparently meaning that he will *never* be permitted to work in the position for which he was hired. Does this result circumvent the careful balancing evident in the noncompete cases discussed above? Does the danger that an ex-employee will disclose trade secrets justify this harsher rule?

4. The judicial debate over the appropriateness of the inevitable disclosure doctrine has continued. Several recent cases have followed *PepsiCo* and enjoined employment absent either proof of trade secret misappropriation or an enforceable noncompetition agreement. See Uncle B's Bakery v. O'Rourke, 920 F. Supp. 1405 (N.D. Iowa 1996) (enjoining former plant manager at a bagel manufacturer from working for any competing business within a 500-mile radius); National Starch & Chem. Corp. v. Parker Chem. Corp., 530 A.2d 31 (N.J. App. 1997) (citing *PepsiCo* with approval). By contrast, one court has adopted what might be called a "partial inevitable disclosure" injunction. In Merck & Co. v. Lyon, 941 F. Supp. 1443 (M.D.N.C. 1996), the court enjoined a pharmaceutical marketing director from discussing his former employer's products or pricing for a period of two years, but refused to enjoin him from competing employment altogether absent a "showing of bad faith." Another court has accepted an inevitable disclosure theory, but issued an injunction limited to nine months, reasoning that the secrets likely to be disclosed would turn stale over time. Novell Inc. v. Timpanogos Research Group, 46 U.S.P.Q.2d 1197 (D. Utah 1998).

c. Can Employees Take Their Own Inventions?

Wexler v. Greenberg
Supreme Court of Pennsylvania
399 Pa. 569, 160 A.2d 430 (1960)

Buckingham Wax Company is engaged in the manufacture, compounding and blending of sanitation and maintenance chemicals. In March, 1949, appellant Greenberg, a qualified chemist in the sanitation and maintenance field, entered the employ of Buckingham as its chief chemist and continued there until April 28, 1957. In the performance of his duties, Greenberg consumed half of his working time in Buckingham's laboratory where he would analyze and duplicate competitors' products and then use the resulting information to develop various new formulas. He would change or modify these formulas for color, odor or viscosity in order that greater commercial use could be made of Buckingham's products. . . . As a result of his activities Greenberg was not only familiar with Buckingham's formulas, he was also fully conversant with the costs of the products and the most efficient method of producing them. . . .

. . . In August, 1957, Greenberg left Buckingham and went to work for Brite[, who up until that time had purchased its chemicals exclusively from Buckingham]. At no time during Greenberg's employment with Buckingham did there exist between them a written or oral contract of employment or any restrictive agreement.

Prior to Greenberg's association with Brite, the corporation's business consisted solely of selling a complete line of maintenance and sanitation chemicals, including liquid soap cleaners, wax base cleaners, disinfectants, and floor finishes. Upon Greenberg's arrival, however, the corporation purchased equipment and machinery and, under the guidance and supervision of Greenberg, embarked on a full-scale program for the manufacture of a cleaner, floor finish and disinfectant, products previously purchased from Buckingham. The formulas in issue in this litigation are the formulas for each of these respective products. The appellants dispute the chancellor's findings as to the identity of their formulas with those of Buckingham, but there was evidence that a spectrophotometer examination of the respective products of the parties revealed that the formulas used in making these products are substantially identical. Appellants cannot deny that they thought the products sufficiently similar as to continue delivery of their own products to their customers in the same cans and drums and with the same labels attached which they had previously used in distributing the products manufactured by Buckingham, and to continue using the identical promotional advertising material. Appellees' formulas had been developed during the tenure of Greenberg as chief chemist and are unquestionably known to him.

The Chancellor found that Greenberg did not develop the formulas for Brite's products after he left Buckingham, but rather that he had appropriated them by carrying over the knowledge of them which he had acquired in

Buckingham's employ. The Chancellor went on to find that the formulas constituted trade secrets and that their appropriation was in violation of the duty that Greenberg owed to Buckingham by virtue of his employment and the trust reposed in him. Accordingly, the relief outlined above was ordered [enjoining Greenberg and Brite from disclosing the secret formulas or making or selling any products using the formulas].

We are initially concerned with the fact that the final formulations claimed to be trade secrets were not disclosed to Greenberg by the appellees during his service or because of his position. Rather, the fact is that these formulas had been developed by Greenberg himself, while in the pursuit of his duties as Buckingham's chief chemist, or under Greenberg's direct supervision. We are thus faced with the problem of determining the extent to which a former employer, without the aid of any express covenant, can restrict his ex-employee, a highly skilled chemist, in the uses to which this employee can put his knowledge of formulas and methods he himself developed during the course of his former employment because this employer claims these same formulas, as against the rest of the world, as his trade secrets. This problem becomes particularly significant when one recognizes that Greenberg's situation is not uncommon. In this era of electronic, chemical, missile and atomic development, many skilled technicians and expert employees are currently in the process of developing potential trade secrets. Competition for personnel of this caliber is exceptionally keen, and the interchange of employment is commonplace. . . . We must therefore be particularly mindful of any effect our decision in this case might have in disrupting this pattern of employee mobility, both in view of possible restraints upon an individual in the pursuit of his livelihood and the harm to the public in general in forestalling, to any extent, widespread technological advances.

The principles outlining this area of the law are clear. A court of equity will protect an employer from the unlicensed disclosure or use of his trade secrets by an ex-employee provided the employee entered into an enforceable covenant so restricting his use, Fralich v. Despar, 165 Pa. 24, 30 Atl. 521 (1894), or was bound to secrecy by virtue of a confidential relationship existing between the employer and employee, Pittsburgh Cut Wire Co. v. Suffrin, 350 Pa. 31, 38 A.2d 33 (1944). Where, however, an employer has no legally protectible trade secret, an employee's "aptitude, his skill, his dexterity, his manual and mental ability, and such other subjective knowledge as he obtains while in the course of his employment, are not the property of his employer and the right to use and expand these powers remains his property unless curtailed through some restrictive covenant entered into with the employer." Id. at 35. The employer thus has the burden of showing two things: (1) a legally protectable trade secret; and (2) a legal basis, either a covenant or a confidential relationship, upon which to predicate relief. . . .

The usual situation involving misappropriation of trade secrets in violation of a confidential relationship is one in which an employer *discloses to his employee* a pre-existing trade secret (one already developed or formulated) so that the employee may duly perform his work. In such a case, the trust and confidence upon which legal relief is predicated stems from the instance of the employer's *turning over to the employee* the pre-existing trade secret. It is then that a pledge of secrecy is impliedly extracted from the employee, a pledge which he carries with him even beyond the ties of his employment relationship. Since it is conceptually impossible, however, to elicit an implied pledge of secrecy from the sole act of an employee turning over to his employer a trade secret which he, the employee, has developed, as occurred in the present case, the appellees must show a different manner in which the present circumstances support the permanent cloak of confidence cast upon Greenberg by the chancellor. The only avenue open to the appellees is to show that the nature of the employment relationship itself gave rise to a duty of nondisclosure.

The burden the appellees must thus meet brings to the fore a problem of accommodating competing policies in our law: the right of a businessman to be protected against unfair competition stemming from the usurpation of his trade secrets and the right of an individual to the unhampered pursuit of the occupations and livelihoods for which he is best suited. There are cogent socio-economic arguments in favor of either position. Society as a whole greatly benefits from technological improvements. Without some means of post-employment protection to assure that valuable developments or improvements are exclusively those of the employer, the businessman could not afford to subsidize research or improve current methods. In addition, it must be recognized that modern economic growth and development has pushed the business venture beyond the size of the one-man firm, forcing the businessman to a much greater degree to entrust confidential business information relating to technological development to appropriate employees. While recognizing the utility in the dispersion of responsibilities in larger firms, the optimum amount of "entrusting" will not occur unless the risk of loss to the businessman through a breach of trust can be held to a minimum.

On the other hand, any form of post-employment restraint reduces the economic mobility of employees and limits their personal freedom to pursue a preferred course of livelihood. The employee's bargaining position is weakened because he is potentially shackled by the acquisition of alleged trade secrets; and thus, paradoxically, he is restrained, because of his increased expertise, from advancing further in the industry in which he is most productive. Moreover, as previously mentioned, society suffers because competition is diminished by slackening the dissemination of ideas, processes and methods.

Were we to measure the sentiment of the law by the weight of both English and American decisions in order to determine whether it favors protecting a businessman from certain forms of competition or protecting an individual in his unrestricted pursuit of a livelihood, the balance would heavily favor the latter. . . .

[The court examined cases from other jurisdictions that found a confidential relationship to exist when the employer disclosed trade secrets to the employee.] Upon our examination of the record here, however, we find that the instant circumstances fall far short of such a relationship. The Chancellor's finding that Greenberg, while in the employ of Buckingham, never engaged in research nor conducted any experiments nor created or invented any formula was undisputed. There is nothing in the record to indicate that the formulas in issue were specific projects of great concern and concentration by Buckingham; instead it appears that they were merely the result of Greenberg's routine work of changing and modifying formulas derived from competitors. Since there was no experimentation or research, the developments by change and modification were fruits of Greenberg's own skill as a chemist without any appreciable assistance by way of information or great expense or supervision by Buckingham. . . .

COMMENTS AND QUESTIONS

1. *Wexler* is explicitly based on the absence of a written agreement between employer and employee.[17] Suppose that Greenberg had signed a contract when he was hired which stated that "all discoveries, inventions, or intellectual property rights stemming from employee's work at Buckingham are the exclusive property of Buckingham." Would such a contract be enforceable? Does it matter whether Greenberg's "discoveries" were new formulas or ways to copy competitors' formulas? Cf. Shamrock Technology, Inc. v. Medical Sterilization, Inc., 6 F.3d 788 (Fed. Cir. 1993) (unpub.) (co-inventor of patented process who disclosed invention to new employer prior to issuance of patent held liable for misappropriation of trade secrets in light of agreement to assign inventions to former employer and not to disclose former employer's trade secrets).

An example of the more common situation, in which the court finds either an express or implied agreement to assign an employee's own inventions, is Winston Research Corp. v. 3M Co., 350 F.2d 134 (9th Cir. 1965). There, the court dismissed an argument akin to Greenberg's as follows:

> Winston argues that information is protected from disclosure only if communicated to the employee by the employer who is seeking protection, and that the information involved in this case was not disclosed by [3M] to the employees subsequently hired by Winston, but rather was developed by these employees themselves, albeit while employed by [3M].
>
> We need not examine the soundness of the rule for which Winston contends, or its applicability to a case such as this in which a group of specialists engaged in related facets of a single development project change their employer. * * * [A]n obligation not to disclose may arise from circumstances other than

---

17. The absence of such an agreement can be significant even in situations in which it might be assumed that the "employer" had a reasonable claim to the employee's invention. Cf. Lariscey v. United States, 20 U.S.P.Q.2d 1845 (Fed. Cir. 1991) (prison inmate owns rights to a secret process he developed while working in a prison shop).

communication in confidence by the employer. It may also rest upon an express or implied agreement. In the present case, an agreement not to disclose might be implied from [3M]'s elaborate efforts to maintain the secrecy of its development program, and the employees' knowledge of those efforts and participation in them. In any event, [3M] and its employees entered into express written agreements binding the latter not to disclose confidential information, and these agreements did not exclude information which the employee himself contributed.

Id. at 140.

In addition, even in Pennsylvania *Wexler* is not always followed. See Healthcare Affiliated Services v. Lippancy, 703 F. Supp. 1142, 1155 (W.D. Pa. 1988) (rejecting *Wexler* analysis, emphasizing that although defendant developed inventions on his own, he did so using knowledge and information made available by the plaintiff employer). But see Fidelity Fund v. DiSanto, 500 A.2d 431 (Sup. Ct. Pa. 1985) (denying recovery against ex-employee salesman partly on the basis that he developed client contacts himself during employment).

2. Does the *Wexler* court confuse an employee's *general* knowledge and skills with specific inventions developed by the employee while at work *based* on that knowledge and skills? If an employee's job involves two tasks, working on established products (covered by trade secrets) and developing new ones, what effect will the *Wexler* rule have on the employee's allocation of time between these tasks?

3. A variant on the departing employee who wishes to take her own inventions is the departing sales representative who wishes to "take" a list of customers (either a written list or one that they have memorized) in order to call on those customers for a competitor.[18] Customer lists are generally protectable as trade secrets, but enjoining employees from calling on customers with whom they have had long-standing relationships raises serious concerns about employee mobility. Two California cases frame the dispute. In Moss, Adams & Co. v. Shilling, 179 Cal. App. 3d 124 (1986), the court drew a line between an employee announcing her departure to start a competing company and actively soliciting old clients to follow her. The former was permissible, but the latter was not. This compromise was thrown into doubt by the decision in Morlife Inc. v. Perry, 66 Cal. Rptr. 2d 731 (Ct. App. 1997), where the court held that the passage of the Uniform Trade Secrets Act in California superseded *Moss, Adams* and that continuing personal relationships with former customers could constitute misappropriation of a trade secret.

Where should the line be drawn between permissible work and impermissible use of "secret" lists of customers? To whom does the value inherent in personal relationships belong?

18. Merely memorizing a trade secret than taking physical documents will not preclude a finding of misappropriate, though it may make misappropriation harder to detect. See Ed Nowogroski Ins. Inc. v. Rucker, — P. 2d —(Wash. 1999).

Note on the Common Law Obligation to Assign Inventions

One important dimension of the *Wexler* case is that none of Greenberg's research yielded a patentable invention. If it had, ownership would have been determined according to Greenberg's status under a long line of common law employee invention cases. In general, employees such as Greenberg fall into one of three categories: (1) employees "hired to invent," which results in employer ownership of the invention; (2) employees who invent in the employer's shop, which results in a limited, nonexclusive "shop right" on the part of the employer to practice the invention; and (3) an employee's "independent invention," in which case the employee owns the invention. See generally United States v. Dubilier Condenser Corp., 289 U.S. 178 (1933); John C. Stedman, Employer-Employee Relations, in Fredrik Neumeyer, The Employed Inventor in the United States 30, 40-41 (1971).

The first category is relatively straightforward. It seems logical to extend this treatment to consultants and others who are not "employees" in the strict sense. Cf. McElmurry v. Arkansas Power & Light Co., 27 U.S.P.Q.2d 1129, 1135 n.15 (Fed. Cir. 1993) (upholding shop right in employer where inventor/patentee was a consultant); Robert P. Merges, Intellectual Property and the Costs of Commercial Exchange: A Review Essay, 93 Mich. L. Rev. 1570 (1995) (highlighting the role of intellectual property in structuring non-employment-based organizations, such as consulting companies and joint ventures). Category (2) reflects situations where employers have less than a complete claim to the invention, but where the employer's facilities or resources are combined with the inventor's talent and industry to produce the invention. The employee owns it, but the employer is compensated by receiving a limited right to practice the invention. Category (3) covers cases where the employee invents on his or her own time, outside the field of employment. See *Dubilier,* supra.

All the above rules apply only in the absence of an agreement. Of course, most companies generally require their employees to sign an employment agreement, usually on the company's terms. In an attempt to alleviate the disparity in bargaining power between the employer and the inventor/employee, several states have adopted "freedom to create" statutes restricting the instances in which employers may compel assignment through contract. See Minn. Stat. Ann. §181.78 (1980); N.C. Gen. Stat. §66-57.1 to 57.2 (1981); Wash. Rev. Code Ann. §49.44.140 (1987); Cal. Labor Code §2870 (West 1987).

These statutes provide that any employee invention assignment agreement that purports to give employers greater rights than they have under the statute is against public policy and unenforceable. The California statute is a good example. It provides that contracts may not require assignment of "invention[s] that the employee developed entirely on his or her own time without using the employer's equipment, supplies, facilities, or trade secret information" unless the invention relates to the employer's current or demonstrably anticipated business. Cal. Labor Code §2870.

One commentator has argued that the United States is slipping in technological innovation because this country's inventors (the vast majority of whom are employed) have lost their incentive to create partially due to assignment agreements. After labelling the state legislation described above as "a step forward," the author writes: "Comprehensive federal legislation is necessary to provide employed inventors greater control over, and interests in, their inventions in order to create the psychological and financial incentives required to increase technological innovation in the United States. . . . Federal legislation would obviate the need for state reform and would bring uniformity that would solve the potential problem of inventors employed by multistate corporations." Henrik D. Parker, Reform for Rights of Employed Inventors, 57 S. Cal. L. Rev. 603 (1984).

PROBLEM

Problem 2-11. Helen, an employee of CarTech, is assigned to work on solving a particular problem in car design. She works for two years on that problem for CarTech, without success. Two months after Helen leaves CarTech to start her own company, she puts out a product that incorporates a solution to the same problem. CarTech sues Helen for misappropriation of trade secrets. At trial, CarTech proves that Helen came up with her system while at CarTech and that she decided to start her own company to exploit her invention rather than disclose it to CarTech. As a result, CarTech had no knowledge of the idea until Helen came out with her product. Should CarTech prevail in its trade secret suit, assuming that there was no employment or invention contract between the parties?

Note on Trailer Clauses

The common law rules just described apply in the absence of contracts, as stated; but they also have some relevance even when written agreements are in place. This is so because, under principles similar to those applied in the law of noncompetition agreements, the law limits the ability of employers to claim ownership by contract of all employee inventions no matter how tenuously related to the employment relationship. A case study in the limits of invention assignment agreements involves the "trailer clause," a contractual provision that requires employees to assign their rights not only in inventions made during the period of employment but also for a certain time thereafter.

Trailer clauses developed in response to a number of cases which held that, absent contractual provisions to the contrary, an employee's ideas did not belong to the employer unless they were written down in tangible form (such as a patent disclosure statement) before the employee left work. See Jamesbury Corp. v. Worcester Valve Co., 443 F.2d 205 (1st Cir. 1971).

Of course, even if an employer uses a trailer clause, there is always the risk that the former employee will simply wait out the duration of the term and then conveniently make the discovery upon the trailer clause's expiration. Such a strategy was attempted by an inventor in at least one reported case. Fortunately for the employer, the court did not believe the defendant inventor's story that the conception date for his flow meter was five days after the expiration of the six months specified in the trailer clause:

> The perfection of a flow meter proved to be a painstakingly intricate process involving extensive testing. It is therefore difficult to believe that after a long and distinguished career with Plaintiff, Mr. Halmi in his musing five days after the trailer clause expired for the first time came up with the idea for the NTV. Although the word "Eureka!" has allegedly been uttered by more than one inventor over the years, the concept at issue does not lend itself to such sudden discovery. The court finds that the concept of the '434 patent must have existed in Mr. Halmi's mind before his employment with GSC ended. Mr. Halmi therefore violated his agreement with GSC.

General Signal Corp. v. Primary Flow Signal, Inc., 1987 U.S. Dist. LEXIS 6929 at 10 (D.R.I. Jul. 27, 1987).

Generally, trailer clauses are subject to the same treatment as noncompetition agreements. That is, they are enforceable only to the extent that they are "reasonable." Although trailer clauses are generally enforceable, clauses of particularly long or indefinite duration may run afoul of the antitrust laws as well as being unenforceable. See United Shoe Machinery Co. v. La Chapelle, 212 Mass. 467, 99 N.E. 289 (1912). One court expressed the requirement of reasonableness as follows:

> Hold-over clauses are simply a recognition of the fact of business life that employees sometimes carry with them to new employers inventions or ideas so related to work done for a former employer that in equity and good conscience the fruits of that work should belong to that former employer. In construing and applying hold-over clauses, the courts have held that they must be limited to *reasonable times . . .* and to *subject matter* which an employee worked on or had knowledge of during his employment. . . . Unless expressly agreed otherwise, an employer has no right under a holdover clause to inventions made outside the scope of the employee's former activities, and made on and with a subsequent employer's time and funds.

Dorr-Oliver, Inc. v. United States, 432 F.2d 447, 452 (Ct. Cl. 1970). Corporations may have trouble enforcing restrictions that are broader in scope.

This is particularly true of large conglomerates that attempt to require the assignment of any invention related to their (diverse) fields of business. See Ingersoll-Rand Co. v. Ciavatta, 110 N.J. 609, 542 A.2d 879, 896 n.6 (1988).

Several proposals designed to improve the rights of the employed inventor have been introduced into Congress. In 1981, for example, Rep. Robert Kastenmeier introduced two bills, one of which would have totally eliminated trailer clauses. Do you see any problems with doing so? Who should bear the burden of proof with respect to inventions arguably made during the period of employment?

PROBLEM

Problem 2-12. You have been offered a position with a high-technology start-up company. They ask you to sign the following agreement. Do you sign it? Is it enforceable?

EMPLOYMENT, CONFIDENTIAL INFORMATION, AND INVENTION ASSIGNMENT AGREEMENT

As a condition of my employment with Science Company, its subsidiaries, affiliates, successors, or assigns (together the "Company"), and in consideration of my employment with the Company and my receipt of the compensation now and hereafter paid to me by the Company, I agree to the following:

1. *Confidential Information*

(a) *Company Information.* I agree at all times during the term of my employment and thereafter, to hold in strictest confidence, and not to use, except for the benefit of the Company, or to disclose to any person, firm, or corporation without written authorization of the Board of Directors of the Company, any Confidential Information of the Company. I understand that **"Confidential Information"** means any Company proprietary information, technical data, trade secrets or know-how, including, but not limited to, research, product plans, products, services, customer lists and customers (including, but not limited to, customers of the Company on whom I called or with whom I became acquainted during the term of my employment), markets, software, developments, inventions, processes, formulas, technology, designs, drawings, engineering, hardware configuration information, marketing, finances, or other business information disclosed to me by the Company either directly or indirectly in writing, orally, or by drawings or observation of parts or equipment. I further understand that Confidential Information does not include any of the foregoing items which has become publicly known and made generally available through no wrongful act of mine or of others who were under confidentiality obligations as to the item or items involved.

(b) *Third Party Information.* I recognize that the Company has received and in the future will receive from third parties their confidential or proprietary information subject to a duty on the Company's part to maintain the confidentiality of such information and to use it only for certain limited purposes. I agree to hold all such confidential or proprietary information in the strictest confidence and not to disclose it to any person, firm, or corporation or to use it except as necessary in carrying out my work for the Company consistent with the Company's agreement with such third party.

2. Inventions

I agree that I will promptly make full written disclosure to the Company, will hold in trust for the sole right and benefit of the Company, and hereby assign to the Company, or its designee, all my right, title, and interest in and to any and all inventions, original works of authorship, developments, concepts, improvements or trade secrets, whether or not patentable or registrable under copyright or similar laws, which I may solely or jointly conceive or develop or reduce to practice, or cause to be conceived or developed or reduced to practice (collectively referred to as "Inventions"), during the period of time I am in the employ of the Company and for three months thereafter, except as provided below. I further acknowledge that all original works of authorship which are made by me (solely or jointly with others) within the scope of and during the period of my employment with the Company and which are protectable by copyright are "works made for hire," as that term is defined in the United States Copyright Act.

3. Conflicting Employment

I agree that, during the term of my employment with the Company and for a period of one year thereafter, I will not engage in any other employment, occupation, consulting, or other business activity in competition with or directly related to the business in which the Company is now involved or becomes involved during the term of my employment, nor will I engage in any other activities that conflict with my obligations to the Company.

4. Returning Company Documents

I agree that at the time of leaving the employ of the Company, I will deliver to the Company (and will not keep in my possession, recreate or deliver to anyone else) any and all devices, records, data, notes, reports, proposals, lists, correspondence, specifications, drawings, blueprints, sketches, materials, equipment, other documents or property, or reproductions of any aforementioned items developed by me pursuant to my employment with the Company or otherwise belonging to the Company, its successors or assigns.

5. *Notification to New Employer*

In the event that I leave the employ of the Company, I hereby grant consent to notification by the Company to my new employer about my rights and obligations under this Agreement.

6. *Solicitation of Employees*

I agree that for a period of twelve (12) months immediately following the termination of my relationship with the Company for any reason, whether with or without cause, I shall not either directly or indirectly solicit, induce, recruit or encourage any of the Company's employees to leave their employment, or take away such employees, or attempt to solicit, induce, recruit, encourage, or take away employees of the Company, either for myself or for any other person or entity.

D. AGREEMENTS TO KEEP SECRETS

As we have seen, trade secret law imposes certain limitations on the owner of a trade secret. To qualify for trade secret protection, information must not be generally known, must be valuable, and must not be disclosed. But can an owner of information avoid those restrictions by requiring others to *agree* to keep the information secret, whether or not the information meets the requirements for protection? The question is fundamental in intellectual property law.

Licenses are generally considered good from an economic standpoint because they promote efficiency. The company (or individual) that develops a new product may not be in the best position to market it, particularly if the invention has uses in several different fields. Absent licenses, that company would either have to sell the product outright, use it incompletely or inefficiently, enter a new field itself, or not use it at all. Private contract remedies allow the market to reorder itself efficiently and still determine the appropriate reward for invention. But we cannot always rely on the market to determine an efficient outcome. Indeed, intellectual property in general is an example of pervasive market failure, a failure that may color the role of contract law.

Warner-Lambert Pharmaceutical Co. v. John J. Reynolds, Inc.
United States District Court for the Southern District of
New York
178 F. Supp. 655 (S.D.N.Y. 1959)

BRYAN, District Judge.

Plaintiff sues under the Federal Declaratory Judgment Act, 28 U.S.C. §§2201 and 2202, for a judgment declaring that it is no longer obligated to make periodic payments to defendants based on its manufacture or sale of the well known product "Listerine," under agreements made between Dr. J. J. Lawrence and J. W. Lambert in 1881, and between Dr. Lawrence and Lambert Pharmacal Company in 1885. Plaintiff also seeks to recover the payments made to defendants pursuant to these agreements since the commencement of the action.

Plaintiff is a Delaware corporation which manufactures and sells Listerine, among other pharmaceutical products. It is the successor in interest to Lambert and Lambert Pharmacal Company which acquired the formula for Listerine from Dr. Lawrence under the agreements in question. Defendants are the successors in interest to Dr. Lawrence.

Jurisdiction is based on diversity of citizenship.

For some seventy-five years plaintiff and its predecessors have been making the periodic payments based on the quantity of Listerine manufactured or sold which are called for by the agreements in suit. The payments have totalled more than twenty-two million dollars and are presently in excess of one million five hundred thousand dollars yearly. . . .

[J. J. Lawrence developed the formula for Listerine in 1880. He licensed the secret formula exclusively to Lambert (later Warner-Lambert) in 1881 under a contract which provided that

> I, Jordan Lambert, hereby agree for myself, my heirs, executors and assigns to pay monthly to Dr. Lawrence, his heirs, executors or assigns, the sum of twenty dollars for each and every gross of said Listerine hereafter sold by myself, my heirs, executors or assigns.

The amount was reduced by subsequent agreement to $6.00 per gross.]

The agreements between the parties contemplated, it is alleged, "the periodic payment of royalties to Lawrence for the use of a trade secret, to wit, the secret formula for" Listerine. After some modifications made with Lawrence's knowledge and approval, the formula was introduced on the market. The composition of the compound has remained the same since then and it is still being manufactured and sold by the plaintiff.

It is then alleged that the "trade secret" (the formula for Listerine) has gradually become a matter of public knowledge through the years following

1881 and prior to 1949, and has been published in the United States Pharmacopo[e]ia, the National Formulary and the Journal of the American Medical Association, and also as a result of proceedings brought against plaintiff's predecessor by the Federal Trade Commission. Such publications were not the fault of plaintiff or its predecessors. . . .

(1)

The plaintiff seems to feel that the 1881 and 1885 agreements are indefinite and unclear, at least as to the length of time during which they would continue in effect. I do not find them to be so. These agreements seem to me to be plain and unambiguous. . . .

The obligation to pay on each and every gross of Listerine continues as long as this preparation is manufactured or sold by Lambert and his successors. It comes to an end when they cease to manufacture or sell the preparation. . . . The plain meaning of the language used in these agreements is simply that Lambert's obligation to pay is co-extensive with manufacture or sale of Listerine by him and his successors.

(3)

However, plaintiff urges with vigor that the agreement must be differently construed because it involved the conveyance of a secret formula. The main thrust of its argument is that despite the language which the parties used the court must imply a limitation upon Lambert's obligation to pay measured by the length of time that the Listerine formula remained secret.

To sustain this theory plaintiff relies upon a number of cases involving the obligations of licensees of copyrights or patents to make continuing payments to the owner or licensor, and argues that these cases are controlling here. . . .

. . . [A]ll [these cases hold] is that when parties agree upon a license under a patent or copyright the court will assume, in the absence of express language to the contrary, that their actual intention as to the term is measured by the definite term of the underlying grant fixed by statute.

It is quite plain that were it not for the patent and copyright features of such license agreements the term would be measured by use. . . .

In the patent and copyright cases the parties are dealing with a fixed statutory term and the monopoly granted by that term. This monopoly, created by Congress, is designed to preserve exclusivity in the grantee during the statutory term and to release the patented or copyrighted material to the general public for general use thereafter. This is the public policy of the statutes in reference to which such contracts are made and it is against this background that the parties to patent and copyright license agreements contract.

Here, however, there is no such public policy. The parties are free to contract with respect to a secret formula or trade secret in any manner which they determine for their own best interests. A secret formula or trade secret may remain secret indefinitely. It may be discovered by someone else almost immediately after the agreement is entered into. Whoever discovers it for himself by legitimate means is entitled to its use.

But that does not mean that one who acquires a secret formula or a trade secret through a valid and binding contract is then enabled to escape from an obligation to which he bound himself simply because the secret is discovered by a third party or by the general public. I see no reason why the court should imply such a term or condition in a contract providing on its face that payment shall be co-extensive with use. To do so here would be to rewrite the contract for the parties without any indication that they intended such a result. . . .

One who acquires a trade secret or secret formula takes it subject to the risk that there be a disclosure. The inventor makes no representation that the secret is non-discoverable. All the inventor does is to convey the knowledge of the formula or process which is unknown to the purchaser and which in so far as both parties then know is unknown to any one else. The terms upon which they contract with reference to this subject matter are purely up to them and are governed by what the contract they enter into provides.

If they desire the payments or royalties should continue only until the secret is disclosed to the public it is easy enough for them to say so. But there is no justification for implying such a provision if the parties do not include it in their contract, particularly where the language which they use by fair intendment provides otherwise. . . .

COMMENTS AND QUESTIONS

1. The scope of trade secret protection is limited, not by a fixed term of years but by the length of time the information remains secret. Indeed, trade secrets become a part of the public domain once secrecy is lost, and their owner cannot prevent even direct copying. It seems odd that contract law would require royalty payments from Warner-Lambert for eighty years (or presumably 115 or more years, as Listerine is still sold commercially), while every other competitor can copy the formula for free. Are there sound reasons for such a rule?

One such reason may be the economic value of freedom of contract. If a party chooses to contract to pay royalties for as long as it uses a product, it is free to do so. Presumably the amount of the royalty payments as well as their duration will reflect the value of the secret to the buyer, discounted by the likelihood of public disclosure. But doesn't this argument also apply to extensions of the patent or copyright term as well? Compare Meehan v. PPG Indus., 802 F.2d 881, 886 (7th Cir. 1986), *cert. denied*, 479 U.S. 1091 (1987) and Boggild v. Kenner Prods., 776 F.2d 1315, 1320-1321 (6th Cir.

1985), *cert. denied*, 477 U.S. 908 (1986), which hold that it is illegal patent or copyright misuse to agree to extend a patent or copyright beyond its term. Why does contract law appear to override trade secret law but not patent or copyright law?

Could a licensee evade its royalty obligations by acquiring the now-public secret from an independent source? The answer seems to depend on the way in which the particular license is written. But cf. Aronson v. Quick Point Pencil Co., 440 U.S. 257 (1979) (holding that the contract obligated licensee to continue paying royalties on an invention even though licensor's patent application had been rejected).

2. The Warner-Lambert result is controversial. The Restatement (Third) of Unfair Competition takes the position that nondisclosure agreements which purport to protect information in the public domain may be unenforceable as an unreasonable restraint on trade. See §41, Comment *d*, at 472. Further, the Restatement notes that "because of the public interest in preserving access to information that is in the public domain, such an agreement will not ordinarily estop a defendant from contesting the existence of a trade secret." §39, Comment *d*, at 430. A number of cases support this view, which seems at odds with *Warner-Lambert*. See, e.g., Gary Van Zeeland Talent, Inc. v. Sandas, 267 N.W.2d 242 (Wisc. 1978); Sarkes Tarzian, Inc. v. Audio Devices, Inc., 166 F. Supp. 250 (S.D. Cal. 1958), *aff'd*, 283 F.2d 695 (9th Cir. 1960), *cert. denied*, 364 U.S. 869 (1961).

On the other hand, the Federal Circuit has seemingly endorsed the *Warner-Lambert* approach, holding in a recent case that the issuance of a patent did not extinguish the confidentiality obligation imposed by a nondisclosure agreement, even though the issuance of the patent destroyed the trade secret that was the basis for the agreement. Celeritas Technologies v. Rockwell Intl. Corp., 150 F.3d 1354 (Fed. Cir. 1998). The equities in that case might be thought to favor the plaintiff: the patent was held invalid, and so offered the plaintiff no relief against a theft of its technology. But isn't that the risk a trade secret owner takes in deciding to patent (and therefore disclose) her invention?

3. Should such restrictions be governed by intellectual property law — making them unenforceable — or by contract law, which presumably would enforce them? The answer to that question will determine whether intellectual property laws (including trade secret laws) constitute binding governmental rules balancing competing interests or merely "default" rules that parties may opt to change. In practice, the courts have walked a hazily defined middle line, refusing to hold that intellectual property statues preempt contracts which alter their terms, but also refusing to enforce certain contracts which go "too far" in upsetting the balance the intellectual property laws have struck.

4. Many licensing agreements are between competitors in an industry, since firms in the same industry are generally those who will be interested in a particular invention. Agreements between competitors, though, are generally considered suspect under the antitrust laws. So too are common license

restrictions that limit the use the licensee can make of the technology. We consider the restrictions that antitrust law places on private agreements of this sort in Chapter 8.

E. REMEDIES

As might be expected given the fluidity of doctrine in trade secret law, there are disparate remedies for trade secret misappropriation stemming from different remedial concerns. The Uniform Trade Secrets Act sets forth the following remedies for misappropriation:

Section 2

(a) Actual or threatened misappropriation may be enjoined. Upon application to the court, an injunction shall be terminated when the trade secret has ceased to exist, but the injunction may be continued for an additional reasonable period of time in order to eliminate commercial advantage that otherwise would be derived from the misappropriation.

(b) If the court determines that it would be unreasonable to prohibit future use, an injunction may condition future use upon payment of a reasonable royalty for no longer than the period of time the use could have been prohibited.

(c) In appropriate circumstances, affirmative acts to protect a trade secret may be compelled by court order.

Section 3

(a) In addition to or in lieu of injunctive relief, a complainant may recover damages for the actual loss caused by misappropriation. A complainant also may recover for the unjust enrichment caused by misappropriation that is not taken into account in computing damages for actual loss.

(b) If willful and malicious misappropriation exists, the court may award exemplary damages in an amount not exceeding twice any award made under subsection (a).

Section 4

If (i) a claim of misappropriation is made in bad faith, (ii) a motion to terminate an injunction is made or resisted in bad faith, or (iii) willful and malicious misappropriation exists, the court may award reasonable attorney's fees to the prevailing party.

What motivates this hodgepodge of remedial measures? As we shall see in the chapters to come, most intellectual property statutes operate on the basis of "property rules." That is, as in cases involving real property, the owner of the intellectual property right is entitled to judicial assistance in protecting the right from future interference. Normally this assistance comes in the form of injunctive relief. By contrast, most tort and contract cases do not involve injunctive relief but rather damages designed to make the plaintiff "whole" in the sense of restoring her to the position she occupied before the tort, or to the position she expected to occupy if the contract had been performed.

Section 2 of the Uniform Trade Secrets Act seems to entitle trade secret plaintiffs to property-like protection, as least so long as their secret remains a secret. But section 2(b) holds open the possibility that courts may refuse to grant such an injunction, settling instead for a "reasonable royalty" (presumably a court's attempt to approximate what the parties might have agreed to pay in a licensing transaction). This provision casts some doubt on the "property entitlement" a trade secret owner might expect.

Similar doubt pervades the provisions on damages. Concepts like "reasonable royalty," "lost profits," and limited-time injunctions designed to "eliminate commercial advantage" all sound like restitutionary measures, aimed at making the plaintiff whole after a loss without necessarily punishing or deterring the defendant. But further provisions permit trade secret plaintiffs to recover for "unjust enrichment" on the part of defendants, and to recover treble damages and attorney fees in the case of willful misappropriation. And in some circumstances misappropriation of trade secrets can be a criminal offense, an idea that is certainly more consistent with a property entitlement rule than a tort or contract rule.

Generally, injunctions are available as a remedy in trade secret cases. Since injunctions offer only prospective relief, however, damages for preinjunction activities may also be collected. Since, unlike patents and copyrights, trade secrets have no definite term, the length of the injunction is often a difficult issue. The following case illustrates the use of one important measure of trade secret injunctions, the "head start" theory.

K-2 Ski Company v. Head Ski Co., Inc.
United States Court of Appeals for the Ninth Circuit
506 F.2d 471 (9th Cir. 1974)

WALLACE, Circuit Judge:

K-2 Ski Company (K-2), a Washington corporation with its principal place of business in Washington, brought this action based upon diversity jurisdiction against Head Ski Company (Head), a Delaware corporation with its principal place of business in Maryland, and William Crocker, a citizen of

Maryland, seeking damages and injunctive relief. K-2 alleged that subsequent to his employment by K-2, Crocker began working for Head and disclosed trade secrets. The district court issued a preliminary injunction against Head and Crocker in April, 1970. That action was reversed by us in July 1972. In the interim, the district court had appointed a special master to hear this case. The master made detailed findings of fact and concluded that Head had unlawfully used K-2 trade secrets. Accordingly, the district court issued a permanent injunction barring Head from using one of the K-2 trade secrets for one year and another secret for two years.

Head appeals. . . .

I. THE FACTS

Head and K-2 are competing ski manufacturers. In the early 1960's, Head's metal-laminated skis dominated the quality ski market. By 1967, the new K-2 fiberglass skis had been marketed and the demand for the K-2 skis had grown significantly. Crocker was employed by K-2 from May, 1967, to February 13, 1970. Prior to Crocker's employment with K-2, he had no knowledge of the construction or production of skis nor had he had any background in engineering, manufacturing, shop practice, or purchasing. Crocker had majored in political science in college and had been primarily involved in appliance and furniture retailing prior to joining K-2. By January 1, 1968, Crocker had advanced to the position of general superintendent of the K-2 manufacturing operations. In that position, he supervised all aspects of the production of skis, purchased all the materials, and hired and fired employees. In July, 1969, Crocker's responsibilities were transferred to another employee and by the end of 1969, Crocker had become dissatisfied with his job at K-2 and contacted Head concerning employment. On January 26, 1970, after visiting the Head plant in Maryland, Crocker was offered a job at Head. Coupled with the offer of a base salary was a bonus which was contingent upon production of 5,000 skis with a wet-wrap process by the end of 1970. Crocker joined Head on February 16, 1970, but did not disclose to K-2 the name of his new employer. This action was filed on March 23, 1970. . . .

III. THE INJUNCTION

The district court found that K-2 had established two trade secrets and that Head had unlawfully utilized them in its production of skis. Relying upon Winston Research Corp. v. Minnesota Mining & Manufacturing Co., 350 F.2d 134, 141-42 (9th Cir. 1965), and Plant Industries, Inc. v. Coleman, 287 F. Supp. 636, 645 (C.D. Cal. 1968), the district court enjoined Head from using the base subassembly trade secret for two years and the surfacing veil secret for one year. We are satisfied that the appropriate duration for the injunction should be the period of time it would have taken Head, either by reverse engineering or by independent development, to develop its ski legitimately without use of the K-2 trade secrets. The district court properly determined the period for injunctive relief but, in issuing its permanent injunctions, apparently did not take into account the period of time that Head had already been under the preliminary injunction.

Head argues that since the preliminary injunction lasted for 27 months, the two- and one-year injunctions are barred because Head has already been enjoined beyond the appropriate period of time necessary to deprive it of the benefits of using the K-2 trade secrets. This issue was not raised before the trial judge. . . . Because the record does not contain any transcript of proceedings before the special master, we do not know whether Head argued there that the permanent injunctions should not issue because the preliminary injunction had already been in effect for over two years. The record does reflect that the trial judge read the master's report.

In Nuelsen, we stated:

> Rather than consider the matter sua sponte, of course, the appellate court may note the existence of the unargued, undecided question and remand the case to the lower court. This makes the decision on the matter one reflecting the consideration of a trial court and the counsel in the case.

293 F.2d at 462 (footnote omitted). We remand this issue to the trial court to consider the effect of the twenty-seven-month preliminary injunction on the one- and two-year permanent injunctions. In its determination, the trial court should consider whether the preliminary injunction served the same purpose as the permanent injunctions and whether it properly deprived Head of the benefits reaped by the use of the K-2 trade secrets. If the preliminary injunction had the same effect or accomplished the same result as the permanent injunctions would have, then the permanent injunctions may have been improper. On the other hand, if the preliminary injunction did not serve the same purpose and did not deprive Head of any benefits, then the permanent injunctions may have been the proper way to compensate K-2 for the unlawful use by Head of the K-2 trade secrets. This issue is more appropriate for a trial court determination because K-2 has argued that Head did not comply with the preliminary injunction and the trial judge is in a better position to resolve this conflict and the question of whether the preliminary injunction accomplished the same result as the permanent injunctions.

COMMENTS AND QUESTIONS

1. "Head-start injunctions" are available to plaintiffs who have published or otherwise disclosed their secret at some point after it was misappro-

priated. Suppose Anne possesses a secret that she is in the process of commercializing. Suppose further that it takes Anne two years after developing the secret to bring the product to market, at which point the secret is disclosed. If Benjamin steals Anne's idea during the development process (say, at month 12), Benjamin will be able to get to market one year earlier than if he had waited until the information became public. In such a case, courts will issue a "head-start" injunction for a period of one year, putting Benjamin in the same position he would have been in without the secret.

Even if such an injunction is impossible (for example, because Benjamin has already entered the market), courts may allow him to continue using the former secret but require him to pay a "reasonable royalty" to Anne. The reasonable royalty is presumably set by the court, in an effort to approximate the royalty Anne might have charged Benjamin in a voluntary transaction.

Is a reasonable royalty a fair solution in such a situation? At least one commentator has suggested that such a remedy "is peculiarly inappropriate to redress a situation where injunctive relief ought to be applied." See 12 Roger Milgrim, Milgrim on Trade Secrets §1.01[2][a], at 1-36 n.20 (citing a district court decision concluding that limiting relief to a reasonable royalty invites misappropriation).

2. In Kewanee Oil Co. v. Bicron Corp., 416 U.S. 470, 473 (1974), the former employees had signed confidentiality agreements. The Court upheld the district court's granting of a permanent injunction against the disclosure or use by respondents of 20 of the 40 claimed trade secrets until such time as the trade secrets had been released to the public, had otherwise generally become available to the public, or had been obtained by respondents from sources having the legal right to convey the information. Id. at 473-74. See also Henry Hope X-Ray Prods., Inc. v. Marron Carrel, Inc., 674 F.2d 1336, 1342 (9th Cir. 1982) (the limitation on confidential information contains the implicit temporal limitation that information may be disclosed when it ceases to be confidential). But see Howard Schultz & Assoc. v. Broniec, 239 Ga. 181, 236 S.E.2d 265, 270 (1977) ("The nondisclosure covenant here contains no time limitation and hence it is unenforceable"); Gary Van Zeeland Talent, Inc. v. Sandas, 54 Wis. 2d 202, 267 N.W.2d 242, 250 (1978) (unlimited duration of agreement not to disclose trade secret customer list makes the agreement per se void).

The Eighth Circuit in Sigma Chemical Co. v. Harris, 794 F.2d 371, 375 (8th Cir. 1986) explained, in the course of rejecting a "temporally unlimited" injunction, the rationale for limiting injunctions in time:

> [E]xtending the injunction beyond the time needed for independent development would give the employer "a windfall protection and would subvert the public interest in fostering competition and in allowing employees to make full use of their knowledge and ability."
> We believe the part of the injunction prohibiting disclosure of trade secrets must be limited in duration and, accordingly, reverse in part and remand the

case to the district court for consideration of the time it would take a "legitimate competitor" to independently reproduce the information contained in the product and vendor files. On remand, the district court should also modify the language of the injunction to expressly state that Harris may use that information which is already in the public domain.

3. Should there be a presumption that injunctive relief is appropriate? Such a presumption is typical of cases involving real property, but no such rule exists in most tort cases, and certainly not in typical contract cases. Are damages sufficient to protect trade secret owners? Perhaps not, since trade secrets are often hard to value. On the other hand, the parties could contract for injunctive relief, at least in employment cases, using noncompetition or confidentiality agreements. Should the law impose obligations the parties have not undertaken voluntarily? Should it refuse to enforce contract terms that the parties have agreed to?

One problem with such "automatic" injunctive relief is that it is difficult — and costly — to enforce. Is it an appropriate use of judicial resources to supervise employer-employee relationships on an ongoing basis? Is there a more cost-effective alternative to such supervision?

Note on Criminal Trade Secret Statutes

Misappropriation of trade secrets is not only a tort; in some circumstances, it is a crime. See generally Eli Lederman, Criminal Liability for Breach of Confidential Commercial Information, 38 Emory L.J. 921 (1989) (summarizing criminal laws governing misappropriation of confidential information). A series of well-publicized criminal prosecutions of computer executives accused of taking trade secrets to their new employers have raised the consciousness of industry professionals about trade secrets. The prosecutions have also raised ethical and political questions about the propriety of trade secret prosecutions "engineered" by the real parties in interest, often major companies such as Intel or Borland.

Criminal trade secret cases differ from civil ones in several respects. Obviously, the complaining party is the government, rather than the injured company. However, the injured companies are the "real parties in interest" and usually have some presence in the case. Even though they are not parties to the criminal proceeding, they at least supply a significant number of witnesses and enjoy a high level of communication with the district attorney.

Just as obviously, the burden of proof is higher than in a civil case. Some cases that could be won by the plaintiffs as civil cases will be lost in criminal court. This situation is even more likely because several states have definitions of trade secrets in their criminal laws that are more limited than their civil counterparts. For example, Cal. Penal Code §499c, which governs theft of trade secrets, historically limited the definition of a trade secret to "scientific or technical" information. (The constitutionality of this definition of trade secrets was upheld against a vagueness attack in People v. Serrata, 133 Cal. Rptr. 144 (Ct. App. 1976).) But in 1996, the California legislature amended its criminal trade secret statute to be coextensive with the definition of trade secrets in the Uniform Trade Secrets Act.

Defendants accused of stealing trade secrets may be charged with other offenses as well. For example, defendants who acquire a secret through improper means, as opposed to acquisition in a confidential relationship, may be guilty of larceny, receiving stolen property, or a host of similar crimes. See People v. Gopal, 217 Cal. Rptr. 487, 493-94 (Ct. App. 1985), *cert. denied,* 476 U.S. 1105 (1986). Further, the growth of computer technology has expanded the federal role in prosecuting theft of trade secrets, since data taken over a computer network is considered to cross state lines. See United States v. Riggs, 739 F. Supp. 414 (N.D. Ill. 1990) (allowing indictment of computer "hackers" who published data from a Bell South computer text file for wire fraud and interstate transportation of stolen property).

The prosecution of a criminal (rather than civil) trade secret case has other effects on the parties involved. First, criminal trade secret courtrooms are the scene of constant battles over the publication of information. The real parties in interest will naturally oppose the disclosure in a public courtroom of the very secrets the defendant is accused of stealing. This concern runs headlong into the defendant's constitutional right to a public trial.[19] Second, civil cases are generally stayed pending the outcome of a criminal prosecution. Thus a criminal prosecution may actually delay injunctive relief — the kind of remedy a civil plaintiff is often most interested in.

COMMENTS AND QUESTIONS

1. Given the stay imposed on a parallel civil action and the higher burden of proof in a criminal case, why would a civil plaintiff ever seek a criminal prosecution?

2. Should theft of trade secrets be a criminal offense? Does the presence of criminal sanctions have any effect on the optimal level of deterrence provided in damages suits, and therefore on the damages that should be awarded in a civil suit?

Do you see any problems with California's recent inclusion of business information in the criminal trade secret statute? Are there reasons to

---

19. There is no such right in civil cases. To avoid the very real danger that a misappropriation action will result in disclosure of the very secrets the plaintiff seeks to protect, civil trade secret actions will almost invariably include protective orders limiting the disclosure of information produced in discovery. Such orders are usually agreed to by the parties but may sometimes be imposed by the court. They will sometimes go so far as to prevent the parties themselves (as opposed to the attorneys and hired experts) from reviewing the other side's documents. In such a case, should in-house counsel be given access to discovery documents? See Brown Bag Software v. Symantec Corp., 960 F.2d 1465, 1470 (9th Cir. 1992).

treat theft of scientific information more harshly than theft of business information?

Note on Federal Criminal Liability for Trade Secret Misappropriation[20]

Prior to the passage of the Economic Espionage Act of 1996, there was only a single, very limited federal statute that directly prohibited the misappropriation of a trade secret. When confronted with an allegation that an individual had misappropriated a valuable trade secret, federal prosecutors were forced to turn to a number of federal statutes that were clearly not designed to cover trade secrets. In particular, prosecutors attempted to use the depression-era Interstate Transportation of Stolen Property Act (ITSP), and the wire fraud and mail fraud statutes. However a recent court decision severely curtails the use of ITSP, and the use of the mail and wire fraud statutes has always been limited because the theft often does not involve the use of mail or wire. In addition, since a thief usually merely copies information and does not necessarily "defraud" the victim permanently of the data, prosecutions are further limited. These shortcomings of federal law, coupled with testimony about the threat of economic espionage sponsored by foreign governments, underlie Congress's decision to enact the Economic Espionage Act of 1996 (EEA), 18 U.S.C. §1831 et seq.

The structure of the EEA reflects its rather disparate origins and the confusion that surrounded major changes made in the last days of the 1996 Congressional session. The EEA contains a definition of trade secrets that is taken, with only minor modifications, from the definition in the Uniform Trade Secrets Act. At the same time, the sections defining misappropriation are entirely new, and have no parallels in existing trade secret law.

The sine qua non of an action under the EEA is the existence of a "trade secret." Section 1839(3) defines that term as follows:

> (3) the term "trade secret" means all forms and types of financial, business, scientific, technical, economic or engineering information, including patterns, plans, compilations, program devices, formulas, designs, prototypes, methods, techniques, processes, procedures, programs, or codes, whether tangible or intangible, and whether or how stored, compiled, or memorialized physically, electronically, graphically, photographically, or in writing if
>
> (A) the owner thereof has taken reasonable measures to keep such information secret; and
>
> (B) the information derives independent economic value, actual or potential, from not being generally known to, and not being readily ascertainable through proper means by the public.

---

20. This section is adapted from James H.A. Pooley et al., Understanding the Economic Espionage Act of 1996, 5 Tex. Intell. Prop. L.J. 177 (1997).

This definition, which was added late in the Congressional session, generally tracks the definition of trade secrets in the Uniform Act. However, there are some differences between the language of the two statutes.

First, the list of potential types of secrets is much more expansive in the EEA than in the Uniform Act, though it has been narrowed some from earlier versions of the statute.[21] It is not clear why Congress chose to expand upon the representative list in the Uniform Act. Because of the expansive interpretation already given to the Uniform Act definition,[22] it is likely that the EEA will apply to all the same types of information that qualify as trade secrets under the current civil standard.

Second, the EEA expressly extends the definition of trade secrets to encompass information in any form, "whether tangible or intangible, and whether or how stored, compiled, or memorialized physically, electronically, graphically, photographically, or in writing." The references to intangible information and the "whether or how" language strongly suggest that not only information stored in electronic form, but also information "stored" only in an individual's memory, can be the subject of a prosecution for trade secret theft.

Sections 1831(a) and 1832(a) contain identical language regarding acts of misappropriation, punishing a specified individual who:

> (1) steals, or without authorization appropriates, takes, carries away, or conceals, or by fraud, artifice or deception obtains a trade secret;
> (2) without authorization copies, duplicates, sketches, draws, photographs, downloads, uploads, alters, destroys, photocopies, replicates, transmits, delivers, sends, mails, communicates, or conveys a trade secret;
> (3) receives, buys, or possesses a trade secret, knowing the same to have been stolen or appropriated, obtained, or converted without authorization . . .

These provisions are, in some respects, significantly broader than corresponding civil trade secret laws such as the Uniform Act.

Section (a)(1) lists a number of means of improper acquisition of a trade secret. In general, the provisions in this section seem to track the "improper means" of acquisition punished in civil cases. In particular, the references to stealing, concealment, fraud, artifice, and deception are reminiscent of the civil law's prohibition against both illegal and immoral business conduct. However, the provision also makes it a crime to "appropriate" or "take" a secret without authorization from the trade secret owner. These terms might be thought to encompass the sort of lawful business espionage that has long been permitted by civil trade secret law — conduct on the order of observing

---

21. Specifically, the terms "data," "tools," "mechanisms," "compounds," and "commercial strategies" were removed from earlier version of the Senate bill.

22. See, e.g., Religious Technology Center v. Lerma, 908 F. Supp. 1362, 1368-69 (E.D. Va. 1995) (religious scriptures); Thermodyne Food Service Prods. v. McDonald's Corp., 940 F. Supp. 1300 (N.D. Ill. 1996) (interrelationship of publicly known elements); Allen v. Johar, Inc., 823 S.W.2d 824 (Ark. 1992) (customer lists); ABBA Rubber Co. v. Seaquist, 286 Cal. Rptr. 518 (Ct. App. 1991) (customer lists); Restatement (Third) Unfair Comp. L. §39, cmt.

a competitor's property from across the street. There is some suggestion in the legislative history that the EEA is not intended to inhibit robust competition.[23] While this might be enough to declare broad categories of competitive intelligence gathering proper and therefore lawful, the ambiguity of some of the terms in section (a)(1) is troubling.

Broader still is section (2), which gives the trade secret owner the right to control a whole host of activities on the order of duplication, transportation, or destruction of a trade secret. This provision encompasses more than its civil counterparts. Section (2) is not by its terms limited to secrets acquired by "improper means" like those listed in section (1). Even legally acquired secrets can be misappropriated under the EEA if they are analyzed or duplicated in one of the ways listed in section (2). The civil law does prohibit some such uses of a lawfully acquired trade secret, but limits its reach to disclosure or use of a secret in violation of a confidential relationship. By contrast, there is no requirement of such a confidential relationship in the EEA. So long as the requirements of secrecy are satisfied, the trade secret owner is apparently entitled to seek the aid of the Justice Department in preventing the world from engaging in any of the enumerated acts in section (2), even if the secret is contained in a product that the defendant lawfully acquired.[24] As a result, the EEA effectively implies a confidential relationship between the trade secret owner and the world at large, something trade secret law has never before attempted and an indication that Congress believes trade secrets law is based in property theory, not tort or contract. Further, some of the specific provisions in section (2), such as the prohibition against "altering" or "destroying" a trade secret, are outside the normal reach of trade secret law.[25]

The most troubling thing about 18 U.S.C. §1831(a)(2) is that it arguably prohibits many forms of heretofore lawful reverse engineering activity. While reverse engineering is not expressly prohibited under this section, neither is it expressly permitted. Rather, its legality appears to be judged according to whether the reverse engineer engages in any of the prohibited acts.[26] To be sure, some types of reverse engineering, such as looking at or tasting a lawfully acquired product in order to determine what is in it, will not be illegal under the EEA. On the other hand, several of the restrictions in section (2) will, if read literally, encompass some forms of reverse engineering. For example, reverse engineering of computer software by "decompilation" almost always involves the making of a prohibited "copy" of the

---

23. S. Rep. No. 142-12212 ("Other companies can and must have the ability to determine the elements of a trade secret through their own inventiveness, creativity and hard work. . . . [P]arallel development of a trade secret cannot and should not constitute a violation of this statute.").

24. See H.R. Rep. No. 104-788 ("The concept of control also includes the mere possession of the information, regardless of the manner by which the non-owner gained possession of the information.").

25. "Destroying" a secret might mean "destroying its value by publishing it to the world," in which case civil law does provide a remedy. Alternatively, it might mean the physical destruction of valuable information, which is not covered by the Uniform Act.

26. See S. Rep. No. 142-12212.

program. Reverse engineering of mechanical devices and computer hardware may well involve prohibited "sketching, drawing, or photographing" of the trade secret contained in the publicly sold device. It is even possible that the prohibition against "altering" a trade secret will be interpreted to prevent chemical analysis of a trade secret product if such analysis involves the use of chemical reactants that bond to secret chemicals or that precipitate out certain elements from the formula.

The legislative history of the EEA is not encouraging to reverse engineers. The Senate report suggests that "the important thing is to focus on whether the accused has committed one of the prohibited acts of this statute rather than whether he or she has 'reverse engineered.' If someone has lawfully gained access to a trade secret and can replicate it without violating copyright, patent *or this law*, then that form of 'reverse engineering' should be fine."[27] The report goes on to suggest that observing a lawfully purchased product, or drinking a Coca-Cola, are legitimate activities.[28] On the other hand, there appears to be no reverse engineering defense protecting any of the forms of analysis suggested above. A computer programmer has the right to decompile a software program in certain circumstances under the Uniform Act, copyright law, and the common law without fear of civil liability. Under the EEA, though, that programmer has arguably committed a felony.

The Justice Department is actively prosecuting cases under the EEA. As of this writing, nearly a dozen cases had been filed. One potentially significant problem with the Act, and criminal trade secret statutes in general, is that the Act's provisions for protecting the confidentiality of the trade secrets during trial may be in conflict with the constitutional rights of criminal defendants, particularly the right to a public trial.

In United States v. Hsu, 155 F.3d 189 (3d Cir. 1998), the first appellate decision interpreting the Act, the court concluded that the defendants (who were the subject of a sting operation) could be found guilty of conspiring and attempting to misappropriate information that was not in fact a trade secret, because the crime of attempt did not require proof that the information the defendants tried to steal was in fact a secret.

27. S. Rep. No. 142-12212, at 12-13 (   ) (emphasis added).
28. Id. at 12213. Cf. *Mason*, 518 So. 2d at 130.

3

Patent Law

A. INTRODUCTION

This chapter introduces the basics of patent law. After a brief survey of the historical origins of our modern patent system and an overview of the elements of patentability, we turn to a substantive treatment of each of those elements. Then we discuss the rules and doctrines concerning patent infringement. Finally we consider remedies in patent cases.[1]

1. Historical Background

The first recorded reference to patents seems to be in Aristotle's *Politics*, composed in the fourth century B.C. In the course of a discussion of rival descriptions of a good constitution, Aristotle mentions a proposal by one Hippodamus of Miletos. According to Aristotle, Hippodamus called for a system of rewards to those who discover things useful to the state. Aristotle condemned the proposal as likely to lead to instability.

After this isolated classical reference, the history of patents skips several historical epochs. Although in recent years scholars have begun to revise the orthodox view of the Middle Ages as a period devoid of innovation,[2] this era does not appear to have been conducive to the idea of patents, at least in the

---

1. The Statutory Supplement to this book contains a brief introduction to the science of biotechnology. We strongly recommend that those without a background in the science review that material at this time.
2. See Donald Hill, A History of Engineering in Classical and Medieval Times (1984).

West.[3] Perhaps this situation prevailed because rigid social hierarchies — best exemplified by the guild system — discouraged the recognition of an individual inventor's genius.[4] Or perhaps recent historians are wrong, and not much in the way of innovation was occurring. Whatever the reason, one must look to the early Renaissance to find the first references to a real patent system.

The first regular administrative apparatus for granting patents — the first real patent "system" — arose in Venice in the late fifteenth century. Isolated grants in Venice and elsewhere were made earlier: in Venice in the early fourteenth century (for corn mill designs), and in Florence to the celebrated architect Brunelleschi for his 1421 invention of a barge with a hoist for transporting marble. The term *patent* — from the Latin *patere* (to be open), referring to an open letter of privilege from the sovereign — originated in this period. But not until the Venetian Senate's 1474 Act was the practice of granting patents regularized:

> Be it enacted that, by the authority of this Council, every person who shall build any new and ingenious device in this City, not previously made in this Commonwealth, shall give notice of it to the office of our General Welfare Board when it has been reduced to perfection so that it can be used and operated. It being forbidden to every other person in any of our territories and towns to make any further device conforming with and similar to said one, without the consent and license of the author, for the term of 10 years. And if anybody builds it in violation hereof, the aforesaid author and inventor shall be entitled to have him summoned before any Magistrate of this City, by which Magistrate the said infringer shall be constrained to pay him one hundred ducats; and the device shall be destroyed at once.[5]

The Venetian Act lays out all the essential features of a modern patent statute. It covers "devices"; states that they must be registered with a specific

---

3. One historian, however, claims that the modern patent system has its origins in the Byzantine Empire:

> It is likely that the modern monopoly originated in Byzantium, and became the invention patent at the Renaissance when numbers of inventions appeared. A traveller of the twelfth century, Benjamin of Tudela, mentions an exclusive privilege for dyeing cloth in the semi-Byzantine kingdom of Jerusalem.

Frumkin, Early History of Patents for Invention, 26 Trans. Newcomen Soc. 47, 47 (1947). Frumkin notes several other isolated instances of protective grants: in Bordeaux, in 1236, a fifteen-year monopoly for the manufacture of cloth; and in 1331, by King Edward III of Great Britain, a nonexclusive grant to export woollen cloth. Id. at 48.

4. The uneasy relationship between the craft guilds and early Renaissance patents suggests that the guilds, either directly or indirectly, may have had a hand in the suppression of patents granted to individuals. See Christine MacLeod, Inventing the Industrial Revolution: The English Patent System, 1660-1800 13 (1988) (describing successful efforts by Cutler's guild to block issuance of a patent on new knife handle design). Yet the culture of "collective invention" common to guilds did originate many important technologies. On the guilds generally, see Carlo M. Cipolla, Before the Industrial Revolution 256 et seq. (2d ed. 1980). There is some evidence that early exclusive privileges were granted in various German principalities. See Pohlmann, The Inventor's Right in Early German Law: Materials of the Time from 1531-1700, 43 J. Pat. and Trademk. Off. Society 121, 122 (1961) ("proto-patents were issued as early as 1378").

5. Mandich, Venetian Patents (1450-1550), 30 J. Pat. & Trademk. Off. Society 166, 177 (1948).

administrative agency; says that they must meet the requirements of being "new and ingenious," "reduced to perfection," and "not previously made in this Commonwealth"; provides a fixed term of ten years; and sets forth a procedure to determine infringement, as well as a remedy. Interestingly, the Venetian Act reserved to the Republic the right to use any invention without compensating the inventor.[6] This is an early attempt to reconcile individual interest with the good of the community, a recurring problem in patent law.

The opening of trade in Europe ensured that the new Venetian concept would spread. As Italian craftsmen — particularly glassworkers — fanned out across Europe, they brought with them the idea of legal protection for inventions.

Patents came to Great Britain by this route, sometime in the middle of the sixteenth century. The chief minister under Elizabeth I, William Cecil (Lord Burghley), used patent grants to induce foreign artisans to introduce continental technologies into England.[7] Thus what later became the Anglo-American patent system was ushered in as a mercantilist instrument — what today would be called a "strategic international trade" policy. The idea was to lure immigrants who had desirable skills and know-how with the promise of an exclusive privilege. Faint glimmers of this early policy survive in certain odd corners of today's patent laws.[8] Note that this policy reflects another attempt to balance the rights of the community against the individual interests of inventors. By luring skilled artisans with the reward of exclusive rights, it was hoped the community would gain the fruits of this skilled labor.[9] Ironically, by the mid-eighteenth century, Britain began to show concern over the reverse problem — leakage of its technical prowess to overseas rivals, including the American colonies.[10]

With the accession of James I in England in the early seventeenth century, patents became less an incentive for inventors of new arts and more a royal favor dispensed to well-placed courtiers. Under this rubric, "patents" were granted on such enterprises as running ale-houses. Parliament, whose members represented many trades injured by these special privileges, was

6. Id.

7. For a detailed discussion of the early development of patent law in England and on the continent, see Edward C. Walterscheid, The Early Evolution of the United States Patent Law: Antecedents, 76 J. Pat. & Trademk. Off. Society 697 (1994) (Part I); 76 J. Pat. & Trademk. Off. Society 849 (1994) (Part II); 77 J. Pat. & Trademk. Off. Society 771 and 847 (1995) (Part III); 78 J. Pat. & Trademk. Off. Society 77 (1996) (Part IV).

8. See Donald Chisum, Foreign Activity: Its Effect on Patentability Under United States Law, 11 Intl. Rev. of Indus. Prop. & Copyright L. 26 (1980). One such provision, which favored U.S. inventors in priority disputes with foreigners, was partially abolished in 1995 when the United States ratified GATT. See infra section B.3.e.

9. Thus the first patents were described as "passports" which allowed their holders to move about freely and practice their trade. Walton H. Hamilton, The Politics of Industry 68 (1957).

10. See, e.g., David Jeremy, Transatlantic Industrial Revolution 36-49 (1981) (describing the key role of English emigr'ants in transfer of technology to the colonies, and the lack of impact of the prohibitory English export laws due to the fact that it was not necessary to transfer physical embodiments of technology when people could simply memorize the necessary information).

displeased. Thus arose the Statute of Monopolies of 1624, which forbade all grants of exclusive privilege except those described in the famous Section 6:

> [B]e it declared and enacted that any declaration before mentioned shall not extend to any letters patent and grants of privilege for the term of fourteen years or under, hereafter to be made, of the sole working or making of any manner of new manufactures within this realm, to the true and first inventor and inventors of such manufacture, which others at the time of making such letters patent shall not use, so as also they be not contrary to law, nor mischievous to the State, by raising prices of commodities at home, or hurt of trade, or generally inconvenient; the said fourteen years to be accounted from the date of the first letters patents, or grant of such privilege hereafter to be made. . . .[11]

This statute called on the common law courts to review all privileges granted by the crown and outlawed all but those based on true inventions. The Statute of Monopolies, with its general ban on exclusive rights to manufacture and sell goods, and its limited exception for the purpose of fostering new inventions, is an early example of both antitrust laws and the complex economic interaction between a desire for competition and a desire for new inventions.

Even with the Statute of Monopolies in effect, the British patent system remained a largely informal administrative apparatus. After the vicissitudes of the Civil War period, during which the Cromwell government "called in" all extant patents and privileges, the *status quo ante* was for the most part restored. Court influence was still helpful in obtaining a patent until the latter part of the eighteenth century. Patent applications were registered rather than examined. And, most tellingly, very few patents were granted.

But as the Industrial Revolution picked up steam (so to speak), attention began to focus on patents once again. An important change at this time was the increasingly stringent requirement that the applicant for a patent describe his or her invention clearly and completely, a development most often associated with the 1778 opinion of the well-respected Judge Mansfield in *Liardet v. Johnson*. The importance of the specification requirement is that it reflected a changed perception about what the inventor was contributing to society in exchange for the patent grant. Under the original patent systems, society's benefit was the introduction of a new art or technology into the country. By the late eighteenth century, the primary benefit was seen as the technological know-how behind the inventor's patent. The beneficiaries on this view were not just the public at large, but also others skilled in the technical arts who could learn something from the patentee's invention. This was a major change in the economic role of patents, for it shifted the emphasis from the introduction of finished products into commerce to the introduction of new and useful *information* to the technical arts. While it is difficult to estimate the significance of this transition, it does seem to address a complaint voiced by

---

11. Great Britain, Statutes at Large, 21 Jac. I, c. 3 (1624).

Lord Burghley over the original patent system — its dismal success rate in introducing new industries to the country. Perhaps paradoxically, the emphasis on technical specifications, while recognizing that not every invention will lead to a new industry, may have more efficiently fostered the growth of industry as a whole, by ensuring that up-to-date technical information was disseminated rapidly after its creation.

Although the overall contribution of the patent system to the Industrial Revolution has been a matter of debate in historical circles, it seems no coincidence that the patent system matured alongside the early industrial technologies. One historian, H. I. Dutton, noted that the British patent system of this period was less than water-tight from the inventor's point of view. But Dutton argues that this actually redounded to the benefit of the economy as a whole, since "leaks" in the grant to one inventor benefitted other inventors.[12] In this, too, the early British experience foreshadows problems that courts struggle with today — how to encourage invention through the use of exclusive rights, without at the same time stifling the creativity of inventors other than the patentee.

Patents were among the many British legal concepts introduced to the American colonies between 1640 and 1776. State patents were granted in most of the original thirteen colonies, beginning with a Massachusetts patent in 1641. Even after the Revolution, under the Articles of Confederation, the individual states continued to issue patents.

Perhaps inevitably, however, conflicts began to arise between the states — most notably over steamboat patents, which were issued to two different inventors during this period. This led to a great deal of confusion over who was actually the inventor of the steamboat, which created an obstacle to the successful operation of interstate steam lines. With this problem (among others) in mind, the Constitutional Convention of 1789 resolved to create a national patent system rooted in the Constitution itself. Thus the provision

12. In his thorough review of the role of the patent system in the "first" Industrial Revolution in Great Britain, Dutton concludes that the system was instrumental in fostering almost all of the key technologies of the era. In addition, in chapters on "Trade in Invention" and "Investment in Patents," he documents the historical connections between patents and the financing of invention, thus illustrating that the early patent system did not reward innovation directly but instead played much the same role it does today, i.e., fostering *invention* (creation of new technology) and thus indirectly encouraging innovation (introduction of new products embodying that technology on the market). H. I. Dutton, The Patent System and Inventive Activity During the Industrial Revolution 1750-1852, 103-48 (1984). And in his conclusion Dutton argues that the patent system's inefficiencies actually made it close to an ideal system, since it encouraged invention but did not protect new technology too much from those who would try to improve it. Id. at 204-205. See also MacLeod, Accident or Design? George Ravenscroft's Patent and the Invention of Lead-Crystal Glass, 28 Tech. & Culture 776-780 (1987) (describing long time lag between invention of lead crystal glass and introduction of final product with "bugs" all worked out); F. M. Scherer, Invention and Innovation in the Watt-Boulton Steam-Engine Venture, 6 Tech. & Culture 184 (1965) (role of the patent system in Watt's seminal steam engine invention). But cf. J. Mokyr, The Industrial Revolution and the New Economic History, in The Economics of the Industrial Revolution 1, 28 (Joel Mokyr, ed. 1989) (arguing that "[p]roperty rights in new techniques were protected, albeit imperfectly by British patent law," yet "[t]he cumulative effect of small improvements made by mostly anonymous workers and technicians was often more important than most of the great inventions.").

of Article I, Section 8, authorizing Congress to award exclusive rights for a limited time to authors and inventors "for their respective writings and discoveries." One historical footnote is worth mentioning in this connection: an early draft of this provision, set out in James Madison's notes to the Convention, called for both exclusive rights *and* outright subsidies for new inventions. But this was rejected in favor of exclusive rights only.[13] In any event the first U.S. patent statute was passed in May 1790,[14] in the very early days of the first Congress (reflecting the importance of this matter), and the first patent was issued shortly thereafter — to Samuel Hopkins of Pittsford, Vermont, for a process for making potash from wood ashes.

The story of Thomas Jefferson's involvement in the early national patent system has often been told; he was a significant contributor to the original statute, and he helped to administer the patent system established in 1790. But while the patent system got on its feet under Jefferson,[15] it did not grow to its full stature until the 1836 revision, when a formal system of examination, with professional examiners, was substituted for the pro forma registration system of the 1793 Act, which had itself been substituted for the original (1790) procedure involving three high-level government officials (including Jefferson as Secretary of State).

Since 1836 the patent system has grown dramatically by any standard — number of patents issued, number of cases litigated, number of significant inventions patented, and so on. As greater demands were placed on it, the patent system developed new rules. For example, the requirement that an invention be more than novel, that it reveal an "inventive leap," or what is now called nonobviousness, developed in the mid-nineteenth century to limit the number of patents that were being issued. Late in the nineteenth century the bureaucratic structure of the Patent Office as we know it began to take shape as well.

In Europe, the nineteenth century was a time when a new generation of analytical economists questioned the economic foundations of the patent system.[16] Indeed, Switzerland and the Netherlands had no patent systems for

13. Bruce Bugbee, The Genesis of American Patent and Copyright Law 126, 143 (1967) (describing Madison and Pinckney proposals, both of which included some form of subsidies; and noting that the Senate proposed a compulsory licensing provision in the first Patent Act of 1790, modeled on similar provisions in state copyright acts, but it was rejected by the House).

14. Patent Act of 1790, Ch. 7, 1 Stat. 109-112 (April 10, 1790).

15. There is some evidence that the early patent acts were a significant stimulus to invention in the new nation. See, e.g., Michael B. Folsom & Steven D. Lubar, Introduction, in The Philosophy of Manufacturers: Early Debates over Industrialization in the United States xxvii-xxviii (Folsom & Lubar eds. 1982) ("Given the fact that the primary inducement and reward for industrial development was money, new industrial interests exerted pressure to establish safeguards for the investment of time and capital in technological innovation. The first great American industrial corporation, the Boston Manufacturing Company, made a significant early profit selling other textile companies rights to the machine patents it held.").

Whether the patent laws spurred invention or not, a patent bar was quick to form to help inventors deal with the new act. The first patent treatise, by the eminent Fessenden, appeared in Boston very early in the nineteenth century. See Thomas Fessenden, An Essay on the Law of Patents for Inventions (1st ed. 1810).

16. See Fritz Machlup & Edith Penrose, The Patent Controversy in the Nineteenth Cen-

more than fifty years during this period.[17] Despite the period of transition and questioning, the patent system was a well-accepted feature of the economic landscape by the beginning of the twentieth century. Key patents on the lightbulb, the telephone system, the basic design of the automobile, and the first airplanes symbolized the technical virtuosity and dynamism of the age. As the scale of industry grew, research and development departments began to appear in the larger companies.[18] Patents were not only a valuable output of these departments; they helped measure the productivity of the departments and served to justify their importance.

The history of the U.S. patent system in this century reflects swings between greater and lesser protection. By the 1920s and 1930s, a number of people began to believe that large companies with patent portfolios were a little *too* powerful. Spurred in part by a series of anticompetitive acts by large companies whose patents dominated their respective industries, courts became less willing to enforce patent rights and more willing to punish patentees for exceeding the scope of their patent grant. The pendulum swung back towards patentability during the 1940s. As the nation threw all available resources into the war effort, the armed forces called on engineers and scientists to perfect a vast array of new technologies in short order. Companies during this era had neither the time nor the ability to try to exclude their competitors: government-directed industrial development and "mandatory cooperation" replaced exclusionary acts. By the time the war was over, there was a consensus in Congress in favor of a strong patent system. In fact the 1952 Patent Act, the first major revision of the patent code since the nineteenth century, marked a return to the principles of that century in many respects.

The patent system by general consensus reached a low-water mark during the 1960s, in the wake of this period of strong protection. The patent office issued patents rather freely, without particularly rigorous examination in many circumstances. On the other hand, it was difficult to get a patent upheld in many federal circuit courts, and the circuits diverged widely both as to doctrine and basic attitudes toward patents. As a consequence, industry downplayed the significance of patents.

In 1982 Congress passed the Federal Courts Improvement Act, creating the new Court of Appeals for the Federal Circuit (or CAFC, as it is now

---

tury, 10 J. Econ. Hist. 1 (1950); Fritz Machlup, Patents, in 2 Intl. Encyclopedia of the Social Sciences 461 (1968).

17. E. Schiff, Industrialization Without National Patents: The Netherlands 1869-1912; Switzerland 1850-1907 (1971). Schiff's book is rather inconclusive on the effect of this non-patent era on the economic development of the two countries. He states in regards to the Netherlands, for instance, that "it seems unlikely that the overall rate of progress in industry would have been markedly different if a patent system . . . had been in operation." Id. at 40. On the other hand, he notes statistical evidence "that the reintroduction of a patent system in 1912 has given an extra spur to Dutch inventive activity." Id.

18. See Thomas P. Hughes, American Genesis: A Century of Invention and Technological Enthusiasm 1870-1970, 150-180 (1989).

widely known).[19] The CAFC handles several important types of cases, but from the beginning one of its primary functions has been to hear all appeals involving patents. While the CAFC was ostensibly formed strictly to unify patent doctrine, it was no doubt hoped by some (and expected by others) that the new court would make subtle alterations in the doctrinal fabric, with an eye toward expanding the scope of patent protection.

To judge by results, that is exactly what happened. Patents are more likely to be held valid now than in previous decades.[20] It is much easier to get an injunction against an infringer. And money damages have soared, both on average and in the highest-visibility cases. Whether intentional or not, the creation of the CAFC will surely be seen as a watershed event by future historians of the patent system.

At the same time the environment in which the CAFC operates has been changing as well. Exports have grown considerably as a source of national wealth. As a result, policymakers in the government now pay much closer attention to the legal and economic features of our major trading partners. This naturally includes much closer scrutiny of the intellectual property systems in these other countries. At the same time, it has become clear that other countries have identified technology-intensive industries as keys to economic growth in the future. This awareness has led to a new focus on economic policy instruments that can be used to foster these industries — including protection of intellectual property. Finally, because intellectual property legislation has no direct, immediate cost to the government, it seems to many to be a relatively cheap aid to industry. The result of these factors has been a growing trend in world trade policy towards exporting or "internationalizing" patent law. A number of countries (including the European Union) have created a reciprocal system in which applicants need not file for protection separately in every country. International treaties contain similar arrangements. If recent changes wrought by the Uruguay Round of GATT, GATT-TRIPS, are any indication, the patent laws of most major countries will come into ever-closer harmony.

19. See Rochelle Dreyfuss, The Federal Circuit: A Case Study in Specialized Courts, 64 N.Y.U. L. Rev. 1, 25-26 (1989); Jordan, Specialized Courts: A Choice?, 76 Nw. U. L. Rev. 745 (1981) (describing competing arguments over specialized courts).

20. Several recent studies have provided statistical support for the proposition (widely accepted in the patent bar) that the Federal Circuit is at least moderately "pro-patent." See John R. Allison & Mark A. Lemley, Empirical Evidence on the Validity of Litigated Patents, 26 Am. Intell. Prop. L. Assn. Q.J. 185 (1998); Donald R. Dunner, J. Michael Jakes, and Jeffrey D. Kerceski, A Statistical Look at the Federal Circuit's Patent Decisions: 1982-1994, 5 Fed. Circuit B.J. 151, 154-155 (1995). See Robert P. Merges, Commercial Success and Patent Standards: Economic Perspectives on Innovation, 76 Cal. L. Rev. 803, 820-821 (1988) (comparing pre- and post-Federal Circuit era statistics); Mark A. Lemley, An Empirical Study of the Twenty-Year Patent Term, 22 AIPLA Q.J. 369 (1994).

2. An Overview of the Patent Laws

It is important to appreciate some basic attributes of the patent system before reading the following cases. This is a complex body of law with its own terminology and tradition, and one must become acquainted with the rudiments before diving into the details.

a. *Requirements for Patentability*

The Patent and Trademark Office (PTO) grants a patent when an inventor can show five things: an invention fits one of the general categories of patentable subject matter; it has not been preceded in identical form in the public prior art; it is useful; it represents a nontrivial extension of what was known; and it is disclosed and described by the applicant in such a way as to enable others to make and use the invention. 35 U.S.C. §§101 (utility), 102 (novelty), 103 (nonobviousness), and 112 (enablement) (1982). See generally Graham v. John Deere, 383 U.S. 1 (1966).

Patent lawyers call the no–identical-prior-invention requirement "novelty." In practice, novelty is established by applying a set of technical rules to determine if a patent applicant was really the first to make the invention she is claiming. The novelty test determines whether the claimed invention is unpatentable because it was made before, sold more than a year before a patent application was filed, or otherwise disqualified by prior use or knowledge.

Utility, the second requirement, has devolved over the years into a rather minimal obstacle to obtaining a patent. Section 101 is the source of this requirement. 35 U.S.C. §101 (1982) ("Whoever invents any new *and useful* process, machine, manufacture, or composition of matter, may obtain a patent therefor. . . ." (emphasis added)). Today, a patent will not be withheld even though the invention works only in an experimental setting and has no proven use in the field or factory. Only if an invention has absolutely no "practical utility" will a patent be denied. The only exception is inventions pertaining to pharmaceuticals, where some cases question whether laboratory promise is enough to establish utility in treating human patients.

The next requirement, nontriviality, is known to patent lawyers as "nonobviousness." 35 U.S.C §103. This is the most important requirement; it has been called "the ultimate condition of patentability." The reason is that nonobviousness attempts to measure an even more abstract quality than novelty or utility: the *technical accomplishment* reflected in an invention. This requirement asks whether an invention is a big enough technical advance over the prior art. Even if an invention is new and useful, it will still not merit a patent if it represents merely a trivial step forward in the art.

Finally, 35 U.S.C. §112 requires a patentee to give a sufficiently good description of her invention that "one of ordinary skill in the art" would be able to make and use the invention. The concern here is not so much with whether the inventor has developed something worth patenting as it is with the benefit the public obtains from the patent "bargain." The disclosure and "enablement" requirements of section 112 ensure that those "skilled in the art" of the invention can read and understand the inventor's contribution, and that after the patent expires they will be able to make and use the invention themselves.

b. Rights Conferred by a Patent

"Claims" are the heart of patent law. Claims define the boundaries of the property right that the patent confers. (Innumerable patent cases therefore analogize claims to the "metes and bounds" of a real property deed.) Claims must always be accompanied by a specification, and in addition most patents also have one or more drawings.

The specification describes the invention. It names all the parts or components of the invention, describes how they work, and illustrates how they work together to perform the invention's function. Only at the end of the specification does the inventor (or, more usually, her patent lawyer) state the precise legal definition of the invention. These are the claims. Here are a few (fanciful) examples of claims:

1. Element 95.
2. A composition comprising
    (a) a solid selected from the group consisting of
        (1) sodium chloride,
        (2) potassium chloride and
        (3) lithium chloride;
    (b) a liquid selected from the group consisting of
        (1) sulfuric acid,
        (2) nitric acid. . . . [21]
3. The material wrought tungsten, having a specific gravity of approximately 19 or greater, and capable of being forged and worked.
4. A windmill comprising a wind-catching device, directed to face the oncoming wind force, which turns a shaft, which acts on gears or

---

21. This is an example of a claim to a "genus" that includes several "species," a common construct in patent law. A special form of genus claim, termed a "Markush expression," is often encountered. Such claims take the form of an expression such as "An X selected from the group consisting of A, B, and C," where X describes the functional class to which A, B and C belong. These claims are often, but certainly not always, used in patents to chemical compositions. There is, for example, a famous patent claiming a cigarette filter made of cheese, whose claim 2 reads: "A cigarette filter according to claim 1, in which the cheese comprises grated particles of cheese selected from the group consisting of Parmesan, Romano, Swiss and cheddar cheeses." U.S. Patent No. 3,234,948, to Stebbings, cited in Robert C. Faber, Landis on Mechanics of Patent Claim Drafting §50 (3rd ed. 1990).

another device to change the direction of said wind force, so as to operate a pump that pumps water.

5. A windmill according to claim 4 wherein the force-changing device is a set of gears.

6. A method for treating baldness, which comprises applying to the scalp an aqueous solution of the compound minoxidil.

7. An apparatus for playing record albums comprising a cartridge or stylus made from at least 40 percent graphite by weight, a tone arm on which said cartridge or stylus is mounted, a turntable on which said record albums are placed for playing, and means for turning said turntable at appropriate speeds for the playing of said record albums.

Note that all the claims except claim 5 are examples of *independent claims;* they do not refer to any other claim or claims. This is in contrast to a dependent claim, such as claim 5 above. You can tell it is a dependent claim because it begins with the phrase "A windmill according to claim 4. . . ." A dependent claim incorporates all the limitations of the independent claim on which it depends.

Note also that claim 6 is a claim to a process or method rather than a device, and that the last element of claim 7 is stated in "means plus function" format: the element is not described in detail but is merely listed as "a means" for accomplishing some goal. The importance of this specialized claim format will be discussed *infra.*

A patent confers the right to exclude others from making, using, selling, offering for sale, or importing the claimed invention for a specific term of years. Until 1995, that "patent term" was 17 years from the date the patent issued. With United States adherence to GATT-TRIPS, the patent term was changed to extend for 20 years from the date the patent application was filed, rather than the old term of 17 years from the date it was issued by the Patent Office. The length of a patent term under the new law therefore varies from case to case — a patent is in force for 20 years minus the amount of time the patent spent in the application (or "prosecution") process.[22]

---

22. Uruguay Round Agreements Act of 1994, P.L. 103-465, 108 Stat. 4809 (1994), at §532, codified at 35 U.S.C. §154(a)(2). For an analysis of the likely effects of this change in patent term, see Mark A. Lemley, An Empirical Study of the Twenty Year Patent Term, 22 AIPLA Q.J. 369 (1994).

The history of the patent term in the United States prior to the 1994 Act is interesting in its own right. The first patent act, Act of April 10, 1790, ch. 7, 1 Stat. 109, adopted the same patent term as the British Statute of Monopolies of 1624, 21 Jac. 1, ch. 3 — namely, 14 years. This was based on the notion that a patent should protect an inventor for a period equal to two terms of a standard British trade apprenticeship of seven years. See White, Why a Seventeen Year Patent?, 38 J. Pat. & Trademk. Off. Society 839 (1956). In the patent act of 1836 — which reinstituted the examination system in the original 1790 Act — a seven-year renewal term was added, making the total possible term 21 years. The present 17-year term is the result of a compromise between the House, which wanted to retain the 14 plus 7 term, and the Senate, which wanted to eliminate the renewal term. Economists have long noted that a uniform term sometimes rewards inventors too much and sometimes gives too little incentive; some have pro-

The exclusionary right is in a sense a negative right, for two reasons. First, a patent does not automatically grant an affirmative right to do anything; patented pharmaceuticals, for instance, must still pass regulatory review at the Food and Drug Administration to be sold legally. Second, a patented invention may itself be covered by a preexisting patent. For instance, a broad "pioneering" patent on a product or process may cover later-developed inventions, themselves patented as improvements. In such a case the holder of an improvement patent has the right to exclude everyone from her improvement — including the holder of the broad patent — while at the same time being barred from use of the improvement herself unless the holder of the broad patent authorizes such use. Patents so related are said to be "blocking patents."

Note on the Procedures for Obtaining a Patent

The process of obtaining a patent from the Patent Office is known as "prosecution." The time and effort required to prosecute a patent varies immensely from case to case. A few applications are reviewed quickly and issued within a year of the application date. Others languish in the PTO for years and, especially when several inventors claim they were the first to produce a particular invention, even decades. The "average" prosecution takes approximately two to three years. See Mark A. Lemley, An Empirical Study of the Twenty Year Patent Term, 22 AIPLA Q.J. 369 (1994) (average time for patent prosecution in 1994 was 864 days, or approximately 2 years, 4 months).[23] During this time the examiner and the inventor or inventor's attorney[24] engage in a series of negotiations. First, the examiner may reject the application as deficient under a number of sections of the Patent Act. Then the examiner will normally attempt to narrow the wording of the claims. The process is helped immensely by the fact that the examiners are specialists; they concentrate only on particular technologies, or even, commonly, a precise corner of a particular technology.

A patent specification describes the problem the inventor faced and the steps she took to solve it. It also provides a precise characterization of the "best mode" of solving the problem contemplated by the inventor. The battle over the precise verbal boundary to the patentee's invention — i.e., the claims — is at the heart of patent prosecution, and the specification must *support* the claims.

---

posed to "craft" individual terms to fit individual inventions. See, e.g., William D. Nordhaus, Invention, Growth, and Welfare (1969) (economic model deriving optimal patent term under different circumstances).

23. Lemley predicts that this time will fall, since under the new 20-year term patent applicants who were formerly willing to delay responding to the PTO will be more inclined to act quickly.

24. Nonlawyer "patent agents" are permitted to practice before the PTO, but most who prosecute patent applications are patent lawyers.

Prosecution is a give-and-take affair. In the course of responding to an examiner's objections, an applicant will often amend her specification and claims. She can then file a "continuation" of the original application, changing only the claims, or she can change her specification, and refile her patent application as a so-called continuation-in-part or "C-I-P" application. The original filing date is preserved so long as no "new matter" is added. In either case, the original application is referred to as the parent application. If two continuations are filed, the original is the grandparent, and so on.

An application that the examiner believes contains more than one invention will be subject to a "restriction" requirement. The applicant must then decide which of the multiple "inventions" in the application she wishes to pursue; the others will have to be placed in a new and separate application. The claims corresponding to the invention she decides to pursue in the original application are called the elected claims.

Often the claims section of a patent begins with the broadest claim, which is then "qualified" in a series of dependent claims. This is followed perhaps by a narrower independent claim, which may itself be qualified by a series of dependent claims. In this way, the general structure of a patent often resembles an inverted pyramid: the broadest claims are first, the narrowest last, and the scope of the claims generally "tapers" from the first to the last. For more on the types of claims, see supra.

Prosecution ends when the patent is granted. If this happens, the application — which will have been kept secret during prosecution — will mature into a full-fledged patent. It is then summarized in the Patent Gazette and made available to the public.[25] Prosecution can in a sense continue after a patent has issued. A patentee who comes to believe that her patent claims are either too broad or too narrow can seek a reissue of the patent, so long as the deficiency in the original patent is the result of a bona fide error or omission. Reissues to broaden the scope of claims must be initiated within two years of the original issuance, however.

In a related proceeding called a reexamination, *anyone* (including the patentee) can seek a reexamination of a patent if a substantial new basis for questioning the patentability of the invention arises after issuance. At present, third parties are strictly limited in their ability to participate in a reexamination proceeding, even if they initiated it. Legislation pending in Congress at this writing would allow greater third-party participation.

Of course, sometimes the patent examiner cannot be persuaded to issue a patent. In that case, the Examiner can issue what is known as a "final rejection" of the application. Such a rejection is not actually final, however. Patent applicants faced with such a rejection can (a) negotiate with the examiner to allow an amended version of the patent, usually with narrower

---

25. A bill introduced in Congress in 1999, but which did not pass, would have required that patent applications be published 18 months after they are filed, regardless of whether the patent has yet issued. H.R. 1907, 106th Cong., 1st Sess. This proposal is consistent with the practice in virtually all countries outside the United States, and it is required by a 1994 executive agreement between the United States and Japan.

claims than the original application; (b) abandon their application and refile a continuation application, hoping to convince the examiner with further argument and amendment (or to get a different examiner!); or (c) appeal the rejection. An appeal of a rejected application follows a procedure not unlike that of many other administrative agency appeals. There is an internal Board of Appeals within the Patent Office which hears cases where an application has been the subject of a "final rejection." If an applicant is dissatisfied with the opinion of the Board of Appeals, he or she may appeal again, this time to the Court of Appeals for the Federal Circuit.

Court disputes involving patents also arise when a patentee brings a lawsuit against someone accused of infringing the patent. There are typically two defenses to such a suit: first, the accused infringer will argue that the patentee's patent is invalid[26] for any number of reasons; an invalid patent, of course, cannot be infringed. Second, the accused infringer will usually argue that even if the patent is valid, the products being made or sold by the accused do not infringe the patentee's patent. Because invalidity is often argued as a defense in an infringement case, all the issues of patent validity dealt with by the Patent Office in granting a patent may also be considered by a court. As we shall see, the standard of review of the Patent Office's decision to issue a patent is a contentious question. See, e.g., Dickinson v. Zurko, 119 S.Ct. 1816 (1999) (adopting the deferential "substantial evidence" standard for judicial review of PTO decisions). But there is no doubt that a patent is "born valid": the patent code contains a provision stating that there is a "presumption of validity" that accompanies any issued patent. 35 U.S.C. §282. Appeals in patent infringement cases go automatically to the Court of Appeals for the Federal Circuit, regardless of the district in which the case was filed.

Three other forms of patent litigation deserve brief mention here.

- A unique form of patent proceeding is a priority dispute between two or more inventors, all of whom claim to have been the first inventor of a particular invention. These are known as interference proceedings. They are an outgrowth of the fact that the United States awards a patent to the first inventor, unlike almost every other country, which awards the patent to the first to file. Interference proceedings are handled by the PTO internally through the Board of Patent Appeals and Interferences, again with a right of appeal to the CAFC.
- People who reasonably fear being sued for patent infringement (for example, because the patentee has threatened them with suit) can file suit themselves for a declaratory judgment that the patent at issue is invalid, or that their conduct does not infringe the patent. Such de-

---

26. Technically, some defenses in patent cases result in a finding that a patent is unenforceable, rather than invalid. This is the case with the defense of inequitable conduct, for instance, which is concerned with improper disclosure or nondisclosure by a patent attorney during prosecution of the patent application. The antitrust-related defense of patent misuse is another example; when patent misuse is proven the affected patents are found unenforceable (at least until the misuse can be "purged" or cured).

claratory judgment actions are typically met with counterclaims charging infringement; these proceed much as normal patent infringement suits would.

- The International Trade Commission has jurisdiction to block the importation of products into the United States if they infringe U.S. patents and to block the importation of products made abroad by processes that are patented in the United States. Patent owners can complain to the ITC, which will bring "section 337" actions against likely infringers. These actions have their own set of procedural rules that operate parallel to district court actions for patent infringement.

3. Theories of Patent Law

By contrast with trade secret law, which draws on a number of different (and sometimes contradictory) theoretical bases, the central theory behind patent law is relatively straightforward. This theory posits that inventions are public goods that are costly to make and that are difficult to control once they are released into the world. As a result, absent patent protection inventors will not have sufficient incentive to invest in creating, developing, and marketing new products. Patent law provides a market-driven incentive to invest in innovation, by allowing the inventor to appropriate the full economic rewards of her invention.

Other theories often advanced to explain intellectual property law — natural rights and personhood theories, for example — play a much less significant role in patent law than in other areas of intellectual property.[27] In part, this is because of the broad nature of the patent grant. Patents give the inventor the right to sue not only those who "steal" his invention, but those who reverse engineer it and even those who develop the same invention independently. The broad nature of this grant makes it difficult to speak of a "moral entitlement" to a patent. And in part the focus on utilitarian theory mirrors the subject matter of patents, which revolves around mechanical devices, chemical formulae, and the like. It would be anomalous to most to speak of the "personality" invested in stamping machinery or pesticides.

---

27. That is not to say such theories do not exist. For a discussion of reward-based and even natural law theories of scientific invention, see, e.g., A. Samuel Oddi, Un-Unified Economic Theories of Patents — the Not-Quite-Holy Grail, 71 Notre Dame L. Rev. 267, 274-277 (1996); Kevin Rhodes, Comment, The Federal Circuit's Patent Nonobviousness Standards: Theoretical Perspectives on Recent Doctrinal Changes, 85 Nw. U.L. Rev. 1051 (1991); cf. Lawrence C. Becker, Deserving to Own Intellectual Property, 68 Chicago-Kent L. Rev. 609 (1993) (arguing that entitlement-based arguments for patent law are intuitively appealing, but do not necessarily justify the scope of current patent doctrine). An alternative to classical incentive theory is the prospect theory of patents, advanced by Edmund Kitch in an important article twenty years ago. See Edmund Kitch, The Nature and Function of the Patent System, 30 J.L. & Econ. 265 (1977). Kitch offers a property-based vision of patents as entitlements to innovate within a particular field, granted to those who have already started such innovation. For a refinement of Kitch's approach that takes account of rent-seeking, see Mark F. Grady & Jay I. Alexander, Patent Law and Rent Dissipation, 78 Va. L. Rev. 305 (1992).

Although the economic incentive story is straightforward, it does not tell us very much about how to design a patent system to provide optimal incentives. For the reasons described in Chapter 1, overprotecting is as bad as underprotecting in many ways. Thus designing the proper economic incentive requires the policymaker to balance the length of the patent term, the appropriate standard of invention, and the nature of the rights granted to patentees. Resolving these conflicts has occupied courts and Congress since the passage of the first patent statute in 1790. We discuss them in some detail below.

B. THE ELEMENTS OF PATENTABILITY

In the sections that follow we investigate the five primary requirements of patentability: (1) patentable subject matter, (2) novelty, (3) utility, (4) nonobviousness, and (5) enablement.

1. Patentable Subject Matter

This section deals with what is known as patentable subject matter under §101 of the patent code: that is, the issue of which *types* of inventions will be considered for patent protection. Section 101 determines, for instance, whether living things can be patented (as in the case that follows), or whether mathematical algorithms are proper patent subject matter. (Cases specifically dealing with computer-related patents are discussed in Chapter 7.) We are not concerned here with the technical requirements of patentability (novelty, utility, nonobviousness, etc.); these requirements are the subject of subsequent sections.

Section 101 of the statute defines the categories of patentable invention broadly: "any . . . process, machine, manufacture, . . . composition of matter, or . . . improvement thereof . . . ." Patent lawyers often employ a working distinction between process claims and "product" claims, which are simply claims to any of the other three types of invention, i.e, machines, manufactures, or compositions of matter.

≡ *Diamond v. Chakrabarty*
Supreme Court of the United States
447 U.S. 303, 206 U.S.P.Q. (BNA) 193 (1980)

BURGER, C.J.

We granted certiorari to determine whether a live, human-made microorganism is patentable subject matter under 35 U.S.C. §101.

I

In 1972, respondent Chakrabarty, a microbiologist, filed a patent application, assigned to the General Electric Co. The application asserted 36 claims related to Chakrabarty's invention of "a bacterium from the genus *Pseudomonas* containing therein at least two stable energy-generating plasmids, each of said plasmids providing a separate hydrocarbon degradative pathway."[1] This human-made, genetically engineered bacterium is capable of breaking down multiple components of crude oil. Because of this property, which is possessed by no naturally occurring bacteria, Chakrabarty's invention is believed to have significant value for the treatment of oil spills.

Chakrabarty's patent claims were of three types: first, process claims for the method of producing the bacteria; second, claims for an inoculum comprised of a carrier material floating on water, such as straw, and the new bacteria; and third, claims to the bacteria themselves. The patent examiner allowed the claims falling into the first two categories, but rejected claims for the bacteria. His decision rested on two grounds: (1) that micro-organisms are "products of nature," and (2) that as living things they are not patentable subject matter under 35 U.S.C. §101.

Chakrabarty appealed the rejection of these claims to the Patent Office Board of Appeals, and the Board affirmed the examiner on the second ground.[3] Relying on the legislative history of the 1930 Plant Patent Act, in which Congress extended patent protection to certain asexually reproduced plants, the Board concluded that §101 was not intended to cover living things such as these laboratory created micro-organisms. [After the Supreme Court vacated its original opinion, the Court of Customs and Patent Appeals vacated its judgment in *Chakrabarty* for reconsideration in light of *Parker v. Flook,* a recently decided case involving the patentability of computer software. After reexamining the case in the light of *Flook,* the appellate court reaffirmed its earlier opinion.]

The Commissioner of Patents and Trademarks again sought certiorari, and we granted the writ. . . .

II

The Constitution grants Congress broad power to legislate to "promote the Progress of Science and useful Arts." . . . The patent laws promote this

---

1. Plasmids are hereditary units physically separate from the chromosomes of the cell. In prior research, Chakrabarty and an associate discovered that plasmids control the oil degradation abilities of certain bacteria. In particular, the two researchers discovered plasmids capable of degrading camphor and octane, two components of crude oil. In the work represented by the patent application at issue here, Chakrabarty discovered a process by which four different plasmids, capable of degrading four different oil components, could be transferred to and maintained stably in a single *Pseudomonas* bacterium, which itself has no capacity for degrading oil.

3. The Board concluded that the new bacteria were not "products of nature," because *Pseudomonas* bacteria containing two or more different energy-generating plasmids are not naturally occurring.

progress by offering inventors exclusive rights for a limited period as an incentive for their inventiveness and research efforts. The authority of Congress is exercised in the hope that "[t]he productive effort thereby fostered will have a positive effect on society through the introduction of new products and processes of manufacture into the economy, and the emanations by way of increased employment and better lives for our citizens."

The question before us in this case is a narrow one of statutory interpretation requiring us to construe 35 U.S.C. §101. Specifically, we must determine whether respondent's micro-organism constitutes a "manufacture" or "composition of matter" within the meaning of the statute.

III

[T]his Court has read the term "manufacture" in §101 in accordance with its dictionary definition to mean "the production of articles for use from raw or prepared materials by giving to these materials new forms, qualities, properties, or combinations, whether by hand-labor or by machinery." American Fruit Growers, Inc. v. Brogdex Co., 283 U.S. 1, 11 (1931). Similarly, "composition of matter" has been construed consistent with its common usage to include "all compositions of two or more substances and . . . all composite articles, whether they be the results of chemical union, or of mechanical mixture, or whether they be gases, fluids, powders or solids." Shell Development Co. v. Watson, 149 F. Supp. 279, 280 (D.C. 1957). In choosing such expansive terms as "manufacture" and "composition of matter," modified by the comprehensive "any," Congress plainly contemplated that the patent laws would be given wide scope.

The relevant legislative history also supports a broad construction. The Patent Act of 1793, authored by Thomas Jefferson, defined statutory subject matter as "any new and useful art, machine, manufacture, or composition of matter, or any new or useful improvement [thereof]." Act of Feb. 21, 1793, §1, 1 Stat. 319. The Act embodied Jefferson's philosophy that "ingenuity should receive a liberal encouragement." 5 Writings of Thomas Jefferson 75-76 (Washington ed. 1871). Subsequent patent statutes in 1836, 1870, and 1874 employed this same broad language. In 1952, when the patent laws were recodified, Congress replaced the word "art" with "process," but otherwise left Jefferson's language intact. The Committee Reports accompanying the 1952 Act inform us that Congress intended statutory subject matter to "include anything under the sun that is made by man." S. Rep. No. 1979, 82d Cong., 2d Sess., 5 (1952); H.R. Rep. No. 1923, 82d Cong., 2d Sess., 6 (1952).

This is not to suggest that §101 has no limits or that it embraces every discovery. *The laws of nature, physical phenomena, and abstract ideas have been held not patentable.* Thus, a new mineral discovered in the earth or a new plant found in the wild is not patentable subject matter. Likewise, Einstein could not patent his celebrated law that $E=mc^2$; nor could Newton have patented the law of gravity. Such discoveries are "manifestations of . . . nature, free to all men and reserved exclusively to none."

Judged in this light, respondent's micro-organism plainly qualifies as patentable subject matter. His claim is not to a hitherto unknown natural phenomenon, but to a nonnaturally occurring manufacture or composition of matter — a product of human ingenuity "having a distinctive name, character [and] use." The point is underscored dramatically by comparison of the invention here with that in *Funk* [Bros. Seed Co. v. Kalo Inoculant Co., 333 U.S. 127 (1948)]. There, the patentee had discovered that there existed in nature certain species of root-nodule bacteria which did not exert a mutually inhibitive effect on each other. He used that discovery to produce a mixed culture capable of inoculating the seeds of leguminous plants. Concluding that the patentee had discovered "only some of the handiwork of nature," the Court ruled the product nonpatentable:

> Each of the species of root-nodule bacteria contained in the package infects the same group of leguminous plants which it always infected. No species acquires a different use. The combination of species produces no new bacteria, no change in the six species of bacteria, and no enlargement of the range of their utility. Each species has the same effect it always had. The bacteria perform in their natural way. Their use in combination does not improve in any way their natural functioning. They serve the ends nature originally provided and act quite independently of any effort of the patentee.

333 U.S., at 131.

Here, by contrast, the patentee has produced a new bacterium with markedly different characteristics from any found in nature and one having the potential for significant utility. His discovery is not nature's handiwork, but his own; accordingly it is patentable subject matter under §101.

IV

Two contrary arguments are advanced, neither of which we find persuasive.

The petitioner's first argument rests on the enactment of the 1930 Plant Patent Act, which afforded patent protection to certain asexually reproduced plants, and the 1970 Plant Variety Protection Act, which authorized protection for certain sexually reproduced plants but excluded bacteria from its protection.[7] In the petitioner's view, the passage of these Acts evidences con-

___

7. The Plant Patent Act of 1930, 35 U.S.C. §161, provides in relevant part:

Whoever invents or discovers and asexually reproduces any distinct and new variety of plant, including cultivated sports, mutants, hybrids, and newly found seedlings, other than a tuber propagated plant or a plant found in an uncultivated state, may obtain a patent therefor. . . .

gressional understanding that the terms "manufacture" or "composition of matter" do not include living things; if they did, the petitioner argues, neither Act would have been necessary.

We reject this argument. Prior to 1930, two factors were thought to remove plants from patent protection. The first was the belief that plants, even those artificially bred, were products of nature for purposes of the patent law. . . . The second obstacle to patent protection for plants was the fact that plants were thought not amenable to the "written description" requirement of the patent law. See 35 U.S.C. §112. Because new plants may differ from old only in color or perfume, differentiation by written description was often impossible.

In enacting the Plant Patent Act, Congress addressed both of these concerns. . . . Moreover, there is language in the House and Senate Committee Reports suggesting that to the extent Congress considered the matter it found the Secretary's dichotomy unpersuasive. The Reports observe:

> There is a clear and logical distinction *between the discovery of a new variety of plant and of certain inanimate things*, such, for example, as a new and useful natural mineral. The mineral is created wholly by nature unassisted by man. . . . On the other hand, a plant discovery resulting from cultivation is unique, isolated, and is not repeated by nature, nor can it be reproduced by nature unaided by man. . . .

S. Rep. No. 315, supra, at 6; H.R. Rep. No. 1129, supra, at 7 (emphasis added).

Congress thus recognized that the relevant distinction was not between living and inanimate things, but between products of nature, whether living or not, and human-made inventions. Here, respondent's micro-organism is the result of human ingenuity and research. Hence, the passage of the Plant Patent Act affords the Government no support.

Nor does the passage of the 1970 Plant Variety Protection Act support the Government's position. As the Government acknowledges, sexually reproduced plants were not included under the 1930 Act because new varieties could not be reproduced true-to-type through seedlings. By 1970, however, it was generally recognized that true-to-type reproduction was possible and that plant patent protection was therefore appropriate. The 1970 Act extended that protection. There is nothing in its language or history to suggest that it was enacted because §101 did not include living things. . . .

The Plant Variety Protection Act of 1970, provides in relevant part:

> The breeder of any novel variety of sexually reproduced plant (other than fungi, bacteria, or first generation hybrids) who has so reproduced the variety, or his successor in interest, shall be entitled to plant variety protection therefor. . . .

84 Stat. 1547, 7 U.S.C. §2402(a). See generally, 3 A. Deller, Walker on Patents, ch. IX (2d ed. 1964); R. Allyn, The First Plant Patents (1934).

The petitioner's second argument is that micro-organisms cannot qualify as patentable subject matter until Congress expressly authorizes such protection. His position rests on the fact that genetic technology was unforeseen when Congress enacted §101. From this it is argued that resolution of the patentability of inventions such as respondent's should be left to Congress. The legislative process, the petitioner argues, is best equipped to weigh the competing economic, social, and scientific considerations involved, and to determine whether living organisms produced by genetic engineering should receive patent protection. In support of this position, the petitioner relies on our recent holding in Parker v. Flook, 437 U.S. 584 (1978), and the statement that the judiciary "must proceed cautiously when . . . asked to extend patent rights into areas wholly unforeseen by Congress." Id., at 596.

It is, of course, correct that Congress, not the courts, must define the limits of patentability; but it is equally true that once Congress has spoken it is "the province and duty of the judicial department to say what the law is." Marbury v. Madison, 1 Cranch 137, 177 (1803). Congress has performed its constitutional role in defining patentable subject matter in §101; we perform ours in construing the language Congress has employed. In so doing, our obligation is to take statutes as we find them, guided, if ambiguity appears, by the legislative history and statutory purpose. Here, we perceive no ambiguity. The subject-matter provisions of the patent law have been cast in broad terms to fulfill the constitutional and statutory goal of promoting "the Progress of Science and the useful Arts" with all that means for the social and economic benefits envisioned by Jefferson. Broad general language is not necessarily ambiguous when congressional objectives require broad terms.

Nothing in *Flook* is to the contrary. The Court carefully scrutinized the claim at issue to determine whether it was precluded from patent protection under "the principles underlying the prohibition against patents for 'ideas' or phenomena of nature." Id., at 593. We have done that here. *Flook* did not announce a new principle that inventions in areas not contemplated by Congress when the patent laws were enacted are unpatentable per se.

To read that concept into *Flook* would frustrate the purposes of the patent law. This is especially true in the field of patent law. A rule that unanticipated inventions are without protection would conflict with the core concept of the patent law that anticipation undermines patentability. Mr. Justice Douglas reminded that the inventions most benefiting mankind are those that "push back the frontiers of chemistry, physics, and the like." Great A.&P. Tea Co. v. Supermarket Corp., 340 U.S. 147, 154 (1950) (concurring opinion). Congress employed broad general language in drafting §101 precisely because such inventions are often unforeseeable.[10]

---

10. Even an abbreviated list of patented inventions underscores the point: telegraph (Morse, No. 1,647); telephone (Bell, No. 174,465); electric lamp (Edison, No. 223,898); airplane (the Wrights, No. 821,393); transistor (Bardeen & Brattain, No. 2,524,035); neutronic reactor (Fermi & Szilard, No. 2,708,656); laser (Schawlow & Townes, No. 2,929,922). See generally Revolutionary Ideas, Patents & Progress in America, United States Patent and Trademark Office (1976).

To buttress his argument, the petitioner, with the support of *amicus*, points to grave risks that may be generated by research endeavors such as respondent's. The briefs present a gruesome parade of horribles. Scientists, among them Nobel laureates, are quoted suggesting that genetic research may pose a serious threat to the human race, or, at the very least, that the dangers are far too substantial to permit such research to proceed apace at this time. We are told that genetic research and related technological developments may spread pollution and disease, that it may result in a loss of genetic diversity, and that its practice may tend to depreciate the value of human life. These arguments are forcefully, even passionately, presented; they remind us that, at times, human ingenuity seems unable to control fully the forces it creates — that, with Hamlet, it is sometimes better "to bear those ills we have than fly to others that we know not of."

It is argued that this Court should weigh these potential hazards in considering whether respondent's invention is patentable subject matter under §101. We disagree. The grant or denial of patents on micro-organisms is not likely to put an end to genetic research or to its attendant risks. The large amount of research that has already occurred when no researcher had sure knowledge that patent protection would be available suggests that legislative or judicial fiat as to patentability will not deter the scientific mind from probing into the unknown any more than Canute could command the tides. Whether respondent's claims are patentable may determine whether research efforts are accelerated by the hope of reward or slowed by want of incentives, but that is all.

What is more important is that we are without competence to entertain these arguments — either to brush them aside as fantasies generated by fear of the unknown, or to act on them. The choice we are urged to make is a matter of high policy for resolution within the legislative process after the kind of investigation, examination, and study that legislative bodies can provide and courts cannot. That process involves the balancing of competing values and interests, which in our democratic system is the business of elected representatives. Whatever their validity, the contentions now pressed on us should be addressed to the political branches of the Government, the Congress and the Executive, and not to the courts.

Accordingly, the judgment of the Court of Customs and Patent Appeals is [a]ffirmed.

Mr. Justice BRENNAN, with whom Mr. Justice WHITE, Mr. Justice MARSHALL, and Mr. Justice POWELL join, dissenting.

I agree with the Court that the question before us is a narrow one. Neither the future of scientific research, nor even the ability of respondent Chakrabarty to reap some monopoly profits from his pioneering work, is at stake. Patents on the processes by which he has produced and employed the new living organism are not contested. The only question we need decide is whether Congress intended that he be able to secure a monopoly on the living organism itself, no matter how produced or how used. Because I believe the Court has misread the applicable legislation, I dissent. . . .

The patent laws attempt to reconcile th[e] Nation's deep-seated antipathy to monopolies with the need to encourage progress. Given the complexity and legislative nature of this delicate task, we must be careful to extend patent protection no further than Congress has provided. In particular, were there an absence of legislative direction, the courts should leave to Congress the decisions whether and how far to extend the patent privilege into areas where the common understanding has been that patents are not available.

In this case, however, we do not confront a complete legislative vacuum. The sweeping language of the Patent Act of 1793, as re-enacted in 1952, is not the last pronouncement Congress has made in this area. In 1930 Congress enacted the Plant Patent Act affording patent protection to developers of certain asexually reproduced plants. In 1970 Congress enacted the Plant Variety Protection Act to extend protection to certain new plant varieties capable of sexual reproduction. Thus, we are not dealing — as the Court would have it — with the routine problem of "unanticipated inventions." In these two Acts Congress has addressed the general problem of patenting animate inventions and has chosen carefully limited language granting protection to some kinds of discoveries, but specifically excluding others. These Acts strongly evidence a congressional limitation that excludes bacteria from patentability.[2]

First, the Acts evidence Congress' understanding, at least since 1930, that §101 does not include living organisms. If newly developed living organisms not naturally occurring had been patentable under §101, the plants included in the scope of the 1930 and 1970 Acts could have been patented without new legislation. Those plants, like the bacteria involved in this case, were new varieties not naturally occurring.[3] Although the Court rejects this line of argument, it does not explain why the Acts were necessary unless to correct a pre-existing situation. I cannot share the Court's implicit assumption that Congress was engaged in either idle exercises or mere correction of

---

2. But even if I agreed with the Court that the 1930 and 1970 Acts were not dispositive, I would dissent. This case presents cogent reasons not to extend the patent monopoly in the face of uncertainty. At the very least, these Acts are signs of legislative attention to the problems of patenting living organisms, but they give no affirmative indication of congressional intent that bacteria be patentable. The caveat of Parker v. Flook, 437 U.S. 584, 596 (1978), an admonition to "proceed cautiously when we are asked to extend patent rights into areas wholly unforeseen by Congress," therefore becomes pertinent. I should think the necessity for caution is that much greater when we are asked to extend patent rights into areas Congress has foreseen and considered but has not resolved.

3. The Court refers to the logic employed by Congress in choosing not to perpetuate the "dichotomy" suggested by Secretary Hyde. But by this logic the bacteria at issue here are distinguishable from a "mineral . . . created wholly by nature" in exactly the same way as were the new varieties of plants. If a new Act was needed to provide patent protection for the plants, it was equally necessary for bacteria. Yet Congress provided for patents on plants but not on these bacteria. In short, Congress decided to make only a subset of animate "human-made inventions" patentable.

the public record when it enacted the 1930 and 1970 Acts. And Congress certainly thought it was doing something significant. The Committee Reports contain expansive prose about the previously unavailable benefits to be derived from extending patent protection to plants.[5] H.R. Rep. No. 91-1605, pp. 1-3 (1970); S. Rep. No. 315, 71st Cong., 2d Sess., 1-3 (1930). Because Congress thought it had to legislate in order to make agricultural "human-made inventions" patentable and because the legislation Congress enacted is limited, it follows that Congress never meant to make items outside the scope of the legislation patentable. . . .

The Court protests that its holding today is dictated by the broad language of §101, which cannot "be confined to the 'particular application[s] . . . contemplated by the legislators.' " But as I have shown, the Court's decision does not follow the unavoidable implications of the statute. Rather, it extends the patent system to cover living material even though Congress plainly has legislated in the belief that §101 does not encompass living organisms. It is the role of Congress, not this Court, to broaden or narrow the reach of the patent laws. This is especially true where, as here, the composition sought to be patented uniquely implicates matters of public concern.

COMMENTS AND QUESTIONS

1. The Court cites Funk Brothers Seed Co. v. Kalo Inoculant Co., 333 U.S. 127 (1948), which centered around a patent for bacteria that assists in the process of nitrogen fixation in the roots of plants. The invention underlying the patent was said to be the realization that, contrary to accepted wisdom, varieties from different species of bacteria could be mixed together and sold in a single root-treatment formula. Experts had previously believed that this was not possible, and had adopted the practice of selling individual bacterial species. The Supreme Court held that combining the bacteria in this way was a discovery, not invention: "[it] is no more than the discovery of some of the handiwork of nature and hence is not patentable."

2. What distinguishes *Chakrabarty* from *Funk Brothers*? One possible answer is that the Court is drawing a line between *discovery* and *invention*. In *Chakrabarty*, the Court emphasizes the transformations that the inventor

---

5. Secretary Hyde's letter was not the only explicit indication in the legislative history of these Acts that Congress was acting on the assumption that legislation was necessary to make living organisms patentable. The Senate Judiciary Committee Report on the 1970 Act states the Committee's understanding that patent protection extended no further than the explicit provisions of these Acts:

Under the patent law, patent protection is limited to those varieties of plants which reproduce asexually, that is, by such methods as grafting or budding. No protection is available to those varieties of plants which reproduce sexually, that is, generally by seeds. S. Rep. No. 91-1246, p. 3 (1970).

Similarly, Representative Poage, speaking for the 1970 Act, after noting the protection accorded asexually developed plants, stated that "for plants produced from seed, there has been no such protection." 116 Cong. Rec. 40295 (1970).

makes on the admittedly natural raw materials of the invention. But in *Funk Bros.*, the Court emphasizes that the bacteria, though combined in a novel way, still perform their same old natural function. Can the cases be reconciled on this ground? Consider 35 U.S.C. §100, which provides that "invention" as the term is used in the Patent Act means both "invention and discovery."

3. Does *Funk Bros.* suggest that new combinations of known elements can never be patentable? At a minimum, does the case require some sort of "synergy" resulting from a combination — that the elements when combined must do more than they each would have done alone? Would such a requirement make sense?

4. Should human intervention be the touchstone for patentability in this area? Consider the following case.

Parke-Davis & Co. v. H. K. Mulford Co.
United States Circuit Court for the
Southern District of New York
189 F. 95 (C.C.S.D.N.Y. 1911)

L. HAND, District Judge:

[Jokichi Takamine, inventor of the two patents in issue, discovered how to isolate a purified substance of significant medical use from the suprarenal glands of animals. Parke-Davis, assignee of the patents, called the product Adrenalin. Takamine's adrenalin was medically superior to the older isolates of this gland that had been in use, because those products were not nearly as pure, which resulted in the injection into humans of a good deal of extraneous gland tissue left over from incomplete purification techniques. Defendant made and sold a similar product, which it called Adrin.]

The patentee originally attempted to claim the active principle itself. This was in his first application where he claimed process and product; but the examiner would not allow these claims, basing his rejection upon his interpretation of American Wood Paper Co. v. Fibre Disintegrating Co., 23 Wall. 566, 23 L. Ed. 31, that no product is patentable, however it be of the process, which is merely separated by the patentee from its surrounding materials and remains unchanged. . . . In [an] amendment . . . , which was about two months after his first application, he changed all the claims so that they . . . were not limited to the active principle. I think that this effected a substantial change in meaning, and that . . . the claims are now broader than a mere claim for the chemically free base, or active principle, and that they cover any substance which possesses the physiological characteristics of the glands and is substantially pure. . . . [²⁸]

---

28. Claim 1 in U.S. patent 730,176, one of the patents in issue in the case, reads:

1. A substance possessing the herein-described physiological characteristics and reactions of the suprarenal glands in a stable and concentrate form, and practically free from inert and associated gland-tissue. — EDS.

[After deciding that defendant's product infringed several claims of both Takamine patents, Judge Hand turned to the issue of whether the Takamine patents were invalid.]

. . . Nor is the patent only for a degree of purity, and therefore not for a new "composition of matter." As I have already shown, it does not include a salt, and no one had ever isolated a substance which was not in salt form, and which was anything like Takamine's. . . . But, even if it were merely an extracted product without change, there is no rule that such products are not patentable. . . . Takamine was the first to make it available for any use by removing it from the other gland-tissue in which it was found, and, while it is of course possible logically to call this a purification of the principle, it became for every practical purpose a new thing commercially and therapeutically. That was a good ground for a patent. That the change here resulted in ample practical differences is fully proved. Everyone, not already saturated with scholastic distinctions, would recognize that Takamine's crystals were not merely the old dried glands in a purer state, nor would his opinion change if he learned that the crystals were obtained from the glands by a process of eliminating the inactive organic substances. The line between different substances and degrees of the same substance is to be drawn rather from the common usages of men than from nice considerations of dialectic. . . .

COMMENTS AND QUESTIONS

1. To Judge Hand, it is of paramount importance that the patentee has created "for every practical purpose a new thing commercially and therapeutically." But if commercial and therapeutic novelty were enough, wouldn't it be possible to obtain a patent on the unprocessed bark of a rare tree which turns out to be an effective therapy for cancer, and therefore a commercially successful product? Can you think of any reasons why society should be reluctant to grant such a patent to the first person who discovers the therapeutic effects of the bark?

2. A parallel line of cases involves metal compounds found in nature. See, e.g., General Electric Co. v. DeForest Radio Co., 28 F.2d 641 (3d Cir. 1928), *cert. denied*, 278 U.S. 656 (1929) (affirming trial court finding of invalidity of claims to "substantially pure tungsten"). In the *DeForest* case, the court stated:

What [the patentee] produced by his process was natural tungsten in substantially pure form. What he discovered were natural qualities of pure tungsten. Manifestly he did not create pure tungsten, nor did he create its characteristics. These were created by nature. . . .

How was the adrenalin in the *Parke-Davis* case different from the tungsten in *DeForest*? What is the difference between the "discovery" of an "inherent property" and the "invention" of a "new property or use" of a product?

3. By contrast with *Parke-Davis,* see Ex parte Latimer, 1889 Dec. Commr. Pat. 123, 46 Pat. Off. Gazz. 1638 (Commr. Patents 1889), where the Commissioner of Patents affirmed two holdings of the Patent Office: (1) the issuance of process claims to a method of extracting weavable fibers from the *Pinus australis* tree; and (2) the final rejection of the applicant's product claim to the same subject matter, viz., "the cellular tissues of the *Pinus australis* eliminated in full lengths from the silicious, resinous, and pulpy parts of the pine needles and subdivided into long, pliant filaments adapted to be spun and woven," because:

> the mere ascertaining of the character or quality of trees that grow in the forest and the construction of the woody fiber and tissue of which they are composed is not a patentable invention, recognized by the statute, any more than to find a new gem or jewel in the earth would entitle the discoverer to patent all gems which should be subsequently found. . . .

See also In re Bergy, 563 F.2d 1031, 1036 (C.C.P.A. 1977) (distinguishing the patentable, humanly-transformed bacteria claimed in the case from a prior case where "[w]e were thinking of something preexisting and merely plucked from the earth and claimed as such, a far cry from a biologically pure culture produced by great labor in a laboratory and so claimed." (citing In re Mancy, 499 F.2d 1289 (C.C.P.A. 1974)).

4. Consider the following passage written by the noted libertarian philosopher Robert Nozick, in which he sets forth the reasons why one who discovers a new plant is entitled to assert property rights over it:

> He does not worsen the situation of others; if he did not stumble upon the substance no one else would have, and the others would remain without it. However, as time passes, the likelihood increases that others would have come across the substance; upon this fact might be based a limit to his property right in the substance so that others are not below their baseline position; for example, its bequest might be limited.

Robert Nozick, Anarchy, State and Utopia 181 (1974).[29] Does this view, which appears to make sense, imply any particular *content* to the right that follows from discovery? Would trade secret protection — a minimal right protecting the discoverer's information concerning the source of the natural product, for instance — be sufficient? Or is a patent called for, with its protection against independent discovery of the product and its prohibition of any competing makers or sellers of the product?

---

29. The passage comes in the midst of a section discussing John Locke's theory of property; thus the emphasis on not worsening anyone else's position, one of the "Lockean provisos" that must be met under this theory for property rights to be defensible. See supra Chapter 1.

5. It is possible to discern from this patchwork of cases a set of rules for the patenting of living things. First, there is generally no objection to patent claims covering a *process* for extracting a natural product, as opposed to claims to the product itself. See, e.g., Merck & Co. v. Olin Mathieson Chem. Corp., 253 F.2d 156 (4th Cir. 1958). What practical difference would there be between the two types of claims? Consider two cases: (a) a natural product identical to the patentee's but made from a different process, and (b) a natural product identical to the patentee's made from a process whose details are unknown to the patentee.

Should the first person to purify a natural product to a certain degree be entitled to a patent on the purified product itself, however produced? Or only on purified products produced using her method? For a discussion of this issue of patent scope, see Michael D. Davis, The Patenting of Products of Nature, 21 Rutgers Comp. & Tech. L.J. 293, 335-347 (1995).

Second, Nozick's suggestion to the contrary notwithstanding, it does not appear to be enough to discover a previously unknown plant or animal (or, for that matter, a previously unknown mineral). On the other hand, human creation of an entirely new bacterium does qualify for patent protection under *Chakrabarty*. Subsequent decisions have uniformly extended this rule to allow the patenting of "new" plants and animals, including mammals. See, e.g., Ex Parte Hibberd, 227 U.S.P.Q. 443 (Bd. Pat. App. & Int. 1985) (section 101 did not bar the issuance of a utility patent for seeds and plants developed by sexual reproduction in a way not found in nature). See also Ex Parte Allen, 2 U.S.P.Q.2d 1425 (Bd. Pat. App. & Int. 1987) (a genetically developed Pacific Oyster was patentable subject matter but was found "obvious" in view of the prior art); U.S. Patent No. 4,736,866 (Apr. 12, 1988) (issued to Philip Leder and Harvard University for "Transgenic Non-Human Mammals," specifically mice). And patenting human DNA sequences — though not humans themselves, of course — is a booming industry. See S. M. Thomas et al., Ownership of the Human Genome, 380 Nature 387 (April 4, 1996) (1,175 patents on human DNA issued worldwide between 1980 and 1995).

While a number of groups have raised concerns about the patenting of living things, particularly as the living things move up the scale of complexity from bacteria towards humans, these arguments about the morality of patenting life have not found a warm reception in the courts. Patentable subject matter appears to be "morally neutral," at least after *Chakrabarty*. There is a doctrine of moral *utility* that may apply in these cases, however. We consider that doctrine in the next section.

6. Apart from the morality of "patenting life" itself, is there ever a valid concern about allocating exclusive ownership rights to certain types of inventions? Consider the problem of a doctor who has come up with a radically new surgical technique. Should the doctor be allowed to patent his technique? Recall that a patent would allow him not only to prevent others from copying his method, but from using the same method even if they developed it independently. Legislation passed by Congress in 1996 precludes owners of patents on medical and surgical procedures from enforcing those patents,

thus carving out a rare legislative exception to the scope of patentable subject matter. 35 U.S.C. §287(c). Is the doctor's situation any different than that of a pharmaceutical company that wants to patent a new drug it has developed?

PROBLEM

Problem 3-1. Inventor *A* discovers a technique for purifying Hormone *Z* from human tissue and applies for a patent on the purified form of the hormone. Inventor *B* discovers the gene coding for the hormone, clones it, and obtains expression in a mammalian cell culture environment. Are these inventions/discoveries patentable?

Note on Patenting "Abstract Ideas"

While general subject matter limitations are disfavored under section 101 (the "products of nature" rule being a notable exception), there are a number of judicially developed doctrines that limit the subject matter of patents. Of these, one of the most important is the rule against patenting "abstract ideas." This rule appears to derive from the traditional idea that patents are intended to cover "devices" or physical things in the useful arts, not more esoteric matters. As the Supreme Court put it in Rubber-Tip Pencil Co. v. Howard, 87 U.S. (20 Wall.) 498, 507 (1874), "[a]n idea of itself is not patentable, but a new device by which it may be made practically useful is."

In O'Reilly v. Morse, 56 U.S. (15 How.) 62 (1853), the telegraph pioneer Morse (of "Morse code" fame) was allowed a broad patent for a process of using electromagnetism to produce discernible signals over telegraph wires. But the Court denied Morse's famous eighth claim, in which Morse claimed the use of "electro magnetism, however developed for marking or printing intelligible characters, signs, or letters, at any distances." Id. at 112. The Court in disallowing that claim said,

> If this claim can be maintained, it matters not by what process or machinery the result is accomplished. For aught that we now know, some future inventor, in the onward march of science, may discover a mode of writing or printing at a distance by means of the electric or galvanic current, without using any part of the process or combination set forth in the plaintiff's specification. His invention may be less complicated — less liable to get out of order — less expensive in construction, and in its operation. But yet, if it is covered by this patent, the inventor could not use it, nor the public have the benefit of it, without the permission of this patentee.

Id. at 113. Is this truly an objection that Morse's claim does not extend to patentable subject matter, or is the Court really expressing the concern that Morse's claim is *too broad*? That the Court's concern is the latter is suggested by the fact that it allowed Morse's other claims to particular uses of electricity in communication (i.e., the telegraph). It is instructive in this regard to consider The Telephone Cases, 126 U.S. 1, 534 (1887), in which the Court upheld a patent to Alexander Graham Bell for his telephone. The Court explained the *Morse* case as follows:

> The effect of that decision was, therefore, that the use of magnetism as a motive power, without regard to the particular process with which it was connected in the patent, could not be claimed, but that its use in that connection could. Bell's invention was the use of electric current to transmit vocal or other sounds. The claim was not "for the use of a current of electricity in its natural state as it comes from the battery, but for putting a continuous current in a closed circuit into a certain specified condition suited to the transmission of vocal and other sounds, and using it in that condition for that purpose." The claim, in other words, was not "one for the use of electricity distinct from the particular process with which it is connected in his patent." The patent was for that use of electricity "both for the magneto and variable resistance *methods*."

Bell's claim, in other words, was not one for all telephonic use of electricity.

But there are certainly circumstances in which the new idea itself is so abstract that it may run afoul of this doctrine. Consider the hypothetical claim advanced by the Court in *Chakrabarty* — an attempt by Einstein to patent the formula $E=mc^2$. This claim is not overbroad — Einstein has given the world precisely what he claims to own. But there is still an intuitive sense in which this idea is too abstract, too fundamental for society to countenance its ownership by a private individual. Perhaps section 101 is designed in part to exclude from patentability discoveries and inventions so basic and fundamental to future advances that patenting them would unduly burden future inventors.

For an interesting account of a movement from the early part of the twentieth century to protect the findings of basic scientific research with a patent-like property right, see C. J. Hamson, Patent Rights for Scientific Discoveries (1930).

COMMENTS AND QUESTIONS

1. Typically, there is a good deal of development work necessary to turn an invention into a viable commercial product. This first *commercial* version of an invention is usually termed an innovation. The patent systems of today reward such innovations only indirectly, through the granting of patents on inventions. Some observers have argued that this is a fundamental flaw in the patent system. They contend that the original function of patents was to reward innovation directly. In their view, the patent system has strayed from this original function. They propose a new patent regime, one more in keeping with this original function. The new regime would provide *direct* protection of innovation by means of exclusive rights to market newly introduced

products. See Direct Protection of Innovation (W. Kingston, ed. 1987). In this book, Kingston proposes a system of property rights to come into effect only when a new product is actually introduced on the market. See id. at 1-34. For example, under such a system, the first entrepreneur who introduced an overnight package delivery service, or any other new business practice, service, or product, would receive an exclusive property right over that concept for a limited period of time. Can you think of any potential problems with such a system — e.g., administrative or other practical difficulties? Does such a proposal shed any light on why the current patent system focuses on invention rather than innovation? Could you defend the current system, perhaps by arguing that technology, rather than business practices or services, contributes something unique to the economy, or otherwise merits special encouragement? Cf. H. I. Dutton, The Patent System and Inventive Activity During the Industrial Revolution 1750-1852 (1984) (arguing that the early patent system did not reward innovation directly but instead played much the same role it does today). Robert P. Merges, Uncertainty and the Standard of Patentability, 7 High Tech. L. Rev. 1 (1993) (model demonstrating that prospect of receiving patent can have dramatic effect on incentive to develop a basic invention).

Note on Patenting Business Methods and "Printed Matter"

Other artifacts of the traditional focus of patent law on physical machines and devices in the useful arts are the twin rules against patenting "printed matter" and "business methods." Both rules originally developed from a series of cases in which the "inventor" of new printed business forms sought to patent those forms. Courts uniformly rejected both the idea that a printed piece of paper could be a patentable invention and that the new system of conducting business embodied in the paper was patentable. United States Credit System Co. v. American Credit Indemnity Co., 59 F. 139, 143 (2d Cir. 1893) involved a patent for a means of credit insurance; the court struck it down on section 101 grounds:

> The three claims of the patent are concerned solely with the providing of sheets with appropriate headings, adapted to be used in preparing historical records or certain business transactions. There is nothing peculiar or novel in preparing a sheet of paper with headings generally appropriate to classes of facts to be recorded, and whatever peculiarity there may be about the headings in this case is a peculiarity resulting from the transactions themselves. . . . Given a series of transactions, there is no patentable novelty in recording them, where, as in this case, such record consists simply in setting down some of their details in an order or sequence common to each record.

At the same time, courts held that a system of transacting business was not patentable unless what were claimed were the physical means for carrying out the system. See, e.g., Hotel Security Checking Co. v. Lorraine Co., 160 F. 467, 469 (2d Cir. 1908). The result of these two doctrines in combination has been that most business and financial innovations were not considered patentable until recently. (See Comment 1 below.) The printed matter rule has been extended to deny patents for such physical items as games, in which the physical elements used (a board, dice, cards) are not new, but what is printed on those elements was new. See Ex parte Gwinn, 112 U.S.P.Q. 439 (Bd. Pat. App. & Int. 1955).

Printed matter may be part of a patentable invention, however, if the invention as a whole claims a new and useful physical structure, or if the relationship between the printed matter and the physical structure is a new and nonobvious one. How this exception interacts with the printed matter rejection is complex and not always clear. For example, in In re Lowry, 32 F.3d 1579 (Fed. Cir. 1994), the Federal Circuit held that a claim for a novel "data structure" contained entirely within an ordinary computer memory was not printed matter because the data structure necessarily reorganized the electronic components of the computer memory in a way that was not intelligible to humans, but only to the machine itself. The court also noted that broad application of the printed matter doctrine was "disfavored."

COMMENTS AND QUESTIONS

1. The rule that methods of doing business are not patentable was criticized by the dissent in In re Schrader, 22 F.3d 290 (Fed. Cir. 1994) (Newman, J., dissenting):

[T]he Board [of Patent Appeals and Interferences] remarked that the "method of doing business" is a fuzzy concept, observed the inconclusiveness of precedent, and sought guidance from this court. Indeed it is fuzzy; and since it is also an unwarranted encumbrance to the definition of statutory subject matter in section 101, . . . my guidance is that it be discarded as error-prone, redundant, and obsolete. . . .

The decisions that have spoken of "methods of doing business" have, or could have, resolved the issue in each case simply by relying on the statutory requirements of patentability such as novelty and unobviousness. [Judge Newman analyzes a number of business methods cases.]

. . . Any historical distinction between a method of "doing" business and the means of carrying it out blur in the complexity of modern business systems. See Paine, Webber, Jackson and Curtis v. Merrill Lynch, 564 F. Supp. 1358 (D. Del. 1983), wherein a computerized system of cash management was held to be statutory subject matter.

I discern no purpose in perpetuating a poorly defined, redundant, and unnecessary "business methods" exception, indeed enlarging (and enhancing the fuzziness of) that exception by applying it in this case.

The Federal Circuit adopted Judge Newman's view in State Street Bank & Trust Co. v. Signature Financial Group, 149 F.3d 1368 (Fed. Cir. 1998).

The court announced that it had decided "to lay this ill-conceived exception to rest." It further opined that the business-methods exception had been without legal force since the 1952 Patent Act, and the court distinguished cases that, arguably, invoked the exception since that time. In so doing it bolstered the idea that a "process" can be statutory despite the fact that it doesn't act on anything tangible.

In the wake of *State Street*, the Patent office has issued hundreds of patents covering methods of doing business.

2. One justification for the printed matter rule is that it is necessary to "channel" certain creations into the realm of patent law, and other creations (notably those in written form) into copyright law. Does this channeling justification make sense? Does it explain the *Lowry* decision?

PROBLEMS

Problem 3-2. In 1986, an American scientist by the name of John Anderson perfected a process that yielded a new man-made chemical element. The element, named Litigacium (Li), was assigned atomic number 116. Anderson's method of synthesizing the element involves using a neutronic reactor at relative high power (about 200 kilowatts) for approximately 100 days. Anderson's patent application noted that a suitable reactor for producing Litigacium is described in a 1984 patent of Dr. Enrago Firmi.

Although Anderson's process claim met with no resistance from the Patent and Trademark Office, the claim relating to the product Litigacium raised more difficult issues. An examiner notified Anderson that his claimed element was not patentable because it preexists in nature, in some form, since a chemical element can not truly be "man-made."

Dr. Anderson has asked you to represent him in his appeal to the Federal Circuit. The appeal presents the following issues:

a) The Patent Office admits that the level of Litigacium produced under natural circumstances — e.g., in the intense heat and pressure that accompanies the collapse of certain large stars — would be miniscule. For example, the examiner states that using well-known theoretical calculations, star collapse would result in an infinitesimally small amount of the element spread throughout an enormous amount of other matter under tremendous heat and pressure. How can the mere possibility that a star would create an undetectable amount of Litigacium permit rejection of your client's patent application?

b) The examiner concedes that it would not be practical given current technology to "harvest" Litigacium from natural sources such as stars. Should this have a bearing on the outcome of the case? Is it the function of section 101's patentable subject matter requirement to find patentable anything that is cheaper (i.e., more efficient) for humans to make than to harvest in raw form from nature? Or is section 101 an expression of a deeply felt aversion to permitting humans to claim "authorship" over products given to us by nature's bounty? Is there a viable distinction between things we "find" in nature and things we ourselves "invent"? Does the language in the patent clause of the Constitution — i.e., that patents shall be conferred to inventors for their "discoveries" — shed any light on these matters?

Problem 3-3. You are a newly elected U.S. senator. As part of your duties on the legislative committee assigned to introduce intellectual property legislation, you have been asked to co-sponsor a bill that seeks to rewrite section 101 of the patent code. According to the sponsor of the bill, the "archaic" language of the current section 101 was drafted during the era of "quill pens and buggy whips." She claims that it is therefore "out of step in an era of computers and space shuttles."

Several lobbyists have seized on the occasion to introduce their pet revisions into section 101. Environmentalists have asked that "newly discovered natural products" be added to the categories of patentable subject matter; they believe this addition will create an incentive to identify and preserve rare species of potential benefit, e.g., as ingredients in medicines or as sources of raw materials. In addition, academic scientists have testified that they feel "slighted" by the exclusion of natural laws and scientific principles from the coverage of the current patent act. They have stated that they fail to see why the patent system should discriminate against basic research, especially in an era when federal funding of science is shrinking. Even the social scientists have gotten into the act. A famous economist who pioneered a revolutionary technique for measuring the volume of money in circulation testified in hearings on the bill that he would have continued thinking up pioneering ideas like this but "the current system gave me no incentive to do so, at least after I made tenure." And some sports stars have sent letters to your committee, claiming that they wish they had received some form of protection for what everyone refers to as their "patented" backhand or putting style or hurdling technique.

How can you evaluate these claims? Is there any justification in refusing to change the current language — e.g., because it has shown enough flexibility to adapt to computer technology and biotechnology and the like? Is there anything special about traditional technology — machines, manufactures, etc. — that makes it more valuable, or more in need of protection, than other "innovations" such as newly discovered natural products, new "pure" science results, new academic theories, or new techniques for doing business (such as the leveraged buyout, or the multidivisional corporate structure)? Why not grant property rights over all sorts of new things, and let the market figure out which ones are truly worthwhile?

2. Utility

The patent code protects all inventions that are novel, *useful* and non-obvious. This section concentrates on this second requirement, known as the utility requirement.

At first glance it might seem as though this is a simple requirement to apply. After all, whether or not something is useful is normally easy to determine. In a sense patent law reflects this; utility is a relatively rare issue in the Patent Office, or in an infringement suit. But in another sense there is more to utility. Both conceptually and as borne out in the cases, subtle issues lurk within the waters of utility.

Chemists often synthesize compounds that they believe might be useful someday for something but for which no particular use is known. When they apply for patents on these compounds, they sometimes run headlong into the utility requirement, as the following case demonstrates.

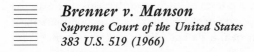

Brenner v. Manson
Supreme Court of the United States
383 U.S. 519 (1966)

Mr. Justice FORTAS delivered the opinion of the Court.

A Patent Office examiner denied Manson's application, and the denial was affirmed by the Board of Appeals within the Patent Office. The ground for rejection was the failure "to disclose any utility for" the chemical compound produced by the process. This omission was not cured, in the opinion of the Patent Office, by Manson's reference to an article in the November 1956 issue of the Journal of Organic Chemistry, 21 J. Org. Chem. 1333-1335, which revealed that steroids of a class which included the compound in question were undergoing screening for possible tumor-inhibiting effects in mice, and that a homologue[3] adjacent to Manson's steroid had proven effective in that role. Said the Board of Appeals, "It is our view that the statutory requirement of usefulness of a product cannot be presumed merely because it happens to be closely related to another compound which is known to be useful."

The Court of Customs and Patent Appeals (hereinafter CCPA) reversed[, stating] "where a claimed process produces a known product it is not necessary to show utility for the product," so long as the product "is not alleged to be detrimental to the public interest."

---

3. "A homologous series is a family of chemically related compounds, the composition of which varies from member to member by CH_2 (one atom of carbon and two atoms of hydrogen). ... Chemists knowing the properties of one member of a series would in general know what to expect in adjacent members."

Our starting point is the proposition, neither disputed nor disputable, that one may patent only that which is "useful." [T]he concept of utility has maintained a central place in all of our patent legislation, beginning with the first patent law in 1790 and culminating in the present [§101]. . . .

Respondent does not — at least in the first instance — rest upon the extreme proposition, advanced by the court below, that a novel chemical process is patentable so long as it yields the intended product and so long as the product is not itself "detrimental." Nor does he commit the outcome of his claim to the slightly more conventional proposition that any process is "useful" within the meaning of §101 if it produces a compound whose potential usefulness is under investigation by serious scientific researchers, although he urges this position, too, as an alternative basis for affirming the decision. Rather, he begins with the much more orthodox argument that his process has a specific utility which would entitle him to a declaration of interference even under the Patent Office's reading of §101. The claim is that the supporting affidavits, by reference to Ringold's 1956 article, reveal that an adjacent homologue of the steroid yielded by his process has been demonstrated to have tumor-inhibiting effects in mice, and that this discloses the requisite utility. We do not accept any of these theories as an adequate basis for overriding the determination of the Patent Office that the "utility" requirement has not been met.

Even on the assumption that the process would be patentable were respondent to show that the steroid produced had a tumor-inhibiting effect in mice, we would not overrule the Patent Office finding that respondent has not made such a showing. The Patent Office held that, despite the reference to the adjacent homologue, respondent's papers did not disclose a sufficient likelihood that the steroid yielded by his process would have similar tumor-inhibiting characteristics. Indeed, respondent himself recognized that the presumption that adjacent homologues have the same utility has been challenged in the steroid field because of "a greater known unpredictability of compounds in that field." In these circumstances and in this technical area, we would not overturn the finding of the Primary Examiner, affirmed by the Board of Appeals and not challenged by the CCPA.

The second and third points of respondent's argument present issues of much importance. Is a chemical process "useful" within the meaning of §101 either (1) because it works — i.e., produces the intended product? or (2) because the compound yielded belongs to a class of compounds now the subject of serious scientific investigation? These contentions present the basic problem for our adjudication. Since we find no specific assistance in the legislative materials underlying §101, we are remitted to an analysis of the problem in light of the general intent of Congress, the purposes of the patent system, and the implications of a decision one way or the other.

In support of his plea that we attenuate the requirement of "utility," respondent relies upon Justice Story's well-known statement that a "useful" invention is one "which may be applied to a beneficial use in society, in contradistinction to an invention injurious to the morals, health, or good order of society, or frivolous and insignificant"[20] — and upon the assertion that to do so would encourage inventors of new processes to publicize the event for the benefit of the entire scientific community, thus widening the search for uses and increasing the fund of scientific knowledge. Justice Story's language sheds little light on our subject. Narrowly read, it does no more than compel us to decide whether the invention in question is "frivolous and insignificant" — a query no easier of application than the one built into the statute. Read more broadly, so as to allow the patenting of any invention not positively harmful to society, it places such a special meaning on the word "useful" that we cannot accept it in the absence of evidence that Congress so intended. There are, after all, many things in this world which may not be considered "useful" but which, nevertheless, are totally without a capacity for harm.

Whatever weight is attached to the value of encouraging disclosure and of inhibiting secrecy, we believe a more compelling consideration is that a process patent in the chemical field, which has not been developed and pointed to the degree of specific utility, creates a monopoly of knowledge which should be granted only if clearly commanded by the statute. Until the process claim has been reduced to production of a product shown to be useful, the metes and bounds of that monopoly are not capable of precise delineation. It may engross a vast, unknown, and perhaps unknowable area. Such a patent may confer power to block off whole areas of scientific development, without compensating benefit to the public. The basic *quid pro quo* contemplated by the Constitution and the Congress for granting a patent monopoly is the benefit derived by the public from an invention with substantial utility. Unless and until a process is refined and developed to this point — where specific benefit exists in currently available form — there is insufficient justification for permitting an applicant to engross what may prove to be a broad field.

These arguments for and against the patentability of a process which either has no known use or is useful only in the sense that it may be an object of scientific research would apply equally to the patenting of the product produced by the process. Respondent appears to concede that with respect to a product, as opposed to a process, Congress has struck the balance on the side of nonpatentability unless "utility" is shown. Indeed, the decisions of the CCPA are in accord with the view that a product may not be patented absent a showing of utility greater than any adduced in the present case. We find absolutely no warrant for the proposition that although Congress intended that no patent be granted on a chemical compound whose sole "utility" consists of its potential role as an object of use-testing, a different set of rules was meant to apply to the process which yielded the unpatentable product. That proposition seems to us little more than an attempt to evade the

---

20. Note on the Patent Laws, 3 Wheat. App. 13, 24. See also Justice Story's decisions on circuit in Lowell v. Lewis, 15 Fed. Cas. 1018 (No. 8568) (C.C.D. Mass.), and Bedford v. Hunt, 3 Fed. Cas. 37 (No. 1217) (C.C.D. Mass.).

impact of the rules which concededly govern patentability of the product itself.

This is not to say that we mean to disparage the importance of contributions to the fund of scientific information short of the invention of something "useful," or that we are blind to the prospect that what now seems without "use" may tomorrow command the grateful attention of the public. But a patent is not a hunting license. It is not a reward for the search, but compensation for its successful conclusion. "[A] patent system must be related to the world of commerce rather than to the realm of philosophy. . . ."

The judgment of the CCPA is Reversed.

Mr. Justice HARLAN, concurring in part and dissenting in part:

. . . Because I believe that the Court's policy arguments are not convincing and that past practice favors the respondent, I would reject the narrow definition of "useful" and uphold the judgment of the CCPA.

The Court's opinion sets out about half a dozen reasons in support of its interpretation. Several of these arguments seem to me to have almost no force. For instance, it is suggested that "[u]ntil the process claim has been reduced to production of a product shown to be useful, the metes and bounds of that monopoly are not capable of precise delineation" and "[i]t may engross a vast, unknown, and perhaps unknowable area." I fail to see the relevance of these assertions; process claims are not disallowed because the products they produce may be of "vast" importance nor, in any event, does advance knowledge of a specific product use provide much safeguard on this score or fix "metes and bounds" precisely since a hundred more uses may be found after a patent is granted and greatly enhance its value. . . .

More to the point, I think, are the Court's remaining, prudential arguments against patentability: namely, that disclosure induced by allowing a patent is partly undercut by patent-application drafting techniques, that disclosure may occur without granting a patent, and that a patent will discourage others from inventing uses for the product. How far opaque drafting may lessen the public benefits resulting from the issuance of a patent is not shown by any evidence in this case but, more important, the argument operates against all patents and gives no reason for singling out the class involved here. The thought that these inventions may be more likely than most to be disclosed even if patents are not allowed may have more force; but while empirical study of the industry might reveal that chemical researchers would behave in this fashion, the abstractly logical choice for them seems to me to [be to] maintain secrecy until a product use can be discovered. As to discouraging the search by others for product uses, there is no doubt this risk exists but the price paid for any patent is that research on other uses or improvements may be hampered because the original patentee will reap much of the reward. From the standpoint of the public interest the Constitution seems to have resolved that choice in favor of patentability.

What I find most troubling about the result reached by the Court is the impact it may have on chemical research. Chemistry is a highly interrelated field and a tangible benefit for society may be the outcome of a number of different discoveries, one discovery building upon the next. To encourage one chemist or research facility to invent and disseminate new processes and products may be vital to progress, although the product or process be without "utility" as the Court defines the term, because that discovery permits someone else to take a further but perhaps less difficult step leading to a commercially useful item. In my view, our awareness in this age of the importance of achieving and publicizing basic research should lead this Court to resolve uncertainties in its favor and uphold the respondent's position in this case.

COMMENTS AND QUESTIONS

1. What is the basis of the majority's claim that until an invention is shown to be useful, the inventor has not supplied the *quid pro quo* for a patent? What harm is there in granting a patent on an invention later shown to be useful? In granting a patent on a useless device? Who would suffer from the grant of such a patent? Would granting the patent help the inventor attract investment, which could be used to establish utility?

2. Would the existence of a market demand for the "intermediate" products at issue in *Brenner* change the result in the case? Companies certainly exist that make money by selling both raw materials and research tools to laboratories. Should we treat patents on these products differently than we do "end" products because of their intended use in the laboratory?

3. The concept of a "use patent" is described in Robert P. Merges & Richard Nelson, On the Complex Economics of Patent Scope, 90 Colum. L. Rev. 839 (1990), where the authors discuss an anomaly alluded to in Justice Harlan's dissent in *Brenner*: an inventor who obtains a patent for a product, e.g., a particular molecule, has the right to exclude all others from making, using or selling that product for *any* and all purposes, including purposes that the inventor did not herself discover or invent. For example, a patented compound created for its use as a leather tanning agent might turn out to be an effective anti-AIDS drug. If so, the patentee would have the right to prevent all others from selling the drug as an AIDS treatment — including the person who discovered that the leather tanning compound had anti-AIDS properties. Note that the utility requirement is met so long as the patentee shows *any* specific utility for the chemical when the patent is first filed — in our example, when the leather tanning property of the compound is discovered. Some have responded to the anomaly by calling for a special type of patent in these situations, a "new use" patent. See id.

New use patents already exist in a limited way. In our example, the one who discovers the anti-AIDS property of the leather tanning agent can obtain a *process* patent for "the process of using [the leather tanning compound] to treat AIDS." See Rohm & Haas v. Roberts Chemical Company, 245 F.2d 693 (4th Cir. 1957) (upholding defendant's patent on a new use of a well-known product as a fungicide); 1 Donald Chisum, Patents §1.03[8] (1978

& Supp. 1996) (collecting other cases on this point). This is in essence only an improvement patent; the discoverer would still have to obtain a license from the patentee to use the compound for treatment of AIDS. But the reverse is also true; unlike the example outlined above, if the one who discovered the leather tanning agent's anti-AIDS properties obtained a process patent, the patentee would have to obtain a license from the improver to have the right to use the compound to treat AIDS.

4. Utility for pharmaceutical products can generally be established by animal testing. See U.S. Patent and Trademark Office, Manual of Patent Examining Procedures §608.01(p) (5th ed. 1983, rev. 1989), "Guidelines for Considering Disclosures of Utility in Drug Cases." In the past, the biotechnology industry sometimes had trouble overcoming utility rejections without clinical data proving that a particular drug was effective *in humans*, despite Federal Circuit precedent holding that a showing of efficacy in a laboratory experiment was sufficient to establish the utility of a new drug. Cross v. Iizuka, 753 F.2d 1040, 1051 (Fed. Cir. 1985).

The Patent Office clarified its policy with new guidelines in 1995, specifying that human clinical trials are not necessary to establish utility. See U.S. P.T.O., Utility Examination Guidelines, 60 Fed. Reg. 36263 (July 14, 1995). The animating concept behind these new guidelines — that the standard for utility in the biotechnology and pharmaceutical industries is the same as it is elsewhere — was strikingly reaffirmed by the Federal Circuit in In re Brana, 51 F.3d 1560 (Fed. Cir. 1995). The court held that the results of in vivo tests in mice were sufficiently probative of efficacy in humans to pass the utility threshold:

> The Commissioner, as did the Board, confuses the requirements under the law for obtaining a patent with the requirements for obtaining government approval to market a particular drug for human consumption. . . . FDA approval is not a prerequisite for finding a compound useful within the meaning of the patent laws. . . . Usefulness in patent law, and in particular in the context of pharmaceutical inventions, necessarily includes the expectation of further research and development. . . . Were we to require Phase II testing in order to prove utility, the associated costs would prevent many companies from obtaining patent protection on promising new inventions, thereby eliminating an incentive to pursue, through research and development, potential cures in many crucial areas such as the treatment of cancer.

Brana, 51 F.3d at 1567. On the other hand, in In re Ziegler, 992 F.2d 1197, 1203 (Fed. Cir. 1993), the Federal Circuit held that a patent applicant was not entitled to claim priority to a 1954 application for "polypropylene," since at the time of the 1954 application "at best, Ziegler was on the way to discovering a practical utility for polypropylene . . . but in that application Ziegler had not yet gotten there." Are these two cases distinguishable?

For an argument that rational drug design strategies may be unfairly subject to harsher treatment under the utility doctrine, see Phillipe Ducor,

New Drug Discovery Techniques and Patents, 22 Rutgers Comp. & Tech. L.J. 369, 431-433 (1996).

5. *Brenner* in many ways represents the "high-water mark" of the utility doctrine. Most applications of the doctrine have been quite limited in the hurdles they place before inventors. Certainly any number of "frivolous" ideas are patented, suggesting that the burden of showing utility cannot be all that high. Thus the reader should not assume that *Brenner* can readily be extended by analogy.

Note on Different Types of Utility

Patent law covers more than one type of utility. We identify three separate types of utility arguments that have been used by the courts in rejecting patent applications. The first type centers on whether an invention is operable or capable of any use. The inquiry here is whether the invention as claimed can really *do* anything. The second major issue is whether the invention works to solve the problem it is designed to solve. The focus here is on the operability of the invention to serve its intended purpose. The third issue, in some ways the most interesting, is whether the intended purpose of the invention has some minimum social benefit, or at least is not completely harmful or deleterious. That is, if the invention does what it is supposed to, is that something that society wants done? The first problem may be thought of as general utility; the second, specific utility, and the third, beneficial or moral utility.

General Utility

In his landmark 1890 treatise, Professor William Robinson wrote that to be patentable, an invention must be more than "a mere curiosity, a scientific process exciting wonder yet not producing physical results, or [a] frivolous or trifling article or operation not aiding in the progress nor increasing the possession of the human race." 1 W. Robinson, Treatise on the Law of Patents for Useful Inventions 463 (1890). In applying this approach, one might perhaps imagine a "machine" with working parts that did not really do anything; perhaps it just spins around, or oscillates back and forth for no particular purpose. Such a machine would fail the test of utility under section 101 of the patent code. Note that machines that serve only to amuse or entertain *are* deemed useful under the patent code, however, so the limitation does not appear to be a very strict one.

Specific Utility

Even if an invention is directed toward a certain function, it must actually perform that function. Otherwise it is not "useful" for achieving that function. A good example of this principle is the case of Newman v. Quigg, 877 F.2d 1575 (Fed. Cir. 1989), *cert. denied*, 495 U.S. 932 (1990), where patent applicant Newman claimed an "Energy Generation System Having Higher Energy Output Than Input" — i.e., a "perpetual motion" machine. The

Federal Circuit, in upholding the denial of Newman's patent, noted that the applicant had not rebutted data from tests performed by the National Bureau of Standards showing that, as feared, the device did not function perpetually. In short, it had no utility because it did not work.

For an argument that the utility requirement diminishes incentives for "revolutionary" inventions, see Samuel Oddi, Beyond Obviousness: Invention Protection in the Twenty-First Century, 38 Am. U.L. Rev. 1097, 1127 (1989) (arguing that utility requirement and statutory classifications under section 101 [i.e., process, machine, manufacture, etc.] discriminate against revolutionary inventions, and calling for special protection of revolutionary inventions to offset these factors).

Beneficial or Moral Utility

The concept of immoral subject matter is thought to have originated in dictum from a Joseph Story opinion. Lowell v. Lewis, 1 Mason 182, 15 F. Cas. 1018 (No. 8568) (C.C.D. Mass. 1817). As examples of nonuseful inventions, he cited patents to "poison people, or to promote debauchery, or to facilitate private assassination." 15 F. Cas. 1018, 1019.

This doctrine was often invoked in the late nineteenth century to deny patents on gambling devices. Interestingly, it was a successful bar to patentability even where inventions appeared to be useful for things other than gambling. See, e.g., Schultz v. Holtz, 82 F. 448 (N.D. Cal. 1897) (patent on coin return device for coin-operated machines denied because it had application to slot machines); National Automatic Device Corp. v. Lloyd, 40 F. 89, 90 (N.D. Ill. 1889) (patent on toy horse race course denied on evidence that toy course was used in bars for betting purposes). Patents were struck down on this basis well into the twentieth century; see, e.g., Meyer v. Buckley Mfg. Co., 15 F. Supp. 640, 641 (N.D. Ill. 1936) (patent denied on "game of chance" vending machine, where user inserted coin and tried to manipulate miniature steam shovel to scoop up a toy), and even as late as 1941, in a pinball machine patent case, the Seventh Circuit was careful to note the distinction between playing pinball and gambling. Chicago Patent v. Genco, 124 F.2d 725, 728 (7th Cir. 1941) (upholding patent on pinball machine). By the 1970s, however, the courts were regularly upholding patents on gambling devices — both because gambling was no longer seen as a major moral issue and because courts had become more wary of denying patents on the basis of an indeterminate moral standard. See, e.g., Ex parte Murphy, 200 U.S.P.Q. (BNA) 801, 803 (Bd. Pat. App. & Int. 1977) (upholding claim for "one-armed bandit").

The fight against immoral inventions was not limited to patents for gambling devices. Another line of cases denied patents for inventions that could be used only to defraud. In one leading case, the Second Circuit invalidated a patent on a process for artificially producing spots on domestic tobacco, finding that the sole use for the process was to make domestic tobacco resemble fine grades of imported tobacco. Richard v. Du Bon, 103 F. 868, 873 (2d Cir. 1900). This was in keeping with other cases holding that patents could not be granted for devices having primarily fraudulent uses. For a more recent, and less expansive notion of "fraudulent," see Juicy Whip, Inc. v. Orange Bang, Inc., 185 F3d 1364 (Fed. Cir. 1999) (Patent for a beverage dispenser that simulated large glass bowl of fresh juice upheld over lack of utility defense).

Cases on medicinal products make up a special class of "fraudulent use" cases. Beginning in the nineteenth century, courts were wary of placing the government's imprimatur on medicines and devices hawked to an unsuspecting public in the free-wheeling days before the establishment of an effective Food and Drug Administration (FDA). The result was a higher standard of utility for health-related inventions, vestiges of which still can be seen in patent cases and Patent Office practices. Brand, Utility in a Pharmaceutical Patent, 39 Food Drug & Cosmetics L.J. 480 (1984).

Now that a powerful FDA has far-reaching powers to regulate drugs and medical devices, however, courts are increasingly willing to focus on functional utility rather than clinical safety when medical patents are at issue. The rationale for this more limited role is to avoid duplication of effort. As one court stated, "[T]o require the Patent Office to make an affirmative finding as to the safety of a drug for human use would work a serious overlapping of the respective jurisdictions of the Patent Office and the FDA." Carter-Wallace, Inc. v. Riverton Labs, Inc., 433 F.2d 1034, 1039 n.7 (2d Cir. 1970).

Recent courts have taken a similar "hands-off" approach to moral utility claims outside the medical area. For example, in Whistler Corp. v. Autotronics, Inc., 14 U.S.P.Q.2d (BNA) 1885 (N.D. Tex. 1988), the court stated:

> This is a patent infringement action brought by Whistler Corporation ("Whistler"), contending defendants are liable for infringing U.S. Patent No. 4,315,261 ("the '261 patent"), entitled "Radar Signal Detector." The case was tried to the court, which now enters judgment in favor of Whistler for the reasons set forth in this memorandum opinion and order.
>
> . . . In denying Whistler's prior preliminary injunction application, the court, in connection with the public interest factor, noted the seeming incongruity of asking a court of law to protect a device used to circumvent the law. Notwithstanding Whistler's evidence that the instant detectors have other uses, the court remains of the view that the primary and almost exclusive purpose for the radar detectors in question is to circumvent law enforcement attempts to detect and apprehend those who violate the law. Although the court seriously questions its being required to referee a contest among entities that manufacture and sell such products, the court concludes that the matter is one for the legislatures of the states, or for the Congress, to decide. Stated another way, only two states have seen fit to prohibit such devices. Unless and until detectors are banned outright, or Congress acts to withdraw patent protection for them, radar detector patentees are entitled to the protection of the patent laws.

COMMENTS AND QUESTIONS

1. What conclusions can be drawn from the attempts of the courts to enforce moral norms by denying patents? First, as in the case of gambling

devices, moral norms — or at least the courts' perceptions of them — change over time. Gambling is not perceived (by most) as a pernicious social ill in the current age, perhaps because other problems now seem so much worse. And other technologies once thought very wrong are now accepted as commonplace, e.g., birth control.

Even conceding that a particular new technology is analogous to gambling or selling fake medicines, another problem remains: what are the limits of the immorality test? How far into the future can the patent challenger look for the immoral effects of an invention, and what consensus version of morality can the courts rely on?

For example, historians and sociologists have long noted the profound social changes that accompanied the invention of the automobile. Some of these changes had unquestionable moral dimensions, such as the impact of automobiles on the incidence of pollution and premarital sex. If these changes could have been foreseen, immoral use might have been raised as a reason not to enforce the patent. A host of other technologies can be thought of in this vein: e.g., cattle prods (sometimes used in torture) and abortion-inducing drugs (safer than other procedures but considered immoral by some). On these points, see Robert Merges, The Patentability of Higher Life Forms: Intellectual Property Rights and Controversial Technologies, 47 Maryland L. Rev. 1051 (1988).

2. In an article by Eric Mirabel, Practical Utility Is a Useless Concept, 36 Am. U.L. Rev. 811 (1987), the author argues that the doctrine of "practical utility" is inconsistent with the history and tradition of the utility requirement.

PROBLEMS

Problem 3-4. Acid Look, Inc. (ALI) is a fabric design company that specializes in designing jeans for the high-end fashion market. ALI develops a process for producing a "random faded effect" on the fabric of new jeans, catering to the market for new jeans that appear to be pre-worn. ALI seeks a patent not only on the process of treating the jeans to produce a random faded effect, but on jeans treated by this process. The patent examiner rejects the claim to pre-faded jeans on the ground that they lack utility. Is this rejection proper? What test of utility is being applied here?

Problem 3-5. Craig Venter, a scientist at the National Institutes of Health, began a project in 1991 to obtain the genetic sequences for many of the natural chemical compounds or proteins that work in the human brain. His plan was to take advantage of the fact that of the many DNA sequences in human genetic material, only those that "code" for operational proteins exist in a certain form — the "cDNA" form. (The "c" in cDNA stands for complementary.) A complementary DNA sequence is one that is constructed at the site of the gene and is transported to the protein-creating site in the cell — the ribosome. (The cDNA sequence forms as chemicals called bases "match up" with the base pairs comprising the DNA sequence on the gene; the "matching up" occurs because the DNA's double helix structure allows it to split apart temporarily like a zipper unzipping.) Scientists believe that only 3 percent of the DNA in the body actually "codes" for some useful protein — the other 97 percent is a mystery.

Venter's project was to get a listing of the sequence of base pairs in the active, protein-encoding DNA sequences in a human brain. Getting all the "raw" sequences together was easy; he used widely available "libraries" of all the active DNA sequences found in various bodily tissues. (These are made by working backward from the cDNA "complementary copies" of the sequences that can be found in the relevant cells.) His contribution was to partially "sequence" the DNA — to obtain a base pair by base pair "listing" for as many cDNAs as he could. Before his work, there were libraries of DNA fragments, but no one had obtained detailed base pair sequences for each fragment.

In practice, he only had to sequence a portion of the DNA segments, and from that portion the entire sequence could be identified. (He called the partial sequences "expressed sequence tags," or ESTs.) This is one part of the larger project to sequence the entire human genome.

In a patent application, Venter claims each of the ESTs he has produced. He also claims fragments of the ESTs, complementary versions of each, and other variants.

Venter claims there is utility in his invention. Specifically, he asserts that the ESTs can be used to (1) map chromosomes, i.e., help give a general idea of the location of the genes on a chromosome; (2) identify tissue types (specifically, tell whether a given sample of tissue comes from the brain or not); and (3) identify gene regions associated with a disease, by drawing cDNA from patients, matching it with its associated EST, and seeing if the EST identifies a particular gene region unique to those with the disease.

Has Venter in fact recited utility? Are the suggested applications "too remote" from known, practical applications? If in fact each application Venter recites is no more than an invitation to further research — e.g., assistance in mapping chromosomes is helpful only when the mapped chromosome is later used to identify and treat a genetic problem — how do you distinguish this case from *Brenner*?

3. Novelty and Statutory Bars

Section 102 of the patent code embodies the principle that only truly new inventions deserve patents. In the technical analysis that has been developed to interpret and apply this principle, two distinct requirements are discernible: novelty and statutory bars. Novelty means "new compared to the prior art"; it states the requirement that, to be patentable, an invention must be somehow different from all published articles, known techniques, and marketed products. Statutory bar means "a bar to patentability based on too long a delay in seeking patent protection." In most cases, the bar arises because of something the inventor herself does. The classic example is publication of a scientific article. If an inventor fails to file a patent application within a year of the article's publication, she is barred from receiving a patent. She has suffered a "loss of right," in the words of the statute.

One key difference between novelty and statutory bars is that generally speaking an inventor's own work cannot destroy the novelty of her invention, but an inventor can create a statutory bar by her own actions (e.g., publication of an article). This distinction follows from the fact that novelty is measured from the date of *invention* whereas statutory bars are a function of the date a patent application is *filed*. It would be anomalous to say that an inventor did not invent first because she herself invented earlier; it is perfectly logical to say that an inventor is barred from receiving a patent because she herself filed an application more than a year after she published her invention.

Major portions of the text of section 102 are printed below, with bracketed insertions. Note that section 102(a) is concerned with novelty, while the statutory bars are set forth in section 102(b).

35 U.S.C. §102. Conditions for Patentability; Novelty and Loss of Right to Patent

A person shall be entitled to a patent unless —

(a) [**Novelty**] the invention was known or used by others in this country, or patented or described in a printed publication in this or a foreign country, before the invention thereof by the applicant for patent, or

(b) [**Statutory Bars**] the invention was patented or described in a printed publication in this or a foreign country or in public use or on sale in this country, more than one year prior to the date of the application for patent in the United States, or

[(c) and (d) omitted]

(e) [**Secret Prior Art: Previously-filed applications**] the invention was described in a patent granted on an application for patent by another filed in the United States before the invention thereof by the applicant for patent . . . or

(f) [**Derivation**] he did not himself invent the subject matter sought to be patented, or

(g) [**Priority; first-to-invent**] before the applicant's invention thereof the invention was made in this country by another who had not abandoned, suppressed, or concealed it. In determining priority of invention there shall be considered not only the respective dates of conception and reduction to practice of the invention, but also the reasonable diligence of one who was first to conceive and last to reduce to practice, from a time prior to conception by the other.

a. The Nature of Novelty

Rosaire v. National Lead Co.
United States Court of Appeals for the Fifth Circuit
218 F.2d 72 (5th Cir. 1955), cert. denied, 349 U.S. 916 (1955)

TUTTLE, Circuit Judge.

In this suit for patent infringement there is presented to us for determination the correctness of the judgment of the trial court, based on findings of fact and conclusions of law, holding that the two patents involved in the litigation were invalid and void and that furthermore there had been no infringement by defendant.

The Rosaire and Horvitz patents relate to methods of prospecting for oil or other hydrocarbons. The inventions are based upon the assumption that gases have emanated from deposits of hydrocarbons which have been trapped in the earth and that these emanations have modified the surrounding rock. The methods claimed involve the steps of taking a number of samples of soil from formations which are not themselves productive of hydrocarbons, either over a horizontal area or vertically down a well bore, treating each sample, as by grinding and heating in a closed vessel, to cause entrained or absorbed hydrocarbons therein to evolve as a gas, quantitatively measuring the amount of hydrocarbon gas so evolved from each sample, and correlating the measurements with the locations from which the samples were taken.

Plaintiff claims that in 1936 he and Horvitz invented this new method of prospecting for oil. In due course the two patents in suit, Nos. 2,192,525 and 2,324,085, were issued thereon. Horvitz assigned his interest to Rosaire.

In view of the fact that the trial court's judgment that the patents were invalid, would of course dispose of [this infringement suit] if correct, we turn our attention to this issue. [Appellee argues] that work carried on by one Teplitz for the Gulf Oil Corporation invalidated both patents by reason of the relevant provisions of the patent laws which state that an invention is not patentable if it "was known or used by others in this country" before the patentee's invention thereof, 35 U.S.C.A. §102(a). Appellee contends that Teplitz and his coworkers knew and extensively used in the field the same alleged inventions before any date asserted by Rosaire and Horvitz.

On this point appellant himself in his brief admits that "Teplitz conceived of the idea of extracting and quantitatively measuring entrained or absorbed gas from the samples of rock, rather than relying upon the free gas in the samples. We do not deny that Teplitz conceived of the methods of the patents in suit." And further appellant makes the following admission: "We admit that the Teplitz-Gulf work was done before Rosaire and Horvitz conceived of the inventions. We will show, however, that Gulf did not apply for patent until 1939, did not publish Teplitz's ideas, and did not otherwise give the public the benefit of the experimental work."

The question as to whether the work of Teplitz was "an unsuccessful experiment," as claimed by appellant, or was a successful trial of the method in question and a reduction of that method to actual practice, as contended by appellee, is, of course, a question of fact. On this point the trial court made the following finding of fact:

> I find as a fact that Abraham J. Teplitz and his coworkers with Gulf Oil Corporation and its Research Department during 1935 and early 1936, before any date claimed by Rosaire, spent more than a year in the oil fields and adjacent territory around Palestine, Texas, taking and analyzing samples both over an area and down drill holes, exactly as called for in the claims of the patents which Rosaire and Horvitz subsequently applied for and which are here in suit. This Teplitz work was a successful and adequate field trial of the prospecting method involved and a reduction to practice of that method. The work was performed in the field under ordinary conditions without any deliberate attempt at concealment or effort to exclude the public and without any instructions of secrecy to the employees performing the work.

As we view it, if the court's findings of fact are correct then under the statute as construed by the courts, we must affirm the finding of the trial court that appellee's patents were invalid.

[T]here was sufficient evidence to sustain the finding of the trial court that there was more here than an unsuccessful or incomplete experiment. It is clear that the work was not carried forward, but that appears to be a result of two things: (1) that the geographical area did not lend itself properly to the test, and (2) that the "entire gas prospecting program was therefore suspended in September of 1936, in order that the accumulated information might be thoroughly reviewed." It will be noted that the program was not suspended to test the worth of the method but to examine the data that was produced by use of the method involved.

With respect to the argument advanced by appellant that the lack of publication of Teplitz's work deprived an alleged infringer of the defense of prior use, we find no case which constrains us to hold that where such work was done openly and in the ordinary course of the activities of the employer, a large producing company in the oil industry, the statute is to be so modified by construction as to require some affirmative act to bring the work to the attention of the public at large.

While there is authority for the proposition that one of the basic principles underlying the patent laws is the enrichment of the art, and that a patent is given to encourage disclosure of inventions, no case we have found requires a holding that, under the circumstances that attended the work of Teplitz, the fact of public knowledge must be shown before it can be urged to invalidate a subsequent patent.

COMMENTS AND QUESTIONS

1. In W. L. Gore & Associates, Inc. v. Garlock, Inc., 721 F.2d 1540, 1548 (Fed. Cir. 1983), *cert. denied,* 469 U.S. 851 (1984), the Federal Circuit had this to say about "secret" prior use, where patentee Gore argued that prior use by another of a machine conforming to the elements of Gore's claim was nonpublic and therefore nonanticipatory:

> The nonsecret use of a claimed process in the usual course of producing articles for commercial purposes is a public use. Electric Storage Battery Co. v. Shimadzu, 307 U.S. 5, 20, 41 U.S.P.Q. 155, 161 (1939). . . . Thus it cannot be said that the district court erred in determining that the invention set forth in claim 1 of the '566 patent was known or used by others under §102(a), as evidenced by . . . operation of the 401 machine before Dr. Gore's asserted date of that invention.

2. The very expansive view taken in *Rosaire* of what it means for a disclosure to be "public" was criticized by one scholar:

> The term "public" . . . seems merely to mean "not secret." It is unnecessary to show that the previous discovery was ever used commercially, that it was in fact observable by the public if such a process or device would not normally be so viewed, or that it was known to more than a few persons. This construction of the term "public" seems questionable, since it may result in the denial of a patent even though the subsequent inventor has conferred a benefit by filing the invention with the public records.

Comment, Prior Art in the Patent Law, 73 Harv. L. Rev. 369, 373 (1959). The author goes on to propose a "higher standard of knowledge or use," arguing that

> an invention should be considered "known or used" only if it was so widely known or used that an ordinary skilled worker exercising reasonable diligence to learn the state of the art would have discovered, recognized, and been able to construct the invention.

Id., at 373. In effect, this proposal would apply a trade secret standard of novelty to the patent law — an invention could be "new" even though it had been made before, as long as it was not in general knowledge or use. Are there any problems with applying this weaker standard of novelty to determine whether a patent should be issued? Can it be reconciled with the broad language of section 102(a), precluding patents on inventions that were previously "known or used by others in this country"?

Would someone exercising "reasonable diligence" have discovered the prior art in *Rosaire*? What if person *A* were located geographically close to an obscure prior art reference, and person *B* was far away. Would the patentability of the same invention be different for the two? If the prior art in question was a device constructed by *C*, would she be barred from continued use of her device in the face of a patent to person *B*, whose invention was only patentable because she *(C)* was so far away that it was not "reasonable" for *B* to know about it?

Even accepting the premise that only "reasonably discoverable" prior art ought to be held against an inventor, how does one measure "reasonableness" here? If inventor *A* is a huge multinational conglomerate and inventor *B* is a home hobbyist with a workbench in the basement, should "reasonable diligence" be defined alike for both? If not, then *A*, the large company, in effect faces a higher standard of patentable novelty than *B*. Is this a good idea? If we adopt it, should the scope of patents issued to the *A*'s and *B*'s of the world be the same, or should *B*'s patent, being easier to get, be limited in some fashion — e.g., by making it effective only in one state or region? Note that the concept of "local novelty" is not unknown in the patent systems of the world.

In Hall v. MacNeale, 107 U.S. (17 Otto) 90 (1883), Hall had in 1866 received a patent on an improved design for the door and walls of burglar-proof safes. The question before the Court was whether the use of certain earlier-model safes before the critical date was *public* enough to create a statutory bar. The Court held:

> The construction and arrangement and purpose and mode of operation and use of the [hidden feature] in the safes were necessarily known to the workmen who put them in. They were, it is true, hidden from view, after the safes were completed, and it required a destruction of the safe to bring them into view. But this was no concealment of them or use of them in secret. They had no more concealment than was inseparable from any legitimate use of them.

3. One issue in *Rosaire* was the patentee's allegation that the Teplitz work was not prior art because it was incomplete, a mere "abandoned experiment." On this issue, consider Picard v. United Aircraft Corp., 128 F.2d 632 (2d Cir. 1942), *cert. denied*, 317 U.S. 651 (1942) (L. Hand, J.):

> It is true that another's experiment, imperfect and never perfected, will not serve either as an anticipation or as part of the prior art, for it has not served to enrich it. The patented invention does not become "known" by such a use or sale, or by anything of which the art cannot take hold and make use as it stands. But the mere fact that an earlier "machine" or "manufacture," sold or used, was an experiment does not prevent its becoming an anticipation or a part of the prior art, provided it was perfected and thereafter became publicly known. Whether it does become so depends upon how far it becomes a part of the stock of

knowledge of the art in question. Judged by that standard, the Curtiss engine [prior art reference] was not an "abandoned experiment"; it had been perfected; it had withstood a severer test than was necessary in use; it had been sold; it remained permanently accessible to the art, a contribution to the sum of knowledge so far as it went.

See also Corona Cord Tire Co. v. Dovan Chemical Corp., 276 U.S. 358, 384-385 (1928), where the Court held that commercial sales were not necessary for a completed experiment to constitute a "public use" under the patent code: "It is not an abandoned experiment because he confines his use of the [product] thus produced to his laboratory or his lecture room."

Note on the Inherency Doctrine

Section 102(a) provides that an applicant is not entitled to a patent if her invention was "known or used by others" prior to the date of the applicant's invention. The meaning of this phrase has been called into question in a series of cases involving unintended, "accidental" anticipation of an invention. Most of these cases involve the inherent, unintended production of a particular physical product in a chemical process. When an inventor later *intentionally* makes the product, presumably because she has some use for it, the prior unintended production of the product may be raised as prior art to the invention.

On the one hand, a venerable line of cases holds that where the first, accidental producer was not aware of the product and did not attempt to produce it, the first production did not bar a patent on the "invention" of the product. Thus, in Tilghman v. Proctor, 102 U.S. 707 (1880), the Supreme Court held that the accidental separation of fat acids from tallow during operation of a steam engine lubricated by tallow did not anticipate Tilghman's patent for a similar separation process. Of the prior production, the Court said:

> They revealed no process for the manufacture of fat acids. If the acids were accidentally and unwittingly produced, whilst the operators were in pursuit of other and different results, without exciting attention and without it even being known what was done or how it had been done, it would be absurd to say that this was an anticipation of Tilghman's discovery.

Id. at 711. On the other hand, an equally venerable line of cases holds that if a product is known in the art already, an inventor cannot obtain a patent *on the product* merely by putting it to a new use, "even if the new result had not before been contemplated." Ansonia Brass & Copper Co. v. Electric Supply Co., 144 U.S. 11 (1892). The inventor could obtain a patent on the new process using that product, however.

COMMENTS AND QUESTIONS

1. Obviously, there is some tension between these lines of cases. Is it possible to reconcile the two, perhaps by distinguishing situations in which the product was unknown (or at least unwanted) from situations in which the product was known and used, albeit for a different purpose? Does such a distinction make sense in light of the purposes of granting patents?

PROBLEM

Problem 3-6. As described in Problem 3-2, physicist John Anderson has isolated and produced a new element in the laboratory — Litigacium, element 116. Although Anderson is the first person to have produced Litigacium in any appreciable amount, and the first to have isolated it, mathematical calculations based on prior experiments show that Dr. Firmi's nuclear reactor inherently produces 3/1,000,000,000 of one gram of Litigacium every year that it operates. Further, this minute amount of Litigacium is spread among 40 tons of highly radioactive uranium fuel, making it impossible to recover or isolate. Nonetheless, Dr. Firmi and his coworkers were aware at the time they ran the reactor that it was producing trace quantities of Litigacium.

Dr. Anderson claims "element 116" in a patent application. Is this claim anticipated by Firmi's prior work?

---

b. Statutory Bars: Publications

Recall from the introduction to this section that one difference between statutory bars (which we cover in this subsection) and novelty is that an inventor can create a statutory bar by her own actions (e.g., publication of an article), whereas she cannot destroy the novelty of her own inventions. In short, an inventor's own work cannot be cited against her (in general) under section 102(a), but it is "fair game" under section 102(b).

Notwithstanding this fact, for many purposes the two provisions are not appreciably different. For example, the definition of "publication" — a term you will recall occurs in both sections 102(a) and (b) — is the same for both. Thus although the following case involves a section 102(b) reference, the reasoning is equally applicable to cases decided under section 102(a).

In re Hall
United States Court of Appeals for the Federal Circuit
781 F.2d 897 (Fed. Cir. 1986)

BALDWIN, Circuit Judge.

This is an appeal from the decision of the U.S. Patent and Trademark Office's (PTO) former Board of Appeals, adhered to on reconsideration by the Board of Patent Appeals and Interferences (board), sustaining the final rejection of claims 1-25 of [Hall's] reissue [a]pplication, based principally on a "printed publication" bar under 35 U.S.C. §102(b). The reference is a doctoral thesis. Because appellant concedes that his claims are unpatentable if the thesis is available as a "printed publication" more than one year prior to the application's effective filing date of February 27, 1979, the only issue is whether the thesis is available as such a printed publication. On the record before us, we affirm the board's decision.

A protest was filed during prosecution of appellant's reissue application which included in an appendix a copy of the dissertation "1,4-α-Glucanglukohydrolase ein amylotylisches Enzym . . ." by Peter Foldi (Foldi thesis or dissertation). The record indicates that in September 1977, Foldi submitted his dissertation to the Department of Chemistry and Pharmacy at Freiburg University in the Federal Republic of Germany, and that Foldi was awarded a doctorate degree on November 2, 1977. . . .

The examiner made a final rejection of the application claims. He said: "On the basis of the instant record it is reasonable to assume that the Foldi thesis was available (accessible) prior to February 27, 1979."[30]

By letter, the PTO's Scientific Library asked Dr. Will whether the Foldi dissertation was made available to the public by being cataloged and placed in the main collection. Dr. Will replied in an October 20, 1983 letter, as translated: "Our dissertations, thus also the Foldi dissertation, are indexed in a special dissertations catalogue, which is part of the general users' catalogue. In the stacks they are likewise set apart in a special dissertation section, which is part of the general stacks."

In response to a further inquiry by the PTO's Scientific Library requesting (1) the exact date of indexing and cataloging of the Foldi dissertation or (2) "the time such procedures normally take," Dr. Will replied in a June 18, 1984 letter: "The Library copies of the Foldi dissertation were sent to us by the faculty on November 4, 1977. Accordingly, the dissertation most probably was available for general use toward the beginning of the month of December, 1977."

The board held that the unrebutted evidence of record was sufficient to conclude that the Foldi dissertation had an effective date as prior art more than one year prior to the filing date of the appellant's initial application.

---

30. The court presumably means February 27, 1978. — EDS.

On appeal, appellant raises two arguments: (1) the §102(b) "printed publication" bar requires that the publication be accessible to the interested public, but there is no evidence that the dissertation was properly indexed in the library catalog prior to the critical date; and (2) even if the Foldi thesis were cataloged prior to the critical date, the presence of a single cataloged thesis in one university library does not constitute sufficient accessibility of the publication's teachings to those interested in the art exercising reasonable diligence.

The [printed publication] bar is grounded on the principle that once an invention is in the public domain, it is no longer patentable by anyone.

The statutory phrase "printed publication" has been interpreted to give effect to ongoing advances in the technologies of data storage, retrieval, and dissemination. Because there are many ways in which a reference may be disseminated to the interested public, "public accessibility" has been called the touchstone in determining whether a reference constitutes a "printed publication" bar under 35 U.S.C. §102(b). The §102 publication bar is a legal determination based on underlying fact issues, and therefore must be approached on a case-by-case basis. The proponent of the publication bar must show that prior to the critical date the reference was sufficiently accessible, at least to the public interested in the art, so that such a one by examining the reference could make the claimed invention without further research or experimentation.

[A]ppellant argues that the Foldi thesis was not shown to be accessible because Dr. Will's affidavits do not say when the thesis was indexed in the library catalog and do not chronicle the procedures for receiving and processing a thesis in the library.

[A]ppellant would have it that accessibility can only be shown by evidence establishing a *specific* date of cataloging and shelving before the critical date. While such evidence would be desirable, in lending greater certainty to the accessibility determination, the realities of routine business practice counsel against requiring such evidence. The probative value of routine business practice to show the performance of a specific act has long been recognized. See, e.g., 1 Wigmore, Evidence § 92 (1940); rule 406, Fed. R. Evid. Therefore, we conclude that competent evidence of the general library practice may be relied upon to establish an approximate time when a thesis became accessible.

We agree with the board that the evidence of record consisting of Dr. Will's affidavits establishes a prima facie case for unpatentability of the claims under the §102(b) publication bar. It is a case which stands unrebutted.

Accordingly, the board's decision sustaining the rejection of appellant's claims is affirmed.

COMMENTS AND QUESTIONS

1. Note that the issue in this case could arise under either section 102(a) or 102(b). If the publication had occurred before the *invention* of the subject

matter by the applicant, section 102(a) would apply. Section 102(b) applied in this case because the publication occurred more than one year before the applicant *filed* for a patent.

2. Compare the "publicness" standard in *Hall* to the rule in trade secret law (information must be "generally known or readily ascertainable" in an industry). Why is it so much easier to lose patent rights than trade secrets?

3. In general, a publication becomes public when it becomes available to at least one member of "the general public." Thus a magazine or technical journal is effective as of its date of publication, i.e., when someone first receives it, rather than the date a manuscript was sent to the publisher or the date the journal or magazine was mailed. As was said in In re Schlittler, 234 F.2d 882 (C.C.P.A. 1956): "[T]he mere placing of a manuscript in the hands of a publisher does not necessarily make it available to the public within the meaning of [the] authorities."

After *Schlittler*, mere receipt of a manuscript by a publisher does not constitute "publication" under section 102(a) or (b). Actual publication — i.e., receipt by subscribers — is required. See, e.g., Protein Found. v. Brenner, 147 U.S.P.Q. (BNA) 429 (1965); U.S. Patent and Trademark Office, Manual of Patent Examining Procedures §706.02 (5th ed. rev. 1991); Ex parte Hudson, 18 U.S.P.Q.2d (BNA) 1322 (Bd. App. & Int. 1990). Is a grant proposal sent to a limited number of expert reviewers a publication under §102(a)? See E. I. du Pont de Nemours & Co. v. Cetus Corp., 19 U.S.P.Q.2d (BNA) 1174 (N.D. Cal. 1990) (no).

PROBLEM

Problem 3-7. In late 1967 the Navy encountered a problem with its new fighter plane. The Navy's aircraft — stationed on aircraft carriers — were more likely to suffer structural damage than Air Force planes of the same type that were positioned on land. Navy scientists concluded that the widespread problem of fuselage cracking, known as "stress corrosion," was attributable to the harsh marine environment. The Navy presented its problem to the aluminum industry, and Algol Corp. was awarded a research contract to develop a new alloy that would be more resistant to the phenomenon.

Pursuant to the terms of the defense contract, Algol would periodically report on its research results in "progress letters." The Navy drew up a distribution list of 33 designees to receive Algol's progress letters, and each letter bore the following "export control" notice: "This Document Is Subject to Special Export Controls and Each Transmittal to Foreign Governments or Foreign Nationals May Be Made Only with Prior Approval of the Naval Air Systems Command."

The designees included aluminum producers (including Algol's competitors), aircraft manufacturers, government agencies, branches of the military, and academic researchers. Several progress letters were mailed, the last on April 5, 1969. As of the last progress letter, Algol had succeeded in developing a high-strength, stress-corrosion-resistant alloy that met the Navy's needs. Algol successfully applied for a patent covering its process for aging and treating its new alloy on February 5, 1971.

When Algol refused to license the patent to its competitor Richards, the latter began to manufacture and sell a high-strength alloy using Algol's patented technology. Algol filed suit alleging infringement.

You are Senior Counsel for Intellectual Property Matters at Richards Aluminum. Your defense strategy is limited by the fact that if the Algol patent is valid, your company is undeniably engaged in infringement. Algol gathers evidence that the 33 designees who received Algol's progress reports treated them as highly confidential although they contained no express limitation on access other than the export control notice. Depositions of Richards' own executives reveal that your company kept its copies of the letters in a protected area, screened even from some of its own employees. Such procedures were typical in defense industry circles when a report bearing an export control notice was received. Nevertheless, the absence of any other access restrictions on the letters supports Richards' argument that Richards could have shared the report's contents with every American citizen.

Is the patent anticipated by Algol's prior "publication"? If so, under section 102(a) or under section 102(b)?

c. Statutory Bars: Public Use

Egbert v. Lippmann
Supreme Court of the United States
104 U.S. 333 (1881)

Mr. Justice WOODS delivered the opinion of the court.

This suit was brought for an alleged infringement of the complainant's reissued letters-patent, No. 5216, dated Jan. 7, 1873, for an improvement in corset-springs.

The original letters were issued to Samuel H. Barnes. The reissue was made to the complainant, under her then name, Frances Lee Barnes, executrix of the original patentee.

The specification for the reissue declares: —
This invention consists in forming the springs of corsets of two or more metallic plates, placed one upon another, and so connected as to prevent them from sliding off each other laterally or edgewise, and at the same time admit of their playing or sliding upon each other, in the direction of their length or

longitudinally, whereby their flexibility and elasticity are greatly increased, while at the same time much strength is obtained.

The bill alleges that Barnes was the original and first inventor of the improvement covered by the reissued letters-patent, and that it had not, at the time of his application for the original letters, been for more than two years in public use or on sale, with his consent or allowance.

The answer takes issue on this averment and also denies infringement. On a final hearing the court dismissed the bill, and the complainant appealed.

We have to consider whether the defence that the patented invention had, with the consent of the inventor, been publicly used for more than two years prior to his application for the original letters, is sustained by the testimony in the record.

[The patent statute] render[s] letters-patent invalid if the invention which they cover was in public use, with the consent and allowance of the inventor, for more than two years prior to his application.[31] Since the passage of the act of 1839 it has been strenuously contended that the public use of an invention for more than two years before such application, even without his consent and allowance, renders the letters-patent therefor void.

It is unnecessary in this case to decide this question, for the alleged use of the invention covered by the letters-patent to Barnes is conceded to have been with his express consent.

The evidence on which the defendants rely to establish a prior public use of the invention consists mainly of the testimony of the complainant.

She testifies that Barnes invented the improvement covered by his patent between January and May, 1855; that between the dates named the witness and her friend Miss Cugier were complaining of the breaking of their corset-steels. Barnes, who was present, and was an intimate friend of the witness, said he thought he could make her a pair that would not break. At their next interview he presented her with a pair of corset-steels which he himself had made. The witness wore these steels a long time. In 1858 Barnes made and presented to her another pair, which she also wore a long time. When the corsets in which these steels were used wore out, the witness ripped them open and took out the steels and put them in new corsets. This was done several times.

It is admitted, and, in fact, is asserted, by complainant, that these steels embodied the invention afterwards patented by Barnes and covered by the reissued letters-patent on which this suit is brought.

Joseph H. Sturgis, another witness for complainant, testifies that in 1863 Barnes spoke to him about two inventions made by himself, one of which was a corset-steel, and that he went to the house of Barnes to see them. Before this time, and after the transactions testified to by the complainant, Barnes and [the complainant] had intermarried. Barnes said his wife had a pair of steels made according to his invention in the corsets which she was then

---

31. This has since been changed to one year. — EDS.

wearing, and if she would take them off he would show them to [Sturgis]. Mrs. Barnes went out, and returned with a pair of corsets and a pair of scissors, and ripped the corsets open and took out the steels. Barnes then explained to the witness how they were made and used.

The question for our decision is, whether this testimony shows a public use within the meaning of the statute.

We observe, in the first place, that to constitute the public use of an invention it is not necessary that more than one of the patented articles should be publicly used. The use of a great number may tend to strengthen the proof, but one well-defined case of such use is just as effectual to annul the patent as many. For instance, if the inventor of a mower, a printing press, or a railway-car makes and sells only one of the articles invented by him, and allows the vendee to use it for two years, without restriction or limitation, the use is just as public as if he had sold and allowed the use of a great number.

We remark, secondly, that, whether the use of an invention is public or private does not necessarily depend upon the number of persons to whom its use is known. If an inventor, having made his device, gives or sells it to another, to be used by the donee or vendee, without limitation or restriction, or injunction of secrecy, and it is so used, such use is public, even though the use and knowledge of the use may be confined to one person.

We say, thirdly, that some inventions are by their very character only capable of being used where they cannot be seen or observed by the public eye. An invention may consist of a lever or spring, hidden in the running gear of a watch, or of a rachet, shaft, or cog-wheel covered from view in the recesses of a machine for spinning or weaving. Nevertheless, if its inventor sells a machine of which his invention forms a part, and allows it to be used without restriction of any kind, the use is a public one. So, on the other hand, a use necessarily open to public view, if made in good faith solely to test the qualities of the invention, and for the purpose of experiment, is not a public use within the meaning of the statute.

Tested by these principles, we think the evidence of the complainant herself shows that for more than two years before the application for the original letters there was, by the consent and allowance of Barnes, a public use of the invention, covered by them. He made and gave to her two pairs of corset-steels, constructed according to his device, one in 1855 and one in 1858. They were presented to her for use. He imposed no obligation of secrecy, nor any condition or restriction whatever. They were not presented for the purpose of experiment, nor to test their qualities. No such claim is set up in her testimony. The invention was at the time complete, and there is no evidence that it was afterwards changed or improved. The donee of the steels used them for years for the purpose and in the manner designed by the inventor. They were not capable of any other use. She might have exhibited them to any person, or made other steels of the same kind, and used or sold them without violating any condition or restriction imposed on her by the inventor.

According to the testimony of the complainant, the invention was completed and put to use in 1855. The inventor slept on his rights for eleven years. Letters-patent were not applied for till March, 1866. In the mean time, the invention had found its way into general, and almost universal, use. A great part of the record is taken up with the testimony of the manufacturers and venders of corset-steels, showing that before he applied for letters the principle of his device was almost universally used in the manufacture of corset-steels. It is fair to presume that having learned from this general use that there was some value in his invention, he attempted to resume, by his application, what by his acts he had clearly dedicated to the public.

We are of opinion that the defence of two years' public use, by the consent and allowance of the inventor, before he made application for letters-patent, is satisfactorily established by the evidence.

Decree affirmed.

Mr. Justice MILLER dissenting.

A private use with consent, which could lead to no copy or reproduction of the machine, which taught the nature of the invention to no one but the party to whom such consent was given, which left the public at large as ignorant of this as it was before the author's discovery, was no abandonment to the public, and did not defeat his claim for a patent. If the little steel spring inserted in a single pair of corsets, and used by only one woman, covered by her outer-clothing, and in a position always withheld from public observation, is a *public* use of that piece of steel, I am at a loss to know the line between a private and a public use.

COMMENTS AND QUESTIONS

1. In what sense was Frances Lee (later Barnes) using the corset steels (i.e., the springs) "publicly"? Why does it not matter how many people use the invention, for purposes of deciding if the use was public? Compare this holding to that of the court in *Rosaire* above.

2. The Court emphasizes that Ms. Barnes never entered into a confidentiality agreement with the inventor. But the Court also relates that the two later married. Does Barnes have a reasonable argument that although no *express* "injunction of secrecy" was made, nonetheless an *implied* requirement of secrecy could be inferred from the surrounding circumstances: i.e., an invention of an intimate nature, by a boyfriend (later husband), disclosed to few people? Justice Miller, in dissent, added: "It may well be imagined that a prohibition to the party so permitted [to use the springs] against her use of the steel spring to public observation, would have been supposed to be a piece of irony." 104 U.S. (14 Otto), at 339. Perhaps the embarrassment of asking was enough to keep Mr. Barnes from requesting that Frances Lee not show the invention to anyone.

3. Subsequent cases have not abandoned *Egbert*, but they have perhaps not read it as broadly as they could. In particular, the Federal Circuit has focused on the nature and purpose of the use in deciding whether it was "public" or not. Use for a commercial purpose is generally a public use, even if it is secret and even if it only occurs once. However, use for personal interest or enjoyment will generally not be considered a "public" use.

A similar rule has developed with respect to the "on sale" prong of the section 102(b) statutory bar. A single sale or offer to sell a product will start the clock running, whether or not it was made in secret. On the other hand, discussions that do not rise to the level of a definite offer to sell the invention will not bar a later patent on that invention.

4. Obviously, the policy behind the statutory bars in section 102(b) is to encourage inventors to file patent applications early. The inventor in *Egbert* did not do so. Was it reasonable to bar him from receiving a patent on his invention for this reason? Is anyone entitled to a patent on this invention?

5. While "public use" and "on sale" bars frequently result from the inventor's own conduct, they do not have to. Indeed, one recent case held that sales of the invention by a third party precluded the inventor from later obtaining a patent, even though the third party had stolen the invention from the inventor. Evans Cooling Systems, Inc. v. General Motors, 125 F.3d 1448 (Fed. Cir. 1997). The case arose out of inventor John Evans' demonstration of his aqueous reverse flow cooling system to GM engineers in the spring of 1989. GM, as early as June 13, 1991, received a binding order for sale of a Corvette model car incorporating the aqueous reverse flow cooling system. The district court agreed with GM that this sale created a statutory bar for Evans' subsequent patent application filed July 1, 1992. Evans appealed on two grounds: that no enforceable contract could have been formed in June of 1991 because GM had yet to clear certain requisite regulatory hurdles routinely required for sale of a new car model and that the theft of a trade secret ought to be considered an exception to the general statutory bar rule. The court disposed of the first ground by stating simply that "the mere fact that the offer for sale was illegal or ineffective does not remove it from the purview of the section 102(b) bar." Id. at 1452.

The court found the second ground — the "theft exception" — equally unpersuasive in light of existing case law. It wrote:

> While such a result may not seem fair, Evans is not without recourse if GM in fact misappropriated his invention. Evans would have an appropriate remedy in state court for misappropriation of a trade secret. We note as well that the facts Evans alleges in support of its misappropriation claim demonstrate that Evans knew GM stole the invention at the very time it was allegedly stolen because during the demonstration GM employees allegedly told Mr. Evans they intended to steal the invention and a sealed room was unsealed during the night between the tests. Evans' patent rights would have nevertheless been protected if Mr. Evans had filed a patent application no more than one year from the date of the demonstration. This he did not do; instead Mr. Evans waited for more

than two years after the demonstration and some six years after it was reduced to practice.

Id. at 1453-54.

6. Must an invention actually be completed and built before it can be sold or "offered for sale?" The Supreme Court said no in a recent case. Pfaff v. Wells Electronics, Inc., 119 S. Ct. 304 (1998). The Court reasoned that "invention" occurred when the inventor conceived of the product and did not require proof that the invention has actually been built. Indeed, it noted, many famous inventions (including Alexander Graham Bell's telephone) were patented before a prototype was ever built. Id. at 308-309. However, the Court did require evidence that the inventive concept itself was "complete, rather than merely . . . substantially complete":

> [T]he on-sale bar applies when two conditions are satisfied before the critical date. First, the product must be the subject of a commercial offer for sale. . . .
>
> Second, the invention must be ready for patenting. That condition may be satisfied in at least two ways: by proof of reduction to practice before the critical date; or by proof that prior to the critical date the inventor had prepared drawings or other descriptions of the invention that were sufficiently specific to enable a person skilled in the art to practice the invention. In this case the second condition of the on-sale bar is satisfied because the drawings Pfaff sent to the manufacturer before the critical date fully disclosed the invention.

Id. at 311-312. See also Weatherchem Corp. v. J.L. Clark, Inc., 163 F.3d 1326, 1332-1334, 49 U.S.P.Q.2d 1001 (Fed.Cir. 1998) (applying *Pfaff* to case where detailed drawings had been made in preparation of an offer for sale). Contrast *Pfaff* with Micro Chem., Inc. v. Great Plains Chem. Co., 103 F.3d 1538 (Fed. Cir. 1997) ("Because Pratt was not close to completion of the invention at the time of the alleged offer and had not demonstrated a high likelihood that the invention would work for its intended purpose upon completion, his December 1984 'offer' could not trigger the on-sale bar.").

Does it make sense to use an enablement standard in determining whether an invention has been made, or should there be some requirement that the invention be physically constructed before it can be placed on sale? Does the 1994 addition of the "offer for sale" language to the list of exclusive rights in 35 U.S.C. §271(a) suggest that completing the invention should not be required?

The late Judge Giles Rich, in one of the last of his many patent law opinions over the years, commented on how complete knowledge of an invention must be under *Pfaff* in Scaltech Inc. v. Retec/Tetra, L.L.C., 178 F.3d 1378, 1383-1384 (Fed. Cir. 1999):

> The district court did not address whether the process that was offered in the 30 March 1988 proposal to Chevron or the 15 November 1988 proposal to Champlin would necessarily have satisfied the claim limitations relating to small

particle size and high solids concentration. We note that there is no requirement that the offer specifically identify these limitations. Nor is there a requirement that Scaltech must have recognized the significance of these limitations at the time of offer. If the process that was offered for sale inherently possessed each of the claim limitations, then the process was on sale, whether or not the seller recognized that his process possessed the claimed characteristics. Inherency may not be established by probabilities or possibilities. The mere fact that a certain thing may result from a given set of circumstances is not sufficient to establish inherency. However, if the natural result flowing from the operation of the process offered for sale would necessarily result in achievement of each of the claim limitations, then claimed invention was offered for sale. The district court must determine if the process offered for sale, in its normal use, inherently satisfies each claim limitation. If so, then the offer creates a §102(b) bar. The district court did not perform this analysis. Therefore, we vacate the district court's holding and remand for a determination as to whether the process on sale inherently satisfies each claim limitation.

PROBLEM

Problem 3-8. Scientists involved in the field of spectroscopy were being haunted by ghosts. Spectroscopy is the study and analysis of materials to determine their components and molecular structure. One form of spectroscopy, magnetic resonance spectroscopy, works by observing a material's reaction to imposed radiation. A spectroscope bombards the sample material (usually contained in a test tube) with pulses of radio frequency radiation and then records and analyzes the material's response. If the signal response given by the material was weak, the spectroscope would often confuse interfering noise and static for the material's response. Scientists attempted to solve the problem of unwanted frequencies (or "ghost" frequencies) by various mathematical manipulations.

One such scientist was Daniel Murray, a doctoral candidate studying at Oxford under the supervision of Sir William Akeroyd. In the course of his experiments, Murray arrived at a means of "ghostbusting" through shifting the phase of the frequency transmitter at 90 degree angles between pulses, thus avoiding sophisticated manipulations of data. A patent application embodying the Murray invention was filed in the United Kingdom on April 8, 1984, and in the United States on December 21, 1986. Murray secured patent protection in both countries by 1989.

In the spring of 1983, Akeroyd attended a scientific meeting in Boulder, Colorado. On a bus ride during the conference, Akeroyd engaged Dr. Christopher Squire in an informal conversation. Squire headed a research team at the U.S.-based Monsanto Company which developed related technology for use in the company's chemical analyses. Akeroyd informed Squire of the promising results of Murray's preliminary research on the phase shifting spectroscope. By 1985, Squire had updated his spectrometer at Monsanto to include a similar phase shifting transmitter.

Squire immediately applied his improved spectrometer technology to assess whether a particular Monsanto herbicide was safe for release into the environment. This was a project Squire's team had been working on for a full year prior to the improvement in the lab's spectrometer. Excited by the improved results, Squire invited several of his superiors and fellow employees into the lab to observe his data.

Is Squire's use a public use under section 102(b)? Does your analysis depend on any of the following hypothetical facts:

- Dr. Squire was the only person who was permitted access to the Monsanto lab where the updated spectrometer was located.
- Anyone *could* have entered the lab, but in fact Squire was the only one who did.
- The only person who set eyes upon the updated spectrometer besides Squire was a Monsanto custodian.
- Squire left the door unlocked regularly at night.
- The phase shifting transmitter is located deep inside the apparatus, and cannot be seen without dismantling the spectrometer.

d. The Experimental Use Exception

City of Elizabeth v. Pavement Company
Supreme Court of the United States
97 U.S. 126 (1877)

Mr. Justice BRADLEY delivered the opinion of the court.

This suit was brought by the American Nicholson Pavement Company against the city of Elizabeth, N. J., upon a patent issued to Samuel Nicholson, dated Aug. 20, 1867, for a new and improved wooden pavement. [I]n the specification, it is declared that the nature and object of the invention consists in providing a process or mode of constructing wooden block pavements upon a foundation along a street or roadway with facility, cheapness, and accuracy, and also in the creation and construction of such a wooden pavement as shall be comparatively permanent and durable, by so uniting and combining all its parts, both superstructure and foundation, as to provide against the slipping of the horses' feet, against noise, against unequal wear, and against rot and consequent sinking away from below.

The bill charges that the defendants infringed this patent by laying down wooden pavements in the city of Elizabeth, N.J., constructed in substantial conformity with the process patented, and prays an account of profits, and an injunction.

The defendants . . . averred that the alleged invention of Nicholson was in public use, with his consent and allowance, for six years before he applied for a patent, on a certain avenue in Boston called the Mill-dam; and contended that said public use worked an abandonment of the pretended invention. . . .

The next question to be considered is, whether Nicholson's invention was in public use or on sale, with his consent and allowance, for more than two years prior to his application for a patent, within the meaning of . . . the acts in force in 1854, when he obtained his patent. It is contended by the appellants that the pavement which Nicholson put down by way of experiment, on Mill-dam Avenue in Boston, in 1848, was publicly used for the space of six years before his application for a patent, and that this was a public use within the meaning of the law.

To determine this question, it is necessary to examine the circumstances under which this pavement was put down, and the object and purpose that Nicholson had in view. It is perfectly clear from the evidence that he did not intend to abandon his right to a patent. He had filed a *caveat* in August, 1847, and he constructed the pavement in question by way of experiment, for the purpose of testing its qualities. The road in which it was put down, though a public road, belonged to the Boston and Roxbury Mill Corporation, which received toll for its use; and Nicholson was a stockholder and treasurer of the corporation. The pavement in question was about seventy-five feet in length, and was laid adjoining to the toll-gate and in front of the toll-house. It was constructed by Nicholson at his own expense, and was placed by him where it was, in order to see the effect upon it of heavily loaded wagons, and of varied and constant use; and also to ascertain its durability, and liability to decay. Joseph L. Lang, who was toll-collector for many years, commencing in 1849, familiar with the road before that time, and with this pavement from the time of its origin, testified as follows:

> Mr. Nicholson was there almost daily, and when he came he would examine the pavement, would often walk over it, cane in hand, striking it with his cane, and making particular examination of its condition. He asked me very often how people liked it, and asked me a great many questions about it. I have heard him say a number of times that this was his first experiment with this pavement, and he thought that it was wearing very well. The circumstances that made this locality desirable for the purpose of obtaining a satisfactory test of the durability and value of the pavement were: that there would be a better chance to lay it there; he would have more room and a better chance than in the city; and, besides, it was a place where most everybody went over it, rich and poor. It was a great thoroughfare out of Boston. It was frequently travelled by teams having a load of five or six tons, and some larger. As these teams usually stopped at the toll-house, and started again, the stopping and starting would make as severe a trial to the pavement as it could be put to.

This evidence is corroborated by that of several other witnesses in the cause; the result of the whole being that Nicholson merely intended this piece of pavement as an experiment, to test its usefulness and durability. Was this a public use, within the meaning of the law?

An abandonment of an invention to the public may be evinced by the conduct of the inventor at any time, even within the two years named in the law. The effect of the law is, that no such consequence will necessarily follow from the invention being in public use or on sale, with the inventor's consent and allowance, at any time within two years before his application; but that, if the invention is in public use or on sale prior to that time, it will be conclusive evidence of abandonment, and the patent will be void.

But, in this case, it becomes important to inquire what is such a public use as will have the effect referred to. That the use of the pavement in question was public in one sense cannot be disputed. But can it be said that the invention was in public use? The use of an invention by the inventor himself, or of any other person under his direction, by way of experiment, and in order to bring the invention to perfection, has never been regarded as such a use. Curtis, Patents, §381; Shaw v. Cooper, [32 U.S.] 7 Pet. 292 [1833].

Now, the nature of a street pavement is such that it cannot be experimented upon satisfactorily except on a highway, which is always public.

When the subject of invention is a machine, it may be tested and tried in a building, either with or without closed doors. In either case, such use is not a public use, within the meaning of the statute, so long as the inventor is engaged, in good faith, in testing its operation. He may see cause to alter it and improve it, or not. His experiments will reveal the fact whether any and what alterations may be necessary. If durability is one of the qualities to be attained, a long period, perhaps years, may be necessary to enable the inventor to discover whether his purpose is accomplished. And though, during all that period, he may not find that any changes are necessary, yet he may be justly said to be using his machine only by way of experiment; and no one would say that such a use, pursued with a *bona fide* intent of testing the qualities of the machine, would be a public use, within the meaning of the statute. So long as he does not voluntarily allow others to make it and use it, and so long as it is not on sale for general use, he keeps the invention under his own control, and does not lose his title to a patent.

It would not be necessary, in such a case, that the machine should be put up and used only in the inventor's own shop or premises. He may have it put up and used in the premises of another, and the use may inure to the benefit of the owner of the establishment. Still, if used under the surveillance of the inventor, and for the purpose of enabling him to test the machine, and ascertain whether it will answer the purpose intended, and make such alterations and improvements as experience demonstrates to be necessary, it will still be a mere experimental use, and not a public use, within the meaning of the statute.

Whilst the supposed machine is in such experimental use, the public may be incidentally deriving a benefit from it. If it be a grist-mill, or a carding-machine, customers from the surrounding country may enjoy the use of it by having their grain made into flour, or their wool into rolls, and still it will not be in public use, within the meaning of the law.

But if the inventor allows his machine to be used by other persons generally, either with or without compensation, or if it is, with his consent, put on sale for such use, then it will be in public use and on public sale, within the meaning of the law.

If, now, we apply the same principles to this case, the anaology will be seen at once. Nicholson wished to experiment on his pavement. He believed it to be a good thing, but he was not sure; and the only mode in which he could test it was to place a specimen of it in a public roadway. He did this at his own expense, and with the consent of the owners of the road. Durability was one of the qualities to be attained. He wanted to know whether his pavement would stand, and whether it would resist decay. Its character for durability could not be ascertained without its being subjected to use for a considerable time. He subjected it to such use, in good faith, for the simple purpose of ascertaining whether it was what he claimed it to be. Did he do any thing more than the inventor of the supposed machine might do, in testing his invention? The public had the incidental use of the pavement, it is true; but was the invention in public use, within the meaning of the statute? We think not. The proprietors of the road alone used the invention, and used it at Nicholson's request, by way of experiment. The only way in which they could use it was by allowing the public to pass over the pavement.

Had the city of Boston, or other parties, used the invention, by laying down the pavement in other streets and places, with Nicholson's consent and allowance, then, indeed, the invention itself would have been in public use, within the meaning of the law; but this was not the case. Nicholson did not sell it, nor allow others to use it or sell it. He did not let it go beyond his control. He did nothing that indicated any intent to do so. He kept it under his own eyes, and never for a moment abandoned the intent to obtain a patent for it. . . .

It is sometimes said that an inventor acquires an undue advantage over the public by delaying to take out a patent, inasmuch as he thereby preserves the monopoly to himself for a longer period than is allowed by the policy of the law; but this cannot be said with justice when the delay is occasioned by a *bona fide* effort to bring his invention to perfection, or to ascertain whether it will answer the purpose intended. His monopoly only continues for the allotted period, in any event; and it is the interest of the public, as well as himself, that the invention should be perfect and properly tested, before a patent is granted for it. Any attempt to use it for a profit, and not by way of experiment, for a longer period than two years before the application, would deprive the inventor of his right to a patent.

COMMENTS AND QUESTIONS

1. Compare the facts in the preceding case to those in Egbert v. Lippmann, supra, the "corset case." How did Nicholson's actions in this case differ from those of Barnes in the *Egbert* case? What evidence did Nicholson

have regarding his six-year prefiling period that Barnes did not have for his comparably long period?

2. Could Nicholson simply have filed a patent application at the end of year one and prosecuted it while conducting his continued test? Would a patent have been granted on such an application? Note that because Nicholson was able to delay filing his application for six years, he received a patent that expired later and was therefore arguably more valuable to him than the patent he would have received if he had applied at the end of one year.

e. Priority Rules and the First to Invent

Having dealt with novelty, we turn now to the closely related topic of *priority*.

35 U.S.C. §102 Novelty and Loss of Right

An inventor shall be entitled to a patent unless —

(g) before the applicant's invention thereof the invention was made in this country by another who had not abandoned, suppressed, or concealed it. In determining priority of invention there shall be considered not only the respective dates of conception and reduction to practice of the invention, but also the reasonable diligence of one who was first to conceive and last to reduce to practice, from a time prior to conception by the other.

Although there are a number of fine points, this section states a basic set of rules. These are: (1) As between two claimants to priority, the first inventor wins, so long as (2) she does not abandon, suppress or conceal. In determining who is the first inventor, generally the first to embody the invention in an actual working version (i.e., the first to "reduce to practice") is the winner. The only exception is stated in the second sentence of section 102(g): if the second to reduce to practice is the first to think up (conceive) the invention, she *may* be permitted to "backdate" her date of invention to the time she conceived of the idea for the invention. Whether or not she can take advantage of her earlier date of conception depends on her diligence and the date of the other inventor's conception. Issues such as these are the common fare of patent priority contests between rival inventors, termed patent "interferences." Recent changes in other sections of the patent code will soon bring a more international flavor to interferences, as explained in the Note on the GATT Amendments that follows the next case.

The following case centers on the second sentence of section 102(g), particularly the "reasonable diligence of one who was first to conceive but last to reduce to practice." For variations on the facts that present other issues under section 102(g), see Note 1 after the case. Also, notice that one of the rival inventors in this priority contest, Kanamaru, introduced only evidence of a patent *filing date* but no evidence regarding dates of conception and reduction to practice. In such cases the Patent Office effectively collapses the entire sequence of inventive events into the single date of patent filing; it is as if Kanamaru conceived, reduced to practice, and filed a patent application on the invention all on the same day. For a likely explanation of *why* Kanamaru relied only on this date, see Note 2 after the case.

═══
═══ **Griffith v. Kanamaru**
═══ *United States Court of Appeals for the Federal Circuit*
═══ *816 F.2d 624 (Fed. Cir. 1987)*
═══

NICHOLS, Senior Circuit Judge.

Owen W. Griffith (Griffith) appeals the decision of the Board of Patent Appeals and Interferences (board) that Griffith failed to establish a prima facie case that he is entitled to an award of priority against the filing date of Tsuneo Kanamaru, et al. (Kanamaru) for a patent on aminocarnitine compounds. We affirm.

Background

This patent interference case involves the application of Griffith, an Associate Professor in the Department of Biochemistry at Cornell University Medical College, for a patent on an aminocarnitine compound, useful in the treatment of diabetes, and a patent issued for the same invention to Kanamaru, an employee of Takeda Chemical Industries. The inventors assigned their rights to the inventions to the Cornell Research Foundation, Inc. (Cornell) and to Takeda Chemical Industries respectively.

Griffith had established conception by June 30, 1981, and reduction to practice on January 11, 1984. Kanamaru filed for a United States patent on November 17, 1982. The board found, however, that Griffith failed to establish reasonable diligence for a prima facie case of prior invention. . . .

The board . . . decided that Griffith failed to establish a prima facie case for priority against Kanamaru's filing date. This result was based on the board's conclusion that Griffith's explanation for inactivity between June 15, 1983, and September 13, 1983, failed to provide a legally sufficient excuse to satisfy the "reasonable diligence" requirement of 35 U.S.C. §102(g). Griffith appeals on the issue of reasonable diligence.

Analysis

This is a case of first impression and presents the novel circumstances of a university suggesting that it is reasonable for the public to wait for disclosure until the most satisfactory funding arrangements are made. The applicable law is the "reasonable diligence" standard contained in 35 U.S.C. §102(g) and we must determine the appropriate role of the courts in construing this exception to the ordinary first-in-time rule. As a preliminary matter we note that, although the board focused on the June 1983 to September 1983 lapse in work, and Griffith's reasons for this lapse, Griffith is burdened with establishing a prima facie case of reasonable diligence from immediately before Kanamaru's filing date of November 17, 1982, until Griffith's reduction to practice on January 11, 1984. 35 U.S.C. §102(g).

On appeal, Griffith presents two grounds intended to justify his inactivity on the aminocarnitine project between June 15, 1983, and September 13, 1983. The first is that . . . it is reasonable, and as a policy matter desirable, for Cornell to require Griffith and other research scientists to obtain funding from outside the university. The second reason Griffith presents is that he reasonably waited for Ms. Debora Jenkins to matriculate in the Fall of 1983 to assist with the project. He had promised her she should have that task which she needed to qualify for her degree. We reject these arguments and conclude that Griffith has failed to establish grounds to excuse his inactivity prior to reduction to practice.

The reasonable diligence standard balances the interest in rewarding and encouraging invention with the public's interest in the earliest possible disclosure of innovation. Griffith must account for the entire period from just before Kanamaru's filing date until his [i.e., Griffith's] reduction to practice. . . .

The board in this case was, but not properly, asked to pass judgment on the reasonableness of Cornell's policy regarding outside funding of research. The correct inquiry is rather whether it is reasonable for Cornell to require the public to wait for the innovation, given the well settled policy in favor of early disclosure. . . . A review of caselaw on excuses for inactivity in reduction to practice reveals a common thread that courts may consider the reasonable everyday problems and limitations encountered by an inventor. See, e.g., . . . Reed v. Tornqvist, 436 F.2d 501, 168 U.S.P.Q. 462 (C.C.P.A. 1971) (concluding it is not unreasonable for inventor to delay completing a patent application until after returning from a three week vacation in Sweden, extended by illness of inventor's father). . . . De Wallace v. Scott, 15 App. D.C. 157 (1899) (where applicant made bona fide attempts to perfect his invention, applicant's poor health, responsibility to feed his family, and daily job demands excused his delay in reducing his invention to practice).

. . . We first note that, in regard to waiting for a graduate student, Griffith does not even suggest that he faced a genuine shortage of personnel. He does not suggest that Ms. Jenkins was the only person capable of carrying on with the aminocarnitine experiment. We can see no application of precedent to suggest that the convenience of the timing of the semester schedule justifies a three-month delay for the purpose of reasonable diligence. Neither do we believe that this excuse, absent even a suggestion by Griffith that Jenkins was uniquely qualified to do his research, is reasonable.

Griffith's second contention that it was reasonable for Cornell to require outside funding, therefore causing a delay in order to apply for such funds, is also insufficient to excuse his inactivity. The crux of Griffith's argument is that outside funding is desirable as a form of peer review, or monitoring of the worthiness of a given project. He also suggests that, as a policy matter, universities should not be treated as businesses, which ultimately would detract from scholarly inquiry. Griffith states that these considerations, if accepted as valid, would fit within the scope of the caselaw excusing inactivity for "reasonable" delays in reduction to practice and filing.

Griffith's excuses sound more in the nature of commercial development, not accepted as an excuse for delay, than the "hardship" cases most commonly found and discussed supra. Delays in reduction to practice caused by an inventor's efforts to refine an invention to the most marketable and profitable form have not been accepted as sufficient excuses for inactivity.

. . . [I]t seems evident that Cornell has consciously chosen to assume the risk that priority in the invention might be lost to an outside inventor, yet, having chosen a noncommercial policy, it asks us to save it the property that would have inured to it if it had acted in single-minded pursuit of gain.

Although we agree with the board's conclusion, it is appropriate to go further and consider other circumstances as they apply to the reasonable diligence analysis of 35 U.S.C. §102(g). The record reveals that from the relevant period of November 17, 1982 (Kanamaru's filing date), to September 13, 1983 (when Griffith renewed his efforts towards reduction to practice), Griffith interrupted and often put aside the aminocarnitine project to work on other experiments. Between June 1982 and June 1983 Griffith admits that, at the request of the chairman of his department, he was primarily engaged in an unrelated research project. . . . Griffith also put aside the aminocarnitine experiment to work on a grant proposal on an unrelated project. . . . Griffith made only minimal efforts to secure funding directly for the aminocarnitine project.

The conclusion we reach from the record is that the aminocarnitine project was second and often third priority in laboratory research as well as the solicitation of funds. We agree that Griffith failed to establish a prima facie case of reasonable diligence or a legally sufficient excuse for inactivity to establish priority over Kanamaru.

COMMENTS AND QUESTIONS

1. What would the result in the interference have been if Professor Griffith had reduced the invention to practice on November 16, 1982? (Recall that Kanamaru's filing date — and therefore, in this case, his effective conception and reduction to practice dates — was November 17, 1982.)

If Griffith had reduced to practice on November 16, 1982, would it matter whether he was diligent between his conception date (June 30, 1981) and his reduction to practice? If not, why not?

Assume that Kanamaru introduced the following evidence: a conception date of January 1, 1982, and reduction to practice on November 1, 1982. Assume also, as actually happened, that Griffith's reduction to practice came after Kanamaru's. With Griffith the first to conceive of the invention, is *Kanamaru's* diligence an issue? *Should* it be?

2. Since Takeda Chemical, Kanamaru's assignee, is based primarily in Japan, it is reasonable to conclude that Kanamaru did the research leading to his patent application in Japan. This would explain the exclusive reliance on the patent filing date in this priority contest. Before 1995, and thus at the time this case was decided, foreign inventors could not introduce evidence of foreign inventive activity (e.g., conception and reduction to practice). In 1994, however, as part of the legislative package implementing the Trade Related Aspects of Intellectual Property (TRIPs) portion of the Uruguay Round negotiations under the GATT, Congress changed 35 U.S.C. §104 to permit evidence of inventive activity taking place in any country that is a member of the World Trade Organization (the successor organization to the GATT) beginning in 1996. See Uruguay Round Agreements Act of 1994, P.L. 103-465, 108 Stat. 4809 (1994), at §531, codified at 35 U.S.C. §104. Congress had made a similar change earlier for members of the North American Free Trade Agreement, or NAFTA.

3. Most countries in the world award patents not to the first person to invent the subject matter but to the first person to file a patent application covering the subject. Who would have won if priority were determined not by a first to invent rule but by a first to *file* rule? Would the outcome in that case be fair? In deciding, keep in mind that such a rule would have saved the cost of the entire interference proceeding, including the appeal that yielded the decision reproduced above. Is the extra cost of this additional "due process" worth it?

4. Griffith and Kanamaru were well on their way to the same invention at roughly the same time. Under the Federal Circuit's decision, only Kanamaru gets a patent on the invention. What happens to Griffith? Not only is he not entitled to the exclusionary power of a patent on his idea, but Kanamaru can exclude *him* from using his own, independently developed idea. This result seems harsh, particularly where (as here) Griffith was working on his version of the invention long before the patent ever issued to Kanamaru.

To ameliorate this problem, a number of bills have been proposed in recent years that would grant "prior user rights" to non-patentees, if they could show that they were using the invention before it was patented by someone else. This proposal is similar to the rule in many European nations, which give "shop rights" to continue doing whatever you were doing before the patent issued. Note that these rights are limited, however, because they do not allow the prior user to expand its sales or improve its product if doing so would infringe on the patent.

Are prior user rights a good idea? Do they unfairly weaken the patent grant? If prior user rights are to be enacted, should the patentee be compensated somehow — say, by receiving a compulsory licensing fee from the prior user? Should such rights be transferable? If so, would a patentee worry that its most aggressive competitor could buy a prior user right from someone else?

Note on the GATT Amendments

As part of its obligations under the Uruguay Round of trade negotiations under the General Agreement on Trade and Tariffs (GATT), concluded in 1994, the United States made the following changes in domestic U.S. patent law.[32] They were the result of the 1994 GATT agreement known as "Trade-Related Aspects of Intellectual Property," or TRIPS.

- The term of a patent is now 20 years, measured from the date the patent application was filed. It had been 17 years, measured from the date the patent was issued by the Patent Office. 35 U.S.C. §154. Under certain circumstances, such as interferences and appealed rejections, this term may be extended for up to five years. Id.
- Members of the World Trade Organization (WTO) may introduce evidence of pre–patent filing inventive acts in their home country for purposes of establishing entitlement to a patent under the U.S. "first to invent" system. See 35 U.S.C. §104. Prior to this amendment, only U.S. inventors were permitted to introduce such evidence, i.e., evidence of conception and reduction to practice; foreign inventors applying for a U.S. patent could only rely on inventive activity *in the U.S.* or on their first patent filing date in the United States or elsewhere. This was an obvious disadvantage to foreign applicants, and in keeping with GATT-TRIPS, it was eliminated. (The same treatment had been extended earlier to members of the North American Free Trade Agreement, or NAFTA, viz., Canada and Mexico.)
- The definition of infringement was expanded to include the acts of unauthorized offering for sale and importing. The old definition specified only making, using, and selling. 35 U.S.C. §271.
- A new section of the patent statute permitting "provisional applications," 35 U.S.C. §111, was added. While a provisional application must be fully enabling under §112, it does not have to include any claims. A brief description and drawing suffice under this new section to establish the applicant's priority, provided that a more complete application is filed within one year. Importantly, filing a provisional application does not begin the 20-year clock for the applicant's patent term; only the filing of a full-blown application, with a claim or claims, can do so.

---

32. For most provisions, the effective date of these amendments was January 1, 1996. Some, however, took effect on June 8, 1995. For details see The Uruguay Round Agreement Act of 1994.

Many believe that these changes have opened the way for several more major amendments. These would include: (1) publication of all U.S. patent applications 18 months after filing; (2) "prior user rights" for independent inventors using an invention later claimed in a U.S. patent by another inventor; and (3) perhaps even a first-to-file priority rule. Legislation that failed in Congress in 1996 would have enacted the first two proposed changes; the third appears to have been put off indefinitely. Only time will tell whether GATT-TRIPS was the first wave in a major revolution or an isolated set of changes that exhausted the U.S. urge to harmonize.

4. Nonobviousness

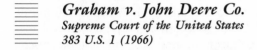

Graham v. John Deere Co.
Supreme Court of the United States
383 U.S. 1 (1966)

Mr. Justice CLARK delivered the opinion of the Court.

After a lapse of 15 years, the Court again focuses its attention on the patentability of inventions under the standard of Art. I, §8, cl. 8, of the Constitution and under the conditions prescribed by the laws of the United States. Since our last expression on patent validity, A. & P. Tea Co. v. Supermarket Corp., 340 U.S. 147 (1950), the Congress has for the first time expressly added a third statutory dimension to the two requirements of novelty and utility that had been the sole statutory test since the Patent Act of 1793. This is the test of obviousness, i.e., whether "the subject matter sought to be patented and the prior art are such that the subject matter as a whole would have been obvious at the time the invention was made to a person having ordinary skill in the art to which said subject matter pertains. Patentability shall not be negatived by the manner in which the invention was made." §103 of the Patent Act of 1952, 35 U.S.C. §103 (1964 ed.).

The questions, involved in each of the companion cases before us, are what effect the 1952 Act had upon traditional statutory and judicial tests of patentability and what definitive tests are now required. We have concluded that the 1952 Act was intended to codify judicial precedents embracing the principle long ago announced by this Court in Hotchkiss v. Greenwood, 11 How. 248 [52 U.S.] (1851), and that, while the clear language of §103 places emphasis on an inquiry into obviousness, the general level of innovation necessary to sustain patentability remains the same.

I

(a). Graham v. John Deere Co., an infringement suit by petitioners, presents a conflict between two Circuits over the validity of a single patent on a "Clamp for vibrating Shank Plows." The invention, a combination of old mechanical elements, involves a device designed to absorb shock from plow shanks as they plow through rocky soil and thus to prevent damage to the plow. We granted certiorari. Although we have determined that neither Circuit applied the correct test, we conclude that the patent is invalid under §103. . . .

II

At the outset it must be remembered that the federal patent power stems from a specific constitutional provision which authorizes the Congress "To promote the Progress of . . . useful Arts, by securing for limited Times to . . . Inventors the exclusive Right to their . . . Discoveries." Art. I, §8, cl. 8. The clause is both a grant of power and a limitation. This qualified authority, unlike the power often exercised in the sixteenth and seventeenth centuries by the English Crown, is limited to the promotion of advances in the "useful arts." It was written against the backdrop of the practices — eventually curtailed by the Statute of Monopolies — of the Crown in granting monopolies to court favorites in goods or businesses which had long before been enjoyed by the public. See Meinhardt, Inventions, Patents and Monopoly, pp. 30-35 (London, 1946). The Congress in the exercise of the patent power may not overreach the restraints imposed by the stated constitutional purpose. Nor may it enlarge the patent monopoly without regard to the innovation, advancement or social benefit gained thereby. Moreover, Congress may not authorize the issuance of patents whose effects are to remove existent knowledge from the public domain, or to restrict free access to materials already available. Innovation, advancement, and things which add to the sum of useful knowledge are inherent requisites in a patent system which by constitutional command must "promote the Progress of . . . useful Arts." This is the standard expressed in the Constitution and it may not be ignored. And it is in this light that patent validity "requires reference to a standard written into the Constitution." A. & P. Tea Co. v. Supermarket Corp., supra, at 154 (concurring opinion).

Within the limits of the constitutional grant, the Congress may, of course, implement the stated purpose of the Framers by selecting the policy which in its judgment best effectuates the constitutional aim. This is but a corollary to the grant to Congress of any Article I power. Gibbons v. Ogden, 9 Wheat. [22 U.S.] 1 [1824]. Within the scope established by the Constitution, Congress may set out conditions and tests for patentability. McClurg v. Kingsland, 1 How. [42 U.S.] 202, 206 [1843]. It is the duty of the Commissioner of Patents and of the courts in the administration of the patent system to give effect to the constitutional standard by appropriate application, in each case, of the statutory scheme of the Congress.

Congress quickly responded to the bidding of the Constitution by enacting the Patent Act of 1790 during the second session of the First Congress. It created an agency in the Department of State headed by the Secretary of

State, the Secretary of the Department of War and the Attorney General, any two of whom could issue a patent for a period not exceeding 14 years to any petitioner that "hath . . . invented or discovered any useful art, manufacture, . . . or device, or any improvement therein not before known or used" if the board found that "the invention or discovery [was] sufficiently useful and important. . . ." 1 Stat. 110. This group, whose members administered the patent system along with their other public duties, was known by its own designation as "Commissioners for the Promotion of Useful Arts."

Thomas Jefferson, who as Secretary of State was a member of the group, was its moving spirit and might well be called the "first administrator of our patent system." See Federico, Operation of the Patent Act of 1790, 18 J. Pat. Off. Soc. 237, 238 (1936). He was not only an administrator of the patent system under the 1790 Act, but was also the author of the 1793 Patent Act. In addition, Jefferson was himself an inventor of great note. His unpatented improvements on plows, to mention but one line of his inventions, won acclaim and recognition on both sides of the Atlantic. Because of his active interest and influence in the early development of the patent system, Jefferson's views on the general nature of the limited patent monopoly under the Constitution, as well as his conclusions as to conditions for patentability under the statutory scheme, are worthy of note.

Jefferson, like other Americans, had an instinctive aversion to monopolies. It was a monopoly on tea that sparked the Revolution and Jefferson certainly did not favor an equivalent form of monopoly under the new government. His abhorrence of monopoly extended initially to patents as well. From France, he wrote to Madison (July 1788) urging a Bill of Rights provision restricting monopoly, and as against the argument that limited monopoly might serve to incite "ingenuity," he argued forcefully that "the benefit even of limited monopolies is too doubtful to be opposed to that of their general suppression," V Writings of Thomas Jefferson, at 47 (Ford ed., 1895).

His views ripened, however, and in another letter to Madison (Aug. 1789) after the drafting of the Bill of Rights, Jefferson stated that he would have been pleased by an express provision in this form: "Art. 9. Monopolies may be allowed to persons for their own productions in literature & their own inventions in the arts, for a term not exceeding ___ years but for no longer term & no other purpose." Id., at 113. And he later wrote: "Certainly an inventor ought to be allowed a right to the benefit of his invention for some certain time. . . . Nobody wishes more than I do that ingenuity should receive a liberal encouragement." Letter to Oliver Evans (May 1807), V Writings of Thomas Jefferson, at 75-76 (Washington ed.).

Jefferson's philosophy on the nature and purpose of the patent monopoly is expressed in a letter to Isaac McPherson (Aug. 1813), a portion of which we set out in the margin.[2] He rejected a natural-rights theory in in-

2. Stable ownership is the gift of social law, and is given late in the progress of society. It would be curious then, if an idea, the fugitive fermentation of an individual brain, could,

tellectual property rights and clearly recognized the social and economic rationale of the patent system. The patent monopoly was not designed to secure to the inventor his natural right in his discoveries. Rather, it was a reward, an inducement, to bring forth new knowledge. The grant of an exclusive right to an invention was the creation of society — at odds with the inherent free nature of disclosed ideas — and was not to be freely given. Only inventions and discoveries which furthered human knowledge, and were new and useful, justified the special inducement of a limited private monopoly. Jefferson did not believe in granting patents for small details, obvious improvements, or frivolous devices. His writings evidence his insistence upon a high level of patentability.

As a member of the patent board for several years, Jefferson saw clearly the difficulty in "drawing a line between the things which are worth to the public the embarrassment of an exclusive patent, and those which are not." The board on which he served sought to draw such a line and formulated several rules which are preserved in Jefferson's correspondence.[3] Despite the board's efforts, Jefferson saw "with what slow progress a system of general rules could be matured." Because of the "abundance" of cases and the fact that the investigations occupied "more time of the members of the board than they could spare from higher duties, the whole was turned over to the judiciary, to be matured into a system, under which every one might know when his actions were safe and lawful." Letter to McPherson, supra,

---

of natural right, be claimed in exclusive and stable property. If nature has made any one thing less susceptible than all others of exclusive property, it is the action of the thinking power called an idea, which an individual may exclusively possess as long as he keeps it to himself; but the moment it is divulged, it forces itself into the possession of every one, and the receiver cannot dispossess himself of it. Its peculiar character, too, is that no one possesses the less, because every other possesses the whole of it. He who receives an idea from me, receives instruction himself without lessening mine; as he who lights his taper at mine, receives light without darkening me. That ideas should freely spread from one to another over the globe, for the moral and mutual instruction of man, and improvement of his condition, seems to have been peculiarly and benevolently designed by nature, when she made them, like fire, expansible over all space, without lessening their density in any point, and like the air in which we breathe, move, and have our physical being, incapable of confinement or exclusive appropriation. Inventions then cannot, in nature, be a subject of property. Society may give an exclusive right to the profits arising from them, as an encouragement to men to pursue ideas which may produce utility, but this may or may not be done, according to the will and convenience of the society, without claim or complaint from any body.
VI Writings of Thomas Jefferson 180-81 (H. A. Washington ed.).

3. "[A] machine of which we are possessed, might be applied by every man to any use of which it is susceptible." Letter to Isaac McPherson, id. at 181.
"[A] change of material should not give title to a patent. As the making a ploughshare of cast rather than of wrought iron; a comb of iron instead of horn or of ivory . . ." Id.
"[A] mere change of form should give no right to a patent, as a high-quartered shoe instead of a low one; a round hat instead of a three-square; or a square bucket instead of a round one." Id. at 181-82.
"[A combined use of old implements.] A man has a right to use a saw, an axe, a plane separately; may he not combine their uses on the same piece of wood?" Letter to Oliver Evans (Jan. 1814), id. at 298.

at 181, 182. Apparently Congress agreed with Jefferson and the board that the courts should develop additional conditions for patentability. Although the Patent Act was amended, revised or codified some 50 times between 1790 and 1950, Congress steered clear of a statutory set of requirements other than the bare novelty and utility tests reformulated in Jefferson's draft of the 1793 Patent Act.

III

The difficulty of formulating conditions for patentability was heightened by the generality of the constitutional grant and the statutes implementing it, together with the underlying policy of the patent system that "the things which are worth to the public the embarrassment of an exclusive patent," as Jefferson put it, must outweigh the restrictive effect of the limited patent monopoly. The inherent problem was to develop some means of weeding out those inventions which would not be disclosed or devised but for the inducement of a patent.

This Court formulated a general condition of patentability in 1851 in Hotchkiss v. Greenwood, 11 How. 248 [52 U.S. (1851)]. The patent involved a mere substitution of materials — porcelain or clay for wood or metal in doorknobs — and the Court condemned it, holding:

> [Unless] more ingenuity and skill . . . were required . . . than were possessed by an ordinary mechanic acquainted with the business, there was an absence of that degree of skill and ingenuity which constitute essential elements of every invention. In other words, the improvement is the work of the skilful mechanic, not that of the inventor. At p. 267.

Hotchkiss, by positing the condition that a patentable invention evidence more ingenuity and skill than that possessed by an ordinary mechanic acquainted with the business, merely distinguished between new and useful innovations that were capable of sustaining a patent and those that were not. The *Hotchkiss* test laid the cornerstone of the judicial evolution suggested by Jefferson and left to the courts by Congress. The language in the case, and in those which followed, gave birth to "invention" as a word of legal art signifying patentable inventions. Yet, as this Court has observed, "[t]he truth is the word [invention] cannot be defined in such manner as to afford any substantial aid in determining whether a particular device involves an exercise of the inventive faculty or not." McClain v. Ortmayer, 141 U.S. 419, 427 (1891); A. & P. Tea Co. v. Supermarket Corp., supra, at 151. Its use as a label brought about a large variety of opinions as to its meaning both in the Patent Office, in the courts, and at the bar. The *Hotchkiss* formulation, however, lies not in any label, but in its functional approach to questions of patentability. In practice, *Hotchkiss* has required a comparison between the subject matter of the patent, or patent application, and the background skill of the calling. It has been from this comparison that patentability was in each case determined.

IV.   The 1952 Patent Act

The Act sets out the conditions of patentability in three sections. An analysis of the structure of these three sections indicates that patentability is dependent upon three explicit conditions: novelty and utility as articulated and defined in §101 and §102, and non-obviousness, the new statutory formulation, as set out in §103. The first two sections, which trace closely the 1874 codification, express the "new and useful" tests which have always existed in the statutory scheme and, for our purposes here, need no clarification. The pivotal section around which the present controversy centers is §103. It provides:

> §103. *Conditions for patentability; non-obvious subject matter*
>
> A patent may not be obtained though the invention is not identically disclosed or described as set forth in section 102 of this title, if the differences between the subject matter sought to be patented and the prior art are such that the subject matter as a whole would have been obvious at the time the invention was made to a person having ordinary skill in the art to which said subject matter pertains. Patentability shall not be negatived by the manner in which the invention was made.

The section is cast in relatively unambiguous terms. Patentability is to depend, in addition to novelty and utility, upon the "non-obvious" nature of the "subject matter sought to be patented" to a person having ordinary skill in the pertinent art. . . .

It is undisputed that this section was, for the first time, a statutory expression of an additional requirement for patentability, originally expressed in *Hotchkiss*. It also seems apparent that Congress intended by the last sentence of §103 to abolish the test it believed this Court announced in the controversial phrase "flash of creative genius," used in Cuno Corp. v. Automatic Devices Corp., 314 U.S. 84 (1941).[7] . . .

---

7. The sentence in which the phrase occurs reads: "[T]he new device, however useful it may be, must reveal the flash of creative genius, not merely the skill of the calling." At p. 91. Although some writers and lower courts found in the language connotations as to the frame of mind of the inventors, none were so intended. The opinion approved *Hotchkiss* specifically, and the reference to "flash of creative genius" was but a rhetorical embellishment of language going back to 1833. Cf. "exercise of genius," Shaw v. Cooper, 7 Pet. 292; "inventive genius," Reckendorfer v. Faber, 92 U.S. 347 (1876); Concrete Appliances Co. v. Gomery, 269 U.S. 177; "flash of thought," Densmore v. Scofield, 102 U.S. 375 (1880); "intuitive genius," Potts v. Creager, 155 U.S. 597 (1895). Rather than establishing a more exacting standard, *Cuno* merely rhetorically restated the requirement that the subject matter sought to be patented must be beyond the skill of the calling. It was the device, not the invention, that had to reveal the "flash of creative genius." See Boyajian, The Flash of Creative Genius, An Alternative Interpretation,

V . . .

While the ultimate question of patent validity is one of law, A. & P. Tea Co. v. Supermarket Corp., supra, at 155, the §103 condition, which is but one of three conditions, each of which must be satisfied, lends itself to several basic factual inquiries. Under §103, the scope and content of the prior art are to be determined; differences between the prior art and the claims at issue are to be ascertained; and the level of ordinary skill in the pertinent art resolved. Against this background, the obviousness or nonobviousness of the subject matter is determined. Such secondary considerations as commercial success, long felt but unsolved needs, failure of others, etc., might be utilized to give light to the circumstances surrounding the origin of the subject matter sought to be patented. As indicia of obviousness or nonobviousness, these inquiries may have relevancy. See Note, Subtests of "Nonobviousness": A Nontechnical Approach to Patent Validity, 112 U. Pa. L. Rev. 1169 (1964).

This is not to say, however, that there will not be difficulties in applying the nonobviousness test. What is obvious is not a question upon which there is likely to be uniformity of thought in every given factual context. The difficulties, however, are comparable to those encountered daily by the courts in such frames of reference as negligence and scienter, and should be amenable to a case-by-case development. We believe that strict observance of the requirements laid down here will result in that uniformity and definiteness which Congress called for in the 1952 Act.

While we have focused attention on the appropriate standard to be applied by the courts, it must be remembered that the primary responsibility for sifting out unpatentable material lies in the Patent Office. To await litigation is — for all practical purposes — to debilitate the patent system. We have observed a notorious difference between the standards applied by the Patent Office and by the courts. While many reasons can be adduced to explain the discrepancy, one may well be the free rein often exercised by Examiners in their use of the concept of "invention." In this connection we note that the Patent Office is confronted with a most difficult task. Almost 100,000 applications for patents are filed each year. Of these, about 50,000 are granted and the backlog now runs well over 200,000. 1965 Annual Report of the Commissioner of Patents 13-14. This is itself a compelling reason for the Commissioner to strictly adhere to the 1952 Act as interpreted here. This would, we believe, not only expedite disposition but bring about a closer concurrence between administrative and judicial precedent.

We have been urged to find in §103 a relaxed standard, supposedly a congressional reaction to the "increased standard" applied by this Court in

---

25 J. Pat. Off. Society 776, 780, 781 (1943); Pacific Contact Laboratories, Inc. v. Solex Laboratories, Inc., 209 F.2d 529, 533; Brown & Sharpe Mfg. Co. v. Kar Engineering Co., 154 F.2d 48, 51-52; In re Shortell, 31 C.C.P.A. (Pat.) 1062, 1069, 142 F.2d 292, 295-96.

its decisions over the last 20 or 30 years. The standard has remained invariable in this Court. Technology, however, has advanced — and with remarkable rapidity in the last 50 years. Moreover, the ambit of applicable art in given fields of science has widened by disciplines unheard of a half century ago. It is but an evenhanded application to require that those persons granted the benefit of a patent monopoly be charged with an awareness of these changed conditions. The same is true of the less technical, but still useful arts. He who seeks to build a better mousetrap today has a long path to tread before reaching the Patent Office.

VI

We now turn to the application of the conditions found necessary for patentability to the cases involved here:

A. The Patent in Issue in Graham v. John Deere Co.

This patent, No. 2,627,798 (hereinafter called the '798 patent) relates to a spring clamp which permits plow shanks to be pushed upward when they hit obstructions in the soil, and then springs the shanks back into normal position when the obstruction is passed over. The device . . . is fixed to the plow frame as a unit. The mechanism around which the controversy centers is basically a hinge. The top half of it, known as the upper plate, is a heavy metal piece clamped to the plow frame and is stationary relative to the plow frame. The lower half of the hinge, known as the hinge plate, is connected to the rear of the upper plate by a hinge pin and rotates downward with respect to it. The shank, which is bolted to the forward end of the hinge plate, runs beneath the plate and parallel to it for about nine inches, passes through a stirrup, and then continues backward for several feet curving down toward the ground. The chisel, which does the actual plowing, is attached to the rear end of the shank. As the plow frame is pulled forward, the chisel rips through the soil, thereby plowing it. In the normal position, the hinge plate and the shank are kept tight against the upper plate by a spring, which is atop the upper plate. A rod runs through the center of the spring, extending down through holes in both plates and the shank. Its upper end is bolted to the top of the spring while its lower end is hooked against the underside of the shank [see Figure 3-1].

When the chisel hits a rock or other obstruction in the soil, the obstruction forces the chisel and the rear portion of the shank to move upward. The shank is pivoted against the rear of the hinge plate and pries open the hinge against the closing tendency of the spring. This closing tendency is caused by the fact that, as the hinge is opened, the connecting rod is pulled downward and the spring is compressed. When the obstruction is passed over, the upward force on the chisel disappears and the spring pulls the shank and hinge plate back into their original position. The lower, rear portion of the hinge plate is constructed in the form of a stirrup which brackets the shank, passing around and beneath it. The shank fits loosely into the stirrup (permitting a slight up and down play). The stirrup is designed to prevent the shank from recoiling away from the hinge plate, and thus prevents excessive strain on the shank near its bolted connection. The stirrup also girds the shank, preventing it from fishtailing from side to side.

In practical use, a number of spring-hinge-shank combinations are clamped to a plow frame, forming a set of ground-working chisels capable of withstanding the shock of rocks and other obstructions in the soil without breaking the shanks.

Background of the Patent

Chisel plows, as they are called, were developed for plowing in areas where the ground is relatively free from rocks or stones. Originally, the shanks were rigidly attached to the plow frames. When such plows were used in the rocky, glacial soils of some of the Northern States, they were found to have serious defects. As the chisels hit buried rocks, a vibratory motion was set up and tremendous forces were transmitted to the shank near its connection to the frame. The shanks would break. Graham, one of the petitioners, sought to meet that problem, and in 1950 obtained a patent, U.S. No. 2,493,811 (hereinafter '811), on a spring clamp which solved some of the difficulties. Graham and his companies manufactured and sold the '811 clamps. In 1950, Graham modified the '811 structure and filed for a patent. That patent, the one in issue, was granted in 1953. This suit against competing plow manufacturers resulted from charges by petitioners that several of respondents' devices infringed the '798 patent.

The Prior Art

Five prior patents indicating the state of the art were cited by the Patent Office in the prosecution of the '798 application. Four of these patents, 10 other United States patents and two prior-use spring-clamp arrangements not of record in the '798 file wrapper were relied upon by respondents as revealing the prior art. The District Court and the Court of Appeals found that the prior art "as a whole in one form or another contains all of the mechanical elements of the '798 Patent." One of the prior-use clamp devices not before the Patent Examiner — Glencoe — was found to have "all of the elements."

We confine our discussion to the prior patent of Graham, '811, and to the Glencoe clamp device, both among the references asserted by respondents. The Graham '811 and '798 patent devices are similar in all elements, save two: (1) the stirrup and the bolted connection of the shank to the hinge

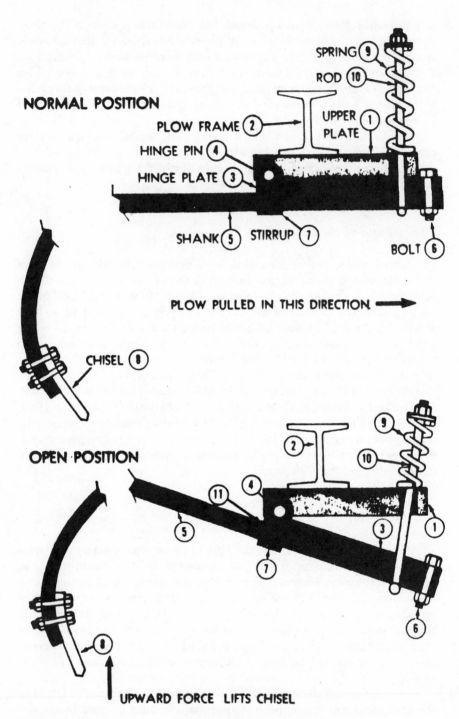

NORMAL POSITION

PLOW FRAME ②

HINGE PIN ④

HINGE PLATE ③

SPRING ⑨

ROD ⑩

UPPER PLATE ①

SHANK ⑤   STIRRUP ⑦

BOLT ⑥

PLOW PULLED IN THIS DIRECTION. ➡

CHISEL ⑧

OPEN POSITION

UPWARD FORCE LIFTS CHISEL

FIGURE 3-1
Patent drawing for the Graham '798 patent.

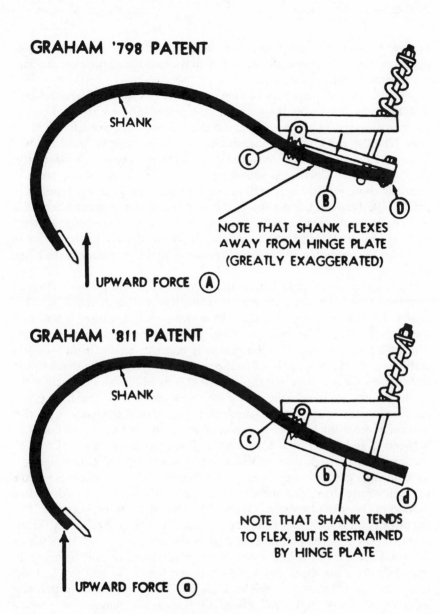

GRAHAM '798 PATENT

SHANK

C

B

D

NOTE THAT SHANK FLEXES
AWAY FROM HINGE PLATE
(GREATLY EXAGGERATED)

UPWARD FORCE (A)

GRAHAM '811 PATENT

SHANK

c

b

d

NOTE THAT SHANK TENDS
TO FLEX, BUT IS RESTRAINED
BY HINGE PLATE

UPWARD FORCE (a)

FIGURE 3-1 cont.
Patent drawing for the Graham '798 patent.

plate do not appear in '811; and (2) the position of the shank is reversed, being placed in patent '811 above the hinge plate, sandwiched between it and the upper plate. The shank is held in place by the spring rod which is hooked against the bottom of the hinge plate passing through a slot in the shank. Other differences are of no consequence to our examination. In practice the '811 patent arrangement permitted the shank to wobble or fishtail because it was not rigidly fixed to the hinge plate; moreover, as the hinge plate was below the shank, the latter caused wear on the upper plate, a member difficult to repair or replace.

Graham's '798 patent application contained 12 claims. All were rejected as not distinguished from the Graham '811 patent. The inverted position of the shank was specifically rejected as was the bolting of the shank to the hinge plate. The Patent Office examiner found these to be "matters of design well within the expected skill of the art and devoid of invention." Graham withdrew the original claims and substituted the two new ones which are substantially those in issue here. His contention was that wear was reduced in patent '798 between the shank and the heel or rear of the upper plate.[11] He also emphasized several new features, the relevant one here being that the bolt used to connect the hinge plate and shank maintained the upper face of the shank in continuing and constant contact with the underface of the hinge plate.

Graham did not urge before the Patent Office the greater "flexing" qualities of the '798 patent arrangement which he so heavily relied on in the courts. The sole element in patent '798 which petitioners argue before us is the interchanging of the shank and hinge plate and the consequences flowing from this arrangement. The contention is that this arrangement — which petitioners claim is not disclosed in the prior art — permits the shank to flex under stress for its *entire* length. [W]hen the chisel hits an obstruction the resultant force (A) pushes the rear of the shank upward and the shank pivots against the rear of the hinge plate at (C). The natural tendency is for that portion of the shank between the pivot point and the bolted connection (i.e., between C and D) to bow downward and away from the hinge plate. The maximum distance (B) that the shank moves away from the plate is slight — for emphasis, greatly exaggerated in the sketches. This is so because of the strength of the shank and the short — nine inches or so — length of that portion of the shank between (C) and (D). On the contrary, in patent '811 the pivot point is the upper plate at point (c); and while the tendency for the shank to bow between points (c) and (d) is the same as in '798, the shank is restricted because of the underlying hinge plate and cannot flex as freely. In practical effect, the shank flexes only between points (a) and (c), and not along the entire length of the shank, as in '798. Petitioners say that this difference in flex, though small, effectively absorbs the tremendous forces of the shock of obstructions whereas prior art arrangements failed.

11. In '811, where the shank was above the hinge plate, an upward movement of the chisel forced the shank up against the underside of the rear of the upper plate. The upper plate thus provided the fulcrum about which the hinge was pried open. Because of this, as well as the location of the hinge pin, the shank rubbed against the heel of the upper plate causing wear both to the plate and to the shank. By relocating the hinge pin and by placing the hinge plate between the shank and the upper plate, as in '798, the rubbing was eliminated and the wear point was changed to the hinge plate, a member more easily removed or replaced for repair.

The Obviousness of the Differences

We cannot agree with petitioners. We assume that the prior art does not disclose such an arrangement as petitioners claim in patent '798. Still we do not believe that the argument on which petitioners' contention is bottomed supports the validity of the patent. The tendency of the shank to flex is the same in all cases. If free-flexing, as petitioners now argue, is the crucial difference above the prior art, then it appears evident that the desired result would be obtainable by not boxing the shank within the confines of the hinge. The only other effective place available in the arrangement was to attach it below the hinge plate and run it through a stirrup or bracket that would not disturb its flexing qualities. Certainly a person having ordinary skill in the prior art, given the fact that the flex in the shank could be utilized more effectively if allowed to run the entire length of the shank, would immediately see that the thing to do was what Graham did, i.e., invert the shank and the hinge plate.

Petitioners' argument basing validity on the free-flex theory raised for the first time on appeal is reminiscent of Lincoln Engineering Co. v. Stewart-Warner Corp., 303 U.S. 545 (1938), where the Court called such an effort "an afterthought. No such function . . . is hinted at in the specifications of the patent. If this were so vital an element in the functioning of the apparatus it is strange that all mention of it was omitted." At p. 550. No "flexing" argument was raised in the Patent Office. Indeed, the trial judge specifically found that "flexing is not a claim of the patent in suit . . ." and would not permit interrogation as to flexing in the accused devices. Moreover, the clear testimony of petitioners' experts shows that the flexing advantages flowing from the '798 arrangement are not, in fact, a significant feature in the patent.

We find no nonobvious facets in the '798 arrangement. The wear and repair claims were sufficient to overcome the patent examiner's original conclusions as to the validity of the patent. However, some of the prior art, notably Glencoe, was not before him. There the hinge plate is below the shank but, as the courts below found, all of the elements in the '798 patent are present in the Glencoe structure. Furthermore, even though the position of the shank and hinge plate appears reversed in Glencoe, the mechanical operation is identical. The shank there pivots about the underside of the stirrup, which in Glencoe is *above* the shank. In other words, the stirrup in Glencoe serves exactly the same function as the heel of the hinge plate in '798. The mere shifting of the wear point to the heel of the '798 hinge plate from the stirrup of Glencoe — itself a part of the hinge plate — presents no operative mechanical distinctions, much less nonobvious differences. . . .

The judgment of the Court of Appeals in [Graham v. John Deere] is affirmed. . . .

COMMENTS AND QUESTIONS

1. The Court states that it is proper in interpreting the Graham '798 claims to refer to the prosecution history (or "file wrapper") of the patent. We will see that this is an important tool in the law of infringement later in this chapter. For now, what is important is that the validity of the patent claims is only one part of a larger picture — "claim interpretation." Patentees not only want to have their patents upheld over the prior art, but they also want those patents to be interpreted broadly, to cover a wide range of potential infringements. Because these goals are in tension, it is important to keep them both in mind whenever you interpret the claims of a patent.

2. Without question, the plow design in *Graham* was new, i.e., novel under section 102 of the patent code. What policy is served by the detailed inquiry into whether it is "new enough" to deserve a patent, i.e., nonobvious? What would be the effect of a patent system that only required novelty? For one view of the matter, see Robert P. Merges, Uncertainty and the Standard of Patentability, 7 High Tech. L.J. 1 (1993). Arguably, section 103 requires an "inventive leap" of some degree over what has been done before as a counterbalance to the strong rights given patentholders.

3. A 1996 amendment to section 103 gives preferential treatment to inventors of a "biotechnological process," a term defined in detail in the act. New section 103(b) provides that where an inventor has developed a new and nonobvious composition of matter produced using a biotechnological process, the process as well as the composition of matter shall be considered nonobvious (and therefore patentable). The new amendment is designed to reverse the Federal Circuit decision in In re Durden, 763 F.2d 1406 (Fed. Cir. 1985), at least in the biotechnology industry. In *Durden*, the court held that a process for making a patentable product was not separately patentable if the steps in the process were themselves well known. *Durden* had previously been limited by the Federal Circuit in In re Pleuddemann, 910 F.2d 823 (Fed. Cir. 1990), where the court drew a distinction between methods of making a new product (which were subject to *Durden*) and methods of *using* a new product (which were not).

What was wrong with the rule in *Durden*? If the process itself was familiar, why should the inventor of a new product be entitled to a process patent *and* a product patent? And even if you think *Durden* was wrong, why change the law only in the biotechnology industry? Note that *Durden* was in any event overruled by the Federal Circuit shortly after the passage of the new statute. In re Ochiai, 71 F.3d 1565 (Fed. Cir. 1996).

The test established in 35 U.S.C. §103 — whether the invention as a whole would be "obvious to one of ordinary skill in the art" — does not itself tell courts very much about how to decide what is obvious. Courts have developed a number of rules to assist in this determination. Two of the most important rules are discussed below.

a. Combining References

To anticipate a patent application under section 102, a single prior art reference must disclose every element of what the patentee claims as his invention. If a prior art reference does not disclose all the parts of an invention, it does not "anticipate" the application. Under section 103, however, a single reference need not disclose the entire invention to bar a patent. Thus section 103 asks whether a researcher who is aware of all the prior art would think to create the claimed invention. In deciding the question of obviousness, it is sometimes permissible to analyze a combination of ideas from different sources of prior art (known as prior art "references").

In re Vaeck
United States Court of Appeals for the Federal Circuit
947 F.2d 488, 20 U.S.P.Q.2d 1438 (Fed. Cir. 1991)

RICH, J.

This appeal is from the September 12, 1990 decision of the Patent and Trademark Office (PTO) Board of Patent Appeals and Interferences (Board), affirming the examiner's rejection of [almost all] claims . . . of [appellant's] application . . . , filed March 4, 1987, . . . as unpatentable under 35 U.S.C. 103. . . . We reverse the §103 rejection.

Background

A. The Invention

The claimed invention is directed to the use of genetic engineering techniques for production of proteins that are toxic to insects such as larvae of mosquitos and black flies. These swamp-dwelling pests are the source of numerous human health problems, including malaria. It is known that certain species of the naturally-occurring Bacillus genus of bacteria produce proteins ("endotoxins") that are toxic to these insects. Prior art methods of combatting the insects involved spreading or spraying crystalline spores of the insecticidal Bacillus proteins over swamps. The spores were environmentally unstable, however, and would often sink to the bottom of a swamp before being consumed, thus rendering this method prohibitively expensive. Hence the need for a lower-cost method of producing the insecticidal Bacillus proteins in high volume, with application in a more stable vehicle.

As described by appellants, the claimed subject matter meets this need by providing for the production of the insecticidal Bacillus proteins within host cyanobacteria. Although both cyanobacteria and bacteria are members of the procaryote kingdom [i.e., they both are organisms lacking a distinct cellular nucleus, as opposed to eukaryotes], the cyanobacteria (which in the past have been referred to as "blue-green algae") are unique among procaryotes in that the cyanobacteria are capable of oxygenic photosynthesis. The cyanobacteria grow on top of swamps where they are consumed by mosquitos and black flies. Thus, when Bacillus proteins are produced within transformed cyanobacterial hosts [i.e., when the Bacillus genes for those proteins have been successfully taken up by the foreign cyanobacteria and the genetic material has been made a permanent part of that host organism, to be replicated when it reproduces] according to the claimed invention, the presence of the insecticide in the food of the targeted insects advantageously guarantees direct uptake by the insects.

More particularly, the subject matter of the application on appeal includes a chimeric (i.e., hybrid) gene comprising (1) a gene derived from a bacterium of the Bacillus genus whose product is an insecticidal protein, united with (2) a DNA promoter effective for expressing the Bacillus gene in a host cyanobacterium, so as to produce the desired insecticidal protein. . . .

D. The Grounds of Rejection

1. The §103 Rejections

Claims 1-6, 16-21, 33-38, 47-48 and 52 (which include all independent claims in the application) were rejected as unpatentable under 35 U.S.C. 103 based upon [the] Dzelzkalns [reference] in view of [the] Sekar I or Sekar II and Ganesan [references]. The examiner stated that Dzelzkalns discloses a chimeric gene capable of being highly expressed in a cyanobacterium, said gene comprising a promoter region effective for expression in a cyanobacterium operably linked to a structural gene encoding [a protein called chloramphenicol acetyl transferase, abbreviated "CAT"]. The examiner acknowledged [the differences between the proteins, but] pointed out [that] Sekar I, Sekar II, and Ganesan teach genes encoding insecticidally active proteins produced by Bacillus, and the advantages of expressing such genes in . . . hosts [from another species] to obtain larger quantities of the protein. The examiner contended that it would have been obvious to one of ordinary skill in the art to substitute the Bacillus genes taught by Sekar I, Sekar II, and Ganesan for the CAT gene in the . . . Dzelzkalns [reference] in order to obtain high level expression of the Bacillus genes in the transformed cyanobacteria. The examiner further contended that it would have been obvious to use cyanobacteria as [a host] for expression of the claimed genes. . . .

Opinion

A. Obviousness

We first address whether the PTO erred in rejecting the claims on appeal as prima facie obvious within the meaning of 35 U.S.C. §103. Obviousness is a legal question which this court independently reviews, though based upon underlying factual findings which we review under the clearly erroneous standard.

Where claimed subject matter has been rejected as obvious in view of a combination of prior art references, a proper analysis under §103 requires, inter alia, consideration of two factors: (1) whether the prior art would have suggested to those of ordinary skill in the art that they should make the claimed composition or device, or carry out the claimed process; and (2) whether the prior art would also have revealed that in so making or carrying out, those of ordinary skill would have a reasonable expectation of success. See In re Dow Chemical Co., 837 F.2d 469, 473, 5 U.S.P.Q.2d 1529, 1531 (Fed. Cir. 1988). Both the suggestion and the reasonable expectation of success must be founded in the prior art, not in the applicant's disclosure. Id.

We agree with appellants that the PTO has not established the prima facie obviousness of the claimed subject matter. The prior art simply does not disclose or suggest the expression in cyanobacteria of a chimeric gene encoding an insecticidally active protein, or convey to those of ordinary skill a reasonable expectation of success in doing so. More particularly, there is no suggestion in Dzelzkalns, the primary reference cited against all claims, of substituting in the disclosed plasmid a structural gene encoding Bacillus insecticidal proteins for the CAT gene utilized for selection purposes. The expression of antibiotic resistance-conferring genes in cyanobacteria, without more, does not render obvious the expression of unrelated genes in cyanobacteria for unrelated purposes.

The PTO argues that the substitution of insecticidal Bacillus genes for CAT marker genes in cyanobacteria is suggested by the secondary references Sekar I, Sekar II, and Ganesan, which collectively disclose expression of genes encoding Bacillus insecticidal proteins in two species of host Bacillus bacteria (B. megaterium and B. subtilis) as well as in the bacterium E. coli. While these references disclose expression of Bacillus genes encoding insecticidal proteins in certain transformed bacterial hosts, nowhere do these references disclose or suggest expression of such genes in transformed cyanobacterial hosts.

To remedy this deficiency, the PTO emphasizes similarity between bacteria and cyanobacteria, namely, that these are both procaryotic organisms, and argues that this fact would suggest to those of ordinary skill the use of cyanobacteria as hosts for expression of the claimed chimeric genes. While it is true that bacteria and cyanobacteria are now both classified as procaryotes, that fact alone is not sufficient to motivate the art worker as the PTO contends. As the PTO concedes, cyanobacteria and bacteria are not identical; they are classified as two separate divisions of the kingdom Procaryotae. Moreover, it is only in recent years that the biology of cyanobacteria has been clarified, as evidenced by references in the prior art to "blue-green algae." Such evidence of recent uncertainty regarding the biology of cyanobacteria tends to rebut, rather than support, the PTO's position that one would consider the cyanobacteria effectively interchangeable with bacteria as hosts for expression of the claimed gene. . . .

The PTO asks us to agree that the prior art would lead those of ordinary skill to conclude that cyanobacteria are attractive hosts for expression of any and all heterologous genes. Again, we can not. The relevant prior art does indicate that cyanobacteria are attractive hosts for expression of both native and heterologous [i.e., foreign] genes involved in photosynthesis (not surprisingly, for the capability of undergoing oxygenic photosynthesis is what makes the cyanobacteria unique among procaryotes). However, these references do not suggest that cyanobacteria would be equally attractive hosts for expression of unrelated heterologous genes, such as the claimed genes encoding Bacillus insecticidal proteins.

In [In re] O'Farrell [853 F.2d 894, 7 U.S.P.Q.2d (BNA) 1673 (Fed. Cir. 1988)], this court affirmed an obviousness rejection of a claim to a method for producing a "predetermined protein in a stable form" in a transformed bacterial host. 853 F.2d at 895, 7 U.S.P.Q.2d at 1674. . . . The main difference between the prior art and the claim at issue was that in [the] Polisky [reference], the heterologous gene was a gene for ribosomal RNA, while the claimed invention substituted a gene coding for a predetermined protein. Id. at 901, 7 U.S.P.Q.2d at 1679. Although, as the appellants therein pointed out, the ribosomal RNA gene is not normally translated into protein, Polisky mentioned preliminary evidence that the transcript of the ribosomal RNA gene was translated into protein, and further predicted that if a gene coding for a protein were to be substituted, extensive translation might result. Id. We thus affirmed, explaining that

> the prior art explicitly suggested the substitution that is the difference between the claimed invention and the prior art, and presented preliminary evidence suggesting that the [claimed] method could be used to make proteins. . . .
> . . . Polisky contained detailed enabling methodology for practicing the claimed invention, a suggestion to modify the prior art to practice the claimed invention, and evidence suggesting that it would be successful.

Id. at 901-02, 7 U.S.P.Q.2d at 1679-80.

In contrast with the situation in O'Farrell, the prior art in this case offers no suggestion, explicit or implicit, of the substitution that is the difference between the claimed invention and the prior art. Moreover, the "reasonable expectation of success" that was present in O'Farrell is not present here. Accordingly, we reverse the §103 rejections.

COMMENTS AND QUESTIONS

1. As the *Vaeck* case makes clear, it is not always permissible to combine two or more references. Only if one of ordinary skill in the art would have some motivation to do so can the court consider the references together. Most often, combining references requires some *suggestion* in the art that would lead one to combine the two references. This was the case in In re O'Farrell, cited in *Vaeck*.

Even though this "suggestion test" has been applied consistently by the courts, it leaves some interesting questions open. For instance, several cases hold that even an "implicit" suggestion to combine references, or to modify the teachings of a single reference, is enough to make the resulting invention obvious. See In re Gorman, 933 F.2d 982 (Fed. Cir. 1991) ("the extent to which such suggestion must be explicit in, or may be fairly inferred from, the references, is decided on the facts of each case"); Gillette Co. v. S. C. Johnson & Son, 919 F.2d 720 (Fed. Cir. 1990) (explicit statement of suggestion not required) Pro-Mold and Tool, Inc. v. Great Lakes Plastics, Inc., 75 F.3d 1568, 1573 (Fed. Cir. 1996) (motivation to combine references may come from knowledge of skilled artisan or nature of problem); Robotic Vision Systems, Inc. v. View Engineering, Inc., 189 F.3d 1370 (Fed. Cir. 1999) (same; dictum). These cases raise the difficult issue of what is enough to constitute an "implicit suggestion," motivation, or incentive — and how such a standard can be kept from undermining the entire "suggestion or motivation" concept. A suggestion that is merely implicit, in other words, seems to carry one away from the attractive idea that references must refer to possible combination or extension *on their face* to make out a case of obviousness. On the other hand, doing away with the implicit suggestion idea would seem to doom accused infringers to lengthy searches for prior art that says what anyone in the field knows — e.g., that a new type of automobile engine would be enhanced by greater gas mileage. In such cases, it seems quite reasonable for courts to maintain that a reference implicitly calls for further optimization along lines that are well-recognized in the art — i.e., that are obvious even *without* an explicit suggestion that they be pursued. On this point, see In re Jones, 958 F.2d 347 (Fed. Cir. 1992) (suggestion to combine references must be "found either in the references themselves *or in the knowledge generally available to one of ordinary skill in the art.*") (emphasis added).

2. An important and controversial issue is the relationship between uncertainty and nonobviousness. If a particular experiment is obvious to try, for example, but there is no guarantee it will succeed, is the resulting invention patentable? Courts have generally said that the invention is obvious if the prior art suggests a motivation to try a particular combination, and that combination once tried is likely to be successful. On the other hand, if reaching the invention would have required "undue experimentation" by one skilled in the art, obtaining those results can still be patentable.

An important case on this issue is In re Bell, 991 F.2d 781 (Fed. Cir. 1993). *Bell* involved a patent application claiming human gene sequences that code for insulin-like growth factors I and II (IGF-I and II). The examiner rejected the claims on grounds of obviousness, citing two publications by Rinderknecht disclosing amino acid sequences for IGF-I and IGF-II, and a patent to Weissman on a "Method for Cloning Genes." After the Board upheld the examiner, the applicants appealed. The Federal Circuit reversed.

The court held that although Rinderknecht provided the structure of the protein, there were an extraordinary number of possible nucleotide sequences that might code for it. Given the large number of possibilities suggested by the prior art, and the failure of the cited prior art to suggest which of those possibilities is the correct human sequence, the claimed sequences would not have been obvious. Further, the court stated, combining Rinderknecht with Weissman does not make the claimed sequences obvious, since Weissmann does not expressly teach nor fairly suggest that its general method for isolating genes should be combined with the disclosed protein of the Rinderknecht references. The Board clearly erred, the court continued, when it held that Weissman teaches toward, rather than away from, the claimed sequences. Therefore, the requisite teaching or suggestion to combine the teachings of the cited prior art references is absent. In the words of the court:

> It may be true that, knowing the structure of the protein, one can use the genetic code to hypothesize possible structures for the corresponding gene and that one thus has the potential for obtaining that gene.
>
> However, because of the degeneracy of the genetic code, there are a vast number of nucleotide sequences that might code for a specific protein. In the case of IGF, Bell has argued without contradiction that the Rinderknecht amino acid sequences could be coded for by more than 10^{36} different nucleotide sequences, only a few of which are the human sequences that Bell now claims. Therefore, given the nearly infinite number of possibilities suggested by the prior art, and the failure of the cited prior art to suggest which of those possibilities is the human sequence, the claimed sequences would not have been obvious.
>
> Bell does not claim all of the 10^{36} nucleic acids that might potentially code for IGF. Neither does Bell claim all nucleic acids coding for a protein having the biological activity of IGF. Rather, Bell claims only the human nucleic acid sequences coding for IGF. Absent anything in the cited prior art suggesting which of the 10^{36} possible sequences suggested by Rinderknecht corresponds to the IGF gene, the PTO has not met its burden of establishing that the prior art would have suggested the claimed sequences.

Accord In re Deuel, 51 F.3d 1552, 1559 (Fed. Cir. 1995). Does the result in *Bell* and *Deuel* make sense? Does it matter how difficult it would be to test all of the 10^{36} possible nucleic acids to see if they code for IGF? Does it matter whether there is some reason to prefer the actual code used in the human body, as opposed to codons producing equivalent proteins?

For an argument that rewarding discovery in the face of uncertainty is appropriate, and that the cost of overcoming the uncertainty is as important as the degree of uncertainty, see Robert P. Merges, Uncertainty and the Standard of Patentability, 7 High Tech. L.J. 1 (1993).

PROBLEM

Problem 3-9. You have drafted a patent application for a client claiming a lollipop in the shape of a human thumb. An examiner rejected the claims as obvious over a combination of numerous prior art references. You must now prepare an argument trying to overturn the examiner's decision by appeal to the Board of Patent Appeals and Interferences. The claimed invention consists of a lollipop filled with a plug of gum, chocolate or food-grade wax. A thumb-shaped elastomeric mold served as the product's wrapper, and after it was peeled from the candy, the user could wear the mold on his or her own thumb. The examiner relied on the following prior art references in rejecting claims to the thumb-shaped lollipop invention:

a) Siciliano shows ice cream in a mold with a stick inserted. The removable mold also serves as the product's wrapper.

b) Copeman shows candy lollipops in elastomeric molds taking "varying shapes, such as fruit or animals." The molds may be used as toy balloons after being removed.

c) Harris shows a hollow, thumb-shaped lollipop into which the user's thumb is inserted.

d) Webster shows a chewing gum entirely enclosing a liquid syrup product. This patent also suggests the greater appeal to consumers of providing two different components in the same confection.

Although some of the references cite at least one other reference, no reference explicitly suggests combining its teaching with that of any other reference. What is your basis for arguing that the invention is patentable?

---

b. "Secondary" Considerations

In *Graham*, the Supreme Court stated that the "secondary factors" of commercial success, long felt need, and so on "may have relevancy." 383 U.S. at 18. The Federal Circuit, by contrast, routinely speaks of these factors — under the rubric "objective evidence" — as a *required* fourth element in the §103 analysis. See, e.g., Greenwood v. Haitori Seiko Co., Ltd., 900 F.2d 238, 241 (Fed. Cir. 1990):

[C]ertain factual predicates are required before the legal conclusion of obviousness or nonobviousness can be reached. . . . The underlying factual determinations to be made are (1) the scope and content of the prior art, (2) the differences between the claimed invention and the prior art, (3) the level of ordinary skill in the art, and (4) *objective evidence of non-obviousness, such as commercial success, long-felt but unsolved need, failure of others, copying, and unexpected results.*

These "secondary considerations" may work as "plus factors," tipping the balance of obviousness one way or the other in a particular case. But the significance of each of these factors has been hotly debated. Consider the following argument from Rochelle Dreyfuss, The Federal Circuit: A Case Study in Specialized Courts, 64 N.Y.U. L. Rev. 1 (1989):

> Th[e] use of secondary considerations is not new to the CAFC. Rather, these considerations were previously accorded little weight because their appearance can sometimes be attributed to factors other than nonobviousness. For instance, commercial success may be due to the dominant market position of the patentee before the introduction of the new invention; the sudden ability to meet long felt need could derive from other technological advances, unrelated to the inventor's contribution; acquiescence may be attributed to the relative cost of obtaining a license, as opposed to challenging the patent. Rather than reject these considerations entirely, the CAFC has recognized their importance in making the law precise and instead has sought to minimize the extent to which they can be misused. Thus, the court has elaborated a "nexus" requirement, which requires that before secondary considerations can be used to demonstrate nonobviousness, a showing must be made that their appearance is attributable to the inventive characteristics of the discovery as claimed in the patent. Secondary considerations do not constitute a complete answer to the problem posed by obviousness. It is, for instance, possible for a nonobvious invention to fail to present secondary considerations. Nonetheless, it is now less probable that a lower court will declare invalid the patent on an invention that, because of the insight of its inventor, met long felt need, enjoyed commercial success, or displayed other objective indicia of having made an important social contribution. Since it is likely that the inconsistent treatment of such inventions was the most destabilizing element of the system, the CAFC has, in this area, made strides in achieving the appearance of precision.

For an economically oriented critique of one particular factor, evidence of commercial success, see Robert P. Merges, Economic Perspectives on Innovation: Patent Standards and Commercial Success, 76 Cal. L. Rev. 803 (1988):

> Commercial success is a poor indicator of patentability because it depends for its effectiveness on a long chain of inferences, and because the links in the chain are often subject to doubt. This was one of the central insights of a seminal article on patentability, Graham v. John Deere Co.: New Standards for Patents, written by Edmund Kitch in 1966.[33] In it Kitch argued that commercial success was an unreliable indicator of nonobviousness. To make his point, Kitch identified four inferences a judge must make to work backward from evidence of market success to a conclusion of patentable invention:

---

33. 1966 Sup. Ct. Rev. 293. — EDS.

First, that the commercial success is due to the innovation. Second, that . . . potential commercial success was perceived before its development. Third, the potential commercial success having been perceived, it is likely that efforts were made [by a number of firms] to develop the improvement. Fourth, the efforts having been made by men of skill in the art, they failed because the patentee was the first to reduce his development to practice.

With only this last event as a starting point, a court is asked to reconstruct a long series of events, and, more importantly, to decide how much of the final success is attributable to each factor introduced along the way. Each inference is weak, because there are almost always several explanations why a product was successful or why other firms missed a market opportunity. Only the *last* piece of the puzzle is indisputably established; the goal of the exercise is reached through a series of inferences that only begins with this last piece. It is an altogether extraordinary job of factual reconstruction, one that reveals the falsity of the term "objective evidence," which is often used by proponents of the secondary considerations.

Merges goes on to argue in favor of another "secondary consideration" — failure of others:

Unlike commercial success, the failure of others to make an invention proves *directly* that parallel research efforts were under way at a number of firms, and that one firm (the patentee) won the race to a common goal. So long as the race was long enough, and so long as there was a clear winner, it is difficult to find fault with such evidence as proof of patentability.[242] In fact, since the failure of others is often one of the inferential steps underlying the commercial success doctrine, it makes sense for courts to adopt a rule of thumb requiring the patentee in most cases to prove failure of others before commercial success will be given substantial weight.

In Richardson-Vicks Inc. v. The Upjohn Co., 122 F.3d 1476 (Fed. Cir. 1997), the Federal Circuit cautioned against giving secondary considerations determinative weight in an obviousness inquiry. It reiterated that courts considering obviousness must evaluate all of the relevant factors in tandem.

5. Describing and Enabling the Invention

The overall goal when drafting patent claims is to make them as broad as the Patent Office will allow. There are essentially two constraints on the

---

242. Many judges have sung the praises of long felt need. Justice William R. Day of the Supreme Court said: "It may be safely said that if those skilled in the mechanical arts are working in a given field and have failed after repeated efforts to discover a certain new and useful improvement, that he who first makes the discovery . . . is entitled to protection as an inventor." Expanded Metal Co. v. Bradford, 214 U.S. 366, 381 (1908). See also Krementz v. S. Cottle Co., 148 U.S. 556, 560 (1892). Recently, the Federal Circuit has shown a willingness to consider such evidence, but has at times appeared to relax one of the two elements conventionally required to establish it — actual parallel research.

breadth of the claims you can draft: (a) the mass of publicly available information on your problem — what patent practitioners call "the prior art"; and (b) the actual work the inventor has done, in the sense that you may not claim anything beyond what the inventor has discovered, i.e., beyond the limits of the principle of the invention. To use a famous example: the inventor of the telegraph, Samuel Morse, was not permitted to claim "all forms of communicating at a distance" using electromagnetic waves, since he had only discovered one — the telegraph. It would be unfair to permit Morse to claim, e.g., microwave communications, since he did not actually discover this. In patent parlance, these "embodiments" were not "enabled" by Morse, and so he may not claim them.

Some inventions are capable of being manifested in a very wide array of embodiments. Think of the Velcro fastener, for instance, present on everything from shoes to spacesuits to huge industrial storage sacks. The point to keep in mind is that all these embodiments share the same inventive principle, i.e., are instances of the same underlying invention. Thus it is reasonable to give the person who invented Velcro fasteners a patent on the entire concept of the new fastener, in whatever context it might be applied.

An invention must be "described" sufficiently well that one of ordinary skill in the art can, relying on the description in the patent, make and use the invention. How much information must be provided to meet this requirement — and how broad the resulting claim will be — are both matters of some controversy. The basic standard is that the written description must do two things: (1) prove to the world that the applicant was in fact in possession of the invention at the time of the application, and (2) enable those skilled in the relevant art to make and use the invention.

An example of the first issue is Fiers v. Revel, 984 F.2d 1164, 1170-1171 (Fed. Cir. 1993). The facts are complex, and not fully relevant here. Basically, Fiers, Revel and Sugano all claimed to be the first to have isolated the DNA sequence coding for human beta-interferon, a protein with potential use in the treatment of cancer. The Federal Circuit had to determine which of the three had the first claim on the invention, since only that party could obtain a patent.[34] The court rejected Revel's claim to priority on the grounds that he had not listed the actual human DNA sequence in his patent application:

> An adequate written description of a DNA requires more than a mere statement that it is part of the invention and reference to a potential method for isolating it; what is required is a description of the DNA itself. Revel's specification does not do that. . . . A bare reference to a DNA with a statement that it can be obtained by reverse transcription is not a description; it does not indicate that Revel was in possession of the DNA. . . . [S]uch a disclosure just represents a wish, or arguably a plan, for obtaining the DNA.

---

34. Indeed, that party would also obtain the right to exclude the other two from practicing the invention, even though all three were researching in the area contemporaneously. See infra section C.

Even in cases where it is clear that one has in fact achieved and described one's invention, that description must enable readers of the patent to make and use the full scope of the invention as claimed by the patentee. Consider the following case.

The Incandescent Lamp Patent
Supreme Court of the United States
159 U.S. 465 (1895)

This was a bill in equity, filed by the consolidated Electric Light Company against the McKeesport Light Company, to recover damages for the infringement of letters patent No. 317,076, issued May 12, 1885, to the Electro-Dynamic Light Company, assignee of Sawyer and Man, for an electric light. The defendants justified [their actions] under certain patents to Thomas A. Edison, particularly No. 223,898, issued January 27, 1880; denied the novelty and utility of the complainants' patent, and averred that the same had been fraudulently and illegally procured. The real defendant was the Edison Electric Light Company, and the case involved a contest between what are known as the Sawyer and Man and the Edison systems of electric lighting.

In their application, Sawyer and Man stated that their invention related to "that class of electric lamps employing an incandescent conductor enclosed in a transparent, hermetically-sealed vessel or chamber, from which oxygen is excluded, and . . . more especially to the incandescing conductor, its substance, its form, and its combination with the other elements composing the lamp. Its object is to secure a cheap and effective apparatus; and our improvement consists, first, of the combination, in a lamp chamber, composed wholly of glass, . . . of an incandescing conductor of carbon made from a vegetable fibrous material, in contradistinction to a similar conductor made from mineral or gas carbon, and also in the form of such conductor so made from such vegetable carbon, and combined in the lighting circuit with the exhausted chamber of the lamp."

The specification further stated that:

> In the practice of our invention we have made use of carbonized paper, and also wood carbon. We have also used such conductors or burners of various shapes, such as pieces with their lower ends secured to their respective supports, and having their upper ends united so as to form an inverted V-shaped burner. We have also used conductors of varying contours — that is, with rectangular bends instead of curvilinear ones; but we prefer the arch shape.
>
> No especial description of making the illuminating carbon conductors, described in this specification and making the subject-matter of this improvement, is thought necessary, as any of the ordinary methods of forming the material to be carbonized to the desired shape and size, and carbonizing it according to the methods in practice before the date of this improvement, may be adopted in the practice thereof by any one skilled in the arts appertaining to the making of

carbons for electric lighting or for other use in the arts. The advantages resulting from the manufacture of the carbon from vegetable fibrous or textile material instead of mineral or gas carbon are many. Among them may be mentioned the convenience afforded for cutting and making the conductor in the desired form and size, the purity and equality of the carbon obtained, its susceptibility to tempering, both as to hardness and resistance, and its toughness and durability.

The claims were as follows:

1. An incandescing conductor for an electric lamp, of carbonized fibrous or textile material and of an arch or horseshoe shape, substantially as hereinbefore set forth.
2. The combination, substantially as hereinbefore set forth, of an electric circuit and an incandescing conductor of carbonized fibrous material, included in and forming part of said circuit, and a transparent hermetically sealed chamber in which the conductor is enclosed.
3. The incandescing conductor for an electric lamp, formed of carbonized paper, substantially as described.

The commercial Edison lamp used by the appellee is composed of a burner made of carbonized bamboo of a peculiar quality discovered by Mr. Edison to be highly useful for the purpose, and having a length of about six inches, a diameter of about five one thousandths of an inch, and an electrical resistance of upwards of 100 ohms. This filament of carbon is bent into the form of a loop, and its ends are secured by good electrical and mechanical connections to two fine platinum wires. . . .

Upon a hearing in the Circuit Court before Mr. Justice Bradley upon pleadings and proofs, the court held the patent to be invalid, and dismissed the bill. 40 Fed. Rep. 21 [C.C.W.D.P.A. (1889)]. Thereupon complainant appealed to this court.

Mr. Justice BROWN, after stating the case as above reported, delivered the opinion of the court.

In order to obtain a complete understanding of the scope of the Sawyer and Man patent, it is desirable to consider briefly the state of the art at the time the application was originally made, which was in January, 1880.

. . . The form of illumination . . . known as the incandescent system . . . consists generally in the passage of a current of electricity through a continuous strip or piece of refractory material, which is a conductor of electricity, but a poor conductor — in other words, a conductor offering a considerable resistance to the flow of the current through it. It was discovered early in this century that various substances might be heated to a white heat by passing a sufficiently strong current of electricity through them. . . .

For many years prior to 1880, experiments had been made by a large number of persons, in various countries, with a view to the production of an incandescent light which could be made available for domestic purposes, and could compete with gas in the matter of expense. Owing partly to a failure to find a proper material, which should burn but not consume, partly to the difficulty of obtaining a perfect vacuum in the globe in which the light was suspended, and partly to a misapprehension of the true principle of incandescent lighting, these experiments had not been attended with success; although it had been demonstrated as early as 1845 that, whatever material was used, the conductor must be enclosed in an air-tight bulb [i.e., vacuum], to prevent it from being consumed by the oxygen in the atmosphere. The chief difficulty was that the carbon burners were subject to a rapid disintegration or evaporation, which electricians assumed was due to the disrupting action of the electric current, and, hence, the conclusion was reached that carbon contained in itself the elements of its own destruction, and was not a suitable material for the burner of an incandescent lamp.

It is admitted that the lamp described in the Sawyer and Man patent is no longer in use, and was never a commercial success; that it does not embody the principle of high resistance with a small illuminating surface; that it does not have the filament burner of the modern incandescent lamp; that the lamp chamber is defective, and that the lamp manufactured by the complainant and put upon the market is substantially the Edison lamp; but it is said that, in the conductor used by Edison (a particular part of the stem of the bamboo lying directly beneath the silicious cuticle, the peculiar fitness for which purpose was undoubtedly discovered by him), he made use of a fibrous or textile material, covered by the patent to Sawyer and Man, and is, therefore, an infringer. It was admitted, however, that the third claim — for a conductor of carbonized paper — was not infringed.

The two main defences to this patent are (1) that it is defective upon its face, in attempting to monopolize the use of all fibrous and textile materials for the purpose of electric illumination; and (2) that Sawyer and Man were not in fact the first to discover that these were better adapted than mineral carbons to such purposes.

Is the complainant entitled to a monopoly of all fibrous and textile materials for incandescent conductors? If the patentees had discovered in fibrous and textile substances a quality common to them all, or to them generally, as distinguishing them from other materials, such as minerals, etc., and such quality or characteristic adapted them peculiarly to incandescent conductors, such claim might not be too broad. If, for instance, minerals or porcelains had always been used for a particular purpose, and a person should take out a patent for a similar article of wood, and woods generally were adapted to that purpose, the claim might not be too broad, though defendant used wood of a different kind from that of the patentee. But if woods generally were not adapted to the purpose, and yet the patentee had discovered a wood possessing certain qualities, which gave it a peculiar fitness for such purpose, it would not constitute an infringement for another to discover and use a different kind of wood, which was found to contain similar or superior qualities. The present case is an apt illustration of this principle. Sawyer and Man supposed they had discovered in carbonized paper the best material for an incandescent conductor. Instead of confining themselves to carbonized paper, as they might properly have done, and in fact did in their third claim, they made a broad claim for every fibrous or textile material, when in fact an examination of over six thousand vegetable growths showed that none of them possessed the peculiar qualities that fitted them for that purpose. Was everybody then precluded by this broad claim from making further investigation? We think not.

The injustice of so holding is manifest in view of the experiments made, and continued for several months, by Mr. Edison and his assistants, among the different species of vegetable growth, for the purpose of ascertaining the one best adapted to an incandescent conductor. Of these he found suitable for his purpose only about three species of bamboo, one species of cane from the Valley of the Amazon, impossible to be procured in quantities on account of the climate, and one or two species of fibres from the agave family. Of the special bamboo, the walls of which have a thickness of about three-eighths of an inch, he used only about twenty-thousandths of an inch in thickness. In this portion of the bamboo the fibres are more nearly parallel, the cell walls are apparently smallest, and the pithy matter between the fibres is at its minimum. It seems that carbon filaments cannot be made of wood — that is, exogenous vegetable growth — because the fibres are not parallel and the longitudinal fibres are intercepted by radial fibres. The cells composing the fibres are all so large that the resulting carbon is very porous and friable. Lamps made of this material proved of no commercial value. After trying as many as thirty or forty different woods of exogenous growth, he gave them up as hopeless. But finally, while experimenting with a bamboo strip which formed the edge of a palmleaf fan, cut into filaments, he obtained surprising results. After microscopic examination of the material, he despatched a man to Japan to make arrangements for securing the bamboo in quantities. It seems that the characteristic of the bamboo which makes it particularly suitable is, that the fibres run more nearly parallel than in other species of wood. Owing to this, it can be cut up into filaments having parallel fibres, running throughout their length, and producing a homogeneous carbon. There is no generic quality, however, in vegetable fibres, because they are fibrous, which adapts them to the purpose. Indeed, the fibres are rather a disadvantage. If the bamboo grew solid without fibres, but had its peculiar cellular formation, it would be a perfect material, and incandescent lamps would last at least six times as long as at present. All vegetable fibrous growths do not have a suitable cellular structure. In some the cells are so large that they are valueless for that purpose. No exogenous, and very few endogenous, growths are suitable. The messenger whom he despatched to different parts of Japan and China sent him about forty different kinds of bamboo, in such quantities as to enable him to make a number of lamps, and from a test of these different species he ascertained which was best for the purpose. From this it appears very clearly that there is no such quality common to fibrous and textile substances generally as makes them suitable for an incandescent conductor, and that the bamboo which was finally pitched upon, and is now generally used, was not selected because it was of vegetable growth, but because it contained certain peculiarities in its fibrous structure which distinguished it from every other fibrous substance. The question really is whether the imperfectly successful experiments of Sawyer and Man, with carbonized paper and wood carbon, conceding all that is claimed for them, authorize them to put under tribute the results of the brilliant discoveries made by others.

It is required by Rev. Stat. §4888 that the application shall contain a written description of the device "and of the manner and process of making, constructing, compounding, and using it in such full, clear, concise, and exact terms as to enable any person, skilled in the art or science to which it appertains or with which it is most nearly connected, to make, construct, compound, and use the same." The object of this is to apprise the public of what the patentee claims as his own, the courts of what they are called upon to construe, and competing manufacturers and dealers of exactly what they are bound to avoid. Grant v. Raymond, 6 Pet. [31 U.S.] 218, 247 [1832]. If the description be so vague and uncertain that no one can tell, except by independent experiments, how to construct the patented device, the patent is void.

It was said by Mr. Chief Justice Taney in Wood v. Underhill, 5 How. [46 U.S.] 1, 5 [1857], with respect to a patented compound for the purpose of making brick or tile, which did not give the relative proportions of the different ingredients:

> But when the specification of a new composition of matter gives only the names of the substances which are to be mixed together, without stating any relative proportion, undoubtedly it would be the duty of the court to declare the patent void. And the same rule would prevail where it was apparent that the proportions were stated ambiguously and vaguely. For in such cases it would be evident, on the face of the specification, that no one could use the invention without first ascertaining, by experiment, the exact proportion of the different ingredients required to produce the result intended to be obtained. . . . And if, from the nature and character of the ingredients to be used, they are not susceptible of such exact description, the inventor is not entitled to a patent.

So in Tyler v. Boston, 7 Wall. [74 U.S.] 327, 330 [1868], wherein the plaintiff professed to have discovered a combination of fuel oil with the mineral and earthy oils, constituting a burning fluid, the patentee stated that the exact quantity of fuel oil, which is necessary to produce the most desirable compound, must be determined by experiment. And the court observed: "Where a patent is claimed for such a discovery it should state the component parts of the new manufacture claimed with clearness and precision, and not leave a person attempting to use the discovery to find it out 'by experiment.' " See also Bene v. Jeantet, 129 U.S. 683 [(1889)]; Howard v. Detroit Stove Works, 150 U.S. 164, 167 [(1893)]; Schneider v. Lovell, 10 Fed. Rep. 666 [C.C.S.D.N.Y. 1882]; Welling v. Crane, 14 Fed. Rep. 571 [C.C.D.N.J. 1882].

If Sawyer and Man had discovered that a certain carbonized paper would answer the purpose, their claim to all carbonized paper would, perhaps, not be extravagant; but the fact that paper happens to belong to the fibrous kingdom did not invest them with sovereignty over this entire kingdom, and thereby practically limit other experimenters to the domain of minerals.

In fact, such a construction of this patent as would exclude competitors from making use of any fibrous or textile material would probably defeat itself, since, if the patent were infringed by the use of any such material, it would be anticipated by proof of the prior use of any such material. In this connection it would appear, not only that wood charcoal had been constantly used since the days of Sir Humphry Davy for arc lighting, but that in the English patent to Greener and Staite of 1846, for an incandescent light, "charcoal, reduced to a state of powder," was one of the materials employed. So also, in the English patent of 1841 to De Moleyns, "a finely pulverized boxwood charcoal or plumbago" was used for an incandescent electric lamp. Indeed, in the experiments of Sir Humphry Davy, early in the century, pieces of well-burned charcoal were heated to a vivid whiteness by the electric current, and other experiments were made which evidently contemplated the use of charcoal heated to the point of incandescence. Mr. Broadnax, the attorney who prepared the application, it seems, was also of opinion that a broad claim for vegetable carbons could not be sustained because charcoal had been used before in incandescent lighting.

We are all agreed that the claims of this patent, with the exception of the third, are too indefinite to be the subject of a valid monopoly. For the reasons above stated the decree of the Circuit Court is Affirmed.

COMMENTS AND QUESTIONS

1. The modern equivalent to old Rev. Stat. §4888 is 35 U.S.C. §112 (1986). The first paragraph of §112 has been interpreted in such a way that three distinct requirements are now said to spring from it: (1) the written description requirement; (2) the clear claim requirement; and (3) the enablement requirement. Which of these requirements was at issue in Incandescent Lamp? More than one?

2. The Court puts forth several rationales for the enablement requirement. For example, the Court asks: "Was everybody then precluded by this broad claim from making further investigation? We think not." What assumption lies behind this statement? Why should the patent system be concerned with "further investigation"?

3. Were *all* claims of the Sawyer and Man patent invalidated? Of what relevance, if any, is it that Edison's research resulted in several patents of his own? Does the fact that inventor *A* received a patent suggest anything about whether *A*'s research results are enabled by the specification of a prior patent issued to *B*?

4. Edison's discovery that only a certain type of bamboo plant would work as a filament is an example of his exhaustive research efforts. Of course, the bamboo itself is not patentable, since it is a product of nature. After

isolating the precise type of bamboo that would work, Edison acted with characteristic speed on a characteristically grand scale: he tried to lock up as many acres of production of the bamboo as he could. See A. Millard, Edison and the Business of Innovation (1990). Cf. Hirshleifer, The Private and Social Value of Information and the Reward to Inventive Activity, 61 Am. Econ. Rev. 561 (1971) (positing reduced need for intellectual property protection, since inventors have "inside information" about their inventions, so they can reap gains by investing in assets that their inventions will make more valuable and selling short assets that their inventions will make less valuable).

If the court was right to invalidate the entire patent, how much research should Sawyer and Man have done to render their broad claims valid? Must they test *every* possible variant known to them?

For more on the early history of the lighting industry, see W. Bright, The Electric-Lamp Industry (1949), and W. MacLaren, The Rise of the Electrical Industry During the Nineteenth Century (1943); A. Millard, supra.

Note on the Written Description Requirement

That one might teach and yet not own — this is what the written description requirement is all about. Like many innovations, it is hard to get used to and may not add much to the sum total of man's knowledge or happiness. But it appears here to stay, at least for a while. So like a hardy new weed, we had better learn something about it.

This newest patent requirement (which might better be deemed "invention description") took root in a judicial desire to rein in the free and easy ways of patent drafters. The particular object of the judges' displeasure is what is called "amendment practice." Patent lawyers are adept at filing a patent application that broadly (if sometimes vaguely) describes an invention. The law has for many years required that at least one formal claim accompany the application. But the original filing has long been viewed merely as the opening bid in a long game of rejections, amendments, "final" rejections (a classic legal misnomer) and assorted wranglings. While thus negotiating and amending, patent lawyers also keep an eye on the inventor's follow-up research and the market into which the invention has found (or will find) its way. As events unfold in these corners, the lawyer makes amendments accordingly. Consider a common example: a firm competing with an inventor may introduce a product containing a variant of the inventor's brainstorm. When the language in the patent application allows, the inventor's patent lawyer adds a claim to the application embracing the new variant. In this manner the competitor's product will infringe the patent if and when it issues. This is standard practice and has been for a long time.

Nevertheless, the Federal Circuit was bothered by several recent cases that arose out of this practice. In an effort to limit the abuses of this practice, the court turned to a classic doctrine which had been largely ignored of late. Some cases — primarily from the 1970s — had cracked down on amendments to first-filed applications, particularly in the chemical arts. See, e.g., In re DiLeone, 436 F.2d 1404, 1405 (C.C.P.A. 1971). In opinions that often sounded in 35 U.S.C. §132's rule against "new matter," these cases had put restrictions on the changes that can be made to the claims in an initial application during prosecution. Dictum from *DiLeone*, supra, provides a good example of the issues bothering these courts:

> [Consider a case] where the specification discusses only compound A and contains no broadening language of any kind. This might very well enable one skilled in the art to make and use compounds B and C; yet the class consisting of A, B and C has not been described.

Id. at 1405 n.1. This is a clear antecedent to the rule developed in recent cases. Yet these early cases reveal a certain amount of judicial trepidation over the new doctrine. The holding is often stated almost in the alternative, with the aforementioned new matter rule, or the traditional enablement requirement, bearing at least part of the decisional weight. Almost every opinion explaining the new rule elicited a strenuous dissent emphasizing either the doctrine's redundancy or its lack of clarity. See, e.g., *DiLeone*, supra, 436 F.2d at 1406 (dissent of Judge Baldwin) (calling for a tighter relationship between enablement and written description). It is therefore no surprise that written description, having met with a cool reception, faded almost completely from subsequent decisions.

However, beginning in the late 1980s, the Federal Circuit resuscitated the doctrine. First sparingly, and with increasing frequency as of late, the court has pointedly deployed "written description" as an independent, and often very stiff, requirement of patentability. Of special import in this connection are two recent decisions: Regents of the University of California v. Eli Lilly & Co., 119 F.3d 1559 (Fed. Cir. 1997) and Gentry Gallery, Inc. v. The Berkline Corp., 134 F.3d 1473 (Fed. Cir. 1998) (excerpted below).

In the *Eli Lilly* case, University of California researchers filed a patent application disclosing the DNA sequence for rat insulin. The final version of the patent claimed rat, mammalian, vertebrate, and human insulin. These are very closely related variants; human insulin differs from the rat version by only twelve (out of more than 300) DNA base pairs that code for the protein. Cf. Michael Delmas Plimier, Genentech, Inc. v. Novo Nordisk & University of California v. Eli Lilly and Co., 13 Berkeley Tech. L.J. 149 (1998). Nonetheless, the court invalidated the claim to human insulin on the grounds that the specification did not meet the written description requirement. The court found it an easy case. It cited contemporary cases holding that disclosure of an amino acid sequence for a protein, together with well-known techniques for identifying a DNA sequence from the amino acids, did not render the resulting sequence obvious under §103. Because even obvious variants of disclosed material might not be within the written description of a patent specification, it was clear to the court that nonobvious variants (under these cases) were also not adequately described. For more background and analysis, see *Plimier*, supra.

For a time it was thought that the "written description" revolution was limited to biotechnology inventions. See, e.g., Janice M. Mueller, The Evolving Application of the Written Description Requirement of 35 U.S.C. §112, ¶1 to Biotechnology Inventions, 13 Berkeley Tech. L.J. 615 (1998). Then came *Gentry Gallery*, our next case.

The Gentry Gallery, Inc., v. The Berkline Corp.
United States Court of Appeals for the Federal Circuit
134 F.3d 1473 (Fed. Cir. 1998)

LOURIE, Circuit Judge.

The Gentry Gallery appeals from the judgment of the United States District Court for the District of Massachusetts holding that the Berkline Corporation does not infringe U.S. Patent 5,064,244, and declining to award attorney fees for Gentry's defense to Berkline's assertion that the patent was unenforceable. . . . Berkline cross-appeals from the decision that the patent was not shown to be invalid. Because the court correctly concluded that the claims were not infringed by Berkline, and that the subject matter of the asserted claims was not shown to have been obvious, and did not abuse its discretion in declining to award attorney fees, we affirm these decisions. However, because the court clearly erred in finding that the written description portion of the specification supported certain of the broader claims asserted by Gentry, we reverse the decision that those claims are not invalid under 35 U.S.C. §112, ¶1 (1994).

Background

Gentry owns the '244 patent, which is directed to a unit of a sectional sofa in which two independent reclining seats ("recliners") face in the same direction. Sectional sofas are typically organized in an L-shape with "arms" at the exposed ends of the linear sections. According to the patent specification, because recliners usually have had adjustment controls on their arms, sectional sofas were able to contain two recliners only if they were located at the exposed ends of the linear sections. Due to the typical L-shaped configuration of sectional sofas, the recliners therefore faced in different directions. See '244 patent; col. 1, ll. 15-19. Such an arrangement was "not usually comfortable when the occupants are watching television because one or both occupants must turn their heads to watch the same [television] set. Furthermore, the separation of the two reclining seats at opposite ends of a sectional sofa is not comfortable or conducive to intimate conversation." Id. at col. 1, ll. 19-25.

The invention of the patent solved this supposed dilemma by, *inter alia*, placing a "console" between two recliners which face in the same direction. This console "accommodates the controls for both reclining seats," thus eliminating the need to position each recliner at an exposed end of a linear section. Id. at col. 1, ll. 36-37. Accordingly, both recliners can then be located on the same linear section allowing two people to recline while watching television and facing in the same direction. Claim 1, which is the broadest claim of the patent, reads in relevant part:

> A sectional sofa comprising:
> a pair of reclining seats disposed in parallel relationship with one another in a
>   double reclining seat sectional sofa section being without an arm at one end
>   . . . ,
> each of said reclining seats having a backrest and seat cushions and movable
>   between upright and reclined positions . . . ,
> *a fixed console* disposed in the double reclining seat sofa section between the
>   pair of reclining seats and with the console and reclining seats together com-
>   prising a unitary structure,
> said console including an armrest portion for each of the reclining seats; said
>   arm rests remaining fixed when the reclining seats move from one to another
>   of their positions,
> and *a pair of control means*, one for each reclining seat; *mounted on the double
>   reclining seat sofa section*. . . .

Id. at col. 4, line 68 to col. 5, ll. 1-27 (emphasis added to most relevant claim language). Claims 9, 10, 12-15, and 19-21 are directed to a sectional sofa in which the control means are specifically located on the console.

In 1991, Gentry filed suit . . . alleging that Berkline infringed the patent by manufacturing and selling sectional sofas having two recliners facing in the same direction. In the allegedly infringing sofas, the recliners were sep-arated by a seat which has a back cushion that may be pivoted down onto the seat, so that the seat back may serve as a tabletop between the recliners. . . . The district court granted Berkline's motion for summary judgment of non-infringement, but denied its motions for summary judgment of invalidity and unenforceability. In construing the language "fixed console," the court

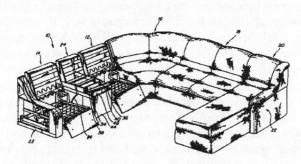

FIGURE 3-2
Patent drawing for the Gentry '244 patent.

relied on, *inter alia*, a statement made by the inventor named in the patent, James Sproule, in a Petition to Make Special (PTMS). See 37 C.F.R. §1.102 (1997). Sproule had attempted to distinguish his invention from a prior art reference by arguing that that reference, U.S. Patent 3,877,747 to Brennan *et al.* ("Brennan"), "shows a complete center seat with a tray in its back." *Gentry I*, 30 U.S.P.Q.2d at 1137. Based on Sproule's argument, the court concluded that, as a matter of law, Berkline's sofas "contain[ ] a drop-down tray identical to the one employed by the Brennan product" and therefore did not have a "fixed console" and did not literally infringe the patent. Id. The court held that Gentry was also "precluded from recovery" under the doctrine of equivalents. Id. at 1138.

*Invalidity * * ***

Berkline . . . argues that claims 1-8, 11, and 16-18 are invalid because they are directed to sectional sofas in which the location of the recliner controls is not limited to the console. According to Berkline, because the patent only describes sofas having controls on the console and an object of the invention is to provide a sectional sofa "with a console . . . that accommodates the controls for both the reclining seats," '244 patent, col. 1, ll. 35-37, the claimed sofas are not described within the meaning of §112, ¶1. Berkline also relies on Sproule's testimony that "locating the controls on the console is definitely the way we solved it [the problem of building sectional sofa with parallel recliners] on the original group [of sofas]." Gentry responds that the disclosure represents only Sproule's preferred embodiment, in which the controls are on the console, and therefore supports claims directed to a sofa in which the controls may be located elsewhere. Gentry relies on Ethicon Endo-Surgery, Inc. v. United States Surgical Corp., 93 F.3d 1572, 1582 n.7, 40 U.S.P.Q.2d 1019, 1027 n.7 (Fed. Cir. 1993), and In re Rasmussen, 650 F.2d 1212, 1214, 211 U.S.P.Q. 323, 326 (CCPA 1981), for the proposition that an applicant need not describe more than one embodiment of a broad claim to adequately support that claim.

We agree with Berkline that the patent's disclosure does not support claims in which the location of the recliner controls is other than on the console. Whether a specification complies with the written description requirement of §112, ¶1, is a question of fact, which we review for clear error on appeal from a bench trial. See Vas-Cath Inc. v. Mahurkar, 935 F.2d 1555, 1563, 19 U.S.P.Q.2d 1111, 1116 (Fed. Cir. 1991). To fulfill the written description requirement, the patent specification "must clearly allow persons of ordinary skill in the art to recognize that [the inventor] invented what is claimed." In re Gosteli, 872 F.2d 1008, 1012, 10 U.S.P.Q.2d 1614, 1618 (Fed. Cir. 1989). An applicant complies with the written description requirement "by describing *the invention*, with all its claimed limitations." Lockwood v. American Airlines, Inc., 107 F.3d 1565, 1572, 41 U.S.P.Q.2d 1961, 1966 (Fed. Cir. 1997).

It is a truism that a claim need not be limited to a preferred embodiment. However, in a given case, the scope of the right to exclude may be limited by a narrow disclosure. For example, as we have recently held, a disclosure of a television set with a keypad, connected to a central computer with a video disk player did not support claims directed to "an individual terminal containing a video disk player." See id. (stating that claims directed to a "distinct invention from that disclosed in the specification" do not satisfy the written description requirement); see also Regents of the Univ. of Cal. v. Eli Lilly & Co., 119 F.3d 1559, 1568, 43 U.S.P.Q.2d 1398, 1405 (Fed. Cir. 1997) (stating that the case law does "not compel the conclusion that a description of a species always constitutes a description of a genus of which it is a part").

In this case, the original disclosure clearly identifies the console as the only possible location for the controls. It provides for only the most minor variation in the location of the controls, noting that the control "may be mounted on top or side surfaces of the console rather than on the front wall . . . without departing from this invention." '244 patent, col. 2, line 68 to col. 3, line 3. No similar variation beyond the console is even suggested. Additionally, the only discernible purpose for the console is to house the controls. As the disclosure states, identifying the only purpose relevant to the console, "[a]nother object of the present invention is to provide . . . a console positioned between [the reclining seats] that accommodates the controls for both of the reclining seats." Id. at col. 1, 11. 33-37. Thus, locating the controls anywhere but on the console is outside the stated purpose of the invention. Moreover, consistent with this disclosure, Sproule's broadest original claim was directed to a sofa comprising, *inter alia*, "control means located upon the center console to enable each of the pair of reclining seats to move separately between the reclined and upright positions." Finally, although not dispositive, because one can add claims to a pending application directed to adequately described subject matter, Sproule admitted at trial that he did not consider placing the controls outside the console until he became aware that some of Gentry's competitors were so locating the recliner controls. Accordingly, when viewed in its entirety, the disclosure is limited to sofas in which the recliner control is located on the console.

Gentry's reliance on *Ethicon* is misplaced. It is true, as Gentry observes, that we noted that "an applicant . . . is generally allowed claims, when the art permits, which cover more than the specific embodiment shown." *Ethicon*, 93 F.3d at 1582 n.7, 40 U.S.P.Q.2d at 1027 n.7 (quoting In re Vickers, 141 F.3d 522, 525, 61 U.S.P.Q.2d 122, 125 (CCPA 1944)). However, we were also careful to point out in that opinion that the applicant "was free to draft claim[s] broadly (within the limits imposed by the prior art) to exclude the lockout precise location as a limitation of the claimed invention" only because he "did not consider the precise location of the lockout to be an element of his invention." Id. Here, as indicated above, it is clear that Sproule considered the location of the recliner controls on the console to be an essential element of his invention. Accordingly, his original disclosure serves to limit the permissible breadth of his later-drafted claims.

Similarly, *In re Rasmussen* does not support Gentry's position. In that case, our predecessor court restated the uncontroversial proposition that "a claim may be broader than the specific embodiment disclosed in a specification." 650 F.2d at 1215, 211 U.S.P.Q. at 326. However, the court also made clear that "[a]n applicant is entitled to claims as broad as the prior art *and his disclosure* will allow." Id. at 1214, 211 U.S.P.Q. at 326 (emphasis added). The claims at issue in *Rasmussen*, which were limited to the generic step of "adheringly applying" one layer to an adjacent layer, satisfied the written description requirement only because "one skilled in the art who read [the] specification would understand that it is unimportant how the layers are adhered, so long as they are adhered." Here, on the contrary, one skilled in the art would clearly understand that it was not only important, but essential to Sproule's invention, for the controls to be on the console.

In sum, the cases on which Gentry relies do not stand for the proposition that an applicant can broaden his claims to the extent that they are effectively bounded only by the prior art. Rather, they make clear that claims may be no broader than the supporting disclosure, and therefore that a narrow disclosure will limit claim breadth. Here, Sproule's disclosure unambiguously limited the location of the controls to the console. Accordingly, the district court clearly erred in finding that he was entitled to claims in which the recliner controls are not located on the console. We therefore reverse the judgment that claims 1-8, 11, and 16-18, were not shown to be invalid.

COMMENTS AND QUESTIONS

1. Berkline's sofa recliners have the reclining controls on arm rests, seat edges, etc. See www.berkline.com. The district court stated that Gentry Gallery's amended claims were " 'broad' but not unsupported." 939 F. Supp. 98, 105 (D. Mass. 1996). The district court stated that the invention had solved the "control quandary" of removing controls from arm rests and underneath seat cushions, and that this location of the controls on this central console was only our way of implementing the invention.

2. *Gentry Gallery* has been described as a case about an "omitted element." Reiffen v. Microsoft Corp., 48 U.S.P.Q.2d 1274 (N.D. Cal. 1998). To those who favor this terminology, the case centers on the *omission* of an originally disclosed element from one or more final claims. There are three things to consider in this analysis. First, who suffers from the inventor's change of heart? In cases such as *Gentry Gallery*, if the amended claims had been allowed and found valid, the Berkline Corporation (the patentee's competitor, and the defendant in the case) would have suffered. Berkline's product (a variant on the sofa design disclosed in Gentry's initial application) would be bounced from the market, perhaps to be replaced by an identical product from Gentry. This would amount to an appropriation of Berkline's variation by the original applicant, Gentry. Before you conclude that this is unfair, however, recall that it is only a trivial variant on Gentry's bona fide invention. Put another way, Gentry had on file in the U.S. Patent Office a

document that taught (enabled) skilled artisans enough to arrive at this variant with minimal effort. Keeping this in mind may temper your sense of the wrongfulness of Gentry's appropriation. As a final consideration, recall that as far as Berkline is concerned, it was purely fortuitous in the actual case that Gentry had not "adequately described" Berkline's variation. There is no way that Berkline could have found out what precisely constituted Gentry's initial disclosure. (Why not?) Thus there is nothing akin to a "reliance" argument in favor of Berkline. But just because there is nothing that Berkline could have done to avoid Gentry's appropriation does not mean that the appropriation is rightful — does it?

3. Does it make sense to have both an enablement requirement and a written description requirement? If the inventor must teach one of ordinary skill in the art how to make and use the invention, why require them to write down the exact invention as well?

4. The *Gentry Gallery* case has been robustly criticized. See, e.g., Harris A. Pitlick, The Mutation on the Description Requirement Gene, 80 J. Pat. & Trademark Off. Soc'y 209, 222 (1998) (recent cases are "an unmitigated disaster"); Laurence H. Pretty, The Recline and Fall of Mechanical Genus Claim Scope Under "Written Description" in the Sofa Case, 80 J. Pat. & Trademark Off. Soc'y 469 (1998). According to Pitlick, commenting on the *Lilly* case,

> The court appears to be concerned with the rush to patenting that accompanied the recombinant DNA revolution, when patents issued, perhaps many, without a sufficiently enabling disclosure. The remedy for such patents is to invalidate them. But the court has lost sight of the real culprit — lack of enablement — and directed its ire at an innocent bystander — the description requirement. Thus, aside from the real world effect of the decisions in *Lilly* (and *Fiers*), the court takes description requirement jurisprudence in an unjustifiably new and reckless direction, freed of any constraints of stare decisis. In *Lilly*, as in *Fiers*, the court mixed up different and unrelated principles of patent law in a manner totally surprising from a tribunal seasoned in complex patent issues.

Pitlick, *supra*, at 222-223. For a contrary view, see Note, Mark J. Stewart, The Written Description Requirement of 35 U.S.C. §112(1): The Standard After Regents of the University of California v. Eli Lilly & Co., 32 Ind. L. Rev. 537, 563-64 (1999):

> [I]t would seem that the holding in *Lilly* . . . avoided a disaster that would have crippled the biotechnology industry. The enormous amount of time and money companies spend to study DNA and protein variants, to clone homologous genes and protein family members, and to mine databases would no longer be justified had the court found the written description in [the] '525 [patent at issue] adequate. Through application of the written description requirement, courts can distinguish between claims to technologies that are too broad or basic to justify patent protection, and those dealing with other types of technologies that are more predictable and may justify broader protection. Thus, the Federal

Circuit has decided that the uniqueness of biotechnology inventions claiming DNA sequences requires the application of a stringent written description requirement to protect the public from inventors seeking to slow the pace of research by preempting future developments before they arrive.

To what extent do these arguments result in a consideration of the relative merits of (and incentives facing) original "pioneer" inventors and subsequent "improvers"? Cf. Jerry R. Green & Suzanne Scotchmer, On the Division of Profit in Sequential Innovation, 26 Rand J. Econ. 20 (1995); and Suzanne Scotchmer, Standing on the Shoulders of Giants: Cumulative Research and the Patent Law, 5 J. Econ. Persp. 29 (1991).

5. One possible benefit of the written description requirement is to preclude patent owners from later claiming what they did not think of at the time they filed their applications. This is a particularly significant issue when the patent does not issue until long after the application is filed. In Hyatt v. Boone, 47 U.S.P.Q.2d 1128 (Fed. Cir. 1998), the court held that Hyatt's patent on a microprocessor did not deserve the benefit of its earliest filing date because the original application did not contain a written description of the invention ultimately claimed. Compare In re Hogan and Banks, 559 F.2d 595, 606 (C.C.P.A. 1977) (refusing to find defect in written description of first-filed application where art continued to develop and applicant updated disclosure accordingly):

> The PTO has not challenged appellants' assertion that their 1953 application enabled those skilled in the art in 1953 to make and use "a solid polymer" as described in claim 13. Appellants disclosed, as the only then existing way to make such a polymer, a method of making the crystalline form. To now say that appellants should have disclosed in 1953 the amorphous form which on this record did not exist until 1962, would be to impose an impossible burden on inventors and thus on the patent system. There cannot, in an effective patent system, be such a burden placed on the right to broad claims. To restrict appellants to the crystalline form disclosed, under such circumstances, would be a poor way to stimulate invention, and particularly to encourage its early disclosure. To demand such restriction is merely to state a policy against broad protection for pioneer inventions, a policy both shortsighted and unsound from the standpoint of promoting progress in the useful arts, the constitutional purpose of the patent laws.

Note on "Analog" Claims in Chemical and Biotechnology Patents: An Exploration of Patent Breadth

Questions of patent breadth continue to be important today, when biotechnology (rather than electric lighting) is one of the key growth fields. In this section, we explore one issue of patent breadth important to the biotechnology industry: the "functional analogs" claim.

Many chemicals have "analogs." Two different molecules with almost identical chemical structure will normally behave in approximately the same way. So, too, similar DNA sequences normally have similar effects. The prevalence of analogs creates problems for those who try to patent new chemical or biological products. If they claim only the precise DNA sequence or chemical structure they have identified, it is relatively easy for others to "design around" their invention by varying an insignificant part of the total structure.

To avoid this problem, many patentees claim not only the precise structure they have produced, but its analogs as well. (This is akin to the claims of Sawyer & Man, in the Incandescent Lamp case, to fibrous and textile materials other than the paper they had actually experimented with.) To what extent a claim should be allowed to preempt such analogs is a difficult problem in patent law. In Amgen, Inc. v. Chugai Pharmaceutical Co., Ltd., 927 F.2d 1200 (Fed. Cir. 1991), cert. denied, 502 U.S. 856 (1991), the plaintiff and defendant each held patents on technology related to the production of erythropoietin (EPO), a critical biological protein that stimulates production of red blood cells and is therefore effective in combating anemia and related conditions. Plaintiff Amgen held a patent on a recombinant DNA version of EPO, while defendant Chugai held a license from codefendant Genetics Institute under a product patent for purified EPO made by concentrating trace amounts of the protein from natural sources. The trial court had held valid certain claims in defendant's patent, and it ruled that plaintiff infringed; it also held that certain claims in both patents were invalid for failure to enable.

Plaintiff's recombinant EPO patent included key claim 7, which reads: "7. A purified and isolated DNA sequence consisting essentially of a DNA sequence encoding a polypeptide having an amino acid sequence sufficiently duplicative of that of erythropoietin to allow possession of the biological property of causing bone marrow cells to increase production of reticulocytes and red blood cells, and to increase hemoglobin synthesis or iron uptake." The "biological property" language, together with the descriptions of the two key functions — blood cell production and iron uptake — was intended to broaden the claim so as to cover any functional substitute or "analog" for the natural EPO protein. The following excerpt comes from the Federal Circuit's discussion of the validity of claim 7 and related claims.

> The essential question here is whether the scope of enablement of claim 7 is as broad as the scope of the claim. See generally In re Fisher. . . .
>   The specification of [Amgen's] patent provides that:
>
>> one may readily design and manufacture genes coding for microbial expression of polypeptides having primary conformations [i.e., proteins with a basic shape] which differ from that herein specified for mature EPO in terms of the identity or location of one or more residues [i.e., amino acids] (e.g., substitutions, terminal and intermediate additions and deletions). . . .
>>   DNA sequences provided by the present invention are thus seen to comprehend all DNA sequences suitable for use in securing expression in a procaryotic or

eucaryotic host cell of a polypeptide product having at least a part of the primary structural conformation and one or more of the biological properties of erythropoietin. . . .

The district court found that over 3,600 different EPO analogs can be made by substituting at only a single amino acid position [in the entire protein], and over a million different analogs can be made by substituting three amino acids. The patent indicates that it embraces means for preparation of "numerous" polypeptide analogs of EPO. Thus, the number of claimed DNA encoding sequences that can produce an EPO-like product is potentially enormous.

In a deposition, Dr. Elliott, who was head of Amgen's EPO analog program, testified that he did not know whether the fifty to eighty EPO analogs Amgen had made "had the biological property of causing bone marrow cells to increase production of reticulocytes and red blood cells, and to increase hemoglobin synthesis or iron uptake" [as required by some of the claims]. Based on this evidence, the trial court [found a lack of enablement]. In making this determination, the court relied in particular on the lack of predictability in the art. . . . After five years of experimentation, the court noted, "Amgen is still unable to specify which analogs have the biological properties set forth in claim 7."

[Although] it is not necessary that a patent applicant test all the embodiments of his invention, In re Angstadt, 537 F.2d 498, 502, 190 U.S.P.Q. 214, 218 (C.C.P.A. 1976)[,] what is necessary is that he provide a disclosure sufficient to enable one skilled in the art to carry out the invention commensurate with the scope of his claims. For DNA sequences, that means disclosing how to make and use enough sequences to justify grant of the claims sought. Amgen has not done that here. It is well established that a patent applicant is entitled to claim his invention generically, when he describes it sufficiently to meet the requirements of Section 112. See Utter v. Hiraga, 845 F.2d 993, 998, 6 U.S.P.Q.2d 1709, 1714 (Fed. Cir. 1988) ("A specification may, within the meaning of 35 U.S.C. §112 ¶1, contain a written description of a broadly claimed invention without describing all species that claim encompasses."). Here, however, despite extensive statements in the specification concerning all the analogs of the EPO gene that can be made, there is little enabling disclosure of particular analogs and how to make them. Details for preparing only a few EPO analog genes are disclosed. Amgen argues that this is sufficient to support its claims; we disagree. This "disclosure" might well justify a generic claim encompassing these and similar analogs, but it represents inadequate support for Amgen's desire to claim all EPO gene analogs. There may be many other genetic sequences that code for EPO-type products. Amgen has told how to make and use only a few of them and is therefore not entitled to claim all of them.

In affirming the district court's [invalidation of these] claims, we do not intend to imply that generic claims to genetic sequences cannot be valid where they are of a scope appropriate to the invention disclosed by an applicant. That is not the case here, where Amgen has claimed every possible analog of a gene containing about 4,000 nucleotides, with a disclosure only of how to make EPO and a very few analogs.

Considering the structural complexity of the EPO gene, the manifold possibilities for change in its structure, with attendant uncertainty as to what utility will be possessed by these analogs, we consider that more is needed concerning identifying the various analogs that are within the scope of the claim, methods for making them, and structural requirements for producing compounds with EPO-like activity. It is not sufficient, having made the gene and a handful of analogs whose activity has not been clearly ascertained, to claim all possible genetic sequences that have EPO-like activity. Under the circumstances, we find no error in the court's conclusion that the generic DNA sequence claims are invalid under Section 112.

COMMENTS AND QUESTIONS

1. To similar effect as *Amgen* is In re Goodman, 11 F.3d 1046, 1052 (Fed. Cir. 1993), where the court held that disclosure of a single working example could not support a broad claim to a method for producing mammalian proteins in the cells of plants.

Consider the interaction between section 103 and section 112 of the act. When we considered nonobviousness in the context of biotechnology, we saw that the courts found the isolation of a human DNA sequence from the protein it creates to be nonobvious, on the grounds that it would take a great deal of effort to identify the relevant DNA sequence. There, the difficulty of working in the field benefitted the patentee. But that same conclusion may make it hard for inventors to enable broad claims in the biotechnology field, since they would have to prove that *the same* "person skilled in the art" would be able to produce other examples covered by the claim without undue experimentation.

2. In addition to the types of broad claims at issue in *Amgen*, recent patents illustrate another expansive claim format. This characteristically reads: "[I claim] the nucleotide sequence in Figure 1, and functionally equivalent sequences having at least x percent sequence homology with said sequence."

Homologous regions are regions with identical nucleotide sequences. Thus the "percent homology" is a measure of the similarity between two strands of DNA; 100 percent homology means the two strands are identical. Homology is normally measured by hybridization, which takes advantage of the complementarity between the two strands of a DNA molecule. One of the two strands from one segment of DNA is joined with the complementary strand from another DNA segment; this forms a *hybrid* segment. If all the nucleotides "match up," or form complementary base pairs, there will be no unmatched regions on the hybrid, and one may conclude that the two original segments are 100 percent homologous. But if there are any mismatches, i.e., if any nucleotide on one strand is not matched with its complementary nucleotide on the other, the hybrid strands will not be bound together at that point.

Does the requirement that the sequence be "functionally equivalent" *and* similar in substantial part provide adequate protection against the concern in *Amgen* that the patentee will control numerous gene sequences that he has not produced or investigated and that may have value in themselves in addition to their use as analogs?

3. To what extent must the invention itself be fully described in the patent specification? In In re Wands, 858 F.2d 731 (Fed. Cir. 1988), the patentee claimed the use of particular monoclonal antibodies to perform a particular immunoassay.[35] The PTO challenged the application under the enablement requirement of 35 U.S.C. §112. The Federal Circuit noted that complete description was not always required:

> Appellants contend that their written specification fully enables the practice of their claimed invention because the monoclonal antibodies needed to perform the immunoassays can be made from readily available starting materials using methods that are well known in the monoclonal antibody art. There is no challenge to their contention that the starting materials (i.e., mice, HBsAg antigen, and myeloma cells) are available to the public. The PTO concedes that the methods used to prepare hybridomas and to screen them for high-affinity IgM antibodies against HBsAg were either well known in the monoclonal antibody art or adequately disclosed in the '145 patent and in the current application. The sole issue is whether, in this particular case, it would require undue experimentation to produce high-affinity IgM monoclonal antibodies.
>
> Enablement is not precluded by the necessity for some experimentation such as routine screening. However, experimentation needed to practice the invention must not be undue experimentation. "The key word is 'undue,' not 'experimentation.' " [In re Angstadt, 537 F.2d 498, 504, 190 U.S.P.Q. (BNA) 214, 219 (C.C.P.A. 1976).]
>
> The determination of what constitutes undue experimentation in a given case requires the application of a standard of reasonableness, having due regard for the nature of the invention and the state of the art.
>
> The test is not merely quantitative, since a considerable amount of experimentation is permissible, if it is merely routine, or if the specification in question provides a reasonable amount of guidance with respect to the direction in which the experimentation should proceed. . . .
>
> Factors to be considered in determining whether a disclosure would require undue experimentation have been summarized by the board in In re Forman. They include (1) the quantity of experimentation necessary, (2) the amount of direction or guidance presented, (3) the presence or absence of working examples, (4) the nature of the invention, (5) the state of the prior art, (6) the relative skill of those in the art, (7) the predictability or unpredictability of the art, and (8) the breadth of the claims. [In re Forman, 230 U.S.P.Q. (BNA) at 547.]

4. In Atlas Powder Co. v. E. I. du Pont De Nemours & Co., 750 F.2d 1569 (Fed. Cir. 1984), the court affirmed a finding that du Pont, the accused infringer, had not proved lack of enablement on the part of the patentee, Atlas Powder. The patent, for explosive compounds, listed in its specification numerous salts, fuels, and emulsifiers that could form thousands of emulsions, but it gave no commensurate information as to which combinations would work. Du Pont had argued that its tests showed a 40 percent failure rate in

---

35. On monoclonal antibody technology, see the Appendix on biotechnology in the statutory supplement.

constructing various embodiments of the claimed invention. The court rejected this "inoperable species" argument:

> Of course, if the number of inoperative combinations becomes significant, and in effect forces one of ordinary skill in the art to experiment unduly in order to practice the claimed invention, the claims might indeed be invalid. See, e.g., In re Cook, 439 F.2d 730, 735, 169 U.S.P.Q. 298, 302 (1971). That, however, has not been shown to be the case here. . . . The district court also found that one skilled in the art would know how to modify slightly many of [the experimental] "failures" to form a better emulsion.

Id. at 1576-1577.

5. Enablement cannot be defined without first determing *who* must be enabled. As one court has noted:

> A patent specification is not addressed to judges or lawyers, but to those skilled in the art; it must be comprehensible to them, even though the unskilled may not be able to gather from it how to use the invention, and even if it is 'all Greek' to the unskilled.

Gould v. Mossinghoff, 229 U.S.P.Q. 1 (D.D.C. 1985), *aff'd in part, vacated in part sub nom.* Gould v. Quigg, 822 F.2d 1074 (Fed. Cir. 1987).

6. In addition to the written description and enablement requirements, section 112 also mandates that the patent disclose the "best mode" of carrying out the invention contemplated by the inventor. This "best mode" requirement is designed to prevent a patentee from "holding back" knowledge from the public, in effect maintaining part of the invention as a trade secret while protecting the whole under patent law. One recent decision suggests that even accidental failure to disclose the best mode can invalidate a patent, though most earlier cases had required some evidence of intent to conceal. See U.S. Gypsum Co. v. National Gypsum Co., 74 F.3d 1209, 1215-1216 (Fed. Cir. 1996). But see Zygo Corp. v. Wyko Corp., 38 U.S.P.Q.2d 1281 (Fed. Cir. 1996) (failure to disclose the commercial embodiment of the patented invention did not violate the best mode rule).

Is disclosure of the "best mode" a reasonable requirement? If a patentee does not have to disclose a particular piece of information in order to tell the public how to make his invention, why require him to disclose it at all?

PROBLEM

Problem 3-10. Nethco, a major oil company, owns the rights to a patent covering the product "crystalline polypropylene." At the time the patent was written, the term referred to a particular substance that the patentee had produced and fully described in the patent specification. After the patent issues, but during its term, another researcher not affiliated with Nethco invents a radically new family of catalysts that for the first time make possible the production of a (crystalline) polypropylene of high molecular weight and intrinsic viscosity — two properties that make the fiber commercially useful. Because this invention occurred after the patent was issued, the new fibers are not described or even alluded to in the specification.

Does Nethco's patent cover these new forms of crystalline polypropylene? How can it be said to "enable" these later-developed versions of polypropylene under §112, ¶1?

C. INFRINGEMENT

1. Claim Interpretation

Patent claims define a patent owner's property right; they have been analogized from time immemorial to the "metes and bounds" of a real property deed. Word meanings determine the precise boundaries of claims. And these boundaries can be crucial: linguistically minor variations in phraseology and meaning are the difference between a finding of infringement (and thus exclusion of the competitor, and perhaps damages) and noninfringement (and thus open entry in at least part of the patentee's market). To the hardheaded businessperson, claim interpretation defines the "shelf space" that belongs exclusively to a patentee. To a patent lawyer, there is deep fascination in the almost scholastic debate over the difference between "therein" and "thereon" in a certain patent claim (see the *Larami* case below). The average businessperson looks straight to the bottom line: can the patentee exclude a competitor (expanding the reach of its exclusive legal franchise) or will it have to share shelf space and profits?

Courts employ a number of rules and procedures to mark off this space, most of which share a common spirit with general canons of legal interpretation applied to statutes, contracts, and the like. Nevertheless, patent interpretation has developed some peculiar details you need to know about. These fall roughly into three categories:

- Interpretive sources: where courts look for guidance regarding the meaning of claim terms;
- Interpretive canons: what general rules of interpretation courts apply in interpreting claims; and
- Interpretive procedures: when in the course of patent litigation a trial court should issue a definitive ruling regarding the meaning of claim terms, and what rules apply to appeals regarding these rulings.

a. Interpretive Sources

Blackletter patent law permits recourse to three primary sources in interpreting claims: the claim language; the patent specification; and the prosecution history of the patent.

i. Claim Language

Foremost is the language of the claim itself. In patent law's version of the "plain meaning rule," courts often resort to common linguistic definitions in deciding between the patentee's and accused infringer's proffered interpretations. (On the broader debate over plain meaning, see William N. Eskridge, Jr., The New Textualism, 37 UCLA L. Rev. 621 (1990)). Everyday dictionaries are often consulted in the process. To pick one of an endless series of examples, we will discuss Virginia Panel Corp. v. MAC Panel Co., 133 F.3d 860 (Fed. Cir. 1997). This case centered around the meaning of "reciprocating" in a claim element reciting the use of "reciprocating slide plates" in a mechanism for testing thousands of tiny electrical connections. The infringer, MAC, argued that reciprocating should be interpreted as *linear* reciprocating. Under this reading of the claim, MAC's product would not infringe. The court refused this interpretation, holding that "reciprocating" should be given its ordinary meaning and not limited to mere linear reciprocation:

> [T]he intrinsic evidence does not require that the word "reciprocating" be interpreted differently from its ordinary meaning . . . That ordinary meaning was established at trial; the uncontroverted evidence, including a recognized treatise, indicated that those skilled in the mechanical arts would have understood "reciprocating" to mean motion in which "a point traverses the same path and reverses its motion at the ends of such a path." Accordingly, because neither the '005 patent nor its prosecution history uses or implies a special definition for "reciprocating," the language "reciprocating slide plates" is not limited to plates that exhibit only linear motion, as opposed to the broader, ordinary meaning of that phrase. Thus, the disputed term literally encompasses rotating motion, such as that exhibited by the MAC mechanism.

Id. at 866. See also Specialty Composites v. Cabot Corp., 845 F.2d 981, 986-987 (Fed. Cir. 1988) (refusing to limit the recited claim term "plasticizer" to external plasticizers where skilled artisans used the term broadly). The court summarized the issue nicely in a succinct dictum from Renishaw PLC v. Marposs Societa' Per Azioni, 158 F.3d 1243 (Fed. Cir. 1998):

> [W]hen a claim term is expressed in general descriptive words, we will not ordinarily limit the term to a numerical range that may appear in the written description or in other claims. . . . Nor may we, in the broader situation, add a

narrowing modifier before an otherwise general term that stands unmodified in a claim. For example, if an apparatus claim recites a general structure (e.g., a noun) without limiting that structure to a specific subset of structures (e.g., with an adjective), we will generally construe the claim to cover all known types of that structure that are supported by the patent disclosure. See, e.g., Sjolund v. Musland, 847 F.2d 1573, 1581-82, 6 U.S.P.Q.2d 2020, 2027 (Fed. Cir.1988) (refusing to limit claim term "baffle" to only rigid baffles and term "panel" to only panels of lattice construction).

Id. at 1249.

ii. Patent Specification

But, as any good lawyer knows, ambiguity can be found in (or insinuated into) almost any source. "Ordinary meanings," and even dictionaries, are no exception. Consider for example the *Renishaw* case cited above. Renishaw held U.S. Patent 5,491,904 for "touch probes," thin mechanical probes whose ends or "sensing tips" precisely measure dimensions in various manufacturing settings. Claim 2 of the patent recited elements of the probe, including the following: "the probe generating a trigger signal when said sensing tip contacts an object. . . ." The issue in the case was deceptively simple: the meaning of "when." To be precise, defendant Marposs argued that "when" means "at the time of contact" — an interpretation, not coincidentally, that would place its test probes outside the scope of the claim. (The Marposs test probes sent a *delayed* signal.) Renishaw by contrast argued essentially that "when" means "at or after," which would bring the Marposs device within the ambit of the claims. Renishaw even unearthed a few dictionary definitions that supported its position. The court, however, was not impressed:

> [A] common meaning, such as one expressed in a relevant dictionary, that flies in the face of the patent disclosure is undeserving of fealty. Thus, where there are several common meanings for a claim term, the patent disclosure serves to point away from the improper meanings and toward the proper meaning. . . . "[W]hen" is not a broad and general term when standing in isolation. Instead, it has several meanings, each of which may prevail based on the context. Here, we have bounteous context. Claim 2 does not exist in rarefied air, but rather is surrounded by a patent disclosure of singular purpose. . . . Replete with references that indicate that the patentee was preeminently concerned with generating a trigger signal as soon as possible after contact, the written description lends precision to the term "when."

158 F.3d at 1251. In many other cases, the specification alone supplies the meaning of a claim term, without recourse to a dictionary. See, e.g., J.T. Eaton & Co. v. Atlantic Paste & Glue Co., 106 F.3d 1563, 1568 (Fed. Cir. 1997) (looking exclusively to specification to construe claim terminology re-

lating to glue-based vermin traps). The interaction between specification and claims is one source of the interpretive "canons" described later in this section.

iii. Prosecution History

An applicant sometimes makes statements during prosecution that limit the meaning of claim terms. (Note the similarity between this general interpretive rule and "prosecution history estoppel," a rule limiting the doctrine of equivalents described later in this chapter.) For example, in Spectrum Intl., Inc. v. Sterilite Corp., 164 F.3d 1372, 1378 (Fed. Cir. 1998), the Federal Circuit encountered a claim to a stackable set of plastic crates that could be used, for example, as containers for items to be recycled. The patent owner, Spectrum, claimed a crate with an opening formed in part by a "front wall" that was, in effect, a ridge along the bottom of the crate's front opening. During reexamination of the patent, Spectrum distinguished a prior art crate by arguing that the bottom of Spectrum's crate was attached to the bottom of the ridge or "front wall" — and thus was different from (and hence patentable over) the prior art reference. Yet during the infringement case Spectrum reversed field, arguing for infringement even when the "front wall" is attached to the floor of a crate along the top of the ridge or wall. (The accused product featured a lip that curled under the front opening in a crate; thus only the *top* of the lip, as a "front wall," was attached to the floor of the crate.) The court sided with the infringer on this issue:

> Spectrum argues for a broad construction [of the claim element]. Spectrum's argument must fail. [Spectrum's interpretation] . . . allows the addition of a downward-projecting appendage to the [front wall]. . . . But this embodiment is in the prior art. Therefore, to have sustained the patentability of claims 2 and 11 over the prior art asserted during reexamination, Spectrum must have relinquished from the language of these claims the possibility that merger of the bottom side also occurs with the top edge of the . . . front wall.

Id. at 1379.

iv. Extrinsic Evidence

More confusion attends the use of "extrinsic evidence" from outside the bounds of the patent itself. One such dispute concerns the role of expert testimony in helping the factfinder construe the claims of the patent and compare them to the accused device. The Federal Circuit has offered inconsistent guidance on this question in several recent opinions. Most notably, in Vitronics Corp. v. Conceptronic, Inc., 90 F.3d 1576 (Fed. Cir. 1996), the court inveighed against the use of expert testimony:

In most situations, an analysis of the intrinsic evidence alone will resolve any ambiguity in a disputed claim term. In such circumstances, it is improper to rely on extrinsic evidence. See, e.g., Pall Corp. v. Micron Separations, Inc., 66 F.3d 1211, 1216 (Fed. Cir. 1995). . . . In those cases where the public record unambiguously describes the scope of the patented invention, reliance on any extrinsic evidence is improper. The claims, specification, and file history, rather than extrinsic evidence, constitute the public record of the patentee's claim, a record on which the public is entitled to rely. In other words, competitors are entitled to review the public record, apply the established rules of claim construction, ascertain the scope of the patentee's claimed invention and, thus, design around the claimed invention. See Markman, 52 F.3d [967] at 978-79. Allowing the public record to be altered or changed by extrinsic evidence introduced at trial, such as expert testimony, would make this right meaningless. The same holds true whether it is the patentee or the alleged infringer who seeks to alter the scope of the claims.

. . . . No doubt there will be instances in which intrinsic evidence is insufficient to enable the court to determine the meaning of the asserted claims, and in those instances, extrinsic evidence, such as that relied on by the district court, may also properly be relied on to understand the technology and to construe the claims. See *Markman*, 52 F.3d at 979. Extrinsic evidence is that evidence which is external to the patent and file history, such as expert testimony, inventor testimony, dictionaries, and technical treatises and articles. Id. at 980. However, as we have recently re-emphasized, extrinsic evidence in general, and expert testimony in particular, may be used only to help the court come to the proper understanding of the claims; it may not be used to vary or contradict the claim language. Id. at 981. Nor may the inventor's subjective intent as to claim scope, when unexpressed in the patent documents, have any effect. Such testimony cannot guide the court to a proper interpretation when the patent documents themselves do so clearly.

Id. at 1583-1584. To similar effect are Bell & Howell Document Mgmt. Prods. v. Altek Sys., 132 F.3d 701 (Fed. Cir. 1997), where the court held that the district court erred by relying on expert testimony to interpret claim language the Federal Circuit considered unambiguous; and J.T. Eaton & Co. v. Atlantic Paste & Glue Co., 106 F.3d 1563 (Fed. Cir. 1997) (same).

On the other hand, the Federal Circuit has backed off considerably from the sweeping rhetoric of *Vitronics*. See Pitney-Bowes, Inc. v. Hewlett-Packard Corp., 51 U.S.P.Q.2d 1161 (Fed. Cir. 1999) (permitting broad leeway to district courts in deciding on outside sources to help in claim construction). Even before *Pitney-Bowes*, the court had approved the use of expert testimony in some cases. Most notably, in Eastman Kodak v. Goodyear Tire & Rubber Co., 114 F.3d 1547 (Fed. Cir. 1997), the court concluded that expert testimony was proper because the documentary record was ambiguous regarding the meaning of the claims. Further, the court was highly deferential to the district court's evaluation of that testimony:

The trial court is best situated to gauge the relevance and need for additional evidence to explicate claim terms. . . . [R]ecognizing both the trial court's trained ability to evaluate expert testimony in relation to the overall structure of the patent and the trial court's better position to ascertain whether an expert's proposed definition fully comports with the specification and claims, this court sustains the trial court's claim interpretation.

Id. at 1555-1556. See also Fromson v. Anitec Printing Plates, 132 F.3d 1437 (Fed. Cir. 1997); Endress & Hauser, Inc. v. Hawk Measurement Sys. Pty., 122 F.3d 1040 (Fed. Cir. 1997) (noting the "wide discretion" the district court has to admit testimony on the issue of claim construction). Cf. Mantech Environmental Corp. v. Hudson Environmental Services, 152 F.3d 1368 (Fed. Cir. 1998) (expert testimony could be admitted for background on the invention, but not relied upon in construing an unambiguous claim).

b. Canons of Construction

i. Claim-Specification Relationship

Renishaw states two important canons of claim construction:

Renishaw . . . alludes to a familiar pair of claim construction canons: (a) one may not read a limitation into a claim from the written description, but (b) one may look to the written description to define a term already in a claim limitation, for a claim must be read in view of the specification of which it is a part. These two rules lay out the general relationship between the claims and the written description. See Vitronics Corp. v. Conceptronic, Inc., 90 F.3d 1576, 1582, 39 U.S.P.Q.2d 1573, 1576 (Fed. Cir. 1996); Markman v. Westview Instruments, Inc., 52 F.3d 967, 979-80, 34 U.S.P.Q.2d 1321, 1329-30 (Fed. Cir. 1995) (en banc), aff'd, 517 U.S. 370, 116 S. Ct. 1384, 134 L.Ed.2d 577, 38 U.S.P.Q.2d 1461 (1996). Although no canon of construction is absolute in its application, these two rules share two underlying propositions. First, it is manifest that a claim must explicitly recite a term in need of definition before a definition may enter the claim from the written description . . . Thus, a party wishing to use statements in the written description to confine or otherwise affect a patent's scope must, at the very least, point to a term or terms in the claim with which to draw in those statements. Without any claim term that is susceptible of clarification by the written description, there is no legitimate way to narrow the property right. . . . The other clear point provided by these two canons covers the situation in which a patent applicant has elected to be a lexicographer by providing an explicit definition in the specification for a claim term. In such a case, the definition selected by the patent applicant controls. . . . If the patentee provides such a clear definition, the two canons require reference to the written description, because only there is the claim term defined as it is used by the patentee. The law provides a patentee with this opportunity because the public may not be schooled in the terminology of the technical art or there may not be an extant term of singular meaning for the structure or concept that is being claimed.

Renishaw PLC v. Marposs Societa' Per Azioni, 158 F.3d at 1249.

ii. Patentee as Lexicographer

Further refinements on this discussion of interpretive canons may be found in another recent opinion, Johnson Worldwide Associates, Inc. v. Zebco Corp., 175 F.3d 985, 990, 50 U.S.P.Q.2d 1607 (Fed. Cir. 1999):

> Our case law demonstrates two situations where a sufficient reason exists to require the entry of a definition of a claim term other than its ordinary and accustomed meaning. The first arises if the patentee has chosen to be his or her own lexicographer by clearly setting forth an explicit definition for a claim term. . . . The second is where the term or terms chosen by the patentee so deprive the claim of clarity that there is no means by which the scope of the claim may be ascertained from the language used. See . . . J.T. Eaton & Co. v. Atlantic Paste & Glue Co., 106 F.3d 1563, 1568, 41 U.S.P.Q.2d 1641, 1646 (Fed. Cir. 1997) (Because "[the disputed claim term] is a term with no previous meaning to those of ordinary skill in the prior art[,] [i]ts meaning, then, must be found [elsewhere] in the patent.") . . . In these two circumstances, a term or terms used in the claim invites — or indeed, requires — reference to intrinsic, or in some cases, extrinsic, evidence. . . .

iii. Claim Differentiation

A time-worn legal maxim is "lex rejicit superflua, pugnantia, incongrua," or "the law rejects superfluous, contradictory, and incongruous things." See Francis Bennion, Statutory Interpretation §316, at 776 (3d ed. 1997). This is commonly applied to the interpretation of statutes. See, e.g., South Carolina v. Catawba Indian Tribe, Inc., 476 U.S. 498, 510 n.22 (1986) ("It is an elementary canon of construction that a statute should be interpreted so as not to render one part inoperative."); Mackey v. Lanier Collection Agency & Service, Inc., 486 U.S. 825, 837 (1988) ("we are hesitant to adopt an interpretation of a congressional enactment which renders superfluous another portion of that same law").

This familiar principle has its application in patent law under the doctrine of "claim differentiation." Construction of a claim that would render another claim in the patent redundant is to be avoided. For example in Transmatic, Inc. v. Gulton Industries, Inc., 53 F.3d 1270 (Fed. Cir. 1995), the Federal Circuit reversed a trial court's construction of claim 1 of a patent that, in effect, imported into the claim various additional structural limitations present in the patent's third claim.

iv. Presumptive Breadth

Two general rules of claim construction amount to a set of presumptions regarding claim breadth. The first is that a claim should be interpreted so as to preserve its validity. Thus in Modine Mfg. Co. v. United States International Trade Commission, 75 F.3d 1545 (Fed. Cir. 1996), the court narrowly interpreted a claim element to "flat side walls" in an air conditioning condenser unit. This preserved the claim's validity over prior art that would have invalidated the claim under the broader meaning. But cf. Rhine v. Casio, Inc., 183 F.3d 1342 (Fed. Cir. 1999) (this axiom does not permit court to engage in wholesale rewrite of claims).

The second, related rule is this: if a claim is subject to two viable alternative interpretations, the narrower one should apply. See Athletic Alternatives, Inc. v. Prince Mfg., Inc., 73 F.3d 1573, 37 U.S.P.Q.2d 1365 (Fed. Cir. 1996). This rule may be thought of as a penalty for unclear drafting, as well as a reasonable "tiebreaker" when viewed from the perspective of public policy. The parallel principle in statutory interpretation has been explained on similar grounds. See generally Cass R. Sunstein, Interpreting Statutes in the Regulatory State, 103 Harv. L. Rev. 405, 457 (1989) ("The 'plain meaning' principle, for example, might be an effort not to discover what Congress meant in the particular case, but instead to tell Congress to be careful with statutory language. The principle warns Congress that courts will not guess about the meaning of statutes or supply remedies for language that leads to absurd results.").

c. Interpretive Procedures

Note on the Proper Role of Judge and Jury in Patent Cases

In patent cases, as elsewhere, the distinction between questions of law and fact is an important issue. As we have seen so far, patent cases often center on scientific or technical details such as how an invention differs from the prior art, or how it works compared to a competitor's product. Whether a judge or a jury decides these key issues may make a significant difference in the outcome of a patent infringement case.

In Markman v. Westview Instruments, 52 F.3d 967 (Fed. Cir. 1995) (en banc), the Federal Circuit put judges squarely in the driver's seat when it comes to interpreting patent claims. Upholding a district judge who had overturned a jury verdict based on improper claim construction, the court ruled (over a vigorous dissent) that claim construction is a matter of law. In addition, the court ruled that the Federal Circuit should review district court findings de novo, notwithstanding such factors as the trial court's proximity to expert witnesses.

The Supreme Court affirmed in a unanimous opinion. Markman v. Westview Instruments, 116 S. Ct. 1384 (1996). The Court evaluated Markman's claim that the Seventh Amendment required that a jury interpret the language of the patent claims by turning to the historical treatment of claim interpretation under English patent law. While the Court noted that "there is no dispute that infringement cases today must be tried to a jury, as their predecessors were more than two centuries ago," it could find no equivalent rule governing the construction of patent claims, since claims per se did not exist in early U.S. patent cases. Those cases the Court did consider did not unambiguously establish that terms of art in patents were historically interpreted by juries. The Court then turned to the policy considerations, which it found to support leaving claim construction in the hands of judges:

> The construction of written instruments is one of those things that judges often do and are likely to do better than jurors unburdened by training in exegesis. Patent construction in particular "is a special occupation, requiring, like all others, special training and practice. The judge, from his training and discipline, is more likely to give a proper interpretation to such instruments than a jury; and he is, therefore, more likely to be right, in performing such a duty, than a jury can be expected to be." Parker v. Hulme, 18 F. Cas. at 1140. Such was the understanding nearly a century and a half ago, and there is no reason to weigh the respective strengths of judge and jury differently in relation to the modern claim; quite the contrary, for "the claims of patents have become highly technical in many respects as the result of special doctrines relating to the proper form and scope of claims that have been developed by the courts and the Patent Office." Woodward, Definiteness and Particularity in Patent Claims, 46 Mich. L. Rev. 755, 765 (1948).
>
> Markman would trump these considerations with his argument that a jury should decide a question of meaning peculiar to a trade or profession simply because the question is a subject of testimony requiring credibility determinations, which are the jury's forte. It is, of course, true that credibility judgments have to be made about the experts who testify in patent cases, and in theory there could be a case in which a simple credibility judgment would suffice to choose between experts whose testimony was equally consistent with a patent's internal logic. But our own experience with document construction leaves us doubtful that trial courts will run into many cases like that. In the main, we expect, any credibility determinations will be subsumed within the necessarily sophisticated analysis of the whole document, required by the standard construction rule that a term can be defined only in a way that comports with the instrument as a whole. Thus, in these cases a jury's capabilities to evaluate demeanor, to sense the "mainsprings of human conduct," or to reflect community standards, are much less significant than a trained ability to evaluate the testimony in relation to the overall structure of the patent. The decisionmaker vested with the task of construing the patent is in the better position to ascertain whether an expert's proposed definition fully comports with the specification and claims and so will preserve the patent's internal coherence. We accordingly think there is sufficient reason to treat construction of terms of art like many other responsibilities that we cede to a judge in the normal course of trial, notwithstanding its evidentiary underpinnings. . . .

Finally, we see the importance of uniformity in the treatment of a given patent as an independent reason to allocate all issues of construction to the court. As we noted in General Elec. Co. v. Wabash Appliance Corp., 304 U.S. 364, 369, 82 L. Ed. 1402, 58 S. Ct. 899 (1938), "the limits of a patent must be known for the protection of the patentee, the encouragement of the inventive genius of others and the assurance that the subject of the patent will be dedicated ultimately to the public." Otherwise, a "zone of uncertainty which enterprise and experimentation may enter only at the risk of infringement claims would discourage invention only a little less than unequivocal foreclosure of the field," United Carbon Co. v. Binney & Smith Co., 317 U.S. 228, 236, 87 L. Ed. 232, 63 S. Ct. 165 (1942), and "the public [would] be deprived of rights supposed to belong to it, without being clearly told what it is that limits these rights." Merrill v. Yeomans, 94 U.S. 568, 573, 24 L. Ed. 235 (1877). It was just for the sake of such desirable uniformity that Congress created the Court of Appeals for the Federal Circuit as an exclusive appellate court for patent cases, observing that increased uniformity would "strengthen the United States patent system in such a way as to foster technological growth and industrial innovation."

On the practical side, *Markman* raises a number of potentially challenging issues for the patent system. Most importantly, it threatens to change a number of routine practices in patent cases. For example, patent cases are usually appealed only after a full trial. The trial will very often include a claim interpretation issue but will of course also cover a host of other issues. *Markman* raises the threat that a full trial on many subordinate issues may become moot depending on how the Federal Circuit interprets a claim. For example, if the claim is interpreted as excluding all of the accused infringer's products, there is no need to consider defenses, such as lack of enablement, failure to disclose best mode, and inequitable conduct. Indeed, to assure that time and money are not spent on pointless issues, the parties in patent cases may well seek to obtain a trial court's legal interpretation of claims and then ask the trial court to certify that finding for interlocutory appeal to the Federal Circuit. If this practice became widespread, the Federal Circuit could well wind up with a heavy caseload.

Another case decided in 1995, and later taken by the Supreme Court, provides a strong counterweight to the "judge-centered" thrust of *Markman*. In Hilton Davis Chemical Co., Inc. v. Warner-Jenkinson Co., Inc., 62 F.3d 1512 (Fed. Cir. 1995) (en banc), the Federal Circuit upheld the power of a jury to resolve questions of infringement under the doctrine of equivalents. Against the argument (supported by some case law and history) that the doctrine of equivalents is an *equitable* issue — and hence one for the court to decide — the Federal Circuit chose to emphasize the close factual questions present in most equivalents cases. The majority dispels a suggestion, made by one of the dissenters, that the issues in the case should have been treated as issues of law, "reviewable de novo on appeal." According to the majority —

Infringement, whether literal or under the doctrine of equivalents, is a question of fact. The Supreme Court made this abundantly clear in *Graver Tank*:

A finding of equivalence is one of fact. . . . Like any other issue of fact, final determination requires a balancing of credibility, persuasiveness and weight of evidence. It is to be decided by the trial court [in a bench trial] and that court's decision, under general principles of appellate review, should not be disturbed unless clearly erroneous. Particularly is this so in a field where so much depends upon familiarity with specific scientific problems and principles not usually contained in the general storehouse of knowledge and experience.

339 U.S. at 609-610. . . .

Hilton Davis, 62 F.3d at 1520-1521. The court dismissed the repeated references by the Supreme Court and the Federal Circuit to the doctrine of equivalents as "equitable," saying that the courts meant the term "in its broadest sense — equity as general fairness." Id.

Most commentators expected the Supreme Court to address this question when it took *Hilton Davis* on certiorari. But the Court declined to do so, rather oddly suggesting that it was not necessary to resolve the issue at the same time that it sent the case back to the district court for further factfinding. Who should do that factfinding is not entirely clear. However, the Court's language strongly hints at a continuing role for the jury. See *Warner-Jenkinson*, 117 S.Ct. at 1053 (suggesting that the use of special verdict forms could solve many of the problems associated with patent jury verdicts).

The changes wrought by *Markman* and *Hilton Davis* have by now thoroughly percolated through the patent system. As predicted, so-called "Markman hearings" have turned into a critical early phase of patent litigation. Special evidentiary and procedural rules have emerged, making these hearings highly specialized "mini-trials" that set the terms of debate for the remaining course of litigation. Of special note are the rules promulgated by the Northern District of California for these hearings (see N.D. Cal. Civ. L.R. §§16(6)-(11)) and a useful handbook aimed at trial court judges preparing for such hearings (James Amend, Patent Law: A Primer for Federal District Court Judges (U.C. Berkeley Center for Law and Technology, 1999)).

Despite the new procedures, some of the patent bar's more dire post-*Markman* predictions have failed to materialize. So far, Markman hearings have not been the subject of interlocutory appeals, for example. The doubling of Federal Circuit caseloads (predicted on the theory that each case would be appealed twice: once after the Markman hearing and again after final resolution) has not happened either — though it is clear that the Federal Circuit spends more time on claim construction than ever before.

COMMENTS AND QUESTIONS

1. Does *Hilton Davis* undo *Markman*? Even if a patentee cannot persuade a court to accept her interpretation of a claim, can't she simply try to persuade the jury of the same interpretation by invoking the doctrine of equivalents? (Consider the standard of review of a jury finding of infringement

under the doctrine of equivalents, which is the same as for any jury finding of fact. *Hilton Davis*, 62 F.3d at 1521.)

2. The logical consequence of the Supreme Court's holding in *Markman* (that claim construction is an issue of law) is that the Federal Circuit reviews it de novo, giving no deference to the trial court's construction of the claims. The Federal Circuit reached this conclusion en banc in Cybor Corp. v. FAS Technologies, 138 F.3d 1448 (Fed. Cir. 1998) (en banc). The Federal Circuit's willingness to reverse claim constructions made by the district courts has caused consternation among district courts that must conduct trials on the basis of claims that they may not fully understand until years later. One study suggests that in the first few years after *Markman*, over 50 percent of the trial court decisions in this area were reversed.

2. Literal Infringement

35 U.S.C. §271 gives the patentee the exclusive right to make, use, sell, offer for sale or import the invention described in the claims of the patent. This right runs from the day the patent is issued until the end of the patent term (formerly 17 years from the day of issue, now 20 years from the day the patent application is filed). Because a patent is defined in terms of its claims, a patent infringement lawsuit is resolved by comparing the claims of the patent to the accused product.

Larami Corp. v. Amron
United States District Court for the Eastern
District of Pennsylvania
27 U.S.P.Q.2d 1280 (E.D. Pa. 1993)

REED, J.

This is a patent case concerning toy water guns manufactured by plaintiff Larami Corporation ("Larami"). Currently before me is Larami's motion for partial summary judgment of noninfringement of United States Patent No. 4,239,129 ("the '129 patent"). . . .

For the reasons discussed below, the motion will be granted.

I. Background

Larami manufactures a line of toy water guns called "SUPER SOAK-ERS." This line includes five models: SUPER SOAKER 20, SUPER SOAKER 30, SUPER SOAKER 50, SUPER SOAKER 100, and SUPER SOAKER 200. All use a hand-operated air pump to pressurize water and a "pinch trigger" valve mechanism for controlling the ejection of the pressurized water. All feature detachable water reservoirs prominently situated outside and above the barrel of the gun. The United States Patent and Trademark Office has issued patents covering four of these models. Larami does not claim to have a patent which covers SUPER SOAKER 20.

Defendants Alan Amron and Talk To Me Products, Inc. (hereinafter referred to collectively as "TTMP") claim that the SUPER SOAKER guns infringe on the '129 patent which TTMP obtained by assignment from Gary Esposito ("Esposito"), the inventor. The '129 patent covers a water gun which, like the SUPER SOAKERS, operates by pressurizing water housed in a tank with an air pump. In the '129 patent, the pressure enables the water to travel out of the tank through a trigger-operated valve into an outlet tube and to squirt through a nozzle. Unlike the SUPER SOAKERS, the '129 patent also contains various electrical features to illuminate the water stream and create noises. Also, the water tank in the '129 patent is not detachable, but is contained within a housing in the body of the water gun.

The "Background of the Invention" contained in the '129 patent reads as follows:

> Children of all ages, especially boys, through the years have exhibited a fascination for water, lights and noise and the subject invention deals with these factors embodied in a toy simulating a pistol.
>
> An appreciable number of U.S. patents have been issued which are directed to water pistols but none appear to disclose a unique assembly of components which can be utilized to simultaneously produce a jet or stream of water, means for illuminating the stream and a noise, or if so desired, one which can be operated without employing the noise and stream illuminating means. A reciprocal pump is employed to obtain sufficient pressure whereby the pistol can eject a stream an appreciable distance in the neighborhood of thirty feet and this stream can be illuminated to more or less simulate a lazer [sic] beam. . . .

Larami has moved for partial summary judgment of noninfringement of the '129 patent . . . and for partial summary judgment on TTMP's counterclaim for infringement of the '129 patent.

II. Discussion . . .

B. *Infringement and Claim Interpretation*

A patent owner's right to exclude others from making, using or selling the patented invention is defined and limited by the language in that patent's claims. Corning Glass Works v. Sumitomo Electric U.S.A., Inc., 868 F.2d 1251, 1257, 9 U.S.P.Q.2d 1962 (Fed. Cir. 1989). Thus, establishing infringement requires the interpretation of the "elements" or "limitations" of the claim and a comparison of the accused product with those elements as so interpreted. Because claim interpretation is a question of law, it is amenable to summary judgment.

The words in a claim should be given their "ordinary or accustomed" meaning. Senmed, Inc. v. Richard-Allan Medical Industries, Inc., 888 F.2d 815, 819 & n.8, 12 U.S.P.Q.2d 1508 (Fed. Cir. 1989). An inventor's interpretations of words in a claim that are proffered after the patent has issued for purposes of litigation are given no weight. . . .

A patent holder can seek to establish patent infringement in either of two ways: by demonstrating that every element of a claim (1) is literally infringed or (2) is infringed under the doctrine of equivalents. To put it a different way, because every element of a claim is essential and material to that claim, a patent owner must, to meet the burden of establishing infringement, "show the presence of every element or its substantial equivalent in the accused device." If even one element of a patent's claim is missing from the accused product, then "[t]here can be no infringement as a matter of law." . . . London v. Carson Pirie Scott & Co., 946 F.2d 1534, 1538-39, 20 U.S.P.Q.2d 1456 (Fed. Cir. 1991).

Larami contends, and TTMP does not dispute, that twenty-eight (28) of the thirty-five (35) claims in the '129 patent are directed to the electrical components that create the light and noise. Larami's SUPER SOAKER water guns have no light or noise components. Larami also contends, again with no rebuttal from TTMP, that claim 28 relates to a "poppet valve" mechanism for controlling the flow of water that is entirely different from Larami's "pinch trigger" mechanism. Thus, according to Larami, the six remaining claims (claims 1, 5, 10, 11, 12 and 16) are the only ones in dispute. Larami admits that these six claims address the one thing that the SUPER SOAKERS and the '129 patent have in common — the use of air pressure created by a hand pump to dispense liquid. Larami argues, however, that the SUPER SOAKERS and the '129 patent go about this task in such fundamentally different ways that no claim of patent infringement is sustainable as a matter of law.

1. Literal Infringement of Claim 1

TTMP claims that SUPER SOAKER 20 literally infringes claim 1 of the '129 patent. Claim 1 describes the water gun as:

[a] toy comprising an elongated housing [case] having a chamber therein for a liquid [tank], a pump including a piston having an exposed rod [piston rod] and extending rearwardly of said toy facilitating manual operation for building up an appreciable amount of pressure in said chamber for ejecting a stream of liquid therefrom an appreciable distance substantially forwardly of said toy, and means for controlling the ejection.

U.S. Patent No. 4,239,129 (bracketed words supplied; see Diagram A, the '129 patent, attached hereto [Figure 3-3]).

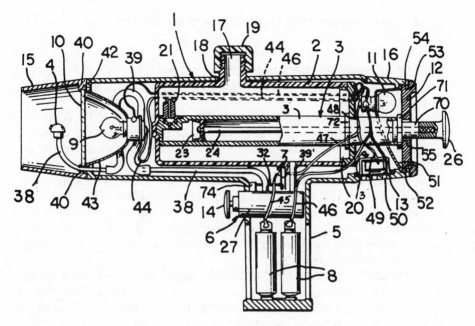

FIGURE 3-3
Patent drawing showing water pistol and/or flashlight structure.

Claim 1 requires, among other things, that the toy gun have "an elongated housing having a chamber therein for a liquid." The SUPER SOAKER 20 water gun, in contrast, has an external water reservoir (chamber) that is detachable from the gun housing, and not contained within the housing. TTMP argues that SUPER SOAKER 20 contains a "chamber therein for a liquid" as well as a detachable water reservoir. It is difficult to discern from TTMP's memorandum of law exactly where it contends the "chamber therein" is located in SUPER SOAKER 20. Furthermore, after having examined SUPER SOAKER 20 . . . , I find that it is plain that there is no "chamber" for liquid contained within the housing of the water gun. The only element of SUPER SOAKER 20 which could be described as a "chamber" for liquid is the external water reservoir located atop the housing. Indeed, liquid is located within the housing only when the trigger causes the liquid to pass from the external water reservoir through the tubing in the housing and out of the nozzle at the front end of the barrel. SUPER SOAKER 20 itself shows that such a transitory avenue for the release of liquid is clearly not a "chamber therein for liquid." Therefore, because the absence of even one element of a patent's claim from the accused product means there can be no finding of literal infringement, London, 946 F.2d at 1538-39, I find that SUPER SOAKER 20 does not infringe claim 1 of the '129 patent as a matter of law. . . .

Accordingly, I conclude that the SUPER SOAKER 20 water gun does not literally infringe claim 1 of the '129 patent.

2. Infringement by Equivalents of Claim 10

[The court further found that defendants TTMP failed to produce evidence which would support a finding that there was a genuine issue of material fact as to whether the Super Soaker guns infringed claim 10 under the doctrine of equivalents.

The court began its analysis by noting that successful use of the doctrine to show infringement requires the patent owner to prove that the accused product has the "substantial equivalent" of every limitation or element of a patent claim. TTMP argued that claim 10 teaches an arrangement of the tank, air pump and outlet nozzle along the same axis. The court, citing a previous decision regarding this same claim, denied that claim 10 required the three components to be located on the same axis; thus the axial placement of these components on the Super Soakers "cannot infringe claim 10 of the '129 patent because there is nothing in the language of claim 10 to which it could be substantially equivalent."

Additionally, Super Soakers' use of an external, detachable water reservoir was found to be such a dramatic improvement over the traditional design — benefitting both the manufacturer and user — that it could not be held to be the "substantial equivalent" of the claim 10 requirement of "a tank in the barrel for a liquid."]

III. Conclusion

In patent cases, summary judgment is appropriate where the accused product does not literally infringe the patent and where the patent owner does not muster evidence that is "sufficient to satisfy the legal standard for infringement under the doctrine of equivalents." London, 946 F.2d at 1538. Thus, and for the foregoing reasons, Larami's motion for partial summary judgment of noninfringement of the '129 patent will be granted.

COMMENTS AND QUESTIONS

1. Suppose TTMP had drafted claim 1 using the following phrase: "an elongated housing having a conjoining chamber for a liquid." Further suppose that the dictionary defines conjoining as "attached to; connected to." If the specification failed to attach any special meaning to this term, would the court's holding regarding literal infringement of claim 1 have been affected? See Casler v. U.S., 9 U.S.P.Q.2d 1753, 1772 (Ct. Cl. 1988); Universal Oil Products Co. v. Globe Oil & Refining Co., 40 F. Supp. 575, 582 (N.D. Ill. 1941), aff'd, 137 F.2d 3 (7th Cir. 1943), aff'd, 322 U.S. 471 (1944).

2. Would the result be different if claim 1 had contained the phrase in question 1 (above) but the word "conjoining" had been defined *in the spec-*

ification to mean "in or on," as opposed to the dictionary meaning in question 1? Which definition would control? See Envirotech Corp. v. Al George, Inc., 730 F.2d 753, 759 (Fed. Cir. 1984).

3. As in question 1 (above), assume the specification was silent as to any particular meaning of the term "conjoining" found in claim 1. But suppose that plaintiff Larami discovers claim 1 was originally rejected by an examiner because the dictionary definition of "conjoining" rendered the claim obvious in light of the prior art. The patent applicant successfully overcame this rejection by disavowing the dictionary definition of "conjoining" and adopting a narrower definition. In the instant litigation, which interpretation of "conjoining" would have been used by the court to determine literal infringement? Recall that the face of the '129 patent would not give a clue to potential infringers such as Larami as to the narrower definition of the claim language conceded during the prosecution history. To find out what the inventor argued during prosecution, one would need to obtain the prosecution history, or "file wrapper," for the patent. See Jonsson v. Stanley Works, 903 F.2d 812, 818 (Fed. Cir. 1990).

4. The elements of a patent claim are of considerable importance in determining its scope. For an accused product to literally infringe a patent, *every* element contained in the patent claim must also be present in the accused product or device. If a claimed apparatus has five parts, or "elements," and the allegedly infringing apparatus has only four of those five, it does not literally infringe. This is true even though the defendant may have copied the four elements exactly, and regardless of how significant or insignificant the missing element is.

The reverse is not always true. If an accused device has all the elements of a patent claim, it will generally be found to infringe. In most cases, that is, it does not matter that the defendant has *added* several new elements — adding new features cannot help a defendant escape infringement. The exact outcome of such a case is determined by the precise claim language and the nature and significance of the "added elements." See, e.g., In re Herz, 537 F.2d 549 (C.C.P.A. 1976).

This rule has an important consequence for the process of innovation. Patentees who have properly claimed a fundamental technology can assert their patent against anyone who uses that technology, even if the defendants have improved it or put it to different use. A broad basic patent therefore gives its owner a great deal of control not only over potential direct competitors, but over a number of derivative or ancillary markets during the term of the patent.

3. The Doctrine of Equivalents

Graver Tank & Mfg. Co. v. Linde Air Products Co.
Supreme Court of the United States
339 U.S. 605 (1950)

Mr. Justice JACKSON delivered the opinion of the Court.

Linde Air Products Co., owner of the Jones patent for an electric welding process and for fluxes to be used therewith, brought an action for infringement against [Graver]. The trial court held four flux claims valid and infringed and certain other flux claims and all process claims invalid. The Court of Appeals affirmed findings of validity and infringement as to the four flux claims but reversed the trial court and held valid the process claims and the remaining contested flux claims. We granted certiorari, and . . . reinstated the District Court decree. 336 U.S. 271. Rehearing was granted, limited to the question of infringement of the four valid flux claims and to the applicability of the doctrine of equivalents to findings of fact in this case. . . .

In determining whether an accused device or composition infringes a valid patent, resort must be had in the first instance to the words of the claim. If accused matter falls clearly within the claim, infringement is made out and that is the end of it.

But courts have also recognized that to permit imitation of a patented invention which does not copy every literal detail would be to convert the protection of the patent grant into a hollow and useless thing. Such a limitation would leave room for — indeed encourage — the unscrupulous copyist to make unimportant and insubstantial changes and substitutions in the patent which, though adding nothing, would be enough to take the copied matter outside the claim, and hence outside the reach of law. One who seeks to pirate an invention, like one who seeks to pirate a copyrighted book or play, may be expected to introduce minor variations to conceal and shelter the piracy. Outright and forthright duplication is a dull and very rare type of infringement. To prohibit no other would place the inventor at the mercy of verbalism and would be subordinating substance to form. It would deprive him of the benefit of his invention and would foster concealment rather than disclosure of inventions, which is one of the primary purposes of the patent system.

The doctrine of equivalents evolved in response to this experience. The essence of the doctrine is that one may not practice a fraud on a patent. Originating almost a century ago in the case of Winans v. Denmead, [56 U.S.] 15 How. 330 [1853], it has been consistently applied by this Court and the lower federal courts, and continues today ready and available for utilization when the proper circumstances for its application arise. "To temper unsparing logic and prevent an infringer from stealing the benefit of the invention" a patentee may invoke this doctrine to proceed against the producer of a device "if it performs substantially the same function in substantially the same way to obtain the same result." Sanitary Refrigerator Co. v. Winters, 280 U.S. 30, 42 (1929). The theory on which it is founded is that "if two devices do the same work in substantially the same way, and accomplish substantially the same result, they are the same, even though they differ in name, form or shape." Union Paper-Bag Machine Co. v. Murphy, 97 U.S. 120, 125 [1877]. The doctrine operates not only in favor of the patentee of a pioneer or primary invention, but also for the patentee of a secondary invention consisting of a combination of old ingredients which produce new and useful results, Imhaeuser v. Buerk, 101 U.S. 647, 655 [1879], although the area of equivalence may vary under the circumstances. The wholesome realism of this doctrine is not always applied in favor of a patentee but is sometimes used against him. Thus, where a device is so far changed in principle from a patented article that it performs the same or a similar function in a substantially different way, but nevertheless falls within the literal words of the claim, the doctrine of equivalents may be used to restrict the claim and defeat the patentee's action for infringement. Westinghouse v. Boyden Power Brake Co., 170 U.S. 537, 568 [1897]. In its early development, the doctrine was usually applied in cases involving devices where there was equivalence in mechanical components. Subsequently, however, the same principles were also applied to compositions, where there was equivalence between chemical ingredients. Today the doctrine is applied to mechanical or chemical equivalents in compositions or devices. Ellis, Patent Claims (1949) §§59-60.

What constitutes equivalency must be determined against the context of the patent, the prior art, and the particular circumstances of the case. Equivalence, in the patent law, is not the prisoner of a formula and is not an absolute to be considered in a vacuum. It does not require complete identity for every purpose and in every respect. In determining equivalents, things equal to the same thing may not be equal to each other and, by the same token, things for most purposes different may sometimes be equivalents. Consideration must be given to the purpose for which an ingredient is used in a patent, the qualities it has when combined with the other ingredients, and the function which it is intended to perform. An important factor is whether persons reasonably skilled in the art would have known of the interchangeability of an ingredient not contained in the patent with one that was.

A finding of equivalence is a determination of fact. Proof can be made in any form: through testimony of experts or others versed in the technology; by documents, including texts and treatises; and, of course, by the disclosures of the prior art. Like any other issue of fact, final determination requires a balancing of credibility, persuasiveness and weight of evidence. It is to be decided by the trial court and that court's decision, under general principles of appellate review, should not be disturbed unless clearly erroneous. Particularly is this so in a field where so much depends upon familiarity with specific scientific problems and principles not usually contained in the general storehouse of knowledge and experience.

In the case before us, we have two electric welding compositions or fluxes: the patented composition, Unionmelt Grade 20, and the accused composition, Lincolnweld 660. . . . Unionmelt actually contains silicates of calcium and magnesium, two alkaline earth metal silicates. Lincolnweld's composition is similar to Unionmelt's, except that it substitutes silicates of calcium and manganese — the latter not an alkaline earth metal — for silicates of calcium and magnesium. In all other respects, the two compositions are alike. The mechanical methods in which these compositions are employed are similar. They are identical in operation and produce the same kind and quality of weld.

The question which thus emerges is whether the substitution of the manganese which is not an alkaline earth metal for the magnesium which is, under the circumstances of this case, and in view of the technology and the prior art, is a change of such substance as to make the doctrine of equivalents inapplicable; or conversely, whether under the circumstances the change was so insubstantial that the trial court's invocation of the doctrine of equivalents was justified.

Without attempting to be all-inclusive, we note the following evidence in the record: Chemists familiar with the two fluxes testified that manganese and magnesium were similar in many of their reactions. There is testimony by a metallurgist that alkaline earth metals are often found in manganese ores in their natural state and that they serve the same purpose in the fluxes; and a chemist testified that "in the sense of the patent" manganese could be included as an alkaline earth metal. Much of this testimony was corroborated by reference to recognized texts on inorganic chemistry. Particularly important, in addition, were the disclosures of the prior art, also contained in the record. The Miller patent, No. 1,754,566, which preceded the patent in suit, taught the use of manganese silicate in welding fluxes. Manganese was similarly disclosed in the Armor patent, No. 1,467,825, which also described a welding composition. And the record contains no evidence of any kind to show that Lincolnweld was developed as the result of independent research or experiments.

It is not for this Court to even essay an independent evaluation of this evidence. This is the function of the trial court. And, as we have heretofore observed, "To no type of case is this . . . more appropriately applicable than to the one before us, where the evidence is largely the testimony of experts as to which a trial court may be enlightened by scientific demonstrations. This trial occupied some three weeks, during which, as the record shows, the trial judge visited laboratories with counsel and experts to observe actual demonstrations of welding as taught by the patent and of the welding accused of infringing it, and of various stages of the prior art. He viewed motion pictures of various welding operations and tests and heard many experts and other witnesses." 336 U.S. 271, 274-275.

The trial judge found on the evidence before him that the Lincolnweld flux and the composition of the patent in suit are substantially identical in operation and in result. He found also that Lincolnweld is in all respects equivalent to Unionmelt for welding purposes. And he concluded that "for all practical purposes, manganese silicate can be efficiently and effectively substituted for calcium and magnesium silicates as the major constituent of the welding composition." These conclusions are adequately supported by the record; certainly they are not clearly erroneous.

It is difficult to conceive of a case more appropriate for application of the doctrine of equivalents. The disclosures of the prior art made clear that manganese silicate was a useful ingredient in welding compositions. Specialists familiar with the problems of welding compositions understood that manganese was equivalent to and could be substituted for magnesium in the composition of the patented flux and their observations were confirmed by the literature of chemistry. Without some explanation or indication that Lincolnweld was developed by independent research, the trial court could properly infer that the accused flux is the result of imitation rather than experimentation or invention. Though infringement was not literal, the changes which avoid literal infringement are colorable only. We conclude that the trial court's judgment of infringement respecting the four flux claims was proper, and we adhere to our prior decision on this aspect of the case.

Affirmed.

Mr. Justice BLACK, with whom Mr. Justice DOUGLAS concurs, dissenting.

I heartily agree with the Court that "fraud" is bad, "piracy" is evil, and "stealing" is reprehensible. But in this case, where petitioners are not charged with any such malevolence, these lofty principles do not justify the Court's sterilization of Acts of Congress and prior decisions, none of which are even mentioned in today's opinion.

The only patent claims involved here describe respondent's product as a flux "containing a major proportion of alkaline earth metal silicate." The trial court found that petitioners used a flux "composed principally of manganese silicate." Finding also that "manganese is not an alkaline earth metal," the trial court admitted that petitioners' flux did not "literally infringe" respondent's patent. Nevertheless it invoked the judicial "doctrine of equivalents" to broaden the claim for "alkaline earth metals" so as to embrace "manganese." On the ground that "the fact that manganese is a proper substitute . . . is fully disclosed in the specification" of respondent's patent, it concluded that "no determination need be made whether it is a known chemical fact outside the teachings of the patent that manganese is an equivalent." . . . Since today's affirmance unquestioningly follows the findings of the trial court, this Court necessarily relies on what the specifications revealed. In so doing, it violates a direct mandate of Congress without even discussing that mandate.

R.S. §4888, as amended, 35 U.S.C. §33 [now code §112], provides that an applicant "shall particularly point out and distinctly claim the part, improvement, or combination which he claims as his invention or discovery." We have held in this very case that this statute precludes invoking the specifications to alter a claim free from ambiguous language, since "it is the claim which measures the grant to the patentee." 336 U.S. 271, 277. What is not specifically claimed is dedicated to the public. For the function of claims under R.S. §4888, as we have frequently reiterated, is to exclude from the patent monopoly field all that is not specifically claimed, whatever may appear in the specifications. Today the Court tacitly rejects those cases. It departs from the underlying principle which, as the Court pointed out in White v. Dunbar, 119 U.S. 47, 51 [1886] forbids treating a patent claim

> like a nose of wax, which may be turned and twisted in any direction, by merely referring to the specification, so as to make it include something more than, or something different from, what its words express. . . . The claim is a statutory requirement, prescribed for the very purpose of making the patentee define precisely what his invention is; and it is unjust to the public, as well as an evasion of the law, to construe it in a manner different from the plain import of its terms.

Giving this patentee the benefit of a grant that it did not precisely claim is no less "unjust to the public" and no less an evasion of R.S. §4888 merely because done in the name of the "doctrine of equivalents."

COMMENTS AND QUESTIONS

1. The dissent makes several other arguments. One is that "the similar use of manganese in prior expired patents" should prevent the patentee from claiming it. As a general matter this is a correct statement of law; the doctrine of equivalents may not be used to expand a claim to include structures described in the prior art. See, e.g., Loctite Corp. v. Ultraseal Ltd., 781 F.2d 861, 870 (Fed. Cir. 1985). In this case, however, the majority apparently concluded that although the accused product's welding flux was known in the prior art, no prior art reference taught the *combination* of this flux with the other elements of the claimed invention. This issue remains unresolved to this day. Compare Baxter Healthcare Corp. v. Spectramed, Inc., 79 F.2d 1165 (Fed. Cir. 1995) (unpublished) (fact that the accused device merely combined references in the prior art did not render it noninfringing) with Laminating Co. of America v. Tri-Star Laminates, Inc., 35 U.S.P.Q.2d 1149 (C.D. Cal. 1995) (patent holder cannot use the doctrine of equivalents to extend the reach of her patent to cover things that would have been obvious at the time the invention was made). For a contemporary statement of the view taken in the dissent, see Marty Adelman & Gary Francione, The Doctrine of Equivalents in Patent Law: Questions that *Pennwalt* Did Not Answer, 137 U. Pa. L. Rev. 673 (1989). See also Paul Janicke, Heat of Passion: What Really Happened in Graver Tank, 24 AIPLA Q.J. 1 (1997) (describing the case in great factual detail and concluding that the defendants should have prevailed).

A recent — and controversial — application of the rule that the patentee cannot claim prior art is Wilson Sporting Goods Co. v. David Geoffrey &

Assoc., 904 F.2d 677 (Fed. Cir. 1990), *cert. denied*, 498 U.S. 992 (1990). In that case, Wilson sued the defendant for infringement of its patent on a golf ball with a particular configuration of "dimples" on it. The defendant's ball had a very similar configuration of dimples, but it lacked the novel feature of Wilson's ball — a group of six "great circles" free of dimples. The court rejected the claim of infringement under the doctrine of equivalents:

> [A] patentee should not be able to obtain, under the doctrine of equivalents, coverage which he could not lawfully have obtained from the PTO by literal claims. The doctrine of equivalents exists to prevent a fraud on a patent, not to give a patentee something which he could not lawfully have obtained from the PTO had he tried. Thus, since prior art always limits what an inventor could have claimed, it limits the range of permissible equivalents of a claim.
>
> . . . To simplify analysis and bring the issue onto familiar turf, it may be helpful to conceptualize the limitation on the scope of equivalents by visualizing a hypothetical patent claim, sufficient in scope to literally cover the accused product. The pertinent question then becomes whether that hypothetical claim could have been allowed by the PTO over the prior art. If not, then it would be improper to permit the patentee to obtain that coverage in an infringement suit under the doctrine of equivalents. If the hypothetical claim could have been allowed, then prior art is not a bar to infringement under the doctrine of equivalents.
>
> Viewing the issue in this manner allows use of traditional patentability rules and permits a more precise analysis than determining whether an accused product (which has no claim limitations on which to focus) would have been obvious in view of the prior art.

But see National Presto Indus. v. The West Bend Co., 76 F.3d 1185 (Fed. Cir. 1996) (no error in failing to conduct a hypothetical claim analysis under *Wilson Sporting Goods*, defendant bears the burden of proving that the accused device is covered by the prior art).

2. The dissent's next argument in *Graver Tank* is that

> Whatever the merits of the doctrine of equivalents where differences between the claims of a patent and the allegedly infringing product are de minimis, colorable only, and without substance, that doctrine should have no application to the facts of this case. For the differences between respondent's welding substance and petitioner's claimed flux were not nearly so slight.

This kind of disagreement — over precisely how much the asserted equivalent varies from the claimed embodiments of the invention — is extremely common in doctrine of equivalents cases. Does the Court in *Graver Tank* provide any guidance in how to determine whether a product is equivalent to a patent claim?

3. A first step in trying to flesh out the meaning of the "function-way-result" test in *Graver Tank* is to apply it to the hypothetical situation considered in the section on literal infringement. Suppose a patent claim has five elements. An accused device copies four of those elements exactly but omits

the (relatively minor) fifth element. We established above that this defendant is not liable for literal infringement. But is her device "equivalent" to the claimed invention?

The answer to this question depends on whether equivalence is tested with reference to the claimed invention as a whole, or element by element. If we think of the doctrine of equivalents as creating a "penumbra" around the exclusive right provided by the claim language itself, the question is whether the whole invention has a penumbra or each element has its own penumbra.

The Federal Circuit addressed this question in 1987 in the case of Pennwalt Corp. v. Durand-Wayland, Inc., 833 F.2d 931 (Fed. Cir. 1987) (en banc), *cert. denied*, 485 U.S. 961 (1988). There, it held that the doctrine of equivalents required an element-by-element analysis of equivalence: each element in the patent claim must have a corresponding part in the accused device — either the identical element or an "equivalent" under *Graver Tank*. Under *Pennwalt*, our hypothetical patentee will lose his suit, because there is nothing in the defendant's device "equivalent" to the missing fifth element.

Judge Bennett dissented in *Pennwalt*:

> The majority facially retains the historical test set forth in *Graver Tank & Mfg. Co.* for infringement under the doctrine of equivalents by stating that infringement in such instances may be found if an accused device performs substantially the same overall function or work, in substantially the same way, to obtain substantially the same overall result as the claimed invention. But in practical effect, the majority has eviscerated the underlying rationale of the *Graver Tank* test by requiring, under the doctrine of equivalents, an exact equivalent for each element of the claimed invention.

Is this a fair criticism?

Several post-*Pennwalt* cases appear implicitly to reject the strict element-by-element approach. See, e.g., Corning Glass Works v. Sumitomo Elec. U.S.A., 868 F.2d 1251 (Fed. Cir. 1989).

≡
≡ *Warner-Jenkinson Company, Inc. v. Hilton*
≡ *Davis Chemical Co.*
≡ Supreme Court of the United States
≡ 520 U.S. 17 (1997)

Mr. Justice Thomas delivered the opinion of the Court.

Nearly 50 years ago, this Court in Graver Tank & Mfg. Co. v. Linde Air Products Co., 339 U.S. 605, 70 S. Ct. 854, 94 L. Ed. 1097 (1950), set out the modern contours of what is known in patent law as the "doctrine of equivalents." Under this doctrine, a product or process that does not literally infringe upon the express terms of a patent claim may nonetheless be found to infringe if there is "equivalence" between the elements of the accused product or process and the claimed elements of the patented invention. *Id.,* at 609, 70 S. Ct., at 856-857. Petitioner, which was found to have infringed upon respondent's patent under the doctrine of equivalents, invites us to speak the death of that doctrine. We decline that invitation. The significant disagreement within the Court of Appeals for the Federal Circuit concerning the application of *Graver Tank* suggests, however, that the doctrine is not free from confusion. We therefore will endeavor to clarify the proper scope of the doctrine.

I

The essential facts of this case are few. Petitioner Warner-Jenkinson Co. and respondent Hilton Davis Chemical Co. manufacture dyes. Impurities in those dyes must be removed. Hilton Davis holds United States Patent No. 4,560,746 ('746 patent), which discloses an improved purification process involving "ultrafiltration." The '746 process filters impure dye through a porous membrane at certain pressures and pH levels, resulting in a high purity dye product.

The '746 patent issued in 1985. As relevant to this case, the patent claims as its invention an improvement in the ultrafiltration process as follows:

> In a process for the purification of a dye . . . the improvement which comprises: subjecting an aqueous solution . . . to ultrafiltration through a membrane having a nominal pore diameter of 5-15 Angstroms under a hydrostatic pressure of approximately 200 to 400 p.s.i.g., *at a pH from approximately 6.0 to 9.0*, to thereby cause separation of said impurities from said dye. . . .

The inventors added the phrase "at a pH from approximately 6.0 to 9.0" during patent prosecution. At a minimum, this phrase was added to distinguish a previous patent (the "Booth" patent) that disclosed an ultrafiltration process operating at a pH above 9.0. The parties disagree as to why the low-end pH limit of 6.0 was included as part of the claim.[2]

In 1986, Warner-Jenkinson developed an ultrafiltration process that operated with membrane pore diameters assumed to be 5-15 Angstroms, at pressures of 200 to nearly 500 p.s.i.g., and at a pH of 5.0. Warner-Jenkinson did not learn of the '746 patent until after it had begun commercial use of its ultrafiltration process. Hilton Davis eventually learned of Warner-Jenkinson's use of ultrafiltration and, in 1991, sued Warner-Jenkinson for patent infringement.

As trial approached, Hilton Davis conceded that there was no literal infringement, and relied solely on the doctrine of equivalents. Over Warner-

---

2. Petitioner contends that the lower limit was added because below a pH of 6.0 the patented process created "foaming" problems in the plant and because the process was not shown to work below that pH level. Respondent counters that the process was successfully tested to pH levels as low as 2.2 with no effect on the process because of foaming, but offers no particular explanation as to why the lower level of 6.0 pH was selected.

Jenkinson's objection that the doctrine of equivalents was an equitable doctrine to be applied by the court, the issue of equivalence was included among those sent to the jury. The jury found that the '746 patent was not invalid and that Warner-Jenkinson infringed upon the patent under the doctrine of equivalents.

The District Court denied Warner-Jenkinson's post-trial motions, and entered a permanent injunction prohibiting Warner-Jenkinson from practicing ultrafiltration below 500 p.s.i.g. and below 9.01 pH. A fractured en banc Court of Appeals for the Federal Circuit affirmed. 62 F.3d 1512 (Fed. Cir.1995). . . .

II . . .

[The court rejected petitioner's argument that the doctrine of equivalents, as set out in *Graver Tank*, did not survive the 1952 revision of the Patent Act.]

III

Understandably reluctant to assume this Court would overrule *Graver Tank*, petitioner has offered alternative arguments in favor of a more restricted doctrine of equivalents than it feels was applied in this case. We address each in turn.

A

Petitioner first argues that *Graver Tank* never purported to supersede a well-established limit on non-literal infringement, known variously as "prosecution history estoppel" and "file wrapper estoppel." See Bayer Aktiengesellschaft v. Duphar Intl. Research B.V., 738 F.2d 1237, 1238 (Fed. Cir. 1984). According to petitioner, any surrender of subject matter during patent prosecution, regardless of the reason for such surrender, precludes recapturing any part of that subject matter, even if it is equivalent to the matter expressly claimed. Because, during patent prosecution, respondent limited the pH element of its claim to pH levels between 6.0 and 9.0, petitioner would have those limits form bright lines beyond which no equivalents may be claimed. Any inquiry into the reasons for a surrender, petitioner claims, would undermine the public's right to clear notice of the scope of the patent as embodied in the patent file.

We can readily agree with petitioner that *Graver Tank* did not dispose of prosecution history estoppel as a legal limitation on the doctrine of equivalents. But petitioner reaches too far in arguing that the reason for an amendment during patent prosecution is irrelevant to any subsequent estoppel. In each of our cases cited by petitioner and by the dissent, prosecution history estoppel was tied to amendments made to avoid the prior art, or otherwise to address a specific concern — such as obviousness — that arguably would have rendered the claimed subject matter unpatentable. . . .

It is telling that in each case this Court probed the reasoning behind the Patent Office's insistence upon a change in the claims. In each instance, a change was demanded because the claim as otherwise written was viewed as not describing a patentable invention at all — typically because what it described was encompassed within the prior art. But, as the United States informs us, there are a variety of other reasons why the PTO may request a change in claim language. And if the PTO has been requesting changes in claim language without the intent to limit equivalents or, indeed, with the expectation that language it required would in many cases allow for a range of equivalents, we should be extremely reluctant to upset the basic assumptions of the PTO without substantial reason for doing so. Our prior cases have consistently applied prosecution history estoppel only where claims have been amended for a limited set of reasons, and we see no substantial cause for requiring a more rigid rule invoking an estoppel regardless of the reasons for a change.

In this case, the patent examiner objected to the patent claim due to a perceived overlap with the Booth patent, which revealed an ultrafiltration process operating at a pH above 9.0. In response to this objection, the phrase "at a pH from approximately 6.0 to 9.0" was added to the claim. While it is undisputed that the upper limit of 9.0 was added in order to distinguish the Booth patent, the reason for adding the lower limit of 6.0 is unclear. The lower limit certainly did not serve to distinguish the Booth patent, which said nothing about pH levels below 6.0. Thus, while a lower limit of 6.0, by its mere inclusion, became a material element of the claim, that did not necessarily preclude the application of the doctrine of equivalents as to that element. Where the reason for the change was not related to avoiding the prior art, the change may introduce a new element, but it does not necessarily preclude infringement by equivalents of that element.

We are left with the problem, however, of what to do in a case like the one at bar, where the record seems not to reveal the reason for including the lower pH limit of 6.0. In our view, holding that certain reasons for a claim amendment may avoid the application of prosecution history estoppel is not tantamount to holding that the absence of a reason for an amendment may similarly avoid such an estoppel. Mindful that claims do indeed serve both a definitional and a notice function, we think the better rule is to place the burden on the patent-holder to establish the reason for an amendment required during patent prosecution. The court then would decide whether that reason is sufficient to overcome prosecution history estoppel as a bar to application of the doctrine of equivalents to the element added by that amendment. Where no explanation is established, however, the court should presume that the PTO had a substantial reason related to patentability for including the limiting element added by amendment. In those circumstances, prosecution history estoppel would bar the application of the doctrine equivalents as to that element. The presumption we have described, one subject to rebuttal if an appropriate reason for a required amendment is established, gives proper deference to the role of claims in defining an invention and providing public notice, and to the primacy of the PTO in ensuring that the claims allowed cover only subject matter that is properly patentable in a proffered patent application. Applied in this fashion, prosecution history estoppel places reasonable limits on the doctrine of equivalents, and further insulates the doctrine from any feared conflict with the Patent Act.

Because respondent has not proffered in this Court a reason for the addition of a lower pH limit, it is impossible to tell whether the reason for that addition could properly avoid an estoppel. Whether a reason in fact exists, but simply was not adequately developed, we cannot say. On remand, the Federal Circuit can consider whether reasons for that portion of the amendment were offered or not and whether further opportunity to establish such reasons would be proper.

B

[Petitioner then argued that infringement should only be found if the accused infringer had actively copied the patented invention. The Federal Circuit had adopted a new test for equivalence that included subjective elements such as intent to copy or alternatively to design around the patent. The Court rejected the use of such subjective elements, stating:] If the essential predicate of the doctrine of equivalents is the notion of identity between a patented invention and its equivalent, there is no basis for treating an infringing equivalent any differently than a device that infringes the express terms of the patent. Application of the doctrine of equivalents, therefore, is akin to determining literal infringement, and neither requires proof of intent. . . .

Although Graver Tank certainly leaves room for petitioner's suggested inclusion of intent-based elements in the doctrine of equivalents, we do not read it as requiring them. The better view, and the one consistent with Graver Tank's predecessors and the objective approach to infringement, is that intent plays no role in the application of the doctrine of equivalents.

C

Finally, petitioner proposes that in order to minimize conflict with the notice function of patent claims, the doctrine of equivalents should be limited to equivalents that are disclosed within the patent itself. A milder version of this argument, which found favor with the dissenters below, is that the doctrine should be limited to equivalents that were known at the time the patent was issued, and should not extend to after-arising equivalents.

As we have noted [above] with regard to the objective nature of the doctrine, a skilled practitioner's knowledge of the interchangeability between claimed and accused elements is not relevant for its own sake, but rather for what it tells the fact-finder about the similarities or differences between those elements. Much as the perspective of the hypothetical "reasonable person" gives content to concepts such as "negligent" behavior, the perspective of a skilled practitioner provides content to, and limits on, the concept of "equivalence." Insofar as the question under the doctrine of equivalents is whether an accused element is equivalent to a claimed element, the proper time for evaluating equivalence and thus knowledge of interchangeability between elements is at the time of infringement, not at the time the patent was issued. And rejecting the milder version of petitioner's argument necessarily rejects the more severe proposition that equivalents must not only be known, but must also be actually disclosed in the patent in order for such equivalents to infringe upon the patent. . . .

V

All that remains is to address the debate regarding the linguistic framework under which "equivalence" is determined. Both the parties and the Federal Circuit spend considerable time arguing whether the so-called "triple identity" test focusing on the function served by a particular claim element, the way that element serves that function, and the result thus obtained by that element is a suitable method for determining equivalence, or whether an "insubstantial differences" approach is better. There seems to be substantial agreement that, while the triple identity test may be suitable for analyzing mechanical devices, it often provides a poor framework for analyzing other products or processes. On the other hand, the insubstantial differences test offers little additional guidance as to what might render any given difference "insubstantial."

In our view, the particular linguistic framework used is less important than whether the test is probative of the essential inquiry: Does the accused product or process contain elements identical or equivalent to each claimed element of the patented invention? Different linguistic frameworks may be more suitable to different cases, depending on their particular facts. A focus on individual elements and a special vigilance against allowing the concept of equivalence to eliminate completely any such elements should reduce considerably the imprecision of whatever language is used. An analysis of the role played by each element in the context of the specific patent claim will thus inform the inquiry as to whether a substitute element matches the function, way, and result of the claimed element, or whether the substitute element plays a role substantially different from the claimed element.

VI

Today we adhere to the doctrine of equivalents. The determination of equivalence should be applied as an objective inquiry on an element-by-element basis. Prosecution history estoppel continues to be available as a defense to infringement, but if the patent-holder demonstrates that an amendment required during prosecution had a purpose unrelated to patentability, a court must consider that purpose in order to decide whether an estoppel is precluded. Where the patentholder is unable to establish such a purpose, a court should presume that the purpose behind the required amendment is such that prosecution history estoppel would apply. Because the Court of Appeals for the Federal Circuit did not consider all of the requirements as described by us today, particularly as related to prosecution history estoppel and the preservation of some meaning for each element in a claim, we reverse and remand for further proceedings consistent with this opinion.

Ms. Justice GINSBURG, with whom Mr. Justice KENNEDY joins, concurring.

I join the opinion of the Court and write separately to add a cautionary note on the rebuttable presumption the Court announces regarding prosecution history estoppel. I address in particular the application of the presumption in this case and others in which patent prosecution has already been completed. The new presumption, if applied woodenly, might in some instances unfairly discount the expectations of a patentee who had no notice at the time of patent prosecution that such a presumption would apply. Such a patentee would have had little incentive to insist that the reasons for all modifications be memorialized in the file wrapper as they were made. Years after the fact, the patentee may find it difficult to establish an evidentiary basis that would overcome the new presumption. The Court's opinion is sensitive to this problem, noting that "the PTO may have relied upon a flexible rule of estoppel when deciding whether to ask for a change" during patent prosecution.

Because respondent has not presented to this Court any explanation for the addition of the lower pH limit, I concur in the decision to remand the matter to the Federal Circuit. On remand, that court can determine — bearing in mind the prior absence of clear rules of the game — whether suitable reasons for including the lower pH limit were earlier offered or, if not, whether they can now be established.

COMMENTS & QUESTIONS

1. On remand, the Federal Circuit held: (1) further remand to the district court was required to determine the reasons the patentee amended the patent claim during prosecution, for the purpose of deciding whether assertion of the claim under doctrine of equivalents was barred by prosecution

history estoppel, (2) if the patentee's assertion of infringement under the doctrine of equivalents was not barred by prosecution history estoppel, the patent was indeed infringed under that doctrine, and (3) the patentee would not be precluded from augmenting the record to show the reason for claiming the disputed claim amendment. Hilton Davis Chemical Co. v. Warner-Jenkinson Company, Inc., 114 F.3d 1161 (Fed. Cir. 1997). On the question of the accused infringer's pH measure during ultrafiltration, the Federal Circuit said:

> Although there is nothing in the written description part of the specification to indicate that the invention extends beyond the specific range given in the claim, there is substantial record evidence to prove that one of ordinary skill in the art would know that performing ultrafiltration at a pH of 5.0 will allow the membrane to perform substantially the same function in substantially the same way to reach substantially the same result as performing ultrafiltration at 6.0. In this regard, Dr. Cook, one of the inventors, testified that the process would work to separate the dye from the impurities at pH values as low as 2.0 (albeit with foaming). Moreover, Warner-Jenkinson's expert testified that the Hilton Davis process would operate at a pH of 5.0. The jury's finding that the accused process with a pH of 5.0 is equivalent to the claimed process with a lower limit of approximately 6.0 does not therefore vitiate the claim limitation. Accordingly, assuming prosecution history estoppel does not preclude such a finding, we reaffirm our prior decision that a pH of 5.0 is equivalent to a pH of "approximately 6.0" in the context of the claimed process.

Id. at 1164.

On the crucial question of implementing a procedure to apply the new Supreme Court presumption on prosecution history estoppel, the Federal Circuit took a flexible approach.

> We hesitate to specify the procedures that the district court can employ to answer the question posed by the newly created presumption of prosecution history estoppel. The better course is to allow the district court to use its discretion to decide whether hearings are necessary or whether the issue can adequately be determined on a written record. If the district court determines that a reason not related to patentability prompted an amendment, the court must then decide if that reason is sufficient to overcome estoppel.

Id. at 1163. In a different case, the Federal Circuit also held that an applicant cannot short-circuit the *Hilton Davis* inquiry by using "boilerplate" language in an amendment, to the effect that the purpose of the amendment is merely to clarify the claim and not to avoid the prior art. Bai v. L&L Wings Inc., 160 F.3d 1350 (Fed. Cir. 1998).

2. In Sage Prod., Inc. v. Devon Indus., Inc. 126 F.3d 1420, 1431 (Fed. Cir. 1997), Devon's patent claimed a receptacle with two baffles, "said first baffle and said second baffle defining a space therebetween sufficient to allow said expendable items to pass through said first opening, between said first

baffle and said second baffle, and through said second opening to be received within said container." The court found that Sage's hazardous waste container only infringed Devon's patent under the doctrine of equivalents if "through" was interpreted as meaning "allowing passage." Devon was estopped from asserting that claim, however, because the prosecution history indicated that the claim had specifically been amended to avoid prior art that had a "flap having a hole therethrough," thereby excluding the "allowing passage" interpretation.

The court in *Sage* offered a strong policy justification for the new presumption of estoppel as well as some cautionary words for patent drafters:

> [A]s between the patentee who had a clear opportunity to negotiate broader claims but did not do so and the public at large, it is the patentee who must bear the cost of its failure to seek protection for this foreseeable alteration of its claimed structure. . . . This court recognizes that such reasoning places a premium on forethought and patent drafting. Indeed, this premium may lead to higher costs of patent prosecution. However, the alternative rule — allowing broad play for the doctrine of equivalents to encompass foreseeable variations, not just of a claim element, but of a patent claim — also leads to higher costs. Society at large would bear these latter costs in the form of virtual foreclosure of competitive activity within the penumbra of each issued patent claim.

Id. at 1425. See also Pharmacia & Upjohn Co. v. Mylan Pharmaceuticals, Inc., 170 F.3d 1373, 1377 (Fed. Cir. 1999) ((1) prosecution estoppel may apply even where there is no amendment made to a patent application's claim(s); and (2) prosecution history should be viewed from the perspective of a competitor of the applicant, skilled in the art of the patent's subject matter).

3. The Supreme Court in *Warner-Jenkinson* held that "the court should presume that the PTO had a substantial reason related to *patentability* for including the limiting element added by amendment." What did the Court mean by patentability? From the factual context of the case, it clearly encompasses amendment to avoid a rejection in light of the prior art (i.e. under 35 U.S.C. §§102 or 103).

Rejections are not all based on prior art, however. Patent applications can be rejected under §112 as well. If one of a patent applicant's claims is rejected for lack of enablement, can that result in a later "enablement estoppel" under the doctrine of equivalents? Ted Apple suggests this hypothetical:

> Consider a patent application for a novel blood diluent, in which the applicant's originally filed claim is to a composition comprising "1% sucrose in an osmotically balanced solution of metallic phosphates." The examiner rejects the claim under §112, first paragraph, as "broader than the enabling disclosure in the recitation of the enormously broad term 'metallic phosphates.'" In response, the applicant amends the claim, replacing the generic term, metallic phosphates, with "sodium phosphate," a specific metallic phosphate for which the specifi-

cation provides ample support. Consequently, the patent issues. When a competitor later markets a blood diluent identical to the plaintiff's except for the use of potassium phosphate in place of sodium phosphate, the applicant (now patentee) sues, asserting infringement under the doctrine of equivalents. In defense, the competitor asserts that, by narrowing the original amendment in order to secure allowance of the claim, the plaintiff has surrendered compositions containing metallic phosphates other than sodium phosphate and is now estopped from claiming them.

Ted Apple, Enablement Estoppel: Should Prosecution History Estoppel Arise When Claims Are Amended to Overcome Enablement Rejections?, 13 Santa Clara Comp. & High Tech L. J. 107, 116 (1997). Apple argues that so-called "enablement estoppel" makes sense because it would align the contribution of the inventor with claim scope that could be asserted under the doctrine of equivalents. See id. at 134.

The Federal Circuit recently called the viability of this theory into doubt, however. In Litton Systems, Inc. v. Honeywell, Inc., 140 F.3d 1449 (Fed. Cir. 1998), the court stated in dicta that the Supreme Court's opinion was limited to estoppel in situations where amendments were made in response to the prior art:

> Thus, from the outset, the Court recognized that the *Warner-Jenkinson* standard "related to patentability" encompassed amendments "made to avoid the prior art."
> The Court then observed:
>> It is telling that in each case this Court probed the reasoning behind the Patent Office's insistence upon a change in the claims. In each instance, a change was demanded because the claim as otherwise written was viewed as not describing a patentable invention at all — typically because what it described was encompassed within the prior art.
>
> [*Warner-Jenkinson*, 117 S. Ct.] at 1050.
> The Supreme Court then proceeded to recognize that "there are a variety of other reasons why the PTO may request a change in claim language."

Litton Systems, 140 F.3d at 1457-1458. How would "enablement estoppel" actually work? Would it make sense to estop a patent holder from asserting any infringement that was not enabled by the patent disclosure? This seems unlikely because the doctrine of equivalents is measured from the date of infringement rather than the date of filing. Under the doctrine of equivalents, infringement can be found even though the equivalent technology was not known at the time the application was filed. An alternate approach is to estop patent holders from asserting infringement under the doctrine of equivalents for embodiments that were rejected as not enabled by the examiner. However, the applicant would not be estopped from asserting her rights to future equivalent embodiments that were neither originally claimed nor enabled.

4. How broad is the estoppel that arises during patent prosecution?

Consider the following, from Hughes Satellite Corp. v. United States, 140 F.3d 1470 (Fed. Cir. 1998):

> The government argues . . . that Warner-Jenkinson further requires prosecution history estoppel to act as an absolute bar, and thus to preclude any equivalents to a claim limitation that was added to overcome a patentability rejection, regardless of what subject matter was surrendered. The patentee thus would be limited to the literal scope of the particular claim limitation. . . .
>
> Thus, the government urges that Hughes should be precluded from asserting any equivalents to paragraphs (e), (f), and (g) because these limitations were added to overcome prior art. . . .
>
> Hughes responds that, aside from the rebuttable presumption, the Supreme Court did not alter this court's prosecution history estoppel analysis, which previously rejected the bright-line rule posited by the government. According to Hughes, such an approach is inconsistent with the requirement that the reasons for an amendment are to be considered.
>
> We reject the government's contention that Warner-Jenkinson requires such a wooden approach to prosecution history estoppel. The Supreme Court has long recognized that the key to prosecution history estoppel is the surrender or disclaimer of subject matter by the patentee, which the patentee is then unable to reclaim through the doctrine of equivalents. See Exhibit Supply Co. v. Ace Patents Corp., 315 U.S. 126, 136, 62 S. Ct. 513, 86 L. Ed. 736 (1942) ("By the amendment he recognized and emphasized the difference between the two phrases and proclaimed his abandonment of all that is embraced in that difference. The difference which he thus disclaimed must be regarded as material, and since the amendment operates as a disclaimer of that difference, it must be strictly construed against him." (citations omitted)); see also Sutter v. Robinson, 119 U.S. 530, 541, 7 S. Ct. 376, 30 L. Ed. 492 (1886) ("He is not at liberty now to insist upon a construction of his patent which will include what he was expressly required to abandon and disavow as a condition of the grant.").
>
> In evaluating the reason behind an amendment, a court must determine what subject matter the patentee actually surrendered. If the accused device wholly fails to meet a limitation to which the patentee has expressly limited the claims, a finding of equivalence is precluded under prosecution history estoppel.
>
> We conclude, however, that no such estoppel arises here. The PTO originally rejected Hughes' claim for obviousness in light of U.S. Patent No, 3,216,674, issued on November 9, 1965 to McLean (the McLean Patent). In response to this rejection, Williams canceled his original claims and inserted, inter alia, the claim that issued as claim 1 and contained [a number of important] limitations. . . . The McLean patent relates to a space vehicle with a self-contained navigation system that maintains a collision course with a target body. As the target body changes its position, the McLean vehicle automatically alters its course relative only to the target. Although the vehicle disclosed in the McLean patent precesses in the same fashion as the Williams invention, it differs from the Williams vehicle by performing precession without two-way communication with an external location and without reference to a fixed, external coordinate system. Williams' amendments thus did not surrender subject matter

covering a device, such as the accused device, which provides two-way communication with an external location (although some calculations are made on-board instead of at the external location) and which uses an external coordinate system.

Because these amendments to the claim language were made to overcome a prior art rejection, they do serve to narrow the range of equivalents; however, they do not preclude all equivalents available to Hughes. See Hughes VII, 717 F.2d at 1363, 219 U.S.P.Q. at 482. Because the accused device does not fall within the range of subject matter surrendered, prosecution history estoppel does not preclude infringement under the doctrine of equivalents. See id. at 1362, 717 F.2d 1351, 219 U.S.P.Q. at 481. * * * *

Id. at 1477.

5. A different problem can arise for the patentee if the specification discloses more than is claimed. According to the Federal Circuit in Maxwell v. J. Baker, Inc., 86 F.3d 1098 (Fed. Cir. 1996), "subject matter disclosed but not claimed in a patent application is dedicated to the public." Id. at 1106. The court held that if a patent discloses but does not claim certain equivalents of an invention, then the patentee cannot later assert an infringement claim under the doctrine of equivalents for those equivalents. To support this holding, the court argued that the alternative would "encourage a patent applicant to present a broad disclosure in the specification and file narrow claims, avoiding examination of broader claims that the applicant could have filed consistent with the specification." Id. at 1107. By contrast, the Federal Circuit seemed to back away from this rule (and create yet another conflict within the circuit) in YBM Magnex Inc. v. International Trade Commn., 145 F.3d 1317, 1322 (Fed. Cir. 1998) (holding that *Maxwell* "did not displace the wealth of precedent that permits determination of equivalency, vel non, as to subject matter included in the written description but not claimed."). The court essentially limited *Maxwell* to its facts and concluded that in the case before it, material that was disclosed but not claimed in a patent could be recaptured under the doctrine of equivalents.

As the court in Vehicular Tech. Corp. v. Titan Wheel Intl. Inc., 141 F.3d 1084 (Fed. Cir. 1998) indicated, the "dedication to the public" rule is akin to prosecution history estoppel: both doctrines limit the scope of a patent under the doctrine of equivalents. The interaction of the written description requirement and the dedication to the public rule in *Maxwell* requires that a patent agent or attorney draft a patent application such that the subject matter recited in the claims has a precise overlap with the subject matter claimed. If the disclosure is not sufficient to support the claims, the claims will be rejected under the written description requirement. On the other hand, if the disclosure is too broad, the patent applicant gives up that part of his invention and "dedicates it to the public." In fact, if a patent is later found invalid under the written description requirement, then the entire invention is effectively disclosed to the public.

Note: What is an Element?

In *Warner Jenkinson*, the Supreme Court tightened application of the doctrine of equivalents. Henceforth, an accused product infringes only if it has "all elements" of a claimed invention. But what exactly is an element? And how precisely must the accused product track each element?

It did not take long for these issues to present themselves in the wake of *Hilton Davis*. The Supreme Court, in the wake of that case, had vacated the Federal Circuit's decision in *Hughes Aircraft Co. v. United States* — further complicating an already tortuous procedural history. On remand, the Federal Circuit was obliged to consider the impact of *Warner Jenkinson* on its prior ruling finding infringement under the doctrine of equivalents. See 140 F.3d 1470 (Fed. Cir. 1998). In its opinion, the Federal Circuit revisited the equivalents issue and in particular the application of the Supreme Court's "all elements" test. The Federal Circuit began with a brief description of the satellite technology at issue in the case:

> The Williams patent relates to an apparatus for control over the orientation, or attitude, of a spacecraft using commands from a ground control station. The relevant limitations of claim 1 of the Williams patent read:
>
> > (e) means disposed on said body for providing an *indication to a location external to said body of the instantaneous spin angle position* of said body about said axis and the orientation of said axis with reference to a fixed external coordinate system;
> > (f) and means disposed on said body for *receiving from said location control signals synchronized* with said indication;
> > (g) said valve being coupled to said last-named means and responsive to said control signals for applying fluid to said fluid expulsion means *in synchronism therewith* for precessing said body to orient said axis into a predetermined desired relationship with said fixed external coordinate system.
>
> (Emphasis added). In order to correct the attitude of the spacecraft, the ground crew must be able to calculate the instantaneous spin angle (ISA) position. The ISA position is the angle between two specific planes. The first plane, the rotating plane, is defined by the location of the precessing jet and the satellite's axis of rotation. The second plane, the reference plane, is defined by a fixed reference point in an external coordinate system (such as the sun or another star) and the spin axis. The angle between these planes at a given moment in time is the ISA position with reference to a fixed external coordinate system. Thus, the ISA position generally measures the location of the precessing jet in its rotational cycle relative to the reference plane.
>
> Two pieces of information are needed to calculate the ISA position: the spin rate of the satellite and the instant in time at which the rotating plane passes by the fixed point in the fixed external coordinate system and at which the jet is closest to the fixed reference point. The invention uses onboard sensors to collect this data and then transmits this information to earth to allow the ground crew to determine the satellite's existing and desired orientations. After making the necessary calculations, the ground crew pulses the attitude jet by radio signal commands to precess, or tip, the spin axis of the satellite to the desired position.
>
> In the accused "store and execute" (S/E) craft, the satellite retrieves the same raw data but calculates the ISA position onboard. The spin rate and in-

formation to determine the orientation of the satellite is transmitted to the ground. In most of the S/E craft, the satellite does not provide information sufficient to calculate the ISA position. After receiving the spin rate, the ground crew performs the necessary calculations to adjust the attitude of the craft. This information is then sent to and stored in the satellite. The precession does not occur, however, until the ground crew sends an execute command to the satellite.

In 1982, the then-Court of Claims originally determined, inter alia, that the accused S/E devices did not infringe the patent literally or under the doctrine of equivalents. See Hughes Aircraft Co. v. United States, 215 U.S.P.Q. 787, 812 (Ct.Cl. Trial Div.1982) (Hughes VI). On appeal, this court reversed the noninfringement judgment, holding that the S/E devices infringe under the doctrine of equivalents, and remanded for a determination of just compensation. See Hughes VII, 717 F.2d at 1366, 219 U.S.P.Q. at 484. After the decision by the Court of Federal Claims on remand, Hughes appealed, challenging the assessment of damages, and the government cross-appealed, again challenging the liability determination of Hughes VII in light of this court's in banc decision in Pennwalt Corp. v. Durand-Wayland, Inc., 833 F.2d 931, 4 U.S.P.Q.2d 1737 (Fed. Cir. 1987) (in banc). See Hughes XIII, 86 F.3d at 1566, 39 U.S.P.Q.2d at 1066 (Fed. Cir. 1996). This court affirmed the damages determination and refused, under the doctrine of law of the case, to reconsider the Hughes VII decision. See id. at 1576, 86 F.3d 1566, 39 U.S.P.Q.2d at 1072. The Supreme Court, however, granted certiorari, vacated the judgment, and remanded the case (GVR order) to this court for reconsideration in light of its decision in Warner-Jenkinson. See Hughes XIV, 117 S.Ct. at 1466. . . .

Turning to the merits, we first address the effect of the "all-elements" rule (sometimes referred to as the "all-limitations" rule) enunciated in Warner-Jenkinson on the Hughes VII decision. The Supreme Court clarified the doctrine of equivalents by noting that:

> [e]ach element contained in a patent claim is deemed material to defining the scope of the patented invention, and thus the doctrine of equivalents must be applied to individual elements of the claim, not to the invention as a whole. It is important to ensure that the application of the doctrine, even as to an individual element, is not allowed such broad play as to effectively eliminate that element in its entirety.

___ U.S. at ___, 117 S.Ct. at 1049. Thus, the test for equivalence is to be applied to the individual claim limitations:

> An analysis of the role played by each element in the context of the specific patent claim will thus inform the inquiry as to whether a substitute element matches the function, way, and result of the claimed element, or whether the substitute element plays a role substantially different from the claimed element.

Id. at 1054; see Pennwalt, 833 F.2d at 935, 4 U.S.P.Q.2d at 1740 (" '[T]he plaintiff must show the presence of every element or its substantial equivalent in the accused device.' To be a 'substantial equivalent,' the element substituted in the accused device for the element set forth in the claim must not be such as would substantially change the way in which the function of the claimed invention is performed." (citations omitted) (quoting Perkin-Elmer Corp. v. Westinghouse Elec. Corp., 822 F.2d 1528, 1532-33, 3 U.S.P.Q.2d 1321, 1324-25 (Fed. Cir. 1987))).

The government argues that the all-elements rule demands that we depart from the reasoning in Hughes VII, in which the court stated that the trial court erred in not "apply[ing] the doctrine of equivalents to the claimed invention as a whole." Hughes VII, 717 F.3d at 1364, 219 U.S.P.Q. at 482. According to the government, to conclude that the claim limitations in paragraphs (e), (f), and (g) are met equivalently by elements in the accused devices would vitiate those claim limitations. The government additionally urges that the arguably corresponding elements of the S/E system differ substantially from the claim limitations by storing the ISA position value onboard in lieu of transmitting an indication of the ISA position to the ground, by not acting in synchronism with the control signals, and by not firing the precession jet within a fixed period of time after receiving the command signal.

Hughes responds that there is no reason to depart from the conclusion reached in Hughes VII because Warner-Jenkinson did not significantly alter the all-elements rule as stated in Pennwalt. Moreover, the Hughes VII court, in Hughes' opinion, did perform the required element-by-element analysis mandated by Warner-Jenkinson.

We conclude that the analysis performed in Hughes VII satisfies the all-elements rule as stated in Warner-Jenkinson. Regarding claim paragraph (e), the court in Hughes VII concluded that the transmission to the ground crew of the spin rate and information sufficient to calculate the sun angle in the S/E vehicles "is the modern day equivalent to providing an indication of the ISA to the ground. . . ." 717 F.2d at 1365, 219 U.S.P.Q.2d at 483. This information, while insufficient to calculate the ISA position, was sufficient to enable the ground crew to control the satellite, which is substantially the same function performed and the identical result achieved by transmitting the indication of the ISA position to the ground. See id.

Furthermore, although the information sent is insufficient to calculate the ISA position, this does not demonstrate a substantial difference in the way the element functions. This is a case in which a "subsequent change in the state of the art, such as later-developed technology, obfuscated the significance of [the] limitation at the time of its incorporation into the claim." Sage Prods., Inc. v. Devon Indus., Inc., 126 F.3d 1420, 1425, 44 U.S.P.Q.2d 1103, 1107 (Fed. Cir. 1997); cf. Warner-Jenkinson, __ U.S. at __, 117 S.Ct. at 1053 ("Insofar as the question under the doctrine of equivalents is whether an accused element is equivalent to a claimed element, the proper time for evaluating equivalency . . . is at the time of infringement, not at the time the patent was issued."); Pennwalt, 833 F.2d at 938, 4 U.S.P.Q.2d at 1742 ("[T]he facts here do not involve later developed computer technology which should be deemed within the scope of the claims to avoid the pirating of an invention."). The court in Hughes VII determined that the change in the S/E devices was the result of a technological advance not available until after the patent issued. See Hughes VII, 717 F.2d at 1365, 219 U.S.P.Q. at 483. Relying on testimony of one of skill in the art at the time of infringement, the court in Hughes VII concluded that this advance resulted in an insubstantial change in the way the element performed its function. See id. (citing testimony that an engineer would realize that transmission of the ISA position was no longer necessary as a result of the change in technology).

The court in Hughes VII also concluded that the "synchronism" limitations in paragraphs (f) and (g) were also equivalently met by the accused devices.

Again, as a result of an advance in technology, the satellite system was able at the time of infringement to store the precession information and to wait to precess the satellite until receipt of the execute command. Thus, the synchronism in the accused device is coordinated by the computer instead of by real-time execution of the command from the ground. As recognized in Hughes VII, "[t]he difference between operation by retention and operation by sending is achieved by relocating the function, making no change in the function performed, or in the basic manner of operation, or in the result obtained." Id. at 1366, 717 F.2d 1351, 219 U.S.P.Q. at 484. The court in Hughes VII correctly performed an analysis of the function, way, and result of the individual elements in the accused devices and concluded that these elements equivalently met the claim limitations at issue.

Accordingly, we conclude that Warner-Jenkinson provides no basis to alter the decision in Hughes VII because the court properly applied the all-elements rule.

Id. at 1472-75.

COMMENTS AND QUESTIONS

1. *An Essential Tension.* In the *Hughes Satellite* case, the Federal Circuit struggled to reconcile two difficult aspects of the *Warner Jenkinson* opinion: (1) the adoption of the "element by element" approach to equivalence and (2) the preservation of the rule that equivalence is measured as of the time of infringement, and not as of the time of invention. The difficulty here — best illustrated by the facts in the *Hughes* case itself — is that "after developed" technologies (those developed after the date of invention) will often challenge the definition of an element in the original invention. Recall that in *Hughes*, the accused embodiments incorporated "on board" computers that were not in existence at the time of the invention of the Hughes satellite. The Federal Circuit, in the original opinion finding infringement under the doctrine of equivalents, found that these after-developed satellites were the equivalent of those claimed in the original patent. This was certainly defensible under prevailing doctrine of equivalents analysis at the time of this opinion. However, the Federal Circuit's subsequent decision in *Pennwalt*, which was endorsed by the Supreme Court in the *Warner Jenkinson* case, called this conclusion into doubt. This is so because the original claims in the Hughes patent include two crucial elements or limitations that are arguably not present in the later-developed satellites: (1) "means for providing . . . [an] *indication to a location external to said body of the instantaneous spin angle position*"; and (2) "means disposed on said body *for receiving from said location control signals synchronized* with said indication."

2. *Hard Questions.* Ultimately, the court in *Hughes* concludes that subsequent developments in the technology "*obfuscate[d] the significance of [the] limitation . . .*," (emphasis added), quoting *Sage Products*. What good is the "all-elements rule" if it is ignored in the important and common case of after-

developed technologies, or applied in such a way that it is difficult to tell what an "element" is? How does the temporal aspect of the post–*Warner Jenkinson* test, i.e., measuring equivalence at the time of infringement, square with the Supreme Court's emphasis on the "notice to competitors" function of patent claims? With these issues in mind, consider the following comments from Judge Clevenger of the Federal Circuit, dissenting from the court's denial of a motion to reconsider its *Hughes* opinion on remand in light of *Warner Jenkinson*:

> We have come a long way from the days when judges frowned on patents as pernicious monopolies deserving scant regard. Today, patents are the backbone of much of the national economy . . . This court has emphasized the requirement of clear and concise drafting of claims, and valid claims that are written properly are now enforced to the hilt. I think the Supreme Court understands this, for, in its *Warner-Jenkinson* opinion, it emphasized the cardinal significance of clear claim language, and highlighted the unwanted tension put on such claim language by a loosely applied doctrine of equivalents. I come away from the Supreme Court's opinion with the impression that the Court understands the mischief that the doctrine of equivalents can do to patent law. To counter such mischief, I think the Supreme Court meant to whittle the doctrine down in size. As I recently explained, however, see Litton Sys., supra, nothing in the Warner-Jenkinson opinion narrows the reach of the doctrine of equivalents, unless one accepts the reading of the opinion asserted by the United States in this case. In short, the United States thus should not give up on this case just because we deny its petition. Instead, the Solicitor General should ask the Supreme Court to grant a writ of certiorari in this case.

Hughes Aircraft Co. v. U.S., 148 F.3d 1384, 1385-86, 47 U.S.P.Q.2d 1542 (Fed. Cir. 1998) (Clevenger, J., dissenting from denial of petition for rehearing and suggestion for rehearing en banc). As this volume went to press, the Federal Circuit had agreed to rehear *en banc* Festo Corp. v. Shoketsu Kinzoku Kogyo Kabushiki Co., 172 F.3d 1361 (Fed. Cir. 1999), to address these issues.

PROBLEM

Problem 3-11. Grimm, the patentee, claimed a mechanical patent consisting of a sealed mechanical piston-cylinder arrangement containing hydraulic fluid and recited one element as "a pair of resilient sealing rings situated near opposite axial ends of the central mounting member and engaging the cylinder to effect a fluid-tight seal therewith." The patent diagram is reprinted in Figure 3-4.

Kobayashi, the accused infringer, built a device that employed a *single* "two-way" sealing ring in the cylinder. The Kobayashi device is depicted in Figure 3-5.

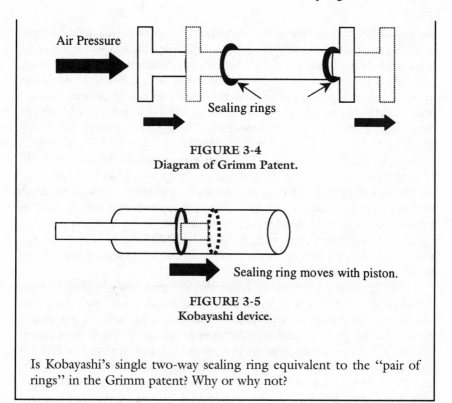

FIGURE 3-4
Diagram of Grimm Patent.

Sealing ring moves with piston.

FIGURE 3-5
Kobayashi device.

Is Kobayashi's single two-way sealing ring equivalent to the "pair of rings" in the Grimm patent? Why or why not?

Note on the Problem of Later-Developed Technologies

Courts have determined how broadly they see "equivalents" based on the degree of advance over the art the original patent represents. When the patent is on a "mere improvement" the courts tend not to consider as "equivalent" a product or process that is even a modest distance beyond the literal terms of the claims. Brill v. Washington Elec. & Ry. Co., 215 U.S. 527 (1910); Kinzenbaw v. Deere & Co., 741 F.2d 383 (Fed. Cir. 1984), *cert. denied*, 470 U.S. 1004 (1985). On the other hand, a patent representing a "pioneer invention" — which the Supreme Court has defined as "a patent concerning a function never before performed, a wholly novel device, or one of such novelty and importance as to make a distinct step in the progress in the art," Boyden Power-Brake Co. v. Westinghouse, 170 U.S. 537, 569 (1898) — is "entitled to a broad range of equivalents." 4 Donald Chisum, Patents §18.04[2] (1998). That is, when a pioneer patent is involved, a court will stretch to find infringement even by a product whose characteristics lie considerably outside the boundaries of the literal claims.

Of course the question of infringement also turns on the precise characteristics of the allegedly infringing device. Following the test laid down by the Supreme Court in *Graver Tank*, courts confronted with a device accused of infringing inquire whether it performs the same function and achieves the same result as the invention in the claims, and whether it does so in the same way. Where the accused device shows only minor or "insubstantial" variations in one of these elements — such as the small movement of one part or a minor change in structure — infringement will be found even if the patentee's invention is a "mere improvement." See, e.g., Tigrett Indus., Inc. v. Standard Indus., Inc., 162 U.S.P.Q. (BNA) 32, 36 (W.D. Tenn. 1967), *aff'd*, 411 F.2d 1218 (6th Cir. 1969), *aff'd by an equally divided court*, 397 U.S. 586 (1970) (claim for playpen calling for "a pair of spaced openings" for two converging drawstrings to adjust side webbing infringed by device with one hole for drawstrings). And even a pioneer patent is not infringed by a device that achieves a different result, or achieves it in a different way. See, e.g., Mead Digital Sys., Inc. v. A. B. Dick Co., 723 F.2d 455, 464 (6th Cir. 1983) (finding that ink-jet printer patent, though a "quantum leap" in the art, was not infringed).

One important set of cases under this doctrine has grappled with the question of whether new technologies, unforeseen at the time the patent was issued, can constitute equivalents. This issue arises when a subsequent device that uses new technology is accused of infringing the original patent. The early cases were split, but the prevailing view now is that new technology can be equivalent. This is true despite the statement in *Graver Tank* that an important determinant in the equivalents inquiry is whether "persons reasonably skilled in the art would have known of the interchangeability of an ingredient not contained in the patent with one that was." Cf. Marty Adelman & Gary Francione, The Doctrine of Equivalents in Patent Law: Questions Pennwalt Did Not Answer, 137 U. Pa. L. Rev. 673, 696 n.103, 697 (1989) (arguing that "this factor [interchangeability] should be used to reject rather than support the application of the doctrine of equivalents," because it signifies that a patentee could have, but mistakenly or intentionally did not, include these interchangeable elements in her original claims).

Notwithstanding the "interchangeability" language in the leading Supreme Court case on the subject, a device performing the same function and achieving the same result in the same way as a patented invention can be found to infringe even if it uses technology developed after the patent was issued. This possibility is subject to two caveats: 1) new technologies can constitute equivalents only so long as they do not perform a different function or cause the device to operate in a substantially different way; and 2) a truly meritorious improvement can escape even *literal* infringement under the "reverse" doctrine of equivalents discussed below.

That these distinctions may not always be easy to make is demonstrated by the case of Hughes Aircraft Co. v. United States, 717 F.2d 1351 (Fed. Cir. 1983), discussed earlier in this seciton. In its 1983 opinion, Federal circuit stated: "Advanced computers and digital communications techniques developed since [the] Williams [patent]," said the Federal Circuit, "permit doing on-board a *part* of what Williams taught as done on the ground." The court concluded: "[P]artial variation in technique, an embellishment made possible by post-Williams technology, does not allow the accused spacecraft to escape the 'web of infringement.' " Id. at 1365 (emphasis in original) (citation omitted). Another case found a patented method for laying pipe, calling for a beam of light to align pipe segments, infringed by the use of later-developed laser beam technology. Laser Alignment, Inc. v. Woodruff & Sons, Inc., 491 F.2d 866 (7th Cir. 1974), *cert. denied*, 419 U.S. 874 (1974).

On the other hand, in Texas Instruments, Inc. v. U.S. Intern. Trade Commn., 805 F.2d 1558 (Fed. Cir. 1986), the Federal Circuit held that major improvements in all the essential elements of hand-held calculators rendered the improved devices non-infringing. 805 F.2d at 1570:

> It is not appropriate in this case, where all of the claimed functions are performed in the accused devices by subsequently developed or improved means, to view each such change as if it were the only change from the disclosed embodiment of the invention. It is the entirety of the technology embodied in the accused devices that must be compared with the patent disclosure. . . .

The specification supporting Texas Instruments' pioneer patent, for instance, described the use of integrated circuits containing bipolar transistors. The improvements all used integrated circuits having metal oxide semiconductor (MOS) transistors. This is an example of improvements in *materials.*

The improved calculators receive input via a device that scans the "matrix" under the keyboard at frequent intervals, whereas the original design had a conductive strip underneath the keypad. This is an example of an improvement that *reduced the number of components* in the invention. Also, the original Texas Instruments display was shown in its specification as a small thermal printer that printed dots on a tape in response to output signals from the processor. The accused devices all use liquid crystal displays (LCDs), the familiar lighted display that does not produce a paper copy, an example of an improvement that *increases the efficiency of an individual component.*

Finally, the internal processing elements of the original calculator were manufactured as discrete components, that were electrically interconnected only in the final design. The newer calculators, in contrast, have all their logic on one integrated circuit, eliminating the necessity for many electrical interconnections. This is an example of enhanced *overall design.* See generally Robert Merges & Richard Nelson, On the Complex Economics of Patent Scope, 90 Colum. L. Rev. 839 (1990).

PROBLEM

Problem 3-12. Mike Molar, a production engineer for Tasty Toothpaste Corp. (Tasty) hit upon a way to make a sanitary, cheap, and small disposable toothbrush. The problem he had been running into was how to keep the toothbrush small while still providing a feature he thought necessary to make it attractive — a built-in supply of toothpaste. The solution: a small reservoir for holding toothpaste in the handle of the toothbrush. Then the user could squeeze the toothpaste from the reservoir to the brush bristles.

He quickly built a prototype of the invention. To build it, he bought a fountain pen, pulled out the ink cartridge and other parts from the interior, tore the bristles off a toothbursh and glued them onto a hollow tube, poked two holes in the top of the tube between two rows of bristles, fitted up a plastic plunger that slid in and out at the back of the fountain pen body, pulled the plunger all the way out, filled the tube with toothpaste, glued the tube inside the fountain pen body, and pushed the plunger. Voila! Toothpaste came out of the holes more or less onto the bristles, ready for brushing. It worked! He drafted a patent specification that included this passage:

> It is one important object of the present invention to provide a portable toothbrush which is easily carried in the pocket and which is thus available whenever it is required for brushing teeth.
>
> The open end of the device includes a reduced section portion which carries the bristles. The cylindrical body shaft includes an interior passage that extends into the reduced section end (i.e., the one with the bristles). The passage includes two termination openings which are in the area at the base of the toothbrush bristles. An open space for placing toothpaste is provided within the interior passage, and a movable plunger is provided for forcing the toothpaste from this space through the narrow part of the passage and out through the termination openings into the toothpaste bristles for use.
>
> In the preferred embodiment, the body shaft is made of plastic. The invention may be made so as to be disposable. Various means for filling the space with toothpaste are envisioned, including pressure-injection through a small hole in the top of the body, which hole can then be sealed. This would make the toothbrush usable only once; it would then be disposed.

He concluded with the following claim.

I claim:
1. A pocket toothbrush having an exterior structure resembling a traditional fountain pen case comprising

    a. a removable cylindrical end cap cover,

    b. a main cylindrical body shaft over at least one end of which said end cap cover fits and having means for engaging the interior of said end cap cover to retain said end cap cover,

    c. said cylindrical body shaft having one end which contains toothbrush bristles [the "bristle end"] extending transversely and capable of being confined within said end cap,

    d. said cylindrical body shaft including an interior passage extending into said bristle end and having at least one termination opening in the area at the base of said bristles,

    e. a movable plunger extending into said cylindrical body shaft in said main cylindrical body shaft,

    f. said body shaft including an interior space for the accommodation of a charge of toothpaste to be fed to said bristles by the operation of said movable plunger, said space being at least big enough to hold a charge for a single application of toothpaste.

Once the patent issues, Tasty begins selling a disposable, portable toothbrush that garners a loyal following. Soon competitors begin entering the market. One, KopyCat Industries, Inc. (KCI), begins selling a portable toothbrush that includes a replaceable toothpaste cartridge so that the brush can be used over and over if the user wishes. The cartridges are designed to have a weak plastic closure that easily breaks when the plunger is pushed against the cartridge. This keeps the toothpaste from hardening in the openings to the bristles. Also, instead of a cap, the KCI design has a telescoping retractable cover that remains attached to the nonbristle end of the brush. The cover is collapsed down, the brush used, and then the cover is pulled back into place. The retractable cover is attached very firmly with two tiny screws. The screws can be taken out and the cover removed, but it takes a tiny jeweler's screwdriver and is difficult.

Tasty has threatened to sue KCI for infringement of the Molar patent. KCI has come to you for advice. Focusing on the *claim language* — and using the specification only to interpret that language — determine (1) whether KCI runs a serious risk of being found liable for infringing the Molar patent; and (2) what changes KCI might make in its product to avoid a future infringement action by Tasty. In particular, pay attention to these issues: Does the KCI product include a "removable end cap cover"? Is it relevant that KCI improved the design of the replaceable toothpaste cartridge?

4. The "Reverse" Doctrine of Equivalents[36]

The doctrine of equivalents helps the patentee by expanding the scope of her claims beyond its literal boundaries. In a roughly symmetrical way, two similar devices are available to the accused infringer: blocking patents and the reverse doctrine of equivalents. Recent cases on biotechnology, as well as recently issued patents in this field, illustrate the relationship between these two legal devices and the way they might be applied in particular cases. The recent case of Scripps Clinic & Research Foundation v. Genentech is a good example; this case is discussed below.

Two patents are said to block each other when one patentee has a broad patent on an invention and another has a narrower patent on some improved feature of that invention. The broad patent is said to "dominate" the narrower one. In such a situation, the holder of the narrower ("subservient") patent cannot practice her invention without a license from the holder of the dominant patent. At the same time, the holder of the dominant patent cannot practice the particular improved feature claimed in the narrower patent without a license.[37]

It is of course preferable for an inventor to have her own patent free and clear of anyone else's claims. An inventor will therefore not often voluntarily characterize her invention as subservient.[38] But a court may do so in the course of litigation. Where the court upholds the validity of an accused infringer's patent on some enhanced feature but nevertheless finds that the accused product infringes a prior, broad patent, it is in effect making the accused infringer's patent subservient to the broad patent.

---

36. Portions of this discussion are taken from Robert P. Merges, A Brief Note on Blocking Patents and the Reverse Doctrine of Equivalents in Biotechnology Cases, 72 J. Pat. & Trademk. Off. Society 870 (1991).

37. Two aspects of this situation may seem counterintuitive: that the narrower (subservient) patent could ever be issued by the Patent Office, given the existence of the broad patent in the prior art; and that once the subservient patent was issued the holder of the dominant patent would be prevented from practicing an invention that clearly falls within the scope of her claims. Subservient patents may be issued, however, when they disclose an improved feature that meets the statutory tests of novelty and nonobviousness. See, e.g., Atlas Powder Co. v. E. I. du Pont & Co., 750 F.2d 1569 (Fed. Cir. 1984) (the fact that the subservient patentee has invented a nonobvious variant of a device covered by a broad patent does not mean that the broad patent is invalid for lack of enabling disclosure under 35 U.S.C. §112). In addition, a subservient patent can prevent a dominant patent holder from practicing the particular improved feature claimed in the subservient patent. This stems from the fact that the patent grant is a right to *exclude,* not an affirmative right to practice an invention. See 35 U.S.C. §154. Thus the dominant patentee can exclude the subservient patentee from practicing her invention at all; and the subservient patentee can exclude the dominant patentee from practicing her specific improved feature. See Atlas Powder, supra; Ziegler v. Phillips Petroleum Co., 483 F.2d 858 (5th Cir. 1973).

38. One example of patents that are so characterized is an improvement patent whose claims are drafted in a special format called "Jepson claims." See, e.g., Pentec, Inc. v. Graphic Controls Corp., 776 F.2d 309 (Fed. Cir. 1985). See generally R. Ellis, Patent Claims §197 (1949). Improvement patents are specifically provided for in the patent code, see 35 U.S.C §101 (1988). A Jepson claim has the same effect as a judicial finding that a patented invention is "dominated" by another invention. Strictly speaking only a patent drafted in Jepson format is an improvement patent. But practitioners often use "improvement patent" more loosely, to describe both consciously drafted improvement claims *and* patents later found to be dominated by an earlier patent.

Even where a court finds a patent subservient to another — thus creating blocking patents — the holder of the subservient patent is still better off than if she had never filed a patent application at all. This is so because she can exclude the holder of the broad patent from practicing her improvement. Although the improver may literally infringe the broad patent, she may gain some bargaining leverage by obtaining the subservient patent.

Although blocking patents are often an effective way to protect improvers, another doctrine can much more effectively mitigate the impact of literal infringement in the rare cases in which it is applied: the "reverse" doctrine of equivalents. Courts have long recognized that, "[c]arried to an extreme, the doctrine of equivalents could undermine the entire patent system." Scope could be enlarged so far beyond the literal language of claims that patents would take on unlimited power. To check the potentially destructive impact of the doctrine and to preserve symmetry in the rules on infringement, the Supreme Court long ago ruled that while

> a charge of infringement is sometimes made out, though the letter of the claims be avoided . . . [t]he converse is equally true. The patentee may bring the defendant within the letter of his claims, but if the latter has so far changed the principle of the device that the claims of the patent, literally construed, have ceased to represent his actual invention, he is as little subject to be adjudged an infringer as one who has violated the letter of a statute has to be convicted, when he has done nothing in conflict with its spirit and intent.

Boyden Power-Brake Co. v. Westinghouse, 170 U.S. 537, 562 (1898). An example drawn from this case may help to illuminate the doctrine. In 1869 George Westinghouse invented a train brake that used a central reservoir of compressed air for stopping power. Further advances in his design, primarily the addition of an air reservoir in each brake cylinder, resulted in a brake that was patented in 1887. An improvement on this 1887 brake, invented by George Boyden, added an ingenious mechanism for pushing compressed air into the brake piston from both the central reservoir *and* from a local reservoir in each brake cylinder. (Westinghouse's brake required a complicated series of passageways to supply air from the two sources.) With the added stopping power of the Boyden brake, engineers could safely operate the increasingly long trains of the late nineteenth century.

The Westinghouse patent included a claim for "the combination of a main air-pipe, an auxiliary reservoir, a brake-cylinder, a triple valve [the device that coordinated the airflows from the main reservoir and the individual brake reservoir] and an auxiliary-valve device, actuated by the piston of the triple-valve . . . for admitting air in the application of the brake." The Court noted that the literal wording of the Westinghouse patent could be read to cover Boyden's brake, since it included what could be described as a "triple valve." But it refused to find infringement, on the ground that Boyden's was a significant contribution that took the invention outside the equitable bounds of the patent:

We are induced to look with more favor upon this device, not only because it is a novel one and a manifest departure from the principle of the Westinghouse patent, but because it solved at once in the simplest manner the problem of quick [braking] action, whereas the Westinghouse patent did not prove to be a success until certain additional members had been incorporated in it.[39]

More recently, a fascinating reverse equivalents issue was raised in Scripps Clinic & Research Foundation v. Genentech, Inc., 927 F.2d 1565, 18 U.S.P.Q.2d (BNA) 1896 (Fed. Cir. 1991), a case typical of an early wave of biotechnology patent actions. Genentech invented and patented the recombinant DNA form of the blood protein Factor VIII:C, a blood clotting agent made by the body and useful in treating patients with clotting disorders. Scripps had previously obtained a patent on purified Factor VIII:C, which it made by isolating and purifying the protein from raw human blood. Scripps sued Genentech for infringement of its product patent, citing the conventional rule that a product patent covers the product no matter how it is made. After attempting to distinguish its recombinant version from Scripps' purified natural protein, Genentech ultimately relied on a pragmatic defense: that the recombinant version was by far cheaper to make, and therefore ought not to be deemed an infringement. The Federal Circuit remanded the case for a determination whether the reverse doctrine of equivalents applied in these circumstances:

> The so-called "reverse doctrine of equivalents" is an equitable doctrine invoked in applying properly construed claims to an accused device. Just as the purpose of the "doctrine of equivalents" is to prevent "pirating" of the patentee's invention, *Graver Tank*, 339 U.S. 605, 607, 608 (1950), so the purpose of the "reverse" doctrine is to prevent unwarranted extension of the claims beyond a fair scope of the patentee's invention.
> . . . Application of the doctrine requires that facts specific to the accused device be determined and weighed against the equitable scope of the claims, which in turn is determined in light of the specification, the prosecution history, and the prior art.
> The record contained evidence of the properties of plasma-derived and recombinantly produced VIII:C, which was presented primarily by Scripps in connection with its proofs of infringement. There was deposition testimony that there were differences between VIII:C from plasma and VIII:C obtained by recombinant techniques; a Scripps' witness described the products as "apples and oranges," referring specifically to stability and formulations. The parties disputed, in connection with the summary judgment motions, the capabilities

---

39. 170 U.S. 537 at 572. On the application of this standard to specific cases, see Jacoby-Bender, Inc. v. Foster Metal Products, Inc., 152 F. Supp. 289 (D. Mass. 1957), *aff'd*, 255 F.2d 869 (1st Cir. 1958) ("I am disposed to regard [the accused] device as . . . an equivalent unless what it accomplished was a marked improvement. . . . In such event it would be appropriate to judge equivalency by the extent of the improvement — the significance of the departure in relation to the remaining basic concept."); Piggott, Equivalents in Reverse, 48 J. Pat. and Trademk. Off. Society 291, 295-299 (1966).

of the respective processes in terms of the purity and specific activities that were enabled for the respective products. The record on this point is extensive.

Genentech argues that its product is equitably seen as changed "in principle," particularly when viewed in the context of the prior art. Genentech asserts that the specific activities and purity that are obtainable by recombinant technology exceed those available by the Scripps process; an assertion disputed by Scripps, but which if found to be correct could provide — depending on the specific facts of similarities and differences — sufficient ground for invoking the reverse doctrine. These aspects were not discussed by the district court.

The principles of patent law must be applied in accordance with the statutory purpose, and the issues raised by new technologies require considered analysis. Genentech has raised questions of scientific and evidentiary fact that are material to the issue of infringement. Consideration of extrinsic evidence is required, and summary judgment is inappropriate. The grant of summary judgment of infringement of [the product] claims is reversed. The issue requires trial.

927 F.2d at 1581.

COMMENTS AND QUESTIONS

1. The *Genentech* case was settled before the district court reached a decision on the reverse doctrine of equivalents issue on remand. See Genentech Litigation Settled, Marketletter, Jan. 24, 1994, p. 1.

2. The importance of the reverse doctrine of equivalents in practice should not be overstated. While it is an important part of the balance in patent law between initial and subsequent inventors, it has only rarely been applied to preclude a finding of infringement. See Jonathan Geld, Back Up! — Using the Reverse Doctrine of Equivalents to Halt the Advance of Functionally Claimed Software (working paper 1996) (reverse doctrine successfully applied in only five cases).

5. Equivalents for Means-Plus-Function Claims

Prior to the Patent Act of 1952, courts invalidated many patent claims — often to processes — on the grounds that they covered entire "functions" rather than specific machines. See, e.g., Halliburton Oil Well Cementing Co. v. Walker, 329 U.S. 1 (1946). Section 112, ¶6 of the 1952 Act reflected the widespread sentiment in favor of some mechanism to claim broad functional features of inventions. Section 112 ¶6 falls far short of permitting claims to an entire function, e.g., "broadcasting" or "cutting," but it does allow functional language in claims to cover claim *elements*. Thus, for example, it allows a patentee to claim as a combination an invention consisting of a novel microprocessor together with "means for inputting data to be processed." Depending, then, on the content of the patent specification, this claim might include the use of the novel microprocessor with data entered from a keyboard, or read in from a disk, or beamed in from a remote source via wireless communication, or perhaps other "means for inputting" as well.

Section 112, ¶6 reads in full as follows:

> An element in a claim for a combination may be expressed as a means or step for performing a specified function without the recital of structure, material, or acts in support thereof, and such claim shall be construed to cover the corresponding structure, material, or acts described in the specification and equivalents thereof.

35 U.S.C. §112 (1986 & Supp. VI 1991). Section 112, ¶6 claims are commonly known as "means plus function" claims. Thus, in our example, the patent applicant could simply include "means for processing data" or the like in his claims. The phrase "means for" is a signal to the PTO and the courts to turn to the specification in order to define the invention. The patentee's claim does not capture *all* means for processing data — only those actually discussed in the specification, and equivalents thereof.

Section 112 ¶6 was enacted in 1952. While it was clear in *infringement* cases that the patentee could only claim to cover the structure disclosed in the specification and equivalents thereof, the Patent Office took the position that in deciding whether to *issue* a patent in the first place, it should check to see whether the prior art disclosed *any* means for performing the specified function. In 1994 the Federal Circuit abolished this long-standing PTO practice:

> The plain and unambiguous meaning of paragraph six is that one construing means-plus-function language in a claim must look to the specification and interpret that language in light of the corresponding structure, material or acts described therein, and equivalents thereof, to the extent that the specification provides such disclosure. Paragraph six does not state or even suggest that the PTO is exempt from this mandate, and there is no legislative history indicating that Congress intended that the PTO should be. Thus, this court must accept the plain and precise language of paragraph six. Accordingly, because no distinction is made in paragraph six between prosecution in the PTO and enforcement in the courts, or between validity and infringement, we hold that paragraph six applies regardless of the context in which the interpretation of means-plus-function language arises, i.e. whether as part of a patentability determination in the PTO or as part of a validity or infringement determination in a court.

In re Donaldson, 16 F.3d 1189, 1193 (Fed. Cir. 1994) (en banc).

COMMENTS AND QUESTIONS

1. As pointed out in commentary written by one of the chief drafters of the 1952 Patent Act:

> The last paragraph of [35 U.S.C. §] 112 relating to so-called functional claims is new. It provides that an element of a claim for a combination . . . may be expressed as a means or step for performing a specified function, without the recital of structure, material or acts in support thereof. It is unquestionable that some measure of greater liberality in the use of functional expressions in combination claims is authorized than had been permitted by some court decisions, and that decisions such as that in Halliburton Oil Well Cementing Co. v. Walker, 329 U.S. 1 (1946[)], are modified or rendered obsolete, but the exact limits of the enlargement remain to be determined. . . . The paragraph ends by stating that such a claim shall be construed to cover the corresponding structure, material, or acts described in the specification and equivalents thereof. This relates primarily to the construction of such claims for the purpose of determining when the claim is infringed (note the use of the word "cover"), and would not appear to have much, if any, applicability in determining the patentability of such claims over the prior art, that is, the Patent Office is not authorized to allow a claim which "reads on" the prior art.

P. J. Federico, Commentary on the New Patent Act, 35 U.S.C.A. p. 1 at pp. 24-25 (1954). Does this passage justify the PTO's position that the specification should not limit the claims for purposes of determining the validity of the patent?

2. Is section 112, ¶6 consistent with the point of having patent claims in the first place? If a potential infringer reads a patent claim that includes an element phrased in "means for *x*" language, how can he be sure he is not infringing on the patent? What if the accused infringer invents a new "means for" performing an old function — is that new means covered by the old patent as well?

3. Congress' use of the phrase "and equivalents thereof" in the statute has given rise to considerable confusion. For ordinary patent claims, the literal scope of the claim is determined by the language of the claim itself. The patentee may also receive broader protection for such claims under the judicially created "doctrine of equivalents."

For means-plus-function claim language, the "literal" scope of the claim itself includes the means described in the specification *and* equivalents to the specification. In addition to this literal scope, the patentee may also receive broader protection under the doctrine of equivalents. Should the patentee who uses "means for" language obtain the benefit of an additional, broader range of equivalents on top of the first range of equivalents spoken of in the statute? See Alpex Computer Corp. v. Nintendo, 34 U.S.P.Q.2d 1167 (S.D.N.Y. 1994) (the doctrine of equivalents is slightly broader than equivalents under section 112, ¶6), *rev'd on other grounds,* 40 U.S.P.Q.2d (BNA) 1667 (Fed. Cir. 1996).

In Chiuminatta Concrete Concepts v. Cardinal Indus. Inc., 46 U.S.P.Q.2d 1752 (Fed. Cir. 1998), the Federal Circuit offered a new explanation of the difference between the two types of equivalents. Not only do the origins of the doctrines differ, but their purposes may differ as well. In particular, the court held that the doctrine of equivalents may encompass later-developed technologies, but section 112 ¶6 equivalents cannot do so. Does this distinction make sense?

4. What claims precisely are subject to section 112, ¶6? In one case, the Federal Circuit rejected the district court's conclusion that particular claim language must be interpreted as creating a "means-plus-function" claim (and thereby limiting the claim). The court pointed out that the patentee could elect whether to proceed under ¶6, and it refused to apply the section to ambiguous claim language. Greenberg v. Ethicon Endo-Surgery Inc., 91 F.3d 1580 (Fed. Cir. 1996). Compare Cole v. Kimberly-Clark Corp., 102 F.3d 1524 (Fed. Cir. 1996) ("perforation means for tearing" language in claim did not invoke section 112, ¶6 because "perforation" disclosed structure) with Unidynamics Corp. v. Automatic Prods. Int'l, 157 F.3d 1311 (Fed. Cir. 1998) ("spring means tending to keep the door closed" did not disclose structure, and therefore invoked section 112, ¶6); see generally York Prods. Inc. v. Central Tractor Farm & Family Center, 99 F.3d 1568 (Fed. Cir. 1996) (whether the word "means" invoked section 112, ¶6 was to be determined on a case-by-case basis by reference to the patent and the prosecution history). Who should be empowered to determine whether a claim is limited under ¶6?

6. Contributory Infringement

C. R. Bard, Inc. v. Advanced Cardiovascular Systems, Inc.
United States Court of Appeals for the Federal Circuit
911 F.2d 670 (Fed. Cir. 1990)

This is a case of claimed infringement of a method patent for a medical treatment. Defendant-Appellant Advanced Cardiovascular Systems, Inc. (ACS) was marketing [a] perfusion catheter for use in coronary angioplasty. Plaintiff-Appellee C.R. Bard, Inc. (Bard) sued ACS for alleged infringement of U.S. Patent No. 4,581,017 ('017), which Bard had purchased all rights to as of December 31, 1986. The '017 patent relates to a method for using a catheter in coronary angioplasty. The district court granted plaintiff Bard summary judgment against ACS, finding infringement of claim 1 of the '017 patent. We reverse the grant of summary judgment and remand the case for further proceedings.

Plaintiff Bard alleges that the ACS catheter is especially adapted for use by a surgeon in the course of administering a coronary angioplasty in a manner that infringes claim 1 of the '017 patent, that therefore ACS is a contributory infringer, and that ACS actively induces infringement. Of course, a finding of induced or contributory infringement must be predicated on a direct infringement of claim 1 by the users of the ACS catheter.

For purposes of this case, the statute requires that ACS sell a catheter for use in practicing the '017 process, which use constitutes a material part of the invention, knowing that the catheter is especially made or adapted for use in infringing the patent, and that the catheter is not a staple article or commodity of commerce suitable for substantial noninfringing use.

In asserting ACS's contributory infringement of claim 1, Bard seeks to establish the requisite direct infringement by arguing that there is no evidence that any angioplasty procedures using the ACS catheter would be noninfringing. Testing this assertion requires a two step analysis. First is a determination of the scope of the claim at issue. Second is an examination of the evidence before the court to ascertain whether, under §271(c), use of the ACS catheter would infringe the claim as interpreted.

Bard argues that [a] prior art patent teaches the use of the catheter with the inlets (side openings) where the blood enters the tube placed only in the aorta, whereas the '017 method in suit involves insertion of the catheter into the coronary artery in such a manner that the openings "immediately adjacent [the] balloon fluidly connect locations within [the] coronary artery surrounding [the] proximal and distal portions of [the] tube." Thus, Bard argues, a surgeon, inserting the ACS catheter into a coronary artery to a point where an inlet at the catheter's proximal end draws blood from the artery, infringes the '017 patent.

[I]t is important to note that the ACS catheter has a series of ten openings in the tube near, and at the proximal end of, the balloon. The first of these openings — the one closest to the balloon — is approximately six millimeters (less than 1/4 inch) from the edge of the proximal end of the balloon. The remainder are located along the main lumen at intervals, the furthest from the balloon being 6.3 centimeters (approximately 2 1/2 inches) away.

It would appear that three possible fact patterns may arise in the course of using the ACS catheter. The first pattern involves positioning the catheter such that all of its side openings are located only in the aorta. This is clearly contemplated by the prior art '725 patent cited by the examiner. In the second of the possible fact patterns, all of the side openings are located within the coronary artery. This situation appears to have been contemplated by the '017 patent, the method patent at issue. In the third fact pattern, some of the side openings are located in the aorta and some are located in the artery.

There is evidence in the record that 40 to 60 percent of the stenoses that require angioplasty are located less than three centimeters from the entrance to the coronary artery. ACS argues that therefore the ACS catheter may be used in such a way that all of the openings are located in the aorta. Even assuming that the trial judge's conclusion is correct that claim 1 is applicable to the third of the fact patterns, it remains true that on this record a reasonable jury could find that, pursuant to the procedure described in the first of the fact patterns (a noninfringing procedure), there are substantial noninfringing uses for the ACS catheter.

Whether the ACS catheter "has no use except through practice of the patented method," Dawson Chemical Co. v. Rohm & Haas Co., is thus a critical issue to be decided in this case. As the Supreme Court recently noted, "[w]hen a charge of contributory infringement is predicated entirely on the sale of an article of commerce that is used by the purchaser [allegedly] to infringe a patent, the public interest in access to that article of commerce is necessarily implicated." Sony Corp. v. Universal City Studios, Inc., 464 U.S. 417, 440 (198[4]) [declining to find contributory copyright infringement in sale of video cassette recorders]. Viewing the evidence in this case in a light most favorable to the nonmoving party, and resolving reasonable inferences in ACS's favor, it cannot be said that Bard is entitled to judgment as a matter of law. The grant of summary judgment finding ACS a contributory infringer under §271(c) is not appropriate.

A person induces infringement under §271(b) by actively and knowingly aiding and abetting another's direct infringement. Bard argues that ACS induced infringement under §271(b) by: 1) providing detailed instructions and other literature on how to use its catheter in a manner which would infringe claim 1; and 2) having positioned the inlets near the balloon's proximal end so as to allow a user of the ACS catheter to infringe claim 1. Because a genuine issue of material fact exists, a grant of summary judgment finding ACS induced infringement is also not appropriate.

COMMENTS AND QUESTIONS

1. While Bard sued ACS, a competing manufacturer of catheters, the "real" infringer in this case is the doctor who completed the catheterization under circumstances that violated Bard's patent. What should doctors do in such a situation? Refuse to perform procedures that infringe a patent? Seek a license? In 1996, Congress amended the patent laws to exempt doctors who perform medical processes from liability for infringement, rendering this problem moot. However, Congress apparently intended to leave device manufacturers liable for contributory infringement. See 35 U.S.C. §287.

2. The Federal Circuit has held that section 271(c) requires a showing that the alleged contributory infringer knew that the combination for which her component was especially designed was both patented and infringing. Trell v. Marlee Electronics Corp., 912 F.2d 1443 (Fed. Cir. 1990). Does this requirement make sense? Is it consistent with the rule for direct infringement, which does not require evidence of knowledge or intent?

If an unpatented product has no other use except in conjunction with the patented machine, process or product, is it reasonable automatically to presume that making, using or selling the *unpatented* product is contributory infringement? Or should companies be allowed to compete in the market for nonstaple products that work with the patented invention? For a discussion of these issues, see Dawson Chemical Co. v. Rohm & Haas, Inc., 448 U.S. 176 (1980).

Note on Inducement

The concepts of contributory infringement and inducement evolved to address infringing activity that somehow lacked the element of a direct making, using or selling of the patented invention. As we have seen, contributory infringement sweeps into the net of infringement the making, use or sale of *less than* the entire patented device. Inducement, on the other hand, involves behavior that omits any making, using or selling but that nevertheless amounts to an attempt to appropriate the value of an invention. It is often described as activity that "aids and abets" infringement. Although inducing infringement commonly involves instructing another to violate a patent, this branch of liability is broad enough to ensnare a host of diverse activities.

A good example of inducement at work is Water Technologies Corp. v. Calco, Ltd., 850 F.2d 660, 669 (Fed. Cir.), *cert. denied*, 488 U.S. 968 (1988). In this case Gartner, one of the defendants, was hired as a consultant by Calco to design a portable water purification system to compete with that sold by the plaintiff/patentee. (The patentee's system involved an advanced purification resin.) Gartner complied by supplying plans for an infringing device. The court called it a classic case of inducement:

> Although section 271(b) does not use the word "knowing," the case law and legislative history uniformly assert such a requirement. Gartner argues that no proof of a specific, knowing intent to induce infringement exists. While proof of intent is necessary, direct evidence is not required; rather, circumstantial evidence may suffice. The requisite intent to induce infringement may be inferred from all of the circumstances. Gartner's activities provide sufficient circumstantial evidence for this court to affirm the district court's finding that he intentionally induced Calco's and the public's direct infringement. Under the facts here, although Gartner's liability as a direct infringer may be de minimis, we see no reason to hold him liable for less than all damages attributable to Calco's infringing sales on the basis of his inducement of direct infringement.

Id. at 668-669.

Water Technologies brings out a number of themes common to cases involving inducement to infringe. First is the importance of intent. As the court states, though the statute does not mention intent, the cases all require it. Notice, however, what evidence the court recites in support of its conclusion that Gartner (one of the defendants in the case) did intend to infringe. Why is extensive circumstantial evidence — whether called intent or something else — required to establish infringement in such a case? Is it relevant that Gartner did not himself make, use, or sell the item that infringed the patent? Why would more extensive evidence be required in such a case?

Second, notice that one item of evidence centers on the existence of a patent. Part of the requisite intent, in other words, is that the one accused of inducing infringement must know of the patent. Beyond this, it is also generally required that the accused infringer know that his or her activities will lead to infringement of the patent. See, e.g., Hewlett-Packard Co. v. Bausch & Lomb, Inc., 909 F.2d 1464 (Fed. Cir. 1990). Since knowledge of infringement depends in part on the accused infringer's understanding of the meaning of the patentee's claims, a number of cases turn on the existence and credibility of a patent attorney's opinion letter regarding whether the accused activities will in fact result in patent infringement. See Manville Sales Corp. v. Paramount Systems, Inc., 917 F.2d 544 (Fed. Cir. 1990) (refusing to find inducement to infringe, partially due to an opinion letter to accused infringer).

COMMENTS AND QUESTIONS

1. Why isn't a consultant who advises a company on an infringing course of action liable for contributory infringement? Was the defendant in *Bard* liable for *both* contributory infringement and inducing infringement? (Recall the instructions to buyers of the defendant's product.)

2. On other acts that may constitute inducement, see, e.g., 4 Donald Chisum, Patents §17.04[2], [3], [4] (1978 & Supp. 1991) (and cases cited therein) (licensing, design, and advertising of infringing product may constitute active inducement). On remedies, see id., at §20.03[7][b][iv] (appropriate relief against one inducing infringement may be same as that against direct infringer). See generally Charles E. Miller, Some Views on the Law of Patent Infringement by Inducement, 53 J. Pat. & Trademk. Off. Society 86 (1971).

PROBLEM

Problem 3-13. Nichols, a scientist who enjoys puzzles, designs a "rotating cube" puzzle in which each face of the cube is composed of a number of smaller cubes, each face is initially of a different color, and the object of the puzzle is to restore the original color scheme once it has been disturbed. Nichols obtains a patent on a method of solving this puzzle, but not on the physical puzzle itself.

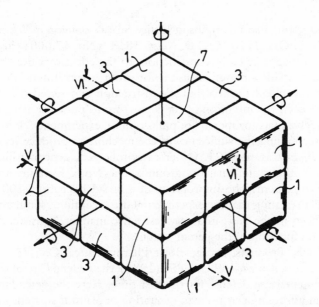

FIGURE 3-6
Patent drawing for Rubik's cube, to which Nichols patented the solutions.

Rubik builds and sells puzzles similar to the ones Nichols has designed. Has Rubik infringed the Nichols patent, either directly or indirectly? Does it matter whether Nichols' patent covers the only known solution to the puzzle, or only one among many possible solutions? [See Figure 3-6.] What if Rubik's product includes a "cheat" sheet advising buyers how to solve the puzzle using a number of methods, including Nichols'? What if Rubik includes a copy of the Nichols patent with each cube sold, ostensibly to advise users how to avoid infringement, but arguably with the intent of encouraging them to use the Nichols method?

D. DEFENSES

1. The "Experimental Use" Defense

Despite the absolute language of section 271, there is a well-recognized, judicially created exception to infringement that is commonly known as the "experimental use" exception. The exception allows for the unlicensed construction and use of a patented invention for purposes of pure scientific inquiry.

The doctrine had its origins in Justice Story's opinion in Whittemore v. Cutter, 29 F. Cas. 1120 (C.C.D. Mass. 1813) (No. 17,600). In this case, the defendant appealed a jury instruction, to the effect that the "making of a machine . . . with a design to use it for profit" constituted infringement. Justice Story upheld the trial judge's instruction and stated that "it could never have been the intention of the legislature to punish a man, who constructed such a machine merely for philosophical experiments, or for the purpose of ascertaining the sufficiency of the machine to produce its described effects." 29 F. Cas. at 1121. Other cases followed, generally limiting the exception to these quite narrow grounds. See Note, Experimental Use as Patent Infringement: The Impropriety of a Broad Exception, 100 Yale L.J. 2169 (1991) (stating that the experimental use exception "should be applied as it has been in the past: in a very restrictive manner, consistent with the purpose and function of the patent system.").

In Roche Products, Inc. v. Bolar Pharmaceutical Co., 733 F.2d 858 (Fed. Cir.), *cert. denied*, 469 U.S. 856 (1984), the Federal Circuit considered the experimental use defense for the first time. Here the defendant (Bolar) engaged in infringing acts prior to expiration of plaintiff's patent in order to facilitate FDA testing, so as to be ready to market the drug as soon as the patent expired. The Federal Circuit overruled the district court's finding of noninfringement, holding the experimental use exception did not include "the limited use of a patented drug for testing and investigation strictly related to FDA drug approval requirements. . . ." 733 F.2d at 861.

> Bolar's intended "experimental" use is solely for business reasons and not for amusement, to satisfy idle curiosity, or for strictly philosophical inquiry. . . . Bolar may intend to perform "experiments," but unlicensed experiments conducted with a view to the adaption of the patented invention to the experimentor's business are a violation of the rights of the patentee to exclude others from using his patented invention.

733 F.2d at 863.

In an excellent article, Rebecca Eisenberg thoroughly reviews the history and rationale behind the doctrine and makes the following recommendations concerning its scope:

(1) Research use of a patented invention to check the adequacy of the specification and the validity of the patent holder's claims about the invention should be exempt from infringement liability.

(2) Research use of a patented invention with a primary or significant market among research users should not be exempt from infringement liability when the research user is an ordinary consumer of the patented invention.

(3) A patent holder should not be entitled to enjoin the use of a patented invention in subsequent research in the field of the invention, which could potentially lead to improvements in the patented technology or to the development of alternative means of achieving the same purpose. However, it might be appropriate in some cases to award a reasonable royalty

after the fact to be sure that the patent holder receives an adequate return on the initial investment in developing the patented invention.

Rebecca Eisenberg, Patents and the Progress of Science: Exclusive Rights and Experimental Use, 56 U. Chi. L. Rev. 1017 (1989).

COMMENTS AND QUESTIONS

1. How would you go about setting the reasonable royalty mentioned by Professor Eisenberg in a case where a subsequent improver infringed a patent in the course of developing a superior alternative that destroys the patentee's market?

2. For an argument that experimental use should usually be extended to patented inventions funded with federal research dollars, see Suzanne T. Michel, The Experimental Use Exception to Infringement Applied to Federally Funded Inventions, 7 High Tech. L.J. 369 (1992).

3. One holding in the *Bolar* case was almost immediately overruled through legislation; see 35 U.S.C. §271(e)(1), enacting a regulatory drug testing exemption from infringement, which permits preexpiration regulatory testing of a patented drug. This provision was introduced as part of the Drug Price Competition Act, Public Law 98-417 (1984) and amended in 1988 to include veterinary drugs. (The trade-off for this exemption was patent term restoration for pharmaceutical patents, which allows patentees to obtain extensions of their patents to partly offset the regulatory review period.) The scope of exempt subject matter under §271(e) has been interpreted to cover medical devices. See Eli Lilly & Co. v. Medtronic, Inc., 872 F.2d 402 (Fed. Cir. 1989), *aff'd*, 496 U.S. 661 (1990).

2. Inequitable Conduct

Kingsdown Medical Consultants, Ltd. v. Hollister Inc.
United States Court of Appeals for the Federal Circuit
863 F.2d 867 (Fed. Cir. 1988), cert. denied, 490 U.S. 1067 (1989)

Kingsdown Medical Consultants, Ltd. and E. R. Squibb & Sons, Inc., (Kingsdown) appeal from a judgment of the United States District Court for the Northern District of Illinois, holding U.S. Patent No. 4,460,363 ('363) unenforceable because of inequitable conduct before the United States Patent and Trademark Office (PTO). We reverse and remand.

Background

Kingsdown sued Hollister Incorporated (Hollister) for infringement of . . . Kingsdown's '363 patent. The district court held the patent unenforceable because of Kingsdown's conduct in respect of claim 9 and reached no other issue.

The invention claimed in the '363 patent is a two-piece ostomy appliance for use by patients with openings in their abdominal walls for release of waste.

The two pieces of the appliance are a pad and a detachable pouch. The pad is secured to the patient's body encircling the abdominal wall opening. Matching coupling rings are attached to the pad and to the pouch. When engaged, the rings provide a water tight seal. Disengaging the rings allows for removal of the pouch.

The Prosecution History

Kingsdown filed its original patent application in February 1978. The '363 patent issued July 17, 1984. The intervening period of more than six-and-a-half years saw a complex prosecution, involving the submission, rejection, amendment, re-numbering, etc., of 118 claims, a continuation application, an appeal, a petition to make special, and citation and discussion of 44 references. . . .

[Certain claims in Kingsdown's original application were allowed by the examiner; others were rejected. Kingsdown appealed the rejection of the latter claims.] While Kingsdown's appeal of other rejected claims was pending, Kingsdown's patent attorney saw a two-piece ostomy appliance manufactured by Hollister. Kingsdown engaged an outside counsel to file a continuation application and withdrew the appeal.

[The continuation application] indicated, incorrectly, that claim 43 in the continuation application corresponded to allowed claim 50 in the parent application. Claim 43 actually corresponded to [one of the other claims] that had been rejected for indefiniteness under §112. Claim 43 was [allowed by the Examiner and] renumbered as the present claim 9 in the '363 patent. . . .

Having examined the prosecution history, the district court found that the examiner could have relied on the representation that claim 43 corresponded to allowable claim 50 and rejected Kingsdown's suggestion that the examiner must have made an independent examination of claim 43. . . .

The district court stated that the narrower language of amended claim 50 gave Hollister, [the defendant accused of infringement here], a possible defense, i.e., that Hollister's coupling member does not encircle the intersection of the aperture and the pad surface because it has an intervening "floating flange" member. The court inferred motive to deceive the PTO because Kingsdown's patent attorney viewed the Hollister appliance after he had amended claim 50 and before the continuation application was filed. The court expressly declined to make any finding on whether the accused device would or would not infringe any claims, but stated that Kingsdown's patent attorney must have perceived that Hollister would have a defense against infringement of the amended version of claim 50 that it would not have against the unamended version.

Issue

Whether the district court's finding of intent to deceive was clearly erroneous, rendering its determination that inequitable conduct occurred an abuse of discretion.[5]

Opinion

We confront a case of first impression, in which inequitable conduct has been held to reside in an incorrect inclusion in a continuation application of a claim that contained allowable subject matter, but had been rejected as indefinite in the parent application.

Inequitable conduct resides in failure to disclose material information, or submission of false material information, with an intent to deceive, and those two elements, materiality and intent, must be proven by clear and convincing evidence. J. P. Stevens & Co., Inc. v. Lex Tex Ltd., Inc., 747 F.2d 1553, 1559, 223 U.S.P.Q. 1089, 1092 (Fed. Cir. 1984), *cert. denied*, 474 U.S. 822 (1985). The findings on materiality and intent are subject to the clearly erroneous standard of Rule 52(a) Fed. R. Civ. P. and are not to be disturbed unless this court has a definite and firm conviction that a mistake has been committed. J. P. Stevens, 747 F.2d at 1562, 223 U.S.P.Q. at 1094.

"To be guilty of inequitable conduct, one must have intended to act inequitably." FMC Corp. v. Manitowoc Co., Inc., 835 F.2d 1411, 1415, 5 U.S.P.Q.2d 1112, 1115 (Fed. Cir. 1987). Kingsdown's attorney testified that he was not aware of the error until Hollister mentioned it in March 1987, and the experts for both parties testified that they saw no evidence of deceptive intent. As above indicated, the district court's finding of Kingsdown's intent to mislead is based on the alternative grounds of: (a) gross negligence; and (b) acts indicating an intent to deceive. Neither ground, however, supports a finding of intent in this case.

a. *Negligence*

The district court inferred intent based on what it perceived to be Kingsdown's gross negligence. Whether the intent element of inequitable conduct is present cannot always be inferred from a pattern of conduct that may be described as gross negligence. That conduct must be sufficient to require a finding of deceitful intent in the light of all the circumstances. We are not convinced that deceitful intent was present in Kingsdown's negligent filing of its continuation application or, in fact, that its conduct even rises to a level that would warrant the description "gross negligence."

5. Because of our decision on intent, it is unnecessary to discuss materiality. Allen Archery, Inc. v. Browning Mfg. Co., 819 F.2d 1087, 1094 (Fed. Cir. 1987).

It is well to be reminded of what actually occurred in this case — a ministerial act involving two claims, which, because both claims contained allowable subject matter, did not result in the patenting of anything anticipated or rendered obvious by anything in the prior art and thus took nothing from the public domain. In preparing and filing the continuation application, a newly-hired counsel for Kingsdown had two versions of "claim 50" in the parent application, an unamended rejected version and an amended allowed version. As is common, counsel renumbered and transferred into the continuation all (here, 22) claims "previously allowed." In filing its claim 43, it copied the "wrong," i.e., the rejected, version of claim 50. That error led to the incorrect listing of claim 43 as corresponding to allowed claim 50 and to incorporation of claim 43 as claim 9 in the patent. In approving the continuation for filing, Kingsdown's regular attorney did not, as the district court said, "catch" the mistake.

In view of the relative ease with which others also overlooked the differences in the claims, Kingsdown's failure to notice that claim 43 did not correspond to the amended and allowed version of claim 50 is insufficient to warrant a finding of an intent to deceive the PTO. Undisputed facts indicating that relative ease are: (1) the similarity in language of the two claims; (2) the use of the same claim number, 50, for the amended and unamended claims; (3) the multiplicity of claims involved in the prosecution of both applications; (4) the examiner's failure to reject claims using "encircled" in the parent application's first and second office actions, making its presence in claim 43 something less than a glaring error; (5) the two-year interval between the rejection/amendment of claim 50 and the filing of the continuation; (6) failure of the examiner to reject claim 43 under §112 or to notice the differences between claim 43 and amended claim 50 during what must be presumed, absent contrary evidence, to have been an examination of the continuation; and (7) the failure of Hollister to notice the lack of correspondence between claim 43 and the amended version of claim 50 during three years of discovery and until after it had carefully and critically reviewed the file history 10 to 15 times with an eye toward litigation. That Kingsdown did not notice its mistake during more than one opportunity of doing so, does not in this case, and in view of Hollister's frequent and focused opportunities, establish that Kingsdown intended to deceive the PTO.

. . . The district court correctly noted that an examiner has a right to expect candor from counsel. Its indication that examiners "must" rely on counsel's candor would be applicable, however, only when the examiner does not have the involved documents or information before him, as the examiner did here. Blind reliance on presumed candor would render examination unnecessary, and nothing in the statute or Manual of Patent Examining Procedure would justify reliance on counsel's candor as a substitute for an examiner's duty to examine the claims.

Thus the first basis for the district court's finding of deceitful intent (what it viewed as "gross negligence") cannot stand.

b. Acts

The district court also based its finding of deceitful intent on the separate and alternative inferences it drew from Kingsdown's acts in viewing the Hollister device, in desiring to obtain a patent that would "cover" that device, and in failing to disclaim or reissue after Hollister charged it with inequitable conduct. The district court limited its analysis here to claim 9 and amended claim 50.

It should be made clear at the outset of the present discussion that there is nothing improper, illegal or inequitable in filing a patent application for the purpose of obtaining a right to exclude a known competitor's product from the market; nor is it in any manner improper to amend or insert claims intended to cover a competitor's product the applicant's attorney has learned about during the prosecution of a patent application. Any such amendment or insertion must comply with all statutes and regulations, of course, but, if it does, its genesis in the marketplace is simply irrelevant and cannot of itself evidence deceitful intent. . . .

Faced with Hollister's assertion that an experienced patent attorney would knowingly and intentionally transfer into a continuing application a claim earlier rejected for indefiniteness, without rearguing that the claim was not indefinite, the district court stated that "how an experienced patent attorney could allow such conduct to take place" gave it "the greatest difficulty." A knowing failure to disclose and knowingly false statements are always difficult to understand. However, a transfer of numerous claims en masse from a parent to a continuing application, as the district court stated, is a ministerial act. As such, it is more vulnerable to errors which by definition result from inattention, and is less likely to result from the scienter involved in the more egregious acts of omission and commission that have been seen as reflecting the deceitful intent element of inequitable conduct in our cases. . . .

Resolution of Conflicting Precedent

"Gross Negligence" and the Intent Element of Inequitable Conduct

Some of our opinions have suggested that a finding of gross negligence compels a finding of an intent to deceive. Others have indicated that gross negligence alone does not mandate a finding of intent to deceive.

"Gross negligence" has been used as a label for various patterns of conduct. It is definable, however, only in terms of a particular act or acts viewed in light of all the circumstances. We adopt the view that a finding that particular conduct amounts to "gross negligence" does not of itself justify an inference of intent to deceive; the involved conduct, viewed in light of all the evidence, including evidence indicative of good faith, must indicate sufficient culpability to require a finding of intent to deceive. . . .

Effect of Inequitable Conduct

When a court has finally determined that inequitable conduct occurred in relation to one or more claims during prosecution of the patent application, the entire patent is rendered unenforceable. We, in banc, reaffirm that rule as set forth in J. P. Stevens & Co. v. Lex Tex Ltd.

Conclusion

Having determined that the district court's finding of intent is clearly erroneous, the panel reverses the judgment based on a conclusion of inequitable conduct before the PTO and remands the case for such further proceedings as the district court may deem appropriate.

COMMENTS AND QUESTIONS

1. Why is the penalty for inequitable conduct so harsh — rendering the patent completely unenforceable? Who has more information about the inventor's work and the precise prior art it relates to — the examiner or the inventor? Who has more to gain from the issuance (or rejection) of a patent?

2. Many inequitable conduct cases involve suppressing a key prior art reference. See, e.g., *J. P. Stevens*, cited in the *Kingsdown* case. Do you think it should matter whether (1) the reference was in a class or group that the Examiner *should* have searched?; (2) that it was disclosed, but in a long "string cite" of less relevant prior art?; (3) that the patentee didn't know of the key reference, but *should have*?

3. The practice of "submarine patenting" — deliberately keeping a secret patent application pending for years in the PTO, only to spring it on an unsuspecting industry — has survived charges of inequitable conduct. See Ford Motor Co. v. Lemelson, 42 U.S.P.Q.2d 1706 (D. Nev. 1997); Advanced Cardiovascular Systems v. Medtronic Inc., No. __, 1996 WL 467293 (N.D. Cal. 1996) (both rejecting inequitable conduct claims). But one court has held that allegations of submarine patenting may state an *antitrust* claim against the patentee. See Discovision Assoc. v. Disc Mfg. Inc., 42 U.S.P.Q.2d 1749 (D. Del. 1997). We discuss the interaction between patent and antitrust law in more detail in Chapter 8.

3. Patent Misuse

Patent misuse is a judicially created doctrine that bars patentees from enforcing their patent against infringers when they have "misused" the patent. It was first adopted by the Supreme Court in 1917.

Motion Picture Patents Company v. Universal Film Manufacturing Company et al.
Supreme Court of the United States
243 U.S. 502 (1917)

Mr. Justice CLARKE delivered the opinion of the court.

In this suit relief is sought against three defendant corporations as joint infringers of claim number seven of United States letters patent No. 707,934 granted to Woodville Latham, assignor, on August 26, 1902, for improvements in Projecting-Kinetoscopes. It is sufficient description of the patent to say that it covers a part of the mechanism used in motion picture exhibiting machines for feeding a film through the machine with a regular, uniform and accurate movement and so as not to expose the film to excessive strain or wear.

The defendants in a joint answer do not dispute the title of the plaintiff to the patent but they deny the validity of it, deny infringement, and claim an implied license to use the patented machine.

Evidence which is undisputed shows that the plaintiff on June 20, 1912, in a paper styled "License Agreement" granted to The Precision Machine Company a right and license to manufacture and sell machines embodying the inventions described and claimed in the patent in suit, and in other patents, throughout the United States, its territories and possessions. This agreement contains a covenant on the part of the grantee that every machine sold by it, except those for export, shall be sold "under the restriction and condition that such exhibiting or projecting machines shall be used solely for exhibiting or projecting motion pictures containing the inventions of reissued letters patent No. 12,192, leased by a licensee of the licensor while it owns said patents, and upon other terms to be fixed by the licensor and complied with by the user while the said machine is in use and while the licensor owns said patents (which other terms shall only be the payment of a royalty or rental to the licensor while in use)." . . .

The agreement further provides that the grantee shall not sell any machine at less than the plaintiff's list price, except to jobbers and others for purposes of resale and that it will require such jobbers and others to sell at not less than plaintiff's list price. . . .

It was admitted at the bar that 40,000 of the plaintiff's machines are now in use in this country and that the mechanism covered by the patent in suit is the only one with which motion picture films can be used successfully.

This state of facts presents two questions for decision:

. . .

Second. May the assignee of a patent, which has licensed another to make and sell the machine covered by it, by a mere notice attached to such machine, limit the use of it by the purchaser or by the purchaser's lessee to terms not stated in the notice but which are to be fixed, after sale, by such assignee in its discretion? . . .

Since Pennock v. Dialogue, 2 Pet. 1, was decided in 1829 this court has consistently held that the primary purpose of our patent laws is not the creation of private fortunes for the owners of patents but is "to promote the progress of science and useful arts" (Constitution, Art. I, §8). . . .

Plainly, this language of the statute and the established rules to which we have referred restrict the patent granted on a machine, such as we have in this case, to the mechanism described in the patent as necessary to produce the described results. It is not concerned with and has nothing to do with the materials with which or on which the machine operates. The grant is of the exclusive right to use the mechanism to produce the result with any appropriate material, and the materials with which the machine is operated are no part of the patented machine or of the combination which produces the patented result. The difference is clear and vital between the exclusive right to use the machine which the law gives to the inventor and the right to use it exclusively with prescribed materials to which such a license notice as we have here seeks to restrict it. . . . Both in form and in substance the notice attempts a restriction upon the use of the supplies only and it cannot with any regard to propriety in the use of language be termed a restriction upon the use of the machine itself.

Whatever right the owner may have to control by restriction the materials to be used in operating the machine must be derived through the general law from the ownership of the property in the machine and it cannot be derived from or protected by the patent law. . . .

This construction gives to the inventor the exclusive use of just what his inventive genius has discovered. It is all that the statute provides shall be given to him and it is all that he should receive, for it is the fair as well as the statutory measure of his reward for his contribution to the public stock of knowledge. If his discovery is an important one his reward under such a construction of the law will be large, as experience has abundantly proved, and if it be unimportant he should not be permitted by legal devices to impose an unjust charge upon the public in return for the use of it. For more than a century this plain meaning of the statute was accepted as its technical meaning, and that it afforded ample incentive to exertion by inventive genius is proved by the fact that under it the greatest inventions of our time, teeming with inventions, were made. . . .

The construction of the patent law which justifies as valid the restriction of patented machines, by notice, to use with unpatented supplies necessary in the operation of them, but which are no part of them, is believed to have originated in Heaton-Peninsular Button-Fastener Co. v. Eureka Specialty

Co., 77 Fed. Rep. 288 (which has come to be widely referred to as the *Button-Fastener Case*), decided by the Circuit Court of Appeals of the Sixth Circuit in 1896. In this case the court, recognizing the pioneer character of the decision it was rendering, speaks of the "novel restrictions" which it is considering and says that it is called upon "to mark another boundary line around the patentee's monopoly, which will debar him from engrossing the market for an article not the subject of a patent," which it declined to do.

This decision proceeds upon the argument that, since the patentee may withhold his patent altogether from public use he must logically and necessarily be permitted to impose any conditions which he chooses upon any use which he may allow of it. The defect in this thinking springs from the substituting of inference and argument for the language of the statute and from failure to distinguish between the rights which are given to the inventor by the patent law and which he may assert against all the world through an infringement proceeding and rights which he may create for himself by private contract which, however, are subject to the rules of general as distinguished from those of the patent law. While it is true that under the statutes as they were (and now are) a patentee might withhold his patented machine from public use, yet if he consented to use it himself or through others, such use immediately fell within the terms of the statute and as we have seen he is thereby restricted to the use of the invention as it is described in the claims of his patent and not as it may be expanded by limitations as to materials and supplies necessary to the operation of it imposed by mere notice to the public.

. . . The perfect instrument of favoritism and oppression which such a system of doing business, if valid, would put into the control of the owner of such a patent should make courts astute, if need be, to defeat its operation. If these restrictions were sustained plainly the plaintiff might, for its own profit or that of its favorites, by the obviously simple expedient of varying its royalty charge, ruin anyone unfortunate enough to be dependent upon its confessedly important improvements for the doing of business.

. . . [F]ollowing the decision of the *Button-Fastener Case*, it was widely contended as obviously sound, that the right existed in the owner of a patent to fix a price at which the patented article might be sold and resold under penalty of patent infringement. But this court, when the question came before it in Bauer v. O'Donnell, 229 U.S. 1, . . . decided that the owner of a patent is not authorized by either the letter or the purpose of the law to fix, by notice, the price at which a patented article must be sold after the first sale of it, declaring that the right to vend is exhausted by a single, unconditional sale, the article sold being thereby carried outside the monopoly of the patent law and rendered free of every restriction which the vendor may attempt to put upon it. The statutory authority to grant the exclusive right to "use" a patented machine is not greater, indeed it is precisely the same, as the authority to grant the exclusive right to "vend," and, looking to that authority, for the reasons stated in this opinion we are convinced that the exclusive right granted in every patent must be limited to the invention described in the claims of the patent and that it is not competent for the owner of a patent by notice attached to its machine to, in effect, extend the scope of its patent monopoly by restricting the use of it to materials necessary in its operation but which are no part of the patented invention, or to send its machines forth into the channels of trade of the country subject to conditions as to use or royalty to be paid to be imposed thereafter at the discretion of such patent owner. The patent law furnishes no warrant for such a practice and the cost, inconvenience and annoyance to the public which the opposite conclusion would occasion forbid it.

It is argued as a merit of this system of sale under a license notice that the public is benefitted by the sale of the machine at what is practically its cost and by the fact that the owner of the patent makes its entire profit from the sale of the supplies with which it is operated. This fact, if it be a fact, instead of commending, is the clearest possible condemnation of, the practice adopted, for it proves that under color of its patent the owner intends to and does derive its profit, not from the invention on which the law gives it a monopoly but from the unpatented supplies with which it is used and which are wholly without the scope of the patent monopoly, thus in effect extending the power to the owner of the patent to fix the price to the public of the unpatented supplies as effectively as he may fix the price on the patented machine. . . .

Coming now to the terms of the notice attached to the machine sold to the Seventy-second Street Amusement Company under the license of the plaintiff and to the first question as we have stated it.

This notice first provides that the machine, which was sold to and paid for by the Amusement Company may be used only with moving picture films containing the invention of reissued patent No. 12,192, so long as the plaintiff continues to own this reissued patent.

Such a restriction is invalid because such a film is obviously not any part of the invention of the patent in suit; because it is an attempt, without statutory warrant, to continue the patent monopoly in this particular character of film after it has expired, and because to enforce it would be to create a monopoly in the manufacture and use of moving picture films, wholly outside of the patent in suit and of the patent law as we have interpreted it. . . .

A restriction which would give to the plaintiff such a potential power for evil over an industry which must be recognized as an important element in the amusement life of the nation, under the conclusions we have stated in this opinion, is plainly void, because wholly without the scope and purpose of our patent laws and because, if sustained, it would be gravely injurious to that public interest, which we have seen is more a favorite of the law than is the promotion of private fortunes. . . .

Mr. Justice HOLMES, dissenting.

I suppose that a patentee has no less property in his patented machine than any other owner, and that in addition to keeping the machine to himself the patent gives him the further right to forbid the rest of the world from making others like it. In short, for whatever motive, he may keep his device wholly out of use. Continental Paper Bag Co. v. Eastern Paper Bag Co., 210 U.S. 405, 422. So much being undisputed, I cannot understand why he may not keep it out of use unless the licensee, or, for the matter of that, the buyer, will use some unpatented thing in connection with it. Generally speaking the measure of a condition is the consequence of a breach, and if that consequence is one that the owner may impose unconditionally, he may impose it conditionally upon a certain event. Ashley v. Ryan, 153 U.S. 436, 443. Lloyd v. Dollison, 194 U.S. 445, 449. . . .

No doubt this principle might be limited or excluded in cases where the condition tends to bring about a state of things that there is a predominant public interest to prevent. But there is no predominant public interest to prevent a patented tea pot or film feeder from being kept from the public, because, as I have said, the patentee may keep them tied up at will while his patent lasts. Neither is there any such interest to prevent the purchase of the tea or films, that is made the condition of the use of the machine. The supposed contravention of public interest sometimes is stated as an attempt to extend the patent law to unpatented articles, which of course it is not, and more accurately as a possible domination to be established by such means. But the domination is one only to the extent of the desire for the tea pot or film feeder, and if the owner prefers to keep the pot or the feeder unless you will buy his tea or films, I cannot see in allowing him the right to do so anything more than an ordinary incident of ownership, or at most, a consequence of the *Paper Bag Case,* on which, as it seems to me, this case ought to turn. See Grant v. Raymond, 6 Pet. 218, 242. . . .

COMMENTS AND QUESTIONS

1. The asserted justification for the patent misuse doctrine lies in the public policy underpinning the patent laws. It is no coincidence, however, that the doctrine first developed around 1900, shortly after the antitrust laws were passed. Patent misuse and antitrust are strikingly similar in the sorts of conduct they prohibit. For example, in *Motion Picture,* the conduct at issue was (among other things) resale price maintenance, which is per se illegal under the antitrust laws. Patent misuse, as originally conceived, was broader than the antitrust laws, however. Neither market power nor actual effect on competition need be proven to show patent misuse, as they would to prove an antitrust violation. We will return to the scope of the doctrine vis-à-vis the antitrust laws shortly.

2. The Court continually expanded the scope of the patent misuse doctrine in the decades that followed *Motion Picture.* That expansion culminated in the cases of Mercoid Corp. v. Mid-Continent Investment Co., 320 U.S. 661 (1944) (*Mercoid I*) and Carbice Corp. v. American Patents Development Corp., 283 U.S. 27 (1931). *Mercoid* held that it was patent misuse to tie a

patented product to a nonstaple product.[40] *Carbice* held that a showing of patent misuse (which does not require proof of market power or impact) was prima facie evidence of an antitrust violation.

Both of these rules were short-lived. Congress amended the patent laws in 1952 by adding 35 U.S.C. §271(d)(1)-(3).

35 U.S.C. §271. Infringement of Patent

. . .

(d) No patent owner otherwise entitled to relief for infringement or contributory infringement of a patent shall be denied relief or deemed guilty of misuse or illegal extension of the patent right by reason of his having done one or more of the following: (1) derived revenue from acts which if performed by another without his consent would constitute contributory infringement of the patent; (2) licensed or authorized another to perform acts which if performed without his consent would constitute contributory infringement of the patent; (3) sought to enforce his patent rights against infringement or contributory infringement; . . .

The import of this statutory limit on the judicially created doctrine of misuse was described in detail by the Supreme Court in the case of Dawson Chemical Co. v. Rohm & Haas Co., 448 U.S. 176 (1980). The Court stated:

Section 271(c) identifies the basic dividing line between contributory infringement and patent misuse. It adopts a restrictive definition of contributory infringement that distinguishes between staple and nonstaple articles of commerce. It also defines the class of nonstaple items narrowly. In essence, this provision places materials like the dry ice of the Carbice case outside the scope of the contributory infringement doctrine. As a result, it is no longer necessary to resort to the doctrine of patent misuse in order to deny patentees control over staple goods used in their inventions.

The limitations on contributory infringement written into §271(c) are counterbalanced by limitations on patent misuse in §271(d). Three species of conduct by patentees are expressly excluded from characterization as misuse. First, the patentee may "deriv[e] revenue" from acts that "would constitute contributory infringement" if "performed by another without his consent." This provision clearly signifies that a patentee may make and sell nonstaple goods used in connection with his invention. Second, the patentee may "licens[e] or authoriz[e] another to perform acts" which without such authorization would constitute contributory infringement. This provision's use in the disjunctive of the term "authoriz[e]" suggests that more than explicit licensing agreements is contemplated. Finally, the patentee may "enforce his patent rights against . . . contributory infringement." This provision plainly means that the patentee may

---

40. A staple product is one which has an existing market, beyond use with the patent. A non-staple product has no commercial use except in connection with the patent.

bring suit without fear that his doing so will be regarded as an unlawful attempt to suppress competition. The statute explicitly states that a patentee may do "one or more" of these permitted acts, and it does not state that he must do any of them.

In our view, the provisions of §271(d) effectively confer upon the patentee, as a lawful adjunct of his patent rights, a limited power to exclude others from competition in nonstaple goods. A patentee may sell a nonstaple article himself while enjoining others from marketing that same good without his authorization. By doing so, he is able to eliminate competitors and thereby to control the market for that product. Moreover, his power to demand royalties from others for the privilege of selling the nonstaple item itself implies that the patentee may control the market for the nonstaple good; otherwise, his "right" to sell licenses for the marketing of the nonstaple good would be meaningless, since no one would be willing to pay him for a superfluous authorization. See Note, 70 Yale. L.J. 649, 659 (1961).

3. Much of the debate over the wisdom of the patent misuse doctrine parallels similar debates over the wisdom of particular antitrust rules. We discuss these rules in the context of intellectual property in more detail in Chapter 8.

4. The patent misuse doctrine is closely connected to the idea of *exhaustion* of patent rights by a first sale in commerce. The exhaustion doctrine provides that once a patent owner has sold or licensed goods that embody the patented product, it cannot exercise further control over those goods. The exhaustion doctrine is venerable and universally accepted.

Recent Federal Circuit precedent casts serious doubt on the modern efficacy of the exhaustion doctrine, however. First, in Mallinckrodt v. MediPart, 976 F.2d 700 (Fed. Cir. 1992), the court held that a patentee could prevent doctors from reusing a patented medical device by putting a "label license" on the device indicating that it was authorized for a single use only. More recently, the Federal Circuit extended *Mallinckrodt* in B. Braun Medical, Inc. v. Abbott Labs., Inc., 124 F.3d 1419 (Fed. Cir. 1997), but also noted the limitations that patent misuse places on such restrictive licenses:

The resolution of this issue is governed by our precedent in *Mallinckrodt*. . . . In that case, we canvassed precedent concerning the legality of restrictions placed upon the post-sale use of patented goods. As a general matter, we explained that an unconditional sale of a patented device exhausts the patentee's right to control the purchaser's use of the device thereafter. . . . The theory behind this rule is that in such a transaction, the patentee has bargained for, and received, an amount equal to the full value of the goods. This exhaustion doctrine, however, does not apply to an expressly conditional sale or license. In such a transaction, it is more reasonable to infer that the parties negotiated a price that reflects only the value of the "use" rights conferred by the patentee. As a result, express conditions accompanying the sale or license of a patented product are generally upheld. . . . Accordingly, conditions that violate some law or equitable consideration are unenforceable. On the other hand, violation of valid conditions entitles the patentee to a remedy for either patent infringement or breach of contract. . . . This, then, is the general framework.

In *Mallinckrodt*, we also outlined the framework for evaluating whether an express condition on the post-sale use of a patented product constitutes patent misuse. The patent misuse doctrine, born from the equitable doctrine of unclean hands, is a method of limiting abuse of patent rights separate from the antitrust laws. The key inquiry under this fact-intensive doctrine is whether, by imposing the condition, the patentee has "impermissibly broadened the 'physical or temporal scope' of the patent grant with anticompetitive effect". . . . Two common examples of such impermissible broadening are using a patent which enjoys market power in the relevant market, see 35 U.S.C. §271(d)(5) (1994), to restrain competition in an unpatented product or employing the patent beyond its 17-year term. In contrast, field of use restrictions (such as those at issue in the present case) are generally upheld, see General Talking Pictures, 305 U.S. at 127, 59 S.Ct. at 117, and any anticompetitive effects they may cause are reviewed in accordance with the rule of reason. See *Mallinckrodt*, 976 F.2d at 708, 24 U.S.P.Q.2d at 1179-80.

Id. at 1426-1427.

Note that this argument runs counter to a bedrock assumption lying behind the first-sale doctrine in all areas of intellectual property law: that at least some of the property rights held by an intellectual property owner are extinguished when an item embodying those rights is sold. To the extent that *Mallinckrodt* and *Braun* assume a different status for tangible items incorporating an intellectual property right, they undercut this basic assumption.

For a discussion of the relationship between price discrimination and the exhaustion doctrine, albeit in the copyright context, see Michael Meurer, Price Discrimination, Personal Use and Piracy: Copyright Protection of Digital Works, 45 Buffalo L. Rev. 845 (1997).

Note on the Scope of the Patent Misuse Doctrine

Congress has shown continued dissatisfaction with the scope of the patent misuse doctrine. In 1988, it passed the Patent Misuse Reform Act (PMRA), which provides:

35 U.S.C. §271.   Infringement of Patent

. . .

(d) No patent owner otherwise entitled to relief for infringement or contributory infringement of a patent shall be denied relief or deemed guilty of misuse or illegal extension of the patent right by reason of his having done one or more of the following: . . .

(4) refused to license or use any rights to the patent; or (5) conditioned the license of any rights to the patent or the sale of the

patented product on the acquisition of a license to rights in another patent or purchase of a separate product, unless, in view of the circumstances, the patent owner has market power in the relevant market for the patent or patented product on which the license or sale is conditioned.

There is obviously a great deal of overlap between the policies of the patent misuse doctrine and those underlying the antitrust laws. A similar overlap existed in scope between the two doctrines, even before (and certainly after) passage of the Patent Misuse Reform Act. In fact, Judge Posner argues that patent misuse and antitrust are actually coextensive — that an act cannot be misuse unless it also violates the antitrust laws. See USM Corp. v. SPS Technologies, 694 F.2d 505 (7th Cir. 1982), *cert. denied*, 462 U.S. 107 (1983):

> The [patent misuse] doctrine arose before there was any significant body of federal antitrust law, and reached maturity long before that law (a product very largely of free interpretation of unclear statutory language) attained its present broad scope. Since the antitrust laws as currently interpreted reach every practice that could impair competition substantially, it is not easy to define a separate role for a doctrine also designed to prevent an anticompetitive practice — the abuse of a patent monopoly. One possibility is that the doctrine of patent misuse, unlike antitrust law, condemns any patent licensing practice that is even trivially anticompetitive, at least if it has no socially beneficial effects. . . .
>
> If misuse claims are not tested by conventional antitrust principles, by what principles shall they be tested? Our law is not rich in alternative concepts of monopolistic abuse; and it is rather late in the day to try to develop one without in the process subjecting the rights of patent holders to debilitating uncertainty. Cf. Hensley Equipment Co. v. Esco Corp., 383 F.2d 252, 261-262 n.19, *amended*, 386 F.2d 442 (5th Cir. 1967).

Contrary to Judge Posner's suggestion, however, there are several differences in scope between misuse and conventional antitrust doctrines. Patent misuse extends (or may extend) to a number of practices that do not fall within the antitrust laws at all. Many of these are practices that arise exclusively or primarily in the context of patents. Several examples follow.

Nonmetered Licenses. Patentees are entitled to charge royalties on the patents they license to others. What form may those license payments take? Hazeltine Research, the consumer electronics patent consortium, charged a royalty based on a flat percentage of a purchaser's sales, regardless of how many patents the purchaser needed or how important they were to the end product. The Court upheld this arrangement in 1950 in Automatic Radio Mfg. Co. v. Hazeltine Research, Inc., 339 U.S. 827, 833-834 (1950). But by 1969 it was of a different view. In Zenith Radio Corp. v. Hazeltine Research, Inc., 395 U.S. 100 (1969), the Court struck down the royalty provision:

We also think patent misuse inheres in a patentee's insistence on a per-centage-of-sales royalty, regardless of use, and his rejection of licensee proposals to pay only for actual use. Unquestionably, a licensee must pay if he uses the patent. Equally, however, he may insist upon paying only for use, and not on the basis of total sales, including products in which he may use a competing patent or in which no patented ideas are used at all. There is nothing in the right granted the patentee to keep others from using, selling, or manufacturing his invention which empowers him to insist on payment not only for use but also for producing products which do not employ his discoveries at all.

What is wrong with nonmetered royalties? Here, Bork's and Posner's arguments in the leveraging debate seem to have some force. See Robert Bork, The Antitrust Paradox (1978); Richard A. Posner, Antitrust Law: An Economic Perspective (1976). If a licensee is only willing to pay a fixed amount for a license, should it matter how that amount is collected? Certainly the "percentage of sales" formula is not itself anticompetitive, and it may be far more convenient than more exact measures of how much a licensee uses a patent. Should the transaction costs of royalty accounting come into play in deciding whether a practice amounts to misuse?

One problem with such provisions is that they also provide a perfect source of information for policing a cartel. A licensee subject to a percentage of sales license must turn its sales and output information over to the licensor. With this information, the licensor can detect attempts to cheat on the cartel by expanding output. This danger is particularly acute when a participant in a market licenses his patent horizontally — that is, to its competitors in the same market. There is therefore an antitrust danger associated with permitting the use of these agreements.

Grantback Clauses. As a condition to a license agreement, a patentee will sometimes require the licensee to grant him rights to any "improvement patents" the licensee is issued while using the licensed patent. Courts have been fairly lenient with respect to such "grantback clauses." See Transparent-Wrap Machine Corp. v. Stokes & Smith Co., 329 U.S. 637 (1947). But grantbacks may sometimes run afoul of the patent misuse doctrine.

Field-of-Use Restrictions. Most patent licenses contain some sort of restriction on how the patent may be used by the licensee. As we discuss in Chapter 8, restrictions that attempt to control the price at which the licensee sells products made using the patented process or equipment are illegal per se under Sherman Act section 1. Other restrictions, however, control not the price but the geographic or product market in which the licensee sells. These agreements have generally been upheld, unless they are part of a tying ar-rangement. See, e.g., U.S. v. Studiengesellschaft Kohle, 670 F.2d 1122 (D.C. Cir. 1981) (upholding field of use restriction in licenses for industrial catalysts).

Patent Suppression. For a variety of reasons, many (indeed, most) patents are never commercialized. The reasons can include lack of adequate financing, absence of commercial value, or simply bad business judgment on the part of an inventor. But there are less benign reasons as well. Inventions may be deliberately patented as a matter of course, in case they turn out to be useful. Many biotechnology companies follow this practice. Inventions may be deliberately patented and then not used in order to deny competitors the opportunity that a new technological advance might present, even though the patentee does not intend to alter its equipment or production process to take advantage of the new invention. Cf. Special Equipment Co. v. Coe, 324 U.S. 370 (1945) (Douglas, J., dissenting) ("The right of suppression of a patent came into the law over a century after the first patent act was passed. . . . I think it is time to be rid of that rule. It is inconsistent with the Constitution and the patent legislation which Congress has enacted.").

Commentators have long argued over whether suppression actually occurs, and whether it is an economically rational strategy. See, e.g., Mark Clark, Suppressing Innovation: Bell Laboratories and Magnetic Recording, 34 Tech. & Cult. 516, 532 (1993) (documenting suppression of Hickman patents covering magnetic recording, a policy that the author argues was pursued by AT&T to preserve markets and for "ideological reasons," i.e., the threat to privacy from recorded phone conversations). Nonuse of patents for benign reasons certainly occurs on a daily basis. Whether true suppression occurs has never been resolved. Advocates of compulsory licensing point to several instances in which companies are supposed to have designed or purchased a patented invention for the sole purpose of denying it to others. Automobile manufacturers are charged with suppressing new auto antipollution technology because they feared that they would be forced by government regulators to install the new devices. And according to a rumor, manufacturers of panty hose are supposed to have suppressed their design for "no-run" hose in order to maintain the lucrative market for disposable hose.

In all these cases (and numerous others), patents for the allegedly suppressed products exist. What is uncertain is *why* the products were never commercialized. The companies involved normally claim that the products do not, in fact, work as promised. Those claims are bolstered by the argument of economists that suppression simply makes no sense. If a new invention is truly superior to current products, they argue, the patentee could sell that invention and more than make up for any losses it might sustain in the market for its current products. Indeed, economists continue, that is the definition of a superior product.

This argument has substantial force when the allegedly suppressed invention is in the same market as the patentee's current products. Thus panty hose manufacturers could presumably switch from selling disposable hose to selling no-run hose and, assuming the no-run hose was really a better product, charge prices high enough to make up for the fact that they would sell fewer pairs of hose. This argument is not completely convincing, however. If the process for producing no-run hose requires different machinery from that which the manufacturers currently use, there will be substantial fixed costs associated with the switch. Manufacturers may prefer to delay introducing the new product until they have to replace their machines anyway; they will suppress the invention until then. By contrast, if a new entrant into the market could use the patented process, he would choose to build the new machines immediately. By suppressing the patent, therefore, the patentee causes society to lose the benefits of the immediate production of the new process.[41] An example of this occurred in the rubber industry, when Standard Oil admitted in a 1942 consent decree to inhibiting the introduction of synthetic rubber. The reason was clearly to preserve its extensive investment in natural rubber production processes. See D. R. B. Ross, Patents and Bureaucrats: U.S. Synthetic Rubber Development Before Pearl Harbor, in Business and Government 119, 120 (J. R. Frese & J. Judd, eds., 1985). See generally Dunford, Suppression of Technology, 32 Admin. Sci. Q. 512 (1987).

This "retooling" problem is even more serious when the patentee is not in the same market as the patented invention but rather in an upstream or downstream market that would be affected by the patent. Consider the rumor that Exxon purchased and buried the design for the "momentum engine," which would tremendously increase automobile engine efficiency (and therefore tremendously decrease the demand for gasoline). It could produce and sell the momentum engine, using the revenues from those sales to offset its loss in gasoline revenues. However, Exxon is not in the engine business and is likely to be less efficient at that business than it is at refining and selling gasoline. Its profit-maximizing course may therefore be to conceal the invention, so that no one else can use it, and to continue to sell gasoline.[42]

In any event, Congress — and the courts before it — has deemed that the costs of determining when nonuse or suppression is actionable, together with potential strategic uses of any remedies for it, outweigh the benefits of rooting out actual cases of suppression. See 35 U.S.C. §271(d)(4), amended in 1988 to protect nonuse.

---

41. Of course, the patentee could always license the patent to the new entrant. However, innovation is associated with strong first-mover advantages, so the first company to manufacture and sell a product is likely to maintain a dominant position even after the patent expires and after further inventions supercede the original one. Thus licensing to competitors may not be an attractive option for many patentees.

42. For reasons similar to those discussed above, licensing to an existing engine manufacturer may not be an attractive option in this situation. Although Exxon would not be in direct competition with the licensee, its business would be adversely affected by licensing the engine. And there is often doubt that a licensor can extract the full value of its invention in the licensing agreement. Exxon would have changed from a continuing enterprise to one whose major asset was licensing revenue for the next 20 years, and which could then expect to go out of business. This option, even if economically rational, is not attractive to most major corporations.

E. INTERNATIONAL PATENT LAW

To obtain international protection for patented technology, lawyers and businesspeople must ultimately rely on the domestic patent law of each country. In two respects, however, international treaties do come into play. First, a number of international conventions streamline the procedures under which patents originating outside the host country are prosecuted. Second, certain international treaties provide a uniform substantive floor of protection, a minimum set of rights below which treaty signatories must not fall in conferring patents. Chief among these is the Trade-Related Aspects of Intellectual Property (TRIPs) Agreement, part of the 1994 Uruguay Round of GATT revisions, which are discussed later in this section.

1. Procedural Rules

The most important set of international procedural rules for patentees involve filing and priority dates. To understand them, however, one must first recognize that in practice they interact with elements of each country's domestic law. Recall, in this regard, the crucial divergence between U.S. and foreign priority rules. As you will recall, the United States is virtually the world's only "first-to-invent" country. All other countries, with trivial exceptions, establish priority based on a "first-to-file" system. If an inventor has an interest in securing patent protection overseas — whether in addition to U.S. protection or in place of it — he or she must file there *as soon after invention as possible*. Unlike the situation in the United States, where one's filing date is not the only significant event, in foreign countries a late filing (relative to a rival) can be fatal. There is no opportunity to prove that although a rival filed first, you in fact invented earlier. The filing date is the *only* relevant event in most foreign countries.

One additional rule in place in many foreign countries is the notion of "absolute priority." This means that an application must be filed before the invention is described or used in a publicly available forum. In other words, there is no one-year statutory grace period for filing in foreign systems. An application must be filed *before* publication of an article describing the same invention, or before any public use or sale — as opposed to within one year of these events as under 35 U.S.C. §102(b).

Thus even if one has solid proof of the date of invention, one must still make haste to the patent office, since any activity — by the inventor or a third party — prior to the date of filing can disqualify the invention from patentability. Contrast this with the statutory bars, whose critical date of one year prior to filing gives the patentee one year from the date of certain activities to file for a patent.

Another difference in domestic patent laws is also relevant. In many foreign countries, most notably those in Europe, "prior user" rights protect first inventors/later filers. Prior inventors are allowed to continue their use of the invention after a patent issues to the first filer/second inventor. See Brownlee, Trade Secret Use of Patentable Inventions, Prior User Rights and Patent Law Harmonization, 72 J. Pat. & Trademk. Off. Society 523 (1990). Would this rule make sense for the United States also? See, e.g., Franklin Pierce Law Center: Third Patent System Major Problems Conference, 32 Idea 1, 41-64 (1991) (discussion of need for prior user rights system in U.S.). Such a bill was pending in Congress at this writing. H.R. 3460, 104th Cong., 2d Sess. Note that at a minimum a system of prior user rights mitigates somewhat the harshness of the first-to-file system. Be aware, however, that this is true only to a limited extent; the prior user right only allows the prior inventor to practice the invention in the manner and to the extent that she was using it when the patent issued to the other inventor. This limitation will often keep the first inventor/prior user right holder from commercializing the patented invention, unless of course efforts to commercialize were already under way when the patent issued. See Brownlee, supra.

Coordinating International Prosecution

Patent lawyers face two problems in coordinating the prosecution of a series of national patents. First, a common priority date must be obtained, to insure that protection will be uniform and unaffected by prior art published (or otherwise having an effective date) before one or more of the national patent applications. Also, a common date will insure that prosecution of a patent in country *A* does not somehow compromise the patentability of the invention in Country *B*. Second, the patent lawyer has to deal with the logistics of international protection; she must oversee multiple filings in diverse languages in numerous countries. The wide variations in national practices and the high cost of conducting a large-scale application barrage make multiple filings one of the more challenging professional tasks in patent law.

Fortunately, two international agreements make these tasks a bit more tolerable. First is the Paris Convention, a longstanding international organization created by treaty in 1886 whose primary function is to guarantee a uniform worldwide priority date across all member countries. An applicant may file in any member country of the Convention up to one year after an initial (typically home-country) filing, without losing the priority date of the initial filing. (This treatment is provided under United States law in 35 U.S.C. §119.) The second international agreement is the Patent Cooperation Treaty, or PCT, which streamlines the filing of multiple national patent applications. Each agreement in its own way is an indispensable tool of the patent trade. Although detailed discussion of the agreements would take up too much room for this volume, a few words about the essential features of each is in order.

a) The Paris Convention

The Paris Convention was signed in 1883, a product of the first true "internationalization" wave in the field of patent law. Paris Convention for the Protection of Industrial Property, as last revised, July 14, 1967, 21 U.S.T. 1583, T.I.A.S. No. 6295, 828 U.N.T.S. 305 (the last revision, sometimes referred to as the "Stockholm" revision, entered into force April 26, 1970). Its primary function is to define a common priority date so that one may file an application in one member state and have the benefit of that same filing date when filing later in another member state. One purpose of this is to prevent interlopers from copying patents applied for or issued in one state and claiming them as their own in another, before the legitimate owner has time to file in the other country.

The key provision in the Convention as regards priority is Article 4. The relevant portions read as follows:

Article 4

A(1) Any person who has duly filed an application for a patent, or for the registration of a utility model, or of an industrial design, or of a trademark, in one of the countries of the Union, or his successor in title, shall enjoy, for the purpose of filing in the other countries, a right of priority in the periods hereinafter fixed.

A(2) Any filing that is equivalent to a regular national filing under the domestic legislation of any country of the Union or under bilateral or multilateral treaties concluded between countries of the Union shall be recognized as giving rise to the right of priority.

A(3) By a regular national filing is meant any filing that is adequate to establish the date on which the application was filed in the country concerned, whatever may be the subsequent fate of the application.

B. Consequently, any subsequent filing in any of the other countries of the Union before the expiration of the periods referred to above shall not be invalidated by reason of any acts accomplished in the interval, in particular, another filing, the publication or exploitation of the invention, the putting on sale of copies of the design, or the use of the mark, and such acts cannot give rise to any third-party right or any right of personal possession. Rights acquired by third parties before the date of the first application that serves as the basis for the right of priority are reserved in accordance with the domestic legislation of each country of the Union.

C(1) The periods of priority referred to above shall be twelve months for patents and utility models, and six months for industrial designs and trademarks.

Thus filing in one country which is a signatory to the Paris convention gives an applicant some "breathing room" — 12 months in which to prepare to file in other signatory nations.

b) The Patent Cooperation Treaty (PCT)

The PCT was signed in 1970. The Patent Cooperation Treaty, opened for signature June 19, 1970, 28 U.S.T. 7645, T.I.A.S. No. 8733 (entered into force Jan. 24, 1978). Its major purpose is to streamline the early prosecution stages of patent applications filed in numerous countries. It is often described as a clearinghouse for international patent applications. As a practical matter, its major advantage is that it gives an inventor (and her patent lawyer) more time, a precious commodity in the prosecution of an application destined for many countries. The signatories to the PCT have agreed to permit an applicant to wait for up to 30 months after the initial filing of a patent application in one country to begin the in-depth prosecution of the application in other countries. This allows the inventor more time, compared to non-PCT prosecution, in which to test the product, decide which countries' protection is worthwhile, and pay the patent office filing fees in the various countries.

There are two main parts of the PCT. Chapter 1 provides that an applicant who files in a national patent office may elect within 12 months to add a PCT filing. The PCT filing is simply an additional filing in any national patent office designated in the PCT. In this case, the applicant has up to 20 months from the initial filing to request that the PCT preliminary prosecution procedure be initiated. At that time, the applicant must also select the PCT member nations in which the applicant wishes to be covered under the PCT filing. Note that Chapter 1 preserves the applicant's priority date (in PCT member countries), without having to begin active prosecution, for eight months longer than the simple Paris Convention priority period.

Chapter 2 of the PCT extends the election period to 30 months. To qualify under Chapter 2, the applicant must make her PCT filing at most five months after the first national filing. Chapter 2 gives an inventor 18 extra months, compared to the Paris Convention, to select countries for coverage and initiate multiple national prosecutions. In other words, so-called "Chapter Two" PCT filings give the inventor up to 30 months to make his or her "national elections."

The extra time is a substantial advantage. Besides simply delaying the expenditure of filing and examination fees, the PCT allows an inventor a significant extra period to assess the technical merits and commercial potential of the invention. This extra time helps the inventor save wasted filing fees for inventions that fail to blossom; for those that show great promise, the various patent applications that grow out of the PCT filing can be tailored to reflect the commercially significant embodiments that have emerged from the extensive testing.

c) Foreign Filing Licenses

The U.S. Patent Act reflects the fundamental fact that technology is important not only to inventors but to the country as a whole. Specifically, the act permits the U.S. government to review all applications for their potential impact on national security. The statute states that before filing a foreign counterpart application for inventions made in the United States, the application must be authorized by a license obtained from the Commissioner of Patents. 35 U.S.C. §184; see Beckman Indus., Inc. v. Coleman Instruments, Inc., 338 F.2d 573 (7th Cir. 1964) (purpose of section 184 is to protect national defense information). The Commissioner's license does not authorize the sale or use of technical data in a foreign country. Separate approvals are required from the appropriate export agency. The license simply provides permission to file a foreign application. Rules promulgated by the Commissioner provide that if a foreign filing license is not issued within six months from the U. S. filing date, foreign counterparts may be filed without penalty unless a secrecy order has been issued by the Patent Office at the time of filing. 37 CFR §5.11 (1990). In practice, foreign filing licenses are routinely sent with the official filing receipt for the patent application.

2. Substantive Harmonization and GATT-TRIPs

As noted earlier, the lasting impact of the Paris Convention was primarily procedural, especially with respect to the uniform, worldwide priority date it made possible. The Paris Convention did, however, contain some minimum substantive standards of protection, as described earlier. Even so, further efforts at harmonizing worldwide standards of patentability, patentees' rights, and infringement bore little fruit for most of the twentieth century.

The slow pace of harmonization was in part due to the fact that the international business community seemed little interested in investing time in the esoteric issues of detailed harmonization. Although it was cumbersome to deal with the divergent standards held by the world's many individual domestic patent jurisdictions, some level of effective protection was available in most of the commercially important countries. Furthermore, patent law harmonization was the domain of a highly specialized affiliate of the United Nations, the World Intellectual Property Organization (WIPO), a forum not thought to be particularly friendly to Western business interests.

Things were changing by the mid-1980s, however. For one thing, the perceived value of intellectual property was increasing; it was beginning to take on a more central role in business planning and strategy. Of special concern was the increasing importance to U.S. businesses of overseas markets in developing countries — countries that had traditionally opposed strong intellectual property protection. Businesses in the United States and Europe, aware both of the increasing importance of intellectual property and of

WIPO's slow progress in harmonization, were thus on the lookout for an alternative forum in which to pursue harmonization. The search ended with the announcement of the Uruguay Round of negotiations to revise the main international trade agreement/organization, the General Agreement on Tariffs and Trade (GATT).

In addition to proposed reforms to the core function and structure of the GATT, the early Uruguay Round agenda soon grew to encompass negotiations on the "Trade-Related Aspects of Intellectual Property," or TRIPs, as it became known. The reference in the title to "Trade-Related" issues was a concession to those who doubted the relevance of intellectual property to the basic GATT mission; it soon became clear, however, that most of the basic elements of intellectual property protection would be up for discussion and potential harmonization in the TRIPs negotiations. By the time the GATT round ended in late 1993, TRIPs had become one of the principal components in the overall package of changes.

Although the post-TRIPs amendments to U.S. law are important, they pale in contrast to the revolutionary changes the agreement makes to the intellectual property regimes of many developing countries. To summarize the highlights, all signatories of the Uruguay Round treaty (who, under the agreement, become members of the newly created World Trade Organization (WTO)) must now:

- Include virtually all important commercial fields within the ambit of patentable subject matter, a major change for countries that, for example, have traditionally refused to enforce pharmaceutical patents on public health/access grounds.
- Test patent applications for (a) the presence of an inventive step, which is defined as precisely synonymous with nonobviousness under section 103 of the U.S. Patent Act, and for (b) "industrial application," similarly defined as coextensive with the U.S. utility requirement.
- Include in the patentees' bundle of exclusive rights the right to control the market for imports of the patented products.
- Eliminate or severely curtail the practice of granting compulsory licenses for patented technology.

See the Final Act Embodying the Results of the Uruguay Round of Multilateral Trade Negotiations, April 15, 1994 2-3 (GATT Secretariat 1994); Annex 1C: Agreement on Trade-Related Aspects of Intellectual Property Rights, id. at 6-19, 365-403. For implementation in the United States, see Uruguay Round Agreements Act, Pub. L. No. 103-465 (H.R. 5110), Dec. 8, 1994. See also J. H. Reichman, Universal Minimum Standards of Intellectual Property Protection Under the TRIPs Component of the WTO Agreement, 29 Intl. Law. 345 (1995) (able summary of provisions and open questions).

We have discussed GATT-related changes to U.S. law earlier in this chapter, in the section on priority, novelty, and statutory bars. By way of summary, the most important provisions:

- Changed the U.S. patent term to 20 years, measured from the date the patent application is filed, rather than 17 years from the date the patent was issued by the Patent Office. 35 U.S.C. §154. (Under certain circumstances, such as interferences and appealed rejections, this term may be extended for up to five years. Id.)
- Opened up the U.S. "first-to-invent" system by allowing members of the WTO to introduce evidence of inventive acts in their home country for purposes of establishing priority. See 35 U.S.C. §104.
- Expanded the definition of infringement to include acts of unauthorized offering for sale and importing. 35 U.S.C. §271.
- Adding a new procedure for filing "provisional applications," 35 U.S.C. §111, which must satisfy section 112 but need not include claims. Such an application does not begin the 20-year clock for the applicant's patent term.

As discussed earlier, many expect a wave of follow-on amendments to U.S. law. Typical candidates include:

- Publication of all U.S. patent applications 18 months after filing.
- "Prior user rights" for independent inventors using an invention later claimed in a U.S. patent by another inventor.
- A first-to-file priority rule, in line with the rest of the world.

These additional changes, if enacted, would further the substantive harmonization of world patent law by bringing the United States into line with the rest of the world on a number of important issues. So long as patent applicants must file, prosecute (and pay fees!) in each country in which they require protection, though, the impact of substantive harmonization will be limited.

F. REMEDIES

As we have seen in the cases in this chapter, the rights conferred by a patent are different from those that grow out of other property rights. On the other hand, there are a number of similarities between patent law and real property law. Like real property, intellectual property is freely alienable, for example. One of the most important similarities comes in the area of remedies. With patents, as with most other legal entitlements that share the label "property," the rule is that the owner of the property right may obtain an injunction to prevent ongoing "trespasses." Indeed, in one important conceptual framework used to describe many types of legal entitlements, it is the injunctive remedy that distinguishes the property right from other entitlements.[43]

The holder of a legal entitlement other than a property right can obtain compensation when the right is violated but cannot prevent violations *ex ante*. Because those who violate such a right must pay damages, this form of entitlement is referred to as a *liability rule*. Classic examples of liability rule entitlements (which we will call "liability rights") are the rights of tort victims and the rights of parties to a contract in the face of a breach. In the latter case, for example, it is generally understood that the normal remedy for breach of contract is damages — a monetary payment — rather than compelling the breaching party to perform the contract. Contracting parties cannot usually obtain injunctions to prevent breach. ("Specific performance" is an exception to this rule.)

Comparing intellectual property rights to the rights of a party to a contract provides insight into why the standard remedies are different in these two areas. Consider first the paradigmatic contract, say for delivery of 100 bushels of wheat at $5.00 per bushel. If the seller breaches, what is the measure of the buyer's injury? Under basic contract law, it is simply the difference between the contract price and the market price at the time of breach. And of course, since wheat is a standard commodity, the market price can be determined easily — found in a newspaper, for example.

Now consider the infringement of a patented invention. What is the measure of the patentee's injury? At first blush, this seems easy to establish: just subtract the patentee's profits after infringement from those she would have made absent the infringement, and extrapolate for the full 15- to 20-year patent term. A moment's reflection should reveal a number of tricky issues, however. What would have happened to the patentee's operations without the infringement? What could competitors have done if the patentee had controlled the market for the patented component by herself? In the field of technology in which the patent exists, what will happen over the next 15 to 20 years?

In most, if not all, cases, these questions will be very difficult to answer. It is hard to put a precise dollar value on a particular piece of patented tech-

---

43. Guido Calabresi & A. Douglas Melamed, Property Rules, Liability Rules, and Inalienability: One View of the Cathedral, 85 Harv. L. Rev. 1089 (1972). According to Calabresi and Melamed, the extent and nature of transaction costs in a particular case dictate whether one of the parties to the Coasian bargain ought to have an absolute property right or simply the right to collect damages caused by the other party's encroachment (i.e., a "liability rule"). This doctrine holds that several factors point toward a property rule: few parties, difficult valuation problems, and otherwise low transaction costs. Other factors indicate that a liability rule might better effectuate the bargain: many parties (especially where any one has the power to "hold up" the whole enterprise), likelihood of strategic bargaining, and otherwise high transaction costs.

Note that the authors refer to entitlements with injunctive remedies as property *rules,* to facilitate discussion of unconventional entitlements not normally thought of as property rights.

nology, especially over a protracted period. This difficulty — which we refer to as the *valuation problem* — is the key to understanding why the injunction, and not damages, is the standard remedy in a patent case. (Of course, with respect to infringement that takes place *prior to* litigation, damages are the only remedy available, so valuation problems cannot be avoided entirely.)

As an aside, you should be aware that the subject matter of copyrights, trademarks, trade secrets, and other forms of intellectual property suffer from similar valuation problems. Indeed, you might say that the various requirements of "uniqueness" in intellectual property law (e.g. novelty and non-obviousness in patent law almost guarantee that its subject matter will not be amenable to treatment as a commodity. As we explore the remedies issues presented below, and in later chapters, keep this in mind. It explains much about the policies behind intellectual property remedies, and it also helps explain why, when they are called for (e.g., in assessing past damages for infringement), intellectual property remedies of the "liability right" variety seem so difficult to quantify and defend.

The purpose of injunctive remedies goes beyond allowing the rightholder to prevent activities of the infringer. To the extent that a rightholder will consider negotiating a license with the infringer, the threat of an injunction will heavily influence the *terms* of the license. Specifically, it allows the rightholder to set her own price for the injury. In intellectual property cases, it allows the *rightholder*, and not a court, to set the terms of a license agreement settling the infringement litigation. This is assumed to be the efficient result, as a court called on to set the terms of the exchange would have a difficult time doing so quickly and cheaply, given the specialized nature of the assets and the varied and complex business environments in which they are deployed.[44] Hence the parties are left to make their own deal.

COMMENTS AND QUESTIONS

1. Is it true that giving a property entitlement entirely to one party to a transaction and allowing both parties to bargain will produce the efficient price? There are at least three potential problems with this approach. First, the parties themselves may not know the value of the technology over which they are negotiating. See David J. Teece, The Market for Know-How and the Efficient International Transfer of Technology, Annals of the Am. Assn. Pol. & Soc. Sci., Nov. 1981, at 81, 86 ("[N]either [the] buyer nor seller of

---

44. This follows from the fact that the parties are participants in the same industry and therefore have more knowledge of the technology at work in their industry. Courts, by contrast, usually have very little technical sophistication; even if they have some, it is of the most generalized variety. Cf. Roberts, Patent Litigation in the United States: Part 3: Remedies and Pretrial Relief, Pat. World, Nov. 1992, at 36, 38 (in discussion of standard a patentee must meet to obtain a preliminary injunction, author states: "The patentee wins on the issue of validity unless the defendant can educate the judge on the relevant elements of science, the history of the technology, the level of the skill, the words of the prior art . . . , all in a relatively few words or minutes.").

technology seems to have a clear idea of the value of the commodity in which they are trading," which means that "royalty rates may simply be a function of [the] negotiating skills of the parties involved."). Second, the transactions costs associated with such licensing arrangements can be significant. See David J. Teece, The Multinational Corporation and the Resource Cost of International Technology Transfer 36 (1976) (transaction costs average 20 percent of total value of technology exchanges studied); Farok J. Contractor, International Technology Licensing: Compensation, Costs, and Negotiation 105 (1981) (transaction costs averaged more than $100,000 for licensing deals studied). Finally, strategic behavior may cause parties to forgo some deals that would be in the best interests of both sides. See Robert Cooter, The Cost of Coase, 11 J. Legal Stud. 1, 19-23 (1982). For a suggestion that the cumulative effect of these problems is that efficient licensing of improvements will not always occur, see Mark A. Lemley, The Economics of Improvement in Intellectual Property Law, 75 Tex. L. Rev. 989 (1997).

Even if you think that bargaining is likely to be inefficient for one of the foregoing reasons, what alternatives are there? Are they any better?

2. As an example of strategic behavior, consider the following: A pioneering inventor patents a basic discovery which leads to a product that yields $100 in profits. An improver comes up with a radical extension of the technology that builds on the insight of the pioneer and produces a product yielding much higher profits: say $1000. The improver may offer to split the cooperative surplus — $1000 (the $1000 that the improver could make selling the improved product, plus the $100 the inventor can continue to make, less the $100 profit the inventor would make absent a bargain) — available in the joint profit-maximizing scenario under which the patentee licenses the basic patent to the improver. That would leave the patentee with $600: her $100 profits from selling her product, plus the royalty of $500, which again is half the cooperative surplus of $1000. Can we expect the patentee to agree to this bargain? Recall that the patentee has a wonderful threat: the right to banish the improver from the market altogether, i.e., to exercise the "property right" conferred by the patent, in the form of a permanent injunction. Will the patentee therefore be inclined to ask for more than half the surplus? At what point will the improver give up on the whole deal? Does this scenario justify replacing the property right with a liability right, at least in some cases? How would you determine which cases would be subject to this modified rule?

3. What role (if any) should culpability play in determining the appropriate remedy for patent infringement? Is it reasonable to treat a willful infringer more harshly than an innocent one? How would such differential treatment square with the rationale for patent remedies discussed above?

1. Injunctions

H. H. Robertson Co. v. United Steel Deck, Inc.
United States Court of Appeals for the Federal Circuit
820 F.2d 384 (Fed. Cir. 1987)

The district court applied to Robertson's motion the Third Circuit standard:

> An applicant for a preliminary injunction against patent infringement must show: . . . (1) a reasonable probability of eventual success in the litigation and (2) that the movant will be irreparably injured pendente lite if relief is not granted. . . . Moreover, while the burden rests upon the moving party to make these two requisite showings, the district court "should take into account, when they are relevant, (3) the possibility of harm to other interested persons from the grant or denial of the injunction, and (4) the public interest."

This is substantially the same standard enunciated by this court. See, for example, Roper Corp. v. Litton Systems, Inc., 757 F.2d 1266, 1270-73, 225 U.S.P.Q. 345, 347-50 (Fed. Cir. 1985), and Atlas Powder, 773 F.2d at 1231-34, 227 U.S.P.Q. at 290-93.

In matters involving patent rights, irreparable harm has been presumed when a clear showing has been made of patent validity and infringement. Smith International, 718 F.2d at 1581, 219 U.S.P.Q. at 692. This presumption derives in part from the finite term of the patent grant, for patent expiration is not suspended during litigation, and the passage of time can work irremediable harm. The opportunity to practice an invention during the notoriously lengthy course of patent litigation may itself tempt infringers.

The nature of the patent grant thus weighs against holding that monetary damages will always suffice to make the patentee whole, for the principal value of a patent is its statutory right to exclude. The presumption of irreparable harm in patent cases is analogous to that applicable to other forms of intellectual property, as discussed in Roper Corp., 757 F.2d at 1271-72, 225 U.S.P.Q. at 348.

The district court held that irreparable injury was presumed because Robertson had established a "strong likelihood of success in establishing validity and infringement." Such presumption of injury was not, however, irrebuttable. During oral argument, in response to an inquiry from the bench, USD and Bouras urged that money damages are an adequate remedy. In response, Robertson emphasized the few remaining years of patent life.

Even when irreparable injury is presumed and not rebutted, it is still necessary to consider the balance of hardships. The magnitude of the threatened injury to the patent owner is weighed, in the light of the strength of the showing of likelihood of success on the merits, against the injury to the accused infringer if the preliminary decision is in error. Results of other litigation involving the same patent may be taken into account, and the public interest is considered. No one element controls the result.

When the movant has shown the likelihood that the acts complained of are unlawful, the preliminary injunction "preserves the status quo if it prevents future trespasses but does not undertake to assess the pecuniary or other consequences of past trespasses." Atlas Powder Co., 773 F.2d at 1232, 227 U.S.P.Q. at 291. The court in *Atlas Powder* thus distinguished between remedies for past infringement, where there is no possibility of other than monetary relief, and prospective infringement, "which may have market effects never fully compensable in money." Id. The cautionary corollary is that a preliminary injunction improvidently granted may impart undeserved value to an unworthy patent.

COMMENTS AND QUESTIONS

1. The district court in its opinion referred to the USD/Bouras portrayal of disruption, loss of business, and loss of jobs, and to Robertson's business needs and patent rights. Observing that "[t]his patent does not have many more years to run," the court held "the equities weigh heavily against the wrongdoer." Id. The court stated that the "protection of patents furthers a strong public policy . . . advanced by granting preliminary injunctive relief when it appears that, absent such relief, patent rights will be flagrantly violated."

To the same effect is the case of Polaroid v. Kodak, where the court was asked to consider the "public interest" in deciding whether to grant a stay of an injunction. The response:

> Not only will an injunction injure Kodak customers and goodwill, it will also, according to Kodak, cause a "major disruption" of business. If and when Kodak is forced to shut down its instant camera production, 800 full-time and 3700 part-time employees will lose their jobs, and the company will lose its $200 million investment in plant and equipment.
>
> I am not unmindful of the hardship an injunction will cause — particularly to Kodak customers and employees. It is worth noting, however, that the harm Kodak will suffer simply mirrors the success it has enjoyed in the field of instant photography. To the extent Kodak has purchased that success at Polaroid's expense, it has taken a "calculated risk" that it might infringe existing patents. . . .
>
> Kodak's characterization of the public interest . . . misconstrues the very concept of public benefit. The public policy at issue in patent cases is the 'protection of rights secured by valid patents.' Courts grant — or refuse to stay — injunctions in order to safeguard that policy, even if those injunctions discommode business and the consuming public.

Polaroid Corp. v. Eastman Kodak Co., Civ. No. 76-1634-Z, D. Mass., October 11, 1985, Slip Op. at 3, 4.

2. Keep in mind that *H. H. Robertson* and *Kodak* were decisions involving *preliminary* injunctions. When a full trial concludes with a finding of

infringement, there is almost never any doubt that a permanent injunction will issue. In very rare cases, however, the court may deny a permanent injunction even after a finding of infringement. In Foster v. American Mach. & Foundry Co., 492 F.2d 1317 (2d Cir. 1974), for example, the court upheld what amounted to a compulsory license: a reasonable royalty damage award (described later in this chapter), but no injunction. This is an example of a rare case in patent law: the treatment of a patent as a liability right. The appellate court opinion is concise on this point:

> We do not find any difficulty in agreeing that an injunction would be an inappropriate remedy in this case. An injunction to protect a patent against infringement, like any other injunction, is an equitable remedy to be determined by the circumstances. It is not intended as a club to be wielded by a patentee to enhance his negotiating stance. Here, as the District Court noted, the appellee manufactures a product; the appellant does not. In the assessment of relative equities, the court could properly conclude that to impose irreparable hardship on the infringer by injunction, without any concomitant benefit to the patentee, would be inequitable.
>
> Instead, the District Court avoided ordering a cessation of business to the benefit of neither party by compensating appellant in the form of a compulsory license with royalties. This Court has approved such a "flexible approach" in patent litigation. Here the compulsory license is a benefit to the patentee who has been unable to prevail in his quest for injunctive relief. To grant him a compulsory royalty is to give him half a loaf. In the circumstance of his utter failure to exploit the patent on his own, that seems fair.

492 F.2d at 1324. What difference, if any, should it make that the patentee (appellant in *Foster*) does not manufacture anything — i.e., is not "practicing" the patent? While in theory the patentee has the right not to practice the patent, other cases have taken this into account in setting remedies as well. See, e.g., E. I. du Pont de Nemours & Co. v. Phillips Petroleum Co., 835 F.2d 277, 278 (Fed. Cir. 1987) (upholding grant of stay of preliminary injunction where patentee du Pont (1) had licensed all who desired entry into the polyethylene business, the subject of the patent, and (2) planned to exit the market; "harm to duPont here is of a different nature than harm to a patentee who is practicing its invention and fully excluding others"); Vitamin Technologists, Inc. v. Wisconsin Alumni Research Foundation, 146 F.2d 941 (9th Cir. 1944), *cert. denied*, 325 U.S. 876 (1945) (describing policy reasons why a permanent injunction should be denied in a case where the patentee had refused to license a health-related patent).

2. Damages: Reasonable Royalty and Lost Profits

Courts have struggled with the problem of setting appropriate measures of damages for past infringement. In Panduit Corp. v. Stahlin Bros. Fibre

Works, Inc., 575 F.2d 1152, 1158 n.5 (6th Cir. 1978), the court adopted an oft-used four-factor test for determining lost profits:

> To obtain as damages the profits on sales he would have made absent the infringement, i. e., the sales made by the infringer, a patent owner must prove: (1) demand for the patented product, (2) absence of acceptable noninfringing substitutes, (3) his manufacturing and marketing capability to exploit the demand, and (4) the amount of the profit he would have made.
>
> When actual damages, e.g., lost profits, cannot be proved, the patent owner is entitled to a reasonable royalty. A reasonable royalty is an amount "which a person, desiring to manufacture and sell a patented article, as a business proposition, would be willing to pay as a royalty and yet be able to make and sell the patented article, in the market, at a reasonable profit." Goodyear Tire and Rubber Co. v. Overman Cushion Tire Co., 95 F.2d 978 at 984, 37 U.S.P.Q. 479 at 484 (6th Cir. 1937) (citing Rockwood v. General Fire Extinguisher Co., 37 F.2d 62 at 66, 4 U.S.P.Q. 299 at 303 (2d Cir. 1930)), *appeal dismissed,* 306 U.S. 665 (1938).

In Polaroid Corp v. Eastman Kodak Co., 16 U.S.P.Q.2d (BNA) 1481 (D. Mass. 1990), the court ultimately assessed total damages of $873,158,971 (corrected from the original figure of $909,457,567). In its discussion of damages, the court said:

> At various times during the trial Polaroid seemed to argue that all evidence about conventional photography should be excluded because conventional photography was not an acceptable substitute for instant photography. Kodak argued that conventional photography was an "economic substitute" for instant. Kodak did not attempt to quantify the number of Kodak instant purchasers who would have turned to conventional products in Kodak's absence; Kodak simply urged that the relative price of instant and conventional photography was a significant variable in the demand formula for instant products and therefore must be considered in any assessment of the market. Instant photography occupied a unique niche in the overall photography market during the infringement period. Consumers sought the emotional "instant experience" of having a picture develop immediately, usually in the presence of the subject. Although instant photography was unique, it did not exist in a vacuum; it also competed with conventional photography for the consumer's photographic dollar. Those who purchased infringing Kodak instant products at Kodak prices would not have considered conventional products as an acceptable substitute at the time of purchase, but the relationship between the relative advantages of conventional and instant photography was an integral part of each consumer's decision. If Polaroid were simply claiming that it could have made Kodak's sales at the same time at similar prices, there would be no question in my mind that consumers would have made the same choice vis-à-vis conventional photography. However, Polaroid's claim is not so simple.
>
> Polaroid's experts spun a scenario in which the prices Polaroid would have charged [if Kodak had not been in the market] were substantially higher and in which, through a complicated massaging of the demand curve, the great bulk of sales would have occurred later than they did historically. In this "but for"

scenario, the effect of conventional photography on the instant photography market must be considered. The evidence of competition between instant and conventional products is overwhelming. Even Polaroid's econometric expert attempted to include the effect of competition from conventional photography in his computation. The relative values of instant and conventional changed throughout the period of infringement. On the facts before me, I must consider that relationship when deciding whether Polaroid's scenario, which differs so substantially from the historical world, is feasible. *Contrary to Polaroid's urging, the law does not require this Court to ignore the effect of competitive forces on the market for patented goods just because there are no non-infringing alternatives. Indeed, the law requires a careful assessment of all market influences when determining lost profit or reasonable royalty damages.* I find that conventional photography was not an acceptable substitute for instant photography during the period of infringement. Competition between conventional and instant photography (which changed throughout the ten years of infringement) did, however, affect the price that consumers were willing to pay for instant photography. Therefore, I have assessed what effect the competition would have had on Polaroid's ability to charge more for its instant products and on the profitability of delaying the introduction of certain lower-priced products. (Emphasis added.)

[In another section of the opinion, the court concludes that in fact this competition had no effect on Polaroid's profit margins or product-introduction-related profits.]

Under *Panduit*, lost profits are available if there are "no adequate substitutes." Since some Federal Circuit lost profits cases suggested that a patent almost insures that there are no adequate substitutes (recall our discussion of "uniqueness" in the introduction to this section), the *Polaroid* court might have simply refused to consider evidence of conventional camera sales altogether. At the other extreme, recent antitrust thinking comes close to arguing that a patent almost never confers "market power" in an economic sense. Under this view, the existence of the patents on elements of instant photography would have been irrelevant in *Polaroid*. Only evidence and testimony on cross-elasticities and the like could establish the absence of substitute products.

Polaroid steers an interesting middle course. While conceding that there are no substitutes in the market for patented products, the court goes on to consider the effect of *economically significant* products on the *profit* the patentee would have made absent infringement, even though those products are not close substitutes in a strict sense. It admits, in other words, that the patented product is unique; but it refuses to equate uniqueness with total insulation from market forces. In a word, there may be no exact substitutes, but at some price (and for some consumers) there are *economic* substitutes. These exert price discipline on the patentee's pricing.

One way to state this in terms of conventional antitrust analysis is to say that although the cross elasticities of demand for the patented item and related items are not high enough to call them identical products, the availability of some (perhaps imperfect) substitutes, coupled with at least some price elasticity in the demand for the patented good, means that the patentee would have faced some price discipline even if she could have made the infringer's sales. One can even imagine a scenario where, if the infringer had not been in the patentee's market, he would have moved into a parallel market from which he could have exerted greater price discipline on the patentee.

COMMENTS AND QUESTIONS

1. Some recent scholarship in the area of antitrust market definition is of special relevance to a consideration of the noninfringing substitute element of the Panduit test. In a paper by R. S. Hartman, D. J. Teece, W. Mitchell and T. M. Jorde, Assessing Market Power in Regimes of Rapid Technological Change, 2 Indus. & Corp. Change 3 (1993), the authors argue that traditional market definition techniques are seriously flawed when applied to markets for highly innovative and technology-intensive products. Specifically, they point out that current Department of Justice Guidelines assume a much too static view, which "biases market definition downwards in industries experiencing rapid technological change, where competition often takes place on performance attributes, and not price." "The more innovative the new product or process," the authors argue, "the greater the conventional market power will appear, because price changes will have little or no influence on demand for a truly innovative product." Id. at 6-7. The authors propose an approach whereby product attributes are included in the assessment of demand and supply. See also Thomas M. Jorde & David J. Teece, Rule of Reason Analysis of Horizontal Arrangements: Agreements Designed to Advance Innovation and Commercialize Technology, 61 Antitrust L.J. 579 (1993).

Note on the Frontiers of Lost Profits Damages

Although the statute establishes the "reasonable royalty" as the floor for patent damages, much of the action in recent years has been in the law of lost damages. In particular, four novel theories have been used to increase damage awards in the past several years.

1. Price Erosion

In a market with only two suppliers, it is obvious that sales made by the infringer would have been made by the patentee if there had been no infringement. Thus it is easy to establish the *number* of units the patentee would have sold without the infringement: the infringer's sales plus the patentee's actual sales. But at what *price* could the patentee have sold? Economic theory advises that a monopolist will charge a higher price than a duopolist, i.e., a seller in a two-firm market. As a result, some patentees have successfully argued that they would have sold the infringer's units plus their own, all at a higher price than they actualy charged, given the presence of the infringer. See, e.g., Lam, Inc. v. Johns-Manville Corp., 718 F.2d 1056 (Fed. Cir. 1983).

Query: What assumption does this theory make about the demand for the patentee's product? Wouldn't you expect the number of units sold to decline as the price increases?

2. The "Market Share" Rule

When more than two sellers share a market and at least one seller is a non-infringing competitor of the patentee, it would appear that one element in the *Panduit* test is missing: the absence of noninfringing substitutes. However, some patentees have overcome this deficiency with a clever argument regarding what would have happened if the infringer had not been in the market. Under this theory, the court is asked to assume that the patentee's market share *relative to the noninfringer* would have remained the same in the absence of the infringer. Consequently, it is assumed that the patentee would have made the same percentage of the infringer's sales as the patentee made in the overall market. See, e.g., State Indus. v. Mor-Flo, Inc., 883 F.2d 1573 (Fed. Cir. 1988), *cert. denied*, 493 U.S. 1022 (1990). With this theory, in other words, the presence or absence of noninfringing substitutes does not prevent the patentee from laying claim to at least some of the infringing sales. Id. at 1578.

Does this make sense to you? Is it possible that the infringer's advertising, customer base, or other assets were responsible for the sales, and that no one else — patentee or competitor — would have made them? Cf. Litton Systems v. Honeywell Inc., 87 F.3d 1559 (Fed. Cir. 1996) ($1.2 billion jury verdict vacated because plaintiff's expert improperly assumed without evidence that all government contracts would have been awarded to plaintiff if the patented component had not been included in defendant's product). In deciding whether the infringer's customers will switch to the patentee or to another competitor, is the relative price of the infringer's goods relevant? What should a court do if the remaining competitors in the industry also sell infringing products, but were not parties to the suit?

3. Lost Sales of Unpatented Components or Products

In yet a third line of cases, patentees have argued that since the infringer's presence cost it sales of unpatented components and products, these losses are compensable. The components cases are analyzed under the "entire market value" rule, which directs the court to ascertain whether the unpatentable components at issue are functionally integral to an overall product whose consumer demand is based on the patented component. See, e.g., TWM Mfg. Co. v. Dura Corp., 789 F.2d 895, 900 (Fed. Cir.), *cert. denied*, 479 U.S. 852 (1986). The idea is that the unpatented component would naturally and normally be sold along with the patented one, and thus the patentee deserves compensation for the lost sales of both. Does this amount to stretching the boundary of the claimed subject matter?[45] Would the patentee price patented components differently if the rule were otherwise?

A related theory was adopted by the Federal Circuit in the 1995 case of Rite-Hite Corp. v. Kelley Co., 56 F.3d 1538 (Fed. Cir. 1995) (en banc), *cert. denied*, 116 S. Ct. 184 (1995). In this case, the patentee, Rite-Hite, argued that the infringer's presence in the market cost it not only sales of its patented product but also sales of an *unpatented* competing product as well. The court accepted this argument, reasoning that "[i]f a particular injury was or should have been reasonably foreseeable by an infringing competitor in the relevant market, broadly defined, that injury is generally compensable absent a persuasive reason to the contrary." 56 F.3d at 1546.

Does this impermissibly shift the focus from the claimed invention to the infringer's behavior? If we accept a counterfactual world when the patentee is alone in the market for the patented item, why not consider an alternative where the infringer sells only the unpatented, competing product?

4. Post-Expiration Sales

It is accepted black-letter law that once a patent expires, others are free to use its teachings and to make, use, and sell competing products. But some ex-patentees have successfully asserted a claim for post-expiration damages on the theory that infringement during the term of the patent gave the infringer a "head start" on post-expiration sales, since the infringer did not have to start up production and marketing after the patent expired. Courts awarding or endorsing damages on such an "accelerated re-entry" theory include TP Orthodontics, Inc. v. Professional Positioners, Inc., 17 U.S.P.Q.2d 1497, 1504-1506 (E.D. Wisc. 1990); Amsted Indus. v. National Castings, Inc., 16 U.S.P.Q.2d 1737 (N.D. Ill. 1990); BIC Leisure Prods. v. Windsurfing Intl., 687 F. Supp. 134 (S.D.N.Y. 1988). In each case, post-expiration damages were part of an award that included damages for infringement during the term of the patent as well.

Compare these cases to the "head-start" injunction rule in trade secret law. If trade secret plaintiffs can get damages and injunctive relief after the secret has been disclosed, is there any reason to deny similar relief to patent plaintiffs?

COMMENTS AND QUESTIONS

The bottom line on patent damages — whether based on lost profits or reasonable royalty — is that they are growing, on average. See Paul M. Janicke, Contemporary Issues in Patent Damages, 42 Am. U.L. Rev. 691

---

45. Compare the patent misuse doctrine, which distinguishes between patentees who attempt to control "staple" products and those who attempt to control "nonstaple" products associated with the invention.

(1993). In this environment, one can expect even more novel arguments to be put forward. Cf. Grain Processing Corp. v. American Maize Prods. Co., 185 F.3d 1341, 1351 (permitting evidence of infringer's likely responses if he or she had known about infringement — e.g., hypothetical, non-infringing alternatives).

G. DESIGN AND PLANT PATENTS

Section 101 is the subject matter provision of what is referred to as the "utility" patent statute. Most patents issued in the United States are utility patents, and utility patents are what most people generally think of when they speak generically of "patent law." Several other provisions of the patent law cover related material, most importantly designs and certain types of plants. Here we review briefly these special statutes.

1. Design Patents

The aesthetic appearance of a product rather than its functional features are protected by the design patent. Design patents have been obtained for a wide range of products including shoes, hats, furniture, tools, packaging, televisions, automobile designs, and computer graphics. Design patent law overlaps both copyright law and trademark and unfair competition law in its coverage of the nonfunctional features of useful objects. 1 Donald S. Chisum, Patents, §1.04 at 1-180 (1992). Due to a lengthy processing time, high application cost, strict requirements that are vague and difficult to apply, and a long history of judicial hostility, the design patent system has been criticized as ineffective and in need of reform. As the courts have recently become more receptive to design patents, interest has been renewed in this form of intellectual property protection.

a. Introduction

The original design patent law in the United States was enacted in 1842 to fill a gap that then existed between copyright protection for authors and patent protection for inventors in the mechanical arts. Chisum, supra, §1.04[1] at 1-180. The intent of the statute, which extended patent protection to "new and original designs for articles of manufacture" was to "give encouragement to the decorative arts." Gorham Mfg. Co. v. White, 81 U.S. (14 Wall.) 511, 524-525 (1871). Appearance, rather than utility, is the crucial factor for consideration in design patent protection. A design may consist of surface ornamentation, configuration, or a combination of both. Id. The Design Patent Act, which was codified in 35 U.S.C. §171, allows a design patent to be obtained for "any new, original and ornamental design for an article of manufacture . . ." and provides that most provisions relating to patents for inventions also apply to design patents. 35 U.S.C. §171 (1988). Design patents are issued for a 14-year term.

b. Requirements for Patentability

A design is patentable if it meets the requirements of novelty, originality, and nonobviousness, is ornamental, and is not dictated by functional considerations. Chisum, supra, §1.04[2] at 1-184. The Patent and Trademark Office defines a design as "the visual characteristics or aspects displayed by the object. It is the appearance presented by the object which creates a visual impact upon the mind of the observer." U.S. Patent and Trademark Office, Manual of Patent Examining Procedure, §1502 (5th ed. Rev. 8, 1988).

An "article of manufacture" has been broadly defined to include silverware, Gorham Mfg. Co. v. White, 81 U.S. (14 Wall.) 511 (1871); cement mixers, In re Koehring, 37 F.2d 421 (C.C.P.A. 1930); furniture, In re Rosen, 673 F.2d 388 (C.C.P.A. 1982); and containers for liquids, Unette Corp. v. Unit Pack Co., 785 F.2d 1026 (Fed. Cir. 1986). In In re Hruby, 373 F.2d 997, 1000 (C.C.P.A. 1967), the Court of Customs and Patent Appeals held that "a manufacture is anything 'made by the hands of man' from raw materials whether literally by hand or by machinery or by art" and found that a design created by the flow of water in a fountain was patentable. While the design must be embodied in an article of manufacture, a design patent may be obtained for only part of an article. In re Zahn, 617 F.2d 261 (C.C.P.A. 1980) (holding that the shank of a drill bit was patentable under section 171). Thus the subject matter of a design patent is not limited to designs for discrete articles.

i. Novelty

Novelty is established if no prior art shows exactly the same design. A design is novel if the "ordinary observer," viewing the new design as a whole, would consider it to be different from, rather than a modification of an already existing design. See, e.g., Clark Equip. Co. v. Keller, 570 F.2d 778, 799 (8th Cir. 1978). Section 102(b) of the Patent Act of 1952 provides for a one-year novelty grace period, allowing a patent for a design as long as the design has not been published anywhere or sold or displayed within the United States more than one year before the application is filed with the Patent and Trademark Office. 35 U.S.C. §102(b) (1988). This gives a designer the opportunity to "test market" a design before incurring the expense of a patent application. J. H. Reichman, Design Protection and the New Technologies: The United States Experience in a Transnational Perspective, 19 U. Balt. L. Rev. 6, 23 (1991).

ii. Nonobviousness

Because section 171 provides that the provisions of Title 35 relating to utility patents also apply to design patents, a design must be nonobvious, which requires the "exercise of the inventive or originative faculty." Smith v. Whitman, 148 U.S. 674, 679 (1893). The test for determining nonobviousness of a utility patent is whether the differences between the invention sought to be patented and the prior art would be obvious at the time of invention to a person with "ordinary skill in the pertinent art." Graham v. John Deere Co., 383 U.S. 1, 14 (1966). See also Chisum, supra, §1.04[2] at 1-200. The *Graham* test requires the court to ascertain the scope and content of the prior art, the differences between the prior art and the claim at issue, and the level of ordinary skill in the art. *Graham*, 383 U.S. at 17. In addition to these subjective inquiries, secondary considerations such as commercial success and long-felt need in the industry are relevant to determine obviousness. Id. at 17-18.

Prior to the adoption of a uniform standard to determine nonobviousness in design patents, the courts were in conflict over whether to use an "ordinary designer" standard or that of an "ordinary intelligent man" when applying the *Graham* test. See In re Laverne, 356 F.2d 1003 (C.C.P.A. 1966). Determining the nonobviousness of a design patent, as opposed to a utility patent, is unpredictable, as it is an inherently subjective inquiry, depending largely on personal taste. In evaluating nonobviousness of utility patents, judges can measure the distance between the prior art and a new invention on the basis of uniform scientific criteria and technical data, while the evaluation of the distance between an appearance design and its predecessors necessarily involves "value judgments that are hard to quantify and unreliable at best." Reichman, supra, at 33 n.164. See In re Bartlett, 300 F.2d 942, 944 (C.C.P.A. 1962) (noting that the determination of patentability in design cases depends on the subjective conclusion of each judge). The nonobviousness requirement has been cited as a primary factor in limiting the availability of design protection in the United States from the 1920s on. Reichman, supra, at 24. Without a clear standard to follow, the appellate courts could apply their own stricter view of nonobviousness, and they routinely invalidated design patents, despite the presumption of validity accorded to a grant by the Patent and Trademark Office. Id.

In 1981, the Court of Customs and Patent Appeals held that the *Graham* test could apply to design patents, and nonobviousness would be measured in terms of a "designer of ordinary capability who designs articles of the type presented in the application." In re Nalbandian, 661 F.2d 1214, 1216 (C.C.P.A. 1981). The new standard allows for objective evidence of expert testimony from designers in the field to be used to prove nonobviousness. Id. at 1217. The Federal Circuit, which was created in 1982 to hear all patent appeals, subsequently adopted this standard. While this approach appears more evenhanded, and has led to more patents being upheld as valid, it may not solve the problem of unpredictability, because the opinions of different designers can vary considerably. William T. Fryer, III, Industrial Design Protection in the United States of America — Present Situation and Plans for Revision, 70 J. Pat. & Trademk. Off. Society 821, 829 (1988).

The Federal Circuit has emphasized the presumptive validity of a design patent, and placed the burden on the challenger to come forward with clear and convincing proof of nonobviousness. See, e.g., Trans-World Mfg. Corp. v. Al Nyman & Sons, Inc., 750 F.2d 1552, 1559-1560 (Fed. Cir. 1984); Avia Group Intl., Inc. v. L.A. Gear Cal., 853 F.2d 1557, 1562 (Fed. Cir. 1988). The court also upheld the notion that to find obviousness, reference must be made to prior art with the same overall appearance of the patented design, rather than to a combination of features from several references. Litton Sys., Inc. v. Whirlpool Corp., 728 F.2d 1423, 1443 (Fed. Cir. 1984); see also In re Rosen, 673 F.2d 388, 390-391 (C.C.P.A. 1982). Most significantly, the Federal Circuit held that objective secondary considerations which apply to utility patents, such as commercial success and copying, are also relevant to determine nonobviousness of design patents. See, e.g., Litton, 728 F.2d at 1441; Avia, 853 F.2d at 1564. The theory supporting the consideration of commercial success is that the purpose of a design patent is to increase salability, so if a design has been a success it "must have been sufficiently novel and superior to attract attention." Robert W. Brown & Co. v. De Bell, 243 F.2d 200, 202 (9th Cir. 1957); see also Chisum, supra, at §1.04[2][f] at 1-208. Evidence of commercial success must be related to the patented design rather than to factors such as functional improvement or advertising. See, e.g., Litton, 728 F.2d at 1443; Avia, 853 F.2d at 1564. The approach adopted by the Federal Circuit with respect to designs challenged for obviousness has been cited as effectively lowering the invalidation rate to only 38 percent from a rate of 75 to 100 percent only a few years earlier. Reichman, supra, at 37.

iii. Ornamentality

A patentable design must be ornamental, creating a more pleasing appearance. To satisfy the requirement of ornamentality, a design "must be the product of aesthetic skill and artistic conception." Blisscraft of Hollywood v. United Plastics Co., 294 F.2d 694, 696 (2d Cir. 1964). This requirement has been met by articles which are outside the realm of traditional "art." See In re Koehring, 37 F.2d 421, 422 (C.C.P.A. 1930) (determining a design for a cement mixer to be ornamental because it "possessed more grace and pleasing appearance" than prior art). A number of cases have denied patentability to designs which are concealed during the normal use of an object, on the basis that ornamentality requires the design to be visible while the object is in its normal and intended use. See Chisum, supra, §1.04[2][c] at 1-190-91.

iv. Functionality

If a design is "primarily functional rather than ornamental," or is "dictated by functional considerations," it is not patentable. Power Controls Corp. v. Hybrinetics, Inc., 806 F.2d 234, 238 (Fed. Cir. 1986). The functionality rule furthers the purpose of the design patent statute, which is to promote the decorative arts. In addition, the rule prevents granting in essence a monopoly to functional features that do not meet the requirements of a utility patent. Chisum, supra, §1.04[2][d] at 1-195. Recognizing that strict application of the functionality rule would invalidate the majority of modern designs, the Federal Circuit validated designs with a higher functionality factor than had been tolerated by the courts previously. Reichman, supra, at 40. This is evidenced in cases upholding design patents for an eyeglass display rack, Trans-World Mfg. Corp. v. Al Nyman & Sons, Inc., 750 F.2d 1552 (Fed. Cir. 1984); fiberglass camper shells, Fiberglass in Motion, Inc. v. Hindelang, No. 83-1266 (Fed. Cir. Apr. 19, 1984); and containers for dispensing liquids, Unette Corp v. Unit Pack Co., 785 F.2d 1026 (Fed. Cir. 1986). The Federal Circuit has held that a design may have functional components as long as the design does not embody a function that is necessary to compete in the market. Avia, 853 F.2d at 1563. If the functional aspect of a design may be achieved by other design techniques, then it is not primarily functional. Id. This more flexible approach reflects a recognition by the court that the majority of valuable industrial designs which should be granted protection in order to stimulate economic growth are a combination of functional and aesthetic features.

c. *Claim Requirements and Procedure*

Two major criticisms of the design patent system in the United States are that it is too expensive and that protection takes too long to obtain. See Fryer, supra, at 834; Reichman, supra, at 24 (procedural requirements make design protection in the United States much "slower and costlier to obtain" than in other countries); Perry J. Saidman, Design Patents — The Whipping Boy Bites Back, 73 J. Pat. & Trademk. Off. Society 859 (1991) (defending against these criticisms).

In 1988, the cost of a design patent application was estimated at $1000. Fryer, supra, at 835. A large part of this cost is the expense of preparing the drawings which constitute the claim. The drawings "must contain a sufficient number of views to constitute a complete disclosure of the appearance of the article." 37 CFR §1.152. All that is required in writing is a very brief description of the drawings. The adequate disclosure and definiteness of the claim required by section 112 are accomplished by the drawings. Chisum, supra, §1.04[3]. Only one claim can be included in a design application. An application may illustrate more than one embodiment of a design only if the embodiments involve a "single inventive concept" and can be protected by a single claim. In re Rubenfield, 270 F.2d 391, 396 (C.C.P.A. 1959), *cert. denied*, 362 U.S. 903 (1960).

The pendency time for a design patent is approximately two to three years. Saidman, supra, at 861. Reasons for this delay include budget and staffing restraints at the Patent and Trademark Office. Id. This leaves a design patent applicant without patent protection from copiers during this long waiting period, as opposed to copyright protection, which requires no initial procedural requirement and takes only a few months for registration to be issued. Fryer, supra, at 840, 835. Of course, it is still possible to protect such designs in appropriate cases under both copyright and trade dress principles. More on that in Chapters 4 and 5.

It has been noted that the current system is "unsuited to the fast-moving but short-lived product cycle characteristic of today's market for mass-produced consumer goods." Reichman, supra, at 24. While there is general agreement that a new form of protection for industrial design is needed, Congress has yet to adopt any of the proposed legislation. An example of proposed legislation that would afford better protection for designs is the Design Copyright Protection Act, which employs a modified copyright form of protection as an alternative to the design patent. See Fryer, supra, at 839-846. Alternatively, some current proposed legislation would protect functional industrial designs of every kind, including those that are neither aesthetically nor technically innovative. Reichman, supra, at 121-122.

d. Infringement

The standard for finding infringement of a design patent was defined in Gorham Mfg. Co. v. White, where the Supreme Court held that "if in the eye of an ordinary observer, giving such attention as a purchaser usually gives, two designs are substantially the same, if the resemblance is such as to deceive such an observer, inducing him to purchase one supposing it to be the other, the first one patented is infringed by the other." 81 U.S. (14 Wall.) 511, 528 (1872). This "eye of the ordinary observer" standard continues to be the rule followed by the Federal Circuit. See Oakley, Inc. v. Intern. Tropic-Cal, Inc., 923 F.2d 167, 169 (Fed. Cir. 1991). The ordinary observer is one who has "reasonable familiarity" with the object in question and is capable of making a comparison to other objects which have preceded it. Applied Arts Corp. v. Grand Rapids Metalcraft Corp., 67 F.2d 428, 430 (6th Cir. 1933). The key factor to determine infringement under the Gorham test is similarity, rather than consumer confusion. Unette, 785 F.2d at 1029 (holding that "likelihood of confusion as to the source of goods is not a necessary or appropriate factor for determining infringement of a design patent").

The second prong of the infringement analysis is the "point of novelty" test, which is distinct from the issue of similarity. Under the "point of novelty" test, the similarity found by the ordinary observer must be attributable to the novel elements of the patented design that distinguish it from prior art. Litton, 728 F.2d at 1444; Avia, 853 F.2d at 1565; FMC Corp. v. Hennessy Indus., Inc., 836 F.2d 521, 527 (Fed. Cir. 1987). Unless the accused design appropriates the novel features of the patented design, there has been no infringement. Avia, 853 F.2d at 1565. The scope of the patented claim, and its points of novelty are determined by examining the field of the prior art. See Litton, 728 F.2d at 1444 (holding that where the field of prior art is "crowded," the scope of a claim will be construed narrowly).

After determining the scope of the patented claim, the infringement inquiry focuses only on the protectable aesthetic components of a patented design. See Lee v. Dayton-Hudson Corp., 838 F.2d 1186, 1188 (Fed. Cir. 1988) (holding that "it is the non-functional, design aspects that are pertinent to determinations of infringement"). Thus this test permits strong similarities to be excused if the defendant can prove that she borrowed "only commonplace or generic ideas, functional features, or other nonprotectable matter" while adding sufficient variation to protectable elements of the design. Reichman, supra, at 44.

Whether a design is infringed when it is used on an entirely different type of article than the patented one has not been settled. Chisum, supra at §1.04[4] at 1-225. In Avia Group Intl. v. L.A. Gear Cal., where the patented design was for an adult athletic shoe and the accused design was for a children's shoe, the court found infringement and held that even in a situation where the patent holder has not put out a product or where the patented design is embodied in a product that does not compete with the patent holder's product, a finding of infringement is not precluded. 853 F.2d at 1565. This decision indicates that the Federal Circuit is willing to extend design patent protection beyond "literal infringement" and protect the design concept itself. Reichman, supra, at 53.

2. Plant Patents

Today, new plants can be protected with utility patents under the general standards discussed earlier in this chapter.[46] In addition to such protection, plants can be protected under two acts specifically designed for them — the Plant Patent Act and the Plant Variety Protection Act.

---

46. The utility patentability of plants was confirmed by the PTO Board of Appeals and Interferences in Ex parte Hibberd, 227 U.S.P.Q. 443 (Bd. Pat. App. & Inter. 1985). Although this decision would seem consistent with the Supreme Court's holding in Diamond v. Chakrabarty, no court of competent jurisdiction has yet to pass on the utility patentability of plants. Nevertheless, patent applicants claiming inventions in plants may currently proceed under §101 or §161.

a. The Plant Patent Act

Prior to passage of the Plant Patent Act of 1930, 35 U.S.C. §§161-164 (1988) (PPA), inventions in plants faced two obstacles to patentability. Like other living organisms, plants were believed to be ineligible for patent protection as they were products of nature. See, e.g., Ex parte Latimer, 1889 Commn. Dec. 123 (1889) (fiber from needle of evergreen tree unpatentable product of nature). The second barrier was the fact that "plants were not thought amenable to the 'written description' requirement" now found at §112 of the Patent Act. Congress addressed these issues in enacting the Plant Patent Act, now embodied in sections 161-64 of Title 35 of the U.S. Code.

The PPA extends patent protection to inventors and discoverers of "any distinct and new" variety of asexually reproducing plant.[47] By judicial interpretation, the term *plant* is used in its lay rather than scientific sense and thus bacteria are not eligible subject matter. In re Arzberger, 112 F.2d 834 (C.C.P.A. 1940).

According to section 161 of the PPA, the provisions of Title 35 that apply to utility patents apply equally to plant patents "except as otherwise provided." However, the elements of plant patentability are not the same as for utility patents. For example, the PPA substitutes the requirement of distinctness for that of utility. Yoder Brothers, Inc. v. California-Florida Plant Corp., 537 F.2d 1347 (5th Cir. 1976), *cert. denied,* 429 U.S. 1094 (1977). Distinctness is measured by examining the characteristics that make the plant clearly distinguishable from other existing plants. As with utility patents, the characteristics which distinguish the plant from others do not have to be superior, merely different.

Although the nonobviousness requirement is theoretically applicable to plant patents, courts have admitted difficulty in applying it in this context. After rephrasing the traditional three-part obviousness test for the plant world, the courts are left with a standard reminiscent of the "distinctness" requirement discussed above. Thus nonobviousness is viewed as a requirement for "invention" resulting in substantial, rather than merely minor, distinctions in the plant.

Perhaps the most significant difference between prosecution of utility patents and plant patents lies in the exemption from compliance with the written description requirement (35 U.S.C. §112) granted plant patent applicants by the PPA. 35 U.S.C. §162. Noncompliance with §112 is permitted provided the applicant's description is as complete as is reasonably possible.

The protection offered by plant patents is the exclusive right to reproduce the plant asexually. 35 U.S.C. §163. Each plant patent application may include but one claim directed specifically to the plant shown and described.

---

47. 35 U.S.C. §161 (1988). Asexual reproduction is reproduction that does not involve the use of seeds. Examples of such methods are budding, grafting, cutting, division, or layering. This limitation was apparently premised on the perception that plants which reproduced other than asexually could not be reproduced reliably true to type. Nicholas O. Seay, Protecting the Seeds of Innovation: Patenting Plants, 16 AIPLA Q.J. 418 (1989).

35 U.S.C. §162. Courts are split as to whether a plant patent covers independent derivation of a plant having the same varietal characteristics or only covers plant material actually derived from the patentee's plant. Cole Nursery Co. v. Youdath Perennial Gardens, Inc., 17 F. Supp. 159 (N.D. Ohio 1936); Yoder Bros., 537 F.2d at 1380; Pan-American Plant Co. v. Matsui, 433 F. Supp. 693, 694 (N.D. Cal. 1977). Regarding the doctrine of equivalents, courts are unanimous in holding that the doctrine is inappropriate in interpreting the claim of a plant patent.

While prosecution under the PPA may be less expensive than obtaining a utility patent, the applicant must be aware of the shortcomings of a plant patent. First, parts of plants are not protected as they would be under a utility patent. Additionally, there may be no doctrine of equivalents for plant patents. Lastly, there is no infringement in the sexual reproduction of patented plant material nor any biological material derived from the sexual reproduction of the plant.

b. The Plant Variety Protection Act

The second special statutory system to protect rights to plant properties is the Plant Variety Protection Act, 7 U.S.C. §2321-2582 (PVPA). Enacted in 1970, the PVPA essentially parallels the PPA for protection of plant varieties that are sexually reproduced by seed. Successful applicants under the PVPA are awarded not a patent, but a "Certificate of Plant Variety Protection."

PVPA certificates are available to the breeder "of any novel variety of sexually reproduced plant," provided the variety possess "distinctness," "uniformity," and "stability." 7 U.S.C. §2402. The "distinctness" requirement is most critical, requiring that the variety "clearly differs by one or more identifiable morphological, physiological or other characteristics. . . ." 7 U.S.C. §2401. Just as the PPA has been interpreted to include "plants" only in the layman's sense, the PVPA explicitly denies protection to fungi, bacteria, or first-generation hybrid plants.[48] In re Bergy, 596 F.2d 952 (C.C.P.A. 1979), *vacated on other grounds,* 444 U.S. 1028 (1980).

Although the PVPA is an example of a registration system, it does share many procedural requirements with the utility patent scheme: a series of statutory bars, content requirements for the application, and a requirement for a seed deposit to be made with the Plant Variety Protection Office. Still, the PVPA features several provisions found in neither the utility patent nor plant patent systems. These provisions include a mandatory license, a series of statutory exemptions for saved seed, an exemption for sales by farmers, and a research exemption.

---

48. First-generation hybrid plants are those produced by mass-breeding two different inbred varieties. First-generation hybrids do not reproduce true to type.

A PVPA certificate grants the breeder the right to exclude others from "selling the variety, or offering it for sale, or reproducing it, or importing it, or exporting it, or using it in producing . . . a hybrid or different variety therefrom. . . ." 7 U.S.C. §2483(a). The protection lasts for 18 years. Although the PVPA does not specifically address the question of independent derivation as a potential defense to infringement, the inclusion of a provision for an interference proceeding suggests that independent derivation is not a defense. Recent technological advances enable scientists to determine a plant's ancestry from its molecular composition, thus making detection of infringement much easier. This development, coupled with the significant monetary investment required to develop novel seed varieties, may account for the recent willingness of firms to vigorously assert their legal claims to specific plant varieties. See A. Hagedorn, Suits Sprout over Rights to Seeds, Wall St. J., Mar. 5, 1990, p. B1 col. 3, B8 cols. 1-2.

The "farmer's exemption" mentioned above permits individual farmers to sell protected varieties to other farmers without liability. By diluting the exclusive rights enjoyed by a PVPA certificate holder, this exemption appears to be at odds with the PVPA's primary purpose of establishing incentives for breeders to develop new strains. After struggling with this inherent contradiction in Delta and Pine Land Company v. People Gin Company, 694 F.2d 1012 (5th Cir. 1983), the Fifth Circuit held that Congress intended the farmer's exemption to be read narrowly; thus the exemption requires that sales be made directly from farmer to farmer — the active participation of a middleman being prohibited.

More recently, defendant farmers accused of engaging in "brown bag sales" invoked the farmer's exemption as a defense to infringement liability. In a recent decision, the United States Supreme Court held that the "farmer's exemption" for saved seed allowed farmers to sell only such seed as was saved to replant their own acreage and not to engage in the business of selling protected seeds. Asgrow Seed Co. v. Winterboer, 115 S. Ct. 788 (1995).

COMMENTS AND QUESTIONS

1. There is no question that intellectual property rights have affected the growth and structure of the seed industry. One recent article observes:

> [Since *Chakrabarty*,] "[n]umerous companies have . . . filed patent applications that cover the genes, the processes of isolating the genes, and making the genetically modified plants and seeds themselves. . . . Although no one disputes that companies that have invested heavily in R&D to isolate, test, and commercialize genes are entitled to protection for their inventions, there is considerable debate within the seed industry concerning how much protection is deserved and what impact patents will have on the cooperative nature of the seed industry itself."

C. S. Gasser & R. T. Fraley, Genetically Engineering Plants for Crop Improvement, 244 Science 1293 (16 June 1989). The authors note that before the passage of the Plant Variety Protection Act in 1970, only three companies sold commercial soybean seeds; now there are "more than 40." Id. See also W. Lesser & R. Masson, An Economic Analysis of the Plant Variety Protection Act 123 (1985) (summarizing positive stimulus to industry from PVPA); Evenson, Intellectual Property Rights and Agribusiness Research and Development: Implications for the Public Agricultural Research System, 65 Am. J. Agric. Econ. 967 (1983) (same).

A good case can be made that the expanded coverage of the conventional patent act (as opposed to the PVPA) will further spur plant-related research in this country. See Lesser, Patenting Seeds in the United States of America: What to Expect, Industrial Prop., Sept. 1986, at 360. There is an inevitable cost, however: more litigation. See, e.g., A. Hagedorn, Suits Sprout over Rights to Seeds, Wall St. J., Mar. 5, 1990, p. B1 col. 3, B8 cols. 1-2 (describing suit under PVPA over "Napolean" celery variety: "With companies spending millions of dollars yearly on biotechnology to create novel seed varieties, the costs of losing the seeds to competitors are greater than ever.").

2. Critics of intellectual property rights in the agriculture sector contend that they help accelerate undesirable trends such as centralization and the loss of economic power by small farmers. See, e.g., J. Kloppenburg, First the Seed: The Political Economy of Plant Biotechnology (1988). Kloppenburg notes that there was a great controversy in the agriculture world when in 1956 a researcher received a patent for hybrid plant breeding techniques. Id. at 113. Kloppenburg uses this as an example of a longstanding rift between the open and public-minded nature of federally funded agriculture research and the orientation toward private gain of private commercial researchers in this sector, whose growing prominence he says contributes to "the commodification of the seed." Id. at 282-284. His general thrust, echoed by others, is that agriculture is a special industry that is not always well-served by competition among private interests.

3. Recall the excerpt from the book by Robert Nozick at the beginning of this chapter. What is the difference between what is protected by plant patents and the protection of "discovered" plants alluded to in the excerpts? What arguments could you make that no special incentives are needed to discover plants, as opposed to the creation of plant-related "inventions" protectable under the patent code and the UPOV? Even if "discovered" plants were given some sort of protection, does that mean that all work on plant-related inventions would cease? See J. Brodovsky, The Mexican Pharmochemical and Pharmaceutical Industries, in The United States and Mexico: Face to Face with New Technology 198 (C. Thorup ed. 1987) (Mexican monopoly on barbasco, a plant that was a good source for making steroids, ended when purchasers developed alternative supply sources after Mexican government imposed higher price for barbasco).

4. Some countries have asserted that they "own" the genetic material from plants that grow inside their borders. Many are poor countries from tropical regions where a great variety of plant species grow. These countries insist that companies from developed countries pay "royalties" for the right to remove genetic material for research or the development of new products.

See John F. Burns, Tradition in India vs. a Patent in the U.S., N.Y. Times, Sept. 15, 1995, p. C4; M. Simons, Poor Nations Seeking Rewards for Contributions to Plant Species, N.Y. Times, May 16, 1989, p. 4, col. 4. Is this a form of intellectual property? Could the U.S. government make such a claim?

Others have contended that for humans to assert any "ownership" over inventions derived essentially from nature is sheer hubris. See Leon R. Kass, Patenting Life, 63 J. Pat. & Trademk. Off. Soc. 571, 599 (1981). The standard reply is to point out that incentives are needed to induce people to perform research in this socially valuable field, an argument that does not directly address the moral objection. Cf. Comment, In His Image: On Patenting Human-Based Bioproducts, 25 U.S.F. L. Rev. 583 (1991). An interesting middle ground is suggested in the following excerpt, which is drawn from a longer discussion of the extent to which property rights ought to be based on how hard someone works to create something — i.e., on a "labor" theory.

> [A]ssuming that labor's fruits are valuable, and that laboring gives the laborer a property right in this value, this would entitle the laborer only to the value she added, and not to the *total* value of the resulting product. Though exceedingly difficult to measure, these two components of value (that attributable to the object labored on and that attributable to the labor) need to be distinguished.

Edwin C. Hettinger, Justifying Intellectual Property, 18 Phil. & Pub. Aff. 31, 37 (1989). Imagine that a company from a developed country "prospects" for genetic material in a poor tropical country, takes some plant specimens back to the lab, and inserts a gene from the "prospected" material into an ordinary domestic plant, thereby producing a very valuable new plant. According to the approach taken in the excerpt, how should the rights be allocated between (1) the poor country's government, which claims ownership of the raw genetic material, and (2) the company that developed the new plant? How would one determine the "value added" by the company's researchers as opposed to the value contributed by the original genetic material?

5. The notion of granting intellectual property rights over the genetic material in native species may seem strange to some, but it might actually serve laudable social purposes. In addition to more fairly distributing the gains from recombinant genetic products based on those species, it would also give developing countries an incentive to protect rainforests and other genetically rich areas. In general, the granting of property rights over a resource can be expected to lead to more efficient use of the resource; at the very least, it will prevent over-exploitation of the resource due to its free (or "public good") quality. See, e.g., Harold Demsetz, Toward a Theory of Property Rights, in Ownership, Control, and the Firm 104 (1988); Charles, Fishery Socioeconomics: A Survey, 64 Land Econ. 276, 279-280 (1988) (describing allocation of fish catches via property rights).

Can a similar argument be made in favor of the attempts of some scientists to patent portions of the human genome? How do the legal and policy considerations differ?

4

Copyright Law

A. INTRODUCTION

1. Brief History of Copyright Protection

The origin of copyright law in Anglo-American jurisprudence parallels the origin of patent law in certain respects. Both laws grew out of original grants of monopoly to merchant guilds. The first "copyright" was granted in England by royal decree in 1556, not long after the introduction of the printing press in England.[1] See Elizabeth L. Eisenstein, The Printing Revolution in Early Modern Europe (Cambridge 1993) (discussing the ties between printing technology and the creation of copyright). For political reasons, the Crown consolidated the new printing business in the hands of the Stationers' Company. It gave to the printers of this company — not to authors — the exclusive right to control the printing and sale of books, forever. Not incidentally, the government conferred these copyrights upon loyal publishers who would not publish books that the Crown considered politi-

---

1. A spirited debate has taken place over the existence of common law copyrights before this time. While it was certainly true that an author had the property right to physical possession of his manuscript at common law, recent scholarship casts serious doubt on the existence of a common law right to prevent copying of words or ideas. See, e.g., Howard B. Abrams, The Historic Foundation of American Copyright Law: Exploding the Myth of Common Law Copyright, 29 Wayne L. Rev. 1119 (1983). Abrams examines in great detail the case of Donaldson v. Becket, 98 Eng. Rep. 257, 262 (H.L. 1774), where the House of Lords concluded that a copyright had *never* existed at common law and therefore could not be said to survive the passage of the Statute of Anne, described below. One author identifies a case of common law copyright infringement much earlier — in the fifth century. See Ernest A. Savage, Old English Libraries (1912) (relating the case of an individual who had secretly copied a manuscript belonging to another and who was forced to turn over the copy).

cally or religiously objectionable, and indeed it subjected the printing business to the oversight of the Star Chamber.[2]

After the exclusive right of the Stationers' Company ended in 1694, members of the company for the first time faced substantial competition in the printing of books. They promptly sought assistance from Parliament. In 1710, Parliament responded by passing the Statute of Anne. The Statute of Anne vested in *authors* of books a monopoly over their works, no doubt to the surprise of the publishers. Unlike the rights granted to publishers by decree, the statutory right was limited to only 14 years, renewable for an additional 14 years by the author. The statute contained a complex system of registration, notice, and deposit requirements; and strict compliance with those requirements was required by the English courts for many years. The Statute of Anne further turned its back on its monopolistic roots by providing that the government could set maximum prices for copyrighted books upon application by disgruntled consumers — what might be described as the first compulsory license in copyright.

Coexisting with the English history of copyright is the continental approach, which treats an author's right in his works of authorship as a fundamental moral right. Justice Breyer's work on copyright quotes a sixteenth-century French lawyer who argued that "as the heavens and the earth belong to God, because they are the work of his word . . . so the author of a book is its complete master, and as such can dispose of it as he chooses." Stephen Breyer, The Uneasy Case for Copyright: A Study of Copyright in Books, Photocopies, and Computer Programs, 84 Harv. L. Rev. 281, 284 (1970) (quoting Marion). Copyright on the continent, and particularly in France, accordingly developed parallel means of protecting authors: a property right similar to the English statutory right, based on a decree by the Revolutionary government in January 1791[3] and on a series of moral rights established in judicial decisions. These rights have only relatively recently been embodied in statutory form. See Raymond Sarraute, Current Theory on the Moral Right of Authors and Artists Under French Law, 16 Am. J. Comp. L. 465 (1968).

Most of the United States passed state copyright laws modeled on the Statute of Anne shortly after gaining their independence. Some of these state statutes went even further than the Statute of Anne in their "pro-consumer" orientation, allowing the government to regulate both the price charged for copyrighted works and the number of such works produced. These state acts also contained complex and often conflicting formal requirements. At the same time, these early state statutes also contained elements of the continental

---

2. The English Crown's desire to control the unlicensed printing of manuscripts manifested itself even earlier, in a 1529 edict directed at dissident speech during a tumultuous time. See David Lange, At Play in the Fields of the Word: Copyright and the Construction of Authorship in the Post-Literate Millennium, 55 L. & Contemp. Probs. 139 (1992).

3. Pre-Revolution French copyright took the form of royal grants of exclusive publishing monopolies, with rights sometimes granted to authors and sometimes to printers. See Jane C. Ginsburg, A Tale of Two Copyrights: Literary Property in Revolutionary France and America, 64 Tulane L. Rev. 991 (1990).

tradition favoring a natural right of authors. For example, the 1783 Massachusetts statute defined the purpose of copyright as *both* the utilitarian goal of producing new creative works, and the securing of "one of the natural rights of all men." See Jane C. Ginsburg, A Tale of Two Copyrights: Literary Property in Revolutionary France and America, 64 Tulane L. Rev. 991 (1990) (arguing that French law is more instrumental, and American law more reward-based, than previously recognized).

Problems with applying these conflicting state laws across state borders led to a general consensus that a national law was necessary, and the Constitution granted explicit power to the federal government to create both patents and copyrights. The patent and copyright clause passed the Convention and was ratified without significant debate.[4] One of the first acts of the new Congress was to pass the Copyright Act of 1790. That Act, like the Statute of Anne, granted protection for authors for 14 years, and allowed renewal for 14 more years. Indeed, this new act was similar in significant respects to the English law that preceded it.[5] The 1790 Act allowed copyrights to be registered with the local district court and notice to be published in local newspapers. Like its predecessors, it was limited to books. However, a series of amendments and court decisions throughout the nineteenth century progressively expanded the scope of the act, adding prints, musical compositions, dramatic works, photographs, artistic works, and sculpture. See, e.g., Bleistein v. Donaldson Lithographing Co., 188 U.S. 239 (1903) (extending copyright protection to chromolithographs). At each turn, copyright's adaptation to a new technology was a rocky one, and there were those who argued that copyright simply could not be adapted to fit the new technology. On the process of technological accretion in copyright law, see generally Paul Goldstein, Copyright's Highway (1995); Jessica Litman, Copyright Legislation and Technological Change, 68 Or. L. Rev. 265 (1989).

The most significant overhaul of the Copyright Act since its founding occurred in 1909. Like the amendments that led up to it, the 1909 Act generally broadened the scope of copyright protection. Protection for literary works was expanded from books to include "all writings," reaching works in progress, and speeches, among other new matter. The duration of copyright protection was extended to 28 years initial duration and 28 more years on renewal. Some protection was provided for foreign works registered in the United States, although the statute discriminated quite bluntly against foreign works. The 1909 Act kept the formalities that the 1790 Act had established, placing the United States at odds with the growing number of countries that had adopted the Berne Convention and abandoned formality — notice and registration — in copyright protection.

---

4. For a discussion of the history of this clause, see Karl Fenning, The Origin of the Patent and Copyright Clause of the Constitution, 17 Geo. L.J. 109 (1929).

5. David Lange argues that "Anglo-American copyright, in all of its contemporary manifestations, can be traced directly to this Statute [of Anne] and to the history that led to its enactment." Lange, supra.

Difficulties in interpretation of the 1909 Act led to numerous attempts to reform it. Significant reform was finally achieved in 1976, when Congress passed the act which (with some modifications) governs most works today. The 1976 Act expanded both the scope and duration of protection. All written works became protectable once they were "fixed in a tangible medium of expression," even if they were unpublished. The duration of copyright was expanded to the life of the author plus fifty years, or 75 years in the case of corporate "authors." Further, the formal notice and registration requirements were loosened, although not discarded.

In other respects, though, the 1976 Act weakened rather than strengthened intellectual property protection. First, the Act explicitly preempted much state and common law copyright, as well as any other statutes or rights that were substantially equivalent to copyright. 17 U.S.C. §301. Second, the 1976 Act codified the existing judicial defense of "fair use" of a copyrighted work. We will return to both of these provisions in some detail.

The 1976 Act was modified in 1980 to expressly incorporate computer programs into the Copyright Act for the first time, as a result of the report of the federal Commission on New Technological Uses of Copyright (CONTU). It was modified again in 1988, when the United States finally ratified the Berne Convention, an international copyright agreement signed by numerous countries. The 1988 amendments abolished the requirement of notice and certain other formalities. The new law provided a two-tier system: foreign copyright owners need not comply with the formalities that remained (thus bringing the United States into compliance with the Berne Convention), but U.S. owners would still have to comply before bringing suit.

The Architectural Works Copyright Protection Act of 1990 also served to bring the United States into compliance with the Berne Convention. In 1992, Congress passed the Audio Home Recording Act to address enhanced concerns about piracy of musical works and sound recordings through the introduction of digital audio tape technology. In 1998, Congress enacted the Sonny Bono Copyright Term Extension Act and the Digital Millennium Copyright Act, which affords protection against the circumvention of copy protection technologies as well as other provisions.

2. An Overview of the Copyright Regime

Although the copyright and patent laws flow from the same constitutional basis and share the same general approach — statutorily created monopolies to foster progress — they feature different elements and rights, reflecting the very different fields of creativity that they seek to encourage. We sketch below the basic elements and rights of copyright. As you review these features, contrast them to the analogous provisions of the patent law. How do you explain the differences?

A protectable copyright has the following elements:

Copyrightable Subject Matter. The subject matter protectable by copy-right spans the broad range of literary and artistic expression — including books, song, dance, computer programs, movies, sculp-ture, painting. Ideas themselves are not copyrightable, but the au-thor's particular expression of an idea is protectable.

Threshold for Protection. A work must exhibit a modicum of originality and be fixed in a "tangible medium of expression."

Formalities. Notice of copyright, in different forms, is required on all works created prior to 1989. Registration of a copyright is not strictly required for its validity, but it is required of U.S. authors prior to instituting an infringement suit. Deposit of copies of the work is required to obtain registration of copyright.

Authorship and Ownership. The work must have been created by the party bringing suit, or rights in the work must have been transferred by the author to the party bringing suit. In the case of works made "for hire," the employer and not the original creator is considered the author and the owner of the work.

Duration of Copyright. A copyright lasts for the life of the author plus 70 years, or 95 years from first publication in the case of entity authors (or 120 years from the year of creation, whichever occurs first).

While the United States Copyright Office registers works, its function is very different than that of the Patent Office. The Copyright Office does not "examine" applications for copyright (other than to ensure that the works contain some original content), and it does not "issue" copyrights. Rather, like trade secrets, a copyright is protectable at the moment the work is created. Further, the range of protectable subject matter is potentially much broader than that of patent law, since the standard of originality is so much lower.

This breadth in copyright protection is balanced, however, by the more limited rights that copyright law confers. Ownership of a valid copyright con-fers the following rights:

Copying. The owner has the exclusive right to make copies of her work. She may sue a copier for infringement if the copying is "material" and "substantial," even if the copy is in a different form or is of only part of the whole.

Derivative Works. The owner has the exclusive right to prepare derivative works, which are works based on the original but in different forms or otherwise altered (such as translations, movies based on books, etc.). These derivative works are themselves copyrightable, to the extent that they contain their own original expression. Note that the right to create derivative works overlaps with the right to make copies.

Distribution. The owner has the right to control the sale and distribution of the original and all copies or derivative works, including licensed copies. However, this right extends only to the first sale of such works. The owner does not have the right to limit resale by purchasers of her works (except in certain limited circumstances).

Performance and Display. The owner has the right to control the public (but not private) performance and display of her works, including both literary and performance-oriented works. This right extends to computer programs and other audiovisual works. The owner generally does not have the right to prevent the public display of a particular original or copy of a work of art, however.

These rights are limited by the fair use doctrine, a balancing test that allows limited use of copyrighted material. In addition, the Copyright Act establishes compulsory licensing for musical compositions and cable television, among others.

There is a fundamental difference between the rights granted by copyright law and those granted by patent law. Unlike patents, copyrights do not give their owner the exclusive right to make or use the thing copyrighted. Rather, they give the author only the right to prevent *unauthorized copying* of their works, as well as the right to prevent some limited types of uses of those works (such as public performances). The independent development of a similar work is perfectly legal.

This means that copyright law must have some mechanism for determining when a work has been copied illegally. While in rare cases direct proof of copying may be available, usually it is not. In its place, courts accept proof that the defendant had *access* to the plaintiff's work, combined with evidence that the two works are *substantially similar.* Only copying — whether proven directly or inferred — constitutes infringement.

3. Philosophical Perspectives on Copyright Protection

In the vast body of court decisions, legislation, and commentaries on copyright law, one can find references to a great many philosophical justifications for its existence. Lord Justice Mansfield, writing in the mid-eighteenth century, states: "From what source, then is the common law drawn, which is admitted to be so clear, in respect of the copy before publication? From this argument — because it is just, that an author should reap the pecuniary profits of his own ingenuity and labor." Millar v. Taylor, 4 Burr. 230, 238 (1769).[6]

For other philosophical views, see generally Alfred Yen, Restoring the Natural Law: Copyright as Labor and Possession, 51 Ohio St. L.J. 517 (1990) (tracing roots of natural law in American copyright law); Justin Hughes, The Philosophy of Intellectual Property, 77 Georgetown L.J. 287, 350-353 (1988) (suggesting various strains of the personhood justification in American copyright law). Immanuel Kant spoke of the "natural obligation" to respect the author's ownership of his works, see Immanuel Kant, Of the Injustice of Counterfeiting Books, 1 Essays and Treatises on Moral, Political, and Various Philosophical Subjects 225, 229-230 (Richardson ed. 1798). And as we have seen, the "natural right" of the author to control the use of his work and to be rewarded for it is one of the significant underpinnings of at least some American copyright jurisprudence. See Benjamin Kaplan, An Unhurried View of Copyright 79 (1967 ed.); James M. Treece, American Law Analogues of the Author's Moral Right, 16 Am. J. Comp. L. 487 (1968).

The predominant philosophical framework undergirding American copyright law, however, is utilitarian. The Constitution grants Congress the power to enact copyright laws in order to "promote the Progress of Science and useful Arts." Art. I, §8, cl. 8. The instrumentalist nature of this provision has not been lost on the courts. In the early decision of Wheaton v. Peters, 33 U.S. (8 Pet.) 591 (1834), the Court treated copyright as a statutory creation designed primarily to enhance the public interest and only secondarily to confer a reward upon authors. Id. at 661. More recently, Justice Stewart described the basic purpose of the Copyright Act as follows

> The limited scope of the copyright holder's statutory monopoly, like the limited duration required by the Constitution, reflects a balance of competing claims upon the public interest: Creative work is to be encouraged and rewarded, but private motivation must ultimately serve the cause of promoting broad public availability of literature, music, and the other arts.[6] The immediate effect of our copyright law is to secure a fair return to an "author's" creative labor. But the ultimate aim is, by this incentive, to stimulate artistic creativity for the general public good. "The sole interest of the United States and the primary object in conferring the monopoly," this Court has said, "lie in the general benefits derived by the public from the labors of authors."

---

6. It is worth noting that this ruling was overturned by the House of Lords five years later, in Donaldson v. Becket, 98 Eng. Rep. 257 (H.L. 1774).

6. Lord Mansfield's statement of the problem almost 200 years ago in Sayre v. Moore, quoted in a footnote to Cary v. Longman, 1 East *358, 362 n.(b), 102 Eng. Rep. 138, 150 n.(b) (1801), bears repeating:

> [W]e must take care to guard against two extremes equally prejudicial; the one, that men of ability, who have employed their time for the service of the community, may not be deprived of their just merits, and the reward of their ingenuity and labour; the other, that the world may not be deprived of improvements, nor the progress of the arts be retarded.

Twentieth Century Music Corp. v. Aiken, 422 U.S. 151, 156 (1975). See also Mazer v. Stein, 347 U.S. 201, 219 (1954). And Justice Stevens has commented that

> The monopoly privileges that Congress may authorize are neither unlimited nor primarily designed to provide a special private benefit. Rather, the limited grant is a means by which an important public purpose may be achieved. It is intended to motivate the creative activity of authors and inventors by the provision of a special reward, and to allow the public access to the products of this genius after the limited period of exclusive control has expired.

Sony Corp. of America v. Universal City Studios, Inc., 464 U.S. 417, 429 (1984).

American copyright law can thus be seen as primarily striving to achieve an optimal balance between fostering incentives for the creation of literary and artistic works and the optimal use and dissemination of such works. Nonetheless, copyright law reflects the other philosophical perspectives as well. Society grants copyrights both because it wants to encourage creation and because it wants to reward authors for their work. Copyright also reflects a moral sense that authors deserve to own the works they have created. The law limits the scope of those copyrights because it wants to make sure that works are freely disseminated and that the next generation of authors can make use of the ideas of the prior generation in creating still more works. As we will see later in this chapter, international pressure and appeals by artists have brought increased recognition of the moral rights of artists. These policies interact in complex ways. In many cases, there is still great controversy over which policy should prevail. We will encounter these controversies throughout this chapter.

COMMENTS AND QUESTIONS

1. Does copyright law appear to strike the proper balance between fostering incentives for the creation of literary and artistic works and the optimal use and dissemination of such works? For example, does the duration of protection strike the right balance? Does the same duration of protection for all copyrightable works — whether books, computer programs, songs, paintings, or choreographic works — make sense? Do other justifications beyond the utilitarian balance better explain copyright's structure and provisions? Cf. Peter S. Menell, Intellectual Property: General Theories, Encyclopedia of Law and Economics (B. Bouckaert and G. De Geest, eds., 2000); Alfred Yen, Restoring the Natural Law: Copyright as Labor and Possession, 51 Ohio St. L.J. 517 (1990); Wendy Gordon, An Inquiry into the Merits of Copyright: The Challenges of Consistency, Consent, and Encouragement Theory, 41 Stan. L. Rev. 1343 (1989); Justin Hughes, The Philosophy of Intellectual Property, 77 Georgetown L.J. 287 (1988).

2. Contrast the way in which copyright law, trade secret law, and patent law vary along the following dimensions:

- threshold for protection
- duration of protection
- rights conferred
- treatment of independent creation
- defenses to infringing use

To what extent can the differences among these legal regimes be explained by differences in the subject of coverage (and the nature of the innovative process in these areas)? differences in the philosophical justifications for these modes of protection? other factors?

3. The term "copyright" reflects well the underlying philosophy of the Anglo-American regime for protecting literary and artistic works — regulation of the right to make *copies* for the purpose of promoting progress in the arts and literature. The emphasis is on the benefit to the public, not the benefits or rights of authors. By contrast, the civil law analog to copyright has a different name and orientation. In France, the comparable body of law is *droit d'auteur,* which translates to "author's rights." The laws in Germany and Spain are similar — *Urheberrecht* and *derecho de autor.* This civil law tradition derives from a Kantian (natural rights) or Hegelian (personhood) justification for legal entitlements, and thus focuses on the rights of *authors.* Thus, the civil law countries have long expressly protected the moral rights of authors — e.g., the right of an author to prevent the mutilation of his or her work after it is sold. See Marshall Leaffer, Understanding Copyright Law 1-3 (3d ed. 1999).

Of what significance is the underlying philosophical perspective — whether utilitarian, natural rights, or personhood — for the structure and content of copyright law? Which perspective is more appropriate as a matter of social justice? public policy? Can these perspectives be effectively harmonized without losing their coherence?

B. REQUIREMENTS

17 U.S.C. §102. Subject Matter of Copyright: In General

(a) Copyright protection subsists, in accordance with this title, in original works of authorship fixed in any tangible medium of expression, now known or later developed, from which they can be perceived, reproduced, or otherwise communicated, either directly or with the aid of a machine or device. . . .

1. Original Works of Authorship

H.R. Rep. No. 94-1476, 94th Cong., 2d Sess.
51 (1976)

The two fundamental criteria of copyright protection — originality and fixation in a tangible form — are restated in the first sentence of this cornerstone provision. The phrase "original works of authorship," which is purposively left undefined, is intended to incorporate without change the standard of originality established by the courts under the present copyright statute. This standard does not include requirements of novelty, ingenuity, or esthetic merit, and there is no intention to enlarge the standard of copyright protection to require them. . . .

As developed by the courts, originality entails *independent creation* of a work featuring a *modicum of creativity.* Independent creation requires only that the author not have copied the work from some other source. As the eminent copyright jurist Judge Learned Hand eloquently observed,

> If by some magic a man who had never known it were to compose anew Keat's *Ode on a Grecian Urn,* he would be an "author" and, if he copyrighted it, others might not copy that poem, though they might of course copy Keats's.

Sheldon v. Metro-Goldwyn Pictures Corp., 81 F.2d 49, 54 (2d Cir. 1951). This highlights an important distinction between patent and copyright law.

> The alleged inventor is chargeable with full knowledge of all the prior art, although in fact he may be utterly ignorant of it. The "author" is entitled to a copyright if he independently contrived a work completely identical with what went before; similarly, although he obtains a valid copyright, he has no right to prevent another from publishing a work identical with his, if not copied from his.

Alfred Bell & Co. v. Catalda Fine Arts, Inc., 191 F.2d 99, 103 (2d Cir. 1951). Courts have set the threshold of creativity necessary to satisfy the originality requirement quite low. Copyright law does not require that a work be

> strikingly unique or novel. . . . All that is needed to satisfy both the Constitution and the statute is that the "author" contributed something more than "merely trivial" variation, something recognizably "his own." Origininality in this context "means little more than a prohibition of actual copying." No matter how poor artistically the "author's" addition, it is enough if it be his own.

Id. at 102-103.

Courts do not judge the artistic merit of a work:

> It would be a dangerous undertaking for persons trained only to the law to constitute themselves final judges of the worth of pictorial illustrations, outside of the narrowest and most obvious limits. At one extreme some works of genius would be sure to miss appreciation. Their very novelty would make them repulsive until the public had learned the new language in which their author spoke. It may be more than doubted, for instance, whether the etchings of Goya or the paintings of Manet would have been sure of protection when seen for the first time. At the other end, copyright would be denied to pictures which appealed to a public less educated than the judge.

Bleistein v. Donaldson Lithographing Co., 188 U.S. 239, 251-252 (1903) (finding a circus advertisement to be sufficiently original). Courts have rarely found literary or artistic works to fall below the *de minimis* originality threshold of copyright law. The few exceptions generally relate to simple slogans and exceedingly modest variations on another work. See 37 C.F.R. §202.1 Material Not Subject to Copyright ("(a) Words and short phrases such as names, titles, and slogans; familiar symbols or designs; mere variations of typographical ornamentation, letter or coloring; mere listing of ingredients or contents . . . (e) Type face as typeface").

A more difficult problem arises when an author creates a work in a mechanical or functional manner. Such assemblage of information can be costly and time consuming (the "sweat of the brow"), but may lack creativity. Should copyright law protect such works?

John Locke proposed a theory of property in which labor over a previously unowned piece of property (intellectual or physical) can vest ownership rights in the laborer. See John Locke, Two Treatises of Government §§27-28. The sweat of the brow theory also commands as adherents a number of venerable court decisions and some modern commentators. See, e.g., International News Service v. Associated Press, 248 U.S. 215, 236 (1918); Blunt v. Patten, 3 F. Cas. 763, 765 (C.C.S.D.N.Y. 1828); Robert Denicola, Copyright in Collections of Facts: A Theory for the Protection of Nonfiction Literary Works, 81 Colum. L. Rev. 516 (1981). Proponents of the sweat of the brow theory advance two basic rationales. The first is economic: compilation is necessary and efficient, and unless it is rewarded there will be no incentive to compile or create factual works. The second rationale is based on fairness: it is unjust to permit one person to steal the hard work of another.

Opponents of the sweat of the brow approach see it as essentially at odds with the rationale of intellectual property protection, because it protects work without regard to creativity. If a list compiled at great effort is protectable, why not a ditch dug with similar effort? Further, they point to the dangers of monopoly (if no one can duplicate or chooses to duplicate the effort of the compiler) and of wasted resources (if several competing companies do the same fact-intensive work to produce the same product). See, e.g., Jerome

H. Reichman and Pamela Samuelson, Intellectual Property Rights in Data?, 50 Vand. L. Rev. 51 (1997); Jane C. Ginsburg, Creation and Commercial Value: Copyright Protection of Works of Information, 90 Colum. L. Rev. 1865 (1990)[7]; Alan Gorman, Fact or Fancy? The Implications for Copyright, 29 J. Copyright Soc'y 560 (1980); Pamela Samuelson, Information as Property: Do Ruckelshaus and Carpenter Signal a Changing Direction in Intellectual Property Law?, 38 Cath. U.L. Rev. 365 (1989).

There are obviously efficiency concerns associated with granting a compiler of data exclusive rights to that compilation. Such a copyright would result either in a monopoly for the copyright holder or would force competitors to duplicate the effort of compiling the data. One commentator has suggested a compromise — protect labor-based works, but only against competing compilers. See Lum, Copyright Protection for Factual Compilations — Reviving the Misappropriation Doctrine, 56 Fordham U.L. Rev. 933, 952 (1988). Does this approach make sense? Or is it the worst of both worlds, permitting the taking of the compiler's property in some cases while allowing her to maintain a monopoly? In particular, consider whether it would promote the optimal investment in databases.

What should be the role of copyright law in protecting the sweat of the brow? The Supreme Court faced this question in 1991.

Feist Publications v. Rural Telephone Service
Supreme Court of the United States
499 U.S. 340 (1991)

Justice O'CONNOR delivered the opinion of the Court.

This case requires us to clarify the extent of copyright protection available to telephone directory white pages.

I

Rural Telephone Service Company is a certified public utility that provides telephone service to several communities in northwest Kansas. It is subject to a state regulation that requires all telephone companies operating in Kansas to issue annually an updated telephone directory. Accordingly, as a condition of its monopoly franchise, Rural publishes a typical telephone directory, consisting of white pages and yellow pages. The white pages list in alphabetical order the names of Rural's subscribers, together with their towns and telephone numbers. The yellow pages list Rural's business subscribers alphabetically by category and feature classified advertisements of various

---

7. Professor Ginsburg does, however, favor a separate statute to protect some "sweat-works." See Jane C. Ginsburg, "No Sweat?" Copyright and Other Protection of Works of Information After Feist v. Rural Telephone, 92 Colum. L. Rev. 338 (1992).

sizes. Rural distributes its directory free of charge to its subscribers, but earns revenue by selling yellow pages advertisements.

Feist Publications, Inc., is a publishing company that specializes in area-wide telephone directories. Unlike a typical directory, which covers only a particular calling area, Feist's area-wide directories cover a much larger geographical range, reducing the need to call directory assistance or consult multiple directories. The Feist directory that is the subject of this litigation covers 11 different telephone service areas in 15 counties and contains 46,878 white pages listings — compared to Rural's approximately 7,700 listings. Like Rural's directory, Feist's is distributed free of charge and includes both white pages and yellow pages. Feist and Rural compete vigorously for yellow pages advertising.

As the sole provider of telephone service in its service area, Rural obtains subscriber information quite easily. Persons desiring telephone service must apply to Rural and provide their names and addresses; Rural then assigns them a telephone number. Feist is not a telephone company, let alone one with monopoly status, and therefore lacks independent access to any subscriber information. To obtain white pages listings for its area-wide directory, Feist approached each of the 11 telephone companies operating in northwest Kansas and offered to pay for the right to use its white pages listings.

Of the 11 telephone companies, only Rural refused to license its listings to Feist. Rural's refusal created a problem for Feist, as omitting these listings would have left a gaping hole in its area-wide directory, rendering it less attractive to potential yellow pages advertisers. . . .

Unable to license Rural's white pages listings, Feist used them without Rural's consent. Feist began by removing several thousand listings that fell outside the geographic range of its area-wide directory, then hired personnel to investigate the 4,935 that remained. These employees verified the data reported by Rural and sought to obtain additional information. As a result, a typical Feist listing includes the individual's street address; most of Rural's listings do not. Notwithstanding these additions, however, 1,309 of the 46,878 listings in Feist's 1983 directory were identical to listings in Rural's 1982-1983 white pages. Four of these were fictitious listings that Rural had inserted into its directory to detect copying.

Rural sued for copyright infringement. . . .

II

A

This case concerns the interaction of two well-established propositions. The first is that facts are not copyrightable; the other, that compilations of facts generally are. Each of these propositions possesses an impeccable pedigree. That there can be no valid copyright in facts is universally understood. The most fundamental axiom of copyright law is that "no author may copyright his ideas or the facts he narrates." Harper & Row, Publishers, Inc. v. Nation Enterprises, 471 U.S. 539, 556 (1985). Rural wisely concedes this point, noting in its brief that "facts and discoveries, of course, are not themselves subject to copyright protection." At the same time, however, it is beyond dispute that compilations of facts are within the subject matter of copyright. Compilations were expressly mentioned in the Copyright Act of 1909, and again in the Copyright Act of 1976.

There is an undeniable tension between these two propositions. Many compilations consist of nothing but raw data — i.e., wholly factual information not accompanied by any original written expression. On what basis may one claim a copyright in such a work? Common sense tells us that 100 uncopyrightable facts do not magically change their status when gathered together in one place. Yet copyright law seems to contemplate that compilations that consist exclusively of facts are potentially within its scope.

The key to resolving the tension lies in understanding why facts are not copyrightable. The *sine qua non* of copyright is originality. To qualify for copyright protection, a work must be original to the author. See *Harper & Row,* supra, at 547-549. Original, as the term is used in copyright, means only that the work was independently created by the author (as opposed to copied from other works), and that it possesses at least some minimal degree of creativity. 1 M. Nimmer & D. Nimmer, Copyright §§2.01[A], [B] (1990) (hereinafter Nimmer). To be sure, the requisite level of creativity is extremely low; even a slight amount will suffice. The vast majority of works make the grade quite easily, as they possess some creative spark, "no matter how crude, humble or obvious" it might be. Id., §1.08[C][1]. Originality does not signify novelty; a work may be original even though it closely resembles other works so long as the similarity is fortuitous, not the result of copying. . . .

It is this bedrock principle of copyright that mandates the law's seemingly disparate treatment of facts and factual compilations. "No one may claim originality as to facts." Id., §2.11[A], p. 2-157. This is because facts do not owe their origin to an act of authorship. The distinction is one between creation and discovery: the first person to find and report a particular fact has not created the fact; he or she has merely discovered its existence. . . .

Factual compilations, on the other hand, may possess the requisite originality. The compilation author typically chooses which facts to include, in what order to place them, and how to arrange the collected data so that they may be used effectively by readers. These choices as to selection and arrangement, so long as they are made independently by the compiler and entail a minimal degree of creativity, are sufficiently original that Congress may protect such compilations through the copyright laws. Nimmer §§2.11[D], 3.03; Denicola 523, n.38. Thus, even a directory that contains absolutely no protectable written expression, only facts, meets the constitutional minimum for copyright protection if it features an original selection or arrangement. See Harper & Row, 471 U.S., at 547. Accord Nimmer §3.03.

This protection is subject to an important limitation. The mere fact that a work is copyrighted does not mean that every element of the work may be protected. Originality remains the sine qua non of copyright; accordingly, copyright protection may extend only to those components of a work that are original to the author. . . .

This inevitably means that the copyright in a factual compilation is thin. Notwithstanding a valid copyright, a subsequent compiler remains free to use the facts contained in an another's publication to aid in preparing a competing work, so long as the competing work does not feature the same selection and arrangement. As one commentator explains it: "[N]o matter how much original authorship the work displays, the facts and ideas it exposes are free for the taking. . . . [T]he very same facts and ideas may be divorced from the context imposed by the author, and restated or reshuffled by second comers, even if the author was the first to discover the facts or to propose the ideas." Ginsburg 186-8.

It may seem unfair that much of the fruit of the compiler's labor may be used by others without compensation. As Justice Brennan has correctly observed, however, this is not "some unforeseen byproduct of a statutory scheme." Harper & Row, 471 U.S., at 589 (dissenting opinion). It is, rather, "the essence of copyright," ibid., and a constitutional requirement. The primary objective of copyright is not to reward the labor of authors, but "to promote the Progress of Science and useful Arts." Art. I, §8, cl. 8. Accord Twentieth Century Music Corp. v. Aiken, 422 U.S. 151, 156 (1975). . . .

III

There is no doubt that Feist took from the white pages of Rural's directory a substantial amount of factual information. At a minimum, Feist copied the names, towns, and telephone numbers of 1,309 of Rural's subscribers. Not all copying, however, is copyright infringement. To establish infringement, two elements must be proven: (1) ownership of a valid copyright, and (2) copying of constituent elements of the work that are original. See Harper & Row, 471 U.S., at 548. The first element is not at issue here; Feist appears to concede that Rural's directory, considered as a whole, is subject to a valid copyright because it contains some foreword text, as well as original material in its yellow pages advertisements.

The question is whether Rural has proved the second element. In other words, did Feist, by taking 1,309 names, towns, and telephone numbers from Rural's white pages, copy anything that was "original" to Rural? Certainly, the raw data does not satisfy the originality requirement. Rural may have been the first to discover and report the names, towns, and telephone numbers of its subscribers, but this data does not " 'owe its origin' " to Rural. Burrow-Giles, 111 U.S., at 58. Rather, these bits of information are uncopyrightable facts; they existed before Rural reported them and would have continued to exist if Rural had never published a telephone directory. The originality requirement "rules out protecting . . . names, addresses, and telephone numbers of which the plaintiff by no stretch of the imagination could be called the author." Patterson & Joyce 776. . . .

The question that remains is whether Rural selected, coordinated, or arranged these uncopyrightable facts in an original way. As mentioned, originality is not a stringent standard; it does not require that facts be presented in an innovative or surprising way. It is equally true, however, that the selection and arrangement of facts cannot be so mechanical or routine as to require no creativity whatsoever. The standard of originality is low, but it does exist. See Patterson & Joyce 760, n.144 ("While this requirement is sometimes characterized as modest, or a low threshold, it is not without effect"). As this Court has explained, the Constitution mandates some minimal degree of creativity, see The Trade-Mark Cases, 100 U.S., at 94; and an author who claims infringement must prove "the existence of . . . intellectual production, of thought, and conception." *Burrow-Giles*, supra, at 59-60.

The selection, coordination, and arrangement of Rural's white pages do not satisfy the minimum constitutional standards for copyright protection. As mentioned at the outset, Rural's white pages are entirely typical. Persons desiring telephone service in Rural's service area fill out an application and Rural issues them a telephone number. In preparing its white pages, Rural simply takes the data provided by its subscribers and lists it alphabetically by surname. The end product is a garden-variety white pages directory, devoid of even the slightest trace of creativity.

Rural's selection of listings could not be more obvious: it publishes the most basic information — name, town, and telephone number — about each person who applies to it for telephone service. This is "selection" of a sort, but it lacks the modicum of creativity necessary to transform mere selection into copyrightable expression. Rural expended sufficient effort to make the white pages directory useful, but insufficient creativity to make it original.

We note in passing that the selection featured in Rural's white pages may also fail the originality requirement for another reason. Feist points out that Rural did not truly "select" to publish the names and telephone numbers of its subscribers; rather, it was required to do so by the Kansas Corporation Commission as part of its monopoly franchise. See 737 F. Supp., at 612. Accordingly, one could plausibly conclude that this selection was dictated by state law, not by Rural.

Nor can Rural claim originality in its coordination and arrangement of facts. The white pages do nothing more than list Rural's subscribers in alphabetical order. This arrangement may, technically speaking, owe its origin to Rural; no one disputes that Rural undertook the task of alphabetizing the names itself. But there is nothing remotely creative about arranging names alphabetically in a white pages directory. It is an age-old practice, firmly rooted in tradition and so commonplace that it has come to be expected as a matter of course. See Brief for Information Industry Association et al. as *Amici Curiae* 10 (alphabetical arrangement "is universally observed in directories published by local exchange telephone companies"). It is not only unoriginal, it is practically inevitable. This time-honored tradition does not possess the minimal creative spark required by the Copyright Act and the Constitution.

We conclude that the names, towns, and telephone numbers copied by Feist were not original to Rural and therefore were not protected by the copyright in Rural's combined white and yellow pages directory. . . .

COMMENTS AND QUESTIONS

1. How can a compiler of telephone directory information obtain copyright protection for its work? Does *Feist* provide any guidelines for determining the precise threshold of originality?

2. As highlighted in Chapter 1, a principal purpose for society in affording protection against copying the work of others is to provide incentives for the creation of new and useful texts, machines, and so on. Property rights in intellectual work would also flow, morally, from the labor of the author. Under *Alfred Bell*, however, copyright protects the most minor variations on material that is already in the public domain. By contrast, hundreds of hours devoted to compiling a valuable telephone directory are for naught, at least under the copyright law. Others can reproduce a telephone directory with impunity. In view of the incentive and labor rationales for intellectual property, don't the principal cases on originality come out exactly wrong?

3. A relatively recent strain of copyright scholarship has applied the lessons of postmodern literary theory to copyright, attempting to challenge the entire concept of originality as a myth. See, e.g., James Boyle, Shamans, Software, and Spleens: Law and the Construction of the Information Society (1996); Peter Jaszi, Toward a Theory of Copyright: The Metamorphoses of "Authorship," 41 Duke L.J. 455 (1991); David Lange, At Play in the Fields of the Word: Copyright and the Construction of Authorship in the Post-Literate Millennium, 55 L. & Contemp. Probs. 139 (1992). Does this approach suggest that the idea of identifying and rewarding "originality" is misguided, because it is falsely premised on a romantic notion of "authorship" that does not in fact exist in the modern world?

4. Is there any room for protection of factual compilations after *Feist*? In particular, a number of authors have expressed concern that *Feist* precluded copyright protection of automated databases, since such databases were built around user search capabilities rather than an original (and copyrightable) organizational structure. See, e.g., John F. Hayden, Copyright Protection of Computer Databases After *Feist*, 5 Harv. J. L. & Tech. 215 (1991); Phillip H. Miller, Life After *Feist:* Facts, the First Amendment, and the Copyright Status of Automated Databases, 60 Fordham L. Rev. 507 (1991); see also Jane Ginsburg, Creation and Commercial Value: Copyright Protection of Works of Information, 90 Colum. L. Rev. 1865 (1990). Would an electronic "phone book" that users could search by entering the name of the desired party be copyrightable under *Feist*? Does it matter how the database itself is organized (something which the user may not see)?

Cases decided after *Feist* have refused to protect a number of compilations similar in nature to the telephone white pages. For example, in Warren Publishing v. Microdos Data Corp., 115 F.3d 1509 (11th Cir. 1997), the court held that a compilation of cable television companies serving each community in the nation was not copyrightable. The arrangement was unprotectable because the communities were listed alphabetically, and the selection was not original because every community in the U.S. served by a cable company was included. See also TransWestern Publishing Co. v. Multimedia Marketing Assoc., 133 F.3d 773 (10th Cir. 1998).

By contrast, in American Dental Assn. v. Delta Dental Plans, 126 F.3d 977 (7th Cir. 1997), Judge Easterbrook restricted the scope of both *Feist* and Baker v. Selden, infra. The court there held that a taxonomy of dental insurance billing codes was original and copyrightable. The court distinguished a taxonomy created by the plaintiff itself from a compilation of pre-existing facts. The court went further than this, however, strongly suggesting that the choice of a single five-digit number itself reflected protectable expression.

If the *American Dental* court is correct, could Rural copyright the phone numbers it has assigned to customers?

5. In the wake of *Feist*, owners of databases and other factual compilations have turned increasingly to contract law to protect their "sweat of the brow" investment. In at least some cases, it appears that a party can protect by contract what she cannot protect by copyright, so that someone who copies a database may be liable for breach of contract even if the database is composed entirely of unprotectable facts. (On this point, compare the discussion of protecting trade secrets by contract in Chapter 2.B.3.) The efficacy of this approach may vary widely, however, depending on whether the factual compilation at issue is widely distributed or sold to a limited number of vendors. We discuss contract protection for databases in more detail in Chapter 6.

An alternative to contract protection is a new state or federal law of intellectual property designed specifically to protect "sweatworks." Certain Scandanavian countries have such a law already, and the European Union directive on database protection has strong "sweat of the brow" elements. In addition, some have argued that there is already a common law restitution-like remedy in the United States when hard work is blatantly appropriated by a copyist. See Wendy J. Gordon, On Owning Information: Intellectual Property and the Restitutionary Impulse, 78 Va. L. Rev. 149 (1992) (corrective justice approach justifies giving authors a right — albeit conditional and defeasible — to a reward for their efforts).

At this writing, several bills are pending in Congress to provide statutory protection for databases under the Commerce Clause, rather than the Copyright Clause. These bills take two different forms: H.R. 354 would create a new intellectual property right in the database itself, which could be infringed not only by wholesale copying of the database but by the use or extraction of individual facts from that database. By contrast, H.R. 1858 would provide a form of unfair competition protection against those who take most or all of a database from a database provider in order to enter into competition with that provider. It seems likely that some form of database protection will be enacted by Congress in the next few years.

Should Congress amend the Copyright Act to protect such "sweat of the brow" works? Is Congress constitutionally empowered to protect such works, given the "limited times" provision in the Copyright Clause? The "author" requirement?

Any *state* law protecting compilations against appropriation might well be preempted by the copyright laws, since it would be directly at odds with *Feist*. See David Rice, Public Goods, Private Contract and Public Policy: Federal Preemption of Software License Prohibitions Against Reverse Engineering, 53 U. Pitt. L. Rev. 543, 595-621 (1992) (arguing that copyright law preempts state laws that contradict copyright policy); but cf. ProCD, Inc. v. Zeidenberg, 86 F.3d 1447 (7th Cir. 1996) (contract law not preempted under one legal theory even where the contract protects telephone white pages; court did not consider alternative grounds of preemption, and left standing district court ruling that state misappropriation and unfair competition laws were preempted). We discuss copyright preemption in more detail in Chapter 6.

6. Copyright protection extends to maps under section 101 (defining "pictorial, graphic and sculptural works"). Can a map satisfy the originality requirement for copyright? Isn't a map by its nature entirely driven by functional considerations (i.e., accurate representation of its subject matter)?

In Mason v. Montgomery Data, Inc., 967 F.2d 135 (5th Cir. 1992), a case decided after *Feist*, the court held that a series of real estate maps of a certain county in Texas were eligible for copyright protection. The court found the maps original in two different respects: (1) Mason had exercised "sufficient creativity in both the selection, coordination and arrangement of the facts that [the maps] depict"; and (2) the graphic artistry of the maps themselves was sufficiently original to qualify for protection.

Are either of these elements truly original in the typical map? How many ways are there to accurately depict the United States and its 50 states, for example? Should copyright depend on something as trivial as the colors chosen for each political subdivision? Cf. Dennis Karjala, Copyright in Electronic Maps, 35 Jurimetrics J. 395 (1995).

PROBLEM

Problem 4-1. Central Bell, the local telephone utility, distributes both a "white pages" telephone book and a "yellow pages," which lists businesses who have chosen to advertise there. Bell's yellow pages are organized alphabetically by subject matter of the business, and alphabetically within each subject. Bell itself created the subject headings with input from its advertisers. Bell also sells larger advertisements to certain companies for more money.

Christopher Publications decides to create and distribute its own yellow pages directory in competition with Bell. Christopher does this by taking a copy of Bell's yellow pages and calling every business that advertises there, asking each if it would like to advertise in Christopher's publication. Christopher places the resulting ads in its own subject matter listings (which do overlap somewhat with Bell's). Many of the advertisements themselves (which are submitted by the advertisers) are identical to those in the Bell directory. Bell sues Christopher for copyright infringement. Who should prevail?

2. Fixation in a Tangible Medium of Expression

H.R. Rep. No. 94-1476, 94th Cong., 2d Sess. 52-53 (1976)

As a basic condition of copyright protection, the bill [the Copyright Act of 1976] perpetuates the existing requirement that a work be fixed in a "tangible medium of expression," and adds that this medium may be one "now known or later developed," and that the fixation is sufficient if the work "can be perceived, reproduced, or otherwise communicated, either directly or with the aid of a machine or device." This broad language is intended to avoid the artificial and largely unjustifiable distinctions, derived from cases such as White-Smith [Music] Publishing Co. v. Apollo Co., 209 U.S. 1 (1908), under which statutory copyrightability in certain cases has been made to depend upon the form or medium in which the work is fixed. Under the bill it makes no difference what the form, manner, or medium of fixation may be whether it is in words, numbers, notes, sounds, pictures, or any other graphic or symbolic indicia, whether embodied in a physical object in written, printed, photographic, sculptural, punched, magnetic, or any other stable form, and whether it is capable of perception directly or by means of any machine or device "now known or later developed."

Under the bill, the concept of fixation is important since it not only determines whether the provisions of the statute apply to a work, but it also represents the dividing line between common law and statutory protection. As will be noted in more detail in connection with section 301, an unfixed work of authorship, such as an improvisation or an unrecorded choreographic work, performance, or broadcast, would continue to be subject to protection under State common law or statute, but would not be eligible for Federal statutory protection under section 102.

The bill seeks to resolve, through the definition of "fixation" in section 101, the status of live broadcasts — sports, news coverage, live performances of music, etc. — that are reaching the public in unfixed form but that are simultaneously being recorded. When a football game is being covered by four television cameras, with a director guiding the activities of the four cameramen and choosing which of their electronic images are sent out to the public and in what order, there is little doubt that what the cameramen and the director are doing constitutes "authorship." The further question to be considered is whether there has been a fixation. If the images and sounds to be broadcast are first recorded (on a video tape, film, etc.) and then transmitted, the recorded work would be considered a "motion picture" subject to statutory protection against unauthorized reproduction or retransmission of the broadcast. If the program content is transmitted live to the public while being recorded at the same time, the case would be treated the same; the copyright owner would not be forced to rely on common law rather than statutory rights in proceeding against an infringing user of the live broadcast.

Thus, assuming it is copyrightable — as a "motion picture" or "sound recording," for example — the content of a live transmission should be accorded statutory protection if it is being recorded simultaneously with its transmission. On the other hand, the definition of "fixation" would exclude from the concept purely evanescent or transient reproductions such as those projected briefly on a screen, shown electronically on a television or other cathode ray tube, or captured momentarily in the "memory" of a computer.

Under the first sentence of the definition of "fixed" in section 101, a work would be considered "fixed in a tangible medium of expression" if there has been an authorized embodiment in a copy of phonorecord and if that embodiment "is sufficiently permanent or stable" to permit the work "to be perceived, reproduced, or otherwise communicated for a period of more than transitory duration." The second sentence makes clear that, in the case of "a work consisting of sounds, images, or both, that are being transmitted," the work is regarded as "fixed" if a fixation is being made at the same time as the transmission.

Under this definition "copies" and "phonorecords" together will comprise all of the material objects in which copyrightable works are capable of being fixed. The definitions of these terms in section 101, together with their usage in section 102 and throughout the bill, reflect a fundamental distinction between the "original work" which is the product of "authorship" and the multitude of material objects in which it can be embodied. Thus, in the sense of the bill, a "book" is not a work of authorship, but is a particular kind of "copy." Instead, the author may write a literary "work," which in turn can be embodied in a wide range of "copies" and "phonorecords," including books, periodicals, computer punch cards, microfilm, tape recordings, and so forth. It is possible to have an "original work of authorship" without having a "copy" or "phono-record" embodying it, and it is also possible to have a "copy" or "phon-orecord" embodying something that does not qualify as an "original work of authorship." The two essential elements — original work and tangible object — must merge through fixation in order to produce subject matter copyrightable under the statute.

COMMENTS AND QUESTIONS

1. The fixation requirement arises in two separate portions of the Copyright Act. First, it is a requirement for copyright protection. Section 102(a) of the Copyright Act provides that "[c]opyright protection subsists . . . in original works of authorship *fixed in any tangible medium of expression,* now known or later developed, from which they can be perceived, reproduced, or otherwise communicated, either directly or with the aid of a machine or device" (emphasis added). Unless and until a work of authorship is so fixed, it does not qualify for copyright protection.

Fixation also plays a role in determining whether a defendant has infringed a copyright. Section 106(1) of the act provides that the copyright owner has the exclusive right to "reproduce the copyrighted work in copies or phonorecords." Section 101 of the act defines "copies" as "material objects . . . in which a work is *fixed.* . . ." (emphasis added). Thus a defendant does not infringe the right to reproduce unless she has reproduced the copyrighted work in fixed form. This rule has been at issue in a number of recent cases concerned with "copies" of computer programs or data made by the computer during its operation. Such "RAM copies" occur every time a computer is turned on or a file is opened; they are necessary for a computer user to view a program or file; and they disappear when the computer is turned off. Does opening a file make a "copy" for copyright purposes? See MAI v. Peak Computing, 991 F.2d 511 (9th Cir. 1993) (yes); see infra Chapter 7 (discussing the issue in detail).

Both meanings of fixation were at issue in White-Smith Music Pub. Co. v. Apollo Co., 209 U.S. 1 (1908). That case involved the protection afforded perforated rolls used in the then-new technology of "player pianos." The case itself raised the question of whether the piano rolls were "copies" of a copyrighted "musical composition." The Court held that the copyright laws extended only to human-readable materials and therefore did not encompass piano rolls. Subsequent interpretations of this decision extended it to suggest that works not in human-readable form were not subject to copyright protection either. Congress overruled *White-Smith* in the 1976 Copyright Act, as noted in the legislative history to the act. And the use of the phrase "musical works" in 17 U.S.C. §102 — in contrast to "musical composition" in prior acts — might be thought to suggest a move away from protection only of works contained in a defined medium of expression.

2. Why have a fixation requirement at all? Historically, the practical effect of the fixation requirement has been to slow copyright's adaption to new technologies. *White-Smith* is an obvious example but not the only one. Some might assert that slowing change in copyright law is itself a persuasive justification for the requirement, but Congress appears largely to have rejected that view in the 1976 Act.

Alternatively, fixation may be constitutionally required. Article I, §8, clause 8 refers to "writings," which could be interpreted to require a tangible fixation of the work. Cf. Goldstein v. California, 412 U.S. 546 (1973). On the other hand, if the constitutional term "writings" is broad enough to encompass sculpture and computer code, can it really be said to place a significant limit on the subject matter of protection?

One explanation for fixation lies in a view of copyright as intended to protect communication. See David G. Luettgen, Functional Usefulness vs. Communicative Usefulness: Thin Copyright Protection for the Nonliteral Elements of Computer Programs, 4 Tex. Intell. Prop. L.J. 233 (1996). Certainly, the original copyright laws — and the Constitution — speak of authors and writings, which we associate with relatively direct communication between writer and audience. On this view, material that does not communicate (directly) to people is undeserving of copyright protection. Professor Dratler analogizes the "communicative function" argument to Count Dracula, because it "keeps emerging from the coffin, despite repeated interment." Jay Dratler, Intellectual Property Law 5.03[1][b], at 5-63. Is there any necessary connection between whether a work is "fixed" and whether it communicates directly with an audience? Compare an extemporaneous public speech (unfixed, but communicative) with an individual's private diary (fixed, but private).

One persuasive argument for the fixation requirement relates to the practical requirements of copyright litigation. Fixation helps in proving authorship. If any expression could be copyrighted, the law might face a large number of frivolous infringement suits that would be virtually impossible to verify — along the lines of "I gave them the idea (or rather the expression) for that book!" Analogous rules exist in other areas of law. Consider the statute of frauds in contract law and the best evidence rule, for example. Could the same result be accomplished more equitably by means of a presumption of noninfringement? By a requirement of access to copyrighted works?

3. Even when the fixation requirement bars federal copyright protection, common law protection may still be available. 17 U.S.C. §301, which preempts state laws with the same scope as federal copyright, expressly retains state laws that protect unfixed expression. 17 U.S.C. §301(b)(1). Many states afford such protection. See, e.g., Estate of Hemingway v. Random House, 244 N.E.2d 250 (N.Y. 1968); Cal. Civil Code §980. We discuss state copyright protection in Chapter 6.

4. In 1994, Congress added section 1101 to the Copyright Act. This provision gives some protection to unfixed musical performances,

by making it illegal to traffic in unauthorized recordings of a live musical performance.

PROBLEM

Problem 4-2. Armstrong, a heavy metal musician, performs an unscheduled jam session in an outdoor concert in 1993, featuring songs never before recorded. Unbeknownst to Armstrong or his producers, Fan records the concert and begins to sell copies of the recording. Armstrong sues Fan for infringement of copyright. Fan replies "What copyright?" Who should prevail?

---

3. Formalities

Copyright "formalities" are requirements imposed on authors by the government that are necessary to obtain copyright protection but that do not relate to the substance of the copyright. The United States has traditionally had four such formal requirements: notice of copyright, publication of the work, registration of the work with the Copyright Office, and deposit of a copy of the work with the Library of Congress. Over the past century, U.S. law has progressed from a regime in which failure to adhere to certain technical requirements resulted in forfeiture of copyright protection to the current regime in which formalities are largely voluntary and failure to comply does not risk forfeiture. The principal reason for this transformation was the decision by the United States to join the Berne Convention, which provides that copyright shall "not be subject to any formality." Nonetheless, copyright formalities have continuing relevance to the practice of copyright law.

a. Notice

U.S. copyright law has experienced three decreasingly restrictive notice regimes since the early part of the twentieth century:

1909 Act. Under the 1909 Act, federal copyright law protected only those works that contained the following information in the appropriate form and location: the year of first publication; the word "Copyright," the abbreviation "Copr." or the symbol ©; and the name of the copyright holder.[8] 17

---

8. Unpublished works were protectable at common law prior to the 1976 Act. See Chapter 6, discussing common law copyright and preemption.

U.S.C. §§19, 20. Failure to satisfy the precise technical requirements of the 1909 Act, such as placing notice on the "title page or the page immediately following" of books or other printed publications, typically resulted in forfeiture of copyright protection and the work's falling into the public domain. Booth v. Haggard, 184 F.2d 470 (8th Cir. 1950). Section 21 of the 1909 Act excused omissions due to "accident or mistake," although this provision has been interpreted narrowly.

1976 Act/Pre-Ratification of Berne Convention (January 1, 1978–March 1, 1989). The 1976 Act provided that copyright protection begins upon the creation of a work, not upon publication. Congress maintained the notice requirement, although it liberalized the rules governing form and location of notice, see 17 U.S.C. §§401-403,[9] and it took much of the harshness out of the requirement, see 17 U.S.C. §§405-406. Failure to give notice on a small number of copies would not result in forfeiture; nor would even large-scale omissions, so long as they were inadvertent and the copyright holder registered the work within five years after publication and made reasonable efforts to give notice after the omission was discovered.

Post-Ratification of the Berne Convention (since March 1, 1989). The Berne Convention Implementation Act, Pub. L. No. 100-568, 102 Stat. 2853 (1988), eliminated the notice requirement of U.S. copyright law prospectively. Thus both the 1909 Act and 1976 Act (pre-Berne) regimes still apply with regard to works publicly distributed without proper notice prior to March 1, 1989. Congress, however, *encouraged* voluntary notice by precluding an alleged infringer from claiming "innocent infringement" in mitigation of actual or statutory damages except in limited circumstances if the copy she had access to contained a proper notice. 17 U.S.C. §401(d).

b. Publication

1909 Act. Federal copyright protection under the 1909 Act was triggered by the act of publishing a work. Unpublished works could be protected under state common law, or "constructively" published by registration with the Copyright Office. See Chapter 6. Despite the importance of "publication" to federal copyright protection, the 1909 Act did not specifically define the term. This gap spawned a rich and complex body of case law defining publication for purposes of copyright law. Because of the grave consequences of publishing a work without proper notice — risk of forfeiture of copyright protection — the courts developed a distinction between *divestive* publica-

---

9. The Copyright Office has issued regulations that have attempted to accommodate different works by requiring only "reasonable" placement of the notice. In some cases, this may be on packaging, particularly where the user would not encounter the notice on the actual work (say, computer software object code) or where affixing the notice to the actual work would interfere with the work (sound recordings, and some forms of visual art).

tion, resulting in forfeiture (divestiture) of common law copyright protection, and *investive* publication, resulting in forfeiture of federal copyright protection if notice is inadequate. In a leading case, the Second Circuit observed that

> . . . courts apply different tests of publication depending on whether plaintiff is claiming protection because he did not publish and hence has a common law claim of infringement — in which case the distribution must be quite large to constitute "publication" — or whether he is claiming under the copyright statute — in which case the requirements for publication are quite narrow. In each case the courts appear so to treat the concept of "publication" as to prevent piracy.

American Visuals Corp. v. Holland, 239 F.2d 740, 744 (2d Cir. 1956). Thus the extent of distribution required to divest common law copyright protection is substantially greater than that required to invest, i.e., require notice for, a federal copyrighted work. In White v. Kimmell, 193 F.2d 744, 746747 (9th Cir. 1952), *cert. denied,* 343 U.S. 957 (1952), the court distinguished between "limited publication" whereby a distribution of copies to "a definitely selected group and for a limited purpose, and without the right of diffusion, reproduction, distribution or sale" did not constitute a publication for purposes of the 1909 Act, and "general publication," which did operate to divest common law protection. In addition, courts generally found that public performance or display of a work did not constitute publication unless tangible copies of the work were distributed to the public. See Ferris v. Frohman, 223 U.S. 424 (1912) (applying pre-1909 Act law); McCarthy v. White, 259 F. 364 (S.D.N.Y. 1919) (public performance of musical work); King v. Mister Maestro, Inc., 224 F. Supp. 101 (S.D.N.Y. 1963) (Martin Luther King, Jr.'s "I Have a Dream" speech carried over radio and TV). Furthermore, courts determined that unauthorized distribution did not constitute publication. In these ways, courts alleviated some of the harsh effects of failure to adhere to the notice requirements.

Is a television broadcast a "publication" within the meaning of the 1909 Act? Does it matter whether permanent copies are made at the time of the broadcast?

1976 Act/Pre-Ratification of Berne Convention (January 1, 1978–March 1, 1989). Under the 1976 Act, federal copyright protection is triggered by the act of creating a work fixed in a tangible medium of expression, and common law copyright is preempted. Hence publication no longer served to distinguish between statutory and common law copyright. Nonetheless, publication still served to determine when notice was required.

The 1976 Act clarified the copyright law by defining "publication" as

> . . . the distribution of copies or phonorecords of a work to the lay public by sale or other transfer of ownership, or by rental, lease, or lending. The offering

to distribute copies or phonorecords to a group of persons for purposes of
further distribution, public performance, or public display, constitutes publication. A public performance or display of a work does not of itself constitute
publication.

17 U.S.C. §101. This definition largely codified the principal considerations that evolved through judicial interpretation of the 1909 Act. See H.R.
Rep. No. 1476, 94th Cong., 2d Sess., 138 (1976). Publication turns on
physical transfer of copies to the public generally without restriction with
respect to disclosure or to a narrower group for purposes of distribution or
dissemination. The 1976 Act expressly excludes public performance or
display.

Post-Ratification of the Berne Convention (since March 1, 1989).
With the elimination of a mandatory notice requirement, the act of publication is no longer a factor in determining the validity of works created after
March 1, 1989. Nonetheless, publication still has relevance for works created
after March 1, 1989 in the following respects:

- *Deposit.* Deposit at the Library of Congress is mandatory only for
  published works, 17 U.S.C. §407.
- *Works of Foreign Authors.* Whereas all unpublished works are protected regardless of nationality or domicile of the author, *published*
  works of foreign authors are protected only under the conditions described in section 104(b).
- *Duration of Copyright Protection.* The term of protection for entity
  owners and works for hire is 95 years from the year of first publication,
  17 U.S.C. §302(c); see also 17 U.S.C. §302(d), (e).
- *Library Photocopying.* The reproduction rights of libraries depend on
  whether the work has been published, see 17 U.S.C. §108(b), (c).
- *Termination of Transfers.* See 17 U.S.C. §203(a)(3).
- *Certain Performance Rights.* See 17 U.S.C. §§110(9), 118(b), (d).
- *Establishing Prima Facie Evidence of Validity of Copyright.* Registration of copyright must occur within five years of first publication, 17
  U.S.C. §410(c).
- *Damages.* Statutory damages and attorney fees are available for published works only if registration preceded the infringement or if the
  work was registered within three months after publication, 17 U.S.C.
  §412.

PROBLEM

Problem 4-3. Which of the following acts would constitute publication under the 1909 Act? the 1976 Act?

a) Penelope Poet brings copies of her latest three poems to the monthly meeting of the Philadelphia Aspiring Poets Society. She distributes copies to the eight people in attendance that month prior to reading the poems.

b) Professor Edgar Edifice assembles a course reader consisting of public domain materials and excerpts written by himself. Students in his Advanced Copyright Theory seminar purchase the reader for the cost of reproduction from the school's reprographics service.

c) Arnold Author recently completed a draft of his first novel. He sends a copy to his agent, whom he authorizes to distributes copies to publishing houses for consideration.

c. Registration

Unlike the notice requirement, registration of a copyrighted work with the Copyright Office has always been "voluntary." However, it is voluntary in only a very limited sense, as we shall see. Also unlike the notice requirement, the registration requirement remains in effect today, in spite of United States adherence to the Berne Convention.

1909 Act. Under the 1909 Act, the term of a copyright ran 28 years, with the author having the right to renew the copyright for an additional 28 years. The act did not require registration in order to obtain a copyright, but it did require registration by the 28th year in order to renew the copyright. Further, registration of the work was a prerequisite to bringing an infringement suit.

1976 Act/Pre-Ratification of Berne Convention (January 1, 1978–March 1, 1989). Under the 1976 Act, which abolished renewal and set a single copyright term, a copyright owner is not required to register her work. However, there are important incentives that encourage her to do so. First, successful registration constitutes prima facie evidence of the validity of the copyright. 17 U.S.C. §410. Second, a copyright holder must register the copyright before bringing an infringement action, 17 U.S.C. §411, so in practice an unregistered copyright is only a potential rather than an actual right.[10] Finally, there is also a powerful incentive for *early* registration: a copyright holder can obtain statutory damages and attorney fees only for infringements that occurred after registration (or which occurred after publication if the work was registered within three months after publication).

---

10. Interestingly, a copyright holder need only *file* for registration of copyright and obtain a response before bringing suit. She is entitled to bring suit even if the Copyright Office rejects her registration application. See 17 U.S.C. §411(a).

Post-Ratification of the Berne Convention (since March 1, 1989). To comply with the Berne Convention, Congress amended section 411 to eliminate the requirement that copyright owners whose country of origin[11] is another Berne member nation must register their works prior to instituting suits. However, Congress retained the requirement of registration prior to suit for domestic works, thereby imposing a greater burden on those who first publish in the United States.

d. Deposit

Section 407 of the Copyright Act requires deposit of two copies of each work published in the United States[12] for which copyright is claimed within three months after publication. (Certain categories of works are exempted from the requirement.) The purpose of this requirement is to enhance the collection of the Library of Congress. Under the 1909 Act, the Register of Copyrights could demand compliance with this deposit requirement, and failure to comply could result in forfeiture of the copyright. Under the 1976 Act, however, the Library deposit requirement — while still mandatory — does not affect the validity of a copyright or the author's right to bring suit. 17 U.S.C. §407. Failure to comply gives rise to a fine, but no other penalty.[13]

A separate deposit with the Copyright Office is also required for copyright registration by section 408. However, this requirement may generally be satisfied by section 407's Library deposit. Unlike the Library of Congress deposit, failure to comply with the section 408 deposit requirement results in a refusal to register the copyright.

The purpose behind registration deposit under section 408 is to enable the Copyright Office to know what it is registering. The section 407 Library of Congress deposit requirement is one of the clearest examples of the dissemination function in the copyright law.

Note on the Restoration of Foreign Copyrighted Works

The system of copyright formalities that prevailed in the United States until quite recently was at odds with the prevailing rule in the rest of the world, which did not generally require such formalities. The existence of such rules meant that a number of foreign works that received copyright protection

---

11. A work's country of origin is the country in which it was first published, or in the case of unpublished works the country in which it was created.

12. The provision limiting the Library deposit requirement to works published in the United States was added by the Berne implementing legislation in 1988.

13. In amending the 1976 Act to adhere to the Berne Convention, Congress concluded that retention of the deposit requirement was not inconsistent with the Berne Convention because failure to comply with the deposit requirement did not result in forfeiture of any copyright protection. H.R. Rep. No. 609, 100th Cong., 2d Sess. 44 (1988).

in their home country (and elsewhere in the world) entered the public domain in the United States because their authors failed to comply with these formalities — for example, if the authors failed to register their work in the United States upon publication. These works have been freely copyable in the United States but still received copyright protection elsewhere in the world.

In 1994 the United States adhered to the Uruguay Round of the General Agreements on Tariffs and Trade (GATT), including the provisions on intellectual property rights (generally referred to as GATT-TRIPs). To comply with TRIPs, Congress in 1994 added section 104A to the Copyright Act. Section 104A provides for the "restoration" of copyright protection for certain foreign works that had lost protection in the United States "due to noncompliance with formalities imposed . . . by United States copyright law." In effect, these works are retrieved from the public domain in the United States and are treated in the same way as any other copyrighted works for purposes of duration, ownership, and so on.

Copyright restoration presents a number of potential problems, and efforts to deal with them have made section 104A a complex provision. Copyright protection in restored works is not retroactive. However, the copyright owner does have the right to "cut off" future uses of the work after a limited grace period by giving the Copyright Office notice of her intention to enforce the copyright. More intractable problems are presented by those who have created "derivative works" of their own based on the newly protected work. Should such authors be deprived of their own creative works as well? The act provides that the authors of such derivative works may continue to exploit them, provided they pay "reasonable compensation" to the original copyright owner.

It is reasonable to expect substantial litigation over the copyright restoration law.

Does restoring copyright in works already in the public domain "promote the progress of science and the useful arts"? How? What additional incentive does such a right give to current authors? If it does not confer such an additional incentive, is it constitutional?

C. COPYRIGHTABLE SUBJECT MATTER

1. Limitations on Copyrightability: Distinguishing Function and Expression

This section considers the fundamental doctrines that operate to channel protection for works between the patent and copyright regimes.

a. The Idea-Expression Dichotomy

The most significant doctrine limiting the copyrightability of works is the "idea-expression" dichotomy, which is codified in section 102(b):

17 U.S.C. §102. Subject Matter of Copyright: In General

(a) Copyright protection subsists, in accordance with this title, in original works of authorship. . . .

(b) In no case does copyright protection for an original work of authorship extend to any idea, procedure, process, system, method of operation, concept, principle, or discovery regardless of the form in which it is described, explained, illustrated, or embodied in such work.

The division between protectable expression and unprotectable ideas was developed in the seminal case of *Baker v. Selden*.

Baker v. Selden
Supreme Court of the United States
101 U.S. 99 (1879)

Mr. Justice BRADLEY delivered the opinion of the court.

Charles Selden, the testator of the complainant in this case, in the year 1859 took the requisite steps for obtaining the copyright of a book, entitled "Selden's Condensed Ledger, or Book-keeping Simplified," the object of which was to exhibit and explain a peculiar system of book-keeping. In 1860 and 1861, he took the copyright of several other books, containing additions to and improvements upon the said system. The bill of complaint was filed against the defendant, Baker, for an alleged infringement of these copyrights. The latter, in his answer, denied that Selden was the author or designer of the books, and denied the infringement charged, and contends on the argument that the matter alleged to be infringed is not a lawful subject of copyright. . . .

The book or series of books of which the complainant claims the copyright consists of an introductory essay explaining the system of book-keeping referred to, to which are annexed certain forms or blanks, consisting of ruled lines, and headings, illustrating the system and showing how it is to be used and carried out in practice. This system effects the same results as book-keeping by double entry; but, by a peculiar arrangement of columns and headings, presents the entire operation, of a day, a week, or a month, on a single page, or on two pages facing each other, in an account-book. The defendant uses a similar plan so far as results are concerned; but makes a different arrangement of the columns, and uses different headings. If the complainant's testator had the exclusive right to the use of the system explained in his book, it would be difficult to contend that the defendant does not infringe it, notwithstanding the difference in his form of arrangement; but if it be assumed that the system is open to public use, it seems to be equally difficult to contend that the books made and sold by the defendant are a violation of the copyright of the complainant's book considered merely as a book explanatory of the system. Where the truths of a science or the methods of an art are the common property of the whole world, and author has the right to express the one, or explain and use the other, in his own way. As an author, Selden explained the system in a particular way. It may be conceded that Baker makes and uses account-books arranged on substantially the same system; but the proof fails to show that he has violated the copyright of Selden's book, regarding the latter merely as an explanatory work; or that he has infringed Selden's right in any way, unless the latter became entitled to an exclusive right in the system.

... [T]he question is, whether the exclusive property in a system of book-keeping can be claimed, under the law of copyright, by means of a book in which that system is explained? ...

There is no doubt that a work on the subject of book-keeping, though only explanatory of well-known systems, may be the subject of a copyright; but, then, it is claimed only as a book. Such a book may be explanatory either of old systems, or of an entirely new system; and, considered as a book, as the work of an author, conveying information on the subject of book-keeping, and containing detailed explanations of the art, it may be a very valuable acquisition to the practical knowledge of the community. But there is a clear distinction between the book, as such, and the art which it is intended to illustrate. The mere statement of the proposition is so evident, that it requires hardly any argument to support it. The same distinction may be predicated of every other art as well as that of book-keeping. A treatise on the composition and use of medicines, be they old or new; on the construction and use of ploughs, or watches, or churns; or on the mixture and application of colors for painting or dyeing; or on the mode of drawing lines to produce the effect of perspective, — would be the subject of copyright; but no one would contend that the copyright of the treatise would give the exclusive right to the art or manufacture described therein. The copyright of the book, if not pirated from other works, would be valid without regard to the novelty, or want of novelty, of its subject-matter. The novelty of the art or thing described or explained has nothing to do with the validity of the copyright. To give to the author of the book an exclusive property in the art described therein, when no examination of its novelty has ever been officially made, would be a surprise and a fraud upon the public. That is the province of letters-patent, not of copyright. The claim to an invention or discovery of an art or manufacture must be subjected to the examination of the Patent Office before an exclusive right therein can be obtained; and it can only be secured by a patent from the government.

The difference between the two things, letters-patent and copyright, may be illustrated by reference to the subjects just enumerated. Take the case of medicines. Certain mixtures are found to be of great value in the healing art. If the discoverer writes and publishes a book on the subject (as regular physicians generally do), he gains no exclusive right to the manufacture and sale of the medicine; he gives that to the public. If he desires to acquire such exclusive right, he must obtain a patent for the mixture as a new art, manufacture, or composition of matter. He may copyright his book, if he pleases; but that only secures to him the exclusive right of printing and publishing his book. So of all other inventions or discoveries. . . .

Of course, these observations are not intended to apply to ornamental designs, or pictorial illustrations addressed to the taste. Of these it may be said, that their form is their essence, and their object, the production of pleasure in their contemplation. This is their final end. They are as much the product of genius and the result of composition, as are the lines of the poet or the historian's periods. On the other hand, the teachings of science and the rules and methods of useful art have their final end in application and use; and this application and use are what the public derive from the publication of a book which teaches them. But as embodied and taught in a literary composition or book, their essence consists only in their statement. This alone is what is secured by the copyright. The use by another of the same methods of statement, whether in words or illustrations, in a book published for teaching the art, would undoubtedly be an infringement of the copyright.

Returning to the case before us, we observe that Charles Selden, by his books, explained and described a peculiar system of book-keeping, and illustrated his method by means of ruled lines and blank columns, with proper headings on a page, or on successive pages. Now, whilst no one has a right to print or publish his book, or any material part thereof, as a book intended to convey instruction in the art, any person may practice and use the art itself which he has described and illustrated therein. The use of the art is a totally different thing from a publication of the book explaining it. The copyright of a book on book-keeping cannot secure the exclusive right to make, sell, and use account-books prepared upon the plan set forth in such book. Whether the art might or might not have been patented, is a question which is not before us. It was not patented, and is open and free to the use of the public. And, of course, in using the art, the ruled lines and headings of accounts must necessarily be used as incident to it [see Figure 4-1].

The plausibility of the claim put forward by the complainant in this case arises from a confusion of ideas produced by the peculiar nature of the art described in the books which have been made the subject of copyright. In describing the art, the illustrations and diagrams employed happen to correspond more closely than usual with the actual work performed by the operator who uses the art. Those illustrations and diagrams consist of ruled lines and headings of accounts; and it is similar ruled lines and headings of accounts which, in the application of the art, the book-keeper makes with his pen, or the stationer with his press; whilst in most other cases the diagrams and illustrations can only be represented in concrete forms of wood, metal, stone, or some other physical embodiment. But the principle is the same in all. The description of the art in a book, though entitled to the benefit of copyright, lays no foundation for an exclusive claim to the art itself. The object of the one is explanation; the object of the other is use. The former may be secured

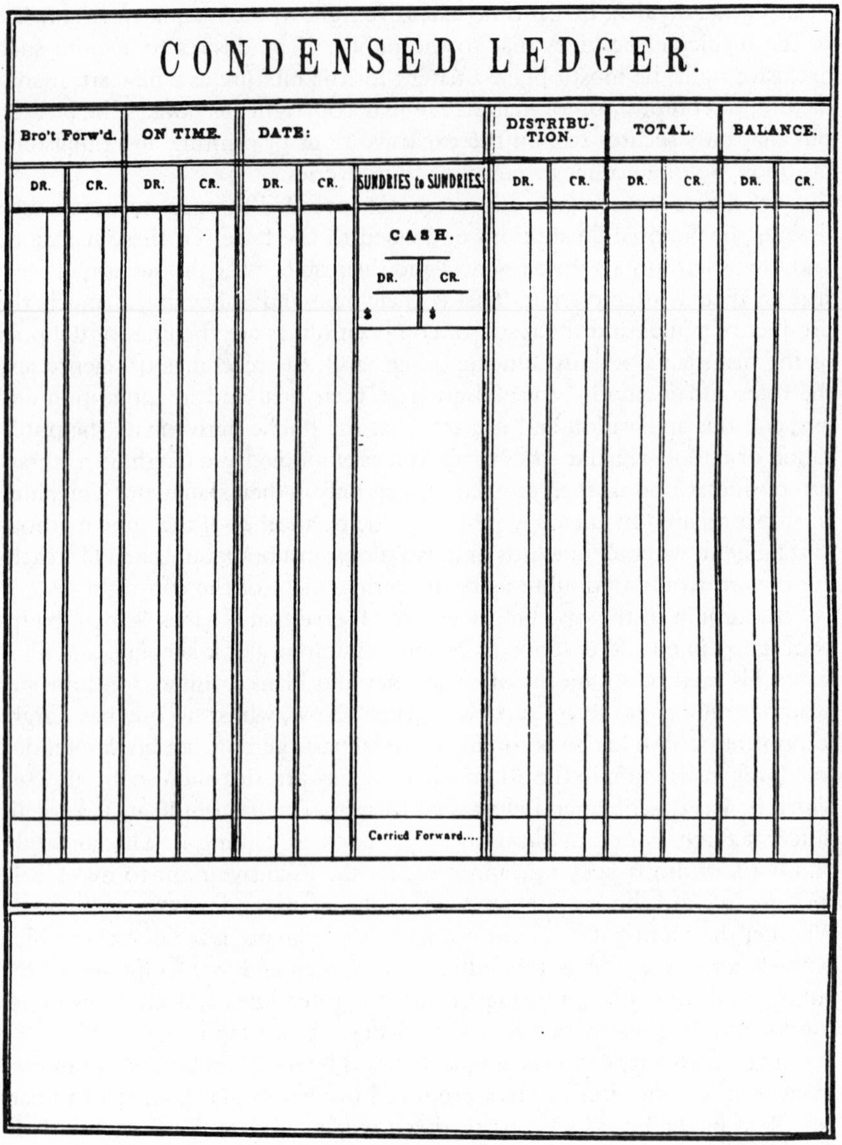

CONDENSED LEDGER.

Bro't Forw'd.		ON TIME.		DATE:		SUNDRIES to SUNDRIES		DISTRIBU-TION.		TOTAL.		BALANCE.	
DR.	CR.	DR.	CR.	DR.	CR.			DR.	CR.	DR.	CR.	DR.	CR.

CASH.

DR. | CR.

$ | $

Carried Forward....

FIGURE 4-1
Selden's blank form for condensed ledger.

by copyright. The latter can only be secured, if it can be secured at all, by letters-patent. . . .

The conclusion to which we have come is, that blank account books are not the subject of copyright; and that the mere copyright of Selden's book did not confer upon him the exclusive right to make and use account-books, ruled and arranged as designated by him and described and illustrated in said book.

The decree of the Circuit Court must be reversed, and the cause remanded with instructions to dismiss the complainant's bill; and it is

So ordered.

COMMENTS AND QUESTIONS

1. What is the Supreme Court's holding in this case? Did the Court rule that Selden's subject matter — accounting forms — are not copyrightable? Or that Selden's particular forms are not copyrightable? Alternatively, did the Court rule that Selden's forms are copyrightable but that the copyright does not prevent Baker's particular use of the forms, because such a result would in effect bestow upon Selden a monopoly over the system in question?

2. What does this case suggest about the relationship between copyright and patent protection? Does it imply that a particular work cannot be protectable by both patent and copyright law? In this regard, consider whether Selden could prevent Baker from reproducing the book explaining the Selden system. Does the doctrine of Baker v. Selden — which establishes an idea-expression dichotomy — coherently channel intellectual property protection between the copyright and patent modes of protection? What differences between patentable subject matter (and the process of innovation) and copyrightable subject matter (and the process of artistic creativity) justify such a doctrine?

In a recent case, the plaintiff claimed to own a copyright in individual recipes for preparing foods using a specific brand of yogurt. The court rejected the argument that the individual recipes were copyrightable:

> The recipes' directions for preparing the assorted dishes fall squarely within the class of subject matter specifically excluded from copyright protection by Section 102(b). . . . The recipes at issue here describe a procedure by which the reader may produce many dishes featuring a specific yogurt. As such, they are excluded from copyright protection as either a "procedure, process, or system" under Section 102(b).
>
> Protection for ideas or processes is the purview of patent. . . .
>
> Nothing in our decision today runs counter to the proposition that recipes may be copyrightable. There are cookbooks in which the authors lace their directions for producing dishes with musings about the spiritual nature of cooking or reminiscences they associate with the wafting odors of certain dishes in various stages or preparation. . . . Other recipes may be accompanied by tales of their historical or ethnic origin.

Publications Intl. v. Meredith Corp., 88 F.3d 473 (7th Cir. 1996).

3. *Distinguishing Idea from Expression.* The most challenging aspect of applying the idea-expression dichotomy is determining where to draw the line between idea and expression. Consider the works in question in *Baker v.*

Selden. What was the "idea" behind Selden's book? Was it to describe a better system of accounting? Was it to describe a system of double-entry bookkeeping? Was it to describe the particular system he had invented? Or was it all of the above?

In the classic statement of how to draw the line between idea and expression, Judge Learned Hand states the problem as follows:

> Upon any work, and especially upon a play, a great number of patterns of increasing generality will fit equally well, as more and more of the incident is left out. The last may perhaps be no more than the most general statement of what the play is about, and at times may consist only of its title; but there is a point in this series of abstractions where they are no longer protected, since otherwise the playwright could prevent the use of his "ideas." . . .

Nichols v. Universal Pictures Corp., 45 F.2d 119, 121 (2d Cir. 1930). Apply this reasoning to *Baker* itself. If the idea of Selden's book is to write a step-by-step guide explaining the use of his new forms (a low level of abstraction), even the detailed structure of the explanatory text may be unprotectable (because it is "necessary" to the idea). At a slightly higher level of abstraction, the Court considered the forms themselves unprotectable but gave Selden considerable latitude to protect his description of his system. At the highest levels of abstraction, had the Supreme Court determined that Selden's idea was to improve accounting, his double-entry bookkeeping forms might well have been considered simply one means of expressing that idea. In that case, Selden would presumably be entitled to copyright the forms.

How do courts determine the appropriate "level of abstraction" for distinguishing idea and expression? Paul Goldstein suggests that there are three categories of unprotectable ideas: the "animating concept" behind the work, the functional principles or "solutions" described or embodied in the work (such as Selden's forms), and the fundamental "building blocks" of creative expression (such as basic plot or character outlines in literary or dramatic works). 1 Paul Goldstein, Copyright §2.3.1.1. Professor Goldstein suggests that courts engage in a rough sort of balancing between the dangers of overprotecting and underprotecting a particular work.

4. *Blank Forms.* The Copyright Office considers the following works not copyrightable and hence ineligible for registration: "Blank forms, such as time cards, graph paper, account books, diaries, bank checks, scorecards, address books, report forms, order forms and the like, which are designed for recording information and do not in themselves convey information." 37 CFR §202(1)(c).

PROBLEM

Problem 4-4. Mediforms designs and sells health care forms to doctors, who submit them to insurance carriers. A sample form is shown in Figure 4-2.

JOHN R. JOHNNSON, M.D.
Type of Practice or Specialty
1000 MAIN STREET, SUITE 10
SOME PLACE, USA 70000

STATE LIC. # 123456789
SOC. SEC. # 000-11-0000

TELEPHONE: (123) 234-5678

☐PRIVATE ☐BLUE CROSS ☐BLUE SHIELD ☐IND. ☐MEDICAID ☐MEDICARE ☐GOV'T.

PATIENT'S LAST NAME		FIRST			INITIAL	BIRTHDATE	SEX [ ]MALE [ ]FEMALE	TODAY'S DATE
ADDRESS		CITY	STATE	ZIP		RELATION TO SUBSCRIBER	REFERRING PHYSICIAN	
SUBSCRIBER OR POLICYHOLDER						INSURANCE CARRIER		
ADDRESS — IF DIFFERENT		CITY	STATE	ZIP		INS. ID	COVERAGE CODE	GROUP

DISABILITY RELATED TO: [ ]ILLNESS [ ]ACCIDENT [ ]IND. [ ]PREGNANCY [ ]
DATE SYMPTOMS APPEARED, INCEPTION OF PREGNANCY, OR ACCIDENT OCCURRED: / /
OTHER HEALTH COVERAGE? [ ]NO [ ]YES · IDENTIFY

ASSIGNMENT: I hereby assign my insurance benefits to be paid directly to the undersigned physician. I am financially responsible for non-covered services.
SIGNED: (Patient or Parent If Minor) Date:

RELEASE: I authorize the undersigned physician to release any information acquired in the course of my examination or treatment.
SIGNED: (Patient or Parent If Minor) Date:

Family Practice

✓ DESCRIPTION		CPT4/MO		FEE	✓ DESCRIPTION		CPT4/MO	FEE	✓ DESCRIPTION		CPT4/MO	FEE
1. OFFICE VISIT	NEW	EST.			**3. HOSP. SERVICES**	NEW	EST.		**9. LABORATORY — IN OFFICE**			
Minimal		90030			Interm.(days)	90215	90260		Urine		81000	
Brief	90000	90040			Extended		90270		Occult Blood		89205	
Limited	90010	90050			Comprehensive		90220		ECG		93000	
Intermediate	90015	90060			Discharge 30 min. - 1 hr.							
Extended		90070			Detention Time 30 min. - 1 hr.		99150					
Comprehensive	90020	90080			Detention Time ___ Hrs.		99151		**10. SURGERY**			
									Anoscopy		46600	
2. INJECTIONS & IMMUNIZATIONS					**4. SPECIAL SERVICES**				Sigmoidoscopy		45355	
Surgical Injection		206			Called to ER during ofc. hrs		99065					
DPT		90701			Night Call - before 10 pm		99050					
DT		90702			Night Call - after 10 pm		99052					
Tetanus		90703			Sundays or Holidays		99054		Surgery Assist		-80	
OPV		90712			**5. EMERGENCY ROOM**							
MMR		90707					905		**11. MISCELLANEOUS**			
					6. HOUSE CALLS				Booklets		99071	
							901		Special Reports		99080	
					7. EXTENDED CARE FACILITY				Supplies, Ace Bandage		99070	
							903		X-Ray			
					8. CONSULTATION							
							906					

DIAGNOSIS	ICD-9						
☐ Abscess		☐ Chenyloma Accuminata	078.1	☐ Hemorrhoids	455.6	☐ Paroxysmal Atrial Tachy.	427.2
☐ Abrasion-sup. Injury	682.9	☐ Conjunctivitis	372	☐ Hypertension	401.9	☐ Pediculosis Pubis-Scabies	133.0
☐ Allergic Reaction	919	☐ Contusion, Hematoma	924.9	☐ Influenza	487.1	☐ Pelvic Congestion	625.5
☐ Amenorrhea	995.3	☐ Coronary Artery Dis.	414.9	☐ Ingrown Toenail	703.0	☐ Pharyngitis, Tonsil.-Acute	462
☐ Anemia	626.0	☐ Cystitis-Pyeloneph.	595.9/590.80	☐ Insomnia	780.52	☐ Pigmented Nevus	M8720/0
☐ Anxiety-Stress-Depression	285.9	☐ Cephalge-Migraine Tension	784.0	☐ Irritable Colon	564.1	☐ Post Nasal Drip	473.9
☐ Arteriosclerosis	309	☐ Dermatitis	692.0	☐ Jaundice-Hepatitis	782.4	☐ Pneumonitis-Pleuritis	486
☐ Arthralgia	440.9	☐ Diabetes Mellitus	250.0	☐ Labyrinth.-Vertigo	386.30/780.4	☐ P.I.D.	614.9
☐ Arthrit.-Osteo Rheum.	719.4	☐ Duodenal Ulcer	532.9	☐ Laryngo-Tracheitis	464	☐ Prostatitis	601.9
☐ Asthma Hayfever	716/714	☐ Duodenitis-Gas	535.6	☐ Lipoma	214.9	☐ Puncture Wound	879.8
☐ Bleeding Internal	493.0	☐ Dysmenorrhea	625.3	☐ Lipid Cholesterol Ab	272.7	☐ Renal Stone	592.0
☐ Bleeding Post Men	626.6	☐ Emphysema-COPD	496	☐ Low Back Pain	724.2	☐ Sebaceous Cyst	706.2
☐ Boil-Carbuncle/Furuncle	627.1	☐ Epicondylitis	726.32	☐ Lymphadenitis	289.3	☐ Seizure Disorder	780.3
☐ Bronchitis-Acute/Chronic	680	☐ Epilepsy	345.9	☐ Lymphangitis	457.2	☐ Sinusitis	473.9
☐ Bronchopneumonitis	490	☐ Eustachian Tube Congest.	381.50	☐ Menorrhagia	626.2	☐ Thyroid Disorder	246.9
☐ Bursitis	485	☐ Exogenous Obesity	278.0	☐ Menopausal Syndrome	627.2	☐ Tinea Corpus-Pedis	110.5
☐ Cellulitis-Impetigo	727	☐ Fatigue	780.7	☐ Myofascitis-Tendonitis	729.1	☐ URI-Viral Syndrome	460
☐ Cerebral Con.	682	☐ Foreign Body	879.8	☐ Muscle Strain/Sprain	848.9	☐ Urethritis-Cystitis	599.0
☐ Cervicitis	850.9	☐ Gastroenteritis	558.9	☐ Otitis-External Cerumen	382.9	☐ Vaginitis	616.10
☐ Cholecyst. & Cholelith.	616.0	☐ Heart Failure	428.0	☐ Otitis-Media Acute	382.9	☐ Weight Loss	783.2
	575/574	☐ Hiatal Hernia	553.3	☐ Pain		☐	

DIAGNOSIS: (IF NOT CHECKED ABOVE)

SERVICES PERFORMED AT: [ ]Office [ ]Johnnson's Hospital [ ]E.R. 100 Main St. [ ]N.H. Some Place, USA 70000

LAB SENT TO: [ ]State Hospital 200 State St. Some Place, USA 7000

DATES DISABLED: FROM: / / TO: / /
OK TO RETURN TO WORK/SCHOOL / /

RETURN APPOINTMENT INFORMATION: 5 - 10 - 15 - 20 - 30 - 45 - 60
DAYS WKS. MOS. PRN PX.

NEXT APPOINTMENT: M — T — W — TH — F — S
DATE: / / TIME: ___ AM PM

DOCTOR'S SIGNATURE/DATE

INSTRUCTIONS TO PATIENT FOR FILING INSURANCE CLAIMS
1. COMPLETE UPPER PORTION OF THIS FORM.
2. SIGN & DATE.
3. MAIL THIS FORM DIRECTLY TO YOUR INSURANCE COMPANY. YOU MAY ATTACH YOUR OWN INSURANCE COMPANY'S FORM IF YOU WISH, ALTHOUGH IT IS NOT NECESSARY.

ACCEPT ASSIGNMENT? [ ]YES [ ]NO
REC'D BY: [ ]CASH [ ]CR. CD [ ]CHECK

TOTAL TODAY'S FEE	
OLD BALANCE	
TOTAL	
AMT. REC'D. TODAY	
NEW BALANCE	

INSUR-A-BILLS • SIBBERO SYSTEMS, INC. • PETALUMA, CA. • © 8/84

FIGURE 4-2
Sample health care form.

The diagnosis checklist contains categories specified by the American Medical Association or government publications (including official code numbers). Mediforms offers a variety of forms, according to specialty, to reflect the illnesses and treatments most relevant to the particular doctor. The forms are personalized to include the doctor's name and address, the nature of the doctor's practice, and the hospitals or clinics at which the doctor performs services. Doctors may use the checklists provided or may customize their own checklists, which most doctors choose to do. The forms also contain brief instructions for filling in each blank and instructions explaining how to obtain insurance reimbursement. The forms were widely praised in the industry and were copied by competitors and individual doctors. Are the forms copyrightable? Should Mediforms be entitled at least to recoup its investment in developing the forms?

≡ ### *Morrissey v. Procter & Gamble*
≡ *United States Court of Appeals for the First Circuit*
≡ *379 F.2d 675 (1st Cir. 1967)*

ALDRICH, Chief Judge.

This is an appeal from a summary judgment for the defendant. The plaintiff, Morrissey, is the copyright owner of a set of rules for a sales promotional contest of the "sweepstakes" type involving the social security numbers of the participants. Plaintiff alleges that the defendant, Procter & Gamble Company, infringed, by copying almost precisely, Rule 1. In its motion for summary judgment, based upon affidavits and depositions, defendant denies that plaintiff's Rule 1 is copyrightable material, and denies access. The district court held for the defendant on both grounds. . . .

The second aspect of the case raises a more difficult question. Before discussing it we recite plaintiff's Rule 1, and defendant's Rule 1, the italicizing in the latter being ours to note the defendant's variations or changes.

> 1. Entrants should print name, address and social security number on a boxtop, or a plain paper. Entries must be accompanied by . . . boxtop or by plain paper on which the name . . . is copied from any source. Official rules are explained on . . . packages or leaflets obtained from dealer. If you do not have a social security number you may use the name and number of any member of your immediate family living with you. Only the person named on the entry will be deemed an entrant and may qualify for prize.
>
> Use the correct social security number belonging to the person named on entry. . . . [A] wrong number will be disqualified.

(Plaintiff's Rule)

> 1. Entrants should print name, address and Social Security number on a Tide boxtop, or *on* [a] plain paper. Entries must be accompanied by Tide boxtop *(any size)* or by plain paper on which the name "Tide" is copied from any source. Official rules are *available* on Tide Sweepstakes packages, or *on* leaflets *at* Tide dealers, *or you can send a stamped, self-addressed envelope to:* Tide "Shopping Fling" Sweepstakes, P.O. Box 4459, Chicago 77, Illinois.

If you do not have a Social Security number, you may use the name and number of any member of your immediate family living with you. Only the person named on the entry will be deemed an entrant and may qualify for a prize.

Use the correct Social Security number, belonging to the person named on *the* entry — wrong numbers will be disqualified.

(Defendant's Rule)

The district court, following an earlier decision, Gaye v. Gillis, D. Mass., 1958, 167 F. Supp. 416, took the position that since the substance of the contest was not copyrightable, which is unquestionably correct, Baker v. Selden, 1879, 101 U.S. 99, 25 L. Ed. 841; Affiliated Enterprises v. Gruber, 1 Cir., 1936, 86 F.2d 958; Chamberlin v. Uris Sales Corp., 2 Cir., 1945, 150 F.2d 512, and the substance was relatively simple, it must follow that plaintiff's rule sprung directly from the substance and "contains no original creative authorship." 262 F. Supp. at 738. This does not follow. Copyright attaches to form of expression, and defendant's own proof, introduced to deluge the court on the issue of access, itself established that there was more than one way of expressing even this simple substance. Nor, in view of the almost precise similarity of the two rules, could defendant successfully invoke the principle of a stringent standard for showing infringement which some courts apply when the subject matter involved admits of little variation in form of expression. . . .

Nonetheless, we must hold for the defendant. When the uncopyrightable subject matter is very narrow, so that "the topic necessarily requires," Sampson & Murdock Co. v. Seaver-Radford Co., 1 Cir., 1905, 140 F. 539, 541; cf. Kaplan, An Unhurried View of Copyright, 64-65 (1967), if not only one form of expression, at best only a limited number, to permit copyrighting would mean that a party or parties, by copyrighting a mere handful of forms, could exhaust all possibilities of future use of the substance. In such circumstances it does not seem accurate to say that any particular form of expression comes from the subject matter. However, it is necessary to say that the subject matter would be appropriated by permitting the copyrighting of its expression. We cannot recognize copyright as a game of chess in which the public can be checkmated. Cf. Baker v. Selden, supra.

Upon examination the matters embraced in Rule 1 are so straightforward and simple that we find this limiting principle to be applicable. Furthermore, its operation need not await an attempt to copyright all possible forms. It cannot be only the last form of expression which is to be condemned, as completing defendant's exclusion from the substance. Rather, in these circumstances, we hold that copyright does not extend to the subject matter at all, and plaintiff cannot complain even if his particular expression was deliberately adopted.

Affirmed.

COMMENTS AND QUESTIONS

1. The *Morrissey* case applies what has come to be known as the "merger" doctrine: When there is only one or but a few ways of expressing an idea, then courts will find that the idea behind the work *merges* with its expression and the work is not copyrightable. The doctrine is an extension of the basic rationale of Baker v. Selden.

2. Some commentators have questioned whether there is any idea, system, or method that cannot be expressed in a variety of ways. *See* 1 Nimmer on Copyright §2.18[D] (1988). Is this true of the system in Baker v. Selden? the sweepstakes in *Morrissey?* Could you express the rules of the sweepstakes in a different manner than the original?

The application of Baker v. Selden and the merger doctrine has become particularly important in computer software cases where subsequent computer manufacturers seek to develop computers offering compatible operating systems and subsequent programmers seek to develop application programs featuring the same menu command structure of popular programs. We study these cases in Chapter 7.

3. Application of the merger doctrine involves the same "levels of abstraction" problem described earlier. Characterizing the "idea" of a work at a lower level of abstraction brings the idea in close alignment with the expression, rendering that expression unprotectable. Logically, the doctrine follows the form, "Categories (or genuses) are unprotectable; instances (or species) are protectable." The game then becomes to define what is the category or genus to which the work belongs. The more fine-grained and precise the definition of the category or genus, the more likely that the "expression" alleged to have been infringed is in fact an "idea."

Some courts have softened the harsh consequences of finding merger (i.e., no protection against copying) by finding that works with limited means of expression are copyrightable but that the scope of copyright protection for such works is "thin." In Continental Casualty Co. v. Beardsley, 253 F.2d 702 (2d Cir. 1958), *cert. denied,* 358 U.S. 816 (1958), the Second Circuit determined that the forms developed by the creator of new type of insurance covering lost securities were copyrightable. The forms in question included explanatory information. The court reasoned that

> in the fields of insurance and commerce the use of specific language in forms and documents may be so essential to accomplish a desired result and so integrated with the use of a legal or commercial conception that the proper standard of infringement is one which will protect as far as possible the copyrighted language and yet allow free use of the thought beneath the language.

Since the competitor copied only the forms and not the explanatory language, however, the court determined that no infringement had occurred. Does this approach better comport with the objectives and structure of copyright law?

Or might it force subsequent authors to use awkward variations in expression in order to make use of an unprotectable idea?

4. *Historical Facts and Research.* Copyright law does not protect historical facts, on the ground that such information is not an original work of authorship. Some courts have extended this doctrine to deny copyright protection for historical research. In Miller v. Universal Studios, 650 F.2d 1365 (5th Cir. 1981), an investigative reporter spent more than 2,500 hours researching a bizarre kidnapping and rescue in which the victim had been buried alive in an underground coffin for 5 days. The researcher and the victim published a book describing the events. After efforts to obtain movie rights for the book failed, Universal Studios proceeded to produce a film based largely upon the book. The Fifth Circuit held "the valuable distinction in copyright between facts and the expression of facts cannot be maintained if research is held to be copyrightable." 650 F.2d at 1365.

Should copyright deny protection for the discoveries of historians? Is this consistent with the incentive basis for copyright protection? What arguments can be made in defense of the doctrine?

To what extent is the selection and arrangement of facts protectable under copyright? What about an original theory interpreting historical research? These questions have been the subject of extensive scholarly debate. See Jane C. Ginsburg, Sabotaging and Reconstructing History: A Comment on the Scope of Copyright Protection in Works of History After Hoehling v. Universal Studios, 29 J. Copyright Soc'y 647 (1982); Robert Denicola, Copyright in Collections of Facts: A Theory for the Protection of Nonfiction Literary Works, 81 Colum. L. Rev. 516 (1981); Robert Gorman, Copyright Protection for the Collection and Representation of Facts, 76 Harv. L. Rev. 1569 (1963).

5. *Scenes à Faire.* Copyright protection does not extend to the "incidents, characters or settings which are as a practical matter indispensable, or at least standard, in the treatment of a given topic." Atari, Inc. v. North American Phillips Consumer Electronics, 672 F.2d 607, 616 (7th Cir. 1982). To allow protection for such aspects of a work would unduly restrict subsequent authors in building their own works within general settings with which their audiences will relate.

PROBLEMS

Problem 4-5. Insect Representations, Inc., sells a highly successful line of jewelry in the shape of various insects. One of its products is a stickpin in the shape of a bumblebee, made of gold veining and encrusted with a number of white gems. Animal Jewelry Corp. designs and markets a jewelled bee pin that also has gold veining and is encrusted with gems. However, Animal uses a different number of total gems, and some of its gems are colors other than white. When Insect sues Animal for copyright infringement, Animal defends on the grounds that it has taken merely the "idea" of a jewelled bee pin. Who should prevail?

Problem 4-6. Harry Historian had always been curious about the cause of the Hindenberg disaster, the explosion of a German zeppelin in 1938. After carefully investigating records and news accounts of the disaster and interviewing witnesses, Harry concluded that the disaster was caused by a disgruntled crew member who sabotaged the dirigible so as to embarrass the Nazi regime. He then wrote a book developing his hypothesis. The book contained rich descriptions of the events leading up to the disaster, including detailed accounts of German beer hall revelry and the passionate patriotism of German nationals (as expressed in their enthusiastic singing of the German national anthem). Without obtaining the movie rights to Harry's book, Capitalistic Studios produced a movie of the disaster which featured a crewman-saboteur and many of the richly detailed scenes in Harry's book. Does Harry have a valid copyright infringement claim? Should he have such a claim?

b. The Useful Article Doctrine

Section 101 of the Copyright Act defines "Pictorial, graphic, and sculptural [PGS] works," one of the categories of works protected under section 102, to include

> two-dimensional and three-dimensional works of fine, graphic, and applied art, photographs, prints and art reproductions, maps, globes, charts, diagrams, models, and technical drawings, including architectual plans. Such works shall include works of artistic craftsmanship insofar as their form but not their mechanical or utilitarian aspects are concerned; the design of a useful article, as defined in this section, shall be considered a pictorial, graphic, or sculptural work only if, and only to the extent that, such design incorporates pictorial, graphic, or sculptural features that can be identified separately from, and are capable of existing independently of, the utilitarian aspects of the article.

The act defines a "useful article" as an "article having an intrinsic utilitarian function that is not merely to portray the appearance of the article or to convey information. An article that is normally a part of a useful article is considered a 'useful article.' " 17 U.S.C. §101.

H.R. Rep. No. 94-1476, 94th Cong., 2d Sess., 47, 54-55 (1976)

[T]he definition of "pictorial, graphic, and sculptural works" carries with it no implied criterion of artistic taste, aesthetic value, or intrinsic quality.

The term is intended to comprise not only "works of art" in the traditional sense but also . . . works of "applied art." . . .

In accordance with the Supreme Court's decision in Mazer v. Stein, 347 U.S. 201 (1954), works of "applied art" encompass all original pictorial, graphic, and sculptural works that are intended to be or have been embodied in useful articles, regardless of factors such as mass production, commercial exploitation, and the potential availability of design patent protection. . . .

The Committee has added language to the definition of "pictorial, graphic, and sculptural works" in an effort to make clearer the distinction between works of applied art protectable under the bill and industrial designs not subject to copyright protection. The declaration that "pictorial, graphic, and sculptural works" include "works of artistic craftsmanship insofar as their form but not their mechanical or utilitarian aspects are concerned" is classic language: it is drawn from Copyright Office regulations promulgated in the 1940's and expressly endorsed by the Supreme Court in the *Mazer* case.

The second part of the amendment states that "the design of a useful article . . . shall be considered a pictorial, graphic, or sculptural work only if, and only to the extent that, such design incorporates pictorial, graphic, or sculptural features that can be identified separately from, and are capable of existing independently of, the utilitarian aspects of the article." . . .

In adopting this amendatory language, the Committee is seeking to draw as clear a line as possible between copyrightable works of applied art and uncopyrighted works of industrial design. A two-dimensional painting, drawing, or graphic work is still capable of being identified as such when it is printed on or applied to utilitarian articles such as textile fabrics, wallpaper, containers, and the like. The same is true when a statue or carving is used to embellish an industrial product or, as in the *Mazer* case, is incorporated into a product without losing its ability to exist independently as a work of art. On the other hand, although the shape of an industrial product may be aesthetically satisfying and valuable, the Committee's intention is not to offer it copyright protection under the bill. Unless the shape of an automobile, airplane, ladies' dress, food processor, television set, or any other industrial product contains some element that, physically or conceptually, can be identified as separable from the utilitarian aspects of that article, the design would not be copyrighted under the bill. The test of separability and independence from "the utilitarian aspects of the article" does not depend upon the nature of the design — that is, even if the appearance of an article is determined by aesthetic (as opposed to functional) considerations, only elements, if any, which can be identified separately from the useful article as such are copyrightable. And, even if the three-dimensional design contains some such element (for example, a carving on the back of a chair or a floral relief design on silver flatware), copyright protection would extend only to that element, and would not cover the over-all configuration of the utilitarian article as such.

Brandir International Inc. v. Cascade Pacific Lumber Co.
United States Court of Appeals for the Second Circuit
834 F.2d 1142 (2d Cir. 1987)

OAKES, Circuit Judge:

In passing the Copyright Act of 1976 Congress attempted to distinguish between protectable "works of applied art" and "industrial designs not subject to copyright protection." See H.R. Rep. No. 1476, 94th Cong., 2d Sess. 54, reprinted in 1976 U.S. Code Cong. & Admin. News 5659, 5667 (hereinafter H.R. Rep. No. 1476). The courts, however, have had difficulty framing tests by which the fine line establishing what is and what is not copyrightable can be drawn. Once again we are called upon to draw such a line, this time in a case involving the "RIBBON Rack," a bicycle rack made of bent tubing that is said to have originated from a wire sculpture. [Figure 4-3] . . . The Register of Copyright, named as a third-party defendant under the statute, 17 U.S.C. §411, but electing not to appear, denied copyrightability. In the subsequent suit brought in the United States District Court for the Southern District of New York, Charles S. Haight, Jr., Judge, the district court granted summary judgment on . . . the copyright claim. . . .

Against the history of copyright protection well set out in the majority opinion in Carol Barnhart Inc. v. Economy Cover Corp., 773 F.2d 411, 415-18 (2d Cir. 1985), and in Denicola, Applied Art and Industrial Design: A Suggested Approach to Copyright in Useful Articles, 67 Minn. L. Rev. 707, 709-17 (1983), Congress adopted the Copyright Act of 1976. . . .

As courts and commentators have come to realize, however, the line Congress attempted to draw between copyrightable art and noncopyrightable design "was neither clear nor new." Denicola, supra, 67 Minn. L. Rev. at 720. One aspect of the distinction that has drawn considerable attention is the reference in the House Report to "physically *or conceptually*" (emphasis added) separable elements. . . .

. . . As Judge Newman's dissent made clear, the *Carol Barnhart* majority did not dispute "that 'conceptual separability' is distinct from 'physical sepability' and, when present, entitles the creator of a useful article to a copyright on its design." 773 F.2d at 420.

"Conceptual separability" is thus alive and well, at least in this circuit. The problem, however, is determining exactly what it is and how it is to be applied. Judge Newman's illuminating discussion in dissent in Carol Barnhart, see 773 F.2d at 419-24, proposed a test that aesthetic features are conceptually separable if "the article . . . stimulate[s] in the mind of the beholder a concept that is separate from the concept evoked by its utilitarian function." Id. at 422. This approach has received favorable endorsement by at least one commentator, W. Patry, Latman's The Copyright Law 43-45 (6th ed. 1986), who calls Judge Newman's test the "temporal displacement" test. It is to be

FIGURE 4-3
Brandir Ribbon bicycle rack.

distinguished from other possible ways in which conceptual separability can be tested, including whether the primary use is as a utilitarian article as opposed to an artistic work, whether the aesthetic aspects of the work can be said to be "primary," and whether the article is marketable as art, none of which is very satisfactory. But Judge Newman's test was rejected outright by the majority as "a standard so ethereal as to amount to a 'nontest' that would be extremely difficult, if not impossible, to administer or apply." 773 F.2d at 419 n.5.

Perhaps the differences between the majority and the dissent in *Carol Barnhart* might have been resolved had they had before them the Denicola article on Applied Art and Industrial Design: A Suggested Approach to Copyright in Useful Articles, supra. There, Professor Denicola points out that although the Copyright Act of 1976 was an effort " 'to draw as clear a line as possible,' " in truth "there is no line, but merely a spectrum of forms and shapes responsive in varying degrees to utilitarian concerns." 67 Minn. L. Rev. at 741. Denicola argues that "the statutory directive requires a distinction between works of industrial design and works whose origins lie outside the design process, despite the utilitarian environment in which they appear." He views the statutory limitation of copyrightability as "an attempt to identify elements whose form and appearance reflect the unconstrained perspective of the artist," such features not being the product of industrial design. Id. at 742. "Copyrightability, therefore, should turn on the relationship between the proffered work and the process of industrial design." Id. at 741. He suggests that "the dominant characteristic of industrial design is the influence of nonaesthetic, utilitarian concerns" and hence concludes that copyrightability "ultimately should depend on the extent to which the work reflects artistic expression uninhibited by functional considerations."[2] Id. To state the Denicola test in the language of conceptual separability, if design elements reflect a merger of aesthetic and functional considerations, the artistic aspects of a work cannot be said to be conceptually separable from the utilitarian elements. Conversely, where design elements can be identified as reflecting the designer's artistic judgment exercised independently of functional influences, conceptual separability exists.

We believe that Professor Denicola's approach provides the best test for conceptual separability and, accordingly, adopt it here for several reasons. First, the approach is consistent with the holdings of our previous cases. . . . Second, the test's emphasis on the influence of utilitarian concerns in the design process may help, as Denicola notes, to "alleviate the de facto discrimination against nonrepresentational art that has regrettably accompanied much of the current analysis." Id. at 745. Finally, and perhaps most importantly, we think Denicola's test will not be too difficult to administer in practice. The work itself will continue to give "mute testimony" of its origins. In addition, the parties will be required to present evidence relating to the design process and the nature of the work, with the trier of fact making the determination whether the aesthetic design elements are significantly influenced by functional considerations.

Turning now to the facts of this case, we note first that Brandir contends, and its chief owner David Levine testified, that the original design of the

2. Professor Denicola rejects the exclusion of all works created with some utilitarian application in view, for that would not only overturn Mazer v. Stein, 347 U.S. 201, 98 L. Ed. 630, 74 S. Ct. 460 (1954), on which much of the legislation is based, but also "a host of other eminently sensible decisions, in favor of an intractable factual inquiry of questionable relevance." 67 Minn. L. Rev. at 741. He adds that "any such categorical approach would also undermine the legislative determination to preserve an artist's ability to exploit utilitarian markets." Id. (citing 17 U.S.C. §113(a) (1976)).

RIBBON Rack stemmed from wire sculptures that Levine had created, each formed from one continuous undulating piece of wire. These sculptures were, he said, created and displayed in his home as a means of personal expression, but apparently were never sold or displayed elsewhere. He also created a wire sculpture in the shape of a bicycle and states that he did not give any thought to the utilitarian application of any of his sculptures until he accidentally juxtaposed the bicycle sculpture with one of the self-standing wire sculptures. It was not until November 1978 that Levine seriously began pursuing the utilitarian application of his sculptures, when a friend, G. Duff Bailey, a bicycle buff and author of numerous articles about urban cycling, was at Levine's home and informed him that the sculptures would make excellent bicycle racks, permitting bicycles to be parked under the overloops as well as on top of the underloops. Following this meeting, Levine met several times with Bailey and others, completing the designs for the RIBBON Rack by the use of a vacuum cleaner hose, and submitting his drawings to a fabricator complete with dimensions. The Brandir RIBBON Rack began being nationally advertised and promoted for sale in September 1979.

In November 1982 Levine discovered that another company, Cascade Pacific Lumber Co., was selling a similar product. Thereafter, beginning in December 1982, a copyright notice was placed on all RIBBON Racks before shipment and on December 10, 1982, five copyright applications for registration were submitted to the Copyright Office. The Copyright Office refused registration by letter, stating that the RIBBON Rack did not contain any element that was "capable of independent existence as a copyrightable pictorial, graphic or sculptural work apart from the shape of the useful article." An appeal to the Copyright Office was denied by letter dated March 23, 1983, refusing registration on the above ground and alternatively on the ground that the design lacked originality, consisting of "nothing more than a familiar public domain symbol." In February 1984, after the denial of the second appeal of the examiner's decision, Brandir sent letters to customers enclosing copyright notices to be placed on racks sold prior to December 1982.

Between September 1979 and August 1982 Brandir spent some $38,500 for advertising and promoting the RIBBON Rack, including some 85,000 pieces of promotional literature to architects and landscape architects. Additionally, since October 1982 Brandir has spent some $66,000, including full-, half-, and quarter-page advertisements in architectural magazines such as Landscape Architecture, Progressive Architecture, and Architectural Record, indeed winning an advertising award from Progressive Architecture in January 1983. The RIBBON Rack has been featured in Popular Science, Art and Architecture, and Design 384 magazines, and it won an Industrial Designers Society of America design award in the spring of 1980. In the spring of 1984 the RIBBON Rack was selected from 200 designs to be included among 77 of the designs exhibited at the Katonah Gallery in an exhibition entitled "The Product of Design: An Exploration of the Industrial Design Process," an exhibition that was written up in the New York Times.

Sales of the RIBBON Rack from September 1979 through January 1985 were in excess of $1,367,000. Prior to the time Cascade Pacific began offering for sale its bicycle rack in August 1982, Brandir's sales were $436,000. The price of the RIBBON Rack ranges from $395 up to $2,025 for a stainless steel model and generally depends on the size of the rack, one of the most popular being the RB-7, selling for $485.

Applying Professor Denicola's test to the RIBBON Rack, we find that the rack is not copyrightable. It seems clear that the form of the rack is influenced in significant measure by utilitarian concerns and thus any aesthetic elements cannot be said to be conceptually separable from the utilitarian elements. This is true even though the sculptures which inspired the RIBBON Rack may well have been — the issue of originality aside — copyrightable.

Brandir argues correctly that a copyrighted work of art does not lose its protected status merely because it subsequently is put to a functional use. The Supreme Court so held in Mazer v. Stein, 347 U.S. 201, 98 L. Ed. 630, 74 S. Ct. 460 (1954), and Congress specifically intended to accept and codify *Mazer* in section 101 of the Copyright Act of 1976. See H.R. Rep. No. 1476 at 54-55. The district court thus erred in ruling that, whatever the RIBBON Rack's origins, Brandir's commercialization of the rack disposed of the issue of its copyrightability.

Had Brandir merely adopted one of the existing sculptures as a bicycle rack, neither the application to a utilitarian end nor commercialization of that use would have caused the object to forfeit its copyrighted status. Comparison of the RIBBON Rack with the earlier sculptures, however, reveals that while the rack may have been derived in part from one or more "works of art," it is in its final form essentially a product of industrial design. In creating the RIBBON Rack, the designer has clearly adapted the original aesthetic elements to accommodate and further a utilitarian purpose. These altered design features of the RIBBON Rack, including the spacesaving, open design achieved by widening the upper loops to permit parking under as well as over the rack's curves, the straightened vertical elements that allow in- and above-ground installation of the rack, the ability to fit all types of bicycles and mopeds, and the heavy-gauged tubular construction of rustproof galvanized steel, are all features that combine to make for a safe, secure, and maintenance-free system of parking bicycles and mopeds. Its undulating shape is said in Progressive Architecture, January 1982, to permit double the storage of conventional bicycle racks. Moreover, the rack is manufactured from 2-3/8-inch standard steam pipe that is bent into form, the six-inch radius of the bends evidently resulting from bending the pipe according to a standard formula that yields bends having a radius equal to three times the nominal internal diameter of the pipe.

Brandir argues that its RIBBON Rack can and should be characterizas a sculptural work of art within the minimalist art movement. Minimalist sculpture's most outstanding feature is said to be its clarity and simplicity, in that it often takes the form of geometric shapes, lines, and forms that are pure and free of ornamentation and void of association. As Brandir's expert put it, "The meaning is to be found in, within, around and outside the work of art, allowing the artistic sensation to be experienced as well as intellectualized." People who use Foley Square in New York City see in the form of minimalist art the "Tilted Arc," which is on the plaza at 26 Federal Plaza. Numerous museums have had exhibitions of such art, and the school of minimalist art has many admirers.[14]

It is unnecessary to determine whether to the art world the RIBBON Rack properly would be considered an example of minimalist sculpture. The result under the copyright statute is not changed. Using the test we have adopted, it is not enough that, to paraphrase Judge Newman, the rack may stimulate in the mind of the reasonable observer a concept separate from the bicycle rack concept. While the RIBBON Rack may be worthy of admiration for its aesthetic qualities alone, it remains nonetheless the product of industrial design. Form and function are inextricably intertwined in the rack, its ultimate design being as much the result of utilitarian pressures as aesthetic choices. Indeed, the visually pleasing proportions and symmetricality of the rack represent design changes made in response to functional concerns. Judging from the awards the rack has received, it would seem in fact that Brandir has achieved with the RIBBON Rack the highest goal of modern industrial design, that is, the harmonious fusion of function and aesthetics. Thus there remains no artistic element of the RIBBON Rack that can be identified as separate and "capable of existing independently of, the utilitarian aspects of the article." Accordingly, we must affirm on the copyright claim. . . .

WINTER, Circuit Judge, concurring in part and dissenting in part:

. . . I respectfully dissent from the majority's discussion and disposition of the copyright claim.

My colleagues, applying an adaptation of Professor Denicola's test, hold that the aesthetic elements of the design of a useful article are not conceptually separable from its utilitarian aspects if "form and function are inextricably intertwined" in the article, and "its ultimate design [is] as much the result of utilitarian pressures as aesthetic choices." Applying that test to the instant matter, they observe that the dispositive fact is that "in creating the Ribbon Rack, [Levine] has clearly adapted the *original* aesthetic elements to accommodate and further a utilitarian purpose." (emphasis added). The grounds of my disagreement are that: (1) my colleagues' adaptation of Professor Denicola's test diminishes the statutory concept of "conceptual separability" to the vanishing point; and (2) their focus on the process or sequence followed by the particular designer makes copyright protection depend upon largely fortuitous circumstances concerning the creation of the design in issue. . . .

[T]he relevant question is whether the design of a useful article, however

14. "Tilted Arc" has itself been the subject of copyright litigation on the issue of "moral rights." See infra. — EDS.

intertwined with the article's utilitarian aspects, causes an ordinary reasonable observer to perceive an aesthetic concept not related to the article's use. The answer to this question is clear in the instant case because any reasonable observer would easily view the Ribbon Rack as an ornamental sculpture.[2] Indeed, there is evidence of actual confusion over whether it is strictly ornamental in the refusal of a building manager to accept delivery until assured by the buyer that the Ribbon Rack was in fact a bicycle rack. Moreover, Brandir has received a request to use the Ribbon Rack as environmental sculpture, and has offered testimony of art experts who claim that the Ribbon Rack may be valued solely for its artistic features. As one of those experts observed: "If one were to place a Ribbon Rack on an island without access, or in a park and surround the work with a barrier, . . . its status as a work of art would be beyond dispute."

My colleagues also allow too much to turn upon the process or sequence of design followed by the designer of the Ribbon Rack. They thus suggest that copyright protection would have been accorded "had Brandir merely adopted . . . as a bicycle rack" an enlarged version of one of David Levine's original sculptures rather than one that had wider upper loops and straightened vertical elements. I cannot agree that copyright protection for the Ribbon Rack turns on whether Levine serendipitously chose the final design of the Ribbon Rack during his initial sculptural musings or whether the original design had to be slightly modified to accommodate bicycles. Copyright protection, which is intended to generate incentives for designers by according property rights in their creations, should not turn on purely fortuitous events. For that reason, the Copyright Act expressly states that the legal test is how the final article is perceived, not how it was developed through various stages. It thus states in pertinent part: "[T]he design of a useful article . . . shall be considered a . . . sculptural work only if, and only to the extent that, such design incorporates . . . *sculptural features that can be identified separately from, and are capable of existing independently of, the utilitarian aspects of the article.*" 17 U.S.C. §101 (1982) (emphasis added). . . .

COMMENTS AND QUESTIONS

1. *Useful Article.* A threshold issue is whether a work is a useful article at all. According to section 101, a useful article is a work "having an intrinsic utilitarian function that is not merely to portray the appearance of the article or to convey information." Thus a drawing of a tablecloth design would not satisfy this definition, but the tablecloth itself would be a useful article. Nonetheless, the definition is difficult to apply as the degree of functionality wanes and the degree of fanciful and artistic expression rises. Consider the following examples:

---

2. The reasonable observer may be forgiven, however, if he or she does not recognize the Ribbon Rack as an example of minimalist art.

- a distinctively decorated toy airplane
- an ornamental fireplace hearth that cannot burn wood
- pet rocks in various shapes fashioned after cute fauna

2. *Physical and/or Conceptual Separability.* In one of the first cases interpreting the useful article doctrine as embodied in the Copyright Act of 1976, the D.C. Circuit took the narrow view that the artistic elements of a useful article (an outdoor lamp) must be physically separable from its overall design to be protectable. The court did not apply any conceptual separability test. Esquire, Inc. v. Ringer, 591 F.2d 796 (D.C. Cir. 1978).[15] Although some courts have followed this approach, see e.g. Norris Indus., Inc. v. International Tel. & Tel. Corp., 212 U.S.P.Q. (BNA) 754 (N.D. Fla. 1981), *aff'd*, 696 F.2d 918 (11th Cir. 1983), *cert. denied*, 464 U.S. 818 (1983), most courts and commentators have rejected such a narrow reading of the statute. As noted in *Brandir*, it is well established in the Second Circuit that physical or conceptual separability is sufficient to establish copyrightability.

The physical separability test is relatively straightforward to apply. An expressive element of a useful article is physically separable if it can stand alone from the article as a whole and if such separation does not impair the utility of the article.

3. *The Conceptual Separability Test.* As the *Brandir* case suggests, the conceptual separability test has been difficult to apply. In addition to the various tests discussed in *Brandir* — the temporal displacement test (Newman dissent in *Barnhart*), Denicola's test (majority in *Brandir*), Winter's test (dissent in *Brandir*) — Paul Goldstein has suggested the following formulation: "[A] pictorial, graphic or sculptural feature incorporated in the design of a useful article is conceptually separable if it can stand on its own as a work of art traditionally conceived, and if the useful article in which it is embodied would be equally useful without it." Paul Goldstein, Copyright §2.5.3.1 (1989). Which test comports best with the statute and its legislative history? Which of these tests will produce the most predictable results? Of the various tests for conceptual separability, which test best promotes innovation in utilitarian works? artistic works?

4. *Relation to Patent and Design Patent Protection.* Contrast the useful article doctrine with the idea-expression dichotomy. Recall that the principal purpose of the idea-expression dichotomy was to channel protection between patent and copyright so as not to undermine the patent system (as the primary mode of protection for utiliarian works, with its higher threshold for protection). Does the useful article doctrine reflect a similar objective? or does it show greater solicitude for the protection of artistic works? If so, what justifies the difference?

---

15. Another curious aspect of the case was the court's reliance upon the provisions and legislative history of the 1976 Act despite the fact that the case involved a work created before the effective date of the act.

How is your analysis affected by Congress' statement in the legislative history to the Copyright Act of 1976 that copyright protection extends to works of "applied art" satisfying the separability requirement regardless of "the potential availability of design patent protection." In re Yardley, 493 F.2d 1389 (C.C.P.A. 1974) (allowing a design patent for a copyrighted work). But cf. 37 CFR §202.10(a) (1989) (Copyright Office regulation disallowing copyright registration for works for which a design patent has issued). Contrast copyright and design patents, supra Chapter 3.G.1, in their protection for ornamental designs. What purpose is served by overlapping protection?

5. *Artistic Value Judgments.* Recall that the Supreme Court warned in *Bleistein* about the dangers of embroiling the courts in artistic value judgments. Is there any way to decide the separability question without rendering such a value judgment? See Alfred C. Yen, Copyright Opinions and Aesthetic Theory, 71 S. Cal. L. Rev. 247 (1998).

6. *Design Protection Legislation.* Congress has considered proposals to protect design by way of copyright at various times during the past century. Title II of the 1976 Copyright Revision Bill would have raised the threshold for originality (by excluding "staple or commonplace" designs) and required registration. Designs would have been protected for a term of ten years. The provision passed the Senate but failed to get out of committee in the House. Is such protection necessary to stimulate innovation in design? If so, what is the appropriate form of protection? See generally J. H. Reichman, Design Protection and the Legislative Agenda, 55 L. & Contemp. Probs. 281 (1992); Ralph Brown, Design Protection: An Overview, 34 U.C.L.A. L. Rev. 1341 (1987); J. H. Reichman, Design Protection in Domestic and Foreign Copyright Law: From the Berne Revision of 1948 to the Copyright Act of 1976, 1983 Duke L.J. 1143.

7. *Sui Generis Protection of Vessel Hull Designs.* Through a plug-molding technique, boat builders can copy hull designs relatively easily. In 1989, the Supreme Court struck down a Florida law protecting boat hull designs as preempted by federal patent law. Bonito Boats, Inc. v. Thunder Craft Boats, Inc., 489 U.S. 141 (1989). Boat designers successfully brought their concerns to Congress. As part of the Digital Millennium Copyright Act of 1998, see infra Chapter 7, Congress afforded designers of original boat hulls ten years of protection. 17 U.S.C. §§1301-32. The Act excludes from such protection designs that are "dictated solely by a utilitarian function of the article that embodies it." §1302(4). In order to secure protection, the designer must register the work within two years of making the design public. Like patent law, the Act affords the owner the exclusive right to make, sell, import, or distribute for sale or any commercial use any hull embodying the design. Like copyright law, §1308 does not impose liability for independently created designs. Should specialized protection be available for other fields of design? Some see the Vessel Hull Design Protection Act as a template for a possible general design protection law.

8. *Other Utilitarian Work.* The utilitarian function exception to copyright protection applies only in the case of a particular type of copyrighted works — "pictorial, graphic and sculptural works." It has no counterpart in literary or musical works, for example. Does this limitation make sense? Do other doctrines, such as the idea-expression dichotomy, serve the same purpose in other contexts?

PROBLEMS

Problem 4-7. The Walt Disney Company markets a line of telephones in the shape of the cartoon characters Mickey and Minnie Mouse. The telephones resemble the characters standing up, with push buttons on the torso and the telephone receiver resting on the hand. Are these designs copyrightable?

Problem 4-8. Armond Artist designs a belt buckle cast in gold and silver which is generally recognized as a work of abstract art. The shape of the buckle is the key artistic component.

The buckles retail for $200 to $6,000 at high-fashion jewelry stores. Some of the designs have been made a part of the permanent collection of the Metropolitan Museum of Art. Is the buckle copyrightable, or is it merely a "useful article"?

Problem 4-9. Versace Fashion Display Inc. has developed a collection of original sculptural forms in the shape of the human torso. They are life-size and anatomically accurate, but without neck, arms, or a back. Versace supplies these forms to clothing retailers for the display of fashion clothing. These forms became popular in the market. Versace customers found the distinctive lines of the forms to be visually attractive and effective in selling blouses, shirts, and sweaters displayed on the forms. Discount Display, Inc., began to notice a decline in their mannequin sales after Versace entered the market. Thereafter, Discount Display expanded its catalog to include forms that it copied from Versace. In response, Versace registers its forms with the Copyright Office and sues Discount Display for copyright infringement.

How should a court resolve this dispute?

---

c. Government Works

Section 105 of the Copyright Act provides that copyright protection is not available "for any work of the United States Government, but the United

States Government is not precluded from receiving and holding copyrights transferred to it by assignment, bequest, or otherwise." Section 101 defines "a work of the United States Government" as "a work prepared by an officer or employee of the United States Government as part of that person's official duties." The House Report notes that this definition should be construed in the same manner as the definition of "work made for hire." We focus on the "work for hire" doctrine in section D.1, infra.

A potential problem arises where a federal government agency commissions a work by an independent contractor. The legislative history states:

> The bill deliberately avoids making any sort of outright, unqualified prohibition against copyright in works prepared under Government contract or grant. There may well be cases where it would be in the public interest to deny copyright in the writings generated by Government research contracts and the like; it can be assumed that, where a government agency commissions a work for its own use merely as an alternative to having one of its own employees prepare the work, the right to secure a private copyright would be withheld. However, there are almost certainly many other cases where the denial of copyright protection would be unfair or would hamper the production and publication of important works. Where, under the particular circumstances, Congress or the agency involved finds that the need to have a work freely available outweighs the need of the private author to secure a copyright, the problem can be dealt with by specific legislation, agency regulations, or contractual restrictions.

H.R. Rep. No. 94-1476, 94th Cong., 2d Sess. 59 (1976).

The Copyright Act does not expressly limit the protectability of works created by state government officers or employees in their offical capacities. Nonetheless courts have held that certain types of government works created by state officials, such as statutes, codes, and judicial opinions, are inherently part of the public domain. These courts have emphasized the inherently public nature of laws and the sense in which the citizens of a state are owners of such works. Limitations on access to such works would undoubtedly raise due process concerns where failure to adhere to such laws or regulations threatens civil or criminal liability. See, e.g., Georgia v. The Harrison Co., 548 F. Supp. 110 (N.D. Ga. 1982), *vacated by agreement of the parties,* 559 F. Supp. 37 (N.D. Ga. 1983) (holding new statutory codification to be in public domain); Building Officials & Code Administrators International v. Code Technology, Inc., 628 F.2d 730 (1st Cir. 1980) (holding building code written by private party and later enacted into law to be in the public domain).

Should works created by the government be copyrightable? What arguments favor affording the government copyright interests in its works? What are the arguments against such protection?

PROBLEM

Problem 4-10. To celebrate the bicentennial of the federal court system, the Administative Office of the United States Courts commissioned KPBS, a public broadcasting station, to create a television series dramatizing famous early federal cases. The production contract assigned all copyright interests in the works to the federal government. The History Channel seeks to air these programs. Must it obtain a license from the federal government? Would your analysis differ if there were no assignment clause? What if a historian employed by the Administrative Office of the United States Courts served as a consultant to the television series? What if federal public officials — including Supreme Court Justices and the Solicitor General — provided commentaries about the famous cases?

---

2. The Domain and Scope of Copyright Protection

17 U.S.C. §102. Subject Matter of Copyright: In General

(a) Copyright protection subsists, in accordance with this title, in original works of authorship fixed in any tangible medium of expression, now known or later developed, from which they can be perceived, reproduced, or otherwise communicated, either directly or with the aid of a machine or device. Works of authorship include the following categories:

    (1)   literary works;
    (2)   musical works, including any accompanying words;
    (3)   dramatic works, including any accompanying music;
    (4)   pantomimes and choreographic works;
    (5)   pictorial, graphic, and sculptural works;
    (6)   motion pictures and other audiovisual works;
    (7)   sound recordings; and
    (8)   architectural works.

H.R. Rep. No. 94-1476, 94th Cong., 2d Sess. (1976)

The second sentence of section 102 lists [eight][16] broad categories which the concept of "works of authorship" is said to "include." The use of the

---

16. The 1976 Act did not separately list architectural works, although the legislative history made clear that copyright extended to an architect's plans and drawings. H.R. Rep. at 55. The category of "architectural works" was added to section 102(a) by the Architectural Works Copyright Protection Act of 1990. — EDS.

"include," as defined in section 101, makes clear that the listing is "illustrative and not limitative," and that the [eight] catgories do not necessarily exhaust the scope of "original works of authorship" that the bill is intended to protect. Rather, the list sets out the general area of copyrightable subject matter, but with sufficient flexibility to free the courts from rigid and outmoded concepts of the scope of particular categories. The items are also overlapping in the sense that a work falling within one class may encompass works coming within some or all of the other categories. . . .

a. Literary Works

Section 101 of the Copyright Act defines "literary works" as "works, other than audiovisual works, expressed in words, numbers, or other verbal or numerical symbols or indicia, regardless of the nature of the material objects, such as books, periodicals, manuscripts, phonorecords, films, tapes, disks, or cards, in which they are embodied."

The legislative history notes that "[t]he term 'literary works' does not connote any criterion of literary merit or qualitative value: it includes catalogs, directories, and similar factual reference, or instructional works and compilations of data. It also includes computer data bases, and computer programs to the extent that they incorporate authorship in the programmer's expression of original ideas, as distinguished from the ideas themselves." Copyright Law Revision, H.R. Rep. No. 94-1476, 94th Cong., 2d Sess. (1976). Although the domain of literary works is quite broad and the originality threshold for protection is low, the Copyright Office regulations state that "words and short phrases such as names, titles, and slogans" are not subject to copyright. 37 CFR §202.1.[17]

The scope of copyright protection for literary works extends not only to the literal text but also to non-literal elements of a work such as its structure, sequence, and organization. Thus a second comer may not circumvent copyright law merely by paraphrasing an original text. As Judge Learned Hand has explained, were the rule otherwise, a "plagiarist would escape liability by immaterial variations" of the copyrighted work. Nichols v. Universal Pictures, 45 F.2d 119, 121 (2d Cir. 1930).

While the words of a story and other expressive elements of its text are clearly protectable, courts have struggled to delineate the scope of protection for other elements of literary works such as fictional characters. As we saw earlier, the scope of copyright protection is limited to expressive content and does not extend to the underlying ideas. To what extent, therefore, should an original literary description of a fictional character, such as James Bond or Superman, limit other authors' use of characters featuring similar attributes?

---

17. Titles, names of characters, and phrases may, however, be protectable under unfair competition law and trademark law. See infra Chapters 5 and 6.

In *Nichols,* Judge Learned Hand provided the following standard for determining the scope of protection for fictional characters:

> [W]e do not doubt that two plays may correspond in plot closely enough for infringement. How far that correspondence must go is another matter. Nor need we hold that the same may not be true as to the characters, quite independently of the "plot" proper, though, as far as we know, such a case has never arisen. If *Twelfth Night* were copyrighted, it is quite possible that a second comer might so closely imitate Sir Toby Belch or Malvolio as to infringe, but it would not be enough that for one of his characters he cast a riotous knight who kept wassail to the discomfort of the household, or a vain and foppish steward who became amorous of his mistress. These would be no more than Shakespeare's "ideas" in the play, as little capable of monopoly as Einstein's Doctrine of Relativity, or Darwin's theory of the Origin of Species. It follows that the less developed the characters, the less they can be copyrighted; that is the penalty an author must bear for marking them too indistinctly.

b. Pictorial, Graphic, and Sculptural Works

Section 101 of the Copyright Act defines "[p]ictorial, graphic, and sculptural works" to include "two-dimensional and three-dimensional works of fine, graphic, and applied art, photographs, prints and art reproductions, maps, globes, charts, diagrams, models, and technical drawings, including architectural plans. . . ." As with literary works, courts are not authorized to judge the artistic merit of the work in deciding whether pictorial, graphic, and sculptural works ought to qualify for copyright protection other than to determine that the threshold of original expression has been attained.

The most significant limitation placed on pictorial, graphic, and sculptural works is the utilitarian function exception. Such works are not protectable to the extent that they have a utilitarian rather than artistic function. Thus the useful article doctrine poses a significant limitation on the scope of protection for sculptural works.

As with literary works, the scope of protection for pictorial, graphic, and sculptural works depends on the degree to which the author has delineated the subjects of the work. In some cases, particularly photographs, drawings and maps, the limited range of expressive choices necessarily limits the scope of protection afforded by copyright law.

PROBLEM

Problem 4-11. The New York Arrows, a professional soccer team, hired Sports Images, an advertising firm, to develop a logo to advertise the team. Sports Images developed the accompanying logo (Figure 4-4).

FIGURE 4-4
Logo for the New York Arrows.

After a dispute with the team about fees, Sports Images submitted an application to the Copyright Office to register the work. Is the work copyrightable?

c. Architectural Works

Under the 1976 Act, achitectural plans in the United States were protected as a species of pictorial, graphic, and sculptural works. The protection afforded architectural works — actual structures — was therefore limited by the useful article doctrine and the idea/expression dichotomy. As provided in the definition of pictorial, graphic, and sculptural works in §101, "the design of a useful article, as defined in this section, shall be considered a pictorial, graphic, or sculptural work only if, and only to the extent that, such design incorporates pictorial, graphic, or sculptural features that can be identified separately from, and are capable of existing independently of, the utilitarian aspects of the article." Hence architectural structures, as opposed to the drawings for them, had little if any protection under copyright law. See Demetriades v. Kaufmann, 690 F. Supp. 658 (S.D.N.Y. 1988) (holding that traced architectural plans infringed the originals, but that the construction of an identical building would not violate a copyright in architectural plans under the principles of *Baker v. Selden*).

The United States became party to the Berne Convention for the Protection of Literary and Artistic Works on March 1, 1989. In order to adhere to Article 2(1) of the convention, Congress passed the Architectural Works Copyright Protection Act of 1990. The act specifically provides protection for "the design of a building as embodied in any tangible medium of expression, including a building, architectural plans, or drawings. The work includes the overall form and elements in the design, but does not include individual standard features." 17 U.S.C. §101. The House Report accompanying the act goes on to state:

. . . By creating a new category of protectable subject matter in new section 102(a)(8), and therefore, by deliberately not encompassing architectural works as pictorial, graphic, or sculptural works in existing section 102(a)(5), the copyrightability of architectural works shall not be evaluated under the separability test applicable to pictorial, graphic, or sculptural works embodied in useful articles. There is considerable scholarly and judicial disagreement over how to apply the separability test, and the principal reason for not treating architectural works as pictorial, graphic, or sculptural works is to avoid entangling architectural works in this disagreement.

The Committee does not suggest, though, that in evaluating the copyrightability or scope of protection for architectural works, the Copyright Office or the courts should ignore functionality. A two-step analysis is envisioned. First, an architectural work should be examined to determine whether there are original design elements present, including overall shape and interior architecture. If such design elements are present, a second step is reached to examine whether the design elements are functionally required. If the design elements are not functionally required, the work is protectable without regard to physical or conceptual separability. As a consequence, contrary to the Committee's report accompanying the 1976 Copyright Act with respect to industrial products, the aesthetically pleasing overall shape of an architectual work would be protected under this bill.

H.R. Rep. No. 101-735, 101st Cong., 2d Sess. 20-21 (1990). Is the distinction that the committee seeks to make between the conceptual separability analysis required for pictorial, graphic, and sculptural works and the two-step functionality limitation for architectural works coherent?

It should be noted that the effective date for protection for architectural works in the United States is December 1, 1990; architectural works produced before that time are governed by the standard for pictorial works. In addition, the protection of section 102(a)(8) is subject to two limitations set forth in section 120:

(a) **Pictorial representations permitted.** — The copyright in an architectural work that has been constructed does not include the right to prevent the making, distributing, or public display of pictures, paintings, photographs, or other pictorial representations of the work, if the building in which the work is embodied is located in or ordinarily visible from a public place.

(b) **Alterations to and destruction of buildings.** — Notwithstanding the provisions of section 106(2), the owners of a building embodying an architectural work may, without the consent of the author or copyright owner of the architectural work, make or authorize the making of alterations to such building, and destroy or authorize the destruction of such building.

COMMENTS AND QUESTIONS

Should an architect be able to copyright a house itself? Obviously, no architect can copyright the myriad *functional* aspects of a house, or of blueprints. Even if she did, it would have little value in many cases, since alleged infringers could persuasively argue that they would have come up with the idea of adding, say, a roof without having seen the copyrighted house. But there are unique aspects of most houses. Does it make sense to protect those specific aspects in the house against those who see the house and decide to imitate it? Has too much protection been extended?

d. Dramatic, Pantomime, and Choreographic Works

The protection afforded these three distinct copyrightable forms is similar, in that each form of protection extends to written or otherwise fixed instructions for performing a work of art. A dramatic work portrays a story by means of dialogue or acting. "It gives direction for performance or actually represents all or a substantial portion of the action as actually occurring rather than merely being narrated or described." U.S. Copyright Office, Compendium II of Copyright Office Practices §431. Distinguishing between literary, musical, and dramatic works can be important in practice. Although the three types of works are often captured in the same form, i.e., written text, performance and display rights may vary depending on whether the work is dramatic or nondramatic. See, e.g., 17 U.S.C. §§110(2), 115.

Pantomime and choreographic works were first brought within copyright law by the 1976 Act. Copyright in such works inheres either in notation — such as Labanotation, a system of symbols for representing movements that can be related to a musical score — or (more commonly) in a film recording. Impromptu, unrecorded dancing is not a protectable work because it is not fixed in a tangible medium of expression.

Many of the rules governing dramatic, pantomime, and choreographic works parallel the law with regard to literary works. For example, copyright protection for choreographic works does not extend to simple dance steps in social dance settings, and protection for pantomimes does not extend to "conventional gestures." Both are akin to the "short phrases" denied protection among literary works. All three forms of work are entitled to protection not only against literal copying, but also against copying of their expressive elements, character, action, and dialogue.

PROBLEM

Problem 4-12. At the 1995 Pan-World Figure Skating Championships in Denver, Colorado, skater Kurt Klutzinovich performed the first series of quadruple lutzes ever seen in competition. ABC sports commentator Bill McKay proclaimed it the "klutz," "Kurt's patented move." At the 1996 Championships, Kurt's coach threatens to bring a copyright infringement action against any other skater who performs a "klutz." Is Kurt's series of moves protectable as a choreographic work? If so, would another skater's performance of a klutz infringe Kurt's copyright?

e. Musical Works and Sound Recordings

Copyright law protects musical works that are written on paper, pressed onto a phonorecord, recorded on audiotape, or otherwise fixed in a tangible medium of expression. The work must be original in its melody, harmony, or rhythm, individually or in combination.

Since 1972, sound recordings have been protectable independently of the musical, dramatic, or literary works which are recorded. Thus the sheet music and lyrics to the song "Hey Jude" are protectable as a *musical composition,* in much the same way a play or short story is protected. But beyond this, the actual recording — the magnetically or electronically recorded version of the song — is protectable as a separate work: *a sound recording.*[18] These two distinct (but related) works might have different owners, or one entity might own both. Importantly, each of these distinct copyrighted works comes with its own set of legal entitlements. Musical compositions, for example, are subject to a compulsory license once they have been released to the public. That is, other recording artists can record their own versions of the copyrighted composition, so long as they pay the statutory fee and follow the statutory procedure laid out in section 115. (When a second musician records a song written and previously released by a first musician, the second is said to "cover" the song; thus this provision is often referred to as the "cover license." You could sell recordings of yourself singing and playing the composition "Hey Jude," so long as you followed the statute.) There is no such compulsory license, however, for the first sound recording. That is, it is illegal for you or anyone else to sell unauthorized copies of the Beatles' original version of "Hey Jude."

---

18. The defintion of "sound recordings" excludes "the sounds accompanying a motion picture or other audiovisual work." 17 U.S.C. §101. Such recordings are protected as part of the motion picture or audiovisual work.

The basis for affording such dual protection, which is a relatively recent addition to the panoply of copyright, is the individuality in expression that inevitably occurs during a particular performance of a prewritten musical work. Thus Elvis Presley, Willie Nelson, and the Pet Shop Boys can obtain separate copyrights in their respective recordings of the copyrighted song "You Are Always on My Mind" because each artist adds individual expression to the way the song is performed. The sound recording copyright protects the individual artist from those who would copy her recorded work, regardless of who owns the copyright to the music that the artist performed. Note, however, that because the sound recording copyright does not extend to the music itself, owners of sound recording copyrights are unable to prevent "nonliteral infringement" (i.e., another group's recording of the same song).

The distinction between musical works and sound recordings is also important with regard to public performance rights. Prior to 1972, sound recordings were not protected by federal copyright law. Any protection that they received was through state statutes or common law. The Sound Recording Act of 1971 extended federal protection to sound recordings. Due to the political might of broadcasters, the Copyright Act does not afford a general public performance right to owners of copyright in sound recordings, although this right does exist for owners of copyright in the underlying work. Recent amendments to the Copyright Act qualify Chapter IV.E.4 of this statement. See infra.

Protection for sound recordings also has the effect of extending copyright protection to impromptu performances if they happen to be recorded and are therefore fixed in a tangible medium of expression.[19]

f. Motion Pictures and Other Audiovisual Works

The Copyright Act defines "audiovisual works" as

> works that consist of a series of related images which are intrinsically intended to be shown by the use of machines or devices such as projectors, viewers, or electronic equipment, together with accompanying sounds, if any, regardless of the nature of the material objects, such as films or tapes, in which the works are embodied.

17 U.S.C. §101. "Motion pictures" are a subset of audiovisual works "consisting of a series of related images, which, when shown in succession, impart an impression of motion, together with accompanying sounds, if any." 17 U.S.C. §101. This definition establishes that for purposes of copyright law sound tracks are an integral part of the motion pictures that they accompany, as opposed to separate musical works.

19. Audiovisual recordings of impromptu performances will have the same effect.

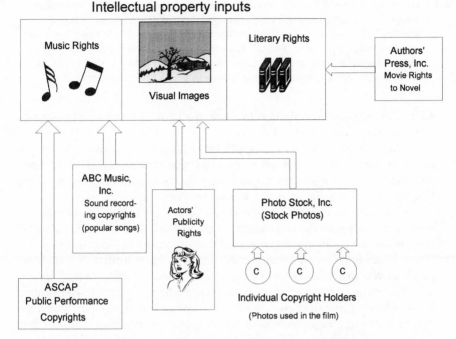

FIGURE 4-5
Intellectual property inputs for a multimedia project.

Protection for audiovisual works is one of the most important issues in intellectual property protection for computer software and multimedia. We discuss this aspect of audiovisual works in detail in Chapter 7.

In some traditional media, and increasingly in the digital era, multiple components are assembled into a single commercial product. As each component is covered by one or more intellectual property rights, this collection of multiple components involves numerous complex transactions. Figure 4-5 illustrates the assembly of a series of "intellectual property inputs" into a commercial multimedia product.

g. Derivative Works and Compilations

Section 103 of the Copyright Act provides protection for derivative works and compilations. The copyright in a derivative work or compilation "extends only to the material contributed by the author of such work, as distinguished from the preexisting material employed in the work." §103(b).

A "derivative work" is a work

based upon one or more preexisting works, such as a translation, musical arrangement, dramatization, fictionalization, motion picture version, sound recording, art reproduction, abridgment, condensation, or any other form in
which a work may be recast, transformed, or adapted. A work consisting of
editorial revisions, annotations, elaborations, or other modifications which, as a
whole, represent an original work of authorship, is a "derivative work."

§101. We explore this definition in detail in section E.2, infra, addressing the
right to make derivative works.

Why should the copyright owner be entitled to a separate copyright in
the derivative work, in addition to the copyright in the original work? There
are at least two possible explanations for granting separate copyright protection to the original elements of a derivative work.

First, derivative works at the end of a "chain" of related works may bear
little resemblance to the original copyrighted work. For example, an author
may produce a children's book, then a movie script based on the book, followed by a movie based on the script, and a series of stuffed animals based
on characters from the movie. The stuffed animals at the end of this "chain"
may bear little if any resemblance to identifiable characters in the original
book. Allowing the derivative works to be copyrighted gives the copyright
holder a stronger argument that each new step in the chain is in fact protectable as a derivative of the prior copyright. A second justification, suggested by Paul Goldstein, is that we may wish to protect new expression
"derived" from works that are already in the public domain.

In addition, because derivative works often capture different markets,
the copyright owner in the original work may wish to license to others the
right to produce derivative works. Just because someone is a successful writer,
for example, does not mean that he will be particularly successful at manufacturing and selling plush toys. Separating the copyright in original elements
of a derivative work may facilitate the division of ownership between the
original author and his licensees. Cf. Gracen v. The Bradford Exchange, 698
F.2d 300 (7th Cir. 1983) (determining whether original copyright owner or
subsequent licensee owns rights to derivative work produced by the licensee).

A "compilation" is a work "formed by the collection and assembling of
preexisting materials or of data that are selected, coordinated, or arranged in
such a way that the resulting work as a whole constitutes an original work of
authorship. The term 'compilation' includes collective works." §101. A "collective work" is a work, "such as a periodical issue, anthology, or encyclopedia, in which a number of contributions, constituting separate and
independent works in themselves, are assembled into a collective whole."
§101.

The level of originality required for a compilation to be copyrightable
has been a contentious area of law. Many fact-based works, such as telephone
directories, require tremendous time, effort, and expense to compile, yet have
little creative content. As we saw in the *Feist* case, however, there must be

"some minimal degree of creativity" to merit copyright protection. Since neither the underlying material being compiled (names, which are factual material) nor the method of arrangement (alphabetical order) were original, the compilation thus created was not copyrightable, regardless of how much effort was involved in producing the directory.

At what point does a collection of uncopyrightable material become copyrightable as a compilation? Consider the following case:

Roth Greeting Cards v. United Card Company
United States Court of Appeals for the Ninth Circuit
429 F.2d 1106 (9th Cir. 1970)

HAMLEY, Circuit Judge:

Roth Greeting Cards (Roth) and United Card Company (United), both corporations, are engaged in the greeting card business. Roth brought this suit against United to recover damages and obtain injunctive relief for copyright infringement of seven studio greeting cards. After a trial to the court without a jury, judgment was entered for defendant. Plaintiff appeals.

Roth's claim involves the production and distribution by United of seven greeting cards which bear a remarkable resemblance to seven of Roth's cards on which copyrights had been granted. Roth employed a writer to develop the textual material for its cards. When Roth's president determined that a textual idea was acceptable, he would integrate that text into a rough layout of a greeting card with his suggested design for the art work. He would then call in the company artist who would make a comprehensive layout of the card. If the card was approved, the artist would do a finished layout and the card would go into production.

During the period just prior to the alleged infringements, United did not have any writers on its payroll. Most of its greeting cards came into fruition primarily through the activities of United's president, Mr. Koenig, and its vice-president, Edward Letwenko.

The source of the art and text of the cards of United, here in question, is unclear. Letwenko was unable to recall the origin of the ideas for most of United's cards. He speculated that the gags used may have come from plant personnel, persons in bars, friends at a party, Koenig, or someone else. He contended that the art work was his own. But he also stated that he visited greeting card stores and gift shows in order to observe what was going on in the greeting card business. Letwenko admitted that he may have seen the Roth cards during these visits or that the Roth cards may have been in his office prior to the time that he did his art work on the United cards. . . .

[T]he trial court found that the art work in plaintiff's greeting cards was copyrightable, but not infringed by defendant. The trial court also found that, although copied by defendant, the wording or textual matter of each of the plaintiff's cards in question consist of common and ordinary English words and phrases which are not original with Roth and were in the public domain prior to first use by plaintiff.

Arguing that the trial court erred in ruling against it on merits, Roth agrees that the textual material involved in their greeting cards may have been in the public domain, but argues that this alone did not end the inquiry into the copyrightability of the entire card. Roth argued that "It is the arrangement of the words, their combination and plan, together with the appropriate art work . . ." which is original, the creation of Roth, and entitled to copyright protection.

In order to be copyrightable, the work must be the original work of the copyright claimant or of his predecessor in interest. M. Nimmer, Copyright (hereafter Nimmer), 10 at 32 (1970). But the originality necessary to support a copyright merely calls for independent creation, not novelty. Alfred Bell & Co., Ltd. v. Catalda Fine Arts, Inc., 191 F.2d 99, 102 (2d Cir. 1951). Cf. Baker v. Selden, 101 U.S. 99, 102-103 (1879).

United argues, and we agree, that there was substantial evidence to support the district court's finding that the textual matter of each card, considered apart from its arrangement on the cards and its association with artistic representations, was not original to Roth and therefore not copyrightable. However, proper analysis of the problem requires that all elements of each card, including text, arrangement of text, art work, and association between art work and text, be considered as a whole.

Considering all of these elements together, the Roth cards are, in our opinion, both original and copyrightable. In reaching this conclusion we recognize that copyright protection is not available for ideas, but only for the tangible expression of ideas. Mazer v. Stein, 347 U.S. 201, 217 (1954). We conclude that each of Roth's cards, considered as a whole, represents a tangible expression of an idea and that such expression was, in totality, created by Roth. See Dorsey v. Old Surety Life Ins. Co., 98 F.2d 872, 873 (10th Cir. 1938).

This brings us to the question of infringement. Greeting cards are protected under 17 U.S.C. 5(a) or (k) as a book, Jackson v. Quickslip Co., Inc., 110 F.2d 731 (2d Cir. 1940), or as a print, 37 C.F.R. 202.14. They are the embodiment of humor, praise, regret or some other message in a pictorial and literary arrangement. As proper subjects of copyright, they are susceptible to infringement in violation of the Act. Detective Comics, Inc. v. Bruns Publications, Inc., 111 F.2d 432 (2d Cir. 1940).

To constitute an infringement under the Act there must be substantial similarity between the infringing work and the work copyrighted; and that similarity must have been caused by the defendant's having copied the copyright holder's creation. The protection is thus against copying — not against any possible infringement caused when an independently created work coincidentally duplicates copyrighted material. Sheldon v. Metro-Goldwyn Pictures Corp., 81 F.2d 49, 54 (2d Cir. 1936).

It appears to us that in total concept and feel the cards of United are the same as the copyrighted cards of Roth. With the possible exception of one

United card (exhibit 6), the characters depicted in the art work, the mood they portrayed, the combination of art work conveying a particular mood with a particular message, and the arrangement of the words on the greeting card are substantially the same as in Roth's cards. In several instances the lettering is also very similar.

It is true, as the trial court found, that each of United's cards employed art work somewhat different from that used in the corresponding Roth cards. However, "The test of infringement is whether the work is recognizable by an ordinary observer as having been taken from the copyrighted source." White-Smith Music Pub. Co. v. Apollo Company, 209 U.S. 1, 17 (1907), Bradbury v. Columbia Broadcasting System, Inc., 287 F.2d 478, 485 (9th Cir. 1961).

The remarkable similarity between the Roth and United cards in issue (with the possible exception of exhibits 5 and 6) is apparent to even a casual observer. For example, one Roth card (exhibit 9) has, on its front, a colored drawing of a cute moppet suppressing a smile and, on the inside, the words "i wuv you." With the exception of minor variations in color and style, defendant's card (exhibit 10) is identical. Likewise, Roth's card entitled "I miss you already," depicts a forlorn boy sitting on a curb weeping, with an inside message reading ". . . and You Haven't even Left . . ." (exhibit 7), is closely paralleled by United's card with the same caption, showing a forlorn and weeping man, and with the identical inside message (exhibit 8). . . .

KILKENNY, Circuit Judge (dissenting).

The majority agrees with a specific finding of the lower court that the words on the cards are not the subject of copyright. By strong implication, it likewise accepts the finding of the trial court that the art work on the cards, although subject to copyright, was not infringed. Thus far, I agree.

I cannot, however, follow the logic of the majority in holding that the uncopyrightable words and the imitated, but not copied art work, constitutes such total composition as to be subject to protection under the copyright laws. The majority concludes that in the overall arrangement of the text, the art work and the association of the art work to the text, the cards were copyrightable and the copyright infringed. This conclusion, as I view it, results in the whole becoming substantially greater than the sum total of its parts. With this conclusion, of course, I cannot agree. . . .

Feeling, as I do, that the copyright act is a grant of limited monopoly to the authors of creative literature and art, I do not think that we should extend a 56-year monopoly in a situation where neither infringement of text, nor infringement of art work can be found. On these facts, we should adhere to our historic philosophy requiring freedom of competition. I would affirm.

COMMENTS AND QUESTIONS

1. Would there be sufficient incentive for authors and publishers to produce greeting cards without the type of protection by the court in *Roth?* On
the contrary, does this decision enable a greeting card company to monopolize most of the stock phrases and images simply by being first?

2. Is *Roth* still good law after *Feist?* Certainly, *Feist* suggests that the
copyright in compilations of facts or other uncopyrightable elements is
"thin." Copyright protection can only extend to the selection or arrangement
of such uncopyrightable material, and then only to the extent such selection
or arrangement was sufficiently creative as to be "original."

PROBLEMS

Problem 4-13. Assume that West Publishing Company, which has
published both official and unofficial reports of federal and state court
decisions for over 100 years, and Mead Data Central are the only competitors in the market for computer legal research databases. Because
cases (especially in the federal courts) are cited by the West's volume
and page number, Mead decides to copy West's pagination in its computer database. West sues, arguing that the arrangement of its cases in
its reporters is copyrighted. Does West have a valid copyright?

Problem 4-14. The Bond News and Investor's World are both
financial reporting services. Each provides to its subscribers a weekly
update of all municipal bonds that have been "called" by the city, and
several pieces of information about the bonds: the bond series, the call
price, the date of the call, and the address and phone number of the
calling agency. Bond obtains this information by having researchers cull
through published notices in 250 newspapers nationwide each day.
Bond, suspecting that Investor's is copying its data rather than conducting a similar search, plants false information in its updates. When
Investor's publishes the same false information in its update, Bond sues.
Investor's admits copying the data but claims that it had a right to do
so because the data were not copyrightable. Who should prevail?

Note on the Rights of Authors and Publishers in
Electronic Compilations

Section 201(c) of the Copyright Act provides that when a copyrighted
work is contributed to a collective work, the copyright in the collective work
(held by its publisher) is separate from the copyright in the component works

(held in the first instance by the authors). The statute provides that unless the parties agree otherwise, "the owner of the copyright in the collective work is presumed to have acquired only the privilege of reproducing and distributing the contribution as part of that particular collective work, any revision of that collective work, and any later collective work in the same series." 17 U.S.C. §201(c).

What happens when a collective work such as a magazine or law review is put on the Internet? At least until recently, most publication contracts didn't discuss this eventuality at all. Nonetheless, publishers have treated the electronic version of their publication as if it were the same as the print version. That is, they have acted as though they had the rights to authorize the online use or reproduction of articles from their magazine.

A number of recent suits filed by authors assert that authors, not publishers, have the right to control the copying and distribution of their articles online. Court decisions have parsed the language of section 201(c) in an effort to find an answer. In Tasini v. New York Times Co., 192 F.3d 356 (2d cir. 1999), the court concluded that putting an entire collective work online was a form of "revision" to the collective work, and therefore the publishers had the right to do it without permission from each author. And in Ryan v. CARL Corp., 23 F. Supp. 2d 1146 (N.D. Cal. 1998), the court held that an information retrieval service was not engaged in revising collective works because it made copies of individual articles, not entire magazine issues. The *Ryan* court concluded that authors — not publishers — controlled the exclusive rights to reproduce particular articles that appeared in the publication, unless the parties agreed to the contrary. It did, however, express some concern whether this result was the most efficient one.

D. OWNERSHIP AND DURATION

Copyrights today are in many ways like other forms of property interests. As we will see in section E, a copyright includes numerous distinct rights. Once a copyright is acquired, the entire bundle of rights may be assigned to others. Alternatively, the owner may divide and transfer particular rights. Nonetheless, there are significant differences between the ownership rules governing copyrights and those governing other property interests in society. The most important difference is that copyrights have limited duration. As required by the U.S. Constitution, copyrights enter the public domain after a statutorily determined "limited time." Furthermore, Congress has differentiated the ownership, transfer, and termination of these rights in other ways as well. As you work through these provisions, consider how copyright ownership rules differ from the ownership rules for other forms of property —intellectual (patents) and real. What justifies these differences?

More specifically, how do these differences promote the objectives of the copyright law?

This section first surveys the law that determines who initially acquires a copyright. This law includes the rules governing works for hire and jointly produced works. We then outline the rules governing the duration of copyright protection. These rules are complicated by the renewal process under the 1909 Act. We conclude with a discussion of the rules governing the division, transfer, and termination of transfers of copyright interests.

1. Initial Ownership of Copyrights

As noted earlier, copyright law since the 1700s has vested copyright in authors and artists who create works satisfying the requirements of copyright law. In the case of an individual who writes, composes, or paints an original work of authorship on her or his own today, the individual acquires the copyright upon creation. 17 U.S.C. §201(a). Many works in our modern and increasingly interconnected economy, however, are not produced in this straightforward manner. Many copyrightable works are produced in the employment context. In addition, many works are the result of collaborations among numerous authors and artists.

a. Works for Hire

Community for Creative Non-Violence et al. v. Reid
Supreme Court of the United States
490 U.S. 730 (1989)

Justice MARSHALL delivered the opinion of the Court.

In this case, an artist and the organization that hired him to produce a sculpture contest the ownership of the copyright in that work. To resolve this dispute, we must construe the "work made for hire" provisions of the Copyright Act of 1976 (Act or 1976 Act), 17 U.S.C. §§101 and 201(b), and in particular, the provision in §101, which defines as a "work made for hire" a "work prepared by an employee within the scope of his or her employment" (hereinafter §101(1)).

Petitioners are the Community for Creative Non-Violence (CCNV), a nonprofit unincorporated association dedicated to eliminating homelessness in America, and Mitch Snyder, a member and trustee of CCNV. In the fall of 1985, CCNV decided to participate in the annual Christmastime Pageant of Peace in Washington, D.C., by sponsoring a display to dramatize the plight of the homeless. As the District Court recounted:

Snyder and fellow CCNV members conceived the idea for the nature of the display: a sculpture of a modern Nativity scene in which, in lieu of the traditional Holy Family, the two adult figures and the infant would appear as contemporary homeless people huddled on a streetside steam grate. The family was to be black (most of the homeless in Washington being black); the figures were to be life-sized, and the steam grate would be positioned atop a platform "pedestal," or base, within which special-effects equipment would be enclosed to emit simulated "steam" through the grid to swirl about the figures. They also settled upon a title for the work — "Third World America" — and a legend for the pedestal: "and still there is no room at the inn." 652 F. Supp. 1453, 1454 (D.C. 1987).

Snyder made inquiries to locate an artist to produce the sculpture. He was referred to respondent James Earl Reid, a Baltimore, Maryland, sculptor. In the course of two telephone calls, Reid agreed to sculpt the three human figures. CCNV agreed to make the steam grate and pedestal for the statue. Reid proposed that the work be cast in bronze, at a total cost of approximately $100,000 and taking six to eight months to complete. Snyder rejected that proposal because CCNV did not have sufficient funds, and because the statue had to be completed by December 12 to be included in the pageant. Reid then suggested, and Snyder agreed, that the sculpture would be made of a material known as "Design Cast 62," a synthetic substance that could meet CCNV's monetary and time constraints, could be tinted to resemble bronze, and could withstand the elements. The parties agreed that the project would cost no more than $15,000, not including Reid's services, which he offered to donate. The parties did not sign a written agreement. Neither party mentioned copyright.

After Reid received an advance of $3,000, he made several sketches of figures in various poses. At Snyder's request, Reid sent CCNV a sketch of a proposed sculpture showing the family in a crechelike setting: the mother seated, cradling a baby in her lap; the father standing behind her, bending over her shoulder to touch the baby's foot. Reid testified that Snyder asked for the sketch to use in raising funds for the sculpture. Snyder testified that it was also for his approval. Reid sought a black family to serve as a model for the sculpture. Upon Snyder's suggestion, Reid visited a family living at CCNV's Washington shelter but decided that only their newly born child was a suitable model. While Reid was in Washington, Snyder took him to see homeless people living on the streets. Snyder pointed out that they tended to recline on steam grates, rather than sit or stand, in order to warm their bodies. From that time on, Reid's sketches contained only reclining figures.

Throughout November and the first two weeks of December 1985, Reid worked exclusively on the statue, assisted at various times by a dozen different people who were paid with funds provided in installments by CCNV. On a number of occasions, CCNV members visited Reid to check on his progress and to coordinate CCNV's construction of the base. CCNV rejected Reid's proposal to use suitcases or shopping bags to hold the family's personal belongings, insisting instead on a shopping cart. Reid and CCNV members did not discuss copyright ownership on any of these visits.

On December 24, 1985, 12 days after the agreed-upon date, Reid delivered the completed statue to Washington. There it was joined to the steam grate and pedestal prepared by CCNV and placed on display near the site of the pageant. Snyder paid Reid the final installment of the $15,000. The statue remained on display for a month. In late January 1986, CCNV members returned it to Reid's studio in Baltimore for minor repairs. Several weeks later, Snyder began making plans to take the statue on a tour of several cities to raise money for the homeless. Reid objected, contending that the Design Cast 62 material was not strong enough to withstand the ambitious itinerary. He urged CCNV to cast the statue in bronze at a cost of $35,000, or to create a master mold at a cost of $5,000. Snyder declined to spend more of CCNV's money on the project.

In March 1986, Snyder asked Reid to return the sculpture. Reid refused. He then filed a certificate of copyright registration for "Third World America" in his name and announced plans to take the sculpture on a more modest tour than the one CCNV had proposed. Snyder, acting in his capacity as CCNV's trustee, immediately filed a competing certificate of copyright registration.

Snyder and CCNV then commenced this action against Reid and his photographer, Ronald Purtee, seeking return of the sculpture and a determination of copyright ownership. The District Court granted a preliminary injunction, ordering the sculpture's return. After a 2-day bench trial, the District Court declared that "Third World America" was a "work made for hire" under §101 of the Copyright Act and that Snyder, as trustee for CCNV, was the exclusive owner of the copyright in the sculpture. 652 F. Supp., at 1457. The court reasoned that Reid had been an "employee" of CCNV within the meaning of §101(1) because CCNV was the motivating force in the statue's production. Snyder and other CCNV members, the court explained, "conceived the idea of a contemporary Nativity scene to contrast with the national celebration of the season," and "directed enough of [Reid's] effort to assure that, in the end, he had produced what they, not he, wanted." Id., at 1456.

The Court of Appeals for the District of Columbia Circuit reversed and remanded.

II

A

The Copyright Act of 1976 provides that copyright ownership "vests initially in the author or authors of the work." 17 U.S.C. §201(a). As a general rule, the author is the party who actually creates the work, that is, the person who translates an idea into a fixed, tangible expression entitled to copyright protection. §102. The Act carves out an important exception, however, for "works made for hire." If the work is for hire, "the employer or other person for whom the work was prepared is considered the author" and owns the copyright, unless there is a written agreement to the contrary. §201(b). Classifying a work as "made for hire" determines not only the initial ownership of its copyright, but also the copyright's duration, §302(c), and the owners' renewal rights, §304(a), termination rights, §203(a), and right to import certain goods bearing the copyright, §601(b)(1). See 1 M. Nimmer & D. Nimmer, Nimmer on Copyright §5.03 [A], pp. 5-10 (1988). The contours of the work for hire doctrine therefore carry profound significance for freelance creators — including artists, writers, photographers, designers, composers, and computer programmers — and for the publishing, advertising, music, and other industries which commission their works.[4]

Section 101 of the 1976 Act provides that a work is "for hire" under two sets of circumstances:

(1)   a work prepared by an employee within the scope of his or her employment; or

(2)   a work specially ordered or commissioned for use as a contribution to a collective work, as a part of a motion picture or other audiovisual work, as a translation, as a supplementary work, as a compilation, as an instructional text, as a test, as answer material for a test, or as an atlas, if the parties expressly agree in a written instrument signed by them that the work shall be considered a work made for hire.

The Petitioners do not claim that the statue satisfies the terms of §101(2). Quite clearly, it does not. Sculpture does not fit within any of the nine categories of "specially ordered or commissioned" works enumerated in that subsection, and no written agreement between the parties establishes "Third World America" as a work for hire.

The dispositive inquiry in this case therefore is whether "Third World America" is "a work prepared by an employee within the scope of his or her employment" under §101(1). The Act does not define these terms. In the absence of such guidance, four interpretations have emerged. The first holds that a work is prepared by an employee whenever the hiring party retains the right to control the product. See Peregrine v. Lauren Corp., 601 F. Supp. 828, 829 (Colo. 1985); Clarkstown v. Reeder, 566 F. Supp. 137, 142

---

4. As of 1955, approximately 40 percent of all copyright registrations were for works for hire, according to a Copyright Office study. See Varmer, Works Made for Hire and On Commission, in Studies Prepared for the Subcommittee on Patents, Trademarks, and Copyrights of the Senate Committee on the Judiciary, Study No. 13, 86th Cong., 2d Sess., 139, (Comm. Print 1960) (hereinafter Varmer, Works Made for Hire). The Copyright Office does not keep more recent statistics on the number of work for hire registrations.

(S.D.N.Y. 1983). Petitioners take this view. Brief for Petitioners 15; Tr. of Oral. Arg. 12. A second, and closely related, view is that a work is prepared by an employee under §101(1) when the hiring party has actually wielded control with respect to the creation of a particular work. This approach was formulated by the Court of Appeals for the Second Circuit, Aldon Accessories Ltd. v. Spiegel, Inc., 738 F.2d 548, *cert. denied,* 469 U.S. 982 (1984), and adopted by the Fourth Circuit, Brunswick Beacon, Inc. v. Schock-Hopchas Publishing Co., 810 F.2d 410 (1987), the Seventh Circuit, Evans Newton, Inc. v. Chicago Systems Software, 793 F.2d 889, *cert. denied,* 479 U.S. 949 (1986), and, at times, by petitioners, Brief for Petitioners 17. A third view is that the term "employee" within §101(1) carries its common-law agency law meaning. This view was endorsed by the Fifth Circuit in Easter Seal Society for Crippled Children and Adults of Louisiana, Inc. v. Playboy Enterprises, 815 F.2d 323 (1987), and by the Court of Appeals below. Finally, respondent and numerous amici curiae contend that the term "employee" only refers to "formal, salaried" employees. See, e.g., Brief for Respondent 23-24; Brief for Register of Copyrights as Amicus Curiae 7. The Court of Appeals for the Ninth Circuit recently adopted this view. See Dumas v. Gommerman, 865 F.2d 1093 (1989).

The starting point for our interpretation of a statute is always its language. Consumer Product Safety Comm'n v. GTE Sylvania, Inc., 447 U.S. 102, 108 (1980). The Act nowhere defines the terms "employee" or "scope of employment." It is, however, well established that "where Congress uses terms that have accumulated settled meaning under . . . the common law, a court must infer, unless the statute otherwise dictates, that Congress means to incorporate the established meaning of these terms." NLRB v. Amax Coal Co., 453 U.S. 322, 329 (1981); see also Perrin v. United States, 444 U.S. 37, 42 (1979). In the past, when Congress has used the term "employee" without defining it, we have concluded that Congress intended to describe the conventional master-servant relationship as understood by common-law agency doctrine. See, e.g., Kelley v. Southern Pacific Co., 419 U.S. 318, 322-323 (1974); Baker v. Texas & Pacific R. Co., 359 U.S. 227, 228 (1959) (per curiam); Robinson v. Baltimore & Ohio R. Co., 237 U.S. 84, 94 (1915). Nothing in the text of the work for hire provisions indicates that Congress used the words "employee" and "employment" to describe anything other than " 'the conventional relation of employer and employee.' " *Kelley,* supra, at 323, quoting *Robinson,* supra, at 94; compare NLRB v. Hearst Publications, Inc., 322 U.S. 111, 124-132 (1944) (rejecting agency law conception of employee for purposes of the National Labor Relations Act where structure and context of statute indicated broader definition). On the contrary, Congress' intent to incorporate the agency law definition is suggested by §101(1)'s use of the term, "scope of employment," a widely used term of art in agency law. See Restatement (Second) of Agency §228 (1958) (hereinafter Restatement).

We thus agree with the Court of Appeals that the term "employee" should be understood in light of the general common law of agency.

In contrast, neither test proposed by petitioners is consistent with the text of the Act. The exclusive focus of the right to control the product test on the relationship between the hiring party and the product clashes with the language of §101(1), which focuses on the relationship between the hired and hiring parties. The right to control the product test also would distort the meaning of the ensuing subsection, §101(2). Section 101 plainly creates two distinct ways in which a work can be deemed for hire: one for works prepared by employees, the other for those specially ordered or commissioned works which fall within one of the nine enumerated categories and are the subject of a written agreement. The right to control the product test ignores this dichotomy by transforming into a work for hire under §101(1) any "specially ordered or commissioned" work that is subject to the supervision and control of the hiring party. Because a party who hires a "specially ordered or commissioned" work by definition has a right to specify the characteristics of the product desired, at the time the commission is accepted, and frequently until it is completed, the right to control the product test would mean that many works that could satisfy §101(2) would already have been deemed works for hire under §101(1). Petitioners' interpretation is particularly hard to square with §101(2)'s enumeration of the nine specific categories of specially ordered or commissioned works eligible to be works for hire, e.g., "a contribution to a collective work," "a part of a motion picture," and "answer material for a test." The unifying feature of these works is that they are usually prepared at the instance, direction, and risk of a publisher or producer. By their very nature, therefore, these types of works would be works by an employee under petitioners' right to control the product test.

We therefore conclude that the language and structure of §101 of the Act do not support either the right to control the product or the actual control approaches.[8] The structure of §101 indicates that a work for hire can arise through one of two mutually exclusive means, one for employees and one for independent contractors, and ordinary canons of statutory interpretation indicate that the classification of a particular hired party should be made with reference to agency law.

This reading of the undefined statutory terms finds considerable support in the Act's legislative history.

---

8. We also reject the suggestion of respondent and amici that the §101(1) term "employee" refers only to formal, salaried employees. While there is some support for such a definition in the legislative history, see Varmer, Works Made for Hire 130; n.11, infra, the language of §101(1) cannot support it. The Act does not say "formal" or "salaried" employee, but simply "employee." Moreover, respondent and those amici who endorse a formal, salaried employee test do not agree upon the content of this test. Compare, e.g., Brief for Respondent 37 (hired party who is on payroll is an employee within §101(1)) with Tr. of Oral Arg. 31 (hired party who receives a salary or commissions regularly is an employee within §101(1)); and Brief for Volunteer Lawyers for the Arts Inc. et al. as Amici Curiae 4 (hired party who receives a salary and is treated as an employee for Social Security and tax purposes is an employee within §101(1)). Even the one Court of Appeals to adopt what it termed a formal, salaried employee test in fact embraced an approach incorporating numerous factors drawn from the agency law definition of employee which we endorse. See Dumas, 865 F.2d, at 1104.

. . . [P]etitioners' construction of the work for hire provisions would impede Congress' paramount goal in revising the 1976 Act of enhancing predictability and certainty of copyright ownership. See H.R. Rep. No. 94-1476, supra, at 129. In a "copyright marketplace," the parties negotiate with an expectation that one of them will own the copyright in the completed work. Dumas, 865 F.2d, at 1104-1105, n.18. With that expectation, the parties at the outset can settle on relevant contractual terms, such as the price for the work and the ownership of reproduction rights.

We turn, finally, to an application of §101 to Reid's production of "Third World America." In determining whether a hired party is an employee under the general common law of agency, we consider the hiring party's right to control the manner and means by which the product is accomplished. Among the other factors relevant to this inquiry are the skill required; the source of the instrumentalities and tools; the location of the work; the duration of the relationship between the parties; whether the hiring party has the right to assign additional projects to the hired party; the extent of the hired party's discretion over when and how long to work; the method of payment; the hired party's role in hiring and paying assistants; whether the work is part of the regular business of the hiring party; whether the hiring party is in business; the provision of employee benefits; and the tax treatment of the hired party. See Restatement §220(2) (setting forth a nonexhaustive list of factors relevant to determining whether a hired party is an employee). No one of these factors is determinative. See Ward, 362 U.S., at 400; Hilton Int'l Co. v. NLRB, 690 F.2d 318, 321 (CA2 1982).

Examining the circumstances of this case in light of these factors, we agree with the Court of Appeals that Reid was not an employee of CCNV but an independent contractor. 270 U.S. App. D.C., at 35, n.11, 846 F.2d, at 1494, n.11. True, CCNV members directed enough of Reid's work to ensure that he produced a sculpture that met their specifications. 652 F. Supp., at 1456. But the extent of control the hiring party exercises over the details of the product is not dispositive. Indeed, all the other circumstances weigh heavily against finding an employment relationship. Reid is a sculptor, a skilled occupation. Reid supplied his own tools. He worked in his own studio in Baltimore, making daily supervision of his activities from Washington practicably impossible. Reid was retained for less than two months, a relatively short period of time. During and after this time, CCNV had no right to assign additional projects to Reid. Apart from the deadline for completing the sculpture, Reid had absolute freedom to decide when and how long to work. CCNV paid Reid $15,000, a sum dependent on "completion of a specific job, a method by which independent contractors are often compensated." Holt v. Winpisinger, 258 U.S. App. D.C. 343, 351, 811 F.2d 1532, 1540 (1987). Reid had total discretion in hiring and paying assistants. "Creating sculptures was hardly 'regular business' for CCNV." 270 U.S. App. D.C., at 35, n.11, 846 F.2d, at 1494, n.11. Indeed, CCNV is not a business at all. Finally, CCNV did not pay payroll or Social Security taxes, provide any employee benefits, or contribute to unemployment insurance or workers' compensation funds.

Because Reid was an independent contractor, whether "Third World America" is a work for hire depends on whether it satisfies the terms of §101(2). This petitioners concede it cannot do. Thus, CCNV is not the author of "Third World America" by virtue of the work for hire provisions of the Act. However, as the Court of Appeals made clear, CCNV nevertheless may be a joint author of the sculpture if, on remand, the District Court determines that CCNV and Reid prepared the work "with the intention that their contributions be merged into inseparable or interdependent parts of a unitary whole." 17 U.S.C. §101. In that case, CCNV and Reid would be co-owners of the copyright in the work. See §201(a).

For the aforestated reasons, we affirm the judgment of the Court of Appeals for the District of Columbia Circuit.

It is so ordered.

COMMENTS AND QUESTIONS

1. What policies underlie the work for hire doctrine? From which philosophical perspectives do these policies flow? Does the court's analysis and holding in *CCNV* comport with these policies?

We learned in the introduction to this chapter that copyright law, while originally designed to protect publishers, has since the early days of English law been intended to benefit the authors themselves. Does the "works for hire" provision in the 1976 Act defeat this purpose? In the case of works for hire, section 101 automatically vests copyright in employers, not the particular employees who author the work. This result is somewhat surprising, especially in light of patent law's approach of considering the individual inventor to be the patentee (at least nominally). Why is this so? Why are inventors always people, whereas authors can be companies?

One justification for the works for hire doctrine may be that it reduces transaction costs. A work for hire is deemed to be a corporate creation, so it need not be assigned. The work for hire idea essentially "preassigns" a work to the employer. This has the important effect of eliminating the costs of negotiating and executing assignment agreements. In addition, as we will see in section 3 infra, the works for hire doctrine circumvents the inalienability of termination of transfers. The work belongs to the employer *ab initio,* rather than by assignment; this may make a significant difference. Automatic assignment may help solve "holdout" problems in compilations and multimedia works, where the publisher must compile the works of a number of authors. (Are the works listed in the second part of section 101's definition of works for hire the types of works in which holdout problems are most likely?)

2. *Works of the U.S. Government.* As we saw earlier, works of the U.S. government are not copyrightable. Section 101 defines U.S. government works as works "prepared by an officer or employee of the United States

government as part of that person's official duties." The legislative history makes clear that Congress intended this provision to parallel the scope of the works for hire doctrine.

3. *Teacher Exception.* There is a significant exception to the works for hire doctrine that protects professors and like employees from having to assign the rights in their works to their universities. This exception is venerable, but it finds little or no textual support in the 1976 Act. Some courts and commentators have concluded that the act did in fact abolish the "teacher exception," but most work hard to find a way to keep the doctrine alive. See Hays v. Sony Corp. of Am., 847 F.2d 412, 416 (7th Cir. 1988) (Judge Posner, a prolific former professor, justified the exception upon the "havoc that [a contrary] conclusion would wreak in the settled practices of academic institutions, the lack of fit between the policy of the work-for-hire doctrine and the conditions of academic production, and the absence of any indication that Congress meant to abolish" the exception.); Weinstein v. University of Illinois, 811 F.2d 1091, 1093-94 (7th Cir. 1987); Rochelle Cooper Dreyfuss, The Creative Employee and the Copyright Act of 1976, 54 U. Chi. L. Rev. 590, 597-98 (1987); but see University of Colorado Found. v. American Cyanamid, 880 F. Supp. 1387 (D. Colo. 1995) (assuming that an academic article written by university professors was a work for hire).

Can you think of a way for the exception to survive the passage of the 1976 Act? Should it?

4. *Role of Contract Law.* Contract law is at least as important as copyright law in determining ownership rights. Not only does the definition of a "work for hire" explicitly incorporate contract law, but many people "commissioning" works will require an assignment of rights before paying for or accepting a work. Should such an agreement be enforceable outside the bounds of the copyright statute? Suppose that *A* creates a work which cannot be "for hire" under section 101 (such as a book commissioned for *B*). Under the Copyright Act, *A* retains the copyright to the work. But can the parties agree to evade this result? For example, can they provide that *A* will pay all the royalties he receives to *B*? Can *A* agree in the same document to sell his rights to *B* for $1? Do such agreements undermine the purpose of the work for hire doctrine? (Recall that we have discussed a similar issue before, in the context of trade secret law.)

Note in this connection that under Playboy Ent., Inc. v. Dumas, 831 F. Supp. 295 (N.D. Ill. 1993), work for hire status cannot be achieved by contract after a work has been created, although subsequent writings can be evidence of a work for hire relationship. Thus there is no way to substitute a work for hire contract for an assignment. Why not allow creators to contract into work for hire status regardless of what type of work they created or contributed to, regardless of their employment status, and regardless of when the contract is entered into?

5. *The Evolving Nature of Employers.* The precise language of the works for hire doctrine is becoming increasingly important in a world of "network organizations." With the proliferation of consultants, "outsourcing,"

"downsizing," and "strategic partnering," businesses are placing much more emphasis on quasi-integration through contracts. This in turn makes for some interesting problems in view of the fact that copyright law was designed not for a complex, "organizationally promiscuous" world but a simple binary (big company vs. solo artist/author) world. How should this shift affect the interpretation of the works for hire doctrine?

b. Joint Works

Section 201(a) states that "[t]he authors of a joint work are co-owners of copyright in the work." This means that they *each* enjoy undivided ownership in the copyrighted work and may exercise independently the "exclusive rights" set out in section 106 (subject to an accounting of profits to the other).

Section 101 defines a "joint work" as "a work prepared by two or more authors with the intention that their contributions be merged into inseparable or interdependent parts of a unitary whole," a requirement that courts have interpreted strictly. See Erickson v. Trinity Theatre, Inc., 13 F.3d 1061, 1071-72 (7th Cir. 1994). Courts require either contemporaneous collaboration or evidence that each author knows at the time the work was created that his or her contribution would at a later time be integrated as an inseparable or interdependent part of a unitary whole. See Oddo v. Reis, 743 F.2d 630 (9th Cir. 1984) (holding that an article subsequently rewritten with a second author was a derivative work of the original article rather than a joint work). Further, the contribution of each author must itself be copyrightable (so that each has contributed something which can be protected under the law). *Erickson*, 13 F.3d at 1070-71. In cases where more than one artist contributes to a given work but the intent to collaborate arises after the creation of the initial work, the resulting collaboration is a derivative work of the initial contribution, rather than a single work of joint authorship.

c. Collective Works

Section 201(c) provides:

Copyright in each separate contribution to a collective work is distinct from copyright in the collective work as a whole, and vests initially in the author of the contribution. In the absence of an express transfer of the copyright or of any rights under it, the owner of copyright in the collective work is presumed to have acquired only the privilege of reproducing and distributing the contribution as part of that particular collective work, any revision of that collective work, and a later collective work in the same series.

Section 101 defines a "collective work" as "a work, such as a periodical issue, anthology, or encyclopedia, in which a number of contributions, constituting separate and independent works in themselves, are assembled into a collective whole."

PROBLEMS

Problem 4-15. Smith, a graphic artist, is employed full time by ADCO to design and flesh out illustrations for advertising campaigns. In his spare time, at home and using no materials taken from work, Smith designs an ad campaign for another company on a freelance basis. Who owns the copyright in Smith's freelance work?

Problem 4-16. Edwards, a playwright, wrote three short plays to be produced by a community theatre company. The plays were written on a tight budget, and Edwards made a number of revisions to the script during rehearsals. A number of these revisions, including the re-construction of two scenes, were made at the suggestion of actors dur-ing rehearsals, and the new scenes were worked out largely by consensus.

After a creative disagreement, the theatre company performs the plays without Edwards' permission. Edwards sues for copyright in-fringement, and the actors claim that they are joint authors with a right to perform the work. Who should prevail?

Problem 4-17. Bable comes up with an idea for a toy car with an integrated circuit that responds to commands as well as speaking and singing songs. Bable founds a company called Up and Running, Inc. to market her "talking car" concept. Bable finds several people to re-cord some new material for the talking car.

- The first is Sally Singer, who agrees to record a children's song she has written called "Red Light Go, Green Light Stop — Woops!" for use on the car. Bable has Sally sign an agreement giving Up and Running "all ownership in the song." The song becomes a hit, and Bable licenses the song to Warner Kids Re-cords for inclusion on an album of children's songs. Sally pro-tests, saying she had planned to release her own album with the song on it.
- The second is Telly Talker, a multilingual kids' storyteller who enters into a "long-term requirements employment" contract with Up and Running. Telly's job is to record translations of the songs and slogans that the toy car says into as many languages as Bable requires. The Japanese version of the toy car becomes a big hit, and Telly informs Bable that he is planning to license his recorded voice to a third party, Toyco, for use in their Japa-nese talking Bear product.

- The third is Gary Guitar, a musician who records guitar music for Up and Running. His practice is to record a snippet of music in whatever genre (bluegrass, jazz, etc.) Bable requests. Bable then sends a check with a standard form legend saying "cashing this check confirms your employment relationship with Up and Running, Inc., and the latter's ownership of the copyright in the music paid for hereby."

Who owns what copyrights in the car's songs and slogans?

2. Duration and Renewal

The duration of copyright protection has evolved significantly over the past century, generally moving in the direction of a longer term of protection.

1909 Act. The 1909 Act employed a dual term of protection, granting a first term of 28 years from the date of first publication (with proper notice) that could be renewed in the final year for a second term of 28 years. Failure to renew registration of copyright in that last year resulted in the work falling into the public domain.

1976 Act. The 1976 Act moved to a unitary term of protection lasting for the life of the author plus 50 years[19] (or, in the case of corporate, anonymous, pseudonymous entities, or works for hire, 75 years from publication or 100 years from creation, whichever occurred first). Under this regime, copyright terms run until the end of the calendar year in which they would otherwise expire, 17 U.S.C. §305, thereby adding an additional period of up to one year for purposes of administrative convenience. The 1976 Act also provided for automatic renewal of works published between the beginning of 1964 and the end of 1977, for an extended term of 47 years. 17 U.S.C. §304(a). Section 106A(d) governs the duration of certain moral rights of visual artists.

Sonny Bono Copyright Term Extension Act of 1998. In 1998, Congress further extended the term of copyright protection by another 20 years. This further extension of copyright duration has the practical effect of keeping proprietary such notable old works as Mickey Mouse and the songs of George and Ira Gershwin. Sections 302-05 of the Act govern copyright duration. You should carefully review these provisions at this time. Table 4-1 summarizes the principal features.

---

19. In the case of joint works, the term of copyright protection is measured from the death of the last surviving author. 17 U.S.C. §302(b).

TABLE 4-1
Duration of Copyright Protection

Date work created	Protected from	Term of protection*
Created January 1, 1978, or thereafter	When the work is fixed in a tangible medium of expression	Life of the author + 70 years (or, if work of corporate, anonymous, pseudonymous entity, or work for hire, 95 years from publication, or 120 years from creation, whichever is less
Published between beginning of 1964 and end of 1977	When published with notice	28 years for first term; automatic extension of 67 years for second term
Published between 1923 and end of 1963	When published with notice	28 years for first term; could be renewed for 67 years; if not so renewed, now in public domain
Created before January 1, 1978, but not yet published	January 1, 1978, the effective date of the 1976 Act which eliminated common law copyright protection	Life of the author plus 70 years or at least until 2003 if the work remains unpublished. If the work is published by 2003, term expires in 2048.
Sound recordings created prior to February 15, 1972	Depends upon treatment under applicable state law. "Any rights or remedies [for such works] under the common law or statutes of any State shall not be annulled or limited by this title until February 15, 2047." 17 U.S.C. §301(c).	

*Copyright terms run until the end of the calendar year in which they would otherwise expire.

COMMENTS AND QUESTIONS

1. *Determining the Optimal Duration of Copyright Protection.* In Article 1, Clause 8 of the U.S. Constitution, Congress is authorized to provide limited terms of protection in order to promote progress in science and the useful arts. Does the duration of copyright protection effectuate the appropriate balance between reward to authors and enrichment of the public domain so as to best promote progress in the arts? Why is the term so much longer than for patents? Determining the appropriate balance requires consideration of a broad range of variables, each of which is difficult to assess and measure.

The optimal duration of copyright protection from a utilitarian perspective requires a balancing of the costs and benefits of lengthening protection. Benefits presumably come in the form of an enhanced incentive for authors and artists to create, while the costs imposed are the limitations on the right of subsequent creators to make use of copyrighted works in their creative efforts and the social cost from monopoly pricing. Unfortunately, there is no good empirical data on this trade-off. Using cost and other data from publishing companies, Professor (now Justice) Stephen Breyer has questioned the need for copyright protection of books in view of the lead time advantages and the threat of retaliation. See Stephen Breyer, The Uneasy Case for Copyright: A Study in Copyright of Books, Photocopies and Computer Programs, 84 Harv. L. Rev. 281 (1970). Although Professor Breyer's conclusions are vulnerable to attack, cf. Barry Tyerman, The Economic Rationale for Copyright Protection for Published Books: A Reply to Professor Breyer, 18 UCLA L. Rev. 1100 (1971); Stephen Breyer, Copyright: A Rejoinder, 20 UCLA L. Rev. 75 (1972), he certainly raises serious qualms about lengthening the duration of copyright protection. A second source of data relating to the appropriate duration of copyright protection is the Copyright Office's records on renewal. Over the period 1927 through 1954, only about 9.5 percent of registered works were renewed prior to the end of their first term. The renewal rate was highest for "published music" (45 percent) and "motion picture photoplays" (43.7 percent). See Guinan, Duration of Copyright, Appendices A, B, Copyr. L, Revision Study No. 30 (1957), in 1 Studies in Copyr. 473 (1963).

Consider the following arguments for significantly limiting copyright protection: At a qualitative level, there would seem little basis for protecting most copyrightable works beyond 10 or 15 years. This observation is supported by the renewal data discussed above. Casebooks, for example, are rarely marketable after 5 years unless they are revised. Similarly, the public's interest in many works of literature and art tends to follow popular waves of a decade or less. Moreover, one can argue that after 25 years, the main interest in most literary works is historical. The public would be served by allowing historians the ability to draw upon such works in creating new works of history and social commentary. With regard to those relatively few works that have enduring commercial value beyond a decade or two, there is little question that such works generate substantial revenue for their authors. Therefore the public would be best served by limiting copyright protection for literary works to 25 years. Are you persuaded by this argument? What counterarguments would you offer? What philosophical basis or bases underlie your arguments?

2. *The Political Economy of Copyright Term Extension.* Who benefits most from the extension of the copyright term? Does extension of the copyright term pose a significant threat to the public? What reasons might explain the lack of public concern about this type of legislation? See William Patry, The Failure of the American Copyright System: Protecting the Idle Rich, Notre Dame L. Rev. 907 (1997). A recent lawsuit contends that the latest term extension is unconstitutional because it violates the "limited times" provision of Article I, Clause 8 (the Copyright and Patent Clause). See Eldred

v. Reno, (D.D.C. amended complaint filed May 10, 1999) <http://cyber.
law.harvard.edu/eldredvreno/complaint.html>.

3. Should the duration of copyright protection be the same for all classes
of subject matter?

PROBLEMS

Problem 4-18. Determine the duration of copyright in the following cases:

- Arnold Author completes his novel *You'll Be Mine 'Til the End
  of Time* on February 28, 1996. The next day, February 29th,
  Arnold is hit by a bus and dies instantly. On what day does his
  copyright expire?
- While working for the New Englander Magazine, Arnold Author
  writes a story entitled "You'll Always Be Mine." The story is
  finally published by The New Englander in 2010. When does
  the copyright expire?
- Arnold Author began work on his greatest novel, *Time Is on My
  Side,* in 1990. In 1991, he completes the first three chapters. In
  1992, he writes the middle three chapters. In 1993, he completes
  the final three chapters. In 1995, he signs an agreement with
  Time/Life Books to publish the novel. The contract assigns all
  copyright interests to Time/Life Books in exchange for 20 percent royalties based on the wholesale price. The book is finally
  published on January 1, 1996. Arnold dies on February 28,
  1996. On what day does the copyright expire? Does your answer
  change if you assume that Arnold entered into a contract with
  Time/Life Books before writing the novel?
- At the time of his death on February 29, 1996, Arnold Author
  has completed three-fourths of his novel *Time Lives On.* His will
  leaves all his property to his spouse, Angela Author, who is also
  a writer. She plans to complete the novel by 1999. Time/Life
  Books has agreed to publish the completed manuscript in the
  year 2000. Assuming that all goes according to plan, when will
  copyright in *Time Lives On* expire?

Problem 4-19. What is the duration of copyright in the following
cases?

- Loretta Wrighter composed and sent a letter to her friend, Emily
  Johnson, in 1961. Emily has saved the letter in her correspondence file since that time. Loretta died in 1970. What is the term
  of protection for this work?
- Stephen Morris published his first novel, *Child's Play,* at the age

of 15 in 1920. He is still alive today. What is the term of protection for this work?

- Anita Author published (with proper notice) her novel entitled *The Winds of Change* in 1970. She died three years later. What is the term of protection for this work?
- Penelope Painter painted her masterpiece entitled "Garden of Wildflowers" in 1953. She distributed copies with notice of copyright in 1955. She renewed the copyright in 1982. She died the next year. When does the copyright expire?

3. Division, Transfer, and Termination of Transfers

The preceding section has explored one important difference between copyright interests and the traditional fee simple absolute in real property law: the limited duration of protection. Copyrights, therefore, can be analogized to a hybrid of a term of years and life estate in that the owner has control of the rights of copyright for a defined and limited period of time (life plus 50 years), after which such rights fall to the public at large.

Another important aspect of copyright ownership is the distinction between ownership of the material object on which the work of authorship is fixed — the book manuscript or oil canvas — and ownership of the copyright interests themselves. Section 202 states:

> Ownership of a copyright, or of any of the exclusive rights under a copyright, is distinct from ownership of any material object in which the work is embodied. Transfer of ownership of any material object, including the copy or phonorecord in which the work is first fixed, does not of itself convey any rights in the copyrighted work embodied in the object; nor, in the absence of an agreement, does transfer of ownership of a copyright or of any exclusive rights under a copyright convey property rights in any material object.

Thus the author of a letter retains her copyright interests in the writing even though she sends the letter to the addressee. The addressee thereby obtains ownership of the material object but may not infringe the copyright interests of the author. The addressee may view the material object and he may show it to others, but he may not make copies, prepare derivative works, distribute the work, or perform or display the work publicly.

Two other elements of real property interests are the rights of property owners and the alienability of such rights. Part E (Rights of Copyright Holders) and Part F (Defenses, including fair use) explore the extent and limitations on copyrights. This section discusses the division, transfer, and termination of the rights of copyright holders. In the domain of real property, the owner of a fee interest may freely divide and alienate the various rights of property ownership. For example, a property owner may divide her lot and sell a portion to another person. Alternatively, she may sell one particular right within the bundle of rights, such as the right to use a path running across the property (an easement). Moreover, such transfers are generally not terminable unless so specified in the transfer agreement. Thus a property owner who creates an easement across her land may not unilaterally terminate such right at a later time.

By contrast, copyright law restricts the alienability of the rights of copyright owners in certain ways. As you study these materials, scrutinize the reasons restricting the alienability of copyright interests.

a. Division and Transfer of Copyright Interests Under the 1909 Act

Courts interpreted the 1909 Act to preclude the formal divisibility of the rights comprising a copyright. A copyright owner could assign the entire copyright to another, but a transfer of any lesser interest was considered a license. This doctrine of indivisibility simplified the notice requirement. As noted earlier, failure to provide proper notice could result in forfeiture of copyright protection. The "owner" of the copyright (or "proprietor"), whether the author or the assignee of the entire copyright, was the appropriate name to be included in the copyright notice. One consequence of the doctrine of indivisibility was that only proprietors had standing to bring suit to enforce the copyright. Thus a licensee would have to join the proprietor in order to protect his or her rights.

In practice, however, court decisions under the 1909 Act limited the effect of the indivisibility rule. See 1 Paul Goldstein, Copyright 4:54 (1995).

b. Division and Transfer of Copyright Interests Under the 1976 Act

The 1976 Act eliminated restrictions on the formal divisibility of copyright interests. Section 201(d) provides:

(1) The ownership of a copyright may be transferred in whole or in part by any means of conveyance or by operation of law, and may be bequeathed by will or pass as personal property by the applicable laws of intestate succession.

(2) Any of the exclusive rights comprised in a copyright, including any subdivision of any of the rights specified by section 106, may be trans-

ferred as provided by clause (1) and owned separately. The owner of any particular exclusive right is entitled, to the extent of that right, to all of the protection and remedies accorded to the copyright owner by this title.

Section 101 defines "transfer of copyright ownership" to include an assignment or an *exclusive* license of any of the exclusive rights comprised in a copyright. The divisibility of copyright ownership enables any owner of an exclusive right to bring an infringement suit without having to join the copyright proprietor.

Section 204 requires transfers of copyright ownership to be executed in writing and signed by the copyright owner. Section 205 provides for the voluntary recordation of transfers with the Copyright Office. It also sets forth priority rules for resolving cases of conflicting transfers.

c. *Termination of Transfers*

The 1976 Act retained one important limitation on the free alienability of copyright interests. Under the 1909 Act, authors could reclaim copyright interests that they had licensed at the time of renewal. Although eliminating the renewal regime, the 1976 Act retains the power for authors and their survivors to reclaim copyright interests at a later time by granting an inalienable power to terminate transfers of copyright between the thirty-fifth and fortieth year from the execution of the grant for works created after 1977, §203(a)(3). With respect to works in their second renewal term prior to 1978, §304(c) allows authors and their families to terminate transfers between the fifty-sixth and sixty-first year of copyright protection for such works so as to allow them to profit for the 19 years of protection for such works added by the 1976 Act. You should carefully review §§203 and 304(c) at this time. Note that this termination of transfer rights may not be assigned in advance.[20] Congress enacted these provisions to better ensure that authors and their families are able to reap a fair portion of the benefits of the author's creative efforts. Congress was concerned that authors had "unequal bargaining power" in negotiating rights with publishers and marketers "resulting in part from the impossibility of determining a work's value until it has been exploited." H.R. Rep. No. 1476, 94th Cong., 2d Sess. 124 (1976).

---

20. Copyright owners may assign contingent renewal rights under the 1909 Act in advance. See Fred Fisher Music Co. v. M. Witmark & Sons, 318 U.S. 643 (1943). This power, however, does not apply to the additional 19 years of protection of the extended renewal term bestowed by the 1976 Act.

Do you agree with Congress' premise that authors are at a serious bargaining disadvantage in negotiating the rights to their works? Is making the power of reverter inalienable necessary to address this concern? What problems might arise as a result of the power to terminate copyright transfers? How has Congress sought to reduce these problems? What alternative means might Congress have used to protect the interests of authors and their families short of an inalienable power of reverter?

E. RIGHTS OF COPYRIGHT OWNERS

17 U.S.C. §106. Exclusive Rights in Copyrighted Works

Subject to sections 107 through 118, the owner of copyright under this title has the exclusive rights to do and to authorize any of the following:

(1) to reproduce the copyrighted work in copies or phonorecords;

(2) to prepare derivative works based on the copyrighted work;

(3) to distribute copies or phonorecords of the copyrighted work to the public by sale or other transfer of ownership, or by rental, lease, or lending;

(4) in the case of literary, musical, dramatic, and choreographic works, pantomimes, and motion pictures and other audiovisual works, to perform the copyrighted work publicly; and

(5) in the case of literary, musical, dramatic, and choreographic works, pantomimes, and pictorial, graphic, or sculptural works, including the individual images of a motion picture or other audiovisual work, to display the copyrighted work publicly.

The 1976 Act states that anyone who "violates any of the exclusive rights of the copyright owner as provided by sections 106 through 121" is liable for copyright infringement. 17 U.S.C. §501(a). Copyright infringement may occur by two distinct sets of actors: (1) those who directly infringe the rights of copyright holders through violation of a copyright owners' rights; and (2) those who encourage or assist a third party to infringe.

1. The Right to Make Copies

There is a close relationship between rights in subsections (1), (2) and (3) of section 106. The right to "reproduce" is, of course, the most fundamental of the rights granted to the copyright owner. It includes all rights to fix a work in a tangible medium of expression. This right includes the right to make phonorecords.

The right to copy granted by section 106(1) is not limited to exact reproduction (such as "copying" a disk or making a photocopy). Rather, it is a broad grant of the right to prevent others from making exact or "substantially similar" reproductions by any means "now known or later developed, and from which the work can be perceived, reproduced, or otherwise communicated, either directly or with the aid of a machine or device," 17 U.S.C. §101 (definition of "copies"). "Phonorecords" includes not only disks that may be played on record players, but all "material objects in which sounds, other than those accompanying a motion picture or other audiovisual works, are fixed by any method now known or later developed, and from which the sounds can be perceived, reproduced, or otherwise communicated, either directly or with the aid of a machine or device." 17 U.S.C. §101.

The doctrines governing direct infringement of copyright address two distinct problems in determining copyright liability. The first problem relates to proving whether someone has actually copied the work of another. As you will recall, copyright law prohibits copying, but it does not prohibit independent creation. The fact that two musicians claim copyright in similar musical compositions does not necessarily mean that one copied from the other. They may have independently composed similar songs. Ideally, a court would like to have direct proof of copying — e.g., eyewitness testimony, records indicating that one author obtained the work from another, distinctive flaws in a work (such as unusual errors that are common to the two works). In many cases, however, such evidence is not available. Nonetheless, the circumstances surrounding the works may point inexorably toward a finding of infringement, for example, where a work has been widely disseminated and a second author claims authorship of an identical or nearly identical work. We will first examine the principal test by which courts assess circumstantial proof of copying.

The second problem that courts have confronted in assessing infringement arises even when the defendant acknowledges having developed his or her work with knowledge of the first comer's work. In these cases, the question is whether the second comer appropriated sufficient material to violate a copyright owner's rights. The legislative history to §106 provides the following guidance:

> [A] copyrighted work would be infringed by reproducing it *in whole or in any substantial part*, and by duplicating it exactly or by imitation or simulation. Wide departures or variations from the copyrighted works would still be an infringement as long as the author's "expression" rather [than] merely the author's "ideas" are taken.

H.R. Rep. No. 94-1476, 94th Cong., 2d Sess. 61 (1976) (emphasis added). As suggested by this passage, the inquiry into whether the defendant has violated a plaintiff's copyright is often complicated by the fact that many copyrightable works intermingle original expression with public domain materials, ideas, facts, stock literary elements, scenes à faire, and other nonprotectable elements. Hence courts have had to develop an infringement filter that adequately protects the interests of copyright owners but at the same time does not extend the scope of copyright protection beyond its statutory scope.

As you will see, courts use the term "substantial similarity" in discussing both problems. This can be confusing. As you study these materials, pay special attention to the context in which the courts are applying this term. "Substantial similarity" has a different meaning depending on whether it is being used as an aid in determining proof of copying, on the one hand, or whether the appropriation of protectable material was improper, on the other.

a. Copying

Arnstein v. Porter
United States Court of Appeals for the Second Circuit
154 F.2d 464 (2d Cir. 1946)

FRANK, Circuit Judge.

[Ira B. Arnstein sued Cole Porter for infringement of copyrights in various of plaintiff's musical compositions. He sought a jury trial. Plaintiff alleged that the defendant's "Begin the Beguine" had been plagiarized from plaintiff's "The Lord Is My Shepherd" and "A Mother's Prayer" and that defendant's "My Heart Belongs to Daddy" had been plagiarized from "A Mother's Prayer." Plaintiff testified in deposition that both works had been published and that about 2,000 copies of "The Lord Is My Shepherd" and over a million copies of "A Mother's Prayer" had been sold. Plaintiff offered no direct proof that defendant saw or heard these compositions. Plaintiff further testified that defendant's "Night and Day" had been plagiarized from plaintiff's "I Love You Madly" and that although the latter composition had not been published, it had been performed publicly over the radio. In addition, plaintiff averred that a copy of the song had been stolen from his room. Plaintiff alleged that some other songs of the defendant had been plagiarized from the plaintiff's unpublished works. He suggested in deposition that the defendant had gained access to these songs either through publishers or a movie producer who were sent copies or through "stooges" who defendant had hired to follow, watch, and live with the plaintiff (and who may have been responsible for the ransacking of his room). When asked how he knew that defendant had anything to do with the "burglaries," plaintiff testified "I don't know that he had to do with it, but I only know that he could have." Defendant testified in depositions that he had never seen nor heard the plaintiff's compositions and that he did not have any connection to the alleged theft of such works.

The district court granted defendant's motion for summary judgment.]

. . . The principal question on this appeal is whether the lower court, under Rule 56, properly deprived plaintiff of a trial of his copyright infringement action. The answer depends on whether "there is the slightest doubt as to the facts." In applying that standard here, it is important to avoid confusing two separate elements essential to a plaintiff's case in such a suit: (a) that defendant copied from plaintiff's copyrighted work and (b) that the copying (assuming it to be proved) went so far as to constitute improper appropriation.

As to the first — copying — the evidence may consist (a) of defendant's admission that he copied or (b) of circumstantial evidence — usually evidence of access — from which the trier of the facts may reasonably infer copying. Of course, if there are no similarities, no amount of evidence of access will suffice to prove copying. If there is evidence of access and similarities exist, then the trier of the facts must determine whether the similarities are sufficient to prove copying. On this issue, analysis ("dissection") is relevant, and the testimony of experts may be received to aid the trier of the facts. If evidence of access is absent, the similarities must be so striking as to preclude the possibility that plaintiff and defendant independently arrived at the same result.

If copying is established, then only does there arise the second issue, that of illicit copying (unlawful appropriation). On that issue (as noted more in detail below) the test is the response of the ordinary lay hearer; accordingly, on that issue, "dissection" and expert testimony are irrelevant.

In some cases, the similarities between the plaintiff's and defendant's work are so extensive and striking as, without more, both to justify an inference of copying and to prove improper appropriation. But such double-purpose evidence is not required; that is, if copying is otherwise shown, proof of improper appropriation need not consist of similarities which, standing alone, would support an inference of copying.

Each of these two issues — copying and improper appropriation — is an issue of fact. If there is a trial, the conclusions on those issues of the trier of the facts — of the judge if he sat without a jury, or of the jury if there was a jury trial — bind this court on appeal, provided the evidence supports those findings, regardless of whether we would ourselves have reached the same conclusions. But a case could occur in which the similarities were so striking that we would reverse a finding of no access, despite weak evidence of access (or no evidence thereof other than the similarities); and similarly as to a finding of no illicit appropriation.

We turn first to the issue of copying. After listening to the compositions as played in the phonograph recordings submitted by defendant, we find similarities; but we hold that unquestionably, standing alone, they do not compel the conclusion, or permit the inference, that defendant copied. The similarities, however, are sufficient so that, if there is enough evidence of access to permit the case to go to the jury, the jury may properly infer that the similarities did not result from coincidence.

Summary judgment was, then, proper if indubitably defendant did not have access to plaintiff's compositions. Plainly that presents an issue of fact. On that issue, the district judge, who heard no oral testimony, had before him the depositions of plaintiff and defendant. The judge characterized plaintiff's story as "fantastic"; and, in the light of the references in his opinion to defendant's deposition, the judge obviously accepted defendant's denial of access and copying. Although part of plaintiff's testimony on deposition (as to "stooges" and the like) does seem "fantastic," yet plaintiff's credibility, even as to those improbabilities, should be left to the jury. If evidence is "of a kind that greatly taxes the credulity of the judge, he can say so, or, if he totally disbelieves it, he may announce that fact, leaving the jury free to believe it or not." If, said Winslow, J., "evidence is to be always disbelieved because the story told seems remarkable or impossible, then a party whose rights depend on the proof of some facts out of the usual course of events will always be denied justice simply because his story is improbable." We should not overlook the shrewd proverbial admonition that sometimes truth is stranger than fiction.

But even if we were to disregard the improbable aspects of plaintiff's story, there remain parts by no means "fantastic." On the record now before us, more than a million copies of one of his compositions were sold; copies of others were sold in smaller quantities or distributed to radio stations or band leaders or publishers, or the pieces were publicly performed. If, after hearing both parties testify, the jury disbelieves defendant's denials, it can, from such facts, reasonably infer access. It follows that, as credibility is unavoidably involved, a genuine issue of material fact presents itself. With credibility a vital factor, plaintiff is entitled to a trial where the jury can observe the witnesses while testifying. . . .

Assuming that adequate proof is made of copying, that is not enough; for there can be "permissible copying," copying which is not illicit. Whether (if he copied) defendant unlawfully appropriated presents, too, an issue of fact. The proper criterion on that issue is not an analytic or other comparison of the respective musical compositions as they appear on paper or in the judgment of trained musicians. The plaintiff's legally protected interest is not, as such, his reputation as a musician but his interest in the potential financial returns from his compositions which derive from the lay public's approbation of his efforts. The question, therefore, is whether defendant took from plaintiff's works so much of what is pleasing to the ears of lay listeners, who comprise the audience for whom such popular music is composed, that defendant wrongfully appropriated something which belongs to the plaintiff.

Surely, then, we have an issue of fact which a jury is peculiarly fitted to determine. Indeed, even if there were to be a trial before a judge, it would be desirable (although not necessary) for him to summon an advisory jury on this question.

We should not be taken as saying that a plagiarism case can never arise in which absence of similarities is so patent that a summary judgment for defendant would be correct. Thus suppose that Ravel's *Bolero* or Shostakovitch's *Fifth Symphony* were alleged to infringe "When Irish Eyes Are Smiling." But this is not such a case. For, after listening to the playing of the respective compositions, we are, at this time, unable to conclude that the likenesses are so trifling that, on the issue of misappropriation, a trial judge could legitimately direct a verdict for defendant.

At the trial, plaintiff may play, or cause to be played, the pieces in such manner that they may seem to a jury to be inexcusably alike, in terms of the way in which lay listeners of such music would be likely to react. The plaintiff may call witnesses whose testimony may aid the jury in reaching its conclusion as to the responses of such audiences. Expert testimony of musicians may also be received, but it will in no way be controlling on the issue of illicit copying, and should be utilized only to assist in determining the reactions of lay auditors. The impression made on the refined ears of musical experts or their views as to the musical excellence of plaintiff's or defendant's works are utterly immaterial on the issue of misappropriation; for the views of such persons are caviar to the general — and plaintiff's and defendant's compositions are not caviar. . . .

CLARK, Circuit Judge (dissenting). . . .

[A]fter repeated hearings of the records, I could not find therein what my brothers found. The only thing definitely mentioned seemed to be the repetitive use of the note e♭ in certain places by both plaintiff and defendant, surely too simple and ordinary a device of composition to be significant. In our former musical plagiarism cases we have, naturally, relied on what seemed the total sound effect; but we have also analyzed the music enough to make sure of an intelligible and intellectual decision. Thus in Arnstein v. Edward B. Marks Music Corp., 2 Cir., 82 F.2d 275, 277, Judge L. Hand made quite an extended comparison of the songs, concluding, inter alia: ". . . the seven notes available do not admit of so many agreeable permutations that we need be amazed at the re-appearance of old themes, even though the identity extend through a sequence of twelve notes." See also the discussion in Marks v. Leo Feist, Inc., 2 Cir., 290 F. 959, and Darrell v. Joe Morris Music Co., 2 Cir., 113 F.2d 80, where the use of six similar bars and of an eight-note sequence frequently repeated were respectively held not to constitute infringement, and Wilkie v. Santly Bros., 2 Cir., 91 F.2d 978, *affirming* D.C.S.D.N.Y., 13 F. Supp. 136, *certiorari denied,* Santly Bros. v. Wilkie, 302 U.S. 735, 58 S. Ct. 120, 82 L. Ed. 568, where use of eight bars with other similarities amounting to over three-quarters of the significant parts was held infringement.

It is true that in Arnstein v. Broadcast Music, Inc., 2 Cir., 137 F.2d 410, 412, we considered "dissection" or "technical analysis" not the proper approach to support a finding of plagiarism, and said that it must be "more ingenuous, more like that of a spectator, who would rely upon the complex of his impressions." But in its context that seems to me clearly sound and in accord with what I have in mind. Thus one may look to the total impression to repulse the charge of plagiarism where a minute "dissection" might dredge up some points of similarity. Hence one cannot use a purely theoretical dis-quisition to supply a tonal resemblance which does not otherwise exist. Certainly, however, that does not suggest or compel the converse — that one must keep his brain in torpor for fear that otherwise it would make clear differences which do exist. Music is a matter of the intellect as well as the emotions; that is why eminent musical scholars insist upon the employment of the intellectual faculties for a just appreciation of music.

Consequently I do not think we should abolish the use of the intellect here even if we could. When, however, we start with an examination of the written and printed material supplied by the plaintiff in his complaint and exhibits, we find at once that he does not and cannot claim extensive copying, measure by measure, of his compositions. He therefore has resorted to a comparative analysis — the "dissection" found unpersuasive in the earlier cases — to support his claim of plagiarism of small detached portions here and there, the musical fillers between the better known parts of the melody. And plaintiff's compositions, as pointed out in the cases cited above, are of the simple and trite character where small repetitive sequences are not hard to discover. It is as though we found Shakespeare a plagiarist on the basis of his use of articles, pronouns, prepositions, and adjectives also used by others. The surprising thing, however, is to note the small amount of even this type of reproduction which plaintiff by dint of extreme dissection has been able to find.

Though it is most instructive, it will serve no good purpose for me to restate here this showing as to each of the pieces in issue. As an example of the rest, we may take plaintiff's first cause of action. This involves his "A Modern Messiah" with defendant's "Don't Fence Me In." The first is written in 6/8 time, the second in common or 4/4 time; and there is only one place where there is a common sequence of as many as five consecutive notes, and these without the same values. Thus it goes. The usual claim seems to be rested upon a sequence of three, of four, or of five — never more than five — identical notes, usually of different rhythmical values. Nowhere is there any-thing approaching the twelve-note sequence of the *Marks* case, supra. Inter-esting is the fact that the closest tonal resemblance is to be found between a piece by defendant written back in 1930 and an uncopyrighted waltz by plaintiff (rejected here by my brothers because it is uncopyrighted) which was never published, but, according to his statement, was publicly performed as early as 1923, 1924, and 1925.

In the light of these utmost claims of the plaintiff, I do not see a legal basis for the claim of plagiarism. So far as I have been able to discover, no earlier case approaches the holding that a simple and trite sequence of this type, even if copying may seem indicated, constitutes proof either of access or of plagiarism. . . .

COMMENTS AND QUESTIONS

1. *Infringement Analysis: Copying + Improper Appropriation.* The *Arnstein* majority articulates both elements needed to establish infringement: (a) copying and (b) improper (or unlawful) appropriation ("whether defendant took from plaintiff's works so much of what is pleasing to the ears of lay listeners, who comprise the audience for whom such popular music is composed, that defendant wrongfully appropriated something which belongs to the plaintiff").[21] We explore the former stage in these comments and problems and the latter stage in the next section.

2. *Circumstantial Proof of Copying: Access + Similarity.* The majority opinion in *Arnstein* articulates the leading test for establishing copying by way of circumstantial evidence. The court suggests a sliding scale: "[I]f there are no similarities, no amount of evidence of access will suffice to prove copying. If there is evidence of access and similarities exist, then the trier of the facts must determine whether the similarities are sufficient to prove copying. . . . If evidence of access is absent, the similarities must be so striking as to preclude the possibility that plaintiff and defendant independently arrived at the same result."[22] Courts have differed in their application of this sliding scale. In the Second Circuit, a plaintiff need not adduce any evidence of access "if a copyrighted work and an allegedly infringing work are strikingly similar." Gaste v. Kaiserman, 863 F.2d 1061 (2d Cir. 1988). Thus the trier of fact may infer from evidence of striking similarity that the defendant could not have independently created his or her work. A recent application of the Second Circuit's rule is Repp v. Lloyd Webber, 132 F.3d 882 (2d Cir. 1997). There, a songwriter accused Andrew Lloyd Webber of copying one of his songs in Phantom's Song, part of Webber's *Phantom of the Opera.* Webber submitted his own affidavit and several supporting affidavits indicating that he had never heard Repp's song and that he had written Phantom's Song without outside influence. In response, Repp submitted expert declarations that the two songs were so "strikingly similar" as to preclude the possibility of independent creation. The Second Circuit reversed summary judgment for Webber, holding that the expert testimony about similarity was sufficient to create a genuine issue of fact regarding access.

---

21. The third element necessary to prove infringement — ownership of a valid copyright in the allegedly infringed work — was not contested in *Arnstein.*

22. Courts often loosely use the term "substantial similarity" in referring to both the comparison of similarity in assessing circumstantial proof of copying and the comparison of substantial similarity in assessing improper appropriation. Cf. Amy Cohen, Masking Copyright Decisionmaking: The Meaninglessness of Substantial Similarity, 20 U.C. Davis L. Rev. 719 (1987). As reflected in *Arnstein,* the degree of similarity required to establish circumstantial proof of copying is gauged along a sliding scale and incorporates objective criteria. By contrast, the "substantial similarity" standard applied in assessing improper appropriation is completely subjective and is premised upon a finding that copying has occurred. The issue at this stage is determining whether the defendant's work reproduces so much of the plaintiff's original expression as to violate the plaintiff's copyright. See infra, subsection *b.*

In a 1984 decision, the Seventh Circuit required the plaintiff to "present sufficient evidence to support a reasonable possibility of access because the jury cannot draw an inference of access based upon speculation and conjecture alone." Selle v. Gibb, 741 F.2d 896, 901 (7th Cir. 1984). In that case, a local band in Chicago alleged that the Bee Gees' 1978 hit tune "How Deep Is Your Love" infringed Selle's 1975 composition "Let It End." The only evidence that the Bee Gees might have had access to Selle's song was that it had been played two or three times by Selle's band in the Chicago area and that Selle had sent a recording to eleven recording and publishing companies, eight of whom returned it and three of whom did not respond. Notwithstanding the jury's finding of liability, the court upheld the lower court's judgment notwithstanding the verdict in favor of the Bee Gees on the ground that there was insufficient proof of access and failure to show "striking similarity" as a legal matter to establish copying.

In Ty, Inc. v. GMA Accessories, Inc., 132 F.3d 1167 (7th Cir. 1994), Judge Posner sought to harmonize the 7th Circuit and the 2nd Circuit tests for proving copying through circumstantial evidence by characterizing *Selle* as dealing with a situation in which "two works may be strikingly similar — may in fact be identical — not because one is copied from the other but because both are copies of the same thing in the public domain." Id. at 1170. "[T]he tension between *Gaste* and *Selle* can be resolved and the true relation between similarity and access expressed. Access (and copying) may be inferred when two works are so similar to each other and not to anything in the public domain that it is likely that the creator of the second work copied the first, but the inference can be rebutted by disproving access or otherwise showing independent creation." Id. at 1171.

Does it make sense to dispense with the need for *any* proof of access in some cases? Or would a burden-shifting presumption be more appropriate?

3. *Subconscious Copying of a Work.* Robert Mack composed the song "He's So Fine," which The Chiffons recorded in 1962. It enjoyed popular success, rising to No. 1 on the U.S. billboard charts for five weeks in 1963; it was among the top hits in England for about seven weeks in 1963 as well. In 1970, George Harrison, formerly of The Beatles, wrote the song "My Sweet Lord." A recording of the song by Billy Preston became a popular hit. Both songs consisted of four repetitions of a very short basic musical phrase, "sol-me-ri," followed by four repetitions of another short basic musical phrase, "sol-la-do-la-do." While neither phrase is novel (or uncommon), experts at trial agreed that the pattern of juxtaposing four repetitions of each phrase is highly unique. George Harrison testified at trial that he had composed the song while in Copenhagen, Denmark, while on a gig. He recalled "vamping" some guitar chords while singing "Hallelujah" and "Hare Krishna." The song developed further as he improvised with the other musicians in his entourage. In assessing copying, the court determined that George Harrison had not deliberately reproduced Mack's work. Nonetheless, the court found infringement on the grounds that " 'My Sweet Lord' is the

very same song as 'He's So Fine' with different words, and Harrison had access to 'He's So Fine.' This is, under the law, infringement of copyright, and is no less so even though subconsciously accomplished." Bright Tunes Music Corp. v. Harrisongs Music, Ltd., 420 F. Supp. 177 (S.D.N.Y. 1976).

This case raises an important point — copying may occur (and may rise to the level of infringement) even though it is completely unintentional. For example, in Lipton v. Nature Co., 71 F.3d 464 (2d Cir. 1995), the Nature Company had licensed a literary work from an author named Michael Wein. Wein in turn had copied the work from a scarf he obtained years before. As it turned out, whoever created the scarf had unlawfully copied this work from Lipton. Lipton sued Wein and the Nature Company for copyright infringement. While the Second Circuit accepted Wein's explanation that he did not copy the work from Lipton, it decided that the issue was irrelevant: Wein (and presumably the Nature Company as well) was liable for copyright infringement if he copied material that turned out to be protected by copyright.

4. *Proof of Copying by Deliberate Error or Common Mistake.* Creators of fact works often deliberately plant minor errors in their works to trap copyists. We saw an example of this in *Feist,* supra section B.1, where a telephone company placed false names in a telephone directory. Map writers also frequently bury erroneous or misspelled place names in their works for this reason. In addition, accidental misspellings and other mistakes common to both works have aided copyright owners in establishing copying.

5. *Techniques for Reducing the Risk of Infringement.* Potential copyright defendants sometimes go to great lengths to avoid having access to works that might influence them. For example, movie studios and television producers return unsolicited scripts unopened. In the computer industry, program developers occasionally prepare new programs in "clean rooms," in which procedures are established to regulate the entry and exit of material from the location where the work is being created and the creative process is carefully documented.

PROBLEM

Problem 4-20. Scooter, a ventriloquist, performs a travelling show with a dummy that vocalizes the catchphrase "You Got the Right One, Uh-Huh." Scooter has performed this show since 1984. His performances have primarily been at elementary schools and Job Corps camps, but he did have a pavilion at the 1984 World's Fair in which he used the phrase. Scooter also attempted to promote his show by mailing unsolicited information packets to corporate executives. Included in these packets were letters that referred to his catchphrase. He mailed one such packet to a Pepsi executive in Baltimore in 1988, but the executive cannot recall ever receiving it.

In 1991, Pepsi starts a massive advertising campaign using Ray Charles singing "You Got the Right One Baby, Uh-Huh" with similar voice inflections. Scooter sues for copyright infringement. Does he have a case?

b. Improper Appropriation

The second problem that arises in assessing infringement is determining whether the defendant has copied sufficient expression to violate the plaintiff's copyright interests. This requires careful assessment of the extent to which the defendant has copied "protected" expression. Judge Learned Hand's opinion in Nichols v. Universal Pictures, Corp., decided seventy years ago, remains the starting point for this inquiry. As you study this opinion, pay close attention to the manner in which the court distinguishes protected and unprotected expression and how it determines whether the defendant has improperly appropriated the plaintiff's work.

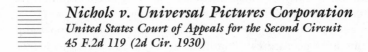

Nichols v. Universal Pictures Corporation
United States Court of Appeals for the Second Circuit
45 F.2d 119 (2d Cir. 1930)

L. HAND, Circuit Judge.

The plaintiff is the author of a play, "Abie's Irish Rose," which it may be assumed was properly copyrighted under section five, subdivision (d), of the Copyright Act, 17 USCA §5(d). The defendant produced publicly a motion picture play, "The Cohens and The Kellys," which the plaintiff alleges was taken from it. As we think the defendant's play too unlike the plaintiff's to be an infringement, we may assume, arguendo, that in some details the defendant used the plaintiff's play, as will subsequently appear, though we do not so decide. It therefore becomes necessary to give an outline of the two plays.

"Abie's Irish Rose" presents a Jewish family living in prosperous circumstances in New York. The father, a widower, is in business as a merchant, in which his son and only child helps him. The boy has philandered with young women, who to his father's great disgust have always been Gentiles, for he is obsessed with a passion that his daughter-in-law shall be an orthodox Jewess. When the play opens the son, who has been courting a young Irish Catholic girl, has already married her secretly before a Protestant minister, and is concerned to soften the blow for his father, by securing a favorable impression of his bride, while concealing her faith and race. To accomplish this he introduces her to his father at his home as a Jewess, and lets it appear that he is interested in her, though he conceals the marriage. The girl somewhat reluctantly falls in with the plan; the father takes the bait, becomes infatuated with the girl, concludes that they must marry, and assumes that of course they will, if he so decides. He calls in a rabbi, and prepares for the wedding according to the Jewish rite.

Meanwhile the girl's father, also a widower, who lives in California, and is as intense in his own religious antagonism as the Jew, has been called to New York, supposing that his daughter is to marry an Irishman and a Catholic. Accompanied by a priest, he arrives at the house at the moment when the marriage is being celebrated, but too late to prevent it and the two fathers, each infuriated by the proposed union of his child to a heretic, fall into unseemly and grotesque antics. The priest and the rabbi become friendly, exchange trite sentiments about religion, and agree that the match is good. Apparently out of abundant caution, the priest celebrates the marriage for a third time, while the girl's father is inveigled away. The second act closes with each father, still outraged, seeking to find some way by which the union, thus trebly insured, may be dissolved.

The last act takes place about a year later, the young couple having meanwhile been abjured by each father, and left to their own resources. They have had twins, a boy and a girl, but their fathers know no more than that a child has been born. At Christmas each, led by his craving to see his grandchild, goes separately to the young folks' home, where they encounter each other, each laden with gifts, one for a boy, the other for a girl. After some slapstick comedy, depending upon the insistence of each that he is right about the sex of the grandchild, they become reconciled when they learn the truth, and that each child is to bear the given name of a grandparent. The curtain falls as the fathers are exchanging amenities, and the Jew giving evidence of an abatement in the strictness of his orthodoxy.

"The Cohens and The Kellys" presents two families, Jewish and Irish, living side by side in the poorer quarters of New York in a state of perpetual enmity. The wives in both cases are still living, and share in the mutual animosity, as do two small sons, and even the respective dogs. The Jews have a daughter, the Irish a son; the Jewish father is in the clothing business; the Irishman is a policeman. The children are in love with each other, and secretly marry, apparently after the play opens. The Jew, being in great financial straits, learns from a lawyer that he has fallen heir to a large fortune from a great-aunt, and moves into a great house, fitted luxuriously. Here he and his family live in vulgar ostentation, and here the Irish boy seeks out his Jewish bride, and is chased away by the angry father. The Jew then abuses the Irishman over the telephone, and both become hysterically excited. The extremity of his feelings make[s] the Jew sick, so that he must go to Florida for a rest, just before which the daughter discloses her marriage to her mother.

On his return the Jew finds that his daughter has borne a child; at first he suspects the lawyer, but eventually learns the truth and is overcome with anger at such a low alliance. Meanwhile, the Irish family who have been forbidden to see the grandchild, go to the Jew's house, and after a violent scene between the two fathers in which the Jew disowns his daughter, who decides to go back with her husband, the Irishman takes her back with her baby to his own poor lodgings. The lawyer, who had hoped to marry the Jew's daughter, seeing his plan foiled, tells the Jew that his fortune really belongs to the Irishman, who was also related to the dead woman, but offers to conceal his knowledge, if the Jew will share the loot. This the Jew repudiates, and, leaving the astonished lawyer, walks through the rain to his enemy's house to surrender the property. He arrives in great defection, tells the truth, and abjectly turns to leave. A reconciliation ensues, the Irishman agreeing to share with him equally. The Jew shows some interest in his grandchild, though this is at most a minor motive in the reconciliation, and the curtain falls while the two are in their cups, the Jew insisting that in the firm name for the business, which they are to carry on jointly, his name shall stand first.

It is of course essential to any protection of literary property, whether at common-law or under the statute, that the right cannot be limited literally to the text, else a plagiarist would escape by immaterial variations. That has never been the law, but, as soon as literal appropriation ceases to be the test, the whole matter is necessarily at large, so that, as was recently well said by a distinguished judge, the decisions cannot help much in a new case. Fendler v. Morosco, 253 N.Y. 281, 292, 171 N.E. 56. When plays are concerned, the plagiarist may excise a separate scene; or he may appropriate part of the dialogue. Then the question is whether the part so taken is "substantial," and therefore not a "fair use" of the copyrighted work; it is the same question as arises in the case of any other copyrighted work. But when the plagiarist does not take out a block in suit, but an abstract of the whole, decision is more troublesome. Upon any work, and especially upon a play, a great number of patterns of increasing generality will fit equally well, as more and more of the incident is left out. The last may perhaps be no more than the most general statement of what the play is about, and at times might consist only of its title; but there is a point in this series of abstractions where they are no longer protected, since otherwise the playwright could prevent the use of his "ideas," to which, apart from their expression, his property is never extended. Nobody has ever been able to fix that boundary, and nobody ever can. In some cases the question has been treated as though it were analogous to lifting a portion out of the copyrighted work; but the analogy is not a good one, because, though the skeleton is a part of the body, it pervades and supports the whole. In such cases we are rather concerned with the line between expression and what is expressed. As respects plays, the controversy chiefly centers upon the characters and sequence of incident, these being the substance.

We did not in Dymow v. Bolton, 11 F.2d 690, hold that a plagiarist was never liable for stealing a plot; that would have been flatly against our ruling in Dam v. Kirk La Shelle Co., 175 F. 902, 41 L.R.A.(N.S.) 1002, 20 Ann.Cas. 1173, and Stodart v. Mutual Film Co., 249 F. 513, affirming my decision in (D.C.) 249 F. 507; neither of which we meant to overrule. We found the plot of the second play was too different to infringe, because the most detailed pattern, common to both, eliminated so much from each that its content went into the public domain; and for this reason we said, "this mere subsection of a plot was not susceptible of copyright." But we do not doubt that two plays may correspond in plot closely enough for infringement. How far that correspondence must go is another matter. Nor need we hold that the same may not be true as to the characters, quite independently of the "plot" proper, though, as far as we know such a case has never arisen. If Twelfth Night were copyrighted, it is quite possible that a second comer might so closely imitate Sir Toby Belch or Malvolio as to infringe, but it would not be enough that for one of his characters he cast a riotous knight who kept wassail to the discomfort of the household, or a vain and foppish steward who became amorous of his mistress. These would be no more than Shakespeare's "ideas" in the play, as little capable of monopoly as Einstein's Doctrine of Relativity, or Darwin's theory of the Origin of Species. It follows that the less developed the characters, the less they can be copyrighted; that is the penalty an author must bear for marking them too indistinctly.

In the two plays at bar we think both as to incident and character, the defendant took no more — assuming that it took anything at all — than the law allowed. The stories are quite different. One is of a religious zealot who insists upon his child's marrying no one outside his faith; opposed by another who is in this respect just like him, and is his foil. Their difference in race is merely an obligato to the main theme, religion. They sink their differences through grandparental pride and affection. In the other, zealotry is wholly absent; religion does not even appear. It is true that the parents are hostile to each other in part because they differ in race; but the marriage of their son to a Jew does no[t] apparently offend the Irish family at all, and it exacerbates the existing animosity of the Jew, principally because he has become rich, when he learns it. They are reconciled through the honesty of the Jew and the generosity of the Irishman; the grandchild has nothing whatever to do with it. The only matter common to the two is a quarrel between a Jewish and an Irish father, the marriage of their children, the birth of grandchildren and a reconciliation.

If the defendant took so much from the plaintiff, it may well have been because her amazing success seemed to prove that this was a subject of enduring popularity. Even so, granting that the plaintiff's play was wholly original, and assuming that novelty is not essential to a copyright, there is no monopoly in such a background. Though the plaintiff discovered the vein, she could not keep it to herself; so defined, the theme was too generalized an abstraction from what she wrote. It was only a part of her "ideas."

Nor does she fare better as to her characters. It is indeed scarcely credible that she should not have been aware of those stock figures, the low comedy Jew and Irishman. The defendant has not taken from her more than their prototypes have contained for many decades. If so, obviously so to generalize her copyright, would allow her to cover what was not original with her. But we need not hold this as matter of fact, much as we might be justified. Even though we take it that she devised her figures out of her brain de novo, still the defendant was within its rights.

There are but four characters common to both plays, the lovers and the fathers. The lovers are so faintly indicated as to be no more than stage properties. They are loving and fertile; that is really all that can be said of them, and anyone else is quite within his rights if he puts loving and fertile lovers in a play of his own, wherever he gets the cue. The Plaintiff's Jew is quite unlike the defendant's. His obsession in his religion, on which depends such racial animosity as he has. He is affectionate, warm and patriarchal. None of these fit the defendant's Jew, who shows affection for his daughter only once, and who has none but the most superficial interest in his grandchild. He is tricky, ostentatious and vulgar, only by misfortune redeemed into honesty. Both are grotesque, extravagant and quarrelsome; both are fond of display; but these common qualities make up only a small part of their simple pictures, no more than any one might lift if he chose. The Irish fathers are even more unlike; the plaintiff's a mere symbol for religious fanaticism and patriarchal pride, scarcely a character at all. Neither quality appears in the defendant's, for while he goes to get his grandchild, it is rather out of a truculent determination not to be forbidden, than from pride in his progeny. For the rest he is only a grotesque hobbledehoy, used for low comedy of the most conventional sort, which any one might borrow, if he chanced not to know the exemplar.

The defendant argues that the case is controlled by my decision in Fisher v. Dillingham (D.C.) 298 F. 145. Neither my brothers nor I wish to throw doubt upon the doctrine of that case, but it is not applicable here. We assume that the plaintiff's play is altogether original, even to an extent that in fact it is hard to believe. We assume further that, so far as it has been anticipated by earlier plays of which she knew nothing, that fact is immaterial. Still, as we have already said, her copyright did not cover everything that might be drawn from her play; its content went to some extent into the public domain. We have to decide how much, and while we are as aware as any one that the line, wherever it is drawn, will seem arbitrary, that is no excuse for not drawing it; it is a question such as courts must answer in nearly all cases. Whatever may be the difficulties a priori, we have no question on which side of the line this case falls. A comedy based upon conflicts between Irish and Jews, into which the marriage of their children enters, is no more susceptible of copyright than the outline of Romeo and Juliet.

The plaintiff has prepared an elaborate analysis of the two plays, showing a "quadrangle" of the common characters, in which each is represented by the emotions which he discovers. She presents the resulting parallelism as proof of infringement, but the adjectives employed are so general as to be quite useless. Take for example the attribute of "love" ascribed to both Jews. The plaintiff has depicted her father as deeply attached to his son, who is his hope and joy; not so, the defendant, whose father's conduct is throughout not actuated by any affection for his daughter, and who is merely once overcome for the moment by her distress when he has violently dismissed her lover. "Anger" covers emotions aroused by quite different occasions in each case; so do "anxiety," "despondency" and "disgust." It is unnecessary to go through the catalogue for emotions are too much colored by their causes to be a test when used so broadly. This is not the proper approach to a solution; it must be more ingenuous, more like that of a spectator, who would rely upon the complex of his impressions of each character. . . .

Decree affirmed.

COMMENTS AND QUESTIONS

1. *Test for Improper Appropriation.* Some decisions in the Ninth Circuit have bifurcated the analysis into an objective test and a subjective test. See, e.g., Shaw v. Lindheim, 919 F.2d 1353 (9th Cir. 1990). The objective test analytically dissects the objective manifestations of creativity (plots, themes, dialogue, mood, setting, pace, sequence, characters) in the plaintiff's work in order to determine those elements that are protectable under copyright law. In the second stage of analysis the trier of fact, applying a purely subjective perspective, determines whether the defendant's work improperly appropriates the plaintiff's protected expression.

Does this bifurcated approach make sense? How can the trier of fact compare the two works without relying on the objective approach to determine what is protectable?

Does it matter what sort of work is at issue? Are juries likely to be able to determine the similarities between two movies more easily than between two computer programs?

2. *Framing the Subjective Analysis Comparison.* A key issue in applying the intrinsic or subjective stage of analysis is delineating what the fact finder compares in deciding whether two works are substantially similar. Does the fact finder compare the two works as a whole or only those elements that are protectable? Courts have differed in their treatment of this critical issue. Some courts have held that the fact-finder shall compare the entirety of the two works, including the "unprotectable" elements. See, e.g., Roth Greeting Cards v. United Card Co., 429 F.2d 1106 (9th Cir. 1970); Sheldon v. Metro-Goldwyn Pictures Corp., 81 F.2d 49 (2d Cir. 1936). Other courts have excluded nonprotectable elements from the comparison. See, e.g., Hoehling v. Universal City Studios, Inc., 618 F.2d 972 (2d Cir.), *cert. denied,* 449 U.S. 841 (1980); Computer Associates International v. Altai, Inc., 982 F.2d 693 (2d Cir. 1992) (addressing the protection of computer code), infra Chapter 7. Which view comports best with copyright principles?

3. *How Much Must Be Taken to Constitute Improper Appropriation?* A copyright owner need not prove that all or nearly all of his or her work has been appropriated to establish infringement. Although the quantum necessary depends on the nature of the work, recall that the legislative history to §106 provides that "a copyrighted work would be infringed by reproducing it *in whole or in any substantial part,* and by duplicating it exactly or by

imitation or simulation. Wide departures or variations from the copyrighted works would still be an infringement as long as the author's 'expression' rather than merely the author's 'ideas' are taken." H.R. Rep. No. 94-1476, 94th Cong., 2d Sess. 61 (1976) (emphasis added). Thus courts have held that "[e]ven a small amount of the original, if it is qualitatively significant, may be sufficient to be an infringement. . . ." Horgan v. Macmillan, Inc., 789 F.2d 157, 162 (2d Cir. 1986).

Determining the threshold for infringement is particularly difficult in those cases in which a defendant has copied distinct literal elements of the plaintiff's work and incorporated them into a larger work of his or her own. This class of cases has been referred to as *fragmented literal similarity*. See Melville Nimmer & David Nimmer, on Copyright ¶13.03[A][2] at 13-46. The Nimmer treatise states:

> The question in each case is whether the similarity relates to matter which constitutes a substantial portion of plaintiff's work — not whether such material constitutes a substantial portion of defendant's work. The quantitative relation of the similar material to the total material contained in plaintiff's work is certainly of importance. However, even if the similar material is quantitatively small, if it is qualitatively important the trier of fact may properly find substantial similarity. In such circumstances the defendant may not claim immunity on the ground the infringement "is such a little one." If, however, the similarity is only as to nonessential matters, then a finding of no substantial similarity should result.

4. *The Role of Expert Testimony in Determining Improper Appropriation.* The court in *Arnstein* held that expert opinion is "utterly immaterial" to the determination of improper appropriation. Does this limitation on evidence make sense with regard to all works? Expert testimony would seem essential in assessing appropriation with regard to technically complex material written for specialized audiences. The issue arises frequently in the context of computer software copyright cases. See infra Chapter 7. Is it desirable to assess similarities in two database programs from the standpoint of the ordinary person in the street rather than the ordinary user of database programs? Isn't expert testimony on the extent to which programming elements are common in the trade essential to determining improper appropriation?

5. *The Appropriate Perspective for Assessing Substantial Similarity: The Ordinary Observer.* The Second Circuit has defined "substantial similarity" as whether the "ordinary observer, unless he set out to detect the disparities [between two works], would be disposed to overlook them, and regard their aesthetic appeal as the same." Peter Pan Fabrics, Inc. v. Martin Weiner Corp., 274 F.2d 487, 489 (2d Cir. 1960).

A number of cases have narrowed the "ordinary observer" perspective by focusing on the impressions of the target audience for the work in question. For example, in Original Appalachian Artworks, Inc. v. Blue Box Factory (USA) Ltd., 577 F. Supp. 625 (S.D.N.Y. 1983), involving copyright protection for a popular line of dolls called "Cabbage Patch Kids," the court allowed expert evidence about how the works would be perceived by children. In Data East USA, Inc. v. Epyx, Inc., 862 F.2d 204 (9th Cir. 1988), the court assessed substantial similarity of two karate video games from the perspective of a "discerning 17.5 year-old boy," based on the district court's finding that "the average age of individuals purchasing 'Karate Champ' is 17.5 years, that the purchasers are predominantly male, and comprise a knowledgeable, critical, and discerning group."

Should the "ordinary observer" test be tailored to the *target audience* for the works? Is the "ordinary observer" perspective, even if tailored to reflect the target audience for the work, likely to distinguish between the protectable and nonprotectable elements of a work in assessing infringement?

Steinberg v. Columbia Pictures Industries, Inc.
United States District Court for the Southern District of New York
663 F. Supp. 796 (S.D.N.Y. 1987)

STANTON, District Judge.

In these actions for copyright infringement, plaintiff Saul Steinberg is suing the producers, promoters, distributors and advertisers of the movie "Moscow on the Hudson" ("Moscow"). Steinberg is an artist whose fame derives in part from cartoons and illustrations he has drawn for *The New Yorker* magazine. Defendant Columbia Pictures Industries, Inc. (Columbia) is in the business of producing, promoting and distributing motion pictures, including "Moscow." . . .

Plaintiff alleges that defendants' promotional poster for "Moscow" infringes his copyright on an illustration that he drew for *The New Yorker* and that appeared on the cover of the March 29, 1976 issue of the magazine, in violation of 17 U.S.C. §§101-810. Defendants deny this allegation and assert the affirmative defenses of fair use as a parody, estoppel and laches.

Defendants have moved, and plaintiff has cross-moved, for summary judgment. For the reasons set forth below, this court rejects defendants' asserted defenses and grants summary judgment on the issue of copying to plaintiff. . . .

II

The essential facts are not disputed by the parties despite their disagreements on nonessential matters. On March 29, 1976, *The New Yorker* published as a cover illustration the work at issue in this suit, widely known as a parochial New Yorker's view of the world. The magazine registered this illustration with the United States Copyright Office and subsequently assigned the copyright to Steinberg. Approximately three months later, plaintiff and *The New Yorker* entered into an agreement to print and sell a certain number of posters of the cover illustration.

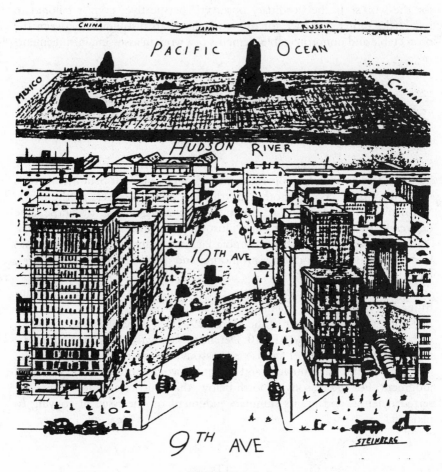

FIGURE 4-6

FIGURE 4-7
Moscow on the Hudson poster.

It is undisputed that unauthorized duplications of the poster were made and distributed by unknown persons, although the parties disagree on the extent to which plaintiff attempted to prevent the distribution of those counterfeits. Plaintiff has also conceded that numerous posters have been created and published depicting other localities in the same manner that he depicted New York in his illustration. These facts, however, are irrelevant to the merits of this case, which concerns only the relationship between plaintiff's and defendants' illustrations.

Defendants' illustration was created to advertise the movie "Moscow on the Hudson," which recounts the adventures of a Muscovite who defects in New York. In designing this illustration, Columbia's executive art director, Kevin Nolan, has admitted that he specifically referred to Steinberg's poster, and indeed, that he purchased it and hung it, among others, in his office. Furthermore, Nolan explicitly directed the outside artist whom he retained to execute his design, Craig Nelson, to use Steinberg's poster to achieve a more recognizably New York look. Indeed, Nelson acknowledged having used the facade of one particular edifice, at Nolan's suggestion that it would render his drawing more "New York-ish." While the two buildings are not identical, they are so similar that it is impossible, especially in view of the artist's testimony, not to find that defendants' impermissibly copied plaintiff's.

To decide the issue of infringement, it is necessary to consider the posters themselves. Steinberg's illustration presents a bird's eye view across a portion of the western edge of Manhattan, past the Hudson River and a telescoped version of the rest of the United States and the Pacific Ocean, to a red strip of horizon, beneath which are three flat land masses labeled China, Japan and Russia. The name of the magazine, in *The New Yorker*'s usual typeface, occupies the top fifth of the poster, beneath a thin band of blue wash representing a stylized sky.

The parts of the poster beyond New York are minimalized, to symbolize a New Yorker's myopic view of the centrality of his city to the world. The entire United States west of the Hudson River, for example, is reduced to a brown strip labeled "Jersey," together with a light green trapezoid with a few rudimentary rock outcroppings and the names of only seven cities and two states scattered across it. The few blocks of Manhattan, by contrast, are depicted and colored in detail. The four square blocks of the city, which occupy the whole lower half of the poster, include numerous buildings, pedestrians and cars, as well as parking lots and lamp posts, with water towers atop a few of the buildings. The whimsical, sketchy style and spiky lettering are recognizable as Steinberg's.

The "Moscow" illustration depicts the three main characters of the film on the lower third of their poster, superimposed on a bird's eye view of New York City, and continues eastward across Manhattan and the Atlantic Ocean, past a rudimentary evocation of Europe, to a clump of recognizably Russian-styled buildings on the horizon, labeled "Moscow." The movie credits appear over the lower portion of the characters. The central part of the poster depicts

approximately four New York city blocks, with fairly detailed buildings, pedestrians and vehicles, a parking lot, and some water towers and lamp posts.
Columbia's artist added a few New York landmarks at apparently random
places in his illustration, apparently to render the locale more easily recognizable. Beyond the blue strip labeled "Atlantic Ocean," Europe is represented by London, Paris and Rome, each anchored by a single landmark
(although the landmark used for Rome is the Leaning Tower of Pisa).

The horizon behind Moscow is delineated by a red crayoned strip, above
which are the title of the movie and a brief textual introduction to the plot.
The poster is crowned by a thin strip of blue wash, apparently a stylization
of the sky. This poster is executed in a blend of styles: the three characters,
whose likenesses were copied from a photograph, have realistic faces and
somewhat sketchy clothing, and the city blocks are drawn in a fairly detailed
but sketchy style. The lettering on the drawing is spiky, in block-printed
handwritten capital letters substantially identical to plaintiff's, while the
printed texts at the top and bottom of the poster are in the typeface commonly associated with *The New Yorker* magazine.[13]

III

To succeed in a copyright infringement action, a plaintiff must prove
ownership of the copyright and copying by the defendant. Reyher v. Children's Television Workshop, 533 F.2d 87, 90 (2d Cir. 1976). There is no
substantial dispute concerning plaintiff's ownership of a valid copyright in his
illustration. Therefore, in order to prevail on liability, plaintiff need establish
only the second element of the cause of action.

"Because of the inherent difficulty in obtaining direct evidence of copying, it is usually proved by circumstantial evidence of access to the copyrighted
work and substantial similarities as to protectible material in the two works."
Reyher, 533 F.2d at 90, citing Arnstein v. Porter, 154 F.2d 464, 468 (2d
Cir. 1946). "Of course, if there are no similarities, no amount of evidence of
access will suffice to prove copying." Arnstein v. Porter, 154 F.2d at 468.

Defendants' access to plaintiff's illustration is established beyond per-
adventure. Therefore, the sole issue remaining with respect to liability is
whether there is such substantial similarity between the copyrighted and accused works as to establish a violation of plaintiff's copyright. The central
issue of "substantial similarity," which can be considered a close question of
fact, may also validly be decided as a question of law.

"Substantial similarity" is an elusive concept. This circuit has recently
recognized that

---

13. The typeface is not a subject of copyright, but the similarity reinforces the impression
that defendants copied plaintiff's illustration.

[t]he "substantial similarity" that supports an inference of copying sufficient to establish infringement of a copyright is not a concept familiar to the public at large. It is a term to be used in a courtroom to strike a delicate balance between the protection to which authors are entitled under an act of Congress and the freedom that exists for all others to create their works outside the area protected by infringement.

Warner Bros., 720 F.2d at 245.

The definition of "substantial similarity" in this circuit is "whether an average lay observer would recognize the alleged copy as having been appropriated from the copyrighted work." Ideal Toy Corp. v. Fab-Lu Ltd., 360 F.2d 1021, 1022 (2d Cir. 1966); Silverman v. CBS, Inc., 632 F. Supp. at 1351-52. A plaintiff need no longer meet the severe "ordinary observer" test established by Judge Learned Hand in Peter Pan Fabrics, Inc. v. Martin Weiner Corp., 274 F.2d 487 (2d Cir. 1960). Uneeda Doll Co., Inc. v. Regent Baby Products Corp., 355 F.Supp. 438, 450 (E.D.N.Y. 1972). Under Judge Hand's formulation, there would be substantial similarity only where "the ordinary observer, unless he set out to detect the disparities, would be disposed to overlook them, and regard their aesthetic appeal as the same." 274 F.2d at 489.

Moreover, it is now recognized that "[t]he copying need not be of every detail so long as the copy is substantially similar to the copyrighted work." Comptone Co. v. Rayex Corp., 251 F.2d 487, 488 (2d Cir. 1958).

In determining whether there is substantial similarity between two works, it is crucial to distinguish between an idea and its expression. It is an axiom of copyright law, established in the case law and since codified at 17 U.S.C. §102(b), that only the particular expression of an idea is protectible, while the idea itself is not.

"The idea/expression distinction, although an imprecise tool, has not been abandoned because we have as yet discovered no better way to reconcile the two competing societal interests that provide the rationale for the granting of and restrictions on copyright protection," namely, both rewarding individual ingenuity, and nevertheless allowing progress and improvements based on the same subject matter by others than the original author. *Durham Industries*, 630 F.2d at 912, quoting *Reyher*, 533 F.2d at 90.

There is no dispute that defendants cannot be held liable for using the idea of a map of the world from an egocentrically myopic perspective. No rigid principle has been developed, however, to ascertain when one has gone beyond the idea to the expression, and "[d]ecisions must therefore inevitably be ad hoc." Peter Pan Fabrics, Inc. v. Martin Weiner Corp., 274 F.2d 487, 489 (2d Cir. 1960) (L. Hand, J.). As Judge Frankel once observed, "Good eyes and common sense may be as useful as deep study of reported and unreported cases, which themselves are tied to highly particularized facts." Couleur International Ltd. v. Opulent Fabrics, Inc., 330 F. Supp. 152, 153 (S.D.N.Y. 1971).

Even at first glance, one can see the striking stylistic relationship between the posters, and since style is one ingredient of "expression," this relationship is significant. Defendants' illustration was executed in the sketchy, whimsical style that has become one of Steinberg's hallmarks. Both illustrations represent a bird's eye view across the edge of Manhattan and a river bordering New York City to the world beyond. Both depict approximately four city blocks in detail and become increasingly minimalist as the design recedes into the background. Both use the device of a narrow band of blue wash across the top of the poster to represent the sky, and both delineate the horizon with a band of primary red.

The strongest similarity is evident in the rendering of the New York City blocks. Both artists chose a vantage point that looks directly down a wide two-way cross street that intersects two avenues before reaching a river. Despite defendants' protestations, this is not an inevitable way of depicting blocks in a city with a grid-like street system, particularly since most New York City cross streets are one-way. Since even a photograph may be copyrighted because "no photograph, however simple, can be unaffected by the personal influence of the author," Time Inc. v. Bernard Geis Assoc., 293 F. Supp. 130, 141 (S.D.N.Y. 1968), quoting *Bleistein*, supra, one can hardly gainsay the right of an artist to protect his choice of perspective and lay-out in a drawing, especially in conjunction with the overall concept and individual details. Indeed, the fact that defendants changed the names of the streets while retaining the same graphic depiction weakens their case: had they intended their illustration realistically to depict the streets labeled on the poster, their four city blocks would not so closely resemble plaintiff's four city blocks. Moreover, their argument that they intended the jumble of streets and landmarks and buildings to symbolize their Muscovite protagonist's confusion in a new city does not detract from the strong similarity between their poster and Steinberg's.

While not all of the details are identical, many of them could be mistaken for one another; for example, the depiction of the water towers, and the cars, and the red sign above a parking lot, and even many of the individual buildings. The shapes, windows, and configurations of various edifices are substantially similar. The ornaments, facades and details of Steinberg's buildings appear in defendants', although occasionally at other locations. In this context, it is significant that Steinberg did not depict any buildings actually erected in New York; rather, he was inspired by the general appearance of the structures on the West Side of Manhattan to create his own New Yorkish structures. Thus, the similarity between the buildings depicted in the "Moscow" and Steinberg posters cannot be explained by an assertion that the artists happened to choose the same buildings to draw. The close similarity can be explained only by the defendants' artist having copied the plaintiff's work. Similarly, the locations and size, the errors and anomalies of Steinberg's shadows and streetlight, are meticulously imitated.

In addition, the Columbia artist's use of the childlike, spiky block print that has become one of Steinberg's hallmarks to letter the names of the streets in the "Moscow" poster can be explained only as copying. There is no inherent justification for using this style of lettering to label New York City streets as it is associated with New York only through Steinberg's poster.

While defendants' poster shows the city of Moscow on the horizon in far greater detail than anything is depicted in the background of plaintiff's illustration, this fact alone cannot alter the conclusion. "Substantial similarity" does not require identity, and "duplication or near identity is not necessary to establish infringement." *Krofft*, 562 F.2d at 1167. Neither the depiction of Moscow, nor the eastward perspective, nor the presence of randomly scattered New York City landmarks in defendants' poster suffices to eliminate the substantial similarity between the posters. As Judge Learned Hand wrote, "no plagiarist can excuse the wrong by showing how much of his work he did not pirate." Sheldon v. Metro-Goldwyn Pictures Corp., 81 F.2d 49, 56 (2d Cir.), cert. denied, 298 U.S. 669, 56 S. Ct. 835, 80 L. Ed. 1392 (1936).

Defendants argue that their poster could not infringe plaintiff's copyright because only a small proportion of its design could possibly be considered similar. This argument is both factually and legally without merit. "[A] copyright infringement may occur by reason of a substantial similarity that involves only a small portion of each work." Burroughs v. Metro-Goldwyn-Mayer, Inc., 683 F.2d 610, 624 n.14 (2d Cir. 1982). Moreover, this case involves the entire protected work and an iconographically, as well as proportionately, significant portion of the allegedly infringing work.

The process by which defendants' poster was created also undermines this argument. The "map," that is, the portion about which plaintiff is complaining, was designed separately from the rest of the poster. The likenesses of the three main characters, which were copied from a photograph, and the blocks of text were superimposed on the completed map.

I also reject defendants' argument that any similarities between the works are unprotectible *scenes a faire*, or "incidents, characters or settings which, as a practical matter, are indispensable or standard in the treatment of a given topic." *Walker*, 615 F. Supp. at 436. See also *Reyher*, 533 F.2d at 92. It is undeniable that a drawing of New York City blocks could be expected to include buildings, pedestrians, vehicles, lampposts and water towers. Plaintiff, however, does not complain of defendants' mere use of these elements in their poster; rather, his complaint is that defendants copied his expression of those elements of a street scene.

While evidence of independent creation by the defendants would rebut plaintiff's prima facie case, "the absence of any countervailing evidence of creation independent of the copyrighted source may well render clearly erroneous a finding that there was not copying." Roth Greeting Cards v. United Card Co., 429 F.2d 1106, 1110 (9th Cir. 1970).

Moreover, it is generally recognized that ". . . since a very high degree of similarity is required in order to dispense with proof of access, it must logically follow that where proof of access is offered, the required degree of similarity may be somewhat less than would be necessary in the absence of such proof." 2 Nimmer § 143.4 at 634, quoted in *Krofft*, 562 F.2d at 1172. As defendants have conceded access to plaintiff's copyrighted illustration, a somewhat lesser degree of similarity suffices to establish a copyright infringement than might otherwise be required. Here, however, the demonstrable similarities are such that proof of access, although in fact conceded, is almost unnecessary. . . .

VI

For the reasons set out above, summary judgment is granted to plaintiffs as to copying. . . .

COMMENTS AND QUESTIONS

1. *Proof of Copying.* Many copyright decisions, like this one, do not clearly distinguish between the analysis of copying and improper appropriation. Was there sufficient evidence to prove copying?

2. *Improper Appropriation.* Do you agree with the court's assessment of improper appropriation? Which elements of Steinberg's work are protected under copyright law? Did Columbia Pictures copy those elements? Did they copy Steinberg's compilation of otherwise unprotectable elements? Suppose you were arguing this case on appeal. What arguments would you make on behalf of the defendant? What counterarguments would you make on behalf of the plaintiff?

PROBLEMS

Problem 4-21. Dinopets markets a line of stuffed animal toys for children. The line includes five popular dinosaurs with exaggerated facial features (e.g., large droopy eyes, long teeth, rounded noses), cheerful pastel colors (pink, lemon, lime), distinctive stitching, and a soft cuddly cotton texture. About a year after Dinopets were on the market, Gigatoys, Inc., a leading toy manufacturer, developed a line of stuffed dinosaur toys. Its line, the Dinomites, featured the five dinosaurs in the Dinopets line as well as three others. Dinomites are about 25 percent larger than comparable Dinopets. Dinomites feature cute facial features (including droopy eyes and long teeth) and come in earth-tone colors (light brown, clay, sand, and stone). They are made of a suede-like material (somewhat coarser than the Dinopets). Dinopets sues Gigatoys, alleging copyright infringement. How would the analysis be conducted? What result?

Problem 4-22. The Stunted Intellectuals, a psychedelic punk rap jazz fusion band, write songs by taking digital "samples" from previous compositions of a number of different genres, juxtaposing or super-imposing them, and laying an electronic beat under and continuous rap commentary over the mixture. Some of the "samples" are from works in the public domain; of the remainder, some are identifiable and un-altered, some are identifiable but altered, and some are unrecognizable as a prior song. No sample lasts for longer than four seconds in any song, but some are repeated within the same song. Have the Stunted Intellectuals infringed the copyright on any of the songs they sample? On all of them?

Problem 4-23. In preparing a biography of the reclusive author J. D. Salinger, Ian Hamilton gained access to letters Salinger wrote to a number of notable people that had been donated to university librar-ies. Through these letters and other sources, Hamilton constructed his biography of Salinger's life. Out of concern for copyright infringement, Hamilton quotes barely more than 200 words from the letters through-out the entire biography. Nonetheless, the letters are quoted or oth-erwise drawn upon in approximately 40 percent of the 192-page biography. To accurately describe events and emotions, impart some of Salinger's distinctive style, and avoid "pedestrian" reporting, Hamilton follows some of the passages from the letters closely.

The following examples illustrate Hamilton's use of the letters to present Salinger's life. In a 1943 letter to Whit Burnett, Salinger's friend, teacher, and editor at *Story* magazine, Salinger expressed his disapproval of the marriage of Oona O'Neil, with whom Salinger had been romantically involved, and Charlie Chaplin, the silent screen film star.

Salinger's Letter	*Hamilton's Biography*
I can see them at home evenings. Chaplin squatting grey and nude, atop his chiffonier, swinging his thy-roid around his head by his bamboo cane, like a dead rat. Oona in an aqua-marine gown, applauding madly from the bathroom. Agnes (her mother) in a Jantzen bathing suit, passing between them with cocktails. I'm facetious, but I'm sorry. Sorry for anyone with a profile as young and lovely as Oona's.	At one point in a letter to Whit Bur-nett, he provides a pen portrait of the Happy Hour Chez Chaplin: the co-median, ancient and unclothed, is brandishing his walking stick — at-tached to the stick, and horribly re-sembling a lifeless rodent, is one of Chaplin's vital organs. Oona claps her hands in appreciation and Agnes, togged out in a bathing suit, pours drinks. Salinger goes on to say he's sorry — sorry not for what he has just written, but for Oona: far too youthful and exquisite for such a dreadful fate.

In another letter to Burnett, Salinger expresses his disfavor of presidential candidate Wendell Wilkie.

Salinger's Letter	*Hamilton's Biography*
He looks to me like a guy who makes his wife keep a scrapbook for him.	[Salinger] had fingered [Wilkie] as the sort of fellow who makes his wife keep an album of his press cuttings.

In another letter describing Parisians' adulation of American soldiers at the liberation of Paris, Salinger writes that the Parisians would have said "What a charming custom!" if:

Salinger's Letter	*Hamilton's Biography*
we had stood on top of the jeep and taken a leak.	the conquerors had chosen to urinate from the roofs of their vehicles.

Has Hamilton infringed Salinger's copyright in his letters?

Limitations on the Exclusive Right to Copy. With certain exceptions (discussed below), the right to prevent copying is absolute. The copyright laws do not apply only to copying for a commercial purpose, or to large-scale copying, or to copying by businesses. Any individual act of copying of protectable expression, for any purpose, presumptively violates the Copyright Act. There are several statutory exceptions to this general rule:

- Section 108 exempts from copyright liability a public "library or archives" which makes only one copy of a work at a time, assuming the copy is made for specified purposes. Copies may be made for the preservation and replacement of existing works, but only if the work cannot be replaced by purchase at a "fair price." Libraries can also make single copies for noncommercial users, as long as the library does not engage in the "systematic reproduction or distribution" of such copies. Finally, libraries are not liable for copyright violations by their patrons (even those using on-site photocopiers), as long as the library posts conspicuous warnings notifying users of the copyright laws.
- Sections 112 and 118(d) permit broadcasters to make "ephemeral" or "ancillary" copies of certain performances and displays during the course of broadcasting. For example, broadcasters are permitted to make a copy during the course of retransmitting a program. We discuss these exceptions in more detail when we consider performance and display rights.
- Section 115 of the act permits anyone to record a musical composition upon the payment of a royalty determined by a formula specified in the statute. (The Copyright Royalty Tribunal, which awarded royalties under this compulsory licensing provision, was abolished by Congress

in 1993.) However, artists cannot record another's musical compositions if they have not been recorded with permission at least once before. Further, section 115 does not permit new artists to make substantial changes in the work recorded, except to the extent necessary to adapt the work to a new genre.

- Section 1008, added by the Audio Home Recording Act of 1992 (AHRA), authorizes "consumers" to make copies of sound recordings for "noncommercial use." The immunity for home taping is part of a broader compromise that resolved issues surrounding the use of "digital audio tape" (DAT) technology. In return for this immunity from suit, manufacturers of DAT decks and tapes must pay a royalty to the Copyright Office for distribution to copyright owners. Further, the act outlaws the sale of DAT decks that can copy copies. We discuss the AHRA in greater detail below.

In addition to these limitations, the fair use doctrine provides a defense in many instances of copying. We consider fair use in detail later.

2. The Right to Prepare Derivative Works

Paul Goldstein, Derivative Rights and
Derivative Works in Copyright
30 J. Copr. Society 209 (1983)

There has been a quiet revolution in copyright law and the copyright industries. Copyright, which once protected only against the production of substantially similar copies in the same medium as the copyrighted work, today protects against uses and media that often lie far afield from the original. Copyright's subject matter has grown, too, making many of these uses and media themselves copyrightable. Both developments reflect the growth of new copyright industries. Hardcover book sales, which once represented the principal measure of a novel's popular success, are today dwarfed by the income from motion pictures, television series, sequels and merchandise derived from the novel. One current, popular motion picture, selling about $3,000,000 in tickets a day, will reportedly earn even more from sales of dolls, sheets, posters, books and a full range of character merchandise.

The 1976 Copyright Act, like the 1909 Act which preceded it, consolidates and advances these expansionary trends in protected rights and protectable subject matter. The 1976 Act gives the copyright owner not only the exclusive right to reproduce its work in copies, but also the exclusive right "to prepare derivative works based upon the copyrighted work." (The Act [§101] defines derivative work as "a work based upon one or more preexisting works" and cites as examples, "a translation, musical arrangement, dramatization, fictionalization, motion picture version, sound recording, art reproduction, abridgment, condensation, or any other form in which a work may be recast, transformed, or adapted.") The 1976 Act also leaves no doubt that derivative works are themselves independently copyrightable and that the derivative author's transformation of the underlying work need not be extensive: "A work consisting of editorial revisions, annotations, elaborations, or other modifications which, as a whole, represent an original work of authorship, is a 'derivative work.'"

Although the Act's commitment to derivative rights and derivative works is clear, judicial acceptance has been uneven. One reason — traced in the first part of this article — is historical. Copyright law was first shaped around the technology of the printing press and around the assumption that the law's proper concern was with literal copies rendered in the same medium. While the subsequent growth in legitimate theaters, motion pictures and television opened vast new markets for derivative uses, impelling Congress to grant derivative rights to copyrighted works and to grant copyright protection to the derivative works created, courts either lagged or overreacted in accommodating the statute to the new rights and subject matter. In many unfortunate respects, copyright doctrine today remains wedded to the economics of the printing press and is divorced from the revolutionary realities of twentieth century markets.

Another problem — to be addressed in the remainder of this article — is that courts have had no general theory to guide them in the evenhanded resolution of cases involving derivative works. Copyright's competing philosophies — biases, really — offer little help in resolving the more difficult issues. Neither "high protectionism," which favors an expansive construction of copyright, nor "low protectionism," which favors a constricted reading, can satisfactorily resolve the paradox that every infringer of a derivative right is, by definition, itself the potential copyright owner of a derivative work, with an equal claim on copyright's system of investment incentives. The fact that the Copyright Act aims to encourage investment in original and derivative works alike seriously complicates the determination whether a particular derivative work infringes an original work. . . .

Copyright's underlying economic principles offer a helpful starting point for answering these questions. The purpose of copyright is to attract private investment to the production of original expression. Copyright seeks to achieve this purpose by giving copyright owners the exclusive right to reap the profits taken from reproductions, performances and other specifically prohibited uses of their original expression. This method implies a floor for investment by requiring that, to be protected, a work be original with the author and not copied from some other source. The method also implies a ceiling on investment by protecting expressions but not their underlying ideas. The incentive structure of copyright is to channel investment to the production and to avoid infringing other copyrighted works.

Copyright's seminal design for regulating the production of copies provides the clearest example of this incentive structure at work. Section 106(1)'s grant of the exclusive reproduction right and section 102(a)'s declaration that copyright subsists in original works of authorship together establish one set of upper and lower limits to copyright investment. Margaret Mitchell and her publisher will invest time and money in writing, editing, producing and promoting the popular novel, *Gone With the Wind,* knowing that no one may copy the work's expressive content without their consent. They are also presumably aware of the floor and ceiling to their protection. They know that others will be free to use any of the novel's underlying ideas — basic plot, theme and character elements — just as they themselves were free in producing the novel to borrow ideas from other works.

The incentive structure of the Act's provisions respecting derivative rights and derivative works differs in two ways. First, section 106(2)'s grant of the exclusive right "to prepare derivative works based upon the copyrighted work" enables prospective copyright owners to proportion their investment in a work's expression to the returns expected not only from the market in which the copyrighted work is first published, but from other derivative markets as well. The copyright owners of *Gone With the Wind* can hope to monopolize not only the sale of the novel's hardcover and paperback editions, but also the use of the novel's expressive elements in translations, motion pictures and countless other derivative formats. Second, just as these owners had a copyright incentive to originate the expression for the novel, *Gone With the Wind,* section 103 — which extends copyright protection to the original elements of derivative works — gives them and their licensees an incentive to add original expression to each derivative work in order to qualify it for copyright protection of its own.

Taken together, sections 102(a) and 103, and sections 106(1) and 106(2), give a prospective copyright owner the incentive to make an original, underlying work, the exclusive right to make new, successive works incorporating expressive elements from the underlying work, and the incentive and exclusive right to make still newer successive works based on these. The continuum may stretch from an underlying novel or story to the work's adaptation into a motion picture, its transformation into a television series, and the eventual embodiment of its characters in dolls, games and other merchandise. The works at the outer reaches of this continuum, and some intermediate works as well, will frequently bear scant resemblance to the expression *or* the ideas of the seminal work and will often be connected only by a license authorizing use of a title or character name.

This analysis offers some help identifying the point at which the right "to reproduce the copyrighted work in copies" leaves off and the right "to prepare derivative works based upon the copyrighted work" begins: It is that point at which the contribution of independent expression to an existing work effectively creates a new work for a different market. The infringer who copies a novel verbatim violates only the right to reproduce, for he has created neither independent expression nor a new market. An infringing novel that borrows expressive elements from the original, but adds new expressive elements of its own, also violated the right to reproduce since, treating two novels with overlapping expression and essentially the same themes and characters as close substitutes, no new market has been created. A paperback edition of a hard-cover novel will also violate the right to reproduce since, though aimed at an arguably different market, it adds nothing expressively different to the original. By contrast, motion pictures, translations and comic strips based on the novel will all infringe the derivative right because they add new expressive elements and serve markets that differ from the market in which the original was first introduced.[29]

Having determined that a derivative right is in issue, it is far more difficult and consequential to draw the line that separates infringing from non-infringing derivative users. The central problem is that *all* works are to some extent based on works that precede them. Ravel's orchestration for *Pictures at an Exhibition* clearly derived from Moussorgsky's *Suite for Piano*. But it is no less true that Moussorgsky derived the inspiration for his work from Victor Hartmann's sketches and drawings displayed in 1874 in the rooms of the St. Petersburg Society of Architects, and the Hartmann sketches and drawings derived from subjects and from compositional, stylistic and thematic elements appearing in earlier works. The example, admittedly stretched, should at least suggest the nature of the problem. . . .

II. Derivative Works

Copyright offers both incentives and deterrents to derivative uses, protecting derivative works as well as prohibiting their unauthorized production. Since the decision to grant protection in any case implies that the derivative work will be given the same array of rights against reproduction and derivative uses as are given to other copyrighted works, the standards to be applied are in many respects identical to those that are applied to original works generally. Yet, the fact that the derivative work rests to some degree on a preexisting work also implies significant differences.

The principal difference in the structure of incentives for the production of derivative works stems from the work's connection to an underlying work, of which the derivative work's proprietor may be the owner, the owner's licensee, or an infringer. The nature of the connection is clearest when the owner of the derivative work is also the owner of the underlying work. Having once produced the expression in the underlying work, the copyright owner should gain no more rights when the same expression appears in an author-

---

29. Has an infringer who strings together excerpts from the novel violated the reproduction right, the derivative right, or both? Although the infringer has added no expression of his own, he has contributed the arguably expressive effort of editing; and although the market for abridgements and condensations substantially overlaps the market for complete novels, the overlap is by no means perfect. Cases like these underscore the fact that the difference between the right to reproduce and the right to prepare derivative works will often be one of degree rather than kind and that, in these cases, the distinction is best resolved according to the comparative degree to which the infringing work belongs to one category rather than the other: the degree to which it adds expressive elements to the underlying work and the degree to which it serves a market different from the market for the underlying work.

ized reproduction. The copyright owner will, however, obtain new rights in the derivative work's original expressive content, just as if it had created an entirely original work. Thus, if the underlying work's expression has, for some reason, fallen into the public domain, so should the same expression in the derivative work, while the derivative work's original expression will be unaffected. And, if the derivative work's expression falls into the public domain, so should the same expression in the underlying work.

The rules should be no different when the derivative work is produced under license from the owner of the underlying work, for a license means only that the owner, who in the previous example elected to act as an integrated producer of underlying works and derivative works, has here chosen to disintegrate, dividing functions among presumably more efficient licensees. Further, licensor and licensee can provide in the license agreement for any assurances and indemnifications needed to protect the licensor against the possibility that the licensee's noncompliance with copyright formalities will place the command expressive elements of the two works in the public domain.

The rules should, however, differ when it is an infringer rather than a licensee who produces the derivative work. The absence of any consensual arrangement between owner and infringer mean, of course, that the owner could not have contracted against acts of the infringer that might divest copyright protection for the underlying expression. As a result, divesting acts by the infringer . . . should not defeat the copyright owner's underlying rights. Should the fact that the derivative work is an infringing work also deprive it of copyright protection for its original, expressive elements? What if the borrowed, expressive elements pervade the entire derivative work? . . . [T]he 1976 Act has not been particularly discriminating in its answer. . . .

More explicit and systematic attention to the economics of copyright's incentive system will produce better legislative and judicial decisions on derivative rights and derivative works. . . . In reaching decisions affecting derivative rights and derivative works, courts and the Congress should attend more closely to copyright's incentive system and to the system's special implications for this growing corner of copyright law.

≡≡≡ *Midway Mfg. Co. v. Artic International, Inc.*
United States Court of Appeals for the Seventh Circuit
≡≡≡ *704 F.2d 1009 (7th Cir. 1983),* cert. denied, *464 U.S. 923 (1983)*

CUMMINGS, Chief Judge. . . .

Plaintiff manufactures video game machines. Inside these machines are printed circuit boards capable of causing images to appear on a television picture screen and sounds to emanate from a speaker when an electric current is passed through them. On the outside of each machine are a picture screen, sound speaker, and a lever or button that allows a person using the machine to alter the images appearing on the machine's picture screen and the sounds emanating from its speaker. Each machine can produce a large number of related images and sounds. These sounds and images are stored on the machine's circuit boards — how the circuits are arranged and connected determines the set of sounds and images the machine is capable of making. When a person touches the control lever or button on the outside of the machine he sends a signal to the circuit boards inside the machine which causes them to retrieve and display one of the sounds and images stored in them. Playing a video game involves manipulating the controls on the machine so that some of the images stored in the machine's circuitry appear on its picture screen and some of its sounds emanate from its speaker.

Defendant sells printed circuit boards for use inside video game machines. One of the circuit boards defendant sells speeds up the rate of play — how fast the sounds and images change — of "Galaxian," one of plaintiff's video games, when inserted in place of one of the "Galaxian" machine's circuit boards. Another of defendant's circuit boards stores a set of images and sounds almost identical to that stored in the circuit boards of plaintiff's "Pac- Man" video game machine so that the video game people play on machines containing defendant's circuit board looks and sounds virtually the same as plaintiff's "Pac-Man" game.

Plaintiff sued defendant alleging that defendant's sale of these two circuit boards infringes its copyrights in its "Galaxian" and "Pac-Man" video games. . . .

The final argument of defendant's that we address is that selling plaintiff's licensees circuit boards that speed up the rate of play of plaintiff's video games is not an infringement of plaintiff's copyrights. Speeding up the rate of play of a video game is a little like playing at 45 or 78 revolutions per minute ("RPM's") a phonograph record recorded at 33 RPM's. If a discotheque licensee did that, it would probably not be an infringement of the record company's copyright in the record. One might argue by analogy that it is not a copyright infringement for video game licensees to speed up the rate of play of video games, and that it is not a contributory infringement for the defendant to sell licensees circuit boards that enable them to do that.

There is this critical difference between playing records at a faster than recorded speed and playing video games at a faster than manufactured rate: there is an enormous demand for speeded-up video games but there is little if any demand for speeded-up records. Not many people want to hear 33 RPM records played at 45 and 78 RPM's so that record licensors would not care if their licensees play them at that speed. But there is a big demand for speeded-up video games. Speeding up a video game's action makes the game more challenging and exciting and increases the licensee's revenue per game. Speeded-up games end sooner than normal games and consequently if players are willing to pay an additional price-per-minute in exchange for the challenge and excitement of a faster game, licensees will take in greater total revenues. Video game copyright owners would undoubtedly like to lay their hands on some of that extra revenue and therefore it cannot be assumed that licensees are implicitly authorized to use speeded-up circuit boards in the machines plaintiff supplies.

Among a copyright owner's exclusive rights is the right "to prepare derivative works based upon the copyrighted work." 17 U.S.C. §106(2). If, as we hold, the speeded-up "Galaxian" game that a licensee creates with a circuit board supplied by the defendant is a derivative work based upon "Galaxian," a licensee who lacks the plaintiff's authorization to create a derivative work is a direct infringer and the defendant is a contributory infringer through its sale of the speeded-up circuit board. See, e.g., Gershwin Publishing Corp. v. Columbia Artists Mgmt., Inc., 443 F.2d 1159, 1162 (2d Cir. 1971); Universal City Studios, Inc. v. Sony Corp. of America, 659 F.2d 963, 975 (9th Cir. 1981), *certiorari granted,* 457 U.S. 1116, 102 S. Ct. 2926, 73 L. Ed. 2d 1326 (1982).

Section 101 of the 1976 Copyright Act defines a derivative work as "a work based upon one or more preexisting works, such as a translation, musical arrangement, dramatization, fictionalization, motion picture version, sound recording, art reproduction, abridgment, condensation, or any other form in which a work may be recast, transformed, or adapted." It is not obvious from this language whether a speeded-up video game is a derivative work. A speeded-up phonograph record probably is not. Cf. Shapiro, Bernstein & Co. v. Jerry Vogel Music Co., 73 F. Supp. 165, 167 (S.D.N.Y. 1947) ("The change in time of the added chorus, and the slight variation in the base of the accompaniment, there being no change in the tune or lyrics, would not be 'new work' "); 1 Nimmer on Copyright §3.03 (1982). But that is because the additional value to the copyright owner of having the right to market separately the speeded-up version of the recorded performance is too trivial to warrant legal protection for that right. A speeded-up video game is a substantially different product from the original game. As noted, it is more exciting to play and it requires some creative effort to produce. For that reason, the owner of the copyright on the game should be entitled to monopolize it on the same theory that he is entitled to monopolize the derivative works specifically listed in Section 101. The current rage for video games was not anticipated in 1976, and like any new technology the video game does not fit with complete ease the definition of derivative work in Section 101 of the 1976 Act. But the amount by which the language of Section 101 must be stretched to accommodate speeded-up video games is, we believe, within the limits within which Congress wanted the new Act to operate. Cf. WGN Continental Broadcasting Co., supra, 693 F.2d at 627; Williams Electronics, Inc., supra, 685 F.2d at 873-874; Atari, supra, 672 F.2d at 614-620. . . .

Litchfield v. Spielberg, 736 F.2d 1352 (9th Cir. 1984). Litchfield claimed that a screenplay she had written and submitted to the defendants called "Lokey From Maldomar" was infringed by defendants' enormously popular movie "E.T." The Ninth Circuit considered the two works in detail, and affirmed the district court's determination that defendants had not violated Litchfield's exclusive right to make copies of her work because the two works were not "substantially similar." The Court then considered Litch-field's alternative contention that E.T. was a derivative work of Lokey From Maldomar:

> Litchfield argues that defendants violated section 106(2) of the Copyright Act, 17 U.S.C. §106(2) (1982), by preparing a derivative work based on Lokey. The district court did not address this issue.
>
> Section 106(2) grants the exclusive rights to prepare a derivative work to the holder of the copyright. 17 U.S.C. §106(2). To constitute a violation of section 106(2) the infringing work must incorporate in some form a portion of the copyrighted work. Notes of the Committee on the Judiciary, House Report No. 94-1476, reprinted in 17 U.S.C. §106(2) (1982).
>
> Litchfield argues that section 106(2) was intended to expand the definition of derivative works to include any work based on a copyrighted work. She argues that the similarities between her play and the movie constitute the required "incorporation" and show that E.T. was based on Lokey for purposes of the statute. Litchfield apparently believes that she does not have to show substantial similarity to show that E.T. is a derivative work.
>
> Litchfield cites no authority to support this novel proposition. The little available authority suggests that a work is not derivative unless it has been sub-stantially copied from the prior work. See Harry Fox Agency, Inc. v. Mills Music, Inc., 543 F. Supp. 844, 849 (S.D.N.Y. 1982), *rev'd on other grounds,* 720 F.2d 733 (2d Cir. 1983), *cert. granted sub nom.* Mills Music, Inc. v. Snyder, 466 U.S. 903, 104 S. Ct. 1676, 80 L. Ed. 2d 151 (1984); Reyher v. Children's Television Workshop, 533 F.2d 87, 90 (2d Cir.), *cert. denied,* 429 U.S. 980, 97 S. Ct. 492, 50 L. Ed. 2d 588 (1976); 1 M. Nimmer, The Law of Copyright, §3.01 at 3-3 (1983 ed).
>
> We have stated that "[a] work will be considered a derivative work only if it would be considered an infringing work if the material which it has derived from a prior work had been taken without the consent of a copyright proprietor of such prior work." United States v. Taxe, 540 F.2d 961, 965 n.2 (9th Cir. 1976).
>
> To prove infringement, one must show substantial similarity. See v. Dur-ang, 711 F.2d at 143, Krofft, 562 F.2d at 1172. Litchfield's arguments that section 106(2) radically altered the protection afforded by the law of copyright are frivolous.

COMMENTS AND QUESTIONS

1. Why protect derivative works under section 106(2) at all? The Nim-mer treatise refers to section 106(2)'s right to prepare derivative works as "completely superfluous." Melville Nimmer & David Nimmer, 2 Nimmer on Copyright ¶8.09[A], at 8-114. It reasons that infringement of the right to prepare derivative works necessarily also infringes either the right to make copies or the right to perform works, so there is no reason to have both. If derivative works must be "substantially similar" to the underlying work to infringe the section 106(2) right, does it add anything to the protections granted elsewhere in the Copyright Act?

Paul Goldstein suggests that protecting derivative works is necessary in some instances to ensure that adequate incentives are given to copyright holders to develop new works. He argues that the author of a book should own the rights to a movie made out of that book, for example. But why are derivative rights necessary to accomplish this? In most such situations, the movie (or sequel or whatever) will of necessity copy places and characters exactly from the original. This seems to be a clear case of direct infringement on protected expression. If on the other hand expression from the original is *not* copied, but only general ideas or themes, is copyright protection really desirable?

2. *Economic Incentives.* Does a broad right to prepare derivative works comport with the economic incentive approach to intellectual property protection? Proponents of an incentive-based view of copyright might well challenge the assumption that authors should receive royalties from derivative works. Do authors really create under the assumption that their works will be translated into different forms? (Some do, certainly; Disney markets its animated films with an eye towards selling T-shirts and stuffed animals as well as movie tickets.) From a strict incentive perspective, should we reward authors in markets they did not originally enter? To what extent does your answer depend on assumptions about the capacity of authors and artists to develop other applications of their work? To what extent does your answer depend on assumptions about the transaction costs of licensing?

The economic rationale for derivative works may break down where the derivative right is used to preclude defendants from developing their own creative works in a market the plaintiff has not herself exploited, but which depends somehow on the plaintiff's work. Something of this sort may have happened in Castle Rock Entertainment v. Carol Publishing Group, 150 F.3d 132 (2d Cir. 1998). There the court enjoined the publication of a Seinfeld trivia book called the *Seinfeld Aptitude Test.* The court reasoned that the book infringed the copyright in the Seinfeld television series because it took numerous facts from the episodes created by the copyright owner. These facts constituted copyrightable expression the court concluded, and the defendants could not prevail on a fair use defense.

Do you find the result in the Seinfeld case problematic? What if the book in question had been an unauthorized biography of Jerry Seinfeld? A news report about a particularly controversial episode? What light, if any, does the case shed on the appropriate limits of the derivative works right?

3. *Comparison to Patent Scope.* Copyright law appears to afford the copyright proprietor broad control of all extensions of their original expression. Section 103 provides that new expression contained in a derivative work is separately copyrightable (and thus gives its owner all the exclusive rights contained in §106). However, §103(b) extends such copyrights only to new expression, and not to preexisting material included in the derivative work. Interestingly, §103(a) provides that only the original author or a licensee is entitled to a copyright in the derivative work. This means that if an infringer makes a movie out of a copyrighted book, adding substantial expression of

her own in the process, *she is not entitled to a copyright in any portion of the movie in which infringing material appears.*

Does this result make sense? If only new expression is copyrightable as part of a derivative work, why assign that new expression (as distinguished from the original expression) exclusively to the original author?

Compare this result with the "blocking patents" situation in patent law. As Merges explains, this doctrine permits a subservient inventor to obtain a patent on his improvement even though that improvement also infringes another patent. Robert P. Merges, Intellectual Property Rights and Bargaining Breakdown: The Case of Blocking Patents, 62 Tenn. L. Rev. 74 (1994). Historically this circumstance has led to enhanced bargaining, but the social costs of occasional bargaining breakdown justify a sort of "patent fair use" principle (the "reverse doctrine of equivalents"). Fortunately, although situations are serious given that they often involve significant new technologies, they are relatively rare. In the great run of cases, the ingenious institution of blocking patents balances the rights of original creators and subsequent improvers rather nicely.

Interestingly, no such institution exists in the law of copyrights. Copyright doctrine prohibits a follow-on creator from appropriating and adding to the copyrighted material of an original creator. Goldstein has argued that the cases provide no compelling reason to deny the derivative artist a copyright in his additional creative efforts. See Paul Goldstein, Derivative Rights and Derivative Works in Copyright, 30 J. Copyright Soc'y 209 (1982) ("The rule [denying copyrightability for unauthorized derivative works] is, however, hard to justify when applied to derivative works such as the motion picture in Sheldon v. MGM [309 U.S. 390 (1940)] in which the underlying work represents only a small part of the value of the derivative work but, because it underlies the whole, will defeat copyright protection for the entire derivative work. Just as an injunction against the motion picture gave plaintiff there a greater return than was needed to induce his investment in the underlying work, so depriving the motion picture owner of all protection against others will give it far less return than is needed to justify investment in the derivative work."); Wendy Gordon, Toward a Jurisprudence of Benefits: The Norms of Copyright and the Problem of Private Censorship, 57 U. Chi. L. Rev. 1009 (1990). Perhaps the best explanation for the lack of a doctrine of "blocking copyrights" is copyright law's policy favoring the reputational interest of authors: by requiring *ex ante* licensing of anyone who wishes to incorporate a copyrighted work into another work, the law ensures that the owner of the copyrighted work will completely control all manifestations of it. The law of derivative works is one way in which a personality or "moral rights" aspect creeps into U.S. copyright law. In effect, the strong protection given derivative works affords the original copyright owner control over alterations or "improvements" upon her work.

Because there is no blocking copyrights doctrine, copyright law is left with a vacuum in certain cases. What should be done with the hypothetical infringer who creates otherwise protectable new expression? Should that new expression be unprotectable because it derives from an infringement? Should it be in the public domain? Should it be deemed "captured" by the original copyright holder?

PROBLEMS

Problem 4-24. Garamon, a French author, wrote a successful novel in French. The novel is copyrighted in France in 1954. Garamon authorized an English translation of his novel but failed to comply with the formalities then required under U.S. law to obtain a U.S. copyright in the translation. Thus the translation fell into the public domain. Subsequently, Oaktree Press photocopies and distributes the English translation of the novel in the United States. Garamon sues for infringement, not of the translation, but of the copyright on the underlying French novel. Who should prevail?

Problem 4-25. A graphic artist for World Enquirer magazine is asked to produce a seamless integration of two photographs, so that it appears that two figures from separate photographs were in the same picture. To accomplish this, he uses digital imaging technology. He feeds both photographs into an optical character reader, or "scanner," and stores both images separately in the computer. Using a commercially available graphics program, he merges the pictures.

Assume one picture was copyrighted, and the other was in the public domain. What rights do the photo owner and World Enquirer have in the resulting image?

---

3. The Distribution Right

Section 106(3) grants copyright owners the right to distribute, through sale or other means, either the original or subsequent copies of a copyrighted work. This right is closely allied with the right to copy, since copying is normally the sine qua non of commercial copyright infringement. Thus, copying and selling a copyright owner's work without authorization violates both the right to copy and the right to distribute. As a corollary, both the copier who never does anything with his or her copies and the unknowing distributor of unauthorized copies are guilty of copyright infringement.

An important limitation on the exclusive right to distribute is the "first sale doctrine." 17 U.S.C. §109(a). This doctrine provides that a copyright holder cannot restrict what a purchaser of a particular lawful copy does *with that copy.* The purchaser may not copy it, but may resell it without restriction or liability.[23]

Quality King Distributors, Inc. v. L'anza Research International, Inc.
Supreme Court of the United States
118 S. Ct. 1125 (1998)

Justice STEVENS delivered the opinion of the Court.

Section 106(3) of the Copyright Act of 1976 (Act), 17 U.S.C. §106(3), gives the owner of a copyright the exclusive right to distribute copies of a copyrighted work. That exclusive right is expressly limited, however, by the provisions of §§107 through 120. Section 602(a) gives the copyright owner the right to prohibit the unauthorized importation of copies. The question presented by this case is whether the right granted by §602(a) is also limited by §§107 through 120. More narrowly, the question is whether the "first sale" doctrine endorsed in §109(a) is applicable to imported copies.

I

Respondent, L'anza Research International, Inc. (L'anza), is a California corporation engaged in the business of manufacturing and selling shampoos, conditioners, and other hair care products. L'anza has copyrighted the labels that are affixed to those products. In the United States, L'anza sells exclusively to domestic distributors who have agreed to resell within limited geographic areas and then only to authorized retailers such as barber shops, beauty salons, and professional hair care colleges. L'anza has found that the American "public is generally unwilling to pay the price charged for high quality products, such as L'anza's products, when they are sold along with the less expensive lower quality products that are generally carried by supermarkets and drug stores." L'anza promotes the domestic sales of its products with extensive advertising in various trade magazines and at point of sale, and by providing special training to authorized retailers.

L'anza also sells its products in foreign markets. In those markets, however, it does not engage in comparable advertising or promotion; its prices to foreign distributors are 35% to 40% lower than the prices charged to domestic distributors. In 1992 and 1993, L'anza's distributor in the United Kingdom arranged the sale of three shipments to a distributor in Malta; each

---

23. Congress created an exception to this rule when it passed §109(b) in 1984. That section prevents the rental of phonorecords or computer programs for profit. It was passed in response to the growth of record rental stores, which were centers for illegal copying of records and tapes. We will return to this issue again when we consider contributory infringement.

shipment contained several tons of L'anza products with copyrighted labels affixed. The record does not establish whether the initial purchaser was the distributor in the United Kingdom or the distributor in Malta, or whether title passed when the goods were delivered to the carrier or when they arrived at their destination, but it is undisputed that the goods were manufactured by L'anza and first sold by L'anza to a foreign purchaser.

It is also undisputed that the goods found their way back to the United States without the permission of L'anza and were sold in California by unauthorized retailers who had purchased them at discounted prices from Quality King Distributors, Inc. (petitioner). . . .

II

This is an unusual copyright case because L'anza does not claim that anyone has made unauthorized copies of its copyrighted labels. Instead, L'anza is primarily interested in protecting the integrity of its method of marketing the products to which the labels are affixed. Although the labels themselves have only a limited creative component, our interpretation of the relevant statutory provisions would apply equally to a case involving more familiar copyrighted materials such as sound recordings or books. Indeed, we first endorsed the first sale doctrine in a case involving a claim by a publisher that the resale of its books at discounted prices infringed its copyright on the books. Bobbs-Merrill Co. v. Straus, 210 U.S. 339, 28 S.Ct. 722, 52 L.Ed. 1086 (1908).

In that case, the publisher, Bobbs-Merrill, had inserted a notice in its books that any retail sale at a price under $1.00 would constitute an infringement of its copyright. The defendants, who owned Macy's department store, disregarded the notice and sold the books at a lower price without Bobbs-Merrill's consent. We held that the exclusive statutory right to "vend" applied only to the first sale of the copyrighted work . . .

Congress subsequently codified our holding in Bobbs-Merrill that the exclusive right to "vend" was limited to first sales of the work. Under the 1976 Act, the comparable exclusive right granted in 17 U.S.C. §106(3) is the right "to distribute copies . . . by sale or other transfer of ownership." The comparable limitation on that right is provided not by judicial interpretation, but by an express statutory provision.

The Bobbs-Merrill opinion emphasized the critical distinction between statutory rights and contract rights. In this case, L'anza relies on the terms of its contracts with its domestic distributors to limit their sales to authorized retail outlets. Because the basic holding in Bobbs-Merrill is now codified in §109(a) of the Act, and because those domestic distributors are owners of the products that they purchased from L'anza (the labels of which were "lawfully made under this title"), L'anza does not, and could not, claim that the statute would enable L'anza to treat unauthorized resales by its domestic distributors as an infringement of its exclusive right to distribute copies of its labels. L'anza does claim, however, that contractual provisions are inadequate to protect it from the actions of foreign distributors who may resell L'anza's products to American vendors unable to buy from L'anza's domestic distributors, and that §602(a) of the Act, properly construed, prohibits such unauthorized competition. To evaluate that submission, we must, of course, consider the text of §602(a). . . .

III

The most relevant portion of §602(a) provides:

> Importation into the United States, without the authority of the owner of copyright under this title, of copies or phonorecords of a work that have been acquired outside the United States is an infringement of the exclusive right to distribute copies or phonorecords under section 106, actionable under section 501. . . .

It is significant that this provision does not categorically prohibit the unauthorized importation of copyrighted materials. Instead, it provides that such importation is an infringement of the exclusive right to distribute copies "under section 106." Like the exclusive right to "vend" that was construed in Bobbs-Merrill, the exclusive right to distribute is a limited right. The introductory language in §106 expressly states that all of the exclusive rights granted by that section — including, of course, the distribution right granted by subsection (3) — are limited by the provisions of §§107 through 120. One of those limitations, as we have noted, is provided by the terms of §109(a), which expressly permit the owner of a lawfully made copy to sell that copy "[n]otwithstanding the provisions of section 106(3)."

After the first sale of a copyrighted item "lawfully made under this title," any subsequent purchaser, whether from a domestic or from a foreign reseller, is obviously an "owner" of that item. Read literally, §109(a) unambiguously states that such an owner "is entitled, without the authority of the copyright owner, to sell" that item. Moreover, since §602(a) merely provides that unauthorized importation is an infringement of an exclusive right "under section 106," and since that limited right does not encompass resales by lawful owners, the literal text of §602(a) is simply inapplicable to both domestic and foreign owners of L'anza's products who decide to import them and resell them in the United States.[14]

Notwithstanding the clarity of the text of §§106(3), 109(a), and 602(a), L'anza argues that the language of the Act supports a construction of the right granted by §602(a) as "distinct from the right under Section 106(3)

---

14. Despite L'anza's contention to the contrary, the owner of goods lawfully made under the Act is entitled to the protection of the first sale doctrine in an action in a United States court even if the first sale occurred abroad. Such protection does not require the extraterritorial application of the Act any more than §602(a)'s "acquired abroad" language does.

standing alone," and thus not subject to §109(a). Otherwise, L'anza argues, both the §602(a) right itself and its exceptions would be superfluous. Moreover, supported by various amici curiae, including the Solicitor General of the United States, L'anza contends that its construction is supported by important policy considerations. We consider these arguments separately.

IV

L'anza advances two primary arguments based on the text of the Act: (1) that §602(a), and particularly its three exceptions, are superfluous if limited by the first sale doctrine; and (2) that the text of §501 defining an "infringer" refers separately to violations of §106, on the one hand, and to imports in violation of §602. The short answer to both of these arguments is that neither adequately explains why the words "under section 106" appear in §602(a). . . .

The Coverage of §602(a)

[The Court rejected the argument that its reading renders §602 superfluous:]
First, even if §602(a) did apply only to piratical copies, it at least would provide the copyright holder with a private remedy against the importer, whereas the enforcement of §602(b) is vested in the Customs Service. Second, because the protection afforded by §109(a) is available only to the "owner" of a lawfully made copy (or someone authorized by the owner), the first sale doctrine would not provide a defense to a §602(a) action against any non-owner such as a bailee, a licensee, a consignee, or one whose possession of the copy was unlawful. Third, §602(a) applies to a category of copies that are neither piratical nor "lawfully made under this title." That category encompasses copies that were "lawfully made" not under the United States Copyright Act, but instead, under the law of some other country.

. . . Even in the absence of a market allocation agreement between, for example, a publisher of the U.S. edition and a publisher of the British edition of the same work, each such publisher could make lawful copies. If the author of the work gave the exclusive U.S. distribution rights — enforceable under the Act — to the publisher of the U.S. edition and the exclusive British distribution rights to the publisher of the British edition, however, presumably only those made by the publisher of the U.S. edition would be "lawfully made under this title" within the meaning of §109(a). The first sale doctrine would not provide the publisher of the British edition who decided to sell in the American market with a defense to an action under §602(a) (or, for that matter, to an action under §106(3), if there was a distribution of the copies).

The argument that the statutory exceptions to §602(a) are superfluous if the first sale doctrine is applicable rests on the assumption that the coverage of that section is co-extensive with the coverage of §109(a). But since it is, in fact, broader because it encompasses copies that are not subject to the first sale doctrine — e.g., copies that are lawfully made under the law of another country — the exceptions do protect the traveler who may have made an isolated purchase of a copy of a work that could not be imported in bulk for purposes of resale. As we read the Act, although both the first sale doctrine embodied in §109(a) and the exceptions in §602(a) may be applicable in some situations, the former does not subsume the latter; those provisions retain significant independent meaning.

[The Court rejected other statutory construction arguments offered by L'Anza].

V

The parties and their amici have debated at length the wisdom or unwisdom of governmental restraints on what is sometimes described as either the "gray market" or the practice of "parallel importation." In K Mart Corp. v. Cartier, Inc., 486 U.S. 281, 108 S.Ct. 1811, 100 L.Ed.2d 313 (1988), we used those terms to refer to the importation of foreign-manufactured goods bearing a valid United States trademark without the consent of the trademark holder. Id., at 285-286, 108 S.Ct., at 1814-1815. We are not at all sure that those terms appropriately describe the consequences of an American manufacturer's decision to limit its promotional efforts to the domestic market and to sell its products abroad at discounted prices that are so low that its foreign distributors can compete in the domestic market. But even if they do, whether or not we think it would be wise policy to provide statutory protection for such price discrimination is not a matter that is relevant to our duty to interpret the text of the Copyright Act.

Equally irrelevant is the fact that the Executive Branch of the Government has entered into at least five international trade agreements that are apparently intended to protect domestic copyright owners from the unauthorized importation of copies of their works sold in those five countries. The earliest of those agreements was made in 1991; none has been ratified by the Senate. Even though they are of course consistent with the position taken by the Solicitor General in this litigation, they shed no light on the proper interpretation of a statute that was enacted in 1976.

The judgment of the Court of Appeals is reversed.

Justice GINSBURG, concurring.

This case involves a "round trip" journey, travel of the copies in question from the United States to places abroad, then back again. I join the Court's opinion recognizing that we do not today resolve cases in which the allegedly infringing imports were manufactured abroad. . . .

Mirage Editions, Inc. v. Albuquerque A.R.T., 856 F.2d 1341 (9th Cir. 1988). Patrick Nagel was a popular 20th Century artist who died in 1984. His widow, Jennifer Dumas, owns the copyrights to his works. Nagel and Dumas published the Nagel works exclusively through Mirage Editions, which sells (among other things) coffee-table books of Nagel's art. A.R.T. purchased numerous copies of these art books, physically cut works out of them, glued the works onto ceramic tile, and offered the tile for sale in retail shops.

Mirage sued A.R.T. for copyright infringement. The parties agreed that A.R.T. had not made any copies of Nagel's works; it merely mounted pictures it purchased from Mirage onto tiles and resold them. The Ninth Circuit found copyright infringement. It reasoned that the process of attaching art from a book onto ceramic tiles constituted the preparation of a derivative work which infringed Mirage's copyright. The court held that the first sale doctrine did not protect A.R.T.:

> We recognize that, under the "first sale" doctrine as enunciated at 17 U.S.C. §109(a) and as discussed in [United States v. Wise, 550 F.2d 1180 (9th Cir. 1977)], appellant can purchase a copy of the Nagel book and subsequently alienate its ownership in that book. However, the right to transfer applies only to the particular copy of the book which appellant has purchased and nothing else. The mere sale of the book to the appellant without a specific transfer by the copyright holder of its exclusive right to prepare derivative works, does not transfer that right to appellant. The derivative works right, remains unimpaired and with the copyright proprietors. . . .

Id. at 1344.

COMMENTS AND QUESTIONS

1. The *Mirage* decision has proved controversial. While it has been followed and even extended in the Ninth Circuit, see Munoz v. Albuquerque A.R.T. Co., 829 F. Supp. 309 (D. Alaska 1993), other courts and most commentators have rejected the Ninth Circuit's holding. See Lee v. A.R.T., Inc., 125 F.3d 580 (7th Cir. 1997); see also Precious Moments v. La Infantil, 971 F. Supp. 66 (D.P.R. 1997); Allison v. Vintage Sports Plaques, 40 U.S.P.Q.2d 1465 (N.D. Ala. 1996); 1 Melville Nimmer & David Nimmer, Nimmer on Copyright §3.03; Edward G. Black & Michael H. Page, Add-On Infringements: When Computer Add-Ons and Peripherals Should (and Should Not) Be Considered Infringing Derivative Works Under Lewis Galoob Toys Inc. v. Nintendo of America Inc. and Other Recent Decisions, 15 Hastings Comm./Ent. L.J. 615, 628-29 (1993). The debate is not over the first sale doctrine but rather over whether what A.R.T. does creates a "derivative work" at all. The *Lee* court reasons that tile-mounting, like framing a picture, is merely a means of displaying the original work. It reads the definition of "derivative works" in section 101 to require the contribution of *copyrightable* new material to the preexisting work. Since framing or mounting aren't activities that add original, copyrightable material, the resulting work is not a derivative work, according to the court.

Is the *Lee* court's approach correct? Is it possible to envision circumstances under which framing a work of art involves original, creative expression?

2. What is the point of having a "right to distribute" if it is limited by the first sale doctrine? Is there any right granted in section 106(3) that does not already exist in the other parts of section 106? One purpose of the statute may be to control the right of first publication of a work. For example, a copyright owner may be able to prevent a library from lending her work to the public without permission, even though the library does not make copies of the work. But if the library has purchased the work from the copyright owner, should she have the right to limit what it does with the work?

Granting copyright owners the right to control distribution may also make it easier for them to find and sue infringers. Distributors may be easier to identify than the original copiers, and (because they operate on a larger scale) there may be fewer to go after. Suing distributors may also be more palatable to copyright owners in some cases than suing actual infringers. For example, textbook publishers may be willing to sue commercial copy shops who sell class readers, but unwilling to sue the university professors who authorized making the copies. After all, most of their customers are the same university professors.

Finally, under the reasoning of some cases, the first sale doctrine does not apply at all to an important new class of works — works made or transmitted by computer. As discussed supra, some cases have held that loading a file into computer memory makes a copy of that file. Under those decisions, is it ever possible to "dispose of" a computer file by e-mailing it to a friend? Do you see why such an act would not fall within the first sale doctrine?

3. What is the significance of Justice Ginsburg's concurrence in *L'Anza?* Suppose that a U.S. copyright owner authorizes the production of copies of his work in a foreign country. Why aren't those copies "lawfully made" within the meaning of the Copyright Act?

4. The *L'Anza* Court is careful to distinguish cases in which the parties have agreed not to reimport goods; in those cases, the Court implies, the contractual restriction may be enforced. If this is true, of what practical import is *L'Anza?* Won't all exporters henceforth include a contractual provision barring reimportation? Do you see any difficulties with such an approach?

PROBLEM

Problem 4-26. Lee, a law student, attends an expensive private school. After buying the books for his first-year classes, he underlines key passages in the books and takes some notes in the margins. At the end of the first year of law school, Lee sells his books back to the law school bookstore. The bookstore in turn sells them to incoming law

> students the next fall as "used books." Under what theory might Lee
> or the bookstore be liable for copyright infringement? Should they be?

4. Public Performance and Display Rights

In addition to the rights to make and distribute copies and adaptations
of a work, copyright owners possess the exclusive right to perform or display
their works publicly. (A former requirement that the performance or display
be made "for profit" was abolished in the 1976 Act.) 17 U.S.C.
§106(4), (5).

The performance and display rights roughly parallel each other, and most
types of copyrighted works are covered by one or the other of the two rights.
Basically, if it moves, it's a performance, and if it stays still, it's a display. See
17 U.S.C. §101 (definitions of "display" and "perform"). Thus paintings,
sculptures, photographs, single photo frames from movies or video games,
and physical copies of books are displayed. Plays, dances, movies, copyrighted
combinations of still photographs (such as slide shows), and readings of books
are performed publicly. Importantly, no public display right exists in architectural works, and only a limited public performance right (discussed below)
exists in sound recordings.[24] This means, for instance, that it is not copyright
infringement to show a house where people can see it, and playing music
over the airwaves does not infringe the copyright *in the sound recordings
played*.[25] However, there are limits on the definition of a performance. For
example, one court has held that playing a board game in a national games
tournament was not a "performance" of the game. The court restricted the
word "play" in section 101 to the playing of sound recordings. Allen v.
Academic Games League of America, 89 F.3d 614 (9th Cir. 1996).

The distinction between performances and displays is important because
the scope of the exclusive protections is quite different. Any physical act taken
to make a work perceivable to the viewer or listener, or cause a work to be
reproduced (even in a transient form that does not create a "copy" under
section 106(1)) is a performance. Playing a CD or a videotape is a performance. Reading a book aloud (but not silently) is a performance. Dancing a
work of choreography is a performance. See 17 U.S.C. §101. By contrast,
the definition of a display is limited by section 109(c), which provides that
the owner of a particular copy of a work is entitled to display it "to viewers
present at the place where the copy is located." This limitation allows most
common displays of a work — including those works fixed at a single location
from which they are visible to the public. Showing artwork in a theater or

---

24. This right covers only "digital" performance of sound recordings, for example performances transmitted over the Internet. The right was added by Congress in 1995. 17 U.S.C.
§106(6).
25. There is also generally a copyright in the musical composition, which is infringed by
public performance unless a royalty is paid to the composer. See below.

gallery without authorization is not copyright infringement, but showing a movie in the same location is infringement. Showing the same piece of art through a television broadcast or over a computer network, by contrast, would infringe the display right.

The scope of the definitions of both performance and display are quite broad. The limitation to "public" performances and displays, however, prevents many commonplace activities from infringing performance and display rights. The definition of a public performance or display is set forth in section 101:

> To perform or display a work "publicly" means —
>
> (1) to perform or display it at a place open to the public or at any place where a substantial number of persons outside of a normal circle of a family and its social acquaintances is gathered; or
>
> (2) to transmit or otherwise communicate a performance or display of the work to a place specified by clause (1) or to the public, by means of any device or process, whether the members of the public capable of receiving the performance or display receive it in the same place or in separate places and at the same time or at different times.

Under this definition, you can probably have a (small) party at which you play a record or video. Large parties — if they go beyond the amorphous definition of your "social acquaintances" — may pose problems. Further, if the place of performance or display is "open to the public," it does not appear to matter how many people actually view the performance. Clause (2) appears to cover all broadcasts to the "public" even if members of the public do not view the copyrighted work at the same place or time. Thus a television broadcast is a performance, even though no one may watch it at all, or if people only watch it in the privacy of their own homes.

A number of courts have grappled with the problem of when a performance becomes public. In two separate cases, the Third Circuit has held that video rental stores cannot provide viewing rooms for customers, because the performance of a rented movie in such a room is "public" (even though the room is rented only to one group at a time). The Third Circuit explained its rationale in terms that interpret the phrase "public performance" very broadly:

> The Copyright Act speaks of performances at a place open to the public. It does not require that the public place be actually crowded with people. A telephone booth, a taxi cab, and even a pay toilet are commonly regarded as "open to the public," even though they are usually occupied only by one party at a time.

Columbia Pictures Indus. v. Aveco, Inc., 800 F.2d 59, 63 (3d Cir. 1986); accord Columbia Pictures Indus. v. Redd Horne, 749 F.2d 154, 158 (3d Cir. 1984).

Sound Recordings

As noted earlier, sound recordings did not come under the protection of federal copyright law until 1972. Even then, the political strength of broadcasters persuaded Congress to deny public performance rights to the owners of sound recordings. Broadcasters did not want to pay royalties to sound recording owners as well as musical composition owners. (Musical compositions copyright owners, and their collecting societies (ASCAP, BMI, and SESAC) were also concerned that public performance royalties paid to sound recording owners would cut into their share of licensing revenues paid by broadcasters).

The introduction of digital transmission technology in the 1990s altered this balance. Recording artists feared that transmission of near perfect quality recordings that could be recorded on digital audio devices posed a tremendous risk of piracy. Traditional broadcasters partially aligned with sound recording owners in favoring some protection for digital performance rights in sound recordings because of the competitive threat posed by new broadcasting entities such as digital audio subscription and interactive services. The resulting interaction of the various interests produced the Digital Performance Right in Sound Recordings Act (DPRSRA) of 1995, codified at 17 U.S.C. §§106(6), 114. The net effect of this statute was to create a new exclusive right "in the case of sound recordings to perform the copyrighted work publicly by means of a digital audio transmission." 17 U.S.C. §106(6). This right is much more circumscribed than the public performance right applicable to other copyrightable works. See §106(4). The complex series of rights and exceptions to the DPRSRA are detailed in 17 U.S.C. §114.

Statutory Limits on Performance and Display Rights

Besides the general fair use provision of section 107, which we will discuss in Part F, there are a number of specific statutory exceptions that limit the scope of the performance and display rights. We summarize each only briefly here; you are encouraged to study the statutory provisions at issue. The provisions are of three basic types.

Public Interest

Section 110 of the Copyright Act exempts many "public interest" performances and displays from the reach of sections 106(4) and (5). Thus most live educational performances and displays are exempt under section 110(1), as are transmissions to classrooms or for other "laudable" purposes. 17 U.S.C. §110(2). Religious performances and displays are exempt from the act under section 110(3). Face-to-face performances of "nondramatic literary or musical works" for free or for charitable purposes are exempt, reviving in part the "for profit" requirement of the 1909 Act. 17 U.S.C. §110(4). Record stores may play records without charge to promote their sale under section 110(7), although the closely analogous performance of videos in video stores appears to be prohibited.

After many years of tense relations between small business and restaurant owners and performing rights societies over the collection of public performance royalties for broadcast and other recorded music played in these establishments, Congress passed the Fairness in Music Licensing Act in 1998 (as part of the Sonny Bono Copyright Term Extension Act). Due to the political strength of retailers and restaurant owners, Congress substantially broadened an exemption for home listening of transmitted performances to extend to small businesses (less than 2,000 square feet), restaurants (less than 3,750 square feet), and larger establishments conforming to limitations on the number of loudspeakers and television screen size. §110(5). These establishments would still require a public performance license from the musical composition owner to host live or taped (e.g., prerecorded music on CD players) performances.[26]

Compulsory Licenses

Section 111 authorizes television broadcast relays, or "secondary transmissions," under a variety of circumstances where they are not for profit and are not content-controlled. For example, the owner of an apartment building with a single reception antenna may relay its signal to residents of the building without charge. Cable systems are also entitled to transmit television broadcasts over their networks and charge a fee for the service, provided that the cable network registers its intent to do so and pays a royalty based on the revenues it receives from subscribers. The royalty rate is set forth in the statute. Until recently, the actual royalties were calculated and collected under the supervision of the Copyright Royalty Tribunal, a government office; now disputes are resolved privately or through ad hoc arbitration panels. (The Copyright Royalty Tribunal was abolished by Congress in 1993.) Cable companies cannot, however, delay or alter the programming they relay from television broadcasts.

Section 119 provides a similar right to a compulsory license for satellite transmission to "unserved households" — that is, households that do not receive the normal transmission signal from a particular network or other station, either through broadcast or cable. Satellites may broadcast the signal of those stations to subscribing recipients upon the payment of a royalty. Unlike section 111, however, section 119 does not specify the royalty rate.

---

26. The Fair Music Licensing Act also provides a new process by which establishment owners can challenge the royalty rates set by performing rights societies. These provisions are contained in section 512 of the Act (Public Law 105-298). Due to a drafting error, there are currently now two sections designated as 512 in the Act, the other corresponding to "Limitations on Liability Relating to Material Online."

Rather, the rate is subject to voluntary negotiation or compulsory arbitration between the satellite owner and the individual stations or networks.

Section 115 entitles musicians and record companies to make and sell their own recordings of copyrighted musical works that have previously been recorded with permission by another artist, as long as they pay a set fee to the copyright owner and do not change the "basic melody or fundamental character of the work." This is the "cover" license referred to above.

Section 116 authorizes owners of jukeboxes ("coin-operated phonorecord players," in the words of the statute) to publicly perform the musical works contained in the jukebox subject to a compulsory license. This compulsory license, like that in section 111, is fixed by statute and was administered until recently by the Copyright Royalty Tribunal.

Section 118 authorizes public broadcasting stations to transmit musical and artistic (but not literary or audiovisual) works upon payment of a compulsory license. This section does not, however, set royalty rates. Rather, it requires public broadcasters and the owners of such works to negotiate a rate every five years under the supervision of the Librarian of Congress.

Section 114(d)-(j) limits the newly created right of performance in digital sound recordings, both restricting the scope of the new right and creating a compulsory license mechanism to be invoked should the parties fail to agree on an appropriate royalty.

Exclusions

Under section 112, entities authorized to transmit a performance or display are allowed to make and keep a small number of copies for archival and security purposes in some (but not all) cases, but may not otherwise use or claim derivative rights in those copies.

PROBLEMS

Problem 4-27. Ralston Hotels, a national hotel chain, offers guests an "in-room video rental" service. A menu is displayed on the guest's interactive television screen, and the guest can select both a movie and a starting time by using his remote control. Portland Pictures, a major movie producer that owns the video rental rights to its movies, sues Ralston, alleging that each selected movie is a "public performance" and demanding royalties. Who should prevail?

Problem 4-28. Buford, the owner of the Pumpwell service station on a busy commercial street, receives approximately 20 calls each day from customers. Because of the volume of repair work that Pumpwell does, it frequently must put callers on hold. Buford decides to provide those callers with music while they are holding, so he patches his radio into the telephone. When a call is placed on hold, the caller hears a station selected by Buford until the call is picked up again. ASCAP sues Pumpwell for copyright infringement, claiming that Pumpwell is publicly performing its songs. Does ASCAP have a meritorious case?

5. Moral Rights

Moral rights are basic rights of artists that extend beyond ownership of economic control of works of authorship, to encompass protections of the "personality" of the author. They normally include (and often go beyond) the right to have one's name associated with one's work (right of attribution); and the right to protect one's works from mutilation or distortion (right of integrity). Moral rights derive from the continental European or "author's rights" tradition in intellectual property law, which differs in many ways from the Anglo-American tradition emphasizing the economic function of intellectual property. See, e.g., Roeder, The Doctrine of Moral Right: A Study in the Law of Artists, Authors and Creators, 53 Harv. L. Rev. 554 (1940); Merryman & Elsen, Law, Ethics and the Visual Arts, ch. IV passim (1979).

In the United States, protection of moral rights has historically been sporadic. The central vehicle for moral rights on the Continent — copyright law — did not grant any explicit moral rights protection in the United States until 1990. Instead, what moral rights protection existed resulted from a combination of state statutes and isolated judicial readings of both the federal copyright and trademark statutes.

In 1990 the United States recognized, for the first time, a limited set of moral rights for a limited class of works. These were set out in the Visual Artists Rights Act of 1990, and are now embodied in §106A of the Copyright Act of 1976. The rights in section 106A include the artist's right to claim authorship of the work, to prevent the use of her name on works she did not create, and to prevent the use of her name on works that have been modified or distorted. In addition, artists get limited rights to object to the modification or destruction of their original works. Section 106A and section 113(d) place important limitations on these rights, however.

COMMENTS AND QUESTIONS

1. Litigation under section 106A has begun. See Carter v. Helmsley-Spear, Inc., 71 F.3d 77 (2d Cir. 1995) (reversing district court injunction in favor of plaintiff sculptors who were creating a sculpture in a building lobby, on the grounds that the work was made for hire). Future litigation will have to resolve some of the detailed questions left open by section 106A.

2. Article 6[bis](1) of the Berne Convention provides

Independently of the author's economic rights, and even after the transfer of said rights, the author shall have the right to claim authorship of the work, and to object to any distortion, mutilation or other modification of, or other derogatory action in relation to, the said work, which would be prejudicial to his honor or reputation.

Upon ratifying the Berne Convention, Congress concluded that existing federal and state protections satisfied Article 6[bis] obligations. In particular, the legislative history of the Berne Convention Implementation Act of 1988 noted rights under section 106; section 115(a)(2) (relating to distortions of musical works under the compulsory license); section 203 (termination of transfers); section 43(a) of the Lanham Act (relating to false designations of origins and false descriptions); state and local laws relating to publicity, contractual violations, fraud and misrepresentation, unfair competition, defamation, and invasion of privacy; and new state statutes in eight states protecting rights of integrity and paternity in certain works of art. H.R. Rep. No. 609, 100th Cong., 2d Sess. 33-34 (1988). For a discussion of state moral rights statutes, see Chapter 6, section A.

Does this collection of protections fully discharge U.S. obligations under the Berne Convention?

3. Significant limits on section 106A make it a much more limited right than exists in Europe. For example, a classic "moral rights" case involves Picasso's painting "Trois Femmes." The painting was cut into one-inch squares by two art investors and sold as "original Picassos." Daniel Grant, Before You Cut Up That Picasso . . . , World Monitor, Feb. 1992, at 58-59. Because Picasso was dead when this occurred, however, section 106A would not prohibit it. 17 U.S.C. §106A(d)(1). In addition, section 106A does not extend to a host of works such as films that are protected under European moral rights regimes. Cf. John Huston–Asphalt Jungle Case, reported at 22 Intl. Rev. Ind. Prop. & Copyrt. L. 121 (1991) (describing French court's decision to bar showing of "colorized" version of film made in black and white by director John Huston).

Moral rights in the continental tradition are normally thought to be "inalienable." Certainly, the Berne Convention speaks of an author's retaining such rights even after relinquishing the copyright. And Sarraute refers to the "inalienable, unbarrable, and perpetual nature of the French moral right." Raymond Sarraute, Current Theory on the Moral Right of Authors and Artists Under French Law, 16 Am. J. Comp. L. 465, 485 (1968). But see Neil W. Netanel, Alienability Restrictions and the Enhancement of Author Autonomy in United States and Continental Copyright Law, 12 Cardozo Arts & Ent. L.J. 1 (1994) (suggesting that the extent of inalienability in continental moral rights law has been overstated). Should moral rights be subject to sale or waiver by contract? If so, what good are they?

4. Artists have also sought moral rights protection under a variety of legal theories other than section 106A. In Serra v. General Services Administration, 667 F. Supp. 1042 (S.D.N.Y. 1987), *aff'd*, 847 F.2d 1045 (2d Cir.

1988), artist Richard Serra, a prominent American sculptor, sought to prevent the United States General Services Administration (GSA) from removing his controversial sculpture "Tilted Arc" from the Federal Plaza in lower Manhattan. The court denied Serra's request for an injunction under various state and federal theories. On the other hand, in Gilliam v. American Broadcasting Companies, 538 F.2d 14 (2d Cir. 1976), the British comedy group Monty Python sued to prevent ABC from broadcasting edited versions of their television series "Monty Python's Flying Circus." On a motion for preliminary injunction, the Second Circuit held that

> It also seems likely that appellants will succeed on the theory that, regardless of the right ABC had to broadcast an edited program, the cuts made constituted an actionable mutilation of Monty Python's work. This cause of action, which seeks redress for defamation of an artist's work, finds its roots in the continental concept of droit moral, or moral right, which may generally be summarized as including the right of the artist to have his work attributed to him in the form in which he created it.
>
> American copyright law, as presently written, does not recognize moral rights or provide a cause of action for their violation, since the law seeks to vindicate the economic, rather than the personal, rights of authors. Nevertheless, the economic incentive for artistic and intellectual creation that serves as the foundation for American copyright law cannot be reconciled with the inability of artists to obtain relief for mutilation or misrepresentation of their work to the public. . . .

The court went on to conlude that this "mutilation" was actionable under the Lanham Act (discussed in Chapter 5).

> It is sufficient to violate the Act that a representation of a product, although technically true, creates a false impression of the product's origin. [citing cases] . . .
>
> These cases cannot be distinguished from the situation in which a television network broadcasts a program properly designated as having been written and performed by a group, but which has been edited, without the writer's consent, into a form that departs substantially from the original work. . . . Thus, an allegation that a defendant has presented to the public a "garbled," distorted version of plaintiff's work seeks to redress the very rights sought to be protected by the Lanham Act.

Id. at 23-24. Other recent cases have held that the publication of public domain fairy tales edited in a form substantially similar to plaintiff's books violated plaintiff's "right of attribution" under the Lanham Act, Waldman Publishing Corp. v. Landoll, Inc., 43 F.3d 775 (2d Cir. 1994), and that the failure to attribute quotes taken from the plaintiff's work to the plaintiff doomed a fair use defense under copyright law. Robinson v. Random House, 877 F. Supp. 830 (S.D.N.Y. 1995).

Recall *Mirage Editions v. Albuquerque A.R.T.*, discussed supra section E.3. The court there held that altering a copyrighted work created an infringing derivative work, even though the alteration did not involve the making of a new copy. Does this rule serve to protect the moral rights of artists by allowing them to prevent alterations of their works even after the artists have sold the work? Does it matter that the copyright owner, rather than the original author, is the one with the power to exercise this right?

5. Moral rights are quite controversial in the United States. They raise a host of difficult philosophical and implementation questions: Should United States law offer general protections to authors and artists against alteration or misattribution of their works? Do such rights interfere with the free licensing of works of intellectual property by giving the creator a continual "veto power" over editing and publication?

Should moral rights extend beyond the first sale of a book or work of art, preventing its owner from (for example) destroying or mutilating that particular copy?

Is it desirable for an artist to be able to control what the owner of a piece of art does with it? In the privacy of her own home? The statutes speak of "defacement, mutilation, or destruction" of works of art, but also prevent mere "alteration." Who should make this determination? The artist? The courts? What might constitute actionable alteration of a work of art? Moving it? See Eric Brooks, Comment, Tilted Justice: Site-Specific Art and Moral Rights After U.S. Adherence to the Berne Convention, 77 Cal. L. Rev. 1431 (1989).

6. Contributory Infringement

Infringement of a copyright is not limited to direct violations of the rights of copyright owners. In some circumstances, a defendant may be liable for encouraging or assisting a third party to infringe a copyright. This is called "inducement" or "contributory infringement."

A number of circumstances can give rise to liability for contributory infringement by one who is not herself an infringer. At one extreme, consider the role of a movie producer who gives an screenwriter a prior film, is present while the screenwriter creates a substantially similar film, and who then produces and distributes that film. The law seems fairly clear that such an individual is liable for "vicarious" (or alternatively, contributory) infringement. See Gershwin Publishing Corp. v. Columbia Artists Management, Inc., 443 F.2d 1150 (2d Cir. 1971) (concert promoter who distributed artists' playlists but who deliberately did not enquire into whether the artists had obtained copyright permissions held liable for an artist's infringement of a performance right).

On the other hand, consider the position of a bookstore that, without knowledge of infringement, sells a book distributed by a reputable publisher that is later determined to be substantially similar to another work. The rule seems fairly clear that the bookstore owner cannot be held liable as a con-

tributory infringer.[27] Knowledge, actual or at least constructive, is a requirement in contributory infringement cases. Defendants are held liable if they
know of or have reason to know of the infringement, but not otherwise. See
Screen Gems v. Mark-Fi Records, Inc., 256 F. Supp. 399 (S.D.N.Y. 1966)
(advertising agency that placed ads selling 20 hit records for $2.98 should
have been on notice due to the low price that the records were pirated).

Should distributors, including retail stores, have a duty to inquire into
the copyright status of the works they sell? How could such a duty be discharged? Should a party's status as a contributory infringer differ depending
on whether an injunction or damages are sought? This issue arises with particular force in the case of Internet service providers (ISPs). ISPs provide
access to the Internet or to online services for a fee. While the ISPs are not
directly responsible for the content of the Internet messages they transmit,
they have been sued in several noteworthy cases by copyright owners upset
about infringement of their works on the Internet. In one such case, Religious
Tech. Center v. Netcom On-line Comm. Corp., 907 F. Supp. 1361, 1373-
74 (N.D. Cal. 1995), the court held that ISPs are not liable for postings they
have no reason to know are infringing, but that they may be liable if they
leave a message on their system after being notified of infringement. Does
this rule unfairly burden ISPs (and their subscribers), by giving ISPs an incentive to remove questionable postings immediately upon complaint, regardless of whether the complaint ultimately has merit?

Distributors or producers are not the only people with reason to be
concerned about contributory infringement. The manufacturers of machines
that help people make copies are also potentially liable for contributory infringement. Certainly Xerox has "assisted" (knowingly or not) countless
copyright infringements by selling photocopiers. Should Xerox be liable for
photocopies made on its machine? Should libraries be liable for providing
photocopiers to patrons? The answer to these questions depends not only on
the nature of the particular technology at issue but also on whether we are
willing to retard the development of copying technologies in the interest of
protecting authors from infringement. We will address these issues when we
consider the "fair use" defense.

COMMENTS AND QUESTIONS

In 1998, Congress passed the Digital Millenium Copyright Act, which
makes it a crime in most circumstances to circumvent so-called "copy protection" schemes or to build or sell a device that helps others to do so. 17
U.S.C. §1201. The DMCA also addresses the liability of "online service pro-

---

27. Consider whether the bookstore owner is *directly* infringing the distribution right in
section 106(3). Cf. Kalem Co. v. Harper Bros., 222 U.S. 55, 63 (1911).

viders." Because the law applies to digital information and computer programs, it is discussed in more detail in Chapter 7.

PROBLEM

Problem 4-29. Industrial Music Co. (IM) develops and produces a device known as a digital music filter for home and personal use. The filter allows individuals with access to audio tapes to "remix" those tapes themselves, and create new works in which sounds from existing works are speeded up, slowed down, added to or subtracted from, or combined with other preexisting works. IM markets its filter as a device for amateur musicians who want to edit their own works, but IM knows that approximately half of the filters it sells are used for "sampling" or otherwise copying copyrighted works without authorization. Is IM guilty of contributory infringement? If so, with every sale, or only with some sales? Can the sale of the filter be enjoined?

---

7. New Restrictions and Rights Aimed at Combating Digital Piracy

Audio Home Recording Rights Act of 1992

By the mid- to late 1980s, consumer electronics companies sought to introduce a host of new products (digital audio tape (DAT), mini-disc (DCC)) that would enable consumers to make digital copies of audio recordings. These technologies would enable ordinary, as well as sophisticated manufacturing entities, to produce identical copies of copyrighted works without any significant degradation of quality. By contrast, analog tape technology could not produce identical reproductions, and the quality of such reproductions degraded across generations. The music industry (comprising music publishers, record companies, broadcasters, performing rights societies, song writers, and recording artists) feared that digital recording technologies would significantly reduce its revenue stream. As occurred with the introduction of video cassette recording technology in the early 1980s, copyright owners and electronics companies prepared for litigation over the introduction of this new recording technology. See Sony Corporation of America v. Universal Studios, Inc., 464 U.S. 417 (1984) (infra §F(1)(a)).

In the shadow of costly and uncertain litigation, see Cahn v. Sony Corp., 90 Civ. 4537 (S.D.N.Y. filed July 9, 1990), the various interests resolved their differences through negotiations, which culminated in Congress' passage of the Audio Home Recording Rights Act of 1992 (codified at 17

U.S.C. §§1001-1010). This compromise legislation paved the way for the introduction of digital home recording technologies by creating a new set of restrictions, rights to compensation, and immunities that go significantly beyond the traditional copyright model:

- *Restrictions on Digital Audio Recording Devices.* Section 1002(a) prohibits the importation, manufacture, and distribution of any digital audio recording device that does not incorporate technological controls (Serial Copy Management System or functional equivalents) that block second-generation digital copying.
- *Compensation of Copyright Owners.* As a means to compensate copyright owners for the inevitable unauthorized copying resulting from these new technologies, the Act requires manufacturers of and importers of digital audio recording equipment and blank tapes, disks, or other storage media, to deposit a percentage of their transfer prices (two percent for digital audio devices and three percent for storage media) into a royalty pool. The royalty pool is distributed to owners of musical works (one-third) and sound recordings (two-thirds) based on the prior year's sales and airtime. §§1003-1007. This compensation mechanism is administered by the Register of Copyright, with provisions for arbitration of disputes.
- *Limited Immunity for Noncommercial Home Taping.* Section 1008 affords immunity for noncommercial home taping by consumers.

Violations of the AHRA are not copyright violations. Rather, the AHRA contains its own enforcement, remedy, and dispute resolution provisions. 17 U.S.C. §§1009-1010.

Digital Millennium Copyright Act of 1998

Somewhat analogous concerns prompted computer software companies to lobby national and international authorities (World Intellectual Property Organization) for greater protections against digital piracy in the mid-1990s. Congress passed a wide-ranging series of amendments and extensions of copyright law in the Digital Millennium Copyright Act of 1998 (DMCA). Somewhat like the AHRA, the DMCA goes beyond traditional copyright law by restricting technologies that facilitate piracy. The DMCA seeks to combat piracy of digital works by prohibiting the circumvention of copy protection systems (or the manufacture or distribution of devices with the primary purpose or effect of circumventing copy protection systems) and the removal or alteration of "copyright management information" conveyed along with a copyrighted work. This latter provision is designed to discourage counterfeiters from stripping identifying information from a work or falsely identifying the author of a work.

F. DEFENSES

1. Fair Use

17 U.S.C. §107. Limitations on Exclusive Rights: Fair Use

Notwithstanding the provisions of section 106, the fair use of a copyrighted work, including such use by reproduction in copies or phonorecords or by any other means specified by that section, for purposes such as criticism, comment, news reporting, teaching (including multiple copies for classroom use), scholarship, or research, is not an infringement of copyright. In determining whether the use made of a work in any particular case is a fair use the factors to be considered shall include —

(1) the purpose and character of the use, including whether such use is of a commercial nature or is for nonprofit educational purposes;

(2) the nature of the copyrighted work;

(3) the amount and substantiality of the portion used in relation to the copyrighted work as a whole; and

(4) the effect of the use upon the potential market for or value of the copyrighted work.

The fact that a work is unpublished shall not itself bar a finding of fair use if such finding is made upon consideration of all the above factors.

It is well established that both the list of potentially fair uses and the factors to be considered in determining fair use are illustrative rather than exhaustive. The House Report accompanying section 107 describes the four factors — and the list of possible fair uses — as "the result of a process of accretion" during the judicial development of the fair use doctrine at common law. According to the report, section 107 is designed to "offer some guidance" in determining fair use, not to formulate "exact rules." H.R. No. 94-1476, at 66 (1976).

Section 107 represents the first attempt to codify what had developed as a common law doctrine. The first reference to fair use (albeit by a different name) is thought to be Folsom v. Marsh, 9 F. Cas. 342 (C.C. Mass. 1841). The court there focused on the loss in value to the plaintiff from copying to determine whether the defendant had engaged in what it called a "justifiable use" of plaintiff's copyrighted work. However, more recent cases have also looked to the purpose of the use and the "public benefit" conferred thereby. See Rosemont Enterprises v. Random House, Inc., 366 F.2d 303 (2d Cir. 1966). Indeed, *Folsom* itself involved the publication of George Washington's private letters, and therefore contained a strong public interest component as well.

The inability of the courts or Congress even to articulate a complete list of the factors to be applied in determining fair use is evidence of the difficulty courts have had in applying the doctrine. Judge Learned Hand described the fair use doctrine as "the most troublesome in the whole law of copyright." Dellar v. Samuel Goldwyn, Inc., 104 F.2d 661 (2d Cir. 1939).

Harper & Row, Publishers, Inc., et al. v. Nation Enterprises et al.
Supreme Court of the United States
471 U.S. 539 (1985)

Justice O'CONNOR delivered the opinion of the Court.

This case requires us to consider to what extent the "fair use" provision of the Copyright Revision Act of 1976 (hereinafter the Copyright Act), 17 U.S.C. §107, sanctions the unauthorized use of quotations from a public figure's unpublished manuscript. In March 1979, an undisclosed source provided The Nation Magazine with the unpublished manuscript of "A Time to Heal: The Autobiography of Gerald R. Ford." Working directly from the purloined manuscript, an editor of The Nation produced a short piece entitled "The Ford Memoirs — Behind the Nixon Pardon." The piece was timed to "scoop" an article scheduled shortly to appear in Time Magazine. Time had agreed to purchase the exclusive right to print prepublication excerpts from the copyright holders, Harper & Row Publishers, Inc. (hereinafter Harper & Row), and Reader's Digest Association, Inc. (hereinafter Reader's Digest). As a result of The Nation article, Time canceled its agreement. Petitioners brought a successful copyright action against The Nation. On appeal, the Second Circuit reversed the lower court's finding of infringement, holding that The Nation's act was sanctioned as a "fair use" of the copyrighted material. We granted certiorari, 467 U.S. 1214 (1984), and we now reverse.

. . . The [Time] issue featuring the excerpts was timed to appear approximately one week before shipment of the full length book version to bookstores. Exclusivity was an important consideration; Harper & Row instituted procedures designed to maintain the confidentiality of the manuscript, and Time retained the right to renegotiate the second payment should the material appear in print prior to its release of the excerpts.

Two to three weeks before the Time article's scheduled release, an unidentified person secretly brought a copy of the Ford manuscript to Victor Navasky, editor of The Nation, a political commentary magazine. Mr. Navasky knew that his possession of the manuscript was not authorized and that the manuscript must be returned quickly to his "source" to avoid discovery. 557 F. Supp. 1067, 1069 (S.D.N.Y. 1983). He hastily put together what he believed was "a real hot news story" composed of quotes, paraphrases, and facts drawn exclusively from the manuscript. Ibid. Mr. Navasky attempted no independent commentary, research or criticism, in part because of the need for speed if he was to "make news" by "publish[ing] in advance of publication of the Ford book." The 2,250-word article, reprinted in the Appendix to this opinion, appeared on April 3, 1979. As a result of The Nation's article, Time canceled its piece and refused to pay the remaining $12,500. . . .

II

We agree with the Court of Appeals that copyright is intended to increase and not to impede the harvest of knowledge. But we believe the Second Circuit gave insufficient deference to the scheme established by the Copyright Act for fostering the original works that provide the seed and substance of this harvest. The rights conferred by copyright are designed to assure contributors to the store of knowledge a fair return for their labors. Twentieth Century Music Corp. v. Aiken, 422 U.S. 151, 156 (1975). . . .

III

A . . .

"[T]he author's consent to a reasonable use of his copyrighted works ha[d] always been implied by the courts as a necessary incident of the constitutional policy of promoting the progress of science and the useful arts, since a prohibition of such use would inhibit subsequent writers from attempting to improve upon prior works and thus . . . frustrate the very ends sought to be attained." Ball 260. Professor Latman, in a study of the doctrine of fair use commissioned by Congress for the revision effort, see Sony Corp. of America v. Universal City Studios, Inc., 464 U.S., at 462-463, n.9 (dissenting opinion), summarized prior law as turning on the "importance of the material copied or performed from the point of view of the reasonable copyright owner. In other words, would the reasonable copyright owner have consented to the use?" Latman 15. . . .

Perhaps because the fair use doctrine was predicated on the author's implied consent to "reasonable and customary" use when he released his work for public consumption, fair use traditionally was not recognized as a defense to charges of copying from an author's as yet unpublished works. Under common-law copyright, "the property of the author . . . in his intellectual creation [was] absolute until he voluntarily part[ed] with the same." American Tobacco Co. v. Werckmeister, 207 U.S. 284, 299 (1907); 2 Nimmer §8.23, at 8-273. This absolute rule, however, was tempered in practice by the equitable nature of the fair use doctrine. In a given case, factors such as implied consent through de facto publication or performance or dissemination of a work may tip the balance of equities in favor of prepublication use. See Copyright Law Revision — Part 2: Discussion and Comments on Report of the Register of Copyrights on General Revision of the U.S. Copyright Law, 88th Cong., 1st Sess., 27 (H.R. Comm. Print 1963) (discussion suggesting works disseminated to the public in a form not constituting a technical "publication" should nevertheless be subject to fair use); 3 Nimmer §13.05, at 13-62, n.2. But it has never been seriously disputed that "the fact that the plaintiff's work is unpublished . . . is a factor tending to negate the defense of fair use." Ibid. Publication of an author's expression before he has authorized its dissemination seriously infringes the author's right to decide when and whether it will be made public, a factor not present in fair use of published works. Respondents contend, however, that Congress, in including first publication among the rights enumerated in §106, which are expressly subject to fair use under §107, intended that fair use would apply in pari materia to published and unpublished works. The Copyright Act does not support this proposition.

We also find unpersuasive respondents' argument that fair use may be made of a soon-to-be-published manuscript on the ground that the author has demonstrated he has no interest in nonpublication. This argument assumes that the unpublished nature of copyrighted material is only relevant to letters or other confidential writings not intended for dissemination. It is true that common-law copyright was often enlisted in the service of personal privacy. See Brandeis & Warren, The Right to Privacy, 4 Harv. L. Rev. 193, 198-199 (1890). In its commercial guise, however, an author's right to choose when he will publish is no less deserving of protection. The period encompassing the work's initiation, its preparation, and its grooming for public dissemination is a crucial one for any literary endeavor. The Copyright Act, which accords the copyright owner the "right to control the first public distribution" of his work, House Report, at 62, echos the common law's concern that the author or copyright owner retain control throughout this critical stage. See generally Comment, The Stage of Publication as a "Fair Use" Factor: Harper & Row, Publishers v. Nation Enterprises, 58 St. John's L. Rev. 597 (1984). The obvious benefit to author and public alike of assuring authors the leisure to develop their ideas free from fear of expropriation outweighs any short-term "news value" to be gained from premature publication of the author's expression. See Goldstein, Copyright and the First Amendment, 70 Colum. L. Rev. 983, 1004-1006 (1970) (The absolute protection the common law accorded to soon-to-be published works "[was] justified by [its] brevity and expedience"). The author's control of first public distribution implicates not only his personal interest in creative control but his property interest in exploitation of prepublication rights, which are valuable in themselves and serve as a valuable adjunct to publicity and marketing. See Belushi v. Woodward, 598 F. Supp. 36 (D.C. 1984) (successful marketing depends on coordination of serialization and release to public); Marks, Subsidiary Rights and Permissions, in What Happens in Book Publishing 230

(C. Grannis ed. 1967) (exploitation of subsidiary rights is necessary to financial success of new books). Under ordinary circumstances, the author's right to control the first public appearance of his undisseminated expression will outweigh a claim of fair use. . . .

IV . . .

The four factors identified by Congress as especially relevant in determining whether the use was fair are: (1) the purpose and character of the use; (2) the nature of the copyrighted work; (3) the substantiality of the portion used in relation to the copyrighted work as a whole; (4) the effect on the potential market for or value of the copyrighted work. We address each one separately.

Purpose of the Use. The Second Circuit correctly identified news reporting as the general purpose of The Nation's use. News reporting is one of the examples enumerated in §107 to "give some idea of the sort of activities the courts might regard as fair use under the circumstances." Senate Report, at 61. This listing was not intended to be exhaustive, see ibid.; §101 (definition of "including" and "such as"), or to single out any particular use as presumptively a "fair" use. The drafters resisted pressures from special interest groups to create presumptive categories of fair use, but structured the provision as an affirmative defense requiring a case-by-case analysis. See H.R. Rep. No. 83, 90th Cong., 1st Sess., 37 (1967); Patry 477, n.4. "[W]hether a use referred to in the first sentence of section 107 is a fair use in a particular case will depend upon the application of the determinative factors, including those mentioned in the second sentence." Senate Report, at 62. The fact that an article arguably is "news" and therefore a productive use is simply one factor in a fair use analysis.

We agree with the Second Circuit that the trial court erred in fixing on whether the information contained in the memoirs was actually new to the public. As Judge Meskill wisely noted, "[c]ourts should be chary of deciding what is and what is not news." 723 F.2d, at 215 (dissenting). Cf. Gertz v. Robert Welch, Inc., 418 U.S. 323, 345-346 (1974). "The issue is not what constitutes 'news,' but whether a claim of news reporting is a valid fair use defense to an infringement of copyrightable expression." Patry 119. The Nation has every right to seek to be the first to publish information. But The Nation went beyond simply reporting uncopyrightable information and actively sought to exploit the headline value of its infringement, making a "news event" out of its unauthorized first publication of a noted figure's copyrighted expression.

The fact that a publication was commercial as opposed to nonprofit is a separate factor that tends to weigh against a finding of fair use. "[E]very commercial use of copyrighted material is presumptively an unfair exploitation of the monopoly privilege that belongs to the owner of the copyright."

Sony Corp. of America v. Universal City Studios, Inc., 464 U.S., at 451. In arguing that the purpose of news reporting is not purely commercial, The Nation misses the point entirely. The crux of the profit/nonprofit distinction is not whether the sole motive of the use is monetary gain but whether the user stands to profit from exploitation of the copyrighted material without paying the customary price. See Roy Export Co. Establishment v. Columbia Broadcasting System, Inc., 503 F. Supp., at 1144; 3 Nimmer §13.05[A][1], at 13-71, n.25.3.

In evaluating character and purpose we cannot ignore The Nation's stated purpose of scooping the forthcoming hardcover and Time abstracts. The Nation's use had not merely the incidental effect but the intended purpose of supplanting the copyright holder's commercially valuable right of first publication. See Meredith Corp. v. Harper & Row, Publishers, Inc., 378 F. Supp. 686, 690 (S.D.N.Y.) (purpose of text was to compete with original), *aff'd*, 500 F.2d 1221 (CA2 1974). Also relevant to the "character" of the use is "the propriety of the defendant's conduct." 3 Nimmer §13.05[A], at 13-72. "Fair use presupposes 'good faith' and 'fair dealing.' " Time Inc. v. Bernard Geis Associates, 293 F. Supp. 130, 146 (S.D.N.Y. 1968), quoting Schulman, Fair Use and the Revision of the Copyright Act, 53 Iowa L. Rev. 832 (1968). The trial court found that The Nation knowingly exploited a purloined manuscript. Unlike the typical claim of fair use, The Nation cannot offer up even the fiction of consent as justification. Like its competitor news-weekly, it was free to bid for the right of abstracting excerpts from "A Time to Heal." Fair use "distinguishes between 'a true scholar and a chiseler who infringes a work for personal profit.' " Wainwright Securities Inc. v. Wall Street Transcript Corp., 558 F.2d, at 94, quoting from Hearings on Bills for the General Revision of the Copyright Law before the House Committee on the Judiciary, 89th Cong., 1st Sess., ser. 8, pt. 3, p. 1706 (1966) (statement of John Schulman).

Nature of the Copyrighted Work.   Second, the Act directs attention to the nature of the copyrighted work. "A Time to Heal" may be characterized as an unpublished historical narrative or autobiography. The law generally recognizes a greater need to disseminate factual works than works of fiction or fantasy. See Gorman, Fact or Fancy? The Implications for Copyright, 29 J. Copyright Soc. 560, 561 (1982).

> [E]ven within the field of fact works, there are gradations as to the relative proportion of fact and fancy. One may move from sparsely embellished maps and directories to elegantly written biography. The extent to which one must permit expressive language to be copied, in order to assure dissemination of the underlying facts, will thus vary from case to case. Id., at 563.

Some of the briefer quotes from the memoirs are arguably necessary adequately to convey the facts; for example, Mr. Ford's characterization of the White House tapes as the "smoking gun" is perhaps so integral to the idea expressed as to be inseparable from it. Cf. 1 Nimmer §1.10[C]. But The Nation did not stop at isolated phrases and instead excerpted subjective descriptions and portraits of public figures whose power lies in the author's individualized expression. Such use, focusing on the most expressive elements of the work, exceeds that necessary to disseminate the facts.

The fact that a work is unpublished is a critical element of its "nature." 3 Nimmer §13.05[A]; Comment, 58 St. John's L. Rev., at 613. Our prior discussion establishes that the scope of fair use is narrower with respect to unpublished works. While even substantial quotations might qualify as fair use in a review of a published work or a news account of a speech that had been delivered to the public or disseminated to the press, see House Report, at 65, the author's right to control the first public appearance of his expression weighs against such use of the work before its release. The right of first publication encompasses not only the choice whether to publish at all, but also the choices of when, where, and in what form first to publish a work.

In the case of Mr. Ford's manuscript, the copyright holders' interest in confidentiality is irrefutable; the copyright holders had entered into a contractual undertaking to "keep the manuscript confidential" and required that all those to whom the manuscript was shown also "sign an agreement to keep the manuscript confidential." While the copyright holders' contract with Time required Time to submit its proposed article seven days before publication, The Nation's clandestine publication afforded no such opportunity for creative or quality control. It was hastily patched together and contained "a number of inaccuracies." A use that so clearly infringes the copyright holder's interests in confidentiality and creative control is difficult to characterize as "fair."

Amount and Substantiality of the Portion Used. Next, the Act directs us to examine the amount and substantiality of the portion used in relation to the copyrighted work as a whole. In absolute terms, the words actually quoted were an insubstantial portion of "A Time to Heal." The District Court, however, found that "[T]he Nation took what was essentially the heart of the book." 557 F. Supp., at 1072. We believe the Court of Appeals erred in overruling the District Judge's evaluation of the qualitative nature of the taking. See, e.g., Roy Export Co. Establishment v. Columbia Broadcasting System, Inc., 503 F. Supp., at 1145 (taking of 55 seconds out of 1 hour and 29-minute film deemed qualitatively substantial). A Time editor described the chapters on the pardon as "the most interesting and moving parts of the entire manuscript." The portions actually quoted were selected by Mr. Navasky as among the most powerful passages in those chapters. He testified that he used verbatim excerpts because simply reciting the information could not adequately convey the "absolute certainty with which [Ford] expressed himself," or show that "this comes from President Ford," or carry the "definitive quality" of the original. In short, he quoted these passages precisely because they qualitatively embodied Ford's distinctive expression.

As the statutory language indicates, a taking may not be excused merely because it is insubstantial with respect to the infringing work. As Judge Learned Hand cogently remarked, "no plagiarist can excuse the wrong by showing how much of his work he did not pirate." Sheldon v. Metro-Goldwyn Pictures Corp., 81 F.2d 49, 56 (CA2), *cert. denied*, 298 U.S. 669 (1936). Conversely, the fact that a substantial portion of the infringing work was copied verbatim is evidence of the qualitative value of the copied material, both to the originator and to the plagiarist who seeks to profit from marketing someone else's copyrighted expression.

Stripped to the verbatim quotes, the direct takings from the unpublished manuscript constitute at least 13% of the infringing article. See Meeropol v. Nizer, 560 F.2d 1061, 1071 (CA2 1977) (copyrighted letters constituted less than 1% of infringing work but were prominently featured). The Nation article is structured around the quoted excerpts which serve as its dramatic focal points. In view of the expressive value of the excerpts and their key role in the infringing work, we cannot agree with the Second Circuit that the "magazine took a meager, indeed an infinitesimal amount of Ford's original language." 723 F.2d, at 209.

Effect on the Market. Finally, the Act focuses on "the effect of the use upon the potential market for or value of the copyrighted work." This last factor is undoubtedly the single most important element of fair use.[9] See 3 Nimmer §13.05[A], at 13-76, and cases cited therein. "Fair use, when properly applied, is limited to copying by others which does not materially impair the marketability of the work which is copied." 1 Nimmer §1.10[D], at 1-87. The trial court found not merely a potential but an actual effect on the market. Time's cancellation of its projected serialization and its refusal to pay the $12,500 were the direct effect of the infringement. . . .

More important, to negate fair use one need only show that if the challenged use "should become widespread, it would adversely affect the potential market for the copyrighted work." Sony Corp. of America v. Universal City Studios, Inc., 464 U.S., at 451 (emphasis added); id., at 484, and n. 36 (collecting cases) (dissenting opinion). This inquiry must take account not only of harm to the original but also of harm to the market for derivative works. See Iowa State University Research Foundation, Inc. v. American Broadcasting Cos., 621 F.2d 57 (CA2 1980); Meeropol v. Nizer, supra, at 1070; Roy Export v. Columbia Broadcasting System, Inc., 503 F. Supp., at 1146. "If the defendant's work adversely affects the value of any of the rights in the copyrighted work (in this case the adaptation [and serialization] right)

---

9. Economists who have addressed the issue believe the fair use exception should come into play only in those situations in which the market fails or the price the copyright holder would ask is near zero. See, e.g., T. Brennan, Harper & Row v. The Nation, Copyrightability and Fair Use, Dept. of Justice Economic Policy Office Discussion Paper 13-17 (1984); Gordon, Fair Use as Market Failure: A Structural and Economic Analysis of the Betamax Case and its Predecessors, 82 Colum. L. Rev. 1600, 1615 (1982). As the facts here demonstrate, there is a fully functioning market that encourages the creation and dissemination of memoirs of public figures. In the economists' view, permitting "fair use" to displace normal copyright channels disrupts the copyright market without a commensurate public benefit.

the use is not fair." 3 Nimmer §13.05[B], at 13-77 — 13-78 (footnote omitted).

It is undisputed that the factual material in the balance of The Nation's article, besides the verbatim quotes at issue here, was drawn exclusively from the chapters on the pardon. The excerpts were employed as featured episodes in a story about the Nixon pardon — precisely the use petitioners had licensed to Time. The borrowing of these verbatim quotes from the unpublished manuscript lent The Nation's piece a special air of authenticity — as Navasky expressed it, the reader would know it was Ford speaking and not The Nation. Thus it directly competed for a share of the market for prepublication excerpts. The Senate Report states:

> "With certain special exceptions . . . a use that supplants any part of the normal market for a copyrighted work would ordinarily be considered an infringement."

Senate Report, at 65. Placed in a broader perspective, a fair use doctrine that permits extensive prepublication quotations from an unreleased manuscript without the copyright owner's consent poses substantial potential for damage to the marketability of first serialization rights in general.

> "Isolated instances of minor infringements, when multiplied many times, become in the aggregate a major inroad on copyright that must be prevented."

Ibid.

COMMENTS AND QUESTIONS

1. A spirited dissent in *Harper & Row* (omitted here) argued that *The Nation* had taken no more than was necessary to report the story, and that most of what was taken reflected ideas rather than expression.

2. One common category of "fair use" involves a minor, noncommercial use of a copyrighted work, often on behalf of a non-profit organization such as an educational institution. There are two important differences between *Harper & Row* and the traditional "public interest" concept of fair use: the use in *Harper & Row* was for profit, and it had a direct market impact. *The Nation* took significant excerpts from Ford's book, and those excerpts constituted the bulk of the *Nation* article. These factors weigh against a finding of fair use. Is there a strong "public interest" in allowing *The Nation* to publish its story that should outweigh those factors?

Suppose *The Nation* had published its article two weeks *after Time's* article. Would the court reach a different result? Should it?

3. As with other intellectual property regimes, the basic theory behind copyright law is that protection will promote more creation and the public will therefore benefit. What does this suggest for copyrighted works that are kept private? Should they be entitled to more or less protection? The *Harper & Row* court seems to believe that such works are entitled to more protection

because the privacy of the author is implicated. Can an argument be made to the contrary, particularly where as here the work is about to be released? Do we really want the monopoly provided to copyright holders to be extended to "monopoly reporting" of news events by the media?

For that matter, are there circumstances in which the "public interest" is so strong that it will outweigh the privacy interests of those who have no intention of *ever* publishing a copyrighted work? Consider the Pentagon Papers case (New York Times v. United States, 403 U.S. 713 (1971)). In that case, the *New York Times* was prosecuted for appropriating and publishing written government secrets about the Vietnam War. The Supreme Court concluded that the government could not prevent the *New York Times* from publishing the Papers. But suppose instead that the author of the Pentagon Papers had sued for copyright infringement. What result? Is publication of a document intended to be kept private a "fair use"? The *New York Times* acted "for profit" in the same way *The Nation* did, although the element of market loss is missing. Would the plaintiff in such a case be entitled to damages? To injunctive relief?

4. One of the factors the *Harper & Row* court cites as militating against allowing *The Nation* to publish excerpts from Ford's memoir is that the author of the book will lose "creative control" over the excerpts. Is this a valid consideration? Must fair use be use within the control or intent of the original author? Courts in general have said no and have allowed parody or negative criticism of a work to use excerpts from the work. We discuss parodies later in this section.

5. The Court refers to the four factors in section 107 as "nonexclusive." Can you think of other factors that are (or should be) relevant to the fair use inquiry?

One factor the Court relies on that is not identified in section 107 is the percentage of the *potentially infringing* work composed of copied material. (The statute mentions only the amount of the *copyrighted work* that is taken.) The Court thus makes much of the fact that copied material constituted 13 percent of the infringing work, even though it was well less than 1 percent of the copyrighted work. Should this be a factor in fair use analysis? If so, how should book reviews be analyzed for fair use purposes?

A particularly striking example of this phenomenon is the recent spate of suits against the makers of movies and television shows, alleging that the movie or show infringes the copyright of a work that is shown briefly in the background of the film. In these cases, the copyrighted work (a chair, a sculpture, a poster, or a song) may be "taken" in its entirety, but it is generally a miniscule portion of the defendant's work. Courts dealing with these cases have come to different conclusions under the fair use doctrine. For example, in Woods v. Universal City Studios, 920 F. Supp. 62 (S.D.N.Y. 1996), the court enjoined the distribution of the film *12 Monkeys* during its run because three scenes in the film featured a chair strikingly similar to the plaintiff's copyrighted chair. And in Ringgold v. Black Entertainment TV Inc., 126 F.3d 70 (2d Cir. 1997), the court held that the out-of-focus background appearance of a copyrighted poster in a television show was not fair use as a matter of law. By contrast, the depiction of background art for sixty seconds was held to be fair use in Jackson v. Warner Bros., 993 F. Supp. 585 (E.D. Mich. 1997). See also Sandoval v. New Line Cinema, 973 F. Supp. 409 (S.D.N.Y. 1997) (unpublished photographs appearing in the background of the movie *Seven* for less than 30 seconds constituted fair use).

How would the court that decided *Harper & Row* approach such a case? Does it make sense to allow the copying of most of a work simply because the defendant has contributed so much noninfringing material? For an argument to excuse copying in some such circumstances, see Mark A. Lemley, The Economics of Improvement in Intellectual Property Law, 75 Tex. L. Rev. 989 (1997).

6. The copyrighted work in *Harper* was an autobiography, which is a factual work. As we discovered earlier, the copyright protection afforded factual works is much "thinner" than for works of fiction, because the facts themselves cannot be protected. Should the Court have distinguished between factual and protectable material in determining the extent of copying by *The Nation?* Would such an analysis have changed the result? The dissent in the case argued that there could be no copyright infringement because *The Nation* took only ideas, and not protectable expression, from Ford's book.

7. In footnote 9, the Court rejects claims of market failure because, in its view, "there is a fully functioning market that encourages the creation and dissemination of memoirs of public figures." Is this the right question? Is the relevant market here the one for memoirs, or for articles like *The Nation*'s? If the latter, is there a "fully functioning market" for excerpts of this nature? For critical reviews of memoirs?

PROBLEMS

Problem 4-30. A home movie taken by a witness named Zales captures the shooting of President John F. Kennedy in Dallas in 1963. Only days after the shooting, Earth magazine buys the exclusive rights to the Zales film. It subsequently publishes some Zales frames in a magazine special on the assassination. The Zales frames are also appended to a government report by the Whitewash Commission on the assassination. Stone, a writer who is convinced that the Whitewash Report is flawed, unsuccessfully seeks permission from Earth to reprint the Zales pictures in his book alleging a conspiracy to kill the president. Undaunted by Earth's refusal, Stone breaks into Earth's offices and photocopies the pictures, which he then publishes in his book. Earth sues for copyright infringement. Stone defends on grounds of fair use, and offers to turn over all profits from his book to Earth. What result?

Problem 4-31. Garrison, a scholar who believes that Lee Harvey Oswald acted alone, is incensed by Stone's book. Garrison publishes a book that he styles *A Rebuttal to Stone*. In it, Garrison follows Stone's organization in detail, presenting and refuting each of Stone's arguments. In doing so, Garrison quotes liberally from Stone's work. Garrison does not use the Zales pictures, however. Stone sues for copyright infringement. Is Garrison's work fair use?

Problem 4-32. Refer back to the Salinger-Hamilton case (problem 4-25 above). Can Hamilton successfully argue a fair use defense?

a. Videotaping

═══ **Sony Corporation of America v. Universal City Studios, Inc.**
Supreme Court of the United States
464 U.S. 417 (1984)

Justice STEVENS delivered the opinion of the Court.

[The respondents, a group of movie studios, sued the makers of video cassette recorders ("VCR's," or "VTR's" in the opinion), alleging that they were liable for contributory copyright infringement because consumers bought VCRs and used them to tape movies and other television shows. The court identified the predominant purpose behind such taping as "time-shifting" — that is, taping a show when you were not at home in order to watch it later.]

When a charge of contributory infringement is predicated entirely on the sale of an article of commerce that is used by the purchaser to infringe a patent, the public interest in access to that article of commerce is necessarily implicated. A finding of contributory infringement does not, of course, remove the article from the market altogether; it does, however, give the patentee effective control over the sale of that item. Indeed, a finding of contributory infringement is normally the functional equivalent of holding that the disputed article is within the monopoly granted to the patentee.[21]

For that reason, in contributory infringement cases arising under the patent laws the Court has always recognized the critical importance of not allowing the patentee to extend his monopoly beyond the limits of his specific grant. . . .

We recognize there are substantial differences between the patent and

---

21. It seems extraordinary to suggest that the Copyright Act confers upon all copyright owners collectively, much less the two respondents in this case, the exclusive right to distribute VTR's simply because they may be used to infringe copyrights. That, however, is the logical implication of their claim. The request for an injunction below indicates that respondents seek, in effect, to declare VTR's contraband. Their suggestion in this Court that a continuing royalty pursuant to a judicially created compulsory license would be an acceptable remedy merely indicates that respondents, for their part, would be willing to license their claimed monopoly interest in VTR's to Sony in return for a royalty.

copyright laws. But in both areas the contributory infringement doctrine is grounded on the recognition that adequate protection of a monopoly may require the courts to look beyond actual duplication of a device or publication to the products or activities that make such duplication possible. The staple article of commerce doctrine must strike a balance between a copyright holder's legitimate demand for effective — not merely symbolic — protection of the statutory monopoly, and the rights of others freely to engage in substantially unrelated areas of commerce. Accordingly, the sale of copying equipment, like the sale of other articles of commerce, does not constitute contributory infringement if the product is widely used for legitimate, unobjectionable purposes. Indeed, it need merely be capable of substantial noninfringing uses.

IV

The question is thus whether the Betamax is capable of commercially significant noninfringing uses. In order to resolve that question, we need not explore *all* the different potential uses of the machine and determine whether or not they would constitute infringement. Rather, we need only consider whether on the basis of the facts as found by the District Court a significant number of them would be noninfringing. Moreover, in order to resolve this case we need not give precise content to the question of how much use is commercially significant. For one potential use of the Betamax plainly satisfies this standard, however it is understood: private, noncommercial time-shifting in the home. It does so both (A) because respondents have no right to prevent other copyright holders from authorizing it for their programs, and (B) because the District Court's factual findings reveal that even the unauthorized home time-shifting of respondents' programs is legitimate fair use.

A. *Authorized Time-Shifting*

Each of the respondents owns a large inventory of valuable copyrights, but in the total spectrum of television programming their combined market share is small. The exact percentage is not specified, but it is well below 10%. If they were to prevail, the outcome of this litigation would have a significant impact on both the producers and the viewers of the remaining 90% of the programming in the Nation. No doubt, many other producers share respondents' concern about the possible consequences of unrestricted copying. Nevertheless the findings of the District Court make it clear that time-shifting may enlarge the total viewing audience and that many producers are willing to allow private time-shifting to continue, at least for an experimental time period.[23]

---

23. The District Court did not make any explicit findings with regard to how much broadcasting is wholly uncopyrighted. The record does include testimony that at least one movie —

The District Court found:

> Even if it were deemed that home-use recording of copyrighted material con-
> stituted infringement, the Betamax could still legally be used to record nonco-
> pyrighted material or material whose owners consented to the copying. An
> injunction would deprive the public of the ability to use the Betamax for this
> noninfringing off-the-air recording.

If there are millions of owners of VTR's who make copies of televised
sports events, religious broadcasts, and educational programs such as Mister
Rogers' Neighborhood, and if the proprietors of those programs welcome
the practice, the business of supplying the equipment that makes such copying
feasible should not be stifled simply because the equipment is used by some
individuals to make unauthorized reproductions of respondents' works. The
respondents do not represent a class composed of all copyright holders. Yet
a finding of contributory infringement would inevitably frustrate the interests
of broadcasters in reaching the portion of their audience that is available only
through time-shifting.

Of course, the fact that other copyright holders may welcome the practice of time-shifting does not mean that respondents should be deemed to
have granted a license to copy their programs. Third-party conduct would be
wholly irrelevant in an action for direct infringement of respondents' copyrights. But in an action for contributory infringement against the seller of
copying equipment, the copyright holder may not prevail unless the relief
that he seeks affects only his programs, or unless he speaks for virtually all
copyright holders with an interest in the outcome. In this case, the record
makes it perfectly clear that there are many important producers of national
and local television programs who find nothing objectionable about the en-
largement in the size of the television audience that results from the practice
of time-shifting for private home use.[28] The seller of the equipment that

---

My Man Godfrey — falls within that category, id., at 2300-2301, and certain broadcasts produced by the Federal Government are also uncopyrighted. See 17 U.S.C. §105 (1982 ed.). Cf.
Schnapper v. Foley, 215 U.S. App. D.C. 59, 667 F.2d 102 (1981) (explaining distinction between work produced by the Government and work commissioned by the Government). To the
extent such broadcasting is now significant, it further bolsters our conclusion. Moreover, since
copyright protection is not perpetual, the number of audiovisual works in the public domain
necessarily increases each year.

28. It may be rare for large numbers of copyright owners to authorize duplication of their
works without demanding a fee from the copier. In the context of public broadcasting, however,
the user of the copyrighted work is not required to pay a fee for access to the underlying work.
The traditional method by which copyright owners capitalize upon the television medium —
commercially sponsored free public broadcast over the public airwaves — is predicated upon the
assumption that compensation for the value of displaying the works will be received in the form
of advertising revenues.

In the context of television programming, some producers evidently believe that permitting
home viewers to make copies of their works off the air actually enhances the value of their
copyrights. Irrespective of their reasons for authorizing the practice, they do so, and in significant
enough numbers to create a substantial market for a noninfringing use of the Sony VTR's. No
one could dispute the legitimacy of that market if the producers had authorized home taping of
their programs in exchange for a license fee paid directly by the home user. The legitimacy of

expands those producers' audiences cannot be a contributory infringer if, as is true in this case, it has had no direct involvement with any infringing activity.

B. *Unauthorized Time-Shifting*

Even unauthorized uses of a copyrighted work are not necessarily infringing. An unlicensed use of the copyright is not an infringement unless it conflicts with one of the specific exclusive rights conferred by the copyright statute. Twentieth Century Music Corp. v. Aiken, 422 U.S., at 154-155. Moreover, the definition of exclusive rights in §106 of the present Act is prefaced by the words "subject to sections 107 through 118." Those sections describe a variety of uses of copyrighted material that "are not infringements of copyright" "notwithstanding the provisions of section 106." The most pertinent in this case is §107, the legislative endorsement of the doctrine of "fair use."

That section identifies various factors that enable a Court to apply an "equitable rule of reason" analysis to particular claims of infringement. Although not conclusive, the first factor requires that "the commercial or nonprofit character of an activity" be weighed in any fair use decision. If the Betamax were used to make copies for a commercial or profit-making purpose, such use would presumptively be unfair. The contrary presumption is appropriate here, however, because the District Court's findings plainly establish that time-shifting for private home use must be characterized as a noncommercial, nonprofit activity. Moreover, when one considers the nature of a televised copyrighted audiovisual work, see 17 U.S.C. §107(2) (1982 ed.), and that time-shifting merely enables a viewer to see such a work which he had been invited to witness in its entirety free of charge, the fact that the entire work is reproduced, see §107(3), does not have its ordinary effect of militating against a finding of fair use.

This is not, however, the end of the inquiry because Congress has also directed us to consider "the effect of the use upon the potential market for or value of the copyrighted work." §107(4). The purpose of copyright is to create incentives for creative effort. Even copying for noncommercial purposes may impair the copyright holder's ability to obtain the rewards that Congress intended him to have. But a use that has no demonstrable effect

---

that market is not compromised simply because these producers have authorized home taping of their programs without demanding a fee from the home user. The copyright law does not require a copyright owner to charge a fee for the use of his works, and as this record clearly demonstrates, the owner of a copyright may well have economic or noneconomic reasons for permitting certain kinds of copying to occur without receiving direct compensation from the copier. It is not the role of the courts to tell copyright holders the best way for them to exploit their copyrights: even if respondents' competitors were ill-advised in authorizing home videotaping, that would not change the fact that they have created a substantial market for a paradigmatic noninfringing use of Sony's product.

upon the potential market for, or the value of, the copyrighted work need not be prohibited in order to protect the author's incentive to create. The prohibition of such noncommercial uses would merely inhibit access to ideas without any countervailing benefit. . . .

. . . [T]o the extent time-shifting expands public access to freely broadcast television programs, it yields societal benefits. In Community Television of Southern California v. Gottfried, 459 U.S. 498, 508, n.12 (1983), we acknowledged the public interest in making television broadcasting more available. Concededly, that interest is not unlimited. But it supports an interpretation of the concept of "fair use" that requires the copyright holder to demonstrate some likelihood of harm before he may condemn a private act of time-shifting as a violation of federal law.

COMMENTS AND QUESTIONS

1. Did the Court get it right? Four Justices dissented in *Sony*. The four dissenters would have applied a rule in contributory infringement cases that held device manufacturers liable if the "primary purpose or effect" of their product was to enable others to infringe copyright, even if the devices were capable of a substantial noninfringing use. The dissent would also have concluded that copying audiovisual works in their entirety was not a fair use.

The issue seems more complex than either the majority or the dissent are willing to recognize. On the one hand, take another look at footnote 21 of the Court's opinion. The argument that copyright owners should not get control over the VCR market seems persuasive. But why shouldn't they? The argument makes sense only if VCRs are a separate, staple commodity from television programs. This in turn depends on whether they have noninfringing uses, an issue over which the Court splits.

Sony was based on the state of the VCR market in its infancy in 1979.[28] Consideration of the issue today is complicated (as it is in the audiotape market) by the fact that VCRs have two distinct uses: they can record programs and they can play them. It seems clear that the use of a VCR to watch movies rented from a licensed supplier — the predominant use of VCRs today — is noninfringing. But does this justify the recording function of a VCR?

Further developments also make it clear that Hollywood was wrong to predict imminent disaster if Sony were allowed to continue making VCRs. Indeed, the movie industry has discovered ways to make a tidy profit from video rentals and sales. Does this suggest that courts should be cautious about using copyright law to stifle a developing new market for technology?

2. What is a "substantial noninfringing use"? The Court defines it as a "commercially significant noninfringing use." Is this a reasonable definition?

---

28. On the development of this industry and the legal issues raised, see James Lardner, Fast Forward: Hollywood, the Japanese and the VCR Wars (1987).

Can you think of circumstances in which a noninfringing use might be substantial enough to justify marketing a product even though it is not "commercially significant"?

Note that the dissent would hold a defendant liable for contributory infringement unless use of its product was "predominantly" noninfringing. Legislation introduced in Congress in 1996, but which did not pass, would have adopted a version of the dissent's test for judging contributory infringement (whether the "primary purpose or effect" of a device was to encourage infringement), at least in the context of the Internet. Similar legislation passed in 1998 under the title Digital Millenium Copyright Act, though with significant changes and limitations. The Act is discussed in detail in Chapter 7. Is this a better standard? Why or why not?

3. *Sony* focused on the nature of the "public benefit" conferred by home taping. Public benefits are not expressly listed as a factor in section 107, although the "purpose and character" of the use is important. To constitute fair use, must a use benefit the public? If so, does this factor effectively eclipse those identified in the statute?

Is there a public benefit to home taping? If so, what is it?

4. Recall our discussion of contributory infringement in the last section. Of what significance is the defendant's knowledge of infringing activity? The *Sony* court distinguishes this case from the trademark cases on the grounds that Sony was not selling to "identified individuals known to be engaging in continued infringement." The careful phrasing is apparently intentional; Sony was surely aware that a large number of its products were being used to make copies, and that at least some of those copies infringed the original works. Should such generalized knowledge be sufficient?

5. *Fair Use as a Remedy for Market Failure.* Wendy Gordon argues that the fair use doctrine can and does serve to remedy market failures. Wendy Gordon, Fair Use as Market Failure: A Structural and Economics Analysis of the Betamax Case and Its Predecessors, 82 Colum. L. Rev. 1600 (1982).

> . . . Though the copyright law has provided a means for excluding non-purchasers and thus has attempted to cure the public goods problem, and though it has provided mechanisms to facilitate consensual transfers, at times bargaining may be exceedingly expensive or it may be impractical to obtain enforcement against nonpurchasers, or other market flaws might preclude achievement of desirable consensual exchanges. In those cases, the market cannot be relied on to mediate public interests in dissemination and private interests in remuneration. . . .
>
> Fair use should be awarded to the defendant in a copyright infringement action when (1) market failure is present; (2) transfer of the use to defendant is socially desirable; and (3) an award of fair use would not cause substantial injury to the incentives of the plaintiff copyright owner. The first element of this test ensures that market bypass will not be approved without good cause. The second element of the test ensures that the transfer of a license to use from the copyright holder to the unauthorized user effects a net gain in social value. The third

element ensures that the grant of fair use will not undermine the incentive-creating purpose of the copyright law.

More recently, Robert Merges has suggested that the fair use doctrine and other liability-type rules in intellectual property doctrine might undermine economic efficiency. He suggests that copiers' reliance on the fair use doctrine may actually prevent the development of innovative cooperative market clearing mechanisms, such as ASCAP. Such licensing societies are a reasonable way to reduce the transactions costs associated with bulk licensing of copyrights. See Robert Merges, Contracting into Liability Rules: Intellectual Property Rights and Collective Rights Organizations, 84 Cal. L. Rev. 1293 (1996).

Does a focus on market failure and licensing unreasonably limit the scope of the fair use doctrine? Should some copying be allowed without the copier having to pay a royalty, even if the copyright owner would have demanded one?

PROBLEM

Problem 4-33. Part of the Court's analysis of fair use in *Sony* was premised on its conclusion that time-shifting does not have a market impact because television programs are offered "free" to consumers, their costs being paid by advertisers who broadcast commercials during programs. How is this analysis affected by the following facts:

a) Fast-forward buttons on VCRs allow consumers watching taped shows to move quickly through certain parts of those shows. A sizeable number of consumers watching taped shows use this feature to bypass commercials. Further, companies have begun to market VCRs that automatically edit out commercials.

b) Consumers who tape particularly popular or "classic" shows, especially movies, sometimes keep the tapes in a video library and watch them repeatedly.

c) The majority of television programming today is not broadcast without charge, but is transmitted over ground cable. Cable television companies charge a monthly fee for access to their stations. Further, some "premium" channels do not use commercials, instead recouping their costs by charging an additional fee per channel (in some cases) or a fee per view (in others).

b. Photocopying

American Geophysical Union, et al. v. Texaco Inc.
United States Court of Appeals for the Second Circuit
60 F.3d 913 (2d Cir. 1994)

JON O. NEWMAN., Chief Judge:

This interlocutory appeal presents the issue of whether, under the particular circumstances of this case, the fair use defense to copyright infringement applies to the photocopying of articles in a scientific journal. This issue arises on the appeal of defendant Texaco Inc. from the July 23, 1992, order of the United States District Court for the Southern District of New York (Pierre N. Leval, Judge) holding, after a limited-issue bench trial, that the photocopying of eight articles from the Journal of Catalysis for use by one of Texaco's researchers was not fair use. See American Geophysical Union v. Texaco Inc., 802 F. Supp. 1 (S.D.N.Y. 1992). Though not for precisely the same reasons, we agree with the District Court's conclusion that this particular copying was not fair use and therefore affirm.

Background

The District Court Proceedings. Plaintiffs American Geophysical Union and 82 other publishers of scientific and technical journals (the "publishers") brought a class action claiming that Texaco's unauthorized photocopying of articles from their journals constituted copyright infringement. Among other defenses, Texaco claimed that its copying was fair use under section 107 of the Copyright Act, 17 U.S.C. §107 (1988). Since it appeared likely that the litigation could be resolved once the fair use defense was adjudicated, the parties agreed that an initial trial should be limited to whether Texaco's copying was fair use, and further agreed that this issue would be submitted for decision on a written record.

Although Texaco employs 400 to 500 research scientists, of whom all or most presumably photocopy scientific journal articles to support their Texaco research, the parties stipulated — in order to spare the enormous expense of exploring the photocopying practices of each of them — that one scientist would be chosen at random as the representative of the entire group. The scientist chosen was Dr. Donald H. Chickering, II, a scientist at Texaco's research center in Beacon, New York. For consideration at trial, the publishers selected from Chickering's files photocopies of eight particular articles from the Journal of Catalysis. . . .

Chickering, a chemical engineer at the Beacon research facility, has worked for Texaco since 1981 conducting research in the field of catalysis, which concerns changes in the rates of chemical reactions. To keep abreast of developments in his field, Chickering must review works published in various scientific and technical journals related to his area of research. Texaco assists in this endeavor by having its library circulate current issues of relevant journals to Chickering when he places his name on the appropriate routing list.

The copies of the eight articles from Catalysis found in Chickering's files that the parties have made the exclusive focus of the fair use trial were photocopied in their entirety by Chickering or by other Texaco employees at Chickering's request. Chickering apparently believed that the material and data found within these articles would facilitate his current or future professional research. The evidence developed at trial indicated that Chickering did not generally use the Catalysis articles in his research immediately upon copying, but placed the photocopied articles in his files to have them available for later reference as needed. Chickering became aware of six of the photocopied articles when the original issues of Catalysis containing the articles were circulated to him. He learned of the other two articles upon seeing a reference to them in another published article. As it turned out, Chickering did not have occasion to make use of five of the articles that were copied. . . .

I. The Nature of the Dispute . . .

A. *Fair Use and Photocopying . . .*

As with the development of other easy and accessible means of mechanical reproduction of documents, the invention and widespread availability of photocopying technology threatens to disrupt the delicate balances established by the Copyright Act. See 3 Melville B. Nimmer & David Nimmer, Nimmer on Copyright ¶13.05[E][1], at 13-225 to 13-226 (1994) [hereinafter Nimmer on Copyright] (noting that "unrestricted photocopying practices could largely undercut the entire law of copyright"); see also Sony, 464 U.S. at 467-68 n.16 (Blackmun, J., dissenting) (recognizing that the "advent of inexpensive and readily available copying machines . . . has changed the dimensions" of the legal issues concerning the practice of making personal copies of copyrighted materials). As a leading commentator astutely notes, the advent of modern photocopying technology creates a pressing need for the law "to strike an appropriate balance between the authors' interest in preserving the integrity of copyright, and the public's right to enjoy the benefits that photocopying technology offers." 3 Nimmer on Copyright ¶13.05[E][1], at 13-226.

Indeed, if the issue were open, we would seriously question whether-the fair use analysis that has developed with respect to works of authorship alleged to use portions of copyrighted material is precisely applicable to copies produced by mechanical means. The traditional fair use analysis, now codified in section 107, developed in an effort to adjust the competing interests of authors — the author of the original copyrighted work and the author of the secondary work that "copies" a portion of the original work in the course of producing what is claimed to be a new work. Mechanical "copying" of an entire document, made readily feasible and economical by the advent of xerography, see SCM Corp. v. Xerox Corp., 463 F. Supp. 983, 991-94 (D. Conn. 1978), *aff'd,* 645 F.2d 1195 (2d Cir. 1981), *cert. denied,* 455 U.S. 1016, 72 L. Ed. 2d 132, 102 S. Ct. 1708 (1982), is obviously an activity entirely different from creating a work of authorship. Whatever social utility copying of this sort achieves, it is not concerned with creative authorship. . . .

II. The Enumerated Fair Use Factors of Section 107 . . .

A. *First Factor: Purpose and Character of Use*

The first factor listed in section 107 is "the purpose and character of the use, including whether such use is of a commercial nature or is for nonprofit educational purposes." 17 U.S.C. §107(1). Especially pertinent to an assessment of the first fair use factor are the precise circumstances under which copies of the eight Catalysis articles were made. After noticing six of these articles when the original copy of the journal issue containing each of them was circulated to him, Chickering had them photocopied, at least initially, for the same basic purpose that one would normally seek to obtain the original — to have it available on his shelf for ready reference if and when he needed to look at it. The library circulated one copy and invited all the researchers to make their own photocopies. It is a reasonable inference that the library staff wanted each journal issue moved around the building quickly and returned to the library so that it would be available for others to look at. Making copies enabled all researchers who might one day be interested in examining the contents of an article in the issue to have the article readily available in their own offices. In Chickering's own words, the copies of the articles were made for "my personal convenience," since it is "far more convenient to have access in my office to a photocopy of an article than to have to go to the library each time I wanted to refer to it." Significantly, Chickering did not even have occasion to use five of the photocopied articles at all, further revealing that the photocopies of the eight Catalysis articles were primarily made just for "future retrieval and reference."

It is true that photocopying these articles also served other purposes. The most favorable for Texaco is the purpose of enabling Chickering, if the need should arise, to go into the lab with pieces of paper that (a) were not as bulky as the entire issue or a bound volume of a year's issues, and (b) presented no risk of damaging the original by exposure to chemicals. And these purposes might suffice to tilt the first fair use factor in favor of Texaco if these purposes were dominant. For example, if Chickering had asked the library to buy him a copy of the pertinent issue of Catalysis and had placed it on his shelf, and one day while reading it had noticed a chart, formula, or other material that he wanted to take right into the lab, it might be a fair use for him to make a photocopy, and use that copy in the lab (especially if he did not retain it and build up a mini-library of photocopied articles). This is the sort of "spontaneous" copying that is part of the test for permissible nonprofit classroom copying. See Agreement on Guidelines for Classroom Copying in Not-For-Profit Educational Institutions, quoted in Patry, The Fair Use Privilege, at 308. But that is not what happened here as to the six items copied from the circulated issues.

As to the other two articles, the circumstances are not quite as clear, but they too appear more to serve the purpose of being additions to Chickering's office "library" than to be spontaneous copying of a critical page that he was reading on his way to the lab. One was copied apparently when he saw a reference to it in another article, which was in an issue circulated to him. The most likely inference is that he decided that he ought to have copies of both items — again for placement on his shelf for later use if the need arose. The last article was copied, according to his affidavit, when he saw a reference to it "elsewhere." What is clear is that this item too was simply placed "on the shelf." As he testified, "I kept a copy to refer to in case I became more involved in support effects research."

The photocopying of these eight Catalysis articles may be characterized as "archival" — i.e., done for the primary purpose of providing numerous Texaco scientists (for whom Chickering served as an example) each with his or her own personal copy of each article without Texaco's having to purchase another original journal. The photocopying "merely 'supersedes the objects' of the original creation," Campbell, 114 S. Ct. at 1171 (quoting Folsom v. Marsh, 9 F. Cas. 342, 348 (C.C.D. Mass. 1841) (No. 4,901)), and tilts the first fair use factor against Texaco. We do not mean to suggest that no instance of archival copying would be fair use, but the first factor tilts against Texaco in this case because the making of copies to be placed on the shelf in Chickering's office is part of a systematic process of encouraging employee researchers to copy articles so as to multiply available copies while avoiding payment.

Texaco criticizes three aspects of the District Court's analysis of the first factor. . . . We consider these three lines of attack separately.

1. Commercial use. We generally agree with Texaco's contention that the District Court placed undue emphasis on the fact that Texaco is a for-profit corporation conducting research primarily for commercial gain. Since many, if not most, secondary users seek at least some measure of commercial gain from their use, unduly emphasizing the commercial motivation of a copier will lead to an overly restrictive view of fair use. . . .

Indeed, Campbell warns against 'elevating . . . to a per se rule' Sony's language about a presumption against fair use arising from commercial use. 114 S. Ct. at 1174. Campbell discards that language in favor of a more subtle, sophisticated approach, which recognizes that 'the more transformative the new work, the less will be the significance of other factors, like commercialism, that may weigh against a finding of fair use.' Id. at 1171. The Court states that 'the commercial or nonprofit educational purpose of a work is only one element of the first factor enquiry' id. at 1174, and points out that 'if, indeed, commerciality carried presumptive force against a finding of fairness, the presumption would swallow nearly all of the illustrative uses listed in the preamble paragraph of §107. . . . ' Id. . . .

We do not consider Texaco's status as a for-profit company irrelevant to the fair use analysis. Though Texaco properly contends that a court's focus should be on the use of the copyrighted material and not simply on the user, it is overly simplistic to suggest that the "purpose and character of the use" can be fully discerned without considering the nature and objectives of the user.

Ultimately, the somewhat cryptic suggestion in section 107(1) to consider whether the secondary use "is of a commercial nature or is for nonprofit educational purposes" connotes that a court should examine, among other factors, the value obtained by the secondary user from the use of the copyrighted material. See Rogers, 960 F.2d at 309 ("The first factor . . . asks whether the original was copied in good faith to benefit the public or primarily for the commercial interests of the infringer."); MCA, Inc. v. Wilson, 677 F.2d 180, 182 (2d Cir. 1981) (court is to consider "whether the alleged infringing use was primarily for public benefit or for private commercial gain"). The commercial/nonprofit dichotomy concerns the unfairness that arises when a secondary user makes unauthorized use of copyrighted material to capture significant revenues as a direct consequence of copying the original work. See Harper & Row, 471 U.S. at 562 ("The crux of the profit/nonprofit distinction is . . . whether the user stands to profit from exploitation of the copyrighted material without paying the customary price.").

Consistent with these principles, courts will not sustain a claimed defense of fair use when the secondary use can fairly be characterized as a form of "commercial exploitation," i.e., when the copier directly and exclusively acquires conspicuous financial rewards from its use of the copyrighted material. . . . Conversely, courts are more willing to find a secondary use fair when it produces a value that benefits the broader public interest. . . . The greater the private economic rewards reaped by the secondary user (to the exclusion of broader public benefits), the more likely the first factor will favor the copyright holder and the less likely the use will be considered fair.

As noted before, in this particular case the link between Texaco's commercial gain and its copying is somewhat attenuated: the copying, at most, merely facilitated Chickering's research that might have led to the production of commercially valuable products. Thus, it would not be accurate to conclude that Texaco's copying of eight particular Catalysis articles amounted to "commercial exploitation," especially since the immediate goal of Texaco's copying was to facilitate Chickering's research in the sciences, an objective that might well serve a broader public purpose. See Twin Peaks, 996 F.2d at 1375; Sega Enterprises, 977 F.2d at 1522. Still, we need not ignore the for-profit nature of Texaco's enterprise, especially since we can confidently conclude that Texaco reaps at least some indirect economic advantage from its photocopying. As the publishers emphasize, Texaco's photocopying for Chickering could be regarded simply as another "factor of production" utilized in Texaco's efforts to develop profitable products. Conceptualized in this way, it is not obvious why it is fair for Texaco to avoid having to pay at least some price to copyright holders for the right to photocopy the original articles.

2. Transformative Use. The District Court properly emphasized that Texaco's photocopying was not "transformative." After the District Court issued its opinion, the Supreme Court explicitly ruled that the concept of a "transformative use" is central to a proper analysis under the first factor, see Campbell, 114 S. Ct. at 1171-73. The Court explained that though a "transformative use is not absolutely necessary for a finding of fair use, . . . the more transformative the new work, the less will be the significance of other factors, like commercialism, that may weigh against a finding of fair use." Id. at 1171.

The "transformative use" concept is pertinent to a court's investigation under the first factor because it assesses the value generated by the secondary use and the means by which such value is generated. To the extent that the secondary use involves merely an untransformed duplication, the value generated by the secondary use is little or nothing more than the value that inheres in the original. Rather than making some contribution of new intellectual value and thereby fostering the advancement of the arts and sciences, an untransformed copy is likely to be used simply for the same intrinsic purpose as the original, thereby providing limited justification for a finding of fair use. See Weissmann v. Freeman, 868 F.2d 1313, 1324 (2d Cir.) (explaining that a use merely for the same "intrinsic purpose" as original "moves the balance of the calibration on the first factor against" secondary user and "seriously weakens a claimed fair use"), *cert. denied,* 493 U.S. 883, 107 L. Ed. 2d 172, 110 S. Ct. 219 (1989).[9]

In contrast, to the extent that the secondary use "adds something new, with a further purpose or different character," the value generated goes beyond the value that inheres in the original and "the goal of copyright, to promote science and the arts, is generally furthered." Campbell, 114 S. Ct. at 1171; see also Pierre N. Leval, Toward a Fair Use Standard, 103 Harv. L. Rev. 1105, 1111 (1990) [hereinafter Leval, Toward a Fair Use Standard]. It is therefore not surprising that the "preferred" uses illustrated in the preamble to section 107, such as criticism and comment, generally involve some transformative use of the original work. See 3 Nimmer on Copyright ¶13.05[A][1][b], at 13-160.

---

9. See also Marcus v. Rowley, 695 F.2d at 1175 (emphasizing that "a finding that the alleged infringers copied the material to use it for the same intrinsic purpose for which the copyright owner intended it to be used is strong indicia of no fair use."). See generally Leon E. Seltzer, Exemptions and Fair Use in Copyright 24 (1978) (noting traditional limit on applicability of fair use doctrine when reproduction of original work is done "in order to use it for its intrinsic purpose — to make what might be called the 'ordinary' use of it").

Texaco suggests that its conversion of the individual Catalysis articles through photocopying into a form more easily used in a laboratory might constitute a transformative use. However, Texaco's photocopying merely transforms the material object embodying the intangible article that is the copyrighted original work. See 17 U.S.C. §§101, 102 (explaining that copyright protection in literary works subsists in the original work of authorship "regardless of the nature of the material objects . . . in which they are embodied"). Texaco's making of copies cannot properly be regarded as a transformative use of the copyrighted material. See Steven D. Smit, "Make a Copy for the File . . .": Copyright Infringement by Attorneys, 46 Baylor L. Rev. 1, 15 & n.58 (1994); see also Basic Books, 758 F. Supp. at 1530-31 (repackaging in anthology form of excerpts from copyrighted books not a transformative use).

Even though Texaco's photocopying is not technically a transformative use of the copyrighted material, we should not overlook the significant independent value that can stem from conversion of original journal articles into a format different from their normal appearance. See generally Sony, 464 U.S. at 454, 455 n.40 (acknowledging possible benefits from copying that might otherwise seem to serve "no productive purpose"); Weinreb, Fair's Fair, at 1143 & n.29 (discussing potential value from nontransformative copying). As previously explained, Texaco's photocopying converts the individual Catalysis articles into a useful format. Before modern photocopying, Chickering probably would have converted the original article into a more serviceable form by taking notes, whether cursory or extended;[10] today he can do so with a photocopying machine. Nevertheless, whatever independent value derives from the more usable format of the photocopy does not mean that every instance of photocopying wins on the first factor. In this case, the predominant archival purpose of the copying tips the first factor against the copier, despite the benefit of a more usable format. . . .

On balance, we agree with the District Court that the first factor favors the publishers, primarily because the dominant purpose of the use is "archival." . . .

B. Second Factor: Nature of Copyrighted Work

The second statutory fair use factor is "the nature of the copyrighted work." 17 U.S.C. §107(2). . . . Ultimately the manifestly factual character of

---

10. In stating that a handwritten copy would have been made, we do not mean to imply that such copying would necessarily have been a fair use. Despite the 1973 dictum in Williams & Wilkins asserting that "it is almost unanimously accepted that a scholar can make a handwritten copy of an entire copyrighted article for his own use. . . ," 487 F.2d at 1350, the current edition of the Nimmer treatise reports that "there is no reported case on the question of whether a single handwritten copy of all or substantially all of a book or other protected work made for the copier's own private use is an infringement or fair use." 3 Nimmer on Copyright ¶1305[E][4][a], at 13-229.

the eight articles precludes us from considering the articles as "within the core of the copyright's protective purposes," Campbell, 114 S. Ct. at 1175; see also Harper & Row, 471 U.S. at 563 ("The law generally recognizes a greater need to disseminate factual works than works of fiction or fantasy."). Thus, in agreement with the District Court, we conclude that the second factor favors Texaco.

C. Third Factor: Amount and Substantiality of Portion Used

The third statutory fair use factor is "the amount and substantiality of the portion used in relation to the copyrighted work as a whole." 17 U.S.C. §107(3). The District Court concluded that this factor clearly favors the publishers because Texaco copied the eight articles from Catalysis in their entirety. [The Court of Appeals agreed].

D. Fourth Factor: Effect upon Potential Market or Value

The fourth statutory fair use factor is "the effect of the use upon the potential market for or value of the copyrighted work." 17 U.S.C. §107(4). Assessing this factor, the District Court detailed the range of procedures Texaco could use to obtain authorized copies of the articles that it photocopied and found that "whatever combination of procedures Texaco used, the publishers' revenues would grow significantly." 802 F. Supp. at 19. The Court concluded that the publishers "powerfully demonstrated entitlement to prevail as to the fourth factor," since they had shown "a substantial harm to the value of their copyrights" as the consequence of Texaco's copying. See id. at 18-21. . . .

In analyzing the fourth factor, it is important (1) to bear in mind the precise copyrighted works, namely the eight journal articles, and (2) to recognize the distinctive nature and history of "the potential market for or value of" these particular works. Specifically, though there is a traditional market for, and hence a clearly defined value of, journal issues and volumes, in the form of per-issue purchases and journal subscriptions, there is neither a traditional market for, nor a clearly defined value of, individual journal articles. As a result, analysis of the fourth factor cannot proceed as simply as would have been the case if Texaco had copied a work that carries a stated or negotiated selling price in the market.

Like most authors, writers of journal articles do not directly seek to capture the potential financial rewards that stem from their copyrights by personally marketing copies of their writings. Rather, like other creators of literary works, the author of a journal article "commonly sells his rights to publishers who offer royalties in exchange for their services in producing and marketing the author's work." Harper & Row, 471 U.S. at 547. In the distinctive realm of academic and scientific articles, however, the only form of royalty paid by a publisher is often just the reward of being published, publication being a key to professional advancement and prestige for the author, see Weissmann, 868 F.2d at 1324 (noting that "in an academic setting, profit is ill-measured in dollars. Instead, what is valuable is recognition because it so often influences professional advancement and academic tenure."). The publishers in turn incur the costs and labor of producing and marketing authors' articles, driven by the prospect of capturing the economic value stemming from the copyrights in the original works, which the authors have transferred to them. Ultimately, the monopoly privileges conferred by copyright protection and the potential financial rewards therefrom are not directly serving to motivate authors to write individual articles; rather, they serve to motivate publishers to produce journals, which provide the conventional and often exclusive means for disseminating these individual articles. It is the prospect of such dissemination that contributes to the motivation of these authors. . . .

1. Sales of Additional Journal Subscriptions, Back Issues, and Back Volumes. Since we are concerned with the claim of fair use in copying the eight individual articles from Catalysis, the analysis under the fourth factor must focus on the effect of Texaco's photocopying upon the potential market for or value of these individual articles. Yet, in their respective discussions of the fourth statutory factor, the parties initially focus on the impact of Texaco's photocopying of individual journal articles upon the market for Catalysis journals through sales of Catalysis subscriptions, back issues, or back volumes.

As a general matter, examining the effect on the marketability of the composite work containing a particular individual copyrighted work serves as a useful means to gauge the impact of a secondary use "upon the potential market for or value of" that individual work, since the effect on the marketability of the composite work will frequently be directly relevant to the effect on the market for or value of that individual work. Quite significantly, though, in the unique world of academic and scientific articles, the effect on the marketability of the composite work in which individual articles appear is not obviously related to the effect on the market for or value of the individual articles. Since (1) articles are submitted unsolicited to journals, (2) publishers do not make any payment to authors for the right to publish their articles or to acquire their copyrights, and (3) there is no evidence in the record suggesting that publishers seek to reprint particular articles in new composite works, we cannot readily conclude that evidence concerning the effect of Texaco's use on the marketability of journals provides an effective means to appraise the effect of Texaco's use on the market for or value of individual journal articles.

These considerations persuade us that evidence concerning the effect of Texaco's photocopying of individual articles within Catalysis on the traditional market for Catalysis subscriptions is of somewhat limited significance in determining and evaluating the effect of Texaco's photocopying "upon the potential market for or value of" the individual articles. We do not mean to suggest that we believe the effect on the marketability of journal subscriptions is completely irrelevant to gauging the effect on the market for and value of individual articles. Were the publishers able to demonstrate that Texaco's type of photocopying, if widespread, would impair the marketability of journals, then they might have a strong claim under the fourth factor. Likewise, were Texaco able to demonstrate that its type of photocopying, even if widespread, would have virtually no effect on the marketability of journals, then it might have a strong claim under this fourth factor.

On this record, however, the evidence is not resounding for either side. The District Court specifically found that, in the absence of photocopying, (1) "Texaco would not ordinarily fill the need now being supplied by photocopies through the purchase of back issues or back volumes . . . [or] by enormously enlarging the number of its subscriptions," but (2) Texaco still "would increase the number of subscriptions somewhat." 802 F. Supp. at 19. This moderate conclusion concerning the actual effect on the marketability of journals, combined with the uncertain relationship between the market for journals and the market for and value of individual articles, leads us to conclude that the evidence concerning sales of additional journal subscriptions, back issues, and back volumes does not strongly support either side with regard to the fourth factor. Cf. Sony, 464 U.S. at 451-55 (rejecting various predictions of harm to value of copyrighted work based on speculation about possible consequences of secondary use). At best, the loss of a few journal subscriptions tips the fourth factor only slightly toward the publishers because evidence of such loss is weak evidence that the copied articles themselves have lost any value.

2. Licensing Revenues and Fees. The District Court, however, went beyond discussing the sales of additional journal subscriptions in holding that Texaco's photocopying affected the value of the publishers' copyrights. Specifically, the Court pointed out that, if Texaco's unauthorized photocopying was not permitted as fair use, the publishers' revenues would increase significantly since Texaco would (1) obtain articles from document delivery services (which pay royalties to publishers for the right to photocopy articles), (2) negotiate photocopying licenses directly with individual publishers, and/or (3) acquire some form of photocopying license from the Copyright Clearance Center Inc. ("CCC"). See 802 F. Supp. at 19. Texaco claims that the District Court's reasoning is faulty because, in determining that the value of the publishers' copyrights was affected, the Court assumed that the publishers were entitled to demand and receive licensing royalties and fees for photocopying. Yet, continues Texaco, whether the publishers can demand a fee for permission to make photocopies is the very question that the fair use trial is supposed to answer.

It is indisputable that, as a general matter, a copyright holder is entitled to demand a royalty for licensing others to use its copyrighted work, see 17 U.S.C. §106 (copyright owner has exclusive right "to authorize" certain uses), and that the impact on potential licensing revenues is a proper subject for consideration in assessing the fourth factor, see, e.g., Campbell, 114 S. Ct. at 1178; Harper & Row, 471 U.S. at 568-69; Twin Peaks, 996 F.2d at 1377; DC Comics Inc. v. Reel Fantasy, Inc., 696 F.2d 24, 28 (2d Cir. 1982); United Telephone Co. of Missouri v. Johnson Publishing Co., Inc., 855 F.2d 604, 610 (8th Cir. 1988).

However, not every effect on potential licensing revenues enters the analysis under the fourth factor.[17] Specifically, courts have recognized limits on the concept of "potential licensing revenues" by considering only traditional, reasonable, or likely to be developed markets when examining and assessing a secondary use's "effect upon the potential market for or value of the copyrighted work." See Campbell, 114 S. Ct. at 1178 ("The market for potential derivative uses includes only those that creators of original works would in general develop or license others to develop."); Harper & Row, 471 U.S. at 568 (fourth factor concerned with "use that supplants any part of the normal market for a copyrighted work") (emphasis added) (quoting S. Rep. No. 473, 94th Cong., 1st Sess. 65 (1975)); see also Mathieson v. Associated Press, 1992 U.S. Dist. LEXIS 9269, 23 U.S.P.Q.2d 1685, 1690-91 (S.D.N.Y. 1992) (refusing to find fourth factor in favor of copyright holder because secondary use did not affect any aspect of the normal market for copyrighted work).

. . . Texaco is correct, at least as a general matter, when it contends that it is not always appropriate for a court to be swayed on the fourth factor by the effects on potential licensing revenues. Only an impact on potential licensing revenues for traditional, reasonable, or likely to be developed markets should be legally cognizable when evaluating a secondary use's "effect upon the potential market for or value of the copyrighted work."

Though the publishers still have not established a conventional market for the direct sale and distribution of individual articles, they have created, primarily through the CCC, a workable market for institutional users to obtain licenses for the right to produce their own copies of individual articles via photocopying. The District Court found that many major corporations now subscribe to the CCC systems for photocopying licenses. 802 F. Supp. at 25. Indeed, it appears from the pleadings, especially Texaco's counterclaim, that Texaco itself has been paying royalties to the CCC. Since the Copyright Act explicitly provides that copyright holders have the "exclusive rights" to

---

17. As Texaco notes and others have recognized, a copyright holder can always assert some degree of adverse [e]ffect on its potential licensing revenues as a consequence of the secondary use at issue simply because the copyright holder has not been paid a fee to permit that particular use. See Leval, Toward a Fair Use Standard, at 1124 ("By definition every fair use involves some loss of royalty revenue because the secondary user has not paid royalties."); Fisher, Reconstructing Fair Use, at 1671 (noting that in almost every case "there will be some material adverse impact on a 'potential market' " since the secondary user has not paid for the use). Thus, were a court automatically to conclude in every case that potential licensing revenues were impermissibly impaired simply because the secondary user did not pay a fee for the right to engage in the use, the fourth fair use factor would always favor the copyright holder. See Leval, Toward a Fair Use Standard, at 1125; Fisher, Reconstructing Fair Use, at 1672.

"reproduce" and "distribute copies" of their works, see 17 U.S.C. §106(1) & (3), and since there currently exists a viable market for licensing these rights for individual journal articles, it is appropriate that potential licensing revenues for photocopying be considered in a fair use analysis.

Despite Texaco's claims to the contrary, it is not unsound to conclude that the right to seek payment for a particular use tends to become legally cognizable under the fourth fair use factor when the means for paying for such a use is made easier. This notion is not inherently troubling: it is sensible that a particular unauthorized use should be considered "more fair" when there is no ready market or means to pay for the use, while such an unauthorized use should be considered "less fair" when there is a ready market or means to pay for the use. The vice of circular reasoning arises only if the availability of payment is conclusive against fair use. Whatever the situation may have been previously, before the development of a market for institutional users to obtain licenses to photocopy articles, see Williams & Wilkins, 487 F.2d at 1357-59, it is now appropriate to consider the loss of licensing revenues in evaluating "the effect of the use upon the potential market for or value of" journal articles. It is especially appropriate to do so with respect to copying of articles from Catalysis, a publication as to which a photocopying license is now available. We do not decide how the fair use balance would be resolved if a photocopying license for Catalysis articles were not currently available. . . .

Primarily because of lost licensing revenue, and to a minor extent because of lost subscription revenue, we agree with the District Court that "the publishers have demonstrated a substantial harm to the value of their copyrights through [Texaco's] copying," 802 F. Supp. at 21, and thus conclude that the fourth statutory factor favors the publishers. . . .

The order of the District Court is affirmed.[19]

JACOBS, Circuit Judge, dissenting:

The stipulated facts crisply present the fair use issues that govern the photocopying of entire journal articles for a scientist's own use, either in the laboratory or as part of a personal file assisting that scientist's particular inquiries. I agree with much in the majority's admirable review of the facts and the law. Specifically, I agree that, of the four nonexclusive considerations bearing on fair use enumerated in section 107, the second factor (the nature of the copyrighted work) tends to support a conclusion of fair use, and the third factor (the ratio of the copied portion to the whole copyrighted work) militates against it. I respectfully dissent, however, in respect of the first and fourth factors. As to the first factor: the purpose and character of Dr. Chick-

___

19. Though neither the limited trial nor this appeal requires consideration of the publishers' remedy if infringement is ultimately found, we note that the context of this dispute appears to make ill-advised an injunction, which, in any event, has not been sought. If the dispute is not now settled, this appears to be an appropriate case for exploration of the possibility of a court-imposed compulsory license. See Campbell, 114 S. Ct. at 1171 n.10; 3 Nimmer on Copyright ¶13.05[E][4][e], at 13-241 to 13-242.

ering's use is integral to transformative and productive ends of scientific research. As to the fourth factor: the adverse effect of Dr. Chickering's use upon the potential market for the work, or upon its value, is illusory. For these reasons, and in light of certain equitable considerations and the overarching purpose of the copyright laws, I conclude that Dr. Chickering's photocopying of the Catalysis articles was fair use.

A. Purpose and Character of the Use . . .

A use that is reasonable and customary is likely to be a fair one. See Harper & Row Publishers, Inc. v. Nation Enterprises, 471 U.S. 539, 550, 85 L. Ed. 2d 588, 105 S. Ct. 2218 (1985) ("the fair use doctrine was predicated on the author's implied consent to 'reasonable and customary' use"). The district court, the majority and I start from the same place in assessing whether Dr. Chickering's photocopying is a reasonable and customary use of the material: making single photocopies for research and scholarly purposes has been considered both reasonable and customary for as long as photocopying technology has been in existence. See Williams & Wilkins Co. v. United States, 487 F.2d 1345, 1355-56, 203 Ct. Cl. 74 (Ct. Cl. 1973), *aff'd by an equally divided court,* 420 U.S. 376, 43 L. Ed. 2d 264, 95 S. Ct. 1344 (1976). The majority quotes the district court's short answer to this important insight: "To the extent the copying practice was 'reasonable' in 1973 [when *Williams v. Wilkins* was decided], it has ceased to be 'reasonable' as the reasons that justified it before [photocopying licensing] have ceased to exist." 802 F. Supp. at 25. I do not agree at all that a reasonable and customary use becomes unfair when the copyright holder develops a way to exact an additional price for the same product. Moreover, I view the advent of the CCC as an event that bears analytically upon the distinct question of whether Dr. Chickering's use supersedes the original (the fourth factor). I therefore reach an issue — reasonable and customary use — not explored by the district court or by the majority.

Consider what Dr. Chickering actually does with scientific journals. As a research scientist, he routinely sifts through the latest research done by his peers, much of which is printed in journals such as Catalysis. He determines which articles potentially assist his specific trains of thought and lines of inquiry, and he photocopies them. Relative to the volume of articles in each issue, his photocopying is insubstantial. He then files the articles for possible future use or study. As the majority observes, "before modern photocopying, Chickering probably would have converted the original article into a more serviceable form by taking notes, whether cursory or extended; today he can do so with a photocopying machine." The majority's footnote 10, appended to this passage, questions whether or not a scholar's handwritten copy of a full work is "necessarily" a fair use. As the majority adds, however, *Williams & Wilkins* says: "It is almost unanimously accepted that a scholar can make a handwritten copy of an entire copyrighted article for his own use, and in the era before photoduplication it was not uncommon (and not seriously questioned) that he could have his secretary make a typed copy for his personal use and files. These customary facts of copyright-life are among our givens." Williams & Wilkins, 487 F.2d at 1350. What Dr. Chickering does is simply a technologically assisted form of note-taking, such as has long been customary among researchers: the photocopy machine saves Dr. Chickering the toil and time of recording notes on index cards or in notebooks, and improves the accuracy and range of the data, charts, and formulas he can extract from the passing stream of information; but the note-taking purpose remains the same. . . .

The majority emphasizes that, as it happened, Dr. Chickering did not "use" the photocopied articles because, in five out of eight instances, he filed them away. There is nothing odd about making notes one does not immediately use, or that one may never consult again. Photocopies, which to Dr. Chickering are the functional counterpart of notes, are used (or not, as the case may be) in the same way. Dr. Chickering's filing away of these photocopies does not subvert his claim of fair use. Like the majority, I am convinced that his deposit of the photocopied articles in his personal file, pending his personal use of them in the future, is an important fact bearing upon fair use; but the dominant significance of that fact, under the first factor of section 107, is that (whether he "uses" them or files them) the articles are not resold or retailed in any way. If the copies were sold by Dr. Chickering, that would be a telling — possibly determinative — fact. What Dr. Chickering has done reinforces the view that his photocopying was not commercial in purpose or character. . . .

The majority emphasizes passim that the photocopying condemned here is "systematic" and "institutional." These terms furnish a ground for distinguishing this case from the case that the majority expressly does not reach: the copying of journal articles by an individual researcher outside an institutional framework. For all the reasons adduced above, I conclude that the institutional environment in which Dr. Chickering works does not alter the character of the copying done by him or at his instance, and that the selection by an individual scientist of the articles useful to that scientist's own inquiries is not systematic copying, and does not become systematic because some number of other scientists in the same institution — four hundred or four — are doing the same thing.

First, the majority's reliance on Texaco's institutional framework does not limit the potentially uncontrolled ramifications of the result. Research is largely an institutional endeavor nowadays, conducted by employees pursuing the overall goals of corporations, university laboratories, courts and law firms, governments and their agencies, think-tanks, publishers of newspapers and magazines, and other kinds of institutions. The majority's limitation of its holding to institutional environments may give comfort to inventors in bicycle shops, scientists in garage laboratories, freelance book reviewers, and solo conspiracy theorists, but it is not otherwise meaningful.

The majority's reliance on the systematic character of the photocopying here also seems to me erroneous. The majority deems Texaco's photocopying systematic because Texaco uses circulation lists to route a copy of each journal issue to the scientists interested in the field. The majority, however, ignores the one determinative issue: whether the decision to photocopy individual articles is made by the individual researcher, as Dr. Chickering did here. Journal issues may be systematically circulated to all scientists in a given group, rather than (say) at random, but the circulation of journal issues is not photocopying, systematic or otherwise. The journal issues circulated by Texaco are procured by subscription. Once Texaco receives the subscription copies from the publisher, Texaco is free to circulate them in-house so that they can be seen by as many scientists as can lay eyes on them. This circulation of copies allows individual scientists to select individual articles for copying. The majority opinion, which leaves open the idea that this practice may comport with copyright law if done by an individual scientist, does not explain why it is impermissible when done by more than one. . . .

B. Effect upon Potential Market or Value

In gauging the effect of Dr. Chickering's photocopying on the potential market or value of the copyrighted work, the majority properly considers two separate means of marketing: (1) journal subscriptions and sales, and (2) licensing revenues and fees.

(1) Subscriptions and Sales. . . . As to the individual articles photocopied by Dr. Chickering, I agree with the majority — as I read the opinion — that one cannot put a finger on any loss suffered by the publisher in the value of the individual articles or in the traditional market for subscriptions and back issues. The district court found that Texaco would not purchase back-issues or back volumes in the numbers needed to supply individual copies of articles to individual scientists.

Finally, the circulation of Catalysis among a number of Texaco scientists can come as no surprise to the publisher of Catalysis, which charges double the normal subscription rate to institutional subscribers. The publisher must therefore assume that, unless they are reading Catalysis for pleasure or committing it to memory, the scientists will extract what they need and arrange to copy it for personal use before passing along the institutional copies.

(2) Licensing Revenues and Fees. . . . In this case the only harm to a market is to the supposed market in photocopy licenses. The CCC scheme is neither traditional nor reasonable; and its development into a real market is subject to substantial impediments. There is a circularity to the problem: the market will not crystallize unless courts reject the fair use argument that Texaco presents; but, under the statutory test, we cannot declare a use to be an infringement unless (assuming other factors also weigh in favor of the secondary user) there is a market to be harmed. At present, only a fraction of journal publishers have sought to exact these fees. I would hold that this fourth factor decisively weighs in favor of Texaco, because there is no normal market in photocopy licenses, and no real consensus among publishers that there ought to be one.

The majority holds that photocopying journal articles without a license is an infringement. Yet it is stipulated that (a) institutions such as Texaco subscribe to numerous journals, only 30 percent of which are covered by a CCC license; (b) not all publications of each CCC member are covered by the CCC licenses; and (c) not all the articles in publications covered by the CCC are copyrighted. It follows that no CCC license can assure a scientist that photocopying any given article is legal. . . .

Under a transactional license, the user must undertake copyright research every time an article is photocopied. First, one must consult a directory to determine whether or not the publisher of the journal is a member of the CCC. If it is, one must ascertain whether the particular publication is one that is covered by the CCC arrangement, because not all publications of participating publishers are covered. Then one must somehow determine whether the actual article is one in which the publisher actually holds a copyright, since there are many articles that, for such reasons as government sponsorship of the research, are not subject to copyright. The production director of plaintiff Springer-Verlag testified at trial that it is almost impossible to tell which articles might be covered by a copyright. Since even an expert has difficulty making such a determination, the transactional scheme would seem to require that an intellectual property lawyer be posted at each copy machine. Finally, once it is determined that the specific article is covered, the copyist will need to record in a log the date, name of publication, publisher, title and author of article, and number of pages copied.

It may be easier to hand copy the material. The transactions costs alone would compel users to purchase a blanket license. However, if (as the majority holds) three of the fair use factors tip in favor of the publishers even without considering the market for license fees, a blanket license offers Texaco no safe harbor. Individual publishers remain free to stand upon the rights conferred in this Court's opinion, and negotiate separate licenses with separate terms, or sell offprints and refuse any license at all. Unless each publisher's licensing rights are made to depend upon whether or not that publisher participates in the CCC, we have the beginnings of a total market failure: with many thousands of scientific publications in circulation, a user cannot negotiate licensing fees individually with numerous publishers — unless it does nothing else. For many publications, licenses are simply not available. As to those, Dr. Chickering has the choice of hand copying, typescript, or the photocopying of selected pages only. . . .

It is hard to escape the conclusion that the existence of the CCC — or the perception that the CCC and other schemes for collecting license fees are or may become "administratively tolerable" — is the chief support for the idea that photocopying scholarly articles is unfair in the first place. The majority finds it "sensible" that a use "should be considered 'less fair' when there is a ready market or means to pay for the use." That view is sensible only to a point. There is no technological or commercial impediment to imposing a fee for use of a work in a parody, or for the quotation of a paragraph in a review or biography. Many publishers could probably unite to fund a bureaucracy that would collect such fees. The majority is sensitive to this problem, but concludes that "the vice of circular reasoning arises only if the availability of payment is conclusive against fair use." That vice is not avoided here. The majority expressly declines to "decide how the fair use balance would be resolved if a photocopying license for Catalysis articles were not currently available." Moreover, the "important" fourth factor tips in favor of the publishers (according to the majority) "primarily because of lost licensing revenue" and only "to a minor extent" on the basis of journal sales and subscriptions. . . .

COMMENTS AND QUESTIONS

1. Is the majority's reasoning in *Texaco* circular? Isn't the issue to be resolved whether or not Texaco has to pay a license fee in order to photocopy journal articles? The answer to that question depends on whether Texaco's copies constitute a "fair use." But the court hinges the fair use inquiry on whether or not the copyright owners have lost licensing revenue from Texaco. Does this make sense?

2. Most people would agree that a single photocopy of a single academic article for research purposes — whether by the NIH or by a commercial group — should constitute fair use. But at what point do such individual fair uses become a collective infringement of copyright? Louise Weinberg stated the problem as follows:

> Copyright proprietors claim that even if each individual act of library photocopying constitutes a 'fair use,' the problem is so great in the aggregate as to effect a shift from fair use to infringement. But why should a use that is fair in an individual case be held unfair simply because so many individuals make the use? This was a stumbling block for the Court of Claims. . . . The language [of section 107] cannot mean that photocopying for the enumerated purposes can never be infringement, in view of obvious contingencies which would require a contrary interpretation: the wholesale reprography of textbooks for distribution to schoolchildren, for example. Thus, the proposed enactment does little more than set the stage for judicial analysis. The fundamental difficulty remains: What may be fair use in the individual case may seem less so when advanced technology can multiply the transaction endlessly.

Louise Weinberg, The Photocopying Revolution and the Copyright Crisis, Pub. Interest 99, 108 (1975).

3. How does *Texaco* apply in the academic context? Copyright owners have brought a number of cases against "copy shops" that reproduce reading material for classes. In Basic Books v. Kinko's Graphics Corp., 758 F. Supp. 1522 (S.D.N.Y. 1991), the court rejected Kinko's fair use defense, largely on the grounds that the copy shop is a for-profit operation. A case decided before

an en banc panel of the Sixth Circuit, Princeton University Press v. Michigan Document Servs., 99 F.3d 1381 (6th Cir. 1996) (en banc), provides a striking reaffirmation of this holding. The court held, as a matter of law, that the for-profit copying of academic readers (or "coursepacks") could not be a fair use. Five of the 13 judges dissented.

In *Basic Books*, Kinko's argued that it is the professors, not the copy shop, that have perpetrated and authorized the copyright infringement. The court dismissed the argument fairly summarily, but it is worthy of further consideration. Were the staff at Kinko's or MDS in a position to know whether professors who submitted packets had obtained the proper copyright permissions?[29] Should all copy shops be required to verify the copyright status of any works copied under their auspices? Should such checking extend even to "self-serve" copiers? Isn't the truly responsible entity here the professor who took 110 pages from a work and put it into a reader? Could Kinko's absolve itself from liability by having the professors sign a statement that they take the responsibility for complying with copyright law?

As a practical matter, copyright owners in academic books are unlikely to sue college professors for copyright infringement. First, most of their authors are college professors. Second, college professors are responsible (directly or indirectly) for most of the purchases of academic books. Thus they have a powerful influence on the market. Finally, it is much more difficult to sue many individual college professors than a few large national copy shops.

It is noteworthy that public libraries are not subject to the requirements the *Kinko's* decision seems to impose on commercial shops. 17 U.S.C. §108 expressly exempts libraries from liability for copies made on their premises as long as they post a warning notice specified in 37 CFR §201.14. Is there a justification for this difference in treatment? Does it stem from the presumed public, noncommercial nature of libraries?

4. Recall that a major policy underlying the copyright law is to encourage the development and dissemination of new works. Is that policy disserved by decisions such as *Texaco* and *Kinko's*? Judge Leval noted that the effect of requiring permissions would be that Texaco would have to buy more copies of journals it wished to circulate (or at least would have to pay more money to the CCC for permission to make photocopies). The corollary of this is that Texaco will buy fewer different journals, since it must buy more copies of each journal. Does this promote "the progress of science and the useful arts"? Is it a result the authors of journal articles would applaud? At the least, isn't the loss of information transfer that results from these decisions a factor to be considered in determining the issue of fair use?

5. In a thoughtful analysis of the fair use doctrine, Lloyd Weinreb has suggested that the "fairness" in fair use is an important factor, because it "gives effect to the community's established practices and understandings and allows the location of copyright within the framework of property gen-

---

29. The answer to this question is apparently yes. There is language in the opinion which suggests that Kinko's undertook to obtain the necessary copyright permissions for the professors.

erally." Lloyd Weinreb, Fair's Fair: A Comment on the Fair Use Doctrine, 103 Harv. L. Rev. 1137, 1161 (1990). If we apply that standard, what does it suggest about decisions such as *Kinko's?* Is there something troubling about a ruling that virtually every academic institution in the United States is regularly committing copyright infringement? Does this suggest that society is out of step with the law, or vice versa?

Robert Ellickson, Order Without Law 258-264 (1991) documents how informal norms among professors "trump" the formal law of copyright as regards photocopying articles for course packets. Ellickson ties this practice into his general thesis that formal law is not nearly as important as social norms in structuring behavior. See also Robert Merges, Among the Tribes of Shasta County, 18 L. & Soc. Inquiry 299 (1993) (reviewing Ellickson's book). Should such private law arrangements be respected (i.e., left alone)? Is there any reason to favor the CCC's social norm over the professorial norm?

c. Parodies

One circumstance in which the fair use defense has repeatedly arisen is in the analysis of parodies. A parody makes fun of an original work by at once imitating and distorting the work. For the parody to be effective, the audience must recognize the connection between the parody and the original work. This necessarily involves some deliberate copying of the original. Not surprisingly, authors unhappy at being parodied have turned to copyright law to defend themselves.

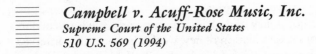

Campbell v. Acuff-Rose Music, Inc.
Supreme Court of the United States
510 U.S. 569 (1994)

Justice SOUTER delivered the opinion of the Court.

We are called upon to decide whether 2 Live Crew's commercial parody of Roy Orbison's song, "Oh, Pretty Woman," may be a fair use within the meaning of the Copyright Act of 1976, 17 U.S.C. §107 (1988 ed. and Supp. IV). Although the District Court granted summary judgment for 2 Live Crew, the Court of Appeals reversed, holding the defense of fair use barred by the song's commercial character and excessive borrowing. Because we hold that a parody's commercial character is only one element to be weighed in a fair use enquiry, and that insufficient consideration was given to the nature of parody in weighing the degree of copying, we reverse and remand.

I

In 1964, Roy Orbison and William Dees wrote a rock ballad called "Oh, Pretty Woman" and assigned their rights in it to respondent Acuff-Rose Music, Inc. See Appendix A. Acuff-Rose registered the song for copyright protection.

Petitioners Luther R. Campbell, Christopher Wongwon, Mark Ross, and David Hobbs, are collectively known as 2 Live Crew, a popular rap music group. In 1989, Campbell wrote a song entitled "Pretty Woman," which he later described in an affidavit as intended, "through comical lyrics, to satirize the original work. . . ." On July 5, 1989, 2 Live Crew's manager informed Acuff-Rose that 2 Live Crew had written a parody of "Oh, Pretty Woman," that they would afford all credit for ownership and authorship of the original song to Acuff-Rose, Dees, and Orbison, and that they were willing to pay a fee for the use they wished to make of it. Enclosed with the letter were a copy of the lyrics and a recording of 2 Live Crew's song. Acuff-Rose's agent refused permission, stating that "I am aware of the success enjoyed by 'The 2 Live Crews,' but I must inform you that we cannot permit the use of a parody of 'Oh, Pretty Woman.' " Nonetheless, in June or July 1989, 2 Live Crew released records, cassette tapes, and compact discs of "Pretty Woman" in a collection of songs entitled "As Clean As They Wanna Be." The albums and compact discs identify the authors of "Pretty Woman" as Orbison and Dees and its publisher as Acuff-Rose.

Almost a year later, after nearly a quarter of a million copies of the recording had been sold, Acuff-Rose sued 2 Live Crew and its record company, Luke Skyywalker Records, for copyright infringement. The District Court granted summary judgment for 2 Live Crew, reasoning that the commercial purpose of 2 Live Crew's song was no bar to fair use; that 2 Live Crew's version was a parody, which "quickly degenerates into a play on words, substituting predictable lyrics with shocking ones" to show "how bland and banal the Orbison song" is; that 2 Live Crew had taken no more than was necessary to "conjure up" the original in order to parody it; and that it was "extremely unlikely that 2 Live Crew's song could adversely affect the market for the original." 754 F. Supp. 1150, 1154-1155, 1157-1158 (MD Tenn. 1991). The District Court weighed these factors and held that 2 Live Crew's song made fair use of Orbison's original. Id., at 1158-1159.

The Court of Appeals for the Sixth Circuit reversed and remanded. 972 F.2d 1429, 1439 (1992). Although it assumed for the purpose of its opinion that 2 Live Crew's song was a parody of the Orbison original, . . . the court concluded that its "blatantly commercial purpose . . . prevents this parody from being a fair use." Id., at 1439.

We granted certiorari, 507 U.S. 1003 (1993), to determine whether 2 Live Crew's commercial parody could be a fair use.

II

It is uncontested here that 2 Live Crew's song would be an infringement of Acuff-Rose's rights in "Oh, Pretty Woman," under the Copyright Act of 1976, 17 U.S.C. §106 (1988 ed. and Supp. IV), but for a finding of fair use through parody. From the infancy of copyright protection, some opportunity for fair use of copyrighted materials has been thought necessary to fulfill copyright's very purpose, "to promote the Progress of Science and useful Arts. . . ." U.S. Const., Art. I, §8, cl. 8. For as Justice Story explained, "in truth, in literature, in science and in art, there are, and can be, few, if any, things, which in an abstract sense, are strictly new and original throughout. Every book in literature, science and art, borrows, and must necessarily borrow, and use much which was well known and used before." Emerson v. Davies, 8 F. Cas. 615, 619 (No. 4,436) (CCD Mass. 1845). Similarly, Lord Ellenborough expressed the inherent tension in the need simultaneously to protect copyrighted material and to allow others to build upon it when he wrote, "while I shall think myself bound to secure every man in the enjoyment of his copy-right, one must not put manacles upon science." Carey v. Kearsley, 4 Esp. 168, 170, 170 Eng. Rep. 679, 681 (K.B. 1803). In copyright cases brought under the Statute of Anne of 1710, English courts held that in some instances "fair abridgements" would not infringe an author's rights, see W. Patry, The Fair Use Privilege in Copyright Law 6-17 (1985) (hereinafter Patry); Leval, Toward a Fair Use Standard, 103 Harv. L. Rev. 1105, 1105 (1990) (hereinafter Leval), and although the First Congress enacted our initial copyright statute, Act of May 31, 1790, 1 Stat. 124, without any explicit reference to "fair use," as it later came to be known, the doctrine was recognized by the American courts nonetheless. . . . [10]

A

The first factor in a fair use enquiry is "the purpose and character of the use, including whether such use is of a commercial nature or is for nonprofit educational purposes." §107(1). This factor draws on Justice Story's formulation, "the nature and objects of the selections made." Folsom v. Marsh, 9 F. Cas., at 348. The enquiry here may be guided by the examples given in the preamble to §107, looking to whether the use is for criticism, or comment, or news reporting, and the like, see §107. The central purpose of this investigation is to see, in Justice Story's words, whether the new work merely "supersedes the objects" of the original creation, Folsom v. Marsh, supra, at

---

10. Because the fair use inquiry often requires close questions of judgment as to the extent of permissible borrowing in cases involving parodies (or other critical works), courts may also wish to bear in mind that the goals of the copyright law, "to stimulate the creation and publication of edifying matter," Leval 1134, are not always best served by automatically granting injunctive relief when parodists are found to have gone beyond the bounds of fair use. See §17 U.S.C. 502(a) (court "may . . . grant . . . injunctions on such terms as it may deem reasonable to prevent or restrain infringement") (emphasis added); Leval 1132 (while in the "vast majority of cases, [an injunctive] remedy is justified because most infringements are simple piracy," such cases are "worlds apart from many of those raising reasonable contentions of fair use" where "there may be a strong public interest in the publication of the secondary work [and] the copyright owner's interest may be adequately protected by an award of damages for whatever infringement is found"); Abend v. MCA, Inc., 863 F.2d 1465, 1479 (CA9 1988) (finding "special circumstances" that would cause "great injustice" to defendants and "public injury" were injunction to issue).

348; accord, *Harper & Row,* supra, at 562 ("supplanting" the original), or instead adds something new, with a further purpose or different character, altering the first with new expression, meaning, or message; it asks, in other words, whether and to what extent the new work is "transformative." Leval 1111. Although such transformative use is not absolutely necessary for a finding of fair use, *Sony,* supra, at 455, n.40,[11] the goal of copyright, to promote science and the arts, is generally furthered by the creation of transformative works. Such works thus lie at the heart of the fair use doctrine's guarantee of breathing space within the confines of copyright, see, e.g., *Sony,* supra, at 478-480 (Blackmun, J., dissenting), and the more transformative the new work, the less will be the significance of other factors, like commercialism, that may weigh against a finding of fair use.

This Court has only once before even considered whether parody may be fair use, and that time issued no opinion because of the Court's equal division. Benny v. Loew's Inc., 239 F.2d 532 (CA9 1956), *aff'd sub nom.* Columbia Broadcasting System, Inc. v. Loew's Inc., 356 U.S. 43 (1958). Suffice it to say now that parody has an obvious claim to transformative value, as Acuff-Rose itself does not deny. Like less ostensibly humorous forms of criticism, it can provide social benefit, by shedding light on an earlier work, and, in the process, creating a new one. We thus line up with the courts that have held that parody, like other comment or criticism, may claim fair use under §107.

The germ of parody lies in the definition of the Greek parodeia, quoted in Judge Nelson's Court of Appeals dissent, as "a song sung alongside another." 972 F.2d, at 1440, quoting 7 Encyclopedia Britannica 768 (15th ed. 1975). Modern dictionaries accordingly describe a parody as a "literary or artistic work that imitates the characteristic style of an author or a work for comic effect or ridicule," or as a "composition in prose or verse in which the characteristic turns of thought and phrase in an author or class of authors are imitated in such a way as to make them appear ridiculous." For the purposes of copyright law, the nub of the definitions, and the heart of any parodist's claim to quote from existing material, is the use of some elements of a prior author's composition to create a new one that, at least in part, comments on that author's works. See, e.g., Fisher v. Dees, supra, at 437; MCA, Inc. v. Wilson, 677 F.2d 180, 185 (CA2 1981). If, on the contrary, the commentary has no critical bearing on the substance or style of the original composition, which the alleged infringer merely uses to get attention or to avoid the drudgery in working up something fresh, the claim to fairness in borrowing from another's work diminishes accordingly (if it does not vanish), and other factors, like the extent of its commerciality, loom larger.[14] Parody needs to mimic

---

11. The obvious statutory exception to this focus on transformative uses is the straight reproduction of multiple copies for classroom distribution.

14. A parody that more loosely targets an original than the parody presented here may still be sufficiently aimed at an original work to come within our analysis of parody. If a parody whose wide dissemination in the market runs the risk of serving as a substitute for the original or licensed derivatives (see infra, discussing factor four), it is more incumbent on one claiming fair use to

an original to make its point, and so has some claim to use the creation of its victim's (or collective victims') imagination, whereas satire can stand on its own two feet and so requires justification for the very act of borrowing. See Ibid.; Bisceglia, Parody and Copyright Protection: Turning the Balancing Act Into a Juggling Act, in ASCAP, Copyright Law Symposium, No. 34, p. 25 (1987).

The fact that parody can claim legitimacy for some appropriation does not, of course, tell either parodist or judge much about where to draw the line. Like a book review quoting the copyrighted material criticized, parody may or may not be fair use, and petitioner's suggestion that any parodic use is presumptively fair has no more justification in law or fact than the equally hopeful claim that any use for news reporting should be presumed fair, see Harper & Row, 471 U.S., at 561. The Act has no hint of an evidentiary preference for parodists over their victims, and no workable presumption for parody could take account of the fact that parody often shades into satire when society is lampooned through its creative artifacts, or that a work may contain both parodic and non-parodic elements. Accordingly, parody, like any other use, has to work its way through the relevant factors, and be judged case by case, in light of the ends of the copyright law.

Here, the District Court held, and the Court of Appeals assumed, that 2 Live Crew's "Pretty Woman" contains parody, commenting on and criticizing the original work, whatever it may have to say about society at large. . . .

We have less difficulty in finding that critical element in 2 Live Crew's song than the Court of Appeals did, although having found it we will not take the further step of evaluating its quality. The threshold question when fair use is raised in defense of parody is whether a parodic character may reasonably be perceived.[16] Whether, going beyond that, parody is in good taste or bad does not and should not matter to fair use. As Justice Holmes explained, "it would be a dangerous undertaking for persons trained only to the law to constitute themselves final judges of the worth of [a work], outside of the narrowest and most obvious limits. At the one extreme some works of genius would be sure to miss appreciation. Their very novelty would make them repulsive until the public had learned the new language in which their author spoke." Bleistein v. Donaldson Lithographing Co., 188 U.S. 239, 251 (1903) (circus posters have copyright protection); cf. Yankee Publishing Inc. v. News America Publishing, Inc., 809 F. Supp. 267, 280 (S.D.N.Y.

---

establish the extent of transformation and the parody's critical relationship to the original. By contrast, when there is little or no risk of market substitution, whether because of the large extent of transformation of the earlier work, the new work's minimal distribution in the market, the small extent to which it borrows from an original, or other factors, taking parodic aim at an original is a less critical factor in the analysis, and looser forms of parody may be found to be fair use, as may satire with lesser justification for the borrowing than would otherwise be required.

16. The only further judgment, indeed, that a court may pass on a work goes to an assessment of whether the parodic element is slight or great, and the copying small or extensive in relation to the parodic element, for a work with slight parodic element and extensive copying will be more likely to merely "supersede the objects" of the original.

1992) (Leval, J.) ("First Amendment protections do not apply only to those who speak clearly, whose jokes are funny, and whose parodies succeed") (trademark case).

While we might not assign a high rank to the parodic element here, we think it fair to say that 2 Live Crew's song reasonably could be perceived as commenting on the original or criticizing it, to some degree. 2 Live Crew juxtaposes the romantic musings of a man whose fantasy comes true, with degrading taunts, a bawdy demand for sex, and a sigh of relief from paternal responsibility. The later words can be taken as a comment on the naivete of the original of an earlier day, as a rejection of its sentiment that ignores the ugliness of street life and the debasement that it signifies. It is this joinder of reference and ridicule that marks off the author's choice of parody from the other types of comment and criticism that traditionally have had a claim to fair use protection as transformative works.[17]

The Court of Appeals, however, immediately cut short the enquiry into 2 Live Crew's fair use claim by confining its treatment of the first factor essentially to one relevant fact, the commercial nature of the use. The court then inflated the significance of this fact by applying a presumption ostensibly culled from Sony, that "every commercial use of copyrighted material is presumptively . . . unfair . . . ." Sony, 464 U.S., at 451. In giving virtually dispositive weight to the commercial nature of the parody, the Court of Appeals erred.

The language of the statute makes clear that the commercial or non profit educational purpose of a work is only one element of the first factor enquiry into its purpose and character. . . .

B

The second statutory factor, "the nature of the copyrighted work," §107(2), draws on Justice Story's expression, the "value of the materials used." Folsom v. Marsh, 9 F. Cas., at 348. This factor calls for recognition that some works are closer to the core of intended copyright protection than others, with the consequence that fair use is more difficult to establish when the former works are copied. See, e.g., Stewart v. Abend, 495 U.S., at 237-238 (contrasting fictional short story with factual works); Harper & Row, 471 U.S., at 563-564 (contrasting soon-to-be-published memoir with published speech); Sony, 464 U.S., at 455, n.40 (contrasting motion pictures with news broadcasts); Feist, 499 U.S., at 348-351 (contrasting creative works with bare factual compilations); 3 M. Nimmer & D. Nimmer, Nimmer on Copyright ¶13.05[A][2] (1993) (hereinafter Nimmer); Leval 1116. We

---

17. We note in passing that 2 Live Crew need not label its whole album, or even this song, a parody in order to claim fair use protection, nor should 2 Live Crew be penalized for this being its first parodic essay. Parody serves its goals whether labeled or not, and there is no reason to require parody to state the obvious (or even the reasonably perceived). See Patry & Perlmutter 716-717.

agree with both the District Court and the Court of Appeals that the Orbison original's creative expression for public dissemination falls within the core of the copyright's protective purposes. 754 F. Supp., at 1155-1156; 972 F.2d, at 1437. This fact, however, is not much help in this case, or ever likely to help much in separating the fair use sheep from the infringing goats in a parody case, since parodies almost invariably copy publicly known, expressive works.

C

The third factor asks whether "the amount and substantiality of the portion used in relation to the copyrighted work as a whole," §107(3) (or, in Justice Story's words, "the quantity and value of the materials used," Folsom v. Marsh, supra, at 348) are reasonable in relation to the purpose of the copying. Here, attention turns to the persuasiveness of a parodist's justification for the particular copying done, and the enquiry will harken back to the first of the statutory factors, for, as in prior cases, we recognize that the extent of permissible copying varies with the purpose and character of the use. See Sony, 464 U.S., at 449-450 (reproduction of entire work "does not have its ordinary effect of militating against a finding of fair use" as to home videotaping of television programs); Harper & Row, 471 U.S., at 564 ("Even substantial quotations might qualify as fair use in a review of a published work or a news account of a speech" but not in a scoop of a soon-to-be-published memoir). The facts bearing on this factor will also tend to address the fourth, by revealing the degree to which the parody may serve as a market substitute for the original or potentially licensed derivatives. See Leval 1123.

The District Court considered the song's parodic purpose in finding that 2 Live Crew had not helped themselves overmuch. 754 F. Supp., at 1156-1157. The Court of Appeals disagreed, stating that "while it may not be inappropriate to find that no more was taken than necessary, the copying was qualitatively substantial. . . . We conclude that taking the heart of the original and making it the heart of a new work was to purloin a substantial portion of the essence of the original." 972 F.2d, at 1438. . . .

Where we part company with the court below is in applying [this factor] to parody, and in particular to parody in the song before us. Parody presents a difficult case. Parody's humor, or in any event its comment, necessarily springs from recognizable allusion to its object through distorted imitation. Its art lies in the tension between a known original and its parodic twin. When parody takes aim at a particular original work, the parody must be able to "conjure up" at least enough of that original to make the object of its critical wit recognizable. See, e.g., Elsmere Music, 623 F.2d, at 253, n.1; Fisher v. Dees, 794 F.2d, at 438-439. What makes for this recognition is quotation of the original's most distinctive or memorable features, which the parodist can be sure the audience will know. Once enough has been taken to assure identification, how much more is reasonable will depend, say, on the extent to which the song's overriding purpose and character is to parody the original or, in contrast, the likelihood that the parody may serve as a market substitute for the original. But using some characteristic features cannot be avoided.

We think the Court of Appeals was insufficiently appreciative of parody's need for the recognizable sight or sound when it ruled 2 Live Crew's use unreasonable as a matter of law. It is true, of course, that 2 Live Crew copied the characteristic opening bass riff (or musical phrase) of the original, and true that the words of the first line copy the Orbison lyrics. But if quotation of the opening riff and the first line may be said to go to the "heart" of the original, the heart is also what most readily conjures up the song for parody, and it is the heart at which parody takes aim. Copying does not become excessive in relation to parodic purpose merely because the portion taken was the original's heart. If 2 Live Crew had copied a significantly less memorable part of the original, it is difficult to see how its parodic character would have come through. See Fisher v. Dees, 794 F.2d, at 439.

This is not, of course, to say that anyone who calls himself a parodist can skim the cream and get away scot free. In parody, as in news reporting, see *Harper & Row,* supra, context is everything, and the question of fairness asks what else the parodist did besides go to the heart of the original. It is significant that 2 Live Crew not only copied the first line of the original, but thereafter departed markedly from the Orbison lyrics for its own ends. 2 Live Crew not only copied the bass riff and repeated it, but also produced otherwise distinctive sounds, interposing "scraper" noise, overlaying the music with solos in different keys, and altering the drum beat. See 754 F. Supp., at 1155. This is not a case, then, where "a substantial portion" of the parody itself is composed of a "verbatim" copying of the original. It is not, that is, a case where the parody is so insubstantial, as compared to the copying, that the third factor must be resolved as a matter of law against the parodists.

Suffice it to say here that, as to the lyrics, we think the Court of Appeals correctly suggested that "no more was taken than necessary," 972 F.2d, at 1438, but just for that reason, we fail to see how the copying can be excessive in relation to its parodic purpose, even if the portion taken is the original's "heart." As to the music, we express no opinion whether repetition of the bass riff is excessive copying, and we remand to permit evaluation of the amount taken, in light of the song's parodic purpose and character, its transformative elements, and considerations of the potential for market substitution sketched more fully below.

D

The fourth fair use factor is "the effect of the use upon the potential market for or value of the copyrighted work." §107(4). It requires courts to consider not only the extent of market harm caused by the particular actions of the alleged infringer, but also "whether unrestricted and widespread conduct of the sort engaged in by the defendant . . . would result in a substan-

tially adverse impact on the potential market" for the original. Nimmer
¶13.05[A][4], p. 13-102.61 (footnote omitted); accord Harper & Row, 471
U.S., at 569; Senate Report, p. 65; Folsom v. Marsh, 9 F. Cas., at 349. The
enquiry "must take account not only of harm to the original but also of harm
to the market for derivative works." Harper & Row, supra, at 568.

Since fair use is an affirmative defense, its proponent would have difficulty carrying the burden of demonstrating fair use without favorable evidence about relevant markets. In moving for summary judgment, 2 Live Crew
left themselves at just such a disadvantage when they failed to address the
effect on the market for rap derivatives, and confined themselves to uncontroverted submissions that there was no likely effect on the market for the
original. They did not, however, thereby subject themselves to the evidentiary
presumption applied by the Court of Appeals. In assessing the likelihood of
significant market harm, the Court of Appeals quoted from language in *Sony*
that " 'if the intended use is for commercial gain, that likelihood may be
presumed. But if it is for a noncommercial purpose, the likelihood must be
demonstrated.' " 972 F.2d, at 1438, quoting Sony, 464 U.S., at 451. The
court reasoned that because "the use of the copyrighted work is wholly commercial, . . . we presume a likelihood of future harm to Acuff-Rose exists."
972 F.2d, at 1438. In so doing, the court resolved the fourth factor against
2 Live Crew, just as it had the first, by applying a presumption about the
effect of commercial use, a presumption which as applied here we hold to be
error.

No "presumption" or inference of market harm that might find support
in *Sony* is applicable to a case involving something beyond mere duplication
for commercial purposes. *Sony*'s discussion of a presumption contrasts a context of verbatim copying of the original in its entirety for commercial purposes, with the non-commercial context of *Sony* itself (home copying of
television programming). In the former circumstances, what *Sony* said simply
makes common sense: when a commercial use amounts to mere duplication
of the entirety of an original, it clearly "supersedes the objects," Folsom v.
Marsh, 9 F. Cas., at 348, of the original and serves as a market replacement
for it, making it likely that cognizable market harm to the original will occur.
Sony, 464 U.S., at 451. But when, on the contrary, the second use is transformative, market substitution is at least less certain, and market harm may
not be so readily inferred. Indeed, as to parody pure and simple, it is more
likely that the new work will not affect the market for the original in a way
cognizable under this factor, that is, by acting as a substitute for it ("superseding [its] objects"). See Leval 1125; Patry & Perlmutter 692, 697-698.
This is so because the parody and the original usually serve different market
functions. Bisceglia, ASCAP, Copyright Law Symposium, No. 34, p. 23.

We do not, of course, suggest that a parody may not harm the market
at all, but when a lethal parody, like a scathing theater review, kills demand
for the original, it does not produce a harm cognizable under the Copyright
Act. Because "parody may quite legitimately aim at garroting the original,
destroying it commercially as well as artistically," B. Kaplan, An Unhurried

View of Copyright 69 (1967), the role of the courts is to distinguish between "biting criticism [that merely] suppresses demand [and] copyright infringement[, which] usurps it." Fisher v. Dees, 794 F.2d, at 438.

This distinction between potentially remediable displacement and unremediable disparagement is reflected in the rule that there is no protectable derivative market for criticism. The market for potential derivative uses includes only those that creators of original works would in general develop or license others to develop. Yet the unlikelihood that creators of imaginative works will license critical reviews or lampoons of their own productions removes such uses from the very notion of a potential licensing market. "People ask . . . for criticism, but they only want praise." S. Maugham, Of Human Bondage 241 (Penguin ed. 1992). Thus, to the extent that the opinion below may be read to have considered harm to the market for parodies of "Oh, Pretty Woman," see 972 F.2d, at 1439, the court erred. Accord, Fisher v. Dees, 794 F.2d, at 437; Leval 1125; Patry & Perlmutter 688-691. n.22.

In explaining why the law recognizes no derivative market for critical works, including parody, we have, of course, been speaking of the later work as if it had nothing but a critical aspect (i.e., "parody pure and simple," supra, at 22). But the later work may have a more complex character, with effects not only in the arena of criticism but also in protectable markets for derivative works, too. In that sort of case, the law looks beyond the criticism to the other elements of the work, as it does here. 2 Live Crew's song comprises not only parody but also rap music, and the derivative market for rap music is a proper focus of enquiry, see Harper & Row, 471 U.S., at 568; Nimmer §13.05[B]. Evidence of substantial harm to it would weigh against a finding of fair use, because the licensing of derivatives is an important economic incentive to the creation of originals. See 17 U.S.C. §106(2) (copyright owner has rights to derivative works). Of course, the only harm to derivatives that need concern us, as discussed above, is the harm of market substitution. The fact that a parody may impair the market for derivative uses by the very effectiveness of its critical commentary is no more relevant under copyright than the like threat to the original market.

Although 2 Live Crew submitted uncontroverted affidavits on the question of market harm to the original, neither they, nor Acuff-Rose, introduced evidence or affidavits addressing the likely effect of 2 Live Crew's parodic rap song on the market for a non-parody, rap version of "Oh, Pretty Woman." . . . It is impossible to deal with the fourth factor except by recognizing that a silent record on an important factor bearing on fair use disentitled the proponent of the defense, 2 Live Crew, to summary judgment. The evidentiary hole will doubtless be plugged on remand.

III

It was error for the Court of Appeals to conclude that the commercial nature of 2 Live Crew's parody of "Oh, Pretty Woman" rendered it presumptively unfair. No such evidentiary presumption is available to address either the first factor, the character and purpose of the use, or the fourth, market harm, in determining whether a transformative use, such as parody, is a fair one. The court also erred in holding that 2 Live Crew had necessarily copied excessively from the Orbison original, considering the parodic purpose of the use. We therefore reverse the judgment of the Court of Appeals and remand for further proceedings consistent with this opinion.

It is so ordered.

<table>
<tr><td>Appendix A
"Oh, Pretty Woman" by
Roy Orbison and William Dees</td><td>Appendix B
"Pretty Woman" as Recorded by
2 Live Crew</td></tr>
</table>

Pretty Woman, walking down the street,	Pretty woman walkin' down the street
Pretty Woman, the kind I like to meet,	Pretty woman girl you look so sweet
Pretty Woman, I don't believe you,	Pretty woman you bring me down to that knee
you're not the truth,	Pretty woman you make me wanna beg please
No one could look as good as you	Oh, pretty woman
Mercy	
	Big hairy woman you need to shave that stuff
Pretty Woman, won't you pardon me,	Big hairy woman you know I bet it's tough
Pretty Woman, I couldn't help but see,	Big hairy woman all that hair it ain't legit
Pretty Woman, that you look lovely as can be	'Cause you look like 'Cousin It'
Are you lonely just like me?	Big hairy woman
Pretty Woman, stop a while,	Bald headed woman girl your hair won't grow
Pretty Woman, talk a while,	Bald headed woman you got a teeny weeny afro
Pretty Woman give your smile to me	Bald headed woman you know your hair could look nice
Pretty Woman, yeah, yeah, yeah	Bald headed woman first you got to roll it with rice
Pretty Woman, look my way,	
Pretty Woman, say you'll stay with me	Bald headed woman here, let me get this hunk of biz for ya
'Cause I need you, I'll treat you right	Ya know what I'm saying you look better than rice a roni
Come to me baby, Be mine tonight	Oh bald headed woman

Pretty Woman, don't walk on by,	Big hairy woman come on in
Pretty Woman, don't make me cry,	And don't forget your bald headed friend
Pretty Woman, don't walk away,	Hey pretty woman let the boys
Hey, O. K.	Jump in
If that's the way it must be, O. K.	
I guess I'll go on home, it's late	Two timin' woman girl you
There'll be tomorrow night, but wait!	know you ain't right
	Two timin' woman you's out with my boy last night
What do I see	Two timin' woman that takes a
Is she walking back to me?	load off my mind
Yeah, she's walking back to me!	Two timin' woman now I know
Oh, Pretty Woman.	the baby ain't mine
	Oh, two timin' woman
	Oh pretty woman

COMMENTS AND QUESTIONS

1. *Burden of Proof.* 2 Live Crew was successful in persuading the Supreme Court to reverse the Court of Appeals and reduce the significance of the presumption that commercial uses are unfair. But the Supreme Court did not simply affirm the district court's ruling that 2 Live Crew was entitled to summary judgment. Instead, the Court remanded the case for a determination of the parody's effect on the market for (non-parodic) rap derivatives of "Pretty Woman." Further, the Court obviously assumed that it was 2 Live Crew's burden to show that its parody had no such effect. Is 2 Live Crew —or any defendant — likely to be able to present such evidence and thus to prevail on summary judgment? Is it appropriate to assign the burden of proof to defendants on this issue?

2. *Scope of the Derivative Works Right.* The debate over parodies reflects a larger copyright issue that the Court does not directly address. Does the copyright owner have the right to prevent derivative works from being created, or only the right to receive royalties from derivative works produced by others? Traditionally, copyright law has assumed that the copyright owner has the right to prevent others from publishing a work at all without his or her permission. But the Court in *Acuff-Rose* seems to assume the reverse — that copyright owners are expected to license their rights in order to promote the widespread distribution of information (a key policy underlying copyright law). In particular, see footnote 10 of the Court's opinion.

Are copyright owners likely to allow parodies voluntarily? Because many parodies lampoon the original, the authors of the original are often unwilling to permit a parody to be prepared at any price. Robert Merges has described this reluctance as the "bargaining breakdown" problem — noneconomic factors prevent the parties from agreeing to what might be an efficient license (in purely economic terms). See Robert Merges, Are You Making Fun of Me? Notes on Market Failure and the Parody Defense in Copyright, 21 Am. Intellectual Prop. L. Assn. Q.J. 305 (1993). Cf. Richard Posner, When Is Parody Fair Use?, 21 J. Legal Stud. 79 (1992). In particular, in this case, Acuff-Rose refused to negotiate a mutually advantageous transfer because they didn't like the defendant's song. Does the breakdown of bargaining justify fair use treatment of parodies? Does the policy of encouraging dissemination do so?

3. Must the alleged infringer's primary intent be to parody the work? Arguably, 2 Live Crew simply wanted to engage in satire. After all, their song is only obliquely directed at Orbison's. Can a work ever be a parody if the author did not intend it to be? (Note that 2 Live Crew originally claimed that their song was a *cover* of the original, and was therefore protected under section 115. This argument failed because 2 Live Crew substantially changed the song from its original version.)

4. *Parody or Satire?* The problems with the parody/satire distinction were put into sharp relief by the Ninth Circuit's decision in Dr. Seuss Enterprises v. Penguin Books, 109 F.3d 1394 (9th Cir. 1997). In that case, the defendant published a poem making fun of the verdict in the O.J. Simpson case. Called "Cat NOT In the Hat," the poem copied the lyric structure of Theodore Geisel's famous children's book *The Cat In the Hat*. The Ninth Circuit rejected the defendant's claim of fair use. It held that the defendant's work was satire, not parody, because his intent was to poke fun at the O.J. Simpson verdict and not at the Dr. Seuss book. The case is troubling because the defendant's work seems precisely the sort of transformative use that *Campbell* seemed to encourage and because there is no way to create such a work without relying on Geisel's preexisting work.

PROBLEMS

Problem 4-34. Greenwich Systems is a software company that sells the world's most popular "screen saver." (A screen saver is a computer program that blanks a computer screen after a certain period of inactivity and then replaces the old screen with another. Screen savers are highly creative and are designed to entertain the viewer as well as to protect the integrity of the screen.) Greenwich's screen saver, called "flying toasters," shows pop-up toasters with wings flying across the screen.

Delaware, a small start-up company, licenses the popular "Bloom County/Outland" characters Opus and Bill for use in its own screen saver. Its program contains a number of humorous replacement screens, including one screen in which Opus shoots down flying toasters with a shotgun. The flying toasters look very similar to Greenwich's. Greenwich sues for copyright infringement, and Delaware defends on grounds of parody. Who should prevail?

Problem 4-35. Air Pirates, a publisher of counter-culture adult comic books, developed a series of comic books cynically depicting popular children's cartoon characters, including Mickey and Minnie Mouse, Donald Duck, and Goofy, as promiscuous, drug dealing rogues. One of the magazine covers had the accompanying image (Figure 4-8).
Should Air Pirates' use be considered fair use?

2. Other Defenses

In addition to statutory defenses (17 U.S.C. §§108-118) and the fair use doctrine (17 U.S.C. §107), there are a number of other defenses against a claim of copyright infringement. Many of these are straightforward, and the reader has no doubt encountered similar defenses elsewhere in the law. The most significant defenses include:

Independent Creation. A defendant may present evidence to prove that he or she created the work independently.

Consent/License. A defendant may defend a copyright action on the ground that he or she has the copyright owner's permission to make use of the protected material. Note that section 204 requires transfers of copyright ownership to be executed in writing and signed by the copyright owner. Nonexclusive licenses need not be in writing, however.

Inequitable Conduct. Closely related to copyright invalidity, the inequitable conduct doctrine parallels the patent law defense of inequitable conduct discussed in Chapter 3. Inequitable conduct occurs when a copyright owner obtains a copyright through fraud or other deceptive conduct on the Copyright Office — for example, by failing to disclose the owner's own plagiarism of a prior work. Circumstances that give rise to inequitable conduct generally also render copyrights invalid, but the defense is distinct because some of the consequences may be different.

Copyright Misuse. There has been substantial uncertainty whether copyright law recognizes a defense analogous to patent misuse. See, e.g., Timothy H. Fine, Misuse and Antitrust Defenses to Copyright Infringement Actions, 17 Hastings L.J. 315 (1965); Frank Gibbs, Copyright Misuse: Thirty Years Waiting for the Other Shoe, 23 Copyrt. L. Symp. (ASCAP) 31 (1977). The Fourth Circuit was the first to apply the copyright misuse doctrine. The court held that a copyright defendant may defend an infringement action on the

FIGURE 4-8
Air Pirates cover.

ground the plaintiff has misused its copyright in an anticompetitive manner. See Lasercomb America, Inc. v. Reynolds, 911 F.2d 970 (4th Cir. 1990). The Fifth Circuit has followed *Lasercomb*. DSC, Inc. v. DGI, Inc., 81 F.3d 597 (5th Cir. 1996); see also Alcatel v. DGI Technologies, 166 F.3d 772 (5th Cir. 1999). This defense will likely became more common as copyright law is increasingly used to protect information technologies. Cf. Ramsey Hanna, Comment, Misusing Antitrust: The Search for Functional Copyright Misuse Standards, 46 Stan. L. Rev. 401 (1994). Indeed, in 1997 the Ninth Circuit held that the American Medical Association committed copyright misuse by selling its copyrighted reference work to the government only on the condition that they would not buy similar works from a competitor. Practice Management Info. Corp. v. American Medical Assn., 121 F.3d 516 (9th Cir. 1997). While misuse doctrine is historically tied to antitrust, what is notable about these cases is that they have relied as much on copyright policy as antitrust to justify their misuse findings. See Mark A. Lemley, Beyond Preemption: The Law and Policy of Intellectual Property Licensing, 87 Cal. L. Rev. 111 (1999). We focus on the application of antitrust doctrine to intellectual property disputes in Chapter 8.

First Amendment. The argument is often made that an accused infringer cannot be prevented from publishing an infringing work because to do so would violate her rights under the First Amendment to the United States Constitution.[30] Courts generally reject this argument on the grounds that the Constitution also contains an explicit provision allowing Congress to protect copyrights and that this explicit provision has been squared with the First Amendment by the fair use and idea-expression limits on copyright. See Kalem Co. v. Harper Bros., 222 U.S. 55, 63 (1911); Sid & Marty Krofft Television v. McDonald's Corp., 562 F.2d 1157 (9th Cir. 1977). These courts thus implicitly identify the idea-expression dichotomy and the fair use doctrine as the appropriate places to draw the line between copyrightable works and protected speech.

Are certain statements on matters of public concern so important that a copyright owner should not be allowed to appropriate them to his own use? Do statements by certain government officials, acting in their official capacity, fall in this category? Should there be a general constitutional protection against copyright infringement for the press? Cf. Melville Nimmer, Does Copyright Abridge the First Amendment Guarantees of Free Speech and Press?, 17 UCLA L. Rev. 1180, 1199 (1970) (suggesting a compulsory licensing

---

30. This alleged conflict is not too surprising in view of the fact, noted by Goldstein, that "copyright is the stepchild of censorship." 2 Paul Goldstein, Copyright sec. 10.3, at 239. See section A of this chapter.

scheme for "news photographs"). The Ninth Circuit has rejected this view, concluding that unedited videotapes of news events (in this case an airplane crash and a train wreck) were copyrightable. The court did, however, leave open the possibility that copyright owners would be forced to release their footage if there was no other avenue of access to the news contained in the film. See Los Angeles News Service v. Tullo, 24 U.S.P.Q.2d 1026 (9th Cir. 1992). Compare Los Angeles News Service v. KCAL-TV, 108 F.3d 1119 (9th Cir. 1997). There, the court held that LANS' videotape of the Reginald Denny beating could not be broadcast by a television news program that did not pay for a license. The court noted that in that case LANS was, in fact, offering to license the news footage to various news stations for a fee.

For a different First Amendment argument — that preliminary injunctions in copyright cases are presumptively unconstitutional prior restraints on speech — see Mark A. Lemley & Eugene Volokh, Freedom of Speech and Injunctions in Intellectual Property Cases, 48 Duke L.J. 147 (1998).

Surely Congress cannot insulate a statute from First Amendment scrutiny merely by enacting it under the copyright laws. See H.R. 3883, 104th Cong., 2d Sess. (a bill declaring the U.S. flag to be copyrighted, and making its destruction by burning a criminal copyright offense).

Immoral/Illegal/Obscene Works. Although some early cases refused to enforce copyrights in obscene or otherwise antisocial works, see William Patry, Latman's The Copyright Law 50 (6th ed. 1986), the modern trend rejects such a defense. See Mitchell Bros. Film Group v. Adult Cinema Theater, 604 F.2d 852 (9th Cir. 1980) (pornography), *cert. denied,* 445 U.S. 917 (1980); Belcher v. Tarbox, 486 F.2d 1087 (9th Cir. 1973) (racing forms).

Statute of Limitations. The statute of limitations for copyright infringement is three years "after the claim accrued." §507. There is some question whether a continuing infringement gives rise to one or many claims and, if it is only a single claim, when it accrues. To see the difference, consider the position of an accused infringer who sold two copies of the same painting, one four years ago and one two years ago. If each sale constitutes a separate infringement, she can be held liable for the recent sale but not for the older sale. If selling the same painting constitutes a single act of infringement, however, whether she can be held liable at all depends on whether the infringement occurred two years ago or four years ago.

G. INTERNATIONAL ISSUES IN COPYRIGHT

Great strides have been made in the last century in harmonizing international copyright laws. The drive towards international uniformity is the result of several factors, notably the increasingly international nature of markets for music, movies, and books. The historical failure of governments to treat copyrights consistently is particularly important for copyrights in modern technological markets, because innovation travels very quickly beyond national boundaries. The perceived failure of some countries to give sufficient respect to intellectual property was a major sticking point in trade negotiations like the General Agreements on Tariffs and Trade (GATT) and the North American Free Trade Agreement (NAFTA).

The GATT agreement, as well as regional copyright harmonization through forums such as NAFTA, the European Union, and the Andean Pact, have made it easier to protect copyrights across national boundaries. The agreements have also established substantive minimum standards of copyright protection in member countries, continuing a trend begun with the Berne Convention in the late nineteenth century. Both Berne and the Uruguay Round of GATT are adhered to by over half the nations in the world; there is substantial but not complete overlap in membership between the two.[31] Both treaties contain choice of law provisions, providing for nondiscrimination in the application of national copyright privileges. Further, both the Berne Convention and the GATT reject the principle of reciprocity in favor of one of national treatment — that is, the level of copyright protection afforded depends on the nation in which infringement takes place, not the nationality of the author.

Berne and GATT also provide a statutory minima of protection that all member nations must afford. These include the term of protection (life of the author plus 50 years for both single and joint works; there is no provision for works for hire. Berne Article 7, 7[bis]); the scope of exclusive rights afforded to copyright holders (literary and artistic works, including works of applied art and audiovisual works. Berne Article 2(1)); the rights conferred (rights of reproduction, Berne Article 9; rights of public performance, Berne Article 11, 11[bis], 11[ter]; moral rights of authors, Berne Article 6[bis]; and the right to prepare derivative works, Berne Article 2(3), 12).

Beyond these substantive minima, each country is free to provide its own specialized form of protection. Because of this, international copyright protection remains less than uniform. For example, countries such as France provide authors and artists with much stronger moral and resale rights than

---

31. A third convention, the Universal Copyright Convention, was written shortly after World War II in large part because of the refusal of the United States to adhere to the Berne Convention. It is similar in many respects to the Berne Convention, and it has lost much of its significance since the United States ratified the Berne Convention.

does the United States. Significantly, the Berne Convention requires that each country grant to citizens of other Berne signatories the same rights that it would grant to its own citizens, *including at least the substantive protections specified above*. Article 5(1). Thus nations that grant rights beyond the Berne minima must provide such protection to foreign citizens as well, even though the foreign country does not provide similar protection to authors of any nationality.

Interestingly, Berne does not require that a signatory adopt each of the minimum substantive protections for *its own* citizens. It does require that foreign nationals be granted those protections, however. Thus a country can offer less protection to its own citizens than to foreign nationals, but not more. The United States has used this provision to maintain some of its copyright formalities (such as registration) even after its acceptance of Berne in 1989. United States nationals must comply with those formalities in order to obtain copyright protection, but citizens of other countries need not do so.

The growth of multilateral copyright treaties is an important step toward global protection for the rights of authors, and toward preventing protectionism. Such treaties may be less effective, however, as a means of preventing international piracy of works in nations that historically have lacked strong copyright protection. The problem in those nations is generally one of inadequate enforcement rather than inadequate statutory protection.

COMMENTS AND QUESTIONS

1. Consider a number of different possible ways of protecting copyrights across national boundaries.

- First, a world body (or regional bodies such as the European Union) could impose (or all countries could agree to) a single, uniform copyright law. In either case, there would be no question of differential treatment between nations covered by the law, because the substantive laws of all the states would be identical.
- Second, the nations of the world or of a particular region could agree to respect each other's copyright laws by signing a multilateral non-discrimination treaty. Such a treaty could take one of two basic forms — granting to each copyright holder the same protection he would get in his home state, or granting to each copyright holder the same protection local copyright holders would get. The first option would ensure that each author received uniform protection throughout the world, but it would require nations to apply different legal rules to different authors. It might also encourage nations to enhance the copyright protection they afford to their citizens, since their citizens will benefit abroad without the nation having to grant a similar benefit to foreign nationals. The second option would ensure that a

nation's copyright laws remain uniform regardless of the nationality of the plaintiff, but it would require giving different protection to the same work in different countries.

- Third, copyright may be protected by bilateral arrangements between individual nations. This approach was predominant until the end of the nineteenth century, when the Berne Convention and a number of multilateral treaties in the Western Hemisphere rendered such arrangements less important. Bilateral arrangements serve much the same purpose as multilateral arrangements, but they are considerably less efficient.

- Finally, international copyright protection may be left entirely to the control of the country in which protection is sought. The problem with this approach is that nations generally take a short-term, protectionist view of allowing protection to foreign nationals. The United States limited foreign access to its copyright laws until 1986 by requiring certain proof of "domestic manufacturing" of a foreign copyright before an infringement suit could be brought. 17 U.S.C. §601.

Is there a reason to prefer one of these models over others? Does your answer depend on how much faith you put in public international law? See Anthony D'Amato & Doris Estelle Long, International Intellectual Property Anthology 199-200 (1996).

2. *WIPO.* While GATT is the highest-profile example of international treaties that affect the content of laws in member nations, it is not the only one. The World Intellectual Property Organization (WIPO), an arm of the United Nations, also develops treaties governing copyright. In 1996, WIPO adopted two new treaties dealing with copyright law. WIPO Copyright Treaty, adopted December 20, 1996, WIPO Doc. CRNR/DC/94; WIPO Performances and Phonograms Treaty, adopted December 20, 1996, CRNR/DC/95. The WIPO Copyright Treaty seeks to promote modernization of national copyright laws to address new issues raised by digital technology. Thus, in addition to a range of largely uncontroversial provisions reaffirming traditional copyright principles, the WIPO Copyright Treaty also addresses such controversial issues as the treatment of temporary copies within a computer's memory, online service provider liability, and circumvention of security devices designed to prevent privacy. See generally, Julie S. Sheinblatt, The WIPO Copyright Treaty, 13 Berkeley Tech. L.J. 535 (1998); Pamela Samuelson, The U.S. Digital Agenda at WIPO, 37 Va. J. Intl. L. 369 (1997). These issues have been subsequently addressed in the U.S. Digital Millennium Copyright Act.

There has been a long-standing "turf battle" between GATT and WIPO over the "proper" forum for establishing international intellectual property standards. See Bal Gopal Das, Intellectual Property Dispute, GATT, WIPO: Of Playing by the Game Rules and Rules of the Game, 35 Idea 149 (1994); Doris Estelle Long, Copyright and the Uruguay Round Agreements: A New Era of Protection or an Illusory Promise?, 22 AIPLA Q.J. 531 (1994). It is reasonable to assume that GATT's creation of a new forum for dispute resolution will exacerbate this tension.

3. *Foreign Law.* U.S. copyright owners have an interest in foreign law to the extent that they intend to distribute their works or otherwise compete in foreign markets. In some respects, foreign countries provide different and stronger rights than U.S. copyright law. For example, the European Union approved Directive 96/9/EC on the Legal Protection of Databases in 1996. This directive establishes a sui generis right against unauthorized extraction of the contents of database for a term of 15 years from the creation of the database. See generally, Mark Schneider, The European Union Database Directive, 13 Berkeley Tech. L.J. 551 (1998). In essence, the EU has developed an intellectual property right that protects "sweat of the brow," thereby filling the potential gap in protection reflected in Feist Publications, Inc. v. Rural Tel. Serv. Co., Inc., 499 U.S. 340 (1991). The EU Database Directive affords rights only to database makers or rightholders who are nationals of a member state or who have their residence in the territory of the European Community. Art. 11(1).

4. Does the whole concept of "internationalizing" copyright protection improperly assume the existence of a shared set of cultural values concerning works to be protected? Paul Geller has argued that the "transplanting" of copyright rules across national boundaries is based on an assumption of "universal, permanent cultural values." Paul Edward Geller, Legal Transplants in International Copyright: Some Problems of Method, 13 UCLA Pac. Basin L.J. 199 (1994). Geller contends that those values are frequently lacking in practice:

> Bear in mind that the notion of a "work" is understood against the background of aesthetic sensibilities that vary from culture to culture. For example, Brad Sherman describes the reluctance of the Anglo-Australians to dignify graphic creations by native Australians as "artistic works," much less find them to be "original" or "creative." . . . To take another example, the People's Republic of China, in its Copyright Act of 1990, introduced new categories of "works," for example, quyi works "based on traditional forms created mainly for performance through recitation, music, or both." As indicated above, some commentators might argue that works in this new category, if it is construed to fall outside the list in Article 2 of the Berne Convention, do not benefit from Berne minimum rights. . . . [T]he "common core" meaning of the Berne notion of "work," historically the European meaning, is made the standard for non-European works.

See also Cathryn A. Berryman, Toward More Universal Protection of Intangible Cultural Property, 1 J. Intell. Prop. L. 293 (1994) (criticizing the failure of intellectual property regimes to protect works of folklore and oral culture). Do these criticisms, if valid, doom the harmonization enterprise? Or are the benefits of harmonization sufficiently great that we should put up with some disagreement on subject matter? One might also ask whether there are certain levels of protection that all nations could agree on. If so, should protection be limited to the areas of universal agreement, or should it on the contrary be extended to any type of work for which an authorial interest is claimed?

5. One version of the internationalization of copyright is exemplified in 19 U.S.C. §2411 et seq., commonly referred to as "Special 301." This is a provision of United States law that permits the U.S. Trade Representative to impose trade sanctions on another nation if he or she concludes that that nation unfairly restricts commerce with the United States. Recent administrations have used this provision to put countries on a "priority watch list" if they do not adequately enforce intellectual property rights and to obtain a series of bilateral treaty concessions on intellectual property rights from those countries. Some scholars argue that Special 301 is a highly efficient means of enforcing intellectual property rights. Tara Kalagher Giunta & Lily H. Shang, Ownership of Information in a Global Economy, 27 Geo. Wash. J. Intl. L. & Econ. 327 (1994). However, many nations have criticized the procedure as discriminatory and at odds with multilateral dispute resolution mechanisms such as the World Trade Organization. See Theodore H. Davis, Combating Piracy of Intellectual Property in International Markets: A Proposed Modification of the Special 301 Action, 24 Vand. J. Transnatl. L. 505 (1991) (noting such criticism). Davis himself proposes that the provision be applied only in carefully limited circumstances.

6. Establishing a more or less uniform set of legal rules is only part of the problem facing international intellectual property. Once legal rules are established, they must also be enforced. By and large, copyright enforcement is lax in developing nations. The reasons for this are complex. In some cases, the government may turn a blind eye to piracy because it believes that the local economy will benefit from the efforts of the pirates. In other cases, enforcement efforts are genuine but hampered by inadequate resources. Finally, in some countries the problem is what might be called the "cycle of piracy" — legitimate copyright owners refuse to sell in the country because of the piracy problem, which means that the only way the public can obtain the goods it wants is to turn to piracy. This in turn only strengthens the resolve of copyright owners not to do business in the country. For a discussion of the enforcement problem, see Marshall A. Leaffer, Protecting United States Intellectual Property Abroad: Toward a New Multilateralism, 76 Iowa L. Rev. 273 (1991).

7. International copyright law concerns not only treaties and agreements but also the application of domestic laws to conduct occuring, in part, outside national boundaries. Several recent cases have highlighted the limits of the territoriality principle in copyright.

First, the Ninth Circuit held in a controversial decision that a defendant cannot be held liable under U.S. law for authorizing conduct that occurred overseas even though the same conduct would have been infringing had it occurred in the U.S. Subafilms v. M.G.M.-Pathe Communications, 24 F.3d 1088 (9th Cir. 1994) (en banc). This decision is consistent with the Copyright Act and the territoriality principle, but it does rely heavily on the assumption that the copyright owner can enforce his rights against foreign infringement in foreign courts.

Second, the court in Computer Associates Intl. v. Altai, 126 F.3d 365 (2d Cir. 1997) enforced the flip side of this territoriality principle. In that case, Computer Associates had sued Altai for copyright infringement in the U.S. and lost. It then filed suit in France, asserting the same claim against the same set of products. The Second Circuit refused to enjoin the French suit under principles of res judicata or collateral estoppel, reasoning that the application of French law to conduct occurring in France presented a separate issue from the one litigated in the U.S. case. Whether this result is right or not (it potentially raises the spectre of having to bring or defend suits separately in every country in the world) is presumably a question of remedy. The court's rationale suggests that copyright plaintiffs should never be entitled to a remedy that applies to infringement beyond the borders of the U.S. If that is correct, then Computer Associates should be free to sue in other countries. If it is incorrect though, the court's decision unfairly gives Computer Associates a second (and third, and fourth . . .) bite at the apple.

Finally, numerous courts have litigated the issue of parallel imports in copyright law. Cases such as *L'Anza*, reprinted supra, suggest that the territoriality principle may not be used to enforce barriers to trade. The decision of the European Free Trade Association Court in Mag Instrument v. California Trading Co., Case E-2/97 (EFTA 1997) permits parallel imports after the first sale of a good. By contrast, the decision of the European Court of Justice in Silhouette Intl. v. Hartlauer Handelsgesellschaft mbH, Case C-355/96 (E.C.J. 1998) allows countries to preclude parallel imports from outside the European Union. We return to the parallel importation issue in more detail in the next chapter.

H. REMEDIES

Economists often classify legal rules into one of two types, depending on the remedies available. The first type of legal regime is known as the "liability rule." Contract law is a good example. In essence, the defendant in a liability rule system has the "right" to violate the law — for example, the right to breach a contract — as long as he pays the plaintiff an amount sufficient to compensate her for her loss. Damages are characteristically the sole remedy under a liability rule; injunctions are rare.

The second type of legal regime is the "property rule." Not surprisingly, property law is a good example. Here, the plaintiff holds the "right" to prevent improper behavior by the defendant — for example, the right to prevent a trespass on her property. While damages may well be awarded as compensation in a property rule regime, other remedies are also available to the plaintiff (or at least against the defendant). These other remedies may include punitive damages, injunctions, and in some cases criminal sanctions.

For a more detailed discussion of property and liability rules, see A. Mitchell Polinsky, An Introduction to Law and Economics (1983); Robert Cooter & Thomas Ulen, Law and Economics 88-324 (1988); John P. Dwyer and Peter S. Menell, Property Law and Policy: A Comparative Institutional Perspective 281-287 (1998).

For better or for worse, the copyright laws are largely set up as property rules. Copyright infringement is not made acceptable simply by the infringer's payment of damages. Indeed, intentional infringement of copyright for financial gain is a criminal offense. 17 U.S.C. §506(a).[32] In civil cases, infringement can be — and normally is — enjoined under 17 U.S.C. §502. Further, courts will often order the seizure or impoundment of allegedly infringing articles while a copyright case is pending, under the authority of 17 U.S.C. §503(a) (in civil cases) or 509(a) (in criminal cases). Those articles may be destroyed or forfeited to the United States after judgment. Finally, in addition to the above remedies, the copyright holder may collect as damages either "actual damages" (defined as the loss to the copyright holder plus additional profits made by the infringer) or statutory damages of up to $20,000 for innocent infringement and $100,000 for willful infringement. 17 U.S.C. §504. Prevailing parties (plaintiffs *and* defendants) are also entitled to costs and attorney fees in some cases. 17 U.S.C. §505.

To be sure, not all aspects of copyright law are reminiscent of a property rule. The fair use doctrine, for example, has strong liability rule components, as Judge Laval has argued. Pierre Laval, Toward a Fair Use Standard, 103 Harv. L. Rev. 1105 (1990). Unlike most liability rules, however, the fair use doctrine actually allows the infringer to avoid making any payment. As we have discussed, there are also a number of other provisions for compulsory licensing in the copyright laws, provisions that are normally found in liability rule regimes. Finally, a close analogy to contract law can be found in the provision for statutory damages, which may be seen as a sort of liquidated damages clause designed to avoid market failure in cases in which injury to the copyright holder is hard to quantify.

As is abundantly clear from even cursory exposure to intellectual property right valuation problems, including licensing negotiations, these assets are exceedingly difficult to value. This is partly because, by statutory definition, each of these rights is in some sense unique. And when an administrative tribunal — consisting of strangers to the industry and neophytes to the technology or creative work covered by the intellectual property right — is involved, the costs rise even more. The damages phase of copyright infringement litigation speaks for itself on this score. Even Mitch Polinsky, who shows

---

32. Criminal penalties for copyright infringement are limited to those set forth in that section. In 1985 the Supreme Court rejected an attempt by the government to prosecute copyright infringement as transportation of stolen property. The Court reasoned that the specific statutory scheme of section 506 was meant to replace, not supplement, general laws that were not written with intellectual property in mind. Dowling v. United States, 473 U.S. 207 (1985).

the interchangeability of property and liability rules under many conditions, found that a property rule is superior to a liability rule when "the court lacks information about both damages and benefits." A. Mitchell Polinksy, Resolving Nuisance Disputes: The Simple Economics of Injunctive and Damage Remedies, 32 Stan. L. Rev. 1075, 1112 (1980). The unique features of intellectual property rights create just such circumstances. Similarly, Alan Schwartz argues in favor of enforcement of contractual provisions stipulating a specific performance remedy. See Alan Schwartz, The Myth That Promisees Prefer Supracompensatory Remedies: An Analysis of Contracting for Damage Measures, 100 Yale L.J. 369, 371 (1990). Schwartz presumes that the presence of such a term in a contract is a sure sign that the parties have had difficulty valuing the potential loss from nonperformance of the contract; while this might be questionable in the general case, the statutorily defined status of intellectual property rights as unique assures that it is realistic in many copyright cases. The *Sheldon* case which follows is a good example of these issues in practice.

Sheldon et al. v. Metro-Goldwyn Pictures Corp. et al.
Supreme Court of the United States
309 U.S. 390 (1940)

Mr. Chief Justice HUGHES delivered the opinion of the Court.

The questions presented are whether, in computing an award of profits against an infringer of a copyright, there may be an apportionment so as to give to the owner of the copyright only that part of the profits found to be attributable to the use of the copyrighted material as distinguished from what the infringer himself has supplied, and, if so, whether the evidence affords a proper basis for the apportionment decreed in this case.

Petitioners' complaint charged infringement of their play "Dishonored Lady" by respondents' motion picture "Letty Lynton," and sought an injunction and an accounting of profits. The Circuit Court of Appeals, reversing the District Court, found and enjoined the infringement and directed an accounting. 81 F.2d 49. Thereupon the District Court confirmed with slight modifications the report of a special master which awarded to petitioners all the net profits made by respondents from their exhibitions of the motion picture, amounting to $587,604.37. 26 F. Supp. 134, 136. The Circuit Court of Appeals reversed, holding that there should be an apportionment and fixing petitioners' share of the net profits at one-fifth. 106 F.2d 45, 51. In view of the importance of the question, which appears to be one of first impression in the application of the copyright law, we granted certiorari. December 4, 1939.

Petitioners' play "Dishonored Lady" was based upon the trial in Scotland, in 1857, of Madeleine Smith for the murder of her lover, — a cause celebre included in the series of "Notable British Trials" which was published in 1927. The play was copyrighted as an unpublished work in 1930, and was produced here and abroad. Respondents took the title of their motion picture "Letty Lynton" from a novel of that name written by an English author, Mrs. Belloc Lowndes, and published in 1930. That novel was also based upon the story of Madeleine Smith and the motion picture rights were bought by respondents. There had been negotiations for the motion picture rights in petitioners' play, and the price had been fixed at $30,000, but these negotiations fell through.

As the Court of Appeals found, respondents in producing the motion picture in question worked over old material; "the general skeleton was already in the public defense. A wanton girl kills her lover to free herself for a better match; she is brought to trial for the murder and escapes." [106 F.2d 50.] But not content with the mere use of that basic plot, respondents resorted to petitioners' copyrighted play. They were not innocent offenders. From comparison and analysis, the Court of Appeals concluded that they had "deliberately lifted the play"; their "borrowing was a deliberate plagiarism." It is from that standpoint that we approach the questions now raised.

Respondents contend that the material taken by infringement contributed in but a small measure to the production and success of the motion picture. They say that they themselves contributed the main factors in producing the large net profits; that is, the popular actors, the scenery, and the expert producers and directors. Both courts below have sustained this contention.

The District Court thought it "punitive and unjust" to award all the net profits to petitioners. The court said that, if that were done, petitioners would receive the profits that the "motion picture stars" had made for the picture "by their dramatic talent and the drawing power of their reputations." "The directors who supervised the production of the picture and the experts who filmed it also contributed in piling up these tremendous net profits." The court thought an allowance to petitioners of 25 percent of these profits "could be justly fixed as a limit beyond which complainants would be receiving profits in no way attributable to the use of their play in the production of the picture." But, though holding these views, the District Court awarded all the net profits to petitioners, feeling bound by the decision of the Court of Appeals in Dam v. Kirk La Shelle Co., 175 F. 902, 903, a decision which the Court of Appeals has now overruled.

The Court of Appeals was satisfied that but a small part of the net profits was attributable to the infringement, and, fully recognizing the difficulty in finding a satisfactory standard, the court decided that there should be an apportionment and that it could fairly be made. The court was resolved "to avoid the one certainly unjust course of giving the plaintiffs everything, because the defendants cannot with certainty compute their own share." The court would not deny "the one fact that stands undoubted," and, making the best estimate it could, it fixed petitioners' share at one-fifth of the net profits, considering that to be a figure "which will favor the plaintiffs in every reasonable chance of error."

Petitioners stress the provision for recovery of "all" the profits, but this is plainly qualified by the words "which the infringer shall have made from such infringement." This provision in purpose is cognate to that for the recovery of "such damages as the copyright proprietor may have suffered due to the infringement." The purpose is thus to provide just compensation for the wrong, not to impose a penalty by giving to the copyright proprietor profits which are not attributable to the infringement.

Prior to the Copyright Act of 1909, there had been no statutory provision for the recovery of profits, but that recovery had been allowed in equity both in copyright and patent cases as appropriate equitable relief incident to a decree for an injunction. Stevens v. Gladding, 17 How. 447, 455. That relief had been given in accordance with the principles governing equity jurisdiction, not to inflict punishment but to prevent an unjust enrichment by allowing injured complainants to claim "that which, ex aequo et bono, is theirs, and nothing beyond this." Livingston v. Woodworth, 15 How. 546, 560. See Root v. Railway Co., 105 U.S. 189, 194, 195. Statutory provision for the recovery of profits in patent cases was enacted in 1870. The principle which was applied both prior to this statute and later was thus stated in the leading case of Tilghman v. Proctor, 125 U.S. 136, 146:

> The infringer is liable for actual, not for possible gains. The profits, therefore, which he must account for, are not those which he might reasonably have made, but those which he did make, by the use of the plaintiff's invention; or, in other words, the fruits of the advantage which he derived from the use of that invention, over what he would have had in using other means then open to the public and adequate to enable him to obtain an equally beneficial result. If there was no such advantage in his use of the plaintiff's invention, there can be no decree for profits, and the plaintiff's only remedy is by an action at law for damages. . . .

Petitioners stress the point that respondents have been found guilty of deliberate plagiarism, but we perceive no ground for saying that in awarding profits to the copyright proprietor as a means of compensation, the court may make an award of profits which have been shown not to be due to the infringement. That would be not to do equity but to inflict an unauthorized penalty. To call the infringer a trustee ex maleficio merely indicates "a mode of approach and an imperfect analogy by which the wrongdoer will be made to hand over the proceeds of his wrong." Larson Co. v. Wrigley Co., 277 U.S. 97, 99, 100. He is in the position of one who has confused his own gains with those which belong to another. Westinghouse Co. v. Wagner Co., supra, p. 618. He "must yield the gains begotten of his wrong." Duplate Corp. v. Triplex Co., 298 U.S. 448, 457. Where there is a commingling of gains, he must abide the consequences, unless he can make a separation of the profits so as to assure to the injured party all that justly belongs to him. When such an apportionment has been fairly made, the copyright proprietor receives all the profits which have been gained through the use of the infringing material and that is all that the statute authorizes and equity sanctions.

Both courts below have held in this case that but a small part of the profits were due to the infringement, and, accepting that fact and the principle that an apportionment may be had if the evidence justifies it, we pass to the consideration of the basis of the actual apportionment which has been allowed.

Third. The controlling fact in the determination of the apportionment was that the profits had been derived, not from the mere performance of a copyrighted play, but from the exhibition of a motion picture which had its distinctive profit-making features, apart from the use of any infringing material, by reason of the expert and creative operations involved in its production and direction. In that aspect the case has a certain resemblance to that of a patent infringement, where the infringer has created profits by the addition of non-infringing and valuable improvements. And, in this instance, it plainly appeared that what respondents had contributed accounted for by far the larger part of their gains. . . .

COMMENTS AND QUESTIONS

1. Damages limited to actual profits gained or lost is a remedy normally found in contract cases. Does *Sheldon* unfairly prevent copyright owners from prohibiting access to their works? Consider the case of private diaries that are published. Might not the injury to the author in some cases exceed the profits of the publisher? 17 U.S.C. §504(a) attempts to take care of this problem by allowing the copyright owner to recover both his actual losses and any additional profits incurred by the infringer, so long as the copyright owner does not thereby obtain a "double recovery."

2. *Attorneys' Fees.* Section 505 of the Copyright Act authorizes the award of attorney fees in the discretion of the court to "prevailing parties." The Supreme Court recently interpreted this provision as allowing fees to be awarded to prevailing plaintiffs or defendants on an equal basis. Fogarty v. Fantasy, Inc., 114 S. Ct. 1023 (1994). The Court's reasoning in rejecting a "dual standard" that favored plaintiffs is instructive on the issue of appropriate copyright incentives:

> [T]he policies served by the Copyright Act are more complex, more measured, than simply maximizing the number of meritorious suits for copyright infringement. . . .
>
> Because copyright law ultimately serves the purpose of enriching the general public through access to creative works, it is peculiarly important that the boundaries of copyright law be demarcated as clearly as possible. To that end, defendants who seek to advance a variety of meritorious copyright defenses should be encouraged to litigate them to the same extent that plaintiffs are encouraged to litigate meritorious claims of infringement. In the case before us, the successful defense of "The Old Man Down the Road" increased public exposure to a musical work that could, as a result, lead to further creative pieces. Thus a successful defense of a copyright infringement action may further the

policies of the Copyright Act every bit as much as a successful prosecution of an infringement claim by the holder of a copyright.

Id. at 2584-85. Are statutory damages, injunctions and criminal penalties consistent with the Court's statement of copyright policy?

3. Is it reasonable to draw a distinction between "counterfeiting" a work — that is, copying it in its entirety for profit — and infringing the copyright through another means (say, by writing a substantially similar screenplay)? Should counterfeiting be punished more severely than other forms of infringement? In 1982, Congress passed the Piracy and Counterfeiting Amendments Act, 18 U.S.C. §§2318-19. The act makes it a felony to "knowingly traffic" in counterfeit phonorecords or audiovisual works, and it sets fines of up to $250,000 and prison terms of up to five years.

4. In Feltner v. Columbia Pictures Television, 118 S. Ct. 1279 (1998), the Supreme Court held that 17 U.S.C. §504(c), which provides that statutory damages are to be awarded in the discretion of the Court, violated the Seventh Amendment right to a jury trial. The Court concluded that the Seventh Amendment required both that liability issues be tried before a jury even if the plaintiff seeks only statutory damages and that the amount of the statutory damages itself must be determined by a jury. How this will interact with the Copyright Act's provision allowing plaintiffs to elect statutory damages at any time before judgment is unclear. It is possible that if a copyright plaintiff is dissatisfied with a jury's award of actual damages, he can then reconvene the jury and ask them to determine statutory damages.

Note on Injunctive Relief

In addition to damage remedies, injunctive relief is freely granted in copyright cases, even when the relief is sought before trial on the merits. See, e.g., Concrete Machinery Co. v. Classic Lawn Ornaments Inc., 843 F.2d 600 (1st Cir. 1988); Wainwright Securities, Inc. v. Wall Street Transcript Corp., 558 F.2d 91, 94 (2d Cir. 1977); Conrad Fabrics, Inc. v. Marcus Bros. Textile Corp., 409 F.2d 1315, 1316-17 (2d Cir. 1969). These preliminary injunctions are generally granted as a matter of course where a plaintiff can convince the court that a finding of infringement is likely. Further, courts are often willing to presume irreparable harm to the copyright owner, on the theory that it is hard to "close the door" after an infringing work has been publicly distributed.

After trial, the "entitlement" to injunctive relief is even more firmly established. Indeed, one commentator goes so far as to suggest that there is no question as to the plaintiff's entitlement to permanent injunctive relief. See Jay Dratler Jr., Intellectual Property Law §13.01[3], at 13-46 to 13-47. In some cases, however, particularly those involving derivative works, a permanent injunction against infringement may appear inequitable. For example, in Abend v. MCA, Inc., 863 F.2d 1465 (9th Cir. 1988), Abend, the licensee of the original owner of the copyright in the story *It Had to Be Murder,* sued the owners of the Alfred Hitchcock film *Rear Window,* which was based on the story. While the copyright in the story had been validly licensed at the time the movie was produced, the Ninth Circuit held that Abend owned the renewal right in the copyright, and that continued exploitation of the film *Rear Window* without permission constituted infringement.[33] However, the court denied Abend injunctive relief:

> . . . We are mindful that this case presents compelling equitable considerations which should be taken into account by the district court in fashioning an appropriate remedy. . . . Defendants invested substantial money, effort, and talent in creating the "Rear Window" film. Clearly the tremendous success of that venture initially and upon re-release is attributable in significant measure to, inter alia, the outstanding performances of its stars — Grace Kelly and James Stewart — and the brilliant directing of Alfred Hitchcock. The district court must recognize this contribution in determining Abend's remedy.
>
> The district court may choose from several available remedies for the infringement. Abend seeks first an injunction against the continued exploitation of the "Rear Window" film. 17 U.S.C. sec. 502(a) provides that the court "*may* . . . grant temporary and final injunctions on such terms as it may deem reasonable to prevent or restrain infringement of a copyright." Defendants argue . . . that a finding of infringement presumptively entitles the plaintiff to an injunction, citing Professor Nimmer. See 3 M. Nimmer, Nimmer on Copyright sec. 14.06[B] at 14-55 to 14-56.2 (1988). However, Professor Nimmer also states that "where great public injury would be worked by an injunction, the courts might . . . award damages or a continuing royalty instead of an injunction in such special circumstances."
>
> We believe such special circumstances exist here. The "Rear Window" film resulted from the collaborative efforts of many talented individuals other than Cornell Woolrich, the author of the underlying story. The success of the movie resulted in large part from factors completely unrelated to the underlying story, "It Had to be Murder." It would cause a great injustice for the owners of the film if the court enjoined them from further exhibition of the movie. An injunction would also effectively foreclose defendants from enjoying legitimate profits derived from exploitation of the "new matter" comprising the derivative work, which is given express copyright protection by section 7 of the 1909 Act. Since defendants could not possibly separate out the "new matter" from the underlying work, their right to enjoy the renewal copyright *in the derivative work* would be rendered meaningless by the grant of an injunction. We also note that an injunction would cause public injury by denying the public the opportunity to view a classic film for many years to come.
>
> This is not the first time that we have recognized that an injunction may be an inappropriate remedy for copyright infringement. In Universal City Studios v. Sony Corp., we stated that Professor Nimmer's suggestion of damages or a continuing royalty would constitute an acceptable resolution for infringement caused by in-home taping of television programs by VCR — "time-shifting."

---

33. The Supreme Court affirmed this holding. Stewart v. Abend, 495 U.S. 207 (1990).

As the district court pointed out in the *Sony* case, an injunction is a "harsh and drastic" discretionary remedy, never an absolute right. Abend argues nonetheless that defendants' attempts to interfere with his production of new derivative works can only be remedied by an injunction. We disagree. Abend has not shown irreparable injury which would justify imposing the severe remedy of an injunction on defendants. Abend can be compensated adequately for the infringement by monetary compensation. 17 U.S.C. sec. 504(b) provides that the copyright owner can recover actual damages and "any profits of the infringement that are *attributable to the infringement* and are not taken into account in computing the actual damages." (Emphasis added by court.)

The district court is capable of calculating damages caused to the fair market value of plaintiff's story by the re-release of the film. . . .

COMMENTS AND QUESTIONS

1. Why shouldn't Abend be entitled to an injunction? Given the high public value of the *Rear Window* film, isn't it reasonable to expect that he would simply license MCA to continue showing the film? Should the court really be in the business of determining the relative value of Abend's contribution to the film?

2. Protecting the moral rights of the artists offers one possible justification for injunctive relief. Would the result in *Abend* be different if the author objected to the use that was made of his work? Should it?

3. The fair use doctrine may serve as a limit on the strong form of the property rule in copyright law. This is particularly true given the Supreme Court's distinction between superseding and transformative works. The copyright owner's theoretical right to "hold up" all progress based on the copyright will not prevent transformative (and therefore fair) uses, at least in some instances. Dictum in the Supreme Court's most recent pronouncement on the issue supports this view (and cites the Ninth Circuit decision in *Abend* with approval):

Because the fair use enquiry often requires close questions of judgment as to the extent of permissible borrowing in cases involving parodies (or other critical works), courts may also wish to bear in mind that the goals of the copyright law, to stimulate the creation and publication of edifying matter, are not always best served by automatically granting injunctive relief when parodists are found to have gone beyond the bounds of fair use.

Campbell v. Acuff-Rose Music Co., 114 S. Ct. 1164, 1171 n.10 (1994).

5

Trademarks and Trade Dress

A. INTRODUCTION

1. Background

Trademarks have been in existence for almost as long as trade itself. Once human economies progressed to the point where a merchant class specialized in making goods for others, the people who made and sold clothing or pottery began to "mark" their wares with a word or symbol to identify the maker. Such marks — often no more than the name of the maker — have been discovered on goods from China, India, Persia, Egypt, Rome, Greece, and elsewhere, and date back as much as 4000 years.[1]

These early marks served several purposes. First, they were a form of advertising, allowing makers to get their name in front of potential customers. Second, they may have been used to prove that the goods were sold by a particular merchant, thus helping to resolve ownership disputes. Third, the marks served as a guarantee of quality, since a merchant who identifies herself with her goods puts her reputation on the line.

These functions coalesced in modern practice, where trademarks are widely viewed as devices that help to reduce information and transaction costs by allowing customers to estimate the nature and quality of goods before purchase.

Consumers rely most on trademarks when it is difficult to inspect a product quickly and cheaply to determine its quality. Many products fit this description: cars, computers, electronic equipment, even food and toys. In

---

1. See, e.g., William H. Browne, A Treatise on the Law of Trademarks 1-14 (1885) (tracing the history of trademarks).

precisely these cases, unscrupulous competitors may be tempted to copy the trademark of a rival producer known for superior quality. After all, it is easier to copy a trademark than to duplicate production techniques, quality assurance programs, and the like.

Even the earliest trademark law cases reflect an awareness of the need to provide a legal remedy against counterfeiting.[2] Under English common law, a party who used a trademark was entitled to prevent subsequent use of the same mark by others selling the same types of goods. See, e.g., Sykes v. Sykes, 107 Eng. Rep. 834 (1824).

In the United States, statutory trademark law appeared late on the scene, particularly given the early legislative interest in patents and copyrights.[3] This was no doubt a function of the low volume of trade in finished goods early in the history of the Republic. As Lawrence Friedman has written,

> [T]rademark law . . . was relatively undeveloped in [the early nineteenth century]. No trademark infringement case was decided in the United States before 1825. Joseph Story granted the first injunction for trademark infringement, in 1844, to protect the makers of "Taylor's Persian Thread." Congress provided neither guidance nor any machinery for registration. Legal protection for designers of trademarks had to be forged in the rough mills of the courts. The economy was still deeply rooted in land and its produce. Intellectual property, despite the name, was not valued for intellectual reasons at all, but because of mercantile and industrial applications. As such, this property was not a central concern of the law until the full-blown factory age.

Trademarks were protected at common law in the United States until 1870, when Congress enacted the first federal trademark statute. That statute, which grounded protection for trademark rights in the patent and copyright clause of the Constitution, was struck down by the Supreme Court as beyond the powers of Congress. The Court reasoned that the patent and copyright clause of the Constitution could not support the statute, since it protected all marks regardless of any novelty or originality. Trade-Mark Cases, 100 U.S. 82, 94 (1879).

Congress reenacted limited federal trademark protection in the Act of 1881, this time basing the statute in the Commerce Clause. The trademark statute was significantly modified in the Act of 1905 and further changed by subsequent amendment in 1920. Today, trademarks are protected by the Lanham Act, 15 U.S.C. §1051 et seq., which was enacted in 1946. With few exceptions, the history of these statutory changes has been one of expansion of the rights of trademark owners.

---

2. See Frank Schechter, The Historical Foundations of the Law Relating to Trademarks (1925) (reviewing medieval origins of modern trademark law).

3. One commentator has traced the earliest common law American trademark decision as far back as 1584, to the decision in *Sandforth's Case*.

2. A Brief Overview of Trademark Theory

Trademarks differ in fundamental ways from the other types of intellectual property protection we have studied so far. Patents, copyrights, and trade secrets are designed to protect and/or reward something new, inventive, or creative, whether it be an idea, a physical creation, or an expression. A trademark, by contrast, does not "depend upon novelty, invention, discovery, or any work of the brain. It requires no fancy or imagination, no genius, no laborious thought." Trade-Mark Cases, 100 U.S. at 94. Rather, trademark protection is awarded merely to those who were the first to use a distinctive mark in commerce.[4] In trademark parlance, the senior (that is, first) user of a mark may prevent junior (subsequent) users from employing the same or a similar mark, where there is a "likelihood of confusion" between the two marks.

Traditionally, there has been nothing in trademark law analogous to the desire to encourage invention or creation that underlies (at least in part) patent and copyright law. There is no explicit federal policy to encourage the creation of more trademarks.[5] Rather, the fundamental principles of trademark law have essentially been ones of tort: the tort of misappropriation of the goodwill of the trademark owner, and the tort of deception of the consumer.[6] In this sense, trademarks may not be thought of as analogous to "property rights" at all. See, e.g., Hanover Star Milling Co. v. Metcalf, 240 U.S. 403 (1916) and cases cited therein. Rather, they are rights acquired with the use of a trademark in commerce, and they continue only so long as that use continues.[7]

Early trademark cases (and statutes) took a restrictive view of these rights. Trademark infringement originally was limited to the use of a name or mark identical to the trademark in the sale of identical goods, where the infringer's use was intended to deceive consumers. These cases were essentially an extension of common law misrepresentation principles. Normally, if seller *A*

---

4. Certain marks are unprotectable because they are generic terms describing an entire class of goods. Allowing one user to appropriate these marks would give that user an unfair degree of control over the product the generic term describes. See infra.

5. Some commentators discern in federal trademark law a purpose to promote interbrand competition by strengthening brands, thereby providing information to consumers and potential competitors. See William P. Kratzke, Normative Economic Analysis of Trademark Law, 21 Memphis St. U.L. Rev. 199, 212-219 (1991). To the extent that this is true, trademarks may be thought of as comporting with federal competition policy, as embodied in the antitrust laws and elsewhere. See Continental T.V., Inc. v. GTE Sylvania, 433 U.S. 36 (1977) (policy of the antitrust laws is to promote brand strength in order to promote interbrand competition).

Further, many commentators believe that recent decisions are moving the Court in the direction of an incentive-based theory of trademark law. We discuss this trend infra.

6. These are two very different interests held by different parties, both protected by a trademark granted to one of the parties. This combination of public and private interests in trademark protection is significant, and we will return to it later in this chapter.

7. The Trademark Law Revision Act of 1988 changed this general principle in an important respect. Under that act, it is now possible to register and protect a trademark based on an *intention* to use that mark in commerce at some time within the next three years. 15 U.S.C. §1051(b) (1988). But the right remains inchoate until use actually occurs.

misrepresents to buyer *B* that *A*'s goods are made by seller *C,* then buyer *B* has a cause of action for misrepresentation against *A*. This would void any contract between *A* and *B*. Seller *C* might also have a common law cause of action against *A*, but the *consumer's* interests can be directly protected at common law only through the misrepresentation cause of action.

One reading of these old common law unfair competition cases is that they conflated *B*'s and *C*'s discrete harms into a single cause of action. They in some sense allowed *C* to stand in *B*'s shoes and sue *A* for misrepresentation. As Kenneth J. Vandevelde, The New Property of the Nineteenth Century: The Development of the Modern Concept of Property, 29 Buffalo L. Rev. 325, 341 (1980) points out, trademarks in the eighteenth century were protected only by the common law of fraud. Beginning with Millington v. Fox, 3 My. & Cr. 338, 40 Eng. Rep. 956 (Ch. 1838), where a tradesman was permanently enjoined from using another's mark, dicta began to appear suggesting that marks were a form of property. Further support for the notion that trademark law served primarily to prevent consumer fraud can be found in Mira Wilkins, The Neglected Intangible Asset: The Influence of the Trademark on the Rise of the Modern Corporation, 34 Bus. & Hist. 66, 72 (1992), where the author writes: "[I]n the American colonies laws passed to maintain the quality of manufactured articles came to form the basis of the country's subsequent trademark legislation. . . . [B]ut not until the late 1840s was the first state law passed to 'prevent fraud in the use of false stamps and labels.' . . ."

The Act of 1905 eliminated the requirements of identicality and intention to deceive, substituting instead the more fluid test of likelihood of confusion. The Lanham Act further liberalized trademark law by providing advantages to registration of trademarks and introducing a separate statutory prohibition against "unfair methods of competition" that afforded protection even to unregistered marks. 15 U.S.C. §1125(a). The result is that a broad class of "marks" now qualify for Lanham Act protection.

One way to analyze this "propertization" trend is to say that it vested rights in competitors in order to increase the likelihood that these causes of action would be brought in the first place. Individual consumers are isolated, and each generally purchases only a few of a given trademarked item. If we also consider the United States' elaborate system of interstate transportation, an individual consumer has very little incentive to police trademark infringements. The difficulty of forming a class for purposes of class action remedies is simply too great.

Competitors — *C* in the story here — have a much greater incentive to police misuse of their marks. In this view, the legitimate trademark user's lower transaction costs in policing the mark are harnessed to the original, fundamental consumer protection rationale to obtain the modern trademark infringement suit. Trademark "ownership," in this view, essentially begins as something like a legal fiction that gives the trademark user a cause of action he would not otherwise have, in order to benefit consumers and the competitive process.

Giving the originator of a mark the right to police counterfeiting also serves to protect three types of investment: (1) investment in the creation of the mark; (2) investment in advertising and promoting the product in association with the mark; and (3) product-related investments such as high-quality raw materials, production equipment, and quality assurance techniques.

3. The Basic Economics of Trademarks and Advertising

Economists have studied advertising and promotion extensively. Despite the fact that developed economies spend more than 2 percent of their GNPs on advertising, economists have not reached consensus regarding the economic function of trademarks. See Richard Schmalensee, Advertising and Market Structure, in New Developments in the Analysis of Market Structure 373 (Joseph Stiglitz and G. Frank Matathewson eds. 1991). In an earlier era, analysis centered on the notion that advertising "unnaturally" stimulated demand and perpetuated oligopoly through "artificial" product differentiation. See, e.g., Joan Robinson, The Economics of Imperfect Competition 89 (1933); Sherwin Rosen, Advertising, Information, and Product Differentiation, in Issues in Advertising: The Economics of Persuasion 161-91 (David G. Tuerck ed. 1978) (summarizing these arguments). See generally Beverly W. Pattishall, Trademarks and the Monopoly Phobia, 50 Mich. L. Rev. 967 (1952) (dismissing these arguments).

There is something to this concern. Certainly, advertising may have the effect of differentiating in the minds of consumers products that are in fact similar or identical. The result of this brand differentiation may be that the trademark owner obtains some power over price. An example is over-the-counter drugs, where brand-name drugs regularly sell for twice the price of their "generic" equivalents, even though the two drugs are chemically identical. Thus one could argue that advertising actually hurts rather than helps consumers. See Ralph S. Brown Jr., Advertising and the Public Interest: Legal Protection of Trade Symbols, 57 Yale L.J. 1165 (1948).

Starting in the 1960s and 1970s, however, economists have viewed advertising in a much more positive light. See George Stigler, The Economics of Information, 69 J. Pol. Econ. 213 (1961). The consensus view now is that advertising cheaply conveys information to consumers. See Phillip Nelson, Advertising as Information, 82 J. Pol. Econ. 729 (1974). Cf. Thomas T. Nagle, Do Advertising-Profitability Studies Really Show That Advertising Creates a Barrier to Entry?, 24 J.L. & Econ. 333 (1981) (brand loyalty demonstrates the costliness of information; advertising and brand promotion lower these costs somewhat).

Economists studying the informational role of advertising make an important distinction, one worth keeping in mind when studying trademark law. They distinguish between a product's simple "search characteristics" (price, color, shape, size, and product category) and its more complex "experience characteristics," such as taste or long-term durability. The former are aspects of product quality that consumers can verify by inspecting the product before they buy it. It is easy to see why firms would invest in advertising these qualities: they are basic aspects of a product, and consumers can easily verify whether the claims are true.

Experience characteristics, by contrast, cannot be cheaply investigated before purchase. They will be discovered only through use. The taste of fruit or wine is an obvious example, but many other goods share the same characteristic: motor oil, house paint, cough suppressant, and computer software, to name a few.

Since consumers cannot easily verify advertising about these aspects of a product, economists wondered why advertisers spend so much money on such advertising. They have come to believe that it serves a "signaling function." Jack Hirshleifer, Where Are We in the Theory of Information?, 63 Am. Econ. Rev. 31, 35-38 (1973) (interpreting work of Phillip Nelson, supra); Paul Milgrom and John Roberts, Price and Advertising Signals of Product Quality, 94 J. Pol. Econ. 796 (1986). That is, "[t]he primary information content of advertisements for experience goods is the information that the brand advertises." Phillip Nelson, supra, at 745. The idea is that advertising only pays when consumers become repeat purchasers; consumers know this about advertising; hence advertising signals a product that the producer believes will attract repeat buyers. Cf. Benjamin Klein and Keith Leffler, The Role of Market Forces in Assuring Contractual Performance, 89 J. Pol. Econ. 615, 630-631 (1981) (advertising signals high "firm-specific selling costs," and a "high premium," hence signals presence of high-quality goods). The basic idea has borne up under recent empirical investigation. See I. P. L. Png & David Reitman, Why Are Some Products Branded and Others Not?, 38 J. L. & Econ. 207 (1995) (summarizing empirical research on branded versus unbranded sales of gasoline and other products at gasoline stations, which shows branding is more common where goods are hard to inspect and where consumer search costs are higher); Lee Benham, The Effect of Advertising on the Price of Eyeglasses, 15 J.L. & Econ. 337, 344-345 (1972) (advertising correlates with lower prices). But see John A. Rizzo, Advertising and Competition in the Ethical Pharmaceutical Industry: The Case of Antihypertensive Drugs, 42 J.L. Econ. 89 (1999) (primary effect of advertising drugs is to reduce consumer price sensitivity).

Obviously, no amount of advertising makes sense unless consumers remember the name of the product. This is where trademarks come in: they are essential shorthand. When they are effective, consumers strongly associate the trademark with the producer's product. The trademark comes to embody all of the firm's informational investments.

Often, courts and commentators speak of the economic value of consumers' associations with a firm and its trademark as the firm's *goodwill*. Goodwill can be thought of as the residual benefit the firm receives from making the three types of investments described above. The direct benefit is the "good feeling" consumers have when they see, hear, or think of the firm and/or its trademark; in economic terms, it is the probability that, based on this good feeling, customers will come back in the future. For many firms — as for Wall Street — the value of goodwill is considerable indeed. By one estimate, for example, the value of Coca Cola's trademark independent of any of its tangible assets is $24 billion. See Industry Calls for Stiffer Enforcement of Anti-Counterfeiting Laws Abroad, 44 Pat., Trademark & Copyright J. (BNA) 585, 586, Oct. 1, 1992. See generally Russell L. Parr, The Value of Trademarks, in ALI-ABA Trademarks, Copyrights, and Unfair Competition for the General Practitioner, April 14, 1994 (Pub. No. C913 ALI-ABA 229), at 235 (using Marlboro brand as case study for trademark valuation technique; estimated value at $65 billion); Jerre B. Swann & Theodore H. Davis, Jr., Dilution, An Idea Whose Time Has Gone: Brand Equity as Protectable Property, The New/Old Paradigm, 1 J. Intell. Prop. L. 219, 229 (1994) (documenting key role played by famous trademarks — Nabisco, Winston, Miracle Whip, Burger King — in raising acquisition prices up to five times book value in recent corporate mergers).

In general, then, the "consumer protection" and "producer incentive" theories of trademark law often seem to be flip sides of the same coin. Yet, as we will see, the two theories may lead to different outcomes in certain cases. You can get a preliminary sense of this by considering a simple question: what if a producer has spent a great deal of money creating and advertising a mark, but the public has not yet begun to associate it with the product? Can a competitor use the mark on identical or similar goods? What incentives are created by a rule either way?

What if instead an accused infringer could demonstrate conclusively in a particular case that (i) the trademark owner invested nothing in the creation of the mark; and (ii) she has spent next to nothing promoting the mark or the product it is attached to. Her success, in other words, has been entirely through repeat purchases and word of mouth. If an incentive/reward theory is at the heart of trademark, what would we be rewarding if we enforced the trademark in this case — *choice* of an effective trademark? (Note that the trademark chosen would appear to be highly successful in one sense: it fosters repeat business with little promotional spending.)

COMMENTS AND QUESTIONS

1. Note how the two economic approaches to trademarks — the older product differentiation theory (trademarks bad) and the newer product information theory (trademarks good) — follow from differing views of the economic role of advertising and promotional expenditure in general. Do you think advertising (1) communicates valuable price/quality information; (2) artifically creates demand for nonessential product features and product "image"; or (3) some combination of the two?

2. Why is a legal remedy necessary for false representations about the origins of consumer goods? If you hire a carpenter to fix your roof and he does a bad job, you will not hire him again. What are the differences between

this scenario and products purchased less often from more diverse sources? See George A. Akerlof, The Market for Lemons: Quality Uncertainty and the Market Mechanism, 84 Q.J. Econ. 488 (1970), reprinted in An Economic Theorist's Book of Tales 7 (1984) (observing that under some conditions markets for goods such as used cars may not function effectively because buyers find it difficult to test the quality of the goods offered, giving sellers an incentive to sell poor quality items, with a resultant diminution of activity across the entire market). Akerlof concludes that one way to stop the "market for lemons" dynamic, where bad (low quality) sellers drive out good ones, is through the use of brand names.

> Brand names not only indicate quality but also give the consumer means of retaliation if the quality does not meet expectations. For the consumer will then curtail future purchases. Often too, new products are associated with old brand names. This ensures the prospective consumer of the quality of the product.

Id., in An Economic Theorist's Book of Tales, at 21. Akerlof identifies other institutions that serve the same purpose: guarantees; chain stores; and government licensing, as of doctors. Could all trademark law be eliminated by mandatory warranty terms? Would consumers necessarily trust them? Would such terms lower search costs as much as brand names? How about government certification programs — e.g., "U.S. Grade A Refrigerators"? If the idea is that consumers need certification of quality levels, why wouldn't they prefer independent, third-party certification? Under what circumstances do consumers demand just that? Do they do so, e.g., for airlines, doctors, lawyers? Why in some cases and not others? (Consider the costs of a bad product choice.)

Alternatively, private organizations might be expected to spring up to provide unbiased evaluation of products for a fee. Such organizations exist. Groups like Consumer's Union sell their evaluations of products, and their reputation depends on continued accuracy and integrity in product investigation. Other examples of private, third-party certification include Good Housekeeping Magazine and Underwriter's Laboratories, both of which give "seals of approval" to certain products that meet their standards for quality.

3. As you will see as you work your way through this chapter, trademark law has been expanding in recent years. Trade dress, antidilution protection, and other developments are all part of the trend. This expansion has moved trademark's conceptual center of gravity well beyond its traditional moorings in "consumer confusion." For cogent critiques of this movement — often called the "propertization" of trademark — see Mark A. Lemley, The Modern Lanham Act and the Death of Common Sense, 108 Yale L.J. 1687 (1999) and Glynn S. Lunney, Jr., Trademark Monopolies, 48 Emory L.J. 367 (1999). Lunney summarizes things this way:

[M]any courts and commentators succumbed to "property mania" — the belief that expanded trademark protection was necessarily desirable so long as the result could be characterized as "property." The result has been a radical and ongoing expansion of trademark protection, both in terms of what can be owned as a trademark and in terms of what trademark ownership entails. This expansion, and its associated reinterpretation of trademark's underlying policies, presents a serious threat to social welfare and has placed at risk the competitive balance that deception-based trademark law originally established. Like deception-based trademark protection, property-based trademark protection can enable a trademark owner to differentiate her product and exclude others from using the differentiating feature. It can thereby cede control over distinct product markets to individual producers and generate for a trademark owner the downward sloping demand curve of a monopolist. However, unlike deception-based trademark, property-based trademark has only a tenuous relationship to consumer deception, and therefore lacks the offsetting efficiency advantages associated with deception-based trademark's quality control and certification functions. As a result, property-based trademark appears presumptively anticompetitive — it generates market power and associated efficiency losses without the offsetting efficiency gains that are thought to justify deception-based trademark.

Id. at 372. Critics of expansive trademark law address one form of trademark expansion: the strengthening of the bundle of rights associated with a particular trademark. However, trademark owners often engage in another type of expansion: once a trademark is established for one type of product, they try to "leverage the brand" by selling other products under the same trademark. This has become especially popular in recent years as businesspeople and consultants awake to the economic power that accompanies established brands. See, e.g., Brands: The New Wealth Creators (Susannah Hart & John Murphy, eds. 1998).

B. WHAT CAN BE PROTECTED AS A TRADEMARK?

What exactly is a trademark? As noted above, the first trademarks were simply names or identifying symbols attached to goods. Names, symbols, and logos remain important trademarks, but they have been joined by a host of other sorts of trademarks. Company names now exist alongside product names. Slogans or phrases qualify for protection as trademarks. The design of a product itself or its packaging may be distinctive "trade dress" entitled to protection under the Lanham Act.

Of course, not all identifying marks are entitled to protection. If all a party needed to do to gain exclusive use of a mark or name was to use it, product comparisons would be virtually impossible because the very words that describe a product would be appropriated to a specific company. Suppose, for example, that Ford owned the exclusive right to describe its products as "cars" or "automobiles." Customers looking for automated means of ground transportation offered by other companies might encounter difficulty in knowing what to ask for. See Kenneth L. Port, Foreword: Symposium on Intellectual Property Law Theory, 68 Chicago-Kent L. Rev. 585, 596-598 (1993).

To avoid this problem, only certain trademarks are entitled to legal protection. Whether an identifying name or phrase may be trademarked at all, and the degree of protection accorded to it, both depend on the "strength" of the mark. This in turn depends on, among other things, the "classification" of the mark as either (1) arbitrary, (2) suggestive, (3) descriptive, or (4) generic. An arbitrary or fanciful mark is a word or phrase that bears no relationship whatsoever to the product it describes. "Exxon" is a good example of a fanciful mark. Arbitrary and fanciful marks are the strongest, because any value they possess in terms of name recognition obviously comes from the corporate use of the name, rather than the natural association in people's minds between a name and a product. The other three categories decrease in strength as they increase in natural association: "suggestive" marks suggest a product in people's minds; "descriptive" marks describe the product or service offered; "generic" marks are so associated with a particular product class that they have become the natural way to refer to that type of product.

1. Trademarks, Trade Names, and Service Marks

To the layperson, trademarks are often thought of as the public name of a producer or other business. In fact, however, the Lanham Act distinguishes in form between several different types of marks. "Trademarks" are the words, phrases, logos, and symbols that producers use to identify their goods:

> **Trademark.** The term "trademark" includes any word, name, symbol, or device, or any combination thereof —
> (1) used by a person, or
> (2) which a person has a bona fide intention to use in commerce and applies to register on the principal register established by this Act, to identify and distinguish his or her goods, including a unique product, from those manufactured or sold by others and to indicate the source of the goods, even if that source is unknown.

Lanham Act §45, 15 U.S.C. §1127.

The term *trademark* does not cover another closely associated type of business identifier, the service mark. Service marks serve the same purposes as trademarks, but they are used to identify services rather than goods. The Act defines service marks as follows:

Service Mark: The term "service mark" means any word, name, symbol, or device, or any combination thereof —

    (1)   used by a person, or

    (2)   which a person has a bona fide intention to use in commerce

and applies to register on the principal register established by this Act, to identify and distinguish the services of one person, including a unique service, from the services of others and to indicate the source of the services, even if that source is unknown. Titles, character names, and other distinctive features of radio and television programs may be registered as service marks notwithstanding that they, or the programs, may advertise the goods of the sponsor.

Lanham Act §45, 15 U.S.C. §1127. In general, service marks are subject to many of the same rules as trademarks, e.g., the rules on establishing priority of use. See, e.g., Martahus v. Video Duplication Services, Inc., 3 F.3d 417 (Fed. Cir. 1993) (cancellation of service mark on grounds that similar service mark had priority).

One issue that sometimes arises involves attempts to register service marks for services that are closely related to the sale of goods. In general, where the services are "expected or routine" in connection with the goods, such registrations are rejected. The theory is that, otherwise, closely related registrations will proliferate, clogging the register. See, e.g., In re Dr. Pepper Co., 836 F.2d 508 (Fed. Cir. 1987) (affirming trademark office rejection of service mark for conducting contests in connection with sale of soft drinks).

A third category of marks is trade names. Rather than goods or services, trade names identify the company itself. Unlike trademarks and service marks, trade names cannot be registered under the Lanham Act unless they actually function to identify the source of particular goods or services, rather than merely identifying a company. See Bell v. Streetwise Records, Ltd., 761 F.2d 67, 75 (1st Cir. 1985). However, trade names are generally registrable in state offices, and state and federal common law may provide protection against confusingly similar company names.

COMMENTS AND QUESTIONS

Although the Lanham Act defines both trade names and service marks, it allows registration of only the latter. Why?

2. Color, Fragrance, and Sounds

≣≣≣ *Qualitex Co. v. Jacobson Products Co., Inc.*
Supreme Court of the United States
115 S. Ct. 1300, 34 U.S.P.Q.2d 1161 (1995)

Justice BREYER delivered the opinion of the Court.

The question in this case is whether the Lanham Trademark Act . . . permits the registration of a trademark that consists, purely and simply, of a color. We conclude that, sometimes, a color will meet ordinary legal trademark requirements. And, when it does so, no special legal rule prevents color alone from serving as a trademark.

I

The case before us grows out of petitioner Qualitex Company's use (since the 1950's) of a special shade of green-gold color on the pads that it makes and sells to dry cleaning firms for use on dry cleaning presses. In 1989 respondent Jacobson Products (a Qualitex rival) began to sell its own press pads to dry cleaning firms; and it colored those pads a similar green-gold. In 1991 Qualitex registered the special green-gold color on press pads with the Patent and Trademark Office as a trademark. Registration No. 1,633,711 (Feb. 5, 1991). Qualitex subsequently added a trademark infringement count . . . in a lawsuit it had already filed challenging Jacobson's use of the green-gold color.

Qualitex won the lawsuit in the District Court. 21 U.S.P.Q.2d 1457, 1991 WL 318798 (CD Cal. 1991). But, the Court of Appeals for the Ninth Circuit set aside the judgment in Qualitex's favor on the trademark infringement claim because, in that Circuit's view, the Lanham Act does not permit Qualitex, or anyone else, to register "color alone" as a trademark. 13 F.3d 1297, 1300, 1302 (1994).

The courts of appeals have differed as to whether or not the law recognizes the use of color alone as a trademark. Compare NutraSweet Co. v. Stadt Corp., 917 F.2d 1024, 1028 (CA7 1990) (absolute prohibition against protection of color alone), with In re Owens-Corning Fiberglas Corp., 774 F.2d 1116, 1128 (CA Fed. 1985) (allowing registration of color pink for fiberglass insulation). . . . Therefore, this Court granted certiorari. . . . We now hold that there is no rule absolutely barring the use of color alone, and we reverse the judgment of the Ninth Circuit.

II

The Lanham Act gives a seller or producer the exclusive right to "register" a trademark . . . and to prevent his or her competitors from using that trademark. . . . Both the language of the Act and the basic underlying principles of trademark law would seem to include color within the universe of things that can qualify as a trademark. The language of the Lanham Act describes that universe in the broadest of terms. It says that trademarks "includ[e] any word, name, symbol, or device, or any combination thereof." §1127. Since human beings might use as a "symbol" or "device" almost anything at all that is capable of carrying meaning, this language, read literally, is not restrictive. The courts and the Patent and Trademark Office have authorized for use as a mark a particular shape (of a Coca-Cola bottle), a particular sound (of NBC's three chimes), and even a particular scent (of plumeria blossoms on sewing thread). See, e.g., Registration No. 696,147 (Apr. 12, 1960); Registration Nos. 523,616 (Apr. 4, 1950) and 916,522 (July 13, 1971); In re Clarke, 17 U.S.P.Q.2d 1238, 1240 (TTAB 1990). If a shape, a sound, and a fragrance can act as symbols why, one might ask, can a color not do the same?

. . . True, a product's color is unlike "fanciful," "arbitrary," or "suggestive" words or designs, which almost automatically tell a customer that they refer to a brand. . . . [S]ee Two Pesos, Inc. v. Taco Cabana, Inc., 112 S. Ct. 2753, 2757 (1992). The imaginary word "Suntost," or the words "Suntost Marmalade," on a jar of orange jam immediately would signal a brand or a product "source"; the jam's orange color does not do so. But, over time, customers may come to treat a particular color on a product or its packaging (say, a color that in context seems unusual, such as pink on a firm's insulating material or red on the head of a large industrial bolt) as signifying a brand. And, if so, that color would have come to identify and distinguish the goods — i.e. to "indicate" their "source" — much in the way that descriptive words on a product (say, "Trim" on nail clippers or "Car-Freshner" on deodorizer) can come to indicate a product's origin. . . . In this circumstance, trademark law says that the word (e.g., "Trim"), although not inherently distinctive, has developed "secondary meaning." See Inwood Laboratories, Inc. v. Ives Laboratories, Inc., 456 U.S. 844, 851, n.11, 102 S. Ct. 2182, 2187, n.11, 72 L. Ed. 2d 606 (1982) ("secondary meaning" is acquired when "in the minds of the public, the primary significance of a product feature . . . is to identify the source of the product rather than the product itself"). Again, one might ask, if trademark law permits a descriptive word with secondary meaning to act as a mark, why would it not permit a color, under similar circumstances, to do the same?

We cannot find in the basic objectives of trademark law any obvious theoretical objection to the use of color alone as a trademark, where that color has attained "secondary meaning" and therefore identifies and distinguishes a particular brand (and thus indicates its "source"). In principle, trademark law, by preventing others from copying a source-identifying mark, "reduce[s] the customer's costs of shopping and making purchasing decisions," 1 J. McCarthy, McCarthy on Trademarks and Unfair Competition §2.01[2], p. 2-3 (3d ed. 1994) (hereinafter McCarthy), for it quickly and easily assures a potential customer that this item — the item with this mark — is made by the same producer as other similarly marked items that he or she liked (or disliked) in the past. At the same time, the law helps assure a producer that it (and not an imitating competitor) will reap the financial, reputation-related rewards associated with a desirable product. The law thereby "encourage[s] the production of quality products," ibid., and simultaneously discourages those who hope to sell inferior products by capitalizing on a consumer's inability quickly to evaluate the quality of an item offered for sale. . . . It is the source-distinguishing ability of a mark — not its ontological status as color, shape, fragrance, word, or sign — that permits it to serve these basic purposes. See Landes & Posner, Trademark Law: An Economic Perspective, 30 J. Law & Econ. 265, 290 (1987). And, for that reason, it is difficult to find, in basic trademark objectives, a reason to disqualify absolutely the use of a color as a mark.

Neither can we find a principled objection to the use of color as a mark in the important "functionality" doctrine of trademark law. The functionality doctrine prevents trademark law, which seeks to promote competition by protecting a firm's reputation, from instead inhibiting legitimate competition by allowing a producer to control a useful product feature. It is the province of patent law, not trademark law, to encourage invention by granting inventors a monopoly over new product designs or functions for a limited time, 35 U.S.C. §§154, 173, after which competitors are free to use the innovation. If a product's functional features could be used as trademarks, however, a monopoly over such features could be obtained without regard to whether they qualify as patents and could be extended forever (because trademarks may be renewed in perpetuity). See Kellogg Co. v. National Biscuit Co., 305 U.S. 111, 119-120, 59 S. Ct. 109, 113-114 . . . (1938) (Brandeis, J.); Inwood Laboratories, Inc., supra, 456 U.S., at 863, 102 S. Ct., at 2193 (White, J., concurring in result) ("A functional characteristic is 'an important ingredient in the commercial success of the product,' and, after expiration of a patent, it is no more the property of the originator than the product itself") (citation omitted). Functionality doctrine therefore would require, to take an imaginary example, that even if customers have come to identify the special illumination-enhancing shape of a new patented light bulb with a particular manufacturer, the manufacturer may not use that shape as a trademark, for doing so, after the patent had expired, would impede competition — not by protecting the reputation of the original bulb maker, but by frustrating competitors' legitimate efforts to produce an equivalent illuminationenhancing bulb. See, e.g., Kellogg Co., supra, 305 U.S., at 119-120, 59 S. Ct., at 113-114 (trademark law cannot be used to extend monopoly over "pillow" shape of shredded wheat biscuit after the patent for that shape had expired). This Court consequently has explained that, "[i]n general terms, a product feature is functional," and cannot serve as a trademark, "if it is essential to the use or purpose of the article or if it affects the cost or quality of the article," that is, if exclusive use of the feature would put competitors at a significant non-reputation-related disadvantage. Inwood Laboratories, Inc., 456 U.S., at 850, n.10, 102 S. Ct., at 2186, n.10. Although sometimes color plays an important role (unrelated to source identification) in making a product more desirable, sometimes it does not. And, this latter fact — the fact that sometimes color is not essential to a product's use or purpose and does not affect cost or quality — indicates that the doctrine of "functionality" does not create an absolute bar to the use of color alone as a mark. See Owens-Corning, 774 F.2d, at 1123 (pink color of insulation in wall "performs no nontrademark function").

It would seem, then, that color alone, at least sometimes, can meet the basic legal requirements for use as a trademark. It can act as a symbol that distinguishes a firm's goods and identifies their source, without serving any other significant function. See U.S. Dept. of Commerce, Patent and Trademark Office, Trademark Manual of Examining Procedure s 1202.04(e), p. 1202-13 (2d ed. May, 1993) (hereinafter PTO Manual) (approving trademark registration of color alone where it "has become distinctive of the applicant's goods in commerce," provided that "there is [no] competitive need for colors to remain available in the industry" and the color is not "functional"); see also 1 McCarthy §§3.01[1], 7.26 ("requirements for qualification of a word or symbol as a trademark" are that it be (1) a "symbol," (2) "use[d] . . . as a mark," (3) "to identify and distinguish the seller's goods from goods made or sold by others," but that it not be "functional"). Indeed, the District Court, in this case, entered findings (accepted by the Ninth Circuit) that show Qualitex's green-gold press pad color has met these requirements. The green-gold color acts as a symbol. Having developed secondary meaning (for customers identified the green-gold color as Qualitex's), it identifies the press pads' source. And, the green-gold color serves no other function. (Although it is important to use some color on press pads to avoid noticeable stains, the court found "no competitive need in the press pad industry for the green-gold color, since other colors are equally usable." 21 U.S.P.Q.2d, at 1460, 1991 WL 318798.) Accordingly, unless there is some special reason that convincingly militates against the use of color alone as a trademark, trademark law would protect Qualitex's use of the green-gold color on its press pads.

III

Respondent Jacobson Products says that there are four special reasons why the law should forbid the use of color alone as a trademark. We shall explain, in turn, why we, ultimately, find them unpersuasive.

First, Jacobson says that, if the law permits the use of color as a trademark, it will produce uncertainty and unresolvable court disputes about what shades of a color a competitor may lawfully use. Because lighting (morning sun, twilight mist) will affect perceptions of protected color, competitors and courts will suffer from "shade confusion" as they try to decide whether use of a similar color on a similar product does, or does not, confuse customers and thereby infringe a trademark. Jacobson adds that the "shade confusion" problem is "more difficult" and "far different from" the "determination of the similarity of words or symbols." . . .

We do not believe, however, that color, in this respect, is special. Courts traditionally decide quite difficult questions about whether two words or phrases or symbols are sufficiently similar, in context, to confuse buyers. They have had to compare, for example, such words as "Bonamine" and "Dramamine" (motion-sickness remedies); "Huggies" and "Dougies" (diapers); "Cheracol" and "Syrocol" (cough syrup); "Cyclone" and "Tornado" (wire fences); and "Mattres" and "1-800-Mattres" (mattress franchisor telephone numbers). . . . Legal standards exist to guide courts in making such comparisons. See, e.g., 2 McCarthy §15.08; 1 McCarthy §§11.24-11.25 ("[S]trong" marks, with greater secondary meaning, receive broader protection than "weak" marks). We do not see why courts could not apply those standards to a color, replicating, if necessary, lighting conditions under which a colored product is normally sold. . . .

Second, Jacobson argues, as have others, that colors are in limited supply. See, e.g., NutraSweet Co., 917 F.2d, at 1028; Campbell Soup Co. v. Armour & Co., 175 F.2d 795, 798 (CA3 1949). Jacobson claims that, if one of many competitors can appropriate a particular color for use as a trademark, and each competitor then tries to do the same, the supply of colors will soon be depleted. Put in its strongest form, this argument would concede that "[h]undreds of color pigments are manufactured and thousands of colors can be obtained by mixing." L. Cheskin, Colors: What They Can Do For You 47 (1947). But, it would add that, in the context of a particular product, only some colors are usable. By the time one discards colors that, say, for reasons of customer appeal, are not usable, and adds the shades that competitors cannot use lest they risk infringing a similar, registered shade, then one is left with only a handful of possible colors. And, under these circumstances, to permit one, or a few, producers to use colors as trademarks will "deplete" the supply of usable colors to the point where a competitor's inability to find a suitable color will put that competitor at a significant disadvantage.

This argument is unpersuasive, however, largely because it relies on an occasional problem to justify a blanket prohibition. When a color serves as a mark, normally alternative colors will likely be available for similar use by others. See, e.g., Owens-Corning, 774 F.2d, at 1121 (pink insulation). Moreover, if that is not so — if a "color depletion" or "color scarcity" problem does arise — the trademark doctrine of "functionality" normally would seem available to prevent the anticompetitive consequences that Jacobson's argument posits, thereby minimizing that argument's practical force.

The functionality doctrine, as we have said, forbids the use of a product's feature as a trademark where doing so will put a competitor at a significant disadvantage because the feature is "essential to the use or purpose of the article" or "affects [its] cost or quality." Inwood Laboratories, Inc., 456 U.S., at 850, n.10, 102 S. Ct., at 2186, n.10. The functionality doctrine thus protects competitors against a disadvantage (unrelated to recognition or reputation) that trademark protection might otherwise impose, namely their inability reasonably to replicate important non-reputation-related product features. For example, this Court has written that competitors might be free to copy the color of a medical pill where that color serves to identify the kind of medication (e.g., a type of blood medicine) in addition to its source. See id., at 853, 858, n.20, 102 S. Ct., at 2188, 2190, n.20 ("[S]ome patients commingle medications in a container and rely on color to differentiate one from another"); see also J. Ginsburg, D. Goldberg, & A. Greenbaum, Trademark and Unfair Competition Law 194-195 (1991) (noting that drug color cases "have more to do with public health policy" regarding generic drug substitution "than with trademark law"). And, the federal courts have demonstrated that they can apply this doctrine in a careful and reasoned manner, with sensitivity to the effect on competition. Although we need not comment on the merits of specific cases, we note that lower courts have permitted competitors to copy the green color of farm machinery (because customers wanted their farm equipment to match) and have barred the use of black as a trademark on outboard boat motors (because black has the special functional attributes of decreasing the apparent size of the motor and ensuring compatibility with many different boat colors). . . . The Restatement (Third) of Unfair Competition adds that, if a design's "aesthetic value" lies in its ability to "confe[r] a significant benefit that cannot practically be duplicated by the use of alternative designs," then the design is "functional." Restatement (Third) of Unfair Competition §17, Comment c, pp. 175-176 (1995). The "ultimate test of aesthetic functionality," it explains, "is whether the recognition of trademark rights would significantly hinder competition." Id., at 176.

The upshot is that, where a color serves a significant nontrademark function — whether to distinguish a heart pill from a digestive medicine or to satisfy the "noble instinct for giving the right touch of beauty to common and necessary things," G.K. Chesterton, Simplicity and Tolstoy 61 (1912) — courts will examine whether its use as a mark would permit one competitor (or a group) to interfere with legitimate (nontrademark-related) competition through actual or potential exclusive use of an important product ingredient. That examination should not discourage firms from creating aesthetically pleasing mark designs, for it is open to their competitors to do the same. See, e.g., W. T. Rogers Co. v. Keene, 778 F.2d 334, 343 (CA7 1985) (Posner, J.). But, ordinarily, it should prevent the anticompetitive consequences of Jacobson's hypothetical "color depletion" argument, when, and if, the circumstances of a particular case threaten "color depletion."

Third, Jacobson points to many older cases — including Supreme Court cases — in support of its position. [The Court distinguishes these as arising before the Lanham Act of 1946.]

Fourth, Jacobson argues that there is no need to permit color alone to function as a trademark because a firm already may use color as part of a trademark, say, as a colored circle or colored letter or colored word, and may rely upon "trade dress" protection, under §43(a) of the Lanham Act, if a competitor copies its color and thereby causes consumer confusion regarding the overall appearance of the competing products or their packaging, see 15 U.S.C. §1125(a) (1988 ed., Supp. V). The first part of this argument begs the question. One can understand why a firm might find it difficult to place a usable symbol or word on a product (say, a large industrial bolt that customers normally see from a distance); and, in such instances, a firm might want to use color, pure and simple, instead of color as part of a design. Neither is the second portion of the argument convincing. Trademark law helps the holder of a mark in many ways that "trade dress" protection does not. See 15 U.S.C. §1124 (ability to prevent importation of confusingly similar goods); §1072 (constructive notice of ownership); §1065 (incontestible status); §1057(b) (prima facie evidence of validity and ownership). Thus, one can easily find reasons why the law might provide trademark protection in addition to trade dress protection.

IV

Having determined that a color may sometimes meet the basic legal requirements for use as a trademark and that respondent Jacobson's arguments do not justify a special legal rule preventing color alone from serving as a trademark (and, in light of the District Court's here undisputed findings that Qualitex's use of the green-gold color on its press pads meets the basic trademark requirements), we conclude that the Ninth Circuit erred in barring Qualitex's use of color as a trademark. For these reasons, the judgment of the Ninth Circuit is

Reversed.

COMMENTS AND QUESTIONS

1. The Court mentions "NBC's three chimes" as an example of a sound that is registered as a trademark under the Lanham Act. Another famous example is MGM's "lion's roar," usually heard at the beginning of an MGM film. See also Harley Wants Roar of Engine Protected by a Trademark, Sacramento Bee, Mar. 27, 1996, p. D6 (describing Harley-Davidson trademark application for engine sound opposed by competitors). Compare this with section 43(a) cases alleging imitation of famous voices, e.g., Midler v. Ford Motor Co., 849 F.2d 460 (9th Cir. 1988), *cert. denied*, 503 U.S. 951 (1992); see Chapter 6.

2. Justice Breyer's opinion for a unanimous Court in *Qualitex* is quite important not only for resolution of the color issue but also as a statement of current Court thinking on the cutting-edge issue of the functionality defense. For more on this topic, see infra.

3. Certification and Collective Marks

For the most part, the "source" identified by a trademark is a single company or individual. But for two special types of marks — certification and collective marks — this is not the case.

Lanham Act §45 (15 U.S.C. §1127) defines a certification mark as follows.

> The term "certification mark" means any word, name, symbol, or device, or any combination thereof —
> (1)   used by a person other than its owner, or
> (2)   which its owner has a bona fide intention to permit a person other than the owner to use in commerce and files an application to register on the principal register established by this chapter,
> to certify regional or other origin, material, mode of manufacture, quality, accuracy, or other characteristics of such person's goods or services or that the work or labor on the goods or services was performed by members of a union or other organization.

Certification marks are generally used by trade associations or other commercial groups to identify a particular type of goods. For example, the French winemakers hold a certification mark in "Champagne." Only sparkling wine made within the geographic region of Champagne, in France, and complying with applicable French law governing winemaking can legally call itself "Champagne." Certification marks cannot be limited to a single producer; they must be open to anyone who meets the standards set forth for certification.

Certification marks serve to certify conformity with centralized standards. See, e.g., Levy v. Kosher Overseers Association of America Inc., 36 U.S.P.Q.2d 1724 (S.D.N.Y. 1995), rev'd 104 F.3d 38 (2d Cir. 1997) (case involving plaintiff, Organized Kashruth Laboratories, suing another for infringement of its well-known kosher certification mark, the "circle K" found on many kosher foods). Because trademarks are thought by many to have grown out of trade guilds, which had much the same quality-control function, it could be argued that certification marks were the first true trademarks. See Frank Schechter, The Historical Foundations of the Law Relating to Trademarks 47 (1925).

Certification marks are meant to bear the "seal of approval" of a central organization, so they can be cancelled on the ground that the organization no longer exercises sufficient control over its members to assure consistent product standards. See 15 U.S.C. §1064 (providing that a certification mark may be cancelled if not policed effectively); see American Angus Association v. Sysco Corp., 865 F. Supp. 1180 (W.D.N.C. 1993) (denying cancellation standing to trademark infringement defendant who sought to cancel plaintiff's beef quality certification mark).

Collective marks are defined as follows in §45 of the Lanham Act (15 U.S.C. §1127):

> The term "collective mark" means a trademark or service mark —
>   (1)   used by the members of a cooperative, an association, or other collective group or organization, or
>   (2)   which such cooperative, association, or other collective group or organization has a bona fide intention to use in commerce and applies to register on the principal register established by this chapter, and includes marks indicating membership in a union, an association, or other organization.

See also Aloe Creme Laboratories, Inc. v. American Society for Aesthetic Plastic Surgery, Inc., 192 U.S.P.Q. 170, (TTAB 1976), a decision denying an opposition to registration of a collective mark:

> There are two basic types of collective marks. A collective trademark or collective service mark is a mark adopted by a "collective" (i.e., an association, union, cooperative, fraternal organization, or other organized collective group) for use only by its members, who in turn use the mark to identify their goods or services and distinguish them from those of nonmembers. The "collective" itself neither sells goods nor performs services under a collective trademark or collective service mark, but the collective may advertise or otherwise promote the goods or services sold or rendered by its members under the mark. A collective membership mark is a mark adopted for the purpose of indicating membership in an organized collective group, such as a union, an association, or other organization. Neither the collective nor its members uses the collective membership mark to identify and distinguish goods or services; rather, the sole function of such a mark is to indicate that the person displaying the mark is a member of the organized collective group. For example, if the collective group is a fraternal organization, members may display the mark by wearing pins or rings upon which the mark appears, by carrying membership cards bearing the mark, etc. Cf. Ex parte The Supreme Shrine of the Order of the White Shrine of Jerusalem, 109 U.S.P.Q. 248 (Comr., 1956); and Notes From the Patent Office (United States Trademark Association, 1965), 3, Part 2, Note 1-3 through 1-5. Of course, a collective group may itself be engaged in the marketing of its own goods or services under a particular mark, in which case the mark is not a collective mark but is rather a trademark for the collective's goods or service mark for the collective's services.
>   In the instant case, applicant, a professional association of plastic surgeons, is as noted above seeking to register the mark shown in its application as a collective membership mark. Thus the mark is not used, either by applicant or its members, to identify and distinguish any goods or services; rather it simply serves to indicate that plastic surgeons who display the mark (as, for example, by hanging upon the walls of their offices the membership certificates bearing the mark which were submitted as specimens in applicant's application) are members of applicant.

192 U.S.P.Q. 170, 173-174.

Collective marks of the first type — those attached to goods and services — are useful in franchising and related arrangements where individual stores or outlets are at least somewhat independent from the central organization holding the collective mark. For the most part, these types of collective marks are treated the same as ordinary trademarks. See, e.g., *Sebastian Intl. v. Long's Drug Stores Corp.*, 53 F.3d 1073 (9th Cir. 1995), *cert. denied,* 116 S. Ct. 302 (1995) (holding trademark "first sale" doctrine applicable to sales under a collective mark).

4. Trade Dress and Product Configurations

Words or phrases that serve to identify a product are not all that the Lanham Act protects. The act also protects "trade dress," the design and packaging of materials, and even the design and shape of a product itself, if the packaging or the product configuration serve the same source-identifying function as trademarks. It is possible to register both trade dress and product configurations as "trademarks" under the Lanham Act. (Indeed, such a registration was at issue in the *Qualitex* case discussed above.) However, because of their complex and changing nature, most trade dress and product configurations are protected without registration under section 43(a) of the Lanham Act, 15 U.S.C. §1125(a).

§1125 [Lanham Act §43]. False Designations of Origin and False Descriptions Forbidden

(a)(1) Any person who, on or in conection with any goods or services, or any container for goods, uses in commerce any word, term, name, symbol, or device, or any combination thereof, or any false designation of origin, false or misleading description of fact, or false or misleading representation of fact, which —

    (A) Is likely to cause confusion, or to cause mistake, or to deceive as to the affiliation, connection, or association of such person with another person, or as to the origin, sponsorship, or approval of his or her goods, services, or commerial activities by another person, or

    (B) in commercial advertising or promotion, misrepresents the nature, characteristics, qualities, or geographic origin of his or her or another person's goods, services, or commercial activities,

shall be liable in a civil action by any person who believes that he or she is or is likely to be damaged by such act.

Section 43(a) is commonly referred to as providing "federal common law" protection for trademarks and related source identifiers. In the next section, we discuss the requirements for protecting a trademark, either under the Lanham Act's registration procedures or under section 43(a).

C. ESTABLISHMENT AND EXTENSION OF TRADEMARK RIGHTS

1. Distinctiveness

a. *Classification of Marks and Requirements for Protection*

When a trademark is immediately capable of identifying a unique product source, rights to the mark are determined solely by priority of use. (See for example the *Zasu Designs* case at the beginning of section 2.) Marks such as these are labeled "inherently distinctive," though for analytical completeness they are further subdivided into arbitrary, fanciful, and suggestive marks. For all other trademarks — those deemed not inherently distinctive — the Lanham Act requires proof of an additional element to secure trademark rights: secondary meaning.

The *Zatarain's* case that follows explores the hierarchy of trademark classifications and illustrates why classifications matter.

The most important type of word or symbol requiring proof of secondary meaning is the *descriptive* trademark. A descriptive mark is defined as "[a] word, picture, or other symbol that directly describes something about the goods or services in connection with which it is used as a mark." J. Thomas McCarthy, McCarthy's Desk Encyclopedia of Intellectual Property 119 (2d ed. 1995). Examples include: Tender Vittles for cat food, Arthriticare for arthritis treatment, and Investacorp for financial services.

In addition to descriptive marks, several other categories of marks require secondary meaning to acquire legal protection: most notably *geographic marks* (such as Nantucket soft drinks), and *personal name marks* (such as O'Malley's beer).[8]

Secondary meaning exists when buyers associate a product with a single source. Thus when consumers recognize the Tender Vittles brand of cat food — when they expect the can so labeled to be of that brand — this descriptive term is functioning as a trademark. To be sure, Tender Vittles retains its primary meaning as a product descriptor. But proof that it has acquired a secondary meaning as a source identifier elevates it to trademark status.

It is important to understand the nature of this secondary meaning. It does not mean that buyers need to know the *identity* of the source, only that the product or service comes from a *single* source. The phrase "single source" may thus be understood to mean "single though anonymous source." See A. J. Canfield Co. v. Honickman, 808 F.2d 291 (3d Cir. 1986).

---

8. In addition, the following require proof of secondary meaning: "Titles of Single Literary Works; Descriptive Titles of Literary Series; Non-inherently Distinctive Designs and Symbols; Non-inherently Distinctive Trade Dress and Packaging; and Non-inherently Distinctive Product and Container Shapes." McCarthy §15.01[2], p. 15-26.

The next case is a leading example of the application of these and related concepts.

▤ *Zatarain's, Inc. v. Oak Grove Smokehouse, Inc.*
   United States Court of Appeals for the Fifth Circuit
   698 F.2d 786 (5th Cir. 1983)

GOLDBERG, Circuit Judge:

This appeal of a trademark dispute presents us with a menu of edible delights sure to tempt connoisseurs of fish and fowl alike. At issue is the alleged infringement of two trademarks, "Fish-Fri" and "Chick-Fri," held by appellant Zatarain's, Inc. ("Zatarain's"). The district court held that the alleged infringers had a "fair use" defense to any asserted infringement of the term "Fish-Fri" and that the registration of the term "Chick-Fri" should be cancelled. We affirm.

I.   Facts and Proceedings Below

A.   *The Tale of the Town Frier*

Zatarain's is the manufacturer and distributor of a line of over one hundred food products. Two of these products, "Fish-Fri" and "Chick-Fri," are coatings or batter mixes used to fry foods. These marks serve as the entree in the present litigation.

Zatarain's "Fish-Fri" consists of 100% corn flour and is used to fry fish and other seafood. "Fish-Fri" is packaged in rectangular cardboard boxes containing twelve or twenty-four ounces of coating mix. The legend "Wonderful FISH-FRI®" is displayed prominently on the front panel, along with the block Z used to identify all Zatarain's products. The term "Fish-Fri" has been used by Zatarain's or its predecessor since 1950 and has been registered as a trademark since 1962.

Zatarain's "Chick-Fri" is a seasoned corn flour batter mix used for frying chicken and other foods. The "Chick-Fri" package, which is very similar to that used for "Fish-Fri," is a rectangular cardboard container labelled "Wonderful CHICK-FRI." Zatarain's began to use the term "Chick-Fri" in 1968 and registered the term as a trademark in 1976.

Zatarain's products are not alone in the marketplace. At least four other companies market coatings for fried foods that are denominated "fish fry" or "chicken fry." Two of these competing companies are the appellees here, and therein hangs this fish tale.

Appellee Oak Grove Smokehouse, Inc. ("Oak Grove") began marketing a "fish fry" and a "chicken fry" in March 1979. Both products are packaged in clear glassine packets that contain a quantity of coating mix sufficient to fry enough food for one meal. The packets are labelled with Oak Grove's name and emblem, along with the words "FISH FRY" or "CHICKEN FRY." Oak Grove's "FISH FRY" has a corn flour base seasoned with various spices; Oak Grove's "CHICKEN FRY" is a seasoned coating with a wheat flour base.

B. Out of the Frying Pan, Into the Fire

Zatarain's first claimed foul play in its original complaint filed against Oak Grove on June 19, 1979, in the United States District Court for the Eastern District of Louisiana. The complaint alleged trademark infringement and unfair competition under the Lanham Act §§32(1), 43(a), 15 U.S.C. §§1114(1), 1125(a) (1976), and La. Rev. Stat. Ann. §51:1405(A) (West Supp. 1982).

The district court found that Zatarain's trademark "Fish-Fri" was a descriptive term with an established secondary meaning, but held that Oak Grove and Visko's had a "fair use" defense to their asserted infringement of the mark. The court further found that Zatarain's trademark "Chick-Fri" was a descriptive term that lacked secondary meaning, and accordingly ordered the trademark registration cancelled.

Battered, but not fried, Zatarain's appeals from the adverse judgment on several grounds. First, Zatarain's argues that its trademark "Fish-Fri" is a suggestive term and therefore not subject to the "fair use" defense. Second, Zatarain's asserts that even if the "fair use" defense is applicable in this case, appellees cannot invoke the doctrine because their use of Zatarain's trademarks is not a good faith attempt to describe their products. Third, Zatarain's urges that the district court erred in cancelling the trademark registration for the term "Chick-Fri" because Zatarain's presented sufficient evidence to establish a secondary meaning for the term. For these reasons, Zatarain's argues that the district court should be reversed. . . .

III.   The Trademark Claims

A.   Basic Principles

1.   Classifications of Marks

The threshold issue in any action for trademark infringement is whether the work or phrase is initially registerable or protectable. Vision Center v. Opticks, Inc., 596 F.2d 111, 115 (5th Cir. 1980); American Heritage Life Insurance Co. v. Heritage Life Insurance Co., 494 F.2d 3, 10 (5th Cir. 1974). Courts and commentators have traditionally divided potential trademarks into four categories. A potential trademark may be classified as (1) generic, (2) descriptive, (3) suggestive, or (4) arbitrary or fanciful. These categories, like the tones in a spectrum, tend to blur at the edges and merge together. The labels are more advisory than definitional, more like guidelines than pigeonholes. Not surprisingly, they are somewhat difficult to articulate and to apply. Soweco, Inc. v. Shell Oil Co., 617 F.2d 1178, 1183 (5th Cir. 1980); Vision Center, 596 F.2d at 115.

A generic term is "the name of a particular genus or class of which an individual article or service is but a member." Vision Center, 596 F.2d at 115; Abercrombie & Fitch Co. v. Hunting World, Inc., 537 F.2d 4, 9 (2d Cir. 1976). A generic term connotes the "basic nature of articles or services" rather than the more individualized characteristics of a particular product. American Heritage, 494 F.2d at 11. Generic terms can never attain trademark protection. William R. Warner & Co. v. Eli Lilly & Co., 265 U.S. 526, 528, 44 S. Ct. 615, 616, 68 L. Ed. 1161 (1924); Soweco, 617 F.2d at 1183; Vision Center, 596 F.2d at 115. Furthermore, if at any time a registered trademark becomes generic as to a particular product or service, the mark's registration is subject to cancellation. Lanham Act §14, 15 U.S.C. §1064(c) (1976). Such terms as aspirin and cellophane have been held generic and therefore unprotectable as trademarks. See Bayer Co. v. United Drug Co., 272 F. 505 (S.D.N.Y. 1921) (aspirin); DuPont Cellophane Co. v. Waxed Products Co., 85 F.2d 75 (2d Cir. 1936) (cellophane).

A descriptive term "identifies a characteristic or quality of an article or service," Vision Center, 596 F.2d at 115, such as its color, odor, function, dimensions, or ingredients. American Heritage, 494 F.2d at 11. Descriptive terms ordinarily are not protectable as trademarks, Lanham Act §2(e)(1), 15 U.S.C. §1052(e)(1) (1976); they may become valid marks, however, by acquiring a secondary meaning in the minds of the consuming public. See id. §2(f), 15 U.S.C. §1052(f). Examples of descriptive marks would include "Alo" with reference to products containing gel of the aloe vera plant, Aloe Creme Laboratories, Inc. v. Milsan, Inc., 423 F.2d 845 (5th Cir. 1970), and "Vision Center" in reference to a business offering optical goods and services, Vision Center, 596 F.2d at 117. As this court has often noted, the distinction between descriptive and generic terms is one of degree. Soweco, 617 F.2d at 1184; Vision Center, 596 F.2d at 115 n.11 (citing 3 R. Callman, The Law of Unfair Competition, Trademarks and Monopolies §70.4 (3d ed. 1969)); American Heritage, 494 F.2d at 11. The distinction has important practical consequences, however; while a descriptive term may be elevated to trademark status with proof of secondary meaning, a generic term may never achieve trademark protection. Vision Center, 596 F.2d at 115 n.11.

A suggestive term suggests, rather than describes, some particular characteristic of the goods or services to which it applies and requires the consumer to exercise the imagination in order to draw a conclusion as to the nature of the goods and services. Soweco, 617 F.2d at 1184; Vision Center, 596 F.2d at 115-116. A suggestive mark is protected without the necessity for proof of secondary meaning. The term "Coppertone" has been held suggestive in regard to sun tanning products. See Douglas Laboratories, Inc. v. Copper Tan, Inc., 210 F.2d 453 (2d Cir. 1954).

Arbitrary or fanciful terms bear no relationship to the products or services to which they are applied. Like suggestive terms, arbitrary and fanciful marks are protectable without proof of secondary meaning. The term "Kodak" is properly classified as a fanciful term for photographic supplies, see Eastman Kodak Co. v. Weil, 137 Misc. 506, 243 N.Y.S. 319 (1930) ("Kodak"); "Ivory" is an arbitrary term as applied to soap. Abercrombie & Fitch, 537 F.2d at 9 n.6.

2. Secondary Meaning

As noted earlier, descriptive terms are ordinarily not protectable as trademarks. They may be protected, however, if they have acquired a secondary meaning for the consuming public. The concept of secondary meaning recognizes that words with an ordinary and primary meaning of their own "may [after] long use with a particular product, come to be known by the public as specifically designating that product." Volkswagenwerk Aktiengesellschaft v. Rickard, 492 F.2d 474, 477 (5th Cir. 1974). In order to establish a secondary meaning for a term, a plaintiff "must show that the primary significance of the term in the minds of the consuming public is not the product but the producer." Kellogg Co. v. National Biscuit Co., 305 U.S. 111, 118, 59 S. Ct. 109, 113, 83 L. Ed. 73 (1938). The burden of proof to establish secondary meaning rests at all times with the plaintiff; this burden is not an easy one to satisfy, for "[a] high degree of proof is necessary to establish secondary meaning for a descriptive term." Vision Center, 596 F.2d at 118 (quoting 3 R. Callman, supra, §77.3, at 359). Proof of secondary meaning is an issue only with respect to descriptive marks; suggestive and arbitrary or fanciful marks are automatically protected upon registration, and generic terms are unprotectable even if they have acquired secondary meaning. See Soweco, 617 F.2d at 1185 n.20.

3. The "Fair Use" Defense

Even when a descriptive term has acquired a secondary meaning sufficient to warrant trademark protection, others may be entitled to use the mark without incurring liability for trademark infringement. When the allegedly infringing term is "used fairly and in good faith only to describe to users the goods or services of [a] party, or their geographic origin," Lanham Act §33(b)(4), 15 U.S.C. §1115(b)(4) (1976), a defendant in a trademark infringement action may assert the "fair use" defense. The defense is available only in actions involving descriptive terms and only when the term is used in its descriptive sense rather than its trademark sense. Soweco, 617 F.2d at 1185; see Venetianaire Corp. v. A & P Import Co., 429 F.2d 1079, 1081-1082 (2d Cir. 1970). In essence, the fair use defense prevents a trademark registrant from appropriating a descriptive term for its own use to the exclusion of others, who may be prevented thereby from accurately describing their own goods. Soweco, 617 F.2d at 1185. The holder of a protectable descriptive mark has no legal claim to an exclusive right in the primary, descriptive meaning of the term; consequently, anyone is free to use the term in its primary, descriptive sense so long as such use does not lead to customer confusion as to the source of the goods or services. See 1 J. McCarthy, Trademarks and Unfair Competition §11.17, at 379 (1973).

4. Cancellation of Trademarks

Section 37 of the Lanham Act, 15 U.S.C. §1119 (1976), provides as follows:

> In any action involving a registered mark the court may determine the right to registration, order the cancellation of registrations, in whole or in part, restore cancelled registrations, and otherwise rectify the register with respect to the registrations of any party to the action. Decrees and orders shall be certified by the court to the Commissioner, who shall make appropriate entry upon the records of the Patent Office, and shall be controlled thereby.

This circuit has held that when a court determines that a mark is either a generic term or a descriptive term lacking secondary meaning, the purposes of the Lanham Act are well served by an order cancelling the mark's registration. American Heritage, 494 F.2d at 14.

We now turn to the facts of the instant case.

B. "FISH-FRI"[3]

1. Classification

Throughout this litigation, Zatarain's has maintained that the term "Fish-Fri" is a suggestive mark automatically protected from infringing uses by virtue of its registration in 1962. Oak Grove and Visko's assert that "fish fry" is a generic term identifying a class of foodstuffs used to fry fish; alternatively, Oak Grove and Visko's argue that "fish fry" is merely descriptive of the characteristics of the product. The district court found that "Fish-Fri" was a descriptive term identifying a function of the product being sold. Having reviewed this finding under the appropriate "clearly erroneous" standard, we affirm. See Vision Center, 596 F.2d at 113.

---

3. We note at the outset that Zatarain's use of the phonetic equivalent of the words "fish fry" — that is, misspelling it — does not render the mark protectable. Soweco, 617 F.2d at 1186 n.24.

We are mindful that "[t]he concept of descriptiveness must be construed rather broadly." 3 R. Callman, supra, §70.2. Whenever a word or phrase conveys an immediate idea of the qualities, characteristics, effect, purpose, or ingredients of a product or service, it is classified as descriptive and cannot be claimed as an exclusive trademark. Id. §71.1; see Stix Products, Inc. v. United Merchants & Manufacturers, Inc., 295 F.Supp. 479, 488 (S.D.N.Y. 1968). Courts and commentators have formulated a number of tests to be used in classifying a mark as descriptive.

A suitable starting place is the dictionary, for "[t]he dictionary definition of the word is an appropriate and relevant indication 'of the ordinary significance and meaning of words' to the public." American Heritage, 494 F.2d at 11 n.5; see also Vision Center, 596 F.2d at 116. Webster's Third New International Dictionary 858 (1966) lists the following definitions for the term "fish fry": "1. a picnic at which fish are caught, fried, and eaten; . . . 2. fried fish." Thus, the basic dictionary definitions of the term refer to the preparation and consumption of fried fish. This is at least preliminary evidence that the term "Fish-Fri" is descriptive of Zatarain's product in the sense that the words naturally direct attention to the purpose or function of the product.

The "imagination test" is a second standard used by the courts to identify descriptive terms. This test seeks to measure the relationship between the actual words of the mark and the product to which they are applied. If a term "requires imagination, thought and perception to reach a conclusion as to the nature of goods," Stix Products, 295 F. Supp. at 488, it is considered a suggestive term. Alternatively, a term is descriptive if standing alone it conveys information as to the characteristics of the product. In this case, mere observation compels the conclusion that a product branded "Fish-Fri" is a prepackaged coating or batter mix applied to fish prior to cooking. The connection between this merchandise and its identifying terminology is so close and direct that even a consumer unfamiliar with the product would doubtless have an idea of its purpose or function. It simply does not require an exercise of the imagination to deduce that "Fish-Fri" is used to fry fish. See Vision Center, 596 F.2d at 116-17; Stix Products, 295 F. Supp. at 487-88. Accordingly, the term "Fish-Fri" must be considered descriptive when examined under the "imagination test."

A third test used by courts and commentators to classify descriptive marks is "whether competitors would be likely to need the terms used in the trademark in describing their products." Union Carbide Corp. v. Ever-Ready, Inc., 531 F.2d 366, 379 (7th Cir. 1976). A descriptive term generally relates so closely and directly to a product or service that other merchants marketing similar goods would find the term useful in identifying their own goods. Vision Center, 596 F.2d at 116-17; Stix Products, 295 F. Supp. at 488. Common sense indicates that in this case merchants other than Zatarain's might find the term "fish fry" useful in describing their own particular batter mixes. While Zatarain's has argued strenuously that Visko's and Oak Grove could have chosen from dozens of other possible terms in naming their coating mix, we find this position to be without merit. As this court has held, the fact that a term is not the only or even the most common name for a product is not determinative, for there is no legal foundation that a product can be described in only one fashion. Vision Center, 596 F.2d at 117 n.17. There are many edible fish in the sea, and as many ways to prepare them as there are varieties to be prepared. Even piscatorial gastronomes would agree, however, that frying is a form of preparation accepted virtually around the world, at restaurants starred and unstarred. The paucity of synonyms for the words "fish" and "fry" suggests that a merchant whose batter mix is specially spiced for frying fish is likely to find "fish fry" a useful term for describing his product.

A final barometer of the descriptiveness of a particular term examines the extent to which a term actually has been used by others marketing a similar service or product. Vision Center, 596 F.2d at 117; Shoe Corp. of America v. Juvenile Shoe Corp., 266 F.2d 793, 796 (C.C.P.A. 1959). This final test is closely related to the question whether competitors are likely to find a mark useful in describing their products. As noted above, a number of companies other than Zatarain's have chosen the word combination "fish fry" to identify their batter mixes. Arnaud's product, "Oyster Shrimp and Fish Fry," has been in competition with Zatarain's "Fish-Fri" for some ten to twenty years. When companies from A to Z, from Arnaud to Zatarain's, select the same term to describe their similar products, the term in question is most likely a descriptive one.

The correct categorization of a given term is a factual issue, Soweco, 617 F.2d at 1183 n.12; consequently, we review the district court's findings under the "clearly erroneous" standard of Fed. R. Civ. P. 52. See Vision Center, 596 F.2d at 113; Volkswagenwerk, 492 F.2d at 478. The district court in this case found that Zatarain's trademark "Fish-Fri" was descriptive of the function of the product being sold. Having applied the four prevailing tests of descriptiveness to the term "Fish-Fri," we are convinced that the district court's judgment in this matter is not only not clearly erroneous, but clearly correct. . . .

2. Secondary Meaning

Descriptive terms are not protectable by trademark absent a showing of secondary meaning in the minds of the consuming public.[5] To prevail in its trademark infringement action, therefore, Zatarain's must prove that its mark "Fish-Fri" has acquired a secondary meaning and thus warrants trademark

___

5. A mark that has become "incontestable" under section 15 of the Lanham Act, 15 U.S.C. §1065 (1976), cannot be challenged as lacking secondary meaning, although it is subject to seven statutory defenses. See id. §33(b), 15 U.S.C. §1115(b). In order for a registrant's mark to be deemed "incontestable," the registrant must use the mark for five continuous years following the registration date and must file certain affidavits with the Commissioner of Patents. Id. §15, 15 U.S.C. §1065. No evidence in the record indicates that Zatarain's has satisfied the requirements of "incontestability"; consequently, we must determine whether proof of secondary meaning otherwise exists.

protection. The district court found that Zatarain's evidence established a secondary meaning for the term "Fish-Fri" in the New Orleans area. We affirm.

The existence of secondary meaning presents a question for the trier of fact, and a district court's finding on the issue will not be disturbed unless clearly erroneous. American Heritage, 494 F.2d at 13; Volkswagenwerk, 492 F.2d at 477. The burden of proof rests with the party seeking to establish legal protection for the mark — the plaintiff in an infringement suit. Vision Center, 596 F.2d at 118. The evidentiary burden necessary to establish secondary meaning for a descriptive term is substantial. Id.; American Heritage, 494 F.2d at 12; 3 R. Callman, supra, §77.3, at 359.

In assessing a claim of secondary meaning, the major inquiry is the consumer's attitude toward the mark. The mark must denote to the consumer "a single thing coming from a single source," Coca-Cola Co. v. Koke Co., 254 U.S. 143, 146, 41 S. Ct. 113, 114, 65 L. Ed. 189 (1920); Aloe Creme Laboratories, 423 F.2d at 849, to support a finding of secondary meaning. Both direct and circumstantial evidence may be relevant and persuasive on the issue.

Factors such as amount and manner of advertising, volume of sales, and length and manner of use may serve as circumstantial evidence relevant to the issue of secondary meaning. See, e.g., Vision Center, 596 F.2d at 119; Union Carbide Corp., 531 F.2d at 380; Aloe Creme Laboratories, 423 F.2d at 849-50. While none of these factors alone will prove secondary meaning, in combination they may establish the necessary link in the minds of consumers between a product and its source. It must be remembered, however, that "the question is not the extent of the promotional efforts, but their effectiveness in altering the meaning of [the term] to the consuming public." Aloe Creme Laboratories, 423 F.2d at 850.

Since 1950, Zatarain's and its predecessor have continuously used the term "Fish-Fri" to identify this particular batter mix. Through the expenditure of over $400,000 for advertising during the period from 1976 through 1981, Zatarain's has promoted its name and its product to the buying public. Sales of twelve-ounce boxes of "Fish-Fri" increased from 37,265 cases in 1969 to 59,439 cases in 1979. From 1964 through 1979, Zatarian's sold a total of 916,385 cases of "Fish-Fri." The district court considered this circumstantial evidence of secondary meaning to weigh heavily in Zatarain's favor.

In addition to these circumstantial factors, Zatarain's introduced at trial two surveys conducted by its expert witness, Allen Rosenzweig. In one survey, telephone interviewers questioned 100 women in the New Orleans area who fry fish or other seafood three or more times per month. Of the women surveyed, twenty-three percent specified Zatarain's "Fish-Fri" as a product they "would buy at the grocery to use as a coating" or a "product on the market that is especially made for frying fish." In a similar survey conducted in person at a New Orleans area mall, twenty-eight of the 100 respondents answered "Zatarain's 'Fish-Fri' " to the same questions. . . .

The authorities are in agreement that survey evidence is the most direct and persuasive way of establishing secondary meaning. Vision Center, 596 F.2d at 119; Aloe Creme Laboratories, 423 F.2d at 849; 1 J. McCarthy, supra, §15.12(D). The district court believed that the survey evidence produced by Zatarain's, when coupled with the circumstantial evidence of advertising and usage, tipped the scales in favor of a finding of secondary meaning. Were we considering the question of secondary meaning de novo, we might reach a different conclusion than did the district court, for the issue is close.

Mindful, however, that there is evidence in the record to support the finding below, we cannot say that the district court's conclusion was clearly erroneous. Accordingly, the finding of secondary meaning in the New Orleans area for Zatarain's descriptive term "Fish-Fri" must be affirmed.

3. The "Fair Use" Defense

Although Zatarain's term "Fish-Fri" has acquired a secondary meaning in the New Orleans geographical area, Zatarain's does not now prevail automatically on its trademark infringement claim, for it cannot prevent the fair use of the term by Oak Grove and Visko's. The "fair use" defense applies only to descriptive terms and requires that the term be "used fairly and in good faith only to describe to users the goods or services of such party, or their geographic origin." Lanham Act §33(b), 15 U.S.C. §1115(b)(4) (1976). The district court determined that Oak Grove and Visko's were entitled to fair use of the term "fish fry" to describe a characteristic of their goods; we affirm that conclusion.

Zatarain's term "Fish-Fri" is a descriptive term that has acquired a secondary meaning in the New Orleans area. Although the trademark is valid by virtue of having acquired a secondary meaning, only that penumbra or fringe of secondary meaning is given legal protection. Zatarain's has no legal claim to an exclusive right in the original, descriptive sense of the term; therefore, Oak Grove and Visko's are still free to use the words "fish fry" in their ordinary, descriptive sense, so long as such use will not tend to confuse customers as to the source of the goods. See 1 J. McCarthy, supra, §11.17.

The record contains ample evidence to support the district court's determination that Oak Grove's and Visko's use of the words "fish fry" was fair and in good faith. Testimony at trial indicated that the appellees did not intend to use the term in a trademark sense and had never attempted to register the words as a trademark. Oak Grove and Visko's apparently believed "fish fry" was a generic name for the type of coating mix they manufactured. In addition, Oak Grove and Visko's consciously packaged and labelled their products in such a way as to minimize any potential confusion in the minds of consumers. The dissimilar trade dress of these products prompted the district court to observe that confusion at the point of purchase — the grocery shelves — would be virtually impossible. Our review of the record convinces us that the district court's determinations are correct. We hold, therefore, that Oak Grove and Visko's are entitled to fair use of the term "fish fry" to describe their products; accordingly, Zatarain's claim of trademark infringement must fail.

C. "CHICK-FRI"

1. Classification

Most of what has been said about "Fish-Fri" applies with equal force to Zatarain's other culinary concoction, "Chick-Fri." "Chick-Fri" is at least as descriptive of the act of frying chicken as "Fish-Fri" is descriptive of frying fish. It takes no effort of the imagination to associate the term "Chick-Fri" with Southern fried chicken. Other merchants are likely to want to use the words "chicken fry" to describe similar products, and others have in fact done so. Sufficient evidence exists to support the district court's finding that "Chick-Fri" is a descriptive term; accordingly, we affirm.

2. Secondary Meaning

The district court concluded that Zatarain's had failed to establish a secondary meaning for the term "Chick-Fri." We affirm this finding. The mark "Chick-Fri" has been in use only since 1968; it was registered even more recently, in 1976. In sharp contrast to its promotions with regard to "Fish-Fri," Zatarain's advertising expenditures for "Chick-Fri" were mere chickenfeed; in fact, Zatarain's conducted no direct advertising campaign to publicize the product. Thus the circumstantial evidence presented in support of a secondary meaning for the term "Chick-Fri" was paltry.

Allen Rosenzweig's survey evidence regarding a secondary meaning for "Chick-Fri" also "lays an egg." The initial survey question was a "qualifier": "Approximately how many times in an average month do you, yourself, fry fish or other seafood?" Only if respondents replied "three or more times a month" were they asked to continue the survey. This qualifier, which may have been perfectly adequate for purposes of the "Fish-Fri" questions, seems highly unlikely to provide an adequate sample of potential consumers of "Chick-Fri." This survey provides us with nothing more than some data regarding fish friers' perceptions about products used for frying chicken. As such, it is entitled to little evidentiary weight.[10]

---

10. Even were we to accept the results of the survey as relevant, the result would not change. In the New Orleans area, only 11 of the 100 respondents in the telephone survey named "Chick-Fri," "chicken fry," or Zatarain's "Chick-Fri" as a product used as a coating for frying chicken. Rosenzweig himself testified that this number was inconclusive for sampling purposes. Thus the survey evidence cannot be said to establish a secondary meaning for the term "Chick-Fri."

It is well settled that Zatarain's, the original plaintiff in this trademark infringement action, has the burden of proof to establish secondary meaning for its term. Vision Center, 596 F.2d at 118; American Heritage, 494 F.2d at 12. This it has failed to do. The district court's finding that the term "Chick-Fri" lacks secondary meaning is affirmed.

COMMENTS AND QUESTIONS

1. Do not be misled by *Zatarain's* reference to the "fair use" of a trademark. The trademark "fair use" defense is a totally different and much more limited defense than its counterpart in copyright law. Fair use in the trademark context establishes that a junior user will not be liable for using a mark in its descriptive sense, as opposed to trading on the senior user's established trademark meaning. While many argue that "fair use" applies only to descriptive marks that have acquired secondary meaning, the leading trademark commentator has argued for a broader role for the doctrine. See 1 J. Mc-Carthy, Trademarks and Unfair Competition §11.17[1], at 11-80 (3d ed. 1992 & Supp. 1996). Professor McCarthy's position was endorsed by the Second Circuit in Car-Freshner Corp. v. S. C. Johnson & Son, 70 F.3d 267 (2d Cir. 1995).

An example of the application of trademark's fair use doctrine can be found in Cosmetically Sealed Industries, Inc. v. Chesebrough-Pond's USA, 125 F.3d 28 (2d Cir. 1997). There the plaintiff registered the trademark "Sealed With a Kiss" for a brand of long-lasting lipstick. Plaintiff sued after the defendant began an advertising campaign for its own brand of lipstick that encouraged users to place a lipstick "kiss" on a postcard and mail it to someone. The defendant's campaign used the phrase "Seal it with a Kiss!!" The court held that the phrase "sealed with a kiss" was in common use and that the defendant was not liable because it merely used that common phrase in its descriptive (rather than its trademark) sense.

2. In the course of its opinion, the court states: "Oak Grove and Visko's apparently believed 'fish fry' was a generic name for the type of coating mix they manufactured." A generic term is a term that indicates what *type* or *category* a product belongs to, rather than indicating the producer or other source of the product. A generic term for a product "tell[s] the buyer what it is, not where it came from." 2 J. McCarthy, Trademarks and Unfair Competition §12.01[1] (3d ed. 1992 & Supp. 1996). Examples would include "bed," "chair," and "pants." (Contrast these with "Eas-a-bed," "ComfortMax Chairs," and "Aunt's Pants" — all at least capable of indicating the source of a product.) A generic term has no distinctiveness vis-à-vis source, so it cannot serve as a trademark. For more on generic trademarks, including the process by which formerly distinctive terms become generic ("genericide"), see section E.2 below.

3. Note that in the case, the survey sample seems to be drawn from a very small area — the New Orleans metropolitan region. Should secondary

meaning in this area permit a descriptive mark to be enforced anywhere, or only in this area? Is this in effect a geographic limitation on the scope of a trademark that is not apparent from the statute — which provides ostensibly national protection for registered marks? How does this geographic limitation interact with the doctrines governing the conflict between junior common law users of a mark and a senior, federally registered mark owner?

A recent case discusses geographical issues related to secondary meaning. Adray v. Adry-Mart, Inc., 76 F.3d 984 (9th Cir. 1996), *amending* 68 F.3d 362 (9th Cir. 1995), centered around rival operators of electronics stores in Southern California (Lou Adray and Adry-Mart), both of whom used the name "Adray's." In 1979 Adry-Mart acquired from Lou Adray's relatives the right to use the Adray's trademark in stores in Los Angeles County. Lou Adray continued to operate his Orange County store, as he had since 1968. In 1989, however, Adry-Mart began to expand to areas first near, and then inside, Orange County.

In this appeal from Lou Adray's trademark infringement suit, the Ninth Circuit held that the district court was correct in instructing the jury that for Lou to receive damages he must show secondary meaning in the market areas in which Adry-Mart operated its Orange County stores. The court observed that there was no possibility that Lou could show national secondary meaning. Also, the court noted, the alleged infringer had not adopted its mark in bad faith in this case. The court went on to hold, however, that it was clear error for the district court to find that Adry-Mart's market included all of Los Angeles County; the evidence established that in some parts of the county Lou Adray had a bigger market share than Adry-Mart. Given the overlap of sales and advertising in these areas, it is likely that neither Lou nor Adry-Mart had secondary meaning there. On remand, the court ordered, the district court should reexamine the issue of where each party had established secondary meaning.

In addition, the Trademark Office has generally required applicants for federal registration (which confers presumptively nationwide protection) to show more than secondary meaning in a limited area. See Phillip Morris, Inc. v. Liggett & Myers Tobacco Co., 139 U.S.P.Q. 240 (TTAB 1963).

Compare *Adray* to another fairly recent case, Fuddruckers, Inc. v. Doc's B.R. Others, Inc., 826 F.2d 837 (9th Cir. 1987). Here defendant Doc's negotiated with Fuddruckers for a Fuddruckers franchise, but the negotiations fell through and Fuddruckers found another franchisee. In mid-June of 1983, Fuddruckers announced plans to open a restaurant in Phoenix in December of that year. Defendants knew of Fuddruckers's plans but proceeded to open a restaurant in Phoenix with trade dress quite similar to that of Fuddruckers. As Fuddruckers's trade dress (restaurant decor) was not inherently distinctive, Fuddruckers was required to show that its trade dress had acquired secondary meaning in its action under §43(a) of the Lanham Act. The court ruled, however, that Fuddruckers did not have to show secondary meaning *in Arizona* prior to Doc's opening its restaurant there: "Fuddruckers is a national restaurant chain, and restaurant customers travel. Fuddruckers should be permitted to show that its trade dress had acquired secondary meaning among some substantial portion of consumers nationally." 826 F.2d at 844.

4. What is the proper role of "circumstantial evidence" of secondary meaning, such as advertising expenditures, the commercial success of the product, and attempts at imitation? Such evidence is generally allowed by courts in cases where secondary meaning is at issue. But should it be? If it is clear, for example, that advertising expenditures have been completely ineffective in swaying the public, is the fact of such expenditures relevant? The answer may depend on your views as to why we are protecting trademarks. If our goal is to provide incentives for businesses to invest in marks (and therefore in quality control), we may want to encourage such expenditures directly. But cf. Robert Merges, Economic Perspectives on Innovation: Commercial Success and Patent Standards, 76 Cal. L. Rev. 803 (1988) (criticizing the analogous role that "secondary considerations of nonobviousness" play in patent law).

Note that some courts (particularly the Second Circuit) have explicitly done away with the need to prove secondary meaning where plaintiffs can point to some other equitable factor in their favor, such as intentional deception or "palming off" by the defendants. This view is known as the New York rule, although there is some question as to its continued vitality in New York state courts. For a detailed discussion of the New York rule, see 2 J. Thomas McCarthy, Trademarks and Unfair Competition §15.04. What does the New York rule suggest about the rationale for trademark protection? Is it incentive-based, property-based, or tort-based?

5. *Foreign Descriptive Terms.* In some circumstances, it may not even be clear what a descriptive term is. Consider the problem of terms that are descriptive in a language other than English. The doctrine of foreign equivalents holds that foreign words must be translated into English for purposes of determining their protectability. But application of this doctrine has been uneven. Is "La Posada" (Spanish for "inn") descriptive of lodging services? Does it matter whether a substantial portion of the clientele speaks Spanish? See In re Pan Tex Hotel Corp., 190 U.S.P.Q. 109 (TTAB 1976) (La Posada is not descriptive because it is unlikely that consumers will translate the name into English). Compare In re Hag Aktiengesellschaft, 155 U.S.P.Q. 598 (TTAB 1967) ("Kaba," meaning coffee, is descriptive of coffee). Similarity of meaning in translation is important, but not determinative, in deciding the issue of descriptiveness. The relation in sight and sound between the English and foreign terms is also important; "a much closer approximation [between the meaning of the foreign term and the English equivalent] is necessary to justify a refusal to register on that basis alone where the marks otherwise are totally dissimilar." In re Sarkli, Ltd., 220 U.S.P.Q. 111, 113 (Fed. Cir. 1983).

6. *Acronyms.* One possible way around a finding that a term is descriptive (or generic) is to alter the term, either by misspelling it or by using an acronym. This, and not an intrinsic aversion to proper spelling, explains the

profusion of product names with words like "Evr," "EZ" and "Klear." Acronyms are occasionally, but not normally, effective in creating a distinctive mark. The relevant question for the courts is whether the misspelling or acronym has the same connotation as the original descriptive or generic mark. If it does, the acronym is not entitled to protection. Thus, ROM is generic because it conveys the same meaning to the listener as "read-only memory." Intel Corp. v. Radiation, Inc., 184 U.S.P.Q. 54 (TTAB 1984). By contrast, "L.A." was not merely descriptive of "low alcohol" beer. Anheuser-Busch, Inc. v. Stroh Brewery Co., 750 F.2d 631 (8th Cir. 1984) ("if some operation of the imagination is required to connect the initials with the product, the initials cannot be equated with the generic phrase, but are suggestive in nature, thereby rendering them protectable."). See generally 1 J. Thomas McCarthy, McCarthy on Trademarks and Unfair Competition 12-72 to 12-75 (1993).

7. The test of "secondary meaning" is very fact-specific, and relies on the reactions of consumers to a mark, generally as tested through consumer surveys. We will again see the use of consumer reaction as a test, when we consider trademark infringement. Does it make sense to rely on such empirical evidence to establish a party's intellectual property rights, particularly given the problems with most consumer surveys? Is there a better alternative?

b. Distinctiveness of Trade Dress and Product Configuration

Two Pesos, Inc. v. Taco Cabana, Inc.
Supreme Court of the United States
505 U.S. 763 (1992)

Justice WHITE delivered the opinion of the Court.

The issue in this case is whether the trade dress[1] of a restaurant may be protected under §43(a) of the Trademark Act of 1946 (Lanham Act), 60 Stat. 441, 15 U.S.C. §1125(a) (1982 ed.), based on a finding of inherent distinctiveness, without proof that the trade dress has secondary meaning.

---

1. The District Court instructed the jury: " 'Trade dress' is the total image of the business. Taco Cabana's trade dress may include the shape and general appearance of the exterior of the restaurant, the identifying sign, the interior kitchen floor plan, the decor, the menu, the equipment used to serve food, the servers' uniforms and other features reflecting on the total image of the restaurant." The Court of Appeals accepted this definition and quoted from Blue Bell Bio-Medical v. Cin-Bad, Inc., 864 F.2d 1253, 1256 (CA5 1989): "The 'trade dress' of a product is essentially its total image and overall appearance." See 932 F.2d 1113, 1118 (CA5 1991). It "involves the total image of a product and may include features such as size, shape, color or color combinations, texture, graphics, or even particular sales techniques." John H. Harland Co. v. Clarke Checks, Inc., 711 F.2d 966, 980 (CA11 1983). Restatement (Third) of Unfair Competition §16, Comment a (Tent. Draft No. 2, Mar. 23, 1990).

I

Respondent Taco Cabana, Inc., operates a chain of fast-food restaurants in Texas. The restaurants serve Mexican food. The first Taco Cabana restaurant was opened in San Antonio in September 1978, and five more restaurants had been opened in San Antonio by 1985. Taco Cabana describes its Mexican trade dress as "a festive eating atmosphere having interior dining and patio areas decorated with artifacts, bright colors, paintings and murals. The patio includes interior and exterior areas with the interior patio capable of being sealed off from the outside patio by overhead garage doors. The stepped exterior of the building is a festive and vivid color scheme using top border paint and neon stripes. Bright awnings and umbrellas continue the theme." 932 F.2d 1113, 1117 (CA5 1991).

In December 1985, a Two Pesos, Inc., restaurant was opened in Houston. Two Pesos adopted a motif very similar to the foregoing description of Taco Cabana's trade dress. Two Pesos restaurants expanded rapidly in Houston and other markets, but did not enter San Antonio. In 1986, Taco Cabana entered the Houston and Austin markets and expanded into other Texas cities, including Dallas and El Paso where Two Pesos was also doing business.

In 1987, Taco Cabana sued Two Pesos in the United States District Court for the Southern District of Texas for trade dress infringement under §43(a) of the Lanham Act, 15 U.S.C. §1125(a) (1982 ed.), and for theft of trade secrets under Texas common law. The case was tried to a jury, which was instructed to return its verdict in the form of answers to five questions propounded by the trial judge. The jury's answers were: Taco Cabana has a trade dress; taken as a whole, the trade dress is nonfunctional; the trade dress is inherently distinctive; the trade dress has not acquired a secondary meaning in the Texas market; and the alleged infringement creates a likelihood of confusion on the part of ordinary customers as to the source or association of the restaurant's goods or services. Because, as the jury was told, Taco Cabana's trade dress was protected if it either was inherently distinctive or had acquired a secondary meaning, judgment was entered awarding damages to Taco Cabana. In the course of calculating damages, the trial court held that Two Pesos had intentionally and deliberately infringed Taco Cabana's trade dress.

The Court of Appeals ruled that the instructions adequately stated the applicable law and that the evidence supported the jury's findings. In particular, the Court of Appeals rejected petitioner's argument that a finding of no secondary meaning contradicted a finding of inherent distinctiveness.

II

The Lanham Act was intended to make "actionable the deceptive and misleading use of marks" and "to protect persons engaged in . . . commerce against unfair competition." §45, 15 U.S.C. §1127. Section 43(a) "prohibits a broader range of practices than does §32," which applies to registered marks, Inwood Laboratories, Inc. v. Ives Laboratories, Inc., 456 U.S. 844, 858 (1982), but it is common ground that §43(a) protects qualifying unregistered trademarks and that the general principles qualifying a mark for registration under §2 of the Lanham Act are for the most part applicable in determining whether an unregistered mark is entitled to protection under §43(a). See A. J. Canfield Co., v. Honickman, 808 F.2d 291, 299, n.9 (CA3 1986); Thompson Medical Co. v. Pfizer Inc., 753 F.2d 208, 215-216 (CA2 1985).

The Court of Appeals determined that the District Court's instructions were consistent with the foregoing principles and that the evidence supported the jury's verdict. Both courts thus ruled that Taco Cabana's trade dress was not descriptive but rather inherently distinctive, and that it was not functional. None of these rulings is before us in this case, and for present purposes we assume, without deciding, that each of them is correct. In going on to affirm the judgment for respondent, the Court of Appeals, following its prior decision in *Chevron*, held that Taco Cabana's inherently distinctive trade dress was entitled to protection despite the lack of proof of secondary meaning. It is this issue that is before us for decision, and we agree with its resolution by the Court of Appeals. There is no persuasive reason to apply to trade dress a general requirement of secondary meaning which is at odds with the principles generally applicable to infringement suits under §43(a). Petitioner devotes much of its briefing to arguing issues that are not before us, and we address only its arguments relevant to whether proof of secondary meaning is essential to qualify an inherently distinctive trade dress for protection under §43(a). Petitioner argues that the jury's finding that the trade dress has not acquired a secondary meaning shows conclusively that the trade dress is not inherently distinctive. The Court of Appeals' disposition of this issue was sound:

> Two Pesos' argument — that the jury finding of inherent distinctiveness contradicts its finding of no secondary meaning in the Texas market — ignores the law in this circuit. While the necessarily imperfect (and often prohibitively difficult) methods for assessing secondary meaning address the empirical question of current consumer association, the legal recognition of an inherently distinctive trademark or trade dress acknowledges the owner's legitimate proprietary interest in its unique and valuable informational device, regardless of whether substantial consumer association yet bestows the additional empirical protection of secondary meaning.

932 F.2d, at 1120, n.7.

This brings us to the line of decisions by the Court of Appeals for the Second Circuit that would find protection for trade dress unavailable absent proof of secondary meaning, a position that petitioner concedes would have to be modified if the temporary protection that it suggests is to be recognized. In Vibrant Sales, Inc. v. New Body Boutique, Inc., 652 F.2d 299 (1981), the plaintiff claimed protection under §43(a) for a product whose features the defendant had allegedly copied. The Court of Appeals held that unregistered marks did not enjoy the "presumptive source association" enjoyed by registered marks and hence could not qualify for protection under §43(a) without proof of secondary meaning. Id. at 303, 304. The court's rationale seemingly denied protection for unregistered but inherently distinctive marks of all kinds, whether the claimed mark used distinctive words or symbols or distinctive product design. The court thus did not accept the arguments that an unregistered mark was capable of identifying a source and that copying such a mark could be making any kind of a false statement or representation under §43(a).

This holding is in considerable tension with the provisions of the Act. If a verbal or symbolic mark or the features of a product design may be registered under §2, it necessarily is a mark "by which the goods of the applicant may be distinguished from the goods of others," 60 Stat. 428, and must be registered unless otherwise disqualified. Since §2 requires secondary meaning only as a condition to registering descriptive marks, there are plainly marks that are registrable without showing secondary meaning. These same marks, even if not registered, remain inherently capable of distinguishing the goods of the users of these marks. Furthermore, the copier of such a mark may be seen as falsely claiming that his products may for some reason be thought of as originating from the plaintiff.

Some years after *Vibrant*, the Second Circuit announced in Thompson Medical Co. v. Pfizer Inc., 753 F.2d 208 (CA2 1985), that in deciding whether an unregistered mark is eligible for protection under §43(a), it would follow the classification of marks set out by Judge Friendly in Abercrombie & Fitch, 537 F.2d, at 9. Hence, if an unregistered mark is deemed merely descriptive, which the verbal mark before the court proved to be, proof of secondary meaning is required; however, "suggestive marks are eligible for protection without any proof of secondary meaning, since the connection between the mark and the source is presumed." 753 F.2d, at 216. The Second Circuit has nevertheless continued to deny protection for trade dress under §43(a) absent proof of secondary meaning, despite the fact that §43(a) provides no basis for distinguishing between trademark and trade dress. See, e.g., Stormy Clime Ltd. v. ProGroup, Inc., 809 F.2d, at 974; Union Mfg. Co. v. Han Baek Trading Co., 763 F.2d 42, 48 (1985); Le Sportsac, Inc. v. K Mart Corp., 754 F.2d 71, 75 (1985).

The Fifth Circuit was quite right in *Chevron*, and in this case, to follow the *Abercrombie* classifications consistently and to inquire whether trade dress for which protection is claimed under §43(a) is inherently distinctive. If it is, it is capable of identifying products or services as coming from a specific source and secondary meaning is not required. This is the rule generally applicable to trademark, and the protection of trademarks and trade dress under §43(a) serves the same statutory purpose of preventing deception and unfair competition. There is no persuasive reason to apply different analysis to the two. The "proposition that secondary meaning must be shown even if the trade dress is a distinctive, identifying mark, [is] wrong, for the reasons explained by Judge Rubin for the Fifth Circuit in Chevron." Blau Plumbing, Inc. v. S.O.S. Fix-it, Inc., 781 F.2d 604, 608 (CA7 1986). The Court of Appeals for the Eleventh Circuit also follows *Chevron*, Ambrit, Inc. v. Kraft, Inc., 805 F.2d 974, 979 (1986), and the Court of Appeals for the Ninth Circuit appears to think that proof of secondary meaning is superfluous if a trade dress is inherently distinctive. Fuddruckers, Inc. v. Doc's B. R. Others, Inc., 826 F.2d 837, 843 (1987).

It would be a different matter if there were textual basis in §43(a) for treating inherently distinctive verbal or symbolic trademarks differently from inherently distinctive trade dress. But there is none. The section does not mention trademarks or trade dress, whether they be called generic, descriptive, suggestive, arbitrary, fanciful, or functional. Nor does the concept of secondary meaning appear in the text of §43(a). Where secondary meaning does appear in the statute, 15 U.S.C. §1052 (1982 ed.), it is a requirement that applies only to merely descriptive marks and not to inherently distinctive ones. We see no basis for requiring secondary meaning for inherently distinctive trade dress protection under §43(a) but not for other distinctive words, symbols, or devices capable of identifying a producer's product.

Engrafting onto §43(a) a requirement of secondary meaning for inherently distinctive trade dress also would undermine the purposes of the Lanham Act. Protection of trade dress, no less than of trademarks, serves the Act's purpose to "secure to the owner of the mark the goodwill of his business and to protect the ability of consumers to distinguish among competing producers. National protection of trademarks is desirable, Congress concluded, because trademarks foster competition and the maintenance of quality by securing to the producer the benefits of good reputation." *Park 'N Fly*, 469 U.S., at 198, citing S. Rep. No. 1333, 79th Cong., 2d Sess., 3-5 (1946) (citations omitted). By making more difficult the identification of a producer with its product, a secondary meaning requirement for a nondescriptive trade dress would hinder improving or maintaining the producer's competitive position.

Suggestions that under the Fifth Circuit's law, the initial user of any shape or design would cut off competition from products of like design and shape are not persuasive. Only nonfunctional, distinctive trade dress is protected under §43(a). The Fifth Circuit holds that a design is legally functional, and thus unprotectable, if it is one of a limited number of equally efficient options available to competitors and free competition would be unduly hindered by according the design trademark protection. See Sicilia Di R. Biebow & Co. v. Cox, 732 F.2d 417, 426 (CA5 1984). This serves to assure that competition will not be stifled by the exhaustion of a limited number of trade dresses.

On the other hand, adding a secondary meaning requirement could have anticompetitive effects, creating particular burdens on the start-up of small companies. It would present special difficulties for a business, such as respondent, that seeks to start a new product in a limited area and then expand into new markets. Denying protection for inherently distinctive nonfunctional trade dress until after secondary meaning has been established would allow a competitor, which has not adopted a distinctive trade dress of its own, to appropriate the originator's dress in other markets and to deter the originator from expanding into and competing in these areas.

As noted above, petitioner concedes that protecting an inherently distinctive trade dress from its inception may be critical to new entrants to the market and that withholding protection until secondary meaning has been established would be contrary to the goals of the Lanham Act. Petitioner specifically suggests, however, that the solution is to dispense with the requirement of secondary meaning for a reasonable, but brief period at the outset of the use of a trade dress. If §43(a) does not require secondary meaning at the outset of a business' adoption of trade dress, there is no basis in the statute to support the suggestion that such a requirement comes into being after some unspecified time.

III

We agree with the Court of Appeals that proof of secondary meaning is not required to prevail on a claim under §43(a) of the Lanham Act where the trade dress at issue is inherently distinctive, and accordingly the judgment of that court is affirmed.

COMMENTS AND QUESTIONS

1. *Two Pesos* reverses the prior rule of a number of circuit courts, which had held that secondary meaning was a required element of a section 43(a) action for trade dress infringement, even if the trade dress was inherently distinctive. See, e.g., Ferrari S.P.A. v. Roberts, 944 F.2d 1235, 1239 (6th Cir. 1991), *cert. denied*, 505 U.S. 1219 (1992). But see the case that follows these comments, throwing the issue into some doubt.

2. Focus for a moment on the policy justifications the Court offers in support of its ruling. Two such justifications appear: that protecting distinctive trade dress will enable companies to appropriate their own goodwill, and that small companies may be unable to protect their trade dress from larger infringers if they are required to establish secondary meaning. Are these justifications persuasive? Should the Court focus on the rights of the trademark holder to protect its goodwill, or on the rights of the public to accurate product information? Are these two goals likely to come into conflict here?

3. Does the Court adopt either the "incentive" or property-rights theory of trademark law? See Stephen L. Carter, Does It Matter Whether Intellectual Property Is Property?, 68 Chicago-Kent L. Rev. 715, 721-22 (1993) (*Two Pesos* is indicative of a trend to strengthen the "property" nature of intellectual property). Certainly, the Court expresses a concern that seems

consistent with the incentive theory: "protecting an inherently distinctive trade dress from its inception may be critical to new entrants to the market." Early protection encourages the development of new marks, in the Court's view.

A moral argument can also be made in favor of protecting inherently distinctive trade dress, at least against "knock-off" copying. Such an argument asserts that the infringer can make no valid moral claim to copy the trade dress, so a right to it ought to reside in its creator. This is different from saying that the originator of the trade dress has earned it; it says that he gets it by default, because no one else has a superior claim. See Lawrence Becker, Property Rights: Philosophic Foundations 41 (1977) ("[i]t is not so much that the producers deserve the produce of their labors. It is rather that no one else does. . . ."); see also Robert C. Denicola, Institutional Publicity Rights: An Analysis of the Merchandising of Famous Trade Symbols, 62 N.C. L. Rev. 603, 640-41 (1984) (arguing no one has better claim to trademark's commercial value than its producer). This moral view — that knock-offs are bad — appears to be behind the "New York rule" discussed above. Arguably, though, this approach misses the point. Isn't the question not *who* should own the right to a mark, but whether *anyone* should have exclusive rights to it?

Judge Posner gives an economic answer to this question; he says that in some situations we declare property rights merely to prevent wasteful dissipation of the value of the asset covered by the right. He explicitly applies this to the right of publicity. See Richard Posner, Economic Analysis of Law 43 (4th ed. 1992):

> [W]hatever information value a celebrity's endorsement has to consumers would be lost if every advertiser can use the celebrity's name and picture. . . . The existence of a congestion externality provides an argument that rights of publicity should be perpetual and thus inheritable (a matter of legal controversy today). We don't want this form of information or expression to be in the public domain, because it will be less valuable there, whether the celebrity is dead or alive.

Is this such a situation? What waste would occur if other restaurants were permitted to duplicate the Taco Cabana décor?

4. Should trade dress protection extend to the décor of a restaurant? Does *Two Pesos* open the door to a wide variety of claims that a particular style of doing business is protectable "trade dress?" An extreme example of a trade dress claim allowed in the wake of *Two Pesos* is Toy Mfgrs. of America v. Helmsley-Spear, Inc., 960 F. Supp. 673 (S.D.N.Y. 1997), where the court held that the plaintiff's "unique" registration process, forms, and location for a toy fair (taken together) constituted protectable trade dress which the defendants had infringed.

Knitwaves, Inc. v. Lollytogs, Ltd.
United States Court of Appeals for the Second Circuit
71 F.3d 996 (2d Cir. 1995)

OAKES, Senior Circuit Judge: . . .

In 1990, Knitwaves, Inc., a manufacturer of children's knitwear, introduced its "Ecology Group" collection of sweaters, consisting of various styles of girls' sweaters and accompanying skirts and pants, presenting "ecology" themes in "fall" colors. It obtained copyrights for the designs of the two sweaters at issue in this litigation — its "Leaf Sweater," a multicolored striped sweater with puffy leaf appliques, and its "Squirrel Cardigan," which has a squirrel and leaves appliqued onto its multipaneled front. In 1992, Lollytogs Ltd., a larger manufacturer which sells children's clothing under the nationally advertised French Toast label, introduced a competing line of fall sweaters, including a similar-looking Leaf Sweater and Squirrel Cardigan admittedly copied from Knitwaves' sweaters. Knitwaves sued Lollytogs in the Southern District of New York for copyright infringement and for unfair competition under the Lanham Act and New York law.

[Plaintiff obtained a preliminary injunction, and later won at trial. Defendant appealed. The court, in an omitted portion of the opinion, upheld the trial court's finding of copyright infringement. It then proceeded to plaintiff's Lanham Act claims.]

II.   Knitwaves' Lanham Act Claim

Lollytogs advances two reasons why we should reverse the [district] court's finding that Lollytogs violated the Lanham Act. It contends, first, that the trade dress of Knitwaves' sweaters is not protectible, and, second, that the district court erred in finding a likelihood of confusion between Knitwaves' and Lollytogs' sweaters. As discussed below, we agree that the sweaters' trade dress is not protectible, and so we vacate that part of the court's order finding a Lanham Act violation, without addressing the issue of likelihood of confusion.

While "trade dress" at one time "referred only to the manner in which a product was 'dressed up' to go to market with a label, package, display card, and similar packaging elements," the concept "has taken on a more expansive meaning and includes the design and appearance of the product as well as that of the container and all elements making up the total visual image by which the product is presented to customers." Jeffrey Milstein, Inc. v. Greger, Lawlor, Roth, Inc., 58 F.3d 27, 31 (2d Cir. 1995). Trade dress, thus, is " 'essentially [a product's] total image and overall appearance.' " Id. (quoting Two Pesos, Inc., v. Taco Cabana, Inc., 505 U.S. 763, 764 n.1 . . . (1992)). See also Restatement (Third) of Unfair Competition §16 & cmt. a (1995).

In contending that Knitwaves' sweaters are not protectible under the Lanham Act, Lollytogs relies primarily on what has been dubbed the "functionality" (or, more precisely in this case, the "aesthetic functionality") defense. [The court rejects this defense; for detailed discussion, see infra.]

. . . [T]o prevail in an action for trade dress infringement under §43(a) of the Lanham Act, a plaintiff must prove (1) that its dress is distinctive of the source and (2) that a likelihood of confusion exists between its product and defendant's product. Jeffrey Milstein, 58 F.3d at 31. To establish that a trade dress is distinctive of a particular source, a plaintiff must demonstrate either that it is "inherently distinctive" or that it has become distinctive through acquiring "secondary meaning" to the consuming public. Villeroy, 999 F.2d at 620.

Prior to the Supreme Court's decision in Two Pesos, Inc. v. Taco Cabana, Inc., 505 U.S. 763 . . . (1992), this circuit had held that trade dress, unlike trademarks, could never be inherently distinctive, and thus we required plaintiffs seeking §43(a) protection of trade dress to establish distinctiveness by proving that the trade dress had acquired secondary meaning. See, e.g., . . . Stormy Clime, 809 F.2d at 974. In *Two Pesos*, however, the Supreme Court rejected this circuit's approach, finding no "textual basis in §43(a) for treating inherently distinctive verbal or symbolic trademarks differently from inherently distinctive trade dress." 505 U.S. at 774. . . . Plaintiffs in trade dress infringement cases, just as in trademark cases, should be given a chance, the Court reasoned, to demonstrate that their trade dress is "capable of identifying products or services as coming from a specific source." Id. . . .

The *Two Pesos* decision left this circuit with the task of determining what it means for trade dress to be "inherently distinctive." In Paddington Corp. v. Attiki Importers & Distribs., Inc., 996 F.2d 577 (2d Cir. 1993), we addressed this issue and concluded that the test set out by Judge Friendly in Abercrombie & Fitch Co. v. Hunting World, Inc., 537 F.2d 4, 9 (2d Cir. 1976), for evaluating the inherent distinctiveness of trademarks was also applicable in the trade dress context. 996 F.2d at 583-84. . . . Under the Abercrombie test, marks are classified as either (1) generic, (2) descriptive, (3) suggestive, or (4) arbitrary or fanciful. *Paddington*, 996 F.2d at 583. While generic marks can never serve to distinguish a source, and descriptive marks require a demonstration of secondary meaning in order to become distinctive, suggestive and arbitrary or fanciful marks are considered to be inherently distinctive. Id. Applying this reasoning, we examined the trade dress of a liquor bottle and found the bottle's label arbitrary in that "[t]he tone and layout of the colors, the style and size of the lettering, and, most important, the overall appearance of the bottle's labeling . . . were selected from an almost limitless supply of patterns, colors and designs." Id. at 584. Finding the bottle's trade dress arbitrary, we concluded that it was per se distinctive. Id.

We do not find the analysis of *Paddington* appropriate in the case before us, in which the trade dress at issue consists of a product's features — the artwork on a sweater — rather than, as in *Paddington*, a product's packaging. Not only does the classification of marks into "generic," "descriptive," "suggestive," or "arbitrary or fanciful" make little sense when applied to product features, but it would have the unwelcome, and likely unintended, result of treating a class of product features as "inherently distinctive," and thus eligible for trade dress protection, even though they were never intended to serve a source-identifying function. The Third Circuit has made this point persuasively in a recent case:

> The difficulty is that . . . a product configuration differs fundamentally from a product's trademark, insofar as it is not a symbol according to which one can relate the signifier (the trademark, or perhaps the packaging) to the signified (the product). Being constitutive of the product itself and thus having no such dialectical relationship to the product, the product's configuration cannot be said to be "suggestive" or "descriptive" of the product, or "arbitrary" or "fanciful" in relation to it. The very basis for the trademark taxonomy — the descriptive relationship between the mark and the product, along with the degree to which the mark describes the product — is unsuited for application to the product itself.

Duraco Prods., Inc. v. Joy Plastic Enterprises, Ltd., 40 F.3d 1431, 1440-41 (3d Cir. 1994) (citations omitted). . . .

The Supreme Court made a similar point recently when it addressed whether the Lanham Act permits registration of a trademark which consists solely of a color. A color, it noted, is "capable of satisfying the more important part of the statutory definition of a trademark, which requires that a person 'us[e]' or 'inten[d] to use' the mark 'to identify and distinguish his or her goods . . . from those manufactured or sold by others and to indicate the source of the goods. . . .'" Qualitex, 115 S. Ct. at 1303 (quoting 15 U.S.C. §1127). On the other hand, the Court noted, in an observation equally applicable to product features:

> [A] product's color is unlike "fanciful," "arbitrary," or "suggestive" words or designs, which almost automatically tell a customer that they refer to a brand. The imaginary word "Suntost," or the words "Suntost Marmalade," on a jar of orange jam immediately would signal a brand or a product "source"; the jam's orange color does not do so.

Id. (citations omitted). Thus, the Court concluded, a product's color is capable of indicating a product's source, and thereby becoming eligible for trademark protection, only when it acquires secondary meaning as, over time, customers come to associate that color with the particular product. Id.

As noted above, in *Two Pesos* the Court concluded that trade dress — unlike color alone, see *Qualitex* — may be inherently distinctive even without the addition of secondary meaning. We do not think, however, that the Court thereby intended to nullify what it called, in *Qualitex*, "the more important part of the statutory definition of a trademark" — the requirement "that a person 'us[e]' or 'inten[d] to use' the mark 'to identify and distinguish his or her goods . . . from those manufactured or sold by others and to indicate the source of the goods.' " Qualitex, 115 S. Ct. at 1303 (quoting 15 U.S.C. §1127). . . . While "arbitrary," "fanciful," or "suggestive" packaging of a product may be presumed to serve this source-identifying function, and thus may be deemed per se distinctive of the source, see Paddington, 996 F.2d at 583, the same presumption may not be made with regard to product features or designs whose primary purposes are likely to be functional or aesthetic.

Accordingly, in determining whether each of Knitwaves' sweater designs can be protected as a trademark, we do not ask whether it is "generic," "descriptive," "suggestive," or "arbitrary or fanciful" — categorizations which we find inapplicable to product features. Rather, we ask whether it is "likely to serve primarily as a designator of origin of the product," Duraco, 40 F.3d at 1449. Cf. Restatement (Third), supra, §13(a) (whether "because of the nature of the [design] and the context in which it is used, prospective purchasers are likely to perceive it as a [design] that . . . identifies goods or services produced or sponsored by [Knitwaves]"); and see id. §16; Imagineering, Inc. v. Van Klassens, Inc., 53 F.3d 1260, 1263-64 (Fed. Cir. 1995) ("Trade dress is inherently distinctive when, by its 'intrinsic nature,' it identifies the particular source of a product.") (quoting *Two Pesos*, 505 U.S. at 768, 112 S. Ct. at 2757). . . .

As Knitwaves' objective in the two sweater designs was primarily aesthetic, the designs were not primarily intended as source identification. Those sweater designs therefore fail to qualify for protection of trade dress inherent in product design. Accordingly, the judgment in favor of Knitwaves on the Lanham Act claim is reversed. Judgment should be entered for Lollytogs on this claim.

COMMENTS AND QUESTIONS

1. Some courts at least seem relieved by the fact that *Knitwaves* removed the burden of determining such classification-related chores as determining what "suggestive" trade dress might be. See, e.g., EFS Marketing Inc. v. Russ Berrie & Co., 76 F.3d 487 (2d Cir. 1996) ("No longer need we decide whether a product-configuration trade dress falls within a particular distinctiveness category; we need determine only whether it is likely to serve primarily as a designator of origin of the product. . . . EFS's trade dress also fails to meet this newer standard because, as already pointed out, EFS's dolls are so similar to the many other troll dolls on the market that they cannot be said to identify EFS as their particular source."). But see Fun-Damental Too Ltd. v. Gemmy Indus. Corp., 111 F.3d 993 (2d Cir. 1997) (traditional classification for word marks was appropriate for product packaging).

Other Second Circuit cases have taken the rule even farther. See Fabrication Enterprises v. Hygenic Corp., 64 F.3d 53, 58 n.3 (2d Cir. 1995); Mana Prods. v. Columbia Cosmetics Mfg., 65 F.3d 1063, 1071 (2d Cir. 1995) (concluding that under *Qualitex* a color can be protected *only* on a showing of secondary meaning). Is this approach consistent with *Two Pesos?* If the trade dress of a restaurant can be inherently distinctive, why can't color? See Michael B. Landau, Reconciling *Qualitex* with *Two Pesos*: Ambiguity and Inconsistency from the Supreme Court, 3 UCLA Ent. L. Rev. 219 (1996) (identifying the problem and criticizing the Second Circuit approach).

2. While *Knitwaves* held that the design of a sweater was not primarily intended as an indicator of source, the Second Circuit concluded in Samara Brothers v. Wal-Mart Stores, 49 U.S.P.Q.2d 1260 (2d Cir. 1998) that the overall look of an *entire line* of clothing was entitled to trade dress protection where the line of clothes was identified with the plaintiff over a period of years. The court distinguished *Knitwaves* by emphasizing (1) the fact that it was reviewing a jury verdict, and (2) that on several substantive grounds the case for protection was more compelling:

> We find the record in this case entirely distinguishable [from *Knitwaves*]. Here, the jury heard the testimony of Samara's Vice-President for sales that Samara chose to design its line of spring/summer seersucker children's clothes using consistent design elements so that the look would be identified with Samara, building brand loyalty. Samara has produced this very same product line for years, and it represents, according to the witness's testimony, "the core [of Samara's] business" and the "lifeblood of the company." When Samara has produced other lines of clothes reflecting different designs, its sales have suffered. In *Knitwaves*, by contrast, the two sweaters were part of a line of fall clothes that had never before been manufactured by Knitwaves nor did the evidence reveal any intent on the part of Knitwaves to establish the "fall motif" as its core product by which it would be recognized in the marketplace. . . .
>
> We agree that ample evidence supports the finding that Wal-Mart's marketing of the knock-offs was willful piracy with an intent to deceive consumers as to the source. That evidence includes the fact that a Wal-Mart representative specifically requested that JPI create seersucker garments based on Samara samples (among other samples); the garments Wal-Mart subsequently ordered from JPI were, as admitted by a Wal-Mart witness, "strikingly similar" to the Samara garments; the breadth of copying was significant as Wal-Mart sold not one but sixteen garments that were practically identical to Samara clothes; and Wal-Mart buyers knew that Samara clothes were being copied.

165 F.3d 120, 125, 127-128. Judge Newman, dissenting, argued that the basic dress design protected by the court was not entitled to copyright protection and that Samara should not be entitled to obtain equivalent protection through the "back door" of trade dress law. The case illustrates a common tension in this area between punishing what the majority opinion brands "willful piracy" and upholding important principles of open competition. The Supreme Court agreed in 1999 to decide the case.

3. Several recent cases seem to point out the futility of applying a classification scheme designed for word marks to product configurations. In Rock and Roll Hall of Fame and Museum, Inc. v. Gentile Productions, 134 F.3d 749 (6th Cir. 1998), the owner of the usual Rock and Roll Hall of Fame

building in Cleveland, Ohio, sued a photographer who marketed pictures of the building set against the Cleveland skyline, alleging that the photographer was infringing the building's trade dress. The district court issued a preliminary injunction, but the Sixth Circuit reversed. The majority and the dissent battled over the proper classification of the museum. The majority agreed that the design of the museum was "fanciful" but denied that it was "fanciful in a trademark sense." It concluded that the design of the museum itself did not function as a trademark and that the public did not recognize it as such. Important to the court's determination was the trademark owner's inconsistent use of the museum's design as a source-identifying function:

> [A]lthough the Museum has used drawings or pictures of its building design on various goods, it has not done so with any consistency. As museum director Robert Bosak stated in his affidavit, "the Museum has used versions of the building shape trademark on . . . a wide variety of products." Several items marketed by the Museum display only the rear of the Museum's building, which looks dramatically different from the front. Drawings of the front of the Museum on the two T-shirts in the record are similar, but they are quite different from the photograph featured in the Museum's poster. And, although the photograph from the poster is also used on a postcard, another postcard displays various close-up photographs of the Museum which, individually and perhaps even collectively, are not even immediately recognizable as photographs of the Museum. In this regard, this case is similar to those in which a party has claimed trademark rights in a famous person's likeness. In Estate of Presley [v. Russen, 513 F.Supp. 1339, 1363-64 (D.N.J. 1981)], the court concluded that, although one particular image of [Elvis] Presley had been used consistently as a mark, "the available evidence [did] not support [the estate's] broad position" that all images of Presley served such a function.

134 F.3d 749, 754-755.

4. In abandoning one classification problem, however, courts like *Knitwaves* may simply have jumped out of the frying pan and into the fire. For if the court is to treat product configurations differently from "ordinary" trade dress such as product packaging, it must figure out how to tell which is which. Consider whether the shape of the classic Coca-Cola bottle is product configuration or trade dress. How would you classify the restaurant décor in *Two Pesos?*

5. Courts, beginning with the Supreme Court in *Qualitex*, have drawn on the recent Restatement (Third) of Unfair Competition for guidance. Section 16, on trade dress protection, attempts to penetrate "classificatory" issues and get to the heart of the issues at stake in trade dress cases.

§16. Configurations of Packaging and Products: Trade Dress and Product Designs

The design of elements that constitute the appearance or image of goods or services as presented to prospective purchasers, including the design of packaging, labels, containers, displays, decor, or the design of a product, a product feature, or a combination of product features, is eligible for protection as a mark under the rules stated in this Chapter if:

(a)   the design is distinctive under the rule stated in §13; and

(b)   the design is not functional under the rule stated in §17.

Comments: . . .

b.   Distinctiveness. The freedom to copy product and packaging features is limited by the law of trademarks only when the copying is likely to confuse prospective purchasers as to the source or sponsorship of goods or services. The imitation or even complete duplication of another's product or packaging will not create a risk of confusion unless some aspect of the duplicated appearance is identified with a particular source. Thus, unless a feature is distinctive under the rule stated in §13, it is ineligible for protection as a trademark. A further restriction that prohibits protection for functional features is discussed in Comment *d* and in §17. Rigorous application of the requirements of distinctiveness and nonfunctionality is necessary in order to avoid undermining the carefully circumscribed statutory regimes for the protection of useful and ornamental designs under federal patent and copyright law.

Many of the cases adjudicating trademark rights in product and packaging designs recite that proof of secondary meaning is a prerequisite for protection. At least with respect to packaging and related features, however, it is now recognized that trade dress can be inherently distinctive. This Section, through reference to the rules stated in §13, recognizes both inherent and acquired distinctiveness, thus assimilating trade dress within the general rules applicable to other trademarks. If the trade dress used by a particular seller differs in significant respects from that employed by others, consumers may be expected to utilize the trade dress as an indication of source. The wide range of designs available for labels and packaging generally permits the recognition of exclusive rights without significantly hindering competition. Trade dress that is unique and prominent can thus be inherently distinctive. If the trade dress is descriptive (see §14), or inconspicuous, or not sufficiently different from that used by others to justify a conclusion of inherent distinctiveness, trademark rights will depend upon proof of distinctiveness through evidence of secondary meaning.

This Section is also applicable to the recognition of trademark rights in the distinctive design features of the goods themselves. As a practical matter, however, it is less common for consumers to recognize the design of a product or product feature as an indication of source. Product designs are more likely to be seen merely as utilitarian or ornamental aspects of the goods. In addition, the competitive interest in copying product designs is more substantial than in the case of packaging, containers, labels, and related subject matter. Product designs are therefore not ordinarily considered inherently distinctive and are thus normally protected only upon proof of secondary meaning. . . .

Note on Product Configurations and Overlapping Intellectual Property Rights

There remains some disagreement among the courts concerning precisely how to treat product configurations. In Duraco Prods. v. Joy Plastic Enterprises, 40 F.3d 1431 (3d Cir. 1994), the Third Circuit rejected the proposition (arguably mandated by *Two Pesos*) that traditional trade dress doctrine applies to product configurations. Instead, the Third Circuit developed a new, stricter test for protecting product configurations. Under the *Duraco* test, a product configuration qualifies as inherently distinctive (and therefore protectable without proof of secondary meaning) only if it is (1) unusual and memorable, (2) capable of being conceptually separated from the product itself, and (3) likely to serve primarily a source-identifying function rather than a utilitarian function. By contrast, the Fourth and Eighth Circuit have specifically rejected the *Duraco* test, holding that product configurations need not be "memorable" or "striking in appearance" to qualify as inherently distinctive. Stuart Hall Co. v. Ampad Corp., 51 F.3d 780 (8th Cir. 1995); Ashley Furniture Indus. v. San Giacomo N.A., 187 F.3rd 363 (4th cir. 1999).

Is there any reason to treat product configurations differently from product names or product packaging for trademark purposes? Protecting product configurations directly means that trademark overlaps not only with copyright law but also with design patent law. Should such overlap be permissible? Note that some courts are not willing to permit product configurations that are the subject of a utility patent to be protected under trade dress law as well, perhaps out of a concern that the owner of the design will be able to extend protection beyond the term of the patent. See Vornado Air Circulation Systems v. Duracraft Corp., 58 F.3d 1498 (10th Cir. 1995). At the same time, trademark law has been extended to protect design configurations that would otherwise be protected under design patent law. In Kohler Co. v. Moen, Inc., 12 F.3d 632 (7th Cir. 1993), the Seventh Circuit affirmed the trademark registration of a faucet design and a faucet handle design.[9] The court relied in part on the legislative history of the Trademark Law Reform Act of 1988. The Senate Report on that act interpreted the terms "symbol or device" as permitting the registration of "colors, shapes, sounds or configurations where they function as trademarks." S. Rep. 100-515, 100th Cong. 2d Sess. at 44, 1988 U.S.C.C.A.N. at 5607.

The *Kohler* court acknowledged that granting trademark protection to product design configurations created an overlap with design patents, which are also available to protect such configurations. The court rejected the arguments that the patent laws by implication preempt overlapping trademark protection and that such preemption is necessary to prevent the grant of a perpetual trademark monopoly in place of the limited 14-year design patent monopoly. The court relied on the "fundamentally different" legal protection offered by trademarks and design patents. Accord In re Yardley, 493 F.2d 1389 (C.C.P.A. 1974).

---

9. Product design configurations are also protectable without registration under section 43(a). See Ferrari S.P.A. v. Roberts, 944 F.2d 1235 (6th Cir. 1991).

Trademark protection under section 43(a) has been extended by at least one court to protect the "artistic style" of an artist from copying. In Romm Art Creations Ltd. v. Simcha Intl., Inc., 786 F. Supp. 1126 (E.D.N.Y. 1992), the court concluded that the style of a series of paintings by Itzchak Tarkay entitled "Women and Cafes" could not be imitated by defendants because the style of the paintings qualified for trade dress protection. The court concluded that the Tarkay paintings were arbitrary or fanciful, and therefore protectable under *Two Pesos* as inherently distinctive works.

The overlap between trade dress and copyright protection in this case is obvious. Artistic works like the Tarkay paintings fall squarely within the protection of the Copyright Act. And copyright's test for infringement — proof of access plus substantial similarity between the works — seems designed precisely for cases such as this, in which intangible stylistic elements are being copied. Note that, unlike copyright law, trademark law does not provide an "independent creation" defense.

COMMENTS AND QUESTIONS

1. Is there anything wrong with overlapping intellectual property protection? After all, patent and copyright themselves overlap in certain areas, trade secrets may overlap with copyright as well, and contract law may overlap or displace all other forms of protection. One concern noted in the text is that in both the examples presented, trademark law not only duplicates but actually expands the protection that would be granted by patent or copyright. Does this more expansive protection suggest that a balance is being upset in these cases?

An old rule of intellectual property law called the "principle of election" limited an author to one type of intellectual property protection. That doctrine was resoundingly rejected in In re Yardley, 493 F.2d 1389 (C.C.P.A. 1974). *Yardley* involved a design patent application for a watch that featured a caricature of Spiro Agnew. The patent examiner rejected the application on the ground that Yardley had already received several copyright registrations for the watch. The Court of Customs and Patent Appeals reversed, concluding that the principle of election was in direct conflict with the patent, copyright, and trademark statutes. The court reasoned that because each statute strikes its own "coverage" balance, a work that qualifies for protection under more than one statute is entitled to the protection of each.

PROBLEM

Problem 5-1. Alice Richland, a gourmet chef world-renowned for her chocolate desserts, designs a new line of chocolate products. Called "Chocolate Shells," these products are made of dark chocolate flavored in special ways with a combination of ingredients Alice hit upon after months of work in her kitchens. The use of the special ingredients imparts a unique flavor to the Shells, and has the additional property of making the usually soft chocolate feel sandy or grainy to the touch. The Shells are made in the shapes of different seashells native to the Florida coast, and each shell is colored in a different, unusual pattern. On the advice of her lawyer, Alice files a trademark application seeking to register each of her shell designs. What aspects of the Shells are entitled to registration? To section 43(a) protection?

2. Priority

As in patent law, many trademark cases turn on issues of priority. Section 45(a) of the Lanham Act, which defines a trademark, requires that the mark either be (1) "used in commerce" or (2) registered with a bona fide intention to use it in commerce. Both at common law and under the traditional Lanham Act registration procedures, determining who owned a trademark meant determining who was first to use it to identify her goods.[10]

The requirement of "use in commerce" is an historical result of the constitutional basis for the trademark laws, which (unlike the patent and copyright statutes) rely on the congressional power to regulate interstate commerce. This requirement also goes hand in hand with the basic trademark theory elaborated in the introduction to this chapter; recall that this theory emphasizes the protection of consumer associations of a brand with a particular product, which can arise only after a trademark is placed on goods sold in commerce.

But just what constitutes use of a trademark? And how much use is enough to secure legal protection for the mark? The case and notes that follow provide some guidance.

≡≡≡ *Zazu Designs v. L'Oreal, S.A.*
United States Court of Appeals for the Seventh Circuit
979 F.2d 499 (7th Cir. 1992)

EASTERBROOK, Circuit Judge.

In 1985 Cosmair, Inc., concluded that young women craved pink and blue hair. To meet the anticipated demand, Cosmair developed a line of "hair cosmetics" — hair coloring that is easily washed out. These inexpensive prod-

---

10. Section (2), defining so-called "intent to use" trademarks, is relatively new; it dates from 1989. This change added to, but did not displace, the traditional touchstone for legal recognition of a trademark: *use*, section (1). And, as we shall see, even intent-to-use registrations e no more than inchoate reservations of rights that, to mature into actual trademarks, must be 'owed by actual use in commerce. The requirement of actual use is thus still the trigger for emark protection.

ucts, under the name ZAZU, were sold in the cosmetic sections of mass merchandise stores. Apparently the teenagers of the late 1980s had better taste than Cosmair's marketing staff thought. The product flopped, but its name gave rise to this trademark suit. Cosmair is the United States licensee of L'Oreal, S.A., a French firm specializing in perfumes, beauty aids, and related products. Cosmair placed L'Oreal's marks on the bottles and ads. . . .

L'Oreal hired Wordmark, a consulting firm, to help it find a name for the new line of hair cosmetics. After checking the United States Trademark Register for conflicts, Wordmark suggested 250 names. L'Oreal narrowed this field to three, including ZAZU, and investigated their availability. This investigation turned up one federal registration of ZAZU as a mark for clothing and two state service mark registrations including that word. One of these is Zazu Hair Designs; the other was defunct.

Zazu Hair Designs is a hair salon in Hinsdale, Illinois, a suburb of Chicago. We call it "ZHD" to avoid confusion with the ZAZU mark. . . . The salon is a partnership between Raymond R. Koubek and Salvatore J. Segretto, hairstylists who joined forces in 1979. ZHD registered ZAZU with Illinois in 1980 as a trade name for its salon. L'Oreal called the salon to find out if ZHD was selling its own products. The employee who answered reported that the salon was not but added, "we're working on it." L'Oreal called again; this time it was told that ZHD had no products available under the name ZAZU.

L'Oreal took the sole federal registration, held by Riviera Slacks, Inc., as a serious obstacle. Some apparel makers have migrated to cosmetics, and if Riviera were about to follow Ralph Lauren (which makes perfumes in addition to shirts and skirts) it might have a legitimate complaint against a competing use of the mark. Sands, Taylor & Wood Co. v. Quaker Oats Co., 24 U.S.P.Q.2d 1001, 1011 (7th Cir. 1992). Riviera charged L'Oreal $125,000 for a covenant not to sue if L'Oreal used the ZAZU mark on cosmetics. In April 1986, covenant in hand and satisfied that ZHD's state trade name did not prevent the introduction of a national product, L'Oreal made a small interstate shipment of hair cosmetics under the ZAZU name. It used this shipment as the basis of an application for federal registration, filed on June 12, 1986. By August L'Oreal had advertised and sold its products nationally.

Unknown to L'Oreal, Koubek and Segretto had for some time aspired to emulate Vidal Sassoon by marketing shampoos and conditioners under their salon's trade name. In 1985 Koubek began meeting with chemists to develop ZHD's products. Early efforts were unsuccessful; no one offered a product that satisfied ZHD. Eventually ZHD received acceptable samples from Gift Cosmetics, some of which Segretto sold to customers of the salon in plain bottles to which he taped the salon's business card. Between November 1985 and February 1986 ZHD made a few other sales. Koubek shipped two bottles to a friend in Texas, who paid $13. He also made two shipments to a hair stylist friend in Florida — 40 bottles of shampoo for $78.58. These were designed to interest the Floridian in the future marketing of the product line. These bottles could not have been sold to the public, because they lacked labels listing the ingredients and weight. See 21 U.S.C. §362(b); 15 U.S.C. §§1452, 1453(a); 21 CFR §§701.3, 701.13(a). After L'Oreal's national marketing was under way, its representatives thrice visited ZHD and found that the salon still had no products for sale under the ZAZU name. Which is not to say that ZHD was supine. Late in 1985 ZHD had ordered 25,000 bottles silkscreened with the name ZAZU. Later it ordered stick-on labels listing the ingredients of its products. In September 1986 ZHD began to sell small quantities of shampoo in bottles filled (and labeled) by hand in the salon. After the turn of the year ZHD directed the supplier of the shampoo and conditioner to fill some bottles; the record does not reveal how many.

After a bench trial the district court held that ZHD's sales gave it an exclusive right to use the ZAZU name nationally for hair products. 9 U.S.P.Q.2d 1972 (N.D. Ill. 1988). The court enjoined L'Oreal from using the mark (a gesture, since the product had bombed and L'Oreal disclaimed any interest in using ZAZU again). It also awarded ZHD $100,000 in damages on account of lost profits and $1 million more to pay for corrective advertising to restore luster to the ZAZU mark. [The final judgment included a $1 million punitive damage award for "oppressive and deceitful" tactics used in the litigation.]

Federal law permits the registration of trademarks and the enforcement of registered marks. Through §43(a) of the Lanham Act, 15 U.S.C. §1125(a), a provision addressed to deceit, it also indirectly allows the enforcement of unregistered marks. But until 1988 federal law did not specify how one acquired the rights that could be registered or enforced without registration. That subject fell into the domain of state law, plus federal common law elaborating on the word "use" in §43(a). . . . At common law, "use" meant sales to the public of a product with the mark attached. Trade-Mark Cases, 100 U.S. 82, 94-95 (1879). See also Hanover Star Milling Co. v. Metcalf, 240 U.S. 403, 414 (1916); United Drug Co. v. Theodore Rectanus Co., 248 U.S. 90, 97 (1918).

"Use" is neither a glitch in the Lanham Act nor a historical relic. By insisting that firms use marks to obtain rights in them, the law prevents entrepreneurs from reserving brand names in order to make their rivals' marketing more costly. Public sales let others know that they should not invest resources to develop a mark similar to one already used in the trade. Blue Bell, Inc. v. Farah Manufacturing Co., 508 F.2d 1260, 1264-65 (5th Cir. 1975); see also William M. Landes and Richard A. Posner, Trademark Law: An Economic Perspective, 30 J.L. & Econ. 265, 281-84 (1987). Only active use allows consumers to associate a mark with particular goods and notifies other firms that the mark is so associated.

Under the common law, one must win the race to the marketplace to establish the exclusive right to a mark. Blue Bell v. Farah; La Societe Anonyme des Parfums LeGalion v. Jean Patou, Inc., 495 F.2d 1265, 1271-74 (2d Cir. 1974). Registration modifies this system slightly, allowing slight sales plus notice in the register to substitute for substantial sales without notice. 15 U.S.C. §1051(a). (The legislation in 1988 modifies the use requirement further, but we disregard this.) ZHD's sales of its product are insufficient use to establish priority over L'Oreal. A few bottles sold over the counter in Hinsdale, and a few more mailed to friends in Texas and Florida, neither link the ZAZU mark with ZHD's product in the minds of consumers nor put other producers on notice. As a practical matter ZHD had no product, period, until months after L'Oreal had embarked on its doomed campaign.

In finding that ZHD's few sales secured rights against the world, the district court relied on cases such as Department of Justice v. Calspan Corp., 578 F.2d 295 (C.C.P.A. 1978), which hold that a single sale, combined with proof of intent to go on selling, permit the vendor to register the mark. See also Axton-Fisher Tobacco Co. v. Fortune Tobacco Co., 82 F.2d 295 (C.C.P.A. 1936); Maternally Yours, Inc. v. Your Maternity Shop, Inc., 234 F.2d 538, 542 (2d Cir. 1956). . . . But use sufficient to register a mark that soon is widely distributed is not necessarily enough to acquire rights in the absence of registration. The Lanham Act allows only trademarks "used in commerce" to be registered. 15 U.S.C. §1051(a). Courts have read "used" in a way that allows firms to seek protection for a mark before investing substantial sums in promotion. See Fort Howard Paper Co. v. Kimberly-Clark Corp., 390 F.2d 1015 (C.C.P.A. 1968); cf. Jim Dandy Co. v. Martha White Foods, Inc., 458 F.2d 1397, 1399 (C.C.P.A. 1972) (party may rely on advertising to show superior registration rights); but see Weight Watchers International, Inc. v. I. Rokeach & Sons, Inc., 211 U.S.P.Q. 700, 709 (T.M.T.A.B. 1981) (more than minimal use is required to register because the statute allows only "owner[s]" to register, and ownership of a mark depends on commercial use). Liberality in registering marks is not problematic, because the registration gives notice to latecomers, which token use alone does not. Firms need only search the register before embarking on development. Had ZHD registered ZAZU, the parties could have negotiated before L'Oreal committed large sums to marketing.

ZHD applied for registration of ZAZU after L'Oreal not only had applied to register the mark but also had put its product on the market nationwide. Efforts to register came too late. At oral argument ZHD suggested that L'Oreal's knowledge of ZHD's plan to enter the hair care market using ZAZU establishes ZHD's superior right to the name. Such an argument is unavailing. Intent to use a mark, like a naked registration, establishes no rights at all. Hydro-Dynamics, Inc. v. George Putnam & Co., 811 F.2d 1470, 1472 (Fed. Cir. 1987). Even under the 1988 amendments (see note), which allow registration in advance of contemplated use, an unregistered plan to use a mark creates no rights. Just as an intent to buy a choice parcel of land does not prevent a rival from closing the deal first, so an intent to use a mark creates no rights a competitor is bound to respect. A statute granting no rights in bare registrations cannot plausibly be understood to grant rights in "intents" divorced from either sales or registrations. Registration itself establishes only a rebuttable presumption of use as of the filing date. Rolley, Inc. v. Younghusband, 204 F.2d 209, 211 (9th Cir. 1953). ZHD made first use of ZAZU in connection with hair services in Illinois, but this does not translate to a protectable right to market hair products nationally. The district court construed L'Oreal's knowledge of ZHD's use of ZAZU for salon services as knowledge "of [ZHD's] superior rights in the mark." 9 U.S.P.Q.2d at 1978. ZHD did not, however, have superior rights in the mark as applied to hair products, because it neither marketed such nor registered the mark before L'Oreal's use. Because the mark was not registered for use in conjunction with hair products, any knowledge L'Oreal may have had of ZHD's plans is irrelevant. Cf. Weiner King, Inc. v. Wiener King Corp., 615 F.2d 512 (C.C.P.A. 1980).

Imagine the consequences of ZHD's approach. Businesses that knew of an intended use would not be entitled to the mark even if they made the first significant use of it. Businesses with their heads in the sand, however, could stand on the actual date they introduced their products, and so would have priority over firms that intended to use a mark but had not done so. Ignorance would be rewarded — and knowledgeable firms might back off even though the rivals' "plans" or "intent" were unlikely to come to fruition. Yet investigations of the sort L'Oreal undertook prevent costly duplication in the development of trademarks and protect consumers from the confusion resulting from two products being sold under the same mark. See Natural Footwear Ltd. v. Hart, Shaffner & Marx, 760 F.2d 1383, 1395 (3d Cir. 1985). L'Oreal should not be worse off because it made inquiries and found that, although no one had yet used the mark for hair products, ZHD intended to do so. Nor should a potential user have to bide its time until it learns whether other firms are serious about marketing a product. The use requirement rewards those who act quickly in getting new products in the hands of consumers. Had L'Oreal discovered that ZHD had a product on the market under the ZAZU mark or that ZHD had registered ZAZU for hair products, L'Oreal could have chosen another mark before committing extensive marketing resources. Knowledge that ZHD planned to use the ZAZU mark in the future does not present an obstacle to L'Oreal's adopting it today. Selfway, Inc. v. Travelers Petroleum, Inc., 579 F.2d 75, 79 (C.C.P.A. 1978).

Occasionally courts suggest that "bad faith" adoption of a mark defeats a claim to priority. See California Cedar Products Co. v. Pine Mountain Corp., 724 F.2d 827, 830 (9th Cir. 1984); Stern Electronics, Inc. v. Kaufman, 669 F.2d 852, 857 (2d Cir. 1982); Blue Bell v. Farah, 508 F.2d at 1267. Although ZHD equates L'Oreal's knowledge of its impending use with "bad faith," the cases use the term differently. In each instance the court applied the label "bad faith" to transactions designed merely to reserve a mark, not to link the name to a product ready to be sold to the public. In California Cedar Products, for example, two firms sprinted to acquire the abandoned DURAFLAME mark. One shipped some of its goods in the abandoning company's wrapper with a new name pasted over it. Two days later the other commenced bona fide sales under the DURAFLAME mark. The court disregarded the first shipment, calling it "both premature and in bad faith," 724 F.2d at 830, and held that the first firm to make bona fide sales to customers was the prior user. "Bad faith" was no more than an epithet stapled to the basic conclusion: that reserving a mark is forbidden, so that the first producer to make genuine sales gets the rights. If these cases find a parallel in our dispute, ZHD occupies the place of the firm trying to reserve a mark for "intended" exploitation. ZHD doled out a few samples in bottles lacking labeling necessary for sale to the public. Such transactions are the sort of pre-marketing maneuvers that these cases hold insufficient to establish rights in a trademark.

The district court erred in equating a use sufficient to support registration with a use sufficient to generate nationwide rights in the absence of registration. Although whether ZHD's use is sufficient to grant it rights in the ZAZU mark is a question of fact on which appellate review is deferential, California Cedar Products, 724 F.2d at 830 . . . , the extent to which ZHD used the mark is not disputed. ZHD's sales of hair care products were insufficient as a matter of law to establish national trademark rights at the time L'Oreal put its electric hair colors on the market.

[In a forcefully stated section of the opinion, the court also reversed the punitive damages holding.]

Reversed and remanded.

CUDAHY, C.J., dissenting:

On the important issue of good faith, L'Oreal's conduct here merits a very hard look. In the case of Riviera, a men's clothing retailer, L'Oreal was careful to pay $125,000 for an agreement not to sue. Yet men's clothing and hair cosmetics marketed to women hardly seem related at all. On the other hand, a women's hair salon developing a line of hair care products is a purveyor of goods and services that seem closely related to hair cosmetics. Therefore, L'Oreal's knowledge of ZHD's use defeats any claim L'Oreal may have to priority.

One of the keys here seems to be the use of ZAZU as a service mark connected with the provision of salon services by ZHD. A service mark can be infringed by its use on a closely related product.[1] . . . [See] 2 J. Thomas McCarthy, Trademark and Unfair Competition §24:6, at 71 (2d ed. 1984 & Supp. 1991) (stating that "[w]here the services consist of retail sales services, likelihood of confusion is found when another mark is used on goods which are commonly sold through such a retail outlet"). A service and a product are related if buyers are likely to assume a common source or sponsorship. . . . The salon services and hair products at issue in this case, which are nearly as kindred as a service and product can be, offer the paradigmatic illustration

---

1. Our recent opinion in Sands, Taylor & Wood Co. v. Quaker Oats Co., 24 U.S.P.Q.2d 1001 (7th Cir. 1992), states explicitly that modern trademark law prohibits the use of a senior user's mark on products that are closely related to the senior user's, as well as those products in direct competition. See 24 U.S.P.Q.2d at 1010; International Kennel Club, Inc. v. Mighty Star, Inc., 846 F.2d 1079, 1089 (7th Cir. 1988). This "protects the owner's ability to enter product markets . . . into which it might reasonably be expected to expand in the future." Sands, Taylor & Wood, 24 U.S.P.Q.2d at 1011; 2 J. Thomas McCarthy, Trademark and Unfair Competition §24:5, at 177 (2d ed. 1984).

of things that are closely related. Thus the majority's disregard for ZHD's substantial use of ZAZU in connection with salon services is unfounded.

As the majority correctly notes, the standard for granting federal registration is somewhat less exacting than that for establishing common law trademark rights. . . . But that distinction is so slight as to be inconsequential here. While any bona fide transaction that is more than a "mere sham" will, when combined with intent to continue use, suffice to support federal registration, Avakoff v. Southern Pacific Co., 765 F.2d 1097, 1098 (Fed. Cir. 1985) . . . , any use greater than de minimis will still warrant trademark protection in the absence of registration. Thus, bona fide test marketing or small experimental sales — indeed, any use that is not nominal or token — can satisfy the test. . . .

In this case, ZHD's use of the ZAZU mark, both in its highly successful salon service business, which drew some out-of-state clients, and in its local and interstate product sales to customers and to a potential marketer, surely is more than de minimis. The extensive evidence of ZHD's intent to step up hair product sales — such as its order for 25,000 ZAZU-emblazoned bottles and its inquiry about advertising rates in a national magazine — bolsters this assessment. Even if ZHD did fail to demonstrate more than a de minimis market penetration nationally, at the very least it successfully established exclusive rights within its primary area of operation. The salon's substantial advertising, increasing revenue and staff and preliminary product sales indicate sufficient market penetration to afford trademark protection in that region. See Natural Footwear Ltd. v. Hart, Schaffner & Marx, 760 F.2d 1383 (3d Cir.) (senior user can establish common law rights in geographic areas where it achieved market penetration), *cert. denied,* 474 U.S. 920, 106 S. Ct. 249, 88 L. Ed. 2d 257 (1985). . . .

L'Oreal concedes that ZHD has exclusive rights to use ZAZU for salon services in the Hinsdale area. Those exclusive rights also preclude L'Oreal from using the mark on hair products in the local area because of the likelihood of confusion between those products and ZHD's salon services, even apart from any confusion between the two parties' products. Given the deferential standard of review on the factual question of use, therefore, I think it clear that ZHD has achieved market penetration and exclusive rights to the ZAZU mark at the very least in the Chicago area.

ZHD's contention that its rights in the ZAZU mark extend beyond the local area is enhanced by evidence that L'Oreal did not, as we have noted, act in good faith. The majority's consideration of the good faith issue minimizes the important role good faith plays in trademark disputes, particularly disputes involving unregistered marks. . . . See, e.g., A. J. Canfield Co. v. Honickman, 808 F.2d 291 (3d Cir. 1986) (stating the doctrine that a senior user "has enforceable rights against any junior user who adopted the mark with knowledge of its senior use"). . . . Contrary to the majority's narrow characterization of bad faith as a concept employed solely to deter attempts to reserve marks prior to genuine sales, courts have examined junior users' good faith in a variety of contexts. In fact, this court has held that a good faith junior user is simply one that begins using a mark without knowledge that another party already is using it. The Money Store v. Harriscorp Finance, Inc., 689 F.2d 666, 674 (7th Cir. 1982); see 2 McCarthy, supra, §26:4 at 292 (equating good faith to "the junior user's lack of knowledge"). And while such knowledge may not automatically negate good faith, only the most unusual situations encompass both knowledge and good faith. . . .

COMMENTS AND QUESTIONS

1. The majority opinion mentions that Zazu Hair Design, "ZHD," had registered its "trade name" as a "service mark" in Illinois in 1980. But how does prior use of a trade name or "service mark" such as Zazu Hair Design, used on a hair salon, affect subsequent use of the same term as a *trademark?* See Malcolm Nicol & Co. v. Witco Corp., 881 F.2d 1063, 1065 (Fed. Cir. 1989) (a trade name, even one that lacks any independent trademark or service mark significance, may bar registration of a trademark or service mark that is confusingly similar to that trade name).

2. The *Zazu* case was decided under pre-1989 federal trademark law. As noted earlier, Congress in 1989 made several important changes in trademark law, including the adoption of "intent to use" registration under section 1 of the Lanham Act. See Trademark Law Revision Act of 1989 (hereinafter "TLRA"), Pub. L. No. 100-667, 102 Stat. 3935, codified at 15 U.S.C. §1051. Under pre-1989 law, *actual* use in commerce, prior to application for registration, was a requirement for registration. This requirement spurred prospective applicants to ship and/or sell a small batch of goods in order to secure trademark protection, a practice that came to be known as "token use."

Among other things, the TLRA amended section 1 of the Lanham Act, 15 U.S.C. §1051, to provide that "[a] person who has a bona fide intention, under circumstances showing the good faith of such person, to use a trademark in commerce may apply to register the trademark . . . on the principal register." Assuming an application based on an intention to use a mark is otherwise allowable, the Trademark Office will issue a "notice of allowance" to the trademark owner (rather than simply registering the mark on the Principal Register). 15 U.S.C. §1063(b)(2). After the notice of allowance is granted, the trademark owner has six months (extendable to one year automatically and to three years for good cause shown) to submit a verified statement that the trademark has in fact been used in commerce, at which point it is entered on the Principal Register. If the trademark owner does not submit such a statement, the trademark is considered abandoned. 15 U.S.C. §1051(d). Assuming that the intent-to-use registrant does eventually use the mark, however, the initial application will be considered "constructive use," entitling the registrant to nationwide priority from the date of the application. 15 U.S.C. §1057(c).

The TLRA was passed largely to bring United States law into harmony with the law in the rest of the world. No country other than the United States required actual use before a mark could be registered. Because international treaties required the United States to give full credit to foreign applicants who had registered marks outside the United States, the United States use requirement in practice meant that *only* U.S. citizens were required to show use in commerce in order to register a mark in the United States.

Is the intent-to-use provision constitutional? Recall that the original United States trademark statute was struck down as unsupported by the patent and copyright clause, and that subsequent trademark statutes (including the Lanham Act) were then enacted under the Commerce Clause. The old requirement in the Lanham Act that registered marks be used in interstate commerce was thought necessary to invoke federal jurisdiction under the Commerce Clause. Does an intent to use a product in interstate commerce confer federal jurisdiction? Compare Charles J. Vinicombe, The Constitutionality of an Intent to Use Amendment to the Lanham Act, 78 Trademark Rep. 361, 369-373 (1988) (courts likely to defer to Congress by assuming that an intention to use a mark means that it is part of the "flow of commerce") with Kenneth L. Port, Foreword: Symposium on Intellectual Property Law Theory, 68 Chicago-Kent L. Rev. 585, 595 n.63 (1993) (Trademark Law Revision Act "should fail constitutional review based on the Commerce Clause").

3. It is widely believed that the TLRA has eliminated "token use" by raising the standard for determining when a mark has been "use[d] in commerce." Consider for example the recent case of Paramount Pictures Corp. v. White, 31 U.S.P.Q.2d 1768 (TTAB 1994) aff'd 108 F.3d 1392 (Fed Cir. 1997) (unpub.), where Paramount opposed registration of "The Romulans" for a connect-the-dots game distributed by White, leader of a rock group called The Romulans. One ground for the opposition was that the mark had not been used in commerce; in particular, that the distribution of connect-the-dots games on promotional fliers for the band was not a statutory use in commerce justifying registration. In commenting on the magnitude of use, the Trademark Trial and Appeal Board stated: "The legislative history of the Trademark Law Revision Act reveals that the purpose of the amendment was to eliminate 'token use' as a basis for registration, and that the new stricter standard contemplates instead commercial use of the type common to the particular industry in question." 31 U.S.P.Q.2d at 1774. Footnote 8 of the Board's opinion quotes from the Congressional Record of November 19, 1987, p. 196-197:

> Amendment of the definition of "use in commerce" is one of the most far-reaching changes the legislation contains. Revised to eliminate the commercially-transparent practice of token use, which becomes unnecessary with the legislation's provision for an intent-to-use application system, it will have a measurable effect on improving the accuracy of the register. . . . The committee intends that the revised definition of "use in commerce" be interpreted to mean commercial use which is typical in a particular industry.

Id. *But cf.* Allard Ent., Inc. v. Advanced Programming Resources, Inc., 146 F.3d 350 (6th Cir. 1998) (plaintiff's somewhat extensive "word-of-mouth" campaign to popularize "APR" mark for computer professional placement service established priority under post-TLRA law). For an example of the pre-1989 rule, see Fort Howard Paper v. Kimberly Clark Corp., 390 F.2d 1015 (C.C.P.A. 1968), *cert. denied,* 393 U.S. 831 (1968) (very limited use sufficient to establish priority). But cf. La Societe Anonyme des Parfums LeGalion v. Jean Patou, Inc., 495 F.2d 1265 (2d Cir. 1974) (token sales program not sufficient use to avoid abandonment); Procter and Gamble v. Johnson & Johnson, Inc., 485 F. Supp. 1185 (S.D.N.Y. 1979), *aff'd without opinion,* 636 F.2d 1203 (2d Cir. 1980) ("minor brands program" not sufficient use).

Could it be argued in the *Zazu* case that ZHD's shipments of hair products were only a "token use"? How might the TLRA have affected the dissent in *Zazu?* Would L'Oreal's good faith still be an issue under the TLRA?

4. Consider this passage from Judge Easterbrook's opinion: "Liberality in registering marks is not problematic, because the registration gives notice to latecomers, which token use alone does not. Firms need only search the register before embarking on development. Had ZHD registered ZAZU, the parties could have negotiated before L'Oreal committed large sums to marketing."

But ZHD *did* register Zazu — as a business (trade name) in the State of Illinois. Why is this any different from the *national* registration contemplated by Easterbrook? Does it provide less opportunity for negotiating prior to large investments? Recall the evidence in the case, which established not only that L'Oreal did find the ZHD trade name in its pre-product introduction search, but also that L'Oreal did contact ZHD. Standard trademark search services generally find all state and federal registrations, together with many "common law" (i.e., nonregistered) uses. Given this evidence, why was state registration any less of a basis for negotiation than federal registration?

5. Judge Easterbrook says that through the use requirement, "the law prevents entrepreneurs from reserving brand names in order to make their rivals' marketing more costly." Could a rival reserve all the potential trademarks that would allow a firm to identify its products? Cf. Stephen L. Carter, The Trouble With Trademarks, 99 Yale L.J. 759, 760 (1990):

> The traditional economic justification for trademark law rests on the premise that the set of available marks is virtually infinite and, in consequence, that the actual mark chosen by a firm to represent its goods is irrelevant. If that assumption turns out to be false — if even before the public comes to associate a mark with any particular goods or services, some marks are more desirable than others — then allowing protection of marks devoid of market significance may raise substantial barriers to entry by competitors. For this reason, I . . . argue, the theoretical case is clearer for the common law model than for the Federal scheme.

Note that another rationale for the use requirement stems not from rivals' costs, but from the desire not to encourage firms to specialize in identifying and registering potential trademarks. See the note below on "Priority and Trademark Theory."

6. Judge Easterbrook provides another explanation for the use requirement: "Public sales let others know that they should not invest resources to develop a mark similar to one already used in the trade." Is there a better and more accurate way to do this? After all, public sales may be quite limited. Does the new federal intent-to-use registration process serve this purpose more effectively?

7. An important case, Blue Bell, Inc. v. Farah Mfg. Co., 508 F.2d 1260 (5th Cir. 1975), teaches that although the race is to the swift in trademark law, the race must be run cleanly if a firm wants to prevail. In the case, Farah's management settled on the name "Time Out" for a new line of blue jeans on May 16. Sample tags using the new name were drawn up on June 27, and on July 3 Farah sent out 12 pair of jeans bearing the new mark to its regional sales managers. More extensive shipments occurred on July 11. Meanwhile Blue Bell management decided on the name "Time Out" for *its* new line of jeans on June 18. Blue Bell commissioned several hundred sample tags (bearing the new logo) that were attached over the top of existing tags on a large shipment of jeans sent out on July 5. By October both firms had received substantial orders for their respective new line of jeans. The court ruled (1) that Blue Bell's "secondary" use of the new logo was in "bad faith" and therefore its July 5 shipment did not establish priority and (2) that Farah's minimal shipments on July 3 also did not establish priority. This left Farah's July 11 shipment as the first substantial use of the new mark in commerce — so Farah won.

When Blue Bell slapped new labels on old jeans, the court condemned it as "a bad faith attempt to reserve a mark." What if subsequent to the relabelling, but before Farah's first shipments, Blue Bell had made actual sales to consumers; would their prior bad faith deprive them of priority? Should it? Why did the court not characterize Farah's sales of pants to employees as an attempt to "reserve a mark"? For more on the myriad roles that bad faith plays in trademark cases, see below, sections D.1, Likelihood of Consumer Confusion, and E.2, Abandonment.

8. Both Blue Bell and Farah began advertising prior to their first sales to customers, yet ultimately the respective sales dates to consumers determined priority. Can advertising or other "promotional activities" ever form the basis for priority?

Maryland Stadium Authority v. Becker, 806 F. Supp. 1236 (D. Md. 1992), *aff'd*, 36 F.3d 1093 (4th Cir. 1994), is instructive on this point. As early as 1987, the Maryland Stadium Authority (MSA) used the name "Camden Yards" in promotional materials (widely distributed to the press and public) describing the new stadium for the Baltimore Orioles baseball team. A number of other names were under consideration at the time, however. In 1989, MSA distributed to the press and sold to the public photographic renditions of the proposed ballpark labeled with the name Camden Yards — again, before the name had been finalized. Defendant Becker, meanwhile, began selling baseball T-shirts under the "Camden Yards" mark in July 1991, shortly after the ballpark's name was made final and official. In an action by

MSA against Becker for trademark infringement under Maryland law, the court found that Becker had intentionally adopted MSA's mark, creating a presumption of the likelihood of confusion that Becker could not overcome. The totality of MSA's activities constituted adoption and use of the mark, generated secondary meaning, and was sufficient to establish priority over Becker. Compare Paramount Pictures Corp. v. White, 31 U.S.P.Q. 2d 1768 (TTAB 1994) (use of name The Romulans on a connect-the-dot game distributed by a rock band held mere "promotional use" and hence not use in commerce under §1 of the Lanham Act). But cf. Societe de Developments et D'Innovations des Marches Agricoles et Aliminetaries-Sodima Union de Cooperatives Agricoles v. International Yogurt, 662 F. Supp. 839, 847 (D. Ore. 1987) ("An axiom of trademark law is: no trade, no trademark."). Note the emphasis on "bad faith" in the *Maryland Stadium* case. Does this distinguish it from cases such as *Paramount*? Or is it another instance of Judge Easterbrook's observation in the *Zazu* opinion that bad faith can be merely "an epithet stapled to . . . [a] conclusion"?

Another use-based limitation on trademark rights is highlighted by Buti v. Impressa Perosa S.R.L., 139 F.3d 98 (2d Cir. 1998). In that case, a restaurant located in Milan, Italy, named "Fashion Café" advertised its business in the U.S. just before Buti began using the same mark for restaurants in the U.S. The court held that the Italian restaurant could not obtain U.S. trademark rights through its advertising because the services it was providing were not themselves offered in commerce in the U.S.

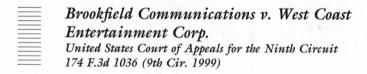

Brookfield Communications v. West Coast Entertainment Corp.
United States Court of Appeals for the Ninth Circuit
174 F.3d 1036 (9th Cir. 1999)

O'SCANNLAIN, Circuit Judge:

We must venture into cyberspace to determine whether federal trademark and unfair competition laws prohibit a video rental store chain from using an entertainment-industry information provider's trademark in the domain name of its web site and in its web site's metatags.

I

Brookfield Communications, Inc. ("Brookfield") appeals the district court's denial of its motion for a preliminary injunction prohibiting West Coast Entertainment Corporation ("West Coast") from using in commerce terms confusingly similar to Brookfield's trademark, "MovieBuff." Brookfield gathers and sells information about the entertainment industry. (A summary of the sequence of events in this case is provided in Figure 5-1.)

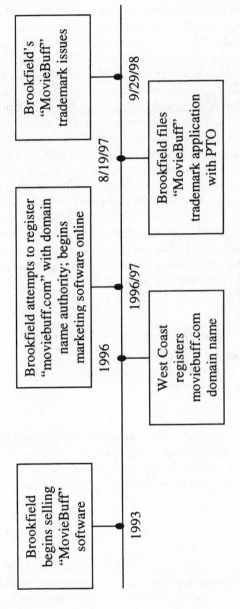

FIGURE 5-1
Brookfield Timeline

Founded in 1987 for the purpose of creating and marketing software and services for professionals in the entertainment industry, Brookfield initially offered software applications featuring information such as recent film submissions, industry credits, professional contacts, and future projects. These offerings targeted major Hollywood film studios, independent production companies, agents, actors, directors, and producers.

Brookfield expanded into the broader consumer market with computer software featuring a searchable database containing entertainment-industry related information marketed under the "MovieBuff" mark around December 1993. Brookfield's "MovieBuff" software now targets smaller companies and individual consumers who are not interested in purchasing Brookfield's professional level alternative, The Studio System, and includes comprehensive, searchable, entertainment-industry databases and related software applications containing information such as movie credits, box office receipts, films in development, film release schedules, entertainment news, and listings of executives, agents, actors, and directors. This "MovieBuff" software comes in three versions — (1) the MovieBuff Pro Bundle, (2) the MovieBuff Pro, and (3) MovieBuff — and is sold through various retail stores, such as Borders, Virgin Megastores, Nobody Beats the Wiz, The Writer's Computer Store, Book City, and Samuel French Bookstores.

Sometime in 1996, Brookfield attempted to register the World Wide Web ("the Web") domain name "moviebuff.com" with Network Solutions, Inc. ("Network Solutions"), but was informed that the requested domain name had already been registered by West Coast. Brookfield subsequently registered "brookfieldcomm.com" in May 1996 and "moviebuffonline.com" in September 1996. Sometime in 1996 or 1997, Brookfield began using its web sites to sell its "MovieBuff" computer software and to offer an Internet-based searchable database marketed under the "MovieBuff" mark. Brookfield sells its "MovieBuff" computer software through its "brookfieldcomm.com" and "moviebuffonline.com" web sites and offers subscribers online access to the MovieBuff database itself at its "inhollywood.com" web site.

On August 19, 1997, Brookfield applied to the Patent and Trademark Office (PTO) for federal registration of "MovieBuff" as a mark to designate both goods and services. Its trademark application describes its product as "computer software providing data and information in the field of the motion picture and television industries." Its service mark application describes its service as "providing multiple-user access to an on-line network database offering data and information in the field of the motion picture and television industries." Both federal trademark registrations issued on September 29, 1998. Brookfield had previously obtained a California state trademark registration for the mark "MovieBuff" covering "computer software" in 1994.

In October 1998, Brookfield learned that West Coast — one of the nation's largest video rental store chains with over 500 stores — intended to launch a web site at "moviebuff.com" containing, inter alia, a searchable entertainment database similar to "MovieBuff." West Coast had registered "moviebuff.com" with Network Solutions on February 6, 1996 and claims that it chose the domain name because the term "Movie Buff" is part of its service mark, "The Movie Buff's Movie Store," on which a federal registration issued in 1991 covering "retail store services featuring video cassettes and video game cartridges" and "rental of video cassettes and video game cartridges." West Coast notes further that, since at least 1988, it has also used various phrases including the term "Movie Buff" . . . On November 10, Brookfield delivered to West Coast a cease-and-desist letter alleging that West Coast's planned use of the "moviebuff.com" would violate Brookfield's trademark rights; as a "courtesy" Brookfield attached a copy of a complaint that it threatened to file if West Coast did not desist.

The next day, West Coast issued a press release announcing the imminent launch of its web site full of "movie reviews, Hollywood news and gossip, provocative commentary, and coverage of the independent film scene and films in production." The press release declared that the site would feature "an extensive database, which aids consumers in making educated decisions about the rental and purchase of" movies and would also allow customers to purchase movies, accessories, and other entertainment-related merchandise on the web site.

Brookfield fired back immediately with a visit to the United States District Court for the Central District of California, and this lawsuit was born. In its first amended complaint filed on November 18, 1998, Brookfield alleged principally that West Coast's proposed offering of online services at "moviebuff.com" would constitute trademark infringement and unfair competition in violation of sections 32 and 43(a) of the Lanham Act, 15 U.S.C. §§1114, 1125(a). Soon thereafter, Brookfield applied ex parte for a temporary restraining order ("TRO") enjoining West Coast "from using . . . in any manner . . . the mark MOVIEBUFF, or any other term or terms likely to cause confusion therewith, including moviebuff.com, as West Coast's domain name, . . . as the name of West Coast's website service, in buried code or metatags on their home page or web pages, or in connection with the retrieval of data or information on other goods or services."

On November 27, West Coast filed an opposition brief in which it argued first that Brookfield could not prevent West Coast from using "moviebuff.com" in commerce because West Coast was the senior user. West Coast claimed that it was the first user of "MovieBuff" because it had used its federally registered trademark, "The Movie Buff's Movie Store,"[5] since 1986 in advertisements, promotions, and letterhead in connection with retail services featuring videocassettes and video game cartridges. Alternatively, West Coast claimed seniority on the basis that it had garnered common-law rights in the domain name by using "moviebuff.com" before Brookfield began offering its "MovieBuff" Internet-based searchable database on the Web. In addition to asserting seniority, West Coast contended that its planned use of

---

5. West Coast applied for a federal trademark registration for this term in 1989, which issued in 1991 and became incontestable in 1996. West Coast purports to have spent over $15,000,000 on advertisements and promotions featuring this mark.

"moviebuff.com" would not cause a likelihood of confusion with Brookfield's trademark "MovieBuff" and thus would not violate the Lanham Act. . . .

II

[The Domain Name System is the hierarchical method used to identify different computers so communications — in the form of data packets — may be properly routed between computers on the Internet. Hierarchical databases store tables of information mapping a domain name (e.g., "berkeley.edu") to a unique server (e.g., 128.32.136.9).]

III

To establish a trademark infringement claim under section 32 of the Lanham Act or an unfair competition claim under section 43(a) of the Lanham Act, Brookfield must establish that West Coast is using a mark confusingly similar to a valid, protectable trademark of Brookfield's. The district court denied Brookfield's motion for preliminary injunctive relief because it concluded that Brookfield had failed to establish that it was the senior user of the "MovieBuff" mark or that West Coast's use of the "moviebuff.com" domain name created a likelihood of confusion. . . .

IV

To resolve whether West Coast's use of "moviebuff.com" constitutes trademark infringement or unfair competition, we must first determine whether Brookfield has a valid, protectable trademark interest in the "MovieBuff" mark. Brookfield's registration of the mark on the Principal Register in the Patent and Trademark Office constitutes prima facie evidence of the validity of the registered mark and of Brookfield's exclusive right to use the mark on the goods and services specified in the registration. See 15 U.S.C. §§1057(b); 1115(a). Nevertheless, West Coast can rebut this presumption by showing that it used the mark in commerce first, since a fundamental tenet of trademark law is that ownership of an inherently distinctive mark such as "MovieBuff" is governed by priority of use. See Sengoku Works Ltd. v. RMC Intl., Ltd., 96 F.3d 1217, 1219 (9th Cir. 1996), *cert. denied*, 521 U.S. 1103, 138 L. Ed. 2d 987, 117 S. Ct. 2478 (1997). The first to use a mark is deemed the "senior" user and has the right to enjoin "junior" users from using confusingly similar marks in the same industry and market or within the senior user's natural zone of expansion. It is uncontested that Brookfield began selling "MovieBuff" software in 1993 and that West Coast did not use "moviebuff.com" until 1996. According to West Coast, however, the fact that it has used "The Movie Buff's Movie Store" as a trademark since 1986 makes it the first user for purposes of trademark priority. In the alternative, West Coast claims priority on the basis that it used "moviebuff.com" in commerce before Brookfield began offering its "MovieBuff" searchable database on the Internet. We analyze these contentions in turn.

A

Conceding that the first time that it actually used "moviebuff.com" was in 1996, West Coast argues that its earlier use of "The Movie Buff's Movie Store" constitutes use of "moviebuff.com." [O]ur sister circuits have explicitly recognized the ability of a trademark owner to claim priority in a mark based on the first use date of a similar, but technically distinct, mark — but only in the exceptionally narrow instance where "the previously used mark is 'the legal equivalent of the mark in question or indistinguishable therefrom' such that consumers 'consider both as the same mark.' " Data Concepts, Inc. v. Digital Consulting, Inc., 150 F.3d 620, 623 (6th Cir. 1998) (quoting Van Dyne-Crotty, Inc. v. Wear-Guard Corp., 926 F.2d 1156, 1159 (Fed. Cir. 1991)); accord *Van Dyne-Crotty*, 926 F.2d at 1159.

This constructive use theory is known as "tacking," as the trademark holder essentially seeks to "tack" his first use date in the earlier mark onto the subsequent mark. See generally 2 J. Thomas McCarthy, McCarthy on Trademarks & Unfair Competition §17:25-27 (4th ed. 1998) [hereafter "McCarthy"].

We agree that tacking should be allowed if two marks are so similar that consumers generally would regard them as essentially the same. Where such is the case, the new mark serves the same identificatory function as the old mark. Giving the trademark owner the same rights in the new mark as he has in the old helps to protect source-identifying trademarks from appropriation by competitors and thus furthers the trademark law's objective of reducing the costs that customers incur in shopping and making purchasing decisions. See Qualitex Co. v. Jacobson Prods. Co., 514 U.S. 159, 163-64, 131 L. Ed. 2d 248, 115 S. Ct. 1300 (1995); Falcon Rice Mill, Inc. v. Community Rice Mill, Inc., 725 F.2d 336, 348 (5th Cir. 1984).

Without tacking, a trademark owner's priority in his mark would be reduced each time he made the slightest alteration to the mark, which would discourage him from altering the mark in response to changing consumer preferences, evolving aesthetic developments, or new advertising and marketing styles. . . .

The standard for "tacking," however, is exceedingly strict: "The marks must create the same, continuing commercial impression, and the later mark should not materially differ from or alter the character of the mark attempted to be tacked." *Van Dyne-Crotty*, 926 F.2d at 1159. This standard is considerably higher than the standard for "likelihood of confusion," which we discuss infra.

[T]he present case is clear cut: "The Movie Buff's Movie Store" and "moviebuff.com" are very different, in that the latter contains three fewer words, drops the possessive, omits a space, and adds ".com" to the end. Because West Coast failed to make the slightest showing that consumers view these terms as identical, we must conclude that West Coast cannot tack its priority in "The Movie Buff's Movie Store" onto "moviebuff.com." As the Federal Circuit explained, "it would be clearly contrary to well-established principles of trademark law to sanction the tacking of a mark with a narrow commercial impression onto one with a broader commercial impression." *Van Dyne-Crotty*, 926 F.2d at 1160 (noting that prior use of "SHAPE UP" could not be tacked onto "EGO," [sic] that prior use of "ALTER EGO" could not be tacked onto "EGO," and that prior use of "Marco Polo" could not be tacked onto "Polo").

Since tacking does not apply, we must therefore conclude that Brookfield is the senior user because it marketed "MovieBuff" products well before West Coast began using "moviebuff.com" in commerce: West Coast's use of "The Movie Buff's Movie Store" is simply irrelevant.

Our conclusion comports with the position of the PTO, which effectively announced its finding of no likelihood of confusion between "The Movie Buff's Movie Store" and "MovieBuff" when it placed the latter on the principal register despite West Coast's prior registration of "The Movie Buff's Movie Store." Priority is accordingly to be determined on the basis of whether Brookfield used "MovieBuff" or West Coast used "moviebuff.com" first.

B

West Coast argues that we are mixing apples and oranges when we compare its first use date of "moviebuff.com" with the first sale date of "MovieBuff" software. West Coast reminds us that Brookfield uses the "MovieBuff" mark with both computer software and the provision of an Internet database; according to West Coast, its use of "moviebuff.com" can cause confusion only with respect to the latter. West Coast asserts that we should accordingly determine seniority by comparing West Coast's first use date of "moviebuff.com" not with when Brookfield first sold software, but with when it first offered its database online.

As an initial matter, we note that West Coast's argument is premised on the assumption that its use of "moviebuff.com" does not cause confusion between its web site and Brookfield's "MovieBuff" software products. Even though Brookfield's computer software and West Coast's offerings on its web site are not identical products, likelihood of confusion can still result where, for example, there is a likelihood of expansion in product lines. See *Official Airline Guides, Inc. v. Goss*, 6 F.3d 1385, 1394 (9th Cir. 1993). As the leading trademark commentator explains: "When a senior user of a mark on product line A expands later into product line B and finds an intervening user, priority in product line B is determined by whether the expansion is 'natural' in that customers would have been confused as to source or affiliation at the time of the intervening user's appearance." 2 McCarthy §16:5. We need not, however, decide whether the Web was within Brookfield's natural zone of expansion, because we conclude that Brookfield's use of "MovieBuff" as a service mark preceded West Coast's use.

Brookfield first used "MovieBuff" on its Internet-based products and services in August 1997, so West Coast can prevail only if it establishes first use earlier than that. In the literal sense of the word, West Coast "used" the term "moviebuff.com" when it registered that domain address in February 1996. Registration with Network Solutions, however, does not in itself constitute "use" for purposes of acquiring trademark priority. See Panavision [Intl., L.P. v. Toeppen 141 F.3d 1316, 1324-25 (9th Cir. 1998)]. The Lanham Act grants trademark protection only to marks that are used to identify and to distinguish goods or services in commerce — which typically occurs when a mark is used in conjunction with the actual sale of goods or services. The purpose of a trademark is to help consumers identify the source, but a mark cannot serve a source-identifying function if the public has never seen the mark and thus is not meritorious of trademark protection until it is used in public in a manner that creates an association among consumers between the mark and the mark's owner. . . .

The district court, while recognizing that mere registration of a domain name was not sufficient to constitute commercial use for purposes of the Lanham Act, nevertheless held that registration of a domain name with the intent to use it commercially was sufficient to convey trademark rights. This analysis, however, contradicts both the express statutory language and the case law which firmly establishes that trademark rights are not conveyed through mere intent to use a mark commercially, see, e.g., Allard Enters. v. Advanced Programming Resources, Inc., 146 F.3d 350, 356 (6th Cir. 1998); Zazu Designs v. L'Oreal, S.A., 979 F.2d 499, 504 (7th Cir. 1992) ("An intent to use a mark creates no rights a competitor is bound to respect."), nor through mere preparation to use a term as a trademark, see, e.g., Hydro-Dynamics, Inc. v. George Putnam & Co., 811 F.2d 1470, 1473-74 (Fed. Cir. 1987); Computer Food Stores, Inc. v. Corner Store Franchises, 176 U.S.P.Q. 535, 538 (T.T.A.B. 1973).

West Coast no longer disputes that its use — for purposes of the Lanham Act — of "moviebuff.com" did not commence until after February 1996. It instead relies on the alternate argument that its rights vested when it began using "moviebuff.com" in e-mail correspondence with lawyers and customers sometime in mid-1996. West Coast's argument is not without support in our case law — we have indeed held that trademark rights can vest even before any goods or services are actually sold if "the totality of [one's] prior actions, taken together, [can] establish a right to use the trademark." New West [Corp. v. NYM Co. of California, Inc., 595 F.2d 1194, 1200 (1979).] Under *New West*, however, West Coast must establish that its e-mail correspondence constituted " 'use in a way sufficiently public to identify or distinguish the marked goods in an appropriate segment of the public mind as those of the adopter of the mark.' " Id. (quoting New England Duplicating Co. v. Mendes, 190 F.2d 415, 418 (1st Cir. 1951)); see also Marvel Comics Ltd. v. Defiant, 837 F. Supp. 546, 550 (S.D.N.Y. 1993) ("The talismanic test is whether or not the use was sufficiently public to identify or distinguish the marked goods in an appropriate segment of the public mind as those of the adopter of the mark.").

West Coast fails to meet this standard. Its purported "use" is akin to putting one's mark "on a business office door sign, letterheads, architectural drawings, etc." or on a prototype displayed to a potential buyer, both of which have been held to be insufficient to establish trademark rights. See Steer Inn Sys., Inc. v. Laughner's Drive-In, Inc., 56 C.C.P.A. 911, 405 F.2d 1401, 1402 (C.C.P.A. 1969); Walt Disney Prods. v. Kusan, Inc., 1979 U.S. Dist. LEXIS 9542, 204 U.S.P.Q. 284, 288 (C.D. Cal. 1979). Although widespread publicity of a company's mark, such as Marvel Comics's announcement to 13 million comic book readers that "Plasma" would be the title of a new comic book, see *Marvel Comics*, 837 F. Supp. at 550, or the mailing of 430,000 solicitation letters with one's mark to potential subscribers of a magazine, see *New West*, 595 F.2d at 1200, may be sufficient to create an association among the public between the mark and West Coast, mere use in limited e-mail correspondence with lawyers and a few customers is not.

West Coast first announced its web site at "moviebuff.com" in a public and widespread manner in a press release of November 11, 1998, and thus it is not until at least that date that it first used the "moviebuff.com" mark for purposes of the Lanham Act. Accordingly, West Coast's argument that it has seniority because it used "moviebuff.com" before Brookfield used "MovieBuff" as a service mark fails on its own terms. West Coast's first use date was neither February 1996 when it registered its domain name with Network Solutions as the district court had concluded, nor April 1996 when it first used "moviebuff.com" in e-mail communications, but rather November 1998 when it first made a widespread and public announcement about the imminent launch of its web site. Thus, West Coast's first use of "moviebuff.com" was preceded by Brookfield's first use of "MovieBuff" in conjunction with its online database, making Brookfield the senior user.

For the foregoing reasons, we conclude that the district court erred in concluding that Brookfield failed to establish a likelihood of success on its claim of being the senior user. [In the remainder of its opinion, the court concluded that West Coast's mark had created a likelihood of confusion under the standard eight-factor *Sleekcraft* test described later in this chapter.]

COMMENTS AND QUESTIONS

1. The court states that it need not decide whether an online database is within the "natural zone of expansion" of West Coast's earlier video rental-

related uses of "Movie Buff." (Recall why not: because Brookfield's use of the mark as a service mark made it the senior use, so West Coast's zone of expansion was irrelevant.) In many other cases, it may well turn out that the first user of a mark will not be the first to take it online by registering its mark (or some variant) in a ".com" format. In today's climate, with "electronic commerce" and Web sites receiving so much attention, when will it *not* be within a firm's natural zone of expansion to take a mark online? Should the zone of expansion concept be applied differently for media and software marks, as compared to marks for industrial products or consumer items?

2. The notion of "tacking" is crucial to the court's analysis. Recall that West Coast argued that its prior use of "Movie Buff" in the video market should be tacked onto its registration of the term with the domain name authority (NSI), thereby bestowing priority on West Coast. Why was West Coast's argument in this regard rejected?

3. The court says that the registration of the domain name "movie-buff.com" with the online registration authority known as NSI merely represented West Coast's "intent to use" the domain name. It did not by itself establish first online use. Do not confuse this aspect of the case with "intent to use" registration under the Lanham Act — an entirely different matter (see section C2, this chapter). In this and other aspects of the case, important guidance is given regarding the interrelationship between the domain name registration system and the traditional statutory registration system. Notice that the former has far less effect in and of itself than the latter and that savvy trademark counsel must learn to coordinate both to carry out clients' strategic objectives regarding "branding" and electronic commerce.

Brookfield bears some resemblance to "cybersquatter" cases. A cybersquatter is someone who registers one or more domain names for the sole purpose of selling them to their "natural" owner. For example, Intermatic, Inc. v. Toeppen, 947 F. Supp. 1227 (N.D. Ill. 1996), Toeppen, a cybersquatter who had registered well-known business names like "eddie-bauer.com," "deltaairlines.com" and, "neiman-marcus.com," was restrained from preventing Intermatic from obtaining the domain name "intermatic.com." Intermatic was not required to compensate Toeppen for the loss of his rights over the domain name. Contrast this with the lucrative (and as of yet uncontested) enterprise of brokering and auctioning domain names. A recent auction of "drugs.com" resulted in a sale of the domain name for $823,456. This type of transfer seems to be appropriate where there are several potential buyers, and the domain name is a generic one that has not acquired any secondary meaning.

The interaction between domain names and trademarks is complex, and we return to it in more detail in Chapter 7.

Note on Geographic Limitations on Trademark Use

At common law (and today for unregistered marks), ownership of a trademark does not necessarily confer nationwide protection. Rather, common law trademarks are protected only in the areas where the marked products are sold or advertised. Thus the owner of an unregistered trademark for goods sold in Oregon and Washington, but not elsewhere, is entitled to prevent others from using that mark for similar goods only in Oregon and Washington. The rationale is that trademarks are not intended to confer a broad property right but merely to protect the goodwill the trademark owner has invested in the mark. Because no one outside Oregon or Washington could associate the mark with the owner, there is no reason to protect it elsewhere. Thus a seller of similar goods in New York can use the same name for the goods without conflict.

There are two exceptions to this common law rule, both based on concerns that the trademark owner's goodwill will be unfairly taken. First, a trademark owner is entitled to the exclusive use of her mark in any geographic area in which the mark's reputation has been established, even if the product is not sold in that geographic area. Such a broader geographic reputation might be established, for example, by national advertising or media coverage of a local business such as a restaurant. Further, the trademark owner is entitled to protect the mark in a territory which he is expected to reach in the normal expansion of his business, even if there is no current likelihood of confusion in that area. See Hanover Star Milling Co. v. Metcalf, 240 U.S. 403 (1916). Second, a trademark owner is entitled to prevent anyone from intentionally trading on her goodwill, even outside her established geographic area. Only innocent (or "good faith") use of the same mark is protected.

One of the principal advantages of trademark registration is that it automatically confers nationwide protection of the mark, retroactive to the date of the trademark application, even if the goods for which the mark is used are sold or advertised in only a small part of the country. Thus trademark registration is vital to protect businesses that plan to expand geographically, as well as those that fear a large national company might use the same name.

What happens, then, if two parties use the same mark for the same goods? If neither party registers its mark, then the common law rule applies. Each party is entitled to exclusive use of the mark in the areas where it has established goodwill. Should the two marks come into conflict in a particular geographic area, the conflict will be resolved in favor of the earliest user *in that area*. If the earliest user anywhere in the country registers her mark, the registrant will generally be entitled to the exclusive right to use the mark throughout the country. However, the non-registering party may assert a "limited area" defense. This defense allows the non-registering party to claim priority in those geographic areas where he has made continuous use of the mark since *before* the registering party filed her application. The non-registering party is "frozen" in the use of his mark, however, and cannot expand it outside his existing territory or a natural "zone of expansion."[11]

***

11. Courts have also allowed junior users to continue sales in certain areas not used by registered trademark owners. Thus in Dawn Donut Co. v. Hart's Food Stores, 267 F.2d 358 (2d Cir. 1959), the court held that Hart's could continue to sell donuts under the trademark

Finally, if both parties file for registration of marks that were in use in different areas at the same time, the Trademark Office will generally declare an "interference" between the two applications. But if the parties agree, or if the Trademark Board determines that registration of both marks is unlikely to cause confusion, it is possible that both marks may be registered for "concurrent use."[12] If two or more marks are registered concurrently, however, the Trademark Office will impose whatever restrictions on the use of the marks are necessary to prevent confusion among consumers.

COMMENTS AND QUESTIONS

1. Compare the current U.S. trademark system to the various schemes for filing patent applications. You will recall that the United States has long used a "first to invent" rule for determining who is entitled to own a patent, while the rest of the world awards a patent to the "first to file." Is there a similar distinction in trademark law, between "first to use" and "first to file"? If so, does the Trademark Law Revision Act turn the United States into a "first to file" system?

2. Why should the user of an arbitrary or suggestive, but unregistered, trademark be limited to protection in a particular geographic area? The asserted justification is that the trademark owner has only established goodwill in that limited area. But why should that matter? Distinctive marks, unlike descriptive marks, are entitled to automatic protection under trademark law without proof that the public associates them with a particular product. See supra section C.1. Isn't it inconsistent to limit the scope of that protection to geographic areas in which the public has formed such an association?

3. Is the *Dawn Donut* rule (discussed in footnote 10) consistent with a consumer protection rationale for trademark law? Consider the likely effect on consumers if Dawn decides to expand into Rochester. For 30 years, they have bought "Dawn" donuts exclusively from Hart. Now, according to the Second Circuit, they will be able to buy "Dawn" donuts only from Dawn. Hart may still sell donuts in Rochester, but it will have to change its product name. How exactly will this keep consumers in Rochester from being confused? What incentives does it give Hart and Dawn to invest in product quality?

"Dawn" in the Rochester, New York, area, even though the plaintiff had registered the Dawn trademark. The court reasoned that because Dawn did not sell in the Rochester area and had no current plans to expand there, Hart's use of the mark was not likely to confuse consumers. However, the court made it clear that Dawn could enter the Rochester market at any time, and that if it did so Hart would have to stop using the mark. Note that the *Dawn Donuts* rule has been questioned in recent years. *See* Circuit City Stores, Inc. v. Car Max, Inc., 165 F.3d 1047 (6th Cir. 1999); Members First Federal Credit Union v. Members 1st Federal Credit Union, 54 F. Supp. 2d 393, 402 (M.D. Pa. 1999) (questioning applicability of *Dawn Donuts* rule in a "highly mobile, technologically-driven society").

12. We discuss interferences in more detail in the next section. Concurrent use applies only to two marks that are actually in use, not to applications based on an intent to use.

4. Priority disputes are often resolved by the Trademark Trial and Appeal Board through oppositions or interferences. Section 13 of the Lanham Act, 15 U.S.C. §1063, provides that "[a]ny person who believes that he would be damaged by the registration of a mark upon the principal register may . . . file an opposition in the Patent and Trademark Office, stating the grounds therefor." Further, 15 U.S.C. §1062(a) expressly provides that trademark applications be published before issuance, so that interested parties may have the opportunity to search for and oppose potentially damaging applications. Applications may be opposed by showing that the mark is not entitled to registration, for example because others had made use of it before the applicant did.

Often the party objecting to an application will herself have filed an application for the same or a similar mark. In that circumstance, the Trademark Commissioner may declare an interference between the two applications. Under section 16 of the Lanham Act, 15 U.S.C. §1066, interferences may also be declared between pending applications and registered trademarks (unless they have become "incontestable"). Interferences are heard by the Trademark Trial and Appeal Board, which may register neither, either, or both marks, and may impose restrictions or conditions on the registration of the mark. 15 U.S.C. §§1067, 1068. Both oppositions and interference decisions may be appealed to the Court of Appeals for the Federal Circuit.

Note on Priority and Trademark Theory

Do any of the tests for priority discussed so far make sense in terms of lowering consumer search costs? Consider: if a second user can freely appropriate a mark, then consumers who have begun to rely on the association between the mark and the first user's product will be thwarted. This situation not only destroys that particular association with the first user's product; it may also make consumers less likely to establish such associations in the future. The upshot is that unless we protect the rights of the first user, more consumers will spend more time searching for goods.

On the other hand, many priority cases involve very limited uses in commerce; these cases typically occur early in the life of a new product or marketing campaign. As a consequence, very few consumers will, at the time of the litigation, have come to associate the mark in question with any goods. Where this is so, we might consider issues other than absolute priority to be important. For example, we might ask whether the first or second user was better positioned to distribute the goods bearing the mark. If the second user was in a better position — e.g., was larger, had more money to spend on advertising, etc. — why not let it use the mark? If consumer search costs are the key, why not take into account the interests of *future* consumers, who may be better served by allowing a search-cost-reducing trademark to fall into the hands of a large company that can make best use of it?

Then again, if the second user can really make better use of the mark, wouldn't it buy the trademark from the first user? When this is so, an award of priority to the first user will give it a share of the value of the market that will ultimately be served by the second user. Perhaps the payment to the first user will compensate it to some extent for the effort expended in coming up with a mark that, in the second user's hands, has real commercial value. (Notice that if this is so, it is a far cry from the consumer protection theory of trademark protection; it comes closer to the property theory described in the Introduction.)

Priority by Contract: Covenants Not to Sue and the Like. Notice in this connection the $125,000 paid by L'Oreal to Riviera Slacks, Inc. in *Zazu*. Recall that Rivera was an early user of the Zazu mark on related goods (clothes). (Note also that the court characterizes this agreement as a "covenant not to sue." For an explanation of this characterization — as opposed to "trademark license agreement" — see below, section C.5.) This payment is interesting for several reasons. First, it suggests that priority is not a once-and-for-all determination. Since the first user can license the trademark to the second-comer, an award of priority does not determine for all time who can use the mark. In some cases, it merely sets the scene for a subsequent contract. Where the second-comer is in a better position to exploit the mark, an award of priority to the first user may simply result in an exchange of money for the right to use the mark. On the other hand, where the first user can more efficiently exploit the mark, economic theory suggests that it will continue to do so.

Priority and the Prevention of Trademark Races. Why should the law grant rights to the first user of a mark? Consider the alternative: if two entities have to "race" to establish nationwide recognition for their mark, will they spend more money widely distributing their new product — and advertising it — than they would otherwise? Would such extensive, early promotional efforts be wasteful? Is the role of trademark priority to forestall such wasteful expenditures and instead promote a more rational product "rollout" nationwide? Should the parties ever be able to divide up the nation?

Some guidance may be gleaned from a prominent justification for secure title in real property holdings. It has been argued that without secure property rights, assigned in advance, those seeking use of a resource will make wasteful expenditures seeking to claim it. The idea is that the orderly, rational development of the resource will be distorted by the absence of property rights. For example, consider "land rushes," such as the famous Oklahoma Land Rush. See Terry L. Anderson & Peter J. Hill, The Race for Property Rights, 33 J.L. & Econ. 177 (1990); R. Taylor Dennen, Some Efficiency Effects of Nineteenth-Century Federal Land Policy: A Dynamic Analysis, 51 Agric. Hist. 718 (1977); David D. Haddock, First Possession Versus Optimal Timing: Limiting the Dissipation of Economic Value, 64 Wash. U. L.Q. 775 (1986). Economic historians have detected evidence of wasteful spending by prospective claimants. See generally Gary D. Libecap, Contracting for Property Rights (1989). Instead of allocating early expenditures rationally — e.g., some money for land, some for seeds, fertilizer, and building materials — people put all their money into the pursuit of land claims. The upshot was that the land was not utilized in an efficient manner. (Compare this line of reasoning to the "prospect theory" of patents described in Chapter 3.)

If the analogy between land development and trademark investment make sense, why not go all the way to a pure registration system, under which virtually no expenditures need be made to secure trademark rights?

As another branch of economic theory would predict, "pure" registration systems — those where broad rights can be acquired without actual use in commerce — have been known to give rise to the scattershot acquisition of numerous trademarks solely for their value to real prospective users. In short, pure registration also invites "rent seeking." Cf. William M. Landes & Richard A. Posner, Trademark Law: An Economic Perspective, 30 J.L. & Econ. 265, 275 (1987). Consider the case of Robert Aries, who in 1965 had the foresight to register over 100 valuable American trademarks including Pan American, NBC, Texaco, Monsanto, and Goodrich with the National Trademark Office of Monaco. After registering the trademarks, Mr. Aries forced the American companies to buy their own marks back from him. Gerald D. O'Brien, The Madrid Agreement Adherence Question, 56 Trademark Rep. 326, 328 (1966). Similar practices were well known under French trademark law, until recent reforms. See, e.g., Andre Armengaud, The New French Law on Trademarks, 56 Trademark Rep. 430, 435-36 (1966):

> In France, during the last few years before the enactment of the 1964 Act, trademark registrations were becoming more and more numerous. This was due to the fact that many persons to whom a fancy name would come to mind, would register the name with the ulterior motive of obtaining some financial return from a possible subsequent user should the occasion arise. As a result, the area of choice for marks for new products was becoming narrower every day. Another drawback was that a merchant, or a manufacturer engaged in a particular trade area like hosiery for instance, would register his mark in all thirty-four classes. The relatively low cost of trademark registration in France, negligible as compared to the high costs a large company usually bears for advertising, made such practices possible. . . . [T]hese two practices . . . were responsible for the tremendous volume of trademark registration.

On the end of these practices in France, see Gerard Dassas, Survey of Experience Under the French Trademark Law, 66 Trademark Rep. 485, 491 (1976).

A similar phenomenon has occurred more recently with Internet domain names. The Internet is run on a decentralized basis, but a private company called Network Solutions, Inc. (NSI) is in charge of assigning e-mail and Web addresses to those who request them. Historically, NSI has given those names out on a first-come, first-serve basis. During the mid-1990s, however, this approach ran into problems. Private individuals registered names such as mcdonald's.com and postit.com before the companies who owned the relevant trademarks could do so. Some companies even registered the marks of their competitors, and used the sites to put up comparative product information. See, e.g., Joshua Quittner, Making a Name on the Internet, Newsday, Oct. 7, 1994, at A4 ("It['s] . . . like a gold rush: Two thousand requests a month are coming in to stake claim to a name on the Internet, nearly 10 times as many as a year ago."). To the extent that NSI's first-come, first-serve policy can be considered an experiment in pure registration, it arguably did not work very well. NSI has modified the policy repeatedly since 1995. Courts have also caught on to the rent-seeking game of Internet domain name registrations solely for sale to "real" users of the name. Intermatic v. Toeppen, 40 U.S.P.Q.2d 1412 (N.D. Ill. 1996). These cases are discussed in more detail in Chapter 7.

There seems to be good reason to allow some sort of early claiming system (i.e., registration with national effect after minimal use) without going all the way to "pure" registration (which invites equally wasteful rent seeking). An alternative would be to allow pure registration with lapse for nonuse after some period of time. Notice that this quite adequately describes the current system of "intent to use" registration.

Incentives to Create Trademarks. Recall the respective efforts of Zazu Hair Design (ZHD), the plaintiff in the case that begins this section, and Wordmark, the consulting firm that L'Oreal hired to help identify a catchy name for its new hair coloring products. Why should ZHD potentially receive rights to the trademark when it appears that Wordmark thought to use Zazu as a product name first? (ZHD had only been using it as a trade name.) In other words, why doesn't the law grant rights to firms that specialize in creating new trademarks? Why must a mark be "used in commerce," as the statute phrases it, to receive legal protection? An argument for abolishing this requirement was made by Stephen Carter in his article, Does It Matter Whether Intellectual Property Is Property?, 68 Chicago-Kent L. Rev. 715, 721 (1993): "But perhaps we should say, So what? How simple and elegant it would be to conclude that secondary meaning is unnecessary because the first to appropriate the mark owns it; owns it not because of its representational nature, but because it is a product of the mind."

Consider some counterarguments:

> Promotional value, good will, popularity, and similar elements of value are joint products of both the public and the creator to a greater extent than are intellectual products themselves. True, even standard intellectual products — collocations of words, music, scenes — will be beneficial only if someone appreciates them; labor is never the only source of value, even for Locke. But with standard intellectual products the active role of the producer and the comparatively passive role of the public makes it easier to assign the resulting value primarily to the laborer. By contrast, with products such as popularity and "commercial magnetism," the chain of causality and responsibility is much harder to trace. . . . Usually a trademark is not purposely created for its own sake; the "benefit" purposely created is the good will of the owning entity (such as its

reputation for manufacturing high-quality products), which the trademark merely happens to represent.

Wendy J. Gordon, A Property Right in Self-Expression: Equality and Individualism in the Natural Law of Intellectual Property, 102 Yale L.J. 1533, 1588 n.277 (1993); cf. Hal Morgan, Symbols of America (1987) (noting broad cultural impact of well-known trademarks). Note the skepticism Professor Gordon apparently feels regarding the claim that trademarks are brought about by legal incentives. This skepticism is shared by Rochelle Dreyfuss. See Rochelle C. Dreyfuss, Expressive Genericity: Trademarks as Language in the Pepsi Generation, 65 Notre Dame L. Rev. 397, 399 (1990) ("[T]here is little need to create economic incentives to encourage businesses to develop a vocabulary with which to conduct commerce. . . .").

What do you think would happen if the law did give rights to firms such as Wordmark? Would it deplete the "stock" of potential trademarks available to actual manufacturers and sellers? Would it be burdensome if many firms who wanted to start a business had to go first to a firm like Wordmark to "shop for a trademark"? How would the parties agree on a fair price for a trademark?

Keep in mind when thinking about these issues that even though firms such as Wordmark are not granted a legal right under the Lanham Act, such firms exist anyway. Indeed, from published reports, it would appear that the image/identity industry is thriving. Is this a good argument against a system that granted rights to these firms?

Do you suppose ZHD would have a cause of action against Wordmark? How about L'Oreal, on the grounds that Wordmark chose a trademark that wound up in costly litigation? Perhaps in the contract specifying Wordmark's services, Wordmark expressly disclaims liability for subsequent litigation.

COMMENTS AND QUESTIONS

1. The *Zazu* case and its ilk center on what constitutes "use" for trademark purposes. But what happens when the "use" asserted by a plaintiff originates not with the seller of goods itself, but with the common parlance of consumers? Consider, for example, Volkswagenwerk A.G. v. Advanced Welding & Mfg. Co., 193 U.S.P.Q. 673 (TTAB 1976), which concerned trademark rights in the word "Bug." To Volkswagen (VW), the official moniker for its classic economy car was "Type I," or, later, "Beetle." But in common parlance, this car was universally referred to as a "Bug." Importantly, it was the consuming public, and not VW, that originated this usage. Should trademark law protect the association between VW and "Bug," even though "Bug" originated with consumers themselves? If so, it is a reflection of the "consumer protection" rationale for trademarks, as opposed to the incentive/property rationale. (Do you see why? How much did VW invest in

creating the "Bug" mark?) Courts tend toward the view that consumer associations should be protected in this context. Id. See also National Cable Television Assn. v. American Cinema Editors, Inc., 937 F.2d 1572 (Fed. Cir. 1991) (ACE used in common parlance as an acronym for association of film (cinema) editors).

2. Difficult priority problems arise when a well-known trademark is abandoned by its original owner. The priority question involves a race among rivals to capture the mark. See California Cedar Products Co. v. Pine Mountain Corp., 724 F.2d 827, 830 (9th Cir. 1984) (bad faith of first rival to claim abandoned Duraflame mark for ersatz fireplace logs negates its claim, leaving second rival to claim the mark with priority). Are the issues the same as when firms race to obtain rights to a new mark? Under a strict consumer protection rationale for trademarks, is there an argument that abandoned marks should be off-limits to rival firms, at least for a number of years? Why would a firm abandon a mark when it has value to rivals — why not sell it? (See the discussion below on Assignments in Gross, in the section on Licensing, section C.5.b.ii.)

Note on Secondary Meaning in the Making

Consider how the doctrine of priority interacts with the doctrine of secondary meaning. Under trademark priority rules, the trademark is presumptively owned by the first person to use it in commerce (barring a federal registration). But the secondary meaning doctrine provides that *descriptive* marks are not entitled to protection until their owner can prove secondary meaning in the minds of consumers. So when does a trademark owner obtain priority of use in a descriptive mark? When she first uses the mark? Or only after she can establish secondary meaning?

This issue has arisen in a number of cases where the defendant is accused of quickly adopting a plaintiff's descriptive mark before the plaintiff can establish secondary meaning. In Laureyssens v. Idea Group, Inc., 964 F.2d 131 (2d Cir. 1992), the court considered a trade dress infringement suit by the makers of "Happy Cube" 3-D puzzles against the makers of "Snafooz" puzzles:

> The district court found that there was no serious question whether actual secondary meaning exists in the HAPPY CUBE trade dress. We think this conclusion is sound given the weak sales of the HAPPY CUBE puzzles, low expenditures for advertising and promotion, minimal unsolicited media coverage, and the brief period of exclusive use of the HAPPY CUBE trade dress.[4] And, while there was evidence of intentional imitation as to the puzzles themselves, there was no evidence of copying of the trade dress.

---

4. The record indicates that the HAPPY CUBE trade dress was adopted in 1990, and for some of that time Idea Group was marketing its puzzles in the allegedly infringing trade dress.

The district court concluded, however, that Laureyssens satisfied the requirement of secondary meaning by raising a serious question whether the flat-form, shrink-wrapped HAPPY CUBE trade dress should be protected under the doctrine of secondary meaning in the making.

In Metro Kane Imports, Ltd. v. Federated Dept. Stores, Inc., 625 F. Supp. 313, 316 (S.D.N.Y. 1985), *aff'd*, 800 F.2d 1128 (2d Cir. 1986), Judge Sweet explained that a trade dress will be "protected against intentional, deliberate attempts to capitalize on a distinctive product" where "secondary meaning is 'in the making' but not yet fully developed." See also Jolly Good Industries, Inc. v. Elegra, Inc., 690 F. Supp. 227, 230-31 (S.D.N.Y. 1988) (indicating that the theory has been "well-received by commentators" and citing as an example, 3 R. Callman, The Law of Unfair Competition, Trademarks and Monopolies §77.3, at 356 (3d ed. 1971)).[5] The supposed doctrine seeks to prevent pirates from intentionally siphoning off another's nascent consumer recognition and goodwill. See, e.g., Jolly Good, 690 F. Supp. at 230-31.

In this case, Judge Sweet found that although Idea Group offered evidence that the SNAFOOZ packaging was developed without prior knowledge of the HAPPY CUBE trade dress, "the evidence does not indicate when [Idea Group] developed its flat-form shrink wrapped package, and therefore Plaintiffs have barely established a serious question of [Idea Group's] 'intentional deliberate attempts' to copy their trade dress." 768 F. Supp. at 1048.

We are, then, squarely presented for the first time with the question whether the doctrine of secondary meaning in the making should be recognized under the Lanham Act. . . .

. . . "The doctrine, if taken literally, is inimical to the purpose of the secondary meaning requirement." Restatement, supra, §13 reporter's note, comment *e* at 53. The secondary meaning requirement exists to insure that something worth protecting exists — an association that has developed in the purchasing public's mind between a distinctive trade dress and its producer — before trademark law applies to limit the freedom of a competitor to compete by copying. As the drafters of the Restatement, supra, §17 comment b at 104-05, explain:

> The freedom to copy product and packaging features is limited by the law of trademarks only when the copying is likely to confuse prospective purchasers as to the source or sponsorship of the goods. The imitation or even complete duplication of another's product or packaging creates no risk of confusion unless some aspect of the duplicated appearance is identified with a particular source. Unless a design is distinctive . . . and thus distinguishes the goods of one producer from those of others, it is ineligible for protection as a trademark.

See also Norwich Pharmacal Co. v. Sterling Drug, Inc., 271 F.2d 569, 572 (2d Cir. 1959) ("Absent confusion, imitation of certain successful features in another's product is not unlawful and to that extent a 'free ride' is permitted."), *cert. denied*, 362 U.S. 919, 4 L. Ed. 2d 739, 80 S. Ct. 671 (1960); Perfect Fit Industries, Inc. v. Acme Quilting Co., 618 F.2d 950, 952-53 (2d Cir. 1980). The so-called doctrine of secondary meaning in the making, by affording protection before prospective purchasers are likely to associate the

---

5. The subsequent edition of Callman's treatise, however, contains no language expressing approval for the doctrine of secondary meaning in the making. See 3 R. Callman, The Law of Unfair Competition, Trademarks, and Monopolies §19.27 (4th ed. 1989).

trade dress with a particular sponsor, constrains unnecessarily the freedom to copy and compete.

The Eighth Circuit previously recognized the improper focus of the concept of secondary meaning in the making in Black & Decker Mfg. v. Ever-Ready Appliance Mfg., 684 F.2d 546, 550 (8th Cir. 1982): "Such a theory focuses solely upon the intent and actions of the seller of the product to the exclusion of the consuming public; but the very essence of secondary meaning is the association in the mind of the public of particular aspects of trade dress with a particular product and producer." See also Scagnelli, Dawn of a New Doctrine? — Trademark Protection for Incipient Secondary Meaning, 71 Trademark Rep. 527, 542-43 (1981).

The argument in favor of permitting development of a doctrine of secondary meaning in the making, offered by Laureyssens, rests principally on the supposition that, without such a doctrine, there will be strong incentives for pirates to capitalize on products that have not yet developed secondary meaning. This argument, however, underestimates the level of protection afforded under existing law to prevent piracy in the early stages of product development. See Scagnelli, supra, at 543-49. For example, intentional copying is "persuasive evidence" of secondary meaning. Coach Leatherware, 933 F.2d at 169 (noting, however, that "conscious replication alone does not establish secondary meaning"); see also Restatement, supra, §17 comment b at 106. Furthermore, secondary meaning can develop quickly to preclude knock-off artists from infringing. See Maternally Yours, Inc. v. Your Maternity Shop, Inc., 234 F.2d 538, 541 (2d Cir. 1956) (secondary meaning acquired in mark MATERNALLY YOURS for maternity apparel store in the 11 months preceding defendant's opening of store named YOUR MATERNITY SHOP). Finally, under New York's common law of unfair competition, a producer's trade dress is protected without proof of secondary meaning against practices imbued with an odor of bad faith. See Saratoga Vichy Spring Co. v. Lehman, 625 F.2d 1037, 1044 (2d Cir. 1980). These practices include palming off, actual deception, appropriation of another's property, see Norwich Pharmacal, 271 F.2d at 570-71; Upjohn Co. v. Schwartz, 246 F.2d 254, 261-62 (2d Cir. 1957), or deliberate copying. See Morex S.P.A. v. Design Inst. of Am., Inc., 779 F.2d 799, 801-02 (2d Cir. 1985); Perfect Fit Indus., 618 F.2d at 952-54; Restatement, supra, §16 reporter's note at 115. Therefore, true innovators, at least under New York law, have adequate means of recourse against free-riders.

For these reasons, we reject the doctrine of secondary meaning in the making under section 43(a) of the Lanham Act. Accordingly, we reverse the district court's decision to grant a preliminary injunction against Idea Group based on section 43(a) of the Lanham Act. Given our holding, we need not address the likelihood of confusion which may have been created by the SNAFOOZ trade dress.

COMMENTS AND QUESTIONS

1. Courts have been virtually unanimous in rejecting the idea of protecting "secondary meaning in the making" as inconsistent with the idea that descriptive marks are not protected. Indeed, many courts expressly require

that a trademark plaintiff obtain secondary meaning before the defendant begins any use of the term. See 1 J. Thomas McCarthy, Trademarks and Unfair Competition §16.12, at 16-40 to 16-43 (collecting cases). Cf. Fuddruckers, Inc. v. Doc's B.R. Others, Inc., 826 F.2d 837 (9th Cir. 1987) (plaintiff was entitled to protect restaurant with trade dress nationally recognized among travellers against infringement by restaurant in Arizona, even though plaintiff had not established secondary meaning in Arizona directly). The Trademark Office, however, subscribes to a different theory of priority of marks with secondary meaning. The Board's position is that where two parties seek to register the same descriptive marks, registration should be granted to whichever mark first acquired secondary meaning, regardless of when the two marks were first used.

Which position makes more sense? If you favor the stricter test of Professor McCarthy and the courts, what happens to trademarks that acquire secondary meaning too late? Are they wholly unprotectable? Or can they be asserted against a third party who uses the mark after secondary meaning is acquired? If you favor the "race to priority" test advocated by the Board, must you necessarily conclude that secondary meaning, once acquired, is retroactive? Is it fair to prevent the use of a mark by someone who had a perfect right to use it at the outset, merely because another party has since acquired secondary meaning?

2. What is wrong with protecting secondary meaning "in the making"? Suppose that in *Laureyssens*, the plaintiff had proven that Idea Group copied its descriptive packaging exactly because it knew that Laureyssens was advertising its product heavily and that Idea Group intended to trade on Laureyssens' expected success in making its packaging distinctive. Is it really fair to permit Idea Group to borrow Laureyssens' future goodwill merely because Laureyssens hasn't yet succeeded in establishing that goodwill? The court acknowledges this argument but contends that the doctrine of secondary meaning adequately protects against "piracy in the early stages of product development." Is the court's argument persuasive?

PROBLEMS

Problem 5-2. Preco Industries began using the term "Porcelaincote" for its porcelain resurfacing material in 1966. Preco concedes that, at the time it began use, the "Porcelaincote" mark was unprotectable because it was descriptive. Preco continued to use the mark for a relatively minor product line. In 1977, Ceramco began to use the identical mark for identical goods. [Both companies sell their products on a nationwide basis.] In 1979, Ceramco began a nationwide advertising campaign using the "Porcelaincote" mark. Shortly thereafter, in 1980, Preco began a similar campaign. In 1981 the parties filed complaints against each other for trademark infringement. The court determined at trial that Ceramco established secondary meaning in 1979, and that Preco established it in 1980.

Assume that neither party ever registered its mark. Who should prevail in the lawsuits? Does either party have rights against a third company that began using the name in 1990? Does your answer to either of these questions change if one or both parties registered its mark on the Principal Register?

Problem 5-3. In May 1989, Shalom Children's Wear begins advertising and planning a line of clothes to be called "Body Gear." In November 1989, In-Wear Corp. files an intent-to-use application for the mark "Body Gear." Shalom files an intent-to-use application for the same mark in December 1989 and begins selling its products in February 1990. Is either party entitled to register the Body Gear mark? (Assume that the mark is otherwise protectable.) Does the result change if In-Wear fails to begin selling its clothes by November 1992?

3. Trademark Office Procedures

Administrative proceedings concerning trademarks in the United States are among the most stringent in the world. The U.S. Patent and Trademark Office (PTO) actively examines applications and — with the help of the courts — polices the trademark registers as well. In this section we review these administrative procedures.

a. Principal vs. Supplemental Register

Although registration is not a prerequisite to trademark protection, trademarks registered on the Principal Register enjoy a number of significant advantages. The primary advantages are: (1) nationwide constructive use and constructive notice, which cut off rights of other users of the same or similar marks, Lanham Act §22 (15 U.S.C. §1072) and Lanham Act §7(c) (15 U.S.C. §1057(c)); (2) the possibility of achieving incontestable status after five years, which greatly enhances rights by eliminating a number of defenses, Lanham Act §15 (15 U.S.C. §1065); and (3) the right to bring a federal cause of action without regard to diversity or minimum amounts in controversy, Lanham Act §39 (15 U.S.C. §1121).[13]

Trademark applications are maintained in an index at the PTO and made

---

13. Other advantages include: (1) the right to request customs officials to bar the importation of goods bearing infringing trademarks, Lanham Act §42 (15 U.S.C. §1124), and (2) provisions for treble damages, attorney fees, and certain other remedies in civil infringement actions, Lanham Act §§34-38 (15 U.S.C. §§1116-1120).

available for public scrutiny soon after filing. This procedure is different from patent applications, the contents of which are kept secret until the patent issues. See Chapter 3.

The Supplemental Register was established by the 1946 Lanham Act "to enable persons in this country to domestically register trademarks so that they might obtain registration under the laws of foreign countries." 2 McCarthy on Trademarks §19.09[1] (1996), at p. 19-68. Under the Paris Convention, foreign registration could not be granted in the absence of domestic registration. Because there are countries where trademark registration is granted to marks that would not qualify for the U.S. Principal Register, the Supplemental Register was created. Thus, even if a U.S. mark cannot gain the advantages of registration on the Principal Register, it may obtain protection in foreign countries.

To be eligible for the Supplemental Register a mark need only be capable of distinguishing goods or services. There is no need to prove that it actually functions in that capacity. The Supplemental Register is not available for clearly generic names, but it is available for the registration of trade dress.

Unlike the Principal Register, registration on the Supplemental Register confers no substantive trademark rights "beyond common law." See, e.g., Clairol, Inc. v. Gillette Co., 389 F.2d 264 (2d Cir. 1968). Registration on the Supplemental Register has no evidentiary effects, it does not provide constructive notice of ownership, the mark cannot become incontestable, and it cannot be used as a basis for the Treasury Department to prevent the importation of infringing goods. However, a mark on the Supplemental Register may be litigated in federal court; may be cited by the PTO against a later applicant; and may provide notice to others that the mark is in use. See In re Clorox Co., 578 F.2d 305 (C.C.P.A. 1978). Marks registered on the Supplemental Register are not subject to intent-to-use filings, interference proceedings, or opposition challenges, but may be canceled at any time by a court.

b. Grounds for Refusing Registration

Section 2 of the Lanham Act provides the basis for many of the grounds for refusing registration on the Principal Register:

No trademark by which the goods of the applicant may be distinguished from the goods of others shall be refused registration on the principal register on account of its nature unless it —

    (a) Consists of or comprises immoral, deceptive, or scandalous matter; or matter which may disparage or falsely suggest a connection with persons, living or dead, institutions, beliefs, or national symbols, or bring them into contempt, or disrepute.

(b) Consists of or comprises the flag or coat of arms or other insignia of the United States, or of any State or municipality, or of any foreign nation, or any simulation thereof.

(c) Consists of or comprises a name, portrait, or signature identifying a particular living individual except by his written consent, or the name, signature, or portrait, of a deceased President of the United States during the life of his widow, if any, except by the written consent of the widow.

(d) Consists of or comprises a mark which so resembles a mark registered in the Patent and Trademark Office, or a mark or trade name previously used in the United States by another and not abandoned, as to be likely, when used on or in connection with the goods of the applicant, to cause confusion, or to cause mistake, or to deceive: *Provided*, That, if the Commissioner determines [accordingly, concurrent registrations may be possible; see below].

(e) Consists of a mark which, (1) when used on or in connection with the goods of the applicant is merely descriptive or deceptively misdescriptive of them, (2) when used on or in connection with the goods of the applicant is primarily geographically descriptive of them, except as indications of regional origin may be registrable under section 1054 of this title, (3) when used on or in connection with the goods of the applicant is primarily geographically deceptively misdescriptive of them, (4) is primarily merely a surname, or (5) comprises any matter that, as a whole, is functional.

Lanham Act §2, 15 U.S.C. §1052. Marks rejected under subsection (e) may be registered if the applicant can demonstrate secondary meaning, however.

i. Immoral or Scandalous Marks

Harjo v. Pro-Football Inc.
Trademark Trial and Appeal Board
50 U.S.P.Q.2d 1705 (TTAB 1999)

Walters, administrative trademark judge.

Suzan Shown Harjo [and others] filed their petition to cancel the registrations of the marks identified below, all owned by Pro-Football, Inc.:

THE WASHINGTON REDSKINS and REDSKINS for "entertainment services — namely, presentations of professional football contests"; REDSKIN-ETTES for "entertainment services, namely, cheerleaders who perform dance routines at professional football games and exhibitions and other personal appearances". . . .

Petitioners allege that they are Native American persons and enrolled members of federally recognized Indian tribes. As grounds for cancellation,

FIGURE 5-2
Trademark Registration Number 986,668

petitioners assert that the word "redskin(s)" or a form of that word appears in the mark in each of the registrations sought to be canceled; that the word "redskin(s)" "was and is a pejorative, derogatory, denigrating, offensive, scandalous, contemptuous, disreputable, disparaging and racist designation for a Native American person"; that the marks in Registration Nos. 986,668 and 987,127 "also include additional matter that, in the context used by registrant, is offensive, disparaging and scandalous"; and that registrant's use of the marks in the identified registrations "offends" petitioners and other Native Americans. Petitioners assert, further, that the marks in the identified registrations "consist of or comprise matter which disparages Native American persons, and brings them into contempt, ridicule, and disrepute" and "consist of or comprise scandalous matter"; and that, therefore, under Section 2(a) of the Trademark Act, 15 U.S.C. 1052(a), the identified registrations should be canceled.

Respondent, in its answer, denies the salient allegations of the petition to cancel and asserts that "through long, substantial and widespread use, advertising and promotion in support thereof and media coverage, said marks have acquired a strong secondary meaning identifying the entertainment services provided by respondent in the form of professional games in the National Football League"; and that "the marks sought to be canceled herein cannot reasonably be understood to refer to the Petitioners or to any of the groups or organizations to which they belong [as] the marks refer to the

Washington Redskins football team which is owned by Respondent and thus cannot be interpreted as disparaging any of the Petitioners or as bringing them into contempt or disrepute."

Petitioners contend that the subject registrations are void ab initio and that the word "redskin(s)" "is today and always has been a deeply offensive, humiliating, and degrading racial slur." Petitioners contend that "a substantial composite of the general public considers 'redskin(s)' to be offensive" and that "the inherent nature of the word 'redskin(s)' and Respondent's use of [its marks involved herein] perpetuate the devastating and harmful effects of negative ethnic stereotyping." Petitioners contend, further, that Native Americans "have understood and still understand" the word "redskin(s)" to be a disparaging "racial epithet" that brings them into contempt, ridicule and disrepute.

Petitioners contend that the Board must consider "the historical setting in which the word 'redskin(s)' has been used." In this regard, petitioners allege that "the history of the relationship between Euro-Americans and Native Americans in the United States has generally been one of conflict and domination by the Euro-Americans"; that " [b]eneath this socioeconomic system lay an important cultural belief, namely, that Indians were 'savages' who must be separated from the Anglo-American colonies and that Anglo-American expansion would come at the expense of Native Americans"; that, in the 1930's, government policies towards Native Americans began to be more respectful of Native American culture; that, however, these policies were not reflected in the activities and attitudes of the general public, who continued to view and portray Native Americans as "simple 'savages' whose culture was treated mainly as a source of amusement for white culture"; and that it was during this time that respondent first adopted the name "Redskins" for its football team.

Petitioners presented the testimony of its linguistics expert, Dr. Geoffrey Nunberg, regarding the usage of the word "redskin(s)." Petitioners contend that the primary denotation of "redskin(s)" is Native American people; that, only with the addition of the word "Washington," has "redskin(s)" acquired a secondary denotation in the sports world, denoting the NFL Football club; that the "offensive and disparaging qualities" of "redskin(s)" arise from its connotations; and that these negative connotations pertain to the word "redskin(s)" in the context of the team name "Washington Redskins." Regarding whether the negative connotations of "redskin(s)" are inherent or arise from the context of its usage, petitioners contend that "redskin(s)" is inherently offensive and disparaging. . . .

Respondent contends that the word "redskin(s) has throughout history, been a purely denotative term, used interchangeably with 'Indian'." In this regard, respondent argues that "redskin(s)" is "an entirely neutral and ordinary term of reference" from the relevant time period to the present; and that, as such, "redskin(s)" is " [synonymous] with ethnic identifiers such as 'American Indian,' 'Indian,' and 'Native American'." Respondent also states that, through its long and extensive use of "Redskins" in connection with professional football, the word has developed a meaning, "separate and distinct from the core, ethnic meaning" of the word "redskin(s)," denoting the "Washington Redskins" football team; and that such use by respondent "has absolutely no negative effect on the word's neutrality — and, indeed, serves to enhance the word's already positive associations — as football is neither of questionable morality nor per se offensive to or prohibited by American Indian religious or cultural practices."

Respondent states that while "the term 'redskin,' used in singular, lower case form references an ethnic group, [this] does not automatically render it disparaging when employed as a proper noun in the context of sports." . . .

Respondent concludes that its marks "do not rise to the level of crudeness and vulgarity that the Board has required before deeming the marks scandalous," nor do its marks disparage or bring Native Americans into contempt or disrepute. Respondent argues that disparagement requires intent on the part of the speaker and that its "intent in adopting the team name was entirely positive" as the team name has, over its history, "reflected positive attributes of the American Indian such as dedication, courage and pride." Similarly, respondent notes that third-party registrations portraying Native Americans and the United States nickel, previously in circulation for many years, portraying a Native American are similar to respondent's "respectful depiction in the team's logo"; and that petitioners have not established that this logo is scandalous, disparaging, or brings Native Americans into contempt or disrepute.

[The Board next presents summaries of testimony by an historian and several social scientists, all of whom described "redskin" as a derogatory term that perpetuated a violent, warlike image of Native Americans, to their continuing detriment.] . . .

The parties' linguistics experts principally disagree over whether a word can be intrinsically negative in connotation, as posited by Dr. Nunberg, or whether, as respondent's witnesses posit, one must always look to the context in which a word is used to determine its connotation and whether that connotation is neutral, positive or negative. However, it is unnecessary for us to determine whether "redskin(s)" is intrinsically positive, negative or neutral, as the record includes numerous examples of the use of the term " redskin(s)," all of them in a "context." Further, as we indicate infra, Section 2(a) requires us to consider the term or other matter at issue in the context of the marks in their entireties, the services identified in the challenged registrations, and the manner of use of the marks in the marketplace. Thus, we consider the meaning of the word "redskin(s)" in this context. [Petitioners' expert] Dr. Nunberg testified that, throughout its approximately 300 years of use, "redskin(s)" has been and is "a connotative term that evokes negative associations, or negative stereotypes, with American Indians. On the other hand, considering the same historical and contemporary material in the record, respondent's experts disagree with Dr. Nunberg's conclusion that the word "redskin(s)" has always been a connotative word of disparagement, or that the evidence of use of the word "redskin(s)" to refer to Native Americans reflects a negative connotation. Rather, Mr. Barnhart described several of the same passages discussed by Dr. Nunberg as connotatively neutral, or even positive, uses of the word "redskin(s)" and concluded that the word "Indian" could easily be substituted therefor without changing the connotation. Dr. Butters, while agreeing that much of the quoted language disparages Native Americans, concluded that it is not the word "redskin(s)" alone that is disparaging. Rather, he concludes that it is the context in which the word appears that portrays Native Americans in a disparaging manner, and that the word "Indian" could be easily substituted in each instance. Dr. Butters states that "Native American," "Indian," and "redskin" are all acceptable words, but that "redskin" is the least formal of the three words and is "only a respectful minor variant alternative for "American Indian.' "

[Petitioners' survey of 301 people showed that 46.6 percent of all respondents, and 36.6 percent of Native Americans, found "redskin" offensive. After listing respondent's objections to the survey, the court concluded:] [W]e find that the survey adequately represents the views of the two populations sampled. While certainly far from dispositive of the question before us in this case, it is relevant and we have accorded some probative value to this survey, as discussed in our legal analysis, infra.

The Board must decide whether, at the times respondent was issued each of its challenged registrations, the respondent's registered marks consisted of or comprised scandalous matter, or matter which may disparage Native American persons, or matter which may bring Native American persons into contempt or disrepute.

The vast majority of the relevant reported cases involving that part of Section 2(a) with which we are concerned in this case were decided principally on the basis of whether the marks consisted of scandalous matter. . . .

While not often articulated as such, determining whether matter is scandalous involves, essentially, a two-step process. First, the Court or Board determines the likely meaning of the matter in question and, second, whether, in view of the likely meaning, the matter is scandalous to a substantial composite of the general public.

[W]hile the decisional law may suggest that intent, or lack thereof, to shock or to ensure that the scandalous connotation of a mark is perceived by a substantial composite of the general public is one factor to consider in determining whether a mark is scandalous, there is no support in the case law for concluding that such intent, or a lack thereof, is dispositive of the issue of scandalousness. See, In re Old Glory Condom Corp., 26 U.S.P.Q.2d 1216 (TTAB 1993). . . .

The plain language of the statute makes clear that disparagement is a separate and distinct ground for refusing or canceling the registration of a mark under Section 2(a). However, there is relatively little published precedent or legislative history to offer us guidance in interpreting the disparagement provision in Section 2(a). As with scandalousness, the determination of whether matter may be disparaging is highly subjective and, thus, general rules are difficult to postulate.

In determining whether or not a mark is disparaging, the perceptions of the general public are irrelevant. Rather, because the portion of Section 2(a) proscribing disparaging marks targets certain persons, institutions or beliefs, only the perceptions of those referred to, identified or implicated in some recognizable manner by the involved mark are relevant to this determination.

Who comprises the targeted, or relevant, group must be determined on the basis of the facts in each case. For example, if the alleged disparagement is of a religious group or its iconography, the relevant group may be the members and clergy of that religion; if the alleged disparagement is of an academic institution, the relevant group may be the students, faculty, administration, and alumni; if the alleged disparagement is of a national symbol, the relevant group may be citizens of that country. See also, In re Reemtsma Cigarettenfabriken G.m.b.H., 122 U.S.P.Q. 339 (TTAB 1959). We distinguish . . . the case herein from the case of Greyhound Corp. v. Both Worlds, Inc., 6 U.S.P.Q.2d 1635 (TTAB 1988). In Greyhound, on summary judgment, the Board sustained the opposition on the grounds of scandalousness, disparagement, and likelihood of confusion. The mark in question was a design of a defecating greyhound dog, for polo shirts and T-shirts. The disparagement in the Greyhound case involved an "offensive" design that disparages a commercial corporate entity and, thus, is akin to the commercial disparagement of property. . . . The disparaging trademark casts doubt upon the quality of opposer's corporate goodwill, as embodied in its running greyhound dog trademarks.

We turn, finally, to the Section 2(a) provisions regarding contempt or disrepute. We find no guidance in the legislative history for interpreting this provision and note that this provision is addressed in the case law, generally, in a conclusory manner with few, if any, guidelines. In view of the "ordinary and common" meanings of the words "contempt" and "disrepute," as they were defined in 1947 and more recently, we believe that the guidelines enunciated herein in connection with determining whether matter in a mark may be disparaging are equally applicable to determining whether such matter brings "persons, living or dead, institutions, beliefs, or national symbols into contempt or disrepute." In all of the reported cases discussed above, the issue was whether the involved marks were scandalous or may be disparaging because of the marks' sexual explicitness or innuendo, vulgarity, religious significance, or reference to illicit activity. The case before us differs factually from the aforementioned types of cases in that petitioners contend, principally, that the word REDSKINS in the marks in question is "a deeply offensive, humiliating, and degrading racial slur" in connection with Native Americans.

We find petitioners have clearly established, by at least a preponderance of the evidence, that, as of the dates the challenged registrations issued, the word "redskin(s)," as it appears in respondent's marks in those registrations and as used in connection with the identified services, may disparage Native Americans, as perceived by a substantial composite of Native Americans. No single item of evidence or testimony alone brings us to this conclusion; rather, we reach our conclusion based on the cumulative effect of the entire record. In particular, the evidence herein shows a portrayal in various media of Native Americans, unrelated to respondent's football team, as uncivilized and, often, buffoon-like characters from, at least, the beginning of this century through the middle to late 1950's. As we move through the 1960's to the present, the evidence shows increasingly respectful portrayal of Native Americans. This is reflected, also, in the decreased use of "redskin(s)," as a term of reference for Native Americans, as society in general became aware of, and sensitive to, the disparaging nature of that word as so used.

[CANCELLATION Ordered.]

COMMENTS AND QUESTIONS

1. If this decision is ultimately upheld on appeal, it would affect only the football team's ownership of federal registrations. What effect will it have on enforcement of the REDSKINS trademarks under Lanham Act §43(a) or under state statutory and common law? Under federal antidilution protection? Later in this chapter it is explained that a mark must be found to be famous to be protected against dilution under the federal statute. Can a mark be disparaging yet still famous? Is it consistent with trademark law to protect consumer associations even though a mark has been declared disparaging?

2. Is it relevant that only 37 percent of Native Americans found "redskin" offensive? Is there a difference between use of the term among members of the community (perhaps in an ironic, perversely proud sense) and use of the same term by non-members of the community? Under the standard for "disparaging" marks announced by the Board, is it relevant that more members of the non-Native American community found "redskin" offensive?

3. Trademarks are a short form of corporate speech. What First Amendment issues are at stake in the case?

4. The court accepted that the opposers had standing in the *Harjo* case given their immediate association with the disparaging marks. Another recent case, Ritchie v. Simpson, 170 F.3d 1092 (Fed. Cir. 1999), presented a more difficult standing issue. In this case, William B. Ritchie opposed registration of various trademarks associated with O.J. Simpson. Ritchie argued that the trademarks recited immoral or scandalous material given the well-known past of the registrant (i.e., spousal abuse, accusation of murder, etc.). The Federal Circuit upheld Ritchie's standing to oppose the marks, holding that he had shown a sufficiently real and personal interest in the outcome of the opposition proceeding (in the form of a deep personal belief in non-abusive relationships) to defeat a motion to dismiss for lack of standing.

5. In re Old Glory Condom Corp., 26 U.S.P.Q.2d 1216 (TTAB 1993), is a good example of a case involving allegedly "immoral or scandalous matter" under §2(a). The applicant was attempting to register the following mark to denote its condom product:

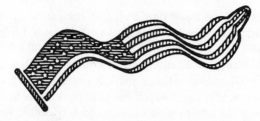

OLD GLORY CONDOM CORP

FIGURE 5-3
Logo for Old Glory Condom Corp.

The Board reversed rejection of the application by the examining attorney; in the course of reviewing older cases finding scandalous marks such as "QUEEN MARY" for women's underwear and "BUBBY TRAP" for brassieres, the court said:

> What was once considered scandalous as a trademark or service mark twenty, thirty or fifty years ago may no longer be considered so. Marks one thought scandalous may now be thought merely humorous (or even quaint). . . . The point to be made here is that, in deciding whether a mark is scandalous under Section 2(a), we must consider that mark in the context of contemporary attitudes. . . .

Id. at 1218–19.

ii. Geographic Marks

In re Nantucket, Inc.
United States Court of Customs and Patent Appeals
677 F.2d 95 (C.C.P.A. 1982)

MARKEY, Chief Judge.

Nantucket, Inc. (Nantucket) appeals from a decision of the Trademark Trial and Appeal Board (board) affirming a refusal to register the mark NANTUCKET for men's shirts on the ground that it is "primarily geographically deceptively misdescriptive." In re Nantucket, Inc., 209 U.S.P.Q. 868 (TTAB 1981). We reverse.

Background

. . . Nantucket, based in North Carolina, filed [an] application . . . for registration of NANTUCKET for men's shirts on the principal register in the Patent and Trademark Office (PTO). . . .

Refusal to register was based on §2(e)(2) of the Lanham Act, 15 U.S.C. §1052(e)(2), as interpreted by the board in In re Charles S. Loeb Pipes, Inc., 190 U.S.P.Q. 238 (TTAB 1975), and in §§1208.02, 1208.05 and 1208.06 of the Trademark Manual of Examining Procedure (TMEP). . . .

TMEP §1208.02 indicates that a mark is primarily geographical, inter alia, if it "is the name of a place which has general renown to the public at large and which is a place from which goods and services are known to emanate as a result of commercial activity."

The examiner, citing a dictionary definition of "Nantucket" as an island in the Atlantic Ocean south of Massachusetts, concluded that the mark NANTUCKET was either primarily geographically descriptive or primarily geographically deceptively misdescriptive, depending upon whether Nantucket's shirts did or did not come from Nantucket Island.

Nantucket informed the PTO that its shirts "do not originate from Nantucket Island," and insisted that the mark would not be understood by purchasers as representing that the shirts were produced there because the island has no market place significance vis-à-vis men's shirts. . . . As applied to shirts, it was argued, NANTUCKET is arbitrary and nondescriptive, because there is no association in the public mind of men's shirts with Nantucket Island.

The examiner's final refusal was based on the view that, because the shirts did not come from Nantucket Island, NANTUCKET is "primarily geographically deceptively misdescriptive."

Before the board, Nantucket relied upon a number of cases, of which In re Circus Ices, Inc., 158 U.S.P.Q. 64 (TTAB 1968), is representative, for its asserted "public association" or "noted for" test. In that case, the board said:

> The term "HAWAIIAN," meaning of or pertaining to Hawaii or the Hawaiian Islands, possesses an obvious geographical connotation, but it does not necessarily follow therefrom that it is primarily geographically descriptive of applicant's product within the meaning of Section 2(e). In determining whether or not a geographical term is primarily geographically descriptive of a product, of primary consideration is whether or not there is an association in the public mind of the product with the particular geographical area, as for example perfumes and wines with France, potatoes with Idaho, rum with Puerto Rico, and beef with Argentina. . . . In the present case, it has not been made to appear that Hawaii or the Hawaiian Islands are noted for flavored-ice products or that the term "HAWAIIAN" is used by anyone to denote the geographical origin of such products. [. . .]

In referring to *Amerise*, the board viewed the "noted for" test, mentioned in *Circus Ices*, as relevant only to whether a geographic term is deceptive under §2(a) ["No trademark . . . shall be refused registration . . . unless it (a) consists of or comprises immoral, *deceptive*, or scandalous matter. . . ." (emphasis added).] Regarding §2(e)(2), the board said:

> (A)ny term which, when applied to the goods or services of the applicant, conveys to consumers primarily or immediately a geographical connotation is pre-

cluded from registration under Section 2(e)(2) notwithstanding the fact that the area or place named is not "noted for" goods or services of that type. . . .

Id. at 871.

The board concluded that the "term NANTUCKET" has a readily recognizable geographic meaning, and . . . no alternative non-geographic significance . . . and hence falls within the proscription of Section 2(e)(2)." Id.

The Board's Test

The board correctly notes that its test for registrability of geographic terms is "easy to administer" and "minimizes subjective determinations by eliminating any need to make unnecessary inquiry into the nebulous question of whether the public associates particular goods with a particular geographical area in applying Section 2(e)(2)." . . . Ease-of-administration considerations aside, the board's approach does raise the question of whether public association of goods with an area must be considered in applying §2(e)(2). That question is one of first impression in this court. We answer in the affirmative.

The board's test rests mechanistically on the one question of whether the mark is recognizable, at least to some large segment of the public, as the name of a geographical area. NANTUCKET is such. That ends the board's test. Once it is found that the mark is the name of a known place, i.e., that it has "a readily recognizable geographic meaning," the next question, whether applicant's goods do or do not come from that place, becomes irrelevant under the board's test, for if they do, the mark is "primarily geographically descriptive"; if they don't, the mark is "primarily geographically deceptively misdescriptive." Either way, the result is the same, for the mark must be denied registration on the principal register unless resort can be had to §2(f).[3]

The Statute

One flaw in the board's test resides in its factoring out the nature of applicant's goods, in contravention of §2(e)(2)'s requirement that the mark be evaluated "when applied to the goods of the applicant," and that regis-

---

3. Section 2(f) provides: "Except as expressly excluded in paragraphs (a), (b), (c), and (d) of this section, nothing in this chapter shall prevent the registration of a mark used by the applicant which has become distinctive of the applicant's goods in commerce. The Commissioner may accept as prima facie evidence that the mark has become distinctive, as applied to the applicant's goods in commerce, proof of substantially exclusive and continuous use thereof as a mark by the applicant in commerce for the five years next preceding the date of the filing of the application for its registration."

tration be denied only when the mark is geographically descriptive or deceptively misdescriptive "of them" (the goods).

Another flaw in the board's test lies in its failure to give appropriate weight to the presence of "deceptively" in §2(e)(2). If the goods do not originate in the geographic area denoted by the mark, the mark might in a vacuum be characterized as geographically misdescriptive, but the statutory characterization required for denial of registration is "geographically *deceptively* misdescriptive." (Emphasis supplied.) Before that statutory characterization may be properly applied, there must be a reasonable basis for believing that purchasers are likely to be deceived.[5] . . .

Geographic terms are merely a specific kind of potential trademark, subject to characterization as having a particular kind of descriptiveness or misdescriptiveness. Registration of marks that would be perceived by potential purchasers as describing or deceptively misdescribing the goods themselves may be denied under §2(e)(1). Registration of marks that would be perceived by potential purchasers as describing or deceptively misdescribing the geographic origin of the goods may be denied under §2(e)(2). In either case, the mark must be judged on the basis of its role in the marketplace.

As the courts have made plain, geographically deceptive misdescriptiveness cannot be determined without considering whether the public associates the goods with the place which the mark names. If the goods do not come from the place named, and the public makes no goods-place association, the public is not deceived and the mark is accordingly not geographically deceptively misdescriptive. . . .

In National Lead Co. v. Wolfe, 223 F.2d 195, 199, 105 U.S.P.Q. 462, 465 (CA 9), *cert. denied,* 350 U.S. 883, 76 S. Ct. 135, 100 L. Ed. 778, 107 U.S.P.Q. 362 (1955), the court held that neither DUTCH, nor DUTCH BOY, as applied to paint, was used "otherwise than in a fictitious, arbitrary and fanciful manner," and noted that "there is no likelihood that the use of the name 'Dutch' or 'Dutch Boy' in connection with the appellant's goods would be understood by purchasers as representing that the goods or their constituent materials were produced or processed in Holland or that they are of the same distinctive kind or quality as those produced, processed or used in that place."

. . . There is no indication that the purchasing public would expect men's shirts to have their origin in Nantucket when seen in the market place with NANTUCKET on them. Hence buyers are not likely to be deceived, and registration cannot be refused on the ground that the mark is "primarily geographically deceptively misdescriptive."

Accordingly, the decision of the board is reversed.

---

5. Reasonable persons are unlikely to believe that bananas labeled ALASKA originated or were grown in Alaska. On the other hand, reasonable persons are quite likely to believe that salmon labeled ALASKA originated in the waters of that state. The board recognized this decisional parameter in its references to French wine, Idaho potatoes, etc., in *Circus Ices,* 158 U.S.P.Q. at 64.

COMMENTS AND QUESTIONS

1. In some cases prospective trademark registrants may be trying to seize on the descriptiveness of a term of geographic origin; "Napa" for wine would be an example. Another motivation is to seize on the good feelings engendered by a place name. See, e.g., Singer Mfg. Co. v. Birginal-Bigsby Corp., 319 F.2d 273 (C.C.P.A. 1963) (American Beauty is primarily geographically deceptively misdescriptive when applied to sewing machines of Japanese origin). For example, a trademark applicant might try to register "Cuban" for cigars made in New Hampshire; this would very likely lead to a finding of deceptiveness (section 2(a)), and the application would be barred completely.

2. In many countries, geographic terms — usually for food or wine — serve also to certify composition, traditional preparation techniques, quality, and actual taste. In Europe, for example, so-called "appellations of origin" for wine — such as "Champagne," "Chablis," and "Chianti" — are protected by statute and administered by national authorities. See Institut National des Appellations D'Origine v. Vintners Intl. Co., 958 F.2d 1574 (Fed. Cir. 1992) (affirming denial of opposition by plaintiff, French national authority for establishing and policing appellations of origin, for registration of "Chablis with a Twist" for citrus flavored wine drink).

As part of the GATT amendments in 1994, Congress (as a concession to European interests) adopted limited recognition of the appellations of origin concept.

Lanham Act §2(a) [15 U.S.C. §1052(a)]

No trademark . . . shall be refused registration on the principal register on account of its nature unless it —

    (a) Consists of or comprises . . . a geographical indication which, when used on or in connection with wines or spirits, identifies a place other than the origin of the goods and is first used on or in connection with wines or spirits by the applicant on or after one year after the date on which the WTO [World Trade Organization] Agreement (as defined in section 3501(9) of Title 19) enters into force with respect to the United States.

Section 2(e) has also long provided that appellations of origin registered under Lanham Act §4 (15 U.S.C §1054) (collective and certification marks), are an exception to the prohibition on geographically descriptive trademarks. A certification mark is a guarantee of authenticity — "Belgian chocolates" must come from Belgium, for example. Certification marks may be registered without proof of secondary meaning, but must be made available in a nondiscriminatory fashion to anyone who complies with the terms of the certification. See Community of Roquefort v. William Faehndrich, Inc., 303 F.2d 494 (2d Cir. 1962) (French city of Roquefort was entitled to prevent use of the term "Roquefort" to describe cheese made in Hungary, based on United

States certification mark). Note that group certification marks serve much the same quality assurance function as individual trademarks, but (assuming collective action problems are overcome) they may effectively protect small businesses who could not establish a well-known mark on their own.

For a spirited debate on appellations of origin, see Jim Chen, A Sober Second Look at Appellations of Origin: How the United States Will Crash France's Wine and Cheese Party, 5 Minn. J. Global Trade 29 (1996) (arguing that appellations of origin are artificial market segmentation devices that detract from consumer welfare); Louis Lorvellec, You've Got to Fight for Your Right to Party, 5 Minn. J. Global Trade 65 (1996) (begging to differ, from French perspective).

Europe has a detailed set of rules regulating the use of geographically descriptive terms, particularly certification marks. Generally, these rules are more restrictive than United States law has been. As part of United States compliance with the North American Free Trade Association (NAFTA) and General Agreements on Tariffs and Trade — Trade-Related Aspects of Intellectual Property (GATT-TRIPS), sections 2(e) and (f) of the Lanham Act have been amended to provide that terms which are primarily geographically deceptively misdescriptive can never be registered, whether or not they acquire secondary meaning. See Final Act Embodying the Results of the Uruguay Round of Multilateral Trade Negotiations, done at Marrakech, Morocco, April 15, 1994, Annex 1C: Agreement on Trade-Related Aspects of Intellectual Property Rights ["TRIPs"], reprinted in 33 I.L.M. 81 (1994). For congressional approval, see Uruguay Round Agreements Act, Pub. L. No. 103-465 [H.R. 5110], Dec. 8, 1994.

iii. Marks Which Are "Primarily Merely a Surname"

From In re Garan, Inc., 3 U.S.P.Q.2d 1537, 1539-1540 (TTAB 1987):

That there are no other meanings of the name in the English language will not support refusal of registration of the surname under the "primarily merely a surname" statutory language unless the average member of the purchasing public would, upon seeing it used as a trademark, recognize it as a surname. This rule was first announced by the late Assistant Commissioner Leeds in her landmark decision, Ex parte Rivera Watch Corporation, 106 U.S.P.Q. 145, 149 (Comr. Pats. 1955), as follows:

> There are some names which by their very nature have only a surname significance even though they are rare surnames. "Seidenberg," if rare, would be in this class. And there are others which have no meaning — well known or otherwise — and are in fact surnames which do not, when applied to goods as trademarks, create the impression of being surnames. It seems to me that the test to be applied in the administration of this provision in the Act is not the rarity of the name, nor whether it is the applicant's name, nor whether it appears in one or more telephone directories, nor whether it is coupled with a baptismal name or initials. The test should be: What is the primary significance to the purchasing public? . . .

Some twenty years later, in In re Kahan & Weisz Jewelry Manufacturing Corp., 508 F.2d 831, 184 U.S.P.Q. 421 (C.C.P.A. 1975), a case involving facts strikingly similar to those in the one now before us, the predecessor of the Court above expressly adopted the Rivera rule and it has since been consistently followed by the Board. See, e.g., In re Pillsbury Co., 174 U.S.P.Q. 318 (TTAB 1972). . . .

Clearly, based on the above, we cannot stop with the directory listings and absence of other meanings, but must evaluate all of the relevant factors.

The six (or seven) directory and NEXIS listings of "Garan" as a surname have limited persuasive impact in view of the fact these were the only ones found in an enormous NEXIS database and some 43 directories of major population centers. In this context, we conclude that "Garan" is an extremely rare surname. While rare surnames may nevertheless be considered to be "primarily merely" surnames, e.g., In re Possis Medical, Inc., 230 U.S.P.Q. 72, 74 (TTAB 1986), and cases cited therein, we agree with appellant that the degree of a surname's rareness should have material impact on the weight given the directory evidence. Here, since it appears that the directory and NEXIS evidence shows "Garan" to be an extremely rare surname, we conclude that the directory and NEXIS evidence only slightly supports the Office's position that GARAN is "primarily merely a surname."

On analysis, we find all of the other factors either neutral or supportive of appellant's position that GARAN would be perceived by purchasers as an arbitrary term or as the trademark and trade name of appellant.

COMMENTS AND QUESTIONS

1. The statutory hesitance to register surnames springs in part from an old common law policy in favor of the right to use one's name, a policy that still has some life. Basile, S.p.A. v. Basile, 899 F.2d 35 (D.C. Cir. 1990), for example, is a case where an Italian watchmaker by the name of Basile sought to use that name on watches in the United States — against the wishes of an established company, well known for its watches sold under the same name. In his opinion reviewing the scope of the injunction granted by the district court in the case, Judge Williams of the D.C. Circuit summarized the common law rule and its more recent interpretation:

A seller's right to use his family name might have carried the day against a risk of buyer confusion in an era when the role of personal and localized reputation gave the right a more exalted status. In Burgess v. Burgess, 3 De G.M. & G. 896, 903-904, 43 Eng. Rep. 351, 354 (1853), the plaintiff's son had followed him into the trade of making anchovy paste; Lord Justice Bruce said:

All the Queen's subjects have a right, if they will, to manufacture and sell pickles and sauces, and not the less that their fathers have done so before them. All the Queen's subjects have a right to sell these articles in their own names, and not the less so that they bear the same name as their fathers.

See also Brown Chemical Co. v. Meyer, 139 U.S. 540, 544, 11 S. Ct. 625, 627, 35 L. Ed. 247 (1891) ("[a] man's name is his own property, and he has the same right to its use and enjoyment as he has to that of any other species of property"); Stix, Baer & Fuller Dry Goods Co. v. American Piano Co., 211 F. 271, 274 (8th Cir. 1913). But even quite old decisions have enjoined a second comer's use of his name where necessary to prevent confusion. See, e.g., Thaddeus Davids Co. v. Davids Manufacturing Co., 233 U.S. 461, 472, 34 S. Ct. 648, 652, 58 L. Ed. 1046 (1914); Hat Corp. of America v. D. L. Davis Corp., 4 F. Supp. 613, 623 (D. Conn. 1933). The commentators, moreover, have been scornful of protecting the second comer's right to use his name at the expense of customer confusion, anathematizing it as the "sacred rights" theory. See Milton Handler & Charles Pickett, Trade-Marks and Trade Names — An Analysis and Synthesis, 30 Colum. L. Rev. 168, 197-200 (1930). . . . Case Note, 38 Harv. L. Rev. 405, 406 (1925).

True, even recent decisions have invoked the right to use one's name, at least as an interest against which the senior user's are balanced. See Taylor Wine [Co., Inc. v. Bully Hill Vineyards, Inc., 569 F.2d 731, 734 (2d Cir. 1978)], . . . at 735. . . . But its weight has decidedly diminished. The courts are now consistent in imposing tighter restrictions on the second comer in the face of possible confusion . . . ; any residual protection of the second comer's use of his own name seems amply explained by the more general principle that an equitable remedy should be no broader than necessary to correct the wrong. . . . Where the second comer had established no reputation under his own name, one court did not even purport to "balance."

This trend in the law unsurprisingly reflects trends in the marketplace. In a world of primarily local trade, the goodwill of an anchovy paste seller may well have depended on his individual reputation within the community. Indeed, one elderly decision tells us that an entrepreneur's failure to use his family name once risked "the reproach of doing business under false colors." Hat Corp., 4 F. Supp. at 623. By contrast, the court went on, "[i]n an age when by corporate activity, mass production, and national distribution, the truly personal element has been so largely squeezed out of business, there is naturally less legitimate pecuniary value in a family name." Id.; see also Handler & Pickett, 30 Colum. L. Rev. at 199. That was in 1933 and the point is more obvious today. Other than understandable pride and sense of identity, the modern businessman loses nothing by losing the name. A junior user's right to use his name thus must yield to the extent its exercise causes confusion with the senior user's mark. . . .

Here Francesco Basile's interest in the use of his name is peculiarly weak. He has no reputation in the United States as a watchmaker. . . . There is no suggestion that the watch industry is one where an individual proprietor's personal presentation of his wares plays a key commercial role, as may still be true of high fashion designers such as Yves St. Laurent and Christian Dior. Nor is it a business that has remained largely local in scope. Although none of these conditions would allow a second comer to use his own name at a serious cost of customer confusion, their absence means that Francesco's interest in his own name here is scarcely greater than his interest in the name Bulova. . . . [W]e think the only plausible motivation for his fight here is a wish to free-ride on Basile's goodwill, precisely what the law means to suppress.

2. Should it matter whether the surname sought to be registered is common or rare? That the senior trademark user is well known or obscure? See S. Smith, Primarily Merely, 63 Trademark Rep. 24 (1973).

3. The scope of injunctions shows the remnants of the common law concern with the right to use one's name. Courts are likely to permit the junior use as long as the user disclaims any connection with the more famous senior user. See Taylor Wine Co. v. Bully Hill Vineyards, 569 F.2d 731 (2d Cir. 1978).

PROBLEM

Problem 5-4. Ernest & Julio Gallo are the nation's largest seller of wine. They sell their wine under the "Gallo" label, and have advertised that trademark heavily. Ernest & Julio's younger brother Joseph Gallo runs a company called Gallo Cattle Co. Gallo Cattle begins selling cheese and salami under the name "Gallo Cheese" and "Gallo Salame." If Ernest & Julio Gallo wish to stop their brother from marketing meat and cheese under the Gallo name, what must they be able to show at trial? If they win, how should a court craft the injunction granting relief; what should it forbid and permit?

---

c. Opposition

Section 13(a) of the Lanham Act reads in part:

(a) Any person who believes that he would be damaged by the registration of a mark upon the principal register may, upon payment of the prescribed fee, file an opposition in the Patent and Trademark Office, stating the grounds therefor, within thirty days after the publication . . . of the mark sought to be registered.

Because potential opposers must become aware of the contested mark's future registration, opposition is only meaningful protection for firms possessing the resources to conduct frequent searches of the Official Gazette, where prospective trademarks are published.

An opposer must plead and prove that: (1) it is likely to be damaged by registration of the applicant's mark (this is the standing requirement); and (2) that there are valid legal grounds why the applicant is not entitled to register its claimed mark. The opposer is in the plaintiff position, and the applicant is in the defendant position. The opposer has the burden of proving that the applicant has no right to register the contested mark. In general, there has been a trend toward liberalization of the standing requirement. 3 McCarthy §20.02[1][a] (1996). "To establish standing to oppose, the opposer need only be something more than a gratuitous interloper or a vicarious avenger of someone else's rights." Id., at p. 20-16.

Once the standing threshold has been crossed, the opposer may rely on any legal ground that negates the applicant's right to registration. See, e.g., Estate of Biro v. Bic Corp., 18 U.S.P.Q.2d 1382 (TTAB 1991). For example, an opposer might argue that the applicant did not make sufficient use of the mark in interstate commerce to receive a use-based registration. Generally, opposers rely on one of the bars to registration described in Lanham Act §2, with section 2(d) being the most common. Section 2(d) prohibits registration where the applicant's mark so resembles either (1) opposer's registered mark or (2) opposer's prior common law mark or trade name, as to be likely to cause confusion. The test for likelihood of confusion is the same used in general litigation. 3 McCarthy, supra, at p. 20-24. It is described in detail below in section D.1.

d. Cancellation

If opposition is the first "backstop" to ex parte examination of trademarks, cancellation may be thought of as the second. Even after a trademark examiner is satisfied that a mark meets the requirements for registration and the mark is in fact registered, it may still be challenged in an inter partes proceeding. The Lanham Act allows one who "believes that he is or will be damaged by registration" to petition for cancellation of marks on either the Principal or Supplemental Register. Lanham Act §18. Cancellation petitions are heard by the Trademark Trial and Appeal Board (TTAB). In civil suits where a federally registered mark is at issue, such as suits under Lanham Act §2, the court may order cancellation of the registration.

The standing requirement for cancellation proceedings is quite similar to that for opposition. In both cases the statute speaks in terms of the plaintiff's belief that he or she will be damaged.

Even after cancellation of the registration, a mark may still enjoy common law rights. See, e.g., National Trailways Bus System v. Trailway Van Lines, 269 F. Supp. 352 (E.D.N.Y. 1965).

e. Concurrent Registration

Concurrent use registration is provided for in Lanham Act §2(d):

No trademark . . . shall be refused registration on the principal register . . . unless it — . . .

    (d)   Consists of or comprises a mark which so resembles a mark . . . as to be likely . . . to cause confusion . . . : *Provided*, That if the Commissioner determines that confusion, mistake, or deception is not likely to result from the continued use by more than one person of the same or similar marks under conditions and limitations as to the mode or place of use of the marks or the goods on or in connection with which such marks are used, concurrent registrations may be issued to such persons when they have become entitled to use such marks as a result of their concurrent lawful use in commerce prior to (1) the earliest of the filing dates of the applications pending or of any registration issued under this Act. . . .

Lanham Act §2(d), 15 U.S.C. §1052(d). Section 2(d) goes on to say that prior use may be waived by agreement of the parties seeking concurrent registration, and that the commissioner may also issue concurrent use registrations when ordered to do so by a court. Id.

If an applicant's mark appears registrable except for the question of concurrent territorial rights, then the application is published for opposition. In the absence of any successful opposition, the Board sends notices to other parties named in the application. Registered owners of the mark, or applicants, who receive notice may file a "statement" within forty days. Mark owners who are not registered or are not applicants must file an answer within forty days. The forum of first resort for concurrent use proceedings is the TTAB.

The most important condition for approval of concurrent use registration is that such registration cannot be likely to cause confusion of buyers or others. Application of Beatrice Foods Co., 429 F.2d 466 (C.C.P.A. 1970). In *Beatrice*, the senior user (Beatrice) had used a mark in 23 states and the junior user (Fairway) had used a mark in five states by the time the registration hearing commenced. Both had filed registration applications, and a concurrent use proceeding was instituted. The court established the general rule that the senior user can be awarded registration covering all parts of the United States except those regions where the subsequent (junior) user can establish existing rights in its actual area of use or zones of natural expansion. The junior user must show that confusion is not likely to result from the concurrent registration. The court then recognized three exceptions to this general rule: (1) the PTO is not required to grant registration contrary to an agreement between the parties that leaves some territory open; (2) where the junior user is the first to obtain federal registration, the junior user obtains nationwide rights subject only to the territorial limitations of the senior user (see, e.g., Weiner King, Inc. v. Weiner King Corp., 615 F.2d 512 (C.C.P.A. 1980) (junior user who registered first and expanded after discovering the senior user was entitled to registration covering the entire United States with the exception of a small enclave encompassing the senior user's territory); (3) areas of mutual nonuse may be maintained if the mark, goods, and territories are such that this is the only way to avoid the likelihood of confusion.

COMMENTS AND QUESTIONS

Should an agreement settling a dispute between parties seeking concurrent registration — hence carving up the nation into exclusive territories —

be scrutinized for antitrust issues? See VMG Enterprises Inc. v. F. Quesada & Franco Inc., 788 F. Supp. 648 (D.P.R. 1992). Absent intellectual property rights, an agreement among competitiors to divide up markets would be illegal per se. See Chapter 8.

4. Incontestability

Park 'N Fly, Inc. v. Dollar Park and Fly, Inc.
Supreme Court of the United States
469 U.S. 189 (1985)

Justice O'CONNOR delivered the opinion of the Court.

In this case we consider whether an action to enjoin the infringement of an incontestable trade or service mark may be defended on the grounds that the mark is merely descriptive. We conclude that neither the language of the relevant statutes nor the legislative history supports such a defense.

I

Petitioner operates long-term parking lots near airports. After starting business in St. Louis in 1967, petitioner subsequently opened facilities in Cleveland, Houston, Boston, Memphis, and San Francisco. Petitioner applied in 1969 to the United States Patent and Trademark Office (Patent Office) to register a service mark consisting of the logo of an airplane and the words "Park 'N Fly." The registration issued in August 1971. Nearly six years later, petitioner filed an affidavit with the Patent Office to establish the incontestable status of the mark. As required by §15 of the Trademark Act of 1946 (Lanham Act), 60 Stat. 433, as amended, 15 U.S.C. §1065, the affidavit stated that the mark had been registered and in continuous use for five consecutive years, that there had been no final adverse decision to petitioner's claim of ownership or right to registration, and that no proceedings involving such rights were pending. Incontestable status provides, subject to the provisions of §15 and §33(b) of the Lanham Act, "conclusive evidence of the registrant's exclusive right to use the registered mark. . . ." §33(b), 15 U.S.C. §1115(b).

Respondent also provides long-term airport parking services, but only has operations in Portland, Oregon. Respondent calls its business "Dollar Park and Fly." Petitioner filed this infringement action in 1978 in the United States District Court for the District of Oregon and requested the court permanently to enjoin respondent from using the words "Park and Fly" in connection with its business. Respondent counterclaimed and sought cancellation of petitioner's mark on the grounds that it is a generic term. See

§14(c), 15 U.S.C. §1064(c). Respondent also argued that petitioner's mark is unenforceable because it is merely descriptive. See §2(e), 15 U.S.C. §1052(e). As two additional defenses, respondent maintained that it is in privity with a Seattle corporation that has used the expression "Park and Fly" since a date prior to the registration of petitioner's mark, see §33(b)(5), 15 U.S.C. §1115(b)(5), and that it has not infringed because there is no likelihood of confusion. See §32(1), 15 U.S.C. §1114(1). . . .

II . . .

This case requires us to consider the effect of the incontestability provisions of the Lanham Act in the context of an infringement action defended on the grounds that the mark is merely descriptive. Statutory construction must begin with the language employed by Congress and the assumption that the ordinary meaning of that language accurately expresses the legislative purpose. See American Tobacco Co. v. Patterson, 456 U.S. 63, 68 (1982). With respect to incontestable trade or service marks, §33(b) of the Lanham Act states that "registration shall be conclusive evidence of the registrant's exclusive right to use the registered mark" subject to the conditions of §15 and certain enumerated defenses. Section 15 incorporates by reference subsections (c) and (e) of §14, 15 U.S.C. §1064. An incontestable mark that becomes generic may be canceled at any time pursuant to §14(c). That section also allows cancellation of an incontestable mark at any time if it has been abandoned, if it is being used to misrepresent the source of the goods or services in connection with which it is used, or if it was obtained fraudulently or contrary to the provisions of §4, 15 U.S.C. §1054, or §§2(a)-(c), 15 U.S.C. §§1052(a)-(c). . . .

The language of the Lanham Act also refutes any conclusion that an incontestable mark may be challenged as merely descriptive. A mark that is merely descriptive of an applicant's goods or services is not registrable unless the mark has secondary meaning. Before a mark achieves incontestable status, registration provides prima facie evidence of the registrant's exclusive right to use the mark in commerce. §33(a), 15 U.S.C. §1115(a). The Lanham Act expressly provides that before a mark becomes incontestable an opposing party may prove any legal or equitable defense which might have been asserted if the mark had not been registered. Ibid. Thus, §33(a) would have allowed respondent to challenge petitioner's mark as merely descriptive if the mark had not become incontestable. With respect to incontestable marks, however, §33(b) provides that registration is conclusive evidence of the registrant's exclusive right to use the mark, subject to the conditions of §15 and the seven defenses enumerated in §33(b) itself. Mere descriptiveness is not recognized by either §15 or §33(b) as a basis for challenging an incontestable mark.

The statutory provisions that prohibit registration of a merely descriptive mark but do not allow an incontestable mark to be challenged on this ground cannot be attributed to inadvertence by Congress. The Conference Committee rejected an amendment that would have denied registration to any descriptive mark, and instead retained the provisions allowing registration of a merely descriptive mark that has acquired secondary meaning. See H.R. Conf. Rep. No. 2322, 79th Cong., 2d Sess., 4 (1946) (explanatory statement of House managers). The Conference Committee agreed to an amendment providing that no incontestable right can be acquired in a mark that is a common descriptive, i. e., generic, term. Id., at 5. Congress could easily have denied incontestability to merely descriptive marks as well as to generic marks had that been its intention.

The Court of Appeals in discussing the offensive/defensive distinction observed that incontestability protects a registrant against cancellation of his mark. 718 F.2d, at 331. This observation is incorrect with respect to marks that become generic or which otherwise may be canceled at any time pursuant to §§14(c) and (e). Moreover, as applied to marks that are merely descriptive, the approach of the Court of Appeals makes incontestable status superfluous. Without regard to its incontestable status, a mark that has been registered five years is protected from cancellation except on the grounds stated in §§14(c) and (e). Pursuant to §14, a mark may be canceled on the grounds that it is merely descriptive only if the petition to cancel is filed within five years of the date of registration. §14(a), 15 U.S.C. §1064(a). The approach adopted by the Court of Appeals implies that incontestability adds nothing to the protections against cancellation already provided in §14. The decision below not only lacks support in the words of the statute; it effectively emasculates §33(b) under the circumstances of this case.

III

Nothing in the legislative history of the Lanham Act supports a departure from the plain language of the statutory provisions concerning incontestability. Indeed, a conclusion that incontestable status can provide the basis for enforcement of the registrant's exclusive right to use a trade or service mark promotes the goals of the statute. The Lanham Act provides national protection of trademarks in order to secure to the owner of the mark the goodwill of his business and to protect the ability of consumers to distinguish among competing producers. See S. Rep. No. 1333, at 3, 5. National protection of trademarks is desirable, Congress concluded, because trademarks foster competition and the maintenance of quality by securing to the producer the benefits of good reputation. Id., at 4. The incontestability provisions, as the proponents of the Lanham Act emphasized, provide a means for the registrant to quiet title in the ownership of his mark. See Hearings on H.R. 82 before the Subcommittee of the Senate Committee on Patents, 78th Cong., 2d Sess., 21 (1944) (remarks of Rep. Lanham); id., at 21, 113 (testimony of Daphne Robert, ABA Committee on Trade Mark Legislation); Hearings on H.R. 102 et al. before the Subcommittee on Trade-Marks of the House

Committee on Patents, 77th Cong., 1st Sess., 73 (1941) (remarks of Rep. Lanham). The opportunity to obtain incontestable status by satisfying the requirements of §15 thus encourages producers to cultivate the goodwill associated with a particular mark. This function of the incontestability provisions would be utterly frustrated if the holder of an incontestable mark could not enjoin infringement by others so long as they established that the mark would not be registrable but for its incontestable status.

Respondent argues, however, that enforcing petitioner's mark would conflict with the goals of the Lanham Act because the mark is merely descriptive and should never have been registered in the first place. . . .

The alternative of refusing to provide incontestable status for descriptive marks with secondary meaning was expressly noted in the hearings on the Lanham Act. Id., at 64, 69 (testimony of Robert Byerley, New York Patent Law Assn.); Hearings on S. 895 before the Subcommittee of the Senate Committee on Patents, 77th Cong., 2d Sess., 42 (1942) (testimony of Elliot Moyer, Special Assistant to the Attorney General). Also mentioned was the possibility of including as a defense to infringement of an incontestable mark the "fact that a mark is a descriptive, generic, or geographical term or device." Id., at 45, 47. Congress, however, did not adopt either of these alternatives. Instead, Congress expressly provided in §§33(b) and 15 that an incontestable mark could be challenged on specified grounds, and the grounds identified by Congress do not include mere descriptiveness.

The dissent echoes arguments made by opponents of the Lanham Act that the incontestable status of a descriptive mark might take from the public domain language that is merely descriptive. As we have explained, Congress has already addressed concerns to prevent the "commercial monopolization" of descriptive language. The Lanham Act allows a mark to be challenged at any time if it becomes generic, and, under certain circumstances, permits the nontrademark use of descriptive terms contained in an incontestable mark. Finally, if "monopolization" of an incontestable mark threatens economic competition, §33(b)(7), 15 U.S.C. §1115(b)(7), provides a defense on the grounds that the mark is being used to violate federal antitrust laws. At bottom, the dissent simply disagrees with the balance struck by Congress in determining the protection to be given to incontestable marks. . . .

VI

We conclude that the holder of a registered mark may rely on incontestability to enjoin infringement and that such an action may not be defended on the grounds that the mark is merely descriptive. Respondent urges that we nevertheless affirm the decision below based on the "prior use" defense recognized by §33(b)(5) of the Lanham Act. Alternatively, respondent argues that there is no likelihood of confusion and therefore no infringement justifying injunctive relief. The District Court rejected each of these arguments, but they were not addressed by the Court of Appeals. 718 F.2d, at

331-332, n.4. That court may consider them on remand. The judgment of the Court of Appeals is reversed, and the case is remanded for further proceedings consistent with this opinion.

It is so ordered.

Justice STEVENS, dissenting.

In trademark law, the term "incontestable" is itself somewhat confusing and misleading because the Lanham Act expressly identifies over 20 situations in which infringement of an allegedly incontestable mark is permitted.[1] Moreover, in §37 of the Act, Congress unambiguously authorized judicial review of the validity of the registration "in any action involving a registered mark."[2] The problem in this case arises because of petitioner's attempt to enforce as "incontestable" a mark that Congress has plainly stated is inherently unregistrable.

The mark "Park 'N Fly" is at best merely descriptive in the context of airport parking.[3] Section 2 of the Lanham Act plainly prohibits the registration of such a mark unless the applicant proves to the Commissioner of the Patent and Trademark Office that the mark "has become distinctive of the applicant's goods in commerce," or to use the accepted shorthand, that it has acquired a "secondary meaning." See 15 U.S.C. §§1052(e), (f). Petitioner never submitted any such proof to the Commissioner, or indeed to the District Court in this case. Thus, the registration plainly violated the Act.

The violation of the literal wording of the Act also contravened the central purpose of the entire legislative scheme. Statutory protection for trademarks was granted in order to safeguard the goodwill that is associated with

1. Section 33(b) enumerates seven categories of defenses to an action to enforce an incontestable mark. See 15 U.S.C. §1115(b), quoted ante, at 3, n.1. In addition, a defendant is free to argue that a mark should never have become incontestable for any of the four reasons enumerated in §15. 15 U.S.C. §1065. Moreover, §15 expressly provides that an incontestable mark may be challenged on any of the grounds set forth in subsections (c) and (e) of §14, 15 U.S.C. §1064, and those sections, in turn, incorporate the objections to registrability that are defined in §§2(a), 2(b), and 2(c) of the Act. 15 U.S.C. §§1052(a), (b), and (c).

2. Section 37, in pertinent part, provides: "In any action involving a registered mark the court may determine the right to registration, order the cancellation of registrations in whole or in part, restore canceled registrations, and otherwise rectify the register with respect to registrations of any party to the action." 15 U.S.C. §1119.

3. In the Court of Appeals petitioner argued that its mark was suggestive with respect to airport parking lots. The Court of Appeals responded: "We are unpersuaded. Given the clarity of its first word, Park 'N Fly's mark seen in context can be understood readily by consumers as an offering of airport parking — imagination, thought, or perception is not needed. Simply understood, 'park and fly' is a clear and concise description of a characteristic or ingredient of the service offered — the customer parks his car and flies from the airport. We conclude that Park 'N Fly's mark used in the context of airport parking is, at best, a merely descriptive mark." 718 F.2d 327, 331 (CA9 1983).

Although the Court appears to speculate that even though the mark is now merely descriptive it might not have been merely descriptive in 1971 when it was first registered. . . . I find such speculation totally unpersuasive. But even if the Court's speculation were valid, the entire rationale of its opinion is based on the assumption that the mark is in the "merely descriptive" category.

particular enterprises. A mark must perform the function of distinguishing the producer or provider of a good or service in order to have any legitimate claim to protection. A merely descriptive mark that has not acquired secondary meaning does not perform that function because it simply "describes the qualities or characteristics of a good or service." No legislative purpose is served by granting anyone a monopoly in the use of such a mark.

Instead of confronting the question whether an inherently unregistrable mark can provide the basis for an injunction against alleged infringement, the Court treats the case as though it presented the same question as Union Carbide Corp. v. Ever-Ready, Inc., 531 F.2d 366 (CA7), *cert. denied*, 429 U.S. 830 (1976), a case in which the merely descriptive mark had an obvious and well-established secondary meaning. In such a case, I would agree with the Court that the descriptive character of the mark does not provide an infringer with a defense. In this case, however, the provisions of the Act dealing with incontestable marks do not support the result the Court has reached.

I

The word "incontestable" is not defined in the Act. Nor, surprisingly, is the concept explained in the Committee Reports on the bill that was enacted in 1946. The word itself implies that it was intended to resolve potential contests between rival claimants to a particular mark. And, indeed, the testimony of the proponents of the concept in the Committee hearings that occurred from time to time during the period when this legislation was being considered reveals that they were primarily concerned with the problem that potential contests over the ownership of registrable marks might present. No one ever suggested that any public purpose would be served by granting incontestable status to a mark that should never have been accepted for registration in the first instance.

In those hearings the witnesses frequently referred to incontestability as comparable to a decree quieting title to real property. Such a decree forecloses any further contest over ownership of the property, but it cannot create the property itself. Similarly the incontestability of a trademark precludes any competitor from contesting the registrant's ownership, but cannot convert unregistrable subject matter into a valid mark. Such a claim would be clearly unenforceable.

. . . The testimony in the Committee hearings concerning the public interest in preventing the grant of monopoly privileges in the use of merely descriptive phrases expressly relied on the administrative practice that was incorporated into §2(f), 15 U.S.C. §1052(f), as a protection against the improper registration of merely descriptive marks. . . .

. . . In light of this legislative history, it is apparent that Congress could not have intended that incontestability should preserve a merely descriptive trademark from challenge when the statutory procedure for establishing secondary meaning was not followed and when the record still contains no evidence that the mark has ever acquired a secondary meaning.

If the registrant of a merely descriptive mark complies with the statutory requirement that prima facie evidence of secondary meaning must be submitted to the Patent and Trademark Office, it is entirely consistent with the policy of the Act to accord the mark incontestable status after an additional five years of continued use. For if no rival contests the registration in that period, it is reasonable to presume that the initial prima facie showing of distinctiveness could not be rebutted. But if no proof of secondary meaning is ever presented, either to the Patent and Trademark Office or to a court, there is simply no rational basis for leaping to the conclusion that the passage of time has transformed an inherently defective mark into an incontestable mark.

Congress enacted the Lanham Act "to secure trade-mark owners in the goodwill which they have built up." But without a showing of secondary meaning, there is no basis upon which to conclude that petitioner has built up any goodwill that is secured by the mark "Park 'N Fly." In fact, without a showing of secondary meaning, we should presume that petitioner's business appears to the consuming public to be just another anonymous, indistinguishable parking lot. When enacting the Lanham Act, Congress also wanted to "protect the public from imposition by the use of counterfeit and imitated marks and false trade descriptions." Upon this record there appears no danger of this occurrence, and as a practical matter, without any showing that the public can specifically identify petitioner's service, it seems difficult to believe that anyone would imitate petitioner's marks, or that such imitation, even if it occurred, would be likely to confuse anybody.

The Court suggests that my reading of the Act "effectively emasculates §33(b) under the circumstances of this case." But my reading would simply require the owner of a merely descriptive mark to prove secondary meaning before obtaining any benefit from incontestability. If a mark is in fact "distinctive of the applicant's goods in commerce" as §2(f) requires, that burden should not be onerous. If the mark does not have any such secondary meaning, the burden of course could not be met. But if that be the case, the purposes of the Act are served, not frustrated, by requiring adherence to the statutory procedure mandated by Congress.[28]

---

28. Moreover, even if the owner of a registered mark may not enjoin infringement, it is not true that the registration has become "meaningless." . . . A registration may be used to prevent the importation of goods bearing infringing marks into this country. See 15 U.S.C. §1124, 19 U.S.C. §1526, and 19 U.S.C. §1337(a). Additionally, registration in this country is a prerequisite to registration in some foreign countries. A. Seidel, What the General Practitioner Should Know About Trademarks and Copyrights 26 (4th ed. 1979); E. Vandenburgh, Trademark Law and Procedure 58 (2d ed. 1968). Further, the United States Court of Customs and Patent Appeals, in an opinion recognizing that Congress had expressed its desire that scandalous matter not be registered, wrote the following regarding the benefits of registration:

> Once a registration is granted, the responsibilities of the government with respect to a mark are not ended. The benefits of registration, in part with government assistance,

In sum, if petitioner had complied with §2(f) at the time of its initial registration, or if it had been able to prove secondary meaning in this case, I would agree with the Court's disposition. I cannot, however, subscribe to its conclusion that the holder of a mark which was registered in violation of an unambiguous statutory command "may rely on incontestability to enjoin infringement." . . . Accordingly, I respectfully dissent.

COMMENTS AND QUESTIONS

1. Why should marks be allowed to become incontestable? Does this privilege reflect a judgment that judicial review of trademarks is unnecessary after a certain time? That the Trademark Office can be trusted to make the right decision in an ex parte proceeding? Certainly, no similar right is afforded the holders of patents, even though they are in some sense a "stronger" intellectual property right. At least one commentator has suggested that incontestability ought to be abolished. Kenneth L. Port, The Illegitimacy of Trademark Incontestability, 26 Ind. L. Rev. 519 (1993). Professor Port reasons that trademark incontestability creates what is in effect a property right in trademarks, since 15 U.S.C. §1115(b) speaks of a conclusive presumption of the trademark holder's "ownership of the mark." (Certainly, the Court's "quiet title" rationale smacks of real property law.) This property right, he asserts, is inconsistent with general principles of trademark law. He explains:

> Patent and copyright owners enjoy the "bundle of rights" notion of property. Their rights are divisible, freely alienable, and exclusive for the duration of statutory protection.
>
> Trademarks, on the other hand, enjoy none of the "bundle of rights" that other forms of property enjoy. Trademark holders possess only the right to exclude others from using that specific trademark on similar goods. Holders of marks possess the right to protect the sphere of interest in which they are using the mark by excluding others, but nothing more. Mark holders do not possess a property right in the mark itself, because trademarks are nothing when devoid of the goodwill they have come to represent or the product on which they are used.

Id. Read section 1115(b) carefully. Is this a fair criticism?

---

include public notice of the mark in an official government publication and in official records which are distributed throughout the world, maintenance of permanent public records concerning the mark, availability of Customs Service for blocking importation of infringing goods, access to federal courts where there is a presumption of validity of the registration . . . , notices to the registrant concerning maintenance of the registration and, to some extent, direct government protection of the mark in that the PTO searches its records and refuses registration to others of conflicting marks. Apart from nominal fees, these costs are underwritten by public funds. In re Robert L. McGinley, 660 F.2d 481, 486 (1981).

2. How does the dissent in *Park 'N Fly* deal with the majority's argument that "incontestability" would be worthless if a trademark holder could not use it to enforce his rights against an infringer? Is Justice Stevens's argument persuasive? If the owner of a descriptive mark improperly registered is not entitled to judicial enforcement of his "incontestable" trademark, why should he be entitled to Customs assistance in stopping the importation of goods? To registration of the invalid trademark in other countries? Further, why have incontestability at all? Wouldn't the legal system be better served by a rule that improperly issued trademarks could be cancelled at any time?

This debate was resurrected in Shakespeare Co. v. Silstar Corp., 9 F.3d 1091 (4th Cir. 1993). There, the District Court cancelled plaintiff's registration for a color scheme on a fishing rod on the grounds that it was functional (apparently because it helped to catch fish). The Fourth Circuit reversed, noting that the mark had become incontestable and holding that incontestable marks cannot be cancelled for functionality because functionality is not specifically enumerated in section 1064. The dissent argued that federal courts should not enforce the mark, even if they could not cancel it.

The functionality problem raises an interesting question about incontestability. What happens when trade dress which is initially not functional *becomes* functional over time? (This may happen because of widespread adoption of the owner's trade dress. For example, a computer graphical user interface (GUI) may become so familiar to so many users that it becomes a "standard" in the industry, and it is a commercial advantage to use the standard product.) Would trademark defendants lose *any* opportunity to defeat the mark on functionality grounds? Apparently concerned by the result in *Shakespeare*, Congress revised the Lanham Act in 1998 to provide that functionality may be asserted as a defense even against an "incontestable" mark. 15 U.S.C. §1064(3).

3. Note that the presumptions afforded registered trademarks differ in several ways from those given to issued patents. Registered trademarks, while entitled to a presumption of validity, are not entitled to the benefit of the "clear and convincing evidence" rule applied in patent cases. At the same time, there is no provision for incontestability of patents after a certain number of years. Is there any reason for these differing presumptions?

4. Patent law gives inventors no rights at all unless the Patent Office issues them a patent. By contrast, copyright law gives full protection to unregistered works as soon as they are created, and there are current legislative efforts to abolish the Copyright Office entirely. Trademark law appears to fall somewhere in between. Trademark owners are entitled to protection without registration, but there are still substantial benefits to registering a trademark. Can you think of any reason for the different administrative schemes? Which is preferable?

5. Extension Overseas: International Trademark Practice

National borders are less of a barrier to economic exchange now than at almost any time in history. As economic activity continues its relentless drive toward world-wide scope, trademarks become even more important.[14] Just as the growth of the national economy in the United States set the stage for trademark law of national scope, the growth of the worldwide economy shows the importance of some sort of international trademark system.

U.S. Trademarks Abroad: The Failure (So Far) of Internationalization. Despite the obvious desirability of international trademark registration, the United States has not joined an international convention that goes beyond the rather minimal protections afforded by the Paris Union. Trademark owners still have to prosecute trademark applications in each country where they want their rights protected.[15] Doing this is expensive; the various national filing fees, translations, and local counsel fees add up.[16] But trademark law in the United States is completely outside the structure of the Madrid Trademark Agreement (MTA), the main international trademark convention.[17] (In 1980, the United States refused to join the Trademark Registration Treaty, a minor international agreement;[18] and although the United States in 1994 did sign the Trademark Law Treaty (TLT), this is a limited agreement among many countries relating mostly to subject matter and prosecution procedures.[19]) Now, with the advent of a new variant on the MTA — called the "Madrid Protocol" — there is renewed interest in trademark internationalization in the United States.[20]

---

14. Shari Caudron, Strike Up the Brand, Indus. Wk., Mar. 7, 1994, at 64 ("[I]n a world ruled by mass production and product standardization, the brand is the only remaining distinction among many products and services.").

15. A U.S. corporation may, however, obtain the benefits of international registration under the Madrid Trademark Agreement in the name of a subsidiary corporation domiciled in one of the Madrid Agreement member states. See MTA, infra, arts. 1(2), (3), 828 U.N.T.S. at 393.

16. Daniel C. Schulte, The Madrid Trademark Agreement's Basis in Registration-based Systems: Does the Protocol Overcome Past Biases? (Part I), 77 J. Pat. & Trademk. Off. Society 595, 597 (1995).

17. Madrid Agreement Concerning the International Registration of Marks, April 14, 1891, 175 Consol. T.S. 57. This is an agreement under the Paris Convention for the Protection of Industrial Property, Stockholm Revision, done July 14, 1967, art. 2(1), 21 U.S.T. 1629, 1631, 828 U.N.T.S. 305, 313. There are approximately 40 members, including China and most of the European nations. See World Intellectual Property Org., World Intellectual Property Organization General Information 30 (1993). Various other international agreements exist, but for the most part they only cover relations between pairs of countries. See, e.g., 3 J. Thomas McCarthy, Trademarks and Unfair Competition §29.10[2] (describing bilateral treaties signed by the U.S.).

18. See Trademark Registration Treaty, done at Vienna, June 12, 1973, and amended on Sept. 26, 1980, reprinted in World Intellectual Property Organization, Trademark Registration Treaty (1989).

19. See Trademark Law Treaty, October 28, 1994. The Trademark Law Treaty was ratified by the Senate in June 1998.

20. Protocol Relating to the Madrid Agreement Concerning the International Registration of Marks, June 28, 1989, reprinted in World Intellectual Property Organization, Protocol, Re-

The MTA does not itself protect any trademark rights; it simply facilitates trademark prosecution in member states. It is effective in this regard, however: over 290,000 international Madrid registrations are now in force. On average, each international registration is extended to ten countries.

Under the MTA, a trademark owner files a single registration with its domestic trademark office.[21] The owner may then extend this "basic registration" by filing, within six months of the original filing, an international application with the domestic trademark office designating other MTA member countries in which protection is sought.[22] The international application is then forwarded to the international trademark bureau of the World Intellectual Property Organization (WIPO), which issues an international registration for the mark and publishes it in "Les Marques Internationales," the gazette of the international trademark community. WIPO then forwards the application to the designated countries for prosecution under each country's national law.[23] Unless the trademark office or an opponent takes action to deny protection, the mark is registered in each of the designated countries.[24]

Several objections have prevented the United States from joining the MTA over the years. First, the MTA follows the international norm in its priority rule, which turns on registration rather than use.[25] Second, because of this, and under the MTA's reciprocity clause, foreign trademark registrants might well flood the U.S. trademark register with unused marks. Thus by staying out of the MTA, the U.S. trademark community is preventing clutter on the trademark register — clutter that has a real cost, in the form of higher search, clearance, and litigation costs. Third, the Madrid Agreement includes a provision known as "central attack," which provides that if the home country registration is successfully opposed or cancelled during the first five years after international registration, all related registrations fall with it. This would harm U.S. trademark owners disproportionately because of the many bases on which U.S. registrations may be challenged under the Lanham Act relative to other countries. Finally, the MTA has only one official language: French.

---

lating to the Madrid Agreement Concerning the International Registration of Marks [hereinafter Madrid Protocol]; S. 977, 103d Cong., 1st Sess. (1993); H.R. 2129, 103d Cong., 2d Sess. (1994). In early 1996, the Madrid Protocol had about thirty signatories but was not yet in force. For a general discussion, see Monique L. Cordray, GATT v. WIPO, 76 J. Pat. & Trademk. Off. Society 121, 124 (1994).

21. Madrid Agreement, art. 1(2), 828 U.N.T.S. at 393.

22. See arts. 3(1), 3ᵗᵉʳ(1), 828 U.N.T.S. at 395, 397; Art. 4(2), 828 U.N.T.S. at 399; Paris Convention, supra, Art. 4(C), 21 U.S.T. at 1632, 828 U.N.T.S. at 315 (six month priority "grace period" for trademark registration filings).

23. Arts. 3(4), 5, 828 U.N.T.S. at 395-397, 399.

24. Art. 5(1), (2), (5), 828 U.N.T.S. at 399-401.

25. See Gerald D. O'Brien, The Madrid Agreement Adherence Question, 56 Trademark Rep. 326 (1966). See generally Jeffrey M. Samuels & Linda B. Samuels, The Changing Landscape of International Trademark Law, 27 Geo. Wash. J. Intl. L. & Econ. 433 (1994).

Some knowledgeable observers believe that many of the traditional objections to U.S. adherence to the MTA no longer apply in light of the new Madrid Protocol (MP).[26] Indeed, the MP was drafted with an eye to overcoming these very objections. The MP thus considerably softens some of the harsher impacts of the MTA, especially on the United States.[27] Specifically, recent legislation to implement the MP under U.S. law (1) allows for filing in English or French, in the U.S. PTO or any other member country trademark office; (2) allows the filing of an international application based not only on a "basic" (or home country) registration but also on a basic *application* — a substantial advantage given the relatively stringent prosecution standards in the United States, which can lead to long gaps between applications and registrations, and especially now in light of intent-to-use registrations under the Lanham Act; and (3) substantially lessens the impact of "central attack" by allowing an international registrant to convert a successfully attacked registration into a bundle of individual national registrations if the attack occurs during the first five years of the international registration.[28]

Although some problems remain,[29] and it had not yet passed in the United States when this book went to press, the MP faces better prospects than any trademark harmonization initiative in the recent past. Indeed, some degree of harmonization would seem to be necessary, if only to keep up with trading partners. In Europe, for example, 1995 saw the advent of the Community Trademark Convention, a treaty providing for a single, Community

26. See U.S. House of Representatives, Committee on the Judiciary, Subcommittee on Courts and Intellectual Property, Hearings on Trademarks Legislation, July 19, 1995 (Statement of Nils Victor Montan, Trademark Counsel, Warner Bros., Inc.) (supporting passage of bill proposing U.S. version of MP).

27. Technically, members of the MP join the Madrid Union without signing the MTA. See MTA art. 1, 828 U.N.T.S. at 391 (states adhering to the Madrid Agreement "constitute a Special Union for the international registration of marks," known as the "Madrid Union."); Madrid Agreement Protocol, art. 1, at 9 (states adhering to the Madrid Agreement Protocol are also members of the Madrid Union).

28. MP, art. 9, at 43-45.

29. See, e.g., Allan Zelnick, The Madrid Protocol — Some Reflections, 82 Trademark Rep. 651 (1992). Zelnick argues that U.S. companies might still prefer to pursue foreign trademark rights by filing individual foreign applications because of the liberality of most foreign trademark systems regarding the classes of goods that can be identified with the mark in the application. (In many countries, a trademark application can specify *all* classes of goods — giving applicants substantially broader rights than under U.S. law.) According to Zelnick,

> [I]t seems likely that on the whole, most American trademark owners who file abroad today are unlikely to make use of the Madrid Protocol. The reason for this conclusion is that under our practice the specification of goods or services that will appear in domestic application, on which the international application will be based, must be narrowly drawn to the specific goods or services in respect of which the mark shall have been used in interstate or foreign commerce of the United States (in the case of applications based on use) or restricted to the particular items or services identified by their common trade names in respect of which the applicant shall have a bona fide intention to use the mark. Since the international application under the Madrid Protocol will have the same specification of goods or services as the United States application upon which it is based, we will have, in effect, transferred our domestic practice in this regard to United States trademark owners' international filings.

Id. at 652.

Trade Mark (CTM) based on a single European trademark registration.[30] If further proof were needed, the CTC shows convincingly that the logic of trademark harmonization is compelling in this era of global trade and instantaneous communication.

Foreign Trademarks in the United States: Limited Internationalization. Because the United States does not yet adhere to any international trademark treaty, foreign trademark owners cannot avail themselves of a centralized prosecution system such as that provided by the Trademark Registration Treaty or the Madrid Protocol. Even so, certain provisions of U.S. law do make some slight concessions to the need for foreign trademark owners to register their marks in the United States. Primarily, section 44 of the Lanham Act allows foreign trademark owners to register their marks in the United States. Lanham Act §44, 15 U.S.C. §1126.

Although §44(d)(2) requires a statement of bona fide intent to use,[31] no subsequent proof of actual use in commerce in the United States is required. This allows foreign trademark owners to register their marks in the United States on terms slightly more favorable than those extended to domestic applicants under the Lanham Act.

One court has held that the Lanham Act incorporates the substantive provisions of the Paris Convention governing trademarks and unfair competition, providing a cause of action for violation of those provisions. General Motors Corp. v. Lopez, 948 F. Supp. 684 (E.D. Mich. 1996).

---

30. Council Regulation (EC) no. 40/94 of 20 December 1993 on the Community Trademark, OJ L 11/1. The CTM Regulation built on a 1980 Directive that substantially harmonized the national trademark laws of the EC member states. Primary features of the recently added CTM are:

- Trademarks with renewable ten-year term, to be registered in the new European Trademark Office in Alicante, Spain.
- Marks valid in all member states but invalid everywhere if revoked.
- Substantive requirements for registration and grounds for opposition that would largely be familiar to American trademark practitioners, e.g., registration of nondistinctive marks upon proof of secondary meaning, and for a limited class of goods or services.
- The working languages are German, English, Spanish, French and Italian. Applications can be filed in any Community language and are translated for free; opposition proceedings, however, are only conducted in one of the official languages, which must be selected at the time of filing.

See Eric P. Raciti, The Harmonization of Trademarks in the European Community: The Harmonization Directive and the Community Trademark, 78 J. Pat. & Trademark Off. Society 51 (1996).

31. This provision was added in 1988 (effective 1989) in response to the decision of Trademark Trial and Appeal Board's decision in Crocker National Bank v. Canadian Imperial Bank of Commerce, 223 U.S.P.Q. (BNA) 909 (TTAB 1984), which held valid a trademark registration application filed by a Canadian bank, despite the absence of an allegation of use anywhere in the world.

Note on the "Gray Market"

United States law prohibits the importation of any goods bearing a registered United States trademark without the consent of the domestic trademark owner, even if the goods are genuine and even if they are imported by a licensed manufacturer (or the original trademark owner itself). 19 U.S.C. §1526. The so-called gray market or "parallel imports" problem is a troubling one for courts and commentators alike. The Supreme Court has described the problem as follows:

> The gray market arises in any of three general contexts. The prototypical gray-market victim (case 1) is a domestic firm that purchases from an independent foreign firm the rights to register and use the latter's trademark as a United States trademark and to sell its foreign-manufactured products here. Especially where the foreign firm has already registered the trademark in the United States or where the product has already earned a reputation for quality, the right to use that trademark can be very valuable. If the foreign manufacturer could import the trademarked goods and distribute them here, despite having sold the trademark to a domestic firm, the domestic firm would be forced into sharp intrabrand competition involving the very trademark it purchased. Similar intrabrand competition could arise if the foreign manufacturer markets its wares outside the United States, as is often the case, and a third party who purchases them abroad could legally import them. In either event, the parallel importation, if permitted to proceed, would create a gray market that could jeopardize the trademark holder's investment.
>
> The second context (case 2) is a situation in which a domestic firm registers the United States trademark for goods that are manufactured abroad by an affiliated manufacturer. In its most common variation (case 2a), a foreign firm wishes to control distribution of its wares in this country by incorporating a subsidiary here. The subsidiary then registers under its own name (or the manufacturer assigns to the subsidiary's name) a United States trademark that is identical to its parent's foreign trademark. The parallel importation by a third party who buys the goods abroad (or conceivably even by the affiliated foreign manufacturer itself) creates a gray market. Two other variations on this theme occur when an American-based firm establishes abroad a manufacturing subsidiary corporation (case 2b) or its own unincorporated manufacturing division (case 2c) to produce its United States trademarked goods, and then imports them for domestic distribution. If the trademark holder or its foreign subsidiary sells the trademarked goods abroad, the parallel importation of the goods competes on the gray market with the holder's domestic sales.
>
> In the third context (case 3), the domestic holder of a United States trademark authorizes an independent foreign manufacturer to use it. Usually the holder sells to the foreign manufacturer an exclusive right to use the trademark in a particular foreign location, but conditions the right on the foreign manufacturer's promise not to import its trademarked goods into the United States. Once again, if the foreign manufacturer or a third party imports into the United States, the foreign-manufactured goods will compete on the gray market with the holder's domestic goods.

K-Mart Corp. v. Cartier, Inc., 486 U.S. 281, 286-287 (1988).

What is wrong with importing gray market goods? In this case, unlike the typical trademark dispute, there is no question that the goods are genuine and are being sold as precisely what they are. Further, there is no question that they originated with the trademark owner (at least at some point). If a gray market transaction occurred entirely within the United States, would it constitute trademark infringement? If not, is there any reason to treat importation differently?

One obvious rationale for prohibiting gray market imports is that the United States trademark owner is losing the benefit of what it thought was an "exclusive" right to sell the trademarked goods in the United States. But does it really have such an exclusive right? Certainly, the right to use the Ford trademark on cars in the United States does not prevent bona fide purchasers of Ford cars from identifying them as such upon resale. And even if the United States trademark owner did obtain the exclusive rights to the first sale of goods in the United States from the manufacturer, isn't the problem here really one of breach of contract?

Section 1526 of 19 U.S.C., which regulates gray market goods, may violate Article 2 of the Paris Convention because it treats United States trademark holders differently than foreign trademark holders, and United States infringers differently than foreign infringers. See Raimund Steiner & Robert Sabath, Intellectual Property and Trade Law Approaches to Gray Market Importation, and the Restructuring of Transnational Entities to Permit Blockage of Gray Goods in the United States, 15 Wm. Mitchell L. Rev. 433, 441 (1989) (after *K-Mart*, to obtain benefits of Section 526 of the Tariff Act of 1930 and the Lanham Act, companies must separate ownership of foreign and domestic trademarks); see generally Note, The Use of Copyright Law to Block the Importation of Gray-Market Goods: The Black and White of It All, 23 Loyola (LA) L. Rev. 645 (1990).

There is a debate in the economics literature about the wisdom of permitting gray market imports. The argument generally runs along these lines: those favoring restrictions on the gray market cite the importance of encouraging investments in brand quality by retailers, coupled with differing international product quality standards. In essence, the argument is that retailers will not engage in optimal expenditures to promote the product (advertising, clean showrooms, knowledgeable sales staff, etc.) if gray market imports will undercut the retailers' prices. On the opposite side, those who support the gray market say it undercuts blatant price discrimination. See Robert J. Staaf, International Price Discrimination and the Gray Market, 4 Intellectual Prop. J. 301 (1989) (gray market presents an opportunity for arbitrage, allowing competitors to circumvent attempted price discrimination and therefore benefitting consumers).

The Problem of Worldwide Famous Marks. Worldwide trademark protection is based on proof of either registration or use in each country in which protection is sought. In theory this creates a significant burden for companies with marks that are famous worldwide: they must register their mark in every country in the world or risk losing rights to that mark to a competing registrant. In fact a number of opportunistic individuals have registered famous marks such as "Coke" in countries where Coca-Cola does not yet do business.

In an effort to deal with this problem, courts around the world have increasingly been willing to recognize goodwill in a world-famous mark even if the mark is not actually used in the country see Paris Convention art. 6bis. Courts have held that K-Mart was entitled to prevent a competing use in Jamaica (see Kmart Corp. v. Kay Mart Ltd., Suit No. C.L. 1993, K066) and that McDonald's was permitted to "take back" rights in South Africa to marks it had ceased using there during the world embargo of that country. Ironically, the United States seems less willing than other nations to act against domestic citizens who trade on the goodwill of foreign marks. See Buti v. Impressa Perosa S.R.L., 139 F.3d 98 (2d Cir. 1998); Person's Co. v. Christman, 900 F.2d 1565 (Fed. Cir. 1990).

6. Extension by Contract: Licensing and Franchising

"Turning Logos into Profit Centers." Evidence is everywhere that a boom is under way in the licensing of trademarks. From sports team logos to university names to designer symbols, badges of affiliation and prestige are ever more common on products of all kinds. And, importantly, these badges come at a premium. As Robert Denicola has written,

> At any sporting goods store one can find plain, unadorned shirts, shorts, and jackets in assorted styles and colors. They are usually near the rear. Closer to the front are items apparently similar in all respects except one — they are prominently decorated with a variety of words and symbols. Some bear the names of athletic equipment manufacturers. Others display the names or insignia of professional sports teams, the name and seal of the state university, or the nickname and mascot of the local high school. They frequently cost significantly more than the items in the rear, yet they sell.

Robert C. Denicola, Institutional Publicity Rights: An Analysis of the Merchandising of Famous Trade Symbols, 62 N.C. L. Rev. 603 (1983).

Empirical evidence confirms the observation. A licensing industry newsletter placed the volume of licensing activity at $13.15 billion in 1994. See Licensing Letter (EPM Communications, Inc.), April 1, 1995, at p. 1. As for future growth prospects, the same newsletter quotes an industry source as a representative voice: "[O]ne of the trends in licensing last year [1994] was the tendency for corporations to examine the potential 'of anything marketable to turn logos into profit centers.'"

Turning logos into profit centers may make sense as a business strategy, but it poses problems for the legal system. The cases in this chapter highlight the fundamental problem: traditionally, trademarks were thought of as symbols representing products, rather than as products in and of themselves. Thus traditional trademark law protects a trademark only because, and only insofar as, it is emblematic of the goodwill behind a product. The mark itself is not the point; it is simply a vehicle to convey useful information regarding a product's quality, prestige, and so on. The trademark guides the consumer to the transaction; sale of the underlying product is the "profit center."

All this changes when the trademark becomes the *subject* of the transaction rather than an adjunct to it. In such a transaction, the mark does not represent the product: it *is* the product.

Consider for example the licensing of a sports team logo for use on a T-shirt or hat. The logo does not summarize information about the quality of the hat; it demonstrates loyalty to a team. Indeed, many if not most sports logos are licensed to a broad array of products, many of differing degrees of quality. The "high end" Red Sox cap ("just like the pros wear!") is a far different product — qua hat — than the cheap synthetic cap costing a few dollars and sold in discount stores. The same for sweatshirts and T-shirts emblazoned with the logos of college and professional teams. Similar examples can be drawn from other avenues of commerce. Thus outdoor equipment companies with a certain consumer cachet have been known to lend their logos for application to rugged four-wheel drive vehicles. Yet no one, or very few, can suppose that companies specializing in backpacks, long underwear, and hiking boots have suddenly taken up truck production.

In these cases, the consumer is buying an image. The trademark owner possesses rights in a symbol associated with certain qualities. Lending that symbol to diverse products connects those products to the feelings the symbol evokes. (The same thing happens outside commerce as well; consider the difference when the "stars and stripes" image is added to a plain object such as a flag or a tombstone.)

Trademark law has begun to recognize the importance of the logo as a product. For instance, a 1975 case held that unauthorized sales of team emblems violated the Lanham Act:

> The certain knowledge of the buyer that the source and origin of the trademark symbols were the plaintiffs satisfies the requirements of the Act. The argument that confusion must be as to source of the manufacture of the emblem itself is unpersuasive, where the trademark, originated by the team, is the triggering mechanism for the sale of the emblem.

Boston Professional Hockey Assn. v. Dallas Cap & Emblem Mfg., Inc., 510 F.2d 1004, 1012 (5th Cir. 1975) (permanent injunction), *cert. denied*, 423 U.S. 868, *reh'g denied*, 423 U.S. 991 (1975).

This holding seems to follow a felt sense that it would be improper to permit an anonymous third party to profit from the plaintiff's logo. However equitable it may appear, though, such a holding is arguably at odds with traditional trademark theory. How does the logo lower consumer search costs? If the logo does not summarize product attributes such as quality, why protect it? Tribunals such as the *Boston Professional Hockey* court have ignored these anomalies and protected logos against use by another. Thus, although trademark owners must still clear some formidable doctrinal hurdles (see below, section on Abandonment and Assignments in Gross), courts have shown a willingness to expand doctrine beyond traditional limits in the face of these new commercial trends.

Trademarks and Organizational Forms: The Growth of Franchising. From its modern founding, trademark law has been at the service of emerging patterns in the organization of industry. In the beginning, that meant protecting emerging channels of trade in relatively local settings. Next came the great nationalization of the economy with the growth of large retail empires in the late nineteenth and early twentieth centuries. Today trademarks form an integral part of the variegated economic landcape of the industrialized world.

Business historian Mira Wilkins has written:

> The legally-backed trade marks . . . became essential intangible assets, providing the basis for the rise of the modern enterprise. . . . The trade mark's fundamental contribution to the modern corporation was that it generated efficiency gains by creating for the firm the opportunity for large sales over long periods. . . . Without the trade mark, the introduction and acceptance by buyers of modern products, produced with economies of scale or scope, and marketed over long distances, would have been impossible. . . . The trade mark by reducing the costs of information led to efficiencies in production and distribution.

Mira Wilkins, The Neglected Intangible Asset: The Influence of the Trademark on the Rise of the Modern Corporation, 34 Bus. & Hist. 66, 87-88 (1992). Interestingly, Wilkins also points out the role that trademarks play in facilitating organizational diversity, or making possible various alternative forms of production. The clearest and most important recent example is franchising.

Trademarks are the "cornerstone of a franchise system." Susser v. Carvel Corp., 206 F. Supp. 636, 640 (S.D.N.Y. 1962), *aff'd*, 332 F.2d 505, 141 U.S.P.Q. 609, *cert. granted*, 379 U.S. 885, *cert. dismissed*, 381 U.S. 125 (1965). The trademark of the franchisor is the identifiable symbol of continuity; it indicates the presence of the national brand at each location. Thus, whatever the precise nature of the franchise, the franchisor and franchisee are very likely to be parties to a trademark license agreement.

Here again we see the stretching of traditional theory. The individual source from which the national brand emanates is the franchisor. This is often a remote corporate entity, whereas one could argue that the franchisee is the real "source" (at the local level) of the goods. To maintain uniformity (and stay on the good side of the abandonment issue; see below), the franchisor almost invariably imposes certain contractual requirements on the franchisee.

Yet it is still the franchisee, in the last instance, who actually runs the establishment where the goods are sold. See James M. Treece, Trademark Licensing and Vertical Restraints in Franchising Arrangements, 116 U. Pa. L. Rev. 435 (1968).

In economic terms, franchising is an interesting mix of contractual and integrated governance characteristics — a kind of hybrid organization in which the franchisee is neither an employee of the franchisor nor an arm's-length buyer of the franchisor's goods. See James A. Brickley & Frederick H. Dark, The Choice of Organizational Form: The Case of Franchising, 18 J. Fin. Econ. 401, 403-407 (1987); Gillian K. Hadfield, Problematic Relations: Franchising and the Law of Incomplete Contracts, 42 Stan. L. Rev. 927 (1990). The franchise trademark plays an interesting role in this relationship. It is one of the franchisor's great assets, one of the things it can sell to franchisees. Yet once a franchisee begins to use the trademark, he or she is in a position to harm the franchise's reputation by selling inferior quality goods. (Note that doing so hurts other franchisees as well.) Indeed, if many of the franchise's customers are travelers who come from afar and do not pass through often, a franchisee might be tempted to "free ride" off the quality investments of the franchisor and other franchisees, by selling inferior goods. Many of the provisions in franchise agreements are directed at preventing such an outcome, e.g., agreements to purchase ingredients and other inputs from the franchisor, stipulations to frequent inspections, and profit-sharing arrangements. See Paul H. Rubin, The Theory of the Firm and the Structure of the Franchise Contract, 21 J. Law & Econ. 223 (1978). Cf. Patrick J. Kaufman & Francine Lafontaine, Costs of Control: The Source of Economic Rents for McDonald's Franchisees, 37 J.L. & Econ. 417 (1994) (finding substantial "rents" or profits left to the franchisee after the franchisor's "cut," indicating that incentive system in franchise contracts was working). In this complicated relationship, trademarks are a key strategic asset. Note also that in many states various aspects of franchise contracts — especially termination provisions — are closely regulated. See, e.g., Thomas M. Pitegoff, Franchise Relationship Laws: A Minefield for Franchisors, 45 Bus. Law 289 (1989).

What significance does the growth of franchising have for trademark doctrine? Should we be concerned about the disaggregation of trademarks from the goods they represent? See the discussion in the previous section, on trademarks sold as valuable items in and of themselves. Does the selling of trademarks without goods attached to them make trademark owners less able to use the mark as a guarantee of quality? Alternatively, does it suggest that the trademark owner will invest even more in quality assurance, since the value of the mark is all it has to offer?

PROBLEM

Problem 5-5. A popular "cult" movie includes characters who have tattooed famous brand logos on various parts of their bodies. The practice quickly catches on, first among "avant garde" artists, and then among college and high school students.

Post-Modern Concepts, Inc. (PMC), seeing an opportunity, opens a storefront tattoo parlor in a popular retail mall in the Midwest. Here PMC applies popular logos as tattoos. At first, customers are primarily interested in the ironic use of logos, as in the film; the logos are primarily used as a spoof or commentary on consumerism. One popular tattoo, for instance, includes the famous "Calvin Klein" wordmark and logo, drawn inside a red circle with a slash. Soon, however — perhaps predictably — the ironic intent of the original practice is lost as it becomes very chic to have a logo tattoo. Thus many people begin to request tattoos without the red slash and circle. The most popular trademark logo tattoos are "Rolls Royce," "Nike," the original "Calvin Klein," and "Harley-Davidson."

Success is instantaneous. In response to torrid demand, PMC sets up franchises all over the country.

Owners of the trademarks begin to notice PMC. They are concerned primarily about lost revenues, because PMC has never taken out any trademark licenses. But some trademark owners are also concerned that pictures of people with logo tattoos are beginning to appear in the press. Even some famous celebrities, such as popular musicians and sports stars, have tattoos in prominent places. When their picture appears in the newspaper, the logo — or part of it — does too. Other trademark owners are concerned that logo tattoos are being applied on parts of the body traditionally covered by clothing. These owners fear that when pictures of people with logo tattoos in these places appear — sometimes in press outlets not fit for family viewing — the trademarks are denigrated.

What hurdles will the trademark owners have to clear to enforce their rights? (Are these marks attached to a product? Also, PMC will undoubtedly argue that sports team fanatics have long painted their faces in team colors, or drawn the team logo on their faces; and that even before PMC, some permanent tattoos consisting of corporate logos were popular, such as "Chevy," "Red Sox Forever," or, perhaps most common, "Harley Davidson.") Are there any risks to letting the practice continue without any enforcement efforts? Does PMC, or its customers, have a defense not based in trademark law?

D. INFRINGEMENT

1. Likelihood of Consumer Confusion

AMF Incorporated v. Sleekcraft Boats
United States Court of Appeals for the Ninth Circuit
599 F.2d 341, 204 U.S.P.Q. 808 (9th Cir. 1979)

ANDERSON, Circuit Judge.

In this trademark infringement action, the district court, after a brief non-jury trial, found appellant AMF's trademark was valid, but not infringed, and denied AMF's request for injunctive relief.

AMF and appellee Nescher both manufacture recreational boats. AMF uses the mark Slickcraft, and Nescher uses Sleekcraft. The crux of this appeal is whether concurrent use of the two marks is likely to confuse the public. The district judge held that confusion was unlikely. We disagree and remand for entry of a limited injunction.

I. Facts

AMF's predecessor used the name Slickcraft Boat Company from 1954 to 1969 when it became a division of AMF. The mark Slickcraft was federally registered on April 1, 1969, and has been continuously used since then as a trademark for this line of recreational boats.

Slickcraft boats are distributed and advertised nationally. AMF has authorized over one hundred retail outlets to sell the Slickcraft line. For the years 1966-1974, promotional expenditures for the Slickcraft line averaged approximately $200,000 annually. Gross sales for the same period approached $50,000,000.

After several years in the boatbuilding business, appellee Nescher organized a sole proprietorship, Nescher Boats, in 1962. This venture failed in 1967. In late 1968 Nescher began anew and adopted the name Sleekcraft. Since then Sleekcraft has been the Nescher trademark. The name Sleekcraft was selected without knowledge of appellant's use. After AMF notified him of the alleged trademark infringement, Nescher adopted a distinctive logo and added the identifying phrase "Boats by Nescher" on plaques affixed to the boat and in much of its advertising. The Sleekcraft mark still appears alone on some of appellee's stationery, signs, trucks, and advertisements.

The Sleekcraft venture succeeded. Expenditures for promotion increased from $6,800 in 1970 to $126,000 in 1974. Gross sales rose from $331,000 in 1970 to over $6,000,000 in 1975. Like AMF, Nescher sells his boats through authorized local dealers.

Slickcraft boats are advertised primarily in magazines of general circulation. Nescher advertises primarily in publications for boat racing enthusiasts. Both parties exhibit their product line at boat shows, sometimes the same show. . . .

IV. Likelihood of Confusion

When the goods produced by the alleged infringer compete for sales with those of the trademark owner, infringement usually will be found if the marks are sufficiently similar that confusion can be expected.[9] When the goods are related,[10] but not competitive, several other factors are added to the calculus. If the goods are totally unrelated, there can be no infringement because confusion is unlikely.

AMF contends these boat lines are competitive. Both lines are comprised of sporty, fiberglass boats often used for water skiing; the sizes of the boats are similar as are the prices. Nescher contends his boats are not competitive with Slickcraft boats because his are true high performance boats intended for racing enthusiasts.

The district court found that although there was some overlap in potential customers for the two product lines, the boats "appeal to separate submarkets." Slickcraft boats are for general family recreation, and Sleekcraft boats are for persons who want high speed recreation; thus, the district court concluded, competition between the lines is negligible. Our research has led us to only one case in which a similarly fine distinction in markets has been recognized, Sleeper Lounge Co. v. Bell Manufacturing Co., 253 F.2d 720, 117 U.S.P.Q. 117 (CA 9 1958). Yet, after careful review of all the exhibits introduced at trial, we are convinced the district court's finding was warranted by the evidence.

The Slickcraft line is designed for a variety of activities: fishing, water skiing, pleasure cruises, and sunbathing. The promotional literature emphasizes family fun. Sleekcraft boats are not for families. They are low-profile racing boats designed for racing, high speed cruises, and water skiing. Seating capacity and luxury are secondary. Unlike the Slickcraft line, handling capability is emphasized. The promotional literature projects an alluring, perhaps flashier, racing image; absent from the pictures are the small children prominently displayed in the Slickcraft brochures.

Even though both boats are designed for towing water skiers, only the highly skilled enthusiast would require the higher speeds the Sleekcraft promises. We therefore affirm the district court's finding that, despite the potential

9. The alleged infringer's intent in adopting the mark is weighed, both as probative evidence of the likelihood of confusion and as an equitable consideration.

10. Related goods are those "products which would be reasonably thought by the buying public to come from the same source if sold under the same mark." Standard Brands, Inc. v. Smidler, 151 F.2d 34, 37, 66 U.S.P.Q. 337, 340-341 (CA 2 1945). See Yale Electric Corp. v. Robertson, 26 F.2d 972 (CA 2 1928).

market overlap, the two lines are not competitive. Accordingly, we must consider all the relevant circumstances in assessing the likelihood of confusion. See Durox Co. v. Duron Paint Manufacturing Co., 320 F.2d 882, 885, 138 U.S.P.Q. 353, 355 (CA 4 1963).

V. Factors Relevant to Likelihood of Confusion

In determining whether confusion between related goods is likely, the following factors are relevant:[11]

1. strength of the mark;
2. proximity of the goods;
3. similarity of the marks;
4. evidence of actual confusion;
5. marketing channels used;
6. type of goods and the degree of care likely to be exercised by the purchaser;
7. defendant's intent in selecting the mark; and
8. likelihood of expansion of the product lines.

1. *Strength of the Mark . . .*

[W]e hold that Slickcraft is a suggestive mark when applied to boats. . . .

Although appellant's mark is protectable and may have been strengthened by advertising, . . . it is a weak mark entitled to a restricted range of protection. Thus, only if the marks are quite similar, and the goods closely related, will infringement be found. . . .

2. *Proximity of the Goods*

For related goods, the danger presented is that the public will mistakenly assume there is an association between the producers of the related goods, though no such association exists. . . . The more likely the public is to make such an association, the less similarity in the marks is requisite to a finding of likelihood of confusion. . . . Thus, less similarity between the marks will suffice when the goods are complementary, . . . the products are sold to the same class of purchasers, . . . or the goods are similar in use and function. . . .

---

11. The list is not exhaustive. Other variables may come into play depending on the particular facts presented. Triumph Hosiery Mills, Inc. v. Triumph International Corp., 308 F.2d 196, 198, 135 U.S.P.Q. 45, 46-47 (CA 2 1962); Restatement of Torts §729, Comment a (1938).

Although these product lines are non-competing, they are extremely close in use and function. In fact, their uses overlap. Both are for recreational boating on bays and lakes. Both are designed for water skiing and speedy cruises. Their functional features, for the most part, are also similar: fiberglass bodies, outboard motors, and open seating for a handful of people. Although the Sleekcraft boat is for higher speed recreation and its refinements support the market distinction the district court made, they are so closely related that a diminished standard of similarity must be applied when comparing the two marks. . . .

3. Similarity of the Marks

The district court found that "the two marks are easily distinguishable in use either when written or spoken." Again, there is confusion among the cases as to whether review of this finding is subject to the clearly erroneous standard. . . .

Similarity of the marks is tested on three levels: sight, sound, and meaning. . . . Each must be considered as they are encountered in the marketplace. Although similarity is measured by the marks as entities, similarities weigh more heavily than differences. . . .

Standing alone the words Sleekcraft and Slickcraft are the same except for two inconspicuous letters in the middle of the first syllable. . . . To the eye, the words are similar.

In support of the district court's finding, Nescher points out that the distinctive logo on his boats and brochures negates the similarity of the words. We agree: the names appear dissimilar when viewed in conjunction with the logo, but the logo is often absent. The exhibits show that the word Sleekcraft is frequently found alone in trade journals, company stationery, and various advertisements.

Nescher also points out that the Slickcraft name is usually accompanied by the additional trademark AMF. As a result of this consistent use, Nescher argues, AMF has become the salient part of the mark indicative of the product's origin. . . .

Although Nescher is correct in asserting that use of a housemark can reduce the likelihood of confusion, . . . the effect is negligible here even though AMF is a well-known house name for recreational equipment. The exhibits show that the AMF mark is down-played in the brochures and advertisements; the letters AMF are smaller and skewed to one side. Throughout the promotional materials, the emphasis is on the Slickcraft name. Accordingly, we find that Slickcraft is the more conspicuous mark and serves to indicate the source of origin to the public. . . .

Sound is also important because reputation is often conveyed word-of-mouth. We recognize that the two sounds can be distinguishable, but the difference is only in a small part of one syllable. In G. D. Searle & Co. v. Chas. Pfizer & Co., 265 F.2d 385, 121 U.S.P.Q. 74 (CA 7 1959), *cert.*

denied, 361 U.S. 819, 123 U.S.P.Q. 590 (1959), the court reversed the trial court's finding that Bonamine sounded "unlike" Dramamine, stating that: "Slight differences in the sound of trademarks will not protect the infringer." Id. at 387, 121 U.S.P.Q. at 75-76. The difference here is even slighter. . . .

Neither expert testimony nor survey evidence was introduced below to support the trial court's finding that the marks were easily distinguishable to the eye and the ear. . . . The district judge based his conclusion on a comparison of the marks. After making the same comparison, we are left with a definite and firm conviction that his conclusion is incorrect. . . .

The final criterion reinforces our conclusion. Closeness in meaning can itself substantiate a claim of similarity of trademarks. See, e.g., S. C. Johnson & Son, Inc. v. Drop Dead Co., 210 F. Supp. 816, 135 U.S.P.Q. 292 (S.D. Cal. 1962), *aff'd*, 326 F.2d 87, 139 U.S.P.Q. 465 (1963) (Pledge and Promise). Nescher contends the words are sharply different in meaning. This contention is not convincing; the words are virtual synonyms. Webster's New World Dictionary of the American Language 1371 (1966).

Despite the trial court's findings, we hold that the marks are quite similar on all three levels.

4. Evidence of Actual Confusion

Evidence that use of the two marks has already led to confusion is persuasive proof that future confusion is likely. . . . Proving actual confusion is difficult, however, . . . and the courts have often discounted such evidence because it was unclear or insubstantial. . . .

AMF introduced evidence that confusion had occurred both in the trade and in the mind of the buying public. A substantial showing of confusion among either group might have convinced the trial court that continued use would lead to further confusion. . . .

The district judge found that in light of the number of sales and the extent of the parties' advertising, the amount of past confusion was negligible. We cannot say this finding is clearly erroneous though we might have viewed the evidence more generously.

5. Marketing Channels

Convergent marketing channels increase the likelihood of confusion. . . . There is no evidence in the record that both lines were sold under the same roof except at boat shows; the normal marketing channels used by both AMF and Nescher are, however, parallel. Each sells through authorized retail dealers in diverse localities. The same sales methods are employed. The price ranges are almost identical. Each line is advertised extensively though different national magazines are used; the retail dealers also promote the lines, by participating in smaller boat shows and by advertising in local newspapers and classified telephone directories. Although different submarkets are involved, the general class of boat purchasers exposed to the products overlap.

6. Type of Goods and Purchaser Care

Both parties produce high quality, expensive goods. According to the findings of fact, the boats "are purchased only after thoughtful, careful evaluation of the product and the performance the purchaser expects."

In assessing the likelihood of confusion to the public, the standard used by the courts is the typical buyer exercising ordinary caution. . . . Although the wholly indifferent may be excluded, . . . the standard includes the ignorant and the credulous. . . . When the buyer has expertise in the field, a higher standard is proper though it will not preclude a finding that confusion is likely. . . . Similarly, when the goods are expensive, the buyer can be expected to exercise greater care in his purchases; again, though, confusion may still be likely. . . .

The parties vigorously dispute the validity of the trial court's finding on how discriminating the average buyer actually is. Although AMF presented expert testimony to the contrary, the court's finding is amply supported by the record. The care exercised by the typical purchaser, though it might virtually eliminate mistaken purchases, does not guarantee that confusion as to association or sponsorship is unlikely.

The district court also found that trademarks are unimportant to the average boat buyer. Common sense and the evidence indicate this is not the type of purchase made only on "general impressions." . . . This inattention to trade symbols does reduce the possibilities for confusion. . . .

The high quality of defendant's boats is also relevant in another way. The hallmark of a trademark owner's interest in preventing use of his mark on related goods is the threat such use poses to the reputation of his own goods. . . . When the alleged infringer's goods are of equal quality, there is little harm to the reputation earned by the trademarked goods. Yet this is no defense, for present quality is no assurance of continued quality. . . . The wrong inheres in involuntarily entrusting one's business reputation to another business. . . . AMF, of course, cannot control the quality of Sleekcraft boats. . . . [Indeed, e]quivalence in quality may actually contribute to the assumption of a common connection.

7. Intent

The district judge found that Nescher was unaware of appellant's use of the Slickcraft mark when he adopted the Sleekcraft name. There was no evidence that anyone attempted to palm off the latter boats for the former. And after notification of the purported infringement, Nescher designed a distinctive logo. . . . We agree with the district judge: appellee's good faith cannot be questioned.

When the alleged infringer knowingly adopts a mark similar to another's, reviewing courts presume that the defendant can accomplish his purpose: that is, that the public will be deceived. . . . Good faith is less probative of the likelihood of confusion, yet may be given considerable weight in fashioning a remedy.

8. *Likelihood of Expansion*

Inasmuch as a trademark owner is afforded greater protection against competing goods, a "strong possibility" that either party may expand his business to compete with the other will weigh in favor of finding that the present use is infringing. . . . When goods are closely related, any expansion is likely to result in direct competition. . . . The evidence shows that both parties are diversifying their model lines. The potential that one or both of the parties will enter the other's submarket with a competing model is strong.

Remedy

[A] limited mandatory injunction is warranted. Upon remand the district court should consider the above interests in structuring appropriate relief.

COMMENTS AND QUESTIONS

1. An interesting case on proximity of goods (factor 2) is Death Tobacco, Inc. v. Black Death USA, 31 U.S.P.Q.2d 1899, 1903 (C.D. Cal. 1993):

> In this case, defendant's ["Black Death"] vodka and plaintiff's ["Death"] cigarettes are related products. They are, to some extent, complementary in that smoking and drinking are related vices that are often undertaken together. They are also somewhat similar in use and function. Cigarettes and vodka both have a mood altering effect and both are used for recreation and relaxation. . . .
>
> The most important factor in the determination of relatedness, here, is that the two items are sold to the same class of purchasers. Defendant introduced evidence that Black Death vodka and Death cigarettes are sold in the same package stores in both San Francisco and Los Angeles. Several store owners display the two products adjacent to one another on the shelf.

2. For an older case listing factors to consider in determining likelihood of confusion, see Polaroid Corp. v. Polarad Electronics Corp., 287 F.2d 492,

495 (2d Cir.), *cert. denied*, 368 U.S. 820 (1961). The actual factors considered vary in each circuit, but in substance the different tests are similar.

3. Are all of these factors (however many there may in fact be) equally important? Strength of the mark may obviously be significant for validity purposes as well as for showing likelihood of confusion, but its use in an infringement proceeding presumes that the mark has already been determined valid. Must a plaintiff prove all the factors? Only some of them? Are there some that must be proven in all cases? Conversely, can proof of one factor ever be sufficient by itself?

4. Proof of actual confusion among consumers is obviously an important step towards showing a "likelihood of confusion." Such proof can sometimes be offered anecdotally, particularly where both products have already been in the market for a significant period of time. But confused consumers are difficult to find, and if the infringement is challenged early enough there may not be very many of them at all. In those circumstances, courts generally allow the results of consumer surveys as evidence of "actual" confusion. See, e.g., Union Carbide v. Ever-Ready, Inc., 531 F.2d 366, 387-388 (7th Cir. 1976); Mutual of Omaha Ins. Co. v. Novak, 836 F.2d 397, 400-401 (8th Cir. 1987). Because of the importance of actual confusion in proving likelihood of confusion, however, and because of the potential for abuse of the survey process, courts are relatively strict about the surveys they allow, routinely rejecting or discounting surveys that are improperly designed or ask ambiguous or leading questions.

5. Who must be likely to be confused? Obviously, the strictness with which this test is applied will determine the chances that a trademark plaintiff has of proving infringement. The consumer who is tested is the "reasonable purchaser" of the products at issue. This standard allows for a great deal of flexibility in testing marks used with different products. For example, the "reasonable purchaser" of fleets of commercial airplanes may be expected both to be more sophisticated and to pay more attention to the decision than the reasonable purchaser of pencils for home use. In light of this, should trademark law offer more protection to the owner of a trademark for pencils (and thus indirectly to the consumers of pencils) than to the owners of airplane trademarks?

A related question is how many consumers must be likely to be confused. The fact that the "reasonable consumer" is at issue might suggest that at least half of the consuming public must be confused in order to constitute trademark infringement. After all, if the median consumer is not "reasonable," who is? But courts have not been willing to require such a showing from plaintiffs. Instead, likelihood of confusion is regularly found if as few as 10 to 15 percent of the consumers surveyed were confused. Is it reasonable to test infringement on the basis of the reactions of a small minority of the population? If not, how much confusion should be required? See Mushroom Makers, Inc. v. R. G. Barry Corp., 580 F.2d 44, 47 (2d Cir. 1978) (testing the "likelihood that *an appreciable number of ordinarily prudent purchasers* will be misled, or indeed simply confused, as to the source of the goods in question.") (emphasis added).

6. What does it mean for two marks to be similar? The courts generally consider three kinds of similarity — sight, sound, and meaning. (Note that the second, and possibly the third, kind of similarity have little or no relevance to trade dress.) Similarity is generally tested by comparing the marks as a whole, rather than by dissecting them. This approach makes sense, doesn't it, since consumers are likely to pay attention to the whole mark in context. On the other hand, dissection is appropriate if the aim is to prevent trademark owners from exercising control over generic, functional, or disclaimed portions of a trademark or trade dress.

Because the perception of the consumer when exposed to the whole mark in context is the linchpin of trademark infringement, a defendant's use of a similar or even identical trademark on similar products may be ameliorated by other differences between the mark and the packaging. For example, similarities in trade dress may not confuse consumers if the packages contain very different product names in large, obvious letters on the front. Disclaimers may also be effective in reducing confusion, if consumers notice them. If consumers would nonetheless be confused, however, efforts to ameliorate the effects of similar marks will not avoid trademark infringement.

7. What does the defendant's intent have to do with the likelihood that consumers will be confused by the two marks? Certainly, intentional copying is likely to be closer than accidental similarity, and therefore quite possibly more confusing, but isn't that adequately tested by the other factors that the court employs? Has the court slipped in a "fairness" factor to create what is in effect a presumption that counterfeiting (that is, intentional copying of a trademark) is illegal? Is such a presumption appropriate?

8. Finally, note that we are assuming that consumers are the ones confused. Suppose that consumers do not confuse two products but that, because of similarities in design, third parties who see the product from a distance (or stripped of identifying trademarks) will assume that the products are the same. This was the situation in Lois Sportswear v. Levi Strauss & Co., 799 F.2d 867 (2d Cir. 1986). There, Levi Strauss had registered not only its trade name and its jean labels, but also the pattern of stitching on the back pockets of its jeans. Lois Sportswear sold jeans with clearly different labels, but with an identical stitching pattern. The trial court found that there was no evidence of actual confusion by purchasers but that nonpurchasers, seeing the jeans "worn by a passer-by," would likely be confused. Should such "post-sale confusion" be actionable? The Second Circuit held that post-sale confusion constituted trademark infringement: "The confusion the Act seeks to prevent in this context is that a consumer seeing the familiar stitching pattern will associate the jeans with [Levi's] and that association will influence his buying decisions."

The district court also concluded that the Lois jeans were not of inferior quality. Given that, how is Levi Strauss & Co. hurt by any confusion that does arise? It will not lose sales, since customers themselves are not confused.

Further, any confusion inures to the company's benefit, doesn't it? Whose interests are being protected by this decision?

9. The fact that a trademark has been made incontestable precludes a defendant from challenging it on the basis of descriptiveness. See *Park 'N Fly v. Dollar Park 'N Fly*, discussed supra. An incontestable mark is not necessarily a *strong* mark for purposes of determining likelihood of confusion, however. In Petro Stopping Centers v. James River Petroleum Inc., 130 F.3d 88 (4th Cir. 1997), the Fourth Circuit affirmed a district court conclusion that plaintiff's Petro mark was merely descriptive and therefore weak, despite the fact that it was incontestable. The court noted that validity and infringement are separate inquiries and that incontestability applies only to the validity of a trademark. Accord Sports Authority Inc. v. Abercrombie & Fitch, 965 F. Supp. 925 (E.D. Mich. 1997).

Note on Other Types of Confusion

Our discussion of infringement so far has focused on only one type of consumer confusion — that consumers will believe the infringer's product is the same as the trademark owner's product because of the similarity between the marks. Presumably, this confusion will allow the infringer to take sales away from the trademark owner by trading on the latter's goodwill. Although such confusion is the central problem at which trademark law is aimed, it is only one of many different ways in which consumers may be confused by different marks. Consider the following types of consumer confusion actionable under the Lanham Act.

Confusion as to Source. Two products bearing similar trademarks are likely to be confused only if both companies actually sell the same or similar products. Suppose, however, that an accused infringer uses a trademark on a group of products that the trademark owner does not sell. In that case, presumably, consumers cannot buy the infringer's products *instead of* the trademark owner's, because the trademark owner does not sell the products at all. For this reason, early courts sometimes found no infringement in situations where the plaintiff's and defendant's products did not directly compete. See, e.g., Borden Ice Cream v. Borden's Condensed Milk Co., 201 F. 510 (7th Cir. 1912) (use of identical trademark for different milk products does not constitute trademark infringement).

But courts have gradually come to recognize that identical trademarks can sometimes cause confusion *as to the source of the products* even in the absence of direct product competition. For example, "Prell" is a trademark for a common brand of shampoo and conditioner. If another company uses the term "Prell" for its hair coloring product, consumers may well believe that they are buying a product sold by the same company that sells Prell shampoo, particularly given the proximity of the two goods. While the infringer will not take sales of hair colors away from the trademark owner in this instance, the trademark owner may be hurt by this confusion in at least two ways. First, if the quality of the hair color is inferior, consumers may blame the maker of shampoo and stop buying Prell products altogether. Second, it is possible that the trademark owner may wish to expand into the market for hair colors. If it does, confusion between the products will almost certainly result if both parties use the Prell mark.

Confusion as to Sponsorship. Even in situations in which consumers will not believe that the trademark owner is the one selling the product, the use of a similar trademark may still confuse them by causing them to believe that the trademark owner is affiliated with or sponsors the infringer's products. For example, suppose that a company that sells soup uses the trade symbol of the United States Olympic Committee on its soup cans. Presumably, the USOC does not sell soup, and no reasonable consumer would be likely to conclude that she was in fact buying USOC soup. But consumers might well conclude that the infringer was somehow affiliated with the USOC, or that one group had agreed to sponsor the other. This confusion as to affiliation or sponsorship is actionable under the Lanham Act, assuming the other requirements for protecting a mark are met.[32]

In cases of confusion either as to source or as to sponsorship, the essential question remains whether consumers are likely to be confused by the similarity of the marks. Because of this, many of the same factors listed in *AMF* apply with equal force to these inquiries. For example, proximity between the products sold by the parties, while not required, does tend to increase both the likelihood of confusion as to source and the chance that the parties will eventually be in direct competition. A similar analysis of infringement can be conducted to determine whether any of these types of confusion is likely.

Initial Interest Confusion. What happens when confusion is dispelled before a product is ever purchased? For example, suppose that a competing fast-food restaurant put a large McDonald's sign by a highway exit. There is no McDonald's at the exit, so consumers won't end up thinking they are buying from one. However, they might be lured into the competitor's establishment once they have left the highway. Should this sort of deliberate effort to confuse consumers, but not at the point of purchase, be actionable? At least some courts have said yes. See Mobil Oil Corp. v. Pegasus Petroleum Corp., 818 F.2d 254 (2d Cir. 1987); J. Thomas McCarthy, Trademarks and Unfair Competition §23:6.

A similar problem has come up more recently on the Internet. Web pages frequently use "metatags," which are words on the page invisible to the user

---

32. Confusion as to affiliation or sponsorship is only expressly addressed in section 43(a) of the Lanham Act, which applies to unregistered marks. But at least one commentator has quite reasonably suggested that the protections afforded to unregistered marks under section 43(a) also apply to registered marks, whether under section 32 of the Lanham Act or under section 43(a). See Jay Dratler, Jr., Intellectual Property Law: Commercial, Creative, and Industrial Property §10.01[1][i], at 10-5 to 10-6.

but which are read by computer search engines. Some companies have begun putting the names of their competitors' products (or even just popular trademarks such as "Playboy") into their metatags in an effort to draw unsuspecting consumers to their site. Is the use of a competitor's trademark in a metatag infringement? There are only a few reported decisions so far, but the cases seem to turn on whether there was a legitimate reason to make reference to the trademark on the Web page. Compare Brookfield Communications v. West Coast Entertainment Corp., 174 F.3d 1036 (9th Cir. 1999) (use of plaintiff's trademark in a Web site "metatag" was trademark infringement where it contributed to customer confusion) with Playboy Enterprises v. Welles, 7 F. Supp. 2d 1098 (S.D. Cal. 1998) (use of "Playmate of the Year" in a metatag to accurately describe defendant's resume was not illegal).

Reverse Confusion. Ordinary trademark confusion occurs when the junior user trades on the reputation of the trademark owner, confusing the public into thinking that his goods are associated with those of the senior trademark owner. At times, however, a large company will adopt the mark of a smaller trademark owner. In this case, the danger is presumably not that the junior user will trade on the smaller company's goodwill. The risk is, instead, that the public will come to associate the mark not with its true owner but with the infringer (who has spent a great deal of money to advertise it). Several courts have made it clear that reverse confusion is trademark infringement, and that the relative size of the companies doesn't matter. Indeed, companies have periodically been forced to halt or even retract major advertising campaigns because of reverse confusion problems. See, e.g., Big O Tire Dealers, Inc. v. Goodyear Tire & Rubber Co., 561 F.2d 1365 (10th Cir. 1977); Sands, Taylor & Wood v. Quaker Oaks, 34 F.3d 1340 (7th Cir. 1994).

In a recent case, Judge Posner took a remarkably different approach. In Illinois High School Assn. v. GTE Vantage, 99 F.3d 244 (7th Cir. 1996), the plaintiffs were the owners of the trademark "March Madness" for their annual high school basketball tournament. After the media began using the term to describe the NCAA *college* basketball tournament, the NCAA and its various licensees adopted the term and used it in advertising their own products and services. The Illinois High School Association sued alleging reverse confusion — that their mark had been appropriated by the NCAA. The Seventh Circuit rejected this claim, creating a new trademark classification: the "dual use trademark." The court reasoned that where *the public*, not the defendant, had appropriated a trademark for use to describe another product or service, trademark law should not stand in the way.

Does this result make sense? Why isn't this an ordinary case of reverse confusion?

Under the Seventh Circuit's opinion, presumably *both* IHSA and NCAA now have rights to the mark. Does this comport with the rationales for trademark protection? Who now has the right to sue infringers? To license the use of the mark?

PROBLEMS

Problem 5-6. Bristol-Myers, a major pharmaceutical company, markets "Excedrin" pain reliever. Since 1968, B-M has marketed "Excedrin PM," which is a pain reliever that does not interfere with sleep. Excedrin PM tablets are sold in a solid blue box whose color fades from dark at the top to light at the bottom. The box contains the words "Excedrin PM" in large white letters across the top third of the box. In the bottom right-hand corner of the box is a depiction of two tablets labeled "PM." B-M also sells Excedrin PM capsules, which are packaged identically except that the background is green and the two capsules in the picture read "Excedrin PM." Both the mark Excedrin PM and the dress of both boxes are registered with the Trademark Office.

In 1991, McNeil Pharmaceuticals introduced "Tylenol PM," a pain reliever chemically identical to Excedrin PM. Tylenol PM tablets are sold in a solid green box whose color fades from dark at the top to light at the bottom. The box contains the words "Tylenol PM" in large white and yellow letters across the top third of the box. In the bottom right-hand corner of the box is a depiction of two tablets, one labeled "Tylenol" and the other labelled "PM." McNeil also sells Tylenol PM capsules, which are packaged identically except that the background is blue and the two capsules in the picture both read "Tylenol PM."

B-M sues McNeil, alleging that both its use of the term PM and its trade dress are likely to confuse consumers. Who should prevail?

Problem 5-7. Ivory Soap is sold for approximately $1.00 a bar. It is a heavily advertised brand name, which identifies itself as "99.44% pure." Whitewash Soap Co. makes and sells counterfeit Ivory Soap bars, which are packaged identically to Ivory (using the Ivory name and trade dress) and which contain chemically identical bars of soap. These counterfeit "Ivory" bars are sold for $.50 each. If a consumer buys the Whitewash soap thinking it is Ivory, how has he been injured? Hasn't he benefited? Is there any reason to protect Ivory's trademark against infringement at the expense of consumer welfare?

2. Dilution

Likelihood of confusion is the traditional benchmark of trademark infringement. Yet in some cases courts have found trademark infringement even where consumers were not likely to be confused as to source. The canonical case involved Kodak bicycles. Despite the fact — well recognized by the court — that consumers were not likely to assume a connection between

Kodak, the famous maker of film and cameras, and the seller of bicycles, the court issued an injunction. It did so because it found that the use of the Kodak name on bicycles would harm Kodak, Inc., even in the absence of confusion as to source. See Eastman Photographic Materials Co. v. Kodak Cycle Co., 15 [British] R.P.C. 105 (1898). See generally Beverly W. Pattishall, Dawning Acceptance of the Dilution Rationale for Trademark–Trade Identity Protection, 74 Trademark Rep. 289, 289 n.2 (1984) (citing early trademark "dilution" cases).

What is the nature of this harm? The legislative history of a recently enacted federal statute on dilution, the Federal Trademark Dilution Act of 1995, Pub. L. 104-374, codified at 15 U.S.C. §1125(c), which added a new section 43(c) to the Lanham Act, contained this instructive summary:

> [This bill would] create a federal cause of action to protect famous marks from unauthorized users that attempt to trade upon the goodwill and established reknown of such marks and, thereby, dilute their distinctive quality. The provision is intended to protect famous marks where the subsequent, unauthorized commercial use of such marks by others dilutes the distinctiveness of the mark. The bill defines the term "dilution" to mean "the lessening of the capacity of a famous mark to identify and distinguish goods or services regardless of the presence or absence of (a) competition between the parties, or (b) likelihood of confusion, mistakes, or deception." Thus, for example, the use of DUPONT shoes, BUICK aspirin, and KODAK pianos would be actionable under this legislation. The protection of marks from dilution differs from the protection accorded marks from trademark infringement. Dilution does not rely upon the standard test of infringement, that is, likelihood of confusion, deception or mistake. Rather, it applies when the unauthorized use of a famous mark reduces the public's perception that the mark signifies something unique, singular, or particular. As summarized in one decision:
>
>> Dilution is an injury that differs materially from that arising out of the orthodox confusion. Even in the absence of confusion, the potency of a mark may be debilitated by another's use. This is the essence of dilution. Confusion leads to immediate injury, while dilution is an infection, which if allowed to spread, will inevitably destroy the advertising value of the mark.
>
> Mortellito v. Nina of California, Inc., 335 F. Supp. 1288, 1296 (S.D.N.Y. 1972).

a. The New Federal Regime

On January 16, 1996, section 43 of the Lanham Act was amended as follows:

(c)(1) The owner of a famous mark shall be entitled, subject to the principles of equity and upon such terms as the court deems reasonable, to an injunction against another person's commercial use in commerce of a mark or trade name, if such use begins after the mark has become famous and causes dilution of the distinctive quality of the mark, and to obtain such other relief as is provided in this subsection. In determining whether a mark is distinctive and famous, a court may consider factors such as, but not limited to —

(A) the degree of inherent or acquired distinctiveness of the mark;

(B) the duration and extent of use of the mark in connection with the goods or services with which the mark is used;

(C) the duration and extent of advertising and publicity of the mark;

(D) the geographical extent of the trading area in which the mark is used;

(E) the channels of trade for the goods or services with which the mark is used;

(F) the degree of recognition of the mark in the trading areas and channels of trade used by the marks' owner and the person against whom the injunction is sought;

(G) the nature and extent of use of the same or similar marks by third parties; and

(H) whether the mark was registered under the Act of March 3, 1881, or the Act of February 20, 1905, or on the principal register.

(2) In an action brought under this subsection, the owner of the famous mark shall be entitled only to injunctive relief unless the person against whom the injunction is sought willfully intended to trade on the owner's reputation or to cause dilution of the famous mark. If such willful intent is proven, the owner of the famous mark shall also be entitled to the remedies set forth in sections 35(a) and 36 [i.e., 15 U.S.C. §§1117, 1118], subject to the discretion of the court and the principles of equity.

(3) The ownership by a person of a valid registration under the Act of March 3, 1881, or the Act of February 20, 1905, or on the principal register shall be a complete bar to an action against that person, with respect to that mark, that is brought by another person under the common law or a statute of a State and that seeks to prevent dilution of the distinctiveness of a mark, label, or form of advertisement.

(4) The following shall not be actionable under this section:

(A) Fair use of a famous mark by another person in comparative commercial advertising or promotion to identify the competing goods or services of the owner of the famous mark.

(B) Noncommercial use of a mark.

(C) All forms of news reporting and news commentary.

Federal Trademark Dilution Act of 1995, §3, P.L. 104-374, codified at 15 U.S.C. §1125(c).

The House Report, quoted in the introduction to this section, goes on to describe the effect of the act.

H.R. Rep. 104-374, 104th Cong., 1st Sess. (1995)

A federal dilution statute is necessary because famous marks ordinarily are used on a nationwide basis and dilution protection is currently only available on a patch-quilt system of protection, in that only approximately 25 states have laws that prohibit trademark dilution. Further, court decisions have been inconsistent and some courts are reluctant to grant nationwide injuctions for violation of state law where half of the states have no dilution law.[1] Protection for famous marks should not depend on whether the forum where suit is filed has a dilution statute. This simply encourages forum-shopping and increases the amount of litigation.

Moreover, the recently concluded Agreement on Trade-Related Aspects of Intellectual Property Rights, including Trade in Counterfeit Goods ("TRIPS") which was part of the Uruguay Round of the GATT agreement includes a provision designed to provide dilution protection to famous marks. Thus, enactment of this bill will be consistent with the terms of the agreement, as well as the Paris Convention, of which the U.S. is also a member. Passage of a federal dilution statute would also assist the executive branch in its bilateral and multilateral negotiations with other countries to secure greater protection for the famous marks owned by U.S. companies. Foreign countries are reluctant to change their laws to protect famous U.S. marks if the U.S. itself does not afford special protection for such marks.

It should be noted that as originally introduced, H.R. 1295 only applied to famous registered marks. However, based on testimony by the Patent and Trademark Office, Congresswoman Patricia Schroeder offered an amendment in the nature of a substitute to H.R. 1295, that was adopted by the Subcommittee, to include all famous marks within the scope of the bill. The Patent and Trademark Office made a compelling case that limiting the federal remedy against trademark dilution to those famous marks that are registered is not within the spirit of the United States' position as a leader setting the standards for strong worldwide protection of intellectual property. Such a limitation would undercut the United States position with our trading partners, which is that famous marks should be protected regardless of whether the marks are registered in the country where protection is sought.

The proposal adequately addresses legitimate First Amendment concerns espoused by the broadcasting industry and the media. The bill will not prohibit or threaten "noncommercial" expression, as that term has been defined by the courts. Nothing in this bill is intended to alter existing case law on the subject of what constitutes "commercial" speech. The bill includes specific language exempting from liability the "fair use" of a mark in the context of comparative commercial advertising or promotion as well as all forms of news reporting and news commentary. The latter provision which was added to

---

1. Blue Ribbon Feed Co., Inc. v. Farmers Union Central Exchange, Inc., 731 F.2d 415, 422 (7th Cir. 1984); Deere & Co. v. MTD Products Inc., 34 U.S.P.Q. 2d 1706 (S.D.N.Y. 1995).

[the act] as a result of an amendment offered by Congressman Moorhead that was adopted by the Committee, recognizes the heightened First Amendment protection afforded the news industry.

It is important to note that H.R. 1295 would not pre-empt existing state dilution statutes. State laws could continue to be applied in cases involving locally famous or distinctive marks.[2] Unlike patent and copyright laws, federal trademark law presently coexists with state trademark law, and it is to be expected that a federal dilution statute should similarly coexist with state dilution law. The ownership of valid federal registration would act as a complete bar to a dilution action brought under state law.

With respect to remedies, the bill limits the relief a court could award to an injunction unless the wrongdoer willfully intended to trade on the trademark owner's reputation or to cause dilution of the famous mark, in which case the remedies under sections 35(a) and 36 of the Trademark Act become available.

COMMENTS AND QUESTIONS

1. Dilution was born in a law review article by trademark lawyer Frank I. Schechter entitled The Rational Basis of Trademark Protection, 40 Harv. L. Rev. 813 (1927). Schechter's proposal applied only to coined, fanciful, or arbitrary marks, only to situations in which the junior user's mark was identical to that of the senior user, and only to use of identical marks on non-competing goods. Id. at 825-830. Courts and commentators have since expanded the doctrine to include marks that have acquired secondary meaning (see discussion in Howard Shire, Dilution Versus Deception — Are State Antidilution Laws an Appropriate Alternative to the Law of Infringement?, 77 Trademark Rep. 273, 275-278 (1987)), situations where the junior user's mark has substantial similarity to the senior user's mark (see, e.g., *Mead Data Central*, below), and to use of similar marks on competing goods (see, e.g., Deere & Co. v. MTD Products, Inc., 41 F.3d 39 (2d Cir. 1994)). For an overview of state dilution doctrines see Elliot Staffin, The Dilution Doctrine: Towards a Reconciliation with the Lanham Act, 6 Fordham Intell. Prop. Media & Entertainment L.J. 105 (1995).

2. At least some commentators believe that antidilution law does not go far enough in protecting trademark owners. See Jerre B. Swann & Theodore H. Davis, Jr., Dilution, An Idea Whose Time Has Gone: Brand Equity as Protectable Property, The New/Old Paradigm, 1 J. Intell. Prop. L. 219 (1994) (arguing for explicit property rights in brand equity, going beyond antidilution law). On the other hand, some commentators suggest the opposite. See Shire, supra, at 296: "Dilution, however, could effectively grant the owner of a 'distinctive' mark a trademark in gross, because the doctrine,

2. See, e.g., Wedgewood Homes, Inc. v. Lund, 659 P.2d 377, 222 U.S.P.Q. 446 (Or. 1983).

when properly applied, bars any junior user from using the mark in connection with any goods or services, regardless of whether the junior user's mark is likely to cause any confusion, mistake or deception"; Megan Gray, Defending Against a Dilution Claim: A Practitioner's Guide, 4 Tex. Intell. Prop. L.J. 205 (1996).

b. Basic Principles

Nabisco, Inc. v. PF Brands, Inc.
United States Court of Appeals for the Second Circuit.
191 F.3d 208 (2d Cir. 1999)

Leval, Circuit Judge:

Nabisco, Inc. and Nabisco Brands Company (collectively "Nabisco") appeal from the preliminary injunction entered by the United States District Court for the Southern District of New York (Shira A. Scheindlin, District Judge) upon the motion of Pepperidge Farm, Inc. and PF Brands, Inc. (collectively "Pepperidge Farm" or "Pepperidge"). The district court found that Nabisco's use of an orange, bite-sized, cheddar cheese-flavored, goldfish-shaped cracker (as part of a tie-in promotion of a Nickelodeon Television Network television production) would dilute the distinctive quality of Pepperidge Farm's mark consisting of an orange, bite-sized, cheddar cheese-flavored, goldfish-shaped cracker, in violation of the Federal Trademark Dilution Act ("FTDA"), section 43(c) of the Lanham Act, 15 U.S.C. §1125(c), and New York's antidilution statute, N.Y. Gen. Bus. Law §360-l. The court entered an order requiring Nabisco to recall and cease selling its goldfish cracker. We affirm.

Background

1. Facts

Pepperidge Farm has produced small crackers in the shape of a goldfish (the "Goldfish" cracker) continuously since 1962. Although the Goldfish line of products includes crackers in various flavors and mixes, the primary product is the orange, cheddar cheese-flavored, fish-shaped cracker, sold in a bag or box under the trade name "Goldfish" and exhibiting a picture of the cracker on the exterior. The company has obtained numerous trademark registrations for the Goldfish design and name. In 1994, it launched an aggressive marketing campaign directed at children, who make up about half of Goldfish consumers, and between 1995 and 1998, it spent more than $120 million marketing the Goldfish line nationwide. The cracker has also been the subject of substantial media coverage, including a feature on "The Today Show" and an episode on "Friends." From 1995 to 1998, net sales of Goldfish crackers more than doubled, to $200 million per year. Measured by sales volume, Pepperidge Farm's Goldfish is the second-largest selling cheese snack cracker in America today. Measured in sales dollars, Goldfish ranks number one.

Occasionally, companies other than Pepperidge Farm have produced cheese crackers shaped like sea creatures, including "Guppies," "Dolphins & Friends," and "Whales." Only one of these products has included a cracker shaped like a goldfish — Nabisco's "Snorkels." Snorkels obtained only a small market share, and it is no longer on the market.

In spring 1998, Nickelodeon Television Network approached Nabisco to explore a possible joint promotion for Nickelodeon's new cartoon program, "CatDog." In August 1998, Nabisco and Nickelodeon entered a Joint Promotion Agreement ("JPA"), giving Nabisco the right to produce cheese crackers in shapes based on the CatDog cartoon. The agreement required Nabisco to print on its packages that "CatDog and related titles, logos and characters are trademarks of" Nickelodeon's parent, Viacom International, Inc. Nabisco's CatDog product was intended to compete with other animal-shaped cheese crackers marketed to children.

The star of the CatDog cartoon program is the CatDog — a two-headed creature that is half cat and half dog. Each half of the CatDog has a distinct personality. The fish is the favorite food and the symbol for the cat half; the bone is the preferred meal and emblem for the dog half. Other characters that are featured on the cartoon include a mouse, a rabbit, a squirrel, and several dogs. In its first three months, the CatDog show garnered a 3.9 Nielsen rating, making it close to the most widely watched program for children.

Pursuant to its agreement with Nickelodeon, Nabisco developed a CatDog snack that consists of small orange crackers in three shapes: half the crackers in a package are in the shape of the two-headed CatDog character, one-quarter in the shape of a bone, and one-quarter in the shape of a fish. The fish-shaped cracker closely resembles Pepperidge Farm's Goldfish cracker in color, shape, and size, and taste, although the CatDog fish is somewhat larger and flatter, and has markings on one side. The CatDog product was to be sold in boxes featuring the CatDog and showing fish and bones in the background. The launch of the CatDog product was set for February 1, 1999.

[The district court found no likelihood of consumer confusion, but did find likely dilution of the Goldfish shape].

Discussion

1. *Whether Pepperidge Farm Showed Likelihood of Success in Proving Dilution. . . .*

We understand the FTDA to establish five necessary elements to a claim of dilution: (1) the senior mark must be famous; (2) it must be distinctive;

(3) the junior use must be a commercial use in commerce; (4) it must begin after the senior mark has become famous; and (5) it must cause dilution of the distinctive quality of the senior mark.

The elements numbered (1) and (4) — fame of the senior mark, and junior use beginning after the senior mark has become famous — use terms in their ordinary English language sense. Their meaning is clear. It is not disputed, furthermore, that Pepperidge Farm's Goldfish constitutes a famous mark, and that Nabisco's first use of its goldfish cracker would not occur until Pepperidge Farm's Goldfish had become famous. Nor is it disputed that Nabisco's sale of its goldfish cracker would involve a "commercial use in commerce." The third element is met.

The only elements that require discussion are numbers (2) and (5) — that the senior mark be "distinctive" and that the junior use "dilute its distinctive quality." Here the statute invokes a term of art in trademark law.

Distinctiveness in a mark is a characteristic quite different from fame. Distinctiveness is a crucial trademark concept, which places marks on a ladder reflecting their inherent strength or weakness. The degree of distinctiveness of a mark governs in part the breadth of the protection it can command. [The court reviews the hierarchy of trademark distinctiveness.]

It is quite clear that the statute intends distinctiveness, in addition to fame, as an essential element. The operative language defining the tort requires that "the [junior] person's . . . use . . . cause[ ] dilution of the distinctive quality of the [senior] mark." 15 U.S.C. §1225(c)(1). There can be no dilution of a mark's distinctive quality unless the mark is distinctive. Furthermore, the statute lists factors to be considered in determining "whether the mark is distinctive and famous." Id. Clearly both qualities are required.

The requirement of distinctiveness is furthermore an important limitation. A mark that, notwithstanding its fame, has no distinctiveness is lacking the very attribute that the antidilution statute seeks to protect. The antidilution statute seeks to guarantee exclusivity not only in cases where confusion would occur but throughout the realms of commerce. Many famous marks are of the common or quality-claiming or prominence-claiming type — such as American, National, Federal, Federated, First, United, Acme, Merit or Ace. It seems most unlikely that the statute contemplates allowing the holders of such common, albeit famous, marks to exclude all new entrants. That is why the statute grants that privilege only to holders of distinctive marks.[2] As ex-

---

2. McCarthy's treatise contends the statute does not include an independent requirement of distinctiveness. It argues that the statute's three repetitions of the need for distinctiveness were included only as the result of accidental failure to delete vestigial remnants of an eliminated registration requirement and that "the word 'distinctive' was left floating in the statute, unmoored to either any statutory requirement or underlying policy goal." 3 J. Thomas McCarthy, McCarthy on Trademarks and Unfair Competition §24.91, at 24-147 (4th ed.1998). McCarthy dismisses "distinctiveness" as a synonym for "fame," that would be "redundant" as a separate requirement. Id. at 24-148. We disagree. We think the inclusion of the requirement of distinctiveness was intended, for good reason, to deny the protection of the statute to non-distinctive marks. See Restatement (Third) of Unfair Competition §25(1)(a) (1995) (requiring that the mark be "highly distinctive" and that the use of the junior mark be likely to "cause a reduction in that distinctiveness").

plained below, we believe that the Pepperidge Farm Goldfish mark is neither near the top nor the bottom of the ladder of distinctiveness, but is reasonably distinctive — certainly sufficiently so to qualify for the statute's protection.

The fifth element, "dilution of the distinctive quality of the mark" is the key operative element of the statute. The statute explains that dilution is "the lessening of the capacity of a famous mark to identify and distinguish goods or services." 15 U.S.C. §1127. We have likewise described dilution under the New York statute as the loss of the "ability to clearly and unmistakably distinguish one source." Hormel Foods Corp. v. Jim Henson Prods., Inc., 73 F.3d 497, 506 (2d Cir. 1996) (internal citations and quotation marks omitted).

The antidilution statutes rest on a judgment that the "stimulant effect" of a distinctive and well-known mark is a "powerful selling tool" that deserves legal protection. Restatement (Third) of Unfair Competition §25 cmt. c (1995). This power derives not only from "the merit of the goods upon which [the mark] is used, but equally [from the mark's] own uniqueness and singularity." Frank I. Schechter, The Rational Basis of Trademark Protection, 40 Harv. L.Rev. 813, 831 (1927). Even when an unauthorized use of the mark does not cause consumer confusion, it can "reduce[ ] the public's perception that the mark signifies something unique, singular, or particular." H. Rep. 104-374, at 3 (1995), reprinted in 1995 U.S.C.C.A.N. 1029, 1030. The junior use thereby diminishes the "selling power that a distinctive mark or name with favorable associations has engendered for a product in the mind of the consuming public." Sally Gee, Inc. v. Myra Hogan, Inc., 699 F.2d 621, 624-25 (2d Cir. 1983).

It is not yet entirely clear how courts should determine whether a junior use causes a senior mark to suffer dilution. See, e.g., *Sally Gee*, 699 F.2d 621, 625 ("dilution remains a somewhat nebulous concept") (applying New York statute). In dealing with the related question of infringement by reason of likelihood of consumer confusion, this court through study of the individual facts of the cases, and the factors suggested by the facts of those cases, came to develop gradually over time a nonexclusive list of factors that can help courts in making this determination. See Polaroid Corp. v. Polarad Electronics Corp., 287 F.2d 492, 495 (2d Cir. 1961) (Friendly, J.). We adopt a similarly cautious and gradual approach.

Considering the factors that appear pertinent on these particular facts, which we discuss below, we agree with the district court that Pepperidge Farm is likely to succeed in establishing that Nabisco's use of its goldfish shape in an orange, cheddar-cheese-flavored, bite-sized cracker dilutes the distinctive quality of Pepperidge Farm's previously famous mark, consisting of a goldfish-shaped orange cheddar-cheese-flavored, bite-sized cracker.

(a) Distinctiveness

In our view, distinctiveness plays a dual role. First, as discussed above, it is a statutory element. A mark cannot qualify for protection unless it is distinctive. Second, the degree of distinctiveness of the senior mark has a considerable bearing on the question whether a junior use will have a diluting effect. As the distinctiveness of the mark is the quality that the statute endeavors to protect, the more distinctiveness the mark possesses, the greater the interest to be protected. And conversely, the more the senior mark tends toward the weak, common, quality-claiming, or prominence-claiming type, the more strongly that weakness would argue against a finding of dilution, especially if the senior use is in a distinctly different field. See Restatement §25 cmt. f ("The degree of distinctiveness of the prior user's mark is also relevant because it measures the likelihood that the mark evokes the type of exclusive images and associations protected under the antidilution statutes. A very distinctive mark is thus more likely to suffer dilution of distinctiveness than is a less distinctive mark.").

We believe the goldfish shape of Pepperidge Farm's Goldfish crackers exhibits a moderate degree of distinctiveness. The fish shape has no logical relationship to a cheese cracker. However, we recognize it is not in the highest category of distinctiveness that belongs to a purely fanciful made-up pattern or design, or a made-up word like Kodak. Merchants often use an animal's likeness as a trademark. Furthermore, we recognize that it is not uncommon for children's cookies or crackers to be made in animal shapes. Indeed, there have been other fish-shaped crackers on the market, including one goldfish-shaped cracker (Nabisco's "Snorkels"). But Nabisco's "Snorkels" were not successful and have been withdrawn, and the other fish-shaped crackers were not similar to Pepperidge Farm's Goldfish; these crackers have not significantly affected the distinctiveness of Pepperidge Farm's Goldfish. In sum, because the use of the goldfish shape has no logical relationship to a bite-sized cheese cracker and for the other reasons discussed above, we believe that Pepperidge Farm's senior mark is reasonably distinctive.

(b) Similarity of the Marks

The degree of similarity of the junior mark to the senior is an obvious factor bearing on a finding of dilution. The marks must be of sufficient similarity so that, in the mind of the consumer, the junior mark will conjure an association with the senior. In that manner the junior mark will lessen the distinctiveness of the senior mark. See *Mead Data*, 875 F.2d at 1029 (majority opinion) ("We hold . . . that the marks must be 'very' or 'substantially' similar and that, absent such similarity, there can be no viable claim of dilution.").

Although we recognize that Nabisco's fish is not identical to Pepperidge Farm's, the similarity is sufficient to lessen the distinctive aspect of the Pepperidge Farm Goldfish in the eyes of consumers. Both fish are presented arbitrarily in the form of a cracker. Notwithstanding slight differences in shape, size and marking, Nabisco's crackers are essentially the same color, shape, size, and taste. The markings that appear on Nabisco's but not on Pepperidge Farm's crackers are quite small and faint so that they would not be easily noticed, except on a close inspection of a sort that is not likely to be performed by one who is intent on popping the crackers into his mouth. Furthermore, the markings appear on only one side of Nabisco's cracker.

Nabisco contends that the court's finding of similarity was in error because of the "context" in which Nabisco is using the goldfish. Nabisco emphasizes that it has packaged the CatDog product in a box quite unlike the Goldfish package, and that the CatDog product is a three-shape mix that is only one-fourth goldfish. Nabisco asserts that in these contexts, Nabisco's goldfish cracker will evoke an association only with the CatDog cartoon, not with the Goldfish cracker.

Like the district court, we are not persuaded by Nabisco's argument. As for Nabisco's box, many consumers of its crackers will not see the box; they will find goldfish-shaped cheddar cheese crackers served in a dish at a bar or restaurant or friend's house, looking very much like the familiar Pepperidge Farm Goldfish product. Infringement cases have consistently held post-sale confusion as well point-of-sale confusion to be actionable under the Lanham Act. See, e.g., Lois Sportswear, U.S.A., Inc. v. Levi-Strauss & Co., 799 F.2d 867, 872-73 (2d Cir. 1986); I.P. Lund Trading ApS v. Kohler Co., 163 F.3d 27, 44-45 (1st Cir. 1998). We recognize that dilution can occur as well in a post-sale as in a point-of-sale context.

The fact that only one quarter of the crackers in Nabisco's mix are fish and that the other two shapes reinforce the CatDog theme is helpful to Nabisco, but in our view insufficiently so. Many consumers seeing the crackers in a dish will not recognize the relationship of the fish to the Nickelodeon CatDog story. They will recognize a fish reminiscent of Pepperidge Farm's fish.

(c) Proximity of the Products and Likelihood of Bridging the Gap

The legislative history of the antidilution statutes shows that the legislatures were largely concerned with junior uses of famous marks on products unrelated to the senior area of commerce — as in the hypothetical cases of Buick aspirin, Schlitz varnish, or Kodak pianos. See H. Rep. 104-374, at 3, reprinted in 1995 U.S.C.C.A.N. 1029, 1030; *Mead Data*, 875 F.2d at 1031. Some courts and commentators have questioned the relevance of similarity of products because the law's "primary purpose was to apply in cases of widely differing goods." *I.P. Lund*, 163 F.3d at 49; see also 3 McCarthy §24:94.1, at 24-164. Nabisco also contends the statutes should not apply to use on competing products, which should be governed exclusively by the infringement laws.

We disagree with this view; or at least find it overstated. While the antidilution statutes aim at a different harm than the infringement statute and dilution undoubtedly can occur among non-competing products, we see no reason why dilution cannot occur as well where the products are competing.

See Restatement §25 cmt. f ("The nature of the respective goods is . . . relevant to dilution."). The closer the junior user comes to the senior's area of commerce, the more likely it is that dilution will result from the use of a similar mark.

Consumer confusion — the nub of an action for infringement — is, of course, unnecessary to show the actionable dilution of a famous mark. It does not follow, however, that dilution cannot be found in circumstances that would also support an action for infringement. Consumer confusion would undoubtedly dilute the distinctive selling power of a trademark. There can be little doubt that a junior use of Buick, Schlitz or Kodak on automotive equipment, drinks, and photographic products would lessen the distinctiveness of the senior mark at least as much as if the junior use were on unrelated products, such as aspirin, varnish and pianos. We can see no reason not to apply the antidilution statute to use on competing or closely related products, where likelihood of confusion, and thus infringement, might also be found.

Indeed, there may well be cases in which proximity may be necessary to dilution. It is easy to imagine instances where because of the low level of distinctiveness of the senior mark, or insufficient similarity between the two, the use of the junior mark in a remote area of commerce would have little tendency to remind consumers of the senior mark and thus little capacity to dilute its effectiveness, but where use of the same junior mark in a closely related area would bring about the harm the statute was designed to avoid. A junior use that confuses consumers as to which mark is which surely dilutes the distinctiveness of the senior mark.

For example, as noted above, a fish shape, while reasonably distinctive, is not as distinctive as a made-up word like Kodak. If Pepperidge Farm were seeking to enjoin the use of a fish-shape as the logo for an automobile, a newspaper, or a line of sports clothes, much less as the mark of a fish merchant or fish restaurant, it seems questionable whether dilution should be found. But here the junior user's area of commerce is identical to the senior's. A second major seller of goldfish-shaped, orange-colored, cheddar-flavored, bite-sized crackers can hardly fail, in our view, to dilute the distinctiveness in the eyes of the consumers of the senior mark in a goldfish-shaped, orange-colored, cheddar-flavored, bite-sized cracker.

Where the junior mark is used on a different product, courts examine whether the senior is likely to enter the junior's market, or "bridge the gap." *Polaroid*, 287 F.2d at 495. This factor recognizes "the senior's interest in preserving avenues of expansion and entering into related fields." *Hormel*, 73 F.3d at 504 (citations omitted). Here, because the junior use is in the same segment of commerce as the senior, there is no need to consider the likelihood that either might bridge the gap between them. There is no gap.

(d) Interrelationship Among the Distinctiveness of the Senior Mark, the Similarity of the Junior Mark, and the Proximity of the Products

The foregoing discussion suggests that there is a close interdependent relationship among these factors. The weaker any of the three factors may be, the stronger the others must be to make a case of dilution. To choose one of many possible hypothetical examples to illustrate this interdependence: with a highly distinctive senior mark, like Chevrolet for cars, even an only moderately similar junior mark — such as Chevremont — might dilute the distinctive quality of the senior if it were used in the automotive industry, but probably not if the junior were used in a distant area like perfumes.

Considering these three factors in relationship to one another, in this case we have a moderately distinctive senior mark, a substantial degree of similarity between the senior and junior mark, and the use of the junior mark in exactly the same area of commerce as the senior. In view of the substantial similarity of the marks and their use in precisely the same area of commerce, we believe that Pepperidge Farm made a strong case for the likelihood of dilution, notwithstanding that its mark is only moderately distinctive.

(e) Shared Consumers and Geographic Limitations

Another relevant factor is the extent of overlap among consumers of the senior user's products and the junior user's products. This factor is meaningful because dilution requires that a mark become less distinctive to consumers. If the consumers who buy the products of the senior user never see the junior user's products or publicity, then those consumers will continue to perceive the senior user's mark as unique, notwithstanding the junior use. See Restatement §25 cmt. f ("If the goods are marketed in different stores to different buyers . . . a connection between the prior and subsequent use may be unlikely.").

This factor strongly favors Pepperidge Farm. The two products will be in direct competition with one another, each being a substitute for the other. Both are to be marketed nationally in grocery stores. Children are also target audiences for both products. The consumers of Pepperidge Farm's Goldfish will see Nabisco's product.

(f) Sophistication of Consumers

Courts examining questions of infringement, as well as dilution, have looked at the sophistication of consumers as a relevant factor. See *Mead Data*, 875 F.2d at 1031-32; *Sally Gee*, 699 F.2d at 626; *Polaroid*, 287 F.2d at 495. Consumers who are highly familiar with the particular market segment are less likely to be confused by similar marks and may discern quite subtle distinctions. Conversely, unsophisticated customers lack this discrimination and are more vulnerable to the confusion, mistake and misassociations against which the trademark law protects. "Purchasers of relatively inexpensive goods such as ordinary grocery store foods are held to a lesser standard of purchasing care." 3 McCarthy §23:95, at 23-188.

Nabisco argues that this factor strongly supports its case, because children will have no difficulty recognizing the Nabisco product as a reference to the CatDog and will thus keep the two marks separate and distinct. Even if Nabisco is correct in that surmise, it seems to us to have only moderate importance, for two reasons. First, while children may be the primary ultimate consumers of the crackers, they are generally not the purchasers. Adult purchasers of crackers may be less sophisticated than children in recognizing the differences between the two fish. Even if, in the minds of children, the addition of Nabisco's CatDog family to the cheese cracker landscape does not lessen the distinctiveness of Pepperidge Farm's mark in its Goldfish, it is likely to do so among adults who will have less awareness of Nickelodeon's CatDog and of the differences between the two competing crackers.

Furthermore, even though children maybe the primary target of Pepperidge's marketing, they make up only one half the Goldfish market. Bite-sized cheese crackers are frequently served to adults as snacks accompanying cocktails. Once removed from their packaging, we think Nabisco's new goldfish cheese-crackers may well dilute the distinctiveness of Pepperidge Farm's mark in its goldfish-shaped, cheese cracker.

(g) Actual Confusion

Consumers' actual confusion of the junior and the senior mark may also be a factor in finding dilution. While we recognize that neither actual confusion nor likelihood of confusion is necessary to sustain an action for dilution, it does not follow that actual confusion cannot be highly probative of dilution. Confusion lessens distinction. When consumers confuse the junior mark with the senior, blurring has occurred.

Except as surveys are used to test the likelihood of consumer confusion between existing and yet-to-be marketed products, actual confusion can arise only where the junior and the senior mark have coexisted in the marketplace. Here the CatDog product has not yet been launched on the market. There has been no opportunity for actual confusion to arise. The absence of evidence of confusion of consumers, therefore, has no probative value.

(h) Adjectival or Referential Quality of the Junior Use

Unlike the remedy of trademark infringement, which generally protects the senior user's mark only within the senior's areas of commerce, the antidilution statute guarantees exclusivity to the senior in unrelated areas of commerce — as exhibited by the examples from the legislative history of Buick aspirin, Schlitz varnish or Kodak pianos. This raises a problem. The senior's mark might be arbitrary or fanciful in the senior area of commerce but highly adjectival — or even generic — in the junior's area. For example, the shape of Pepperidge's Goldfish crackers is arbitrary as a cheese cracker, but would be adjectival if used on a sign by a fish market or fish restaurant. Cf. *Abercrombie*, 537 F.2d at 9 n. 6 (noting that "Ivory" is arbitrary as applied to soap, but generic when used to describe products made from elephant tusks).

The stronger the adjectival association between the junior use and the junior area of commerce, the less likelihood there is that the junior's use will dilute the strength of the senior's mark. The logical association between a fish image and a fish business would lead consumers to understand the fish sign as descriptive of the junior's business, regardless whether it is also being used as a mark. Consumers would be unlikely to draw a diluting association between the junior mark and the Goldfish cracker. Similarly, if the hypothetical "Chevremont" mark in our earlier example were used on cheese, rather than automobile equipment, it would be less likely to dilute the distinctive quality of Chevrolet's mark.

It is a generally accepted principle of the trademark law, furthermore, that a senior claim to a mark does not bar a junior from using the same words (or symbols) comprising the mark in their descriptive sense. See 15 U.S.C. §1115(b)(4); Restatement §28. While the principle has been recognized primarily in connection with infringement actions, see, e.g., Car-Freshner Corp. v. S.C. Johnson & Son, Inc., 70 F.3d 267 (2d Cir. 1995), we see no reason why it should have any less application to actions for dilution.

Nabisco claims some protection from the fact that its fish shape is not arbitrary but acts as a reference to the fish in Nickelodeon's CatDog story. The weakness in its argument lies in the fact that when the Nabisco crackers are served in a bowl, consumers who are not familiar with the Nickelodeon entertainment and its cross licensing with Nabisco will see simply crackers very similar to Pepperidge Farm's fish (together with other shapes) and will not know that it celebrates Nickelodeon's CatDog entertainment. . . .

Considering the reasonable distinctiveness of the Goldfish mark, the very close proximity of the products, the degree of similarity between the two goldfish crackers, the low level of sophistication of many consumers, the occurrence of adjudication at the start of the junior use (and consequent absence of injury to the junior user's accumulated goodwill in its mark), we conclude that Pepperidge Farm has demonstrated a high likelihood of success in proving that Nabisco's commercial use of its goldfish shape will dilute the distinctiveness of Pepperidge Farm's nearly identical famous senior mark. We conclude the district court committed no error in granting a preliminary injunction.

2. Application of Antidilution Statutes to Competing Products.

Nabisco contends that the dilution statute has no application to competing products. We disagree. For the reasons discussed above, we believe that dilution can occur where the junior mark's use competes directly with the senior's as well as where the junior use is in a non-competing market. In general, the closer the products are to one another, the greater the likelihood of both confusion and dilution. The senior user has a right to the antidilution law's remedy in either case. Here we further note that the text of the federal statute directly contradicts Nabisco's position. The act expressly states that dilution may occur "regardless of the presence or absence of . . . competition between the owner of the famous mark and other parties." 15 U.S.C. §1127(c) (emphasis added).

We are not persuaded by the argument that it is unnecessary to apply the antidilution statute to a junior use on a competitive product because such a competing use would already be forbidden by the infringement statutes. In the absence of contrary legislative command, the fact that other remedies may be available to prevent a perceived ill does not seem to be sufficient reason to construe a statute as not reaching circumstances that fall squarely within its words. The fact that injured senior users may thus be given a choice of remedies is not sufficient reason to read into the antidilution statute limitations that Congress did not write. Moreover, failure to construe the antidilution statutes as reaching competing products may lead to a gap in coverage; the products might be found too far apart to support a finding of likelihood of confusion — (and therefore an infringement action) — yet too close together to permit a finding of dilution.

We have already held that New York's antidilution statute applies to "competitors as well as noncompetitors," Nikon Inc. v. Ikon Corp., 987 F.2d 91, 96 (2d Cir. 1993), and we now so hold under the FTDA. . . .

4. The Need for Proof of Actual Dilution.

Relying on a recent decision by the Fourth Circuit, Nabisco also asserts that proof of dilution under the FTDA requires proof of an "actual, consummated harm." Ringling Bros.-Barnum & Bailey Combined Shows, Inc. v. Utah Division of Travel Dev., 170 F.3d 449, 464 (4th Cir. 1999). We reject the argument because we disagree with the Fourth Circuit's interpretation of the statute.

It is not clear which of two positions the Fourth Circuit adopted by its requirement of proof of "actual dilution." Id. The narrower position would be that courts may not infer dilution from "contextual factors (degree of mark and product similarity, etc.)," but must instead rely on evidence of "actual loss of revenues" or the "skillfully constructed consumer survey." Id.

at 457, 464-65. This strikes us as an arbitrary and unwarranted limitation on the methods of proof.

To require proof of actual loss of revenue seems inappropriate. If the famous senior mark were being exploited with continually growing success, the senior user might never be able to show diminished revenues, no matter how obvious it was that the junior use diluted the distinctiveness of the senior. Even if diminished revenue could be shown, it would be extraordinarily speculative and difficult to prove that the loss was due to the dilution of the mark. And as to consumer surveys, they are quite subject to manipulation and are therefore not highly reliable. If a junior user began to market Buick aspirin or Schlitz shellac, we see no reason why the senior users could not rely on persuasive circumstantial evidence of dilution of the distinctiveness of their marks without being obligated to show lost revenue or engage in an expensive battle of unreliable surveys. Plaintiffs are ordinarily free to make their case through circumstantial evidence that will justify an ultimate inference of injury. "[C]ontextual factors" have long been used to establish infringement. We see no reason why they should not be used to prove dilution.

The broader reading of the Fourth Circuit's "actual, consummated" dilution element would require not only that dilution be proved by a showing of lost revenues or surveys but also that the junior be already established in the marketplace before the senior could seek an injunction. We recognize that the language of the statute gives some support to this reading, in that it uses the formulation, "causes dilution," rather than referring to "likelihood of dilution." *Ringling Bros.*, 170 F.3d at 464. In our view, however, such a reading depends on excessive literalism to defeat the intent of the statute. Notwithstanding the use of the present tense in "causes dilution," it seems plausibly within Congress's meaning to understand the statute as intending to provide for an injunction to prevent the harm before it occurs.

To read the statute as suggested by the Ringling opinion would subject the senior user to uncompensable injury. The statute could not be invoked until injury had occurred. And, because the statute provides only for an injunction and no damages (absent willfulness), see 15 U.S.C. §1125(c)(2), such injury would never be compensated. The Ringling reading is also disastrously disadvantageous for the junior user. In many instances the junior user would wish to know whether it will be permitted to use a newly contemplated mark before the mark is launched rather than after. That is why Nabisco sought declaratory relief as soon as Pepperidge Farm objected. If the statute is interpreted to mean that no adjudication can be made until the junior mark has been launched and has caused actual dilution, businesses in Nabisco's position will be unable to seek declaratory relief before going to market. They will be obligated to spend the huge sums involved in a product launch without the ability to seek prior judicial assurance that their mark will not be enjoined.

We are not at all sure that the Ringling opinion intends to limit the application of the statute to dilution that has actually occurred, as opposed to the narrower reading discussed above relating to manner of proof.[6] In any event, we read the statute to permit adjudication granting or denying an injunction, whether at the instance of the senior user or the junior seeking declaratory relief, before the dilution has actually occurred. . . .

7. The Standards Employed by the District Court. . . .

We think it would be a serious mistake at the outset of our consideration of the new federal antidilution statute to limit ourselves to these six factors or to any other putatively definitive list. At this stage, even the best-designed test would likely omit some important factors (and perhaps include others that will at times be irrelevant). Rather, in considering a new federal statutory right, it seems to us that courts would do better to feel their way from case to case, setting forth in each those factors that seem to bear on the resolution of that case, and, only eventually to arrive at a consensus of relevant factors on the basis of this accumulated experience. . . .

The promulgation of such a list has a tendency to quash open-minded, constructive thinking about a new statutory right. We believe it is by far premature for federal courts to declare and close the list of factors that will be deemed pertinent in cases under the new federal act.

Conclusion

Pepperidge Farm has demonstrated likelihood of success in proving that Nabisco's use of its goldfish-shaped cheddar cheese cracker will dilute Pepperidge Farm's mark in its similar, famous, goldfish-shaped cheddar cheese cracker. We affirm the court's order preliminarily enjoining the distribution of Nabisco's goldfish cracker.

COMMENTS AND QUESTIONS

1. The district court held that Pepperidge Farm could not prove that consumers would be confused by Nabisco's goldfish. Assuming that this is correct, why doesn't it also dispose of the dilution case? Are you persuaded

---

6. The Fourth Circuit also expressed concern that the federal antidilution statute not be read to create a radical "property right in gross," or a right in the senior user to prohibit any substantially replicating mark. *Ringling Bros.*, 170 F.3d at 454, 459. We agree that the antidilution statutes do not "prohibit all uses of a distinctive mark that the owner prefers not be made." *Deere*, 41 F.3d at 44. As we discussed above, there are many instances in which a junior use of a famous mark might not reduce the capacity of that mark to identify and distinguish products under the dilution statutes. Thus, we agree with the Fourth Circuit that the dilution statutes do not create a "property right in gross."

that dilution should apply to marks in the same market, even if they are not confusingly similar? If so, what role is there left for a trademark infringement claim in such a case? Is dilution a form of "super-trademark law"?

2. The Fourth Circuit's decision in *Ringling Brothers* (discussed in *Nabisco*) arguably has the plain meaning of the statute on its side. After all, in contrast to trademark's "likelihood of confusion" inquiry, the Federal Trademark Dilution Act does not punish "likelihood of dilution," but rather "dilution" itself.

After *Ringling Brothers*, how would you go about proving a case of federal trademark dilution in the Fourth Circuit? Would you be more likely to bring a state cause of action instead? If so, would this partially undermine the purpose of the federal statute? What questions would you include on a consumer survey meant to be used as evidence of actionable dilution under the Fourth Circuit test? How would you phrase a question about actual economic harm to the trademark owner's market(s)?

3. Though federal antidilution protection is still new, and the *Ringling Brothers* case very recent, at least some courts have shown a willingness to follow the "direct economic harm" approach it lays out. See, e.g., World Gym Licensing, Ltd. v. Fitness World, Inc., 47 F. Supp. 2d 614 (D. Md. 1999) ("Plaintiff has not submitted the hard proof of actual economic harm that *Ringling Brothers* demands. . . ."); American Cyanamid Co. v. Nutraceutical Corp., 1999 U.S. Dist. LEXIS 9156 (D.N.J. May 28, 1999) (labels incorporating colors of the visual spectrum incapable of triggering an "instinctive mental association" with Centrum's color-spectrum trademark); Playboy Enterprises v. Netscape Comm. Corp., 1999 U.S. Dist. LEXIS 9638 (C.D. Cal. June 24, 1999) (banner ads displayed by search engine operators did not cause blurring of Playboy's marks). As a result, the dispute highlighed in *Nabisco* is likely to be quite important.

4. Some cases decided before the 1996 federal antidilution provision had held that even "locally famous" marks were deserving of protection under state antidilution laws. See, e.g., Wedgewood Homes, Inc. v. Lund, 294 Or. 493, 659 P.2d 377, 222 U.S.P.Q. 446 (Or. 1983) (neither nationally famous nor arbitrary, "coined" business name is required for protection under Oregon antidilution statute; name of home builder is "locally famous" in eastern Washington County, Oregon, and that suffices for protection). The 1996 Act's legislative history makes clear that section 43(c)(1) of the new federal law does not protect locally famous marks, though state statutes may continue to do so:

Section 3 [of the Act, now §43(c)(1)] identifies a list of nonexclusive factors that a court may consider in determining whether a mark qualifies for protection. These factors include: (1) the degree of distinctiveness of the mark; (2) the duration and extent of use of the mark; (3) the geographical extent of the trading area in which the mark is used; and (4) whether the mark is federally registered. The first factor makes it clear that a mark may be deemed "famous" even if not inherently distinctive, that is, even if the mark is not arbitrary, fanciful, or coined. With respect to the duration and extent of use, generally a famous mark will have been in use for some time. *The geographic fame of the mark must extend throughout a substantial portion of the U.S.* Finally, although a mark need not be federally registered in order to be eligible for dilution protection, the fact that the mark is registered with the U.S. Patent and Trademark Office may be considered by a court in determining whether a mark is distinctive.

Federal Trademark Dilution Act of 1995, H. Rep. 104-374 (accompanying H.R. 1295), November 30, 1995 (emphasis added). Congress has not provided guidance regarding what constitutes a "substantial portion" of the United States.

Unfortunately, a number of courts seem willing to protect locally known marks that lack national fame. See, e.g., Gazette Newspapers, Inc. v. The New Paper, 934 F. Supp. 688 (D. Md. 1996) ("Gazette" famous for local Maryland paper); Wawa, Inc. v. Haaf, 40 U.S.P.Q.2d 1629 (E.D. Pa. 1996) ("Wawa" famous for convenience store in Pennsylvania and surrounding states); but see Star Markets Ltd. v. Texaco, Inc., 950 F. Supp. 1030 (D. Hawaii 1996) ("Star Market" not famous for grocery store that exists only in Hawaii). Cf. Mejia & Assocs. v. IBM, 920 F. Supp. 540 (S.D.N.Y. 1996) (noting that the test for fame under federal law is less rigorous than that under New York law). For criticism of this lax test for fame, see Mark A. Lemley, The Modern Lanham Act and the Death of Common Sense, 108 Yale L.J. 1687 (1999).

5. Dilution doctrine, as developed by the courts, embraces several theories of harm. The two most common are "dilution by blurring" and "dilution by tarnishment." The *Nabisco* case is an example of the former; the *L. L. Bean* case, in section E.5 below on "Parody," is an example of the latter. The dilution by tarnishment theory explicitly seeks to protect a senior user's quality connotations and goodwill against negative connotations or dissonant associations generated by a junior's use of the same or similar mark. See Coca-Cola v. Alma-Leo U.S.A., Inc., 719 F. Supp. 725, 728 (N.D. Ill. 1989) (dilution for a candy company to sell powdered bubble gum which closely resembled cocaine in a container strikingly similar to a Coca-Cola bottle, because the association of "such a noxious substance as cocaine with plaintiff's wholesome beverage" would tend to impugn Coca-Cola's business reputation); Eastman Kodak Co. v. D. B. Rakow, 739 F. Supp. 116 (W.D.N.Y. 1989) (dilution for a stand-up comedian to use "Kodak" as his stage name because his crude jokes would impair Kodak's reputation); Dallas Cowboys Cheerleaders, Inc. v. Pussycat Cinema, Ltd., 604 F.2d 200 (2d Cir. 1979) (dilution to promote a pornographic movie by suggesting that Dallas Cowboys cheerleaders were participants and to use actresses whose costumes closely resembled those of the Dallas Cowboys cheerleaders). See also Deere & Co. v. MTD Productions, Inc., 41 F.3d 39, 44-45 (2d Cir. 1994) (humorous use of competitors' mark found to dilute it). Some also argue that

dilution by genericization can be found in some cases. See Elliot Staffin, The Dilution Doctrine: Towards a Reconciliation with the Lanham Act, 6 Fordham Intell. Prop. Media & Entertainment L.J. 105, 139-142 (1995).

6. The Paris Convention, including Article 6[bis], is silent on the protection of trademarks against dilution. GATT-TRIPs signatory countries, however, must now provide some form of protection against dilution of a mark, at least if it is famous:

> Article 6*bis* of the Paris Convention (1967) shall apply, mutatis mutandis, to goods or services which are not similar to those in respect of which a trademark is registered, provided that use of that trademark in relation to those goods or services would indicate a connection between those goods and services and the owner of the registered trademark and provided that the interests of the owner of the registered trademark are likely to be damaged by such use.

Final Act Embodying the Results of the Uruguay Round of Multilateral Trade Negotiations, done at Marrakech, Morocco, April 15, 1994, Annex 1C: Agreement on Trade-Related Aspects of Intellectual Property Rights ("TRIPs"), at Article 16(3), reprinted in 33 I.L.M. 81, 89 (1994). For congressional approval, see Uruguay Round Agreements Act, Pub. L. No. 103-465 (H.R. 5110), Dec. 8, 1994. To what extent is the extension of trademark rights to "nonsimilar" goods or services mitigated by other requirements in this section?

Why can a mark not currently used in a country be protected against dilution in that country? Does it matter whether the mark is registered? Should foreign marks be protected against dilution in anticipation of their possible entry into the market?

7. The new federal dilution statute does not only protect word marks; it can also protect famous trade dress and product configurations. Sunbeam Prods. v. West Bend Co., 123 F.3d 246 (5th Cir. 1998). It does not protect only against identical marks. Similarity between the marks is sufficient if it causes blurring or tarnishment. See Ringling Bros.-Barnum & Bailey Combined Shows, Inc. v. Utah Div. of Travel Dev., 170 F.3d 449 (4th Cir. 1999); Toys "R" Us, Inc. v. Akkaoui, 40 U.S.P.Q.2d 1836 (N.D. Cal. 1996). For criticism of these expansions, see Paul Heald, Exposing the Malign Application of the Federal Dilution Statute to Product Configurations, 5 J. Intell. Prop. L. 415 (1998).

Note on Dilution and "Search Theory"

How would you state the harm that comes with dilution in terms of the consumer search rationale described earlier in this chapter? As a consumer, is locating Kodak film in the store or picking out the Kodak film ad in a magazine more difficult if there are many unassociated products or ads that use the Kodak name? As an experiment, try walking down the "generic" aisle in a supermarket, where all the boxes are plain black and white, trying to pick out the macaroni among the spaghetti, rigatoni, and so on. Would distinctive labels — all brightly colored, with striking designs — make a difference? From a certain distance, to take a different example, it is difficult to pick out a single marathon runner from the crowd of other runners, even though — or because — each runner is dressed in day-glo warmups, running shoes, and so on. The point is simply this: if many products share a similar label, the fact that the label is striking when viewed in isolation is irrelevant. It is as difficult to discern a single blazing orange poppy in a field full of them as it is to identify a lone grey pebble out of a rock pile.

If, as suggested earlier, attention spans are truly taxed in the current era, with its barrage of information and images, dilution may be an even more important adjunct to traditional "consumer confusion" rationales for trademarks. Indeed, the harm that dilution seeks to address might best be described as a loss of consumer *attention* due to the proliferation of similar or identical symbols of trade. Trademark owners thus seek to protect the distinctiveness of their marks — first, to insure a moment of unfettered attention, and then to indicate a unique connection between the mark and the seller's product.

PROBLEM

Problem 5-8. Mead Data Central is the owner of the registered trademark LEXIS for a legal and business-related electronic database. The service is extremely well-known among lawyers, but not among the general public.

A decade after the first use of the Lexis mark, Toyota names its new luxury car the "Lexus." Mead sues for dilution of its Lexis mark. What result? On what facts does your decision depend?

---

3. Contributory Infringement

Both patent law and copyright law have well-developed doctrines of contributory infringement. Defendants are liable for contributory infringement if, although they did not themselves infringe the patent or copyright, they assisted or encouraged others to infringe. Liability for contributory infringement extends to the makers and vendors of machines on which infringements are performed, but only if the machines are not capable of a substantial noninfringing use.

Is there contributory infringement in trademark law? At least at first blush, it is hard to see what acts might be comparable to selling or making available a machine used in producing copies.

Nonetheless, a number of courts have held that contributory infringement of trademarks is illegal. For example, in Polo Ralph Lauren Corp. v. Chinatown Gift Shop, 855 F. Supp. 648 (S.D.N.Y. 1994), the District Court for the Southern District of New York held that a landlord who knew that a commercial tenant was engaging in trademark infringement by selling counterfeit goods and who did nothing to prevent the activity could be sued for contributory infringement.

Is this result defensible? Does it extend to newspapers that print advertisements by counterfeiters? To graphics and print shops that print ads? To those who sell furniture or office supplies to counterfeiters?

A related problem is the extent to which private organizations are required to help enforce the trademark laws. This problem arose most recently on the Internet, the quasi-public but largely unregulated computer network that links millions of users worldwide. InterNIC, the private registry of Internet domain names, cannot possibly research domain names to ensure that they are not infringing prior trademarks. Instead, domain names are assigned on a first-come, first-served basis and are listed in the Internet Registry. Beginning in 1994, a number of individuals claimed domain names that had analogues in the physical world that were already trademarked by others. One columnist claimed the name mcdonald's.com, for example. Is InterNIC liable for contributory infringement if it assigns the name McDonald's to the first requestor and later denies the same name to McDonald's Corp.? Cf. Lockheed v. Network Solutions, Inc., 985 F. Supp. 949 (C.D. Cal. 1997) (NSI not liable for direct infringement merely for registering a domain name). Does it matter that the Internet is a worldwide organization and that the domain name registrant might be outside the United States?

4. False Advertising

Section 43(a) of the Lanham Act includes a specific prohibition on false or misleading advertising:

> (a) Any person who, on or in connection with any goods or services, or any container for goods, uses in commerce any word, term, name, symbol, or device, or any combination thereof, or any false designation of origin, false or misleading description of fact, or false or misleading representation of fact, which —
>> (1) . . .
>> (2) in commerce advertising or promotion, misrepresents the nature, characteristics, qualities, or geographic origin of his or her or another person's goods, services, or commercial activities,
> shall be liable in a civil action by any person who believes that he or she is or is likely to be damaged by such act.

*Johnson & Johnson*Merck Consumer
Pharmaceuticals Co. v. Smithkline Beecham Corp.*
*Johnson & Johnson*Merck Consumer Pharmaceuticals Co. v. Smithkline Beecham Corp.*
United States Court of Appeals for the Second Circuit
960 F.2d 294 (2d Cir. 1992)

WALKER, Circuit Judge:

Johnson & Johnson*Merck Consumer Pharmaceuticals Company ("J&J*Merck") appeals from the final judgment of the United States District Court for the Southern District of New York, Honorable Miriam Goldman Cedarbaum, Judge, denying J&J*Merck injunctive relief and dismissing its complaint against defendants Smithkline Beecham Corporation ("Smithkline") and Jordan, McGrath, Case & Taylor, Inc. ("Jordan"). J&J*Merck, the manufacturer of MYLANTA, a nonprescription antacid, instituted this action in order to restrain Smithkline and Jordan, the manufacturer and advertiser of TUMS, respectively, a competing brand of antacid, from continuing to air certain television commercials that J&J*Merck claims are false and misleading in violation of §43(a) of the Lanham Act, 15 U.S.C. §1125(a), and §§349 and 350 of the New York General Business Law.

Because we agree with Judge Cedarbaum's legal analysis, and because we conclude that her factual findings were not clearly erroneous, we affirm the judgment of the district court.

Background

J&J*Merck manufactures and markets MYLANTA, a popular nonprescription antacid product that is used to reduce heartburn by dyspepsia sufferers across the nation. MYLANTA contains aluminum hydroxide and magnesium hydroxide as its acid neutralizing agents. Smithkline produces the antacid TUMS. While TUMS is also extremely popular among victims of gastric distress, its formula relies upon calcium carbonate, rather than aluminum or magnesium salts, in treating indigestion and related ailments.

In September of 1990, Smithkline and its advertising agency, Jordan, instituted a television advertising campaign that sought to promote TUMS by comparing the ingredients contained in TUMS to those contained in other leading nonprescription antacids. The first commercial in this media campaign was entitled "Ingredients." Ingredients shows a man and a woman seated at a lunch counter along with other patrons, all of whom are eating such marginally digestible items as grease-laden french fries and chili. A background announcer explains that this couple "ate the same lunch" and "got the same heartburn," but that "his antacid [TUMS] is very different." A series of competing products then appear on the screen — ROLAIDS, MAALOX, and MYLANTA. At the same time, the ingredients of each of these antacids are visually superimposed on the screen and listed by the announcer: ROLAIDS — "Aluminum Salt"; MAALOX — "Aluminum and Magnesium"; MYLANTA — "Aluminum and Magnesium." Immediately following this remedial menu, the ad continues with a close-up image of a roll of TUMS. The voice-over states that TUMS is "aluminum-free," and that "only TUMS helps wipe out heartburn and gives you calcium you need every day." After the woman is depicted pondering the virtues of a calcium-based antacid, the commercial ends with a visual and verbal statement of the advertisement's slogan: "Calcium rich, aluminum free TUMS."

The makers of ROLAIDS, MAALOX and MYLANTA found the Ingredients advertisement impossible to swallow. After it was first aired, they immediately complained to Smithkline that Ingredients falsely implied that their antacids were harmful. In early October, 1990, they filed formal protests with each of the three major television networks, objecting to the Ingredients broadcast on the grounds that it was false and misleading. Representatives of Smithkline and Jordan met with network officials to discuss the protests lodged by TUMS' competitors, at which time they advised the networks that they would revise the Ingredients commercial. Thereafter, Smithkline and Jordan voluntarily withdrew Ingredients from the air. Although necessary background, the Ingredients advertisement is not the subject of this action.

In late October, 1990, Smithkline and Jordan released a second version of the TUMS commercial, entitled "Ingredients-Revised." The difference between Ingredients and Ingredients-Revised was that the second advertisement deleted all references to TUMS as an "aluminum-free" product. Rather, Ingredients-Revised emphasized the fact that TUMS contains calcium, ROLAIDS, MAALOX and MYLANTA did not, and that calcium is good for you. In addition to the full version of Ingredients-Revised, Smithkline and Jordan began showing a shortened one. This commercial singled out MYLANTA as the sole target of comparison, while keeping the rest of the commercial substantially the same.

. . . J&J*Merck sued in the district court to enjoin appellees from continuing to broadcast Ingredients-Revised and its shortened version, on the grounds that the commercials violated §43(a) of the Lanham Act, 15 U.S.C §1125(a), and §§349 and 350 of the New York General Business Law. Specifically, J&J*Merck alleged that both Ingredients-Revised, and the shortened version, falsely represented that: 1) occasional ingestion of TUMS, in the manner directed for antacid relief, results in nutritional benefit to the consumer; and 2) the magnesium and aluminum contained in MYLANTA are unsafe for human consumption.

Initially, J&J*Merck moved for a preliminary injunction. Instead of granting the motion, the district court consolidated the application with an expedited trial on the merits. After a five day bench trial, Judge Cedarbaum concluded that J&J*Merck had "not shown that the message that occasional Tums users will benefit from calcium is false or misleading." Furthermore, she found "that the challenged commercials do not communicate the message that aluminum or magnesium are harmful or unsafe." On the basis of this, Judge Cedarbaum concluded that J&J*Merck failed to demonstrate "that 'Ingredients-Revised' is either false or misleading." Accordingly, she dismissed both J&J*Merck's federal and state law claims. This appeal followed.

Discussion

On appeal, J&J*Merck has abandoned its claims relating to calcium and magnesium. Thus, the only issue before us now is whether the challenged commercials communicated a false or misleading message with respect to the safety of aluminum-based antacids. The district court explicitly found that "if the challenged commercials indicated that . . . aluminum [is] dangerous, they would be false and misleading."

J&J*Merck contends that, even though the content of the challenged commercials is literally true, Ingredients-Revised preys upon a publicly held misperception that the ingestion of aluminum causes Alzheimer's disease. According to J&J*Merck, the commercials accomplish this by repeatedly juxtaposing the absence of aluminum in TUMS with its presence in MYLANTA. In turn, this repetition supposedly links MYLANTA with an allegedly popularly held, yet unsubstantiated concern that aluminum is associated with Alzheimer's. Since the aluminum/Alzheimer's connection has not been scientifically established, J&J*Merck argues that Ingredients-Revised purposefully taps into a preexisting body of public misinformation in order to communicate the false and misleading message that aluminum-based antacids are harmful.

The gravamen of J&J*Merck's claim is that advertisers may be held liable for the knowing exploitation of public misperception. While this argument presents a novel theory of Lanham Act liability — one which we neither reject nor embrace — we note that, in any event, it would be unavailing in this case. Because J&J*Merck has failed to show that it has suffered any injury as a result of the challenged TUMS commercials, it cannot obtain relief under any theory of Lanham Act liability that is premised upon an implied falsehood. . . .

I. Liability for Implied Falsehoods . . .

The law governing false advertising claims under the Lanham Act is well settled in this circuit. In order to recover damages or obtain equitable relief, a plaintiff must show that either: 1) the challenged advertisement is literally false, or 2) while the advertisement is literally true it is nevertheless likely to mislead or confuse consumers. See McNeil-P.C.C., Inc. v. Bristol-Myers Squibb Co., 938 F.2d 1544, 1549, 19 U.S.P.Q.2d (BNA) 1525 (2d Cir. 1991); Johnson & Johnson v. GAC Intl., Inc., 862 F.2d 975, 977, 9 U.S.P.Q.2d (BNA) 1316 (2d Cir. 1988); Coca-Cola Co. v. Tropicana Products Inc., 690 F.2d 312, 317, 216 U.S.P.Q. (BNA) 272 (2d Cir. 1982); American Home Products Corp. v. Johnson & Johnson, 577 F.2d 160, 165, 198 U.S.P.Q. (BNA) 132 (2d Cir. 1978).

Where, as here, a plaintiff's theory of recovery is premised upon a claim of implied falsehood, a plaintiff must demonstrate, by extrinsic evidence, that the challenged commercials tend to mislead or confuse consumers. See GAC Intl., Inc., 862 F.2d at 977; American Home Products Corp., 577 F.2d at 165; Upjohn Co. v. American Home Products Corp., 598 F. Supp. 550, 556, 225 U.S.P.Q. (BNA) 109 (S.D.N.Y. 1984). It is not for the judge to determine, based solely upon his or her own intuitive reaction, whether the advertisement is deceptive. Rather, as we have reiterated in the past, "the question in such cases is — what does the person to whom the advertisement is addressed find to be the message?" American Home Products Corp., 577 F.2d at 166 (quoting American Brands, Inc. v. R. J. Reynolds Tobacco Co., 413 F. Supp. 1352, 1357 (S.D.N.Y. 1976)). That is, what does the public perceive the message to be?

The answer to this question is pivotal because, where the advertisement is literally true, it is often the only measure by which a court can determine whether a commercial's net communicative effect is misleading. Thus, the success of a plaintiff's implied falsity claim usually turns on the persuasiveness of a consumer survey. See Coca-Cola Co., 690 F.2d at 317 ("When the challenged advertisement is implicitly rather than explicitly false, its tendency to violate the Lanham Act by misleading, confusing or deceiving should be tested by public reaction."); American Home Products Corp., 577 F.2d at 165; Upjohn Co., supra.

Where a district court's factual findings regarding an advertisement's tendency to mislead or confuse consumers are based "on its analysis of . . . surveys [and] on the corroborating evidence of the market research experts," those findings are entitled to the clearly erroneous standard of review. American Home Products Corp., 577 F.2d at 167. J&J*Merck does not dispute this. However, in attempting to show that the district court's findings on this point were clearly erroneous, J&J*Merck argues that Judge Cedarbaum failed sufficiently to consider factors other than consumer survey evidence, such as: 1) the general "commercial context" or sea of information in which consumers are immersed; 2) the defendant's intent to harness public misperception; 3) the defendant's prior advertising history; and 4) the sophistication of the advertising audience. This argument causes us some discomfort.

Appellant's criticism of the district court's findings misconstrues the proper role of consumer survey evidence in the analysis of implied falsehood claims. Generally, before a court can determine the truth or falsity of an advertisement's message, it must first determine what message was actually conveyed to the viewing audience. Consumer surveys supply such information. Id., 577 F.2d at 166. "Once the meaning to the target audience has been determined, the court, as the finder of fact, must then judge whether the evidence establishes that they were likely to be misled." Nikkal Indus., Ltd. v. Salton, Inc., 735 F. Supp. 1227, 1235 (S.D.N.Y. 1991) (citing

McNeilab, Inc. v. American Home Products Corp., 501 F. Supp. 517, 525, 207 U.S.P.Q. (BNA) 573 (S.D.N.Y. 1980)).

Three of the factors listed by J&J*Merck, i.e., commercial context, defendant's prior advertising history, and sophistication of the advertising audience, only come into play, if at all, during the latter part of the court's analysis. In a particular case, these factors may shed some light on whether the challenged advertisement contributed to the meaning that was ultimately gleaned by the target audience. In other words, in determining whether an advertisement is likely to mislead or confuse, the district court may consider these factors after a plaintiff has established "that a not insubstantial number of consumers," Coca-Cola Co., 690 F.2d at 317, hold the false belief allegedly communicated in the ad.

Absent such a threshold showing, an implied falsehood claim must fail. This follows from the obvious fact that the injuries redressed in false advertising cases are the result of public deception. Thus, where the plaintiff cannot demonstrate that a statistically significant part of the commercial audience holds the false belief allegedly communicated by the challenged advertisement, the plaintiff cannot establish that it suffered any injury as a result of the advertisement's message. Without injury there can be no claim, regardless of commercial context, prior advertising history, or audience sophistication.

However, we have held that "where a plaintiff adequately demonstrates that a defendant has intentionally set out to deceive the public," and the defendant's "deliberate conduct" in this regard is of an "egregious nature," a presumption arises "that consumers are, in fact, being deceived." Resource Developers, Inc. v. Statue of Liberty — Ellis Island Foundation, Inc., 926 F.2d 134, 140, 17 U.S.P.Q.2d (BNA) 1842 (2d Cir. 1991). This presumption which may be engendered by the expenditure "of substantial funds in an effort to deceive consumers and influence their purchasing decisions" relieves a plaintiff of the burden of producing consumer survey evidence that supports its claim. Id. (quoting U-Haul Intl., Inc. v. Jartran, Inc., 793 F.2d 1034, 1041 (9th Cir. 1986)). In such a case, once a plaintiff establishes deceptive intent, "the burden shifts to the defendant to demonstrate the absence of consumer confusion." Resource Developers, supra. J&J*Merck argues that this principle applies here. We are not persuaded.

The evidence of deceptive intent introduced at trial by J&J*Merck, which was controverted by defense witnesses, was comprised of certain conflicting Smithkline internal memoranda and related solely to the original Ingredients version of the TUMS commercial — the version that contained the "aluminum-free" slogan. No doubt, this evidence might have been used to establish Smithkline's deceptive intent in broadcasting the Ingredients commercial. However, as stated above, Smithkline withdrew Ingredients from the air, and that version of the commercial is not the subject of this action. In response to complaints lodged with Smithkline and the television networks by TUMS' competitors, the "aluminum-free" reference was dropped from the Ingredients ad. J&J*Merck introduced no additional evidence that the same "deceptive intent," which allegedly infected its production of the Ingredients commercial, informed the editing process that resulted in Ingre-dients-Revised. Judge Cedarbaum acknowledged that J&J*Merck "attempted to show that [a false message] was communicated through evidence of defendants' intent." Apparently, she found its attempt unavailing. In any event, given the indirect and controverted nature of the evidence regarding the intent behind the Ingredients-Revised advertisement, we are unwilling to extend the presumption of consumer confusion to this case.

II. The Bruno and Ridgway Survey

At trial, J&J*Merck introduced the results of a consumer survey that it had conducted in conjunction with Bruno and Ridgway Research Associates, a marketing firm that conducts between 200-300 such surveys each year. According to Mr. Joseph Ridgway, the firm's president and J&J*Merck's expert witness, the survey was designed to assess what messages are communicated to consumers by Ingredients-Revised. In order to gain this information, 150 male and 150 female adult nonprescription antacid users were shown the commercial and interviewed in eight different shopping malls.

The aspects of the survey that are relevant to this appeal are those which concern the commercial's message regarding aluminum. Questions 8a and 8b asked: "Aside from trying to get you to buy the product, what are the main ideas the commercial communicates to you?" and "What other ideas does the commercial communicate to you?" Out of the 300 people surveyed, 18 people generally responded that "other antacids contain ingredients that are bad/harmful to you," only six of which specifically commented that "aluminum is not good for you." Two other responses were listed in the survey's tally as "aluminum is bad for brain/causes alzheimer's," and lastly, one additional response was recorded under the heading "miscellaneous negative aluminum comments." Questions 14a-c were asked of the 220 respondents who recalled Ingredients-Revised stating that MAALOX and MYLANTA contained aluminum and magnesium. They respectively inquired:

> 14a — What, if anything, does the commercial communicate to you about the aluminum and magnesium in Maalox and Mylanta?
> 14b — What else, if anything, does the commercial communicate to you about the aluminum and magnesium in Maalox and Mylanta?
> 14c — Based on the commercial you just saw, how do you feel about taking a product for heartburn that contains aluminum and magnesium?

Of the 220 people who responded to these questions, 83 answered with a comment classified as "not good for you/harmful/detrimental to your health." Three answered that aluminum causes Alzheimer's disease and, in addition, 38 responded that these ingredients are not needed by the body.

By adding up all negative comments made about either aluminum or magnesium in response to any of the survey questions, Mr. Ridgway concluded that Ingredients-Revised communicated to 45% of those surveyed that aluminum is either unhealthful or not good for you.

Smithkline and Jordan called Dr. Yoram Wind, a professor of marketing, as their expert witness to testify on the nature of the messages communicated by Ingredients-Revised. Dr. Wind strongly criticized the Bruno and Ridgway study primarily on two grounds. First, he testified that, in his opinion, the survey was almost wholly comprised of leading questions. Second, he faulted the study for not taking into account the fact that respondents may have brought with them previously acquired information regarding calcium and aluminum, and for failing to adjust the survey accordingly. In his opinion, the study should have contained a control group — people who were asked similar questions that sought to elicit their beliefs regarding the safety of antacid ingredients, but who were not shown the challenged commercials beforehand.

The evidentiary value of a survey's results rests upon the underlying objectivity of the survey itself. See Universal City Studios, Inc. v. Nintendo Co., Ltd., 746 F.2d 112, 118, 223 U.S.P.Q. (BNA) 1000 (2d Cir. 1984). This objectivity, in turn, "depends upon many factors, such as whether [the survey] is properly 'filtered' to screen out those who got no message from the advertisement, whether the questions are directed to the real issues, and whether the questions are leading or suggestive." American Home Products Corp. v. Johnson & Johnson, 654 F. Supp. 568, 590 (S.D.N.Y. 1987); see also Weight Watchers Intl., Inc. v. Stouffer Corp., 744 F. Supp. 1259, 1272, 19 U.S.P.Q.2d (BNA) 1321 (S.D.N.Y. 1990) (listing 7 criteria for survey trustworthiness).

Judge Cedarbaum's analysis of the Bruno and Ridgway survey was wholly in keeping with these principles. After hearing testimony from the parties' experts, and reviewing the results of the study itself, she specifically found that the study "did not show that 'Ingredients-Revised' communicates that the aluminum . . . in Mylanta is harmful or unsafe." In her view, the responses to questions 8a and 8b, which elicited only nine anti-aluminum reactions from the 300 people surveyed, were "the most persuasive evidence of the message communicated by 'Ingredients-Revised.' " She attributed this to the fact that questions 8a and 8b were "open-ended," and, therefore, more objective. . . .

The probative value of any given survey is a fact specific question that is uniquely contextual. While certain types of survey questions may be appropriate to discern the message of one advertisement, they may be completely inapposite with regard to another. Thus, "it [is] in the district court's province as trier of fact to weigh the evidence, and in particular the opinion research." American Home Products, 577 F.2d at 167. After reviewing the record in this case, we conclude that Judge Cedarbaum's evaluation of the survey questions is not clearly erroneous.

Conclusion

Based upon the literal message of the challenged commercials, and on the responses obtained from the consumer survey, the district court found that J&J*Merck failed to establish that commercials were either false or misleading. Upon review, we conclude that the district court's findings were not erroneous. Therefore, we affirm the district court's denial of injunctive relief, and its dismissal of J&J*Merck's complaint.

Affirmed.

COMMENTS AND QUESTIONS

1. What did Smithkline Beecham do wrong in its initial advertisement? Did it say anything that was untrue? Did it say that J&J's product was bad for consumers? The creation of often artificial brand distinctions is an integral part of product advertising. Is there anything wrong with creating such distinctions? Further, how should we test whether the "public" is deceived by a true statement? Should it be enough that 10 percent of the public is misled? 50 percent? Should the percentage required vary with the nature or severity of the deception? The Second Circuit has taken the position that "[w]hether or not the statements made in the advertisements are literally true, section 43(a) of the Lanham Act encompasses more than blatant falsehoods. It embraces innuendo, indirect intimations, and ambiguous suggestions evidenced by the consuming public's misapprehension of the hard facts underlying an advertisement." Vidal Sassoon, Inc. v. Bristol-Myers Co., 661 F.2d 272 (2d Cir. 1981) (comparison ad that misled consumers as to the extent and methodology of the survey conducted was actionable under section 43(a)).

2. Suppose that Smithkline Beecham had never mentioned its own product or compared the two, but had merely questioned the wisdom of consuming aluminum. Would J&J have a cause of action under section 43(a)? Until recently, the answer was no. Before the Trademark Law Reform Act of 1988, section 43(a) prohibited only false or misleading statements about the advertiser's own product. Disparagement of the plaintiff's product, while likely actionable under state law tort theories, did not violate the Lanham Act. See Bernard Food Indus. v. Dietene Co., 415 F.2d 1279 (7th Cir. 1969). The 1988 Amendments to section 43 clearly reverse this rule, at least as to statements made in "commercial advertising or promotion." Should federal law provide a cause of action for disparagement of goods or competitors? Johnson & Johnson also sued under state law, specifically §§349 and 350 of the New York General Business Law. These provisions cover, respectively, unfair trade practices in general and false advertising, and they are representative of state law false advertising statutes.

3. While advertisements comparing products or attacking one product often come from a competitor in the market, this is not necessarily the case. Consumers' rights groups like Consumer's Union (publisher of the popular

magazine *Consumer Reports*) regularly compare competing products against each other. Government agencies do the same on occasion, and irate consumers have been known to vent their frustration by advertising the defects of a product they are particularly unhappy with. Do these comparisons or disparagements — which presumably are made by neutral third parties rather than competitors — fall within the scope of section 43(a)? The legislative history suggests that section 43(a) was not intended to reach "consumer or editorial comment" by a group like Consumer Reports. Cong. Rec. H10420-21 (daily ed. Oct. 19, 1988) (statement of Rep. Kastenmeier). The dilution provisions of section 43(c) expressly exempt news reporting and comparative advertising from the reach of that section.

4. What is "advertising"? Certainly television and print advertisements fall within the scope of the false advertising prohibition. But what about flyers handed out on the street? Sales presentations? Letters to a customer? The general rule is that the Lanham Act does not apply to isolated statements or correspondence to individuals. But in Seven-Up Co. v. Coca-Cola Co., 86 F.3d 1379 (5th Cir. 1996), the court concluded that a series of sales presentations made to bottlers qualified as advertising for purposes of the Lanham Act.

5. The false advertising prohibition in section 43(a) is the most obviously consumer-oriented of the provisions in the Lanham Act. Protecting consumers from false information about products obviously furthers one of the key goals of trademark law — to ensure that consumers have accurate information on which to make purchasing decisions. Given this, it is somewhat surprising that consumers do not have standing to sue for false advertising. Barrus v. Sylvania, 55 F.3d 468 (9th Cir. 1995) (no consumer standing under §43(a)); Serbin v. Ziebart Intl. Corp., 11 F.3d 1163 (3d Cir. 1993) (only parties with a "reasonable commercial interest" may bring suit under section 43(a)). Is this requirement inconsistent with the consumer protection rationale for section 43(a)? If not, is there another explanation? On the role of competitors as consumer champions, see B. Sanfield, Inc. v. Finlay Fine Jewelry, Inc., 857 F. Supp. 1241 (N.D. Ill. 1994) (finding that plaintiff, a competitor jeweler, stated viable cause of action under §43(a) and state unfair trade practices statute for defendant's misleading advertising concerning price discounting, despite fact that plaintiff's store, products, and prices are not mentioned in plaintiff's ads).

6. Is private litigation by competitors necessary to ensure accuracy in advertising? See Lillian R. BeVier, Competitor Suits for False Advertising Under Section 43(a) of the Lanham Act: A Puzzle in the Law of Deception, 78 Va. L. Rev. 1, 2-3 (1992) (describing most false advertising claims as "systematically trivial," and arguing that "competitors' incentives to sue are not correlated with the likelihood of consumer harm."). BeVier argues that for the most part the market will discipline producers whose ads are misleading.

Consider two other possible sources of advertising regulation: FTC complaints and "private advertising codes." See Arthur Best, Controlling False

Advertising: A Comparative Study of Public Regulation, Industry Self-Policing, and Private Litigation, 20 Ga. L. Rev. 1 (1985) (comparing benefits of three different sources of ad regulation, and arguing for expansion of private litigation in various cases).

7. In Ross D. Petty, Supplanting Government Regulation with Competitor Lawsuits: The Case of Controlling False Advertising, 25 Ind. L. Rev. 351 (1991), the author describes industry-level self-regulation:

> Many industry trade associations have advertising codes, as do media and media associations. The National Advertising Division (NAD) of the Council of Better Business Bureaus has actively investigated advertising complaints since 1971. It is funded by dues paid to the Council of Better Business Bureaus by advertisers and advertising agencies. During 1983-85, forty-three percent of these complaints were made by competitors. If the NAD cannot resolve the complaint, the case can be appealed to the National Advertising Review Board. The Board is funded in the same manner as NAD, but has decided only forty-one cases that have been appealed to it out of the more than 2,000 investigated by NAD since 1971. In sixty-six percent of the cases, the Board upheld the NAD decision. In twenty percent of the cases, the Board reversed or modified the NAD decision, and fifteen percent of the cases were dismissed or withdrawn. Thus far, advertisers have always complied with a negative Board decision. NAD standards for reviewing advertising seem comparable to the FTC's standards. For example, in 1984, the NAD took formal action on 105 complaints. Eighty percent of these complaints questioned the adequacy of substantiation and eighty-three percent challenged misleading statements or depictions. In 1984, fifteen of these cases involved explicit comparisons with rival offerings and nearly forty involved implicit comparisons (e.g., "the only lawn fertilizer there is"). Lastly, from 1973 through part of 1982, thirty percent of NAD cases dealt with companies in Advertising Age's 100 Leading National Advertisers. Thus, the NAD appears to be more willing than the FTC to deal with national advertisers and comparative advertising. The cost of complaining to NAD is comparable to complaining to the FTC and is lower than that of bringing a Lanham Act case. Like the Lanham Act courts, NAD acts quickly. It frequently resolves complaints within six months of receipt. NAD examines about one hundred complaints annually. Despite its lack of authority to issue binding orders, it obtains discontinuance or modification in about seventy-five percent of its cases with the remainder vindicating the challenged advertisement.

On the related topic of the post-1989 cause of action under §43(a) for trade libel, and its pre-1989 common law antecedents, see 3 J. Thomas McCarthy, McCarthy on Trademarks and Unfair Competition, §27.09[1]-[3] (3d ed. 1992 & Supp. 1996). For more on "private" industry norms and practices supplanting formal, statutory law, see Robert P. Merges, Contracting into Liability Rules: Institutions for Facilitating Transactions in Intellectual Property Rights, 84 Cal. L. Rev. 1293 (1996); Robert Ellickson, Order Without Law (1991).

8. For the most part, plaintiffs in §43(a) advertising cases seek injunctive relief. In some cases, however, they seek — and obtain — damages. See, e.g.,

U-Haul Intl. v. Jartran, Inc., 793 F.2d 1034 (9th Cir. 1986) (sustaining $40 million award of defendant's profits calculated with reference to defendant's advertising costs). At least one commentator has argued that damages ought to be more liberally awarded. See Arthur Best, Monetary Damages for False Advertising, 49 U. Pitt. L. Rev. 1 (1987) (advocating a presumption that ads cause economic injury, and that courts use a defendant's expenditures on the challenged ads as an approximate measure of the injury they caused plaintiff). Do we need enhanced damages remedies in these cases? Or does the nature of competition provide sufficient incentive for competitors to challenge false advertisements?

9. Suing a defendant for making false statements in the press raises obvious First Amendment concerns. Most of the cases involving independent testing agencies, as well as a number of the direct competitor disparagement cases, are actually litigated as First Amendment cases, in which the plaintiff must prove falsity plus actual malice if she is a public figure. See, e.g., Bose, supra, 466 U.S. at 485. We will discuss the First Amendment limits on a false advertising claim in more detail in section E.5 below, when we discuss the parody defense. For a more extensive treatment, see Edwin C. Baker, Advertising and the First Amendment.

E. DEFENSES

1. Genericness

For a term to serve the purpose of a trademark, it must point to a unique source. When a term refers instead to a general class of products, it is deemed "generic" and cannot serve as a trademark. "Ford" is a source of products, a species; "car" is a class of products — a genus.

Generic trademarks are either "born generic," i.e., refused registration on the Principal Register because they are generic *ab initio*, or they become generic over time through a process called "genericide." For an example of the former, see Bellsouth Corp. v. Datanational Corp., 60 F.3d 1565 (Fed. Cir. 1995) ("Walking Fingers" logo for telephone yellow pages was generic). The case that follows discusses the latter issue.

The Murphy Door Bed Co., Inc. v. Interior Sleep Systems, Inc.
United States Court of Appeals, Second Circuit
874 F.2d 95 (2d Cir. 1989)

MINER, Circuit Judge:

Defendants-appellants appeal from a judgment entered in the United States District Court for the Eastern District of New York [for] . . . plaintiff. . . . The judgment includes a permanent injunction enjoining all defendants from use of the Murphy name. . . .

We hold that "Murphy bed" is a generic term, having been appropriated by the public to designate generally a type of bed. Consequently, defendants could not have infringed on plaintiff's trademark, alleged to be Murphy bed, and the district court erred in finding trademark infringement. . . .

Background

At the turn of this century, William Lawrence Murphy invented and manufactured a bed that when not in use could be concealed in a wall closet. By using a counter-balancing mechanism, the bed could be lowered from or raised to a closet in a wall to which the bed is hinged. In 1918, the United States Patent Office granted Mr. Murphy a patent for a "pivot bed," which was substantially similar to the wall bed. Mr. Murphy incorporated in New York in 1925 as the Murphy Door Bed Company and began to sell the wall bed under the name of "Murphy bed." Since its inception, the Murphy Co. has used the words Murphy and Murphy bed as its trademark for concealed beds. Other manufacturers of wall beds generally describe their products as "wall beds," "concealed beds," "disappearing beds," "authentic adjustable hydraulic beds" and the like, but rarely as Murphy beds. In fact, at least twice, when independent companies marketed their products as Murphy beds, Murphy complained to them and, as a result, the companies refrained from further deliberate use of the term Murphy bed.

On March 23, 1981, and again on November 16, 1982, the Patent and Trademark Office ("PTO") denied the Murphy Co.'s application to register the Murphy bed trademark. The PTO examining attorney explained that the words "Murphy bed" had become generic and that the phrase Murphy bed was "merely descriptive of a characteristic of the goods." In August 1984, the Trademark Trial and Appeal Board ("TTAB") affirmed the denial of registration. See In re Murphy Door Bed Co., Inc., 223 U.S.P.Q. 1030 (TTAB 1984). The TTAB noted that "Murphy bed has for a long period of time been used by a substantial segment of the public as a generic term for a bed which folds into a wall or a closet." Id. at 1033.

In December 1981, [defendants] entered into a distributorship agreement with the Murphy Co. and became the exclusive distributor of the Murphy bed in . . . four Florida counties. . . . The agreement, in the form of a letter signed by both Murphy and Zarcone [on behalf of Interior Sleep Systems, Inc.], provided in part that:

> 5) — Whenever the Murphy name is used, it must be in capital letters and identified by the word trademark or TM adjacent to Murphy. Cabinets or other material not furnished by Murphy will not be promoted or sold as Murphy products. . . .
> 8) — Upon termination of this agreement Interior Sleep Systems, Inc. agrees to discontinue the use of the name "Murphy bed".

After learning of the TTAB's 1984 decision denying Murphy's application for trademark registration, [defendants formed a corporation with the term "Murphy Beds" in the title, and then filled certain orders with both beds of its own making and beds from the Murphy Company]. . . .

[The district] court ruled that the name Murphy was not generic because a "secondary meaning" had been attributed to it by the general public, and that the common law of trademark therefore protected the Murphy Co. . . . Thus, the court held that the burden was on the defendants to show abandonment of the Murphy trademark — i.e. that the trademark had lost its significance. . . . The court then found that defendants did not sustain their burden. . . .

Discussion . . .

A term or phrase is generic when it is commonly used to depict a genus or type of product, rather than a particular product. See Abercrombie & Fitch Co. v. Hunting World, Inc., 537 F.2d 4, 9 (2d Cir. 1976). When a term is generic, "trademark protection will be denied save for those markets where the term still has not become generic and a secondary meaning has been shown to continue." Id. at 10. We have held that "the burden is on plaintiff to prove that its mark is a valid trademark . . . [and] that its unregistered mark is not generic." Reese Publishing Co. v. Hampton Intl. Communications, Inc., 620 F.2d 7, 11 (2d Cir. 1980).

As the Murphy mark is unregistered, *Reese* suggests that the district court erred in shifting the burden of proof to the defendants. However, the words at issue in *Reese*, "Video Buyer's Guide," were of common use before the product developer applied them to his product, whereas here, the term Murphy bed was created for its purpose by the manufacturer and only thereafter was it adopted by the public as a matter of common use. See Gimix v. JS & A Group, Inc., 699 F.2d 901, 905 (7th Cir. 1983) (differentiating between term in common usage before application to product and coinage of term to suit product that is later expropriated). It was this genericness of an "invented" term that Learned Hand addressed when determining whether "aspirin," a coined word, had been so adopted by the lay public as to become generic. See Bayer Co. v. United Drug Co., 272 F. 505, 509 (S.D.N.Y. 1921); see also King-Seeley Thermos Co. v. Aladdin Indus., Inc., 321 F.2d 577, 579 (2d Cir. 1963) (wide-spread use of the word "thermos," despite having been invented by plaintiff for description of vacuum bottle, created genericness); DuPont Cellophane Co. v. Waxed Products Co., 85 F.2d 75, 81 (2d Cir.) (expropriation by public of word "cellophane" created genericness), *cert. denied*, 299 U.S. 601, 57 S. Ct. 194, 81 L. Ed. 443 (1936). We find this distinction important and hold that where the public is said to have expropriated a term established by a product developer, the burden is on the defendant to prove genericness. Thus, critical to a trial court's allocation of proof burdens is a determination of whether the term at issue is claimed to be generic by reason of common usage or by reason of expropriation. This presumption of nongenericness of a product name in the case of apparent public expropriation is justified by the commercial protection a developer of innovations deserves.

The Murphy Co. was the first to employ the word Murphy to describe a bed that could be folded into a wall closet. It is claimed that over time the public adopted, or, rather, expropriated, the term Murphy bed as a synonym for any bed that folds into a closet. Accordingly, the district court was correct in placing the burden of proof of genericness upon the defendants. We find, however, that [defendant] did indeed establish the genericness of the term Murphy bed.

The following factors combined lead us to conclude that [defendant] showed at trial that today the term Murphy bed, in the eyes of "a substantial majority of the public," King-Seeley, 321 F.2d at 579, refers to a species of bed that can fold into a wall enclosure. First, the decision of the PTO, and certainly the TTAB, is to be accorded great weight. . . . The district court explicitly rejected the decisions of the PTO and TTAB finding genericness, despite acknowledging their persuasive force. Second, the term Murphy bed is included in many dictionaries as a standard description of a wall-bed. See, e.g., Webster's Third New International Dictionary 1489 (1981). While dictionary definitions are not conclusive proof of a mars generic nature, they are influential because they reflect the general public's perception of a mark's meaning and implication. . . . Third, [defendant] introduced as evidence numerous examples of newspaper and magazine use of the phrase Murphy bed to describe generally a type of bed. Again, such evidence is not proof positive, but it is a strong indication of the general public's perception that Murphy bed connotes something other than a bed manufactured by the Murphy Co. . . .

In finding a lack of genericness, the district court was influenced by Murphy's efforts at policing its mark. The court noted with approval instances where Murphy complained to those who had used the term Murphy bed to describe beds not necessarily produced by Murphy. . . . However, when, as here, the mark has "entered the public domain beyond recall," policing is of no consequence to a resolution of whether a mark is generic. King-Seeley, 321 F.2d at 579. . . .

We conclude that the defendants adequately demonstrated that the Murphy bed mark is generic. . . .

COMMENTS AND QUESTIONS

1. Companies often fight vigorously to prevent their trademarks from becoming generic through casual usage. Consider the strong message sent by Xerox Corporation in the accompanying advertisement (Figure 5-4).

Apart from general advertising such as this, firms also police the uses of their marks via lawsuits. See, e.g., Selchow & Righter Co. v. McGraw-Hill Book Co., 580 F.2d 25 (2d Cir. 1978) (granting Scrabble trademark holder preliminary injunction against publisher of "The Scrabble Dictionary," on grounds that publication would cause irreparable injury by possibly rendering trademark generic); Elliot Staffin, The Dilution Doctrine: Towards a Reconciliation with the Lanham Act, 6 Fordham Intell. Prop. Media & Entertainment L.J. 105, 117 (1995) (collecting recent cases finding that dilution causes of action may lie against those employing a trademark in a way that threatens to make it generic). Cf. Ralph H. Folsom and Larry L. Teply, Trademarked Generic Words, 89 Yale L.J. 1323, 1346-1347 n.110 (1980) (describing organized efforts of trademark attorneys to pressure dictionary publishers into excluding trademarked words, and/or including disclaimers that such inclusion should not bear on genericide issue). This latter form of policing may explain a finding in William M. Landes & Richard A. Posner, Trademark Law: An Economic Perspective, 30 J.L. & Econ. 265, 296 (1987) ("Thus, although words held to be generic are more likely to show up in the dictionary than those held not to be generic, the difference in probabilities is small — 54% versus 41%.").

2. Under the law of at least one state, a generic term already in public use later acquires secondary meaning by virtue of a product developer's unique use, thus warranting trademark protection. See, e.g., AntiDefamation League of B'Nai B'rith v. Arab Anti-Defamation League, 72 Misc. 2d 847, 855, 340 N.Y.S.2d 532, 534-544 (N.Y. Sup. Ct. 1972). Thus genericide may be reversible. Does this rule conflict with the federal trademark scheme?

3. In an omitted portion of the opinion, the court held that defendants breached their contractual obligation to refrain from using the term Murphy bed after termination of their distribution agreement with the Murphy Co., and that defendants engaged in unfair competition by passing off beds of their own manufacture as beds of the Murphy Co. Does it make sense to restrict the rights of the defendants — who had signed an agreement — to a greater degree than other competitors? Does this holding suggest that parties can "contract around" trademark genericness? What would be the effect of such agreements?

4. In Kellogg Co. v. National Biscuit Co., 305 U.S. 111, 116-118 (1938), Justice Brandeis set out a classic discussion of "genericide," though

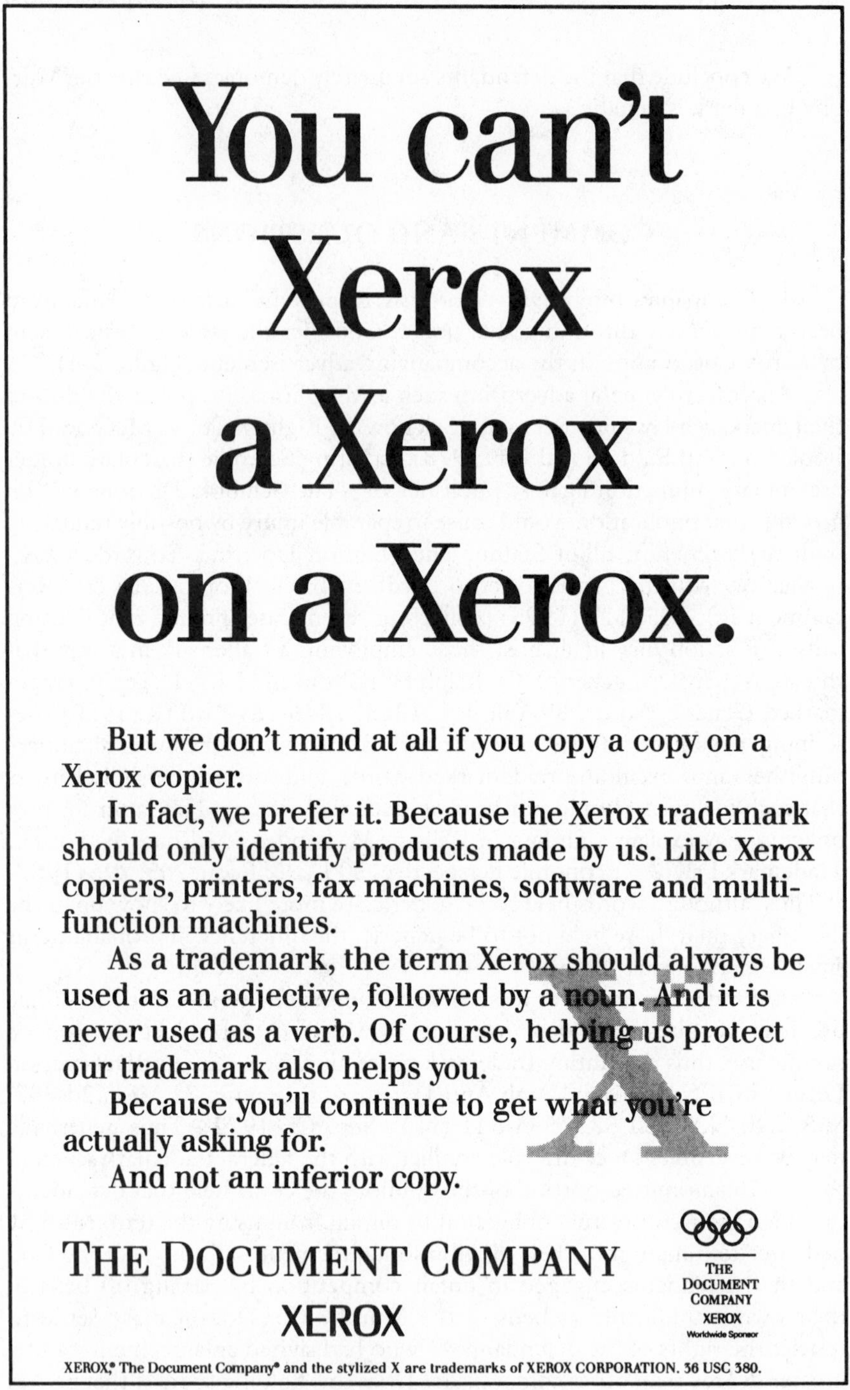
You can't Xerox a Xerox on a Xerox.
But we don't mind at all if you copy a copy on a Xerox copier.
In fact, we prefer it. Because the Xerox trademark should only identify products made by us. Like Xerox copiers, printers, fax machines, software and multi-function machines.
As a trademark, the term Xerox should always be used as an adjective, followed by a noun. And it is never used as a verb. Of course, helping us protect our trademark also helps you.
Because you'll continue to get what you're actually asking for.
And not an inferior copy.
THE DOCUMENT COMPANY
XEROX
THE DOCUMENT COMPANY XEROX
Worldwide Sponsor
XEROX,® The Document Company® and the stylized X are trademarks of XEROX CORPORATION. 36 USC 380.

FIGURE 5-4
Xerox trademark advertisement.

one infused also with elements of functionality and descriptiveness/secondary meaning:

> The plaintiff has no exclusive right to the use of the term "Shredded Wheat" as a trade name. For that is the generic term of the article, which describes it with a fair degree of accuracy; and is the term by which the biscuit in pillow-shaped form is generally known by the public. Since the term is generic, the original maker of the product acquired no exclusive right to use it. As Kellogg Company had the right to make the article, it had, also, the right to use the term by which the public knows it. . . . Ever since 1894 the article has been known to the public as shredded wheat. For many years, there was no attempt to use the term "Shredded Wheat" as a trade-mark. . . .
>
> Moreover, the name "Shredded Wheat," as well as the product, the process and the machinery employed in making it, has been dedicated to the public. The basic patent for the product and for the process of making it, and many other patents for special machinery to be used in making the article, issued to Perky. In those patents the term "shredded" is repeatedly used as descriptive of the product. The basic patent expired October 15, 1912; the others soon after. Since during the life of the patents "Shredded Wheat" was the general designation of the patented product, there passed to the public upon the expiration of the patent, not only the right to make the article as it was made during the patent period, but also the right to apply thereto the name by which it had become known. As was said in Singer Mfg. Co. v. June Mfg. Co., 163 U.S. 169, 185:
>
>> It equally follows from the cessation of the monopoly and the falling of the patented device into the domain of things public, that along with the public ownership of the device there must also necessarily pass to the public the generic designation of the thing which has arisen during the monopoly. . . . To say otherwise would be to hold that, although the public had acquired the device covered by the patent, yet the owner of the patent or the manufacturer of the patented thing had retained the designated name which was essentially necessary to vest the public with the full enjoyment of that which had become theirs by the disappearance of the monopoly.
>
> It is contended that the plaintiff has the exclusive right to the name "Shredded Wheat," because those words acquired the "secondary meaning" of shredded wheat made at Niagara Falls by the plaintiff's predecessor. There is no basis here for applying the doctrine of secondary meaning. The evidence shows only that due to the long period in which the plaintiff or its predecessor was the only manufacturer of the product, many people have come to associate the product, and as a consequence the name by which the product is generally known, with the plaintiff's factory at Niagara Falls. But to establish a trade name in the term "shredded wheat" the plaintiff must show more than a subordinate meaning which applies to it. It must show that the primary significance of the term in the minds of the consuming public is not the product but the producer. This it has not done. The showing which it has made does not entitle it to the exclusive use of the term shredded wheat but merely entitles it to require that the defendant use reasonable care to inform the public of the source of its product.

The Court cites Singer Mfg. Co. v. June Mfg. Co., 163 U.S. 169, 185 (1896) for the proposition that once a patent on a device expires, the name by which that device has been sold also enters the public domain. Surely that is not always the case. Numerous products that once were patented are still sold under the same trademark. And this result — that patent expiration does not automatically end trademark protection — is also consistent with the purposes behind the two laws, which are very different. The key to *Singer*'s holding may lie in its expressed concern that "the owner of the patent . . . had retained the designated name which was essentially necessary to vest the public with the full enjoyment of" the product. Thus it is only when the name itself is "essentially necessary" to sales of the product — that is, when the name is generic — that trademark protection should not survive the expiration of a patent.

5. While most genericness cases involve word marks, trade dress and product configurations can also be generic. See, e.g., Kendall-Jackson Winery v. E. & J. Gallo Winery, 150 F.3d 1042 (9th Cir. 1998), where the court held that an autumnal grape leaf featured on both plaintiff's and defendant's wine bottles was generic in the wine industry. *See also* Sunrise Jewelry Mfg. Corp. v. Fred, S.A. 175 F.3d 1322 (Fed. Cir. 1999) (holding that generic product configuration may be cancelled notwith standing "incontestable" status; remanding for determination of genericness).

6. Lanham Act §14 (15 U.S.C §1064) offers some guidance on how to determine whether a mark has become generic. It provides that "[t]he primary significance of the registered mark to the relevant public rather than purchaser motivation shall be the test for determining whether the registered mark has become generic." This language was added to the statute by the Trademark Clarification Act of 1984, which was passed in order to reverse the holding of Anti-Monopoly, Inc. v. General Mills Fun Group, 684 F.2d 1316 (9th Cir. 1982), *cert. denied*, 459 U.S. 1227 (1983). That case held that a game called "Anti-Monopoly," which used rules and a game board very similar to Parker Brothers' "Monopoly" game, could be sold because "Monopoly" had become generic. The Ninth Circuit reasoned that most consumers did not buy the "Monopoly" game because it was made by Parker Brothers and because they liked the products made by this company. Yet only this motivation for buying, the court said, supports trademark rights in the game's name. Since consumers associated the term "Monopoly" with a single product (the game), *and not a single producer* (at least not one they could name), the court found the mark generic. The court apparently would have limited trademark rights only to those trademarks that point unmistakenly to a single, *known* source. This reasoning was quickly repudiated in the amendments to §14 (15 U.S.C. §1064) just described. Can you see what is wrong with the rationale in *Anti-Monopoly?*

Section 14 provides that a mark may become generic as to some, but not all, goods and that in that case it may be cancelled in part. Cancellation of a mark because it has become generic donates that mark to the public domain, enabling other competitors to use it for the same or similar products (such as Anti-Monopoly).

7. When a court declares a trademark generic, it destroys a property right built up with considerable investment. Is this the same as other instances

where government decrees destroy private property; that is, should "just compensation" be paid for this "taking"? See Stephen L. Carter, Does It Matter Whether Intellectual Property Is Property?, 68 Chicago-Kent L. Rev. 715 (1993) (speculating about the desirability of such an arrangement); compare Ruckelshaus v. Monsanto Co., 467 U.S. 986 (1984) (government required to compensate trade secret owner when it compels disclosure of the secret).

The Second Circuit has periodically suggested that even generic marks are entitled to protection against some forms of unfair competition. See, e.g., Genesee Brewing Co. v. Stroh Brewing Co., 124 F.3d 137, 150 (2d Cir. 1997) (plaintiff's mark for "Honey Brown Ale" was generic, but defendant could still be liable if it did not use "every reasonable means to prevent confusion" as to the source of the products); Forschner Group Inc. v. Arrow Trading Co., 124 F.3d 402 (2d Cir. 1997). This seems contrary to the very purpose of the genericness doctrine, however.

8. Some commentators have noticed that trademarks have come to serve a more expressive role in addition to their original function of identifying a source for a commercial product. Clearly, expressions such as "She looks like a Barbie doll," or "he thinks he is a Marlboro man" make use of trademarks without any attempt to identify a source, or even specifically connect the source to a commercial product. Such expressions have salience precisely because the owners of the mark have blanketed our culture with them. In this sense the communicative content of Barbie and Marlboro — the unique images and associations these terms call to mind — are an unintended offshoot of the advertiser's commercially-motivated saturation campaigns. Even so, these associations are very real.

This is not news to many owners of well-recognized marks. Many of them have instigated aggressive licensing campaigns to take advantage of their unique image in markets far beyond those they originally served. Thus logos such as "NBA" and "NFL," though they may have originally identified unique "brands" (high-quality sports leagues), now appear almost everywhere. Surely no one looks for the NBA logo on toy balls or teddy bears to assure herself of getting a well-made ball or toy. Rather, the buyer is seeking a piece of the NBA image.

Current practice actually reverses the traditional pairing. The logo in some sense is being sold *by itself*; the item to which it is attached is an add-on, a substrate for the real product of value. Certainly this is true of T-shirts, pins, and other logo-bearing items. Instead of logos identifying products, products carry logos. Perhaps it will turn out that these trends represent the final stage in the "propertization" of trademarks — the withering away of the associational role, and the emergence of trademarks as saleable items in themselves.

One perceptive commentator, Rochelle Dreyfuss, has watched these developments, asking whether trademark doctrine has "kept up." Rochelle C. Dreyfuss, Expressive Genericity: Trademarks as Language in the Pepsi Generation, 65 Notre Dame L. Rev. 397 (1990). Her answer: no. This leads Professor Dreyfuss to suggest a number of ways trademark law should be adjusted. Her primary suggestion is to extend traditional concepts of genericness to encompass the realm of expressive (as opposed to commercial, or what she calls "competitive") uses of trademarks. "[C]ourts," she writes,

> would first decide whether there is an expressive component to the challenged use and then consider how central the trademark is to the usage. If the mark is found to be rhetorically unique within its context, it would be considered expressively — but not necessarily competitively — generic, and the trademark owner would not be permitted to suppress its utilization in that context. . . . [T]he suggested analysis would permit courts to drop the fiction of consumer confusion, and instead focus on the primary concern, which is the expressive significance of the mark.

Dreyfuss, at 418-419.

Note on Genericide, Language, and Policing Costs

It might seem that the notion of genericide does not make much sense. Why should a mark pass into the public domain because it has come to signify a category of products rather than a single product/source combination? If a trademark owner has "invented" a trademark, there is a clear sense in which the genericness doctrine takes away some rights that the creator was endowed with because of her creativity. Should trademark protection be permanent, or should we attempt a rough "calibration," along the lines of patent and copyright, to adjust the incentive to call forth the optimum (or at least desirable) level of this activity? Cf. William M. Landes & Richard A. Posner, Trademark Law: An Economic Perspective, 30 J.L. & Econ. 265, 270 (1987) (trademark protection offers a modest incentive to create new commercial terms).

Does it make sense that the very success of a trademark should be its undoing? In some ways, this is analogous to the argument that copyright protection on a software user interface should "lapse" when the interface becomes an industry standard. Cf. Lotus Devel. Corp. v. Borland Intl., Inc., 49 F.3d 807 (1st Cir. 1995), *aff'd by equally divided court*, 116 S. Ct. 804 (1996). The unifying theme is the notion that the success of certain kinds of creative works is largely a result of the need for a single standard, rather than the creative brilliance of the work itself. Economists speak of these works as creating "network externalities": users benefit from the fact that others use the work. Widely used computer protocols, such as the TCP/IP standard allowing exchange of data over the Internet, are advantageous because they make a wide variety of content available to all users. Thus each new person who adopts a protocol benefits all current users of the protocol. See infra, Chapter 7, discussing these markets in detail.

Some courts and legal scholars point out that under such a dynamic, the success of works of this kind is attributable more to the *collective labor* of the users than to the labor of the work's creator. See, e.g., Lotus Devel. Corp. v. Borland Intl., Inc., 49 F.3d 807, 821 (1st Cir. 1995) (Boudin, J., concurring) (accused infringer's product was "merely trying to give former Lotus users an option to exploit their own prior investment in learning. . . . [I]t is hard to see why customers who have learned the Lotus menu and devised macros for it should remain captives of Lotus because of an investment in learning made by the users and not by Lotus."). They argue that the efforts of the users in learning a new work, such as the computer protocol in the example above, rather than the efforts of the creator in designing the work, account for the success of these kinds of works. It follows from this argument that the claims of the work's creator must give way to those of the users in some cases. This logic is applied directly to generic trademarks in the following passage:

> Giving ownership in intellectual products that have come to serve as standards, such as West citations or generic trademarks, would not ordinarily leave "enough, and as good" [in the Lockean sense]. There may be room in the world for only one of a given type of thing, or a long-lived artifice may become a mode of communication. It is the nature of a standard that nothing "as good" is available. For these reasons, the [Lockean] proviso would be violated if the courts gave those who create standards in nonfungible goods a right to prevent people from utilizing them.

Wendy J. Gordon, A Property Right in Self-Expression: Equality and Individualism in the Natural Law of Intellectual Property, 102 Yale L.J. 1533, 1600 (1993) (footnote omitted). See also Justin Hughes, The Philosophy of Intellectual Property, 77 Geo. L.J. 287, 315-323 (1988) (suggesting that ownership of most intellectual products will easily satisfy the Lockean proviso, though "extraordinary ideas" like generic trademarks should remain open for all to use). A recent commentary makes explicit the *public's* contribution to strong trademarks. *See* Steven Wilf, Who Authors Trademarks?, 17 Cardozo Arts & Ent. L.J. 1 (1999).

Economists have observed that human languages exhibit network externalities. Professor Gordon's argument just above in effect extends the network externalities argument to individual words. When a word comes to describe a genus or class of goods, i.e., enters general use, it takes on some of the properties of the computer protocol described earlier. It becomes the standard shorthand descriptor for a complex, multi-attribute product type. Past a certain point, any alternative descriptor becomes clearly second-best. Requiring consumers and competitors to use this second-best descriptor entails costs. See Ralph H. Folsom and Larry L. Teply, Trademarked Generic Words, 89 Yale L.J. 1323, 1340-41 (1980). Though difficult to quantify, these costs may be substantial. See also Landes and Posner, supra, at 280, 292 (discussing costs of having to "describe around" a well-known term).

Consider the example of "plexiglas." The next best alternative to this well-known descriptor might be: "unbreakable clear plastic sheets, or window material." That is not only a mouthful; it is more expensive to advertise (because longer), harder to remember, and more subject to mistakes and confusion. Thus it is not hard to see why "plexiglas" became the preferred shorthand for it. Consequently, although the originator of this term might have put substantial effort into creating it and encouraging its use, there is a good argument that it has become a standard name (or descriptor) — and hence generic. But cf. Rohm & Haas Co. v. Polycast Technology Corp., 172 U.S.P.Q. 167 (D. Del. 1971) (enjoining defendant's use of Plexiglas mark). As another example, consider "Yo-Yo"; how would you describe this kind of toy without using the word *yo-yo*? Cf. Donald F. Duncan, Inc. v. Royal Tops Mfg. Co., 343 F.2d 655 (7th Cir. 1965) (holding *yo-yo* generic, besides being descriptive in Filipino language).

The Role of Policing Costs. Folsom and Teply argue that policing costs (reflected in ads such as the one by Xerox, above) are always wasted and therefore should not be considered in determining whether a trademark has become generic. See Folsom and Teply, supra, at 1354. Do you agree? Can you defend expenditures to keep a mark from going generic on the basis that they are attempts to provide alternative "standards" that will thereby retain the source-indicating function of threatened trademarks? If consumers accept the premise of the ads and faithfully refer to the "copier" instead of the "Xerox," are consumers harmed? And do such expenditures show that the cost of informing the public about alternative product descriptors is worth it to the firm, given its investment in its trademark? Cf. E. I du Pont de Nemours & Co. v. Yoshida Intl., Inc., 393 F. Supp. 502, 523-524 (E.D.N.Y. 1975) ("Teflon" not generic, in part because of du Pont's vigorous education and policing campaign). Of course, it may be impossible for a company to prevent genericide, for example, when a trademark achieves rapid and overwhelming acceptance. Then no amount of policing will work. Cf. Du Pont Cellophane Co. v. Waxed Prods. Co., 85 F.2d 75 (2d Cir.), *cert. denied,* 299 U.S. 601 (1936) ("Cellophane" generic despite vigorous policing efforts). Does such rapid acceptance indicate that the next best alternative descriptor is significantly less effective? Cf. Folsom and Teply, supra, at 1344 (noting that some alternative terms are better than others; comparing "lip balm" as alternative to "Chap Stick"; with "dextro-amphetamine sulphate" as alternative to "Dexadrine"); Comment, Trademarks and Generic Words: An Effect-on-Competition Test, 51 U. Chi. L. Rev. 868, 884-885 (1984) (advocating use of antitrust-like cross-elasticities of demand analysis to determine degree of substitutability between terms).

Can you see a less benign motive for expenditures to maintain the trademark status of a word that has become a widely used name for a product class? See Folsom and Teply, supra, at 1337 (suggesting two: (1) to maintain entry barriers to competition; and (2) to obtain "free advertising" every time someone uses the trademarked word to refer to a product class). This perspective certainly points away from a per se rule that significant policing expenditures alone can preserve the trademark status of a term. And, of course, to the extent that firms attempt to police the noncommercial use of marks, there may be a significant effect on the free flow of information and ideas. The First Amendment, discussed below in the section on parody, comes into play in these situations.

2. Functionality

Stormy Clime Ltd. v. ProGroup, Inc.
United States Court of Appeals for the Second Circuit
809 F.2d 971 (2d Cir. 1987)

JON O. NEWMAN, Circuit Judge:

This appeal concerns the application of the functionality defense in a trade dress infringement suit under section 43(a) of the Lanham Act, 15 U.S.C. §1125(a) (1982), in which a manufacturer alleges that a competitor improperly copied the overall design of its product. The issue arises on an appeal by defendant-appellant ProGroup, Inc. from an order of the District Court for the Southern District of New York (Louis L. Stanton, Judge), entered at the request of plaintiff-appellee Stormy Clime Ltd., which preliminarily enjoined ProGroup from marketing a rainjacket. Because the appropriate legal standard concerning the functionality defense was incompletely applied, we vacate the preliminary injunction and remand for further proceedings.

Background

Stormy Clime, a New York corporation, designs and sells sportswear. Its products are sold primarily in country club proshops. Stormy Clime began operations in 1982 with the introduction of a waterproof rainjacket which it has sold under its "COOL IT" trademark. The rainjacket is made of a high-sheen, waterproof fabric. Its distinctive feature is the employment of three shingles or vents — horizontal slots covered by a flap — designed to facilitate release of perspiration and body heat while keeping the wearer protected from rain and wind. Stormy Clime's COOL IT rainjacket is normally sold as part of a rainsuit comprising a jacket and matching pants. Revenue from sales of the COOL IT line of sportswear has grown from $400,000 in 1982 to approximately $2.1 million in 1985.

ProGroup markets its line of rainsuits under the trademark "DUCKS-TER". After considering Stormy Clime's COOL IT rainjacket, as well as those of several other manufacturers, ProGroup developed its latest DUCKS-TER rainjacket, which features two horizontal shingles/vents, a high-sheen fabric, and a hood. In designing its jacket, ProGroup was careful not to infringe Stormy Clime's United States Patent No. 4,408,356 for tacking the shingles so as to prevent them from being lifted by the wind. The use of ventilation shingles on rainjackets was the subject of U.S. Patent Nos. 1,562,767 and 2,259,460, both of which have long since expired.

At a Professional Golf Association trade show on January 23, 1986, ProGroup introduced its latest DUCKSTER line of rainsuits. Because of their shingled design, high-sheen fabric, and colors, these DUCKSTER rainjackets closely resembled Stormy Clime's COOL IT rainjackets.

On January 24, 1986, the day after ProGroup introduced its DUCKS-TER rainjackets, Stormy Clime filed suit against ProGroup in the Southern District of New York, alleging trademark infringement and unfair competition under the Lanham Act, 15 U.S.C. §§1114, 1125(a) (1982). On April 28, 1986, following initial discovery, Stormy Clime moved for a preliminary injunction.

The exhibits and documentary evidence introduced in the trial court indicate that both the Stormy Clime and ProGroup rainjackets feature horizontal shingles/vents (three in the COOL IT rainjacket, two in the DUCKS-TER rainjacket), a high-sheen fabric, and a hood. The COOL IT rainjacket comes in fourteen colors; the DUCKSTER rainjacket comes in six of these colors. The shoulders of the COOL IT rainjacket are constructed from five pieces of fabric attached by seams; the DUCKSTER rainjacket features a seamless shoulder construction. The COOL IT rainjacket has top-entry, patch pockets; the DUCKSTER rainjacket uses slant-entry, integral pockets. The COOL IT rainjacket has a stand-up collar with snaps; the DUCKSTER rainjacket has a laydown collar without snaps. The COOL IT rainjacket, regardless of fabric color, has a white front zipper, a white drawstring at the waist, and a white zipper on the collar which closes a pocket housing its hood. The DUCKSTER rainjacket has a front zipper matching the color of the jacket fabric and does not have a drawstring. Although the initial version of the DUCKSTER rainjacket featured a white zipper on the collar of lighter-colored jackets and a black zipper on the collar of darker-colored jackets, the current version has a detachable hood and no zipper on the collar. The COOL IT rainjacket does not have any external label. The DUCKSTER rainjacket has a small metal insignia attached to the sliding element of its front zipper. Both rainjackets have labels clearly displaying their respective trademarks sewn just below the inside of the collar.

Stormy Clime alleged that the design of the DUCKSTER rainjacket infringed the "trade dress" of its COOL IT rainjacket. In addition to disputing that the jackets were confusingly similar, ProGroup defended on the ground that the design was functional and hence not protectable under trade dress law. ProGroup emphasized the need for horizontal shingles, a high-sheen waterproof fabric, and a hood in any high-quality, low-cost, waterproof jacket suitable for use by golfers. ProGroup also noted the several differences between the COOL IT and DUCKSTER rainjackets.

On July 2, 1986, Judge Stanton issued a preliminary injunction barring ProGroup from making, marketing, or selling its shingled rainjacket and ordering ProGroup to inform its distributors and customers of the terms of the injunction. Stormy Clime posted a $50,000 bond. ProGroup appeals the preliminary injunction on the sole ground that Judge Stanton applied an incorrect legal standard in analyzing ProGroup's functionality defense.

Discussion . . .

Section 43(a) of the Lanham Act, 15 U.S.C. §1125(a), which prohibits the marketing of a product conveying a "false designation of origin," has been interpreted to entitle the first manufacturer of a product to an unregistered trademark in the "trade dress" of its product. See LeSportsac, 754 F.2d at 75; Warner Bros. Inc. v. Gay Toys, Inc., 658 F.2d 76, 77-78 (2d Cir. 1981) (Warner I). The trade dress of a product "involves the total image of a product and may include features such as size, shape, color or color combinations, texture, [or] graphics," John H. Harland Co. v. Clarke Checks, Inc., 711 F.2d 966, 980 (11th Cir. 1983). Although "trade dress" has traditionally referred to the packaging or labeling of a product, see, id. at 980; LeSportsac, supra, 754 F.2d at 75, this Circuit has recognized that the design of a product may function as its "packaging," thereby entitling the manufacturer to trade dress protection for the appearance of the product. See id.; Warner I, supra, 658 F.2d at 78; Harlequin Enterprises Ltd. v. Gulf & Western Corp., 644 F.2d 946, 949 (2d Cir. 1981).

The typical case in which a first manufacturer seeks to prevent the copying of the appearance of its product under section 43(a) of the Lanham Act involves two stages of inquiry. In the first stage, the plaintiff must show that the trade dress of its product has acquired secondary meaning in the marketplace and that the design of the competitor's product is confusingly similar to that of the plaintiff's product. See LeSportsac, supra, 754 F.2d at 75 (quoting 20th Century Wear, Inc. v. Sanmark-Stardust Inc., 747 F.2d 81, 92 (2d Cir. 1984), *cert. denied*, 470 U.S. 1052, 84 L. Ed. 2d 818, 105 S. Ct. 1755 (1985)); Warner Bros., Inc. v. Gay Toys, Inc., 724 F.2d 327, 332 (2d Cir. 1983) (Warner II); Vibrant Sales, Inc. v. New Body Boutique, Inc., 652 F.2d 299, 303 (2d Cir. 1981), cert. denied, 455 U.S. 909, 71 L. Ed. 2d 448, 102 S. Ct. 1257, 213 U.S.P.Q. (BNA) 1056 (1982); Litton Systems, Inc. v. Whirlpool Corp., 728 F.2d 1423, 1444-45, 221 U.S.P.Q. (BNA) 97 (Fed. Cir. 1984). Even if a manufacturer makes both of these showings, the competitor can prevail in the second stage of inquiry by showing that the similar arrangement of features is functional. See LeSportsac, supra, 754 F.2d at 75-76 (placing the burden of proof of functionality on the alleged infringer); Warner II, supra, 724 F.2d at 330-32.

. . . [T]he sole issue before us now is whether Judge Stanton erred in finding that Stormy Clime has raised serious questions going to the merits of whether ProGroup's functionality defense was available.

Judge Stanton analyzed the functionality question as follows:

> ProGroup asserts that three of the features constituting the overall "look" of the COOL IT jacket, the shingles/vents, the high-sheen material, and the

hood, are functional, and only one (the distinctive white trim, which ProGroup does not use) is nonfunctional. Thus, it argues that the overall "look" of the COOL IT jacket is not entitled to trademark protection. "But by breaking [Stormy Clime's] trade dress into its individual elements and then attacking certain of those elements as functional, [ProGroup] misconceives the scope of the appropriate inquiry." LeSportsac, 754 F.2d at 76. In determining whether the trade dress or "look" is functional one should consider it overall and as a whole, not break the trade dress into its individual features. If, seen as a whole, the design or trade dress "primarily serves a legitimate trademark purpose — identifying the source of the product — . . . [it is] eligible for protection" even though it might also serve functional purposes. Id. at 78. Here the appearance of the Stormy Clime jacket presents a "particular combination and arrangement of design elements that identify" and distinguish it from other jackets. Cf. id. at 76. The "look" of plaintiff's jacket "distinguishes the garment as being a COOL IT jacket specifically from Stormy Clime." The arrangement of the functional features is original with plaintiff's jacket, not required by the function itself, and thus entitled to protection. See LeSportsac, 754 F.2d at 78. At the very least, Stormy Clime has raised serious questions going to the merits of this issue.

Stormy Clime, Ltd. v. ProGroup, Inc., 230 U.S.P.Q. (BNA) 685, 687 (E.D.N.Y. 1986).

Though the District Court drew useful guidance from *LeSportsac*, the analysis of functionality and the application of the defense in this case was incomplete, leaving substantial doubt whether Stormy Clime's objection to ProGroup's defense presents a fair ground of litigation, even at this preliminary stage of the litigation. In *LeSportsac*, Chief Judge Feinberg began an analysis of functionality by noting that a functional feature " 'is essential to the use or purpose of an article or . . . affects the cost or quality of the article.' " LeSportsac, supra, 754 F.2d at 76 (quoting Inwood Laboratories, Inc. v. Ives Laboratories, Inc., 456 U.S. 844, 850 n.10, 72 L. Ed. 2d 606, 102 S. Ct. 2182 (1982)). The *LeSportsac* opinion goes on to quote Judge Oakes' elaboration of this definition in *Warner II*:

> A design feature of a particular article is "essential" only if the feature is dictated by the functions to be performed; a feature that merely accommodates a useful function is not enough. In re Morton-Norwich Products, Inc., 671 F.2d 1332, 1342, 213 U.S.P.Q. (BNA) 9 (Cust. & Pat. App. 1982) (shape of plastic container for spray products not essential to its purpose as a sprayer). And a design feature "affecting the cost or quality of an article" is one which permits the article to be manufactured at a lower cost, e.g., Kellogg Co. v. National Biscuit Co., 305 U.S. 111, 122, 59 S. Ct. 109, 115, 83 L. Ed. 73 (1938) (pillow shape of shredded wheat biscuit functional as cost would be increased and quality lessened by other form), or one which constitutes an improvement in the operation of the goods, e.g., Fisher Stoves, Inc. v. All Nighter Stove Works, Inc., 626 F.2d 193, 195 (1st Cir. 1980) (two-tier design of woodstove functional because improving the operation of the stove in three respects).

LeSportsac, supra, 754 F.2d at 76 (quoting Warner II, supra, 724 F.2d at 331). Even accepting Stormy Clime's claim to an original arrangement of features — shingles/vents, high-sheen waterproof fabric, and hood — on its rainjacket, we fail to see how this arrangement is not "essential" to the purpose of the product — to keep golfers dry, free to move, and comfortable during rain showers, all at low cost. Judge Stanton's assertion that "the arrangement of functional features [on Stormy Clime's COOL IT rainjacket is] not required by the function itself" is neither explained nor justifiable on the record thus far developed. Furthermore, Judge Stanton's opinion does not reckon with ProGroup's substantial contention that the similarities between the DUCKSTER and COOL IT rainjackets reflect cost and quality considerations.

Moreover, the facts in *LeSportsac* provide a contrast with the instant case that points to the availability of the functionality defense. In *LeSportsac*, the plaintiff LeSportsac Inc. began selling a distinctive line of lightweight luggage in 1976. The luggage was made from parachute nylon and trimmed with color-coordinated cotton-webbing straps and color-coordinated zippers with hollow rectangular pulls. The LeSportsac luggage was marked by repetitive printing of the LeSportsac logo on the cotton tape. See LeSportsac, supra, 754 F.2d at 74. In 1984, K mart Corporation began selling a competing line of luggage that resembled the LeSportsac line in substantially every respect including the repetitive logo; K mart used the trademark "di paris sac." Unlike the present case, the District Court in *LeSportsac* made specific findings regarding the functionality of the arrangement of features comprising the plaintiff's product. . . . LeSportsac, Inc. v. K Mart Corp., 607 F. Supp. 183, 185 (E.D.N.Y. 1984). Since extensive evidence established that the arrangement of the features of the LeSportsac luggage was ornamental and nonfunctional, the Court of Appeals could readily rely upon the District Court's analysis and conclude that the District Court acted well within its discretion in issuing a preliminary injunction. See LeSportsac, supra, 754 F.2d at 76-77.

By contrast, the arrangement of the principal features common to both the COOL IT and DUCKSTER rainjackets — shingles/vents, high-sheen waterproof fabric, and hood — appear to be dictated by the purpose of providing a low-cost, [un]encumbering, waterproof jacket for wear while playing golf and other sports. Though the rainjackets are similar in appearance, their purely ornamental features and labeling differ in almost every respect other than color, and Stormy Clime, having elected to manufacture its product in fourteen different colors, is not in a strong position to complain that a competitor's product uses a few of those same colors.

Furthermore, analysis of the functionality defense requires consideration of additional factors, especially the purpose of the functionality defense. Judge Oakes' opinion in *Warner II* points out that the important purpose of the functionality defense is "to protect advances in functional design from being monopolized [so as] to encourage competition and the broadest dissemination of useful design features." Warner II, supra, 724 F.2d at 331. Balancing this purpose with the Lanham Act's purpose of preventing confusion as to the source of products, we believe that the functionality inquiry in the present case should have focused on whether bestowing trade dress protection upon Stormy Clime's arrangement of features " 'will hinder competition or impinge upon the rights of others to compete effectively in the sale of goods.' " Sicilia di R. Biebow & Co. v. Cox, 732 F.2d 417, 429 (5th Cir. 1984) (quoting In re Morton-Norwich Products, Inc., 671 F.2d 1332, 1342, 213 U.S.P.Q. (BNA) 9 (Cust. & Pat. App. 1982)). See Sunbeam Corp. v. Equity Industries Corp., 635 F. Supp. 625, 735-37 (E.D. Va. 1986) (finding that "gumball dispenser configuration" for food processor is functional and hence not protectable under section 43(a) of the Lanham Act); see also LeSportsac, supra, 754 F.2d at 77 (noting that "K mart's ability to compete is not unduly hindered by the determination that LeSportsac's particular configuration of design features is nonfunctional . . .").

We do not mean to suggest that the functionality inquiry is equivalent to the "important ingredient in commercial success" test applied in Industria Arredamenti Fratelli Saporiti v. Charles Craig, Ltd. 725 F.2d 18, 19 (2d Cir. 1984), and limited in LeSportsac, supra, 754 F.2d at 77-78, but cf. Prufrock Ltd., Inc. v. Lasater, 781 F.2d 129, 133 (8th Cir. 1986). As we explain below, a critical aspect in considering hindrance to competition is whether bestowing trade dress protection on a product design prevents potential competitors from entering a market that is not foreclosed by a valid patent. Thus, a distinctive design or arrangement of features that is an important ingredient in the commercial success of a product but is not "essential to the use or purpose" of the product and does not "affect[ ] the cost or quality" of the product could be protectable trade dress. Inwood Laboratories, Inc. v. Ives Laboratories, Inc., 456 U.S. 844, 850 n.10, 72 L. Ed. 2d 606, 102 S. Ct. 2182 (1982).

In conducting its inquiry, the District Court should assess the degree of functionality of the similar features, see In re Morton-Norwich Products, Inc., supra, 671 F.2d 1332, 1340-41, the degree of similarity between the non-functional (ornamental) features of the competing products, see LeSportsac, supra, 754 F.2d at 74, 77 (noting that K mart emulated LeSportsac's logo); cf. Litton Systems, Inc. v. Whirlpool Corp., supra, 728 F.2d at 1446 (discussing labeling), and the feasibility of alternative arrangements of functional features that would not impair the utility of the product, see Sicilia di R. Biebow & Co. v. Cox, supra, 732 F.2d at 429; Sunbeam Corp. v. Equity Industries Corp., supra, 635 F. Supp. at 636-37; see also Sno-Wizard Mfg., Inc. v. Eisemann Products Co., 791 F.2d 423, 426 n.3 (5th Cir. 1986); Prufrock Ltd., Inc. v. Lasater, 781 F.2d 129, 132-33 (8th Cir. 1986). These factors should be considered along a continuum. See 3 R. Callmann The Law of Unfair Competition, Trademarks and Monopolies §19.33 at 132 (L. Altman 4th ed. 1983) ("Functionality is often a matter of degree, rather than a binary yes-or-no matter."). On one end, unique arrangements of purely functional features constitute a functional design. On the other end, distinctive and arbitrary arrangements of predominantly ornamental features that do not hinder potential competitors from entering the same market with differently dressed versions of the product are non-functional and hence eligible for trade dress protection. In between, the case for protection weakens the more clearly the arrangement of allegedly distinctive features serves the purpose of the product (including maintenance of low cost), especially where the competitor copying such features has taken some significant steps to differentiate its product. In this case, for example, ProGroup places its DUCKSTER trademark prominently on a label at the inside of the neck, a traditional means of identifying the manufacturer of clothing. The trademark is also displayed on the zipper pull, on a hang-tab, and on the plastic bag in which the jacket is packaged.

Courts must proceed with caution in assessing claims to unregistered trademark protection in the design of products so as not to undermine the objectives of the patent laws. Patents for inventions and designs, see 35 U.S.C. §§101, 171 (1982), are the principal means by which we protect intellectual property embodied in products. By bestowing limited periods of protection to novel, non-obvious, and useful inventions and new, original, and ornamental designs — seventeen years for invention patents, 35 U.S.C. §154 (1982), and fourteen years for design patents, 35 U.S.C. §173 (1982) — the patent laws encourage progress in science and the useful arts. See U.S. Const. art. 1 §8, cl. 8. Society reaps the rewards of these advances in the short term to the extent that patent holders and their licensees incorporate protected ideas into new and useful products. These rewards are more fully realized in the longer term because novel ideas fall into the public domain upon the expiration of patent protection. Since trademark protection extends for an unlimited period, expansive trade dress protection for the design of products would prevent some functional products from enriching the public domain.

This threat is particularly great when, as in the instant case, a first manufacturer seeks broad trade dress protection for a product on the ground that its arrangement of predominantly functional features is distinctive. Even if Stormy Clime were to have a patent on the use of horizontal shingled vents in rainjackets, it is unlikely that it would have as much monopoly power as an unregistered trademark in the shingled look would provide. . . .

To avoid undermining the purpose of the patent laws to place useful innovations in the public domain after expiration of a limited monopoly, courts must be sensitive to whether a grant of trade dress protection would close all avenues to a market that is otherwise open in the absence of a valid patent. See Prufrock Ltd., Inc. v. Lasater, supra, 781 F.2d at 132-33; Sicilia di R. Biebow & Co. v. Cox, supra, 732 F.2d at 428; Sunbeam Corp. v. Equity Industries Corp., supra, 635 F. Supp. at 636; cf. In re Deister Concentrator Co., Inc., 48 C.C.P.A. 952, 289 F.2d 496, 504, 129 U.S.P.Q. (BNA) 314 (Cust. & Pat. App. 1961) (noting that the reason for refusing to protect functional elements against copying is not that they "cannot or do not indicate source to the purchasing public but that there is an overriding public policy of preventing their monopolization, of preserving the public right to copy"). In the instant case, the District Court does not appear to have considered whether any feasible shingled rainjacket could be manufactured that would not appear similar in key respects to Stormy Clime's COOL IT rainjacket. Cf. LeSportsac, supra, 754 F.2d at 77 (noting that "for example, the cotton carpet tape and carrying straps could be placed differently, contrasted in color with the bag or be made thicker or thinner; zipper pulls could be solid or nonrectangular; the repeating elliptical logo could be changed or placed differently").

By focusing upon hindrances to legitimate competition, the functionality test, carefully applied, can accommodate consumers' somewhat conflicting interests in being assured enough product differentiation to avoid confusion as to source and in being afforded the benefits of competition among producers. Cf. Sicilia de R. Biebow & Co. v. Cox, supra, 732 F.2d at 429-30. Moreover, this test encourages those seeking trade dress protection for their products to select distinctive non-functional identifying marks.

Conclusion

In light of the foregoing, we have substantial doubts whether this record permitted issuance of a preliminary injunction. In any event, we are convinced that the District Court did not apply the governing legal standard with sufficient completeness to demonstrate that preliminary relief was warranted. Consequently, we vacate the preliminary injunction and remand for further proceedings consistent with this opinion. Though the propriety of a preliminary injunction might yet be demonstrated, we strongly urge the parties to proceed directly to the merits so that, on a fully developed record, the District Court can determine whether any relief is warranted.

Preliminary injunction vacated; cause remanded for further proceedings.

COMMENTS AND QUESTIONS

1. Judge Newman suggests that a design or arrangement that is "an important ingredient in the commercial success of a product," but is "not essential" and "does not affect the cost or quality of the product" is not functional. How can such a distinction be drawn? As Landes and Posner suggest, in some sense every worthwhile design feature affects the "cost or quality" of the product. See William Landes and Richard Posner, Trademark Law: An Economic Perspective, 30 J. Law and Econ. 265, 297 (1987). But if this is true, what is left of trademark law? Should all design attributes that succeed in differentiating products therefore be unprotectable?

Furthermore, what should courts do about designs that are partially distinctive and partially functional (a category that includes most, if not all, designs)? See the *Vornado* case following for a discussion of this issue. Should there be a threshold (say, 80 percent functional) below which defendants are free to copy the design? Or must copiers attempt to parse the functional from the nonfunctional? Compare in this regard the copyright rules, which allow the copying of ideas but not expression within a single work. Suppose the two cannot be separated? Ought there to be a "utilitarian function" exception in trademark law, along the lines of the *Carol Barnhart* case in copyright (see Chapter 4)? Cf. Sunbeam Prods. v. West Bend Co., 123 F.3d 246, 257 (5th Cir. 1997) (configuration of mixer was protectable, despite the fact that various aspects were functional, because the "total image" and "overall appearance" were not dictated by function).

One might well wonder after these thought experiments if it is really worthwhile to regulate this aspect of competition at all. Are designs so expensive that they need a special incentive system to call forth the optimal number? Or do design and manufacturing processes provide enough built-in "lead time" to compensate the creators of innovative designs? The prevalence of piracy in many parts of the world suggests that the answer to the last question may be no. See J. H. Reichman, Legal Hybrids Between the Patent and Copyright Paradigms, 94 Colum. L. Rev. 2432 (1994) (arguing that natural lead time has diminished in recent years). Even so, are these the sorts of concerns trademark law should address? Are they more appropriately considered under patent law or some system of design protection?

2. In 1998, Congress codified functionality as a valid reason to refuse registration, as well as a ground for opposition and cancellation. Functional marks may appear on the supplemental register. See Pub. L. 105-330, Title II, §201(a)(4), Title III, §301, 112 Stat. 3070, codified at 15 U.S.C. §§1052, 1064, 1091 (1998).

3. Although the product features in *Stormy Clime* can be described as utilitarian, another line of cases deals with a slippery issue that runs under the banner of "aesthetic functionality." This phrase is sometimes used to describe product features that, while not adopted for utilitarian reasons, are also not adopted specifically to assist buyers in identifying a product's origins — i.e., they are adopted for "non-trademark" purposes. See, e.g., Wallace International Silversmiths, Inc. v. Godinger Silver Art Co., Inc., 916 F.2d 76 (2d Cir. 1990), *cert. denied*, 499 U.S. 976 (1991). In *Wallace*, the court accepted a functionality defense in a case where the disputed design feature was admittedly aesthetic: defendant had copied certain carved features from plaintiff's baroque style silverware. See also American Greetings Corp. v. Dan-Dee Imports, Inc., 807 F.2d 1136 (3d Cir. 1986) (upholding defendant's use of printed words on tummy of teddy bears as functional feature of product); Publications Intl. v. Landoll, 49 U.S.P.Q.2d 1139 (7th Cir. 1998) (book design, including size, cover style, and gilding of page edges, was aesthetically functional because "color plays an important role (unrelated to source identification) in making a product more desirable.").

In Qualitex Co. v. Jacobsen Prods. Co., 115 S. Ct. 1300, 34 U.S.P.Q.2d 1161 (1995), the Supreme Court made some important statements explaining the gist of the contemporary test of functionality, and in so doing fashioned a unifying rubric to embrace the disparate strains of functionality doctrine, including "aesthetic functionality":

> . . . The Restatement (Third) of Unfair Competition adds that, if a design's "aesthetic value" lies in its ability to "confe[r] a significant benefit that cannot practically be duplicated by the use of alternative designs," then the design is "functional." Restatement (Third) of Unfair Competition §17, Comment c, pp. 175-176 (1995). The "ultimate test of aesthetic functionality," it explains, "is whether the recognition of trademark rights would significantly hinder competition." Id., at 176.
>
> This Court . . . has explained that, "[i]n general terms, a product feature is functional," and cannot serve as a trademark, "if it is essential to the use or purpose of the article or if it affects the cost or quality of the article," that is, *if exclusive use of the feature would put competitors at a significant non-reputation-related disadvantage.*

Id. at 1306 (emphasis added). Does this statement differ from Judge Newman's phrasing of the test in *Stormy Clime*, supra? How difficult will it be in practice to determine whether a given product feature is desirable because it conveys reputation, or because it serves some "non-reputation-related" purpose? What if a product feature does both? On this point, see this statement from footnote 19 of the *Vornado* opinion, infra:

> Although a producer may find efficiencies in combining the brand-identifying function with a product's utilitarian function by using a useful product feature as a trademark or trade dress, we accord this type of efficiency little weight. Although the efficient combining of form and function is at the heart of good industrial design, promoting it is not a Lanham Act objective.

For a recent case applying aesthetic functionality to reject trade dress protection, see Herman Miller Inc. v. Palazzetti Imports & Exports, 998 F. Supp. 757 (E.D. Mich. 1998). For a good analysis of the aesthetic functionality doctrine in light of trademark theory, see Mitchell M. Wong, Note, The Aesthetic Functionality Doctrine and the Law of Trade-Dress Protection, 83 Cornell L. Rev. 1116 (1998).

4. While most functionality cases involve trade dress, the functionality doctrine also extends to wordmarks. The precise application of the doctrine to section 32 cases was debated in Inwood Laboratories, Inc. v. Ives Laboratories, Inc., 456 U.S. 844 (1982). An interesting recent case, Kohler Co. v. Moen, Inc., 12 F.3d 632 (7th Cir. 1993), upheld a trademark registration for a product design configuration — the shape of a faucet and handle. The Seventh Circuit in *Kohler* relied in part on the legislative history of the Trademark Law Reform Act of 1988. The Senate Report on that act interpreted the terms "symbol or device" as permitting the registration of "colors, shapes, sounds or configurations where they function as trademarks." S. Rep. 100-515, 100th Cong., 2d Sess. at 44, 1988 U.S.C.C.A.N. at 5607.

The *Kohler* court acknowledged that granting trademark protection to product design configurations created an overlap with design patents, which are also available to protect such configurations. The court rejected the arguments that the patent laws impliedly preempt overlapping trademark protection, and that such preemption was necessary to prevent the grant of a perpetual trademark monopoly in place of the limited 14-year design patent monopoly. The court relied on the "fundamentally different" legal protection offered by trademarks and design patents. Accord In re Yardley, 493 F.2d 1389 (C.C.P.A. 1974).

Can you think of situations in which names, words, or other forms of traditional registered trademarks might be functional? How would a functional name differ from a generic name?

5. Consider the following argument, from William M. Landes & Richard A. Posner, Trademark Law: An Economic Perspective, 30 J.L. & Econ. 265 (1987):

> The concept of functionality, which is mainly important in connection with design features used as trademarks, is a parallel concept to genericness. A functional feature cannot be trademarked, and a trademarked feature loses trademark protection when it becomes functional. The maker of a tire could not trademark its circular shape but could trademark an irregularly shaped hubcap. The maker of a steak knife could not trademark the serrated blade, but could trademark an intricate arabesque carved into the handle. A particular shape for a container might initially be subject to being trademarked, but if technological developments made it much cheaper to manufacture than alternative shapes, it would lose its trademark protection.

Does it make sense for a product feature to be (legally) nonfunctional at time 1 and then become functional later? Shouldn't the intent of the product creator come into play? That is, if a design feature is adopted with the intent to attract consumer associations regarding source, why should the designer be punished when it later turns out to have a cost advantage?

Restatement (Third) of Unfair Competition, §17.
Functional Designs

A design is "functional" for purposes of the rule stated in §16 if the design affords benefits in the manufacturing, marketing, or use of the goods or services with which the design is used, apart from any benefits attributable to the design's significance as an indication of source, that are important to effective competition by others and that are not practically available through the use of alternative designs.

Comment:

a. . . . [I]n determining whether a particular design is "functional" and therefore ineligible for protection as a trademark, the ultimate inquiry is whether a prohibition against copying will significantly hinder competition by others. . . .

b. Functionality. A packaging or product feature is not functional merely because the feature serves a utilitarian purpose. The recognition of trademark rights is precluded only when the particular design affords benefits that are not practically available through alternative designs. A bottle, for example, is a utilitarian element of the packaging for wine, but the design of a particular wine bottle is not functional under the rule of this Section unless the shape or other aspects of the bottle provide significant benefits that are not practically obtainable through the use of alternative bottle designs.

In general, a functional design is one that is costly to do without. Thus, the benefits afforded by a particular design do not themselves determine whether that design is functional; a design is functional only if those benefits cannot practically be duplicated through the use of other designs. The availability of alternative designs that satisfy the utilitarian requirements or that otherwise afford similar advantages is therefore decisive in determining functionality. If a particular design affords benefits that are superior to those of any practical alternative design, it is functional if the benefits are important to effective competition. . . .

The benefit conferred by a particular packaging or product design may take various forms. The benefit may lie, for example, in greater economy in manufacturing, shipping, or handling, in increased utility or durability, or in enhanced effectiveness or ease of use. If the benefit afforded by the design resides solely in its association with a particular source, however, the design is not functional; the extent to which others may use such a design is determined by the rules of this Chapter governing trademark infringement. See §20.

Any evidence that relates to the advantages inherent in a particular design is relevant to the issue of functionality. Functionality may thus be inferred from evidence of advertising or other promotional efforts by the plaintiff that emphasize the benefits of the design. Benefits claimed in pursuit of a utility patent may be particularly persuasive evidence of functionality. Discussions of the design in textbooks or trade literature and engineering analysis of its utilitarian aspects are also relevant. The fact that other manufacturers familiar with the design have rejected it in favor of other designs is evidence that the design is not functional. . . .

A design that is nonfunctional may be protected as a trademark upon proof that it is distinctive under the rule stated in §13. Possible limitations on the protection of product features under state law are discussed in §16, Comment c.

COMMENTS AND QUESTIONS

1. The Restatement, like the cases we have looked at, defines functionality in terms of nonsource (or reputation)-related benefits that are important to consumers. Since we are dealing here with product attributes, however, this test requires courts to make a very difficult inquiry: what attributes of a product does a consumer value, and how much of the total value of the overall product is attributable to the attribute in question? It has been argued that deconstructing consumer utility functions to determine the value of individual attributes — as well as nonproduct attributes such as distribution, service, and advertising — is a very difficult task, and one that courts should avoid in favor of other, more tractable types of evidence.

Note on Policing Backdoor Patents: Functionality and Patent Preemption

Stormy Clime is representative of an increasing number of cases wherein product sellers seek multiple forms of legal protection. To some extent, the law here suffers from a lack of coordination. While there may be the semblance of a coherent policy in the *internal* doctrinal structure of each area of intellectual property law, it is difficult for courts to determine the *aggregate* degree of protection that will be afforded a product across all branches of intellectual property law. Cf. Theodore H. Davis, Jr., Copying in the Shadow of the Constitution: The Rational Limits of Trade Dress Protection, 80 Minn. L. Rev. 595 (1996).

Courts are aware of the possibility that multiple forms of protection will lead to excessive costs from society's point of view. See generally Kellogg Co. v. National Biscuit Co., 305 U.S. 111, 122 (1938) (holding that lapse of patent on shredded wheat cereal, together with lax policing of the mark and widespread use, obviate trademark status of term "Shredded Wheat"). Generally, however, opinions in this sensitive and developing legal area dwell not on issues of aggregate protection but instead on the problem of preserving the exclusive domain of each branch of protection. Primarily, this takes the form of policing against "backdoor patents."

As we saw in Chapter 4, copyright law also confronts the problem of crafting doctrine so as to limit protection of product configurations. The "useful article" doctrine is most often invoked to hold the line against excessive copyright expansion along this dimension. And, as *Baker v. Selden* (Chapter 4) makes so clear, the goal is to prevent the worst of both worlds scenario: patent-like protection over an item extending for the full term of copyright and without any of the strict requirements for patent protection. The social welfare loss of inadvertent "life-plus-fifty-year patents," it is assumed, would be substantial indeed.

But if copyright in this connection sets off alarm bells, current trade dress cases call for screaming sirens and searchlights. The reason is that, unlike copyright, trademark (and trade dress) law has no finite term. The upshot is that, if product configuration protection proceeds unchecked, we will find a myriad of perpetual quasi-property rights protecting products in the marketplace. Seen in this light, the stakes are high indeed when functionality issues arise in trade dress cases.

Courts have struggled to deal with this overlap. In two recent cases, Duraco Prods., Inc. v. Joy Plastic Enterprises, Ltd., 40 F.3d 1431 (3d Cir. 1994), and Versa Prods. Co., Inc. v. BiFold Co. (Mfg.) Ltd., 50 F.3d 189 (3d Cir. 1995), for example, Judge Becker of the Third Circuit has shown an acute awareness of the risks of extending trade dress protection to product configurations. See generally Todd R. Geremia, Protecting the Right to Copy: Trade Dress Claims for Configurations in Expired Utility Patents, 92 Nw. U. L. Rev. 779 (1998) (calling for "bright line" rule prohibiting trade dress protection for any design disclosure in an expired utility patent). Consider as well two other cases at the cutting edge of the patent–trade dress interface.

Vornado Air Circulation Systems, Inc. v. Duracraft Corp., 58 F.3d 1498 (10th Cir. 1995), *cert. denied*, 116 S. Ct. 753 (1996). In this case, the court found that trade dress protection for a nonfunctional product feature was preempted by federal patent law where the product feature was a "significant inventive component" of the plaintiff's patented design.

The patent at issue in the *Vornado* case is U.S. Patent 4,927,324, Re. 34,551, issued May 20, 1990, reissued Feb. 22, 1994, entitled "Ducted Fan." Claim 1 includes the following elements (paraphrased in brackets), including the grill element:

1. A ducted fan comprising:

   [a base with motor;]

   [a funnel-shaped duct and outer cowling;]

   [a tapered, inner cowling inside the outer cowling;]

   a circular grill having an outer radius attached to the discharge end of the inner cowling, the grill including a center hub and a series of arcuate shaped ribs extending outwardly from the hub to said outer radius, each rib having a constant curvature radius and each rib being equally spaced from each other around the hub, the maximum lateral spacing between the ribs is inboard from said outer radius;

   [and a support to hold up the fan].

In addition, the specification describes how the grill functions with the other elements of the invention to produce the superior airflow touted in the patent. (See, e.g., specification p. 6: "The longer length curved ribs [on the grill] as compared with a conventional straight rib, provides a less rigid grill structure which can be desirable under certain circumstances such as impact shocks.")

The fact that the claims include a grill element and that the grill is described at length in the specification led the *Vornado* court to conclude that the grill was "a significant inventive aspect" of the product — and hence that it could not be the subject of trade dress protection. 35 U.S.P.Q.2d 1332, 1342. In summarizing its holding in the case, the court states that "where a disputed product configuration is part of a claim in a utility patent," trade dress protection will be unavailable. Id.

Since the parties stipulated there was no patent infringement, and defendant's fan included a copy of plaintiff's grill design, then Vornado's *precise* grill design must *not* itself be patented. The court therefore affirmed the district court's finding that the fan design is nonfunctional. Does this follow? On the contrary, there would appear to be much unclaimed but disclosed subject matter in patents that is quite functional indeed.

Thomas & Betts Corp. v. Panduit Corp., 65 F.3d 654 (7th Cir. 1995), *cert. denied,* 116 S. Ct. 1044 (1996). A case decided at about the same time as *Vornado* expresses the same concerns. In this case both parties made cable ties, thin plastic straps used to bundle together electrical wires, under the dashboard of a car, for instance. Plaintiff Thomas & Betts, or "T & B", sued competitor Panduit for selling cable ties with similar shaped "heads," the part of the strap into which the other end is fed and pulled tight — somewhat like a miniature belt buckle. In 1965, T & B had obtained a patent on the two-piece cable tie ("the Schwester patent"). Though the Schwester patent expired in 1982 and a related patent also held by T & B expired in 1986, until 1993 T & B remained the sole producer of two-piece cable ties, with annual sales of almost $100 million worldwide.

Defendant Panduit argued that, as in *Vornado*, the fact that the cable tie heads were described in the Schwester patent meant that they could not be protected under trade dress law. The court went on to consider *Vornado* at some length. The court held that T & B's cable ties were not protectable trade dress because they had no secondary meaning to likely purchasers. In the course of its opinion, however, the court commented at length on the *Vornado* opinion:

> The present case is distinguishable from *Vornado* but it is not clear that the differences dictate a different result. The oval shape of the cable tie head was not specifically claimed in the Schwester patent, a point that T & B makes much of, while the spiral grill was a required element in at least one of the claims in the patent at issue in *Vornado*. Therefore, one could infringe T & B's patent without infringing its trade dress while any product which infringed the Vornado patent (at least the claim in which the spiral grill was an element) would also infringe its claimed trade dress.
>
> Whether or not a feature is claimed is not necessarily, however, a good indicator of its relative importance to the invention as a whole. Additional elements in a claim narrow the scope of the patentee's monopoly since an infringing invention must contain each element of a claim or its equivalent. . . . Therefore, a patent applicant seeks to draw his claims as broadly as possible with as few

specific elements as he can get away with. The fact that Vornado included the spiral grill as an element of a claim likely means that it could not have received a patent on the rest of the invention without including the spiral grill limitation — not that the spiral grill itself was a "significant inventive aspect." This is particularly true in *Vornado* where the spiral grill itself was not independently claimed (that is why Duracraft's fan did not infringe the still valid patent) because it was disclosed in the prior art.

Likewise, the fact that the Schwester patent includes the cable tie head as a required element without claiming a particular shape means that T & B was able to obtain a broader monopoly, i.e., they were able to exclude competitors from making metal barbed cable ties with any head shape. There is no reason, therefore, to infer from its absence as an element in a claim that the shape of the head is any less important to the two-piece cable tie than the spiral grill is to Vornado's fan. All one can infer is that because it was not separately claimed, the shape of the head, like the spiral grill, was not independently patentable.

The focus on what is or is not claimed is misguided for another more fundamental reason. In the patent "bargain," the claims define what the patentee receives, the "metes and bounds" from which he can exclude competitors. What the public receives is the entire invention as disclosed in the claims but primarily in the patent specifications which are required to "contain a written description of the invention, and of the manner and process of making and using it, in such full, clear, concise, and exact terms as to enable any person skilled in the art to which it pertains, or with which it is most nearly connected, to make and use the same, and shall set forth the best mode contemplated by the inventor of carrying out his invention." 35 U.S.C. §112.

If one purpose of the functionality defense is to guard against "backdoor patents," then perhaps functionality should be seen as related to the doctrine of federal-state preemption. Cases such as *Vornado* make clear that, in fact, the same concerns that motivate preemption of state statutes apply in some cases *beyond* the limits of trademark's functionality doctrine. Both traditional functionality cases, and "nonfunctional preemption" cases such as *Vornado* are motivated by the policy against backdoor patents, which in this case operates as a sort of federal-federal preemption.

For a discussion of federal preemption of state intellectual property statutes, see Chapter 6, section B.

PROBLEMS

Problem 5-9. Ferrari is a world-famous maker of upscale sports cars. It limits the number of cars it produces, and each car costs approximately $200,000. Ferrari's cars have the same general features as normal cars — wheels, chassis, etc. — but they also have a distinctive look that is easily recognized. They are sleek and low to the ground, a fact that may make them accelerate more quickly and that makes them more attractive to look at.

Roberts sells a fiberglass kit that replicates the exterior features of a Ferrari, though not the engine or performance. When sued for trade dress infringement, Roberts defends on the grounds that the Ferrari design is functional. Is it?

Problem 5-10. Eighteen years ago, Spartan Laboratories invented and patented a new pain-relieving drug called asperol. During the term of the patent, Spartan retained the exclusive right to sell asperol, which it manufactured in bright orange capsules. After the Spartan patent expired, a number of other companies began making generic asperol. Each of these companies sells the generic asperol in the same bright orange capsules as Spartan. Although the orange capsules are not visible inside the manufacturers' boxes (which do not resemble each other), asperol is sold only by prescription, and pharmacists invariably remove the drug from its original packaging and repackage it in their own bottles. The result is that the consumer sees only the name "asperol" and the orange capsules, regardless of who makes the drug.

Spartan filed suit under section 43(a) of the Lanham Act, alleging that the other manufacturers had infringed its trade dress protection by coloring the capsules orange. At trial, Spartan proves that the color orange is protectable because it is distinctive and because, over the seventeen years of the patent, pharmacists and customers had come to equate orange with Spartan's asperol. The generic manufacturers defend the suit on the grounds that the color is functional. In support of this claim, they present survey evidence that patients, particularly elderly patients, may become upset if the color of the drug is changed, and may refuse to believe that the drug is in fact asperol. The generic manufacturers present further evidence that pharmacists rely in part on color in making sure that they have packaged and labelled drugs correctly. Is the color orange functional?

2. Abandonment

Major League Baseball Properties, Inc. v. Sed Non Olet Denarius, Ltd.
United States District Court for the Southern District of New York
817 F. Supp. 1103 (S.D.N.Y. 1993)

MOTLEY, District Judge.

I. Introduction

Plaintiffs, Major League Baseball Properties, Inc. ("Properties") and Los Angeles Dodgers, Inc. ("Los Angeles"), allege in their Amended Complaint that the conduct of the . . . defendants [corporate operators of three restaurants, and their principals] (hereinafter collectively "The Brooklyn Dodger") . . . constitute: a) an infringement upon the rights of plaintiffs' trademarks in violation of 15 U.S.C. §§1114 and 1117; b) a wrongful appropriation of plaintiffs' trademarks in violation of 15 U.S.C. §1125; c) a violation of plaintiffs' common law trademark and property rights; d) a violation of plaintiffs' rights under the New York General Business Law §368-d; e) unfair competition; and f) the intentional use by defendants of a counterfeit mark in violation of 15 U.S.C. §1117(b).

Each of these six causes of action is alleged to flow from defendants' use of the words "The Brooklyn Dodger" as the name and servicemark of the restaurants which defendants have operated in Brooklyn, New York, beginning in March 1988. Plaintiffs initially sought permanent injunctive relief, an accounting of profits, the destruction of physical items containing the allegedly infringing marks, monetary damages, and attorneys' fees.

By their Answer and Amended Answer defendants denied any infringement of plaintiffs' alleged right to use a "Brooklyn Dodger" trademark. Defendants also pleaded the defenses of abandonment by plaintiffs of any "Brooklyn Dodgers" mark which plaintiffs may have owned at one time, as well as laches. The abandonment defense was premised upon the plaintiffs' failure to make any commercial or trademark use of the "Brooklyn Dodgers" name for at least 25 years after Los Angeles left Brooklyn in 1958. The laches defense was premised upon the fact that plaintiffs waited for more than a year and a half after learning of defendants' use of the allegedly infringing trademark before advising defendants of any alleged infringement. During this period defendants expended substantial resources and monies in establishing their restaurants in Brooklyn, New York. . . .

Finally, in their Amended Answer, defendants counterclaimed for the cancellation of various trademark registrations for "Brooklyn Dodgers" filed by plaintiffs after defendants' application to register the "Brooklyn Dodger" servicemark was filed on April 28, 1988. . . .

II. Findings of Facts . . .

Plaintiff Los Angeles is a corporation with offices and its principal place of business in Los Angeles, California. It is the owner of the Los Angeles Dodgers, a professional baseball team which, since 1958, has played baseball in Los Angeles, California under the name the "Los Angeles Dodgers." . . . Prior to 1958 the same professional baseball team played baseball in Brooklyn, New York and were known as the "Brooklyn Dodgers" or the "Dodgers."

In 1958, the team moved the site of its home games from Brooklyn to Los Angeles. . . . It pointedly changed its name to Los Angeles Dodgers, Inc. . . .

By agreement with the Major League Clubs, Properties has been granted the exclusive right to market, license, publish, publicize, promote nationally, and protect the trademarks owned by the Major League Clubs, including those owned by the Los Angeles Dodgers. . . .

. . . By 1991, retail sales of licensed Major League Baseball merchandise were in excess of $2 billion. [Plaintiffs estimate that of the total Dodgers merchandising revenue, only $9 million of 1991 sales carried the *Brooklyn* Dodgers name.]

. . . On March 17, 1988 SNOD began doing business as a restaurant under the name "The Brooklyn Dodger Sports Bar and Restaurant." . . .

It was the individual defendants' decision that their restaurants would emphasize the multiple themes of fun, sports and Brooklyn. Their intention was to create a nostalgic setting where Brooklynites could relax and reminisce about times gone by. . . .

They initially chose to name their establishment "Ebbets Field" after the former baseball park located in Brooklyn, New York in which a baseball team known as the "Brooklyn Dodgers" played baseball until October, 1957. . . .

[Defendants rejected the name "Ebbets Field" because of a conflict with a small restaurant elsewhere in New York.]

The defendants knew that the departure of the "Brooklyn Dodgers" in 1958 had been accompanied by monumental hard feelings in the Borough of Brooklyn. In fact the relocation was one of the most notorious abandonments in the history of sports. . . . At the time defendants selected their logo, they were aware that Los Angeles owned federal trademark registrations for the word "Dodgers." . . . However, at no time during their consideration of the "Brooklyn Dodger" name did the individual defendants have any reason to believe that "The Brooklyn Dodger" mark was being used by Los Angeles, and certainly not for restaurant or tavern services. . . . When considering the use of the "Brooklyn Dodger" mark, at no time was there any discussion among the individual defendants and Brian Boyle about trading on the goodwill of Los Angeles in Brooklyn. . . . Indeed, non-party witness Brian Boyle, a life-long Brooklyn resident, testified that, given the acrimonious abandonment of Brooklyn by Los Angeles, the idea of trading on Los Angeles' "goodwill" in Brooklyn is almost "laughable."

C. Defendants' Use of the Trademark

In connection with each of defendants' [three] "The Brooklyn Dodger" restaurants, defendants make and/or made prominent use of the "Dodger" name and the "Brooklyn Dodger" name, with the word "Dodger" in stylized script, in the color blue, and in blue script.

The defendants' composite design mark consisted of three words: "The," "Brooklyn" and "Dodger" are entwined with one another and with an impish character, . . . which was styled after the Charles Dickens' character, the "Artful Dodger" from the novel Oliver Twist, leaning against the "r" in "Dodger." . . . Defendants, however, make significant use of their logo without the cartoon character to promote their business, including on merchandise such as apparel, in advertisements, on their letterhead and as part of their servicemark.

Defendants' logo is similar to Los Angeles' trademarks. The name "Brooklyn Dodger" is similar to the name "Brooklyn Dodgers" as used by plaintiffs. The script used by the defendants in their logo is similar to that used in Los Angeles' trademarks. The color blue used by defendants is similar to the color blue used by and associated with Los Angeles' [team] in Brooklyn. The swash or tail of the word "Dodger" used by defendants is similar to that used in Los Angeles' trademarks in terms of style and length. . . .

In selecting their logo, defendants intentionally sought to reproduce the Brooklyn Dodgers' trademarks. Indeed, the script for the defendants' logo was intentionally chosen by defendants to track the script used by the Brooklyn Dodgers. . . .

D. Plaintiffs' Use of the Trademark . . .

Plaintiffs' use of the "Brooklyn Dodgers" mark was based upon its physical location, until October 1957, in Brooklyn, New York. However, in 1959, Los Angeles made prominent commercial use and reference to their Brooklyn heritage and trademarks in connection with the promotion of Roy Campanella Night, honoring the former Brooklyn Dodgers player and present employee. . . . Los Angeles made prominent use of their trademarks incorporating the word "Brooklyn" at their annual oldtimers games. . . . Oldtimers games are commercial baseball exhibitions at which former players are honored and perform so that older fans can recall the past and younger fans can learn about the history of the Club.

[The court describes extensive licensing and use of the Los Angeles Dodgers trademarks, including for food services and restaurants.]

While plaintiffs have from time to time made use of their former "Brooklyn Dodgers" mark occasionally and sporadically for historical retrospective[s] such as "Old Timer's Day" festivities, the documentary proof establishes that, following its departure from Brooklyn, Los Angeles' earliest licensing of the "Brooklyn Dodgers" mark occurred on April 6, 1981.

Between 1981 and March 17, 1988, the date of defendants' first use of their mark for restaurant and tavern services, plaintiffs used their "Brooklyn Dodgers" mark for a variety of purposes. Those uses were almost exclusively in the context of T-shirts, jackets, sportswear, sports paraphernalia and on various types of novelty items (i.e. on drinking mugs, cigarette lighters, pens, Christmas tree ornaments, wristbands, etc.). . . . However, none of these uses competes with defendants' use of the mark for restaurant and tavern services.

[The court describes several instances between 1981 and 1988 where businesses sought and received permission to use the "Brooklyn Dodgers" name and logo, including several that were restaurant-related.] . . .

III. Conclusions of Law

[The court concluded that plaintiffs had failed to prove a likelihood of confusion; for completeness, it then considered defenses.]

C. Defendants' Affirmative Defenses

1. Abandonment

Under the Lanham Act a federally registered trademark is considered abandoned if its "use has been discontinued with intent not to resume." Cerveceria Centroamericana, S.A. v. Cerveceria India, Inc., 892 F.2d 1021, 1023 (Fed. Cir. 1989) (quoting Lanham Act, 15 U.S.C. §1127 (1988)).

Abandonment is defined in the Lanham Act: "A mark shall be deemed 'abandoned' — (a) When its use has been discontinued with intent not to resume. Intent not to resume may be inferred from circumstances." 15 U.S.C. §1127 (1988).

. . . The burden of proving abandonment falls upon the party seeking cancellation of a registered mark because a certificate of registration is " 'prima facie evidence of the validity of the registration' and continued use." Cerveceria India, 892 F.2d at 1023 . . . ; Exxon Corp. v. Humble Exploration Co., 695 F.2d 96, 99 (5th Cir.), *reh'g denied*, 701 F.2d 173 (5th Cir. 1983). . . . The party seeking cancellation must establish abandonment by a preponderance of the evidence. See Cerveceria India, 892 F.2d at 1023-24.

The Lanham Act provides that "[n]onuse for two consecutive years shall be prima facie abandonment." 15 U.S.C. §1127. . . . Prima facie abandonment establishes a rebuttable presumption of abandonment. . . . The evidence presented at trial established that between 1958 and 1981 plaintiffs made no commercial trademark use whatsoever of any "Brooklyn Dodgers" mark. Minor changes in a trademark that do not affect the overall commercial impression of the mark do not constitute abandonment. . . . However, Los Angeles' change, in this case, from "Brooklyn Dodgers" to "Los Angeles Dodgers" was not minor; it involved an essential element affecting the public's perception of the mark and the team.[19] . . .

---

19. While Los Angeles ceased commercial use of the trademark because of its relocation from Brooklyn to Los Angeles, its motive, justifiable or not, is irrelevant. *Stetson*, 955 F.2d at 851. Nevertheless, Los Angeles' nonuse in this case was voluntary. Courts have been more reluctant to find an absence of intent to resume where the trademark owner had an excusable reason for nonuse — that is, where nonuse was involuntary. See, e.g., Defiance Button Machine Co. v. C & C Metal Products Corp., 759 F.2d 1053, 1059 (2d Cir.), *cert. denied*, 474 U.S. 844, 106 S. Ct. 131, 88 L. Ed. 2d 108 (1985) (no abandonment where cessation of business was involuntary); American International Group, Inc. v. American International Airways, Inc., 726 F. Supp. 1470 (E.D. Pa. 1989) (where airline declared bankruptcy, remaining goodwill and lack of intent to abandon precluded finding abandonment).

Plaintiffs argue that their "Dodgers" mark without a geographical reference — that is, "Dodgers" alone — is a protected use infringed by defendants actions. However, in this context, "Brooklyn" is more than a geographic designation or appendage to the word "Dodgers." The "Brooklyn Dodgers" was a non-transportable cultural institution separate from the "Los Angeles Dodgers" or the "Dodgers" who play in Los Angeles. It is not simply the "Dodgers," (and certainly not the "Los Angeles Dodgers"), that defendants seek to invoke in their restaurant; rather defendants specifically seek to recall the nostalgia of the cultural institution that was the "Brooklyn Dodgers." It was the "Brooklyn Dodgers" name that had acquired secondary meaning in New York in the early part of this century, prior to 1958. It was that cultural institution that Los Angeles abandoned.

. . . In this case, in order to maintain use of the mark, Los Angeles would have had to continue to use "Brooklyn Dodgers" as the name of its baseball team. Only in this way would the public continue to identify the name with the team. Defiance Button Mach. Co. v. C & C Metal Products Corp., 759 F.2d 1053, 1059 (2d Cir.), *cert. denied*, 474 U.S. 844 . . . (1985). . . . [20]

Rather than using the "Brooklyn Dodgers" mark in the ordinary course of trade, a more accurate description of Los Angeles' use of the mark, at least between 1958 and 1981, was given by its General Counsel in a 1985 letter to someone seeking to use it on a novelty item:

> Since the Dodgers moved to Los Angeles in 1958 the name "Brooklyn Dodgers" has been reserved strictly for use in conjunction with items of historical interest.

. . . Under the law, such warehousing is not permitted. . . . Rights in a trademark are lost when trademarks are "warehoused" as plaintiffs attempted to "warehouse" the "Brooklyn Dodgers" mark for more than two (2) decades. . . . Plaintiffs' failure to use the "Brooklyn Dodgers" trademark between 1958, when Los Angeles left Brooklyn, and 1981 constitutes abandonment of the trademark. . . .

---

20. The outcome in *Defiance*, finding no abandonment, is distinguishable from the case at hand because the court in *Defiance* held that the continued goodwill toward the button company after it stopped producing goods and the fact that the company intended to retain its trademark for some commercial use precluded a finding of abandonment. Similarly, the court in Schenley Industries, 441 F.2d 675 held that continued goodwill and lack of intent to abandon precluded finding of abandonment. In the unique facts of this case, however, plaintiffs have not succeeded in demonstrating that much goodwill in Brooklyn survived Los Angeles' move in 1957. But more importantly, while plaintiff in *Defiance* was at least able to demonstrate an intent not to abandon, plaintiffs here have not even demonstrated an intent not to abandon, much less the statutory requirement of intent to resume. A mark retains "residual" goodwill "if the proponent of a mark stops using it but demonstrates an intent to keep the mark alive for use in resumed business." Pan American World Airways, Inc. v. Panamerican School of Travel, Inc., 648 F.Supp. 1026, 1031 (S.D.N.Y.), *aff'd without op.*, 810 F.2d 1160 (2d Cir. 1986) (citing *Defiance,* supra). The claim to residual goodwill will not preclude a finding of abandonment where, as in this case, the owner unequivocally declares its intention to discontinue use.

Abandonment under the Lanham Act, however, requires both nonuse and intent not to resume use. . . .

Once prima facie abandonment has been proven, the trademark registrants — in this case plaintiffs — must carry their burden of producing evidence that there was an intent to resume use of the trademark. Cerveceria India, 892 F.2d at 1025-26. . . .

. . . Rather than merely proving that it did not intend to abandon its trademark, the trademark registrant must demonstrate that it intended to use or resume use. See Exxon, 695 F.2d at 99, 102-103 ("Stopping at an 'intent not to abandon' [rather than 'intent to resume'] tolerates an owner's protecting a mark with neither commercial use nor plans to resume commercial use. Such a license is not permitted by the Lanham Act"). . . .

. . . Plaintiffs have in no way demonstrated their intent to resume commercial use of the "Brooklyn Dodgers" mark within two years after Los Angeles left Brooklyn in 1958 or at anytime within the ensuing quarter century. . . .

2. Resumption

Having determined that plaintiffs abandoned the "Brooklyn Dodgers" mark, the next inquiry is to determine the effect of that abandonment, given that plaintiffs have recently resumed limited use of the trademark. . . .

. . . [I]f plaintiffs have any interest in a "Brooklyn Dodgers" mark, that interest arose in 1981 when commercial use of the mark resumed after a twenty-three (23) year hiatus. Plaintiffs' preemptive rights in the "Brooklyn Dodgers" mark would extend only to the precise goods on or in connection with which the trademark was used since its resumption (i.e. clothing, jewelry, novelty items).

In other words, the fact that plaintiffs resumed use prior to defendants' use does not mean that plaintiffs may preclude defendants' use of the mark in their restaurant business in Brooklyn. . . .

Accordingly, the court declines to enjoin defendants' very limited use of the "Brooklyn Dodger" mark by defendants for use in connection with its local restaurants directed toward older Brooklyn Dodgers fans in the Brooklyn community in the city of New York. . . .

COMMENTS AND QUESTIONS

1. Section 45 of the Lanham Act, on abandonment, was amended in 1994 (effective Jan. 1, 1996) to change the presumptive abandonment period from two to three years, in accordance with the Uruguay Round amendments. See The Uruguay Round Agreements Act, Pub. L. No. 103-465, 108 Stat. 4809 (1994), at §521, codified at 15 U.S.C. §1127. This provision of the act was the U.S. version of article 19(1) of The Agreement on Trade-

Related Aspects of Intellectual Property Rights, Including Trade in Counterfeit Goods, opened for signature Apr. 15, 1994, 33 I.L.M. 81.

Thus section 45 now reads in part:

> A mark shall be deemed to be "abandoned" when either of the following occurs:
>
> (1) When its use has been discontinued with intent not to resume such use. Intent not to resume may be inferred from circumstances. Nonuse for three consecutive years shall be prima facie evidence of abandonment. "Use" of a mark means the bona fide use of such mark made in the ordinary course of trade, and not made merely to reserve a right in a mark.
>
> (2) When any course of conduct of the owner, including acts of omission as commision, causes the mark to . . . lose its significance as a mark. . . .

Lanham Act §45, 15 U.S.C. §1127. The *Dawn Donuts* case and notes, below, deal with the §45(2) type of abandonment.

2. If trademark rights are essentially about protecting consumers from confusion, why does the legal standard require anything other than a showing that the mark's meaning has faded sufficiently so that it can now be "reclaimed" by another? Why does the statute settle for a presumption; why not a per se rule, perhaps based on a longer period — say ten years? Cf. Note, The Song Is Over but the Melody Lingers On: Persistence of Goodwill and the Intent Factor in Trademark Abandonment, 56 Fordham L. Rev. 1003, 1006-1007 (1988) ("[W]hen a trademark has persisting or residual goodwill, even after a period of nonuse, doubts should be resolved in favor of the trademark owner and against the competitor charging abandonment. . . . [Courts] should . . . make it as difficult as possible to find a trademark abandoned whenever goodwill in the mark persists."). What arguments can you think of in favor of this liberal rule? Against? Should there be an incentive to aggressively "recycle" marks that have shown some usefulness?

Do you think the association between the name "Brooklyn Dodgers" and the major league baseball team now located in Los Angeles had faded sufficiently for someone else to reclaim the name? Do the unusual facts in this case — especially the animosity between ex-Brooklyn Dodger fans and the Los Angeles Dodgers — suggest why a per se rule might not make sense in every case? Is the court right to assume the absence of "goodwill" here, where consumers recognize the Dodger mark but simply have bad associations with it? Or does the continual connection in the public's mind suggest an ongoing role for trademark law?

3. Are there reasons to limit the abandonment doctrine where it may affect names that clearly seem to "belong" to the trademark owner? In Kareem Abdul-Jabaar v. General Motors, 75 F.3d 1361 (9th Cir. 1996), the

court rejected defendant's claim that basketball star Abdul-Jabaar had abandoned his birth name, Lew Alcindor. The court established a per se rule: "A proper name thus cannot be deemed abandoned throughout its possessor's life, despite his failure to use it. . . ."

PROBLEMS

Problem 5-11. Until 1972 the Humble Oil & Refining Co. was one of the largest producers and sellers of gasoline in the world. In that year Humble decided to change its name to Exxon, an arbitrary mark it had selected for the purpose. The company invested an enormous amount of money advertising the new name, and placing its new brand on all its products, services, and correspondence. However, the new Exxon also started a "brand maintenance program" under which it continued to make a few sales every year under the Humble name. It sent certain invoices on Humble letterhead, and bulk oil was sometimes sold in Humble barrels. Exxon continued this program throughout the 1970s.

In 1975, a new oil exploration company decides to call itself the Humble Oil Exploration Co. The new company picked the Humble name because they hoped to trade on residual goodwill left in the name, and because they believed Exxon had abandoned its old name. Can the new company use the Humble name?

Problem 5-12. IBM Corporation is a world-famous maker of computers and other business machines. For decades, people have informally referred to IBM as "Big Blue," both in conversation and in published articles and papers about the company. However, IBM does not itself use or claim the name "Big Blue." It has never registered the name Big Blue, its employees are forbidden by company policy to use the name in referring to the company, and at one point an exasperated company spokesperson is reported to have said "We're not Big Blue — please don't call us that. We want nothing to do with the name." Nonetheless, the nickname persists.

In 1994 a small repair shop that specializes in fixing IBM computers decided to call itself the "Big Blue Repair Shop." After BBRS started using the name, IBM sent it a cease-and-desist letter alleging that IBM owned the rights to the name Big Blue, since people used it to refer to IBM. Who has priority in use of the Big Blue mark?

a. *"Unsupervised Licenses"*

Dawn Donut Company, Inc. v. Hart's Food Stores, Inc.
United States Court of Appeals for the Second Circuit
267 F.2d 358 (2d Cir. 1959)

[Dawn Donut, a Michigan wholesaler of doughnut mix, owned federally registered trademarks Dawn and Dawn Donut. Dawn brought a trademark infringement suit against defendant Hart's Food Stores ("Hart"), which used the mark "Dawn" in connection with its sale of doughnuts and baked goods in stores in and around Rochester, N.Y. Hart filed a counterclaim to cancel plaintiff's registrations, on the ground that plaintiff had abandoned its trademarks due to inadequate quality control and supervision on the part of its licensees — the retail bakeries that sold doughnuts made from Dawn's mix. The district court denied Dawn's request for an injunction and dismissed Hart's counterclaim. This appeal followed. In an omitted (and oft-cited) portion of this opinion, Circuit Judge Lumbard held that plaintiff was not entitled to any relief under the Lanham Act because there was no present likelihood that plaintiff would expand its retail use of its trademarks into defendant's market area. A majority of the panel also found that there was no abandonment by plaintiff of its registrations, and that therefore defendant was not entitled to have plaintiff's registrations of trademarks cancelled.]

LUMBARD, Circuit Judge [dissenting in part]. . . .
The final issue presented is raised by defendant's appeal from the dismissal of its counterclaim for cancellation of plaintiff's registration on the ground that the plaintiff failed to exercise the control required by the Lanham Act over the nature and quality of the goods sold by its licensees.

We are all agreed that the Lanham Act places an affirmative duty upon a licensor of a registered trademark to take reasonable measures to detect and prevent misleading uses of his mark by his licensees or suffer cancellation of his federal registration. The Act, 15 U.S.C.A. §1064, provides that a trademark registration may be cancelled because the trademark has been "abandoned." And "abandoned" is defined in 15 U.S.C.A. §1127 to include any act or omission by the registrant which causes the trademark to lose its significance as an indication of origin.

Prior to the passage of the Lanham Act many courts took the position that the licensing of a trademark separately from the business in connection with which it had been used worked an abandonment. Reddy Kilowatt, Inc. v. MidCarolina Electric Cooperative, Inc., 4 Cir., 1957, 240 F.2d 282, 289; American Broadcasting Co. v. Wahl Co., 2 Cir., 1941, 121 F.2d 412, 413; Everett O. Fisk & Co. v. Fisk Teachers' Agency, Inc., 8 Cir., 1924, 3 F.2d 7, 9. The theory of these cases was that:

A trade-mark is intended to identify the goods of the owner and to safeguard his good will. The designation if employed by a person other than the one whose business it serves to identify would be misleading. Consequently, "a right to the use of a trade-mark or a trade-name cannot be transferred in gross."

American Broadcasting Co. v. Wahl Co., supra, 121 F.2d at page 413.

Other courts were somewhat more liberal and held that a trademark could be licensed separately from the business in connection with which it had been used provided that the licensor retained control over the quality of the goods produced by the licensee. E. I. DuPont de Nemours & Co. v. Celanese Corporation of America, 1948, 167 F.2d 484, 35 CCPA 1061. . . . But even in the DuPont case the court was careful to point out that naked licensing, viz. the grant of licenses without the retention of control, was invalid. E. I. DuPont de Nemours & Co. v. Celanese Corporation of America, supra, 167 F.2d at page 489.

The Lanham Act clearly carries forward the view of these latter cases that controlled licensing does not work an abandonment of the licensor's registration, while a system of naked licensing does. 15 U.S.C.A. §1055 provides:

> Where a registered mark or a mark sought to be registered is or may be used legitimately by related companies, such use shall inure to the benefit of the registrant or applicant for registration, and such use shall not affect the validity of such mark or of its registration, provided such mark is not used in such manner as to deceive the public.

And 15 U.S.C.A. §1127 defines "related company" to mean "any person who legitimately controls or is controlled by the registrant or applicant for registration in respect to the nature and quality of the goods or services in connection with which the mark is used."[33]

Without the requirement of control, the right of a trademark owner to license his mark separately from the business in connection with which it has been used would create the danger that products bearing the same trademark might be of diverse qualities. See American Broadcasting Co. v. Wahl Co., supra; Everett O. Fisk & Co. v. Fisk Teachers' Agency, Inc., supra. If the licensor is not compelled to take some reasonable steps to prevent misuses of his trademark in the hands of others the public will be deprived of its most effective protection against misleading uses of a trademark. The public is hardly in a position to uncover deceptive uses of a trademark before they occur and will be at best slow to detect them after they happen. Thus, unless the licensor exercises supervision and control over the operations of its li-

---

33. Lanham Act §45 now reads in relevant part: "The term 'related company' means any person whose use of a mark is controlled by the owner of the mark with respect to the nature and quality of the goods or services on or in connection with which the mark is used." 15 U.S.C. §1127. The portion of 15 U.S.C. §1055 (Lanham Act §5) quoted just above in the case has not changed. — EDS.

censees the risk that the public will be unwittingly deceived will be increased and this is precisely what the Act is in part designed to prevent. See Sen. Report No. 1333, 79th Cong., 2d Sess. (1946). Clearly the only effective way to protect the public where a trademark is used by licensees is to place on the licensor the affirmative duty of policing in a reasonable manner the activities of his licensees.

The critical question on these facts therefore is whether the plaintiff sufficiently policed and inspected its licensees' operations to guarantee the quality of the products they sold under its trademarks to the public. The trial court found that: "By reason of its contacts with its licensees, plaintiff exercised legitimate control over the nature and quality of the food products on which plaintiff's licensees used the trademark 'Dawn.' Plaintiff and its licensees are related companies within the meaning of Section 45 of the Trademark Act of 1946." It is the position of the majority of this court that the trial judge has the same leeway in determining what constitutes a reasonable degree of supervision and control over licensees under the facts and circumstances of the particular case as he has on other questions of fact; and particularly because it is the defendant who has the burden of proof on this issue they hold the lower court's finding not clearly erroneous.

I dissent from the conclusion of the majority that the district court's findings are not clearly erroneous because while it is true that the trial judge must be given some discretion in determining what constitutes reasonable supervision of licensees under the Lanham Act, it is also true that an appellate court ought not to accept the conclusions of the district court unless they are supported by findings of sufficient facts. It seems to me that the only findings of the district judge regarding supervision are in such general and conclusory terms as to be meaningless. In the absence of supporting findings or of undisputed evidence in the record indicating the kind of supervision and inspection the plaintiff actually made of its licensees, it is impossible for us to pass upon whether there was such supervision as to satisfy the statute. There was evidence before the district court in the matter of supervision, and more detailed findings thereon should have been made.

Plaintiff's licensees fall into two classes: (1) those bakers with whom it made written contracts providing that the baker purchase exclusively plaintiff's mixes and requiring him to adhere to plaintiff's directions in using the mixes; and (2) those bakers whom plaintiff permitted to sell at retail under the 'Dawn' label doughnuts and other baked goods made from its mixes although there was no written agreement governing the quality of the food sold under the Dawn mark.[6]

6. On cross-examination plaintiff's president conceded that during 1949 and 1950 the company in some instances, the number of which is not made clear by his testimony, distributed its advertising and packaging material to bakers with whom it had not reached any agreement relating to the quality of the goods sold in packages bearing the name "Dawn." It also appears from plaintiff's list of the 16 bakers who were operating as exclusive Dawn shops at the time of the trial that plaintiff's contract with 3 of these shops had expired and had not been renewed and

The contracts that plaintiff did conclude, although they provided that the purchaser use the mix as directed and without adulteration, failed to provide for any system of inspection and control. Without such a system plaintiff could not know whether these bakers were adhering to its standards in using the mix or indeed whether they were selling only products made from Dawn mixes under the trademark "Dawn."

The absence, however, of an express contract right to inspect and supervise a licensee's operations does not mean that the plaintiff's method of licensing failed to comply with the requirements of the Lanham Act. Plaintiff may in fact have exercised control in spite of the absence of any express grant by licensees of the right to inspect and supervise.

The question then, with respect to both plaintiff's contract and non-contract licensees, is whether the plaintiff in fact exercised sufficient control.

Here the only evidence in the record relating to the actual supervision of licensees by plaintiff consists of the testimony of two of plaintiff's local sales representatives that they regularly visited their particular customers and the further testimony of one of them, Jesse Cohn, the plaintiff's New York representative, that "in many cases" he did have an opportunity to inspect and observe the operations of his customers. The record does not indicate whether plaintiff's other sales representatives made any similar efforts to observe the operations of licensees.

Moreover, Cohn's testimony fails to make clear the nature of the inspection he made or how often he made one. His testimony indicates that his opportunity to observe a licensee's operations was limited to "those cases where I am able to get into the shop" and even casts some doubt on whether he actually had sufficient technical knowledge in the use of plaintiff's mix to make an adequate inspection of a licensee's operations.

The fact that it was Cohn who failed to report the defendant's use of the mark "Dawn" to the plaintiff casts still further doubt about the extent of the supervision Cohn exercised over the operations of plaintiff's New York licensees.

Thus I do not believe that we can fairly determine on this record whether plaintiff subjected its licensees to periodic and thorough inspections by trained personnel or whether its policing consisted only of chance, cursory examinations of licensees' operations by technically untrained salesmen. The latter system of inspection hardly constitutes a sufficient program of supervision to satisfy the requirements of the Act.

---

that in the case of 2 other such shops the contract had been renewed only after a substantial period of time had elapsed since the expiration of the original agreement. The record indicates that these latter 2 bakers continued to operate under the name "Dawn" and purchase "Dawn" mixes during the period following the expiration of their respective franchise agreements with the plaintiff. Particularly damaging to plaintiff is the fact that one of the 2 bakers whose franchise contracts plaintiff allowed to lapse for a substantial period of time has also been permitted by plaintiff to sell doughnuts made from a mix other than plaintiff's in packaging labeled with plaintiff's trademark.

. . . I would direct the district court to order the cancellation of plaintiff's registrations if it should find that the plaintiff did not adequately police the operations of its licensees. . . .

The district court's denial of an injunction restraining defendant's use of the mark "Dawn" on baked and fried goods and its dismissal of defendant's [abandonment] counterclaim are affirmed.

COMMENTS AND QUESTIONS

1. Despite its age, the discussion in *Dawn Donuts* regarding licensee supervision and abandonment is still the standard in the area. See 3 J. Thomas McCarthy, McCarthy on Trademarks and Unfair Competition §26.14 (3d ed. 1992 & Supp. 1996).

2. Trademark rights are regularly lost because of unsupervised licensing. See, e.g., Stanfield v. Osborne Indus., Inc., 52 F.3d 867 (10th Cir. 1995) (denying plaintiff's advertising-related trademark claim because rights were lost due to unsupervised license).

3. The *Celanese* case cited in *Dawn Donuts*, E. I. du Pont de Nemours & Co. v. Celanese Corporation, 167 F.2d 484 (C.C.P.A. 1948), was one of the first cases to establish the legitimacy of trademark licensing over the objection that licensing necessarily entailed an abandonment. Obviously, the growth of franchising, character merchandising, and related practices depended on such a holding. Much modern business would be impossible if corporations could not expand their brand names in these ways. Cf. the note above on "Extension by Contract: Licensing and Franchising," section B.5.

b. *The Rule Against Assignments in Gross*

The rule that unsupervised or "naked" licenses can amount to abandonment has a logical corollary: that outright assignments of trademarks are invalid. This common law rule was codified in section 10 of the Lanham Act: "A registered mark or a mark for which application to register has been filed shall be assignable with the goodwill of the business in which the mark is used, or with that part of the goodwill of the business connected with the use of and symbolized by the mark." Lanham Act §10, 15 U.S.C. §1060.

Assignments of trademarks alone, without any underlying assets or "goodwill," are called "assignments in gross." They are invalid, see 2 J. Thomas McCarthy, McCarthy on Trademarks and Unfair Competition §18.01 (3d ed. 1992 & Supp. 1996), a result said to follow from the fact that trademarks are only repositories and symbols of goodwill, rather than true property rights. See American Steel Foundries v. Robertson, 269 U.S. 372, 380 (1926). That case reasoned that consumers, having come to rely on a trademark to identify the characteristics of a product as manufactured and sold by Firm *A*, would be harmed if *A* simply sold the mark outright to

Firm *B* without transferring any of the employees or production machinery that Firm *A* had used to make its product. Id. By the same reasoning, courts invalidated trademarks when sold to another firm for use on a different product. See Filkins v. Blackman, 9 Fed. Cas. 50, 51 (No. 4786) (C.C.D. Conn. 1876) ("The right to the use of a trademark cannot be so enjoyed by an assignee that he shall have the right to affix the mark to goods differing in character or species from the article to which it was originally attached.").

As originally applied, the rule against assignments in gross was quite strict; in general, a firm was required to assign tangible assets along with the trademark. See, e.g., Pepsico, Inc. v. Grapette Co., 416 F.2d 285 (8th Cir. 1969) (assignment of trademark Peppy without underlying assets invalid). More recently, however, the traditional rule has been relaxed, partly in recognition of the increased frequency and importance of trademark-related transactions. The contemporary rule can be seen operating in cases involving assignment of "soft" trademark-related assets, such as customer lists, production formulas (as opposed to machinery), and even amorphous "goodwill." See, e.g., In re Roman Cleanser Co., 802 F.2d 207 (6th Cir. 1986) (validating transfer of trademark in satisfaction of security interest in it, together with formulas and customer lists); Money Store v. Harriscorp Finance, Inc., 689 F.2d 666 (7th Cir. 1982) (assignment of Money Store trademark by senior user for $1 not invalid; nominal recitation of "goodwill" in assignment contract, without transfer of any other assets, was enough). Cf. William M. Landes and Richard A. Posner, Trademark Law: An Economic Perspective, 30 J.L. & Econ. 265, 274-275 (1987) (arguing that the "assignment in gross" doctrine makes sense only in "final period" cases, where sellers of goods are leaving the market and hence do not care if consumers are disappointed by the low quality of the assignee's goods).

By contrast, Japan, for example, recognizes private property rights in the trademark itself. Therefore, assignments in gross are valid even if totally divorced from any goodwill. Trademark rights are also severable; they may be assigned by class, providing the goods of the remaining classification would not cause confusion with the goods of the class assigned. See generally Kazuko Matsuo, Trademarks, in 4 Doing Business in Japan (Zentaro Kitagawa ed., 1991).

The rule against assignments in gross makes sense from the point of view of protecting consumer associations between a mark and an underlying product. If the symbol changes hands, and is now used to "refer to" a different product, consumers might be confused. (Imagine if language experts decided to change the meaning of a common word, without telling anyone.) As Landes and Posner, supra, argue, the risk of confusion is greatest when the trademark assignor is leaving the market.

Of course, this rationale for the doctrine assumes that consumers cannot see the lower quality of the assignee's product when they look at the product. Moreoever, the doctrine seems highly questionable as applied to cases where the assignee uses the mark on a completely different type of product altogether. In such a case, any consumers who retain an association between the mark and the old product will quickly grasp the changed circumstances when they see the new product.

With these points in mind, why not encourage the transfer of trademarks that have proven effective? Why restrict transfers to those accompanied by underlying assets? For cogent criticism of the rule against assignments in gross, see Allison Sell McDade, Note, Trading in Trademarks — Why the Anti-Assignment in Gross Doctrine Should Be Abolished When Trademarks Are Used as Collateral, 77 Tex. L. Rev. 465 (1998).

Does (or should) the law similarly prohibit the original trademark owner from significantly decreasing the quality of his or her goods, or from changing the type of product to which the mark is attached? See 2 J. Thomas McCarthy, McCarthy on Trademarks and Unfair Competition §17.09 (3d ed. 1992 & Supp. 1996) (citing cases and arguing that this would amount to deceit under the Lanham Act). The latter may be an issue for registered marks, where the classification of goods is important.

COMMENTS AND QUESTIONS

Article 21 of the General Agreements on Tariffs and Trade, which entered into force in the United States in 1995, provides that "the owner of a registered trademark shall have the right to assign his trademark with or without the transfer of the business to which the trademark belongs." Does this article require the United States to abolish the rule against assignment in gross?

4. Nontrademark or "Nominative" Use

The New Kids on the Block v. News America Publishing, Inc.
United States Court of Appeals for the Ninth Circuit
971 F.2d 302 (9th Cir. 1992)

KOZINSKI, Circuit Judge.

The individual plaintiffs perform professionally as The New Kids on the Block, reputedly one of today's hottest musical acts. This case requires us to weigh their rights in that name against the rights of others to use it in identifying the New Kids as the subjects of public opinion polls.

Background

No longer are entertainers limited to their craft in marketing themselves to the public. This is the age of the multi-media publicity blitzkrieg: Trading on their popularity, many entertainers hawk posters, T-shirts, badges, coffee mugs and the like — handsomely supplementing their incomes while boosting their public images. The New Kids are no exception; the record in this case indicates there are more than 500 products or services bearing the New Kids trademark. Among these are services taking advantage of a recent development in telecommunications: 900 area code numbers, where the caller is charged a fee, a portion of which is paid to the call recipient. Fans can call various New Kids 900 numbers to listen to the New Kids talk about themselves, to listen to other fans talk about the New Kids, or to leave messages for the New Kids and other fans.

The defendants, two newspapers of national circulation, conducted separate polls of their readers seeking an answer to a pressing question: Which one of the New Kids is the most popular? *USA Today*'s announcement contained a picture of the New Kids and asked, "Who's the best on the block?" The announcement listed a 900 number for voting, noted that "any USA Today profits from this phone line will go to charity," and closed with the following: "New Kids on the Block are pop's hottest group. Which of the five is your fave? Or are they a turn off? . . . Each call costs 50 cents. Results in Friday's Life section." The *Star*'s announcement, under a picture of the New Kids, went to the heart of the matter: "Now which kid is the sexiest?" The announcement, which appeared in the middle of a page containing a story on a New Kids concert, also stated: "Which of the New Kids on the Block would you most like to move next door? STAR wants to know which cool New Kid is the hottest with our readers." Readers were directed to a 900 number to register their votes; each call cost 95 cents per minute. [The *USA Today* poll generated less than $300 in revenues, all of which the newspaper donated to the Berklee College of Music. The *Star*'s poll generated about $1600.]

Fearing that the two newspapers were undermining their hegemony over their fans, the New Kids filed a . . . complaint in federal court raising [various state and federal trademark-related claims]. The two papers raised the First Amendment as a defense, on the theory that the polls were part and parcel of their "news-gathering activities." The district court granted summary judgment for defendants. . . .

Discussion

I

A. Since at least the middle ages, trademarks have served primarily to identify the source of goods and services, "to facilitate the tracing of 'false' or defective wares and the punishment of the offending craftsman." F. Schechter, The Historical Foundations of the Law Relating to Trade-marks 47 (1925). The law has protected trademarks since the early seventeenth century, and the primary focus of trademark law has been misappropriation — the problem of one producer's placing his rival's mark on his own goods. . . .

A trademark is a limited property right in a particular word, phrase or symbol. And although English is a language rich in imagery, we need not belabor the point that some words, phrases or symbols better convey their intended meanings than others. . . . Indeed, the primary cost of recognizing property rights in trademarks is the removal of words from (or perhaps non-entrance into) our language. Thus, the holder of a trademark will be denied protection if it is (or becomes) generic, i.e., if it does not relate exclusively to the trademark owner's product. . . . This requirement allays fears that producers will deplete the stock of useful words by asserting exclusive rights in them. When a trademark comes to describe a class of goods rather than an individual product, the courts will hold as a matter of law that use of that mark does not imply sponsorship or endorsement of the product by the original holder. . . .

With many well-known trademarks, such as Jell-O, Scotch tape and Kleenex, there are equally informative non-trademark words describing the products (gelatin, cellophane tape and facial tissue). But sometimes there is no descriptive substitute, and a problem closely related to genericity and descriptiveness is presented when many goods and services are effectively identifiable only by their trademarks. For example, one might refer to "the two-time world champions" or "the professional basketball team from Chicago," but it's far simpler (and more likely to be understood) to refer to the Chicago Bulls. In such cases, use of the trademark does not imply sponsorship or endorsement of the product because the mark is used only to describe the thing, rather than to identify its source.

Indeed, it is often virtually impossible to refer to a particular product for purposes of comparison, criticism, point of reference or any other such purpose without using the mark. For example, reference to a large automobile manufacturer based in Michigan would not differentiate among the Big Three; reference to a large Japanese manufacturer of home electronics would narrow the field to a dozen or more companies. Much useful social and commercial discourse would be all but impossible if speakers were under threat of an infringement lawsuit every time they made reference to a person, company or product by using its trademark. . . .

Cases like these are best understood as involving a non-trademark use of a mark — a use to which the infringement laws simply do not apply, just as videotaping television shows for private home use does not implicate the copyright holder's exclusive right to reproduction. See Sony Corp. v. Universal City Studios, Inc., 464 U.S. 417, 447-51, 104 S. Ct. 774, 791-93, 78 L. Ed. 2d 574 (1984). Indeed, we may generalize a class of cases where the use of the trademark does not attempt to capitalize on consumer confusion or to appropriate the cachet of one product for a different one. Such nominative use of a mark — where the only word reasonably available to describe a particular thing is pressed into service — lies outside the strictures of trademark law. Because it does not implicate the source-identification function that is the purpose of trademark, it does not constitute unfair competition; such use is fair because it does not imply sponsorship or endorsement by the trademark holder. "When the mark is used in a way that does not deceive the public we see no such sanctity in the word as to prevent its being used to tell the truth." Prestonettes, Inc. v. Coty, 264 U.S. 359, 368, 44 S. Ct. 350, 351, 68 L. Ed. 731 (1924) (Holmes, J.).

To be sure, this is not the classic fair use case where the defendant has used the plaintiff's mark to describe the defendant's own product. Here, the New Kids trademark is used to refer to the New Kids themselves. We therefore do not purport to alter the test applicable in the paradigmatic fair use case. If the defendant's use of the plaintiff's trademark refers to something other than the plaintiff's product, the traditional fair use inquiry will continue to govern. But, where the defendant uses a trademark to describe the plaintiff's product, rather than its own, we hold that a commercial user is entitled to a nominative fair use defense provided he meets the following three requirements: First, the product or service in question must be one not readily identifiable without use of the trademark; second, only so much of the mark or marks may be used as is reasonably necessary to identify the product or service; and third, the user must do nothing that would, in conjunction with the mark, suggest sponsorship or endorsement by the trademark holder. . . .

While the New Kids have a limited property right in their name, that right does not entitle them to control their fans' use of their own money. Where, as here, the use does not imply sponsorship or endorsement, the fact that it is carried on for profit and in competition with the trademark holder's business is beside the point. . . . Voting for their favorite New Kid may be, as plaintiffs point out, a way for fans to articulate their loyalty to the group, and this may diminish the resources available for products and services they sponsor. But the trademark laws do not give the New Kids the right to channel their fans' enthusiasm (and dollars) only into items licensed or authorized by them. . . .

The district court's judgment is affirmed.

COMMENTS AND QUESTIONS

1. Compare the holding in this case with (1) cases such as *Zatarain's*, supra, finding fair use of a descriptive trademark by a competitor, and (2) parody cases, such as *L. L. Bean*, infra. Does the doctrine used to explain each of these cases serve essentially the same purpose? Or are the goals separate?

2. Does the use of the "New Kids" name in this fashion imply endorsement of the contest by the band? How would the court rule if the newspaper had used the logo of a famous sports team?

3. In Rock and Roll Hall of Fame and Museum v. Gentile Productions, 134 F.3d 749 (6th Cir. 1998), the plaintiff sought to assert trademark protection in its unusual building design to preclude a photographer from mar-

keting pictures of the building against the Cleveland skyline. Assuming the building is properly the subject of trademark protection, does it follow that the defendant is engaged in trademark infringement? Isn't he making an accurate representation about the "product" itself — that this is what it looks like? Does the fact that he is making money from picture sales give the Museum rights against him? The court did not address this issue because it found the building itself unprotectable. It did note, however, that the photographer was entitled to title his picture "Rock and Roll Hall of Fame and Museum," because that was an accurate description of the subject of the photograph. Id. at 755-756.

Would your answer to these questions change if Coca-Cola brought suit against a company marketing a poster of a Coke bottle? Why or why not?

PROBLEM

Problem 5-13. Toho, Inc. is the owner of the copyrights in several Japanese "Godzilla" movies as well as the registered U.S. trademark "Godzilla." Toho granted a license to a major American movie company to use the name and the monster in its recent high-budget film and various related products and promotional efforts.

Capitalizing on the hype surrounding the new Godzilla film, Edgar publishes a book cataloguing the history of Godzilla in film and print. He titles his book "Godzilla" and places a picture of the monster on the cover. Toho sues for trademark infringement.

How should the court rule? Would it matter if Toho had licensed a different writer to produce an "authorized" history?

---

5. Parody

≡≡≡ *L. L. Bean, Inc. v. Drake Publishers, Inc.*
United States Court of Appeals, First Circuit
811 F.2d 26, 28 (1st Cir. 1987),
cert. denied, *483 U.S. 1013 (1987)*

BOWNES, Circuit Judge.

Imitation may be the highest form of flattery, but plaintiff-appellee L. L. Bean, Inc., was neither flattered nor amused when *High Society* magazine published a prurient parody of Bean's famous catalog. Defendant-appellant Drake Publishers, Inc., owns *High Society*, a monthly periodical featuring adult erotic entertainment. Its October 1984 issue contained a two-page article entitled "L. L. Beam's Back-To-School-Sex Catalog." . . . The article was labelled on the magazine's contents page as "humor" and "parody."

The article displayed a facsimile of Bean's trademark and featured pictures of nude models in sexually explicit positions using "products" that were described in a crudely humorous fashion.

L. L. Bean sought a temporary restraining order to remove the October 1984 issue from circulation. The complaint alleged trademark infringement, unfair competition, trademark dilution, [etc.]. The United States District Court for the District of Maine . . . grant[ed] Bean['s motion for] summary judgment with respect to the trademark dilution claim raised under Maine law.

. . . One need only open a magazine or turn on television to witness the pervasive influence of trademarks in advertising and commerce. Designer labels appear on goods ranging from handbags to chocolates to every possible form of clothing. Commercial advertising slogans, which can be registered as trademarks, have become part of national political campaigns. "Thus, trademarks have become a natural target of satirists who seek to comment on this integral part of the national culture." Dorsen, Satiric Appropriation and the Law of Libel, Trademark and Copyright: Remedies Without Wrongs, 65 B.U. L. Rev. 923, 939 (1986); Note, Trademark Parody: A Fair Use and First Amendment Analysis, 72 Va. L. Rev. 1079 (1986).

The ridicule conveyed by parody inevitably conflicts with one of the underlying purposes of the Maine anti-dilution statute, which is to protect against the tarnishment of the goodwill and reputation associated with a particular trademark. . . . The court below invoked this purpose as the basis for its decision to issue an injunction. The issue before us is whether enjoining the publication of appellant's parody violates the first amendment guarantees of freedom of expression.

II

The district court disposed of the first amendment concerns raised in this matter by relying on the approach taken in Dallas Cowboys Cheerleaders, Inc. v. Pussycat Cinema, Ltd., 604 F.2d 200 (2d Cir. 1979). In rejecting Drake's claim that the first amendment protects the unauthorized use of another's trademark in the process of conveying a message, the district court cited the following language from *Dallas Cowboys Cheerleaders*: "Plaintiff's trademark is in the nature of a property right, . . . and as such it need not 'yield to the exercise of First Amendment rights under circumstances where adequate alternative avenues of communication exist.' " Lloyd Corp. v. Tanner, 407 U.S. 551 [92 S. Ct. 2219, 33 L. Ed. 2d 131] (1972). L. L. Bean v. Drake, 625 F. Supp. at 1537 (quoting Dallas Cowboys Cheerleaders, 604 F.2d at 206).

We do not believe that the first amendment concerns raised here can be resolved as easily as was done in *Dallas Cowboys Cheerleaders*. Cf. Harriet Dorsen, Satiric Appropriation, 65 B.U. L. Rev. at 951; Robert Denicola, Trademarks as Speech: Constitutional Implications of the Emerging Rationales for the Protection of Trade Symbols, 1982 Wis. L. Rev. 158, 206 ("the sweeping rejection of the defendant's first amendment claim in *Dallas Cowboys Cheerleaders* is dangerously simplistic."). Aside from our doubts about whether there are alternative means of parodying plaintiff's catalog, we do not think the court fully assessed the nature of a trademark owner's property rights. A trademark is a form of intellectual property; the Supreme Court case, Lloyd Corp. v. Tanner, 407 U.S. 551, 92 S. Ct. 2219, 33 L. Ed. 2d 131 (1972), relied upon by the *Dallas Cowboys Cheerleaders* court, involved a shopping center. The first amendment issues involved in this case cannot be disposed of by equating the rights of a trademark owner with the rights of an owner of real property:

> [T]rademark is not property in the ordinary sense but only a word or symbol indicating the origin of a commercial product. The owner of the mark acquires the right to prevent the goods to which the mark is applied from being confused with those of others and to prevent his own trade from being diverted to competitors through their use of misleading marks.

Power Test Petroleum Distributors v. Calcu Gas, 754 F.2d 91, 97 (2d Cir. 1985) (quoting Industrial Rayon Corp. v. Dutchess Underwear Corp., 92 F.2d 33, 35 (2d Cir. 1937), *cert. denied*, 303 U.S. 640, 58 S. Ct. 610, 82 L. Ed. 1100 (1938)). . . .

The district court . . . read the anti-dilution statute as granting a trademark owner the unfettered right to suppress the use of its name in any context, commercial or noncommercial, found to be offensive, negative or unwholesome. As one commentator has pointed out, there are serious first amendment implications involved in applying anti-dilution statutes to cover noncommercial uses of a trademark:

> Famous trademarks offer a particularly powerful means of conjuring up the image of their owners, and thus become an important, perhaps at times indispensable, part of the public vocabulary. Rules restricting the use of well-known trademarks may therefore restrict the communication of ideas. . . . If the defendant's speech is particularly unflattering, it is also possible to argue that the trademark has been tarnished by the defendant's use. The constitutional implications of extending the misappropriation or tarnishment rationales to such cases, however, may often be intolerable. Since a trademark may frequently be the most effective means of focusing attention on the trademark owner or its product, the recognition of exclusive rights encompassing such use would permit the stifling of unwelcome discussion.

Denicola, Trademarks As Speech [supra] at 195-96.

The district court's opinion suggests that tarnishment may be found when a trademark is used without authorization in a context which diminishes the positive associations with the mark. Neither the strictures of the first amendment nor the history and theory of anti-dilution law permit a finding of tarnishment based solely on the presence of an unwholesome or negative context in which a trademark is used without authorization. Such a reading of the anti-dilution statute unhinges it from its origins in the marketplace. A trademark is tarnished when consumer capacity to associate it with the appropriate products or services has been diminished. The threat of tarnishment arises when the goodwill and reputation of a plaintiff's trademark is linked to products which are of shoddy quality or which conjure associations that clash with the associations generated by the owner's lawful use of the mark.

. . . The Constitution is not offended when the anti-dilution statute is applied to prevent a defendant from using a trademark without permission in order to merchandise dissimilar products or services. Any residual effect on first amendment freedoms should be balanced against the need to fulfill the legitimate purpose of the anti-dilution statute. . . . The harm occurs when a trademark's identity and integrity — its capacity to command respect in the market — is undermined due to its inappropriate and unauthorized use by other market actors. When presented with such circumstances, courts have found that trademark owners have suffered harm despite the fact that redressing such harm entailed some residual impact on the rights of expression of commercial actors. See, e.g., . . . Chemical Corp. of America v. Anheuser-Busch, Inc., 306 F.2d 433 (5th Cir. 1962), *cert. denied,* 372 U.S. 965, 83 S. Ct. 1089, 10 L. Ed. 2d 129 (1963) (floor wax and insecticide maker's slogan, "Where there's life, there's bugs," harmed strength of defendant's slogan, "Where there's life, there's Bud."); . . . General Electric Co. v. Alumpa Coal Co., 205 U.S.P.Q. (BNA) 1036 (D. Mass. 1979) ("Genital Electric" monogram on underpants and T-shirts harmful to plaintiff's trademark); Gucci Shops, Inc. v. R. H. Macy & Co., 446 F. Supp. 838 (S.D.N.Y. 1977) (defendant's diaper bag labelled "Gucchi Goo" held to injure Gucci's mark); Coca-Cola Co. v. Gemini Rising, Inc., 346 F. Supp. 1183 (E.D.N.Y. 1972) (enjoining the merchandis[ing] of "Enjoy Cocaine" posters bearing logo similar to plaintiff's mark).

While the cases cited above might appear at first glance to be factually analogous to the instant one, they are distinguishable for two reasons. First, they all involved unauthorized commercial uses of another's trademark. Second, none of those cases involved a defendant using a plaintiff's trademark as a vehicle for an editorial or artistic parody. In contrast to the cases cited, the instant defendant used plaintiff's mark solely for noncommercial purposes. Appellant's parody constitutes an editorial or artistic, rather than a commercial, use of plaintiff's mark. The article was labelled as "humor" and "parody" in the magazine's table of contents section; it took up two pages in a one-hundred-page issue; neither the article nor appellant's trademark was featured on the front or back cover of the magazine. Drake did not use Bean's mark to identify or promote goods or services to consumers; it never intended to market the "products" displayed in the parody. . . .

If the anti-dilution statute were construed as permitting a trademark owner to enjoin the use of his mark in a noncommercial context found to be negative or offensive, then a corporation could shield itself from criticism by forbidding the use of its name in commentaries critical of its conduct. . . .

The district court's application of the Maine anti-dilution statute to appellant's noncommercial parody cannot withstand constitutional scrutiny. Drake has not used Bean's mark to identify or market goods or services; it has used the mark solely to identify Bean as the object of its parody. The reading of the anti-dilution provision advanced by the district court would improperly expand the scope of the anti-dilution statute far beyond the frontiers of commerce and deep into the realm of expression. . . .

Our reluctance to apply the anti-dilution statute to the instant case also stems from a recognition of the vital importance of parody. . . . It would be anomalous to diminish the protection afforded parody solely because a parodist chooses a famous trade name, rather than a famous personality, author or creative work, as its object. . . .

Finally, we reject Bean's argument that enjoining the publication of appellant's parody does not violate the first amendment because "there are innumerable alternative ways that Drake could have made a satiric statement concerning 'sex in the outdoors' or 'sex and camping gear' without using plaintiff's name and mark." This argument fails to recognize that appellant is parodying L. L. Bean's catalog, not "sex in the outdoors." The central role which trademarks occupy in public discourse (a role eagerly encouraged by trademark owners), makes them a natural target of parodists. Trademark parodies, even when offensive, do convey a message. The message may be simply that business and product images need not always be taken too seriously; a trademark parody reminds us that we are free to laugh at the images and associations linked with the mark. The message also may be a simple form of entertainment conveyed by juxtaposing the irreverent representation of the trademark with the idealized image created by the mark's owner. . . . While such a message lacks explicit political content, that is no reason to afford it less protection under the first amendment. . . .

Reversed and remanded.

COMMENTS AND QUESTIONS

1. Parody has been asserted as a defense not only against claims grounded in state dilution laws but also against trademark infringement per se. See, e.g., Mutual of Omaha Ins. Co. v. Novak, 836 F.2d 397 (8th Cir. 1987) (affirming district court's finding of likelihood of confusion between plaintiff's famous "Mutual of Omaha" mark and defendant's anti-nuclear T-shirts, bearing funny picture and "Mutant of Omaha" legend), *cert. denied*, 488 U.S. 933, 109 S. Ct. 326, 102 L. Ed. 2d 344 (1988).

2. Do you agree with the court that the defendant did not use L. L. Bean's trademark on a commercial product? Would the case have come out differently if "Beam's Sex Catalog" canvas carry bags were also sold? Was the magazine given away or sold? Was the parody motivated by a desire to make social commentary or to make money? Should it matter?

3. Black Dog Tavern Co., Inc. v. Hall, 823 F. Supp. 48 (D. Mass. 1993), explains the several divergent views of parody that have emerged, and also deals with the "commercial/non-commercial" distinction that forms the basis of the holding in *L. L. Bean*:

> When confronted with trademark parody, some courts have adopted a balancing test, weighing the public interest in avoiding consumer confusion against the public interest in free expression. See, e.g., Cliffs Notes, Inc. v. Bantam Doubleday Dell Publishing Group, Inc., 886 F.2d 490 (2d Cir. 1989) (considering the multifactor likelihood of confusion analysis "at best awkward in the context of parody"). Absent contrary guidance from the Court of Appeals for the First Circuit, this court sees no reason to abandon the present likelihood of confusion analysis, "provided that it is applied with special sensitivity to the purposes of trademark law and the First Amendment rights of [defendant]." Anheuser-Busch, Inc. v. Balducci Publications, 814 F. Supp. 791, 795-96 (E.D. Mo. 1993). "[W]here a party chooses a mark as a parody of an existing mark, the intent is not necessarily to confuse the public but rather to amuse." Jordache [Enterprises, Inc. v. Hogg Wyld, Ltd., 828 F.2d 1482, 1485 (10th Cir. 1987) (large-sized "Lardache" jeans a parody of Jordache mark, so no injunction)]. Of course, the mere fact that a party intends to create a parody does not preclude the possibility of confusion. . . .
>
> As the Court of Appeals for the Second Circuit has observed:
>
>> A parody must convey two simultaneous — and contradictory — messages: that it is the original, but also that it is not the original and is instead a parody. To the extent that it does only the former but not the latter, it is not only a poor parody but also vulnerable under trademark law, since the customer will be confused.
>
> Cliffs Notes, 886 F.2d at 494 . . .
>
> In this case, the court finds that defendant's Black Hog and Dead Dog marks convey just enough of plaintiff's Black Dog marks to allow an ordinarily prudent consumer to appreciate the point of the parody, thereby diminishing the risk of confusion. . . . The fact that defendant's parody appears on products that are sold for profit does not alter this conclusion. This is not a case in which the defendant's spoof is "subservient and only tangentially related" to a primarily commercial message. . . .

823 F. Supp. 48, 57-58.

4. Other cases demonstrate the wide range of approaches to the parody issue. In Hormel Foods Corp. v. Jim Henson Prods., Inc., 73 F.3d 497 (2d Cir. 1996), the court held that defendant's use of a character named "Spa'am" in a movie was an acceptable parody of the plaintiff's famous registered mark "Spam" for meat-related products. The opinion treats the case strictly under "likelihood of confusion" principles. Although no confusion is found to be likely and the parody is permitted, the First Amendment is never mentioned. So too with Lyons Ptshp. v. Giannoulas, 179 F.3d 384 (5th Cir. 1999) (use of character resembling children's show star "Barney," in routines at sporting events, was a parody for purposes of trademark and copyright

law). See also Steven M. Perez, Confronting Biased Treatment of Trademark Parody Under the Lanham Act, 44 Emory L.J. 1451 (1995) (arguing that trademark-based "likelihood of confusion" analysis makes parody cases too unpredictable and inconsistent with free speech interests).

5. The Fifth Circuit recently adapted the parody/satire distinction from copyright law to trademarks. In Elvis Presley Enterprises v. Capece, 141 F.3d 188 (5th Cir. 1998), the court held that defendant's Sixties theme bar could not use the name "Velvet Elvis" because it infringed on the rights of Elvis Presley's estate. The owner of the bar claimed that he was engaged in a legitimate parody of the kitsch associated with certain aspects of the 1960s. The Fifth Circuit concluded that "parody is not a defense to trademark infringement" and that in any event the "Velvet Elvis" was engaged in satire and not parody because its statement did not require the use of the Elvis trademark.

Does it make sense to import the parody/satire distinction into trademark law? Aren't the purposes of the laws different? In any event, how likely is it that consumers will be confused by the use of the "Velvet Elvis" name? Cf. *Lyons Ptshp., supra* (limiting the holding in the *Elvis Presley* case).

PROBLEMS

Problem 5-14. Anheuser-Busch sells beer under a variety of brand names, including Michelob. As part of a general trend of brand proliferation in the beer industry, Busch has recently produced sub-brands of Michelob, including Michelob Dry, Michelob Lite, and Michelob Ice.

After the waters near a Busch plant are polluted in an oil spill, Balducci runs the accompanying advertisement [Figure 5-5] for "Michelob Oily" on the back of a humor magazine. The ad identifies itself as a parody in micro-print along the side. The pictures in the ad are take-offs from those in real Michelob commercials.

Anheuser-Busch sues Balducci for trademark infringement. At trial, Busch can show only a tiny percentage of consumers who thought Michelob Oily was a real product, but it proves that half the people surveyed believed Balducci should have to get permission from Busch to run the ad. Who should prevail?

Problem 5-15. The Coca-Cola Co. maintains an extremely strong and well-recognized trademark in the word "Coke" for its soft drink, and in the phrase "Enjoy Coke," particularly when used in connection with its red and white patterned logo. Gemini Rising, Inc., which distributes commercial posters, designs a poster with a logo identical to Coca-Cola's which reads "Enjoy Cocaine." After passersby who saw the poster complained to the local press, Coca-Cola brings suit against

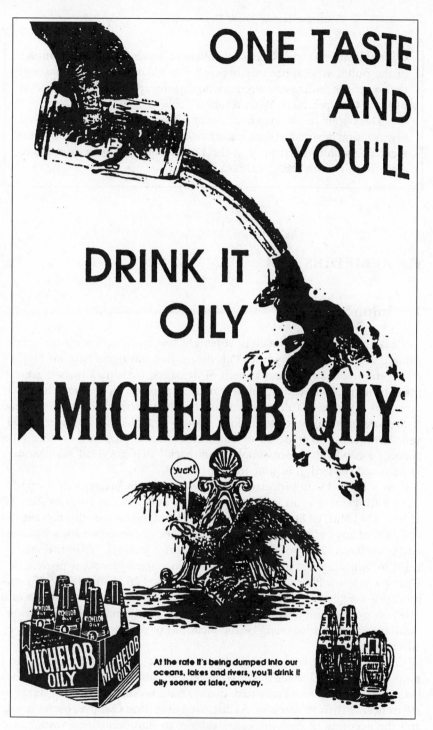

FIGURE 5-5
Parody of Michelob advertisement.

Gemini Rising. At trial, Coca-Cola offers evidence from a few members of the public who apparently believed that Coca-Cola had sponsored the posters, including one woman who threatened to organize a boycott of Coca-Cola products. What result?

Does your result change if Gemini Rising is a nonprofit political organization devoted to drug legalization? If it agrees to include a statement in medium-sized print at the bottom of the poster disclaiming any affiliation with the Coca-Cola Co.?

F. REMEDIES

1. Injunctions

Like patent law and much of copyright law, trademark remedies are organized around a property rule. This means that infringers have no "right" to use the trademark upon payment of damages; trademark owners are entitled to injunctions against infringement as a matter of course. The application of a property rule makes sense in the trademark context, because trademarks serve to protect a unique good — the plaintiff's business goodwill. Infringers who trade on or dilute (and thus appropriate or destroy) a plaintiff's goodwill cannot simply "buy back" that goodwill with money. Once it is dissipated, it is gone forever.

As is normal with property rule regimes, the principal remedy for trademark infringement is an injunction. Section 34(a) of the Lanham Act, 15 U.S.C. §1116(a), gives courts the power to grant injunctions "to prevent the violation of any right of the registrant of a mark . . . or to prevent a violation under section 1125(a) of this title [Lanham Act §43(a)]." Plaintiffs are entitled to injunctions as a matter of course once infringement is proven. See Champion Spark Plug Co. v. Sanders, 331 U.S. 125, 130-131 (1947). Injunctions may include prohibitions against the sale of goods or the use of a mark on goods or in advertising, or they may be affirmative injunctions requiring corrective advertising or the inclusion of disclaimers on products or advertising.

Trademark law is not a typical property rule, however. Owners of real property are entitled to sell it to whomever they wish. They may also let other people rent it for a fee. Trademark owners have no such unfettered rights to sell or license their trademarks. As discussed in section C above, both the sale and the licensing of trademarks are subject to significant legal restrictions. Having given trademark owners a property right to enable them to protect their goodwill, the government appears unwilling to allow trademark owners to do what they see fit with that right. Instead, restrictions on alienation of trademarks are designed to make sure that the trademark is in fact used to promote the goodwill of the associated business.

2. Damages

a. *Infringer's Gain and Mark Owner's Loss*

While injunctions are the fundamental Lanham Act remedy, they are not the only one. Damages are also sometimes available in trademark infringement cases. Section 35(a) of the Lanham Act, 15 U.S.C. §1117(a), provides that plaintiffs are entitled:

> to recover (1) defendant's profits, (2) any damages sustained by the plaintiff, and (3) the costs of the action. . . . In assessing profits the plaintiff shall be required to prove defendant's sales only; defendant must prove all elements of cost or deduction claimed. In assessing damages the court may enter judgment . . . for any sum above the amount found as actual damages, not exceeding three times such amount.

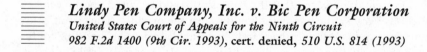

Lindy Pen Company, Inc. v. Bic Pen Corporation
United States Court of Appeals for the Ninth Circuit
982 F.2d 1400 (9th Cir. 1993), cert. denied, *510 U.S. 814 (1993)*

ROLL, District Judge:

This trademark infringement case brought by Lindy Pen Company, Inc. against Bic Pen Corporation returns to this court for the third time seeking review of the district court's damages order upon remand. For the reasons set forth below, we affirm the decision of the district court and deny Bic's cross-appeal.

Introduction

The origins of this case go back to the mid-1960s when Appellant Lindy Pen Company (Lindy) and Appellee Bic Pen Corporation (Bic) were competitors in the production and manufacture of ball point pens. Each marketed a fine-point tip for use by accountants and auditors. In 1965, before Lindy's trademark was perfected and registration issued, Bic used the mark "Auditor's" on its pen barrels. Lindy contacted Bic, making a claim to the mark, and Bic voluntarily agreed to stop using it. On September 20, 1966, United States Trademark Registration No. 815,488 was issued to Lindy for the word "Auditor's." Lindy Pen Co., Inc. v. Bic Pen Corp., 550 F. Supp. 1056, 1059

(C.D. Cal. 1982). Lindy perfected the registration by filing the appropriate documents pursuant to federal statute. Id.

Fourteen years later, Bic adopted the legend "Auditor's Fine Point" to describe a certain pen model. Prior to this use, Bic researched the term "Auditor's" and found that at least three other manufacturers employed a variation of the word "Auditor's" in their marketing materials. This investigation revealed that Lindy also used the term, but that Lindy exerted no proprietary interest over it in its advertising.

Lindy learned of Bic's renewed use of the mark and filed suit in 1980 alleging trademark infringement, unfair competition, breach of contract, and trademark dilution. The district court entered judgment in favor of Bic on all claims. The circuit court upheld the district court's ruling that Bic did not infringe in the major retail markets, but remanded the case to determine whether there was a likelihood of confusion in telephone order sales. Lindy Pen Co., Inc. v. Bic Pen Corp., 725 F.2d 1240 (9th Cir. 1984), *cert. denied*, 469 U.S. 1188, 105 S. Ct. 955, 83 L. Ed. 2d 962 (1985) (Lindy I). The court specifically found that "there is no evidence of actual confusion [and] that Bic adopted the designation 'Auditor's fine point' without intent to capitalize on Lindy's goodwill. . . ." Id. at 1246.

Upon remand, the district court determined that there was a likelihood of confusion in the telephone order market because the purchaser did not have the opportunity to view the pens at the time of purchase. The court went on to find, however, that the confusion could be cured upon receipt of the goods because the purchaser could return the product. The circuit court again disagreed and found that Lindy had established a likelihood of confusion in the telephone order market which post-sale inspection could not cure. Lindy Pen Co., Inc. v. Bic Pen Corp., 796 F.2d 254 (9th Cir. 1986) (Lindy II). In words which now comprise the nub of the current appeal, this court ordered that the case be remanded to district court "with instructions to enter an order enjoining Bic from using the word 'Auditor's' on or in connection with its pens. We additionally instruct the district court to order an accounting and to award damages and other relief as appropriate." Id. . . .

Accounting of Profits

Following the damages proceeding, the district court held that an accounting of profits was inappropriate because Bic's infringement was innocent and accomplished without intent to capitalize on Lindy's trade name. Section 35 of the Lanham Act, 15 U.S.C. §1117(a), governs the award of monetary remedies in trademark infringement cases and provides for an award of defendant's profits, any damages sustained by the plaintiff, and the costs of the action.

The Supreme Court has indicated that an accounting of profits follows as a matter of course after infringement is found by a competitor. Hamilton-Brown Shoe Co. v. Wolf Bros. & Co., 240 U.S. 251, 259, 36 S. Ct. 269,

271, 60 L. Ed. 629, 634 (1916). Nonetheless, an accounting of profits is not automatic and must be granted in light of equitable considerations. Champion Spark Plug Co. v. Sanders, 331 U.S. 125, 131, 67 S. Ct. 1136, 1139, 91 L. Ed. 1386, 1391 (1947). Where trademark infringement is deliberate and willful, this court has found that a remedy no greater than an injunction "slights" the public. Playboy Enterprises, Inc. v. Baccarat Clothing Co., Inc., 692 F.2d 1272, 1274 (9th Cir. 1982). This standard applies, however, only in those cases where the infringement is "willfully calculated to exploit the advantage of an established mark." Id. The intent of the infringer is relevant evidence on the issue of awarding profits and damages and the amount. Maier Brewing Co., 390 F.2d at 123. When awarding profits, the court is cautioned that the "Plaintiff is not . . . entitled to a windfall." Bandag, Inc., 750 F.2d at 918.

Lindy posits that Bic's actions require an accounting as a remedy for willful infringement. The parties agree that as early as 1965, Bic had knowledge of Lindy's claim to trademark rights in the term "Auditor's" when Bic published a catalog advertising pens with the mark stamped on their barrels. Bic voluntarily suspended use of the mark after Lindy informed Bic of its claim to the term. Fourteen years after this exchange, Bic began using the legend "Auditor's Fine Point" on its extra-fine point pens, the subject of the current litigation.

Willful infringement carries a connotation of deliberate intent to deceive. Courts generally apply forceful labels such as "deliberate," "false," "misleading," or "fraudulent" to conduct that meets this standard. . . . Indeed, this court has cautioned that an accounting is proper only where the defendant is "attempting to gain the value of an established name of another." Maier Brewing Co., 390 F.2d at 123.

The present case simply does not involve willful infringement. Evidence taken at the damages proceeding shows that Lindy in general, and its Auditor's line in particular, was experiencing an overall business decline. The trial court also found that any knowledge that Bic may have had of Lindy's mark stemming from the 1965 interchange was conducted by outside counsel, thereby implying that Bic's knowledge of Lindy's interest was attenuated at best. Lindy Pen, 550 F. Supp. at 1059. The district court also determined that Lindy's mark was weak and that there was no evidence of actual confusion. Id. at 1058-60. Based on these facts, it was reasonable for the district court to conclude that Bic's actions were not "willfully calculated to exploit the advantage of an established mark," Bandag, Inc., 750 F.2d at 921, and that Bic's conduct did not rise to the level of willfulness which would have mandated an award.[4]

This circuit has announced a deterrence policy in response to trademark infringement and will grant an accounting of profits in those cases where

---

4. Even assuming that Bic could be termed a willful infringer, "willful infringement may support an award of profits to the plaintiff, but does not require one." Faberge, Inc. v. Saxony Products, Inc., 605 F.2d 426, 429 (9th Cir. 1979) (citing Maier Brewing Co., 390 F.2d at 121).

infringement yields financial rewards. Playboy Enterprises, Inc. v. Baccarat Clothing Co., Inc., 692 F.2d 1272, 1274 (9th Cir. 1982) (plaintiff damaged through defendant's intentional use of counterfeit product labels). However, the policy considerations which clearly justified an award in *Playboy Enterprises* are not present in this case. Unlike the *Playboy Enterprises* plaintiff, Lindy's trademark was weak and Bic's infringement was unintentional. Moreover, Bic's major position in the pen industry makes it clear that it was not trading on Lindy's relatively obscure name. Accordingly, the policy considerations announced by *Playboy Enterprises* would be trivialized by insisting on an accounting in this case. See ALPO Petfoods Inc., 913 F.2d at 969 ("deterrence is too weak and too easily invoked a justification for the severe and often cumbersome remedy of a profits award. . . ."). To award profits in this situation would amount to a punishment in violation of the Lanham Act which clearly stipulates that a remedy "shall constitute compensation not a penalty." 15 U.S.C. §1117(a).

Award of Damages

15 U.S.C. §1117(a) further provides for an award, subject to equitable principles, of "any damages sustained by the plaintiff. . . ." A plaintiff must prove both the fact and the amount of damage. 2 J. Thomas McCarthy, Trademarks and Unfair Competition §30:27, at 511 (2d ed. 1984). Damages are typically measured by any direct injury which a plaintiff can prove, as well as any lost profits which the plaintiff would have earned but for the infringement. Id. at 509. Because proof of actual damage is often difficult, a court may award damages based on defendant's profits on the theory of unjust enrichment. Id. at 511. See also Bandag, Inc., 750 F.2d at 918.

The district court gave Lindy the opportunity to prove its damages under both methods: actual damages in the form of its lost profits, or if that proved too difficult, through proof of Bic's unjust enrichment in the form of Bic's profits. . . .

After final briefing, the court concluded that Lindy had failed to show any actual damage because it did not put forth sufficient proof of its lost profits. To establish damages under the lost profits method, a plaintiff must make a "prima facie showing of reasonably forecast profits." 2 J.T. McCarthy, Trademarks and Unfair Competition §30:27, at 511 (2d ed. 1984). The court found Lindy's calculations to be irreparably flawed because Lindy did not isolate its own telephone order sales from total pen sales. Consequently, Lindy's calculations contained items in which no likelihood of confusion existed and therefore were inappropriately included. The district court reasoned that it had no rational basis upon which to estimate an award as to the infringing items and accordingly denied Lindy's request.

Trademark remedies are guided by tort law principles. Id. at 509 ("Plaintiff's damages should be measured by the tort standard under which the infringer-tortfeasor is liable for all injuries caused to plaintiff by the wrongful act. . . ."). As a general rule, damages which result from a tort must be established with reasonable certainty. Dan B. Dobbs, Remedies §3.3, at 151 (1973). The Supreme Court has held that "damages are not rendered uncertain because they cannot be calculated with absolute exactness," yet, a reasonable basis for computation must exist. Eastman Kodak Co. v. Southern Photo Materials Co., 273 U.S. 359, 379, 47 S. Ct. 400, 405, 71 L. Ed. 684, 691 (1927). . . .

Lindy produced evidence of its total pen sales, as available, for the designated time period. Although it divided its sales into total sales and specific sales of Auditor's, it failed to further subdivide its data into the category of telephone order sales. Lindy was in the best position to identify its own sales, but declined to provide the court with any evidence of its loss caused by Bic's wrong doing. Although Lindy offers excuses for this deficiency, its explanations do not negate the fact that Lindy never furnished the court any reasonable estimate of its own sales. It would have been error for the district court to select an arbitrary percentage of total sales to represent the more narrow submarket of telephone sales. The court was correct, therefore, in finding that Lindy failed to sustain its burden of proving reasonably forecast profits.

Lindy also sought an accounting of Bic's profits based on a theory of unjust enrichment. Lindy assesses as error the district court's requirement that Lindy separate Bic's telephone order sales from total sales of goods. Lindy maintains that this requirement had the effect of shifting the burden of proving infringing sales from the infringer to the trademark holder. However, an accounting is intended to award profits only on sales that are attributable to the infringing conduct. The plaintiff has only the burden of establishing the defendant's gross profits from the infringing activity with reasonable certainty. Once the plaintiff demonstrates gross profits, they are presumed to be the result of the infringing activity. Mishawaka Rubber & Woolen Mfg. Co. v. S. S. Kresge Co., 316 U.S. 203, 206-07, 62 S. Ct. 1022, 1024-25, 86 L. Ed. 1381, 1385-86 (1942). The defendant thereafter bears the burden of showing which, if any, of its total sales are not attributable to the infringing activity, and, additionally, any permissible deductions for overhead. 15 U.S.C. 1117(a).

Lindy failed to come forward with any evidence of sales of the Bic "Auditor's Fine Point" in the infringing market. Lindy instead brought forth proof of Bic's total sales. Lindy averred to the court that a division of Bic's sales into the telephone submarket "is impossible from Bic's records since Bic never separated its pens according to telephone sales. . . ." To the contrary, Lindy had access through discovery to Bic's records from which a reasonable estimate could have been accomplished.

We find that Lindy's appeal on the question of actual damages fails due to a lack of proof at the damages proceeding. . . .

COMMENTS AND QUESTIONS

1. How is the *public* "slighted" by a court's failure to award damages for willful infringement? Does it matter to the public whether an infringer

pays damages to the trademark owner or not? Surely, any confusion that the public has suffered is not ameliorated because the trademark owner receives compensation. Why isn't an injunction, which the courts agree is a sufficient remedy in the case of "innocent infringement," sufficient in cases of intentional infringement as well?

2. Is an accounting for profits ever possible in a dilution case? If so, why? If the plaintiff is compensated for her losses caused by the dilution, does it make any sense to take away the defendant's profits in an unrelated market as well?

3. Who should bear the burden of proof in determining profits? See Mishawaka Rubber & Woolen Mfg. Co. v. S. S. Kresge Co., 316 U.S. 203, 206-207 (1942) (defendant bears the burden of proving that not all profits associated with a sale should be awarded to plaintiff). The Court reasoned: "There may well be a windfall to the trade-mark owner where it is impossible to isolate the profits which are attributable to the use of the infringing mark. But to hold otherwise would give the windfall to the wrongdoer."

Suppose that a defendant has established significant goodwill of its own between the time it adopted a mark and the time it is found to have infringed. If the plaintiff is entitled to an injunction, does the defendant have any way of recovering the value of its investment in goodwill? Is it at least entitled to a setoff against damages for its monetary losses as a result of the injunction? Does it matter if the case is a dilution case and the defendant is in a separate market?

4. In general, many courts follow the holding in Champion Spark Plug v. Sanders, 331 U.S. 125, 130-131, 67 S. Ct. 1136, 1139, 91 L. Ed. 1386 (1947), that an accounting is appropriate only when fraud or palming off is present, and will grant an accounting of defendant's profits only if the defendant acted in bad faith. But in some circuits, lost profits may be awarded even in the absence of proof of willfulness. See, e.g., George Basch Co., Inc. v. Blue Coral, Inc., 968 F.2d 1532, 1540 (2d Cir. 1992) ("[I]n the absence of . . . a showing [of willfulness], a plaintiff is not foreclosed from receiving monetary relief. Upon proof of actual consumer confusion, a plaintiff may still obtain damages — which, in turn, may be inclusive of plaintiff's own lost profits."). For an interesting recent case that explores what constitutes "bad faith" for enhanced damages purposes, see Int'l Star Class Yacht Racing Assoc. v. Tommy Hilfiger U.S.A., Inc., 1999 WL 108739 (S.D. N.Y., Mar. 3, 1999) (on remand from Second Circuit, adhering to earlier opinion finding no bad faith in adoption of plaintiff's mark, despite very limited search of trademark databases). See also Roger D. Blair and Thomas F. Cotter, 39 Wm. & Mary L. Rev. 1585, 1691 (1998) (arguing that the law should require "some level of search activity" for a trademark defendant to escape a damage award). Cf. Banff, Ltd. v. Colberts, Inc., 996 F.2d 33, 35 (2d Cir.), *cert. denied,* 510 U.S. 1010 (1993).

5. On the availability of *defendant's* profits, see Minnesota Pet-Breeders, Inc. v. Schell & Kampeter, Inc., 843 F. Supp. 506, 512-513 (D. Minn. 1993), *aff'd on other grounds,* 41 F.3d 1242 (8th Cir. 1994):

A number of courts have held that an accounting of defendant's profits in cases of trademark infringement or unfair competition is warranted if (1) the plaintiff sustained damage from the infringement, (2) the infringer is unjustly enriched, and (3) necessary to deter a willful infringer from doing so again. At least one court has held that "deterrence alone cannot justify such an award [of infringing defendant's profits]." [Alpo Petfoods, Inc. v. Ralston Purina Co., 913 F.2d 958, 969 (D.C. Cir. 1990).]

Note that in affirming the decision below, the Eigth Circuit added these qualifications: "[I]n this case, in which there was no actual competition between the trademarked and the infringing products, it is most likely that MPB would be entitled to injunctive but not monetary relief, and virtually inconceivable that only monetary relief would be appropriate." Id. at 1247-1248.

6. For a detailed exposition of how to calculate the infringer's profits in an accounting, see Dennis S. Corgill, Measuring the Gains from Trademark Infringement, 65 Fordham L. Rev. 1909 (1997).

PROBLEM

Problem 5-16. A small Vermont company named STW (and its predecessors) has used the mark "Thirst-Aid" for soft drinks since 1921. STW has never sold its products widely, however, and in fact its sales are declining. In 1983, Quaker Oats adopted the slogan "Gatorade is Thirst Aid" for its popular Gatorade beverage. STW sued for trademark infringement. At trial, STW proves that Quaker Oats knew of STW's trademark, but its lawyers advised it that the Gatorade slogan made "fair use" of STW's "descriptive" mark. The trial court disagreed, finding that Quaker Oats had infringed STW's mark in bad faith.

During the period that the "Thirst-Aid" campaign ran, pretax profits on the sale of Gatorade came to $247 million on sales of $2.6 billion. Quaker Oats proves that Gatorade had approximately $475 million in sales before the ad campaign. It also demonstrates that STW had previously offered to license the "Thirst-Aid" mark to another company for 1/3 of 1% of sales, and that STW's total goodwill in 1984 was less than $100,000 and declining. What damages should be awarded in this case?

b.   Corrective Advertising

Big O Tire Dealers, Inc. v. The Goodyear Tire & Rubber Company
United States Court of Appeals for the Tenth Circuit
561 F.2d 1365 (10th Cir. 1977)

Lewis, Chief Judge.

This civil action was brought by Big O Tire Dealers, Inc., ("Big O") asserting claims of unfair competition against The Goodyear Tire & Rubber Co. ("Goodyear") based upon false designation of origin under 15 U.S.C. §1125(a) and common law trademark infringement. After a ten-day trial and three days of deliberation, the jury returned the following verdict:

> We the jury in the above entitled cause, upon our oath do say that we find the following as our verdict herein:
>
>> Upon the claim of liability for trademark infringement we find for Big O Inc.
>> Upon the claim of liability for false designation of origin we find for Goodyear.
>> Upon the claim for trademark disparagement we find for Big O Inc.
>> We find that plaintiff has proven special compensatory damages in the amount of $ None.
>> We find that plaintiff has proven general compensatory damages in the amount of $2,800,000.
>> We assess punitive or exemplary damages in the amount of $16,800,000.
>> Dated September 4, 1975.

Filing a comprehensive post-trial opinion the United States District Court for the District of Colorado entered judgment on the jury's verdict, permanently enjoined Goodyear from infringing on Big O's trademark, and dismissed Goodyear's counterclaim for equitable relief. 408 F. Supp. 1219. Goodyear appeals that judgment.

Big O is a tire-buying organization which provides merchandising techniques, advertising concepts, operating systems, and other aids to approximately 200 independent retail tire dealers in 14 states who identify themselves to the public as Big O dealers. These dealers sell replacement tires using the Big O label on "private brand" tires. They also sell other companies' brands such as B. F. Goodrich and Michelin Tires. At the time of trial Big O's total net worth was approximately $200,000.

Goodyear is the world's largest tire manufacturer. In 1974 Goodyear's net sales totalled more than $5.25 billion and its net income after taxes surpassed $157 million. In the replacement market Goodyear sells through a nationwide network of company-owned stores, franchise dealers, and independent retailers.

In the fall of 1973 Big O decided to identify two of its lines of private brand tires as "Big O Big Foot 60" and "Big O Big Foot 70." These names were placed on the sidewall of the respective tires in raised white letters. The first interstate shipment of these tires occurred in February 1974. Big O dealers began selling these tires to the public in April 1974. Big O did not succeed in registering "Big Foot" as a trademark with the United States Patent and Trademark Office. . . .

In July 1974 Goodyear decided to use the term "Bigfoot" in a nationwide advertising campaign to promote the sale of its new "Custom Polysteel Radial" tire. The name "Custom Polysteel Radial" was molded into the tire's sidewall. Goodyear employed a trademark search firm to conduct a search for "Bigfoot" in connection with tires and related products. This search did not uncover any conflicting trademarks. After this suit was filed Goodyear filed an application to register "Bigfoot" as a trademark for tires but withdrew it in 1975. Goodyear planned to launch its massive, nationwide "Bigfoot" advertising campaign on September 16, 1974.

On August 24, 1974, Goodyear first learned of Big O's "Big Foot" tires. Goodyear informed Big O's president, Norman Affleck, on August 26 of Goodyear's impending "Bigfoot" advertising campaign. Affleck was asked to give Goodyear a letter indicating Big O had no objection to this use of "Bigfoot." When Affleck replied he could not make this decision alone, it was suggested Affleck talk with John Kelley, Goodyear's vice-president for advertising.

Affleck called Kelley and requested more information on Goodyear's impending advertising campaign. A Goodyear employee visited Affleck on August 30 and showed him rough versions of the planned Goodyear "Bigfoot" commercials and other promotional materials. On September 10, Affleck and two Big O directors met in New Orleans, with Kelley and Goodyear's manager of consumer market planning to discuss the problem further. At this time the Big O representatives objected to Goodyear using "Bigfoot" in connection with tires because they believed any such use would severely damage Big O. They made it clear they were not interested in money in exchange for granting Goodyear the right to use the "Bigfoot" trademark, and asked Goodyear to wind down the campaign as soon as possible. Goodyear's response to this request was indefinite and uncertain.

During the trial several Goodyear employees conceded it was technically possible for Goodyear to have deleted the term "Bigfoot" from its television advertising as late as early September. However, on September 16, 1974, Goodyear launched its nationwide "Bigfoot" promotion on ABC's Monday Night Football telecast. By August 31, 1975, Goodyear had spent $9,690,029 on its massive, saturation campaign.

On September 17 [1974] Affleck wrote Kelley a letter setting forth his understanding of the New Orleans meeting that Goodyear would wind up its "Bigfoot" campaign as soon as possible. Kelley replied on September 20, denying any commitment to discontinue use of "Bigfoot" and declaring Goodyear intended to use "Bigfoot" as long as it continued to be a helpful advertising device.

On October 9 Kelley told Affleck he did not have the authority to make the final decision for Goodyear and suggested that Affleck call Charles Eaves, Goodyear's executive vice-president. On October 10 Affleck called Eaves and Eaves indicated the possibility of paying Big O for the use of the term "Bigfoot." When Affleck stated no interest in the possibility Eaves told him Goodyear wished to avoid litigation but that if Big O did sue, the case would be in litigation long enough that Goodyear might obtain all the benefits it desired from the term "Bigfoot."

This was the final communication between the parties until Big O filed suit on November 27, 1974. The district court denied Big O's request for a temporary restraining order and a preliminary injunction. After judgment was entered on the jury's verdict for Big O, Goodyear appealed to this court. Goodyear's allegations of error are discussed below. . . .

IV

. . . Big O does not claim nor was any evidence presented showing Goodyear intended to trade on the goodwill of Big O or to palm off Goodyear products as being those of Big O. Instead, Big O contends Goodyear's use of Big O's trademark created a likelihood of confusion concerning the source of Big O's "Big Foot" tires.

The facts of this case are different from the usual trademark infringement case. As the trial judge stated, the usual trademark infringement case involves a claim by a plaintiff with a substantial investment in a well established trademark. The plaintiff would seek recovery for the loss of income resulting from a second user attempting to trade on the goodwill associated with that established mark by suggesting to the consuming public that his product comes from the same origin as the plaintiff's product. The instant case, however, involves reverse confusion wherein the infringer's use of plaintiff's mark results in confusion as to the origin of plaintiff's product. Only one reported decision involves the issue of reverse confusion. In Westward Coach Mfg. Co. v. Ford Motor Co., 7 Cir., 388 F.2d 627, *cert. denied,* 392 U.S. 927, 88 S. Ct. 2286, 20 L. Ed. 2d 1386, the court held reverse confusion is not actionable as a trademark infringement under Indiana law.

Consequently, Goodyear argues the second use of a trademark is not actionable if it merely creates a likelihood of confusion concerning the source of the first user's product. Since both parties agree Colorado law is controlling in this case, we must decide whether this so-called reverse confusion is actionable under Colorado law. To our knowledge, the Colorado courts have never considered whether a second use creating the likelihood of confusion about the source of the first user's products is actionable. However, the Colorado Court of Appeals in deciding a trade name infringement case involving an issue of first impression, cogently pointed out that the Colorado Supreme Court "has consistently recognized and followed a policy of protecting established trade names and preventing public confusion and the tendency has been to widen the scope of that protection." Wood v. Wood's Homes Inc., 33 Colo. App. 285, 519 P.2d 1212, 1215-16.

Using that language as a guiding light in divining what Colorado law is on this issue of first impression, we hold that the Colorado courts, if given the opportunity, would extend its common law trademark infringement actions to include reverse confusion situations. Such a rule would further Colorado's "policy of protecting trade names and preventing public confusion" as well as having "the tendency [of widening] the scope of that protection."

The district court very persuasively answered Goodyear's argument that liability for trademark infringement cannot be imposed without a showing that Goodyear intended to trade on the goodwill of Big O or to palm off Goodyear products as being those of Big O's when it said

> The logical consequence of accepting Goodyear's position would be the immunization from unfair competition liability of a company with a well established trade name and with the economic power to advertise extensively for a product name taken from a competitor. If the law is to limit recovery to passing off, anyone with adequate size and resources can adopt any trademark and develop a new meaning for that trademark as identification of the second user's products. The activities of Goodyear in this case are unquestionably unfair competition through an improper use of a trademark and that must be actionable.

408 F. Supp. at 1236.

Goodyear further argues there was no credible evidence from which the jury could have found a likelihood of reverse confusion. A review of the record demonstrates the lack of merit in this argument. Big O presented more than a dozen witnesses who testified to actual confusion as to the source of Big O's "Big Foot" tires after watching a Goodyear "Bigfoot" commercial. The jury could have reasonably inferred a likelihood of confusion from these witnesses' testimony of actual confusion. Moreover, two of Goodyear's executive officers, Kelley and Eaves, testified confusion was likely or even inevitable.

[The court ordered Goodyear to perform corrective advertising in the amount of $678,302, a proportionate share of what Goodyear spent advertising the name.]

COMMENTS AND QUESTIONS

1. How was Big O hurt by Goodyear's massive advertising campaign? If the products have the same name, and if Goodyear spends much more on advertising than Big O does, shouldn't sales of Big O Tires increase? One possible answer is that Big O feared that Goodyear's tires would be of lower

quality than its own and that its public image would suffer as a result of being associated with Goodyear. But that doesn't seem to be the case here, since Goodyear was selling premium tires and Big O was selling off-brands. Can you envision any way in which Goodyear's advertising of the brand would reduce Big O's sales?

Is the public hurt by thinking that Big O is affiliated with Goodyear? How? Is Big O hurt? Should that matter?

2. In the *Big O Tires* case, the court ordered as a remedy that Goodyear pay $14 million in punitive damages plus $678,302 — what it would cost in corrective advertising to "undo" the "damage" Goodyear had done in connecting itself with the Big O name.

This punitive damage award raises an interesting point. Is the court justified in ordering compensation — whether in the form of corrective advertising or a cash payment — beyond the point where advertising would have benefited Big O? If Goodyear benefits more from the advertising than Big O would have — as is perhaps shown by the fact that it costs more to restore the status quo for Big O — does this establish that the trademark is better left in Goodyear's hands? What if Goodyear realizes that it can make better use of a trademark currently held by a competitor; should we allow Goodyear to appropriate the mark, by paying damages as set by the court? (Note that this is an example of a "liability rule"; see the discussion on patent remedies in Chapter 3.) What would be the effects of such a rule?

Further, consider that *any* spending is really a windfall to Big O, since it did not have the resources to shell out anything like $14 million in advertising. Is there any consumer interest served by corrective advertising in a reverse confusion case?

The idea of punitive damages for trademark infringement seems to deny the possibility of "efficient" trademark infringement. This in turn suggests that trademarks are property and that courts are willing to grant "specific performance" remedies in infringement cases. Does this view make sense?

3. How should damages be measured in cases of reverse confusion? Presumably, a company like Big O Tires might be able to prove that it lost sales (rather than gained them) as a result of Goodyear's use of its "Bigfoot" trademark. But is Big O also entitled to Goodyear's profits from infringement? Its "unjust enrichment"? Most courts would say yes, at least where Big O can prove intentional or willful infringement.

What unjust profits has Goodyear gained *through infringement*? Must it turn over all its profits from the sale of "Bigfoot" tires, even though it is Goodyear and not Big O that has built a national reputation for the "Bigfoot" name? How about the increase in sales attributable to the use of the name "Bigfoot"? Or are Goodyear's profits from infringement limited to whatever sales it took away from Big O (presumably the same measure as Big O's lost profits)?

Note on the Trademark Counterfeiting Act of 1984

In 1984, Congress substantially increased the penalties for intentional copying of a trademark (called "counterfeiting").[33] 18 U.S.C. §2320 makes it a felony to knowingly use a counterfeit mark in connection with the sale of goods or services. It provides for fines and imprisonment of the offenders, and it permits the destruction of goods bearing counterfeit marks (rather than simply the removal of the marks themselves). The act also added

- section 34(d) of the Lanham Act, 15 U.S.C. §1116(d), which provides for the seizure of counterfeit goods and records of sale before trial upon an ex parte application;
- section 35(b) of the Lanham Act, 15 U.S.C. §1117(b), which provides for the award of treble damages plus attorney fees and prejudgment interest against counterfeiters unless the court finds "extenuating circumstances";
- language in section 36 of the Lanham Act, 15 U.S.C. §1118, concerning the right of the court to destroy counterfeit goods after trial.

An exception to the broad reach of the act applies to those who are authorized to use a trademark "at the time of the manufacture or production" of the goods. See U.S. v. Bohai Trading Co., Inc., 45 F.3d 577 (1st Cir. 1995) (upholding the constitutionality of this provision in the face of a challenge on the grounds of vagueness). Although the ex parte nature of the remedies has brought the 1984 Act under constitutional scrutiny, courts have generally upheld the legality of its remedies and procedures. See, e.g., United States v. McEvoy, 820 F.2d 1170 (11th Cir.), *cert denied*, 484 U.S. 902 (1987). But cf. Time Warner Entertainment Co., L.P. v. Does Nos. 1-2, 876 F. Supp. 407 (E.D.N.Y. 1994) (overturning on Fourth Amendment grounds proposed seizure order drafted by owners of copyrights and trademarks; court cited role of private investigator in conducting seizure and impoundment, failure to provide sufficient particularity for premises to be searched or articles to be seized, inclusion of private residence as site to be searched, and failure to sufficiently describe infringing materials).

COMMENTS AND QUESTIONS

1. The Trademark Counterfeiting Act of 1984 is clearly punitive in nature, providing for both criminal sanctions and treble damages. Is such a

---

33. Section 45 of the Lanham Act, 15 U.S.C. §1127, defines a "counterfeit" as "a spurious mark which is identical with, or substantially indistinguishable from, a registered mark." While this definition does not contain an explicit intent requirement, courts are generally unwilling to find good faith use of an infringing mark to be "spurious."

punitive approach warranted? If so, why not apply these remedies to all trademark infringements?

2. Are consumers hurt by counterfeiting, or is it only trademark owners who are being protected by the Lanham Act? What if the counterfeit goods were not in fact inferior to genuine goods with the same trademark? Have consumers suffered any injury in that case if they mistakenly buy counterfeit goods?

3. Should it be a defense to a counterfeiting claim that the reasonable consumer was aware that the goods were not genuine (for example, because they were purchased from a street vendor without a certificate of authenticity for approximately 10 percent of the retail price of the genuine goods in stores)? Are consumers likely to be "confused" by such sales?

4. Judge Posner suggests one good reason for multiple damages in counterfeiting cases: deterrence. In holding that treble damages were an appropriate remedy even for innocent infringement, he argued:

> [T]he sale of counterfeit merchandise has become endemic — perhaps pandemic. Most of the infringing sellers are small retailers, such as K-Econo. Obtaining an injunction against each and every one of them would be infeasible. Trademark owners cannot hire investigators to shop every retail store in the nation. And even if they could and did, and obtained injunctions against all present violators, this would not stop the counterfeiting. Other infringers would spring up, and would continue infringing until enjoined. . . . Treble damages are a particularly suitable remedy in cases where surreptitious violations are possible, for in such cases simple damages (or profits) will underdeter; the violator will know that he won't be caught every time, and merely confiscating his profits in the cases in which he is caught will leave him with a net profit from infringement. . . . [T]he smaller the violator, the less likely he is to be caught, and the more needful, therefore, is a heavy punishment if he is caught.

Louis Vuitton S.A. v. Lee, 875 F.2d 584 (7th Cir. 1989). Is this rationale persuasive?

6

State Intellectual Property Law and Federal Preemption

As we saw in Chapter 2, state protection of intellectual property — whether rooted in statute or in the common law — has long been a feature of the legal landscape. But state intellectual property law extends far beyond trade secrets. Especially in recent years, state law has reached out to an ever-widening range of issues: from court decisions and legislation on publicity rights to "antidilution" statutes in the states (discussed in Chapter 5) and even to state moral rights and artist's resale royalties, states have become a major force in pushing the frontiers of intellectual property law. Because this new wave of state legislation covers such a wide front, it is impossible to provide a comprehensive survey here. Instead, in section A of this chapter, we consider a sampling of the more important state law initiatives over the years. This survey should be read in the context of other state-law protection discussed elsewhere in this book (notably Chapter 2, regarding trade secrets, and Chapter 5, discussing state trademark and antidilution statutes).

In section B of this chapter we turn to the law of federal preemption. Although there has always been a law of federal preemption, which essentially carves out an exclusive federal sphere in the intellectual property field, preemption has taken on new importance in the current environment of state expansionism. The easiest cases on preemption are those where state law attempts either to duplicate federal protection or to interfere with it; the Supremacy Clause of the Constitution nullifies such attempts. More difficult cases are the norm, however. Most involve state law that does not frontally challenge federal authority but instead directs itself to "interstitial" cases not covered by the federal scheme. The question in these cases is whether the gaps in federal protection are an intentional feature of the federal scheme; whether, in other words, the presence of federal protection in some cases precludes states from offering protection in other cases. This logical structure — which involves a search for the intent behind uncovered cases — is

795

what makes preemption such a doctrinally knotty area. As you will see from the cases, doctrine in this area appears quite unstable, as it is not (yet) supported by an adequate conceptual framework. A few stabs in that direction are attempted in the Note on Preemption Rationales at the end of section B.

A. STATE INTELLECTUAL PROPERTY LAW

1. *The Tort of Misappropriation*

The first category of state law intellectual property rights that we consider is a common law doctrine that began in the famous case of International News Service v. Associated Press.

International News Service v. Associated Press
Supreme Court of the United States
248 U.S. 215 (1918)

Justice PITNEY delivered the opinion of the court.

The parties are competitors in the gathering and distribution of news and its publication for profit in newspapers throughout the United States. The Associated Press, which was complainant in the District Court, is a co-operative organization, incorporated under the Membership Corporations Law of the State of New York, its members being individuals who are either proprietors or representatives of about 950 daily newspapers published in all parts of the United States. . . . Complainant gathers in all parts of the world, by means of various instrumentalities of its own, by exchange with its members, and by other appropriate means, news and intelligence of current and recent events of interest to newspaper readers and distributes it daily to its members for publication in their newspapers. The cost of the service, amounting approximately to $3,500,000 per annum, is assessed upon the members and becomes a part of their costs of operation, to be recouped, presumably with profit, through the publication of their several newspapers. Under complainant's by-laws each member agrees upon assuming membership that news received through complainant's service is received exclusively for publication in a particular newspaper, language, and place specified in the certificate of membership, that no other use of it shall be permitted, and that no member shall furnish or permit anyone in his employ or connected with his newspaper to furnish any of complainant's news in advance of publication to any person not a member. And each member is required to gather the local news of his district and supply it to the Associated Press and to no one else.

Defendant is a corporation organized under the laws of the State of New Jersey, whose business is the gathering and selling of news to its customers and clients, consisting of newspapers published throughout the United States, under contracts by which they pay certain amounts at stated times for defendant's service. It has wide-spread news-gathering agencies; the cost of its operations amounts, it is said, to more than $2,000,000 per annum; and it serves about 400 newspapers located in the various cities of the United States and abroad, a few of which are represented, also, in the membership of the Associated Press.

The parties are in the keenest competition between themselves in the distribution of news throughout the United States; and so, as a rule, are the newspapers that they serve, in their several districts.

Complainant in its bill, defendant in its answer, have set forth in almost identical terms the rather obvious circumstances and conditions under which their business is conducted. The value of the service, and of the news furnished, depends upon the promptness of transmission, as well as upon the accuracy and impartiality of the news; it being essential that the news be transmitted to members or subscribers as early or earlier than similar information can be furnished to competing newspapers by other news services, and that the news furnished by each agency shall not be furnished to newspapers which do not contribute to the expense of gathering it. And further, to quote from the answer:

> Prompt knowledge and publication of world-wide news is essential to the conduct of a modern newspaper, and by reason of the enormous expense incident to the gathering and distribution of such news, the only practical way in which a proprietor of a newspaper can obtain the same is, either through cooperation with a considerable number of other newspaper proprietors in the work of collecting and distributing such news, and the equitable division with them of the expenses thereof, or by the purchase of such news from some existing agency engaged in that business.

The bill was filed to restrain the pirating of complainant's news by defendant in three ways: First, by bribing employees of newspapers published by complainant's members to furnish Associated Press news to defendant before publication, for transmission by telegraph and telephone to defendant's clients for publication by them; Second, by inducing Associated Press members to violate its by-laws and permit defendant to obtain news before publication; and Third, by copying news from bulletin boards and from early editions of complainant's newspapers and selling this, either bodily or after rewriting it, to defendant's customers.

The District Court, upon consideration of the bill and answer, with voluminous affidavits on both sides, granted a preliminary injunction under the first and second heads; but refused at that stage to restrain the systematic practice admittedly pursued by defendant, of taking news bodily from the bulletin boards and early editions of complainant's newspapers and selling it as its own. The court expressed itself as satisfied that this practice amounted to unfair trade, but as the legal question was one of first impression it considered that the allowance of an injunction should await the outcome of an appeal. 240 Fed. Rep. 983, 996. Both parties having appealed, the Circuit Court of Appeals sustained the injunction order so far as it went, and upon complainant's appeal modified it and remanded the cause with directions to issue an injunction also against any bodily taking of the words or substance of complainant's news until its commercial value as news had passed away. 245 Fed. Rep. 244, 253. The present writ of certiorari was then allowed. 245 U.S. 644.

The only matter that has been argued before us is whether defendant may lawfully be restrained from appropriating news taken from bulletins issued by complainant or any of its members, or from newspapers published by them, for the purpose of selling it to defendant's clients. Complainant asserts that defendant's admitted course of conduct in this regard both violates complainant's property right in the news and constitutes unfair competition in business. And notwithstanding the case has proceeded only to the stage of a preliminary injunction, we have deemed it proper to consider the underlying questions, since they go to the very merits of the action and are presented upon facts that are not in dispute. As presented in argument, these questions are: 1. Whether there is any property in news; 2. Whether, if there be property in news collected for the purpose of being published, it survives the instant of its publication in the first newspaper to which it is communicated by the news-gatherer; and 3. Whether defendant's admitted course of conduct in appropriating for commercial use matter taken from bulletins or early editions of Associated Press publications constitutes unfair competition in trade.

The federal jurisdiction was invoked because of diversity of citizenship, not upon the ground that the suit arose under the copyright or other laws of the United States. Complainant's news matter is not copyrighted. It is said that it could not, in practice, be copyrighted, because of the large number of dispatches that are sent daily; and, according to complainant's contention, news is not within the operation of the copyright act. Defendant, while apparently conceding this, nevertheless invokes the analogies of the law of literary property and copyright, insisting as its principal contention that, assuming complainant has a right of property in its news, it can be maintained (unless the copyright act be complied with) only by being kept secret and confidential, and that upon the publication with complainant's consent of uncopyrighted news by any of complainant's members in a newspaper or upon a bulletin board, the right of property is lost, and the subsequent use of the news by the public or by defendant for any purpose whatever becomes lawful. . . .

In considering the general question of property in news matter, it is necessary to recognize its dual character, distinguishing between the substance of the information and the particular form or collocation of words in which the writer has communicated it.

No doubt news articles often possess a literary quality, and are the subject of literary property at the common law; nor do we question that such an article, as a literary production, is the subject of copyright by the terms of the act as it now stands. In an early case at the circuit Mr. Justice Thompson held in effect that a newspaper was not within the protection of the copyright acts of 1790 and 1802 (Clayton v. Stone, 2 Paine, 382; 5 Fed. Cas. No. 2872). But the present act is broader; it provides that the works for which copyright may be secured shall include "all the writings of an author," and specifically mentions "periodicals, including newspapers." Act of March 4, 1909, c. 320, §§4 and 5, 35 Stat. 1075, 1076. Evidently this admits to copyright a contribution to a newspaper, notwithstanding it also may convey news; and such is the practice of the copyright office, as the newspapers of the day bear witness. See Copyright Office Bulletin No. 15 (1917), pp. 7, 14, 16-17.

But the news element — the information respecting current events contained in the literary production — is not the creation of the writer, but is a report of matters that ordinarily are publici juris; it is the history of the day. It is not to be supposed that the framers of the Constitution, when they empowered Congress "to promote the progress of science and useful arts, by securing for limited times to authors and inventors the exclusive right to their respective writings and discoveries" (Const., Art I, §8, par. 8), intended to confer upon one who might happen to be the first to report a historic event the exclusive right for any period to spread the knowledge of it.

We need spend no time, however, upon the general question of property in news matter at common law, or the application of the copyright act, since it seems to us the case must turn upon the question of unfair competition in business. And, in our opinion, this does not depend upon any general right of property analogous to the common-law right of the proprietor of an unpublished work to prevent its publication without his consent; nor is it foreclosed by showing that the benefits of the copyright act have been waived. We are dealing here not with restrictions upon publication but with the very facilities and processes of publication. The peculiar value of news is in the spreading of it while it is fresh; and it is evident that a valuable property interest in the news, as news, cannot be maintained by keeping it secret. Besides, except for matters improperly disclosed, or published in breach of trust or confidence, or in violation of law, none of which is involved in this branch of the case, the news of current events may be regarded as common property. What we are concerned with is the business of making it known to the world, in which both parties to the present suit are engaged. That business consists in maintaining a prompt, sure, steady, and reliable service designed to place the daily events of the world at the breakfast table of the millions at a price that, while of trifling moment to each reader, is sufficient in the aggregate to afford compensation for the cost of gathering and distributing it, with the added profit so necessary as an incentive to effective action in the commercial world. The service thus performed for newspaper readers is not only innocent but extremely useful in itself, and indubitably constitutes a legitimate business. The parties are competitors in this field; and, on fundamental principles, applicable here as elsewhere, when the rights or privileges of the one are liable to conflict with those of the other, each party is under a duty so to conduct its own business as not unnecessarily or unfairly to injure that of the other. Hitchman Coal & Coke Co. v. Mitchell, 245 U.S. 229, 254.

Obviously, the question of what is unfair competition in business must be determined with particular reference to the character and circumstances of the business. The question here is not so much the rights of either party as against the public but their rights as between themselves. See Morison v. Moat, 9 Hare, 241, 258. And although we may and do assume that neither party has any remaining property interest as against the public in uncopyrighted news matter after the moment of its first publication, it by no means follows that there is no remaining property interest in it as between themselves. For, to both of them alike, news matter, however little susceptible of ownership or dominion in the absolute sense, is stock in trade, to be gathered at the cost of enterprise, organization, skill, labor, and money, and to be distributed and sold to those who will pay money for it, as for any other merchandise. Regarding the news, therefore, as but the material out of which both parties are seeking to make profits at the same time and in the same field, we hardly can fail to recognize that for this purpose, and as between them, it must be regarded as quasi property, irrespective of the rights of either as against the public.

In order to sustain the jurisdiction of equity over the controversy, we need not affirm any general and absolute property in the news as such. The rule that a court of equity concerns itself only in the protection of property rights treats any civil right of a pecuniary nature as a property right (In re Sawyer, 124 U.S. 200, 210; In re Debs, 158 U.S. 564, 593); and the right to acquire property by honest labor or the conduct of a lawful business is as much entitled to protection as the right to guard property already acquired. Truax v. Raich, 239 U.S. 33, 37-38; Brennan v. United Hatters, 73 N.J.L. 729, 742; Barr v. Essex Trades Council, 53 N.J. Eq. 101. It is this right that furnishes the basis of the jurisdiction in the ordinary case of unfair competition. . . .

Not only do the acquisition and transmission of news require elaborate organization and a large expenditure of money, skill, and effort; not only has it an exchange value to the gatherer, dependent chiefly upon its novelty and freshness, the regularity of the service, its reputed reliability and thoroughness, and its adaptability to the public needs; but also, as is evident, the news has an exchange value to one who can misappropriate it.

The peculiar features of the case arise from the fact that, while novelty and freshness form so important an element in the success of the business, the very processes of distribution and publication necessarily occupy a good deal of time. Complainant's service, as well as defendant's, is a daily service to daily newspapers; most of the foreign news reaches this country at the Atlantic seaboard, principally at the City of New York, and because of this, and of time differentials due to the earth's rotation, the distribution of news matter throughout the country is principally from east to west; and, since in speed the telegraph and telephone easily outstrip the rotation of the earth, it is a simple matter for defendant to take complainant's news from bulletins or early editions of complainant's members in the eastern cities and at the mere cost of telegraphic transmission cause it to be published in western papers issued at least as early as those served by complainant. Besides this, and irrespective of time differentials, irregularities in telegraphic transmission on different lines, and the normal consumption of time in printing and distributing the newspaper, result in permitting pirated news to be placed in the hands of defendant's readers sometimes simultaneously with the service of competing Associated Press papers, occasionally even earlier.

Defendant insists that when, with the sanction and approval of complainant, and as the result of the use of its news for the very purpose for which it is distributed, a portion of complainant's members communicate it to the general public by posting it upon bulletin boards so that all may read, or by issuing it to newspapers and distributing it indiscriminately, complainant no longer has the right to control the use to be made of it; that when it thus reaches the light of day it becomes the common possession of all to whom it is accessible; and that any purchaser of a newspaper has the right to communicate the intelligence which it contains to anybody and for any purpose, even for the purpose of selling it for profit to newspapers published for profit in competition with complainant's members.

The fault in the reasoning lies in applying as a test the right of the complainant as against the public, instead of considering the rights of complainant and defendant, competitors in business, as between themselves. The right of the purchaser of a single newspaper to spread knowledge of its contents gratuitously, for any legitimate purpose not unreasonably interfering with complainant's right to make merchandise of it, may be admitted; but to transmit that news for commercial use, in competition with complainant — which is what defendant has done and seeks to justify — is a very different matter. In doing this defendant, by its very act, admits that it is taking material that has been acquired by complainant as the result of organization and the expenditure of labor, skill, and money, and which is salable by complainant for money, and that defendant in appropriating it and selling it as its own is endeavoring to reap where it has not sown, and by disposing of it to newspapers that are competitors of complainant's members is appropriating to itself the harvest of those who have sown. Stripped of all disguises, the process amounts to an unauthorized interference with the normal operation of complainant's legitimate business precisely at the point where the profit is to be reaped, in order to divert a material portion of the profit from those who have earned it to those who have not; with special advantage to defendant in the competition because of the fact that it is not burdened with any part of the expense of gathering the news. The transaction speaks for itself, and a court of equity ought not to hesitate long in characterizing it as unfair competition in business.

The underlying principle is much the same as that which lies at the base of the equitable theory of consideration in the law of trusts — that he who has fairly paid the price should have the beneficial use of the property. Pom.

Eq. Jur., §981. It is no answer to say that complainant spends its money for that which is too fugitive or evanescent to be the subject of property. That might, and for the purposes of the discussion we are assuming that it would, furnish an answer in a common-law controversy. But in a court of equity, where the question is one of unfair competition, if that which complainant has acquired fairly at substantial cost may be sold fairly at substantial profit, a competitor who is misappropriating it for the purpose of disposing of it to his own profit and to the disadvantage of complainant cannot be heard to say that it is too fugitive or evanescent to be regarded as property. It has all the attributes of property necessary for determining that a misappropriation of it by a competitor is unfair competition because contrary to good conscience.

The contention that the news is abandoned to the public for all purposes when published in the first newspaper is untenable. Abandonment is a question of intent, and the entire organization of the Associated Press negatives such a purpose. The cost of the service would be prohibitive if the reward were to be so limited. No single newspaper, no small group of newspapers, could sustain the expenditure. Indeed, it is one of the most obvious results of defendant's theory that, by permitting indiscriminate publication by anybody and everybody for purposes of profit in competition with the news-gatherer, it would render publication profitless, or so little profitable as in effect to cut off the service by rendering the cost prohibitive in comparison with the return. The practical needs and requirements of the business are reflected in complainant's by-laws which have been referred to. Their effect is that publication by each member must be deemed not by any means an abandonment of the news to the world for any and all purposes, but a publication for limited purposes; for the benefit of the readers of the bulletin or the newspaper as such; not for the purpose of making merchandise of it as news, with the result of depriving complainant's other members of their reasonable opportunity to obtain just returns for their expenditures.

It is to be observed that the view we adopt does not result in giving to complainant the right to monopolize either the gathering or the distribution of the news, or, without complying with the copyright act, to prevent the reproduction of its news articles; but only postpones participation by complainant's competitor in the processes of distribution and reproduction of news that it has not gathered, and only to the extent necessary to prevent that competitor from reaping the fruits of complainant's efforts and expenditure, to the partial exclusion of complainant, and in violation of the principle that underlies the maxim sic utere tuo, etc.

It is said that the elements of unfair competition are lacking because there is no attempt by defendant to palm off its goods as those of the complainant, characteristic of the most familiar, if not the most typical, cases of unfair competition. Howe Scale Co. v. Wyckoff, Seamans & Benedict, 198 U.S. 118, 140. But we cannot concede that the right to equitable relief is confined to that class of cases. In the present case the fraud upon complainant's rights is more direct and obvious. Regarding news matter as the mere material from which these two competing parties are endeavoring to make money, and treating it, therefore, as quasi property for the purposes of their business because they are both selling it as such, defendant's conduct differs from the ordinary case of unfair competition in trade principally in this that, instead of selling its own goods as those of complainant, it substitutes misappropriation in the place of misrepresentation, and sells complainant's goods as its own.

Besides the misappropriation, there are elements of imitation, of false pretense, in defendant's practices. The device of rewriting complainant's news articles, frequently resorted to, carries its own comment. The habitual failure to give credit to complainant for that which is taken is significant. Indeed, the entire system of appropriating complainant's news and transmitting it as a commercial product to defendant's clients and patrons amounts to a false representation to them and to their newspaper readers that the news transmitted is the result of defendant's own investigation in the field. But these elements, although accentuating the wrong, are not the essence of it. It is something more than the advantage of celebrity of which complainant is being deprived. . . .

The decree of the Circuit Court of Appeals will be Affirmed.

HOLMES, J., concurring:

When an uncopyrighted combination of words is published there is no general right to forbid other people repeating them — in other words there is no property in the combination or in the thoughts or facts that the words express. Property, a creation of law, does not arise from value, although exchangeable — a matter of fact. Many exchangeable values may be destroyed intentionally without compensation. Property depends upon exclusion by law from interference, and a person is not excluded from using any combination of words merely because someone has used it before, even if it took labor and genius to make it. If a given person is to be prohibited from making the use of words that his neighbors are free to make some other ground must be found. One such ground is vaguely expressed in the phrase unfair trade. This means that the words are repeated by a competitor in business in such a way as to convey a misrepresentation that materially injures the person who first used them, by appropriating credit of some kind which the first user has earned. The ordinary case is a representation by device, appearance, or other indirection that the defendant's goods come from the plaintiff. But the only reason why it is actionable to make such a representation is that it tends to give the defendant an advantage in his competition with the plaintiff and that it is thought undesirable that an advantage should be gained in that way. Apart from that the defendant may use such unpatented devices and uncopyrighted combinations of words as he likes. The ordinary case, I say, is palming off the defendant's product as the plaintiff's, but the same evil may follow from the opposite falsehood — from saying, whether in words or by implication, that the plaintiff's product is the defendant's, and that, it seems to me, is what has happened here.

Fresh news is got only by enterprise and expense. To produce such news as it is produced by the defendant represents by implication that it has been acquired by the defendant's enterprise and at its expense. When it comes from one of the great news-collecting agencies like the Associated Press, the source generally is indicated, plainly importing that credit; and that such a representation is implied may be inferred with some confidence from the unwillingness of the defendant to give the credit and tell the truth. If the plaintiff produces the news at the same time that the defendant does, the defendant's presentation impliedly denies to the plaintiff the credit of collecting the facts and assumes that credit to the defendant. If the plaintiff is later in western cities it naturally will be supposed to have obtained its information from the defendant. The falsehood is a little more subtle, the injury a little more indirect, than in ordinary cases of unfair trade, but I think that the principle that condemns the one condemns the other. It is a question of how strong an infusion of fraud is necessary to turn a flavor into a poison. The dose seems to me strong enough here to need a remedy from the law. But as, in my view, the only ground of complaint that can be recognized without legislation is the implied misstatement, it can be corrected by stating the truth; and a suitable acknowledgment of the source is all that the plaintiff can require. I think that within the limits recognized by the decision of the Court the defendant should be enjoined from publishing news obtained from the Associated Press for ___ hours after publication by the plaintiff unless it gives express credit to the Associated Press; the number of hours and the form of acknowledgment to be settled by the District Court. . . .

BRANDEIS, J., dissenting: . . .

News is a report of recent occurrences. The business of the news agency is to gather systematically knowledge of such occurrences of interest and to distribute reports thereof. The Associated Press contended that knowledge so acquired is property, because it costs money and labor to produce and because it has value for which those who have it not are ready to pay; that it remains property and is entitled to protection as long as it has commercial value as news; and that to protect it effectively the defendant must be enjoined from making, or causing to be made, any gainful use of it while it retains such value. An essential element of individual property is the legal right to exclude others from enjoying it. If the property is private, the right of exclusion may be absolute; if the property is affected with a public interest, the right of exclusion is qualified. But the fact that a product of the mind has cost its producer money and labor, and has a value for which others are willing to pay, is not sufficient to ensure to it this legal attribute of property. The general rule of law is, that the noblest of human productions — knowledge, truths ascertained, conceptions, and ideas — become, after voluntary communication to others, free as the air to common use. Upon these incorporeal productions the attribute of property is continued after such communication only in certain classes of cases where public policy has seemed to demand it. These exceptions are confined to productions which, in some degree, involve creation, invention, or discovery. But by no means all such are endowed with this attribute of property. The creations which are recognized as property by the common law are literary, dramatic, musical, and other artistic creations; and these have also protection under the copyright statutes. The inventions and discoveries upon which this attribute of property is conferred only by statute, are the few comprised within the patent law. There are also many other cases in which courts interfere to prevent curtailment of plaintiff's enjoyment of incorporeal productions; and in which the right to relief is often called a property right, but is such only in a special sense. In those cases, the plaintiff has no absolute right to the protection of his production; he has merely the qualified right to be protected as against the defendant's acts, because of the special relation in which the latter stands or the wrongful method or means employed in acquiring the knowledge or the manner in which it is used. Protection of this character is afforded where the suit is based upon breach of contract or of trust or upon unfair competition.

The knowledge for which protection is sought in the case at bar is not of a kind upon which the law has heretofore conferred the attributes of property; nor is the manner of its acquisition or use nor the purpose to which it is applied, such as has heretofore been recognized as entitling a plaintiff to relief. . . .

The means by which the International News Service obtains news gathered by the Associated Press is also clearly unobjectionable. It is taken from papers bought in the open market or from bulletins publicly posted. No breach of contract such as the court considered to exist in Hitchman Coal & Coke Co. v. Mitchell, 245 U.S. 229, 254; or of trust such as was present in Morison v. Moat, 9 Hare, 241; and neither fraud nor force, is involved. The manner of use is likewise unobjectionable. No reference is made by word or by act to the Associated Press, either in transmitting the news to subscribers or by them in publishing it in their papers. Neither the International News Service nor its subscribers is gaining or seeking to gain in its business a benefit from the reputation of the Associated Press. They are merely using its product without making compensation. See Bamforth v. Douglass Post Card & Machine Co., 158 Fed. Rep. 355; Tribune Co. of Chicago v. Associated Press, 116 Fed. Rep. 126. That, they have a legal right to do; because the product is not property, and they do not stand in any relation to the Associated Press, either of contract or of trust, which otherwise precludes such use. The argument is not advanced by characterizing such taking and use a misappropriation.

It is also suggested, that the fact that defendant does not refer to the Associated Press as the source of the news may furnish a basis for the relief. But the defendant and its subscribers, unlike members of the Associated Press, were under no contractual obligation to disclose the source of the news; and there is no rule of law requiring acknowledgment to be made where uncopyrighted matter is reproduced. The International News Service is said to mislead its subscribers into believing that the news transmitted was originally gathered by it and that they in turn mislead their readers. There is, in fact, no representation by either of any kind. Sources of information are sometimes given because required by contract; sometimes because naming the source gives authority to an otherwise incredible statement; and sometimes the source is named because the agency does not wish to take the responsibility itself of giving currency to the news. But no representation can properly be implied from omission to mention the source of information except that the International News Service is transmitting news which it believes to be credible.

. . . The great development of agencies now furnishing country-wide distribution of news, the vastness of our territory, and improvements in the means of transmitting intelligence, have made it possible for a news agency or newspapers to obtain, without paying compensation, the fruit of another's efforts and to use news so obtained gainfully in competition with the original collector. The injustice of such action is obvious. But to give relief against it would involve more than the application of existing rules of law to new facts. It would require the making of a new rule in analogy to existing ones. The unwritten law possesses capacity for growth; and has often satisfied new demands for justice by invoking analogies or by expanding a rule or principle. This process has been in the main wisely applied and should not be discontinued. Where the problem is relatively simple, as it is apt to be when private interests only are involved, it generally proves adequate. But with the increasing complexity of society, the public interest tends to become omnipresent; and the problems presented by new demands for justice cease to be simple. Then the creation or recognition by courts of a new private right may work serious injury to the general public, unless the boundaries of the right are definitely established and wisely guarded. In order to reconcile the new private right with the public interest, it may be necessary to prescribe limitations and rules for its enjoyment; and also to provide administrative machinery for enforcing the rules. It is largely for this reason that, in the effort to meet the many new demands for justice incident to a rapidly changing civilization, resort to legislation has latterly been had with increasing frequency.

The rule for which the plaintiff contends would effect an important extension of property rights and a corresponding curtailment of the free use of knowledge and of ideas; and the facts of this case admonish us of the danger involved in recognizing such a property right in news, without imposing upon news-gatherers corresponding obligations. . . .

Courts are ill-equipped to make the investigations which should precede a determination of the limitations which should be set upon any property right in news or of the circumstances under which news gathered by a private agency should be deemed affected with a public interest. Courts would be powerless to prescribe the detailed regulations essential to full enjoyment of the rights conferred or to introduce the machinery required for enforcement of such regulations. Considerations such as these should lead us to decline to establish a new rule of law in the effort to redress a newly-disclosed wrong, although the propriety of some remedy appears to be clear.

COMMENTS AND QUESTIONS

1. Both the majority and the dissent seem to agree that there is no traditional intellectual property right in published news. Certainly copyright law — the most likely candidate for protecting works of authorship — cannot protect AP's news. See A. A. Hoehling, supra Chapter 4. And the majority's reference to "unfair competition" suggests that its theory is grounded in a tort of wrongful appropriation. But the tort analogy runs into difficulty as well, since (as Justice Brandeis observes) it is hard to fault the conduct at issue in the case. (Note that the Supreme Court did not consider AP's claim that INS had bribed AP newspapers or employees, an allegation which if true seems much more likely to support a tort action.) Instead, the Court seems to settle into what it calls a theory of "quasi-property" that entitles its owner to protection only against direct competitors.

2. How should courts following *International News* determine whether two companies are direct competitors who are then subject to this quasi-property right? For example, suppose that a single newspaper (not a member of AP) had appropriated information from published AP reports. Does the single paper compete with AP in news gathering? Does it matter whether the paper competes directly with one of AP's members, or has its own city monopoly? Or should the test be whether AP could have expected licensing revenues from the paper, whether or not they are in competition? Cf. National Football League v. Delaware, 435 F. Supp. 1372 (D. Del. 1977) (NFL cannot prevent Delaware from implementing a lottery based on NFL games, because the lottery is a "collateral service" rather than one in competition with the NFL).

3. How long does this quasi-property right last? The Court suggests one answer: until AP has appropriated its news value as a return on its news-gathering activities. Justice Holmes's concurrence suggests that this would be a matter of hours. But is that necessarily true? What happens when different news media compete? Should there be different rules for CNN, scheduled television news, daily newspapers, and weekly and monthly news magazines? Is AP entitled to protect its news not only from appropriation by "immediate" news sources, but also from the weeklies and monthlies that seek to use its news reports?

4. What if INS or its newspapers had credited AP with reporting the story? In that case, AP would presumably have received some value for its news — credit for the "scoop." But its member papers would still have faced competition from nonmembers that they would prefer to avoid. Should attribution be enough? Justice Holmes says yes, but neither the majority nor Justice Brandeis' dissent see this as the major issue (although the majority notes that INS's failure to attribute the news to AP "accentuates" the unfairness of its competition). Note that attribution should solve any Lanham Act problem that might otherwise arise, since there is no danger that INS or its member papers will be "passing off" AP stories as their own if the source of the story is clearly identified.

This fact points up a significant difference between the tort of misappropriation and trademark infringement. Trademark law is premised on harm to consumers stemming from a "likelihood of confusion." Absent consumer confusion, trademarks (and trade dress) are generally not protectable. *International News* cuts a much broader swath. If it is not based on consumer protection, what is the rationale for the tort of misappropriation? Is it incentive theory? If so, what exactly is the Court trying to encourage?

5. The *International News* decision has had a checkered subsequent history. Early decisions by lower courts attempted to construe the decision narrowly. Indeed, in 1929 Judge Learned Hand went so far as to say: "While it is of course true that the law ordinarily speaks in general terms, there are cases where the occasion is at once the justification for, and the limit of, what is decided. *[International News]* appears to us such an instance; we think that no more was covered than situations substantially similar to those then at bar." Cheney Bros. v. Doris Silk Corp., 35 F.2d 279 (2d Cir. 1929), *cert. denied*, 281 U.S. 728 (1930). Judge Hand went on to characterize the problems with the Court's decision as "insuperable," and to state that it "flagrantly conflict[ed]" with the federal statutory intellectual property laws. Id.

After Erie R. Co. v. Tompkins, 304 U.S. 64 (1938) abolished federal common law in diversity cases (the grounds on which *International News* was decided), it seemed to many that the *International News* misappropriation doctrine was dead. See, e.g., James Treece, Patent Policy and Preemption: The Stiffel and Compco Cases, 32 U. Chi. L. Rev. 80 (1964). But the doctrine has since reappeared in a number of cases. See, e.g., United States Golf Assn. v. St. Andrews Systems, 749 F.2d 1028 (3d Cir. 1984) (*International News* was based on direct competition between the parties; court refused to apply it absent such direct competition); Imax Corp. v. Cinema Technologies, Inc., 152 F.3d 1161 (9th Cir. 1998) (owner of movie projector equipment could recover on a common law misappropriation theory for the use of information disclosed in confidence, despite the fact that the plaintiff could not prevail on its trade secret claim); Ettore v. Philco Television Broadcasting Corp., 229 F.2d 481 (3d Cir. 1956), *cert. denied*, 351 U.S. 926 (1956) (producer of boxing match could recover damages from television station which broadcast the match without permission). In National Basketball Assn. v. Sports Team Analysis and Tracking Systems, 105 F.3d 841 (2d Cir. 1997), the most recent case involving the misappropriation tort, the court held that *International News* survives today only where

> (i) a plaintiff generates or gathers information at a cost; (ii) the information is time-sensitive; (iii) a defendant's use of the information constitutes free-riding on the plaintiff's efforts; (iv) the defendant is in direct competition with a product or service offered by the plaintiffs; and (v) the ability of other parties to free-ride on the efforts of the plaintiff or others would so reduce the incentive to produce the product that its existence or quality would be substantially threatened.

The court held that these limitations on the misappropriation doctrine made it qualitatively different from a cause of action for copyright infringement and, therefore, saved it from preemption by the copyright laws. We discuss copyright preemption infra.

6. Suppose that the plaintiff in Feist v. Rural Telephone Services, supra Chapter 4, had brought a claim for misappropriation and unfair competition rather than a copyright claim. Should the plaintiff prevail under *International News?* It certainly would be able to claim that it had invested substantial time and effort in putting together a telephone directory, and that defendant (a direct competitor) had merely copied that information as soon as it became public. Assuming this is enough to show unfair competition under *International News*, what does that fact suggest about the interaction between the copyright laws and the unfair competition doctrine? Does it make sense that the rule in *Feist* should coexist with unfair competition law?

7. One solution to the inconsistency between *Feist* and *International News* would be to create an explicit (presumably statutory) property right in factual compilations. Strong arguments in favor of such protection were made in the copyright context both before and after *Feist*. See Note on Sweat of the Brow, supra Chapter 4. Furthermore, doesn't the Court have a valid point about the incentive to invest in a productive activity (collecting news) whose benefits will immediately be dissipated by imitators? Isn't that precisely the justification for the patent system?

On the other hand, what problems do you foresee with such a new federal intellectual property regime? Would it overwhelm patent and copyright as means of protecting intellectual property? What effect would it have on businesses? Would it encourage factual research and compilation, or just encourage monopolization of facts? On this subject, see also Leo J. Raskind, The Misappropriation Doctrine as a Competitive Norm of Intellectual Property Law, 75 Minn. L. Rev. 875, 876-77 (1991), where the author argues that:

> when courts hear patent, copyright, and trademark cases in which statutory protection is inappropriate, but nonetheless the conduct of a party is characterized as "chiseling," "piracy," "unethical," or the like, they should begin their analysis by considering the competitive relationship from which the claim originates. The clear legislative expression of a preference for competition contained in federal antitrust laws warrants this approach. Moreover, courts in these cases should recognize the Supreme Court's continued emphasis on the preemptive effect given federal legislation relating to competition. Ancillary doctrines that impinge on competition, such as misappropriation, should be invoked sparingly. From this perspective, courts should consider allegedly "inappropriate" conduct as an element of behavior in a competitive market context; within that framework, courts should view such conduct as an element of cost that a seller, as a competitor, considers when determining how much of a particular product to offer.

8. To what extent was the Court in *INS* merely enforcing the norms of commercial reasonableness in the newspaper industry? See Richard Epstein, International News Service v. Associated Press: Custom and Law as Sources of Property Rights in News, 78 Va. L. Rev. 85 (1992) (noting the different

customary rules in the news trade under which "lifting" stories was wrong, but following up on another paper's "tips" to report the same story was acceptable; and concluding that, partially unwittingly, Justice Pitney ended up affirming these customary rules in his opinion); Douglas G. Baird, Common Law Intellectual Property and the Legacy of International News Service v. Associated Press, 50 U. Chi. L. Rev. 411 (1983). But see Stephen L. Carter, Custom, Adjudication, and Petrushevsky's Watch: Some Notes from the Intellectual Property Front, 78 Va. L. Rev. 129, 132 (1992) ("Even courts inclined to enforce private orderings might not be very good at anthropology. The judge, after all, is on the outside, looking in. Even assuming — and there is no reason to do so — that the parties tell the whole truth, it will not always be so easy for a court to discern an industry's customs.").

2. State "Common Law" Copyright

Before the American Revolution, copyright was a creature of state law, either by statute or by interpretation of the common law. As federal law has gradually come to occupy more and more of the scope of copyright, the role of state law has been correspondingly reduced. Nonetheless, state common law still provides copyright-style protection in some areas that federal law does not. Consider the following case.

Estate of Hemingway v. Random House, Inc.
Court of Appeals of New York
23 N.Y.2d 341, 244 N.E.2d 250, 296 N.Y.S.2d 771
(Ct. App. N.Y. 1968)

On this appeal — involving an action brought by the estate of the late Ernest Hemingway and his widow against the publisher and author of a book, entitled 'Papa Hemingway' — we are called upon to decide, primarily, whether conversations of a gifted and highly regarded writer may become the subject of common-law copyright, even though the speaker himself has not reduced his words to writing. . . .

Hemingway died in 1961. During the last 13 years of his life, a close friendship existed between him and A. E. Hotchner, a younger and far less well-known writer. Hotchner, who met Hemingway in the course of writing articles about him, became a favored drinking and traveling companion of the famous author, a frequent visitor to his home and the adapter of some of his works for motion pictures and television. During these years, Hemingway's conversation with Hotchner, in which others sometimes took part, was filled with anecdote, reminiscence, literary opinion and revealing comment about actual persons on whom some of Hemingway's fictional characters were based. Hotchner made careful notes of these conversations soon after they occurred, occasionally recording them on a portable tape recorder.

During Hemingway's lifetime, Hotchner wrote and published several articles about his friend in which he quoted some of this talk at length. Hemingway, far from objecting to this practice, approved of it. Indeed, the record reveals that other writers also quoted Hemingway's conversation without any objection from him, even when he was displeased with the articles themselves.

After Hemingway's death, Hotchner wrote "Papa Hemingway," drawing upon his notes and his recollections, and in 1966 it was published by the defendant Random House. Subtitled "a personal memoir", it is a serious and revealing biographical portrait of the world-renowned writer. Woven through the narrative, and giving the book much of its interest and character, are lengthy quotations from Hemingway's talk, as noted or remembered by Hotchner. . . .

[Among the causes of action was] that "Papa Hemingway" consists, in the main, of literary matter composed by Hemingway in which he had a common-law copyright. . . .

Turning to [this] cause of action, we agree with the disposition made below but on a ground more narrow than that articulated by the court at Special Term. It is the position of the plaintiffs (under this count) that Hemingway was entitled to a common-law copyright on the theory that his directly quoted comment, anecdote and opinion were his "literary creations", his "literary property", and that the defendant Hotchner's note-taking only performed the mechanics of recordation. And, in a somewhat different vein, the plaintiffs argue that "(w)hat for Hemingway was oral one day would be or could become his written manuscript the next day", that his speech, constituting not just a statement of his ideas but the very form in which he conceived and expressed them, was as much the subject of common-law copyright as what he might himself have committed to paper.

Common-law copyright is the term applied to an author's proprietary interest in his literary or artistic creations before they have been made generally available to the public. It enables the author to exercise control over the first publication of his work or to prevent publication entirely — hence, its other name, the "right of first publication".[1] No cases deal directly with the question whether it extends to conversational speech and we begin, therefore, with a brief review of some relevant concepts in this area of law.

[A]s a noted scholar in the field has observed, "the underlying rationale for common law copyright (i.e., the recognition that a property status should attach to the fruits of intellectual labor) is applicable regardless of whether such labor assumes tangible form" (Nimmer, Copyright, §11.1, p. 40). The principle that it is not the tangible embodiment of the author's work but the

---

1. Although common-law copyright in an unpublished work lasts indefinitely, it is extinguished immediately upon publication of the work by the author. He must then rely, for his protection, upon Federal statutory copyright. (See Nimmer, Copyright, §11, pp. 38-42 and ch. 4, p. 183 et seq.) Section 2 of the [1909] Copyright Act (U.S. Code, tit. 17) expressly preserves common-law rights in unpublished works against any implication that the field is pre-empted by the Federal statute.

creation of the work itself which is protected finds recognition in a number of ways in copyright law.

One example, with some relevance to the problem before us, is the treatment which the law has accorded to personal letters — a kind of half-conversation in written form. Although the paper upon which the letter is written belongs to the recipient, it is the author who has the right to publish them or to prevent their publication. . . . Nor has speech itself been entirely without protection against reproduction for publication. The public delivery of an address or a lecture or the performance of a play is not deemed a "publication," and, accordingly, it does not deprive the author of his common-law copyright in its contents.

Letters, however — like plays and public addresses, written or not — have distinct, identifiable boundaries and they are, in most cases, only occasional products. Whatever difficulties attend the formulation of suitable rules for the enforcement of rights in such works . . . they are relatively manageable. However, conversational speech, the distinctive behavior of man, is quite another matter, and subjecting any part of it to the restraints of common-law copyright presents unique problems.

One such problem . . . is that of avoiding undue restraints on the freedoms of speech and press and, in particular, on the writers of history and of biographical works of the genre of Boswell's "Life of Johnson". The safeguarding of essential freedoms in this area is, though, not without its complications. The indispensable right of the press to report on what people have *done*, or on what has *happened* to them or on what they have *said* in public does not necessarily imply an unbounded freedom to publish whatever they may have *said* in private conversation, any more than it implies a freedom to copy and publish what people may have put down in private writings. . . .

[S]peech is now easily captured by electronic devices and, consequently, we should be wary about excluding all possibility of protecting a speaker's right to decide when his words, uttered in private dialogue, may or may not be published at large. Conceivably, there may be limited and special situations in which an interlocutor brings forth oral statements from another party which both understand to be the unique intellectual product of the principal speaker, a product which would qualify for common-law copyright if such statements were in writing. Concerning such problems, we express no opinion; we do no more than raise the questions, leaving them open for future consideration in cases which may present them more sharply than this one does.

The defendant Hotchner asserts — without contradiction in the papers before us — that Hemingway never suggested to him or to anyone else that he regarded his conversational remarks to be "literary creations" or that he was of a mind to restrict Hotchner's use of the notes and recordings which Hemingway knew him to be accumulating. On the contrary, as we have already observed, it had become a continuing practice, during Hemingway's lifetime, for Hotchner to write articles about Hemingway, consisting largely of quotations from the latter's conversation — and of all of this Hemingway approved. In these circumstances, authority to publish must be implied, thus negativing the reservation of any common-law copyright.

Assuming, without deciding, that in a proper case a common-law copyright in certain limited kinds of spoken dialogue might be recognized, it would, at the very least, be required that the speaker indicate that he intended to mark off the utterance in question from the ordinary stream of speech, that he meant to adopt it as a unique statement and that he wished to exercise control over its publication. In the conventional common-law copyright situation, this indication is afforded by the creation of the manuscript itself. It would have to be evidenced in some other way if protection were ever to be accorded to some forms of conversational dialogue.

[Affirmed.]

COMMENTS AND QUESTIONS

1. As we saw in Chapter 4, the history of federal copyright law has been one of almost continuous expansion — in protectable subject matter, in the scope of protection afforded, and in the duration of copyright protection. This growth in federal protection has diminished the importance of common law and state statutory copyright, because such state copyright protection is relevant only for works which (for one reason or another) are ineligible for federal protection. (On why this is so, see infra section B.2, discussing federal preemption of state rights equivalent to federal copyright law.)

For example, the 1968 decision in *Hemingway* makes reference to the (pre-1976) rule that federal copyright law did not extend to works until they were published. However, the requirement of publication was abolished in the 1976 Copyright Act. Other formal requirements — registration, notice, and deposit, for example — are also much less significant than they once were.

Today, the primary significance of state copyright is in the area of "unfixed" works, such as the conversations at issue in *Hemingway*. And even in that area, the federal law is encroaching. As we discussed in Chapter 4, and will again in Chapter 7, courts are increasingly inclined to interpret the fixation requirement narrowly, giving copyright protection even to works that are fixed only for a brief period of time. Further, in the most recent set of revisions to the Copyright Act, Congress added section 1101, which provides protection for unfixed musical performances. For a comprehensive discussion of section 1101, see Susan M. Deas, Jazzing Up the Copyright Act? Resolving the Uncertainties of the United States Anti-Bootlegging Law, 20 Hastings Comm. & Ent. L.J. 567 (1998).

2. There are several policies underlying the fixation requirement, including a concern that unscrupulous plaintiffs would falsely claim that an unrecorded conversation had been "copied" by a defendant. Given these

policies, does it make sense to protect "unfixed" works via state copyright law? If we should protect them, is there any reason to prefer doing so at the state rather than the federal level?

3. Are there any First Amendment problems with giving a person the ability to control the reporting of his or her oral statements?

4. For commentary on the *Hemingway* issue, see Vicki L. Ruhga, Note, Ownership of Interviews: A Theory for Protection of Quotations, 67 Neb. L. Rev. 675 (1988): "Some commentators, and the Office of Copyright, indicate that interviews should be protected by a 'dual' copyright. Under the dual theory, both the interviewer and the interviewee can claim copyright in their respective expressions, absent an agreement to the contrary." See also Falwell v. Penthouse Intl., 521 F. Supp. 1204 (W.D. Va. 1981) (rejecting Falwell's claim of common law copyright in interview given by Falwell to two freelance reporters who sold a story based on an interview to Penthouse magazine).

3. Idea Submissions

═══ *Downey v. General Foods Corporation*
Court of Appeals of New York
286 N.E.2d 257, 334 N.Y.S.2d 874 (N.Y. Ct. App. 1972)

FULD, Chief Judge.

The plaintiff, an airline pilot, brought this action against the defendant General Foods Corporation to recover damages [of $2,800,000] for the alleged misappropriation of an idea. It is his claim that he suggested that the defendant's own gelatin product, "Jell-O," be named "Wiggley" or a variation of that word, including "Mr. Wiggle," and that the product be directed towards the children's market; that, although the defendant disclaimed interest in the suggestion, it later offered its product for sale under the name "Mr. Wiggle." The defendant urges — by way of affirmative defense — that the plaintiff's "alleged 'product concept and name' was independently created and developed" by it. The plaintiff moved for partial summary judgment "on the question of liability" on 5 of its 14 causes of action and the defendant cross-moved for summary judgment dismissing the complaint. The court at Special Term denied both motions, and the Appellate Division affirmed, 37 A.D.2d 250, 323 N.Y.S.2d 578, granting leave to appeal to this court on a certified question.

The plaintiff relies chiefly on correspondence between himself and the defendant, or, more precisely, on letters over the signature of a Miss Dunham, vice-president in charge of one of its departments. On February 15, 1965, the plaintiff wrote to the defendant, stating that he had an "excellent idea to increase the sale of your product JELL-O . . . making it available for children". Several days later, the defendant sent the plaintiff an "Idea Submittal

Form" (ISF) which included a form letter and a space for explaining the idea.[2] In that form, the plaintiff suggested, in essence, that the produce "be packaged & distributed to children under the name 'WIG-L-E' (meaning wiggly or wiggley) or 'WIGGLE-E' or 'WIGGLE-EEE' or 'WIGLEY.' " He explained that, although his children did not "get especially excited about the Name JELL-O, or wish to eat it", when referred to by that name, "the kids really took to it fast" when his wife "called it 'wiggle-y,' " noting that they then "associate(d) the name to the 'wiggleing' dessert." Although this is the only recorded proof of his idea, the plaintiff maintains that he sent Miss Dunham two handwritten letters in which he set forth other variations of "Wiggiley," including "Mr. Wiggley, Wiggle, Wiggle-e."[3]

A letter, dated March 8, 1965, over the signature of Miss Dunham, acknowledged the submission of the ISF and informed the plaintiff that it had no interest in promoting his suggestion. However, in July, the defendant introduced into the market a Jell-O product which it called "Mr. Wiggle." The plaintiff instituted the present action some months later. In addition to general denials, the answer contains several affirmative defenses, one of which, as indicated above, recites that the defendant independently created the product's concept and name before the plaintiff's submission to it.

In support of its position, the defendant pointed to depositions taken by the plaintiff from its employees and from employees of Young & Rubicam, the firm which did its advertising. From these it appears that the defendant first began work on a children's gelatin product in May, 1965 — three months after the plaintiff had submitted his suggestion — in response to a threat by Pillsbury Company to enter the children's market with a product named "Jiggly." Those employees of the defendant in charge of the project enlisted the aid of Young & Rubicam which, solely on its own initiative, "came up with the name 'Mr. Wiggle' ". In point of fact, Miss Dunham swore in her deposition that she had had no knowledge whatever of the plaintiff's idea until late in 1966, shortly before commencement of his suit; that ideas submitted by the general public were kept in a file by an assistant of hers "under lock and key"; and that no one from any other of the defendant's departments ever asked to research those files. The assistant, who had alone handled the correspondence with the plaintiff over Miss Dunham's signature — reproduced by means of a signature duplicating machine — deposed that she had no contact whatsoever with Young & Rubicam and had never discussed the name "Wiggle" or "Mr. Wiggle" with any one from that firm.

---

2. The form letter — signed and returned by the plaintiff — recited that "I submit this suggestion with the understanding, which is conclusively evidenced by my use and transmittal to you of this form, that this suggestion is not submitted to you in confidence, that no confidential relationship has been or will be established between us and that the use, if any, to be made of this suggestion by you and the compensation to be paid therefor, if any, if you use it, are matters resting solely in your discretion."

3. Neither of these letters was found in the defendant's files, nor did the plaintiff have the originals or exact copies.

In addition to the depositions of its employees and the employees of its advertising agency, the defendant submitted documentary proof of its prior use of some form of the word "wiggle" in connection with its endeavor to sell Jell-O to children. Thus, it submitted (1) a copy of a report which Young & Rubicam furnished it in June of 1959 proposing "an advertising program directed at children as a means of securing additional sales volume"; (2) a copy of a single dimensional reproduction of a television commercial, prepared in 1959 and used thereafter by the defendant in national and local television broadcasts, which contained the phrase, "ALL THAT WIGGLES IS NOT JELL-O"; and (3) a copy of a newspaper advertisement that appeared in 1960, depicting an Indian "squaw" puppet and her "papoose" preparing Jell-O — the "top favorite in every American tepee" — and suggesting to mothers that they "(m)ake a wigglewam of Jell-O for your tribe tonight!"

The critical issue in this case turns on whether the idea suggested by the plaintiff was original or novel. An idea may be a property right. But, when one submits an idea to another, no promise to pay for its use may be implied, and no asserted agreement enforced, if the elements of novelty and originality are absent, since the property right in an idea is based upon these two elements. (See Soule v. Bon Ami Co., 201 App. Div. 794, 796, 195 N.Y.S. 574, 575, *aff'd*, 235 N.Y. 609, 139 N.E. 754; Bram v. Dannon Milk Prods., 33 A.D.2d 1010, 307 N.Y.S.2d 571. . . .) The *Bram* case is illustrative; in reversing Special Term and granting summary judgment dismissing the complaint, the Appellate Division made it clear that, despite the asserted existence of an agreement, the plaintiff could not recover for his idea if it was not original and had been used before (33 A.D.2d, at p. 1010, 307 N.Y.S.2d 571):

> The idea submitted by the plaintiff to the defendants, the concept of depicting an infant in a highchair eating and enjoying yogurt, was lacking in novelty and had been utilized by the defendants . . . prior to its submission. Lack of novelty in an idea is fatal to any cause of action for its unlawful use. In the circumstances a question of fact as to whether there existed an oral agreement between the parties would not preclude summary judgment.

In the case before us, the record indisputably establishes, first, that the idea submitted — use of a word ("wiggley" or "wiggle") descriptive of the most obvious characteristic of Jell-O, with the prefix "Mr." added — was lacking in novelty and originality and, second, that the defendant had envisaged the idea, indeed had utilized it, years before the plaintiff submitted it. As already noted, it had made use of the word "wiggles" in a 1959 television commercial and the word "wigglewam" in a 1960 newspaper advertisement. It was but natural, then, for the defendant to employ some variation of it to combat Pillsbury's entry into the children's market with its "Jiggly." Having relied on its own previous experience, the defendant was free to make use of "Mr. Wiggle" without being obligated to compensate the plaintiff.

It is only necessary to add that, in light of the complete pretrial disclosure in this case of every one who had any possible connection with the creation of the name, the circumstance, adverted to by the courts below, that the facts surrounding the defendant's development of the name were within the knowledge of the defendant and its advertising agency does not preclude a grant of summary judgment. In the present case, it was shown beyond per-adventure that there was no connection between Miss Dunham's department and the defendant's other employees or the employees of the advertising outfit who took part in the creation of "Mr. Wiggle." In exhaustive discovery proceedings — which included examinations of all parties concerned either with that name or the defendant's idea files — the plaintiff was furnished with every conceivable item of information in the defendant's possession bearing on the privacy and confidentiality of such files and on the absence of access to them by those outside of Miss Dunham's department. The hope, expressed by the plaintiff that he may be able to prove that the witnesses who gave testimony in examinations before trial lied, is clearly insufficient to create an issue of fact requiring a trial or defeat the defendant's motion for summary judgment. . . .

The order appealed from should be reversed. . . .

COMMENTS AND QUESTIONS

1. The case of Soule v. Bon Ami, 139 N.E. 94 (N.Y. 1923) (per curiam) is a much-cited early case on the law of what has come to be called "idea submissions." In this case, Soule suggested essentially that Bon Ami could make more money by squeezing its retailers' profit margin. According to Soule, Bon Ami should raise the price it charged retailers while maintaining the product's price to consumers. Soule's bargain with Bon Ami was that Soule would be compensated only if his pricing strategy increased profits without hurting sales volume. This idea was of course not new, yet Bon Ami subsequently used it. Although Soule turned out to be right — Bon Ami increased profits without reducing sales — Bon Ami refused to pay. New York's highest state court (the Court of Appeals) held that Soule was not entitled to compensation. Most read this as the origin of the "novelty" re-quirement in New York idea submission cases, but at least one commentator claims that a close reading of the opinion supports the conclusion that Soule failed because he did not prove increased profits. See Ronald Caswell, Com-ment, A Comparison and Critique of Idea Protection in California, New York, and Great Britain, 14 Loy. L.A. Intl. & Comp. L.J. 717 (1992).

The novelty requirement was central to the Second Circuit's decision rejecting an idea-submission claim in Murray v. National Broadcasting Co., 844 F.2d 988 (2d Cir. 1988), *cert. denied*, 488 U.S. 955 (1988). In that case, Murray (an NBC employee) alleged that he had given NBC the idea for a television series focusing on the life of a black family and that his idea was eventually used as the basis for "The Cosby Show." The court rejected Murray's claim because it found that his idea was insufficiently novel:

> [W]e believe, as a matter of law, that plaintiff's idea embodied in his "Father's Day" proposal was not novel because it merely represented an "adaptation of existing knowledge" and of "known ingredients" and therefore lacked "genuine novelty and invention." Educational Sales Programs, 317 N.Y.S.2d at 844.
>
> We recognize of course that even novel and original ideas to a greater or lesser extent combine elements that are themselves not novel. Originality does not exist in a vacuum. Nevertheless, where, as here, an idea consists in essence of nothing more than a variation on a basic theme — in this case, the family situation comedy — novelty cannot be found to exist. . . .
>
> Appellant would have us believe that by interpreting New York law as we do, we are in effect condoning the theft of ideas. On the contrary, ideas that reflect "genuine novelty and invention" are fully protected against unauthorized use. But those ideas that are not novel "are in the public domain and may freely be used by anyone with impunity." Since such non-novel ideas are not protectible as property, they cannot be stolen.

Id. at 992-93. Contra Reeves v. Alyeska Pipeline Serv. Co., 926 P.2d 1130 (Alaska 1996) (plaintiff did not need to demonstrate novelty in an idea submission case based on quasi-contract theory).

2. How would you justify the rule that novelty and originality are required for an idea submitter to be entitled to compensation? Does it turn on the courts' characterization of the submitter's interest in the idea as a "property right"? Why is the label "property" a necessary element of the analysis, especially in light of the court's statement that the issue is the enforceability of an "asserted *agreement*"? What contract doctrine is implicated when an idea submitter submits an idea that is already in the "prior art" — consideration? Is there potential value in the act of identifying or locating a preexisting idea and bringing it to the attention of one who might profit from it? Is this at least possibly more than the proverbial peppercorn?

In a recent article, David McGovern articulates five separate rationales for protecting idea submissions: property, unjust enrichment, express contract, implied-in-fact contract, and confidential relationship. David McGovern, What Is Your Pitch? Idea Protection Is Nothing But Curveballs, 15 Loy. (LA) Ent. L.J. 475 (1995). McGovern acknowledges that not all courts accept all theories. Further, some theories are simply not applicable in certain situations (for example, there may be no express contract or confidential relationship between the parties).

3. How does the contract analysis in *Downey* compare to that in the trade secret/breach of confidence cases such as Smith v. Dravo, discussed in Chapter 2? What (if any) differences are there in the relationship of the parties, the duties of the idea/information recipient, and the nature of the idea or information?

4. Are ideas really the subject of property, as the court suggests? Both patent and copyright law go to great lengths to prevent intellectual property

owners from controlling the fundamental ideas resident in their works, even if those ideas are new and nonobvious. Why protect idea submissions at all, if they do not fall within the realm of the intellectual property laws discussed in Chapters 2 to 5? Does protecting ideas that are original or novel undermine the delicate balance in the Copyright Act between the rights of authors and users? After all, it is a fundamental principle of copyright law that ideas, unlike expression, are free for the taking. A similar argument can be made with respect to trade secrecy: what is the point of the secrecy requirement, if original or novel ideas can be protected under state common law even though they are widely known?

Desny v. Wilder
Supreme Court of California
46 Cal. 2d 715, 299 P.2d 257, 110 U.S.P.Q. 433
(Cal. Sup. Ct. 1956)

SCHAUER, Justice.

Plaintiff appeals from a summary judgment rendered against him in this action to recover the reasonable value of a literary composition, or of an idea for a photoplay, a synopsis of which composition, embodying the idea, he asserts he submitted to defendants for sale, and which synopsis and idea, plaintiff alleges, were accepted and used by defendants in producing a photoplay. . . .

[W]e have concluded, for reasons hereinafter stated, that the summary judgment in favor of defendants was erroneously granted and should be reversed. . . .

[I]t appears from the present record that defendant [Billy] Wilder [a famous director] at the times here involved was employed by defendant Paramount Pictures Corporation . . . either as a writer, producer or director, or a combination of the three. In November, 1949, plaintiff telephoned Wilder's office. Wilder's secretary, who was also employed by Paramount, answered, and plaintiff stated that he wished to see Wilder. At the secretary's insistence that plaintiff explain his purpose, plaintiff 'told her about this fantastic unusual story. . . . I told her that it was the life story of Floyd Collins who was trapped and made sensational news for two weeks . . . and I told her the plot.' . . . Two days later plaintiff, after preparing a three or four page outline of the story, telephoned Wilder's office a second time and told the secretary the synopsis was ready. The secretary requested plaintiff to read the synopsis to her over the telephone so that she could take it down in shorthand, and plaintiff did so. . . . Plaintiff on his part told the secretary that defendants could use the story only if they paid him 'the reasonable value of it.' . . . She said that if Billy Wilder of Paramount uses the story, 'naturally we will pay you for it.' . . . Plaintiff's only subsequent contact with the secretary was a telephone call to her in July, 1950, to protest the alleged use of his composition and idea in a photoplay produced and exhibited by defendants. The photoplay, as hereinafter shown in some detail, closely parallels both plaintiff's synopsis and the historical material concerning the life and death of Floyd Collins.

Defendants concede, as they must, that "the act of disclosing an unprotectible idea, if that act is in fact the bargained for exchange for a promise, may be consideration to support the promise." They then add, "But once the idea is disclosed without the protection of a contract, the law says that anyone is free to use it. Therefore, subsequent use of the idea cannot constitute consideration so as to support a promise to pay for such use." And as to the effect of the evidence defendants argue that plaintiff "disclosed his material before . . . (defendants) did or could do anything to indicate their willingness or unwillingness to pay for the disclosure. The act of using the idea, from which appellant attempts to imply a promise to pay, came long after the disclosure." . . .

Generally speaking, ideas are as free as the air. . . . But there can be circumstances when neither air nor ideas may be acquired without cost. The diver who goes deep in the sea, even as the pilot who ascends high in the troposphere, knows full well that for life itself he, or someone on his behalf, must arrange for air. . . . The theatrical producer likewise may be dependent for his business life on the procurement of ideas from other persons as well as the dressing up and portrayal of his self-conceptions; he may not find his own sufficient for survival. As counsel for the Writers Guild aptly say, ideas 'are not freely usable by the entertainment media until the latter are made aware of them.' The producer may think up the idea himself, dress it and portray it; or he may purchase either the conveyance of the idea alone or a manuscript embodying the idea in the author's concept of a literary vehicle giving it form, adaptation and expression. It cannot be doubted that some ideas are of value to a producer.

An idea is usually not regarded as property, because all sentient beings may conceive and evolve ideas throughout the gamut of their powers of cerebration and because our concept of property implies something which may be owned and possessed to the exclusion of all other persons. We quote as an accurate statement of the law in this respect the following language of Mr. Justice Brandeis, dissenting in International News Service v. Associated Press (1918), 248 U.S. 215, 250:

> An essential element of individual property is the legal right to exclude others from enjoying it. [. . .] But the fact that a product of the mind has cost its producer money and labor, and has a value for which others are willing to pay, is not sufficient to ensure to it this legal attribute of property. The general rule of law is, that the noblest of human productions, knowledge, truths ascertained, conceptions, and ideas become, after voluntary communication to others, free as the air to common use.

The principles above stated do not, however, lead to the conclusion that ideas cannot be a subject of contract. As Mr. Justice Traynor stated in his dissenting opinion in Stanley v. Columbia Broadcasting System (1950), 35 Cal. 2d 653, 674, 221 P.2d 73:

> The policy that precludes protection of an abstract idea by copyright does not prevent its protection by contract. Even though an idea is not property subject to exclusive ownership, its disclosure may be of substantial benefit to the person to whom it is disclosed. That disclosure may therefore be consideration for a promise to pay. . . . Even though the idea disclosed may be 'widely known and generally understood' (citation), it may be protected by an express contract providing that it will be paid for regardless of its lack of novelty.

The lawyer or doctor who applies specialized knowledge to a state of facts and gives advice for a fee is selling and conveying an idea. In doing that he is rendering a service. The lawyer and doctor have no property rights in their ideas, as such, but they do not ordinarily convey them without solicitation by client or patient. Usually the parties will expressly contract for the performance of and payment for such services, but, in the absence of an express contract, when the service is requested and rendered the law does not hesitate to infer or imply a promise to compensate for it. In other words the recovery may be based on contract either express or implied. The person who can and does convey a valuable idea to a producer who commercially solicits the service or who voluntarily accepts it knowing that it is tendered for a price should likewise be entitled to recover. In so holding we do not fail to recognize that free-lance writers are not necessarily members of a learned profession and as such bound to the exalted standards to which doctors and lawyers are dedicated. So too we are not oblivious of the hazards with which producers of the class represented here by defendants and their related amici are confronted through the unsolicited submission of numerous scripts on public domain materials in which public materials the producers through their own initiative may well find nuclei for legitimately developing the "stupendous and colossal." The law, however, is dedicated to the proposition that for every wrong there is a remedy. . . . To that end the law of implied contracts assumes particular importance in literary idea and property controversies. . . .

[W]e conclude that conveyance of an idea can constitute valuable consideration and can be bargained for before it is disclosed to the proposed purchaser, but once it is conveyed, i.e., disclosed to him and he has grasped it, it is henceforth his own and he may work with it and use it as he sees fit. In the field of entertainment the producer may properly and validly agree that he will pay for the service of conveying to him ideas which are valuable and which he can put to profitable use. Furthermore, where an idea has been conveyed with the expectation by the purveyor that compensation will be paid if the idea is used, there is no reason why the producer who has been the beneficiary of the conveyance of such an idea, and who finds it valuable and is profiting by it, may not then for the first time, although he is not at that time under any legal obligation so to do, promise to pay a reasonable compensation for that idea, that is, for the past service of furnishing it to him and thus create a valid obligation. . . . But, assuming legality of consideration, the idea purveyor cannot prevail in an action to recover compensation for an abstract idea unless (a) before or after disclosure he has obtained an express promise to pay, or (b) the circumstances preceding and attending disclosure, together with the conduct of the offeree acting with knowledge of the circumstances, show a promise of the type usually referred to as "implied" or "implied-in-fact." . . .

Such inferred or implied promise, if it is to be found at all, must be based on circumstances which were known to the producer at and preceding the time of disclosure of the idea to him and he must voluntarily accept the disclosure, knowing the conditions on which it is tendered. . . . The idea man who blurts out his idea without having first made his bargain has no one but himself to blame for the loss of his bargaining power. . . . So, if the plaintiff here is claiming only for the conveyance of the idea of making a dramatic production out of the life of Floyd Collins he must fail unless in conformity with the above stated rules he can establish a contract to pay.

From plaintiff's testimony, as epitomized above . . . it does not appear that a contract to pay for conveyance of the abstract photoplay idea had been made, or that the basis for inferring such a contract from subsequent related acts of the defendants had been established, at the time plaintiff disclosed his basic idea to the secretary. Defendants, consequently, were at that time and from then on free to use the abstract idea if they saw fit to engage in the necessary research and develop it to the point of a usable script. Whether defendants did that, or whether they actually accepted and used plaintiff's synopsis, is another question.

. . . Literary property which is protectible may be created out of unprotectible material such as historical events. It has been said (and does not appear to have been successfully challenged) that 'There are only thirty-six fundamental dramatic situations, various facets of which form the basis of all human drama.' (Georges Polti, 'The Thirty-Six Dramatic Situations'; see also, Henry Albert Phillips, 'The Universal Plot Catalog'; Eric Heath, 'Story Plotting Simplified.') It is manifest that authors must work with and from ideas or themes which basically are in the public domain. . . . Events from the life of Floyd Collins were avowedly the basic theme of plaintiff's story. . . .

[A]ny literary composition, conceivably, may possess value in someone's estimation and be the subject of contract. . . . Obviously the defendants here used someone's script in preparing and producing their photoplay. That script must have had value to them. As will be hereinafter shown, it closely resembles plaintiff's synopsis. Ergo, plaintiff's synopsis appears to be a valuable literary composition. Defendants had an unassailable right to have their own employes conduct the research into the Floyd Collins tragedy (an historical event in the public domain) and prepare a story based on those facts and to translate it into a script for the play. But equally unassailable (assuming the verity of the facts which plaintiff asserts) is plaintiff's position that defendants had no right except by purchase on the terms he offered to acquire and use the synopsis prepared by him.

[Affirmed in part, reversed in part, and remanded.]

CARTER, Justice [concurring in result]. . . .

When we consider the difference in economic and social backgrounds of those offering such merchandise for sale and those purchasing the same, we are met with the inescapable conclusion that it is the seller who stands in the inferior bargaining position. It should be borne in mind that producers are not easy to contact. . . . It should also be borne in mind that writers have no way of advertising their wares that, as is most graphically illustrated by the present opinion, no producer, publisher, or purchaser for radio or television, is going to buy a pig in a poke. And, when the writer, in an earnest endeavor to sell what he has written, conveys his idea or his different interpretation of an old idea, to such prospective purchaser, he has lost the result of his labor, definitely and irrevocably. And, in addition, there is no way in which he can protect himself. If he says to whomever he is permitted to see, or, as in this case, talk with over the telephone, "I won't tell you what my idea is until you promise to pay me for it," it takes no Sherlock Holmes to figure out what the answer will be! This case is a beautiful example of the practical difficulties besetting a writer with something to sell. . . .

I disagree with the statement in the majority opinion that: "The idea man who blurts out his idea without having first made his bargain has no one but himself to blame for the loss of his bargaining power." It seems to me that in the ordinary situation, when the so-called "idea man" has an opportunity to see, or talk with, the prospective purchaser, or someone in his employ, that it is at that time, without anything being said, known to both parties that the one is there to sell, and the other to buy. This is surely true of a department store when merchandise is displayed on the counter; it is understood by anyone entering the store that the merchandise so displayed is for sale. [I]t is completely unnecessary for the storekeeper, or anyone in his employ, to state to anyone entering the store that all articles there are for sale. I am at a loss to see why any different rules should apply when it is ideas for sale rather than the normal run of merchandise.

COMMENTS AND QUESTIONS

1. The concurring opinion makes an interesting point: "[W]hen the writer, in an earnest endeavor to sell what he has written, conveys his idea or his different interpretation of an old idea, to [a] prospective purchaser, he has lost the result of his labor." This general feature of the "market for information" was noted by Kenneth Arrow, in an article entitled Economic Welfare and the Allocation of Resources for Invention, in The Rate and Direction of Inventive Activity 615 (1962). Indeed, the phenomenon is some-

times called "Arrow's paradox of information." Without a property right, Arrow pointed out, the seller of information is in a pickle: if in trying to strike a deal she discloses the information, she has nothing left to sell, but if she does not disclose anything the buyer has no idea what is for sale. Arrow pointed out that when the information involves an invention, patents protect the seller so she can confidently offer her idea for sale. In other words, patents solve Arrow's Paradox. Does the opinion above prove that property rights are necessary to overcome the information paradox?

One problem with giving property rights to abstract ideas is that it can put the recipients of idea submissions in a bind. They are just as bound by Arrow's information paradox as plaintiffs — if they don't listen to the idea "pitch," they will never know whether the idea was worth paying for. The point of intellectual property protection in this situation is to encourage such idea submissions, by making the plaintiff confident in her ability to protect her idea. But suppose that a defendant hears a "pitch" for an idea that is old, that they have already come up with themselves, or that someone else has already pitched to them. If intellectual property protection extends to idea submissions, the recipient may be forced to pay for an idea it already has! Arguably, therefore, awarding property rights in idea submissions merely changes the nature of Arrow's paradox, rather than eliminating it entirely.

2. Wendy Gordon has explored the notion that a central problem of intellectual property law is to compensate creators of works who bestow benefits on those who follow, up to some socially justifiable point. In this analysis, the basic structure of intellectual property law is closely akin to the law of restitution, which seeks to determine when someone who bestows un–bargained-for benefits deserves compensation. See, e.g., Wendy J. Gordon, On Owning Information: Intellectual Property and the Restitutionary Impulse, 78 Va. L. Rev. 149 (1992); Wendy J. Gordon, Of Harms and Benefits: Torts, Restitution and Intellectual Property, 21 J. Legal Stud. 449 (1992). On restitution generally, see Saul Levmore, Explaining Restitution, 71 Va. L. Rev. 65 (1985). Does this literature help explain the need for a remand in the *Desny* case? Recall that the purpose of the remand is to determine how much Desny's idea contributed to the profits of Wilder's film.

3. The holding in *Desny* has been applied and extended in a number of cases. For instance, in Blaustein v. Burton, 88 Cal. Rptr. 319, 9 Cal. App. 3d 161 (Cal. Ct. App. 1970), the plaintiff Blaustein orally "pitched" the idea of using Richard Burton and Elizabeth Taylor in a film version of Shakespeare's "Taming of the Shrew." Although there was little novel in the pitch, the court held that there were triable issues of fact concerning the enforceability of an implied contract for the idea. The case is notable in that the "pitch" was protectable even though it was never reduced to writing.

PROBLEM

Problem 6-1. In Buchwald v. Paramount Pictures Corp. (Cal. Super. 1990) (unpub.), the Washington humorist and columnist Art Buchwald submitted an eight-page summary of a film idea entitled "It's a Crude, Crude World" to executives at Paramount. The summary described in some detail the storyline, which involved a third-world prince who came to America for "a state visit." Buchwald gave a two-page overview of the plot to Paramount Pictures. Paramount subsequently entered into a contract with Buchwald whereby Paramount bought the rights to Buchwald's story and concept, with the aim of making a movie, starring Eddie Murphy, to be called "King for a Day." Another person would write the actual script, but "King for a Day" was to be based on "It's a Crude, Crude World." Because of various production difficulties, Paramount abandoned the project in March 1985. In May 1986, Buchwald gave an option on his film idea to the Warner Brothers Studio. In the summer of 1987, Paramount began development of a similar script by Eddie Murphy. In this movie — which became the successful film "Coming to America" — Murphy portrayed an African prince who comes to America in search of a suitable woman to marry.

The court described the plots of the two scripts as follows:

> In Buchwald's treatment, a rich, educated, arrogant, extravagant, despotic African potentate comes to America for a state visit. After being taken on a grand tour of the United States, the potentate arrives at the White House. A gaff[e] in remarks made by the President infuriates the African leader. His sexual desires are rebuffed by a black woman State Department officer assigned to him. She is requested by the President to continue to serve as the potentate's United States escort. While in the United States, the potentate is deposed, deserted by his entourage and left destitute. He ends up in the Washington ghetto, is stripped of his clothes, and befriended by a black lady. The potentate experiences a number of incidents in the ghetto, and obtains employment as a waiter. In order to avoid extradition, he marries the black lady who befriended him, becomes the emperor of the ghetto and lives happily ever after.
>
> In "Coming to America" the pampered prince of a mythical African Kingdom (Zamunda) wakes up on his 21st birthday to find that the day for his prearranged marriage has arrived. Discovering his bride to be very subservient and being unhappy about that fact, he convinces his father to permit him to go to America for the ostensible purpose of sewing [sic] his "royal oats." In fact, the prince intends to go to America to find an independent woman to marry. The prince and his friend go to Queens, New York, where their property is stolen and they begin living in a slum area. The prince discovers his true love, Lisa, whose father — McDowell — operates a fast-food restaurant for whom the prince and his friend

begin to work. The prince and Lisa fall in love, but when the King and Queen come to New York and it is disclosed who the prince is, Lisa rejects the prince's marriage invitation. The film ends with Lisa appearing in Zamunda, marrying the prince and apparently living happily ever after.

Because of *Coming to America*, Warner Brothers decided not to pursue Buchwald's story. Buchwald then sued Paramount. The case turned on the contract between Buchwald and Paramount, which provided in pertinent part:

> "Work" means the aforementioned Material and includes all prior, present and future versions, adaptations and translations thereof (whether written by Author or by others), its theme, story, plot, characters and their names, its title or titles and subtitles, if any, . . . and each and every part of all thereof. "Work" does not include any material written or prepared by Purchaser or under Purchaser's Authority. . . .
> ["Contingent consideration":]
> For the first theatrical motion picture (the "Picture"): If, but only if, a feature length theatrical motion picture shall be produced *based upon Author's Work.*

(Emphasis added.) How should the court rule?

COMMENTS AND QUESTIONS

1. For a proposal to deal with the specific situation in *Buchwald*, see Robert M. Winteringham, Note, Stolen from Stardust and Air: Idea Theft in the Entertainment Industry and a Proposal for a Concept Initiator Credit, 46 Fed. Comm. L.J. 373, 377 (1994):

> This Note defines a three-part test for courts to use when deciding whether a concept initiator credit is needed in an entertainment product. First, the idea is eligible for the concept initiator credit if it qualifies as a "narrative crux." A narrative crux is an idea that is so original and highly specific that the idea's expression would receive copyright protection without being invalidated by the merger doctrine. In a narrative crux, the idea underlying the expression does not extend beyond the expression itself. Second, the idea must be used qualitatively and quantitatively in a second work. Finally, the concept initiator credit will not be required unless the idea was knowingly used in another work.

See also King v. Innovation Books, 976 F.2d 824 (2d Cir. 1992) (action by author Stephen King to *prevent* a filmmaker's use of "based upon" credit in the movie "Lawnmower Man." King objected to the movie, which was loosely based on an early short story whose copyright he had assigned. The filmmaker was enjoined from giving full credit to King, but was allowed to use "based upon" credit).

2. If you were an independent scriptwriter, how would you protect your ideas while trying to market them? Under what circumstances would the career damage from obtaining a reputation for litigiousness be worth it? For a description of an institutional response to the theft of idea problem, see the following note on the Hollywood script registry.

3. If you were a film studio or television production company, how could you guard against suits by people like Buchwald, or by people who claim to have submitted ideas that no one in your company ever recalls seeing? Under some cases habitual rejection and return of unsolicited ideas eliminates the prospect of liability. See, e.g., Davis v. General Foods Corp., 21 F. Supp. 445 (S.D.N.Y. 1937) (plaintiff's unsolicited recipe returned with form letter); Whitfield v. Lear, 751 F.2d 90 (2d Cir. 1984) (noting, in decision reversing summary judgment for defendant, plaintiff's evidence that it was customary in the television industry for a studio not desiring outside submissions to say so explicitly and to return scripts so submitted without opening them). Is it desirable for firms to routinely reject good ideas for fear of spurious lawsuits? Is this an example of nuisance lawsuits undermining an otherwise mutually beneficial market? Is it an example of the "market for lemons," where bad idea submissions (i.e., ones that lead to spurious suits) drive out the good ones? Is it enough that studios and production companies can rely on trusted middlepersons such as agents to obtain ideas from proven submitters? See Julie Salamon, Bookshelf: Celluloid Immortals and Literary He-Men, Wall St. J., July 29, 1992, p. A7 col. 1:

> Whatever the outcome, the suit has made Hollywood acutely aware of where it gets its ideas. "Producers are going to have to be very careful what submissions they read, and that makes it harder in the way they conduct business," says producer Howard Rosenman, co-president of Sandollar Productions, adding, "I never accept any unsolicited material. Ever."

Note on Institutions for Effectuating Idea Submissions: The Hollywood Script Registry Example

TV and movie scripts are often written by freelancers not under contract with a studio or production company. These writers must submit the scripts if they are to be evaluated for purchase. Submission of scripts, however, is fraught with peril. Copyright protection does not extend to many basic elements of plot and character. Consequently, it is difficult (and of course, expensive) to enforce rights in these aspects of a script, which creates a serious risk of illicit misappropriation. Firms receiving unsolicited scripts face the opposite problem: the risk of a lawsuit every time a movie or TV show bears even a passing resemblance to an unsolicited script that was received sometime in the past.

The Writer's Guild of America, West, developed a "Script Registry" to lower both these risks. Writers deposit a copy of a script they are going to submit with the Registry, which date-stamps it and stores it for five years. Annual volume runs to approximately 30,000 submissions; a total of 150,000 or more scripts is thus on file at any time. The Registry also operates a proven arbitration service which resolves over 300 cases each year; very few go beyond this tribunal to court.

It should be noted that scriptwriters are part of a fairly close-knit industry with some repeat-play features, factors that make informal dispute resolution, and the Registry institution itself, run more smoothly. See Cynthia Craft, Hollywood's Storehouse of Scripts, Ideas, Movies, Los Angeles Times, March 18, 1992, at F3, col. 1 ("Says one attorney from O'Melveny & Myers, 'Frankly, [these cases often settle confidentially] because everyone works with everyone else. It's almost like mediating a family dispute.' "). See generally Robert Ellickson, Order Without Law (1991), on norms in close-knit groups.

An interesting analogue can be found in the toy industry which, like the TV and film industries, relies on independent "concept creators" for a fair number of new products. As with scriptwriters, at least some toy creators appear to prefer being freelancers. The industry has recognized this and has evolved a transactional mechanism to encourage it: the trusted "pro," or repeat player, who submits many product ideas over a long period of time. One source that discusses product development in the toy industry in detail describes the advantages toy firms have when dealing with trusted agents, or industry "pros." Richard C. Levy and Ronald O. Weingartner, From Workshop to Toy Store (N.Y.: Simon and Schuster, 1990) at 180-181:

> Although pros do not like rejection any more than the amateurs do, they live with the reality that only so much product can be done in any single year. They may not like it that their concept was only a smidge different from an item ultimately selected for market, but they regard that as better execution of a similar concept rather than an infringement. . . . Pros know that their business relationships with manufacturers are based, among other things, on trust and ethics. They are in the business of living off their ideas, not living off legal settlements. As Phil Orbanes at Parker Brothers [a large toy company] says, "There is a greater legal risk [for manufacturers] in having an inventor come in cold."

Reputation in the professional toy creator community keeps manufacturers from violating their end of the "trust bargain":

> [O]n the whole, toy-company executives are honorable professionals who value their relationships with independent developers and do everything possible to maintain the balance in the flow of ideas between in-house and outside development sources. They also know that if they intentionally cheat inventors, the

pipeline would soon be empty. "The last thing we want to do is rip off an inventor [says an R&D executive in the industry]. . . . Word would spread like wildfire and all our sources would dry up instantly."

Id.

In Hollywood, the third party independent "bonding" function of the Script Registry obviously broadens the market for scripts. For certain scripts whose *ex ante* value is low, and which therefore could not justify the higher transaction costs which would attend an individually negotiated submission contract, the Registry can be said to actually *create* a market. Clearly the writers see the advantage of the system (Craft, 1992, at p. F3):

> "You get a little sense of security that your ideas are somewhat protected," said Joe Ferullo, a Westchester writer who estimated this was his 20th trip to the registration office. "At least it provides a bit of ammunition in case questions of authorship come up."
>
> John Bowton . . . [used] the guild's offices to register an episode for Fox TV's "The Simpsons" that he wrote in the hope of selling to the cartoon's producers.
>
> "I always feel that if I put the WGA [Registration] number on it, it's like a little shield, like the sign of the cross keeping the vampire away. . . . This way, they know it's on record and it's the best thing to prove that you came up with the story."

Note that one alternative to the Script Registry, "vertical integration" in the form of hiring script writers as employees, has its own problems. Many writers produce better work as freelancers, since in that capacity they have complete freedom to shape their work environment and creative process. In addition, it is also no doubt very difficult to preidentify potentially productive scriptwriters, manage them, and monitor their output. Finally, the "personal service" contracts necessary to bind such an employee would face serious enforcement problems in the courts. If an employee scriptwriter thought she had hit upon a "blockbuster," she would almost surely quit the company and claim to have developed the idea on her own, rather than allow the employer to reap the lion's share of profits. Surely the costs of such vertical integration — in the form of reduced productivity and less diversity of entertainment products — would be great. For all these reasons the Script Registry may be said to contribute as a net gain the differential between the actual value of scripts created under the Registry system and the value that would be created if all scriptwriters were retained as employees. On these issues generally, see Robert P. Merges, Expanding Boundaries of the Law: Intellectual Property and the Costs of Commercial Exchange: A Review Essay, 93 Mich. L. Rev. 1570 (1995).

Note finally that the Patent Office can be said to serve a similar function. By putting its imprimatur on an invention, the Office certifies that the invention is at least apparently new and nonobvious and that the description is true (since there are penalties for lying about the invention to the Office). Hence the Office serves a "bonding" function in addition to its role of protecting a prospective licensor from the bite of Arrow's information paradox. (In addition, the Patent Office will rule on "interferences" between competing claimants to the same invention.)

COMMENTS AND QUESTIONS

1. Allowing plaintiffs to bring lawsuits claiming that their ideas were improperly taken imposes a certain administrative burden on the courts, both because they must decide more cases and because litigation over abstract ideas is more difficult to resolve than claims that a tangible invention or publication was copied. Private ordering mechanisms like the WGA Script Registry are one way of reducing those adminstrative costs.

2. Writers Guild members agree to submit to Guild arbitration in disputes over movie scripts. Courts are loathe to overturn Guild decisions in these matters. *See* Wellman v. Writers Guild of Am., West, Inc., 146 F.3d 666 (9th Cir. 1998). The same rule applies when film companies disagree with arbitration outcomes. See Orion Pictures Corp. v. Writers Guild of Am., West, Inc., 946 F.2d 722 (9th Cir. 1991).

PROBLEM

Problem 6-2. Lohr, an eccentric engineer, mails an idea she has had for a new invention to several engineering companies. The idea is mailed in a "double envelope." The outer envelope contains a confidentiality agreement, indicating that the contents of a second, sealed envelope are the property of Lohr and may not be used unless the user pays Lohr 10% of any profit that results. If the recipient does not agree to these terms, he is instructed to mail back the sealed, stamped, self-addressed envelope containing the idea. The "agreement" clearly provides that by opening the envelope, the recipient agrees to the terms.

Dupco receives the agreement and opens the inner envelope. The next day, Dupco discovers that Lohr has posted the complete text of her idea on the Internet with no confidentiality restriction. Can Dupco use Lohr's idea without compensating her? Does it matter how novel or original the idea was on the day Dupco opened the envelope?

---

4. Publicity Rights

Midler v. Ford Motor Co.
United States Court of Appeals for the Ninth Circuit
849 F.2d 460 (9th Cir. 1988)

JOHN T. NOONAN, Circuit Judge:

This case centers on the protectibility of the voice of a celebrated chanteuse from commercial exploitation without her consent. Ford Motor Company and its advertising agency, Young & Rubicam, Inc., in 1985 advertised the Ford Lincoln Mercury with a series of nineteen 30 or 60 second television commercials in what the agency called "The Yuppie Campaign." The aim was to make an emotional connection with Yuppies, bringing back memories of when they were in college. Different popular songs of the seventies were sung on each commercial. The agency tried to get "the original people," that is, the singers who had popularized the songs, to sing them. Failing in that endeavor in ten cases the agency had the songs sung by "sound alikes." Bette Midler, the plaintiff and appellant here, was done by a sound alike.

Midler is a nationally known actress and singer. She won a Grammy as early as 1973 as the Best New Artist of that year. Records made by her since then have gone Platinum and Gold. She was nominated in 1979 for an Academy award for Best Female Actress in The Rose, in which she portrayed a pop singer. Newsweek in its June 30, 1986 issue described her as an "outrageously original singer/comedian." Time hailed her in its March 2, 1987 issue as "a legend" and "the most dynamic and poignant singer-actress of her time."

When Young & Rubicam was preparing the Yuppie Campaign it presented the commercial to its client by playing an edited version of Midler singing "Do You Want To Dance," taken from the 1973 Midler album, "The Divine Miss M." After the client accepted the idea and form of the commercial, the agency contacted Midler's manager, Jerry Edelstein. The conversation went as follows: "Hello, I am Craig Hazen from Young and Rubicam. I am calling you to find out if Bette Midler would be interested in doing . . . ?" Edelstein: "Is it a commercial?" "Yes." "We are not interested."

Undeterred, Young & Rubicam sought out Ula Hedwig, whom it knew to have been one of "the Harlettes," a backup singer for Midler for ten years. Hedwig was told by Young & Rubicam that "they wanted someone who could sound like Bette Midler's recording of [Do You Want To Dance]." She was asked to make a "demo" tape of the song if she was interested. She made an a capella demo and got the job.

At the direction of Young & Rubicam, Hedwig then made a record for the commercial. The Midler record of "Do You Want To Dance" was first played to her. She was told to "sound as much as possible like the Bette Midler record," leaving out only a few "aahs" unsuitable for the commercial. Hedwig imitated Midler to the best of her ability.

After the commercial was aired Midler was told by "a number of people" that it "sounded exactly" like her record of "Do You Want To Dance." Hedwig was told by "many personal friends" that they thought it was Midler singing the commercial. Ken Fritz, a personal manager in the entertainment business not associated with Midler, declares by affidavit that he heard the commercial on more than one occasion and thought Midler was doing the singing.

Neither the name nor the picture of Midler was used in the commercial; Young & Rubicam had a license from the copyright holder to use the song. At issue in this case is only the protection of Midler's voice. The district court described the defendants' conduct as that "of the average thief." They decided, "If we can't buy it, we'll take it." The court nonetheless believed there was no legal principle preventing imitation of Midler's voice and so gave summary judgment for the defendants. Midler appeals.

The First Amendment protects much of what the media do in the reproduction of likenesses or sounds. A primary value is freedom of speech and press. Time, Inc. v. Hill, 385 U.S. 374, 388, 17 L. Ed. 2d 456, 87 S. Ct. 534 (1967). The purpose of the media's use of a person's identity is central. If the purpose is "informative or cultural" the use is immune; "if it serves no such function but merely exploits the individual portrayed, immunity will not be granted." Felcher and Rubin, "Privacy, Publicity and the Portrayal of Real People by the Media," 88 Yale L.J. 1577, 1596 (1979). Moreover, federal copyright law preempts much of the area. "Mere imitation of a recorded performance would not constitute a copyright infringement even where one performer deliberately sets out to simulate another's performance as exactly as possible." Notes of Committee on the Judiciary, 17 U.S.C.A. §114(b). It is in the context of these First Amendment and federal copyright distinctions that we address the present appeal.

Nancy Sinatra once sued Goodyear Tire and Rubber Company on the basis of an advertising campaign by Young & Rubicam featuring "These Boots Are Made For Walkin'," a song closely identified with her; the female singers of the commercial were alleged to have imitated her voice and style and to have dressed and looked like her. The basis of Nancy Sinatra's complaint was unfair competition; she claimed that the song and the arrangement had acquired "a secondary meaning" which, under California law, was protectible. This court noted that the defendants "had paid a very substantial sum to the copyright proprietor to obtain the license for the use of the song and all of its arrangements." To give Sinatra damages for their use of the song would clash with federal copyright law. Summary judgment for the defendants was affirmed. Sinatra v. Goodyear Tire & Rubber Co., 435 F.2d 711, 717-718 (9th Cir. 1970), *cert. denied,* 402 U.S. 906, 28 L. Ed. 2d 646, 91 S. Ct. 1376 (1971). If Midler were claiming a secondary meaning to "Do You Want To Dance" or seeking to prevent the defendants from using that song, she would fail like Sinatra. But that is not this case. Midler does not seek damages for Ford's use of "Do You Want To Dance," and thus her claim is not preempted by federal copyright law. Copyright protects "original works of authorship fixed in any tangible medium of expression." 17 U.S.C. at 102(a). A voice is not copyrightable. The sounds are not "fixed." What is put forward as protectible here is more personal than any work of authorship.

Bert Lahr once sued Adell Chemical Co. for selling Lestoil by means of a commercial in which an imitation of Lahr's voice accompanied a cartoon of a duck. Lahr alleged that his style of vocal delivery was distinctive in pitch, accent, inflection, and sounds. The First Circuit held that Lahr had stated a cause of action for unfair competition, that it could be found "that defendant's conduct saturated plaintiff's audience, curtailing his market." Lahr v. Adell Chemical Co., 300 F.2d 256, 259 (1st Cir. 1962). That case is more like this one. But we do not find unfair competition here. One-minute commercials of the sort the defendants put on would not have saturated Midler's audience and curtailed her market. Midler did not do television commercials. The defendants were not in competition with her. See Halicki v. United Artists Communications, Inc., 812 F.2d 1213 (9th Cir. 1987).

California Civil Code section 3344 is also of no aid to Midler. The statute affords damages to a person injured by another who uses the person's "name, voice, signature, photograph or likeness, in any manner." The defendants did not use Midler's name or anything else whose use is prohibited by the statute. The voice they used was Hedwig's, not hers. The term "likeness" refers to a visual image not a vocal imitation. The statute, however, does not preclude Midler from pursuing any cause of action she may have at common law; the statute itself implies that such common law causes of action do exist because it says its remedies are merely "cumulative." Id. §3344(g).

The companion statute protecting the use of a deceased person's name, voice, signature, photograph or likeness states that the rights it recognizes are "property rights." Id. §990(b). By analogy the common law rights are also property rights. Appropriation of such common law rights is a tort in California. Motschenbacher v. R. J. Reynolds Tobacco Co., 498 F.2d 821 (9th Cir. 1974). In that case what the defendants used in their television commercial for Winston cigarettes was a photograph of a famous professional racing driver's racing car. The number of the car was changed and a wing-like device known as a "spoiler" was attached to the car; the car's features of white pinpointing, an oval medallion, and solid red coloring were retained. The driver, Lothar Motschenbacher, was in the car but his features were not visible. Some persons, viewing the commercial, correctly inferred that the car was his and that he was in the car and was therefore endorsing the product. The defendants were held to have invaded a "proprietary interest" of Motschenbacher in his own identity. Id. at 825.

Midler's case is different from Motschenbacher's. He and his car were physically used by the tobacco company's ad; he made part of his living out of giving commercial endorsements. But, as Judge Koelsch expressed it in Motschenbacher, California will recognize an injury from "an appropriation of the attributes of one's identity." Id. at 824. It was irrelevant that Motschenbacher could not be identified in the ad. The ad suggested that it was he. The ad did so by emphasizing signs or symbols associated with him. In the same way the defendants here used an imitation to convey the impression that Midler was singing for them.

Why did the defendants ask Midler to sing if her voice was not of value to them? Why did they studiously acquire the services of a sound-alike and instruct her to imitate Midler if Midler's voice was not of value to them? What they sought was an attribute of Midler's identity. Its value was what the market would have paid for Midler to have sung the commercial in person.

A voice is more distinctive and more personal than the automobile accouterments protected in Motschenbacher. A voice is as distinctive and personal as a face. The human voice is one of the most palpable ways identity is manifested. We are all aware that a friend is at once known by a few words on the phone. At a philosophical level it has been observed that with the sound of a voice, "the other stands before me." D. Ihde, Listening and Voice 77 (1976). A fortiori, these observations hold true of singing, especially singing by a singer of renown. The singer manifests herself in the song. To impersonate her voice is to pirate her identity. See W. Keeton, D. Dobbs, R. Keeton, D. Owen, Prosser & Keeton on Torts 852 (5th ed. 1984).

We need not and do not go so far as to hold that every imitation of a voice to advertise merchandise is actionable. We hold only that when a distinctive voice of a professional singer is widely known and is deliberately imitated in order to sell a product, the sellers have appropriated what is not theirs and have committed a tort in California. Midler has made a showing, sufficient to defeat summary judgment, that the defendants here for their own profit in selling their product did appropriate part of her identity.

Reversed and remanded for trial.

COMMENTS AND QUESTIONS

1. Young and Rubicam paid money to the owner of the copyright in the song "Do You Want to Dance," who permitted them to remake the song. Why wasn't that enough? Should Midler have the additional right to prevent imitation of her voice? Note that the owner of the copyright in sound recordings is often the producer or the studio, rather than the singer who made the original recording. Does it matter whether Young and Rubicam could have purchased the rights to the recording of Midler herself singing "Do You Want to Dance"? from the studio? If an artist doesn't control the rights to her own recording, why should she be able to prevent imitations of that recording?

A related question is whether the right of publicity is subject to an implicit first sale defense. In Allison v. Vintage Sports Plaques, 136 F.3d 1443 (11th Cir. 1998), the Eleventh Circuit held that it was. The defendant had purchased authorized sports trading cards and framed them for resale. The court held that the defendant could lawfully resell the images of celebrities that he had lawfully purchased and that he was not impermissibly using the sports trading cards to sell the associated frames.

2. Both Midler, the imitator, and independent third parties provided evidence that people hearing the Ford commercial were confused — they thought that Midler was the one singing. Is likelihood of confusion (the test for trademark infringement) the relevant question here? Could Midler prevail even if the attempt to imitate her was not very good, so that most people could tell the difference? If a disclaimer at the beginning of the ad had indicated that the song was an impersonation, rather than Midler herself?

A related question is whether the right of publicity applies only to celebrities. Historically, the answer has been no. The right of publicity is derived from the "commercial advantage" wing of the tort of invasion of privacy, and can be invoked by anyone whose name or likeness was appropriated by another for commercial advantage. See Restatement (2d) Torts §652C. Does this lineage suggest that the "likelihood of confusion" test shouldn't limit the right of publicity?

3. Does a right of publicity survive the death of the individual? See, e.g., Estate of Presley v. Russen, 513 F. Supp. 1339 (D.N.J. 1981) (New Jersey right of publicity is descendible). In another New Jersey case, McFarland v. Miller, 14 F.3d 912, 29 U.S.P.Q.2d 1586 (3d Cir. 1994) (involving "Spanky" McFarland of "Our Gang" fame) the court compared the treatment of publicity and defamation actions under the state statute on survival of causes of action: "If the defamation action survives, the publicity action does so *a fortiori* because the right of publicity has become largely proprietary, not personal as in defamation. McFarland's claims did not abate on his death and his personal representative has the right to continue this action." 14 F.2d at 918. Does this result make sense? Whose rights are served by preventing imitation after a singer's death?

If the right of publicity survives death, how long does it last? Does it ever terminate? See, e.g., Cal. Civ. Code §990 (termination of right 50 years after a celebrity's death).

4. New York state law until recently did *not* include "voice" in the list of attributes protectable under its statute protecting commercial use of celebrity attributes. However, the statute was recently amended. See Oliveira v. Frito-Lay Inc., 46 U.S.P.Q.2d 1636 (S.D.N.Y. 1998) (amending earlier opinion in case involving jazz song "The Girl From Ipanema" as sung by Brazilian singer Astrud Gilberto).

5. On the origins of publicity rights out of common law privacy rights, see William Prosser, Privacy, 48 Cal. L. Rev. 383 (1960).

≡≡≡ ### *White v. Samsung Electronics America, Inc.*
≡≡≡ *United States Court of Appeals for the Ninth Circuit*
≡≡≡ *989 F.2d 1512 (9th Cir. 1993)*

Before GOODWIN, PREGERSON and ALARCON, Circuit Judges.

The panel has voted unanimously to deny the petition for rehearing. . . .

KOZINSKI, Circuit Judge, with whom Circuit Judges O'SCANNLAIN and KLEINFELD join, dissenting from the order rejecting the suggestion for rehearing en banc.

I

Saddam Hussein wants to keep advertisers from using his picture in unflattering contexts. Clint Eastwood doesn't want tabloids to write about him. Rudolf Valentino's heirs want to control his film biography. The Girl Scouts don't want their image soiled by association with certain activities. George Lucas wants to keep Strategic Defense Initiative fans from calling it "Star Wars." Pepsico doesn't want singers to use the word "Pepsi" in their songs. Guy Lombardo wants an exclusive property right to ads that show big bands playing on New Year's Eve. Uri Geller thinks he should be paid for ads showing psychics bending metal through telekinesis. Paul Prudhomme, that household name, thinks the same about ads featuring corpulent bearded chefs. And scads of copyright holders see purple when their creations are made fun of.

Something very dangerous is going on here. Private property, including intellectual property, is essential to our way of life. It provides an incentive for investment and innovation; it stimulates the flourishing of our culture; it protects the moral entitlements of people to the fruits of their labors. But reducing too much to private property can be bad medicine. Private land, for instance, is far more useful if separated from other private land by public streets, roads and highways. Public parks, utility rights-of-way and sewers reduce the amount of land in private hands, but vastly enhance the value of the property that remains.

So too it is with intellectual property. Overprotecting intellectual property is as harmful as underprotecting it. Creativity is impossible without a rich public domain. Nothing today, likely nothing since we tamed fire, is genuinely new: Culture, like science and technology, grows by accretion, each new creator building on the works of those who came before. Overprotection stifles the very creative forces it's supposed to nurture.

The panel's opinion is a classic case of overprotection. Concerned about what it sees as a wrong done to Vanna White, the panel majority erects a property right of remarkable and dangerous breadth: Under the majority's opinion, it's now a tort for advertisers to remind the public of a celebrity. Not to use a celebrity's name, voice, signature or likeness; not to imply the celebrity endorses a product; but simply to evoke the celebrity's image in the public's mind. This Orwellian notion withdraws far more from the public domain than prudence and common sense allow. It conflicts with the Copyright Act and the Copyright Clause. It raises serious First Amendment problems. It's bad law, and it deserves a long, hard second look.

II

Samsung ran an ad campaign promoting its consumer electronics. Each ad depicted a Samsung product and a humorous prediction: One showed a raw steak with the caption "Revealed to be health food. 2010 A.D." Another showed [talk show host] Morton Downey, Jr. in front of an American flag with the caption "Presidential candidate. 2008 A.D." The ads were meant to convey — humorously — that Samsung products would still be in use twenty years from now.

The ad that spawned this litigation starred a robot dressed in a wig, gown and jewelry reminiscent of Vanna White's hair and dress; the robot was posed next to a Wheel-of-Fortune-like game board [see Figure 6-1]. The caption read "Longest-running game show. 2012 A.D." The gag here, I take it, was that Samsung would still be around when White had been replaced by a robot.

Perhaps failing to see the humor, White sued, alleging Samsung infringed her right of publicity by "appropriating" her "identity." Under California law, White has the exclusive right to use her name, likeness, signature and voice for commercial purposes. Cal. Civ. Code §3344(a); Eastwood v. Superior Court, 149 Cal. App. 3d 409, 417, 198 Cal. Rptr. 342, 347 (1983). But Samsung didn't use her name, voice or signature, and it certainly didn't use her likeness. The ad just wouldn't have been funny had it depicted White or someone who resembled her — the whole joke was that the game show host(ess) was a robot, not a real person. No one seeing the ad could have thought this was supposed to be White in 2012.

The district judge quite reasonably held that, because Samsung didn't use White's name, likeness, voice or signature, it didn't violate her right of publicity. 971 F.2d at 1396-97. Not so, says the panel majority: The California right of publicity can't possibly be limited to name and likeness. If it were, the majority reasons, a "clever advertising strategist" could avoid using White's name or likeness but nevertheless remind people of her with impunity, "effectively eviscerat[ing]" her rights. To prevent this "evisceration," the panel majority holds that the right of publicity must extend beyond name and likeness, to any "appropriation" of White's "identity" — anything that "evoke[s]" her personality. Id. at 1398-99.

III

But what does "evisceration" mean in intellectual property law? Intellectual property rights aren't like some constitutional rights, absolute guarantees protected against all kinds of interference, subtle as well as blatant. They cast no penumbras, emit no emanations: The very point of intellectual property laws is that they protect only against certain specific kinds of appropriation. I can't publish unauthorized copies of, say, Presumed Innocent; I can't make a movie out of it. But I'm perfectly free to write a book about an idealistic

FIGURE 6-1
Samsung's robot of a game show hostess.

young prosecutor on trial for a crime he didn't commit. So what if I got the idea from Presumed Innocent? So what if it reminds readers of the original? Have I "eviscerated" Scott Turow's intellectual property rights? Certainly not. All creators draw in part on the work of those who came before, referring to it, building on it, poking fun at it; we call this creativity, not piracy.

The majority isn't, in fact, preventing the "evisceration" of Vanna White's existing rights; it's creating a new and much broader property right, a right unknown in California law.[16] It's replacing the existing balance be-

---

16. In fact, in the one California case raising the issue, the three state Supreme Court Justices who discussed this theory expressed serious doubts about it. Guglielmi v. Spelling-Goldberg

tween the interests of the celebrity and those of the public by a different balance, one substantially more favorable to the celebrity. Instead of having an exclusive right in her name, likeness, signature or voice, every famous person now has an exclusive right to anything that reminds the viewer of her. After all, that's all Samsung did: It used an inanimate object to remind people of White, to "evoke [her identity]." 971 F.2d at 1399.[17]

Consider how sweeping this new right is. What is it about the ad that makes people think of White? It's not the robot's wig, clothes or jewelry; there must be ten million blond women (many of them quasi-famous) who wear dresses and jewelry like White's. It's that the robot is posed near the "Wheel of Fortune" game board. Remove the game board from the ad, and no one would think of Vanna White. But once you include the game board, anybody standing beside it — a brunette woman, a man wearing women's clothes, a monkey in a wig and gown — would evoke White's image, precisely the way the robot did. It's the "Wheel of Fortune" set, not the robot's face or dress or jewelry that evokes White's image. The panel is giving White an exclusive right not in what she looks like or who she is, but in what she does for a living.[18]

Prods., 25 Cal. 3d 860, 864 n.5, 160 Cal. Rptr. 352, 355 n.5, 603 P.2d 454, 457 n.5 (1979) (Bird, C. J., concurring) (expressing skepticism about finding a property right to a celebrity's "personality" because it is "difficult to discern any easily applied definition for this amorphous term"). Neither have we previously interpreted California law to cover pure "identity." Midler v. Ford Motor Co., 849 F.2d 460 (9th Cir. 1988), and Waits v. Frito-Lay, Inc., 978 F.2d 1093 (9th Cir. 1992), dealt with appropriation of a celebrity's voice. See id. at 1100-01 (imitation of singing style, rather than voice, doesn't violate the right of publicity). Motschenbacher v. R. J. Reynolds Tobacco Co., 498 F.2d 821 (9th Cir. 1974), stressed that, though the plaintiff's likeness wasn't directly recognizable by itself, the surrounding circumstances would have made viewers think the likeness was the plaintiff's. Id. at 827; see also Moore v. Regents of the Univ. of Cal., 51 Cal. 3d 120, 138, 271 Cal. Rptr. 146, 157, 793 P.2d 479, 490 (1990) (construing Motschenbacher as "hold[ing] that every person has a proprietary interest in his own likeness").

17. Some viewers might have inferred White was endorsing the product, but that's a different story. The right of publicity isn't aimed at or limited to false endorsements, Eastwood v. Superior Court, 149 Cal. App. 3d 409, 419-20, 198 Cal. Rptr. 342, 348 (1983); that's what the Lanham Act is for. Note also that the majority's rule applies even to advertisements that unintentionally remind people of someone. California law is crystal clear that the common-law right of publicity may be violated even by unintentional appropriations. Id. at 417 n.6, 198 Cal. Rptr. at 346 n.6; Fairfield v. American Photocopy Equipment Co., 138 Cal. App. 2d 82, 87, 291 P.2d 194 (1955).

18. Once the right of publicity is extended beyond specific physical characteristics, this will become a recurring problem: Outside name, likeness and voice, the things that most reliably remind the public of celebrities are the actions or roles they're famous for. A commercial with an astronaut setting foot on the moon would evoke the image of Neil Armstrong. Any masked man on horseback would remind people (over a certain age) of Clayton Moore. And any number of songs — "My Way," "Yellow Submarine," "Like a Virgin," "Beat It," "Michael, Row the Boat Ashore," to name only a few — instantly evoke an image of the person or group who made them famous, regardless of who is singing. See also Carlos V. Lozano, West Loses Lawsuit over Batman TV Commercial, L.A. Times, Jan. 18, 1990, at B3 (Adam West sues over Batman-like character in commercial); Nurmi v. Peterson, 10 U.S.P.Q.2d 1775, 1989 WL 407484 (C.D. Cal. 1989) (1950s TV movie hostess "Vampira" sues 1980s TV hostess "Elvira"); text accompanying notes 7-8 (lawsuits brought by Guy Lombardo, claiming big bands playing at New Year's Eve parties remind people of him, and by Uri Geller, claiming psychics who can bend metal remind people of him). Cf. Motschenbacher, where the claim was that viewers would think

This is entirely the wrong place to strike the balance. Intellectual property rights aren't free: They're imposed at the expense of future creators and of the public at large. Where would we be if Charles Lindbergh had an exclusive right in the concept of a heroic solo aviator? If Arthur Conan Doyle had gotten a copyright in the idea of the detective story, or Albert Einstein had patented the theory of relativity? If every author and celebrity had been given the right to keep people from mocking them or their work? Surely this would have made the world poorer, not richer, culturally as well as economically.

This is why intellectual property law is full of careful balances between what's set aside for the owner and what's left in the public domain for the rest of us: The relatively short life of patents; the longer, but finite, life of copyrights; copyright's idea-expression dichotomy; the fair use doctrine; the prohibition on copyrighting facts; the compulsory license of television broadcasts and musical compositions; federal preemption of overbroad state intellectual property laws; the nominative use doctrine in trademark law; the right to make soundalike recordings. All of these diminish an intellectual property owner's rights. All let the public use something created by someone else. But all are necessary to maintain a free environment in which creative genius can flourish.

The intellectual property right created by the panel here has none of these essential limitations: No fair use exception; no right to parody; no idea-expression dichotomy. It impoverishes the public domain, to the detriment of future creators and the public at large. Instead of well-defined, limited characteristics such as name, likeness or voice, advertisers will now have to cope with vague claims of "appropriation of identity," claims often made by people with a wholly exaggerated sense of their own fame and significance. . . . Future Vanna Whites might not get the chance to create their personae, because their employers may fear some celebrity will claim the persona is too similar to her own. The public will be robbed of parodies of celebrities, and our culture will be deprived of the valuable safety valve that parody and mockery create.

Moreover, consider the moral dimension, about which the panel majority seems to have gotten so exercised. Saying Samsung "appropriated" something of White's begs the question: Should White have the exclusive right to something as broad and amorphous as her "identity"? Samsung's ad didn't simply copy White's schtick — like all parody, it created something new. True, Samsung did it to make money, but White does whatever she does to make money, too; the majority talks of "the difference between fun and profit," 971 F.2d at 1401, but in the entertainment industry fun is profit. Why is Vanna White's right to exclusive for-profit use of her persona — a persona that might not even be her own creation, but that of a writer, director or producer — superior to Samsung's right to profit by creating its own in-

---

plaintiff was actually in the commercial, and not merely that the commercial reminded people of him.

ventions? Why should she have such absolute rights to control the conduct of others, unlimited by the idea-expression dichotomy or by the fair use doctrine?

To paraphrase only slightly Feist Publications, Inc. v. Rural Telephone Service Co., 111 S. Ct. 1282, 1289-90 (1991), it may seem unfair that much of the fruit of a creator's labor may be used by others without compensation. But this is not some unforeseen byproduct of our intellectual property system; it is the system's very essence. Intellectual property law assures authors the right to their original expression, but encourages others to build freely on the ideas that underlie it. This result is neither unfair nor unfortunate: It is the means by which intellectual property law advances the progress of science and art. We give authors certain exclusive rights, but in exchange we get a richer public domain. The majority ignores this wise teaching, and all of us are the poorer for it.

IV

The panel, however, does more than misinterpret California law: By refusing to recognize a parody exception to the right of publicity, the panel directly contradicts the federal Copyright Act. Samsung didn't merely parody Vanna White. It parodied Vanna White appearing in "Wheel of Fortune," a copyrighted television show, and parodies of copyrighted works are governed by federal copyright law.

Copyright law specifically gives the world at large the right to make "fair use" parodies, parodies that don't borrow too much of the original. Fisher v. Dees, 794 F.2d 432, 435 (9th Cir. 1986). . . .

The majority's decision decimates this federal scheme. It's impossible to parody a movie or a TV show without at the same time "evok[ing]" the "identit[ies]" of the actors. You can't have a mock Star Wars without a mock Luke Skywalker, Han Solo and Princess Leia, which in turn means a mock Mark Hamill, Harrison Ford and Carrie Fisher. You can't have a mock Batman commercial without a mock Batman, which means someone emulating the mannerisms of Adam West or Michael Keaton. See Carlos V. Lozano, West Loses Lawsuit over Batman TV Commercial, L.A. Times, Jan. 18, 1990, at B3 (describing Adam West's right of publicity lawsuit over a commercial produced under license from DC Comics, owner of the Batman copyright). The public's right to make a fair use parody and the copyright owner's right to license a derivative work are useless if the parodist is held hostage by every actor whose "identity" he might need to "appropriate."

Our court is in a unique position here. State courts are unlikely to be particularly sensitive to federal preemption, which, after all, is a matter of first concern to the federal courts. The Supreme Court is unlikely to consider the issue because the right of publicity seems so much a matter of state law. That leaves us. It's our responsibility to keep the right of publicity from taking away federally granted rights, either from the public at large or from a copyright owner. We must make sure state law doesn't give the Vanna Whites and Adam Wests of the world a veto over fair use parodies of the shows in which they appear, or over copyright holders' exclusive right to license derivative works of those shows. In a case where the copyright owner isn't even a party — where no one has the interests of copyright owners at heart — the majority creates a rule that greatly diminishes the rights of copyright holders in this circuit.

V

The majority's decision also conflicts with the federal copyright system in another, more insidious way. Under the dormant Copyright Clause, state intellectual property laws can stand only so long as they don't "prejudice the interests of other States." Goldstein v. California, 412 U.S. 546, 558, 93 S. Ct. 2303, 2310, 37 L. Ed. 2d 163 (1973). A state law criminalizing record piracy, for instance, is permissible because citizens of other states would "remain free to copy within their borders those works which may be protected elsewhere." Id. But the right of publicity isn't geographically limited. A right of publicity created by one state applies to conduct everywhere, so long as it involves a celebrity domiciled in that state. If a Wyoming resident creates an ad that features a California domiciliary's name or likeness, he'll be subject to California right of publicity law even if he's careful to keep the ad from being shown in California. See Acme Circus Operating Co. v. Kuperstock, 711 F.2d 1538, 1540 (11th Cir. 1983); Groucho Marx Prods. v. Day and Night Co., 689 F.2d 317, 320 (2d Cir. 1982); see also Factors Etc. v. Pro Arts, 652 F.2d 278, 281 (2d Cir. 1981).

The broader and more ill-defined one state's right of publicity, the more it interferes with the legitimate interests of other states. A limited right that applies to unauthorized use of name and likeness probably does not run afoul of the Copyright Clause, but the majority's protection of "identity" is quite another story. Under the majority's approach, any time anybody in the United States — even somebody who lives in a state with a very narrow right of publicity — creates an ad, he takes the risk that it might remind some segment of the public of somebody, perhaps somebody with only a local reputation, somebody the advertiser has never heard of. See note 17 supra (right of publicity is infringed by unintentional appropriations). So you made a commercial in Florida and one of the characters reminds Reno residents of their favorite local TV anchor (a California domiciliary)? Pay up.

This is an intolerable result, as it gives each state far too much control over artists in other states. No California statute, no California court has actually tried to reach this far. It is ironic that it is we who plant this kudzu in the fertile soil of our federal system.

VI

Finally, I can't see how giving White the power to keep others from evoking her image in the public's mind can be squared with the First Amendment. Where does White get this right to control our thoughts? The majority's creation goes way beyond the protection given a trademark or a copyrighted work, or a person's name or likeness. All those things control one particular way of expressing an idea, one way of referring to an object or a person. But not allowing any means of reminding people of someone? That's a speech restriction unparalleled in First Amendment law.

What's more, I doubt even a name-and-likeness-only right of publicity can stand without a parody exception. The First Amendment isn't just about religion or politics — it's also about protecting the free development of our national culture. Parody, humor, irreverence are all vital components of the marketplace of ideas. The last thing we need, the last thing the First Amendment will tolerate, is a law that lets public figures keep people from mocking them, or from "evok[ing]" their images in the mind of the public. 971 F.2d at 1399.

The majority dismisses the First Amendment issue out of hand because Samsung's ad was commercial speech. Id. at 1401 & n. 3. So what? Commercial speech may be less protected by the First Amendment than noncommercial speech, but less protected means protected nonetheless. Central Hudson Gas & Elec. Corp. v. Public Serv. Comm'n, 447 U.S. 557 (1980). And there are very good reasons for this. Commercial speech has a profound effect on our culture and our attitudes. Neutral-seeming ads influence people's social and political attitudes, and themselves arouse political controversy. . . .

In our pop culture, where salesmanship must be entertaining and entertainment must sell, the line between the commercial and noncommercial has not merely blurred; it has disappeared. Is the Samsung parody any different from a parody on Saturday Night Live or in Spy Magazine? Both are equally profit-motivated. Both use a celebrity's identity to sell things — one to sell VCRs, the other to sell advertising. Both mock their subjects. Both try to make people laugh. Both add something, perhaps something worthwhile and memorable, perhaps not, to our culture. Both are things that the people being portrayed might dearly want to suppress. . . .

VII

For better or worse, we are the Court of Appeals for the Hollywood Circuit. Millions of people toil in the shadow of the law we make, and much of their livelihood is made possible by the existence of intellectual property rights. But much of their livelihood — and much of the vibrancy of our culture — also depends on the existence of other intangible rights: The right to draw ideas from a rich and varied public domain, and the right to mock, for profit as well as fun, the cultural icons of our time.

In the name of avoiding the "evisceration" of a celebrity's rights in her image, the majority diminishes the rights of copyright holders and the public at large. In the name of fostering creativity, the majority suppresses it. Vanna White and those like her have been given something they never had before, and they've been given it at our expense. I cannot agree.

COMMENTS AND QUESTIONS

1. The Ninth Circuit seemed to reaffirm its holding in *White* in Wendt v. Host Intl., 125 F.3d 806 (9th Cir. 1997). That case also involved ani-matronic robots. These were representative of characters from the television show Cheers and were placed in licensed airport Cheers bars. The bars had obtained rights from the producers of the television show but not from the actors themselves. The court held that the actors retained publicity rights in their portrayal of the fictional characters under California law and remanded the case for an analysis of the similarities between the plaintiffs and the robots.

If the actors had brought a claim under California law based on the use of their images from the TV show itself, they would have lost. See Fleet v. CBS, 50 Cal. App. 4th 1911 (1996) (California publicity law preempted to the extent it imposes controls on the exploitation of name or likeness through distribution of a motion picture in which actor appeared); Page v. Something Weird Video, 960 F. Supp. 1438 (C.D. Cal. 1996) (First Amendment allows use of drawing of character in film to promote that film). Should the result be any different where a spin-off product is licensed?

2. Cases other than *White* discuss "celebrity attributes" besides appearance and voice. See, e.g., McFarland v. Miller, 14 F.3d 912 (3d Cir. 1994) (granting damages for use of appearance and name of "Our Gang" member Spanky McFarland in restaurant decor). Is there any natural boundary to the right of publicity? If mere "evocation of a celebrity" is required, which of the following could be protected by a right of publicity: (1) a characteristic walk or even running style; (2) a characteristic gesture, such as Clint Eastwood's sneer, or Johnny Carson's musically accompanied golf swing, or Michael Jackson's "moonwalking" dance steps; (3) a "signature" joke, such as Henny Youngman's "take my wife — please," or Joan Rivers' "can we talk?" — notwithstanding that both jokes were well-known when they were adopted as "signatures"; (4) a style of chess opening that has become associated with a particular grand master; (5) a shot, move, or technique in sports that is closely identified with a particular athlete — e.g., Monica Seles' two-handed backhand shot in tennis; Bob Cousy's behind the back basketball pass; Pete Rose's headfirst baseball slide.

On the other hand, is there any natural limit to Judge Kozinski's reasoning? Or does it suggest that the entire concept of a right of publicity is ill-considered? Is there some identifiable reason that Bette Midler's claim seems more plausible than Vanna White's?

3. Judge Kozinski emphasizes throughout his opinion that the right of publicity adds extra burdens to the creators of works that draw on celebrity attributes. For example, he states: "We must make sure state law doesn't give the Vanna Whites and Adam Wests of the world a veto over fair use parodies of the shows in which they appear, or over copyright holders' exclusive right to license derivative works of those shows." Judge Kozinski's point is that the right of publicity creates the need for an entirely new "layer" of transactions *on top of* the traditional copyright license. For example, in many of the "voice-alike" cases such as *Midler*, the defendant in the publicity action is a legitimate licensee of the copyright holder in the song that the defendant used. The right of publicity cases thus implicitly hold that the copyright license does not shield the licensee from liability for using the work, at least under some circumstances. It also means that, at least for copyrighted works assigned prior to the rapid growth of the right of publicity, creators of works can use the new right to extract some extra value. See Eben Shapiro, Rising Caution on Using Celebrity Images, N.Y. Times, Nov. 4, 1992, at D20.

4. Consider Judge Kozinski's last argument, that the First Amendment limits the right of publicity. This point has some force. Certainly, newsworthy figures cannot prevent stories from being published about them by invoking the right of publicity. And this is true even though the paper may be using the name or likeness of the newsworthy figure to gain "commercial advantage" in the form of increased sales. In Montana v. San Jose Mercury News, 34 Cal. App. 4th 790, 40 Cal. Rptr. 2d 639 (1995), the court held that the *San Jose Mercury News* had a right to reprint its news story about and picture of football star Joe Montana on promotional posters "for the purpose of showing the quality and content of the newspaper."

For a case holding that there is a constitutionally grounded parody exception to the right of publicity — which might have applied in the *White* case as well — see Cardtoons v. Major League Baseball Players' Association, 39 U.S.P.Q.2d 1865 (10th Cir. 1996). See also Matthews v. Wozencraft, 15 F.3d 432 (5th Cir. 1994) (Texas right of publicity does not allow plaintiff to block a fictionalized biographical narrative, even though plaintiff can be identified from the book). For an argument that the right of publicity can be reconciled with the parody defense, see Michael E. Hartmann & Daniel R. Kelly, Parody (of Celebrities, in Advertising), Parity (Between Advertising and Other Types of Commercial Speech), and (the Property Right of) Publicity, 17 Hastings Comm. & Ent. L.J. 633 (1995).

5. For interesting theoretical accounts of the right of publicity, see Michael Madow, Private Ownership of Public Image: Popular Culture and Publicity Rights, 81 Cal. L. Rev. 127 (1993); Mark Grady, A Positive Economic Theory of the Right of Publicity, 1 UCLA Entertainment L. Rev. 97 (1994). Madow reviews three major theoretical descriptions of the right of publicity: labor theory, prevention of unjust enrichment, and economic incentive the-

ory. He is generally skeptical of the expansive publicity cases that have come down in recent years.

In the second article, Professor Grady puts forth what he calls a falsifiable theory that accounts for the right of publicity. Grady argues that the right of publicity can be understood as a response to the problem of rent dissipation. If the law does not recognize a property right in celebrity attributes, everyone will be free to use those attributes, and they will be less valuable. Id., at 101. In other words, the general rationale for property rights — that they facilitate efficient exploitation — applies equally well in this context.

Grady argues for instance that "Vanna White's image is so valuable — hence, so vulnerable to dissipation — that it is an unusually strong case for protection." Id., at 118. How do we know how valuable White's image is? Should we require her to introduce evidence of licensing revenue to establish that her image is valuable before we protect it? If we protect it first, on the supposition that it is valuable, what will this do to its value? In general, how does the Grady view of celebrity attributes as a scarce resource differ from that put forward by Judge Kozinski in his opinion in the *White* case? Is it true for all celebrity images that the public can become so oversaturated that the value of the image declines? If so, how do we account for nearly ubiquitous images, such as Mickey Mouse, Michael Jordan, and the Coca-Cola logo? Is it possible that for at least some images more exposure means *higher* value? Should the law try to distinguish between such images and those for which overexposure is more clearly a problem — for instance, the Rolls Royce logo? Is it more efficient to allow a single property holder to determine the optimal exposure strategy? Does the Grady view of publicity rights lead logically to the conclusion that parodies should be prohibited, or at least restricted to a non-dissipatory number? For discussion of a similar theoretical account of some aspects of patent law, compare Mark F. Grady & Jay I. Alexander, Patent Law and Rent Dissipation, 78 Va. L. Rev. 305 (1992), with Robert P. Merges, Rent Control in the Patent District: Observations on the Grady-Alexander Thesis, 78 Va. L. Rev. 359 (1992).

6. For sources of background information, see Frank G. Houdek, Researching the Right of Publicity: A Revised and Comprehensive Bibliography of Law-Related Materials, 16 Hastings Comm. & Ent. L.J. 385 (1994); Note, Right of Publicity Run Riot: The Case for a Federal Statute, 60 S. Cal. L. Rev. 1179 (1987).

California Publicity Statute
Cal. Civ. Code §3344.1 (1999).

In October of 1999, the California legislature passed the Astaire Celebrity Image Protection Act. *See* 1999 Cal. Legis. Serv. Ch. 998 (S.B. 209) (avail. Westlaw), superseding old Cal. Civ. Code §990, codified at new Cal. Civ. Code §3344.1 (1999). This Act made sweeping changes to the law in this area in California, and pushed the frontiers of publicity rights in several important directions, including:

- Extension of the statutory term of protection from 50 to 70 years;
- Creation of Copyright Act-like classes of statutory successors to succeed to the rights when the famous person dies;
- Minimum statutory damages, open-ended punitive damages, and "British Rule"-style (i.e., loser pays) attorney fees;
- Free assignability of the rights, both during and after the famous person's rights; and
- A public registry of rightholders, maintained on the World Wide Web, presumably to lower transaction costs in clearing these now formidable property rights.

In addition, the Act specifically overruled the decision in Astaire v. Best Film & Video Corp., 116 F.3d 1297, amended by 136 F.3d 1208 (9th Cir.), cert. denied, 119 S.Ct. 161 (1998). This case had ruled that the use of images of Fred Astaire in connection with a dance instruction videotape was not actionable under certain exceptions to publicity rights in old Civil Code §990. New §3334.1(a)(3) now sets forth a rule that would effectively reverse the outcome under facts such as those in *Astaire*.

The new section reads as follows:

> 3344.1.(a) (1) Any person who uses a deceased personality's name, voice, signature, photograph, or likeness, in any manner, on or in products, merchandise, or goods, or for purposes of advertising or selling, or soliciting purchases of, products, merchandise, goods, or services, without prior consent from the person or persons specified in subdivision (c), shall be liable for any damages sustained by the person or persons injured as a result thereof. In addition, in any action brought under this section, the person who violated the section shall be liable to the injured party or parties in an amount equal to the greater of seven hundred fifty dollars ($750) or the actual damages suffered by the injured party or parties, as a result of the unauthorized use, and any profits from the unauthorized use that are attributable to the use and are not taken into account in computing the actual damages. In establishing these profits, the injured party or parties shall be required to present proof only of the gross revenue attributable to the use and the person who violated the section is required to prove his or her deductible expenses. Punitive damages may also be awarded to the injured party or parties. The prevailing party or parties in any action under this section shall also be entitled to attorneys' fees and costs.
>
> (2) For purposes of this subdivision, a play, book, magazine, newspaper, musical composition, audiovisual work, radio or television program, single and original work of art, work of political or newsworthy value, or an advertisement or commercial announcement for any of these works, shall not be considered a product, article of merchandise, good, or service if it is fictional or nonfictional entertainment, or a dramatic, literary, or musical work.

(3) If a work that is protected under paragraph (2) includes within it a use in connection with a product, article of merchandise, good, or service, this use shall not be exempt under this subdivision, notwithstanding the unprotected use's inclusion in a work otherwise exempt under this subdivision, if the claimant proves that this use is so directly connected with a product, article of merchandise, good, or service as to constitute an act of advertising, selling, or soliciting purchases of that product, article of merchandise, good, or service by the deceased personality without prior consent from the person or persons specified in subdivision (c).

(b) The rights recognized under this section are property rights, freely transferable, in whole or in part, by contract or by means of trust or testamentary documents, whether the transfer occurs before the death of the deceased personality, by the deceased personality or his or her transferees, or, after the death of the deceased personality, by the person or persons in whom the rights vest under this section or the transferees of that person or persons.

(c) The consent required by this section shall be exercisable by the person or persons to whom the right of consent (or portion thereof) has been transferred in accordance with subdivision (b), or if no transfer has occurred, then by the person or persons to whom the right of consent (or portion thereof) has passed in accordance with subdivision (d).

(d) Subject to subdivisions (b) and (c), after the death of any person, the rights under this section shall belong to the following person or persons and may be exercised, on behalf of and for the benefit of all of those persons, by those persons who, in the aggregate, are entitled to more than a one-half interest in the rights:

(1) The entire interest in those rights belong to the surviving spouse of the deceased personality unless there are any surviving children or grandchildren of the deceased personality, in which case one-half of the entire interest in those rights belong to the surviving spouse.

(2) The entire interest in those rights belong to the surviving children of the deceased personality and to the surviving children of any dead child of the deceased personality unless the deceased personality has a surviving spouse, in which case the ownership of a one-half interest in rights is divided among the surviving children and grandchildren.

(3) If there is no surviving spouse, and no surviving children or grandchildren, then the entire interest in those rights belong to the surviving parent or parents of the deceased personality.

(4) The rights of the deceased personality's children and grandchildren are in all cases divided among them and exercisable in the manner provided in Section 240 of the Probate Code according to the number of the deceased personality's children represented; the share of the children of a dead child of a deceased personality can be exercised only by the action of a majority of them.

(e) If any deceased personality does not transfer his or her rights under this section by contract, or by means of a trust or testamentary document, and there are no surviving persons as described in subdivision (d), then the rights set forth in subdivision (a) shall terminate.

(f) (1) A successor in interest to the rights of a deceased personality under this section or a licensee thereof may not recover damages for a use prohibited by this section that occurs before the successor-in-interest or licensee registers a claim of the rights under paragraph (2).

(2) Any person claiming to be a successor-in-interest to the rights of a deceased personality under this section or a licensee thereof may register that claim with the Secretary of State on a form prescribed by the Secretary of State and upon payment of a fee of ten dollars ($10). The form shall be verified and shall include the name and date of death of the deceased personality, the name and address of the claimant, the basis of the claim, and the rights claimed.

(3) Upon receipt and after filing of any document under this section, the Secretary of State shall post the document along with the entire registry of persons claiming to be a successor in interest to the rights of a deceased personality or a registered licensee under this section upon the World Wide Web, also known as the Internet. The Secretary of State may microfilm or reproduce by other techniques any of the filings or documents and destroy the original filing or document. The microfilm or other reproduction of any document under the provisions of this section shall be admissible in any court of law. The microfilm or other reproduction of any document may be destroyed by the Secretary of State 70 years after the death of the personality named therein.

(4) Claims registered under this subdivision shall be public records.

(g) No action shall be brought under this section by reason of any use of a deceased personality's name, voice, signature, photograph, or likeness occurring after the expiration of 70 years after the death of the deceased personality.

(h) As used in this section, "deceased personality" means any natural person whose name, voice, signature, photograph, or likeness has commercial value at the time of his or her death, whether or not during the lifetime of that natural person the person used his or her name, voice, signature, photograph, or likeness on or in products, merchandise or goods, or for purposes of advertising or selling, or solicitation of purchase of, products, merchandise, goods, or services. A "deceased personality" shall include, without limitation, any such natural person who has died within 70 years prior to January 1, 1985.

(i) As used in this section, "photograph" means any photograph or photographic reproduction, still or moving, or any video tape or live television transmission, of any person, such that the deceased personality is readily identifiable. A deceased personality shall be deemed to be readily identifiable from a photograph when one who views the photograph with the naked eye can reasonably determine who the person depicted in the photograph is.

(j) For purposes of this section, a use of a name, voice, signature, photograph, or likeness in connection with any news, public affairs, or sports broadcast or account, or any political campaign, shall not constitute a use for which consent is required under subdivision (a).

(k) The use of a name, voice, signature, photograph, or likeness in a commercial medium shall not constitute a use for which consent is required under subdivision (a) solely because the material containing the use is commercially sponsored or contains paid advertising. Rather, it shall be a question of fact whether or not the use of the deceased personality's name, voice, signature, photograph, or likeness was so directly connected with the commercial sponsorship or with the paid advertising as to constitute a use for which consent is required under subdivision (a).

(l) Nothing in this section shall apply to the owners or employees of any medium used for advertising, including, but not limited to, newspapers, magazines, radio and television networks and stations, cable television systems, billboards, and transit ads, by whom any advertisement or solicitation in violation of this section is published or disseminated, unless it is established that the owners or employees had knowledge of the unauthorized use of the deceased personality's name, voice, signature, photograph, or likeness as prohibited by this section.

(m) The remedies provided for in this section are cumulative and shall be in addition to any others provided for by law.

(n) This section shall apply to the adjudication of liability and the imposition of any damages or other remedies in cases in which the liability, damages, and other remedies arise from acts occurring directly in this state. For purposes of this section, acts giving rise to liability shall be limited to the use, on or in products, merchandise, goods, or services, or the advertising or selling, or soliciting purchases of, products, merchandise, goods, or services prohibited by this section.

(o) This section shall be known, and may be cited, as the Astaire Celebrity Image Protection Act.

COMMENTS AND QUESTIONS

1. California case law continues to evolve. See Comedy III Productions, Inc. v. Gary Saderup, Inc., 80 Cal. Rptr. 2d 464, 49 U.S.P.Q.2d 1282 (Cal. App. 1998), *review granted and opinion superseded*, 973 P.2d 512 (Cal. 1999) (t-shirts with artist's sketch of "The Three Stooges" held a violation of California publicity statute; case under review by California Supreme Court).

2. Until recently, some states held back from recognizing and extending publicity rights. Publicity rights statutes are expanding, however. See, e.g., J. Thomas McCarthy, The Rights of Publicity and Privacy, (Rev. 1998); O. Yale Lewis, Jr., New Washington Publicity Rights Legislation, Washington State Bar News, June, 1998, <http://www.wsba.org/barnews/archives98/pub-rite.html> (describing legislation recently enacted in Washington State, overturning an earlier case holding that publicity rights are not descendible there).

3. On the application of publicity rights to the Internet and digital media in general, see Michaels v. Internet Entertainment Group Inc., 5 F.Supp.2d 823, 46 U.S.P.Q.2d 1892 (C.D.Cal. 1998) (publicity rights claim brought to prevent dissemination of images from unauthorized videotape ruled distinct from copyright issues; preliminary injunction granted).

PROBLEM

Problem 6-3. In the late 1980s, the "New Kids on the Block" were an enormously successful pop music group, especially among the younger teen market. Capitalizing on this success, the New Kids sold over 500 products or services bearing their trademarked name. Among those services were "900 numbers" that fans could call to learn more about the New Kids, or to talk to the New Kids themselves.

During the height of the New Kids craze, the newspaper *USA Today* conducted a telephone poll that allowed readers to "vote" for their favorite New Kid (or for "none of the above" if they did not like the band at all) by calling a *USA Today* 900 number. As a part of the poll, the paper included captioned pictures of each of the band members. Suppose that the New Kids on the Block sued *USA Today* for infringement of their right of publicity. Do they have a claim under California law?

5. State Moral Rights

Beginning in 1990, federal copyright law (17 U.S.C. §106A) has provided limited protection for visual artists' moral rights. We discussed this protection in some detail in Chapter 4. In some states, this federal protection is augmented by separate state moral rights statutes. One recent summary describes the more expansive nature of state moral rights protection:

> In New Mexico, any original work of art of recognized quality is protected from alteration or destruction if the work is in public view. Rhode Island protects all original works of visual art of any medium, with the exception of film. Connecticut provides an extensive list of protected media, including masters from which copies of an artistic work can be made. Some state laws permit recovery in a broader range of circumstances. For example, in California and Pennsylvania, no intentional defacement, alteration, or destruction is allowed regardless of the artist's ability to prove damage to honor or reputation, so long as the work is of recognized quality. Massachusetts goes further by adding a prohibition against acts of gross negligence.

Amy L. Landers, The Current State of Moral Rights Protection for Visual Artists in the United States, 15 Hastings Comm. & Ent. L.J. 165 (1992), citing 107 N.M. Stat. Ann. §13-4B-2(B), 3(B) (Michie 1991) (work of fine art is "any original work of visual or graphic art of any media . . . of recognized quality."); R.I. Gen. Laws §5-62-2(e) (Michie 1987); Conn. Gen. Stat. Ann. §42-116 §(2) (West Supp. 1991); Cal. Civ. Code §§987-990 (West Supp. 1991); Pa. Stat. Ann. title 73, §§2101-10 (1991); Mass. Gen. Laws Ann. ch. 231, §85S (West Supp. 1991). Most of these state statutes provide greater protection than does the federal statute. The state statutes generally provide more rights to artists, make those rights harder to waive, and apply moral rights to a broader class of works than does the federal law.

COMMENTS AND QUESTIONS

1. In one case construing Cal. Civ. Code §987, a California court held that a mural is a "painting" entitled to protection. Botello v. Shell Oil Co.,

280 Cal. Rptr. 535, 229 Cal. App. 3d 1130 (Ct. App. 1991). Another court held that architectural plans are not works protected by the California Art Preservation Act, since they do not constitute "fine art." Robert H. Jacobs, Inc. v. Westoaks Realtors, Inc., 205 Cal. Rptr. 620, 159 Cal. App. 3d 637 (Ct. App. 1984).

2. The New York moral rights statute takes a different tack. It emphasizes the value of an artist's reputation, rather than the intrinsic value of the work itself. Under this statute, for example, display of a mutilated original art work (or a copy for limited-run works) is prohibited only if the artist's name is associated with it:

> [N]o person other than the artist or a person acting with the artist's consent shall knowingly display in a place accessible to the public or publish a work of fine art or limited edition multiple of not more than three hundred copies by that artist or a reproduction thereof in an altered, defaced, mutilated or modified form if the work is displayed, published or reproduced as being the work of the artist, or under circumstances under which it would reasonably be regarded as being the work of the artist, and damage to the artist's reputation is reasonably likely to result therefrom, except that this section shall not apply to sequential imagery such as that in motion pictures. . . .
>
> [T]he artist shall retain at all times the right to claim authorship, or, for just and valid reason, to disclaim authorship of such work. The right to claim authorship shall include the right of the artist to have his or her name appear on or in connection with such work as the artist. The right to disclaim authorship shall include the right of the artist to prevent his or her name from appearing on or in connection with such work as the artist. Just and valid reason for disclaiming authorship shall include that the work has been altered, defaced, mutilated or modified other than by the artist, without the artist's consent, and damage to the artist's reputation is reasonably likely to result or has resulted therefrom.

N.Y. Arts and Cultural Affairs Law §14.03 (1994). For interpretations of this statute, compare Morita v. Omni Publications Intern., Ltd., 741 F. Supp. 1107 (S.D.N.Y. 1990) (publication of photograph of Hiroshima juxtaposed with pronuclear slogan was not a mutilation under New York statute) with Wojnarowicz v. American Family Assn., 745 F. Supp. 130 (S.D.N.Y. 1990) (pamphlet containing photocopied fragments of only sexually explicit fragments of artist's works, which was distributed in effort to stop public funding of National Endowment for the Arts, violated New York's Artists' Authorship Rights Act). See also Edward J. Damich, The New York Artists' Authorship Rights Act: A Comparative Critique, 84 Colum. L. Rev. 1733, 1735 (1984) ("The New York statute confines its protection of the integrity of an artwork to circumstances damaging to the artist's reputation, leaving the statute susceptible to the interpretation that it does not prohibit complete destruction of art works."). For an argument that the U.S. does not truly protect moral rights and is thus in violation of its Berne Convention obligations, see Brian T. McCartney, "Creepings" and "Glimmers" of the Moral Rights of Artists in American Copyright Law, 6 UCLA Ent. L. Rev. 35 (1998); but see Thomas F. Cotter, Pragmatism, Economics, and the Droit Moral, 76 N.C. L. Rev. 1 (1997) (agreeing, but arguing this is a good thing).

3. For a careful treatment of a potentially tricky issue, see Peter H. Karlen, Joint Ownership of Moral Rights, 38 J. Copyright Soc'y 242 (1991).

4. For a discussion of the preemption issues raised by state moral rights statutes and the Visual Artists Rights Act, see infra section B of this chapter.

Note on Resale Royalties, or the "Droit de Suite"

The "droit de suite" permits an artist to benefit from appreciation in the value of her works by entitling her to a percentage of all subsequent sales. California offers this benefit under the Artist's Resale Royalty Act, which stipulates that a 5 percent royalty shall be paid on each sale of a work. See Morseberg v. Balyon, 621 F.2d 972 (9th Cir. 1980), *cert. denied*, 449 U.S. 983 (1980) (California Resale Royalty Act not preempted by 1909 Copyright Act; no opinion expressed on applicability of analysis to preemption under the 1976 Act). Puerto Rico also has implemented the droit de suite. See P.R. Laws Ann. title 31 (1991) (similar provision under the law of Puerto Rico). See also Larry S. Karp & Jeffrey M. Perloff, Legal Requirements That Artists Receive Resale Royalties, 13 Intl. Rev. L. & Econ. 163, 173 (1993) (arguing, on basis of formal model, that resale royalty mandated by law will almost surely cause more harm than good, and that better policy would be for government to help implement transactional mechanism to make privately bargained resale royalty agreements more attractive).

B. FEDERAL PREEMPTION

1. Patent Preemption

Sears, Roebuck & Co. v. Stiffel Company
Supreme Court of the United States
376 U.S. 225 (1964)

Mr. Justice BLACK delivered the opinion of the Court.

The question in this case is whether a State's unfair competition law can, consistently with the federal patent laws, impose liability for or prohibit the copying of an article which is protected by neither a federal patent nor a copyright. The respondent, Stiffel Company, secured design and mechanical patents on a "pole lamp" — a vertical tube having lamp fixtures along the outside, the tube being made so that it will stand upright between the floor and ceiling of a room. Pole lamps proved a decided commercial success, and soon after Stiffel brought them on the market Sears, Roebuck & Company put on the market a substantially identical lamp, which it sold more cheaply, Sears' retail price being about the same as Stiffel's wholesale price. Stiffel then brought this action against Sears in the United States District Court for the Northern District of Illinois, claiming in its first count that by copying its design Sears had infringed Stiffel's patents and in its second count that by selling copies of Stiffel's lamp Sears had caused confusion in the trade as to the source of the lamps and had thereby engaged in unfair competition under Illinois law. There was evidence that identifying tags were not attached to the Sears lamps although labels appeared on the cartons in which they were delivered to customers, that customers had asked Stiffel whether its lamps differed from Sears', and that in two cases customers who had bought Stiffel lamps had complained to Stiffel on learning that Sears was selling substantially identical lamps at a much lower price.

The District Court, after holding the patents invalid for want of invention, went on to find as a fact that Sears' lamp was "a substantially exact copy" of Stiffel's and that the two lamps were so much alike, both in appearance and in functional details, "that confusion between them is likely, and some confusion has already occurred." On these findings the court held Sears guilty of unfair competition, enjoined Sears "from unfairly competing with [Stiffel] by selling or attempting to sell pole lamps identical to or confusingly similar to" Stiffel's lamp, and ordered an accounting to fix profits and damages resulting from Sears' "unfair competition."

The Court of Appeals affirmed.[1] . . .

The grant of a patent is the grant of a statutory monopoly; indeed, the grant of patents in England was an explicit exception to the statute of James I prohibiting monopolies. Patents are not given as favors, as was the case of monopolies given by the Tudor monarchs, see The Case of Monopolies (Darcy v. Allein), 11 Co. Rep. 84 b., 77 Eng. Rep. 1260 (K.B. 1602), but are meant to encourage invention by rewarding the inventor with the right, limited to a term of years fixed by the patent, to exclude others from the use of his invention. During that period of time no one may make, use, or sell the patented product without the patentee's authority. 35 U.S.C. §271. But in rewarding useful invention, the "rights and welfare of the community must be fairly dealt with and effectually guarded." Kendall v. Winsor, 21 How. 322, 329 (1859). To that end the prerequisites to obtaining a patent are strictly observed, and when the patent has issued the limitations on its exercise are equally strictly enforced. . . .

Thus the patent system is one in which uniform federal standards are carefully used to promote invention while at the same time preserving free competition. Obviously a State could not, consistently with the Supremacy Clause of the Constitution, extend the life of a patent beyond its expiration date or give a patent on an article which lacked the level of invention required

---

1. No review is sought here of the ruling affirming the District Court's holding that the patent is invalid.

for federal patents. To do either would run counter to the policy of Congress of granting patents only to true inventions, and then only for a limited time. Just as a State cannot encroach upon the federal patent laws directly, it cannot, under some other law, such as that forbidding unfair competition, give protection of a kind that clashes with the objectives of the federal patent laws.

In the present case the "pole lamp" sold by Stiffel has been held not to be entitled to the protection of either a mechanical or a design patent. An unpatentable article, like an article on which the patent has expired, is in the public domain and may be made and sold by whoever chooses to do so. What Sears did was to copy Stiffel's design and to sell lamps almost identical to those sold by Stiffel. This it had every right to do under the federal patent laws. That Stiffel originated the pole lamp and made it popular is immaterial. "Sharing in the goodwill of an article unprotected by patent or trade-mark is the exercise of a right possessed by all — and in the free exercise of which the consuming public is deeply interested." Kellogg Co. v. National Biscuit Co., supra, 305 U.S., at 122. To allow a State by use of its law of unfair competition to prevent the copying of an article which represents too slight an advance to be patented would be to permit the State to block off from the public something which federal law has said belongs to the public. The result would be that while federal law grants only 14 or 17 years' protection to genuine inventions, see 35 U.S.C. §§154, 173, States could allow perpetual protection to articles too lacking in novelty to merit any patent at all under federal constitutional standards. This would be too great an encroachment on the federal patent system to be tolerated. . . .

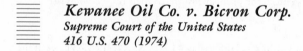

Kewanee Oil Co. v. Bicron Corp.
Supreme Court of the United States
416 U.S. 470 (1974)

Mr. Chief Justice BURGER delivered the opinion of the Court.

We granted certiorari to resolve a question on which there is a conflict in the courts of appeals: whether state trade secret protection is pre-empted by operation of the federal patent law. In the instant case the Court of Appeals for the Sixth Circuit held that there was preemption. The Courts of Appeals for the Second, Fourth, Fifth, and Ninth Circuits have reached the opposite conclusion. . . .

Petitioner brought this diversity action in United States District Court for the Northern District of Ohio seeking injunctive relief and damages for the misappropriation of trade secrets. The district Court, applying Ohio trade secret law, granted a permanent injunction against the disclosure or use by respondents of 20 of the 40 claimed trade secrets until such time as the trade secrets had been released to the public, had otherwise generally become available to the public, or had been obtained by respondents from sources having the legal right to convey the information.

The Court of Appeals for the Sixth Circuit held that the findings of fact by the District Court were not clearly erroneous, and that it was evident from the record that the individual respondents appropriated to the benefit of Bicron secret information on processes obtained while they were employees at Harshaw. Further, the Court of Appeals held that the District Court properly applied Ohio law relating to trade secrets. Nevertheless, the Court of Appeals reversed the District Court, finding Ohio's trade secret law to be in conflict with the patent laws of the United States. The Court of Appeals reasoned that Ohio could not grant monopoly protection to processes and manufacturing techniques that were appropriate subjects for consideration under 35 U.S.C. §101 for a federal patent but which had been in commercial use for over one year and so were no longer eligible for patent protection under 35 U.S.C. §102(b).

We hold that Ohio's law of trade secrets is not preempted by the patent laws of the United States, and accordingly, we reverse. . . .

III

The first issue we deal with is whether the States are forbidden to act at all in the area of protection of the kinds of intellectual property which may make up the subject matter of trade secrets.

Article I, §8, cl. 8, of the Constitution grants to the Congress the power

> [t]o promote the Progress of Science and useful Arts, by securing for limited Times to Authors and Inventors the exclusive Right to their respective Writings and Discoveries. . . .

In the 1972 Term, in Goldstein v. California, 412 U.S. 546 (1973), we held that the cl. 8 grant of power to Congress was not exclusive and that, at least in the case of writings, the States were not prohibited from encouraging and protecting the efforts of those within their borders by appropriate legislation. The States could, therefore, protect against the unauthorized rerecording for sale of performances fixed on records or tapes, even though those performances qualified as "writings" in the constitutional sense and Congress was empowered to legislate regarding such performances and could pre-empt the area if it chose to do so. This determination was premised on the great diversity of interests in our Nation — the essentially nonuniform character of the appreciation of intellectual achievements in the various States. Evidence for this came from patents granted by the States in the 18th century. 412 U.S., at 557.

Just as the States may exercise regulatory power over writings so may the States regulate with respect to discoveries. States may hold diverse viewpoints in protecting intellectual property relating to invention as they do in protecting the intellectual property relating to the subject matter of copyright. The only limitation on the States is that in regulating the area of patents and copyrights they do not conflict with the operation of the laws in this area passed by Congress, and it is to that more difficult question we now turn.

IV

The question of whether the trade secret law of Ohio is void under the Supremacy Clause involves a consideration of whether that law "stands as an obstacle to the accomplishment and execution of the full purposes and objectives of Congress." Hines v. Davidowitz, 312 U.S. 52, 67 (1941). See Florida Avocado Growers v. Paul, 373 U.S. 132, 141 (1963). We stated in Sears, Roebuck & Co. v. Stiffel Co., 376 U.S. 225, 229 (1964), that when state law touches upon the area of federal statutes enacted pursuant to constitutional authority, "it is 'familiar doctrine' that the federal policy 'may not be set at naught, or its benefits denied' by the state law. Sola Elec. Co v. Jefferson Elec. Co., 317 U.S. 173, 176 (1942). This is true, of course, even if the state law is enacted in the exercise of otherwise undoubted state power." . . .

The stated objective of the Constitution in granting the power to Congress to legislate in the area of intellectual property is to "promote the Progress of Science and useful Arts." The patent laws promote this progress by offering a right of exclusion for a limited period as an incentive to inventors to risk the often enormous costs in terms of time, research, and development. . . .

The maintenance of standards of commercial ethics and the encouragement of invention are the broadly stated policies behind trade secret law. "The necessity of good faith and honest, fair dealing, is the very life and spirit of the commercial world." . . .

As we noted earlier, trade secret law protects items which would not be proper subjects for consideration for patent protection under 35 U.S.C. §101. As in the case of the recordings in Goldstein v. California, Congress, with respect to nonpatentable subject matter, "has drawn no balance; rather, it has left the area unattended, and no reason exists why the State should not be free to act." Goldstein v. California, supra, at 570 (footnote omitted).

Since no patent is available for a discovery, however useful, novel, and nonobvious, unless it falls within one of the express categories of patentable subject matter of 35 U.S.C. §101, the holder of such a discovery would have no reason to apply for a patent whether trade secret protection existed or not. Abolition of trade secret protection would, therefore, not result in increased disclosure to the public of discoveries in the area of nonpatentable subject matter. . . .

Congress has spoken in the area of those discoveries which fall within one of the categories of patentable subject matter of 35 U.S.C. §101 and which are, therefore, of a nature that would be subject to consideration for a patent. Processes, machines, manufactures, compositions of matter, and improvements thereof, which meet the tests of utility, novelty, and nonobviousness are entitled to be patented, but those which do not, are not. The question remains whether those items which are proper subjects for consideration for a patent may also have available the alternative protection accorded by trade secret law.

Certainly the patent policy of encouraging invention is not disturbed by the existence of another form of incentive to invention. In this respect the two systems are not and never would be in conflict. Similarly, the policy that matter once in the public domain must remain in the public domain is not incompatible with the existence of trade secret protection. By definition a trade secret has not been placed in the public domain. . . .

. . . Trade secret law will encourage invention in areas where patent law does not reach, and will prompt the independent innovator to proceed with the discovery and exploitation of his invention. Competition is fostered and the public is not deprived of the use of valuable, if not quite patentable, invention. . . .

The final category of patentable subject matter to deal with is the clearly patentable invention, i.e., that invention which the owner believes to meet the standards of patentability. It is here that the federal interest in disclosure is at its peak. . . .

Trade secret law provides far weaker protection in many respects than the patent law. While trade secret law does not forbid the discovery of the trade secret by fair and honest means, e.g., independent creation or reverse engineering, patent law operates "against the world," forbidding any use of the invention for whatever purpose for a significant length of time. The holder of a trade secret also takes a substantial risk that the secret will be passed on to his competitors, by theft or by breach of a confidential relationship, in a manner not easily susceptible of discovery or proof. Painton & Co. v. Bourns, Inc., 442 F.2d, at 224. Where patent law acts as a barrier, trade secret law functions relatively as a sieve. The possibility that an inventor who believes his invention meets the standards of patentability will sit back, rely on trade secret law, and after one year of use forfeit any right to patent protection, 35 U.S.C. §102(b), is remote indeed.

Nor does society face much risk that scientific or technological progress will be impeded by the rare inventor with a patentable invention who chooses trade secret protection over patent protection. The ripeness-of-time concept of invention, developed from the study of the many independent multiple discoveries in history, predicts that if a particular individual had not made a particular discovery others would have, and in probably a relatively short period of time. If something is to be discovered at all very likely it will be discovered by more than one person. . . .

. . . Trade secret law and patent law have co-existed in this country for over one hundred years. Each has its particular role to play, and the operation of one does not take away from the need for the other. . . . Congress, by its silence over these many years, has seen the wisdom of allowing the States to enforce trade secret protection. Until Congress takes affirmative action to the contrary, States should be free to grant protection to trade secrets. . . .

Mr. Justice DOUGLAS, with whom Mr. Justice BRENNAN concurs, dissenting.

Today's decision is at war with the philosophy of Sears, Roebuck & Co. v. Stiffel Co., 376 U.S. 225, and Compco Corp. v. Day-Brite Lighting, Inc., 376 U.S. 234. Those cases involved patents — one of a pole lamp and one of fluorescent lighting fixtures — each of which was declared invalid. The lower courts held, however, that though the patents were invalid the sale of identical or confusingly similar products to the products of the patentees violated state unfair competition laws. We held that when an article is unprotected by a patent, state law may not forbid others to copy it, because every article not covered by a valid patent is in the public domain. Congress in the patent laws decided that where no patent existed, free competition should prevail; that where a patent is rightfully issued, the right to exclude others should obtain for no longer than 17 years, and that the States may not "under some other law, such as that forbidding unfair competition, give protection of a kind that clashes with the objectives of the federal patent laws," 376 U.S., at 231. . . .

The conflict with the patent laws is obvious. The decision of Congress to adopt a patent system was based on the idea that there will be much more innovation if discoveries are disclosed and patented than there will be when everyone works in secret. Society thus fosters a free exchange of technological information at the cost of a limited 17-year monopoly. . . .

A suit to redress theft of a trade secret is grounded in tort damages for breach of a contract — a historic remedy, Cataphote Corp. v. Hudson, 422 F.2d 1290. Damages for breach of a confidential relation are not pre-empted by this patent law, but an injunction against use is pre-empted because the patent law states the only monopoly over trade secrets that is enforceable by specific performance; and that monopoly exacts as a price full disclosure. A trade secret can be protected only by being kept secret. Damages for breach of a contract are one thing; an injunction barring disclosure does service for the protection accorded valid patents and is therefore pre-empted. . . .

COMMENTS AND QUESTIONS

1. The rule set forth by the *Sears* case seems fairly clear. Patent law reflects a compromise between the goal of promoting innovation and the danger of condoning monopoly. Supplementing the scope of patent law may upset that balance, and is therefore prohibited. Supplementing *enforcement* of the federal intellectual property laws was condoned in dictum at the end of *Compco*, a companion case, and in Justice Harlan's concurrence in both cases. After these cases, then, state law served a very limited function in the scheme of intellectual property protection. States could work to further the

goals of federal protection, but they had to work within the parameters set down by federal law.

The *Kewanee* opinion takes a remarkably different tack. Chief Justice Burger's opinion for the Court emphasizes only one of the two policies shaping the patent laws: the goal of promoting innovation. The opinion does not discuss the dangers intellectual property protection poses for free competition. As a result, the *Kewanee* Court finds no problem with trade secret protection which extends beyond the scope of the patent laws. Note that the Court seems to approve not only state laws that protect nonpatentable subject matter (an area in which it could be argued that the federal government has no interest),[1] but also the protection of inventions not patentable for some other reason (i.e., suppression, misuse, lack of novelty, or obviousness).

Does *Kewanee* suggest that an inventor who has lost patent protection because she suppressed her invention may be entitled to trade secret protection for the same invention? Is that the right result?

2. Is *Kewanee* reconcilable with *Sears*? The *Kewanee* court did not overrule *Sears* or *Compco;* indeed, it cited them in support of its holding. Thus state laws preventing copying were treated differently from trade secret laws after *Kewanee*. The latter, although broader in scope (they prevented far more than just outright copying of products), were permissible; the former were not. For an argument that *Kewanee* "closed the circle" on the open-ended preemption analysis of *Sears* and *Compco*, see Paul Goldstein, Kewanee Oil Co. v. Bicron Corp.: Notes on a Closing Circle, 1974 Sup. Ct. Rev. 81 (1974).

3. Is preemption a good idea? That depends on what you think of the balance the federal laws have struck. If you are more concerned about injury to competition by conferring monopoly rights on patentees, you are likely to favor the result in *Sears*. If on the other hand you think that innovation is underrewarded, it is reasonable to oppose federal preemption. One way to reconcile these cases may be to read *Sears* and *Compco* as expressing a federal policy in favor of reverse engineering of products in the public domain. If that is the overarching federal goal, it is logical to strike down the laws in *Sears* and *Compco* but not *Kewanee*, since trade secrets statutes (unlike the unfair competition laws we have discussed) generally allow reverse engineering. This result is also consistent with the reading of trade secret laws as merely an application of tort and contract principles.

4. Recall the Court's statement: "Where patent law acts as a barrier, trade secret law functions relatively as a sieve." Does this square with recent empirical evidence, such as the studies referred to in Chapters 1 and 2 that show trade secret protection to be the preferred form of protection in some industries? For more on the "election" of trade secret vs. patent protection, see the Note on "Electing" Trade Secrets or Patents, following the *Bonito Boats* case.

---

1. Even in this area, though, a federal interest may be discerned. If Congress has declared some subject matter unpatentable, that could reflect a federal determination that that matter is unworthy of protection, a determination that state law should not be allowed to upset.

Bonito Boats, Inc. v. Thunder Craft Boats, Inc.
Supreme Court of the United States
489 U.S. 141 (1989)

Justice O'CONNOR delivered the opinion of the Court. . . .

I

In September 1976, petitioner Bonito Boats, Inc., a Florida corporation, developed a hull design for a fiberglass recreational boat which it marketed under the trade name Bonito Boat Model 5VBR. Designing the boat hull required substantial effort on the part of Bonito. A set of engineering drawings was prepared, from which a hardwood model was created. The hardwood model was then sprayed with fiberglass to create a mold, which then served to produce the finished fiberglass boats for sale. The 5VBR was placed on the market sometime in September 1976. There is no indication in the record that a patent application was ever filed for protection of the utilitarian or design aspects of the hull, or for the process by which the hull was manufactured. The 5VBR was favorably received by the boating public, and "a broad interstate market" developed for its sale.

In May 1983, after the Bonito 5VBR had been available to the public for over six years, the Florida Legislature enacted Fla. Stat. §559.94 (1987). The statute makes "it . . . unlawful for any person to use the direct molding process to duplicate for the purpose of sale any manufactured vessel hull or component part of a vessel made by another without the written permission of that other person." §559.94(2). The statute also makes it unlawful for a person to "knowingly sell a vessel hull or component part of a vessel duplicated in violation of subsection (2)." §559.94(3). Damages, injunctive relief, and attorney's fees are made available to "any person who suffers injury or damage as the result of a violation" of the statute. §559.94(4). The statute was made applicable to vessel hulls or component parts duplicated through the use of direct molding after July 1, 1983. §559.94(5).

On December 21, 1984, Bonito filed this action in the Circuit Court of Orange County, Florida. The complaint alleged that respondent here, Thunder Craft Boats, Inc., a Tennessee corporation, had violated the Florida statute by using the direct molding process to duplicate the Bonito 5VBR fiberglass hull, and had knowingly sold such duplicates in violation of the Florida statute. . . .

III

We believe that the Florida statute at issue in this case so substantially impedes the public use of the otherwise unprotected design and utilitarian ideas embodied in unpatented boat hulls as to run afoul of the teaching of our decisions in *Sears* and *Compco*. It is readily apparent that the Florida statute does not operate to prohibit "unfair competition" in the usual sense that the term is understood. The law of unfair competition has its roots in the common-law tort of deceit: its general concern is with protecting *consumers* from confusion as to source. . . .

In contrast to the operation of unfair competition law, the Florida statute is aimed directly at preventing the exploitation of the design and utilitarian conceptions embodied in the product itself. The sparse legislative history surrounding its enactment indicates that it was intended to create an inducement for the improvement of boat hull designs. See Tr. of Meeting of Transportation Committee, Florida House of Representatives, May 3, 1983 ("There is no inducement for [a] quality boat manufacturer to improve these designs and secondly, if he does, it is immediately copied. This would prevent that and allow him recourse in circuit court"). To accomplish this goal, the Florida statute endows the original boat hull manufacturer with rights against the world, similar in scope and operation to the rights accorded a federal patentee. Like the patentee, the beneficiary of the Florida statute may prevent a competitor from "making" the product in what is evidently the most efficient manner available and from "selling" the product when it is produced in that fashion. Compare 35 U. S. C. §154. The Florida scheme offers this protection for an unlimited number of years to all boat hulls and their component parts, without regard to their ornamental or technological merit. Protection is available for subject matter for which patent protection has been denied or has expired, as well as for designs which have been freely revealed to the consuming public by their creators.

That the Florida statute does not remove all means of reproduction and sale does not eliminate the conflict with the federal scheme. See *Kellogg*, 305 U.S., at 122. In essence, the Florida law prohibits the entire public from engaging in a form of reverse engineering of a product in the public domain. This is clearly one of the rights vested in the federal patent holder, but has never been a part of state protection under the law of unfair competition or trade secrets. . . .

Moreover, as we noted in *Kewanee*, the competitive reality of reverse engineering may act as a spur to the inventor, creating an incentive to develop inventions that meet the rigorous requirements of patentability. 416 U.S., at 489-490. The Florida statute substantially reduces this competitive incentive, thus eroding the general rule of free competition upon which the attractiveness of the federal patent bargain depends. . . .

The Florida statute is aimed directly at the promotion of intellectual creation by substantially restricting the public's ability to exploit ideas that the patent system mandates shall be free for all to use. Like the interpretation of Illinois unfair competition law in Sears and Compco, the Florida statute represents a break with the tradition of peaceful coexistence between state market regulation and federal patent policy. The Florida law substantially restricts the public's ability to exploit an unpatented design in general circulation, raising the specter of state-created monopolies in a host of useful shapes and processes for which patent protection has been denied or is otherwise unobtainable. It thus enters a field of regulation which the patent laws have reserved to Congress. The patent statute's careful balance between public right and private monopoly to promote certain creative activity is a "scheme of federal regulation . . . so pervasive as to make reasonable the inference that Congress left no room for the States to supplement it." Rice v. Santa Fe Elevator Corp., 331 U.S. 218, 230 (1947). . . .

COMMENTS AND QUESTIONS

1. Does this result make sense? It is certainly consistent with *Sears* and *Compco;* less so with *Kewanee. Bonito* can be reconciled with trade secrets statutes, though, if one accepts the reverse engineering rationale described above. Whatever one thinks of the *Bonito* result, the last sentence remains troubling. Characterizing patent law as "pervasive federal regulation" suggests that it might preempt the field, automatically striking down all state laws that attempt to regulate intellectual property. If taken seriously, that approach would leave no room at all for state protection of inventions.

2. Note the similarity between the statutes struck down in *Sears* and *Bonito Boats.* In both cases, what was prohibited was the direct copying of a competitor's design. The *Bonito Boats* statute is more limited than that in *Sears,* since it prohibits only one particular method of copying. Nonetheless, the Court struck it down. Why should the courts be concerned about these comparatively narrow statutes when they allowed the far broader statute in *Kewanee* to pass muster?

3. The *Bonito* decision has been roundly criticized. See., e.g., John S. Wiley, Jr., Bonito Boats: Uninformed but Mandatory Federal Innovation Policy, 1989 Sup. Ct. Rev. 283. Cf. Symposium, Product Simulation: A Right or a Wrong?, 64 Colum. L. Rev. 1178 (1964) (articles criticizing the analogous decisions in *Sears* and *Compco*).

A different form of criticism of the result in *Bonito Boats* was offered by Congress in 1998. As part of the Digital Millenium Copyright Act, Congress created a new federal intellectual property right protecting "original" boat hull designs. 17 U.S.C. §1301 et seq. We discuss this statute in more detail in Chapter 4, supra. Can Congress lawfully accomplish here what the states cannot? Does *Bonito Boats* suggest some sort of constitutional limitation on any form of protection (state or federal) in this area?

4. Some lawyers have tried to apply *Bonito Boats* in contexts beyond the patent-like legislation actually considered in the case. This has not met with much success, however; like other broad preemption decisions before it, *Bonito Boats* has been hemmed in by subsequent qualifications. In Waits v. Frito-Lay, Inc., 978 F.2d 1093 (9th Cir. 1992), *cert. denied,* 506 U.S. 1080

(1993), a right of publicity case with facts remarkably similar to the *Midler* case reproduced above, the Ninth Circuit had the following to say about *Bonito Boats*:

Bonito Boats involved a Florida statute giving perpetual patent-like protection to boat hull designs already on the market, a class of manufactured articles expressly excluded from federal patent protection. The Court ruled that the Florida statute was preempted by federal patent law because it directly conflicted with the comprehensive federal patent scheme. In reaching this conclusion, the Court cited its earlier decisions in Sears Roebuck & Co. v. Stiffel Co., 376 U.S. 225, 84 S. Ct. 784, 11 L. Ed. 2d 661 (1964), and Compco Corp. v. Day-Brite Lighting, 376 U.S. 234, 84 S. Ct. 779, 11 L. Ed. 2d 669 (1964), for the proposition that "publicly known design and utilitarian ideas which were unprotected by patent occupied much the same position as the subject matter of an expired patent," i.e., they are expressly unprotected. Bonito Boats, 489 U.S. at 152, 109 S. Ct. at 978.

The defendants seize upon this citation to Sears and Compco as a reaffirmation of the sweeping preemption principles for which these cases were once read to stand. They argue that Midler was wrongly decided because it ignores these two decisions, an omission that the defendants say indicates an erroneous assumption that Sears and Compco have been "relegated to the constitutional junkyard." Thus, the defendants go on to reason, earlier cases that rejected entertainers' challenges to imitations of their performances based on federal copyright preemption, were correctly decided because they relied on Sears and Compco. See Sinatra v. Goodyear Tire & Rubber Co., 435 F.2d 711, 716-18 (9th Cir. 1970), *cert. denied*, 402 U.S. 906, 91 S. Ct. 1376, 28 L. Ed. 2d 646 (1971); Booth v. Colgate-Palmolive Co., 362 F. Supp. 343, 348 (S.D.N.Y. 1973); Davis v. Trans World Airlines, 297 F. Supp. 1145, 1147 (C.D. Cal. 1969). This reasoning suffers from a number of flaws.

Bonito Boats itself cautions against reading Sears and Compco for a "broad pre-emptive principle" and cites subsequent Supreme Court decisions retreating from such a sweeping interpretation. "[T]he Patent and Copyright Clauses do not, by their own force or by negative implication, deprive the States of the power to adopt rules for the promotion of intellectual creation." Bonito Boats, 489 U.S. at 165, 109 S. Ct. at 985 (citing, inter alia, Goldstein v. California, 412 U.S. 546, 552-61, 93 S. Ct. 2303, 2307-08, 37 L. Ed. 2d 163 (1973) and Kewanee Oil Co. v. Bicron Corp., 416 U.S. 470, 478-79, 94 S. Ct. 1879, 1885, 40 L. Ed. 2d 315 (1974)). Instead, the Court reaffirmed the right of states to "place limited regulations on the use of unpatented designs in order to prevent consumer confusion as to source." Id. Bonito Boats thus cannot be read as endorsing or resurrecting the broad reading of Compco and Sears urged by the defendants, under which Waits' state tort claim arguably would be preempted.

Moreover, the Court itself recognized the authority of states to protect entertainers' "right of publicity" in Zacchini v. Scripps-Howard Broadcasting Co., 433 U.S. 562, 97 S. Ct. 2849, 53 L. Ed. 2d 965 (1977). In Zacchini, the Court endorsed a state right-of-publicity law as in harmony with federal patent and copyright law, holding that an unconsented-to television news broadcast of a commercial entertainer's performance was not protected by the First Amendment. Id. at 573, 576-78, 97 S. Ct. at 2856, 2858-59. The cases Frito asserts were "rightly decided" all predate Zacchini and other Supreme Court precedent narrowing Sears' and Compco's sweeping preemption principles. In sum, our holding in Midler, upon which Waits' voice misappropriation claim rests, has not been eroded by subsequent authority.

5. In a number of cases, restrictions on the content of licensing contracts are said to raise preemption issues. One might question whether contracts are equivalent to state statutes for preemption purposes. On at least one occasion, the Supreme Court has suggested that protection of unpatentable goods (which would be prohibited under state misappropriation statutes) is permissible under contract law. See, e.g., Aronson v. Quick Point Pencil, 440 U.S. 257 (1979). On the other hand, there are a variety of circumstances in which federal patent policy simply precludes the parties from contracting to the contrary. See, e.g., Brulotte v. Thys Co., 379 U.S. 29 (1964) (holding unenforceable patent licensed in agreement that extends beyond patent's then 17-year term); Lear, Inc. v. Adkins, 395 U.S. 653 (1969) (contracts estopping licensee from challenging the validity of a patent are void); Everex Systems Inc. v. Cadtrak Corp., 89 F.3d 673 (9th Cir. 1996) (federal policy precluding assignment of nonexclusive patent licenses prevailed over state doctrine permitting such assignments). Note that in some of these cases, such as *Everex*, federal preemption actually works to the benefit of the intellectual property owner.

Regardless of the particular limitations federal law imposes on licensing agreements, patent licensing in general is a question of state (not federal) law. The Federal Circuit reaffirmed this in Gjerlov v. Schuyler Labs. Inc., 131 F.3d 1016 (Fed. Cir. 1997).

6. What is the prevailing rule regarding patent preemption? Can you articulate a workable guideline for distinguishing permissible from impermissible state laws?

Note on "Electing" Trade Secrets or Patents

The economic significance of the trade secret/patent tradeoff — which might be deemed the "doctrine of quasi-election" — is discussed in Friedman, Landes & Posner, Some Economics of Trade Secret Law, 5 J. Econ. Persp. 61 (1991). The authors argue that trade secret law protects inventions that are not worth patenting, either because they are too trivial (and hence will fail 35 U.S.C. §103) or because they are not worth the cost of patenting. Id. at 65-66. Moreover, they make an interesting point regarding the "race to invent/patent" literature: trade secret protection is better than patent protection in that it does not encourage over-rapid invention in a regime where the cost of invention is expected to fall over time. This takes some of the pressure off the racing/rent dissipation models that have become popular in the economic literature on patents in recent years.

But there is more to the story than a simple election. As Paul Goldstein pointed out some time ago, it is anomalous to speak of state law "diverting" inventors away from patentable research; at the outset of the research, it is impossible to say whether it will be patentable. See Paul Goldstein, Kewanee Oil Co. v. Bicron Corp.: Notes on a Closing Circle, 1974 Sup. Ct. Rev. 81, 91-92 (1974).

This point bears heavily on one of the *Bonito Boats* court's justifications for its holding. According to the Supreme Court, state laws that protect the same subject matter as is protected by federal patent law have the potential to redirect inventive efforts away from potentially patentable research. Presumably this might "distort" research in directions contrary to those desired by Congress when it passed the Patent Act. As a consequence, it is implicit in the scheme of the federal patent system that state protection not get "too close," because this might divert research that would have been directed toward potentially patentable inventions. Here we refer to this as the "crowding out" rationale.

The problem with this thinking is that it assumes that when researchers make decisions about where to invest they choose between two mutually exclusive types of research: potentially patentable, and nonpatentable. The Supreme Court envisions a world where inventors might "slack off," stopping their research efforts short of what is needed to achieve a patent, if they can get "equivalent" protection for less ambitious research. If such protection is available, they will invest in less ambitious, presumably "safer" projects; this will reduce the amount of "patent-oriented" research, and cut into the volume of patents as well. This view of things also assumes that Congress has determined that the extra effort required to make something patentable makes that thing more valuable as well.

Are any of these assumptions true? Only the last one, perhaps; Congress does ask something significant of an inventor who would obtain a patent, whereas state protection schemes often require something more modest, as in *Bonito Boats*. As to the other assumptions, it appears that most researchers do research first and then consider various legal options for protecting it. One reason they do this is clear: *it is impossible in most cases to tell ex ante whether a particular project or project type is of the "patentable" variety or not.* Only when the researcher has the initial experimental results in hand does she turn to the patent attorney for guidance. Only at this point does she know whether she has something patentable or not. Although many researchers take patents into account in deciding which projects to pursue, it is not a major factor. Nor can it be, since patents are granted for results, not plans.

Seen in this light, state intellectual property laws are not a threat for the reasons identified by the Court. In fact, state protection might actually increase the amount of research firms take on — and ultimately the number of patents too! This is because state intellectual property laws may increase the overall incentives faced by the firm trying to decide whether to engage in a research project or not. If the incentive is greater, more research will be undertaken. And if more research is undertaken, more *patentable* research will result.

The preceding point can be restated as follows. The Supreme Court pictures an inventor faced with two urns, one marked "potentially patentable" and the other marked "state protection." The Court's point is that increasing the value of a ball picked from the second urn will decrease the number of balls chosen from the first urn. In fact, in some cases, and perhaps most cases, the inventor *does not know* which type of urn she is facing; it is as though she were presented with one large urn in which some of the balls are marked "patentable" and the others "state protection." Because she does not know at the outset of the game which type of ball she will pick, an increase in the value of *either type* of ball will increase her expected return and thus increase the number of balls she is willing to pick.

This is a more accurate representation of the choices facing many inventors. Consequently one must recognize that state intellectual property laws may have an effect *opposite* of that stated by the Court: they may increase the number of "draws" from the undifferentiated urn marked "potential inventions," thereby increasing the number of patentable as well as state-protectable inventions.

This is not to say that the Court was wrong in its holding in *Bonito Boats*. On the contrary, there may be good reasons to guard against over-encouraging efforts to innovate. In addition, there will often be federalism-based concerns that are implicated by a system of state intellectual property laws. But the "crowding out" theory used by the Court may be wrong, at least in some cases, and therefore should not be used to justify a strong patent preemption position.

2. Copyright Preemption

≡ *Goldstein v. California*
≡ Supreme Court of the United States
≡ 412 U.S. 546 (1973), reh'g denied, 414 U.S. 883 (1973)

Mr. Chief Justice BURGER delivered the opinion of the Court.

We granted certiorari to review petitioners' conviction under a California statute making it a criminal offense to "pirate" recordings produced by others.

In 1971, an information was filed by the State of California, charging petitioners in 140 counts with violating §653h of the California Penal Code. The information charged that, between April 1970 and March 1971, petitioners had copied several musical performances from commercially sold recordings without the permission of the owner of the master record or tape. Petitioners moved to dismiss the complaint on the grounds that §653h was in conflict with Art. I, §8, cl. 8, of the Constitution, the "Copyright Clause," and the federal statutes enacted thereunder. Upon denial of their motion, petitioners entered pleas of nolo contendere to 10 of the 140 counts; the remaining counts were dismissed. On appeal, the Appellate Department of the California Superior Court sustained the validity of the statute. After exhausting other state appellate remedies, petitioners sought review in this Court.

I

Petitioners were engaged in what has commonly been called "record piracy" or "tape piracy" — the unauthorized duplication of recordings of performances by major musical artists. Petitioners would purchase from a retail distributor a single tape or phonograph recording of the popular performances they wished to duplicate. The original recordings were produced and marketed by recording companies with which petitioners had no contractual relationship. At petitioners' plant, the recording was reproduced on blank tapes, which could in turn be used to replay the music on a tape player. . . .

The challenged California statute forbids petitioners to transfer any performance fixed on a tape or record onto other records or tapes with the intention of selling the duplicates, unless they have first received permission from those who, under state law, are the owners of the master recording. Although the protection afforded to each master recording is substantial, lasting for an unlimited time, the scope of the proscribed activities is narrow. No limitation is placed on the use of the music, lyrics, or arrangement employed in making the master recording. Petitioners are not precluded from hiring their own musicians and artists and recording an exact imitation of the performance embodied on the master recording. Petitioners are even free to hire the same artists who made the initial recording in order to duplicate the performance. In essence, the statute thus provides copyright protection solely for the specific expressions which compose the master record or tape.

Petitioners' attack on the constitutionality of §653h has many facets. First, they contend that the statute establishes a state copyright of unlimited duration, and thus conflicts with Art. I, §8, cl. 8, of the Constitution. Second, petitioners claim that the state statute interferes with the implementation of federal policies inherent in the federal copyright statutes. 17 U.S.C. §1 et seq. According to petitioners, it was the intention of Congress, as interpreted by this Court in Sears, Roebuck & Co. v. Stiffel Co., 376 U.S. 225, 84 S. Ct. 784, 11 L. Ed. 2d 661 (1964), and Compco Corp. v. Day-Brite Lighting, 376 U.S. 234, 84 S. Ct. 779, 11 L. Ed. 2d 669 (1964), to establish a uniform law throughout the United States to protect original writings. As part of the federal scheme, it is urged that Congress intended to allow individuals to copy any work which was not protected by a federal copyright. Since §653h effectively prohibits the copying of works which are not entitled to federal protection, petitioners contend that it conflicts directly with congressional policy and must fall under the Supremacy Clause of the Constitution.

We note at the outset that the federal copyright statutes to which petitioners refer were amended by Congress while their case was pending in the state courts. In 1971, Pub. L. 92-140, 85 Stat. 391, 17 U.S.C. §§1(f), 5(n), 19, 20, 26, 101(e), was passed to allow federal copyright protection of recordings. However, §3 of the amendment specifically provides that . . . nothing in Title 17, as amended is to "be applied retroactively or (to) be construed as affecting in any way any rights with respect to sound recordings fixed before" February 15, 1972 . . . [and thus that] the amendments have no application in petitioners' case.

II

Petitioners' first argument rests on the premise that the state statute under which they were convicted lies beyond the powers which the States reserved in our federal system. If this is correct, petitioners must prevail, since the States cannot exercise a sovereign power which, under the Constitution, they have relinquished to the Federal Government for its exclusive exercise. . . .

"Whatever subjects of this power are in their nature national, or admit only of one uniform system, or plan of regulation, may justly be said to be of such a nature as to require exclusive legislation by Congress." 12 How. at 319. The Court's determination that Congress alone may legislate over matters which are necessarily national in import reflects the basic principle of federalism. Mr. Chief Justice Marshall said, "The genius and character of the (federal) government seem to be, that its action is to be applied to all the external concerns of the nation, and to those internal concerns which affect the States generally; but not to those which are completely within a particular State, which do not affect other States, and with which it is not necessary to interfere, for the purpose of executing some of the general powers of the government." Gibbons v. Ogden, 9 Wheat. 1, 195, 6 L. Ed. 23 (1824).

The question whether exclusive federal power must be inferred is not a simple one, for the powers recognized in the Constitution are broad and the nature of their application varied. . . . We must also be careful to distinguish those situations in which the concurrent exercise of a power by the Federal Government and the States or by the States alone may possibly lead to conflicts and those situations where conflicts will necessarily arise. . . .

The objective of the Copyright Clause was clearly to facilitate the granting of rights national in scope. While the debates on the clause at the Constitutional Convention were extremely limited, its purpose was described by James Madison in the Federalist:

The utility of this power will scarcely be questioned. The copyright of authors has been solemnly adjudged, in Great Britain, to be a right of common law. The right to useful inventions seems with equal reason to belong to the inventors.

The public good fully coincides in both cases with the claims of individuals. The States cannot separately make effectual provision for either of the cases, and most of them have anticipated the decision of this point, by laws passed at the instance of Congress.[11]

The difficulty noted by Madison relates to the burden placed on an author or inventor who wishes to achieve protection in all States when no federal system of protection is available. To do so, a separate application is required to each state government; the right which in turn may be granted has effect only within the granting State's borders. The national system which Madison supported eliminates the need for multiple applications and the expense and difficulty involved. In effect, it allows Congress to provide a reward greater in scope than any particular State may grant to promote progress in those fields which Congress determines are worthy of national action.

Although the Copyright Clause thus recognizes the potential benefits of a national system, it does not indicate that all writings are of national interest or that state legislation is, in all cases, unnecessary or precluded. The patents granted by the States in the 18th century show, to the contrary, a willingness on the part of the States to promote those portions of science and the arts which were of local importance. Whatever the diversity of people's backgrounds, origins, and interests, and whatever the variety of business and industry in the 13 Colonies, the range of diversity is obviously far greater today in a country of 210 million people in 50 States. In view of that enormous diversity, it is unlikely that all citizens in all parts of the country place the same importance on works relating to all subjects. Since the subject matter to which the Copyright Clause is addressed may thus be of purely local importance and not worthy of national attention or protection, we cannot discern such an unyielding national interest as to require an inference that state power to grant copyrights has been relinquished to exclusive federal control.

The question to which we next turn is whether, in actual operation, the exercise of the power to grant copyrights by some States will prejudice the interests of other States. As we have noted, a copyright granted by a particular State has effect only within its boundaries. If one State grants such protection, the interests of States which do not are not prejudiced since their citizens remain free to copy within their borders those works which may be protected elsewhere. The interests of a State which grants copyright protection may, however, be adversely affected by other States that do not; individuals who wish to purchase a copy of a work protected in their own State will be able to buy unauthorized copies in other States where no protection exists. However, this conflict is neither so inevitable nor so severe as to compel the conclusion, that state power has been relinquished to the exclusive jurisdiction of the Congress. Obviously when some States do not grant copyright protection — and most do not — that circumstance reduces the economic value of a state copyright, but it will hardly render the copyright worthless. The

11. The Federalist No. 43, p. 309 (B. Wright ed. 1961).

situation is no different from that which may arise in regard to other state monopolies, such as a state lottery, or a food concession in a limited enclosure like a state park; in each case, citizens may escape the effect of one State's monopoly by making purchases in another area or another State. Similarly, in the case of state copyrights, except as to individuals willing to travel across state lines in order to purchase records or other writings protected in their own State, each State's copyrights will still serve to induce new artistic creations within that State — the very objective of the grant of protection. We do not see here the type of prejudicial conflicts which would arise, for example, if each State exercised a sovereign power to impose imposts and tariffs; nor can we discern a need for uniformity such as that which may apply to the regulation of interstate shipments.

Similarly, it is difficult to see how the concurrent exercise of the power to grant copyrights by Congress and the States will necessarily and inevitably lead to difficulty. At any time Congress determines that a particular category of "writing" is worthy of national protection and the incidental expenses of federal administration, federal copyright protection may be authorized. Where the need for free and unrestricted distribution of a writing is thought to be required by the national interest, the Copyright Clause and the Commerce Clause would allow Congress to eschew all protection. In such cases, a conflict would develop if a State attempted to protect that which Congress intended to be free from restraint or to free that which Congress had protected. However, where Congress determines that neither federal protection nor freedom from restraint is required by the national interest, it is at liberty to stay its hand entirely. Since state protection would not then conflict with federal action, total relinquishment of the States' power to grant copyright protection cannot be inferred.

As we have seen, the language of the Constitution neither explicitly precludes the States from granting copyrights nor grants such authority exclusively to the Federal Government. The subject matter to which the Copyright Clause is addressed may at times be of purely local concern. No conflict will necessarily arise from a lack of uniform state regulation, nor will the interest of one State be significantly prejudiced by the actions of another. No reason exists why Congress must take affirmative action either to authorize protection of all categories of writings or to free them from all restraint. We therefore conclude that, under the Constitution, the States have not relinquished all power to grant to authors "the exclusive Right to their respective Writings."

B

Petitioners base an additional argument on the language of the Constitution. The California statute forbids individuals to appropriate recordings at any time after release. From this, petitioners argue that the State has created a copyright of unlimited duration, in violation of that portion of Art. I, §8, cl. 8, which provides that copyrights may only be granted "for limited Times." Read literally, the text of Art. I does not support petitioners' position. Section 8 enumerates those powers which have been granted to Congress; Whatever limitations have been appended to such powers can only be understood as a limit on congressional, and not state, action. Moreover, it is not clear that the dangers to which this limitation was addressed apply with equal force to both the Federal Government and the States. When Congress grants an exclusive right or monopoly, its effects are pervasive; no citizen or State may escape its reach. As we have noted, however, the exclusive right granted by a State is confined to its borders. Consequently, even when the right is unlimited in duration, any tendency to inhibit further progress in science or the arts is narrowly circumscribed. The challenged statute cannot be voided for lack of a durational limitation.

III

Our conclusion that California did not surrender its power to issue copyrights does not end the inquiry. We must proceed to determine whether the challenged state statute is void under the Supremacy Clause. No simple formula can capture the complexities of this determination; the conflicts which may develop between state and federal action are as varied as the fields to which congressional action may apply. . . .

While the area in which Congress may act is broad, the enabling provision of Clause 8 does not require that Congress act in regard to all categories of materials which meet the constitutional definitions. Rather, whether any specific category of "Writings" is to be brought within the purview of the federal statutory scheme is left to the discretion of the Congress. The history of federal copyright statutes indicates that the congressional determination to consider specific classes of writings is dependent, not only on the character of the writing, but also on the commercial importance of the product to the national economy. As our technology has expanded the means available for creative activity and has provided economical means for reproducing manifestations of such activity, new areas of federal protection have been initiated.

Petitioners contend that the actions taken by Congress in establishing federal copyright protection preclude the States from granting similar protection to recordings of musical performances. According to petitioners, Congress addressed the question of whether recordings of performances should be granted protection in 1909; Congress determined that any individual who was entitled to a copyright on an original musical composition should have the right to control to a limited extent the use of that composition on recordings, but that the record itself, and the performance which it was capable of reproducing were not worthy of such protection. [The Court rejects this reading of the legislative history of the 1909 Act.]

Sears and *Compco*, on which petitioners rely, do not support their position. In those cases, the question was whether a State could, under principles of a state unfair competition law, preclude the copying of mechanical configurations which did not possess the qualities required for the granting of a federal design or mechanical patent. . . .

In regard to mechanical configurations, Congress had balanced the need to encourage innovation and originality of invention against the need to insure competition in the sale of identical or substantially identical products. The standards established for granting federal patent protection to machines thus indicated not only which articles in this particular category Congress wished to protect, but which configurations it wished to remain free. The application of state law in these cases to prevent the copying of articles which did not meet the requirements for federal protection disturbed the careful balance which Congress had drawn and thereby necessarily gave way under the Supremacy Clause of the Constitution. No comparable conflict between state law and federal law arises in the case of recordings of musical performances. In regard to this category of 'Writings,' Congress has drawn no balance; rather, it has left the area unattended, and no reason exists why the State should not be free to act.

IV

More than 50 years ago, Mr. Justice Brandeis observed in dissent in International News Service v. Associated Press: "The general rule of law is, that the noblest of human productions — knowledge, truths ascertained, conceptions, and ideas — become, after voluntary communication to others, free as the air to common use." 248 U.S. 215, 250 . . . (1918).

But there is no fixed, immutable line to tell us which "human productions" are private property and which are so general as to become "free as the air." In earlier times, a performing artist's work was largely restricted to the stage; once performed, it remained "recorded' only in the memory of those who had seen or heard it. Today, we can record that performance in precise detail and reproduce it again and again with utmost fidelity. The California statutory scheme evidences a legislative policy to prohibit "tape piracy" and "record piracy," conduct that may adversely affect the continued production of new recordings, a large industry in California. Accordingly, the State has, by statute, given to recordings the attributes of property. . . .

We conclude that the State of California has exercised a power which it retained under the Constitution, and that the challenged statute, as applied in this case, does not intrude into an area which Congress has, up to now, preempted. Until and unless Congress takes further action with respect to recordings fixed prior to February 15, 1972, the California statute may be enforced against acts of piracy such as those which occurred in the present case.

Affirmed.

See, e.g. U.S. ex rel. Berge v. Trustees of the University of Alabama, 104 F.3d 1453, 41 USPQ2d 1481 (4th Cir. 1997) ("[T]he shadow actually cast by the Act's preemption is notably broader than the wing of its protection.").

COMMENTS AND QUESTIONS

1. From the dissent in *Goldstein*:

[Given] the presumption of *Sears* and *Compco* that congressional silence betokens a determination that the benefits of competition outweigh the impediments placed on creativity by the lack of copyright protection, and in the absence of a congressional determination that the opposite is true, we should not let our distaste for "pirates" interfere with our interpretation of the copyright laws. I would therefore hold that, as to sound recordings fixed before February 15, 1972, the States may not enforce laws limiting reproduction.

412 U.S., at 579. Is the "presumption" a fair characterization of *Sears* and *Compco*?

2. Preemption doctrine is rife with troubling distinctions. For example, the state statutes at issue in *Sears* and *Compco* were said to lower certain of the requirements of patentability under the federal patent law; meanwhile, the Court in *Goldstein* said that the California law at issue there applied to *different types of works* than were covered under federal copyright law. Does the distinction between "below requirements" and "unprotected type of work" make sense in light of the fact that many commentators view "proper subject matter" (in both copyright and patent law) as a requirement of protection? Is there a difference in the degree of incursion into the federal sphere between the two cases? Is there any greater presumption that one or the other category of works was intentionally left unprotected by Congress? What about works that fit both categories, e.g., many inventions that would be rejected by the Patent Office under the "printed matter" exception to patentability?

3. The amendment to the 1909 Act referred to in the *Goldstein* opinion was carried over into the 1976 Act. Sound recordings now appear on the list of copyrightable subject matter. But Congress appears to have rejected to some extent the minimal preemption approach of *Goldstein*, by passing §301 of the Act, 17 U.S.C. §301:

17 U.S.C. §301. Preemption with Respect to Other Laws

(a) On and after January 1, 1978, all legal or equitable rights that are equivalent to any of the exclusive rights within the general scope of copyright as specified by section 106 in works of authorship that are fixed in a tangible medium of expression and come within the subject matter of copyright as specified by sections 102 and 103, whether created before or after that date and whether published or unpublished, are governed exclusively by this title. Thereafter, no person is entitled to any such right or equivalent right in any such work under the common law or statutes of any state.

(b) Nothing in this title annuls or limits any rights or remedies under the common law or statutes of any State with respect to —

(1) subject matter that does not come within the subject matter of copyright as specified by sections 102 and 103, including works of authorship not fixed in tangible medium of expression; or

(2) any cause of action arising from undertakings commenced before January 1, 1978; or

(3) activities violating legal or equitable rights that are not equivalent to any of the exclusive rights within the general scope of copyright as specified by section 106.

(c) With respect to sound recordings fixed before February 15, 1972, any rights or remedies under the common law or statutes of any State shall not be annulled or limited by this title until February 15, 2047. The preemptive provisions of subsection (a) shall apply to any such rights and remedies pertaining to any cause of action arising from undertakings commenced on and after February 15, 2047. Notwithstanding the provisions of section 303, no sound recording fixed before February 15, 1972, shall be subject to copyright under this title before, on, or after February 15, 2047.

(d) Nothing in this title annuls or limits any rights or remedies under any other Federal statute.

As a result of section 301, state laws can provide remedies outside copyright, but not "equivalent" to copyright. As one might expect, the meaning of "equivalence" in this context has led to no small degree of confusion. For an example of the problems engendered by section 301, see the *National Car Rental* decision later in this chapter.

4. Several cases have gone beyond section 301 to a general preemption analysis along the lines of the patent preemption cases such as *Kewanee* and *Bonito Boats*. In Associated Film Distribution Corp. v. Thornburgh, 520 F. Supp. 971 (E.D. Pa. 1981), *rev'd and remanded on other grounds*, 683 F.2d 808, 817 (3d Cir. 1982), *cert. denied*, 480 U.S. 933 (1982), the court found preempted a state statute regulating the procedure by which film exhibitors licensed major motion pictures from film distribution arms of the major movie companies. The court went beyond section 301, stating that the "more general question of conflict between the two statutory schemes under the supremacy clause is decisive." 520 F. Supp. at 993. Using this approach, the court held that the statute interfered with the national scheme of copyright protection. The impact of the decision in *Associated* is blunted by the fact that another district court, ruling on a similar state statute in Ohio, found that federal copyright law did not preempt the statute. See Allied Artists Pictures Corp. v. Rhodes, 496 F. Supp. 408 (S.D. Ohio 1980), *aff'd and remanded in part*, 679 F.2d 656, 665 (6th Cir. 1982).

In ASCAP v. Pataki, 930 F. Supp. 873 (S.D.N.Y. 1996), the court invalidated a state statute regulating the activities of performing rights societies such as ASCAP and BMI. The state statute required (among other things) that such groups provide owners of establishments performing music with written notice of an investigation within 72 hours after it is initiated, thus making it difficult for ASCAP and others to conduct "undercover" investigations for violations of the copyright laws. The court's opinion addressed only the issue of "conflict preemption, which occurs either where compliance with both federal and state regulations is a physical impossibility, or where state law stands as an obstacle to the accomplishment and execution of the full purposes and objectives of Congress." Id. The court found that the state statutory provisions "hinder the realization of the federal copyright scheme" for several reasons: the statute made it more difficult for copyright owners to enforce their rights; it effectively established a "statute of limitations" on copyright investigations that was shorter than the federal statute; and it gave copyright defendants a counterclaim that they could use to offset copyright damages. See also College Entrance Examination Board v. Pataki, 889 F. Supp. 554 (N.D.N.Y. 1995) (state law requiring disclosure of standardized test questions and answers preempted by Copyright Act because it conflicted with the rights of copyright owners to restrict distribution of copyrighted material).

In ProCD v. Zeidenberg, 908 F. Supp. 640 (W.D. Wisc. 1996), the district court reasoned in dictum that state contract law could not be used to prevent the copying of telephone white pages, which the Supreme Court had determined in *Feist* were uncopyrightable. The district court applied a constitutional copyright preemption analysis similar to the patent analysis undertaken in *Bonito Boats*. The Seventh Circuit reversed the district court's decision, concluding that under section 301 of the Copyright Act state contract law could not be preempted in this case. ProCD v. Zeidenberg, 86 F.3d 1447 (7th Cir. 1996).[2] Judge Easterbrook's opinion did not mention Supremacy Clause preemption at all, though one can infer from the holding in the case that the court must have intended to reverse on that ground as well. Contra Vault Inc. v. Quaid Corp., 847 F.2d 255 (5th Cir. 1988) (state statute permitting enforcement of shrinkwrap licenses invalid under Supremacy Clause preemption).

In the wake of the Seventh Circuit opinion in *ProCD*, some people have suggested that section 301 replaces the sort of Supremacy Clause analysis undertaken in *Goldstein* and *Bonito Boats*. Is this a fair reading of the statute? Does section 301 cover all possible conflicts between state and federal laws? How should the courts treat a state statute providing, in effect, that federal copyright law did not apply to certain activities within the state?

---

2. This portion of the case is discussed in more detail in the notes following *National Car Rental*, infra.

National Car Rental System, Inc. v. Computer Associates International, Inc.
United States Court of Appeals for the Eighth Circuit
991 F.2d 426 (8th Cir. 1993)

MAGILL, Circuit Judge.

We here deal with the difficult question of the extent to which the Copyright Act preempts state breach of contract actions alleging that the licensee of computer software exceeded limitations on the use of computer software contained in the license agreements. Computer Associates International, Inc., appeals from the district court's order resolving a motion for judgment on the pleadings and dismissing its breach of contract claim against National Car Rental as preempted under the Copyright Act. We conclude that the district court failed to grant Computer Associates all reasonable inferences from its pleadings, and hold that as properly construed, the cause of action as pled is not preempted. We reverse.

I. Background

Computer Associates International, Inc. (CA), creates and licenses computer software. CA licensed its programs to the appellee, National Car Rental Systems, Inc. (National), to process National's data on National's hardware in Bloomington, Minnesota. The 1990 license agreement between CA and National provided, as did earlier licenses, that National may use the licensed programs "only for the internal operations of Licensee and for the processing of its own data." A separate order form, incorporated into the license agreement, similarly provided that "use of the Licensed Program[s] is restricted to the internal operations of Licensee and for the processing of its own data."

Sometime in 1990, National decided to cease its internal computer operations and contract with an independent computer services vendor for computer related information services. Ultimately, National retained Electronic Data Systems Corporation (EDS) to provide these services. In connection with this transaction, National, EDS, and CA entered into a supplement addendum, which provided that EDS could use the licensed programs to process National's data. The supplement addendum provided that EDS would use the programs for the benefit of National subject to the terms and conditions of the 1990 license agreement, and solely "to process data of Licensee and in no event for the processing of data . . . of any third party other than Licensee."

CA subsequently determined that National had been using the programs to process the data of third parties, including Lend Lease Trucks, Inc. (Lend

Lease), and Tilden Car Rental, Inc. (Tilden), in violation of the license agreement, and that such use had continued through EDS under the supplement addendum. CA threatened to sue National if such use did not stop. National then brought a declaratory judgment action in the district court. National admitted in its complaint that it "has used the Licensed Software in its business activities . . . including the activities relating to Tilden and Trucks [Lend Lease]," but requested a declaration that its use of the programs neither breached the license agreement nor infringed CA's copyright. CA asserted two counterclaims. In the first, it claimed that National's use of the programs, either individually or through EDS, for the benefit of Lend Lease and Tilden, breached the license agreement. In the second, CA claimed that National infringed its copyright by making an unauthorized copy of the software.

National moved for judgment on the pleadings . . . , alleging that CA's first counterclaim was preempted under §301(a) of the Copyright Act. In resolving the motion, the district court concluded that CA alleged a lease agreement between National and the third parties: National permitted them to use the software in exchange for payment. The district court concluded that this cause of action, as pled, was "equivalent" to the exclusive copyright right of distribution of copies of the work, and held it was preempted.

The Copyright Act provides the exclusive source of protection for "all legal and equitable rights that are equivalent to any of the exclusive rights within the general scope of copyright as specified by section 106" of the Copyright Act. See 17 U.S.C. §301(a). Concomitantly, all non-equivalent rights are not preempted. A state cause of action is preempted if: (1) the work at issue is within the subject matter of copyright as defined in §§102 and 103 of the Copyright Act, and (2) the state law created right is equivalent to any of the exclusive rights within the general scope of copyright as specified in §106. Harper & Row, Publishers, Inc. v. Nation Enters., 723 F.2d 195, 200 (2d Cir. 1983). . . .

In the first counterclaim, CA alleged, in pertinent part, that:

> The authorization for use granted National [pursuant to the 1990 License Agreement] was for the internal operations of National and for the processing of its own data. Pursuant to the Supplement Addendum . . . it was agreed that the Licensed Programs would be used solely for the benefit of National and subject to the rights, obligations and benefits in all respects of the terms and conditions of the License Agreement. Pursuant to the Supplement Addendum . . . National and EDS further agreed, among other things, to use the Licensed Programs . . . solely to process data of National and in no event for the processing of data of any third party other than National. . . . In none of the License Agreements was [National] granted any authorization to use the licensed programs for the benefit of any company other than itself. . . . National has used and permitted the use of the Licensed Programs for the processing of data for the benefit of third parties. This use of the Licensed Programs for the benefit of third parties includes the use for Lend Lease Trucks, Inc. and Tilden Car Rental, Inc. . . . National has been unjustly enriched by any fees or other compensation received from those third-parties for use of the Licensed Programs for their benefit.

The district court noted that the computer software in question was within the subject matter of copyright, and thus focused on whether CA's breach of contract action sought to protect rights equivalent to the exclusive copyright rights. The court noted that National had not alleged which copyright right was equivalent to CA's action, but concluded that the distribution right was the only right potentially equivalent.

In resolving the preemption issue, the court concluded that "[c]onstrued as true, CA's allegations reflect the existence of a lease arrangement between National and Lend Lease and Tilden: National has permitted Lend Lease and Tilden to use the licensed software in exchange for payment. The distribution right includes specifically the right to lease or lend." Mem. op. at 9-10. The court further concluded that the presence of a contract promise did not create a right qualitatively different from copyright, and stated: "In essence, CA alleges National breached the license agreement by infringing CA's copyrights in the licensed software." Id. at 10.

We believe that in reaching this conclusion, the district court either failed to give CA the benefit of all reasonable inferences from the pleadings or misinterpreted the law of copyright preemption. . . .

The only potential allegation of unauthorized distribution comes in CA's contention that National "permitted the use" of the programs. Because we must give CA the benefit of all reasonable inferences from the pleadings, we cannot conclude that an allegation that National "permitted the use" necessarily amounts to an allegation of the actual distribution of a copy of the program. Rather, we believe that such a pleading can be read, in context, to allege that National permitted EDS to use the programs for the benefit of Lend Lease and Tilden, with no copies ever going to Lend Lease and Tilden. . . .

We thus conclude that CA's pleadings cannot be read to allege that National breached its contract by actually distributing a copy of the licensed program to either Lend Lease or Tilden. . . .

C. Preemption of the Contractual Limitation on Use

The question then becomes whether CA's allegation that National breached their contract by using the program in a fashion not allowed under the contract protects a right equivalent to one of the exclusive copyright rights. We believe it does not.

We agree with the district court that the computer program in question is within the subject matter of copyright. Thus we focus on the second preemption issue: whether the right sought under state law is equivalent to the exclusive rights under copyright. We must consider whether a limitation on the uses to which a licensee may put a licensed work are preempted even though those uses do not involve the exclusive copyright rights. As noted above, courts and commentators have framed this inquiry as whether the right in question is "infringed by the mere act of reproduction, performance, distribution or display." 1 Nimmer on Copyright §1.01[B], at 1-13. Section 301 preempts only those state law rights that " 'may be abridged by an act which, in and of itself, would infringe one of the exclusive rights' provided by federal copyright law." Computer Assocs. Intl., Inc. v. Altai, Inc., 982 F.2d 693, 716 (2d Cir. 1992) (citing Harper & Row, Publishers, Inc. v. Nation Enters., 723 F.2d 195, 200 (2d Cir. 1983), *rev'd on other grounds,* 471 U.S. 539, 105 S. Ct. 2218, 85 L. Ed. 2d 588 (1985)). If an extra element is "required, instead of or in addition to the acts of reproduction, perform-ance, distribution or display, in order to constitute a state-created cause of action, then the right does not lie 'within the general scope of copyright' and there is no preemption." 1 Nimmer on Copyright §1.01[B], at 1-14-15 (footnotes omitted). . . .

We conclude that the alleged contractual restriction on National's use of the licensed programs constitutes an extra element in addition to the copy-right rights making this cause of action qualitatively different from an action for copyright.

National initially contends that any complaint alleging use of a copy-righted work that exceeds the uses allowable under the license must be brought as a copyright infringement claim; contract claims containing such allegations are preempted. In support of this proposition, National cites sev-eral cases finding copyright infringement when the licensee's "use" of a copy-righted and licensed work exceeded the uses allowed under the license. . . .

Because we find no general rule holding breach of contract actions such as this one preempted, we examine specifically whether this cause of action seeks to protect rights equivalent to the exclusive copyright rights. We con-clude that the contractual restriction on use of the programs constitutes an additional element making this cause of action not equivalent to a copyright action.

National disagrees with this characterization and attempts to read the term "use" in the license agreement as synonymous with the rights given to the copyright holder. We believe it is not, as two recent cases make clear.

In a case very similar to this one, involving CA, the court held that a contractual restriction on use of a computer program was distinct from the exclusive copyright rights. In Computer Assocs. v. State St. Bank & Trust, 789 F. Supp. 470 (D. Mass. 1992), the parties had executed a license that provided, inter alia: "Customer agrees to refrain from using the Equipment . . . for other customer-sites or customers on a service basis." Id. at 475. CA argued that State Street violated the provisions of the license agreement by allowing customers direct access to the programs to gain information. Be-cause of that alleged breach, Computer Associates claimed that it could cancel their maintenance contract. State Street moved for a preliminary injunction prohibiting CA from cancelling the maintenance contract. CA then claimed that the violation of the agreements constituted copyright infringement, pre-venting State Street from claiming irreparable harm. In denying this claim, the court stated: infringement results only from the unauthorized copying of copyrighted material. A use of an authorized copy of copyrighted subject matter ordinarily is not infringing. . . . Therefore, applicable limitations on State Street's use of the programs, if any, must be derived initially from the license agreements, not copyright law. Id. at 472.

The Ninth Circuit reached a similar conclusion in G. S. Rasmussen & Assocs. v. Kalitta Flying Serv., 958 F.2d 896 (9th Cir. 1992). In Kalitta, the plaintiff had received from the FAA a certificate (STC) allowing him to modify a plane design. The defendant took the certificate from an existing, modified plane and used it to modify another plane, without paying the plaintiff for doing so. Plaintiff sued for unfair competition, and the defendant raised the defense of copyright preemption. The court first noted that there was no allegation of copying. It went on to hold that:

> Federal copyright law governs only copying. . . . Enforcement of Rasmussen's property right in his STC leaves Kalitta free to make as many copies of the certificate as it wishes; to the extent the manual supplement is not protected by the copyright laws, the same is true of it. That Kalitta is prevented from then using these copies to obtain an airworthiness certificate from the FAA does not interfere in any way with the operation of the copyright laws.

Id. at 904.

In both of these cases, the courts distinguished restrictions on use of a copyrightable work that did not involve "copying" from the exclusive rights in copyright. CA's situation is the same. CA does not claim that National is doing something that the copyright laws reserve exclusively to the copyright holder, or that the use restriction is breached "by the mere act of reproduction, performance, distribution or display." Instead, on this posture, CA must be read to claim that National's or EDS's processing of data for third parties is the prohibited act. None of the exclusive copyright rights grant CA that right of their own force. Absent the parties' agreement, this restriction would not exist. Thus, CA is alleging that the contract creates a right not existing under the copyright law, a right based upon National's promise, and that it is suing to protect that contractual right. The contractual restriction on use of the programs constitutes an extra element that makes this cause of action qualitatively different from one for copyright.

We believe that the legislative history of the Copyright Act supports this conclusion. In elaborating the meaning of the term "equivalent rights" the House committee report to the Copyright Act suggests that breaches of contract were not generally preempted: "nothing in the bill derogates from the rights of parties to contract with each other and to sue for breaches of contract." See H.R. Rep. No. 94-1476, 94th Cong., 2d Sess. 132, reprinted in 1976 U.S.C.C.A.N. 5659, 5748. This is not the end of the inquiry, however.

National contends that while the bill as initially drafted might have excluded breaches of contract from preemption, the bill as passed did not. National notes that §301(b)(3) of the Copyright Act, as initially drafted and reported out of committee in the House, explicitly exempted breach of contract suits from preemption. This provision was then amended on the floor of the House to delete all the specific examples of non-preempted causes of action. National claims this action demonstrated congressional intent to remove the "safe harbor" from preemption for breach of contract actions. See, e.g., Wolff v. Institute of Elec. & Elecs. Eng'rs, Inc., 768 F. Supp. 66 (S.D.N.Y. 1991) (deletion of safe harbor provision for breaches of contracts suggests that Congress did not intend generally to except them from preemption).

We disagree. Although the deletion of a provision from a final bill generally means that Congress intends to disavow what was formerly expressed, we believe in this case the facts surrounding the deletion of §301(b)(3) suggest Congress did not intend to reverse the presumption of non-preemption for the examples initially included in §301(b)(3). Instead, it appears that certain members of the House were concerned about the subsequent addition of the tort of misappropriation to the list of non-preempted causes of action, and suggested deletion of the specific examples in order to prevent confusion about the scope of preemption. . . . [W]e believe, the better view is that the legislative history suggests a congressional intent not to preempt breach of contract actions such as this one. . . .

We do not read CA to allege that National was unjustly enriched as a result of a wrongful exercise of one of the §106 rights. Rather, we read this allegation of damage as a further explanation of the damages CA intends to prove arising from the breach of contract. CA alleges generally that it has been damaged in an amount to be proved at trial, and it will have to prove those damages. In this context, we read its allegations of unjust enrichment as an attempt, albeit inartful, to allege that National received from Lend Lease and Tilden amounts that CA would have received had National not breached their contract. Second, National notes that CA requested return or destruction of any copies of its programs still in National's possession. It notes that the Copyright Act provides precisely that remedy, see 17 U.S.C. §504, and claims that the request for destruction shows the claim is equivalent to a copyright claim. We disagree. The parties' contract specifically provides for the return or destruction of the licensed programs upon any breach of the license agreement. This remedy would apply equally to this asserted breach (improper use) as to an action for breach of an agreement to pay royalties or license fees, which National admits would not be preempted. Furthermore, the copyright remedy of return or destruction applies even absent a preexisting relationship between the parties: it does not have to be stated in a contract or license agreement. We cannot conclude that this action is preempted simply because the parties' contract provides a remedy for breach identical to a remedy provided in copyright. . . .

III. Conclusion

For all the foregoing reasons, we conclude that CA's cause of action, as pled, is not preempted by the Copyright Act. Therefore, we reverse the order of the district court dismissing CA's first counterclaim with prejudice and remand for further proceedings consistent with this opinion.

COMMENTS AND QUESTIONS

1. The court states that since "National's or EDS's processing of data for third parties is the prohibited act," and since nothing in the Copyright Act limits the rights of certain parties to use a copyrighted work, the contract-based cause of action is not preempted. The court distinguishes this behavior — allowing an unauthorized third party to use a work — from the actual distribution of a copy of a work; since the behavior is not equivalent to a protected right under section 106, it is not preempted. What if National had made copies of the program and distributed them to third parties? Would not those third parties still have to use the work? Would this use — which occurs in this hypothetical situation *in addition to* the copyright-protected acts of copying and distribution, rather than *instead* of them, as in the actual case — form the basis for a non-preempted contract cause of action? If not, the case might make sense; it provides independent content to a state cause of action for breach of contract. But if use restrictions based on copying are not preempted, then allegations of "breach through improper use" will *always* be available to form the foundation for an alternative cause of action (breach of contractual use restriction) in what is otherwise a copyright infringement case.

2. Can contract law *ever* be preempted under section 301? At least one court has held that contracts generally are not preempted by copyright and has strongly suggested that no contractual provision would ever be preempted. ProCD v. Zeidenberg, 86 F.3d 1447 (7th Cir. 1996). The court reasoned:

> Rights "equivalent to any of the exclusive rights within the general scope of copyright" are rights established *by law* — rights that restrict the options of persons who are strangers to the author. Copyright law forbids duplication, public performance, and so on, unless the person wishing to copy or perform the work gets permission; silence means a ban on copying. A copyright is a right against the world. Contracts, by contrast, generally affect only their parties; strangers may do as they please, so contracts do not create "exclusive rights." Someone who found a copy of [the plaintiff's software product] on the street would not be affected by the shrinkwrap license. . . .

Id. Is this distinction between judicial enforcement of contracts and other state laws persuasive? It is interesting to note that, whatever the validity of the general distinction, the court in *ProCD* applied it to validate a "shrink-wrap license," a peculiar form of contract that is drafted by the creator of the product and purports to bind to its terms anyone who uses the product. While it is technically true in such a case that only "parties" to the contract are bound by it, anyone who has access to the product will automatically become such a party. See Mark A. Lemley, Intellectual Property and Shrinkwrap Licenses, 68 S. Cal. L. Rev. 1239 (1995) (challenging the assumption that contract terms affect only the parties to the contract).

A number of cases have preempted contract terms that conflict with federal patent and copyright policy. For example, contract terms that purport to extend a patent or copyright beyond its expiration have repeatedly been held unenforceable. One recent decision also preempts a claim for breach of implied contract, on the theory that the "implied contract" for idea protection violated the federal policy in favor of protecting only expression in a copyrighted work. Endemol Entertainment v. Twentieth Television Inc., 48 U.S.P.Q.2d 1524 (C.D. Cal. 1998).

3. Does copyright preemption of state contract law depend on the remedy asserted for breach of contract? Should federal law be more concerned if a party seeks by contract to bring the weapons of copyright law to bear? For example, the House Committee Notes to section 109(a) of the Copyright Act (the "first sale" doctrine) provide that a contract limiting a user's first sale rights may be enforceable by an action for breach of contract but does not give rise to an action for copyright infringement. Contract remedies are ordinarily limited to expectation damages, rather than the consequential damages, statutory damages, attorney fees, and injunctions available under the copyright law. But suppose a license agreement provided that *resale* of a copyrighted work — permissible under copyright law — voids the entire license, rendering the reseller liable for copyright infringement. Should such a contract be preempted by copyright law? If not, is there anything left in practice of the first sale doctrine?

4. The National Conference of Commissioners on Uniform State Laws recently promulgated a controversial new model state law on software licensing called the Uniform Computer Information Transactions Act (UCITA). UCITA would expand the role of contracts in protecting software and would therefore heighten preemption concerns. For detailed discussion of UCITA, including the issue of federal intellectual property preemption of private contracts, see the articles collected in 87 Calif. L. Rev. 1 (1999) and 13 Berkeley Tech. L. J. (1998).

PROBLEM

Problem 6-4. Presco is in the business of printing and selling academic journals to university libraries. When Presco receives a subscription request, it sends a form contract to the requestor. The requesting library is required to sign the contract before Presco will start the subscription. Presco's form contract provides in part that "Independent of and in addition to any provisions of state or federal law, the parties agree that Subscriber will not make, cause to be made, or allow to be made any copies of any Presco journals without the prior express written consent of Presco."

Is the contract provision preempted? Does it matter whether the subscriber's copying would constitute fair use under the copyright laws?

Note on the "Additional Element" Test for Preemption

Section 301 of the 1976 Copyright Act introduced a seemingly straightforward test for federal preemption of state laws in the general domain of copyright: the so-called extra element test. Under this test, if state law provides a cause of action that requires proof of at least one element in addition to those required for a copyright infringement claim, the state law survives. The test tries to state simply, in terms of required pleadings for an effective cause of action, the basic idea that a state can proscribe activities in the general domain of intellectual property as long as those activities are qualitatively different from the ones that are the subject of federal protection.

A detailed treatment of this topic can be found in Patrick McNamara, Note, Copyright Preemption: Effecting the Analysis Prescribed by Section 301, 24 B.C. L. Rev. 963 (1983). In this Note, the author categorizes Supreme Court cases as adopting either a "state presumption" (i.e., presumption that state legislation is not preempted), or a "federal presumption" (i.e., presumption of preemption). McNamara believes a plausible case can be made that Congress rejected the state presumption by prominently mentioning the federal-presumption *Sears* and *Compco* cases as models for guidance. He goes on to point out the deficiencies of the "additional element" test, especially in light of this reading of the legislative history:

> The "extra element" test sometimes works sensible results. More often, however, the test works results contrary to the intent of the statute, due primarily to the ease with which a court can "find" the extra element needed for the state law to survive preemption. The extra element test also allows state legislatures to manipulate preemption analysis by encouraging them to add extra elements to state laws, thereby permitting courts to find non-equivalency due to these extra elements.

Id., at 984-985 (footnotes omitted). The author argues that the extra element test ought to be incorporated into a much more comprehensive test that includes (1) the purpose of the state law; and (2) the impact of the state law on the federal scheme. Id., at 1007-1008.

State publicity rights, considered earlier in this chapter, are an important new test of preemption concepts. For example, in Baltimore Orioles, Inc. v. Major League Baseball Players Association, 805 F.2d 663 (7th Cir. 1986), *cert. denied*, 480 U.S. 941 (1987), the court found that federal copyright law preempted baseball players' assertions of state publicity rights in their images and game performances. Because each game was embodied in a copyrighted telecast and players uniformly assign their copyrights to their teams, the court found that the game performances could not be the subject of independent state publicity rights.

A perceptive student note has criticized the *Orioles* court's approach and reasoning. In Shelley Ross Saxer, Note, Baltimore Orioles, Inc. v. Major League Baseball Players Association: The Right of Publicity in Game Performances and Federal Copyright Preemption, 36 UCLA L. Rev. 861 (1989), the author highlights the cursory nature — and sweeping ramifications — of the court's holding. For example, Saxer observes that the court did not even identify which state's right of publicity would apply. The note then proceeds to criticize the findings underlying the preemption holding:

> If the *Baltimore Orioles* court had used the New York right of publicity in its "additional element" approach to analyzing equivalency [i.e., preemption], it would not have found the rights equivalent. To establish a cause of action in New York, the plaintiff must show: (1) his or her name, portrait or picture has been (2) used for advertising or trade purposes (3) without his or her consent. The New York statute requirement that the person's name, portrait or picture be used for purposes of advertising or trade is an element in addition to the "mere act of reproduction, performance, distribution or display." New York courts have imposed an additional requirement that the person who created the right commercially exploit the name and likeness during their lifetime. Copyright infringement only requires the plaintiff to establish: (1) copyright ownership, and (2) the copying of the copyrighted work by the defendant. Therefore, the advertising or trade purpose element and the commercial exploitation element required for a right of publicity claim are "additional elements" not required for a copyright infringement claim. . . . The "interest created under the right of publicity["] is the person's individual style — his persona — which need not become tangibly fixed in a medium of expression. Copyright protection, in contrast, requires that the author's expression of an idea be fixed in tangible form. The New York right of publicity, and possibly any other state's right of publicity, is qualitatively different from copyright under the "additional element" test.

Id., at 881-882 (footnotes omitted). Saxer goes on to note that

> even if the players . . . impliedly consented to the videotaping [of games], only the telecast itself could be considered copyrightable. The game performance would remain unprotected by federal copyright law. The telecast's copyright protection would prevent unauthorized copying of the telecast, but the players' right of publicity in the game performance would remain unaffected. The players could assign their state rights of publicity to the clubs, or to anyone else. However, *this assignment would be a transaction separate from their consent to the videotaping.*

Id., at 875 (emphasis added). This is a telling point. References to the transactional nightmare that the baseball player's claims would create are scattered throughout the *Orioles* court's opinion. Without doubt, the difficulties of determining compensation for individual players — e.g., measuring how long they appear on the screen each time they are shown — are significant. In addition, another factor which likely influenced the court was the fact that the players were seeking a declaration of their rights as a way of obtaining a better deal in the midst of the then-existing television broadcast contract. In terms of basic intellectual property policy, in other words, the court saw the players' claims as purely an attempt to redistribute the gains from the broadcasting contract. The current group of players certainly did not argue that their new rights would lead them to "produce" more distinctive images. Thus the only potential impact of these rights would have been as a lever to renegotiate the original broadcast rights bargain.

COMMENTS AND QUESTIONS

1. Compare the result in *Orioles* with the Ninth Circuit's approach to copyright preemption of state publicity rights in Waits v. Frito-Lay, 978 F.2d 1093 (9th Cir. 1992), discussed above:

> The defendants ask that we rethink Midler anyway, arguing as the defendants did there that voice misappropriation is preempted by section 114 of the Copyright Act. Under this provision, a state cause of action escapes Copyright Act preemption if its subject matter "does not come within the subject matter of copyright . . . including works or authorship not fixed in any tangible medium of expression." 17 U.S.C. §301(b)(1). We rejected copyright preemption in Midler because voice is not a subject matter of copyright: "A voice is not copyrightable. The sounds are not 'fixed.' " Midler, 849 F.2d at 462. As a three-judge panel, we are not at liberty to reconsider this conclusion, and even if we were, we would decline to disturb it.
>
> Waits' claim, like Bette Midler's, is for infringement of voice, not for infringement of a copyrightable subject such as sound recording or musical composition. Moreover, the legislative history of section 114 indicates the express intent of Congress that "[t]he evolving common law rights of 'privacy,' 'publicity,' and trade secrets . . . remain unaffected [by the preemption provision] as long as the causes of action contain elements, such as an invasion of personal rights . . . that are different in kind from copyright infringement." H.R. Rep. No. 1476, 94th Cong., 2d Sess. 132, reprinted in 1976 U.S.C.C.A.N. 5659, 5748. Waits' voice misappropriation claim is one for invasion of a personal property right: his right of publicity to control the use of his identity as embodied in his voice. See Midler, 849 F.2d at 462-63 ("What is put forward as protectable here is more personal than any work of authorship. . . . A voice is as distinctive and personal as a face."). The trial's focus was on the elements of voice misappropriation, as formulated in Midler: whether the defendants had deliberately imitated Waits' voice rather than simply his style and whether Waits' voice was sufficiently distinctive and widely known to give him a protectable right in its use. These elements are "different in kind" from those in a copyright infringement case challenging the unauthorized use of a song or recording.

Waits' voice misappropriation claim, therefore, is not preempted by federal copyright law.

2. Does section 301 establish a clear rule for determining when state laws are preempted by the Copyright Act? How does copyright preemption compare to patent preemption? Do the two reach the same sort of statutes? Is their approach philosophically consistent?

One way to approach this issue is to consider federal preemption of trade secret laws. In the *Kewanee* opinion reproduced earlier, the Supreme Court held that the patent laws did not preempt state trade secret statutes. Would a copyright preemption analysis reach the same result if the subject matter of the trade secret were also copyrightable? Does it matter whether the trade secret was protected by a confidential relationship, or whether the plaintiff's theory was that the secret was taken by "improper means"? On copyright preemption of trade secret claims, see Computer Associates Intl. v. Altai, Inc., 982 F.2d 693, 715-17 (2d Cir. 1992) (copyright and trade secret claims could coexist regarding the use of the same computer program as long as misappropriation of trade secrets is based on a defendant's breach of a duty of trust or confidence). See also Automated Drawing Systems v. Integrated Network Services, 447 S.E.2d 109 (Ga. Ct. App. 1994) (copyright law did not preempt Georgia Computer Systems Protection Act, which prohibited the "misappropriation" of source code and provided tort damages); Wrench LLC v. Taco Bell Corp., 49 U.S.P.Q.2d 1032, 1038-1039 (W.D. Mich. 1998) (where plaintiffs alleged that defendant Taco Bell misappropiated their concept of a "psycho dog" in its ad campaign for a chain of taco stores, the court permitted misappropriation and conversion claims to go forward: "Because Plaintiffs' misappropriation claim requires the extra element of a legal relationship and Plaintiffs have alleged facts under that claim which, if proven, would support the existence of an implied-in-fact contract, the claim is qualitatively different from a copyright infringement claim and therefore is not preempted by §301.").

Note on the Special Case of Moral Rights Preemption

As noted above, both state and federal statutes protect the moral rights of artists. To what extent are state statutes preempted by the federal law? The Visual Artists Rights Act of 1990 added subsection (f) to § 301, the preemption provision of the Copyright Act:

17 U.S.C. §301 (f). Preemption Provision

(1) On [June 1, 1991], all legal and equitable rights that are equivalent to any of the rights conferred by 106A with respect to works of visual art to which the rights conferred by section 106A apply are governed exclusively by section 106A and section 113(d) and the provisions of this title

relating to such sections. Thereafter, no person is entitled to any such right or equivalent right in any work of visual art under the common law or statutes of any State.

(2) Nothing in paragraph (1) annuls or limits any rights or remedies under the common law or statutes of any State with respect to —

(A) any cause of action from undertakings commenced before [June 1, 1991];

(B) activities violating legal or equitable rights that are not equivalent to any of the rights conferred by section 106A with respect to works of visual art;

(C) activities violating legal or equitable rights which extend beyond the life of the author.

COMMENTS AND QUESTIONS

1. Does this statute preempt the state moral rights statutes discussed earlier in this chapter? On the differences between state and federal moral rights protection, see Amy L. Landers, The Current State of Moral Rights Protection for Visual Artists in the United States, 15 Hastings Comm. & Ent. L.J. 165 (1992). She argues that:

[S]tates should be able to protect media completely unaddressed by the [1990 VARA] Act. For example, a state should be permitted to protect works of applied art. If in the interest of fostering local creativity, Pennsylvania seeks to extend moral rights protection to work made by fine furniture makers, [this should be permitted].

See also Brett Sirota, Visual Artists Rights Act: Federal Versus State Moral Rights, 21 Hofstra L. Rev. 461 (1992). Cf. Jennifer R. Clarke, Note, The California Resale Royalties Act as a Test Case for Preemption Under the 1976 Copyright Law, 81 Colum. L. Rev. 1315, 1316 (1981), which "applies the traditional [preemption] analysis to the California [Resale Royalty] Act . . . and concludes that the California Act does not hinder the purpose of the federal copyright act to a degree that warrants preemption."

One case has indicated that the California moral rights statute, Cal. Civ. Code §987, "appears" to be preempted by this provision. Lubner v. City of Los Angeles, 53 Cal. Rptr. 2d 24 (Ct. App. 1996). The court did not need to reach the issue, however, since it held that negligently damaging a piece of art by losing control of a garbage truck which thereafter rolled into the art did not violate the California statute.

3. Trademark Preemption

There has been surprisingly little litigation over the preemptive effect of the Lanham Act. Because the Lanham Act was passed under the aegis of the

Commerce Clause of the Constitution, rather than the Patents and Copyrights Clause, the analysis developed in the remainder of this chapter is not directly relevant to the problem of trademark preemption. Instead, courts turn to other Commerce Clause cases, which focus on whether Congress intended to "preempt the field" of trademark law and on whether there is an actual or potential conflict between the state and federal statutes.

Two early cases held that the Lanham Act preempted all state trademark laws. Sargen & Co. v. Welco Feed Mfg., 195 F.2d 929 (8th Cir. 1952); Time, Inc. v. T.I.M.E. Inc., 123 F. Supp. 446 (S.D. Cal. 1954). But both cases are in disfavor, largely because Congress clearly intended in passing the Lanham Act to allow it to coexist with some state trademark laws. See 15 U.S.C. §§1065, 1115(b)(5) (both referring to the continued effect of state law). See Richard A. De Sevo, Antidilution Laws: The Unresolved Dilemma of Preemption Under the Lanham Act, 84 Trademark Rptr. 300, 301-04 (1994). However, at least some courts continue to hold that state statutes which are directed at the same types of conduct as the Lanham Act are preempted. See Three Blind Mice Designs Co. v. Cyrk Inc., 892 F. Supp. 303 (D. Mass. 1995) (state antidilution statute "wholly preempted" to the extent that it seeks to regulate directly competitive goods).

A more difficult question is presented where state trademark statutes may conflict with the general federal rule. Sometimes such conflicts are clear. For example, state laws that attempt to grant priority to a party other than the earliest federal registrant are surely preempted under 15 U.S.C. §1127, since the laws "interfere" with the rights granted federal registrants under the Act.

Note on the Preemption of State Antidilution Statutes

Preemption will also occur if the state law is at odds with the "purpose of Congress" in passing the Lanham Act. Depending on the purpose identified, state antidilution laws might fail this test, since they do not require a likelihood of consumer confusion. See De Sevo, supra, at 312-320. Courts and commentators have split on this issue. See David S. Welkowitz, Preemption, Extraterritoriality and the Problem of State Antidilution Laws, 67 Tul. L. Rev. 1, 4 (1992) (arguing that "injunctions like the one issued in Mead Data Central, Inc. v. Toyota Motor Sales, U.S.A., 702 F. Supp. 1031 (S.D.N.Y. 1988), *rev'd on other grounds*, 875 F.2d 1026 (2d Cir. 1989) exceed [constitutional] limits" under the Commerce Clause, though state antidilution statutes are not wholly preempted and in many cases injunctions based on state intellectual property law are enforceable under the Full Faith and Credit Clause). Cf. Milton W. Handler, Are the State Antidilution Laws Compatible with the National Protection of Trademarks?, 75 Trademark Rep. 269 (1985); Joseph P. Bauer, A Federal Law of Unfair Competition: What Should Be the Reach of Section 43(a) of the Lanham Act?, 31 UCLA L. Rev. 671 (1984); Charles Bunn, The National Law of Unfair Competition,

62 Harv. L. Rev. 987 (1949); Paul Heald, Comment, Unfair Competition and Federal Law: Constitutional Restraints on the Scope of State Law, 54 U. Chi. L. Rev. 1411 (1987).

The new federal antidilution law changes the preemption analysis, of course. Many people expected that such a law would preempt state antidilution statutes, replacing the patchwork of inconsistent protections with uniform national protection.[3] Instead, the new section 43(c)(3) of the Lanham Act charts a narrow course around state antidilution laws by allowing them to continue in force but granting owners of federally registered marks immunity from suit under such state laws. Unregistered marks, as well as those registered under state law, receive no such immunity. The legislative history explains:

> Under section 3 of the bill, a new Section 43(c)(3) of the Lanham Act would provide that ownership of a valid federal trademark registration is a complete bar to an action brought against the registrant under state dilution law. This section provides a further incentive for the federal registration of marks and recognizes that to permit a state to regulate the use of federally registered marks is inconsistent with the intent of the Lanham Act "to protect registered marks used in such commerce from interference by state, or territorial legislation." It is important to note that *the proposed federal dilution statute would not preempt state dilution laws.* Unlike patent and copyright laws, federal trademark law coexists with state trademark law, and it is to be expected that the federal dilution statute should similarly coexist with state dilution statutes.

Federal Trademark Dilution Act of 1995, House Report 104-374 (accompanying H.R. 1295), November 30, 1995 (emphasis added).

Thus parallel state antidilution claims will not wither away completely. Indeed, depending on whether state law affords more attractive remedies (note the "injunction only" rule for most cases under the federal dilution provision, §43(c)(2)), parties may well continue to bring at least some antidilution cases under state law. Note that for the most part, however, nationwide injunctions are more difficult to acquire under state antidilution statutes, making federal claims more likely if such an injunction is the trademark owner's major goal.

COMMENTS AND QUESTIONS

1. Is the lack of preemption troubling? Consider the following possibilities:

---

3. Indeed, some saw little purpose for the federal law otherwise. See Kenneth L. Port, The "Unnatural" Expansion of Trademark Rights: Is a Federal Dilution Statute Necessary?, 18 Seton Hall Legis. J. 433 (1994).

- A federal antidilution claim fails because the use is found to be a "fair" one (say, for news reporting purposes). A parallel action is brought under a state statute that contains no such limitation, and the court grants a nationwide injunction against dilution on the basis of the state statute.
- A nationally famous but unregistered product configuration is found to dilute a locally known product configuration under a state statute.

Do the state statutes at issue in these cases interfere with federal trademark policy?

2. Figure 6-2 is a graphic representation of preemption doctrine. The areas inside the circle represent subject matter that is related to an area of federal protection. Areas inside the triangle are eligible for federal statutory protection. Areas inside the circle but not inside the triangle are preempted.

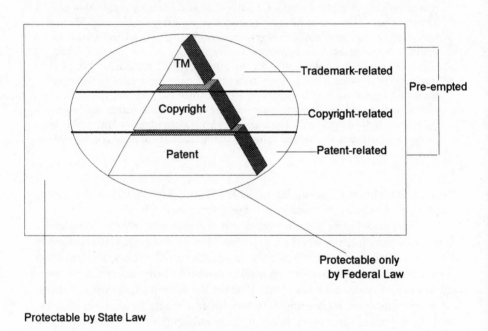

FIGURE 6-2
Areas of federal protection.

7

Protection of Computer Software

The previous chapters have explored the principal modes of intellectual property protection. As we have seen, these statutory and common law regimes comprise a complex system of rights. In some circumstances, the law envisions overlapping layers of protection as in the case of design protection (covered simultaneously by copyright, design patent, and trade dress). In other circumstances, one mode of protection overrides or preempts others, as in the cases of the idea-expression dichotomy and the relationship of federal and state protections. With the advent of new technologies and the evolution of the law, the system of intellectual property protection has increasingly become integrated.

The blurring of traditional doctrinal lines can be seen throughout the practice of intellectual property law. The major transactions and cases today affecting high technology firms typically involve complex technologies, multiple areas of intellectual property law, and antitrust issues. Nowhere is this pattern more evident than in the area of computer and information technologies.

In view of this reality, this chapter explores the manner in which the various modes of intellectual property protection have developed to protect computer technology. As essential background to this study, we begin with an overview of computer technology and the economics of computer markets. We then work through the various modes of protection for computer technology. We begin with trade secret protection, which served as the principal mode of protection during the early and intermediate stages of the computer industry. With the advent and proliferation of microcomputers and mass-marketed software in the mid- to late 1970s, developers turned increasingly toward copyright protection, which has served as the principal mode of protection for software for the past 15 years. Our focus then shifts to patent and trademark, which are becoming increasingly important and will be the prin-

893

cipal battlegrounds in the coming decade. We conclude the chapter with a look at alternative modes of protection for software, which enables us to explore the desirability and practicality of tailoring intellectual property protection to new technologies.

A. THE ECONOMICS OF COMPUTER MARKETS[1]

1. The Market for Computers and Computer Software

At the most basic level, consumers demand "computing services" to meet their data processing needs. As we saw above, these needs can be satisfied completely by hardware, or they can be satisfied by general purpose computer hardware equipped with the appropriate application software. Thus hardware and software may be both substitutes and complements for each other.[2]

Demand for Computer Services. The demand for computer services is driven by the great diversity of entities — businesses, government agencies, research institutions, individuals — with data processing needs. These needs range from simple calculations to complex scientific applications. Consumers also differ in the variety of data processing tasks that they must accomplish. A medium-sized business, for example, might have many data processing tasks for which a computer might prove useful: handling the payroll, record keeping, word processing, and projecting business trends. In contrast, a manufacturing facility might simply need to regulate the temperature of a kiln. A physicist might need a computer to execute high-speed calculations and to maintain complex data bases.

Supply of Computer Services. The hardware sector of the computer industry consists of original equipment manufacturers, semiconductor chip manufacturers, and vendors. The vendors purchase computer components and chips from the other firms and assemble them into computer systems. A few major firms in the industry, like IBM, Digital Equipment Corporation, and Apple, are involved in all aspects of hardware research and development as well as software. Other firms have significant shares of particular hardware

---

1. For those unfamiliar with the basis of computer technology, we have provided an introduction to the technology in an Appendix to the statutory supplement. It is also available on our Web site, http://www.law.berkeley.edu/belt/.

2. In economic terms, goods are "substitutes" if purchase of one makes it unnecessary to buy the other. They are "complements" if they are traditionally purchased or used together.

markets but do not attempt to maintain a presence in vertically related markets (Dell and Compaq in computers, Intel and Advanced Micro Devices in chips, for example).

The software sector offers a wide variety of services and products. Its work includes the design of general operating systems, contract programming, and the development of commercial application packages. Many large hardware systems manufacturers develop operating and application software for their systems. There are also many small, independent firms that specialize in aspects of software services and product development. Recently a few firms, such as Microsoft, have emerged as major forces in the computer software market. As with hardware, many of the largest software firms are vertically integrated, selling both operating systems and application programs. Microsoft is the obvious example of a vertically integrated software company. Other software vendors, such as Borland and WordPerfect, concentrate on particular application program markets. Finally, there is a rapidly growing "middle ground" between operating systems and application programs: the market for client-server and network management software. Significant players in this market include Novell, Sun, and (increasingly) Microsoft.

Evolution of Computer Markets.   The rapid advance of all aspects of computer technology has enabled the computer services market to expand at a blistering pace since the advent of commercial computing in the mid-1950s. Computers, which not too long ago were found only at large corporations, research institutions, and government agencies, are now in a substantial portion of American homes. In the hardware field, the trend has been toward smaller, universal computer systems. In 1965, domestic consumers purchased 260 minicomputers and 5,350 mainframes. Minicomputer unit sales surpassed mainframe unit sales by 1974. And microcomputer unit sales surpassed minicomputer unit sales in their second year of production, 1976. By 1986, sales of microcomputers costing more than $1000 reached approximately 4 million units and revenues of almost $12 billion, giving microcomputers the largest share of computer revenues. In the 1990s, sales of personal computers and notebooks have averaged 10 to 12 million per year. As of this writing, it is estimated that Microsoft's DOS, Windows and Windows 95 operating systems have a collective "installed base" of over 125 million computers, suggesting that an even larger number of personal computers are in use by businesses and by individual households.

These trends in hardware have dramatically changed the structure of the software sector of the industry. As recently as the late 1970s, most software firms produced custom programs for predominantly commercial customers. The advent and proliferation of microcomputers and the increase in flexibility of minicomputers have greatly increased the demand for general purpose application packages. Consequently, firms that produce commercial application packages for a variety of computer systems emerged during the 1980s as the major revenue-generating force in the computer industry. In addition, a major focus of the computer industry in the 1990s has been linking together the vast array of computer and telecommunication systems into computer networks. Computer networks have been a growth area for both hardware and software companies.

2. Market Failures

Public Goods Problem. As discussed in Chapter 1, all markets for goods embodying intellectual property exhibit an externality commonly referred to as the "public goods" problem. Public goods have two distinguishing features: (1) nonexcludability (it is difficult to prevent those who do not pay for the good from consuming it); and (2) nonrivalrous competition (additional consumers of the good do not deplete the supply of the good available to others). Beautiful gardens and military defense are classic examples of public goods — military defense, for example, benefits everyone who is defended, whether they pay for the privilege or not. The result of the public goods problem is well recognized: the private market will undersupply public goods because producers cannot reap the marginal value of their investment in providing such goods.

Innovations in computer software are good (although not perfect) examples of a public good. Given the availability of low-cost copying, it is difficult to exclude nonpurchasers from the benefits of innovative computer programs once they are made commercially available. Even programs written in object code and stored in ROM can be readily copied by inserting the chip into a laboratory device and downloading the information onto another chip or onto paper. Moreover, one person's use of the information does not detract from any other person's use of that same information, since the information can be copied without affecting the original. Since the authors and creators of computer software cannot reap the marginal value of their efforts, economic theory suggests that in the absence of other incentives to innovate they will undersupply technological advances in computer software.

Standardization and Network Externalities. The second principal market failure in the computer software market arises from the presence of network externalities. These exist in markets for products for which the utility or satisfaction that a consumer derives from the product increases with the number of other consumers of the product. The telephone is a classic example of a product with network externalities. The benefits to a person from owning a telephone are a function of the number of other people owning telephones connected to the same telephone network; the more people on the network, the more people each person can call and receive calls from. Another classic network externality flows from standardization. In this case, the value of learning a particular standard (say, how to use a certain word processing program) depends on how many other people use that standard. Consider the prevalence of a "normal" typewriter keyboard.[3] Because almost all English language typewriters feature the same keyboard configuration, commonly referred to as "QWERTY," typists need learn only one keyboard system. This standardization enhances worker mobility and the breadth of products available to those who use the QWERTY keyboard.

Network externalities also inhere to product standards that allow for the interchangeability of complementary products. Examples of products for which this type of network externality is important are videocassette recorders, CD players, and computer operating systems. As discussed above, general purpose computer operating systems allow consumers to use a variety of application software programs on the same system-unit hardware. The only requirement is that the application program be coded to work on the operating system embedded in the general computer system. Thus the operating system serves as a "compatibility nexus" for a particular computer network. Application software producers will develop more programs for systems that are widely used; hardware producers will develop more configurations of disk drives, memory, and other features for popular operating systems. In general, the benefits of a larger computer operating system network include a wider variety of application software that can run on that operating system, lower search costs for consumers seeking particular application programs that run on that operating system, reduced retraining costs, greater labor mobility, and wider availability of compatible hardware configurations and peripherals.

Similarly, the greater the extent to which computer-human interfaces for particular application programs, such as word processing or spreadsheets, are standardized, the easier it will be for computer users to utilize their skills in different working environments. Computer users also want the computer-human interface to be as easy to use as possible — i.e., "user-friendly." In part, user friendliness is a function of what the user has already learned: things that are familiar tend to be easier to use. More generally, user friendliness is determined by the extent to which the design is tailored to achieve human factor goals such as maximizing performance speed, minimizing learning time, minimizing rate of errors, and maximizing retention over time.

Standardization can occur by way of market processes, whereby subsequent entrants to a market adopt the standard of an existing firm. This has been referred to as "bandwagon standardization." See Joseph Farrell, Standardization and Intellectual Property, 30 Jurimetrics J. 35, 41 (1989). In the computer industry, firms often foster this process by freely licensing the use of a standard and by publishing design and interoperability specifications. The UNIX operating system is an example of an open standard promoted by certain private companies, such as Sun Microsystems.

Alternatively, standardization can occur through formal processes, such as government standard-setting and joint development of industry standards

---

3. See Paul A. David, CLIO and the Economics of QWERTY, 75 Am. Econ. Rev. 332 (May 1985).

by a number of firms. This top-down approach to standard-setting has been common in the telecommunications industry, at least until recently. Regulatory agencies, in consultation with the affected companies, set standards for connecting peripheral devices (including computers) to telephone lines. Only devices that comply with these publicly available standards can operate over the telephone lines.

Finally, industries characterized by strong network externalities will sometimes develop "de facto" standards, even though neither the companies in the industry nor the government has made any effort to settle on a particular standard or ensure interoperability. An example is the Microsoft DOS operating system (and the Windows shell later built on top of it). Microsoft did not "open" its operating system to competitors. Nonetheless, once the Microsoft operating system reached a certain "critical mass" of users, its very size began to make it more attractive. More application programmers wrote programs to run on DOS and Windows because of the large user base; more users adopted the Microsoft system in part because of all the programs being written for it. In 1994, DOS and Windows held 80 percent of the personal computer operating system market, despite the closed nature of the system and despite the presence of several arguably superior competing systems.

An important economic consideration in markets with significant network externalities is whether computer firms will have the correct incentives to develop or adopt compatible products, thereby enlarging existing networks. Economists have demonstrated that computer firms might prefer to adopt noncompatible product standards even though their adoption of compatible products would increase net social welfare.[4] The explanation for this behavior is that by adopting a compatible standard, a firm enlarges the size of a network that comprises both the adopter's product and its rivals' products. This will have the effect of increasing the desirability of the rivals' products to consumers, thereby reducing the adopter's market share (although of a larger market) relative to what it would have been had the firm adopted a noncompatible product standard. Although this effect by no means implies that subsequent market entrants will never adopt a compatible operating system or computer-user interface, it does suggest that incentives to adopt compatible standards might be suboptimal.

While a widely adopted product standard can offer important benefits to consumers and firms, it can also "trap" the industry in an obsolete or inferior standard. In essence, the installed base built upon the "old" standard — reflected in durable goods and human capital (training) specific to the old standard — can create an inertia that makes it much more difficult for any one producer to break away from the old standard by introducing a noncompatible product, even if the new standard offers a significant technological

---

4. See Michael L. Katz & Carl Shapiro, Network Externalities, Competition, and Compatibility, 75 Am. Econ. Rev. 424, 435 (May 1985).

improvement over the current standard.[5] In this way, network externalities can retard innovation and slow or prevent adoption of improved product standards. This problem is particularly significant in industries that develop rapidly, such as the computer industry.

As an example of the danger of stagnation, investigators cite the persistence of the standard QWERTY typewriter keyboard despite the apparent availability of a better key configuration developed and patented by August Dvorak and W. L. Dealey in 1932. U.S. Navy studies and world typing speed records suggest that the Dvorak Simplex Keyboard is significantly more efficient than the QWERTY system.[6] Adoption of the better standard appears to have been effectively stymied by "switching costs" — the costs of converting or replacing QWERTY keyboards and retraining those who use them. Because of the fear that national standards would exacerbate the inertia problem, the National Bureau of Standards declined to set interface standards for computers in the early 1970s. Subsequent developments proved that a wise decision.

Nonetheless, standards are essential in many cases. Although the United States government did not adopt computer interface standards in the 1970s, the hardware industry had to develop de facto standards so that computers and peripheral devices could "talk" to each other. The software industry faced similar problems in ensuring compatibility between operating systems and applications programs in the 1980s, and the computer networking industry faces the same problem today. The problem goes away if one company controls the market for all related goods, as AT&T did in the telephone industry until 1984. If AT&T is the only company to sell both telephone services and telephones, there is unlikely to be a problem in ensuring that your telephone works with your phone line. But this form of standardization comes at the expense of competition. Controlling standards and compatibility in dynamic industries is one of the most vexing policy problems the computer industry faces. The problem is exacerbated by the availability of intellectual property protection for many of the potential standards.

COMMENTS AND QUESTIONS

1. What means exist — contractual, intellectual property, business strategy, and otherwise — to appropriate the value of innovation in computer software and hardware? Is the intellectual work embodied in computer technology any more "leaky" than other technologies? Does this leakiness justify special protection — beyond traditional copyright and patent protection —

---

5. See Joseph Farrell & Garth Saloner, Standardization, Compatibility, and Innovation, 16 Rand J. Econ. 70 (1985).

6. However, evidence recently presented by two economists casts doubt on the technological superiority of the Dvorak keyboard. See S. J. Liebowitz & Stephen E. Margolis, Network Externality: An Uncommon Tragedy, 8 J. Econ. Persp. 133, 147 (1994).

for computer technology, especially software, or can this technology fit neatly within existing structures?

2. Which of the following types of firms have greater incentives to adopt an existing standard than to develop a new standard: large firms with significant name recognition in related product markets, or small firms with little name recognition? Why? See Farrell, supra, 30 Jurimetrics J. at 38.

3. Does intellectual property protection for products featuring network externalities increase or decrease the likelihood that firms will seek to develop proprietary standards? How does it affect the likelihood that inefficient standards will arise? that inertia can be broken? How does strong intellectual property protection affect the likelihood that formal standards — either by governmental bodies, private standard setting organizations, or consortia of leading firms — will be adopted? Should formal standardization be encouraged?

4. How does the presence of network externalities with regard to computer software affect the analysis of intellectual property protection? Of what relevance are network externalities to the scope of copyright protection? patent? trademark? Should the network externalities of operating systems be treated in the same manner as network externalities of user interfaces?

5. How should computer software be protected under the intellectual property laws? In particular, consider the implications of the following aspects of computer programming, suggested in Pamela Samuelson et al., A Manifesto Concerning the Legal Protection of Computer Programs, 94 Colum. L. Rev. 2308, 2315-16 (1994):

> Computer programs have a number of important characteristics that have been difficult for legal commentators and decisionmakers to perceive. First, the primary source of value in a program is its behavior, not its text. Second, program text and behavior are independent in the sense that a functionally indistinguishable imitation can be written by a programmer who has never seen the text of the original program. Third, programs are, in fact, machines (entities that bring about useful results, i.e. behavior) that have been constructed in a medium of text (source and object code). The engineering designs embodied in programs could as easily be implemented in hardware as in software, and the user would be unable to distinguish between the two. Fourth, the industrial designs embodied in programs are typically incremental in character, the result of software engineering techniques and a large body of practical know-how.

Does the fact that we are concerned primarily with the behavior of a machine suggest that patent law is the appropriate vehicle for protecting computer software? Does it matter that virtually all of these "machines" are in fact constructed from text? Does the incremental nature of software development suggest the contrary, that patent law is not the appropriate vehicle of protection?

The authors of the "Manifesto" conclude that none of the current intellectual property rules is adequate to protect computer programs. For greater discussion of their proposed alternative, see section F of this chapter.

6. One of the themes of section 1 supra is that programmers focus on efficiency in achieving commercial or technical goals. Should improvements in efficiency be protectable as intellectual property? If so, what is the best way to accomplish this?

7. Paul Goldstein writes:

> Science and technology are centripetal, conducting toward a single optimal result. One water pump can be better than another water pump, and the rule of patent and trade secret is to direct investment toward such improvements. Literature and the arts are centrifugal, aiming at a wide variety of audiences with different tastes. . . . The aim of copyright is to direct investment toward abundant rather than efficient expression.

Paul Goldstein, Infringement of Copyright in Computer Programs, 47 U. Pitt. L. Rev. 1119, 1123 (1986). For all its parallels to literature and art, computer programming is at base engineering — it involves the construction of a machine. Does this fact suggest that copyright and trademark are not well-suited to protecting computer software?

7. Computer programs have a number of different elements — purpose, structure, user interfaces, program interfaces, code, etc. Further, as noted above, different people program in different ways. Do these factors suggest that different types of intellectual property protection might be appropriate for different parts of a computer program? Is it possible that different intellectual property laws could protect the *same* program elements? Do you see any problems with such an overlap? See J. H. Reichman, Legal Hybrids Between the Patent and Copyright Paradigms, 94 Colum. L. Rev. 2432 (1994).

B. TRADE SECRET PROTECTION

Computers are different from other creations subject to intellectual property protection. To a significant degree, they are different because they are "hybrids": part machine, part writing, and part artistry. Because of this, and because of the strangeness of the new technology (it is hard to imagine a "literary work" composed entirely of 1s and 0s), intellectual property law has struggled to protect the intellectual work embodied in computer programs.

The main contours of legal protection for computer technology have emerged in the past decade. Copyright law protects computer software, and it is generally acknowledged that competitors may not make multiple copies of someone else's software without permission, with important exceptions and qualifications discussed later. Patent law protects computer hardware and new processes or "structures" embodied in computer software. But in the late 1970s and early 1980s, things were very different. Whether copyright protected computer programs at all was an open question, debated by the federal Commission on New Technological Uses of Copyrighted Works (CONTU)[7] and not resolved until the enactment of the 1980 Amendments to the Copyright Act. During this period, and even after the 1980 Amendments, courts disagreed over such fundamental issues as the copyrightability of object code and whether copyright protected computer programs against nonliteral infringement.

At the same time, a similar debate was raging in the federal courts over the patentability of computer software. The Supreme Court had held computer programs unpatentable in Gottschalk v. Benson, 409 U.S. 63 (1972), and it did not reopen the door *Gottschalk* closed until 1981. Even after 1981, the patentability of "pure" computer software was open to dispute. The issue was sufficiently clouded in 1994, for example, that the Federal Circuit took In re Alappat, 33 F.3d 1526 (Fed. Cir. 1994) (en banc) in a vain attempt to clarify the standards for issuing software patents. While software patents now appear to be here to stay, their precise scope remains unclear.

Against this backdrop, it was important for software vendors in the late 1970s and early 1980s to establish their rights vis-à-vis users. One way for vendors to prevent users from freely copying their computer programs was to claim those programs as a trade secret. Trade secrecy was a particularly viable alternative in the early days of the computer industry because most computer software was customized and sold to particular customers through detailed licensing agreements. Trade secrets have also been important in the computer industry as a means of protecting projects under development from being taken by departing employees.

The general principles and doctrines of trade secret law were set out in Chapter 2. You should briefly review those materials at this time, taking particular note of the Uniform Trade Secrets Act. The materials below highlight the distinctive manner in which trade secret law has been applied to computer technology.

While it was established early on that the secret elements of a computer program could be protected from misappropriation, computer companies that wanted to protect their intellectual property still had a problem. It is a fundamental principle of trade secret law that the information protected must remain a secret. If the secret is disclosed to the public, protection is forever lost. Computer companies did not want to keep their hardware or software "secret" — they wanted to sell them! An influential early case held that they could do both.

---

7. Final Report of the National Commission on New Technological Uses of Copyrighted Works (U.S. Govt. Printing Office 1978).

Data General Corp. v. Digital Computer Controls, Inc.
Delaware Court of Chancery
297 A.2d 433 (Del. Ct. Chanc. 1971), aff'd, *297 A.2d 437*
(Del. S. Ct. 1972)

MARVEL, Vice Chancellor:

Data General Corporation seeks an order preliminarily enjoining the defendant Digital Computer Controls from making use of claimed trade secrets allegedly contained in design drawings which accompany certain sales of plaintiff's Nova 1200 computer, defendant having acquired such a computer with accompanying drawings from one of plaintiff's customers. This is the decision of the Court on plaintiff's motion for a preliminary injunction as well as on defendants' cross-motion for summary judgment of dismissal of the present action.

The relevant facts thus far adduced are as follows: During the past several years plaintiff has developed and marketed successfully small general-purpose computers to which they have given the name Nova, a large part of plaintiff's research and development budget for the past several years having been allocated to the development of such small computers, which, the parties agree, are the only ones of their type presently being profitably marketed, there having been no device on the market comparable to plaintiff's machines until defendant's entry on the scene.

When a sale of a Nova 1200 computer is made, plaintiff makes available at no extra cost to those customers who wish to do their own maintenance the design or logic drawings of the device sold, in this case a Nova 1200 computer. Such type of maintenance has been found to be desired by some customers in order to avoid periods of unproductive delay while waiting for repairs to be made by plaintiff's trained personnel.

Design drawings made available to customers are furnished subject to the terms of a non-disclosure clause contained in a paper which accompanies a purchase agreement. Furthermore, all drawings bear a legend to the effect that they contain proprietary information of the plaintiff which is not to be used by a purchaser for manufacturing purposes. However, the Nova 1200 is not patented, and its design drawings have not been copyrighted.

In April, 1971, the defendant Ackley, the president of Digital, purchased for his company from a customer of Data General, namely Mini-Computer Systems, Inc., a Nova 1200 computer which the latter had earlier acquired from plaintiff. Although the exact circumstances surrounding such sale are not entirely clear on the present record, the corporate defendant's president, in consummating such purchase, acquired a set of design drawings of the Nova 1200, said drawings having been furnished to MiniComputer by the plaintiff in order to facilitate maintenance by the former of its Nova 1200's.

The corporate defendant thereafter proceeded to use such design drawings as a pattern for the construction of a competing machine which it is now about to market, the corporate defendant's president conceding that only minor changes have been made in the basic design of plaintiff's device in the course of the development of defendant's comparable computer.

[The court held that plaintiff's trade secret claims are not preempted by the federal intellectual property or antitrust laws.]

II

Defendants' other arguments are accordingly based on the law of trade secrets. Thus, in order for the plaintiff to establish its right to relief here, it must demonstrate (1) the existence of a trade secret and that the corporate defendant has either (2) received the information within the confines of a confidential relationship and proposes to misuse the information in violation of such relationship, or (3) that the corporate defendant improperly received the information in question in such a manner that its confidential nature should have been known to it and that it nonetheless proposes to misuse such information.

Trade secrets have been defined in Restatement, Torts §757, comment b. as follows: "A trade secret may consist of any formula, pattern, device or compilation of information which is used in one's business, and which gives him an opportunity to obtain an advantage over competitors who do not know or use it."

Defendants insist, however, that plaintiff has not maintained that degree of secrecy which will preserve its right to relief, either by publicly selling an article alleged to contain a trade secret, or by failing to restrict access to the design drawings for its device, arguing that matters of common knowledge in an industry may not be claimed as trade secrets.

It has been recognized in similar cases that even though an unpatented article, device or machine has been sold to the public, and is therefore subject to examination and copying by anyone, the manner of making the article, device or machine may yet constitute a trade secret until such a copy has in fact been made, Schulenburg v. Signatrol, Inc., 33 Ill. 2d 379, 212 N.E.2d 865, and Tabor v. Hoffman, N.Y., 118 N.Y. 30, 23 N.E. 12.

Defendants contend, however, that the issuance by plaintiff of copies of design drawings to its customers was made without safeguards designed properly to maintain the secrecy requisite to the existence of a trade secret. In other words, it is contended that plaintiff's attempts to maintain secrecy merely consisted of (1) not giving copies of the design drawings to those customers who did not need them for maintenance of their computer, (2) obtaining agreements not to disclose the information from those customers who were given copies of the drawings, and (3) printing a legend on the drawings which contained the allegedly confidential information which identified the drawing as proprietary information, the use of which was restricted.

Plaintiff argues, however, that disclosure of the design drawings to purchasers of the computer is necessary properly to maintain its device, that such disclosure was required by the very nature of the machine, and that reasonable steps were taken to preserve the secrecy of the material released. I conclude at this preliminary stage of the case that it cannot be held as a matter of law that such precautions were inadequate, a factual dispute as to the adequacy of such precautions having clearly been raised. Defendants' motion for summary judgment must accordingly be denied.

Finally, I am satisfied that if plaintiff were to prevail at final hearing, it would only be entitled to injunctive relief during that undetermined period of time which would be required for defendants substantially to reproduce the plaintiff's device without its accompanying drawings. Such period of time would, on the present record, vary according to the number of man hours devoted to so-called reverse engineering. Thus, the granting of a preliminary injunction at this juncture would grant plaintiff all the relief it might hope ultimately to obtain after final hearing. Compare Thomas C. Marshall, Inc. v. Holiday Inn, Inc., 40 Del. Ch. 77, 174 A.2d 27. In addition, while plaintiff has presented evidence of its attempts to preserve the secrecy of the alleged secrets it now seeks to have protected by injunction, it has not adduced sufficient evidence to establish that likelihood of ultimate success on final hearing which would entitle it to the issuance of a preliminary injunction.

On notice, an order may be presented denying defendants' motion for summary judgment as well as plaintiff's motion for a preliminary injunction, which order shall also include a provision that all drawings issued by the corporate defendant in connection with the sale of its Nova-type computer contain a restrictive legend of the type now set forth on plaintiff's drawings here in issue.

COMMENTS AND QUESTIONS

1. Even at a very early stage in the history of computer law, there was no serious question that computer hardware and software *could* be protected by trade secret law. See, e.g., University Computing Co. v. Lykes-Youngstown Corp., 504 F.2d 518 (5th Cir. 1974) (plaintiff's computer system was a protectable trade secret). Cases such as *Lykes-Youngstown* involved a fairly typical trade secret situation — people with access to a secret program still under development attempting to use or sell their own product based in part on the program. See also Cybertek Computer Products v. Whitfield, 203 U.S.P.Q. 1020 (Cal. Super. 1977).

But the contours of trade secret protection of computer programs are not always clear. For example, in Rivendell Forest Prods. v. Georgia-Pacific Corp., 28 F.3d 1042 (10th Cir. 1994), the Tenth Circuit reversed a district court opinion holding that Rivendell was not entitled to protect the basic concepts of organizing a price quote system against misappropriation by others. (The district court opinion is reported at 824 F. Supp. 961 (D. Colo.

1993).) The use of trade secret law to protect such abstract, high-level concepts is somewhat troubling. On the one hand, the District Court is surely correct that if Rivendell's ideas are not in fact secret, Georgia-Pacific is free to copy them. On the other hand, the Court of Appeals points out that basic concepts are protectable under trade secret law in certain combinations, if those combinations are not generally known in the industry. On this issue, compare Integrated Cash Mgmt. Serv. v. Digital Transactions, Inc., 920 F.2d 171 (2d Cir. 1990) (the combination of publicly known utility programs into a specific arrangement is protectable as a trade secret) with Comprehensive Technologies, Inc. v. Software Artisans, Inc., 3 F.3d 730 (4th Cir. 1993) (combination of known elements did not qualify as a trade secret where the particular combination was logically required and common in the industry). Cf. Vermont Microsystems Inc. v. Autodesk Inc., 88 F.3d 142 (2d Cir. 1996) (shading algorithm could be protected as a trade secret even though other developers had independently created similar algorithms, because defendant derived value from using plaintiff's algorithm).

2. The plaintiff in *Data General* sold over 500 Nova computers to the general public. Each purchaser who requested one got a copy of the "confidential" design drawings. Why are these drawings still considered a secret? Does widespread disclosure compromise the secrecy claim at some point, even though all disclosures are made under an agreement of confidentiality? This issue arises frequently today in the software industry. As computers have become ubiquitous in business and quite common in the home, the numbers of "secret" programs in circulation may be counted in the millions rather than the hundreds. *Data General* implicitly concludes that even a relatively widespread disclosure to customers does not compromise the secrecy of the computer design. For cases addressing this issue in the context of computer software, compare Management Science of Am. v. Cyborg Sys., Inc., 1977-1 Trade Cas. (CCH) ¶61,472 (N.D. Ill. 1977) (distribution of 600 copies of a program under a confidentiality agreement did not destroy secrecy) with Young Dental Mfg. Co. v. Q3 Special Prods., Inc., 891 F. Supp. 1345 (E.D. Mo. 1995) (characterizing as "completely frivolous" plaintiff's claim that its publicly sold software was a trade secret).

A closely related question involves attempts by the owners of information to "contract around" the requirement of secrecy. If the parties agree to treat a piece of information as secret, is the licensee bound not to use or disclose the information under contract principles regardless of whether or not it is in fact in the public domain? This issue is a recurring one in intellectual property law, and a problem that has never adequately been addressed.

3. Customers who buy a product on the open market are entitled to break it apart to see how it works. This process is called "reverse engineering" the product. Trade secret law does not protect owners against legitimate purchasers who discover the secret through reverse engineering. But does the possibility that a product might be reverse engineered foreclose *any* trade secret protection? At least one court has said no. In Data General Corp. v. Grumman Systems Support Corp., 825 F. Supp. 340, 359 (D. Mass. 1993),

the court upheld a jury's verdict that Grumman had misappropriated trade secrets contained in object code form in Data General's computer program, despite the fact that many copies of the program had been sold on the open market. The court reasoned: "With the exception of those who lawfully licensed or unlawfully misappropriated MV/ADEX, Data General enjoyed the exclusive use of MV/ADEX. Even those who obtained MV/ADEX and were able to *use* MV/ADEX were unable to discover its trade secrets because MV/ADEX was distributed only in its object code form, which is essentially unintelligible to humans." The court noted that Data General took significant steps to preserve the secrecy of MV/ADEX, requiring that all users of the program sign licenses agreeing not to disclose the program to third parties. Under this decision, a defendant may have to prove that they had some sort of legitimate access to the plaintiff's information — for example, by demonstrating that they reverse-engineered it from a publicly available product — even though the product containing the secret is widely distributed.

The *Data General* cases suggest that reasonable efforts to protect the secrecy of an idea contained in a commercial product — such as locks, black boxes, or the use of unreadable code — may suffice to maintain trade secret protection even after the product itself is widely circulated. Does this result make sense? For a different approach, see Videotronics v. Bend Electronics, 564 F. Supp. 1471, 1476 (D. Nev. 1983) (software cannot be a trade secret if it is publicly distributed and can be readily copied).

4. Why is reverse engineering lawful at all? If one purpose of trade secret law is to promote standards of commercial ethics, doesn't there seem to be something wrong with taking apart a competitor's product in order to figure out how to copy it (or, as in the case that follows, how to render it useless)? Does reverse engineering benefit only those competitors who are not smart enough to develop ideas or products for themselves?

One explanation for reverse engineering may be that it serves to differentiate trade secret law from patent law. As we shall see, patent law grants stronger protection than trade secret law (including protection against reverse engineering), but only if the patent holder meets relatively stringent standards. Reverse engineering may be a legal rule designed to weaken trade secret protection relative to patent protection. Can you think of reasons why we would want to weaken trade secret protection? On this point, see Kewanee Oil Co. v. Bicron Corp., 416 U.S. 470 (1974), in which the Supreme Court held that the defense of reverse engineering in trade secret law was an important part of what allowed patents and trade secrets to coexist.

5. One way in which computer software cases may be distinguished from other trade secret cases has to do with the nature of what is being distributed. While a particular computer program may be widely distributed, in fact all that is sold to the consumer is a disk containing object code. Object code is virtually impossible for humans to read without machine assistance.[8] Because

8. It is possible to "reverse engineer" object code in some cases to create a kind of rough estimate of what must have been in the original source code. The process, however, is demanding

of this, computer software is in some sense unlike a physical product whose design is evident to the casual observer. Even after it is publicly distributed, object code is meaningless to the casual observer. Only a complex process of reverse engineering (sometimes called "disassembly" or "decompilation") can enable the user to decipher the source code that was originally written for the program.

Should it matter that a computer program is distributed only in object code form? Consider the following case, in which the defendant was accused of misappropriating a computer program in object code form:

> The source code can and does qualify as a trade secret. . . .
> Whether the object code is a trade secret is a more difficult question.[8] Atkinson first contends that the object code cannot be a trade secret because it does not derive independent economic value from its secrecy, and therefore fails the first definitional requirement of a trade secret. This argument has no merit. Trandes generates most of its revenues by providing computer services. . . . Armed with a copy of the object code, an individual would have the means to offer much the same engineering services as Trandes. . . .
> Atkinson next argues that the object code cannot be a trade secret because Trandes did not keep it secret. . . . Atkinson asserts that the Tunnel System has been widely disclosed as a mass-marketed product and that its existence and its abilities are not secret. [The court concluded that the object code remained secret because it had only been distributed to two customers, and both of them signed licenses agreeing to keep the program a secret.]

Trandes Corp. v. Guy F. Atkinson Co., 996 F.2d 655, 663-64 (4th Cir. 1993). Consider the court's footnote. Can object code be a trade secret if it can easily be duplicated (whether or not the copier understands what he is copying)? Is the plaintiff in this case really trying to leverage copyright protection out of a trade secret claim?

In both the *CTI* and *ICM* cases, cited above, the alleged trade secret at issue was not the source or object code of the computer program itself but certain high-level design features of the program (its "architecture"). Suppose that, rather than using what they had learned of the architecture of the program while employed by the company, ICM's former employees had copied the object code of the program altogether. (Leave aside for a moment questions of copyright infringement, and consider only the trade secret issue.) Would they be liable for misappropriating the trade secrets contained in the

---

and time-consuming even for expert programmers. See, e.g., Andrew Johnson-Laird, Technical Demonstration of "Decompilation," paper presented at the Computer Law Association's 1993 International Computer Law Retreat, October 21-23, 1993, at 29.

8. This case presents an unusual set of facts. In the ordinary case, the owner of trade secret computer software will maintain the secrecy of the source code but freely distribute the object code. See, e.g., Q-Co Indus. v. Hoffman, 625 F. Supp. 608, 617 (S.D.N.Y. 1985) (program secret where source code secret, even though object code not secret). In such cases, the owner of the software cannot claim trade secret protection for the object code because its disclosure to the public destroyed its secrecy. In this case, however, Trandes maintained the secrecy of the source code and the object code, as we explain below.

program architecture, on the grounds that copying the program in its entirety necessarily copied the architecture? Or would the fact that the object code was publicly disclosed protect them from liability? How would the courts in *Trandes* and the *Data General* cases answer this question? Does the answer suggest a problem with relying on trade secrecy to protect computer programs?

6. While there once was a question as to the availability of copyright and/or patent protection for computer technology, there no longer is. In view of this fact, to what extent should hardware and software manufacturers rely on trade secret protection? Are there disadvantages to trade secret law as a means of protection for innovations in computer technology?

Section 301 of the Copyright Act of 1976, 17 U.S.C. §301, preempts state intellectual property rights that are equivalent to any of the rights granted by the federal copyright law. There has been a good deal of litigation in the computer industry over federal copyright preemption of state trade secret law. Most (but not all) have found no preemption. See, e.g., Computer Associates v. Altai, Inc., 982 F.2d 693 (2d Cir. 1992). If the trade secret laws are interpreted to preclude the copying of object code in the previous example, should they be preempted by copyright law?

PROBLEMS

Problem 7-1. Microsoft Corp.'s Disk Operating System (MS-DOS) became the industry standard for personal computer operating systems. Microsoft sold several million copies of MS-DOS each year; there are estimated to be upwards of 25 million copies in circulation worldwide. Further, MS-DOS was normally "pre-installed" on many new personal computers, so that PC purchasers automatically purchased the operating system as well. Assuming that Microsoft has taken reasonable efforts to require both hardware manufacturers and end users to keep the program confidential, is Microsoft entitled to protect MS-DOS as a trade secret? What efforts would be "reasonable" under these circumstances?

Assume for the moment that Microsoft has succeeded in maintaining the secrecy of its computer operating system and has taken reasonable efforts to protect that secrecy. Are competing application programmers entitled to decompile the operating system to prepare application programs that will run on Microsoft's OS? Is there anything Microsoft can do that will prevent its competitors from legally reverse engineering its product?

Problem 7-2. Software Incorporated (SI), the developer of application software, possesses a valid trade secret on the proprietary computer code in its Calcutec program and has taken reasonable steps to protect its secrecy. Q, a disgruntled employee of User Inc., a licensee of the Calcutec program, wrongfully makes the proprietary code widely available by posting it on BBS, an electronic bulletin board service, from which it can be easily copied. SI learns of this unauthorized disclosure within a few weeks and notifies BBS. BBS promptly removes Calcutec from the bulletin board and posts a notice there of its unauthorized disclosure of Calcutec — but not before Software Distributors, Inc., among other companies, downloads the program and distributes it. Software Distributors sells Calcutec, without designation of source or copyright, to Tech Corp.

- What actions does SI have against Q, User, BBS, Software Distributors, and Tech?
- What difficulties might SI encounter in maintaining such an action?
- Does "wrongful publication" of the Calcutec code on BBS negate its status as a secret, thereby placing it in the public domain?
- What does this problem suggest about the viability of trade secret as a means for protecting widely distributed software products?

Problem 7-3.   Bartles and Vine are graduate students in computer science who regularly log onto a Xanton Corporation computer using a remote, dedicated terminal provided by Xanton to the university. The Xanton computer is not open to the general public, but the students have permission to use it for certain limited purposes in connection with their research. Xanton gives similar permission to other students around the world.

Bartles and Vine are frustrated by the fact that they have to rely on the dedicated Xanton terminal; it is the only one in the computer lab, and it is frequently in use by other students. To solve this problem, they decide to write a program that will allow a PC with a modem to emulate a Xanton terminal so they can access the Xanton computer from home at any time. To accomplish this, the students log onto the computer using the remote terminal and engage in "black box" analysis of the terminal interface — sending particular commands to see how the main computer responds.

Using the information they have thus acquired, Bartles and Vine write an emulation interface program for their PC. They call Xanton Corporation's public relations department, and a receptionist there gives them a telephone modem number. Bartles and Vine then use their PC to log into the Xanton computer.

Have Bartles and Vine misappropriated trade secrets allegedly contained in the Xanton interface? Does the result change if Bartles and Vine decide to market their emulation interface?

C.   COPYRIGHT LAW

Computer software, by its very nature as written work intended to serve utilitarian purposes, defies easy categorization within our intellectual property system. The copyright law has traditionally served as the principal source of legal protection for literary and artistic work, while the patent system and trade secret law have been the primary means for protecting utilitarian works. Faced with the difficult challenge of fitting computer and other new information technologies under the existing umbrella of intellectual property protection, Congress in 1974 established the National Commission on New Technological Uses of Copyrighted Works (CONTU) to study the implications of the new technologies and recommend revisions to federal intellectual property law. After conducting extensive hearings and receiving expert reports, a majority of the blue-ribbon panel of copyright authorities and interest group representatives comprising CONTU concluded in 1978 that the intellectual work embodied in computer software should be protected under copyright law, notwithstanding the fundamental principle that copyright cannot protect "any idea, procedure, process, system, method of operation, concept, principle, or discovery." 17 U.S.C. §102(b). Congress adopted CONTU's recommendations in 1980, passing legislation almost identical to that suggested in the Final Report.

In light of the computer software industry's relative youth and anticipated rapid growth, CONTU's rough empirical judgment that copyright would best promote the invention, development, and diffusion of new and better software products was, by necessity, highly speculative. As CONTU recognized, it was impossible in 1978 to establish a precise line between copyrightable expression of computer programs and the uncopyrightable processes that they implement. Yet the location of this line — the idea/expression dichotomy — was critical to the rough cost-benefit analysis that guided CONTU's recommendation. Drawing the line too liberally in favor of copyright protection would bestow strong monopolies upon those who develop operating systems that become industry standards and upon the first to write programs performing specific applications and would thereby inhibit other creators from developing improved programs and computer systems. Drawing the line too conservatively would allow programmers' efforts to be copied easily, thus discouraging the creation of all but modest incremental advances. The wisdom of Congress's decision to bring computer programs within the scope of copyright law thus depends critically upon where the courts draw this line.

We begin with the first generation of copyright infringement suits under the 1980 Amendments; these cases focus on whether and to what extent literal copying of computer software violates copyright law. Because the coding of operating systems is critically important to hardware system compatibility, the principal economic effect of the first generation of cases was on competition among hardware manufacturers. The following section examines the second generation of copyright infringement suits, focusing on the extent to which nonliteral forms of copying constitute copyright infringement. We then turn to protection for operating elements of programs and program outputs.

1. The Scope of Software Copyright

a. *Protection for Literal Elements of Program Code*

The first copyright cases involving computer programs involved the simplest form of copyright infringement — direct copying of the program code. Because the copying in these early cases was "literal" — that is, the accused infringer took the actual text of the work, not merely the structure, organization, or output of the programs — there is no question of scope of protection or "substantial similarity." Rather, the disputed issue in these early cases was whether computer programs could be protected by copyright at all. Some early cases held that object codes or operating systems were not copyrightable. See Data Cash Systems v. JS & A Group, 480 F. Supp. 1063 (N.D. Ill. 1979). The rationale was that such works were not "communicative."

CONTU Commissioner John Hersey, in his dissent to CONTU's Final Report, highlighted the fact that computer "[p]rograms are profoundly different from the various forms of 'works of authorship' secured under the Constitution by copyright. Works of authorship have always been intended to be circulated to human beings and to be used by them — to be read, heard, or seen, for either pleasurable or practical ends. Computer programs, in their mature phase, are addressed to machines." Hersey argued that copyright protection should "not extend to a computer program in the form in which it is capable of being used to control computer operations."

Do you agree with Commissioner Hersey's critique of bringing computer software, in all its forms, within the copyright law? Does the Constitution limit copyright protection for "works of authorship" solely to works intended to be directly "read, heard, or seen" by humans? Should protection for computer software be limited in this way? Would Hersey's proposed limitation on the scope of copyright protection for software better balance the social interests affected by legal protection for software? What might justify CONTU's decision to accord software full inclusion within the copyright law, irrespective of whether it is directed toward operating machines or communicating with humans? For a detailed discussion of the problems raised by Commissioner Hersey in his dissent, see Pamela Samuelson, CONTU Revisited: The Case Against Copyright Protection for Computer Programs in Machine-Readable Form, 1984 Duke L.J. 663 (1984).

Despite these arguments, it was well settled by the early 1980s that computer programs were copyrightable, just like any other literary work. See, e.g., Apple Computer v. Franklin Computer, 714 F.2d 1240 (3d Cir. 1983) (rejecting argument that operating system is not copyrightable on grounds that it is a process, system, or method of operation, or the idea/expression dichotomy; and observing that achieving compatibility with independently developed application programs "is a commercial and competitive objective which does not enter into the somewhat metaphysical issue of whether particular ideas and expressions have merged"); Williams Electronics v. Artic Intl., 685 F.2d 870 (3d Cir. 1982). No cases since that time have held otherwise.

This argument has some troubling implications. One of the fundamental rules of computing is that anything that can be implemented in software can also, in principle, be implemented in hardware. Indeed, "software" itself is at base nothing more than a temporary (rather than permanent) way of ordering circuit switches in hardware. So if object code is copyrightable, why not more permanent instantiations in hardware — the microcode instructions embedded in computer chips, for example, or the layout of a circuit board itself?

Copyright law does sometimes protect the design elements of utilitarian objects: the sculpture/lamp of *Mazer v. Stein*, discussed in Chapter 4, is a prominent example. While courts have had difficulty at times separating the artistic from the utilitarian, they have steadfastly refused to extend copyright protection to the utilitarian aspects of three-dimensional objects such as lamps, jewelry, and mannequins. Protecting object code and microcode raises many of the same issues. How should courts separate the unprotectable utilitarian aspects of a computer program from the protectable expressive aspects? We will see how the courts have dealt with this issue in the sections that follow.

PROBLEM

Problem 7-4. Letni Corp. produces microprocessors, the central component of microcomputer systems. These "computers on a chip" are integrated circuits capable of performing arithmetic functions, manipulation of data, and other operations in response to object code instructions. Letni's 684 microprocessor currently commands a 70 percent market share of the microcomputer market. Letni's 684 chip utilizes a single "bus" or path for transferring data. It features 512 lines of microcode, each 21 bits long, which enables the microprocessor to interpret 133 assembly instructions.

CEN Corp., a competing microprocessor manufacturer, seeks to sell semiconductor chips capable of running the same software as Letni's 684 microprocessor. A prior patent cross-licensing agreement entitles

CEN to duplicate Letni's 684 microarchitecture and hardware to the extent comprehended by the Letni patent. To increase processing speed and capability relative to Letni's 684 chip, CEN designed its U2 microprocessor with a dual bus system and a larger micro instruction word size (29 bits). The 1024 lines of microcode enable the microprocessor to interpret 156 assembly instructions; 552 of these lines carry out the instruction set of the Letni 684 microprocessor. The remaining lines perform other functions not carried out by the 684 chip. Because of the dual bus architecture and larger word size, CEN was able to write shorter and more efficient microcode.

Nonetheless, there were a number of similarities in the microcodes of the two microprocessors. Letni brought suit, alleging: (1) microprograms are computer programs and hence protected by the Copyright Act; and (2) CEN's U2 microcode infringed the copyright of Letni's 684 chip.

How should this case be resolved? Should microcode be copyrightable? If so, what should be the scope of protection? Can a hardware manufacturer obtain copyright protection for the configuration of the hardwiring of its computer? If so, under what theory? If your answer is different from that for microcode, what explains the distinction?

b. Protection for Nonliteral Elements of Program Code

Apple v. Franklin clearly established that exact copying of computer code infringes the programmer's copyright in the code. This ruling served the important purpose of prohibiting piracy of computer programs. Soon thereafter, a second generation of computer software cases raised the question whether competitors may copy other elements of a computer program, such as the program's underlying structure, sequence, or organization. As we saw in Chapter 4, the scope of copyright protection may extend beyond the literal elements of a work. How far should these doctrines be extended in the context of computer programs?

Whelan Associates, Inc. v. Jaslow Dental Laboratory, Inc.
United States Court of Appeals for the Third Circuit
797 F.2d 1222 (3d Cir. 1986), cert. denied, 479 U.S. 1031 (1987)

BECKER, Circuit Judge.

This appeal involves a computer program for the operation of a dental laboratory, and calls upon us to apply the principles underlying our venerable copyright laws to the relatively new field of computer technology to determine the scope of copyright protection of a computer program. More particularly, in this case of first impression in the courts of appeals, we must determine whether the structure (or sequence and organization) of a computer program is protectible by copyright, or whether the protection of the copyright law extends only as far as the literal computer code. . . .

I. Factual Background

[Jaslow Dental Laboratory, Inc. ("Jaslow Lab") hired Strohl Systems, Inc. ("Strohl"), a custom software firm, to develop a computer program that would organize the bookkeeping and administrative tasks of its business — manufacturing dental prosthetics and devices. The contract provided that Strohl could market the program to other dental laboratories after installation at Jaslow Lab, with Jaslow Lab receiving a 10 percent royalty. Strohl assigned the project to Elaine Whelan, an experienced programmer. After interviewing Jaslow Lab employees and studying the dental laboratory business, Whelan wrote a program called Dentalab in EDL (Event Driven Language), a computer language that ran on Jaslow Lab's IBM Series One computer.

[Elaine Whelan subsequently left Strohl to form Whelan Associates, Inc., in order to pursue the market for Dentalab. Whelan Associates acquired Strohl's interest in Dentalab and entered into an agreement with Jaslow Lab obliging Jaslow Lab to use its "best efforts and to act diligently in the marketing of the Dentalab package" and requiring Whelan Associates to "use its best efforts and to act diligently to improve and augment the previously successfully designed Dentalab package."

[During the period of the agreement, Rand Jaslow, an officer and shareholder of Jaslow Lab, began development of the Dentcom PC program ("Dentcom program"), which performed the same functions as Dentalab but was written in the BASIC computer language, enabling it to run on a broader range of computers. Jaslow Lab subsequently ended the business relationship with Whelan Associates, formed Dentcom, Inc., to market dental laboratory software, and hired a professional computer programmer to complete the Dentcom program. Dentcom subsequently marketed both the Dentalab and Dentcom programs, advertising the Dentcom program as "a new version of the Dentalab computer system."] . . .

V. The Scope of Copyright Protection of Computer Programs

It is well, though recently, established that copyright protection extends to a program's source and object codes. In this case, however, the district court did not find any copying of the source or object codes, nor did the plaintiff allege such copying. Rather, the district court held that the Dentalab copyright was infringed because the overall structure of Dentcom was substantially similar to the overall structure of Dentalab. The question therefore arises whether mere similarity in the overall structure of programs can be the basis for a copyright infringement, or, put differently, whether a program's copyright protection covers the structure of the program or only the program's literal elements, i.e., its source and object codes.

Title 17 U.S.C. §102(a)(1) extends copyright protection to "literary works," and computer programs are classified as literary works for the purposes of copyright. The copyrights of other literary works can be infringed even when there is no substantial similarity between the works' literal elements. One can violate the copyright of a play or book by copying its plot or plot devices. See, e.g., Twentieth Century-Fox Film Corp. v. MCA, Inc., 715 F.2d 1327, 1329 (9th Cir. 1983) (13 alleged distinctive plot similarities between *Battlestar Galactica* and *Star Wars* may be basis for a finding of copyright violation); Sid & Marty Krofft Television Productions, Inc., 562 F.2d at 1167 (similarities between McDonaldland characters and H.R. Pufnstuf characters can be established by " 'total concept and feel' " of the two productions, quoting Roth Greeting Cards v. United Card Co., 429 F.2d 1106, 1110 (9th Cir. 1970)); Sheldon v. Metro-Goldwyn Pictures Corp., 81 F.2d 49, 54-55 (1936); Nichols v. Universal Pictures Corp., 45 F.2d 119, 121 (2d Cir. 1930) (copyright "cannot be limited literally to the text, else a plagiarist would escape by immaterial variations"). By analogy to other literary works, it would thus appear that the copyrights of computer programs can be infringed even absent copying of the literal elements of the program. Defendants contend, however, that what is true of other literary works is not true of computer programs. They assert two principal reasons, which we consider in turn.

A. *Section 102(b) and the dichotomy between idea and expression*

It is axiomatic that copyright does not protect ideas, but only expressions of ideas. . . .

Defendants argue that the structure of a computer program is, by definition, the idea and not the expression of the idea, and therefore that the structure cannot be protected by the program copyright. Under the defendants' approach, any other decision would be contrary to §102(b). We divide our consideration of this argument into two parts. First, we examine the caselaw concerning the distinction between idea and expression, and derive from it a rule for distinguishing idea from expression in the context of computer programs. We then apply that rule to the facts of this case.

1. *A rule for distinguishing idea from expression in computer programs* — It is frequently difficult to distinguish the idea from the expression thereof. No less an authority than Learned Hand, after a career that included writing some of the leading copyright opinions, concluded that the distinction will "inevitably be ad hoc." Peter Pan Fabrics, Inc. v. Martin Weiner Copr., 274

F.2d 487, 489 (2d Cir. 1960). See also Knowles & Palmeri, Dissecting Krofft: An Expression of New Ideas in Copyright?, 8 San. Fern. Val. L. Rev. 109, 126 (1980) (arguing that there can be no meaningful distinction between idea and expression). Although we acknowledge the wisdom of Judge Hand's remark, we feel that a review of relevant copyright precedent will enable us to formulate a rule applicable in this case. In addition, precisely because the line between idea and expression is elusive, we must pay particular attention to the pragmatic considerations that underlie the distinction and copyright law generally. In this regard, we must remember that the purpose of the copyright law is to create the most efficient and productive balance between protection (incentive) and dissemination of information, to promote learning, culture and development. See U.S. Const. Art. I §8 cl. 8 (Copyright Clause) (giving Congress the power to "promote the Progress of Science and useful Arts").[27] . . .

The Court's test in Baker v. Selden suggests a way to distinguish idea from expression. Just as Baker v. Selden focused on the end sought to be achieved by Selden's book, the line between idea and expression may be drawn with reference to the end sought to be achieved by the work in question. In other words, *the purpose or function of a utilitarian work would be the work's idea, and everything that is not necessary to that purpose or function would be part of the expression of the idea.* Cf. Apple Computer, Inc. v. Formula Intl. Inc., 562 F. Supp. 775, 783 (C.D. Ca. 1983) ("Apple seeks here not to protect ideas (i.e. making the machine perform particular functions) but rather to protect their particular *expressions* . . ."), aff'd, 725 F.2d 521 (9th Cir. 1984). Where there are various means of achieving the desired purpose, then the particular means chosen is not necessary to the purpose; hence, there is expression, not idea.[28]

Consideration of copyright doctrines related to *scenes à faire* and fact-intensive works supports our formulation, for they reflect the same underlying

---

27. Achieving the proper incentive has been a longstanding task of courts. See, e.g., Sayre v. Moore, 102 Eng. Rep. 138, 140 n.6 (1785) (Lord Manfield):

> [W]e must take care to guard against two extremes equally prejudicial; the one, that men of ability, who have employed their time for the service of the community, may not be deprived of their just merits, and the reward for their ingenuity and labour; the other, that the world may not be deprived of improvements, nor the progress of the arts be retarded.

See also Twentieth Century Music Corp. v. Aiken, 422 U.S. 151, 156, 45 L. Ed. 2d 84, 95 S. Ct. 2040 (1975); Apple Computer, 714 F.2d at 1253 (Sloviter, J.) ("the line must be a pragmatic one, which also keeps in consideration 'the preservation of the balance between competition and protection reflected in the patent and copyright laws.' " (quoting Herbert Rosenthal Jewelry Corp. v. Kalpakian, 446 F.2d 738, 742 (9th Cir. 1971))).

28. This test is necessarily difficult to state, and it may be difficult to understand in the abstract. It will become more clear as we discuss and explain it in the textual discussion that follows this footnote. See also infra (discussion of the copyrightability of file structures that raise many of the issues considered here). . . . [T]he idea of the Dentalab program was the efficient management of a dental laboratory (which presumably has significantly different requirements from those of other businesses). Because that idea could be accomplished in a number of different ways with a number of different structures, the structure of the Dentalab program is part of the program's expression, not its idea.

principle. *Scenes à faire* are "incidents, characters or settings which are as a practical matter indispensable . . . in the treatment of a given topic." Atari, Inc. v. North American Philips Consumer Elecs. Corp., 672 F.2d 607, 616 (7th Cir.), *cert. denied,* 459 U.S. 880, 74 L. Ed. 2d 145, 103 S. Ct. 176 (1982). See also See v. Durang, 711 F.2d 141, 143 (9th Cir. 1983). It is well-settled doctrine that *scenes à faire* are afforded no copyright protection.

Scenes à faire are afforded no protection because the subject matter represented can be expressed in no other way than through the particular scene à faire. Therefore, granting a copyright "would give the first author a monopoly on the commonplace ideas behind the scenes à faire." Landsberg v. Scrabble Crossword Game Players, Inc., 736 F.2d at 489.[30] This is merely a restatement of the hypothesis advanced above, that the purpose or function of a work or literary device is part of that device's "idea" (unprotectible portion). It follows that anything necessary to effecting that function is also, necessarily, part of the idea, too.

Fact intensive works are given similarly limited copyright coverage. See, e.g., Landsberg, 736 F.2d at 488; Miller v. Universal City Studios, Inc., 650 F.2d 1365, 1372 (5th Cir. 1981). Once again, the reason appears to be that there are only a limited number of ways to express factual material, and therefore the purpose of the literary work — telling a truthful story — can be accomplished only by employing one of a limited number of devices. Landsberg, 736 F.2d at 488. Those devices therefore belong to the idea, not the expression, of the historical or factual work.

Although the economic implications of this rule are necessarily somewhat speculative, we nevertheless believe that the rule would advance the basic purpose underlying the idea/expression distinction, "the preservation of the balance between competition and protection reflected in the patent and copyright laws." Herbert Rosenthal Jewelry Corp. v. Kalpakian, 446 F.2d 738, 742 (9th Cir. 1971); see also Apple Computer, 714 F.2d at 1253 (quoting Kalpakian); supra n.27. As we stated above, . . . among the more significant costs in computer programming are those attributable to developing the structure and logic of the program. The rule proposed here, which allows copyright protection beyond the literal computer code, would provide the proper incentive for programmers by protecting their most valuable efforts, while not giving them a stranglehold over the development of new computer devices that accomplish the same end.

The principal economic argument used against this position — used, that is, in support of the position that programs' literal elements are the only parts of the programs protected by the copyright law — is that computer programs are so intricate, each step so dependent on all of the other steps, that they are almost impossible to copy except literally, and that anyone who attempts to copy the structure of a program without copying its literal ele-

---

30. Hoehling, 618 F.2d at 979 (explaining the *scenes à faire* doctrine as follows: "Because it is virtually impossible to write about a particular historical era or fictional theme without employing certain 'stock' or standard literary devices, we have held that scenes à faire are not copyrightable as a matter of law."). . . .

ments must expend a tremendous amount of effort and creativity. In the works of one commentator: "One cannot simply 'approximate' the entire copyrighted computer program and create a similar operative program without the expenditure of almost the same amount of time as the original programmer expended." Note, 68 Minn. L. Rev. at 1290 (footnote omitted). According to this argument, such work should not be discouraged or penalized. A further argument against our position is not economic but jurisprudential; another commentator argues that the concept of structure in computer programs is too vague to be useful in copyright cases. Radcliffe, Recent Developments in Copyright Law Related to Computer Software, 4 Computer L. Rep. 189, 194-97 (1985). He too would therefore appear to advocate limiting copyright protection to programs' literal codes.

Neither of the two arguments just described is persuasive. The first argument fails for two reasons. In the first place, it is simply not true that "approximation" of a program short of perfect reproduction is valueless. To the contrary, one can approximate a program and thereby gain a significant advantage over competitors even though additional work is needed to complete the program. Second, the fact that it will take a great deal of effort to copy a copyrighted work does not mean that the copier is not a copyright infringer. The issue in a copyrighted case is simply whether the copyright holder's expression has been copied, not how difficult it was to do the copying. Whether an alleged infringer spent significant time and effort to copy an original work is therefore irrelevant to whether he has pirated the expression of an original work.

As to the second argument, it is surely true that limiting copyright protection to computers' literal codes would be simpler and would yield more definite answers than does our answer here. Ease of application is not, however, a sufficient counterweight to the considerations we have adduced on behalf of our position.

Finally, one commentator argues that the process of development and progress in the field of computer programming is significantly different from that in other fields, and therefore requires a particularly restricted application of the copyright law. According to this argument, progress in the area of computer technology is achieved by means of "stepping-stones," a process that "requires plagiarizing in some manner the underlying copyrighted work." Note, 68 Minn. L. Rev. at 1292 (footnote omitted). As a consequence, this commentator argues, giving computer programs too much copyright protection will retard progress in the field.

We are not convinced that progress in computer technology or technique is qualitatively different from progress in other areas of science or the arts. In balancing protection and dissemination, the copyright law has always recognized and tried to accommodate the fact that all intellectual pioneers build on the work of their predecessors.[33] Thus, copyright principles derived from other areas are applicable in the field of computer programs.

---

33. Long before the first computer, Sir Isaac Newton humbly explained that "if [he] had seen farther than other men, it was because [he] had stood on the shoulders of giants."

2. *Application of the general rule to this case* — The rule proposed here is certainly not problem-free. The rule has its greatest force in the analysis of utilitarian or "functional" works, for the purpose of such works is easily stated and identified. By contrast, in cases involving works of literature or "non-functional" visual representations, defining the purpose of the work may be difficult. Since it may be impossible to discuss the purpose or function of a novel, poem, sculpture or painting, the rule may have little or no application to cases involving such works. The present case presents no such difficulties, for it is clear that the purpose of the utilitarian Dentalab program was to aid in the business operations of a dental laboratory.[34] It is equally clear that the structure of the program was not essential to that task; there are other programs on the market, competitors of Dentalab and Dentcom, that perform the same functions but have different structures and designs.

. . . The conclusion is thus inescapable that the detailed structure of the Dentalab program is part of the expression, not the idea, of that program. . . .

Despite the fact that copyright protection extends to sequence and form in the computer context, unless we are able to answer Judge Higginbotham's powerful rhetorical question [in Synercom Technology, Inc. v. University Computing Co., 462 F. Supp. 1003 (N.D. Tex. 1978)] — "if sequencing and ordering [are] expression, what separable idea is being expressed?" — in our own case, we would have to hold that the structure of the Dentalab program is part of its idea and is thus not protectible by copyright. Our answer has already been given, however: the idea is the efficient organization of a dental laboratory (presumably, this poses different problems from the efficient organization of some other kinds of laboratories or businesses). Because there are a variety of program structures through which that idea can be expressed, the structure is not a necessary incident to that idea. . . .

VII. Conclusion

We hold that (1) copyright protection of computer programs may extend beyond the programs' literal code to their structure, sequence, and organization, and (2) the district court's finding of substantial similarity between the Dentalab and Dentcom programs was not clearly erroneous. The judgment of the district court will therefore be affirmed.

---

34. We do not mean to imply that the idea or purpose behind every utilitarian or functional work will be precisely what it accomplishes, and that structure and organization will therefore always be part of the expression of such works. The idea or purpose behind a utilitarian work may be to accomplish a certain function in a certain way, see, e.g., Baker v. Selden, 101 U.S. at 100 (referring to Selden's book as explaining "a peculiar system of book-keeping"), and the structure or function of a program might be essential to that task. There is no suggestion in the record, however, that the purpose of the Dentalab program was anything so refined; it was simply to run a dental laboratory in an efficient way.

COMMENTS AND QUESTIONS

1. The *Whelan* court is concerned with choosing a rule that promotes innovation in the development and diffusion of computer software. In particular, the court is inclined to reward what it describes as the most time- and cost-intensive parts of computer programming. Is the cost of creation a valid factor to consider in affording copyright protection? In this regard, consider Feist v. Rural Telephone Services, 499 U.S. 340 (1991), in which the Supreme Court definitively rejected the "sweat of the brow" theory of copyright law. There, the Court held that factual works that took a great deal of effort to compile are not protectable under copyright if the work that went into them did not meet a threshold standard of originality. If telephone books are not entitled to copyright protection, why should computer programs get protection merely because they took a great deal of effort to write?

Even if you agree that the court properly rewarded the hard work of Whelan's programmers, it is possible to quibble with the court's factual conclusions about programming. What does the court see as the most time-intensive and costly stages in the development of application programs? How do these conclusions influence its interpretation of copyright law? In particular, consider whether the court would have reached a different conclusion if it had focused on the importance to programmers of building on the works of others.

2. What "rule" does the *Whelan* court develop for distinguishing idea from expression in the nonliteral elements of application program code? Applying this rule, what did the court find to be the unprotectable idea of the Dentalab program? What is the protectable expression? Does this test comport with §102(b) and Baker v. Selden?

In applying *Baker* to this case, two points deserve special mention. First, Judge Learned Hand described copyright analysis as relying on a series of "levels of abstraction" of increasing generality. See *Nichols*, cited above. At what "level of abstraction" has *Whelan* defined the idea of the computer program? Is it reasonable to conclude that "to run a dental laboratory efficiently" is the idea of the program? Why not choose a more concrete idea, along the lines of "collecting input on the following five subjects, and directing courses of action to maximize the following three constraints?" On the other hand, can you describe the "idea" of the program as "to make money," and everything else as merely one expression of that idea?

Although the court in *Whelan* purports to derive its rule for distinguishing idea from expression from Baker v. Selden, it is interesting to apply the *Whelan* rule to the facts of Baker v. Selden. Applying the *Whelan* test, what is the idea expressed in Selden's work? What is the protectable expression? Does the result of the case change?

3. The second point about the idea-expression dichotomy is evident from the Second Circuit's decision in Hoehling v. Universal City Studios, 618 F.2d 972 (2d Cir. 1980), *cert. denied*, 449 U.S. 841 (1980), discussed in Chapter 4. Ideas are not the only elements of a work not subject to copy-

right protection. Scenes à faire are mentioned by the court as another unprotectable element. In addition, facts, information, or works in the public domain, and creations of insufficient originality to deserve protection are all unprotectable elements according to copyright law. Does *Whelan*'s analysis take these unprotectable elements into account? If so, how?

PROBLEM

Problem 7-5. The Midwest Corn Cooperative Association (MCCA) is a nonprofit agricultural cooperative comprised of approximately 15,000 corn farmers in the midwestern states. MCCA's purpose is to assist its members in the growing and marketing of corn. During the mid-1970's, MCCA developed a computer software system called CornMart that provided users with information regarding corn prices and availability, accounting services, and a means for electronically completing actual sales. The CornMart system was used by corn farmers and buyers through terminals connected by telephone lines to MCCA's large central computer.

CornMart was developed by a team of programmers that included George Cusher and Ben Fishman. In the early 1980s, after the program had been up and running for a number of years, Cusher and Fishman decided the system would have much greater usefulness if the CornMart program could be operated on personal computers, which were by then increasingly used by MCCA members. Cusher and Fishman took a complete source code version of CornMart and supporting program documentation when they departed MCCA.

In 1985 Cusher and Fishman formed Cornware, Inc., and began work on a personal computer version of CornMart. By early 1986 they succeeded in writing a personal computer version of CornMart that they marketed as Cornware. Soon after they began marketing their new software, MCCA sued alleging copyright infringement.

Expert testimony at trial established that Cornware and CornMart were very similar in terms of functional specification, overall structure, and documentation. The main difference between the systems was that CornMart is designed to run on a mainframe computer and Cornware on a personal computer. There was, however, no literal copying of code. Cusher and Fishman testified that they had not copied or modified the CornMart program but instead had "drawn on their knowledge of the corn industry and expertise in computer programming and design gained over many years." Expert testimony at trial indicated that it would take significant effort to modify a program as complex as Corn Mart to run on a personal computer, perhaps more effort than rewriting the program entirely. Nonetheless, there remained significant structural similarities between the programs, although two experts testified that residual similarity between the programs reflected features of the corn market and that different programmers knowledgeable about the corn industry could conceivably make similar judgments. For example, both programs were designed to present the same information as is contained on a corn recap sheet, a standard form used for corn transactions.

How would you argue the case for MCCA? for Cornware? How would this case be resolved under the *Whelan* test?

≡ *Computer Associates International v. Altai, Inc.*
   United States Court of Appeals for the Second Circuit
≡ *982 F.2d 693 (2d Cir. 1992)*

WALKER, Circuit Judge: . . .

This appeal comes to us from the United States District Court for the Eastern District of New York, the Honorable George C. Pratt, Circuit Judge, sitting by designation. By Memorandum and Order entered August 12, 1991, Judge Pratt found that defendant Altai, Inc.'s ("Altai"), OSCAR 3.4 computer program had infringed plaintiff Computer Associates' ("CA"), copyrighted computer program entitled CA-SCHEDULER. Accordingly, the district court awarded CA $364,444 in actual damages and apportioned profits. Altai has abandoned its appeal from this award. With respect to CA's second claim for copyright infringement, Judge Pratt found that Altai's OSCAR 3.5 program was not substantially similar to a portion of CA-SCHEDULER called ADAPTER, and thus denied relief. . . .

II. Facts . . .

The subject of this litigation originates with one of CA's marketed programs entitled CA-SCHEDULER. CA-SCHEDULER is a job scheduling program designed for IBM mainframe computers. Its primary functions are straightforward: to create a schedule specifying when the computer should run various tasks, and then to control the computer as it executes the schedule. CA-SCHEDULER contains a sub-program entitled ADAPTER, also developed by CA. ADAPTER is not an independently marketed product of CA; it is a wholly integrated component of CA-SCHEDULER and has no capacity for independent use.

Nevertheless, ADAPTER plays an extremely important role. It is an "operating system compatibility component," which means, roughly speaking, it serves as a translator. An "operating system" is itself a program that manages the resources of the computer allocating those resources to other programs as needed. The IBM System 370 family of computers, for which CA-SCHEDULER was created, is, depending upon the computer's size, designed to contain one of three operating systems: DOS/VSE, MVS, or CMS.

As the district court noted, the general rule is that "a program written for one operating system, e.g., DOS/VSE, will not, without modification, run under another operating system such as MVS." Computer Assocs., 775 F. Supp. at 550. ADAPTER's function is to translate the language of a given program into the particular language that the computer's own operating system can understand. . . .

A program like ADAPTER, which allows a computer user to change or use multiple operating systems while maintaining the same software, is highly desirable. It saves the user the costs, both in time and money, that otherwise would be expended in purchasing new programs, modifying existing systems to run them, and gaining familiarity with their operation. The benefits run both ways. The increased compatibility afforded by an ADAPTER-like component, and its resulting popularity among consumers, makes whatever software in which it is incorporated significantly more marketable.

Starting in 1982, Altai began marketing its own job scheduling program entitled ZEKE. The original version of ZEKE was designed for use in conjunction with a VSE operating system. By late 1983, in response to customer demand, Altai decided to rewrite ZEKE so that it could be run in conjunction with an MVS operating system.

[At that time, James P. Williams, then an employee of Altai and now its President, recruited Claude F. Arney, III, a long-standing friend and computer programmer who worked for CA, to assist Altai in designing an MVS version of ZEKE. Unknown to Williams, Arney was intimately familiar with CA's ADAPTER program and he took VSE and MVS source code versions of ADAPTER with him when he left CA to join Altai. Without disclosing his knowledge of ADAPTER, Arney persuaded Williams that the best way to modify ZEKE to run on an MVS operating system was to introduce a "common system interface" component, an approach that stemmed from Arney's familiarity with ADAPTER. Arney subsequently developed a component-program named OSCAR using the ADAPTER source code. Approximately 30% of the first generation of OSCAR was copied from CA's ADAPTER program. In mid 1988, CA discovered the copying from ADAPTER and brought this copyright infringement and trade secret action. Altai learned of the copying from the complaint.]

Upon advice of counsel, Williams initiated OSCAR's rewrite. The project's goal was to save as much of OSCAR 3.4 as legitimately could be used, and to excise those portions which had been copied from ADAPTER. Arney was entirely excluded from the process, and his copy of the ADAPTER code was locked away. Williams put eight other programmers on the project, none of whom had been involved in any way in the development of OSCAR 3.4. Williams provided the programmers with a description of the ZEKE operating system services so that they could rewrite the appropriate code. The rewrite project took about six months to complete and was finished in mid-November 1989. The resulting program was entitled OSCAR 3.5.

From that point on, Altai shipped only OSCAR 3.5 to its new customers. . . .

Discussion

[The district court concluded that version 3.5 was not substantially similar to CA-ADAPTER.]

I. *Copyright Infringement . . .*

As a general matter, and to varying degrees, copyright protection extends beyond a literary work's strictly textual form to its non-literal components. As we have said, "[i]t is of course essential to any protection of literary property that the right cannot be limited literally to the text, else a plagiarist would escape by immaterial variations." Nichols v. Universal Pictures Co., 45 F.2d 119, 121 (2d Cir. 1930) (L. Hand, J.), *cert. denied,* 282 U.S. 902, 51 S. Ct. 216, 75 L. Ed. 795 (1931). Thus, where "the fundamental essence or structure of one work is duplicated in another," 3 Nimmer, §13.03(A][1], at 13-24, courts have found copyright infringement. . . . This black letter proposition is the springboard for our discussion.

A. Copyright Protection for the Non-literal Elements of Computer Programs

It is now well settled that the literal elements of computer programs, i.e., their source and object codes, are the subject of copyright protection. . . . Here, as noted earlier, Altai admits having copied approximately 30% of the OSCAR 3.4 program from CA's ADAPTER source code, and does not challenge the district court's related finding of infringement.

In this case, the hotly contested issues surround OSCAR 3.5. As recounted above, OSCAR 3.5 is the product of Altai's carefully orchestrated rewrite of OSCAR 3.4. After the purge, none of the ADAPTER source code remained in the 3.5 version; thus, Altai made sure that the literal elements of its revamped OSCAR program were no longer substantially similar to the literal elements of CA's ADAPTER.

According to CA, the district court erroneously concluded that Altai's OSCAR 3.5 was not substantially similar to its own ADAPTER program. CA argues that this occurred because the district court "committed legal error in analyzing [its] claims of copyright infringement by failing to find that copyright protects expression contained in the non-literal elements of computer software." We disagree.

CA argues that, despite Altai's rewrite of the OSCAR code, the resulting program remained substantially similar to the structure of its ADAPTER program. As discussed above, a program's structure includes its non-literal components such as general flow charts as well as the more specific organization of inter-modular relationships, parameter lists, and macros. In addition to these aspects, CA contends that OSCAR 3.5 is also substantially similar to

ADAPTER with respect to the list of services that both ADAPTER and OSCAR obtain from their respective operating systems. We must decide whether and to what extent these elements of computer programs are protected by copyright law.

[The court agrees with Whelan v. Jaslow, supra, that the nonliteral elements of computer programs are entitled to copyright protection as literary works.]

1) Idea vs. Expression Dichotomy

It is a fundamental principle of copyright law that a copyright does not protect an idea, but only the expression of the idea. . . .

Drawing the line between idea and expression is a tricky business. Judge Learned Hand noted that "[n]obody has ever been able to fix that boundary, and nobody ever can," Nichols, 45 F.2d at 121. Thirty years later his convictions remained firm. "Obviously, no principle can be stated as to when an imitator has gone beyond copying the 'idea,' and has borrowed its 'expression,' " Judge Hand concluded. "Decisions must therefore inevitably be ad hoc." Peter Pan Fabrics, Inc. v. Martin Weiner Corp., 274 F.2d 487, 489 (2d Cir. 1960).

The essentially utilitarian nature of a computer program further complicates the task of distilling its idea from its expression. See SAS Inst., 605 F. Supp. at 829; cf. Englund, at 893. In order to describe both computational processes and abstract ideas, its content "combines creative and technical expression." See Spivack, at 755. The variations of expression found in purely creative compositions, as opposed to those contained in utilitarian works, are not directed towards practical application. For example, a narration of Humpty Dumpty's demise, which would clearly be a creative composition, does not serve the same ends as, say, a recipe for scrambled eggs — which is a more process oriented text. Thus, compared to aesthetic works, computer programs hover even more closely to the elusive boundary line described in §102(b).

[The court reviewed the facts and holding of Baker v. Selden].

To the extent that an accounting text and a computer program are both "a set of statements or instructions . . . to bring about a certain result," 17 U.S.C. §101, they are roughly analogous. In the former case, the processes are ultimately conducted by human agency; in the latter, by electronic means. In either case, as already stated, the processes themselves are not protectable. But the holding in Baker goes farther. The Court concluded that those aspects of a work, which "must necessarily be used as incident to" the idea, system or process that the work describes, are also not copyrightable. 101 U.S. at 104. Selden's ledger sheets, therefore, enjoyed no copyright protection because they were "necessary incidents to" the system of accounting that he described. Id. at 103. From this reasoning, we conclude that those elements of a computer program that are necessarily incidental to its function are similarly unprotectable.

While Baker v. Selden provides a sound analytical foundation, it offers scant guidance on how to separate idea or process from expression, and moreover, on how to further distinguish protectable expression from that expression which "must necessarily be used as incident to the work's underlying concept." In the context of computer programs, the Third Circuit's noted decision in Whelan has, thus far, been the most thoughtful attempt to accomplish these ends.

[The court quoted from the Whelan v. Jaslow decision reproduced above].

So far, in the courts, the Whelan rule has received a mixed reception. While some decisions have adopted its reasoning, see, e.g., Bull HN Info. Sys., Inc. v. American Express Bank, Ltd., 1990 Copyright Law Dec. (CCH) P 26,555 at 23,278 (S.D.N.Y. 1990); Dynamic Solutions, Inc. v. Planning & Control, Inc., 1987 Copyright Law Dec. (CCH) ¶26,062 at 20,912 (S.D.N.Y. 1987); Broderbund Software Inc. v. Unison World, Inc., 648 F. Supp. 1127, 1133 (N.D. Cal. 1986), others have rejected it. See Plains Cotton Co-op v. Goodpasture Computer Serv. Inc., 807 F.2d 1256, 1262 (5th Cir.), *cert. denied*, 484 U.S. 821, 108 S. Ct. 80, 98 L. Ed. 2d 42 (1987); cf. Synercom Technology, Inc. v. University Computing Co., 462 F. Supp. 1003, 1014 (N.D. Tex. 1978) (concluding that order and sequence of data on computer input formats was idea not expression).

Whelan has fared even more poorly in the academic community, where its standard for distinguishing idea from expression has been widely criticized for being conceptually overbroad. See, e.g., Englund, at 881; Menell, at 1074, 1082; Kretschmer, at 837-39; Spivack, at 747-55; Thomas M. Gage, Note, Whelan Associates v. Jaslow Dental Laboratories: Copyright Protection for Computer Software Structure — What's the Purpose?, 1987 Wis. L. Rev. 859, 860-61 (1987). The leading commentator in the field has stated that, "[t]he crucial flaw in [Whelan's] reasoning is that it assumes that only one 'idea,' in copyright law terms, underlies any computer program, and that once a separable idea can be identified, everything else must be expression." 3 Nimmer §13.03[F], at 13-62.34. This criticism focuses not upon the program's ultimate purpose but upon the reality of its structural design. As we have already noted, a computer program's ultimate function or purpose is the composite result of interacting subroutines. Since each subroutine is itself a program, and thus, may be said to have its own "idea," Whelan's general formulation that a program's overall purpose equates with the program's idea is descriptively inadequate.

Accordingly, we think that Judge Pratt wisely declined to follow Whelan. See Computer Assocs., 775 F. Supp. at 558-60. In addition to noting the weakness in the Whelan definition of "program-idea," mentioned above, Judge Pratt found that Whelan's synonymous use of the terms "structure, sequence, and organization," see Whelan, 797 F.2d at 1224 n.1, demonstrated a flawed understanding of a computer program's method of operation. See Computer Assocs., 775 F. Supp. at 559-60 (discussing the distinction between a program's "static structure" and "dynamic structure"). Rightly, the district court found Whelan's rationale suspect because it is so closely tied to what can now be seen — with the passage of time — as the opinion's somewhat outdated appreciation of computer science.

2) Substantial Similarity Test for Computer Program Structure: Abstraction-Filtration-Comparison

We think that Whelan's approach to separating idea from expression in computer programs relies too heavily on metaphysical distinctions and does not place enough emphasis on practical considerations. Cf. Apple Computer, 714 F.2d at 1253 (rejecting certain commercial constraints on programming as a helpful means of distinguishing idea from expression because they did "not enter into the somewhat metaphysical issue of whether particular ideas and expressions have merged"). As the cases that we shall discuss demonstrate, a satisfactory answer to this problem cannot be reached by resorting, *a priori*, to philosophical first principles.

As discussed herein, we think that district courts would be well-advised to undertake a three-step procedure, based on the abstractions test utilized by the district court, in order to determine whether the non-literal elements of two or more computer programs are substantially similar. This approach breaks no new ground; rather, it draws on such familar copyright doctrines as merger, *scenes à faire*, and public domain. In taking this approach, however, we are cognizant that computer technology is a dynamic field which can quickly outpace judicial decisionmaking. Thus, in cases where the technology in question does not allow for a literal application of the procedure we outline below, our opinion should not be read to foreclose the district courts of our circuit from utilizing a modified version.

In ascertaining substantial similarity under this approach, a court would first break down the allegedly infringed program into its constituent structural parts. Then, by examining each of these parts for such things as incorporated ideas, expression that is necessarily incidental to those ideas, and elements that are taken from the public domain, a court would then be able to sift out all non-protectable material. Left with a kernel, or possibly kernels, of creative expression after following this process of elimination, the court's last step would be to compare this material with the structure of an allegedly infringing program. The result of this comparison will determine whether the protectable elements of the programs at issue are substantially similar so as to warrant a finding of infringement. It will be helpful to elaborate a bit further.

Step One: Abstraction

As the district court appreciated, see Computer Assocs., 775 F. Supp. at 560, the theoretic framework for analyzing substantial similarity expounded by Learned Hand in the Nichols case is helpful in the present context. In

Nichols, we enunciated what has now become known as the "abstractions" test for separating idea from expression:

> Upon any work . . . a great number of patterns of increasing generality will fit equally well, as more and more of the incident is left out. The last may perhaps be no more than the most general statement of what the [work] is about, and at times might consist only of its title; but there is a point in this series of abstractions where they are no longer protected, since otherwise the [author] could prevent the use of his "ideas," to which, apart from their expression, his property is never extended.

Nichols, 45 F.2d at 121.

While the abstractions test was originally applied in relation to literary works such as novels and plays, it is adaptable to computer programs. In contrast to the Whelan approach, the abstractions test "implicitly recognizes that any given work may consist of a mixture of numerous ideas and expressions." 3 Nimmer §13.03[F] at 13-62.34-63.

As applied to computer programs, the abstractions test will comprise the first step in the examination for substantial similarity. Initially, in a manner that resembles reverse engineering on a theoretical plane, a court should dissect the allegedly copied program's structure and isolate each level of abstraction contained within it. This process begins with the code and ends with an articulation of the program's ultimate function. Along the way, it is necessary essentially to retrace and map each of the designer's steps — in the opposite order in which they were taken during the program's creation.

As an anatomical guide to this procedure, the following description is helpful:

> At the lowest level of abstraction, a computer program may be thought of in its entirety as a set of individual instructions organized into a hierarchy of modules. At a higher level of abstraction, the instructions in the lowest-level modules may be replaced conceptually by the functions of those modules. At progressively higher levels of abstraction, the functions of higher-level modules conceptually replace the implementations of those modules in terms of lower-level modules and instructions, until finally, one is left with nothing but the ultimate function of the program. . . . A program has structure at every level of abstraction at which it is viewed. At low levels of abstraction, a program's structure may be quite complex; at the highest level it is trivial.

Englund, at 897-98. Cf. Spivack, at 774.

Step Two: Filtration

Once the program's abstraction levels have been discovered, the substantial similarity inquiry moves from the conceptual to the concrete. Professor Nimmer suggests, and we endorse, a "successive filtering method" for separating protectable expression from non-protectable material. See generally 3 Nimmer §13.03[F]. This process entails examining the structural components at each level of abstraction to determine whether their particular inclusion at that level was "idea" or was dictated by considerations of efficiency, so as to be necessarily incidental to that idea; required by factors external to the program itself; or taken from the public domain and hence is non-protectable expression. See also Kretschmer, at 844-45 (arguing that program features dictated by market externalities or efficiency concerns are unprotectable). The structure of any given program may reflect some, all, or none of these considerations. Each case requires its own fact specific investigation.

Strictly speaking, this filtration serves "the purpose of defining the scope of plaintiff's copyright." Brown Bag Software v. Symantec Corp., 960 F.2d 1465, 1475 (9th Cir.) (endorsing "analytic dissection" of computer programs in order to isolate protectable expression), *cert. denied,* 113 S. Ct. 198 (1992). By applying well developed doctrines of copyright law, it may ultimately leave behind a "core of protectable material." 3 Nimmer §13.03[F](5), at 13-72. Further explication of this second step may be helpful.

(a) Elements Dictated by Efficiency

The portion of Baker v. Selden, discussed earlier, which denies copyright protection to expression necessarily incidental to the idea being expressed, appears to be the cornerstone for what has developed into the doctrine of merger. See Morrissey v. Proctor & Gamble Co., 379 F.2d 675, 678-79 (1st Cir. 1967) (relying on Baker for the proposition that expression embodying the rules of a sweepstakes contest was inseparable from the idea of the contest itself, and therefore were not protectable by copyright); see also Digital Communications, 659 F. Supp. at 457. The doctrine's underlying principle is that "[w]hen there is essentially only one way to express an idea, the idea and its expression are inseparable and copyright is no bar to copying that expression." Concrete Machinery Co. v. Classic Lawn Ornaments. Inc., 843 F.2d 600, 606 (1st Cir. 1988). Under these circumstances, the expression is said to have "merged" with the idea itself. In order not to confer a monopoly of the idea upon the copyright owner, such expression should not be protected. See Herbert Rosenthal Jewelry Corp. v. Kalpakian, 446 F.2d 738, 742 (9th Cir. 1971).

CONTU recognized the applicability of the merger doctrine to computer programs. In its report to Congress it stated that:

> [C]opyrighted language may be copied without infringing when there is but a limited number of ways to express a given idea. . . . In the computer context, this means that when specific instructions, even though previously copyrighted, are the only and essential means of accomplishing a given task, their later use by another will not amount to infringement.

CONTU Report at 20. While this statement directly concerns only the application of merger to program code, that is, the textual aspect of the program, it reasonably suggests that the doctrine fits comfortably within the general context of computer programs.

Furthermore, when one considers the fact that programmers generally strive to create programs "that meet the user's needs in the most efficient manner," Menell, at 1052, the applicability of the merger doctrine to computer programs becomes compelling. In the context of computer program design, the concept of efficiency is akin to deriving the most concise logical proof or formulating the most succinct mathematical computation. Thus, the more efficient a set of modules are, the more closely they approximate the idea or process embodied in that particular aspect of the program's structure.

While, hypothetically, there might be a myriad of ways in which a programmer may effectuate certain functions within a program — i.e., express the idea embodied in a given subroutine — efficiency concerns may so narrow the practical range of choice as to make only one or two forms of expression workable options. See 3 Nimmer §13.03[F](2), at 13-63; see also Whelan, 797 F.2d at 1243 n.43 ("It is true that for certain tasks there are only a very limited number of file structures available, and in such cases the structures might not be copyrightable." . . .) Of course, not all program structure is informed by efficiency concerns. See Menell, at 1052 (besides efficiency, simplicity related to user accommodation has become a programming priority). It follows that, in order to determine whether the merger doctrine precludes copyright protection to an aspect of a program's structure that is so oriented, a court must inquire "whether the use of *this particular set of modules* is necessary efficiently to implement that part of the program's process" being implemented. Englund, at 902. If the answer is yes, then the expression represented by the programmer's choice of a specific module or group of modules has merged with their underlying idea and is unprotected. Id. at 902-03.

Another justification for linking structural economy with the application of the merger doctrine stems from a program's essentially utilitarian nature and the competitive forces that exist in the software marketplace. See Kretschmer, at 842. Working in tandem, these factors give rise to a problem of proof which merger helps to eliminate. Efficiency is an industry-wide goal. Since, as we have already noted, there may be only a limited number of efficient implementations for any given program task, it is quite possible that multiple programmers, working independently, will design the identical method employed in the allegedly infringed work. Of course, if this is the case, there is no copyright infringement. See Roth Greeting Cards v. United Card Co., 429 F.2d 1106, 1110 (9th Cir. 1970); Sheldon, 81 F.2d at 54.

Under these circumstances, the fact that two programs contain the same efficient structure may as likely lead to an inference of independent creation as it does to one of copying. See 3 Nimmer §13.03[F][2], at 13-65; cf. Herbert Rosenthal Jewelry Corp., 446 F.2d at 741 (evidence of independent creation may stem from defendant's standing as a designer of previous similar works). Thus, since evidence of similarly efficient structure is not particularly probative of copying, it should be disregarded in the overall substantial similarity analysis. See 3 Nimmer §13.03[F][2], at 13-65. . . .

(b)   Elements Dictated by External Factors

We have stated that where "it is virtually impossible to write about a particular historical era or fictional theme without employing certain 'stock' or standard literary devices," such expression is not copyrightable. Hoehling v. Universal Studios, Inc., 618 F.2d 972, 979 (2d Cir.), *cert. denied*, 449 U.S. 1 841, 101 S. Ct. 42, 66 L. Ed. 2d 1 (1980). . . .

Professor Nimmer points out that "in many instances it is virtually impossible to write a program to perform particular functions in a specific computing environment without employing standard techniques." 3 Nimmer §13.03[F][3], at 13-65. This is a result of the fact that a programmer's freedom of design choice is often circumscribed by extrinsic considerations such as (1) the mechanical specifications of the computer on which a particular program is intended to run; (2) compatibility requirements of other programs with which a program is designed to operate in conjunction; (3) computer manufacturers' design standards; (4) demands of the industry being serviced; and (5) widely accepted programming practices within the computer industry. Id. at 13-65-71. . . .

(c)   Elements Taken from the Public Domain

Closely related to the non-protectability of *scenes à faire*, is material found in the public domain. Such material is free for the taking and cannot be appropriated by a single author even though it is included in a copyrighted work. See E. F. Johnson Co. v. Uniden Corp. of America, 623 F. Supp. 1485, 1499 (D. Minn. 1985); see also Sheldon, 81 F.2d at 54. We see no reason to make an exception to this rule for elements of a computer program that have entered the public domain by virtue of freely accessible program exchanges and the like. See 3 Nimmer §13.03[F][14]; see also Brown Bag Software, 960 F.2d at 1473 (affirming the district court's finding that "[p]laintiffs may not claim copyright protection of an . . . expression that is, if not standard, then commonplace in the computer software industry."). Thus, a court must also filter out this material from the allegedly infringed program before it makes the final inquiry in its substantial similarity analysis.

Step Three: Comparison

The third and final step of the test for substantial similarity that we believe appropriate for non-literal program components entails a comparison. Once a court has sifted out all elements of the allegedly infringed program which are "ideas" or are dictated by efficiency or external factors, or taken from the public domain, there may remain a core of protectable expression. In terms of a work's copyright value, this is the golden nugget. See *Brown Bag Software*, 960 F.2d at 1475. At this point, the court's substantial similarity inquiry focuses on whether the defendant copied any aspect of this protected expression, as well as an assessment of the copied portion's relative importance with respect to the plaintiff's overall program. See 3 Nimmer §13.03[F][5]; *Data East USA*, 862 F.2d at 208 ("To determine whether similarities result from unprotectable expression, analytic dissection of similarities may be performed. If . . . all similarities in expression arise from use of common ideas, then no substantial similarity can be found.").

3) Policy Considerations

We are satisfied that the three step approach we have just outlined not only comports with, but advances the constitutional policies underlying the Copyright Act. Since any method that tries to distinguish idea from expression ultimately impacts on the scope of copyright protection afforded to a particular type of work, "the line [it draws] must be a pragmatic one, which also keeps in consideration 'the preservation of the balance between competition and protection. . . . ' " Apple Computer, 714 F.2d at 1253 (citation omitted).

CA and some amici argue against the type of approach that we have set forth on the grounds that it will be a disincentive for future computer program research and development. At bottom, they claim that if programmers are not guaranteed broad copyright protection for their work, they will not invest the extensive time, energy and funds required to design and improve program structures. While they have a point, their argument cannot carry the day. The interest of the copyright law is not in simply conferring a monopoly on industrious persons, but in advancing the public welfare through rewarding artistic creativity, in a manner that permits the free use and development of non-protectable ideas and processes.

In this respect, our conclusion is informed by Justice Stewart's concise discussion of the principles that correctly govern the adaptation of the copyright law to new circumstances. In Twentieth Century Music Corp. v. Aiken, he wrote:

> The limited scope of the copyright holder's statutory monopoly, like the limited copyright duration required by the Constitution, reflects a balance of competing claims upon the public interest: Creative work is to be encouraged and rewarded, but private motivation must ultimately serve the cause of promoting broad public availability of literature, music, and the other arts. The immediate effect of our copyright law is to secure a fair return for an "author's" creative labor. But the ultimate aim is, by this incentive, to stimulate artistic creativity for the general public good. . . . When technological change has rendered its literal terms ambiguous, the Copyright Act must be construed in light of this basic purpose.

422 U.S. 151, 156 (1975)(citations and footnotes omitted).

Recently, the Supreme Court has emphatically reiterated that "[t]he primary objective of copyright is not to reward the labor of authors. . . ." Feist Publications, Inc. v. Rural Tel. Serv. Co., 111 S. Ct. 1282, 1290 (1991) (emphasis added). While the Feist decision deals primarily with the copyrightability of purely factual compilations, its underlying tenets apply to much of the work involved in computer programming. Feist put to rest the "sweat of the brow" doctrine in copyright law. Id. 111 S. Ct. at 1295. The rationale of that doctrine "was that copyright was a reward for the hard work that went into compiling facts." Id. 111 S. Ct. at 1291. The Court flatly rejected this justification for extending copyright protection, noting that it "eschewed the most fundamental axiom of copyright law — that no one may copyright facts or ideas." Id.

Feist teaches that substantial effort alone cannot confer copyright status on an otherwise uncopyrightable work. As we have discussed, despite the fact that significant labor and expense often goes into computer program flow-charting and debugging, that process does not always result in inherently protectable expression. Thus, Feist implicitly undercuts the Whelan rationale, "which allow[ed] copyright protection beyond the literal computer code . . . [in order to] provide the proper incentive for programmers by protecting their most valuable efforts. . . ." Whelan, 797 F.2d at 1237 (footnote omitted). We note that Whelan was decided prior to Feist when the "sweat of the brow" doctrine still had vitality. In view of the Supreme Court's recent holding, however, we must reject the legal basis of CA's disincentive argument.

Furthermore, we are unpersuaded that the test we approve today will lead to the dire consequences for the computer program industry that plaintiff and some amici predict. To the contrary, serious students of the industry have been highly critical of the sweeping scope of copyright protection engendered by the Whelan rule, in that it "enables first comers to 'lock up' basic programming techniques as implemented in programs to perform particular tasks." Menell, at 1087; see also Spivack, at 765 (Whelan "results in an inhibition of creation by virtue of the copyright owner's quasi-monopoly power").

To be frank, the exact contours of copyright protection for non-literal program structure are not completely clear. We trust that as future cases are decided, those limits will become better defined. Indeed, it may well be that the Copyright Act serves as a relatively weak barrier against public access to the theoretical interstices behind a program's source and object codes. This results from the hybrid nature of a computer program, which, while it is literary expression, is also a highly functional, utilitarian component in the larger process of computing.

Generally, we think that copyright registration — with its indiscriminating availability — is not ideally suited to deal with the highly dynamic technology of computer science. Thus far, many of the decisions in this area reflect the courts' attempt to fit the proverbial square peg in a round hole. The district court, see Computer Assocs., 775 F. Supp. at 560, and at least one commentator have suggested that patent registration, with its exacting up-front novelty and non-obviousness requirements, might be the more appropriate rubric of protection for intellectual property of this kind. See Randell M. Whitmeyer, Comment, A Plea for Due Processes: Defining the Proper Scope of Patent Protection for Computer Software, 85 Nw. U.L. Rev. 1103, 1123-25 (1991); see also Lotus Dev. Corp. v. Borland Intl., Inc., 788 F. Supp. 78, 91 (D. Mass. 1992) (discussing the potentially supplemental relationship between patent and copyright protection in the context of computer programs). In any event, now that more than 12 years have passed since CONTU issued its final report, the resolution of this specific issue could benefit from further legislative investigation — perhaps a CONTU II.

In the meantime, Congress has made clear that computer programs are literary works entitled to copyright protection. Of course, we shall abide by these instructions, but in so doing we must not impair the overall integrity of copyright law. While incentive based arguments in favor of broad copyright protection are perhaps attractive from a pure policy perspective, see Lotus Dev. Corp., 740 F. Supp. at 58, ultimately, they have a corrosive effect on certain fundamental tenets of copyright doctrine. If the test we have outlined results in narrowing the scope of protection, as we expect it will, that result flows from applying, in accordance with Congressional intent, long-standing principles of copyright law to computer programs. Of course, our decision is also informed by our concern that these fundamental principles remain undistorted.

B. The District Court Decision . . .

2) Evidentiary Analysis

The district court had to determine whether Altai's OSCAR 3.5 program was substantially similar to CA's ADAPTER. . . .

The district court took the first step in the analysis set forth in this opinion when it separated the program by levels of abstraction. The district court stated:

> As applied to computer software programs, this abstractions test would progress in order of "increasing generality" from object code, to source code, to parameter lists, to services required, to general outline. In discussing the particular similarities, therefore, we shall focus on these levels.

Computer Assocs., 775 F. Supp. at 560. While the facts of a different case might require that a district court draw a more particularized blueprint of a program's overall structure, this description is a workable one for the case at hand.

Moving to the district court's evaluation of OSCAR 3.5's structural components, we agree with Judge Pratt's systematic exclusion of non-protectable expression. With respect to code, the district court observed that after the rewrite of OSCAR 3.4 to OSCAR 3.5, "there remained virtually no lines of code that were identical to ADAPTER." Id. at 561. Accordingly, the court found that the code "present[ed] no similarity at all." Id. at 562.

Next, Judge Pratt addressed the issue of similarity between the two programs' parameter lists and macros. He concluded that, viewing the conflicting evidence most favorably to CA, it demonstrated that "only a few of the lists and macros were similar to protected elements in ADAPTER; the others were either in the public domain or dictated by the functional demands of the program." Id. As discussed above, functional elements and elements taken from the public domain do not qualify for copyright protection. With respect to the few remaining parameter lists and macros, the district court could reasonably conclude that they did not warrant a finding of infringement given their relative contribution to the overall program. See Warner Bros., Inc. v. American Broadcasting Cos., Inc., 720 F.2d 231, 242 (2d Cir. 1983) (discussing de minimis exception which allows for literal copying of a small and usually insignificant portion of the plaintiff's work); 3 Nimmer §13.03[F][5], at 13-74. In any event, the district court reasonably found that, for lack of persuasive evidence, CA failed to meet its burden of proof on whether the macros and parameter lists at issue were substantially similar. See Computer Assocs., 775 F. Supp. at 562.

The district court also found that the overlap exhibited between the list of services required for both ADAPTER and OSCAR 3.5 was "determined by the demands of the operating system and of the applications program to which it [was] to be linked through ADAPTER or OSCAR. . . ." Id. In other words, this aspect of the program's structure was dictated by the nature of other programs with which it was designed to interact and, thus, is not protected by copyright.

Finally, in his infringement analysis, Judge Pratt accorded no weight to the similarities between the two programs' organizational charts, "because [the charts were] so simple and obvious to anyone exposed to the operation of the program[s]." Id. CA argues that the district court's action in this regard "is not consistent with copyright law" — that "obvious" expression is protected, and that the district court erroneously failed to realize this. However, to say that elements of a work are "obvious," in the manner in which the district court used the word, is to say that they "follow naturally from the work's theme rather than from the author's creativity." 3 Nimmer §13.03[F][3], at 1365. This is but one formulation of the *scenes à faire* doctrine, which we have already endorsed as a means of weeding out unprotectable expression. . . .

Since we accept Judge Pratt's factual conclusions and the results of his legal analysis, we affirm his dismissal of CA's copyright infringement claim based upon OSCAR 3.5. We emphasize that, like all copyright infringement cases, those that involve computer programs are highly fact specific. The amount of protection due structural elements, in any given case, will vary according to the protectable expression found to exist within the program at issue. . . .

COMMENTS AND QUESTIONS

1. How does the *Altai* court distinguish idea from expression? Compare this method to that proposed and applied in *Whelan*. How would the *Whelan* case have come out if the *Altai* test had been applied? How would Baker v. Selden be decided under the *Altai* approach?

2. Which approach — *Whelan* or *Altai* — comports better with section 102(b) and Baker v. Selden? Is there a more appropriate way of distinguishing idea from expression in the context of application program code?

3. *Altai* seems to ask courts to analytically dissect a computer program to determine what is protectable and copied and compare this to the entire program, rather than to the protectable uncopied elements, to determine substantial similarity. Does this suggest that programs with relatively little protectable material can be freely copied?

4. Compare the manner in which *Altai* and Apple v. Franklin, supra, treat the issue of achieving compatibility between computer programs. Under the *Altai* test, could Franklin legitimately develop a computer program that achieves compatibility with the Apple operating system? If so, how would you advise Franklin to structure its development process so as to minimize the risk of copyright infringement?

5. The *Altai* case has been warmly praised by most commentators. See, e.g., David Bender, Computer Associates v. Altai: Rationality Prevails, Computer Law., Aug. 1992, at 1; Pamela Samuelson et al., The Nature of Copyright Analysis for Computer Program: Copyright Law Professors' Brief Amicus Curiae in Lotus v. Borland, 16 Hastings Comm. & Ent. L.J. 657 (1994) (brief of 24 copyright law professors endorsing *Altai*). However, the decision has also been bitterly attacked by lawyers and scholars representing large computer companies. See Jack E. Brown, Analytical Dissection of Copyrighted Computer Software — Complicating the Simple and Confounding the Complex, 25 Ariz. St. L.J. 801 (1993); Anthony L. Clapes & Jennifer M. Daniels, Revenge of the Luddites: A Closer Look at Computer Associates v. Altai, Computer Law., Nov. 1992, at 11. Professor Arthur Miller has expressed apprehension about the implications of the *Altai* approach:

> [A] court must employ considerable caution in excluding efficient or speedy program expression lest it undermine the effective protection of computer programs. For example, the mere fact that the expression is efficient should not, without more, bar protection for original authorship in the programming context any more than it does in prose work. An uncritical application of *Altai*'s language would penalize the most effective (and in some senses the most artistic) programmers.

Arthur Miller, Copyright Protection for Computer Programs, Databases, and Computer-Generated Works: Is Anything New Since CONTU?, 106 Harv. L. Rev. 977, 1004-05 (1993) (footnotes omitted). Do you agree with Miller's view of the scope of copyright protection for the efficient elements of program structure? Is protection of programming choices dictated by efficiency consistent with copyright cases outside the computer context?

6. The *Altai* test was rapidly adopted by most courts. Judicial convergence on the abstraction-filtration-comparison test has been so complete that every court to confront the issue since 1992 has chosen the *Altai* approach over the *Whelan* approach. See Bateman v. Mnemonics, Inc., 79 F.3d 1532, 1543-45 (11th Cir. 1996); Apple Computer v. Microsoft Corp., 35 F.3d 1435 (9th Cir. 1994), *cert. denied*, 115 S. Ct. 1176 (1995); Engineering Dynamics, Inc. v. Structural Software, Inc., 26 F.3d 1335 (5th Cir. 1994); Kepner-Tregoe, Inc. v. Leadership Software, 12 F.3d 527 (5th Cir. 1994), *cert. denied*, 115 S. Ct. 82 (1994); Gates Rubber Co. v. Bando Chemical Indus., 9 F.3d 823 (10th Cir. 1993); Atari Games Corp. v. Nintendo, 975 F.2d 832 (Fed. Cir. 1992); CMAX/Cleveland, Inc. v. UCR, Inc., 804 F. Supp. 337 (M.D. Ga. 1992); Mark A. Lemley, Convergence in the Law of Software Copyright, 10 High Tech. L.J. 1 (1995). See also Brown Bag Software v. Symantec Corp., 960 F.2d 1465, 1475-76 (9th Cir. 1992), *cert. denied*, 506 U.S. 869 (1992) (endorsing "analytic dissection" of the elements of a computer program prior to the *Altai* decision). Cf. MiTek Holdings Inc. v. Arce Engineering Co., 89 F.3d 1548 (11th Cir. 1996) (upholding *Altai* approach, but allowing plaintiff to skip abstraction step by identifying in advance those elements it considers to be protectable and infringed). In addition, courts in Canada, the United Kingdom, and France have endorsed the *Altai* filtration analysis.

Not all of these courts have approached the abstraction-filtration-comparison analysis in precisely the same way, however. The Tenth Circuit decision in *Gates Rubber* is particularly notable for its elaboration of the test beyond the parameters of *Altai*. In that case, the court acknowledged that "[a]pplication of the abstractions test will necessarily vary from case-to-case and program-to-program. Given the complexity and ever-changing nature of computer technology, we decline to set forth any strict methodology for the abstraction of computer programs." Nonetheless, the court identified six levels of "generally declining abstraction": (1) the main purpose of the computer program, (2) the structure or architecture of a program, generally as represented in a flowchart, (3) "modules" that comprise particular program operations or types of stored data, (4) individual algorithms or data structures employed in each of the modules, (5) the source code that instructs the computer to carry out each necessary operation on each data structure, and (6) the object code that is actually read by the computer. The court used these levels of abstraction to facilitate its analysis of the program at issue.

The *Gates Rubber* court also gave further content to the filtration part of the *Altai* analysis. The court filtered out six unprotectable elements: ideas, the processes or methods of the computer program,[9] facts, material in the

---

9. Id. at 836-37. The court cited the legislative history accompanying section 102(b) of the Copyright Act, which clearly indicates that the actual processes or methods used in a computer program are not entitled to copyright protection. Id. The court noted that processes were most commonly found in the program architecture, module structure, and algorithms used. Id. at 837.

public domain, expression that has "merged" with an idea or process, and expression that is so standard or common as to be a "necessary incident" to an idea or process.[10] Finally, the court indicated that comparison of the protected elements of a program should be done on a case-by-case basis, with an eye toward determining whether a substantial portion of the protectable expression of the original work has been copied.

Is this analysis consistent with *Altai*? Is a court applying *Gates Rubber* likely to give more or less protection to a computer program than would the *Altai* court? Note that applying *Altai* does not mean that the nonliteral elements of computer programs get no protection at all. For example, in the Fifth Circuit, which has expressly adopted the *Gates Rubber* approach, one district court has concluded that a set of threshold values set by a computer manufacturer to indicate a failure threshold for replacement of a hard drive were copyrightable, since the business and engineering decisions that went into setting those values showed sufficient original expression to be protected. Compaq Computer v. Procom Technology, 908 F. Supp. 1409 (S.D. Tex. 1995). The court held that the merger and scenes à faire doctrines did not bar copyrightability, since the precise values Compaq chose were not necessary for competing manufacturers. See also Harbor Software Inc. v. Applied Systems Inc., 925 F. Supp. 1042 (S.D.N.Y. 1996) (finding nonliteral elements of a computer program protectable under the *Altai* test).

7. Note that while both *Whelan* and *Altai* are nominally about the scope of protection afforded computer programs under the copyright laws, both courts merge the analysis of infringement into the protectability analysis. This is particularly evident in *Altai*'s filtration analysis, where the comparison step involves identifying the similarities between the copyrighted program and the accused program.

PROBLEMS

Problem 7-6. Reassess the MCCA problem (Problem 7-5) based on the *Altai* methodology. Do you reach the same conclusion as before?

Problem 7-7. Demento Corporation ("Demento") produces Demento II, the leading home video game system. The Demento II consists of a console, which runs game cartridges and attaches to a monitor or standard television, and a control device, which enables the user to manipulate characters or other images of particular games on the screen. Demento licenses a limited number of games per year.

To prevent unauthorized game producers from manufacturing

---

10. Id. This is the "scenes à faire" doctrine. The court held that such standard devices in computer programs include "those elements of a program that have been dictated by external factors," such as hardware and software standards, specifications, or compatibility requirements, customer design specifications, and basic industry practices. Id.

games to run on the Demento II console, Demento developed a "lock out" device that governs access to the console. The console includes a "master chip" or "lock" that will only run game cartridges that contain an appropriate "slave chip" or "key." When a user inserts an authorized game cartridge into the Demento II, the slave chip transmits an arbitrary data stream "key" that is received by the master chip and unlocks the console, allowing the game to be played. The unlocking device is a sophisticated software program encoded in the master and slave chips. There are a multitude of different programs capable of generating the data stream that unlocks the console, although it would be almost impossible to decipher the "key" by trial and error. It would be like trying to find a needle in a haystack.

The rapid growth of the home video game market eroded the revenue base of Mutant Corporation ("Mutant"), a leading maker of arcade video games. Mutant sought to license the right to produce game cartridges for the Demento II system, but it was put off by the limits on the number of new games that Demento would license per year and the high cost of each license. Mutant decided instead to develop its own slave chip that would unlock the Demento II game cartridge.

Mutant engineers first chemically peeled the layers of the Demento II chips to allow microsopic examination of the circuitry. Through this means, they were able to decipher the object code. After that, they were able to construct the series of pulsating signals that unlocked the master chip. Mutant discovered that only a relatively small portion of the coded message was necessary to unlock the Demento II console. From this information, Mutant software designers built a slave chip that successfully unlocked the Demento II console and enabled Mutant's game cartridges to run. It included the entire coded message for fear that Demento might later alter new versions of the Demento II system in such a way as to make Mutant's cartridges inoperable on newer units seeking the fuller coded message. To differentiate its product, the Mutant slave program was written in a different computer language and employed a different microprocessor than the Demento II system. Since the Mutant microprocessor operated at a faster speed than the Demento II chip, the Mutant program included numerous pauses.

Despite these differences, Demento promptly sued Mutant for copyright infringement after Mutant introduced its first game cartridge for the Demento II system. What is the unprotectable idea of the Demento II system? What is the protectable expression? Is the lockout code functional? How would this case be resolved in a jurisdiction adhering to the *Whelan* test? the *Altai* test?

c. Protection for Functional Elements and Protocols

Many software copyright cases involve the taking of particular pieces of a plaintiff's program. Defendants in these cases may be those who attempted to "clone" the plaintiff's program, writing their own code to create a functionally equivalent program, see Lotus Dev. Corp. v. Paperback Software, 740 F. Supp. 37 (D. Mass. 1990), or those who simply attempt to reproduce the internal program interface necessary to make two programs compatible. Whether functional program elements could be protected under copyright law was tested in a series of cases involving the Lotus 1-2-3 spreadsheet.

Lotus Development Corp. v. Borland International
United States Court of Appeals for the First Circuit
49 F.3d 807 (1st Cir. 1995), aff'd by equally divided Court,
116 S. Ct. 804 (1996)

STAHL, Circuit Judge.

This appeal requires us to decide whether a computer menu command hierarchy is copyrightable subject matter. In particular, we must decide whether, as the district court held, plaintiff-appellee Lotus Development Corporation's copyright in Lotus 1-2-3, a computer spreadsheet program, was infringed by defendant-appellant Borland International, Inc., when Borland copied the Lotus 1-2-3 menu command hierarchy into its Quattro and Quattro Pro computer spreadsheet programs. See Lotus Dev. Corp. v. Borland Int'l, Inc., 788 F. Supp. 78 (D. Mass. 1992) ("Borland I"); Lotus Dev. Corp. v. Borland Int'l, Inc., 799 F. Supp. 203 (D. Mass. 1992) ("Borland II"); Lotus Dev. Corp. v. Borland Int'l, Inc., 831 F. Supp. 202 (D. Mass. 1993) ("Borland III"); Lotus Dev. Corp. v. Borland Int'l, Inc., 831 F. Supp. 223 (D. Mass. 1993) ("Borland IV").

I. Background

Lotus 1-2-3 is a spreadsheet program that enables users to perform accounting functions electronically on a computer. Users manipulate and control the program via a series of menu commands, such as "Copy," "Print," and "Quit." Users choose commands either by highlighting them on the screen or by typing their first letter. In all, Lotus 1-2-3 has 469 commands arranged into more than 50 menus and submenus. [See Figure 7-1.]

Lotus 1-2-3, like many computer programs, allows users to write what are called "macros." By writing a macro, a user can designate a series of

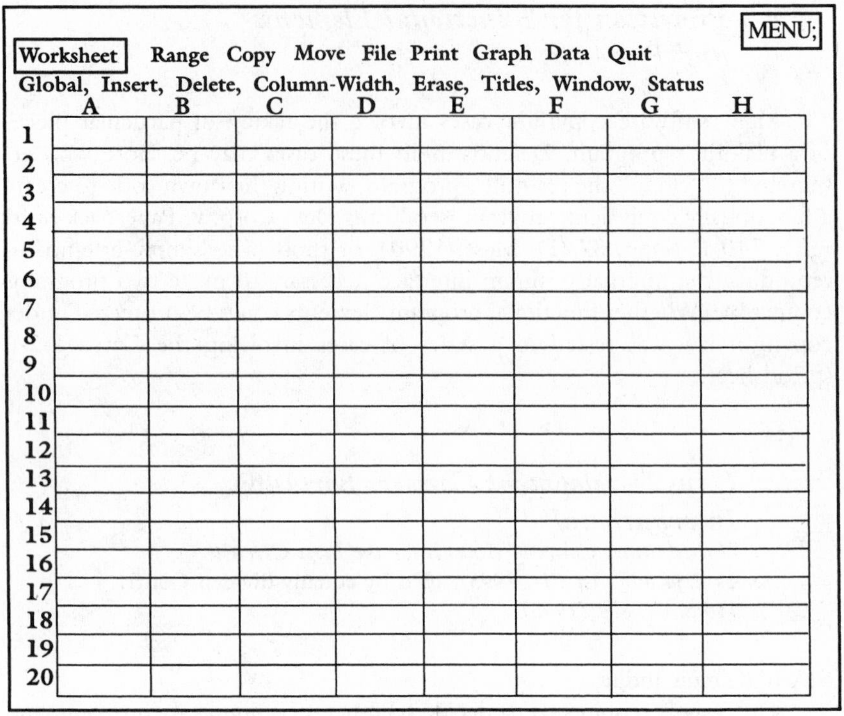

| Worksheet | Range | Copy | Move | File | Print | Graph | Data | Quit | MENU; |

Global, Insert, Delete, Column-Width, Erase, Titles, Window, Status

	A	B	C	D	E	F	G	H
1								
2								
3								
4								
5								
6								
7								
8								
9								
10								
11								
12								
13								
14								
15								
16								
17								
18								
19								
20								

FIGURE 7-1
Facsimile of a Lotus 1-2-3 screen display.

command choices with a single macro keystroke. Then, to execute that series of commands in multiple parts of the spreadsheet, rather than typing the whole series each time, the user only needs to type the single pre-programmed macro keystroke, causing the program to recall and perform the designated series of commands automatically. Thus, Lotus 1-2-3 macros shorten the time needed to set up and operate the program.

Borland released its first Quattro program to the public in 1987, after Borland's engineers had labored over its development for nearly three years. Borland's objective was to develop a spreadsheet program far superior to existing programs, including Lotus 1-2-3. In Borland's words, "from the time of its initial release . . . Quattro included enormous innovations over competing spreadsheet products."

The district court found, and Borland does not now contest, that Borland included in its Quattro and Quattro Pro version 1.0 programs "a virtually identical copy of the entire 1-2-3 menu tree." Borland III, 831 F. Supp. at 212. In so doing, Borland did not copy any of Lotus's underlying computer code; it copied only the words and structure of Lotus's menu command hierarchy. Borland included the Lotus menu command hierarchy in its programs to make them compatible with Lotus 1-2-3 so that spreadsheet users who were already familiar with Lotus 1-2-3 would be able to switch to the Borland programs without having to learn new commands or rewrite their Lotus macros.

In its Quattro and Quattro Pro version 1.0 programs, Borland achieved compatibility with Lotus 1-2-3 by offering its users an alternate user interface, the "Lotus Emulation Interface." By activating the Emulation Interface, Borland users would see the Lotus menu commands on their screens and could interact with Quattro or Quattro Pro as if using Lotus 1-2-3, albeit with a slightly different looking screen and with many Borland options not available on Lotus 1-2-3. In effect, Borland allowed users to choose how they wanted to communicate with Borland's spreadsheet programs: either by using menu commands designed by Borland, or by using the commands and command structure used in Lotus 1-2-3 augmented by Borland-added commands.

Lotus filed this action against Borland in the District of Massachusetts on July 2, 1990, four days after a district court held that the Lotus 1-2-3 "menu structure, taken as a whole — including the choice of command terms [and] the structure and order of those terms," was protected expression covered by Lotus's copyrights. Lotus Dev. Corp. v. Paperback Software Int'l, 740 F. Supp. 37, 68, 70 (D. Mass. 1990) ("Paperback").[1] . . .

On July 31, 1992, the district court denied Borland's motion [for summary judgment] and granted Lotus's motion in part. The district court ruled that the Lotus menu command hierarchy was copyrightable expression because

> [a] very satisfactory spreadsheet menu tree can be constructed using different commands and a different command structure from those of Lotus 1-2-3. In fact, Borland has constructed just such an alternate tree for use in Quattro Pro's native mode. Even if one holds the arrangement of menu commands constant, it is possible to generate literally millions of satisfactory menu trees by varying the menu commands employed.

Borland II, 799 F. Supp. at 217. The district court demonstrated this by offering alternate command words for the ten commands that appear in Lotus's main menu. Id. For example, the district court stated that "the 'Quit' command could be named 'Exit' without any other modifications," and that "the 'Copy' command could be called 'Clone,' 'Ditto,' 'Duplicate,' 'Imitate,' 'Mimic,' 'Replicate,' and 'Reproduce,' among others." Id. Because so many variations were possible, the district court concluded that the Lotus developers' choice and arrangement of command terms, reflected in the Lotus menu command hierarchy, constituted copyrightable expression.

In granting partial summary judgment to Lotus, the district court held that Borland had infringed Lotus's copyright in Lotus 1-2-3 . . . Borland II,

---

1. Judge Keeton presided over both the *Paperback* litigation and this case.

799 F. Supp. at 223. The court nevertheless concluded that while the Quattro and Quattro Pro programs infringed Lotus's copyright, Borland had not copied the entire Lotus 1-2-3 user interface, as Lotus had contended. . . .

Immediately following the district court's summary judgment decision, Borland removed the Lotus Emulation Interface from its products. Thereafter, Borland's spreadsheet programs no longer displayed the Lotus 1-2-3 menus to Borland users, and as a result Borland users could no longer communicate with Borland's programs as if they were using a more sophisticated version of Lotus 1-2-3. Nonetheless, Borland's programs continued to be partially compatible with Lotus 1-2-3, for Borland retained what it called the "Key Reader" in its Quattro Pro programs. Once turned on, the Key Reader allowed Borland's programs to understand and perform some Lotus 1-2-3 macros. With the Key Reader on, the Borland programs used Quattro Pro menus for display, interaction, and macro execution, except when they encountered a slash ("/") key in a macro (the starting key for any Lotus 1-2-3 macro), in which case they interpreted the macro as having been written for Lotus 1-2-3. Accordingly, people who wrote or purchased macros to shorten the time needed to perform an operation in Lotus 1-2-3 could still use those macros in Borland's programs. The district court permitted Lotus to file a supplemental complaint alleging that the Key Reader infringed its copyright.

The parties agreed to try the remaining liability issues without a jury. The district court held two trials, the Phase I trial covering all remaining issues raised in the original complaint (relating to the Emulation Interface) and the Phase II trial covering all issues raised in the supplemental complaint (relating to the Key Reader). . . .

In its Phase I trial decision, the district court found that "each of the Borland emulation interfaces contains a virtually identical copy of the 1-2-3 menu tree and that the 1-2-3 menu tree is capable of a wide variety of expression." Borland III, 831 F. Supp. at 218. The district court also rejected Borland's affirmative defenses of laches and estoppel. Id. at 218-23.

In its Phase II trial decision, the district court found that Borland's Key Reader file included "a virtually identical copy of the Lotus menu tree structure, but represented in a different form and with first letters of menu command names in place of the full menu command names." Borland IV, 831 F. Supp. at 228. In other words, Borland's programs no longer included the Lotus command terms, but only their first letters. The district court held that "the Lotus menu structure, organization, and first letters of the command names . . . constitute part of the protectable expression found in [Lotus 1-2-3]." Id. at 233. Accordingly, the district court held that with its Key Reader, Borland had infringed Lotus's copyright. Id. at 245. The district court also rejected Borland's affirmative defenses of waiver, laches, estoppel, and fair use. Id. at 235-45. The district court then entered a permanent injunction against Borland, id. at 245, from which Borland appeals. . . .

II. Discussion

On appeal, Borland does not dispute that it factually copied the words and arrangement of the Lotus menu command hierarchy. Rather, Borland argues that it "lawfully copied the unprotectable menus of Lotus 1-2-3." Borland contends that the Lotus menu command hierarchy is not copyrightable because it is a system, method of operation, process, or procedure foreclosed from protection by 17 U.S.C. §102(b). Borland also raises a number of affirmative defenses.

A. Copyright Infringement Generally . . .

In this appeal, we are faced only with whether the Lotus menu command hierarchy is copyrightable subject matter in the first instance, for Borland concedes that Lotus has a valid copyright in Lotus 1-2-3 as a whole and admits to factually copying the Lotus menu command hierarchy. As a result, this appeal is in a very different posture from most copyright-infringement cases, for copyright infringement generally turns on whether the defendant has copied protected expression as a factual matter. Because of this different posture, most copyright-infringement cases provide only limited help to us in deciding this appeal. This is true even with respect to those copyright-infringement cases that deal with computers and computer software.

B. Matter of First Impression

Whether a computer menu command hierarchy constitutes copyrightable subject matter is a matter of first impression in this court. While some other courts appear to have touched on it briefly in dicta, see, e.g., Autoskill, Inc. v. National Educ. Support Sys., Inc., 994 F.2d 1476, 1495 n.23 (10th Cir.), *cert. denied*, 126 L. Ed. 2d 254, 114 S. Ct. 307 (1993), we know of no cases that deal with the copyrightability of a menu command hierarchy standing on its own (i.e., without other elements of the user interface, such as screen displays, in issue). Thus we are navigating in uncharted waters.

Borland vigorously argues, however, that the Supreme Court charted our course more than 100 years ago when it decided Baker v. Selden, 101 U.S. 99, 25 L. Ed. 841 (1879). In Baker v. Selden, the Court held that Selden's copyright over the textbook in which he explained his new way to do accounting did not grant him a monopoly on the use of his accounting system. Borland argues: "The facts of Baker v. Selden, and even the arguments advanced by the parties in that case, are identical to those in this case. The only difference is that the 'user interface' of Selden's system was implemented by pen and paper rather than by computer." . . .

We do not think that Baker v. Selden is nearly as analogous to this appeal as Borland claims. Of course, Lotus 1-2-3 is a computer spreadsheet, and as such its grid of horizontal rows and vertical columns certainly resembles an accounting ledger or any other paper spreadsheet. Those grids, however, are not at issue in this appeal for, unlike Selden, Lotus does not claim to have a monopoly over its accounting system. Rather, this appeal involves Lotus's monopoly over the commands it uses to operate the computer. Accordingly, this appeal is not, as Borland contends, "identical" to Baker v. Selden.

C. *Altai*

Before we analyze whether the Lotus menu command hierarchy is a system, method of operation, process, or procedure, we first consider the applicability of the test the Second Circuit set forth in Computer Assoc. Int'l, Inc. v. Altai, Inc., 982 F.2d 693 (2d Cir. 1992). The Second Circuit designed its Altai test to deal with the fact that computer programs, copyrighted as "literary works," can be infringed by what is known as "nonliteral" copying, which is copying that is paraphrased or loosely paraphrased rather than word for word. See id. at 701 (citing nonliteral-copying cases); see also 3 Melville B. Nimmer & David Nimmer, Nimmer on Copyright §13.03[A][1] (1993). When faced with nonliteral-copying cases, courts must determine whether similarities are due merely to the fact that the two works share the same underlying idea or whether they instead indicate that the second author copied the first author's expression. The Second Circuit designed its Altai test to deal with this situation in the computer context, specifically with whether one computer program copied nonliteral expression from another program's code. . . .

In the instant appeal, we are not confronted with alleged nonliteral copying of computer code. Rather, we are faced with Borland's deliberate, literal copying of the Lotus menu command hierarchy. Thus, we must determine not whether nonliteral copying occurred in some amorphous sense, but rather whether the literal copying of the Lotus menu command hierarchy constitutes copyright infringement.

While the Altai test may provide a useful framework for assessing the alleged nonliteral copying of computer code, we find it to be of little help in assessing whether the literal copying of a menu command hierarchy constitutes copyright infringement. In fact, we think that the Altai test in this context may actually be misleading because, in instructing courts to abstract the various levels, it seems to encourage them to find a base level that includes copyrightable subject matter that, if literally copied, would make the copier liable for copyright infringement.[8] While that base (or literal) level would not

---

8. We recognize that Altai never states that every work contains a copyrightable "nugget" of protectable expression. Nonetheless, the implication is that for literal copying, "it is not necessary to determine the level of abstraction at which similarity ceases to consist of an 'expression of ideas,' because literal similarity by definition is always a similarity as to the expression of ideas." 3 Melville B. Nimmer & David Nimmer, Nimmer on Copyright §13.03[A](2) (1993).

be at issue in a nonliteral-copying case like Altai, it is precisely what is at issue in this appeal. We think that abstracting menu command hierarchies down to their individual word and menu levels and then filtering idea from expression at that stage, as both the Altai and the district court tests require, obscures the more fundamental question of whether a menu command hierarchy can be copyrighted at all. The initial inquiry should not be whether individual components of a menu command hierarchy are expressive, but rather whether the menu command hierarchy as a whole can be copyrighted. But see Gates Rubber Co. v. Bando Chem. Indus., Ltd., 9 F.3d 823 (10th Cir. 1993) (endorsing Altai's abstraction-filtration-comparison test as a way of determining whether "menus and sorting criteria" are copyrightable).

D. *The Lotus Menu Command Hierarchy: A "Method of Operation"*

Borland argues that the Lotus menu command hierarchy is uncopyrightable because it is a system, method of operation, process, or procedure foreclosed from copyright protection by 17 U.S.C. §102(b). Section 102(b) states: "In no case does copyright protection for an original work of authorship extend to any idea, procedure, process, system, method of operation, concept, principle, or discovery, regardless of the form in which it is described, explained, illustrated, or embodied in such work." Because we conclude that the Lotus menu command hierarchy is a method of operation, we do not consider whether it could also be a system, process, or procedure.

We think that "method of operation," as that term is used in §102(b), refers to the means by which a person operates something, whether it be a car, a food processor, or a computer. Thus a text describing how to operate something would not extend copyright protection to the method of operation itself; other people would be free to employ that method and to describe it in their own words. Similarly, if a new method of operation is used rather than described, other people would still be free to employ or describe that method.

We hold that the Lotus menu command hierarchy is an uncopyrightable "method of operation." The Lotus menu command hierarchy provides the means by which users control and operate Lotus 1-2-3. If users wish to copy material, for example, they use the "Copy" command. If users wish to print material, they use the "Print" command. Users must use the command terms to tell the computer what to do. Without the menu command hierarchy, users would not be able to access and control, or indeed make use of, Lotus 1-2-3's functional capabilities.

The Lotus menu command hierarchy does not merely explain and present Lotus 1-2-3's functional capabilities to the user; it also serves as the method by which the program is operated and controlled. The Lotus menu command hierarchy is different from the Lotus long prompts, for the long prompts are not necessary to the operation of the program; users could operate Lotus 1-2-3 even if there were no long prompts. The Lotus menu command hierarchy is also different from the Lotus screen displays, for users need not "use" any expressive aspects of the screen displays in order to operate Lotus 1-2-3; because the way the screens look has little bearing on how users control the program, the screen displays are not part of Lotus 1-2-3's "method of operation." The Lotus menu command hierarchy is also different from the underlying computer code, because while code is necessary for the program to work, its precise formulation is not. In other words, to offer the same capabilities as Lotus 1-2-3, Borland did not have to copy Lotus's underlying code (and indeed it did not); to allow users to operate its programs in substantially the same way, however, Borland had to copy the Lotus menu command hierarchy. Thus the Lotus 1-2-3 code is not an uncopyrightable "method of operation."

The district court held that the Lotus menu command hierarchy, with its specific choice and arrangement of command terms, constituted an "expression" of the "idea" of operating a computer program with commands arranged hierarchically into menus and submenus. Borland II, 799 F. Supp. at 216. Under the district court's reasoning, Lotus's decision to employ hierarchically arranged command terms to operate its program could not foreclose its competitors from also employing hierarchically arranged command terms to operate their programs, but it did foreclose them from employing the specific command terms and arrangement that Lotus had used. In effect, the district court limited Lotus 1-2-3's "method of operation" to an abstraction.

Accepting the district court's finding that the Lotus developers made some expressive choices in choosing and arranging the Lotus command terms, we nonetheless hold that that expression is not copyrightable because it is part of Lotus 1-2-3's "method of operation." We do not think that "methods of operation" are limited to abstractions; rather, they are the means by which a user operates something. If specific words are essential to operating something, then they are part of a "method of operation" and, as such, are unprotectable. This is so whether they must be highlighted, typed in, or even spoken, as computer programs no doubt will soon be controlled by spoken words.

The fact that Lotus developers could have designed the Lotus menu command hierarchy differently is immaterial to the question of whether it is a "method of operation." In other words, our initial inquiry is not whether the Lotus menu command hierarchy incorporates any expression. Rather, our initial inquiry is whether the Lotus menu command hierarchy is a "method of operation." Concluding, as we do, that users operate Lotus 1-2-3 by using the Lotus menu command hierarchy, and that the entire Lotus menu command hierarchy is essential to operating Lotus 1-2-3, we do not inquire further whether that method of operation could have been designed differently. The "expressive" choices of what to name the command terms and how to arrange them do not magically change the uncopyrightable menu command hierarchy into copyrightable subject matter.

Our holding that "methods of operation" are not limited to mere abstractions is bolstered by Baker v. Selden. In Baker, the Supreme Court explained that

> the teachings of science and the rules and methods of useful art have their final end in application and use; and this application and use are what the public derive from the publication of a book which teaches them. . . . The description of the art in a book, though entitled to the benefit of copyright, lays no foundation for an exclusive claim to the art itself. The object of the one is explanation; the object of the other is use. The former may be secured by copyright. The latter can only be secured, if it can be secured at all, by letters-patent.

Baker v. Selden, 101 U.S. at 104-05. Lotus wrote its menu command hierarchy so that people could learn it and use it. Accordingly, it falls squarely within the prohibition on copyright protection established in Baker v. Selden and codified by Congress in §102(b).

In many ways, the Lotus menu command hierarchy is like the buttons used to control, say, a video cassette recorder ("VCR"). A VCR is a machine that enables one to watch and record video tapes. Users operate VCRs by pressing a series of buttons that are typically labelled "Record, Play, Reverse, Fast Forward, Pause, Stop/Eject." That the buttons are arranged and labeled does not make them a "literary work," nor does it make them an "expression" of the abstract "method of operating" a VCR via a set of labeled buttons. Instead, the buttons are themselves the "method of operating" the VCR.

When a Lotus 1-2-3 user chooses a command, either by highlighting it on the screen or by typing its first letter, he or she effectively pushes a button. Highlighting the "Print" command on the screen, or typing the letter "P," is analogous to pressing a VCR button labeled "Play."

Just as one could not operate a buttonless VCR, it would be impossible to operate Lotus 1-2-3 without employing its menu command hierarchy. Thus the Lotus command terms are not equivalent to the labels on the VCR's buttons, but are instead equivalent to the buttons themselves. Unlike the labels on a VCR's buttons, which merely make operating a VCR easier by indicating the buttons' functions, the Lotus menu commands are essential to operating Lotus 1-2-3. Without the menu commands, there would be no way to "push" the Lotus buttons, as one could push unlabeled VCR buttons. While Lotus could probably have designed a user interface for which the command terms were mere labels, it did not do so here. Lotus 1-2-3 depends for its operation on use of the precise command terms that make up the Lotus menu command hierarchy. . . .

Computer programs, unlike VCRs, are copyrightable as "literary works." 17 U.S.C. §102(a). Accordingly, one might argue, the "buttons" used to operate a computer program are not like the buttons used to operate a VCR, for they are not subject to a useful-article exception. The response, of course, is that the arrangement of buttons on a VCR would not be copyrightable even without a useful-article exception, because the buttons are an uncopyrightable "method of operation." Similarly, the "buttons" of a computer program are also an uncopyrightable "method of operation."

That the Lotus menu command hierarchy is a "method of operation" becomes clearer when one considers program compatibility. Under Lotus's theory, if a user uses several different programs, he or she must learn how to perform the same operation in a different way for each program used. For example, if the user wanted the computer to print material, then the user would have to learn not just one method of operating the computer such that it prints, but many different methods. We find this absurd. The fact that there may be many different ways to operate a computer program, or even many different ways to operate a computer program using a set of hierarchically arranged command terms, does not make the actual method of operation chosen copyrightable; it still functions as a method for operating the computer and as such is uncopyrightable.

Consider also that users employ the Lotus menu command hierarchy in writing macros. Under the district court's holding, if the user wrote a macro to shorten the time needed to perform a certain operation in Lotus 1-2-3, the user would be unable to use that macro to shorten the time needed to perform that same operation in another program. Rather, the user would have to rewrite his or her macro using that other program's menu command hierarchy. This is despite the fact that the macro is clearly the user's own work product. We think that forcing the user to cause the computer to perform the same operation in a different way ignores Congress's direction in §102(b) that "methods of operation" are not copyrightable. That programs can offer users the ability to write macros in many different ways does not change the fact that, once written, the macro allows the user to perform an operation automatically. As the Lotus menu command hierarchy serves as the basis for Lotus 1-2-3 macros, the Lotus menu command hierarchy is a "method of operation." . . .

We also note that in most contexts, there is no need to "build" upon other people's expression, for the ideas conveyed by that expression can be conveyed by someone else without copying the first author's expression. In the context of methods of operation, however, "building" requires the use of the precise method of operation already employed; otherwise, "building" would require dismantling, too. Original developers are not the only people entitled to build on the methods of operation they create; anyone can. Thus, Borland may build on the method of operation that Lotus designed and may use the Lotus menu command hierarchy in doing so.

Our holding that methods of operation are not limited to abstractions goes against Autoskill, 994 F.2d at 1495 n.23, in which the Tenth Circuit rejected the defendant's argument that the keying procedure used in a computer program was an uncopyrightable "procedure" or "method of operation" under §102(b). The program at issue, which was designed to test and train students with reading deficiencies, id. at 1481, required students to select responses to the program's queries "by pressing the 1, 2, or 3 keys." Id. at 1495 n.23. The Tenth Circuit held that, "for purposes of the preliminary injunction, . . . the record showed that [this] keying procedure reflected at least a minimal degree of creativity," as required by Feist for copyright protection. Id. As an initial matter, we question whether a programmer's decision to have users select a response by pressing the 1, 2, or 3 keys is original. More importantly, however, we fail to see how "a student selecting a response by pressing the 1, 2, or 3 keys," id., can be anything but an unprotectable method of operation.[14] . . .

Reversed.

BOUDIN, Circuit Judge, concurring.

The importance of this case, and a slightly different emphasis in my view of the underlying problem, prompt me to add a few words to the majority's tightly focused discussion.

I

Most of the law of copyright and the "tools" of analysis have developed in the context of literary works such as novels, plays, and films. In this milieu, the principal problem — simply stated, if difficult to resolve — is to stimulate creative expression without unduly limiting access by others to the broader themes and concepts deployed by the author. The middle of the spectrum presents close cases; but a "mistake" in providing too much protection involves a small cost: subsequent authors treating the same themes must take a few more steps away from the original expression.

The problem presented by computer programs is fundamentally different in one respect. The computer program is a means for causing something to happen; it has a mechanical utility, an instrumental role, in accomplishing the world's work. Granting protection, in other words, can have some of the consequences of patent protection in limiting other people's ability to perform a task in the most efficient manner. Utility does not bar copyright (dictionaries may be copyrighted), but it alters the calculus.

Of course, the argument for protection is undiminished, perhaps even enhanced, by utility: if we want more of an intellectual product, a temporary monopoly for the creator provides incentives for others to create other, different items in this class. But the "cost" side of the equation may be different where one places a very high value on public access to a useful innovation

---

14. The Ninth Circuit has also indicated in dicta that "menus, and keystrokes" may be copyrightable. Brown Bag Software v. Symantec Corp., 960 F.2d 1465, 1477 (9th Cir.), *cert. denied*, BB Asset Management, Inc. v. Symantec Corp., 121 L. Ed. 2d 141, 113 S. Ct. 198 (1992). In that case, however, the plaintiff did not show that the defendant had copied the plaintiff's menus or keystrokes, so the court was not directly faced with whether the menus or keystrokes constituted an unprotectable method of operation. Id.

that may be the most efficient means of performing a given task. Thus, the argument for extending protection may be the same; but the stakes on the other side are much higher.

It is no accident that patent protection has preconditions that copyright protection does not — notably, the requirements of novelty and non-obviousness — and that patents are granted for a shorter period than copyrights. This problem of utility has sometimes manifested itself in copyright cases, such as Baker v. Selden, 101 U.S. 99, 25 L. Ed. 841 (1879), and been dealt with through various formulations that limit copyright or create limited rights to copy. But the case law and doctrine addressed to utility in copyright have been brief detours in the general march of copyright law.

Requests for the protection of computer menus present the concern with fencing off access to the commons in an acute form. A new menu may be a creative work, but over time its importance may come to reside more in the investment that has been made by users in learning the menu and in building their own mini-programs — macros — in reliance upon the menu. Better typewriter keyboard layouts may exist, but the familiar QWERTY keyboard dominates the market because that is what everyone has learned to use. See P. David, CLIO and the Economics of QWERTY, 75 Am. Econ. Rev. 332 (1985). The QWERTY keyboard is nothing other than a menu of letters.

Thus, to assume that computer programs are just one more new means of expression, like a filmed play, may be quite wrong. The "form" — the written source code or the menu structure depicted on the screen — look hauntingly like the familiar stuff of copyright; but the "substance" probably has more to do with problems presented in patent law or, as already noted, in those rare cases where copyright law has confronted industrially useful expressions. Applying copyright law to computer programs is like assembling a jigsaw puzzle whose pieces do not quite fit. . . .

II

In this case, the raw facts are mostly, if not entirely, undisputed. Although the inferences to be drawn may be more debatable, it is very hard to see that Borland has shown any interest in the Lotus menu except as a fall-back option for those users already committed to it by prior experience or in order to run their own macros using 1-2-3 commands. At least for the amateur, accessing the Lotus menu in the Borland Quattro or Quattro Pro program takes some effort.

Put differently, it is unlikely that users who value the Lotus menu for its own sake — independent of any investment they have made themselves in learning Lotus' commands or creating macros dependent upon them — would choose the Borland program in order to secure access to the Lotus menu. Borland's success is due primarily to other features. Its rationale for deploying the Lotus menu bears the ring of truth.

Now, any use of the Lotus menu by Borland is a commercial use and deprives Lotus of a portion of its "reward," in the sense that an infringement claim if allowed would increase Lotus' profits. But this is circular reasoning: broadly speaking, every limitation on copyright or privileged use diminishes the reward of the original creator. Yet not every writing is copyrightable or every use an infringement. The provision of reward is one concern of copyright law, but it is not the only one. If it were, copyrights would be perpetual and there would be no exceptions.

The present case is an unattractive one for copyright protection of the menu. The menu commands (e.g., "print," "quit") are largely for standard procedures that Lotus did not invent and are common words that Lotus cannot monopolize. What is left is the particular combination and sub-grouping of commands in a pattern devised by Lotus. This arrangement may have a more appealing logic and ease of use than some other configurations; but there is a certain arbitrariness to many of the choices.

If Lotus is granted a monopoly on this pattern, users who have learned the command structure of Lotus 1-2-3 or devised their own macros are locked into Lotus, just as a typist who has learned the QWERTY keyboard would be the captive of anyone who had a monopoly on the production of such a keyboard. Apparently, for a period Lotus 1-2-3 has had such sway in the market that it has represented the de facto standard for electronic spreadsheet commands. So long as Lotus is the superior spreadsheet — either in quality or in price — there may be nothing wrong with this advantage.

But if a better spreadsheet comes along, it is hard to see why customers who have learned the Lotus menu and devised macros for it should remain captives of Lotus because of an investment in learning made by the users and not by Lotus. Lotus has already reaped a substantial reward for being first; assuming that the Borland program is now better, good reasons exist for freeing it to attract old Lotus customers: to enable the old customers to take advantage of a new advance, and to reward Borland in turn for making a better product. If Borland has not made a better product, then customers will remain with Lotus anyway.

Thus, for me the question is not whether Borland should prevail but on what basis. Various avenues might be traveled, but the main choices are between holding that the menu is not protectable by copyright and devising a new doctrine that Borland's use is privileged. No solution is perfect and no intermediate appellate court can make the final choice.

To call the menu a "method of operation" is, in the common use of those words, a defensible position. After all, the purpose of the menu is not to be admired as a work of literary or pictorial art. It is to transmit directions from the user to the computer, i.e., to operate the computer. The menu is also a "method" in the dictionary sense because it is a "planned way of doing something," an "order or system," and (aptly here) an "orderly or systematic arrangement, sequence or the like." Random House Webster's College Dictionary 853 (1991).

A different approach would be to say that Borland's use is privileged because, in the context already described, it is not seeking to appropriate the advances made by Lotus' menu; rather, having provided an arguably more attractive menu of its own, Borland is merely trying to give former Lotus users an option to exploit their own prior investment in learning or in macros. The difference is that such a privileged use approach would not automatically protect Borland if it had simply copied the Lotus menu (using different codes), contributed nothing of its own, and resold Lotus under the Borland label.

The closest analogue in conventional copyright is the fair use doctrine. E.g., Harper & Row, Publishers, Inc. v. Nation Enters., 471 U.S. 539, 85 L. Ed. 2d 588, 105 S. Ct. 2218 (1985). Although invoked by Borland, it has largely been brushed aside in this case because the Supreme Court has said that it is "presumptively" unavailable where the use is a "commercial" one. See id. at 562. In my view, this is something less than a definitive answer; "presumptively" does not mean "always" and, in any event, the doctrine of fair use was created by the courts and can be adapted to new purposes.

But a privileged use doctrine would certainly involve problems of its own. It might more closely tailor the limits on copyright protection to the reasons for limiting that protection; but it would entail a host of administrative problems that would cause cost and delay, and would also reduce the ability of the industry to predict outcomes. Indeed, to the extent that Lotus' menu is an important standard in the industry, it might be argued that any use ought to be deemed privileged.

In sum, the majority's result persuades me and its formulation is as good, if not better, than any other that occurs to me now as within the reach of courts. Some solutions (e.g., a very short copyright period for menus) are not options at all for courts but might be for Congress. In all events, the choices are important ones of policy, not linguistics, and they should be made with the underlying considerations in view.

COMMENTS AND QUESTIONS

1. The issues raised by the Lotus decision remain unsettled. The Supreme Court granted certiorari in the case, but it deadlocked 4-4 and therefore affirmed the First Circuit, but it produced no precedential opinion. Lotus Dev. Corp. v. Borland Intl., 116 S. Ct. 804 (1996). Note that the First Circuit concedes that its opinion is at odds with the Tenth Circuit's decision in Autoskill, Inc. v. National Educational Support Systems, 994 F.2d 1476, 1495 n.23 (10th Cir. 1993), *cert. denied*, 510 U.S. 916 (1993), and dictum from the Ninth Circuit's decision in Brown Bag Software v. Symantec Corp., 960 F.2d 1465, 1477 (9th Cir. 1992). It is also arguably inconsistent with the Fifth Circuit's decision in Engineering Dynamics, Inc. v. Structural

Software Inc., 26 F.3d 1335 (5th Cir. 1994), which gave broad copyright protection to program interfaces and relied extensively on Judge Keeton's prior decision in Lotus v. Paperback.

On the other hand, the *Lotus* approach seems consistent with the legislative history of 17 U.S.C. §102(b). The House Report accompanying that section states: "Section 102(b) is intended, among other things, to make clear that the expression adopted by the programmer is the copyrightable element in a computer program, and that the actual processes or methods embodied in the program are not within the scope of the copyright law." H.R. Rep. No. 1476, 94th Cong., 2d Sess. 57 (1976), reprinted in 1976 U.S.C.C.A.N. 5659, 5670.

Further, most commentators seem to agree with the ultimate result in *Lotus*, though many are troubled by the court's reasoning. See, e.g., Brief Amicus Curiae of Copyright Law Professors in Support of Respondent in Lotus v. Borland, No. 94-2003 (U.S. 1995) (brief submitted by 34 copyright professors urging affirmance for a variety of reasons); Mark A. Lemley, Convergence in the Law of Software Copyright?, 10 High Tech. L.J. 1 (1995); Glynn S. Lunney Jr., Lotus v. Borland: Copyright and Computer Programs, 70 Tulane L. Rev. 2397 (1996); Peter S. Menell, The Challenges of Reforming Intellectual Property Protection for Computer Software, 94 Colum. L. Rev. 2644, 2653 (1994). Finally, at least one other appellate court has followed the reasoning of *Lotus* in denying protection to computer systems as methods of operation. See MiTek Holdings v. Arce Engineering, 89 F.3d 1548 (11th Cir. 1996) (refusing to protect menu structures under section 102(b) because allowing such protection would infringe on the province of patent law). But see Mitel v. Iqtel, 124 F.3d 1366 (10th Cir. 1997), where the court explicitly rejected the *Lotus* approach, though it did find the communications protocols at issue there unprotectable under *Altai*.

2. Contrast the analysis of section 102(b) and Baker v. Selden in the *Lotus* case with that in Apple v. Franklin. Applying the *Lotus* court's logic, isn't object code a "method of operation" for instructing a CPU how to process information? If so, how can the copyrightability of computer programs be reconciled with the exclusion from copyright of "methods of operation"?

3. Note that the *Lotus* court chose to reject the filtration test adopted by the Second Circuit in Computer Associates v. Altai, and followed by most other circuits. Is the holding in *Lotus* really incompatible with *Altai*'s abstraction-filtration-comparison test? One possible way to reconcile *Lotus* and *Altai* is to treat *Lotus* not as a case involving "literal" copying of a menu structure, but as a case involving nonliteral copying of the program itself. Lotus distributed object code which, in certain combinations, produces on a screen physical images that represent certain words and have certain effects. Borland did not copy that code; rather, it wrote its own code, which produced similar images and had the same effects as the Lotus code. The court could then have applied *Altai*'s abstraction-filtration-comparison analysis to this "non-

literal" copying. This is in effect precisely what the *Lotus* court did — although when it reached the "filtration" step, it determined that the entire program was unprotectable at the menu command hierarchy level. See Mark A. Lemley, Convergence in the Law of Software Copyright?, 10 High Tech. L.J. 1 (1995).

Do the *Altai* and *Lotus* cases really present analogous issues? Is a "filtration" analysis that filters out the entire copyrighted work at one level of abstraction within the contemplation of the *Altai* court?

4. The *Lotus* court compares the 1-2-3 menu command hierarchy to the buttons on a VCR, which it says are uncopyrightable because they are methods of operation. But is this the right analogy? Lotus claimed that Borland had infringed its copyright by copying the structure, sequence, and organization of its menu hierarchy, not the words of the menus themselves. This "structure, sequence and organization" seems the equivalent of the location and arrangement of the VCR buttons, rather than the buttons themselves. Would an original way of arranging the buttons on a VCR be copyrightable? Should it be? Is the question one of originality and the idea-expression dichotomy, or is there some reason not to copyright the arrangement of buttons on a VCR even if it were original and expressive?

5. In the predecessor case to Lotus v. Borland, defendant Paperback Software argued that the command structure of the Lotus program is in essence a language. It can be used to create macros and run the spreadsheet. Is this a fair characterization? The district court attacked this analogy (which it called "strained") on two basic grounds: that Lotus 1-2-3 was not in fact a language, and that languages could be copyrightable. The first conclusion is certainly debatable; one's definition of a language (or alternatively a "system") will go a long way toward answering this question. But consider the court's second premise — that languages might be copyrightable. Is copyrighting "natural" (i.e., human) languages consistent with what you know of copyright law? Is a language functional, or does it express? Does it matter whether the plaintiff is claiming the creation of certain words, on the one hand, or the entire system of grammar and interaction between words? On this point, compare Ronald Johnson & Allen Grogan, Copyright Protection for Command Driven Interfaces, 8:6 Computer Law. 1 (June 1991) (advocating copyright protection for languages) with Pamela Samuelson, How to Interpret the Lotus Decision; And How Not To, 33:11 Communications of the ACM 27 (Nov. 1990) and Elizabeth G. Lowry, Comment, Copyright Protection for Computer Languages: Creative Incentive or Technological Threat, 39 Emory L.J. 1293 (1990).

Does the First Circuit's decision in Lotus v. Borland shed any light on this issue? Are languages "systems" or "methods of operation" which are denied protection under section 102(b)? The Australian case of Data Access Corp. v. Powerflex Services Pty., No. VG473 (Fed. Ct. Aus., Melbourne 1996), may prove instructive. That court held that even a single word serving a menu command function may be copyrightable, since it is a concatenation of letters. Copyright protection therefore clearly extended to the computer language at issue in that case, which consisted of 225 words. Is this result consistent with the goals of copyright?

6. Both the majority and Judge Boudin's concurrence express concern over the harm that would be done to users of Lotus 1-2-3 if they could not copy their macros for use on Quattro Pro. Certainly, users have invested time and effort in using Lotus 1-2-3, and this may make them reluctant to change spreadsheet programs, even if Borland's program really is superior. But is this the sort of problem with which the copyright law should be concerned? Why isn't a customer preference for a known, trusted product part of the reward that the copyright owner is entitled to reap?

Is the QWERTY keyboard protectable by copyright? Why or why not? Does the answer depend on timing — whether the QWERTY keyboard has just been introduced, or whether it is well established as the industry standard for typewriters? How would you apply section 102(b) in each case?

7. In his concurrence, Judge Boudin proposes (though he does not necessarily adopt) an alternative rationale for the decision: that Lotus's menu command hierarchy was copyrightable, but that Borland's use of the menu hierarchy in this case was privileged because it was done for the purpose of making the two programs compatible. The "privileged use" rationale is narrower than the majority's "method of operation" rule. Because the majority holds the menu command hierarchy entirely uncopyrightable, Lotus has no power to prevent anyone from copying its menu structure, regardless of the circumstances in which the copying occurs.

Suppose that, without taking any protected source or object code, a small company managed to acquire an early copy of Lotus 1-2-3. It replicated the menu command hierarchy and began to sell "clone" programs, which looked and worked just like Lotus 1-2-3. Cf. Lotus Dev. Corp. v. Paperback Software, 740 F. Supp. 37 (D. Mass. 1990), a prior decision by Judge Keeton which is apparently overruled by the First Circuit's decision. In such a case, the equities might seem to tilt towards Lotus more than they did in the *Borland* case. Nonetheless, under the majority's rule, the clone manufacturer has the right to make such copies. By contrast, the "privileged use" rationale might not protect the clone manufactuter from a copyright infringement suit. Does Judge Boudin's "privileged use" approach better address this circumstance? For a more detailed discussion of a privilege to copy certain aspects of a computer program in order to achieve compatibility, see infra section C.3 of this chapter. For an argument that the effort that goes into program design should be protected against misappropriation, though not by copyright law, see Pamela Samuelson et al., A Manifesto Concerning the Legal Protection of Computer Programs, 94 Colum. L. Rev. 2308 (1994).

d. Protection for Program Outputs: User Interfaces

Source and object code (as well as the nonliteral elements of program structure) are protectable as literary works. Screen displays (the output of the computer program) are protectable as audiovisual or pictorial works. Given these two sets of doctrines, it might be considered evident that user interfaces — the screen displays that mediate the input to and output from a computer program — should also be protected by copyright.

Acknowledging these facts does not, however, tell us what *portions* of the user interface can be considered copyrightable expression. To a greater extent than other types of screen displays, user interfaces are governed by market standards and functional considerations. As a result, assessing claims of copyright infringement is a challenging task.

Apple Computer, Inc. v. Microsoft Corp.
United States Court of Appeals for the Ninth Circuit
35 F.3d 1435 (9th Cir. 1994)

RYMER, Circuit Judge:

Lisa and Macintosh are Apple computers. Each has a graphical user interface ("GUI") which Apple Computer, Inc. registered for copyright as an audiovisual work. Both GUIs were developed as a user-friendly way for ordinary mortals to communicate with the Apple computer; the Lisa Desktop and the Macintosh Finder are based on a desktop metaphor with windows, icons and pull-down menus which can be manipulated on the screen with a hand-held device called a mouse. When Microsoft Corporation released Windows 1.0, having a similar GUI, Apple complained. As a result, the two agreed to a license giving Microsoft the right to use and sublicense derivative works generated by Windows 1.0 in present and future products. Microsoft released Windows 2.03 and later, Windows 3.0; its licensee, Hewlett-Packard Company (HP), introduced NewWave 1.0 and later, NewWave 3.0, which run in conjunction with Windows to make IBM-compatible computers easier to use. Apple believed that these versions exceed the license, make Windows more "Mac-like," and infringe its copyright. This action followed.

[The district court held that the only triable issue of fact was whether certain minor parts of the Windows interface were "virtually identical" to the Apple GUI. Apple appealed.]

Apple asks us to reverse because of two fundamental errors in the district court's reasoning. First, Apple argues that the court should not have allowed the license for Windows 1.0 to serve as a partial defense. Second, Apple contends that the court went astray by dissecting Apple's works so as to eliminate unprotectable and licensed elements from comparison with Windows 2.03, 3.0 and NewWave as a whole, incorrectly leading it to adopt a standard of virtual identity instead of substantial similarity. We disagree. . . .

III

Apple makes a number of related arguments challenging the district court's copyright analysis. It contends that the district court deprived its works of meaningful protection by dissecting them into individual elements and viewing each element in isolation. Because the Macintosh GUI is a dynamic audiovisual work, Apple argues that the "total concept and feel" of its works — that is, the selection and arrangement of related images and their animation — must be compared with that of the Windows and NewWave GUIs for substantial similarity. Apple further asserts that in this case, the court had no occasion to dissect its works into discrete elements because Microsoft and HP virtually mimicked the composition, organization, arrangement and dynamics of the Macintosh interface, as shown by striking similarities in the animation of overlapping windows and the design, layout and animation of icons. Apple also argues that even if dissection were appropriate, the district court should not have eliminated from jury consideration those elements that are either licensed or unprotected by copyright. Though stated somewhat differently, all of these contentions boil down to the same thing: Apple wants an overall comparison of its works to the accused works for substantial similarity rather than virtual identity.

The fact that Apple licensed the right to copy almost all of its visual displays fundamentally affects the outcome of its infringement claims. Authorized copying accounts for more than 90% of the allegedly infringing features in Windows 2.03 and 3.0, and two-thirds of the features in NewWave. More than that, the 1985 Agreement and negotiations leading up to Microsoft's license left Apple no right to complain that selection and arrangement of licensed elements make the interface as a whole look more "Mac-like" than Windows 1.0.

Thus, we do not start at ground zero in resolving Apple's claims of infringement. Rather, considering the license and the limited number of ways that the basic ideas of the Apple GUI can be expressed differently, we conclude that only "thin" protection, against virtually identical copying, is appropriate. Apple's appeal, which depends on comparing its interface as a whole for substantial similarity, must therefore fail. . . .

Although this litigation has raised difficult and interesting issues about the scope of copyright protection for a graphical user interface, resolving this appeal is a matter of applying well-settled principles. In this, as in other cases, the steps we find helpful to follow are these:

(1) The plaintiff must identify the source(s) of the alleged similarity between his work and the defendant's work.

(2) Using analytic dissection, and, if necessary, expert testimony, the court must determine whether any of the allegedly similar features are protected by copyright. Where, as in this case, a license agreement is involved, the court must also determine which features the defendant was authorized to copy. Once the scope of the license is

determined, unprotectable ideas must be separated from potentially protectable expression; to that expression, the court must then apply the relevant limiting doctrines in the context of the particular medium involved, through the eyes of the ordinary consumer of that product.

(3) Having dissected the alleged similarities and considered the range of possible expression, the court must define the scope of the plaintiff's copyright — that is, decide whether the work is entitled to "broad" or "thin" protection. Depending on the degree of protection, the court must set the appropriate standard for a subjective comparison of the works to determine whether, as a whole, they are sufficiently similar to support a finding of illicit copying. . . .

B

It is not easy to distinguish expression from ideas, particularly in a new medium. However, it must be done, as the district court did in this case. Baker v. Selden, 101 U.S. 99, 25 L. Ed. 841 (1879). As we recognized long ago in the case of competing jeweled bee pins, similarities derived from the use of common ideas cannot be protected; otherwise, the first to come up with an idea will corner the market. Herbert Rosenthal Jewelry Corp. v. Kalpakian, 446 F.2d 738, 742 (9th Cir. 1971). Apple cannot get patent-like protection for the idea of a graphical user interface, or the idea of a desktop metaphor which concededly came from Xerox. It can, and did, put those ideas together creatively with animation, overlapping windows, and well-designed icons; but it licensed the visual displays which resulted.

The district court found that there are five other basic ideas embodied in the desktop metaphor: use of windows to display multiple images on the computer screen and to facilitate user interaction with the information contained in the windows; iconic representation of familiar objects from the office environment; manipulation of icons to convey instructions and to control operation of the computer; use of menus to store information or computer functions in a place that is convenient to reach, but saves screen space for other images; and opening and closing of objects as a means of retrieving, transferring and storing information. Apple V, 799 F. Supp. at 1026. No copyright protection inheres in these ideas. Therefore, substantial similarity of expression in unlicensed elements cannot be based on the fact that the Lisa, the Finder, Windows 2.03, 3.0 and NewWave all have windows, icons representing familiar objects from the office environment that describe functions being performed and that can be moved around the screen to tell the computer what to do, menus which give easy access to information or functions without using space on the screen, or objects that open and close.

Well-recognized precepts guide the process of analytic dissection. First, when an idea and its expression are indistinguishable, or "merged," the expression will only be protected against nearly identical copying. Krofft, 562

F.2d at 1167-68; Kalpakian, 446 F.2d at 742. For example, in this case, the idea of an icon in a desktop metaphor representing a document stored in a computer program can only be expressed in so many ways. An iconic image shaped like a page is an obvious choice.

The doctrine of scenes à faire is closely related. As we explained in Frybarger v. International Business Machines Corp., 812 F.2d 525 (9th Cir. 1987), when similar features in a videogame are " 'as a practical matter indispensable, or at least standard, in the treatment of a given [idea],' " they are treated like ideas and are therefore not protected by copyright. Id. at 530 (quoting Atari, Inc. v. North Am. Philips Consumer Elecs. Corp., 672 F.2d 607, 616 (7th Cir.), *cert. denied*, 459 U.S. 880 (1982)). Furthermore, as Frybarger holds, "the mere indispensable expression of these ideas, based on the technical requirements of the videogame medium, may be protected only against virtually identical copying." Id.; see also Data East, 862 F.2d at 209 (visual displays of karate match conducted by two combatants, one of whom wears red shorts and the other white as in the sport, and who use the same moves, are supervised by a referee and are scored alike as in the sport, are inherent in the sport of karate itself and as such are unprotectable). In this case, for example, use of overlapping windows inheres in the idea of windows. A programmer has only two options for displaying more than one window at a time: either a tiled system, or an overlapping system. As demonstrated by Microsoft's scenes à faire video, overlapping windows have been the clear preference in graphic interfaces. Accordingly, protectable substantial similarity cannot be based on the mere use of overlapping windows, although, of course, Apple's particular expression may be protected.

Apple suggests that scenes à faire should not limit the scope of its audiovisual copyright, or at least that the interactive character of GUIs and their functional purpose should not outweigh their artistry. While user participation may not negate copyrightability of an audiovisual work, see, e.g., Midway Mfg. Co. v. Artic Int'l, Inc., 704 F.2d 1009, 1011-12 (7th Cir.), *cert. denied*, 464 U.S. 823 (1983); Stern Elecs., Inc. v. Kaufman, 669 F.2d 852, 856 (2d Cir. 1982), the district court did not deny protection to any aspect of Apple's works on this basis. In any event, unlike purely artistic works such as novels and plays, graphical user interfaces generated by computer programs are partly artistic and partly functional. They are a tool to facilitate communication between the user and the computer; GUIs do graphically what a character-based interface, which requires a user to type in alphanumeric commands, does manually. Thus, the delete function is engaged by moving an icon on top of a trash can instead of hitting a "delete" key. In Apple's GUI, the ability to move icons to any part of the screen exemplifies an essentially functional process, indispensable to the idea of manipulating icons by a mouse.

To the extent that GUIs are artistic, there is no dispute that creativity in user interfaces is constrained by the power and speed of the computer. See Manufacturers Technologies, Inc. v. Cams, Inc., 706 F. Supp. 984, 994-95 (D. Conn. 1989) (denying protection to formatting style of plaintiff's screen displays because of constraints on viable options available to programmers). For example, hardware constraints limit the number of ways to depict visually the movement of a window on the screen; because many computers do not have enough power to show the entire contents of the window as it is being moved, the illusion of movement must be shown by using the outline of a window or some similar feature. Design alternatives are further limited by the GUI's purpose of making interaction between the user and the computer more "user-friendly." These, and similar environmental and ergonomic factors which limit the range of possible expression in GUIs, properly inform the scope of copyright protection.

Originality is another doctrine which limits the scope of protection. As the Supreme Court recently made clear, protection extends only to those components of a work that are original to the author, although original selection and arrangement of otherwise uncopyrightable components may be protectable. Feist Publications, Inc. v. Rural Tel. Serv. Co., 499 U.S. 340, 348-51 (1991). Apple's argument that components should not be tested for originality because its interface as a whole meets the test, see Roth Greeting Cards v. United Card Co., 429 F.2d 1106, 1109 (9th Cir. 1970) ("The originality necessary to support a copyright merely calls for independent creation, not novelty."), is therefore misplaced. Beyond that, Apple admits that it borrowed heavily from the iconic treatments in the Xerox Star and an IBM Pictureworld research report but disputes several of the district court's individual determinations. For instance, Apple claims that its file folder and page icon designs are original. Even if they are, these particular icons add so little to the mix of protectable material that the outcome could not reasonably be affected.

In sum, the district court's analytic dissection was appropriately conducted under the extrinsic portion of our test for whether sufficient copying to constitute infringement has taken place. We are not persuaded to the contrary by Apple's arguments that the district court shouldn't have dissected at all, or dissected too much; that it "filtered out" unprotectable and licensed elements instead of viewing the Macintosh interface as a whole; and that it should have recognized protectability of arrangements and the "total concept and feel" of the works under a substantial similarity standard.

First, graphical user interface audiovisual works are subject to the same process of analytical dissection as are other works. We have dissected videogames, which are audiovisual works and therefore closely analogous, see, e.g., Data East, 862 F.2d at 208-09 (performing analytic dissection of similarities to determine whether similarities resulted from unprotectable expression); Frybarger, 812 F.2d at 529-30 (district court correctly concluded that similar features in videogames were unprotectable ideas and that no reasonable jury could find expressive elements substantially similar), and we have dissected nonliteral elements of computer programs, which are somewhat analogous, see, e.g., Brown Bag, 960 F.2d at 1475-77 (rejecting argument similar to Apple's about propriety of analytic dissection of computer program components such as screens, menus and keystrokes); Johnson Controls, Inc. v. Phoenix Control Sys., Inc., 886 F.2d 1173, 1176 (9th Cir. 1989) (noting special master's detailed analysis of similarities). Other courts perform the same analysis, although articulated differently. See, e.g., Computer Assocs. Int'l, Inc. v. Altai, Inc., 982 F.2d 693, 706-11 (2d Cir. 1992) (adopting "abstraction-filtration-comparison" test for analyzing nonliteral structure of computer program, relying in part on our own approach); Gates Rubber Co. v. Bando Chem. Indus., 9 F.3d 823, 834, 841 (10th Cir. 1993) (adopting Altai test, but suggesting that comparison of works as a whole may be appropriate as preliminary step before filtering out unprotected elements); Engineering Dynamics, Inc. v. Structural Software, Inc., 26 F.3d 1335, 1342-43 (5th Cir. 1994) (adopting Gates Rubber/Altai test to analyze scope of copyright protection for user interface, input formats and output reports); Lotus Dev. Corp. v. Borland Int'l, Inc., 788 F. Supp. 78, 90, 93 (D. Mass. 1992) (describing similar three-part test); cf. Whelan Assocs. v. Jaslow Dental Lab., Inc., 797 F.2d 1222, 1236 (3d Cir. 1986) (defining idea of utilitarian work as its purpose or function, and everything not necessary to that purpose as expression), *cert. denied*, 479 U.S. 1031, 93 L. Ed. 2d 831, 107 S. Ct. 877 (1987).

Nor did the district court's dissection run afoul of the enjoinder in such cases as Johnson Controls, 886 F.2d at 1176, Krofft, 562 F.2d at 1167, and Roth, 429 F.2d at 1110, to consider the "total concept and feel" of a work. Here, the court did not inappropriately dissect dissimilarities, and so did nothing to distract from subjectively comparing the works as a whole. See Aliotti v. R. Dakin & Co., 831 F.2d 898, 901 (9th Cir. 1987) (indicating that as the concern of Krofft).

As we made clear in Aliotti, the party claiming infringement may place "no reliance upon any similarity in expression resulting from" unprotectable elements. Id. (emphasis added) (similarities between competing stuffed dinosaur toys on account of posture and body design, and being cuddly, stem from the physiognomy of dinosaurs or from the nature of stuffed animals and are thus unprotectable). Otherwise, there would be no point to the extrinsic test, or to distinguishing ideas from expression. In this case, it would also effectively rescind the 1985 Agreement. This does not mean that at the end of the day, when the works are considered under the intrinsic test, they should not be compared as a whole. See McCulloch v. Albert E. Price, Inc., 823 F.2d 316, 321 (9th Cir. 1987) (contrasting artistic work at issue, where decorative plates were substantially similar in more than the one unprotectable element (text), with factual works which have many unprotectable elements and very little protectable expression). Nor does it mean that infringement cannot be based on original selection and arrangement of unprotected elements. However, the unprotectable elements have to be identified, or filtered, before the works can be considered as a whole. See Harper House, 889 F.2d at 207-08 (reversing because "total impact and effect" test of jury instruction did not distinguish between protectable and unprotectable material, thereby improperly making it possible for jury to find copying based on unprotected material instead of selection and arrangement); see also Pasillas, 927 F.2d at

443 (copyright holder could not rely on unprotectable elements to show substantial similarity of expression); Frybarger, 812 F.2d at 529 (to extent that similarities between works were confined to ideas and general concepts, they were noninfringing).

C

The district court's conclusion that the works as a whole are entitled only to limited protection and should be compared for virtual identity follows from its analytic dissection. By virtue of the licensing agreement, Microsoft and HP were entitled to use the vast majority of features that Apple claims were copied. Of those that remain, the district court found no unauthorized, protectable similarities of expression in Windows 2.03 and 3.0, and only a handful in NewWave. Thus, any claim of infringement that Apple may have against Microsoft must rest on the copying of Apple's unique selection and arrangement of all of these features. Under Harper House and Frybarger, there can be no infringement unless the works are virtually identical.

Apple, however, contends that its audiovisual work with animation and icon design cannot be analogized to factual works such as game strategy books, see Landsberg v. Scrabble Crossword Game Players, Inc., 736 F.2d 485, 488 (9th Cir.) ("Similarity of expression may have to amount to verbatim reproduction or very close paraphrasing before a factual work will be deemed infringed."), *cert. denied,* 469 U.S. 1037 (1984), accounting systems, see Selden, 101 U.S. at 104 (copyright in book describing new accounting system not infringed when defendant copied ledger sheets used in system), or organizers, see Harper House, 889 F.2d at 205 (as compilations consisting largely of uncopyrightable elements, plaintiff's organizers entitled only to protection against "bodily appropriation of expression"), which are afforded only "thin" protection because the range of possible expression is narrow. See Feist, 499 U.S. at 349. Rather, it submits that the broader protection accorded artistic works is more appropriate. See, e.g., McCulloch, 823 F.2d at 321 (artistic work like a decorative plate receives broader protection because of endless variations of expression available to artist).

Which end of the continuum a particular work falls on is a call that must be made case by case. We are satisfied that this case is closer to Frybarger than to McCulloch. See also Atari Games Corp. v. Oman, 979 F.2d 242, 245 (D.C. Cir. 1992) (analogizing audiovisual work like a videogame to compilation of facts). Accordingly, since Apple did not contest summary judgment under the virtual identity standard on the merits, judgment was properly entered. . . .

We therefore hold that the district court properly identified the sources of similarity in Windows and NewWave, determined which were licensed, distinguished ideas from expression, and decided the scope of Apple's copyright by dissecting the unauthorized expression and filtering out unprotectable elements. Having correctly found that almost all the similarities spring either from the license or from basic ideas and their obvious expression, it correctly concluded that illicit copying could occur only if the works as a whole are virtually identical.

COMMENTS AND QUESTIONS

1. What is the appropriate standard for testing infringement of the user interfaces of computer programs? The traditional test for evaluating copying has been the "substantial similarity" standard. *Apple*'s "virtual identicality" test comes from a number of cases, including the *Hoehling* case excerpted in Chapter 4, which deal with "thin" copyrights. Is it appropriate to vary the standard for infringement with the scope of copyright protection, or does that "double-count" the results of the filtration analysis?

2. In many nonliteral software infringement cases, as well as in *Apple*, the vast majority of the copied work is determined to be uncopyrightable for one reason or another. If only a few "kernels" of copyrightable expression are mixed in with a large amount of public domain material, how should the courts test the "substantiality" of copying? If only 2 percent of a program is copyrightable, is taking that 2 percent the taking of a *de minimis* amount of the total program, or should it be considered the taking of the *entire* copyrighted work? Can the result possibly depend on whether uncopyrightable elements were also taken?

3. Consider the effect of dissection on copyright incentives. Does the fragmented protection that copyright affords to functional works such as computer programs really encourage creation of more programs? Does it encourage programmers to be more expressive when writing programs? If so, is that a desirable goal?

Before *Apple* and *Lotus*, a number of courts had given protection to relatively small individual elements of a user interface which they considered "artistic." See, e.g., Broderbund Software, Inc. v. Unison World, 648 F. Supp. 1127, 1134 (N.D. Cal. 1986) (choice of typeface on user interface screen and choice of words "Choose a Font" as title for a Print Shop screen were examples of audiovisual displays "dictated primarily by artistic and aesthetic consideration, and not by utilitarian or mechanical ones."); Lotus Dev. Corp. v. Paperback Software, 740 F. Supp. 37 (D. Mass. 1990) (choice of particular words to describe interface commands was expressive and therefore protectable). Is *Apple* inconsistent with protecting such micro-level design choices? Or is it simply harder to do so now?

4. Is Lotus Dev. Corp. v. Borland Intl., reprinted above, really a case about command structures or about user interfaces? If the court approached it as a user interface case, should it reach the same result? Would it? Cf. Mitel Inc. v. Iqtel Inc., 896 F. Supp. 1050 (D. Colo. 1995), where the court followed the First Circuit's decision in *Lotus* and, in addition, rejected the plaintiff's claim that its command codes were protectable as a user interface.

PROBLEM

Problem 7-8. Pueblo Graphics sells software that produces a variety of word processing fonts. (Fonts control the appearance of letters and numbers in typesetting). The software allows the fonts to appear on the screen in exactly the same way they will appear on the printed page. Pueblo's software contains a number of "classic" print fonts, such as Geneva, Helvetica, and Times Roman Bold. In addition, however, Pueblo has designed several new fonts that look significantly different from any fonts previously produced.

Pueblo would like to assert a copyright infringement suit against a competitor who has copied its fonts. The competitor wrote its own program to implement those fonts, so Pueblo wants to know if it can get copyright protection in the fonts themselves. How would you advise Pueblo? Is their claim appropriately characterized as one for a "user interface"? Why or why not?

---

2. Exclusive Rights in Computer Programs

a. The Right to Make "Copies"

Most of the cases considered in this chapter have involved the copying of all or part of a program by a company that sold a competing program. "Copying" has a potentially broader meaning in the computer context, however. The use of any modern computer program necessarily involves the creation of transitory copies in the RAM memory of a computer. Because "running" a computer program necessarily involves copying the program from long-term memory (ROM, CD-ROM, or disk) into the short-term memory of the computer, it is virtually impossible to use any computer program without making a "copy" of the program. However, that copy is not saved; it is erased whenever the application is exited or the computer is turned off. Whether such "copies" are fixed — and therefore potentially violate the copyright owner's rights in section 106(1) — has been a matter of some debate.

In 1993 the Ninth Circuit held that loading a computer program into RAM involved making a copy for purposes of section 106. MAI Systems Corp. v. Peak Computer, Inc., 991 F.2d 511, 518 (9th Cir. 1993). The court reasoned:

> The district court's grant of summary judgment on MAI's claims of copyright infringement reflects its conclusion that a 'copying' for purposes of copyright law occurs when a computer program is transferred from a permanent storage device to a computer's RAM. This conclusion is consistent with its find-

ing, in granting the preliminary injunction, that: 'the loading of copyrighted computer software from a storage medium (hard disk, floppy disk, or read only memory) into the memory of a central processing unit ('CPU') causes a copy to be made. In the absence of ownership of the copyright or express permission by license, such acts constitute copyright infringement.' We find that this conclusion is supported by the record and by the law.

Peak concedes that in maintaining its customer's computers, it uses MAI operating software 'to the extent that the repair and maintenance process necessarily involved turning on the computer to make sure it is functional and thereby running the operating system.' It is also uncontroverted that when the computer is turned on the operating system is loaded into the computer's RAM. As part of diagnosing a computer problem at the customer site, the Peak technician runs the computer's operating system software, allowing the technician to view the systems error log, which is part of the operating system, thereby enabling the technician to diagnose the problem.

Peak argues that this loading of copyrighted software does not constitute a copyright violation because the 'copy' created in RAM is not 'fixed.' However, by showing that Peak loads the software into RAM and is then able to view the system error log and diagnose the problem with the computer, MAI has adequately shown that the representation created in the RAM is 'sufficiently permanent or stable to permit it to be perceived, reproduced, or otherwise communicated for a period of more than transitory duration.'

After reviewing the record, we find no specific facts (and Peak points to none) which indicate that the copy created in the RAM is not fixed. . . .

We have found no case which specifically holds that the copying of software into RAM creates a 'copy' under the Copyright Act. However, it is generally accepted that the loading of software into a computer constitutes the creation of a copy under the Copyright Act. See, e.g., Vault Corp. v. Quaid Software, 847 F.2d 255, 260 (5th Cir. 1988) ('the act of loading a program from a medium of storage into a computer's memory creates a copy of the program'); 2 Nimmer on Copyright, 8.08 at 8-105 (1983) ('Inputting a computer program entails the preparation of a copy.'); Final Report of the National Commission on the New Technological Uses of Copyrighted Works, at 13 (1978) ('the placement of a work into a computer is the preparation of a copy'). We recognize that these authorities are somewhat troubling since they do not specify that a copy is created regardless of whether the software is loaded into the RAM, the hard disk or the read only memory ('ROM'). However, since we find that the copy created in the RAM can be 'perceived, reproduced, or otherwise communicated,' we hold that the loading of software into the RAM creates a copy under the Copyright Act.

COMMENTS AND QUESTIONS

1. If the Ninth Circuit is correct, every use of a computer or any computer program involves the making of at least one (and probably multiple) copies of the computer program. The fact that it is impossible to read or use a computer program without making such "copies" means that the application of the copyright laws to computer software is potentially far broader than

to other types of works. While you can "use" a book without making a copy, under *MAI* any use of a computer program — even turning the computer on — necessarily implicates the copyright laws. But cf. NFLC, Inc. v. Devcom Mid-America, Inc., 45 F.3d 231 (7th Cir. 1995) (suggesting that the use of dumb terminals might not make a RAM copy).

Partly in recognition of this problem, Congress enacted section 117 of the Copyright Act in 1980. Section 117 makes it possible for users of computer programs to make certain copies without fear of liability. We discuss section 117 in detail in the next section of this chapter; for now, keep in mind that making a copy is not always a violation of the copyright laws.

Does the specific enactment by Congress of a statutory provision permitting a user to make copies "necessary" to run a program suggest that Congress agreed that RAM copies are "fixed" for copyright purposes?

2. *MAI* is arguably inconsistent with both prior caselaw and the intent of Congress. See, e.g., British Leland v. Armstrong, 1986 RPC 279 (U.K. House of Lords). In Apple Computer v. Formula Intl., 594 F. Supp. 617, 621-22 (C.D. Cal. 1984), a case cited with apparent approval in *MAI*, the court indicated that copies stored in RAM, unlike those loaded in ROM, were only "temporary." And the House Report discussing the definition of "fixed" in section 101 of the Act had this to say on the subject of computer memory: "[T]he definition of 'fixation' would exclude from the concept purely evanescent or transient reproductions such as those projected briefly on a screen, shown electronically on a television or other cathode ray tube, or captured momentarily in the 'memory' of a computer." H.R. Rep. No. 94-1476, 94th Cong., 2d Sess. 52-53 (1976).

On the other hand, a number of courts have followed *MAI*'s lead in holding that copies loaded in RAM are fixed for purposes of copyright infringement. See Triad Systems v. Southeastern Express Co., 64 F.3d 1330 (9th Cir. 1995), *cert. denied*, 116 S. Ct. 1015 (1996); Stenograph L.L.C. v. Bossard Assocs, 144 F.3d 96 (D.C. Cir. 1998); In re Independent Service Organizations Antitrust Litigation, 910 F. Supp. 1537, 1541 (D. Kan. 1995); ACS v. MAI, 845 F. Supp. 356 (E.D. Va. 1994).

Is RAM memory perceivable for "more than a transitory duration"? Is it more like movies projected on a screen, or like movies stored on videotape?

3. Most commentators have been critical of the *MAI* decision. See, e.g., Niva Elkin-Koren, Copyright Law and Social Dialogue on the Information Superhighway: The Case Against Copyright Liability of Internet Providers, 13 Cardozo Arts & Ent. L.J. 345, 381-82 (1995); Michael E. Johnson, Note, The Uncertain Future of Computer Software Users' Rights in the Aftermath of MAI Systems, 44 Duke L.J. 327 (1994); Mark A. Lemley, Intellectual Property and Shrinkwrap Licenses, 68 S. Cal. L. Rev. 1239, 1280 n.184 (1995); Katrine Levin, Note, MAI v. Peak: Should Loading Operating System Software into RAM Constitute Copyright Infringement? 24 Golden Gate U.L. Rev. 649 (1994); Jessica Litman, The Exclusive Right to Read, 13 Cardozo Arts & Ent. L.J. 29, 41-43 (1994); Carol G. Stovsky, Note, MAI Systems Corp. v. Peak Computer, Inc.: Using Copyright Law to Prevent

Unauthorized Use of Computer Software, 56 Ohio St. L.J. 593 (1995). On the other hand, the Clinton administration has enthusiastically endorsed it in its White Paper on copyright and the Internet. See NII Working Group on Intellectual Property, Intellectual Property and the National Information Infrastructure (August 1995).

b. Copies and Section 117

As noted above, making certain types of "copies" is an essential part of using computer programs. Because of this important difference between computer software and other types of copyrightable material, Congress followed the recommendations of CONTU and passed section 117. That section provides, in relevant part:

> Notwithstanding the provisions of section 106, it is not an infringement for the owner of a copy of a computer program to make or authorize the making of another copy or adaptation of that computer program provided:
>
> > (1) that such a new copy or adaptation is created as an essential step in the utilization of the computer program in conjunction with a machine and that it is used in no other manner, or
> >
> > (2) that such new copy or adaptation is for archival purposes only and that all archival copies are destroyed in the event that continued possession of the computer program should cease to be rightful.

17 U.S.C. §117.

Section 117's right to make archival copies has not been controversial. It allows computer users to make "backup" copies of their programs, as well as to load the programs onto hard drives from a floppy disk. See Vault Corp. v. Quaid Software Ltd., 847 F.2d 255 (5th Cir. 1988); ProCD v. Zeidenberg, 908 F. Supp. 640 (W.D. Wisc. 1996), *rev'd on other grounds,* 86 F.3d 1447 (7th Cir. 1996). The language of the statute also contemplates that multiple archival copies may be made of a single computer program. It clearly does *not* allow users of a program to make copies and distribute them to others for commercial purposes. See, e.g., Allen Myland Inc. v. IBM, 746 F. Supp. 520 (E.D. Pa. 1990); Apple Computer v. Formula Intl., 594 F. Supp. 617 (C.D. Cal. 1984).

The right to make "essential" copies has proven more controversial. The basic intention of the section seems to be to allow copies like those made in MAI v. Peak, supra. Further, the language of section 117(1) extends not only to identical copies but to such "adaptation" of programs as is necessary to use them with a particular computer. In *Vault*, the Fifth Circuit held that such adaptations were not limited to those intended by the software vendor. In that case, Vault sold a computer program called PROLOK, which was used to "copy-protect" software.[18] Quaid reverse engineered Vault's program in order to create its own program, called RAMKEY, which disabled PROLOK's copy protection. In the course of writing RAMKEY, Quaid loaded the PROLOK program into its computer memory. The court rejected Vault's claim of copyright infringement:

> In order to develop RAMKEY, Quaid analyzed Vault's program by copying it into its computer's memory. Vault contends that, by making this unauthorized copy, Quaid directly infringed upon Vault's copyright. The district court held that "Quaid's actions clearly fall within [the §117(1)] exemption. The loading of [Vault's] program into the [memory] of a computer is an 'essential step in the utilization' of [Vault's] program. Therefore, Quaid has not infringed Vault's copyright by loading [Vault's program] into [its computer's memory]." Vault, 655 F. Supp. at 758.
>
> Section 117(1) permits an owner of a program to make a copy of that program provided that the copy "is created as an essential step in the utilization of the computer program in conjunction with a machine and that it is used in no other manner." Congress recognized that a computer program cannot be used unless it is first copied into a computer's memory, and thus provided the §117(1) exception to permit copying for this essential purpose. See CONTU Report at 31. Vault contends that, due to the inclusion of the phrase "and that it is used in no other manner," this exception should be interpreted to permit only the copying of a computer program for the purpose of using it for *its intended purpose*. Because Quaid copied Vault's program into its computer's memory for the express purpose of devising a means of defeating its protective function, Vault contends that §117(1) is not applicable.
>
> We decline to construe §117(1) in this manner. Even though the copy of Vault's program made by Quaid was not used to prevent the copying of the program placed on the PROLOK diskette by one of Vault's customers (which is the purpose of Vault's program), and was, indeed, made for the express purpose of devising a means of defeating its protective function, the copy made by Quaid was "created as an essential step in the utilization" of Vault's program. Section 117(1) contains no language to suggest that the copy it permits must be employed for a use intended by the copyright owner, and, absent clear congressional guidance to the contrary, we refuse to read such limiting language into this exception. We therefore hold that Quaid did not infringe Vault's exclusive right to reproduce its program in copies under §106(1).

Vault Corp. v. Quaid Software Ltd., 847 F.2d 255 (5th Cir. 1988).

To whom does section 117 apply? By its terms, the statute authorizes copies or adaptations made by "the owner of a copy of a computer program" or persons she authorizes to make such copies. In its Final Report, CONTU indicated its intention "that *persons in rightful possession of copies of programs* be able to use them freely without fear of exposure to copyright liability." CONTU Final Report at 13 (emphasis added). The report seems to indicate

---

18. Copy protection, which was common during the early 1980s, refers to programs which attempt to prevent unauthorized copying of software by technical means.

that section 117 applies to anyone who has not obtained the program by improper means.

Without discussion or explanation, Congress changed "persons in rightful possession of copies" to "owners of copies." Does this mean that section 117 doesn't benefit "licensees" — a group which on some readings encompasses all users of a computer program? Should the phrase "owner of a copy of a computer program" be interpreted to refer to the rightful possessor of the physical copy, regardless of who retains the copyright in the program itself? This issue was addressed at length by the Federal Circuit in DSC Communications Corp. v. Pulse Communications Inc., 170 F.3d 1354 (Fed. Cir. 1999):

> The district court concluded that making copies of the POTS-DI software (in the resident memory of POTS cards) was an "essential step in the utilization" of the POTS-DI software and that there was no evidence that the RBOCs used the software in any other manner that would constitute infringement. Accordingly, under the district court's theory of the case there was no direct infringement (and thus no contributory infringement) if the RBOCs were section 117 "owners" of copies of the POTS-DI software.
>
> The district court then held that the RBOCs were "owners" of copies of the POTS-DI software because they obtained the software by making a single payment and obtaining a right to possession of the software for an unlimited period. Those attributes of the transaction, the court concluded, made the transaction a "sale."
>
> DSC challenges the district court's conclusion that, based on the terms of the purchase transactions between DSC and the RBOCs, the RBOCs were "owners" of copies of the POTS-DI software. In order to resolve that issue, we must determine what attributes are necessary to constitute ownership of copies of software in this context.
>
> Unfortunately, ownership is an imprecise concept, and the Copyright Act does not define the term. Nor is there much useful guidance to be obtained from either the legislative history of the statute or the cases that have construed it. The National Commission on New Technological Uses of Copyrighted Works ("CONTU") was created by Congress to recommend changes in the Copyright Act to accommodate advances in computer technology. In its final report, CONTU proposed a version of section 117 that is identical to the one that was ultimately enacted, except for a single change. The proposed CONTU version provided that "it is not an infringement for the rightful possessor of a copy of a computer program to make or authorize the making of another copy or adaptation of that program. . . ." Final Report of the National Commission on New Technological Uses of Copyrighted Works, U.S. Dept. of Commerce, PB-282141, at 30 (July 31, 1978) (emphasis added). Congress, however, substituted the words "owner of a copy" in place of the words "rightful possessor of a copy." See Pub.L. No. 96-517, 96th Cong., 2d Sess. (1980). The legislative history does not explain the reason for the change, see H.R.Rep. No. 96-1307, 96th Cong., 2d Sess., pt. 1, at 23 (1980), but it is clear from the fact of the substitution of the term "owner" for "rightful possessor" that Congress must have meant to require more than "rightful possession" to trigger the section 117 defense.

In the leading case on section 117 ownership, the Ninth Circuit considered an agreement in which MAI, the owner of a software copyright, transferred copies of the copyrighted software to Peak under an agreement that imposed severe restrictions on Peak's rights with respect to those copies. See MAI Sys. Corp. v. Peak Computer, Inc., 991 F.2d 511, 26 USPQ2d 1458, 1462 (9th Cir. 1995). The court held that Peak was not an "owner" of the copies of the software for purposes of section 117 and thus did not enjoy the right to copy conferred on owners by that statute. The Ninth Circuit stated that it reached the conclusion that Peak was not an owner because Peak had licensed the software from MAI. See id. at 518 n. 5. That explanation of the court's decision has been criticized for failing to recognize the distinction between ownership of a copyright, which can be licensed, and ownership of copies of the copyrighted software. See, e.g., 2 Melville B. Nimmer, Nimmer on Copyright ¶8.08[B][1], at 8-119 to 1-121 (3d ed. 1997). Plainly, a party who purchases copies of software from the copyright owner can hold a license under a copyright while still being an "owner" of a copy of the copyrighted software for purposes of section 117. We therefore do not adopt the Ninth Circuit's characterization of all licensees as non-owners. Nonetheless, the MAI case is instructive, because the agreement between MAI and Peak, like the agreements at issue in this case, imposed more severe restrictions on Peak's rights with respect to the software than would be imposed on a party who owned copies of software subject only to the rights of the copyright holder under the Copyright Act. And for that reason, it was proper to hold that Peak was not an "owner" of copies of the copyrighted software for purposes of section 117. See also Advanced Computer Servs. of Mich. v. MAI Sys. Corp., 845 F.Supp. 356, 367, 30 USPQ2d 1443, 1452 (E.D.Va. 1994) ("MAI customers are not 'owners' of the copyrighted software; they possess only the limited rights set forth in their licensing agreements"). We therefore turn to the agreements between DSC and the RBOCs to determine whether those agreements establish that the RBOCs are section 117 "owners" of copies of the copyrighted POTS-DI software.

[The court concluded after a detailed analysis of the facts of the transaction that certain agreements in that case did not transfer ownership of the particular copies of the program at issue, but that other transactions did involve a transfer of ownership].

Id. at 1359-1362. It is worth noting that *DSC* involved a negotiated agreement between sophisticated parties — the buyers were the regional Bell operating companies. By contrast, in the mass-market context, most (though by no means all) cases that have considered the issue have concluded that mass-market, over-the-counter transfers of computer programs at retail stores constitute "sales of goods" subject to the Uniform Commercial Code, rather than license agreements.

The Digital Millenium Copyright Act, passed by Congress in 1998, reversed the holding of *MAI v. Peak*, but in a very narrow way. Rather than declare that section 117 applies to rightful possessors of a computer program, as CONTU originally intended, the DMCA altered section 117 to create a particular exemption from liability for the owner or lessee of a computer to make or authorize the making of a copy of a computer program for purposes of maintenance or repair of the computer hardware. However, the copy thus made must be made solely by turning the machine on, be used for no other purpose, and no other programs can be copied. Further, the new provision does not apply to maintenance or repair of software, just hardware.

Does this solve the problem created by the *MAI* case? Or does it give legislative sanction to the idea that RAM copies are not permitted except in the narrow circumstances covered by the new Act? One might read this new exemption as creating a negative implication that all other copies made by lessees or licensees were illegal even if they fell within sections 117(1) or (2). Thus this change may benefit independent service organizations that service computer hardware, but it does nothing for computer users in general.

Is there a social interest in maintaining a competitive market for hardware and software maintenance? One way of thinking about the issue is to determine whether the copyright in a computer program should permit its owner to control the markets for hardware and software maintenance, or whether using the copyright to prevent competition in that market would expand copyright protection beyond its intended boundaries. For a discussion of the competitive interest in software maintenance, see Pamela Samuelson, Modifying Copyrighted Software: Adjusting Copyright Doctrine to Accommodate a Technology, 28 Jurimetrics J. 179 (1988).

c. The Digital Millenium Copyright Act

In the last days of the 1998 Congressional session, Congress passed sweeping changes to the Copyright Act designed specifically to address a variety of Internet-related copyright issues. The Digital Millenium Copyright Act was nominally intended to bring U.S. law into compliance with the 1996 WIPO treaties on copyright and the Internet, but in fact it went well beyond what those treaties required.[11] The DMCA changed Internet copyright law in three significant ways.

Circumventing Copy Protection. First, the DMCA makes the circumvention of copy protection systems (or the manufacture or distribution of devices with the primary purpose or effect of circumventing copy protection systems) a crime if done without the authority of the copyright owner. 17 U.S.C.§1201(a),(b). Specifically, the statute makes it illegal to "circumvent a technological protection measure" that effectively controls access to a work. Circumvention means to "descramble a scrambled work, to decrypt an encrypted work, or otherwise avoid, bypass, remove, deactivate, or impair a technological protection measure." Id. §1201(a)(3).

This new statute marks a significant change in the balance between copyright owners and users in at least three respects. First, it reflects a rather striking decision to promote innovation by banning one class of innovations

---

11. Indeed, it seems relatively clear that U.S. law did not need to change at all to conform with the WIPO treaties. For detail on this point, see Pamela Samuelson, The U.S. Digital Agenda at WIPO, 37 Va. J. Int'l L. 369 (1997).

entirely — those that involve decryption technologies. If one thinks of the competition between encryption technologies and decryption technologies as a sort of arms race, the DMCA orders unilateral technological disarmament by the decrypters. Second, the statute changes the law of contributory infringement. The production of devices that facilitate infringement has always been illegal under copyright law but only if the device produced was not "capable of a substantial noninfringing use." Sony Corp. v. Universal City Studios, 464 U.S. 417 (1984). The DMCA does not adopt this standard. Instead it adopts a broader definition of illegal conduct, one closer to the dissent's "primary purpose or effect" test in *Sony*. The dissent in *Sony* would have applied that test to declare VCRs illegal; it remains to be seen what sorts of commercial devices will be swept into the DMCA's intermediate standard. Third, and perhaps most notable, the statute itself does not condition the illegality of circumvention on proof that the user violated the copyright laws. Someone who circumvents a copy protection measure in order to make lawful use of a work (to copy the ideas contained therein, for example, or to engage in "fair use") has not violated the copyright laws but has nonetheless committed a crime under the DMCA. Thus while the statute purports not to affect the rights, remedies, limitations, or defenses of copyright law, 17 U.S.C. sec. 1201(c), the protection provided by that section is likely to prove illusory.

The power of the DMCA's prohibitions on circumvention is circumscribed by a number of exceptions that were written into the Act. In particular, section 1201(f) allows software developers to engage in reverse engineering of a computer program for the purpose of achieving interoperability between programs, so long as copyright law would permit the reverse engineering[12] and there is no other readily available means of obtaining that information. Other sections permit circumvention of copy protection systems for purposes of legitimate law enforcement and government intelligence activity, scientific research into encryption, testing of the security of devices, and for various other limited reasons. 17 U.S.C. §1201(e)-(j). Further, the prohibitions on conduct by individuals (as opposed to the creation of circumventing devices) are stayed for two years, pending a report by the Library of Congress.

Copyright Management Information. Section 1202 of the Copyright Act makes it illegal to remove or alter "copyright management information" conveyed along with a copyrighted work or to provide false copyright management information. Copyright management information includes information that identifies the authors, owners, or performers in a work and the terms and conditions under which a work may be used. The intent of this section is to punish those who facilitate counterfeiting by stripping identifying information from a work or who falsely identify themselves as the authors of a work. The literal terms of the Act may go further, however, creating a quasi-moral right of attribution and ensuring that "clickwrap licenses" remain embedded in a computer program.

Online Service Provider Liability. The DMCA limits the liability of "on-

---

12. On that question, see infra.

line service providers" (OSPs) for messages posted to their systems in a variety of circumstances. "Online service provider" is defined broadly to include not only companies in the business of providing Internet access, but also any entity that provides services "such as Internet access, email, chat room and Web page hosting," even if that is not their primary business. 17 U.S.C. §512(k)(1). Thus, universities or companies may fall within the exemption even though the hosting of discussion fora or the provision of e-mail services are incidental to their main mission.

The conditions for exemption are complex. To qualify for the general exemptions, OSPs must accommodate "standard" technological copy protection measures such as encryption and digital watermarking. 17 U.S.C. §512(i)(2). OSPs must also agree to terminate users who repeatedly infringe copyright rights on the system. However, the OSP need not actively monitor messages on its system in an attempt to identify copyright infringements.

For messages posted to an OSP's system, or links provided to infringing material by the OSP, the exemption provides that OSPs generally will not be liable for copyright infringement if they did not know and could not reasonably have known that the material was infringing. However, OSPs that learn of infringement must act expeditiously to remove or block access to the infringing material. In addition the DMCA creates a "notice and take down" rule. Under this rule an OSP that receives written notice that material on their system is infringing must remove the material as soon as practicable. The OSP must also notify the poster of that material that it has been taken down and give the poster an opportunity to respond. If the poster responds by giving "counter notification" that the material is not copyrighted, or is lawfully used, the OSP must put the material back on its system within 10 to 14 days unless the copyright owner files suit against the poster.

The DMCA also creates exemptions for OSPs who do nothing more than automatically route information through their systems without modifying or redirecting it, and creates limited exemptions for certain types of information caching.

COMMENTS AND QUESTIONS

1. Should Congress be in the business of regulating what devices can be developed? Is such regulation necessary to combat piracy on the Internet?

2. Is there anything wrong with prohibiting the alteration or misuse of copyright management information? This provision seems to serve the laudable purposes of preventing misrepresentations as to ownership and combatting counterfeiting. On the other hand at least one commentator is concerned that the provision may encourage the development of "smart CMI" systems that not only include terms and conditions but use technological self-help to enforce those terms and conditions. See Julie E. Cohen, A Right to Read Anonymously: A Closer Look at Copyright Management in Cyberspace, 28 Conn. L. Rev. 981 (1996).

3. Is the DMCA's treatment of OSP liability appropriate? In particular, consider whether the law was already coming to a reasonable accomodation of the competing interests of copyright owners and OSPs.

4. For a good summary of the DMCA, including detailed provisions not discussed here, see Jonathan Band, The Digital Millenium Copyright Act, IP Worldwide, December 1998. ˙

3. Fair Use and Derivative Works

Lewis Galoob Toys, Inc. v. Nintendo of America, Inc.
United States Court of Appeals for the Ninth Circuit
964 F.2d 965 (9th Cir. 1992), cert. denied, 507 U.S. 985 (1993)

FARRIS, Circuit Judge:
Nintendo of America appeals the district court's judgment following a bench trial (1) declaring that Lewis Galoob Toys' Game Genie does not violate any Nintendo copyrights and dissolving a temporary injunction and (2) denying Nintendo's request for a permanent injunction enjoining Galoob from marketing the Game Genie. Lewis Galoob Toys, Inc. v. Nintendo of America, Inc., 780 F. Supp. 1283, 20 U.S.P.Q.2d (BNA) 1662 (N.D. Cal. 1991). We have appellate jurisdiction pursuant to 15 U.S.C. §1121 and 28 U.S.C. §§1291 and 1292(a)(1). We affirm.

Facts

The Nintendo Entertainment System is a home video game system marketed by Nintendo. To use the system, the player inserts a cartridge containing a video game that Nintendo produces or licenses others to produce. By pressing buttons and manipulating a control pad, the player controls one of the game's characters and progresses through the game. The games are protected as audiovisual works under 17 U.S.C. §102(a)(6).

The Game Genie is a device manufactured by Galoob that allows the player to alter up to three features of a Nintendo game. For example, the Game Genie can increase the number of lives of the player's character, increase the speed at which the character moves, and allow the character to float above obstacles. The player controls the changes made by the Game Genie by entering codes provided by the Game Genie Programming Manual and Code Book. The player also can experiment with variations of these codes.

The Game Genie functions by blocking the value for a single data byte sent by the game cartridge to the central processing unit in the Nintendo

Entertainment System and replacing it with a new value. If that value controls the character's strength, for example, then the character can be made invincible by increasing the value sufficiently. The Game Genie is inserted between a game cartridge and the Nintendo Entertainment System. The Game Genie does not alter the data that is stored in the game cartridge. Its effects are temporary.

Discussion . . .

2. *Fair Use*

"The doctrine of fair use allows a holder of the privilege to use copyrighted material in a reasonable manner without the consent of the copyright owner." Narell v. Freeman, 872 F.2d 907, 913, 10 U.S.P.Q.2d (BNA) 1596 (9th Cir. 1989) (citations omitted). The district court concluded that, even if the audiovisual displays created by the Game Genie are derivative works, Galoob is not liable under 17 U.S.C. §107 because the displays are a fair use of Nintendo's copyrighted displays. "Whether a use of copyrighted material is a 'fair use' is a mixed question of law and fact. If the district court found sufficient facts to evaluate each of the statutory factors, the appellate court may decide whether defendants may claim the fair use defense as a matter of law." Abend v. MCA, Inc., 863 F.2d 1465, 1468, 9 U.S.P.Q.2d (BNA) 1337 (9th Cir. 1988), *aff'd sub nom.* Stewart v. Abend, 495 U.S. 207, 109 L. Ed. 2d 184, 110 S. Ct. 1750, 14 U.S.P.Q.2d (BNA) 1614 (1990).

Section 107 codifies the fair use defense. . . .

Much of the parties' dispute regarding the fair use defense concerns the proper focus of the court's inquiry: (1) Galoob or (2) consumers who purchase and use the Game Genie. Nintendo's complaint does not allege direct infringement, nor did it try the case on that theory. The complaint, for example, alleges only that "Galoob's marketing advertising [sic], promoting and selling of Game Genie has and will contribute to the creation of infringing derivatives of Nintendo's copyrighted . . . games." Contributory infringement is a form of third party liability. See Melville B. Nimmer & David Nimmer, 3 Nimmer on Copyright ¶12.04[A]2, at 12-68 (1991). The district court properly focused on whether consumers who purchase and use the Game Genie would be infringing Nintendo's copyrights by creating (what are now assumed to be) derivative works.

Nintendo emphasizes that the district court ultimately addressed its direct infringement by authorization argument. The court concluded that, "because the Game Genie does not create a derivative work when used in conjunction with a copyrighted video game, Galoob does not 'authorize the use of a copyrighted work without the actual authority from the copyright owner.' " Galoob, 780 F. Supp. at 1298 (quoting Sony, 464 U.S. at 435 n.17). Although infringement by authorization is a form of direct infringement, this does not change the proper focus of our inquiry; a party cannot authorize another party to infringe a copyright unless the authorized conduct would itself be unlawful.

The district court concluded that "a family's use of a Game Genie for private home enjoyment must be characterized as a non-commercial, non-profit activity." Galoob, 780 F. Supp. at 1293. Nintendo argues that Game Genie users are supplanting its commercially valuable right to make and sell derivative works. Nintendo's reliance on Harper & Row Publishers, Inc. v. Nation Enters., 471 U.S. 539, 562, 85 L. Ed. 2d 588, 105 S. Ct. 2218, 225 U.S.P.Q. (BNA) 1073 (1985), is misplaced. The commercially valuable right at issue in Harper & Row was the right of first publication; Nation Enterprises intended to publish the copyrighted materials for profit. See id. at 562-63. See also Sony, 464 U.S. at 449 ("If the Betamax were used to make copies for a commercial or profit-making purpose, such use would presumptively be unfair."). Game Genie users are engaged in a non-profit activity. Their use of the Game Genie to create derivative works therefore is presumptively fair. See Sony, 464 U.S. at 449.

The district court also concluded that "the [Nintendo] works' published nature supports the fairness of the use." Galoob, 780 F. Supp. at 1293. Nintendo argues that it has not published the derivative works created by the Game Genie. This argument ignores the plain language of section 107: "the factors to be considered shall include . . . the nature of the copyrighted work." The argument also would make the fair use defense unavailable in all cases of derivative works, including "criticism, comment, news reporting, teaching . . . , scholarship, or research." 17 U.S.C. §107. A commentary that incorporated large portions of *For Whom the Bell Tolls,* for example, would be undeserving of fair use protection because the incorporated portions would constitute an unpublished derivative work. This cannot be the law.

The district court further concluded that the amount of the portion used in relation to the copyrighted work as a whole "cannot assist Nintendo in overcoming the presumption of fair use." Galoob, 780 F. Supp. at 1293. The video tape recorders at issue in Sony allowed consumers to tape copyrighted works in their entirety. The Supreme Court nevertheless held that, "when one considers . . . that [video tape recording] merely enables a viewer to see such a work which he had been invited to witness in its entirety free of charge, the fact that the entire work is reproduced does not have its ordinary effect of militating against a finding of fair use." 464 U.S. 449 at 449-50 (citations omitted). Consumers are not invited to witness Nintendo's audiovisual displays free of charge, but, once they have paid to do so, the fact that the derivative works created by the Game Genie are comprised almost entirely of Nintendo's copyrighted displays does not militate against a finding of fair use.

Nintendo would distinguish Sony because it involved copying copyrighted works rather than creating derivative works based on those works. In other words, the consumers in Sony could lawfully copy the copyrighted works because they were invited to view those works free of charge. Game Genie users, in contrast, are not invited to view derivative works based on

Nintendo's copyrighted works without first paying for that privilege. Sony cannot be read so narrowly. It is difficult to imagine that the Court would have reached a different conclusion if Betamax purchasers were skipping portions of copyrighted works or viewing denouements before climaxes. Sony recognizes that a party who distributes a copyrighted work cannot dictate how that work is to be enjoyed. Consumers may use a Betamax to view copyrighted works at a more convenient time. They similarly may use a Game Genie to enhance a Nintendo Game cartridge's audiovisual display in such a way as to make the experience more enjoyable.

"The fourth factor is the 'most important, and indeed, central fair use factor.'" Stewart, 495 U.S. at 238 (quoting 3 Nimmer on Copyright ¶13.05[A], at 13-81). The district court concluded that "Nintendo has failed to show any harm to the present market for its copyrighted games and has failed to establish the reasonable likelihood of a potential market for slightly altered versions of the games at suit." Galoob, 780 F. Supp. at 1295. Nintendo's main argument on appeal is that the test for market harm encompasses the potential market for derivative works. Because the Game Genie is used for a noncommercial purpose, the likelihood of future harm may not be presumed. See Sony, 464 U.S. at 451. Nintendo must show "by a preponderance of the evidence that some meaningful likelihood of future harm exists." Id.

Nintendo's argument is supported by case law. Although the Copyright Act requires a court to consider "the effect of the use upon the potential market for or value of the copyrighted work," 17 U.S.C. §107(4) (emphasis added), we held in Abend that "although the motion picture will have no adverse effect on bookstore sales of the [underlying] novel — and may in fact have a beneficial effect — it is 'clear that [the film's producer] may not invoke the defense of fair use.'" 863 F.2d at 1482 (quoting 3 Nimmer on Copyright ¶13.05[B], at 13-84). We explained: "'If the defendant's work adversely affects the value of any of the rights in the copyrighted work . . . the use is not fair even if the rights thus affected have not as yet been exercised by the plaintiff.'" Id. (quoting 3 Nimmer on Copyright ¶13.05[B], at 13-84 to 13-85 (footnotes omitted)). The Supreme Court specifically affirmed our finding that the motion picture adaptation "impinged on the ability to market new versions of the story." Stewart, 495 U.S. at 238.

Still, Nintendo's argument is undermined by the facts. The district court considered the potential market for derivative works based on Nintendo game cartridges and found that: (1) "Nintendo has not, to date, issued or considered issuing altered versions of existing games," Galoob, 780 F. Supp. at 1295, and (2) Nintendo "has failed to show the reasonable likelihood of such a market." Id. The record supports the court's findings. According to Stephen Beck, Galoob's expert witness, junior or expert versions of existing Nintendo games would enjoy very little market interest because the original version of each game already has been designed to appeal to the largest number of consumers. Mr. Beck also testified that a new game must include new material or "the game player is going to feel very cheated and robbed, and

[the] product will have a bad reputation and word of mouth will probably kill its sales." Howard Lincoln, Senior Vice President of Nintendo of America, acknowledged that Nintendo has no present plans to market such games.

The district court also noted that Nintendo's assertion that it may wish to re-release altered versions of its game cartridges is contradicted by its position in various other lawsuits:

> In those actions, Nintendo opposes antitrust claims by using the vagaries of the video game industry to rebut the impact and permanence of its market control, if any. Having indoctrinated this Court as to the fast pace and instability of the video game industry, Nintendo may not now, without any data, redefine that market in its request for the extraordinary remedy sought herein. . . . While board games may never die, good video games are mortal.

Galoob, 780 F. Supp. at 1295. The existence of this potential market cannot be presumed. See Sony, 464 U.S. at 451. See also Wright v. Warner Books, Inc., 953 F.2d 731, 739, 20 U.S.P.Q.2d (BNA) 1892 (2d Cir. 1991) (affirming district court's finding of no reasonable likelihood of injury to alleged market because "plaintiff has offered no evidence that the project will go forward"). The fourth and most important fair use factor also favors Galoob.

Nintendo's most persuasive argument is that the creative nature of its audiovisual displays weighs against a finding of fair use. The Supreme Court has acknowledged that "fair use is more likely to be found in factual works than fictional works." Stewart, 495 U.S. at 237. This consideration weighs against a finding of fair use, but it is not dispositive. See Sony, 464 U.S. at 448 (fair use defense is an "equitable rule of reason"). The district court could properly conclude that Game Genie users are making a fair use of Nintendo's displays. . . .

Affirmed.

COMMENTS AND QUESTIONS

1. Suppose that Nintendo had announced its intention to sell add-on devices to its entertainment system that would perform the same function as the Game Genie. Should this fact change the fair use analysis? Certainly, Nintendo could prove that it was losing money to Galoob in this example. Does this fact sway the fourth factor in Nintendo's favor? What if Nintendo could show the existence of a "market" for *licensing* devices like the Game Genie? Is Galoob depriving Nintendo of a royalty payment? On this issue, cf. American Geophysical Union v. Texaco, 37 F.3d 881 (2d Cir. 1994) (existence of "market" for royalty payments of copies undercuts fair use argument).

2. One of the reasons people play video games is for the challenge of overcoming obstacles. Does a device which gives a game character "super powers" or an infinite number of lives defeat this purpose? It is certainly

possible that owners of the Game Genie will grow bored with video games more quickly.[13] If Nintendo can prove that the Game Genie will cut into its sales of games by making the games less challenging, should that affect its copyright argument?

Alternatively, does Nintendo have some "moral right" in the integrity of its audiovisual work that is being violated by Galoob's device?

≡ *Micro Star v. Formgen Inc.*
≡ *United States Court of Appeals for the Ninth Circuit.*
≡ *154 F.3d 1107 (9th Cir. 1998)*

KOZINSKI, Circuit Judge.

Duke Nukem routinely vanquishes Octabrain and the Protozoid Slimer. But what about the dreaded Micro Star?

I

FormGen Inc., GT Interactive Software Corp. and Apogee Software, Ltd. (collectively FormGen) made, distributed and own the rights to Duke Nukem 3D (D/N-3D), an immensely popular (and very cool) computer game. D/N-3D is played from the first-person perspective; the player assumes the personality and point of view of the title character, who is seen on the screen only as a pair of hands and an occasional boot, much as one might see oneself in real life without the aid of a mirror. Players explore a futuristic city infested with evil aliens and other hazards. The goal is to zap them before they zap you, while searching for the hidden passage to the next level. The basic game comes with twenty-nine levels, each with a different combination of scenery, aliens, and other challenges. The game also includes a "Build Editor," a utility that enables players to create their own levels. With FormGen's encouragement, players frequently post levels they have created on the Internet where others can download them. Micro Star, a computer software distributor, did just that: It downloaded 300 user-created levels and stamped them onto a CD, which it then sold commercially as Nuke It (N/I). N/I is packaged in a box decorated with numerous "screen shots," pictures of what the new levels look like when played.

Micro Star filed suit in district court, seeking a declaratory judgment that N/I did not infringe on any of FormGen's copyrights. FormGen counterclaimed, seeking a preliminary injunction barring further production and distribution of N/I. Relying on Lewis Galoob Toys, Inc. v. Nintendo of Am., Inc., 964 F.2d 965 (9th Cir. 1992), the district court held that N/I was not a derivative work and therefore did not infringe FormGen's copyright. The

---

13. Of course, this might induce them to buy more Nintendo games, rather than less. If that is the case, it is hard to see how Nintendo has been injured.

district court did, however, grant a preliminary injunction as to the screen shots, finding that N/I's packaging violated FormGen's copyright by reproducing pictures of D/N-3D characters without a license. The court rejected Micro Star's fair use claims. Both sides appeal their losses. . . .

III

To succeed on the merits of its claim that N/I infringes FormGen's copyright, FormGen must show (1) ownership of the copyright to D/N-3D, and (2) copying of protected expression by Micro Star. See Triad Systems Corp. v. Southeastern Express Co., 64 F.3d 1330, 1335 (9th Cir. 1995). FormGen's copyright registration creates a presumption of ownership, see id., and we are satisfied that FormGen has established its ownership of the copyright. We therefore focus on the latter issue.

FormGen alleges that its copyright is infringed by Micro Star's unauthorized commercial exploitation of user-created game levels. In order to understand FormGen's claims, one must first understand the way D/N-3D works. The game consists of three separate components: the game engine, the source art library and the MAP files. The game engine is the heart of the computer program; in some sense, it is the program. It tells the computer when to read data, save and load games, play sounds and project images onto the screen. In order to create the audiovisual display for a particular level, the game engine invokes the MAP file that corresponds to that level. Each MAP file contains a series of instructions that tell the game engine (and, through it, the computer) what to put where. For instance, the MAP file might say scuba gear goes at the bottom of the screen. The game engine then goes to the source art library, finds the image of the scuba gear, and puts it in just the right place on the screen.[3] The MAP file describes the level in painstaking detail, but it does not actually contain any of the copyrighted art itself; everything that appears on the screen actually comes from the art library. Think of the game's audiovisual display as a paint-by-numbers kit. The MAP file might tell you to put blue paint in section number 565, but it doesn't contain any blue paint itself; the blue paint comes from your palette, which is the low-tech analog of the art library, while you play the role of the game engine. When the player selects one of the N/I levels, the game engine references the N/I MAP files, but still uses the D/N-3D art library to generate the images that make up that level.

FormGen points out that a copyright holder enjoys the exclusive right to prepare derivative works based on D/N-3D. See 17 U.S.C.

3. Actually, this is all a bit metaphorical. Computer programs don't actually go anywhere or fetch anything. Rather, the game engine receives the player's instruction as to which game level to select and instructs the processor to access the MAP file corresponding to that level. The MAP file, in turn, consists of a series of instructions indicating which art images go where. When the MAP file calls for a particular art image, the game engine tells the processor to access the art library for instructions on how each pixel on the screen must be colored in order to paint that image.

§106(2) (1994). According to FormGen, the audiovisual displays generated when D/N-3D is run in conjunction with the N/I CD MAP files are derivative works that infringe this exclusivity. Is FormGen right? The answer is not obvious.

The Copyright Act defines a derivative work as

> a work based upon one or more preexisting works, such as a translation, musical arrangement, dramatization, fictionalization, motion picture version, sound recording, art reproduction, abridgment, condensation, or any other form in which a work may be recast, transformed, or adapted. A work consisting of editorial revisions, annotations, elaborations, or other modifications which, as a whole, represent an original work of authorship, is a "derivative work."

Id. §101. The statutory language is hopelessly overbroad, however, for "[e]very book in literature, science and art, borrows and must necessarily borrow, and use much which was well known and used before." Emerson v. Davies, 8 F. Cas. 615, 619 (C.C.D.Mass. 1845) (No. 4436), quoted in 1 Nimmer on Copyright, §3.01, at 3-2 (1997). To narrow the statute to a manageable level, we have developed certain criteria a work must satisfy in order to qualify as a derivative work. One of these is that a derivative work must exist in a "concrete or permanent form," *Galoob*, 964 F.2d at 967 (internal quotation marks omitted), and must substantially incorporate protected material from the preexisting work, see Litchfield v. Spielberg, 736 F.2d 1352, 1357 (9th Cir.1984). Micro Star argues that N/I is not a derivative work because the audiovisual displays generated when D/N-3D is run with N/I's MAP files are not incorporated in any concrete or permanent form, and the MAP files do not copy any of D/N-3D's protected expression. It is mistaken on both counts.

The requirement that a derivative work must assume a concrete or permanent form was recognized without much discussion in Galoob. There, we noted that all the Copyright Act's examples of derivative works took some definite, physical form and concluded that this was a requirement of the Act. See *Galoob*, 964 F.2d at 967-68; see also Edward G. Black & Michael H. Page, Add-On Infringements, 15 Hastings Comm/Ent. L.J. 615, 625 (1993) (noting that in *Galoob* the Ninth Circuit "re-examined the statutory definition of derivative works offered in section 101 and found an independent fixation requirement of sorts built into the statutory definition of derivative works"). Obviously, N/I's MAP files themselves exist in a concrete or permanent form; they are burned onto a CD-ROM. See ProCD, Inc. v. Zeidenberg, 86 F.3d 1447, 1453 (7th Cir. 1996) (computer files on a CD are fixed in a tangible medium of expression). But what about the audiovisual displays generated when D/N-3D runs the N/I MAP files — i.e., the actual game level as displayed on the screen? Micro Star argues that, because the audiovisual displays in Galoob didn't meet the "concrete or permanent form" requirement, neither do N/I's.

In *Galoob*, we considered audiovisual displays created using a device called the Game Genie, which was sold for use with the Nintendo Entertainment System. The Game Genie allowed players to alter individual features of a game, such as a character's strength or speed, by selectively "blocking the value for a single data byte sent by the game cartridge to the [Nintendo console] and replacing it with a new value." *Galoob*, 964 F.2d at 967. Players chose which data value to replace by entering a code; over a billion different codes were possible. The Game Genie was dumb; it functioned only as a window into the computer program, allowing players to temporarily modify individual aspects of the game. See Lewis Galoob Toys, Inc. v. Nintendo of Am., Inc., 780 F.Supp. 1283, 1289 (N.D.Cal. 1991).

Nintendo sued, claiming that when the Game Genie modified the game system's audiovisual display, it created an infringing derivative work. We rejected this claim because "[a] derivative work must incorporate a protected work in some concrete or permanent form." *Galoob*, 964 F.2d at 967 (internal quotation marks omitted). The audiovisual displays generated by combining the Nintendo System with the Game Genie were not incorporated in any permanent form; when the game was over, they were gone. Of course, they could be reconstructed, but only if the next player chose to reenter the same codes.[4]

Micro Star argues that the MAP files on N/I are a more advanced version of the Game Genie, replacing old values (the MAP files in the original game) with new values (N/I's MAP files). But, whereas the audiovisual displays created by Game Genie were never recorded in any permanent form, the audiovisual displays generated by D/N-3D from the N/I MAP files are in the MAP files themselves. In *Galoob*, the audiovisual display was defined by the original game cartridge, not by the Game Genie; no one could possibly say that the data values inserted by the Game Genie described the audiovisual display. In the present case the audiovisual display that appears on the computer monitor when a N/I level is played is described — in exact detail — by a N/I MAP file.

This raises the interesting question whether an exact, down to the last detail, description of an audiovisual display (and — by definition — we know that MAP files do describe audiovisual displays down to the last detail) counts as a permanent or concrete form for purposes of Galoob. We see no reason it shouldn't. What, after all, does sheet music do but describe in precise detail the way a copyrighted melody sounds? See 1 William F. Patry, Copyright Law and Practice 168 (1994) ("[A] musical composition may be embodied in sheet music. . . ."). To be copyrighted, pantomimes and dances may be "described in sufficient detail to enable the work to be performed from that

---

4. A low-tech example might aid understanding. Imagine a product called the Pink Screener, which consists of a big piece of pink cellophane stretched over a frame. When put in front of a television, it makes everything on the screen look pinker. Someone who manages to record the programs with this pink cast (maybe by filming the screen) would have created an infringing derivative work. But the audiovisual display observed by a person watching television through the Pink Screener is not a derivative work because it does not incorporate the modified image in any permanent or concrete form. The Game Genie might be described as a fancy Pink Screener for video games, changing a value of the game as perceived by the current player, but never incorporating the new audiovisual display into a permanent or concrete form.

description." Id. at 243 (citing Compendium II of Copyright Office Practices §463); see also Horgan v. Macmillan, Inc., 789 F.2d 157, 160 (2d Cir. 1986). Similarly, the N/I MAP files describe the audiovisual display that is to be generated when the player chooses to play D/N-3D using the N/I levels. Because the audiovisual displays assume a concrete or permanent form in the MAP files, *Galoob* stands as no bar to finding that they are derivative works.

In addition, "[a] work will be considered a derivative work only if it would be considered an infringing work if the material which it has derived from a preexisting work had been taken without the consent of a copyright proprietor of such preexisting work." Mirage Editions v. Albuquerque A.R.T. Co., 856 F.2d 1341, 1343 (quoting 1 Nimmer on Copyright §3.01 (1986)) (internal quotation marks omitted). "To prove infringement, [FormGen] must show that [D/N-3D's and N/I's audiovisual displays] are substantially similar in both ideas and expression." Litchfield v. Spielberg, 736 F.2d 1352, 1356 (9th Cir. 1984) (emphasis omitted). Similarity of ideas may be shown by comparing the objective details of the works: plot, theme, dialogue, mood, setting, characters, etc. See id. Similarity of expression focuses on the response of the ordinary reasonable person, and considers the total concept and feel of the works. See id. at 1356-57. FormGen will doubtless succeed in making these showings since the audiovisual displays generated when the player chooses the N/I levels come entirely out of D/N-3D's source art library. Cf. Atari, Inc. v. North Am. Philips Consumer Elec. Corp., 672 F.2d 607, 620 (7th Cir. 1982) (finding two video games substantially similar because they shared the same "total concept and feel").

Micro Star further argues that the MAP files are not derivative works because they do not, in fact, incorporate any of D/N-3D's protected expression. In particular, Micro Star makes much of the fact that the N/I MAP files reference the source art library, but do not actually contain any art files themselves. Therefore, it claims, nothing of D/N-3D's is reproduced in the MAP files. In making this argument, Micro Star misconstrues the protected work. The work that Micro Star infringes is the D/N-3D story itself — a beefy commando type named Duke who wanders around post-Apocalypse Los Angeles, shooting Pig Cops with a gun, lobbing hand grenades, searching for medkits and steroids, using a jetpack to leap over obstacles, blowing up gas tanks, avoiding radioactive slime. A copyright owner holds the right to create sequels, see Trust Co. Bank v. MGM/UA Entertainment Co., 772 F.2d 740 (11th Cir. 1985), and the stories told in the N/I MAP files are surely sequels, telling new (though somewhat repetitive) tales of Duke's fabulous adventures. A book about Duke Nukem would infringe for the same reason, even if it contained no pictures.[5]

---

5. We note that the N/I MAP files can only be used with D/N-3D. If another game could use the MAP files to tell the story of a mousy fellow who travels through a beige maze, killing vicious saltshakers with paper-clips, then the MAP files would not incorporate the protected expression of D/N-3D because they would not be telling a D/N-3D story.

Micro Star nonetheless claims that its use of D/N-3D's protected expression falls within the doctrine of fair use. . . .

As a preliminary matter, Micro Star asks us to focus on the player's use of the N/I CD in evaluating the fair use claim, because-according to Micro Star-the player actually creates the derivative work. In Galoob, after we assumed for purposes of argument that the Game Genie did create derivative works, we went on to consider the fair use defense from the player's point of view. See *Galoob*, 964 F.2d at 970. But the fair use analysis in *Galoob* was not necessary and therefore is clearly dicta. More significantly, Nintendo alleged only contributory infringement — that Galoob was helping consumers create derivative works; FormGen here alleges direct infringement by Micro Star, because the MAP files encompass new Duke stories, which are themselves derivative works.

Our examination of the section 107 factors yields straightforward results. Micro Star's use of FormGen's protected expression was made purely for financial gain. While that does not end our inquiry, see Campbell v. Acuff-Rose Music, Inc., 510 U.S. 569, 584, 114 S.Ct. 1164, 127 L.Ed.2d 500 (1994), "every commercial use of copyrighted material is presumptively an unfair exploitation of the monopoly privilege that belongs to the owner of the copyright." Sony Corp. of Am. v. Universal City Studios, Inc., 464 U.S. 417, 451, 104 S.Ct. 774, 78 L.Ed.2d 574 (1984).[6] The Supreme Court has explained that the second factor, the nature of the copyrighted work, is particularly significant because "some works are closer to the core of intended copyright protection than others, with the consequence that fair use is more difficult to establish when the former works are copied." *Campbell*, 510 U.S. at 586, 114 S.Ct. 1164. The fair use defense will be much less likely to succeed when it is applied to fiction or fantasy creations, as opposed to factual works such as telephone listings. See United Tel. Co. v. Johnson Publ'g Co., 855 F.2d 604, 609 (8th Cir. 1988); see also Stewart v. Abend, 495 U.S. 207, 237, 110 S.Ct. 1750, 109 L.Ed.2d 184 (1990). Duke Nukem's world is made up of aliens, radioactive slime and freezer weapons—clearly fantasies, even by Los Angeles standards. N/I MAP files "expressly use[] the [D/N-3D] story's unique setting, characters, [and] plot," *Stewart*, 495 U.S. at 238, 110 S.Ct. 1750; both the quantity and importance of the material Micro Star used are substantial. Finally, by selling N/I, Micro Star "impinged on [FormGen's] ability to market new versions of the [D/N-3D] story." *Stewart*, 495 U.S. at 238, 110 S.Ct. 1750; see also Twin Peaks Productions, Inc. v. Publications Intl., Ltd., 996 F.2d 1366, 1377 (2d Cir. 1993). Only FormGen has the right to enter that market; whether it chooses to do so is entirely its business. "[N/I] neither falls into any of the categories enumerated in section 107 nor meets the four criteria set forth in section 107." *Stewart*, 495 U.S. at 237, 110 S.Ct. 1750. It is not protected by fair use.

---

6. Of course, transformative works have greater recourse to the fair use defense as they "lie at the heart of the fair use doctrine's guarantee of breathing space within the confines of copyright . . . and the more   transformative the new work, the less will be the significance of other factors, like commercialism, that may weigh against a finding of fair use." Campbell, 510 U.S. at 579, 114 S.Ct. 1164 (citations omitted). N/I can hardly be described as transformative; anything but.

Micro Star also argues that it is the beneficiary of the implicit license FormGen gave to its customers by authorizing them to create new levels. Section 204 of the Copyright Act requires the transfer of the exclusive rights granted to copyright owners (including the right to prepare derivative works) to be in writing. See 17 U.S.C. §204(a); Effects Assocs., Inc. v. Cohen, 908 F.2d 555, 556 (9th Cir. 1990). A nonexclusive license may, however, be granted orally or implied by conduct. See *Effects*, 908 F.2d at 558. Nothing indicates that FormGen granted Micro Star any written license at all; nor is there evidence of a nonexclusive oral license. The only written license FormGen conceivably granted was to players who designed their own new levels, but that license contains a significant limitation: Any new levels the players create "must be offered [to others] solely for free." The parties dispute whether the license is binding, but it doesn't matter. If the license is valid, it clearly prohibits commercial distribution of levels; if it doesn't, FormGen hasn't granted any written licenses at all.

In case FormGen didn't license away its rights, Micro Star argues that, by providing the Build Editor and encouraging players to create their own levels, FormGen abandoned all rights to its protected expression. It is well settled that rights gained under the Copyright Act may be abandoned. But abandonment of a right must be manifested by some overt act indicating an intention to abandon that right. See Hampton v. Paramount Pictures Corp., 279 F.2d 100, 104 (9th Cir. 1960). Given that it overtly encouraged players to make and freely distribute new levels, FormGen may indeed have abandoned its exclusive right to do the same. But abandoning some rights is not the same as abandoning all rights, and FormGen never overtly abandoned its rights to profit commercially from new levels. Indeed, FormGen warned players not to distribute the levels commercially and has actively enforced that limitation by bringing suits such as this one.

IV

Because FormGen will likely succeed at trial in proving that Micro Star has infringed its copyright, we reverse the district court's order denying a preliminary injunction and remand for entry of such an injunction.

COMMENTS & QUESTIONS

1. Exactly what protected expression did Micro Star copy? The MAP files produced by game players themselves contain entirely new program code and contain no code at all copied from Formgen. It is true that these files are designed to run with Formgen's game engine and art library, and indeed would be useless without them. But does it follow from that that the MAP files themselves are "copies" of the original Duke Nukem game?

The real basis of the court's argument seems to be that the MAP files are guides to an infringing work. If a user buys the Micro Star CD and runs the game using the MAP files, he will generate on his screen a new Duke Nukem "level" that incorporates Micro Star's protected art. That might possibly be an infringing act which Micro Star has contributed to or induced. If that is the complaint, the court was wrong to distinguish *Galoob*. Indeed, *Galoob* looks precisely analogous: the defendant makes a product that does not itself contain any copyrighted expression but which end users can employ to generate altered versions of the copyrighted work. The same analytic problem infects the court's analysis of the third fair use factor in which the court concluded that the use copied substantially from Duke Nukem. While the game played by the end user does copy substantially from Formgen's original, it is hard to see how anyone could argue that the MAP files themselves did so.

2. Does the court's analogy to a movie sequel support its result? Why exactly are sequels considered derivative works? If it were possible to imagine a sequel that didn't itself include any copyrighted expression, would it be infringing? Should it be? How does the court square its result with *Litchfield v. Spielberg*, cited in the case, that held a work could not be an infringing derivative work unless it was substantially similar in protected expression to the original?

3. The court's fair use analysis has a number of problems. First, the court disposes of the first factor entirely by reference to the *Sony* "presumption" that commercially motivated uses are unfair. This approach was expressly repudiated by the Supreme Court in Campbell v. Acuff-Rose Music, Inc., 510 U.S. 569, 584 (1994), which flatly rejected the use of any such presumption.

Second, the court suggests in a footnote that there is nothing transformative about the MAP files Micro Star sold. Why not? Aren't these files creative additions to the Duke Nukem universe?

Third, the court disposes of the market-effect factor by asserting that Micro Star's use interferes with Formgen's right to sell additional Duke Nukem levels. Does it matter whether Formgen has made any effort to do so? Whether it plans to do so in the future? Neither fact is considered in the court's opinion. It is also worth considering how Formgen's license to users to distribute their own levels for free on the Internet would affect the market for sales of additional levels.

4. Is this case really about the scope of the license Formgen admittedly granted to the individual users who wrote the MAP files at issue? It appears that Micro Star could have collated and distributed those MAP files for free without running afoul of Formgen's license restriction. If the only basis for the suit is that Micro Star charged money for doing what it had a right to do for free, isn't this case better resolved under contract than copyright law?

5. We discuss one specific type of fair use, reverse engineering, in the next section. For a discussion of a variety of other activities that may qualify for fair use in the digital environment, see Pamela Samuelson, Fair Use for Computer Programs and Other Copyrightable Works in Digital Form: The Implications of Sony, Galoob and Sega, 1 J. Intell. Prop. L. 49 (1993).

Note on Reverse Engineering

In view of the network externalities flowing from many software products and the cumulative nature of innovation in software, two critical issues in the computer industry are the manner by and extent to which competitors can develop compatible or interoperable programs, competitive programs, and enhancements of programs. Of particular relevance is the permissibility of reverse engineering. Reverse engineering is loosely defined as "starting with the known product and working backward to divine the process which aided in its development or manufacture." Kewanee Oil v. Bicron, 416 U.S. 470 (1974). It typically involves two phases: (1) disassembly or decompilation of the program in order to create human-readable source code that may be analyzed, and (2) using the results of this analysis to create a commercially viable program.

Most computer programs are distributed only in object code form, a fact that makes it difficult to discover the ideas and principles contained in a program without reverse engineering. In addition, some programs have special devices designed to prevent inspection. Nonetheless, reverse engineering of software, while difficult, is possible in some limited circumstances. There are a variety of approaches to reverse engineering of computer programs. One is referred to as "black box" testing. By systematically inputting instructions, a computer programmer can learn how the program processes information and can construct a program that produces the same outputs for given instructions. Because of the complexity of most programs today, however, this approach is often infeasible. A second method is to study the technical specifications and user manuals. These materials, however, often do not contain sufficient information to achieve full compatibility and interoperability. In order to feasibly achieve these objectives, it is often necessary to take a machine apart and directly observe its operation or, as is more common with regard to software, to decompile the program. After a program is understood, the competitor will use the results of the analysis to create a commercially viable program. The case and materials that follow consider the validity of such decompilation.

═══
═══ *Sega Enterprises Ltd. v. Accolade, Inc.*
═══ *United States Court of Appeals for the Ninth Circuit*
═══ *977 F.2d 1510 (9th Cir. 1992)*
═══

REINHART, Circuit Judge: . . .

I. Background

Plaintiff-appellee Sega Enterprises, Ltd. ("Sega"), a Japanese corporation, and its subsidiary, Sega of America, develop and market video entertainment systems, including the "Genesis" console (distributed in Asia under the name "Mega-Drive") and video game cartridges. Defendant-appellant Accolade, Inc., is an independent developer, manufacturer, and marketer of computer entertainment software, including game cartridges that are compatible with the Genesis console, as well as game cartridges that are compatible with other computer systems.

Sega licenses its copyrighted computer code and its "SEGA" trademark to a number of independent developers of computer game software. Those licensees develop and sell Genesis-compatible video games in competition with Sega. Accolade is not and never has been a licensee of Sega. Prior to rendering its own games compatible with the Genesis console, Accolade explored the possibility of entering into a licensing agreement with Sega, but abandoned the effort because the agreement would have required that Sega be the exclusive manufacturer of all games produced by Accolade.

Accolade used a two-step process to render its video games compatible with the Genesis console. First, it "reverse engineered" Sega's video game programs in order to discover the requirements for compatibility with the Genesis console. As part of the reverse engineering process, Accolade transformed the machine-readable object code contained in commercially available copies of Sega's game cartridges into human-readable source code using a process called "disassembly" or "decompilation".[2] Accolade purchased a Genesis console and three Sega game cartridges, wired a decompiler into the console circuitry, and generated printouts of the resulting source code. Accolade engineers studied and annotated the printouts in order to identify areas of commonality among the three game programs. They then loaded the disassembled code back into a computer, and experimented to discover the interface specifications for the Genesis console by modifying the programs and studying the results. At the end of the reverse engineering process, Accolade created a development manual that incorporated the information it had discovered about the requirements for a Genesis-compatible game. According to the Accolade employees who created the manual, the manual contained only functional descriptions of the interface requirements and did not include any of Sega's code.

In the second stage, Accolade created its own games for the Genesis. According to Accolade, at this stage it did not copy Sega's programs, but relied only on the information concerning interface specifications for the Genesis that was contained in its development manual. Accolade maintains that

2. Computer programs are written in specialized alphanumeric languages, or "source code". In order to operate a computer, source code must be translated into computer readable form, or "object code". Object code uses only two symbols, 0 and 1, in combinations which represent the alphanumeric characters of the source code. A program written in source code is translated into object code using a computer program called an "assembler" or "compiler", and then imprinted onto a silicon chip for commercial distribution. Devices called "disassemblers" or "decompilers" can reverse this process by "reading" the electronic signals for "0" and "1" that are produced while the program is being run, storing the resulting object code in computer memory, and translating the object code into source code. Both assembly and disassembly devices are commercially available, and both types of devices are widely used within the software industry.

with the exception of the interface specifications, none of the code in its own games is derived in any way from its examination of Sega's code. In 1990, Accolade released "Ishido", a game which it had originally developed and released for use with the Macintosh and IBM personal computer systems, for use with the Genesis console.

[With its Genesis III product, Sega included a new lockout code, called the TMSS system.]

Accolade learned of the impending release of the Genesis III in the United States in January, 1991, when the Genesis III was displayed at a consumer electronics show. When a demonstration at the consumer electronics show revealed that Accolade's "Ishido" game cartridges would not operate on the Genesis III, Accolade returned to the drawing board. During the reverse engineering process, Accolade engineers had discovered a small segment of code — the TMSS initialization code — that was included in the "power-up" sequence of every Sega game, but that had no identifiable function. The games would operate on the original Genesis console even if the code segment was removed. Mike Lorenzen, the Accolade engineer with primary responsibility for reverse engineering the interface procedures for the Genesis console, sent a memo regarding the code segment to Alan Miller, his supervisor and the current president of Accolade, in which he noted that "it is possible that some future Sega peripheral device might require it for proper initialization."

In the second round of reverse engineering, Accolade engineers focused on the code segment identified by Lorenzen. After further study, Accolade added the code to its development manual in the form of a standard header file to be used in all games. The file contains approximately twenty to twenty-five bytes of data. Each of Accolade's games contains a total of 500,000 to 1,500,000 bytes. According to Accolade employees, the header file is the only portion of Sega's code that Accolade copied into its own game programs. In this appeal, Sega does not raise a separate claim of copyright infringement with respect to the header file.

In 1991, Accolade released five more games for use with the Genesis III, "Star Control", "Hardball!", "Onslaught", "Turrican", and "Mike Ditka Power Football." With the exception of "Mike Ditka Power Football", all of those games, like "Ishido", had originally been developed and marketed for use with other hardware systems. All contained the standard header file that included the TMSS initialization code. . . .

[Sega sued Accolade for copyright infringement.]

III. Copyright Issues

Accolade raises four arguments in support of its position that disassembly of the object code in a copyrighted computer program does not constitute copyright infringement. First, it maintains that intermediate copying does not infringe the exclusive rights granted to copyright owners in section 106 of the Copyright Act unless the end product of the copying is substantially similar to the copyrighted work. Second, it argues that disassembly of object code in order to gain an understanding of the ideas and functional concepts embodied in the code is lawful under section 102(b) of the Act, which exempts ideas and functional concepts from copyright protection. Third, it suggests that disassembly is authorized by section 117 of the Act, which entitles the lawful owner of a copy of a computer program to load the program into a computer. Finally, Accolade contends that disassembly of object code in order to gain an understanding of the ideas and functional concepts embodied in the code is a fair use that is privileged by section 107 of the Act.

Neither the language of the Act nor the law of this circuit supports Accolade's first three arguments. Accolade's fourth argument, however, has merit. . . .

A. Intermediate Copying

We have previously held that the Copyright Act does not distinguish between unauthorized copies of a copyrighted work on the basis of what stage of the alleged infringer's work the unauthorized copies represent. Walker v. University Books, 602 F.2d 859, 864 (9th Cir. 1979) ("[T]he fact that an allegedly infringing copy of a protected work may itself be only an inchoate representation of some final product to be marketed commercially does not in itself negate the possibility of infringement."). Our holding in *Walker* was based on the plain language of the Act. Section 106 grants to the copyright owner the exclusive rights "to reproduce the work in copies", "to prepare derivative works based upon the copyrighted work", and to authorize the preparation of copies and derivative works. 17 U.S.C. §106 (1)-(2). Section 501 provides that "anyone who violates any of the exclusive rights of the copyright owner as provided by sections 106 through 118 . . . is an infringer of the copyright." Id. §501(a). On its face, that language unambiguously encompasses and proscribes "intermediate copying". Walker, 602 F.2d at 863-64.

In order to constitute a "copy" for purposes of the Act, the allegedly infringing work must be fixed in some tangible form, "from which the work can be perceived, reproduced, or otherwise communicated, either directly or with the aid of a machine or device." 17 U.S.C. §101. The computer file generated by the disassembly program, the printouts of the disassembled code, and the computer files containing Accolade's modifications of the code that were generated during the reverse engineering process all satisfy that requirement. The intermediate copying done by Accolade therefore falls squarely within the category of acts that are prohibited by the statute.

Accolade points to a number of cases that it argues establish the lawfulness of intermediate copying. Most of the cases involved the alleged copying of books, scripts, or literary characters. See v. Durang, 711 F.2d 141 (9th Cir. 1983); Warner Bros. v. ABC, 654 F.2d 204 (2d Cir. 1981); Miller v. Universal City Studios, Inc., 650 F.2d 1365 (5th Cir. 1981). In each case, however, the eventual lawsuit alleged infringement only as to the final work of the defendants. We conclude that this group of cases does not alter or limit the holding of *Walker*.

The remaining cases cited by Accolade, like the case before us, involved intermediate copying of computer code as an initial step in the development of a competing product. Computer Assoc. Int'l v. Altai, Inc., [supra] (2d Cir. 1992) ("CAI"); NEC Corp. v. Intel Corp., 10 U.S.P.Q.2d (BNA) 1177 (N.D. Cal. 1989); E. F. Johnson Co. v. Uniden Corp., 623 F. Supp. 1485 (D. Minn. 1985). In each case, the court based its determination regarding infringement solely on the degree of similarity between the allegedly infringed work and the defendant's final product. A close reading of those cases, however, reveals that in none of them was the legality of the intermediate copying at issue. Sega cites an equal number of cases involving intermediate copying of copyrighted computer code to support its assertion that such copying is prohibited. Atari Games Corp. v. Nintendo of America, Inc., 18 U.S.P.Q.2d (BNA) 1935 (N.D. Cal. 1991); SAS Institute, Inc. v. S&H Computer Systems, Inc., 605 F. Supp. 816 (M.D. Tenn. 1985); S&H Computer Systems, Inc. v. SAS Institute, Inc., 568 F. Supp. 416 (M.D. Tenn. 1983); Hubco Data Products v. Management Assistance, Inc., 219 U.S.P.Q. (BNA) 450 (D. Idaho 1983). Again, however, it appears that the question of the lawfulness of intermediate copying was not raised in any of those cases.

In summary, the question whether intermediate copying of computer object code infringes the exclusive rights granted to the copyright owner in section 106 of the Copyright Act is a question of first impression. In light of the unambiguous language of the Act, we decline to depart from the rule set forth in Walker for copyrighted works generally. Accordingly, we hold that intermediate copying of computer object code may infringe the exclusive rights granted to the copyright owner in section 106 of the Copyright Act regardless of whether the end product of the copying also infringes those rights. If intermediate copying is permissible under the Act, authority for such copying must be found in one of the statutory provisions to which the rights granted in section 106 are subject.

B. The Idea/Expression Distinction

Accolade next contends that disassembly of computer object code does not violate the Copyright Act because it is necessary in order to gain access to the ideas and functional concepts embodied in the code, which are not protected by copyright. 17 U.S.C. §102(b). Because humans cannot comprehend object code, it reasons, disassembly of a commercially available computer program into human-readable form should not be considered an infringement of the owner's copyright. Insofar as Accolade suggests that disassembly of object code is lawful *per se*, it seeks to overturn settled law.

Accolade's argument regarding access to ideas is, in essence, an argument that object code is not eligible for the full range of copyright protection. Although some scholarly authority supports that view, we have previously rejected it based on the language and legislative history of the Copyright Act. Johnson Controls, Inc. v. Phoenix Control Sys., Inc., 886 F.2d 1173, 1175 (9th Cir. 1989). . . .

Nor does a refusal to recognize a *per se* right to disassemble object code lead to an absurd result. The ideas and functional concepts underlying many types of computer programs, including word processing programs, spreadsheets, and video game displays, are readily discernible without the need for disassembly, because the operation of such programs is visible on the computer screen. The need to disassemble object code arises, if at all, only in connection with operations systems, system interface procedures, and other programs that are not visible to the user when operating — and then only when no alternative means of gaining an understanding of those ideas and functional concepts exists. In our view, consideration of the unique nature of computer object code thus is more appropriate as part of the case-by-case, equitable "fair use" analysis authorized by section 107 of the Act. See infra Part III(D). Accordingly, we reject Accolade's second argument.

C. Section 117

Section 117 of the Copyright Act allows the lawful owner of a copy of a computer program to copy or adapt the program if the new copy or adaptation "is created as an essential step in the utilization of the computer program in conjunction with a machine and . . . is used in no other manner." 17 U.S.C. §117(1). Accolade contends that section 117 authorizes disassembly of the object code in a copyrighted computer program.

Section 117 was enacted on the recommendation of CONTU, which noted that "because the placement of any copyrighted work into a computer is the preparation of a copy [since the program is loaded into the computer's memory], the law should provide that persons in rightful possession of copies of programs be able to use them freely without fear of exposure to copyright liability." CONTU Report at 13. We think it is clear that Accolade's use went far beyond that contemplated by CONTU and authorized by section 117. Section 117 does not purport to protect a user who disassembles object code, converts it from assembly into source code, and makes printouts and photocopies of the refined source code version.

D. Fair Use

Accolade contends, finally, that its disassembly of copyrighted object code as a necessary step in its examination of the unprotected ideas and functional concepts embodied in the code is a fair use that is privileged by section 107 of the Act. Because, in the case before us, disassembly is the only means of gaining access to those unprotected aspects of the program, and because Accolade has a legitimate interest in gaining such access (in order to determine how to make its cartridges compatible with the Genesis console), we agree with Accolade. Where there is good reason for studying or examining the unprotected aspects of a copyrighted computer program, disassembly for purposes of such study or examination constitutes a fair use.

1

As a preliminary matter, we reject Sega's contention that the assertion of a fair use defense in connection with the disassembly of object code is precluded by statute. First, Sega argues that not only does section 117 of the Act *not* authorize disassembly of object code, but it also constitutes a legislative determination that any copying of a computer program *other* than that authorized by section 117 cannot be considered a fair use of that program under section 107. That argument verges on the frivolous. Each of the exclusive rights created by section 106 of the Copyright Act is expressly made subject to all of the limitations contained in sections 107 through 120. 17 U.S.C. §106. Nothing in the language or the legislative history of section 117, or in the CONTU Report, suggests that section 117 was intended to preclude the assertion of a fair use defense with respect to uses of computer programs that are not covered by section 117, nor has section 107 been amended to exclude computer programs from its ambit.

Moreover, sections 107 and 117 serve entirely different functions. Section 117 defines a narrow category of copying that is lawful *per se*. 17 U.S.C. §117. Section 107, by contrast, establishes a *defense* to an otherwise valid claim of copyright infringement. . . .

Second, Sega maintains that the language and legislative history of section 906 of the Semiconductor Chip Protection Act of 1984 (SCPA) establish that Congress did not intend that disassembly of object code be considered a fair use. Section 906 of the SCPA authorizes the copying of the "mask work" on a silicon chip in the course of reverse engineering the chip. 17 U.S.C. §906. The mask work in a standard ROM chip, such as those used in the Genesis console and in Genesis-compatible cartridges, is a physical representation of the computer program that is embedded in the chip. The zeros and ones of binary object code are represented in the circuitry of the mask work by open and closed switches. Sega contends that Congress's express authorization of copying in the particular circumstances set forth in section 906 constitutes a determination that other forms of copying of computer programs are prohibited.

The legislative history of the SCPA reveals, however, that Congress passed a separate statute to protect semiconductor chip products because it believed that semiconductor chips were intrinsically utilitarian articles that were not protected under the Copyright Act. H.R. Rep. No. 781, 98th Cong., 2d Sess. 8-10, reprinted in 1984 U.S.C.C.A.N. 5750, 5757-59. Accordingly, rather than amend the Copyright Act to extend traditional copyright protection to chips, it enacted "a sui generis form of protection, apart from and independent of the copyright laws." Id. at 10, 1984 U.S.C.C.A.N. at 5759. Because Congress did not believe that semiconductor chips were eligible for copyright protection in the first instance, the fact that it included an exception for reverse engineering of mask work in the SCPA says nothing about its intent with respect to the lawfulness of disassembly of computer programs under the Copyright Act. Nor is the fact that Congress did not contemporaneously amend the Copyright Act to permit disassembly significant, since it was focusing on the protection to be afforded to semiconductor chips. Here we are dealing not with an alleged violation of the SCPA, but with the copying of a computer program, which is governed by the Copyright Act. Moreover, Congress expressly stated that it did not intend to "limit, enlarge or otherwise affect the scope, duration, ownership or subsistence of copyright protection . . . in computer programs, data bases, or any other copyrightable works embodied in semiconductor chip products." Id. at 28, 1984 U.S.C.C.A.N. at 5777. Accordingly, Sega's second statutory argument also fails. We proceed to consider Accolade's fair use defense.

2 . . .

In determining that Accolade's disassembly of Sega's object code did not constitute a fair use, the district court treated the first and fourth statutory factors [of 17 U.S.C. §107, reprinted above] as dispositive, and ignored the second factor entirely. Given the nature and characteristics of Accolade's direct use of the copied works, the ultimate use to which Accolade put the functional information it obtained, and the nature of the market for home video entertainment systems, we conclude that neither the first nor the fourth factor weighs in Sega's favor. In fact, we conclude that both factors support Accolade's fair use defense, as does the second factor, a factor which is important to the resolution of cases such as the one before us.

(a)

With respect to the first statutory factor, we observe initially that the fact that copying is for a commercial purpose weighs against a finding of fair use. Harper & Row, 471 U.S. at 562. However, the presumption of unfairness that arises in such cases can be rebutted by the characteristics of a particular commercial use. Hustler Magazine, Inc. v. Moral Majority, Inc., 796 F.2d 1148, 1152 (9th Cir. 1986); see also Maxtone-Graham v. Burtchaell, 803 F.2d 1253, 1262 (2d Cir. 1986), *cert. denied,* 481 U.S. 1059, 95 L. Ed. 2d 856, 107 S. Ct. 2201 (1987). Further "the commercial nature of a use is a matter of degree, not an absolute. . . ." Maxtone-Graham, 803 F.2d at 1262.

Sega argues that because Accolade copied its object code in order to produce a competing product, the Harper & Row presumption applies and precludes a finding of fair use. That analysis is far too simple and ignores a number of important considerations. We must consider other aspects of "the purpose and character of the use" as well. As we have noted, the use at issue was an intermediate one only and thus any commercial "exploitation" was indirect or derivative.

The declarations of Accolade's employees indicate, and the district court found, that Accolade copied Sega's software solely in order to discover the functional requirements for compatibility with the Genesis console — aspects of Sega's programs that are not protected by copyright. 17 U.S.C. §102(b). With respect to the video game programs contained in Accolade's game cartridges, there is no evidence in the record that Accolade sought to avoid performing its own creative work. Indeed, most of the games that Accolade released for use with the Genesis console were originally developed for other hardware systems. Moreover, with respect to the interface procedures for the Genesis console, Accolade did not seek to avoid paying a customarily charged fee for use of those procedures, nor did it simply copy Sega's code; rather, it wrote its own procedures based on what it had learned through disassembly. Taken together, these facts indicate that although Accolade's ultimate purpose was the release of Genesis-compatible games for sale, its direct purpose in copying Sega's code, and thus its direct use of the copyrighted material, was simply to study the functional requirements for Genesis compatibility so that it could modify existing games and make them usable with the Genesis console. Moreover, as we discuss below, no other method of studying those requirements was available to Accolade. On these facts, we conclude that Accolade copied Sega's code for a legitimate, essentially non-exploitative purpose, and that the commercial aspect of its use can best be described as of minimal significance.

We further note that we are free to consider the public benefit resulting from a particular use notwithstanding the fact that the alleged infringer may gain commercially. See Hustler, 796 F.2d at 1153 (quoting MCA, Inc. v. Wilson, 677 F.2d 180, 182 (2d Cir. 1981)). Public benefit need not be direct or tangible, but may arise because the challenged use serves a public interest. Id. In the case before us, Accolade's identification of the functional requirements for Genesis compatibility has led to an increase in the number of independently designed video game programs offered for use with the Genesis console. It is precisely this growth in creative expression, based on the dissemination of other creative works and the unprotected ideas contained in those works, that the Copyright Act was intended to promote. See Feist Publications, Inc. v. Rural Tel. Serv. Co., 111 S. Ct. 1282, 1290, 113 L. Ed. 2d 358 (1991) (citing Harper & Row, 471 U.S. at 556-57). The fact that Genesis-compatible video games are not scholarly works, but works offered for sale on the market, does not alter our judgment in this regard. We conclude that given the purpose and character of Accolade's use of Sega's video game programs, the presumption of unfairness has been overcome and the first statutory factor weighs in favor of Accolade.

(b)

As applied, the fourth statutory factor, effect on the potential market for the copyrighted work, bears a close relationship to the "purpose and character" inquiry in that it, too, accommodates the distinction between the copying of works in order to make independent creative expression possible and the simple exploitation of another's creative efforts. We must, of course, inquire whether, "if [the challenged use] should become widespread, it would adversely affect the potential market for the copyrighted work," Sony Corp. v. Universal City Studios, 464 U.S. 417, 451 (1984), by diminishing potential sales, interfering with marketability, or usurping the market, *Hustler*, 796 F.2d at 1155-56. If the copying resulted in the latter effect, all other considerations might be irrelevant. The Harper & Row Court found a use that effectively usurped the market for the copyrighted work by supplanting that work to be dispositive. 471 U.S. at 567-69. However, the same consequences do not and could not attach to a use which simply enables the copier to enter the market for works of the same type as the copied work.

Unlike the defendant in *Harper & Row*, which printed excerpts from President Ford's memoirs verbatim with the stated purpose of "scooping" a Time magazine review of the book, 471 U.S. at 562, Accolade did not attempt to "scoop" Sega's release of any particular game or games, but sought only to become a legitimate competitor in the field of Genesis-compatible video games. Within that market, it is the characteristics of the game program as experienced by the user that determine the program's commercial success. As we have noted, there is nothing in the record that suggests that Accolade copied any of those elements.

By facilitating the entry of a new competitor, the first lawful one that is not a Sega licensee, Accolade's disassembly of Sega's software undoubtedly "affected" the market for Genesis-compatible games in an indirect fashion. We note, however, that while no consumer except the most avid devotee of President Ford's regime might be expected to buy more than one version of the President's memoirs, video game users typically purchase more than one game. There is no basis for assuming that Accolade's "Ishido" has significantly affected the market for Sega's "Altered Beast", since a consumer might easily purchase both; nor does it seem unlikely that a consumer particularly interested in sports might purchase both Accolade's "Mike Ditka Power Football" and Sega's "Joe Montana Football", particularly if the games are, as Accolade contends, not substantially similar. In any event, an attempt to monopolize the market by making it impossible for others to compete runs counter to the statutory purpose of promoting creative expression and cannot constitute a strong equitable basis for resisting the invocation of the fair use doctrine. Thus, we conclude that the fourth statutory factor weighs in Accolade's, not Sega's, favor, notwithstanding the minor economic loss Sega may suffer.

(c)

The second statutory factor, the nature of the copyrighted work, reflects the fact that not all copyrighted works are entitled to the same level of protection. The protection established by the Copyright Act for original works of authorship does not extend to the ideas underlying a work or to the functional or factual aspects of the work. 17 U.S.C. §102(b). To the extent that a work is functional or factual, it may be copied, Baker v. Selden, 101 U.S. 99, 102-04, (1879), as may those expressive elements of the work that "must necessarily be used as incident to" expression of the underlying ideas, functional concepts, or facts, id. at 104. Works of fiction receive greater protection than works that have strong factual elements, such as historical or biographical works, Maxtone-Graham, 803 F.2d at 1263 (citing Rosemont Enterprises, Inc. v. Random House, Inc., 366 F.2d 303, 307 (2d Cir. 1966), *cert. denied,* 385 U.S. 1009 (1967)), or works that have strong functional elements, such as accounting textbooks, Baker, 101 U.S. at 104. Works that are merely compilations of fact are copyrightable, but the copyright in such a work is "thin." Feist Publications, 111 S. Ct. at 1289.

Computer programs pose unique problems for the application of the "idea/expression distinction" that determines the extent of copyright protection. To the extent that there are many possible ways of accomplishing a given task or fulfilling a particular market demand, the programmer's choice of program structure and design may be highly creative and idiosyncratic. However, computer programs are, in essence, utilitarian articles — articles that accomplish tasks. As such, they contain many logical, structural, and visual display elements that are dictated by the function to be performed, by considerations of efficiency, or by external factors such as compatibility requirements and industry demands. Computer Assoc. Int'l, Inc. v. Altai, Inc. In some circumstances, even the exact set of commands used by the programmer is deemed functional rather than creative for purposes of copyright. "When specific instructions, even though previously copyrighted, are the only and essential means of accomplishing a given task, their later use by another will not amount to infringement." CONTU Report at 20; see CAI, 23 U.S.P.Q. 2d at 1254.

Because of the hybrid nature of computer programs, there is no settled standard for identifying what is protected expression and what is unprotected idea in a case involving the alleged infringement of a copyright in computer software. We are in wholehearted agreement with the Second Circuit's recent observation that "thus far, many of the decisions in this area reflect the courts' attempt to fit the proverbial square peg in a round hole." CAI, 23 U.S.P.Q. 2d at 1257. In 1986, the Third Circuit attempted to resolve the dilemma by suggesting that the idea or function of a computer program is the idea of the program as a whole, and "everything that is not necessary to that purpose or function [is] part of the expression of that idea." Whelan Assoc., Inc. v. Jaslow Dental Laboratory, Inc., 797 F.2d 1222, 1236 (3d Cir. 1986) (emphasis omitted). The Whelan rule, however, has been widely — and soundly — criticized as simplistic and overbroad. See CAI, 23 U.S.P.Q. 2d at 1252 (citing cases, treatises, and articles). In reality, "a computer program's ultimate function or purpose is the composite result of interacting subroutines. Since each subroutine is itself a program, and thus, may be said to have its own 'idea,' Whelan's general formulation . . . is descriptively inadequate." Id. For example, the computer program at issue in the case before us, a video game program, contains at least two such subroutines — the subroutine that allows the user to interact with the video game and the subroutine that allows the game cartridge to interact with the console. Under a test that breaks down a computer program into its component subroutines and sub-subroutines and then identifies the idea or core functional element of each, such as the test recently adopted by the Second Circuit in CAI, 23 U.S.P.Q.2d (BNA) at 1252-53, many aspects of the program are not protected by copyright. In our view, in light of the essentially utilitarian nature of computer programs, the Second Circuit's approach is an appropriate one.

Sega argues that even if many elements of its video game programs are properly characterized as functional and therefore not protected by copyright, Accolade copied protected expression. Sega is correct. The record makes clear that disassembly is wholesale copying. Because computer programs are also unique among copyrighted works in the form in which they are distributed for public use, however, Sega's observation does not bring us much closer to a resolution of the dispute.

The unprotected aspects of most functional works are readily accessible to the human eye. The systems described in accounting textbooks or the basic structural concepts embodied in architectural plans, to give two examples, can be easily copied without also copying any of the protected, expressive aspects of the original works. Computer programs, however, are typically distributed for public use in object code form, embedded in a silicon chip or on a floppy disk. For that reason, humans often cannot gain access to the unprotected ideas and functional concepts contained in object code without disassembling that code — i.e., making copies.[8] Atari Games v. Nintendo of America, 975 F.2d 832 (Fed. Cir. 1992).

Sega argues that the record does not establish that disassembly of its object code is the only available method for gaining access to the interface specifications for the Genesis console, and the district court agreed. An independent examination of the record reveals that Sega misstates its contents, and demonstrates that the district court committed clear error in this respect.

First, the record clearly establishes that humans cannot *read* object code. Sega makes much of Mike Lorenzen's statement that a reverse engineer can work directly from the zeros and ones of object code but "it's not as fun." In full, Lorenzen's statements establish only that the use of an electronic

---

8. We do not intend to suggest that disassembly is always the only available means of access to those aspects of a computer program that are unprotected by copyright. As we noted in Part III(B), supra, in many cases the operation of a program is directly reflected on the screen display and therefore visible to the human eye. In those cases, it is likely that a reverse engineer would not need to examine the code in order to understand what the program does.

decompiler is not absolutely necessary. Trained programmers can disassemble object code by hand. Because even a trained programmer cannot possibly remember the millions of zeros and ones that make up a program, however, he must make a written or computerized copy of the disassembled code in order to keep track of his work. See generally Johnson-Laird, Technical Demonstration of "Decompilation", reprinted in Reverse Engineering: Legal and Business Strategies for Competitive Design in the 1990's 102 (Prentice Hall Law & Business ed. 1992). The relevant fact for purposes of Sega's copyright infringement claim and Accolade's fair use defense is that *translation* of a program from object code into source code cannot be accomplished without making copies of the code.

Second, the record provides no support for a conclusion that a viable alternative to disassembly exists. The district court found that Accolade could have avoided a copyright infringement claim by "peeling" the chips contained in Sega's games or in the Genesis console, as authorized by section 906 of the SCPA, 17 U.S.C. §906. Even Sega's amici agree that this finding was clear error. The declaration of Dr. Harry Tredennick, an expert witness for Accolade, establishes that chip peeling yields only a physical diagram of the *object code* embedded in a ROM chip. It does not obviate the need to translate object code into source code. Atari Games Corp., slip op. at 22.

The district court also suggested that Accolade could have avoided a copyright infringement suit by programming in a "clean room". That finding too is clearly erroneous. A "clean room" is a procedure used in the computer industry in order to prevent direct copying of a competitor's code during the development of a competing product. Programmers in clean rooms are provided only with the functional specifications for the desired program. As Dr. Tredennick explained, the use of a clean room would not have avoided the need for disassembly because disassembly was necessary in order to discover the functional specifications for a Genesis-compatible game.

In summary, the record clearly establishes that disassembly of the object code in Sega's video game cartridges was necessary in order to understand the functional requirements for Genesis compatibility. The interface procedures for the Genesis console are distributed for public use only in object code form, and are not visible to the user during operation of the video game program. Because object code cannot be read by humans, it must be disassembled, either by hand or by machine. Disassembly of object code necessarily entails copying. Those facts dictate our analysis of the second statutory fair use factor. If disassembly of copyrighted object code is *per se* an unfair use, the owner of the copyright gains a *de facto* monopoly over the functional aspects of his work — aspects that were expressly denied copyright protection by Congress. 17 U.S.C. §102(b). In order to enjoy a lawful monopoly over the idea or functional principle underlying a work, the creator of the work must satisfy the more stringent standards imposed by the patent laws. Bonito Boats, Inc. v. Thunder Craft Boats, Inc., 489 U.S. 141, 159-64, 103 L. Ed. 2d 118, 109 S. Ct. 971 (1989). Sega does not hold a patent on the Genesis console. Because Sega's video game programs contain unprotected aspects that cannot be examined without copying, we afford them a lower degree of protection than more traditional literary works. See CAI, 23 U.S.P.Q.2d at 1257. In light of all the considerations discussed above, we conclude that the second statutory factor also weighs in favor of Accolade.

(d)

As to the third statutory factor, Accolade disassembled entire programs written by Sega. Accordingly, the third factor weighs against Accolade. The fact that an entire work was copied does not, however, preclude a finding of fair use. Sony Corp., 464 U.S. at 449-50; Hustler, 795 F.2d at 1155 ("*Sony Corp.* teaches us that the copying of an entire work does not preclude fair use per se."). In fact, where the ultimate (as opposed to direct) use is as limited as it was here, the factor is of very little weight. Cf. Wright v. Warner Books, Inc., 953 F.2d 731, 738 (2d Cir. 1991).

(e)

In summary, careful analysis of the purpose and characteristics of Accolade's use of Sega's video game programs, the nature of the computer programs involved, and the nature of the market for video game cartridges yields the conclusion that the first, second, and fourth statutory fair use factors weigh in favor of Accolade, while only the third weighs in favor of Sega, and even then only slightly. Accordingly, Accolade clearly has by far the better case on the fair use issue.

We are not unaware of the fact that to those used to considering copyright issues in more traditional contexts, our result may seem incongruous at first blush. To oversimplify, the record establishes that Accolade, a commercial competitor of Sega, engaged in wholesale copying of Sega's copyrighted code as a preliminary step in the development of a competing product. However, the key to this case is that we are dealing with computer software, a relatively unexplored area in the world of copyright law. We must avoid the temptation of trying to force "the proverbial square peg into a round hole." CAI, 23 U.S.P.Q.2d at 1257.

In determining whether a challenged use of copyrighted material is fair, a court must keep in mind the public policy underlying the Copyright Act. " 'The immediate effect of our copyright law is to secure a fair return for an author's creative labor. But the ultimate aim is, by this incentive, to stimulate artistic creativity for the general public good.' " Sony Corp., 464 U.S. at 432 (quoting Twentieth Century Music Corp. v. Aiken, 422 U.S. 151, 156 (1975)). When technological change has rendered an aspect or application of the Copyright Act ambiguous, " 'the Copyright Act must be construed in light of this basic purpose.' " Id. As discussed above, the fact that computer programs are distributed for public use in object code form often precludes public access to the ideas and functional concepts contained in those programs, and thus confers on the copyright owner a de facto monopoly over those ideas and functional concepts. That result defeats the fundamental purpose of the Copyright Act — to encourage the production of original works by protecting the expressive elements of those works while leaving the ideas, facts, and functional concepts in the public domain for others to build on. Feist Publications, 111 S. Ct. at 1290.

Sega argues that the considerable time, effort, and money that went into development of the Genesis and Genesis-compatible video games militate against a finding of fair use. Borrowing from antitrust principles, Sega attempts to label Accolade a "free rider" on its product development efforts. In Feist Publications, however, the Court unequivocally rejected the "sweat of the brow" rationale for copyright protection. 111 S. Ct. at 1290-95. Under the Copyright Act, if a work is largely functional, it receives only weak protection. "This result is neither unfair nor unfortunate. It is the means by which copyright advances the progress of science and art." Id. at 1290; see also id. at 1292 ("In truth, 'it is just such wasted effort that the proscription against the copyright of ideas and facts . . . [is] designed to prevent.' ") (quoting Rosemont Enterprises, Inc. v. Random House, Inc., 366 F.2d 303, 310 (2d Cir. 1966), *cert. denied,* 385 U.S. 1009, 87 S. Ct. 714, 17 L. Ed. 2d 546 (1967)); CAI, 23 U.S.P.Q.2d at 1257. Here, while the work may not be largely functional, it incorporates functional elements which do not merit protection. The equitable considerations involved weigh on the side of public access. Accordingly, we reject Sega's argument.

(f)

We conclude that where disassembly is the only way to gain access to the ideas and functional elements embodied in a copyrighted computer program and where there is a legitimate reason for seeking such access, disassembly is a fair use of the copyrighted work, as a matter of law. Our conclusion does not, of course, insulate Accolade from a claim of copyright infringement with respect to its finished products. Sega has reserved the right to raise such a claim, and it may do so on remand.

COMMENTS AND QUESTIONS

1. Commentators have sharply divided on whether compatibility and/ or standardization should justify reverse engineering or copying of parts of a plaintiff's computer program. Virtually all recent courts, as well as most commentators, have endorsed reverse engineering in some circumstances. In addition to *Sega,* see DSC Communications v. DGI Technologies, 81 F.3d 597, 601 (5th Cir. 1996); Bateman v. Mnemonics, Inc., 79 F.3d 1532, 1539 n.18 (11th Cir. 1995); Lotus Dev. Corp. v. Borland Intl., 49 F.3d 807, 817-18 (1st Cir. 1995) (Boudin, J., concurring); Atari Games Corp. v. Nintendo of

America, 975 F.2d 832, 843-44 (Fed. Cir. 1992); Vault v. Quaid, 847 F.2d 255, 270 (5th Cir. 1988); Mitel Inc. v. Iqtel Inc., 896 F. Supp. 1050 (D. Colo. 1995); Jonathan Band & Masanobu Katoh, Interfaces on Trial (1995); Julie Cohen, Reverse Engineering and the Rise of Electronic Vigilantism: Intellectual Property Implications of "Lock-Out" Technologies, 68 S. Cal. L. Rev. 1091 (1995); Lawrence D. Graham & Richard O. Zerbe Jr., Economically Efficient Treatment of Computer Software: Reverse Engineering, Protection, and Disclosure, 22 Rutgers Comp. & Tech. L.J. 61 (1996); Dennis S. Karjala, Copyright Protection of Computer Documents, Reverse Engineering, and Professor Miller, 19 U. Dayton L. Rev. 975, 1016-18 (1994); David A. Rice, Sega and Beyond: A Beacon for Fair Use Analysis . . . At Least as Far as It Goes, 19 U. Dayton L. Rev. 1131, 1168 (1994).

On the other hand, some early decisions rejected compatibility as a justification for copying. See Apple Computer v. Franklin Computer, 714 F.2d 1240 (3d Cir. 1983); Digital Communications Assoc. v. Softklone Distributing Corp., 659 F. Supp. 449 (N.D. Ga. 1987). See also Anthony Clapes, Confessions of an Amicus Curiae: Technophobia, Law and Creativity in the Digital Arts, 19 U. Dayton L. Rev. 903 (1994) (no right to reverse engineer software) and Arthur Miller, Copyright Protection for Computer Programs, Databases, and Computer-Generated Works: Is Anything New Since CONTU?, 106 Harv. L. Rev. 977 (1993) (same). Most of these decisions, unlike *Sega*, involve copied code that appears in the defendant's final product, rather than just intermediate copying. Should the courts treat intermediate copying differently from copying for compatibility in a final product?

In DSC Communications Corp. v. Pulse Communications Inc., 170 F.3d 1354 F.3d 1354 (Fed. Cir. 1999), the court seemed to acknowledge that reverse engineering could be a fair use, but held that it was not always fair. It wrote:

> The district court dismissed DSC's direct copyright infringement claim on the ground that Pulsecom's conduct was excused by the affirmative defense of fair use for reverse engineering, as discussed in Sega Enterprises, Ltd. v. Accolade, Inc., 977 F.2d 1510, 1520, 24 USPQ2d 1561, 1569 (9th Cir. 1992). The *Sega* case, however, does not stand for the proposition that any form of copyright infringement is privileged as long as it is done as part of an effort to explore the operation of a product that uses the copyrighted software. On the basis of DSC's evidence at trial, Pulsecom's activities in creating copies of the POTS-DI software on its POTS cards by using the RBOCs' Litespan systems does [sic] not qualify as "fair use" under the *Sega* analysis. DSC's evidence showed that Pulsecom representatives made copies of the POTS-DI software on Pulsecom POTS cards as part of the ordinary operation of those cards, not as part of an effort to determine how the Litespan system worked. Rather than being part of an attempt at reverse engineering, the copying appears to have been done after Pulsecom had determined how the system functioned and merely to demonstrate the interchangeability of the Pulsecom POTS cards with those made and sold by DSC.

Id. at 1363. Why isn't there a valid social purpose in testing one's system to make sure it is compatible with the plaintiff's system? Does Pulse's use have a market-destroying effect here? If so, what is it? For a result in some tension with *Pulse*, see Alcatel USA Inc. v. DGI Technologies Inc., 166 F.3d 772 (5th Cir. 1999) (holding that it was copyright misuse for DSC to attempt to prevent a competitor from testing the compatibility of its cards with DSC's).

2. Do you agree with the *Sega* court's analysis of the fourth fair use factor? Hasn't Sega been injured because it can no longer control who produces games for its machine? Won't it lose the ability to charge a fee to game writers? If so, is this injury the result of copyright infringement or of Accolade's independent contributions to its games?

3. *Sega* suggests that the "nature" of computer programs entitles them to less protection than other literary works, since programs must be copied in order for the user to have access to the uncopyrightable ideas and facts they contain. Does this reasoning apply only to copying to achieve compatibility, or does it suggest that *any* copying is more likely to constitute fair use?

4. Is standardization a valid rationale for copying? Peter Menell suggests that even arbitrary elements of user interfaces that become de facto industry standards should lose copyright protection in order to foster network externalities. Peter Menell, An Analysis of the Scope of Copyright Protection for Computer Programs, 41 Stan. L. Rev. 1045, 1101 (1989). He argues that the high premium that consumers place on widely learned standards makes compatibility with such standards critical to entering the marketplace. In effect, what began as an arbitrary element has become functional due to widespread consumer learning. See also Frederick Warren-Boulton, Kenneth Baseman, & Glenn Woroch, The Economics of Intellectual Property for Software: The Proper Role for Copyright (working paper June 13, 1994).

On the other hand, not every case requires standardization. The *Sega* court is careful to note that intermediate copying qualifies as fair use only if it is necessary to produce an interoperable program. At least one court has rejected a claim of fair use on the grounds that the copyrighted material taken was not needed for interoperability. Compaq Computer v. Procom Technology, 908 F. Supp. 1409 (S.D. Tex. 1995).

5. Is the fair use doctrine an appropriate means of encouraging compatability or standardization? Does accomodating those policy goals require stretching the doctrine too far? A number of commentators have suggested that encouraging compatability is best accomplished by legislating a right to reverse-engineer, analogous to the right provided in the Semiconductor Chip Protection Act of 1984.[14] See, e.g., Timothy S. Teter, Merger and the Machines: An Analysis of the Pro-Compatibility Trend in Computer Software Copyright Cases, 45 Stan. L. Rev. 1061, 1089-97 (1993).

6. Is Sega entitled to control not only the sales of its copyrighted software-hardware package, but also video games which will run on the Sega

---

14. This statute is discussed in detail infra.

system? The court's decision obliquely suggests that there are antitrust (or possibly copyright misuse) problems with giving Sega such power: "an attempt to monopolize the market by making it impossible for others to compete runs counter to the statutory purpose of promoting creative expression and cannot constitute a strong equitable basis for resisting the invocation of the fair use doctrine." Is this concern properly addressed in the fair use doctrine? Are there other limits on Sega's ability to control the creation of compatible programs?

Possible limits include the antitrust laws and the copyright misuse doctrine. Copyright misuse is an equitable doctrine that has occasionally been applied by the courts to prevent a plaintiff from enforcing a copyright where the effect of that enforcement would be to enlarge the scope of the copyright monopoly beyond its natural bounds. See Lasercomb America, Inc. v. Reynolds, 911 F.2d 970 (4th Cir. 1990); Tamburo v. Calvin, No. 94 C 5206, 1995 U.S. Dist. LEXIS 3399 (N.D. Ill. Mar. 15, 1995) (both holding that overbroad noncompetition clauses attached to copyright licenses prevented enforcement of the copyright). A recent application of the copyright misuse doctrine in an analogous context is DSC Comm. Corp. v. DGI Technologies, 81 F.3d 597, 601 (5th Cir. 1996). In that case, the Fifth Circuit held that a plaintiff was not likely to prevail on a claim of copyright infringement stemming from the creation of RAM copies of its operating system, because the plaintiff's assertion of such a copyright claim was likely to constitute copyright misuse. The court reasoned:

> DSC seems to be attempting to use its copyright to obtain a patent-like monopoly over unpatented microprocessor cards. Any competing microprocessor card developed for use on DSC phone switches must be compatible with DSC's copyrighted operating system software. In order to ensure that its card is compatible, a competitor such as DGI must test the card on a DSC phone switch. Such a test necessarily involves making a copy of DSC's copyrighted operating system, which copy is downloaded into the card's memory when the card is booted up. If DSC is allowed to prevent such copying, then it can prevent anyone from developing a competing microprocessor card, even though it has not patented the card. The defense of copyright misuse "forbids the use of the copyright to secure an exclusive right or limited monopoly not granted by the Copyright Office," including a limited monopoly over microprocessor cards.

Is it reasonable to argue that Sega was misusing its copyright in this case? Would application of the misuse doctrine provide a narrower rationale on which to decide the case? Is such a narrower rationale desirable? See Julie E. Cohen, Reverse Engineering and the Rise of Electronic Vigilantism: Intellectual Property Implications of "Lock-Out" Technologies, 68 S. Cal. L. Rev. 1091 (1995) (suggesting that a misuse-based approach may be warranted but grounding it in patent law).

7. At the time *Sega* was decided, reverse engineering a computer program was a difficult and time-consuming task. If reverse engineering were

simple and flawless, should it still be legal? For an argument that "easy" reverse engineering would be problematic, see Pamela Samuelson et al., A Manifesto Concerning the Legal Protection of Computer Programs, 94 Colum. L. Rev. 2308 (1994).

8. Patent and trade secret laws treat reverse engineering very differently. Reverse engineering a patented product necessarily constitutes infringement if it involves making, using, or selling the patented invention. There is no statutory defense for reverse engineering. By contrast, the Uniform Trade Secrets Act expressly provides that reverse engineering a commercially available product is a legitimate means of discovering a trade secret. What explains the different treatment of reverse engineering among patent, copyright, and trade secret laws?

D. PATENT PROTECTION

1. Is Software Patentable Subject Matter?

With the growth of computer technology came an early crop of patent applications. In the 1950s and early 1960s, the Patent Office met these with a uniform response: whatever software is, it is definitely *not* patentable subject matter. Some programmers persisted in their efforts to have software recognized with the traditional badge of technical achievement, an issued patent. Some even obtained patents, but not for software per se. Overall, the early unwillingness of the PTO and courts to grant patents on software set the stage for the next two decades of legal protection for computer software, bringing trade secrets and copyright to the forefront.

On several notable occasions, early programmer/applicants brought their fight to patent software to the Supreme Court. These early cases — and indeed the vast majority of software patent cases to date — focused on the question of whether an "invention" consisting of the use of a mathematical algorithm could be patentable subject matter under 35 U.S.C. §101. We begin with the first Supreme Court opinion on this issue.

≡ *Gottschalk v. Benson*
≡ **Supreme Court of the United States**
≡ *409 U.S. 63 (1972)*

Mr. Justice DOUGLAS delivered the opinion of the Court.

Respondents filed in the Patent Office an application for an invention which was described as being related "to the processing of data by program and more particularly to the programmed conversion of numerical information" in general-purpose digital computers. They claimed a method for converting binary-coded decimal (BCD) numerals into pure binary numerals. The claims were not limited to any particular art or technology, to any particular apparatus or machinery, or to any particular end use. They purported to cover any use of the claimed method in a general-purpose digital computer of any type. Claims 8 and 13 were rejected by the Patent Office but sustained by the Court of Customs and Patent Appeals.

The question is whether the method described and claimed is a "process" within the meaning of the Patent Act.[2]

A digital computer, as distinguished from an analog computer, operates on data expressed in digits, solving a problem by doing arithmetic as a person would do it by head and hand. Some of the digits are stored as components of the computer. Others are introduced into the computer in a form which it is designed to recognize. The computer operates then upon both new and previously stored data. The general-purpose computer is designed to perform operations under many different programs.

The representation of numbers may be in the form of a time series of electrical impulses, magnetized spots on the surface of tapes, drums, or discs, charged spots on cathode-ray tube screens, the presence or absence of punched holes on paper cards, or other devices. The method or program is a sequence of coded instructions for a digital computer.

The patent sought is on a method of programming a general-purpose digital computer to convert signals from binary-coded decimal form into pure binary form. A procedure for solving a given type of mathematical problem is known as an "algorithm." The procedures set forth in the present claims are of that kind; that is to say, they are a generalized formulation for programs to solve mathematical problems of converting one form of numerical representation to another. From the generic formulation, programs may be developed as specific applications. The decimal system uses as digits the 10 symbols 0, 1, 2, 3, 4, 5, 6, 7, 8, and 9. The value represented by any digit depends, as it does in any positional system of notation, both on its individual value and on its relative position in the numeral. Decimal numerals are written by placing digits in the appropriate positions or columns of the numerical sequence, i.e., "unit" (10^0), "tens" (10^1), "hundreds" (10^2), "thousands" (10^3), etc. Accordingly, the numeral 1492 signifies $(1 \times 10^3) + (4 \times 10^2) + (9 \times 10^1) + (2 \times 10^0)$.

---

2. Title 35 U.S.C. §100(b) provides:

The term 'process' means process, art or method, and includes a new use of a known process, machine, manufacture, composition of matter, or material.

Title 35 U.S.C. §101 provides:

Whoever invents or discovers any new and useful process, machine, manufacture, or composition of matter, or any new and useful improvement thereof, may obtain a patent therefor, subject to the conditions and requirements of this title.

The pure binary system of positional notation uses two symbols as digits — 0 and 1, placed in a numerical sequence with values based on consecutively ascending powers of 2. In pure binary notation, what would be the tens position is the twos position; what would be [the] hundreds position is the fours position; what would be the thousands position is the eights. Any decimal number from 0 to 10 can be represented in the binary system with four digits or positions as indicated in the following table.

Shown as the sum of powers of 2:

Decimal	2^3 (8)		2^2 (4)		2^1 (2)		2^0 (1)	*Pure Binary*
0 =	0	+	0	+	0	+	0	= 0000
1 =	0	+	0	+	0	+	2^0	= 0001
2 =	0	+	0	+	2^1	+	0	= 0010
3 =	0	+	0	+	2^1	+	2^0	= 0011
4 =	0	+	2^2	+	0	+	0	= 0100
5 =	0	+	2^2	+	0	+	2^0	= 0101
6 =	0	+	2^2	+	2^1	+	0	= 0110
7 =	0	+	2^2	+	2^1	+	2^0	= 0111
8 =	2^3	+	0	+	0	+	0	= 1000
9 =	2^3	+	0	+	0	+	2^0	= 1001
10 =	2^3	+	0	+	2^1	+	0	= 1010

The BCD System using decimal numerals replaces the character for each component decimal digit in the decimal numeral with the corresponding four-digit binary numeral, shown in the righthand column of the table. Thus decimal 53 is represented as 0101 0011 in BCD, because decimal 5 is equal to binary 0101 and decimal 3 is equivalent to binary 0011. In pure binary notation, however, decimal 53 equals binary 110101. The conversion of BCD numerals to pure binary numerals can be done mentally through use of the foregoing table. The method sought to be patented varies the ordinary arithmetic steps a human would use by changing the order of the steps, changing the symbolism for writing the multiplier used in some steps, and by taking subtotals after each successive operation. The mathematical procedures can be carried out in existing computers long in use, no new machinery being necessary. And, as noted, they can also be performed without a computer.

The Court stated in Mackay Co. v. Radio Corp., 306 U.S. 86, 94 [1939], that "[w]hile a scientific truth, or the mathematical expression of it, is not a patentable invention, a novel and useful structure created with the aid of knowledge of scientific truth may be." That statement followed the longstanding rule that "[a]n idea of itself is not patentable." "A principle, in the abstract, is a fundamental truth; an original cause; a motive; these cannot be patented, as no one can claim in either of them an exclusive right." Le Roy v. Tatham, 14 How. [55 U.S.] 156, 175 [1853]. Phenomena of nature, though just discovered, mental processes, and abstract intellectual concepts are not patentable, as they are the basic tools of scientific and technological work. As we stated in Funk Bros. Seed Co. v. Kalo Co., 333 U.S. 127, 130 [1948], "He who discovers a hitherto unknown phenomenon of nature has no claim to a monopoly of it which the law recognizes. If there is to be invention from such a discovery, it must come from the application of the law of nature to a new and useful end." We dealt there with a "product" claim, while the present case deals with a "process" claim. But we think the same principle applies.

Here the "process" claim is so abstract and sweeping as to cover both known and unknown uses of the BCD to pure binary conversion. The end use may (1) vary from the operation of a train to verification of drivers' licenses to researching the law books for precedents and (2) be performed through any existing machinery or future-devised machinery or without any apparatus.

In O'Reilly v. Morse, 15 How. [56 U.S.] 62 [1853], Morse was allowed a patent for a process of using electromagnetism to produce distinguishable signs for telegraphy. Id., at 111. But the Court denied the eighth claim in which Morse claimed the use of "electro magnetism, however developed for marking or printing intelligible characters, signs, or letters, at any distances." Id., at 112. The Court in disallowing that claim said, "If this claim can be maintained, it matters not by what process or machinery the result is accomplished. For aught that we now know, some future inventor, in the onward march of science, may discover a mode of writing or printing at a distance by means of the electric or galvanic current, without using any part of the process or combination set forth in the plaintiff's specification. His invention may be less complicated — less liable to get out of order — less expensive in construction, and in its operation. But yet, if it is covered by this patent, the inventor could not use it, nor the public have the benefit of it, without the permission of this patentee." Id., at 113.

In The Telephone Cases, 126 U.S. 1, 534 [1887], the Court explained the *Morse* case as follows: "The effect of that decision was, therefore, that the use of magnetism as a motive power, without regard to the particular process with which it was connected in the patent, could not be claimed, but that its use in that connection could." Bell's invention was the use of electric current to transmit vocal or other sounds. The claim was not "for the use of a current of electricity in its natural state as it comes from the battery, but for putting a continuous current in a closed circuit into a certain specified condition suited to the transmission of vocal and other sounds, and using it in that condition for that purpose." Ibid. The claim, in other words, was not "one for the use of electricity distinct from the particular process with which it is connected in his patent." Id., at 535. The patent was for that use of electricity "both for the magneto and variable resistance methods." Id., at 538. Bell's claim, in other words, was not one for all telephonic use of electricity.

In Corning v. Burden, 15 How. [56 U.S.] 252, 267-268 [1853], the Court said, "One may discover a new and useful improvement in the process of tanning, dyeing, etc., irrespective of any particular form of machinery or mechanical device." The examples given were the "arts of tanning, dyeing, making waterproof cloth, vulcanizing India rubber, smelting ores." Id., at 267. Those are instances, however, where the use of chemical substances or physical acts, such as temperature control, changes articles or materials. The chemical process or the physical acts which transform the raw material are, however, sufficiently definite to confine the patent monopoly within rather definite bounds.

Cochrane v. Deener, 94 U.S. 780 [1876], involved a process for manufacturing flour so as to improve its quality. The process first separated the superfine flour and then removed impurities from the middlings by blasts of air, reground the middlings, and then combined the product with the superfine. Id., at 785. The claim was not limited to any special arrangement of machinery. Ibid. The Court said,

> That a process may be patentable, irrespective of the particular form of the instrumentalities used, cannot be disputed. If one of the steps of a process be that a certain substance is to be reduced to a powder, it may not be at all material what instrument or machinery is used to effect that object, whether a hammer, a pestle and mortar, or a mill. Either may be pointed out; but if the patent is not confined to that particular tool or machine, the use of the others would be an infringement, the general process being the same. A process is a mode of treatment of certain materials to produce a given result. It is an act, or a series of acts, performed upon the subject-matter to be transformed and reduced to a different state or thing.

Id., at 787-788.

It is argued that a process patent must either be tied to a particular machine or apparatus or must operate to change articles or materials to a "different state or thing." We do not hold that no process patent could ever qualify if it did not meet the requirements of our prior precedents. It is said that the decision precludes a patent for any program servicing a computer. We do not so hold. It is said that we have before us a program for a digital computer but extend our holding to programs for analog computers. We have, however, made clear from the start that we deal with a program only for digital computers. It is said we freeze process patents to old technologies, leaving no room for the revelations of the new, onrushing technology. Such is not our purpose. What we come down to in a nutshell is the following.

It is conceded that one may not patent an idea. But in practical effect that would be the result if the formula for converting BCD numerals to pure binary numerals were patented in this case. The mathematical formula involved here has no substantial practical application except in connection with a digital computer, which means that if the judgment below is affirmed, the patent would wholly pre-empt the mathematical formula and in practical effect would be a patent on the algorithm itself.

It may be that the patent laws should be extended to cover these programs, a policy matter to which we are not competent to speak. The President's Commission on the Patent System[4] rejected the proposal that these programs be patentable:

> Uncertainty now exists as to whether the statute permits a valid patent to be granted on programs. Direct attempts to patent programs have been rejected on the ground of nonstatutory subject matter. Indirect attempts to obtain patents and avoid the rejection, by drafting claims as a process, or a machine or components thereof programmed in a given manner, rather than as a program itself, have confused the issue further and should not be permitted. The Patent Office now cannot examine applications for programs because of a lack of a classification technique and the requisite search files. Even if these were available, reliable searches would not be feasible or economic because of the tremendous volume of prior art being generated. Without this search, the patenting of programs would be tantamount to mere registration and the presumption of validity would be all but nonexistent.
>
> It is noted that the creation of programs has undergone substantial and satisfactory growth in the absence of patent protection and that copyright protection for programs is presently available.

If these programs are to be patentable, considerable problems are raised which only committees of Congress can manage, for broad powers of investigation are needed, including hearings which canvass the wide variety of views which those operating in this field entertain. The technological problems tendered in the many briefs before us indicate to us that considered action by the Congress is needed. Reversed.

Appendix to Opinion of the Court

Claim 8 reads:

The method of converting signals from binary coded decimal form into binary which comprises the steps of

(1) storing the binary coded decimal signals in a re-entrant shift register,
(2) shifting the signals to the right by at least three places, until there is a binary 1 in the second position of said register,
(3) masking out said binary 1 in said second position of said register,
(4) adding a binary 1 to the first position of said register,
(5) shifting the signals to the left by two positions,
(6) adding a 1 to said first position, and

---

4. "To Promote the Progress of . . . Useful Arts," Report of the President's Commission on the Patent System (1966).

(7) shifting the signals to the right by at least three positions in preparation for a succeeding binary 1 in the second position of said register.

COMMENTS AND QUESTIONS

1. Recall the idea-expression merger doctrine in copyright law, which is invoked to refuse protection to otherwise copyrightable expression if it will allow the author to effectively control the idea underlying the work. Does *Benson* create a similar "idea-invention" merger doctrine in patent law? Is this approach consistent with other patent cases?

2. Prior to the *Benson* case, the Court of Customs and Patent Appeals (C.C.P.A.) had heard eight appeals from applicants claiming computer program-related inventions. According to an in-depth analysis of these cases and other matters pertaining to *Benson*,

> A curious thing about these eight pre-*Benson* cases is that none of them, not even the CCPA's decision in the Benson case, makes any more than an incidental use of the word "algorithm" in discussing the patentability issue. Hence, none of the analysis contained in these lower court decisions focused on the patentability of "algorithms." It was the Supreme Court's decision in Benson that shifted the focus of attention to "algorithms."

Pamela Samuelson, Benson Revisited: The Case Against Patent Protection for Algorithms and Other Computer Program-Related Inventions, 39 Emory L.J. 1025, 1042-1043 (1990). After *Benson*, Samuelson states, the case law "is focused almost exclusively on algorithms." Id., at 1059.

3. The author of the authoritative patent law treatise, Donald Chisum, has criticized the *Benson* decision and called for it to be overruled. The result in the case, he argues, "stemmed from an antipatent judicial bias that cannot be reconciled with the basic elements of the patent system established by Congress." Donald Chisum, The Future of Software Protection: The Patentability of Algorithms, 47 U. Pitt. L. Rev. 959, 961 (1986). Professor Chisum states that the "awkward distinctions and seemingly irreconcilable results of the case law since *Benson* . . . are the product of the analytical and normative weakness of *Benson* itself." Id., at 961-962. Chisum believes there are strong policy reasons to favor the patentability of computer algorithms.

Consider the section 101 cases outside the computer context, such as Diamond v. Chakrabarty, 447 U.S. 303 (1980). Is Chisum's criticism a fair one? Is the court treating software differently than it does other inventions?

4. Of what value would a patent be if the inventor were limited in her claims to *existing* technology? On the other hand, what about the objection stated in the *Morse* case, that allowing Morse to patent "*any method*" of communicating via electric signals at a distance would preempt much future inventive work? Would this claim have covered microwave communication, space satellites, fiber optics?

5. Justice Douglas writes: "Transformation and reduction of an article 'to a different state or thing' is the clue to the patentability of a process claim that does not include particular machines." What argument would you make to show that Benson's invention did this? The quoted phrase comes from Cochrane v. Deener, a nineteenth century case. This attempt to tie software patentability to some tangible (rather than abstract) transformation presages the direction of the case law after Diamond v. Diehr, excerpted below.

6. Recall this statement from Justice Douglas' opinion in Gottschalk v. Benson:

> Here the "process" claim is so abstract and sweeping as to cover both known and unknown uses of the BCD to pure binary conversion. The end use may (1) vary from the operation of a train to verification of drivers' licenses to researching the law books for precedents and (2) be performed through any existing machinery or future-devised machinery or without any apparatus.

Is this truly a section 101 objection, or is it an objection to the *scope* of the claims sought by the patentee? How are the two related? Is section 101 designed to exclude from patentability discoveries and inventions so basic and fundamental to future advances that patenting them would unduly burden future inventors? Recall the statement by Douglas in *Benson* that "mental processes" (which he holds includes algorithms) represent "the basic tools of scientific and technological work." 409 U.S. at 67. Again, the thought seems to be that these inventions are so basic — have so many applications — that they are not an appropriate subject for patents.

This objection has been made with respect to software patents in general by groups such as the League for Programming Freedom. And yet if this is true, then *any* invention which has a very wide range of applications might be thought to be unpatentable. Yet these are perhaps the most valuable types of inventions!

If section 101 is not the appropriate place to address this problem, what is? Or is it even a problem?

———————————

Attempts to patent software did not end in the wake of *Benson*. Perhaps encouraged by the Court's insistence that it was not precluding computer program patents entirely, computer companies continued to file applications claiming computer software in various forms. In the 1970s, most of these applications were not for what we think of today as "pure software," but instead for the use of special-purpose computers or programs in a particular industrial environment. The Supreme Court addressed the patentability of such programs in Parker v. Flook, 437 U.S. 584 (1978), and again in Diamond v. Diehr, 450 U.S. 175 (1981). In *Flook* the Court held that a patent on a method for calculating an "alarm limit" controlling a temperature was unpatentable because the only novel part of the invention was a mathematical formula. The court held to the contrary on similar facts in *Diehr*, however. There the patent was framed not as a method for calculating a number but as a method for curing synthetic rubber. The fact that the method used a mathematical formula to compute temperature and that the use of that formula in a computer was the only novel part of the invention did not render the process as a whole unpatentable. Rather, as long as the process or machine viewed as a whole was patentable subject matter, it would pass muster under §101.

The court thus differentiated "pure" claims to mathematical formulae from claims to inventions that implement or apply those formulae:

> We recognize, of course, that when a claim recites a mathematical formula (or scientific principle or phenomenon of nature), an inquiry must be made into whether the claim is seeking patent protection for that formula in the abstract. A mathematical formula as such is not accorded the protection of our patent laws, *Gottschalk v. Benson*, and this principle cannot be circumvented by attempting to limit the use of the formula to a particular technological environment. Similarly, insignificant postsolution activity will not transform an unpatentable principle into a patentable process. To hold otherwise would allow a competent draftsman to evade the recognized limitations on the type of subject matter eligible for patent protection. On the other hand, when a claim containing a mathematical formula implements or applies that formula in a structure or process which, when considered as a whole, is performing a function which the patent laws were designed to protect (*e.g.*, transforming or reducing an article to a different state or thing), then the claim satisfies the requirements of §101. Because we do not view respondents' claims as an attempt to patent a mathematical formula, but rather to be drawn to an industrial process for the molding of rubber products, we affirm the judgment of the Court of Customs and Patent Appeals.

Id at 52.

COMMENTS AND QUESTIONS

1. In applying *Diehr* (and, to a lesser extent, *Flook*), the Court of Customs and Patent Appeals and its successor, the United States Court of Appeals for the Federal Circuit, created a two-part test for the patentability of computer programs. Under the decisions in In re Freeman, 573 F.2d 1237 (C.C.P.A. 1978), In re Walter, 618 F.2d 758 (C.C.P.A. 1980), and In re Abele, 684 F.2d 902 (C.C.P.A. 1982), the proper way to evaluate patentability is, first, to determine whether a patent claim recites an algorithm "directly or indirectly." If it does, the second part of the test asks whether the claimed invention as a whole is no more than the algorithm itself. If so, the claim must be rejected under section 101; if not, it is patentable. Under the *Freeman-Walter-Abele* test, a program can be patentable even if the only *new*

idea is the algorithm itself, as long as other process steps or physical structures are included in the patent claims.

Does the *Diehr* approach, as implemented by the C.C.P.A., exalt artful drafting above the substance of the patented invention? Is there any software program that cannot be described and claimed in terms that will meet the *Freeman-Walter-Abele* test? What is left of *Benson* after *Diehr*? See Thomas P. Burke, Note, Software Patent Protection: Debugging the Current System, 69 Notre Dame L. Rev. 1115, 1164 (1994) ("*Benson* has been so limited that a decision explicitly overruling it would not work a great change in the substantive law.").

2. It is not surprising that patent applicants after *Diehr* began to describe their software inventions as being part of a larger machine or process. This shift is well documented in Samuelson, *Benson* Revisited, at 1089-90. These patentees met with considerable success; by 1994, there were an estimated 14,000 issued software patents in the United States. Nonetheless, the Federal Circuit continued to debate the patentability of software.

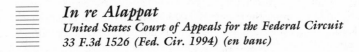

In re Alappat
United States Court of Appeals for the Federal Circuit
33 F.3d 1526 (Fed. Cir. 1994) (en banc)

RICH, J.,

Kuriappan P. Alappat, Edward E. Averill, and James G. Larsen (collectively Alappat) appeal the April 22, 1992, reconsideration decision of the Board of Patent Appeals and Interferences (Board) of the United States Patent and Trademark Office (PTO), Ex Parte Alappat, 23 U.S.P.Q.2d 1340 (BPAI, 1992), which sustained the Examiner's rejection of claims 15-19 of application Serial No. 07/149,792 ('792 application) as being unpatentable under 35 U.S.C. Section 101 (1988). . . .

[In Part I, the Federal Circuit held that the "reconsideration decision" was rendered by a legally constituted panel of the Board of Appeals and Interferences, and therefore the Federal Circuit could properly move to a consideration of the merits.]

II.   The Merits

Our conclusion is that the appealed decision should be reversed because the appealed claims are directed to a "machine" which is one of the categories named in 35 U.S.C. Section 101, as the first panel of the Board held.

A.   Alappat's Invention

Alappat's invention relates generally to a means for creating a smooth waveform display in a digital oscilloscope. The screen of an oscillosope is the front of a cathode-ray tube (CRT), which is like a TV picture tube, whose screen, when in operation, presents an array (or raster) of pixels arranged at intersections of vertical columns and horizontal rows, a pixel being a spot on the screen which may be illuminated by directing an electron beam to that spot, as in TV. Each column in the array represents a different time period, and each row represents a different magnitude. An input signal to the oscilloscope is sampled and digitized to provide a waveform data sequence (vector list), wherein each successive element of the sequence represents the magnitude of the waveform at a successively later time. The waveform data sequence is then processed to provide a bit map, which is a stored data array indicating which pixels are to be illuminated. The waveform ultimately displayed is formed by a group of vectors, wherein each vector has a straight line trajectory between two points on the screen at elevations representing the magnitudes of two successive input signal samples and at horizontal positions representing the timing of the two samples.

Because a CRT screen contains a finite number of pixels, rapidly rising and falling portions of a waveform can appear discontinuous or jagged due to differences in the elevation of horizontally contiguous pixels included in the waveform. In addition, the presence of "noise" in an input signal can cause portions of the waveform to oscillate between contiguous pixel rows when the magnitude of the input signal lies between values represented by the elevations of the two rows. Moreover, the vertical resolution of the display may be limited by the number of rows of pixels on the screen. The noticeability and appearance of these effects is known as aliasing.

To overcome these effects, Alappat's invention employs an anti-aliasing system wherein each vector making up the waveform is represented by modulating the illumination intensity of pixels having center points bounding the trajectory of the vector. The intensity at which each of the pixels is illuminated depends upon the distance of the center point of each pixel from the trajectory of the vector. Pixels lying squarely on the waveform trace receive maximum illumination, whereas pixels lying along an edge of the trace receive illumination decreasing in intensity proportional to the increase in the distance of the center point of the pixel from the vector trajectory. Employing this anti-aliasing technique eliminates any apparent discontinuity, jaggedness, or oscillation in the waveform, thus giving the visual appearance of a smooth continuous waveform. In short, and in lay terms, the invention is an improvement in an oscilloscope comparable to a TV having a clearer picture.

Reference to Fig. 5A of the '792 application, reproduced below [see Figure 7-2], better illustrates the manner in which a smooth appearing waveform is created.

Each square in this figure represents a pixel, and the intensity level at which each pixel is illuminated is indicated in hexadecimal notation by the number or letter found in each square. Hexadecimal notation has sixteen characters, the numbers 0-9 and the letters A-F, wherein A represents 10, B represents 11, C represents 12, D represents 13, E represents 14, and F represents 15. The intensity at which each pixel is illuminated increases from 0 to F. Accordingly, a square with a 0 (zero) in it represents a pixel having no

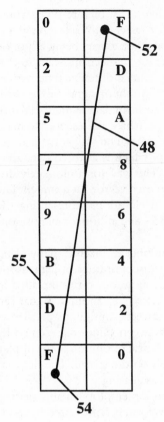

FIGURE 7-2
Facsimile diagram from the '792 application.

illumination, and a square with an F in it represents a pixel having maximum illumination. [The hexadecimal intensity level is stored in a "bit map" in Alappat's system as a binary number.] . . .

Points 54 and 52 in Fig. 5A represent successive observation points on the screen of an oscilloscope. Without the benefit of Alappat's anti-aliasing system, points 54 and 52 would appear on the screen as separate, unconnected spots. [That is, without Alappat's algorithm to "interpolate" between points, only the pixels labelled 54 and 52 would be illuminated on the oscilloscope screen, and none of the points in between would be illuminated.] In Alappat's system, the different intensity level at which each of the pixels is illuminated produces the appearance of the line 48, a so-called vector.

The intensity at which each pixel is to be illuminated is determined as follows, using pixel 55 as an example. First, the vertical distance between the Y coordinates of observation points 54 and 52 (ΔY_i) is determined. In this example, this difference equals 7 units, with one unit representing the center-to-center distance of adjacent pixels. Then, the elevation of pixel 55 above pixel 54 (ΔY_{ij}) is determined, which in this case is 2 units. The Y_i and Y_{ij} values are then "normalized," which Alappat describes as converting these values to larger values which are easier to use in mathematical calculations. In Alappat's example, a barrel shifter is used to shift the binary input to the left by the number of bits required to set the most significant (left-most) bit of its output signal to "1." The ΔY_i and ΔY_{ij} values are then plugged into a mathematical equation for determining the intensity at which the particular pixel is to be illuminated. In this particular example, the equation $I'(i,j)= [1 - (\Delta Y_{ij} / \Delta Y_i)]$ F, wherein F is 15 in hexadecimal notation, suffices. The intensity of pixel 55 in this example would thus be calculated as follows: $[1 - (2/7)]15 = (5/7)15 = 10.71 = 11$ (or B).

Accordingly, pixel 55 is illuminated at 11/15 of the intensity of the pixels in which observation points 54 and 52 lie. Alappat discloses that the particular formula used will vary depending on the shape of the waveform.

B. The Rejected Claims

Claim 15, the only independent claim in issue, reads:

A rasterizer for converting vector list data representing sample magnitudes of an input waveform into anti-aliased pixel illumination intensity data to be displayed on a display means comprising:

    (a)   means for determining the vertical distance between the endpoints of each of the vectors in the data list;

    (b)   means for determining the elevation of a row of pixels that is spanned by the vector;

    (c)   means for normalizing the vertical distance and elevation; and

    (d)   means for outputting illumination intensity data as a predetermined function of the normalized vertical distance and elevation.

Each of claims 16-19 depends directly from claim 15 and more specifically defines an element of the rasterizer claimed therein. Claim 16 recites that means (a) for determining the vertical distance between the endpoints of each of the vectors in the data list, described above, comprises an arithmetic logic circuit configured to perform an absolute value function. Claim 17 recites that means (b) for determining the elevation of a row of pixels that is spanned by the vector, j described above, comprises an arithmetic logic circuit configured to perform an absolute value function. Claim 18 recites that means (c) for normalizing the vertical distance and elevation comprises a pair of barrel shifters. Finally, claim 19 recites that means (d) for outputting comprises a read only memory (ROM) containing illumination intensity data. As the first Board panel found, each of (a)-(d) was a device known in the electronics arts before Alappat made his invention. . . .

D. Analysis

(1) Section 112, Paragraph Six ...

When independent claim 15 is construed in accordance with Section 112 Para. 6, claim 15 reads as follows, the subject matter in brackets representing the structure which Alappat discloses in his specification as corresponding to the respective means language recited in the claims:

> A rasterizer [a "machine"] for converting vector list data representing sample magnitudes of an input waveform into anti-aliased pixel illumination intensity data to be displayed on a display means comprising:
>
> (a) [an arithmetic logic circuit configured to perform an absolute value function, or an equivalent thereof] for determining the vertical distance between the endpoints of each of the vectors in the data list;
>
> (b) [an arithmetic logic circuit configured to perform an absolute value function, or an equivalent thereof] for determining the elevation of a row of pixels that is spanned by the vector;
>
> (c) [a pair of barrel shifters, or equivalents thereof] for normalizing the vertical distance and elevation; and
>
> (d) [a read only memory (ROM) containing illumination intensity data, or an equivalent thereof] for outputting illumination intensity data as a predetermined function of the normalized vertical distance and elevation.

As is evident, claim 15 unquestionably recites a machine, or apparatus, made up of a combination of known electronic circuitry elements.

Despite suggestions by the Commissioner to the contrary, each of dependent claims 16-19 serves to further limit claim 15. Section 112 Para. 6 requires that each of the means recited in independent claim 15 be construed to cover at least the structure disclosed in the specification corresponding to the "means." Each of dependent claims 16-19 is in fact limited to one of the structures disclosed in the specification.

(2) Section 101

The reconsideration Board majority affirmed the Examiner's rejection of claims 15-19 on the basis that these claims are not directed to statutory subject matter as defined in Section 101. ... As discussed ... supra, claim 15, properly construed, claims a machine, namely, a rasterizer "for converting vector list data representing sample magnitudes of an input waveform into anti-aliased pixel illumination intensity data to be displayed on a display means," which machine is made up of, at the very least, the specific structures disclosed in Alappat's specification corresponding to the means-plus-function elements (a)-(d) recited in the claim. According to Alappat, the claimed rasterizer performs the same overall function as prior art rasterizers, but does so in a different way, which is represented by the combination of four elements claimed in means-plus-function terminology. Because claim 15 is directed to a "machine," which is one of the four categories of patentable subject matter enumerated in Section 101, claim 15 appears on its face to be directed to Section 101 subject matter.

This does not quite end the analysis, however, because the Board majority argues that the claimed subject matter falls within a judicially created exception to Section 101 which the majority refers to as the "mathematical algorithm" exception. Although the PTO has failed to support the premise that the "mathematical algorithm" exception applies to true apparatus claims, we recognize that our own precedent suggests that this may be the case. . . . Even if the mathematical subject matter exception to Section 101 does apply to true apparatus claims, the claimed subject matter in this case does not fall within that exception.

(a)

The plain and unambiguous meaning of Section 101 is that any new and useful process, machine, manufacture, or composition of matter, or any new and useful improvement thereof, may be patented if it meets the requirements for patentability set forth in Title 35, such as those found in Sections 102, 103, and 112. The use of the expansive term "any" in Section 101 represents Congress's intent not to place any restrictions on the subject matter for which a patent may be obtained beyond those specifically recited in Section 101 and the other parts of Title 35. Indeed, the Supreme Court has acknowledged that Congress intended Section 101 to extend to "anything under the sun that is made by man." Diamond v. Chakrabarty, 447 U.S. 303, 309 (1980). . . . Thus, it is improper to read into Section 101 limitations as to the subject matter that may be patented where the legislative history does not indicate that Congress clearly intended such limitations. . . .

Despite the apparent sweep of Section 101, the Supreme Court has held that certain categories of subject matter are not entitled to patent protection. In *Diehr*, its most recent case addressing Section 101, the Supreme Court explained that there are three categories of subject matter for which one may not obtain patent protection, namely "laws of nature, natural phenomena, and abstract ideas." Diehr, 450 U.S. at 185. Of relevance to this case, the Supreme Court also has held that certain mathematical subject matter is not, standing alone, entitled to patent protection. . . . A close analysis of *Diehr*, *Flook*, and *Benson* reveals that the Supreme Court never intended to create an overly broad, fourth category of subject matter excluded from Section 101. Rather, at the core of the Court's analysis in each of these cases lies an attempt by the Court to explain a rather straightforward concept, namely, that certain types of mathematical subject matter, standing alone, represent nothing more than abstract ideas until reduced to some type of practical application, and thus that subject matter is not, in and of itself, entitled to patent protection. . . .

(b)

Given the foregoing, the proper inquiry in dealing with the so called mathematical subject matter exception to Section 101 alleged herein is to see whether the claimed subject matter as a whole is a disembodied mathematical concept, whether categorized as a mathematical formula, mathematical equation, mathematical algorithm, or the like, which in essence represents nothing more than a "law of nature," "natural phenomenon," or "abstract idea." If so, *Diehr* precludes the patenting of that subject matter. That is not the case here.

Although many, or arguably even all, of the means elements recited in claim 15 represent circuitry elements that perform mathematical calculations, which is essentially true of all digital electrical circuits, the claimed invention as a whole is directed to a combination of interrelated elements which combine to form a machine for converting discrete waveform data samples into anti-aliased pixel illumination intensity data to be displayed on a display means. This is not a disembodied mathematical concept which may be characterized as an "abstract idea," but rather a specific machine to produce a useful, concrete, and tangible result.

The fact that the four claimed means elements function to transform one set of data to another through what may be viewed as a series of mathematical calculations does not alone justify a holding that the claim as a whole is directed to nonstatutory subject matter. See In re Iwahashi, 888 F.2d at 1375, 12 U.S.P.Q.2d at 1911. Indeed, claim 15 as written is not "so abstract and sweeping" that it would "wholly pre-empt" the use of any apparatus employing the combination of mathematical calculations recited therein. See *Benson*, 409 U.S. at 68-72 (1972). Rather, claim 15 is limited to the use of a particularly claimed combination of elements performing the particularly claimed combination of calculations to transform, i.e., rasterize, digitized waveforms (data) into anti-aliased, pixel illumination data to produce a smooth waveform.

Furthermore, the claim preamble's recitation that the subject matter for which Alappat seeks patent protection is a rasterizer for creating a smooth waveform is not a mere field-of-use label having no significance. Indeed, the preamble specifically recites that the claimed rasterizer converts waveform data into output illumination data for a display, and the means elements recited in the body of the claim make reference not only to the inputted waveform data recited in the preamble but also to the output illumination data also recited in the preamble. Claim 15 thus defines a combination of elements constituting a machine for producing an anti-aliased waveform.

The reconsideration Board majority also erred in its reasoning that claim 15 is unpatentable merely because it "reads on a general purpose digital computer 'means' to perform the various steps under program control." Alappat,

23 U.S.P.Q.2d at 1345. The Board majority stated that it would "not presume that a stored program digital computer is not within the Section 112 Para. 6 range of equivalents of the structure disclosed in the specification." *Alappat*, 23 U.S.P.Q.2d at 1345. Alappat admits that claim 15 would read on a general purpose computer programmed to carry out the claimed invention, but argues that this alone also does not justify holding claim 15 unpatentable as directed to nonstatutory subject matter. We agree. We have held that such programming creates a new machine, because a general purpose computer in effect becomes a special purpose computer once it is programmed to perform particular functions pursuant to instructions from program software. In re Freeman, 573 F.2d 1237, 1247 n.11, 197 U.S.P.Q. 464, 472 n.11 (C.C.P.A. 1978); In re Noll, 545 F.2d 141, 148, 191 U.S.P.Q. 721, 726 (C.C.P.A. 1976); In re Prater, 415 F.2d at 1403 n.29, 162 U.S.P.Q. at 549-50 n.29.

Under the Board majority's reasoning, a programmed general purpose computer could never be viewed as patentable subject matter under Section 101. This reasoning is without basis in the law. The Supreme Court has never held that a programmed computer may never be entitled to patent protection. Indeed, the Benson court specifically stated that its decision therein did not preclude "a patent for any program servicing a computer." *Benson*, 409 U.S. at 71. Consequently, a computer operating pursuant to software may represent patentable subject matter, provided, of course, that the claimed subject matter meets all of the other requirements of Title 35. In any case, a computer, like a rasterizer, is apparatus not mathematics.

Conclusion

For the foregoing reasons, the appealed decision of the Board affirming the examiner's rejection is reversed.

ARCHER, C.J., with whom NIES, J., joins, concurring in part and dissenting in part. . . .

I disagree with the majority's conclusion that Alappat's "rasterizer," which is all that is claimed in the claims at issue, constitutes an invention or discovery within 35 U.S.C. Section 101. I would affirm the board's decision sustaining the examiner's rejection of claims 15-19 to the rasterizer under 35 U.S.C. Section 101 because Alappat has not shown that he invented or discovered a machine within Section 101.

In 1873, George Curtis made certain general observations about patent law, the scope of patentable subject matter being at its heart. He stated them with such force and eloquence, and in my view they have such relevance to the issue we face today, that I repeat them as follows:

> It is necessary . . . to have clear and correct notions of the true scope of a patent right . . . which may be found to assist, in particular cases, the solution

of the question, whether a particular invention or discovery is by law a patentable subject.

In this inquiry it is necessary to commence with the process of exclusion; for although, in their widest acceptation, the terms "invention" and "discovery" include the whole vast variety of objects on which the human intellect may be exercised, so that in poetry, in painting, in music, in astronomy, in metaphysics, and in every department of human thought, men constantly invent or discover, in the highest and the strictest sense, their inventions and discoveries in these departments are not the subjects of the patent law. . . . The patent law relates to a great and comprehensive class of discoveries and inventions of some new and useful effect or result in matter, not referable to the department of the fine arts. The matter of which our globe is composed is the material upon which the creative and inventive faculties of man are exercised, in the production of whatever ministers to his convenience or his wants. Over the existence of matter itself he has no control. . . .

The direct control of man over matter consists, therefore, in placing its particles in new relations. This is all that is actually done, or that can be done, namely, to cause the particles of matter existing in the universe to change their former places, by moving them, by muscular power or some other force. But as soon as they are brought into new relations, it is at once perceived that there are vast latent forces in nature, which come to the aid of man, and enable him to produce effects and results of a wholly new character, far beyond the mere fact of placing the particles in new positions. He moves certain particles of matter into a new juxtaposition, and the chemical agencies and affinities called into action by this new contact produce a substance possessed of new properties and powers, to which has been given the name of gunpowder. He takes a stalk of flax from the ground, splits it into a great number of filaments, twists them together, and laying numbers of the threads thus formed across each other, forms a cloth, which is held together by the tenacity or force of cohesion in the particles, which nature brings to his aid. He moves into new positions and relations certain particles of wood and iron, in various forms, and produces a complicated machine, by which he is able to accomplish a certain purpose, only because the properties of cohesion and the force of gravitation cause it to adhere together and enable the different parts to operate upon each other and to transmit the forces applied to them, according to the laws of motion. It is evident, therefore, that the whole of the act of invention, in the department of useful arts, embraces more than the new arrangement of particles of matter in new relations. The purpose of such new arrangements is to produce some new effect or result, by calling into activity some latent law, or force, or property, by means of which, in a new application, the new effect or result may be accomplished. In every form in which matter is used, in every production of the ingenuity of man, he relies upon the laws of nature and the properties of matter, and seeks for new effects and results through their agency and aid. Merely inert matter alone is not the sole material with which he works. Nature supplies powers, and forces, and active properties, as well as the particles of matter, and these powers, forces, and properties are constantly the subjects of study, inquiry, and experiment, with a view to the production of some new effect or result in matter.

Any definition or description, therefore, of the act of invention, which excludes the application of the natural law, or power, or property of matter, on which the inventor has relied for the production of a new effect, and the object of such application, and confines it to the precise arrangement of the particles of matter which he may have brought together, must be erroneous.

G. Curtis, A Treatise on the Law of Patents for Useful Inventions at xxiii-xxv (4th ed. 1873). . . .

Alappat has arranged known circuit elements to accomplish nothing other than the solving of a particular mathematical equation represented in the mind of the reader of his patent application. Losing sight of the forest for the structure of the trees, the majority today holds that any claim reciting a precise arrangement of structure satisfies 35 U.S.C. Section 101. As I shall demonstrate, the rationale that leads to this conclusion and the majority's holding that Alappat's rasterizer represents the invention of a machine are illogical, inconsistent with precedent and with sound principles of patent law, and will have untold consequences.

B

. . . The terms used in Section 101 have been used for over two hundred years — since the beginnings of American patent law — to define the extent of the subject matter of patentable invention. . . . Coexistent with the usage of these terms has been the rule that a person cannot obtain a patent for the discovery of an abstract idea, principle or force, law of nature, or natural phenomenon, but rather must invent or discover a practical "application" to a useful end. Diamond v. Diehr. . . .

Thus patent law rewards persons for inventing technologically useful applications, instead of for philosophizing unapplied research and theory. . . .

Additionally, unapplied research, abstract ideas, and theory continue to be the "basic tools of scientific and technological work," which persons are free to trade in and to build upon in the pursuit of among other things useful inventions. *Flook*, 437 U.S. at 589, 198 U.S.P.Q. at 197 (quotations omitted). Even after a patent has been awarded for a new, useful, and nonobvious practical application of an idea, others may learn from the underlying ideas, theories, and principles to legitimately "design around" the patentee's useful application. See Slimfold Mfg. Co. v. Kinkead Indus., Inc., 932 F.2d 1453, 1457, 18 U.S.P.Q.2d 1842, 1845-46 (Fed. Cir. 1991). . . .

In addition to the basic principles embodied in the language of Section 101, the section has a pragmatic aspect. That subject matter must be new (Section 102) and nonobvious (Section 103) in order to be patentable is of course a separate requirement for patentability, and does not determine whether the applicant's purported invention or discovery is within Section 101. . . . Section 101 must be satisfied before any of the other provisions apply, and in this way Section 101 lays the predicate for the other provisions of the patent law. . . . When considering that the patent law does not allow patents merely for the discovery of ideas, principles, and laws of nature, ask whether, were it not so, the other provisions of the patent law could be applied at all. If Einstein could have obtained a patent for his discovery that the energy of an object at rest equals its mass times the speed of light squared, how would his discovery be meaningfully judged for nonobviousness, the sine qua non of patentable invention? 35 U.S.C. Section 103. When is the abstract idea "reduced to practice" as opposed to being "conceived"? See id. Section 102(g). What conduct amounts to the "infringement" of another's idea? See id. Section 271. . . .

Consider for example the discovery or creation of music, a new song. Music of course is not patentable subject matter; a composer cannot obtain exclusive patent rights for the original creation of a musical composition. But now suppose the new melody is recorded on a compact disc. In such case, the particular musical composition will define an arrangement of minute pits in the surface of the compact disc material, and therefore will define its specific structure. See D. Macaulay, The Way Things Work 248-49 (Houghton Mifflin 1988). Alternatively suppose the music is recorded on the rolls of a player piano or a music box.

Through the expedient of putting his music on known structure, can a composer now claim as his invention the structure of a compact disc or player piano roll containing the melody he discovered and obtain a patent therefor? The answer must be no. The composer admittedly has invented or discovered nothing but music. The discovery of music does not become patentable subject matter simply because there is an arbitrary claim to some structure.

And if a claim to a compact disc or piano roll containing a newly discovered song were regarded as a "manufacture" and within Section 101 simply because of the specific physical structure of the compact disc, the "practical effect" would be the granting of a patent for a discovery in music. Where the music is new, the precise structure of the disc or roll would be novel under Section 102. Because the patent law cannot examine music for "nonobviousness," the Patent and Trademark Office could not make a showing of obviousness under Section 103. The result could well be the award of a patent for the discovery of music. The majority's simplistic approach of looking only to whether the claim reads on structure and ignoring the claimed invention or discovery for which a patent is sought will result in the awarding of patents for discoveries well beyond the scope of the patent law. . . .

So what did Alappat invent or discover? Alappat's specification clearly distinguishes between an "oscilloscope" and a "rasterizer," and Alappat claims his invention in claims 15-19 to be only the "rasterizer."

The "rasterizer" as claimed is an arrangement of circuitry elements for converting data into other data according to a particular mathematical operation. The rasterizer begins with vector "data" — two numbers. "[I]t does not matter how they are ascertained." Brief for Alappat at 39. The two numbers, as they might to any algebra student, "represent" endpoints of a line.

The claimed "rasterizer" ends with other specific "data" — an array of numbers, as the original and reconsideration panels of the board both expressly agreed. . . .

Alappat admits that each of the circuitry elements of the claimed "ras-terizer" is old. He says they are merely "form." Thus, they are only a con-venient and basic way of electrically representing the mathematical operations to be performed, that is, converting vector data into matrix or raster data. In Alappat's view, it is the new mathematic operation that is the "substance" of the claimed invention or discovery. Claim 15 as a whole thus claims old circuitry elements in an arrangement defined by a mathematical operation, which only performs the very mathematical operation that defines it. Rather than claiming the mathematics itself, which of course Alappat cannot do, Alappat claims the mathematically defined structure. But as a whole, there is no "application" apart from the mathematical operation that is asserted to be the invention or discovery. What is going on here is a charade. . . .

COMMENTS AND QUESTIONS

1. Does Alappat's invention — essentially, interpolating between points to display a line on an oscilloscope's screen — seem straightforward? Does the Federal Circuit say that this invention deserves a patent? What is the precise legal holding of the opinion?

Dictum or not, certain language in *Alappat* suggests an extremely broad reading of section 101. The court acknowledges that Alappat's claims will read on "a general purpose computer programmed to carry out the claimed invention," but it says the claim is nonetheless patentable because "such programming creates a new machine, because a general purpose computer in effect becomes a special purpose computer once it is programmed to perform particular functions pursuant to instructions from programmed software." Is there any program for which this is not true? What remains of the section 101 limitation in light of this language?

2. In one portion of one of the many omitted opinions, Judge Pauline Newman, concurring on the merits, states: "Mathematics is not a monster to be struck down or out of the patent system, but simply another resource whereby technological advance is achieved." In re Alappat, 33 F.3d 1526 (Fed. Cir. 1994) (Newman, J., concurring). Is this sentiment inconsistent with the view taken in *Benson*? How about *Diehr*? What is the difference between (1) pure mathematics; (2) applied mathematics, as that term is used in mathematics department course catalogues; and (3) mathematics applied to solve a problem in the real world?

Note on Post-Alappat Developments

While the en banc nature of the decision might suggest an emerging consensus in support of *Alappat*'s broad reading of section 101, other cases left the outcome in doubt. Two cases decided just after *Alappat* differed over the patentability of data structures, an issue not directly presented in *Alappat* but certainly related to it. Compare In re Lowry, 32 F.3d 1579 (Fed. Cir.

1994) (upholding as patentable abstract claims to a type of software data structure wherein data "objects" are arranged in complex relationships) with In re Warmerdam, 33 F.3d 1354 (Fed. Cir. 1994) (affirming rejection of process claims to data structure for representing proximity and shape of objects in the immediate environment, e.g., for use by a mobile machine with robot vision and based on idea of representing objects in computer memory as increasingly fine-grained "bubbles"). See also In re Schraeder, 22 F.3d 290 (Fed. Cir. 1994) (holding unpatentable claims to method for competitively bidding on plurality of related items, such as contiguous tracts of land, since mathematical algorithm is implicit in claim, and since algorithm is not applied to or limited by physical elements or process steps, in that grouping or regrouping of bids is not physical change, effect, or result, and mere entering of bids in "record" is insufficient to impart patentability under the cases).

In 1995, IBM appealed the PTO's rejection of a claim to "software contained on a floppy disk" to the Federal Circuit. See In re Beauregard, 53 F.3d 1583 (Fed. Cir. 1995). While the appeal was pending, the PTO decided not to oppose the claim, and promised the court that it would shortly issue new examining guidelines for software patents. After several false starts, the PTO did issue final Examination Guidelines for Computer-Implemented Inventions in January of 1996. These guidelines direct examiners as follows:

> The subject matter courts have found to be outside the four statutory categories of invention is limited to abstract ideas, laws of nature and natural phenomena. While this is easily stated, determining whether an applicant is seeking to patent an abstract idea, a law of nature or a natural phenomenon has proven to be challenging. These three exclusions recognize that subject matter that is not a *practical application or use* of an idea, a law of nature or a natural phenomenon is not patentable. . . .
>
> Claims to computer-related inventions that are clearly non-statutory fall into the same general categories as non-statutory claims in other arts, namely natural phenomena such as magnetism, and abstract ideas or laws of nature which constitute "descriptive material." Descriptive material can be characterized as either "functional descriptive material" or "non-functional descriptive material." In this context, "functional descriptive material" consists of data structures and computer programs which impart functionality when encoded on a computer-readable medium. "Non-functional descriptive material" includes but is not limited to music, literary works and a compilation or mere arrangement of data.
>
> Both types of "descriptive material" are non-statutory when claimed as descriptive material per se. When functional descriptive material is recorded on some computer-readable medium it becomes structurally and functionally interrelated to the medium and will be statutory in most cases. When non-functional descriptive material is recorded on some computer-readable medium, it is not structurally and functionally interrelated to the medium but is merely carried by the medium. Merely claiming non-functional descriptive material stored in a computer-readable medium does not make it statutory. Such a result would exalt form over substance. Thus, non-statutory music does not become

statutory by merely recording it on a compact disk. Protection for this type of work is provided under the copyright law. . . .

Data structures not claimed as embodied in computer-readable media are descriptive material per se and are not statutory because they are neither physical "things" nor statutory processes. Such claimed data structures do not define any structural and functional interrelationships between the data structure and other claimed aspects of the invention which permit the data structure's functionality to be realized. In contrast, a claimed computer-readable medium encoded with a data structure defines structural and functional interrelationships between the data structure and the medium which permit the data structure's functionality to be realized, and is thus statutory.

Similarly, computer programs claimed as computer listings per se, i.e., the descriptions or expressions of the programs, are not physical "things," nor are they statutory processes, as they are not "acts" being performed. Such claimed computer programs do not define any structural and functional interrelationships between the computer program and other claimed aspects of the invention which permit the computer program's functionality to be realized. In contrast, a claimed computer-readable medium encoded with a computer program defines structural and functional interrelationships between the computer program and the medium which permit the computer program's functionality to be realized, and is thus statutory. Accordingly, it is important to distinguish claims that define descriptive material per se from claims that define statutory inventions. . . .

Since a computer program is merely a set of instructions capable of being executed by a computer, the computer program itself is not a process and Office personnel should treat a claim for a computer program, without the computer-readable medium needed to realize the computer program's functionality, as non-statutory functional descriptive material. When a computer program is claimed in a process where the computer is executing the computer program's instructions, Office personnel should treat the claim as a process claim. When a computer program is recited in conjunction with a physical structure, such as a computer memory, Office personnel should treat the claim as a product claim. . . .

Where certain types of descriptive material, such as music, literature, art, photographs and mere arrangements or compilations of facts or data, are merely stored so as to be read or outputted by a computer without creating any functional interrelationship, either as part of the stored data or as part of the computing processes performed by the computer, then such descriptive material alone does not impart functionality either to the data as so structured, or to the computer. Such "descriptive material" is not a process, machine, manufacture or composition of matter.

The policy that precludes the patenting of non-functional descriptive material would be easily frustrated if the same descriptive material could be patented when claimed as an article of manufacture. For example, music is commonly sold to consumers in the format of a compact disc. In such cases, the known compact disc acts as nothing more than a carrier for non-functional descriptive material. The purely non-functional descriptive material cannot alone provide the practical application for the manufacture.

Office personnel should be prudent in applying the foregoing guidance. Non-functional descriptive material may be claimed in combination with other functional descriptive material on a computer-readable medium to provide the necessary functional and structural interrelationship to satisfy the requirements of §101. The presence of the claimed non-functional descriptive material is not necessarily determinative of non-statutory subject matter. For example, a computer that recognizes a particular grouping of musical notes read from memory and upon recognizing that particular sequence, causes another defined series of notes to be played, defines a functional interrelationship among that data and the computing processes performed when utilizing that data, and as such is statutory because it implements a statutory process. . . .

Office personnel must treat each claim as a whole. The mere fact that a hardware element is recited in a claim does not necessarily limit the claim to a specific machine or manufacture. If a product claim encompasses *any and every* computer implementation of a process, when read in light of the specification, it should be examined on the basis of the underlying process. Such a claim can be recognized as it will:

- define the physical characteristics of a computer or computer component exclusively as functions or steps to be performed on or by a computer, and
- encompass *any and every* product in the stated class (e.g., computer, computer-readable memory) *configured in any manner* to perform that process. . . .

If a claim is found to encompass any and every product embodiment of the underlying process, and if the underlying process is statutory, the product claim should be classified as a statutory product. By the same token, if the underlying process invention is found to be non-statutory, Office personnel should classify the "product" claim as a "non-statutory product." If the product claim is classified as being a non-statutory product on the basis of the underlying process, Office personnel should emphasize that they have considered all claim limitations and are basing their finding on the analysis of the underlying process. . . .

A claim that requires one or more acts to be performed defines a process. However, not all processes are statutory under §101. To be statutory, a claimed computer-related process must either: (1) result in a physical transformation outside the computer for which a practical application in the technological arts is either disclosed in the specification or would have been known to a skilled artisan (discussed in (i) below), or (2) be limited by the language in the claim to a practical application within the technological arts (discussed in (ii) below). The claimed practical application must be a further limitation upon the claimed subject matter if the process is confined to the internal operations of the computer. If a physical transformation occurs outside the computer, it is not necessary to claim the practical application. A disclosure that permits a skilled artisan to practice the claimed invention, i.e., to put it to a practical use, is sufficient. On the other hand, it is necessary to claim the practical application if there is no physical transformation or if the process merely manipulates concepts or converts one set of numbers into another. . . .

There is always some form of physical transformation within a computer because a computer acts on signals and transforms them during its operation and changes the state of its components during the execution of a process. Even though such a physical transformation occurs within a computer, such activity is not determinative of whether the process is statutory because such transformation alone does not distinguish a statutory computer process from a non-statutory computer process. What is determinative is not how the computer performs the process, but what the computer does to achieve a practical application.

A process that merely manipulates an abstract idea or performs a purely mathematical algorithm is non-statutory despite the fact that it might inherently have some usefulness. For such subject matter to be statutory, the claimed process must be limited to a practical application of the abstract idea or mathematical algorithm in the technological arts. For example, a computer process that simply calculates a mathematical algorithm that models noise is non-statutory. However, a claimed process for digitally filtering noise employing the mathematical algorithm is statutory. . . .

If the "acts" of a claimed process manipulate only numbers, abstract concepts or ideas, or signals representing any of the foregoing, the acts are not being applied to appropriate subject matter. Thus, a process consisting solely of mathematical operations, i.e., converting one set of numbers into another set of numbers, does not manipulate appropriate subject matter and thus cannot constitute a statutory process.

COMMENTS AND QUESTIONS

1. The law of software patents in the last 15 years can be criticized as exalting form over substance, allowing patent attorneys to circumvent the limitations of section 101 through artful drafting. Do the 1996 Guidelines solve this problem? Or are they formalistic as well? In particular, consider what distinguishes a patentable data structure from an unpatentable one under the guidelines. Is it anything other than the language of the claims?

2. Do the guidelines' provisions on patentable subject matter affect the scope of the patents that result? One possible explanation for the formal distinctions drawn in the guidelines is that they are designed to allow narrow patent claims while excluding broad ones. For example, a claim directed to a computer program itself is not statutory, but a claim for a program implemented on a particular medium is statutory. Does this distinction imply that the claim is infringed only by the use of the same program *recorded on the same medium*? Is such a restriction consistent with *Alappat*, supra, where the court acknowledged that various possible computer implementations of a program were functionally equivalent?

3. Patent and copyright protection have coexisted uneasily for some time in the software arena. Do the 1996 Guidelines draw a line between material that is patentable and material that is copyrightable? In particular, is the fact that a particular program element receives copyright protection evidence that it is (or should be) unpatentable? Can you think of a way to patent copyrighted material in a computer-based medium that would pass muster under the guidelines?

4. While the PTO Guidelines for examiners are obviously important to the continued development of the law in this area, they cannot substitute for the judgment of the courts. If the Federal Circuit decides, for example, that

"software on a floppy disk" is not within the ambit of 35 U.S.C. §101, the PTO does not have the power to issue such patents regardless of the guidelines. Similarly, the PTO cannot reject an application under section 101 if the courts have decided that it is patentable subject matter. Thus, while important, the PTO Guidelines do not represent the final word on the section 101 issue.

At the same time, the evidence is growing that the courts may not have the final say on section 101 either. With over 40,000 software patents in force in the United States, and several thousand more being issued every year, the real world seems to have left the courts (and perhaps even the guidelines) behind. Numerous issued patents cover pure data structures (see, e.g., U.S. Patent No. 5,488,717; U.S. Patent No. 5,414,701), methods for performing calculations in a data processor (see, e.g., U.S. Patent No. 5,386,375), data compression algorithms (see, e.g., U.S. Patent No. 5,051,745), and software-based encryption algorithms (see, e.g., U.S. Patent No. 5,530,752) despite the apparently nonstatutory nature of such claims. The "cognitive dissonance" that results between what the law is and what the actual practice has been suggests either that the law can be expected to change or that a large number of issued patents will have to be held invalid.

For an argument that congressional action is necessary to clarify the section 101 issues surrounding software patents, see Maximilian R. Peterson, Note, Now You See It, Now You Don't: Was It a Patentable Machine or an Unpatentable "Algorithm"? On Principle and Expediency in Current Patent Law Doctrines Relating to Computer-Implemented Inventions, 64 Geo. Wash. L. Rev. 90 (1995); Lee Hollaar, Justice Douglas Was Right: The Need for Congressional Action on Software Patents, 24 AIPLA Q.J. — (forthcoming 1997).

5. Despite the ongoing struggle to define the contours of algorithm patentability in the United States, the world trade community in 1994 adopted an expansive patent subject matter provision as part of the Trade-Related Aspects of Intellectual Property (TRIPs) accord in the "Uruguay Round" of agreements under the General Agreement on Tariffs and Trade (GATT). See General Agreement on Tariffs and Trade: Multilateral Trade Negotiations Final Act Embodying the Results of the Uruguay Round of Trade Negotiations, Done at Marrakesh, April 15, 1994, 33 International Legal Materials 1125 (1994), at Article 27, ¶1:

> Subject to the provisions of paragraphs 2 and 3 [which do not deal with software], patents shall be available for any inventions, whether products or processes, in all fields of technology, provided that they are new, involve an inventive step and are capable of industrial application. . . . [P]atents shall be available and patent rights enjoyable without discrimination as to the place of invention, the field of technology and whether products are imported or locally produced.

This language suggests agreement upon a more or less uniform rule worldwide favoring patents for at least some types of software, although there is no evidence that this was intended by the drafters of Article 27. Notwithstanding Article 27, however, most European courts have taken the position that "pure" software is not patentable. However, there is some indication that that attitude may be changing. Article 52 of the European Patent Convention has increasingly been interpreted in ways similar to the old U.S. rule, permitting software inventions to be patented if they are claimed as part of a larger machine.

State Street Bank & Trust v. Signature Financial Services
United States Court of Appeals for the Federal Circuit
149 F.3d 1368 (Fed. Cir. 1998)

RICH, Circuit Judge.

Signature Financial Group, Inc. (Signature) appeals from the decision of the United States District Court for the District of Massachusetts granting a motion for summary judgment in favor of State Street Bank & Trust Co. (State Street), finding U.S. Patent No. 5,193,056 (the '056 patent) invalid on the ground that the claimed subject matter is not encompassed by 35 U.S.C. §101 (1994). See State Street Bank & Trust Co. v. Signature Financial Group, Inc., 927 F.Supp. 502, 38 USPQ2d 1530 (D.Mass. 1996). We reverse and remand because we conclude that the patent claims are directed to statutory subject matter.

Background

Signature is the assignee of the '056 patent which is entitled "Data Processing System for Hub and Spoke Financial Services Configuration." The '056 patent issued to Signature on March 9, 1993, naming R. Todd Boes as the inventor. The '056 patent is generally directed to a data processing system (the system) for implementing an investment structure which was developed for use in Signature's business as an administrator and accounting agent for mutual funds. In essence, the system, identified by the proprietary name Hub and Spoke, facilitates a structure whereby mutual funds (Spokes) pool their assets in an investment portfolio (Hub) organized as a partnership. This investment configuration provides the administrator of a mutual fund with the advantageous combination of economies of scale in administering investments coupled with the tax advantages of a partnership.

State Street and Signature are both in the business of acting as custodians and accounting agents for multi-tiered partnership fund financial services. State Street negotiated with Signature for a license to use its patented data processing system described and claimed in the '056 patent. When negotiations broke down, State Street brought a declaratory judgment action asserting invalidity, unenforceability, and noninfringement in Massachusetts district court, and then filed a motion for partial summary judgment of patent invalidity for failure to claim statutory subject matter under sec. 101. The motion was granted and this appeal followed.

Discussion

. . . The patented invention relates generally to a system that allows an administrator to monitor and record the financial information flow and make all calculations necessary for maintaining a partner fund financial services configuration. As previously mentioned, a partner fund financial services configuration essentially allows several mutual funds, or "Spokes," to pool their investment funds into a single portfolio, or "Hub," allowing for consolidation of, inter alia, the costs of administering the fund combined with the tax advantages of a partnership. In particular, this system provides means for a daily allocation of assets for two or more Spokes that are invested in the same Hub. The system determines the percentage share that each Spoke maintains in the Hub, while taking into consideration daily changes both in the value of the Hub's investment securities and in the concomitant amount of each Spoke's assets.

In determining daily changes, the system also allows for the allocation among the Spokes of the Hub's daily income, expenses, and net realized and unrealized gain or loss, calculating each day's total investments based on the concept of a book capital account. This enables the determination of a true asset value of each Spoke and accurate calculation of allocation ratios between or among the Spokes. The system additionally tracks all the relevant data determined on a daily basis for the Hub and each Spoke, so that aggregate year end income, expenses, and capital gain or loss can be determined for accounting and for tax purposes for the Hub and, as a result, for each publicly traded Spoke.

It is essential that these calculations are quickly and accurately performed. In large part this is required because each Spoke sells shares to the public and the price of those shares is substantially based on the Spoke's percentage interest in the portfolio. In some instances, a mutual fund administrator is required to calculate the value of the shares to the nearest penny within as little as an hour and a half after the market closes. Given the complexity of the calculations, a computer or equivalent device is a virtual necessity to perform the task. . . .

The district court began its analysis by construing the claims to be directed to a process, with each "means" clause merely representing a step in that process. However, "machine" claims having "means" clauses may only be reasonably viewed as process claims if there is no supporting structure in the written description that corresponds to the claimed "means" elements.

See In re Alappat, 33 F.3d 1526, 1540-41, 31 USPQ2d 1545, 1554 (Fed.Cir. 1994) (en banc). This is not the case now before us.

When independent claim 1 is properly construed in accordance with section 112, para. 6, it is directed to a machine, as demonstrated below, where representative claim 1 is set forth, the subject matter in brackets stating the structure the written description discloses as corresponding to the respective "means" recited in the claims.

1. A data processing system for managing a financial services configuration of a portfolio established as a partnership, each partner being one of a plurality of funds, comprising:

(a) computer processor means [a personal computer including a CPU] for processing data;

(b) storage means [a data disk] for storing data on a storage medium;

(c) first means [an arithmetic logic circuit configured to prepare the data disk to magnetically store selected data] for initializing the storage medium;

(d) second means [an arithmetic logic circuit configured to retrieve information from a specific file, calculate incremental increases or decreases based on specific input, allocate the results on a percentage basis, and store the output in a separate file] for processing data regarding assets in the portfolio and each of the funds from a previous day and data regarding increases or decreases in each of the funds, [sic, funds'] assets and for allocating the percentage share that each fund holds in the portfolio;

(e) third means [an arithmetic logic circuit configured to retrieve information from a specific file, calculate incremental increases and decreases based on specific input, allocate the results on a percentage basis and store the output in a separate file] for processing data regarding daily incremental income, expenses, and net realized gain or loss for the portfolio and for allocating such data among each fund;

(f) fourth means [an arithmetic logic circuit configured to retrieve information from a specific file, calculate incremental increases and decreases based on specific input, allocate the results on a percentage basis and store the output in a separate file] for processing data regarding daily net unrealized gain or loss for the portfolio and for allocating such data among each fund; and

(g) fifth means [an arithmetic logic circuit configured to retrieve information from specific files, calculate that information on an aggregate basis and store the output in a separate file] for processing data regarding aggregate year-end income, expenses, and capital gain or loss for the portfolio and each of the funds.

Each claim component, recited as a "means" plus its function, is to be read, of course, pursuant to sec. 112, para. 6, as inclusive of the "equivalents" of the structures disclosed in the written description portion of the specification. Thus, claim 1, properly construed, claims a machine, namely, a data processing system for managing a financial services configuration of a portfolio established as a partnership, which machine is made up of, at the very least, the specific structures disclosed in the written description and corresponding to the means-plus-function elements (a)-(g) recited in the claim. A "machine" is proper statutory subject matter under §101. We note that, for the purposes of a §101 analysis, it is of little relevance whether claim 1 is directed to a "machine" or a "process," as long as it falls within at least one of the four enumerated categories of patentable subject matter, "machine" and "process" being such categories.

This does not end our analysis, however, because the court concluded that the claimed subject matter fell into one of two alternative judicially-created exceptions to statutory subject matter. The court refers to the first exception as the "mathematical algorithm" exception and the second exception as the "business method" exception. . . .

The repetitive use of the expansive term "any" in §101 shows Congress's intent not to place any restrictions on the subject matter for which a patent may be obtained beyond those specifically recited in §101. Indeed, the Supreme Court has acknowledged that Congress intended §101 to extend to "anything under the sun that is made by man." Diamond v. Chakrabarty, 447 U.S. 303, 309, 100 S.Ct. 2204, 65 L.Ed.2d 144 (1980); see also Diamond v. Diehr, 450 U.S. 175, 182, 101 S.Ct. 1048, 67 L.Ed.2d 155 (1981). Thus, it is improper to read limitations into §101 on the subject matter that may be patented where the legislative history indicates that Congress clearly did not intend such limitations. See *Chakrabarty*, 447 U.S. at 308, 100 S.Ct. 2204 ("We have also cautioned that courts 'should not read into the patent laws limitations and conditions which the legislature has not expressed.' " (citations omitted)).

The "Mathematical Algorithm" Exception

The Supreme Court has identified three categories of subject matter that are unpatentable, namely "laws of nature, natural phenomena, and abstract ideas." *Diehr*, 450 U.S. at 185, 101 S.Ct. 1048. Of particular relevance to this case, the Court has held that mathematical algorithms are not patentable subject matter to the extent that they are merely abstract ideas. See *Diehr*, 450 U.S. 175, 101 S.Ct. 1048, *passim*; Parker v. Flook, 437 U.S. 584, 98 S.Ct. 2522, 57 L.Ed.2d 451 (1978); Gottschalk v. Benson, 409 U.S. 63, 93 S.Ct. 253, 34 L.Ed.2d 273 (1972). In *Diehr*, the Court explained that certain types of mathematical subject matter, standing alone, represent nothing more than abstract ideas until reduced to some type of practical application, i.e., "a useful, concrete and tangible result." *Alappat*, 33 F.3d at 1544, 31 USPQ2d at 1557.[4]

Unpatentable mathematical algorithms are identifiable by showing they are merely abstract ideas constituting disembodied concepts or truths that

---

4. This has come to be known as the mathematical algorithm exception. This designation has led to some confusion, especially given the *Freeman-Walter-Abele* analysis. By keeping in mind that the mathematical algorithm is unpatentable only to the extent that it represents an abstract idea, this confusion may be ameliorated.

are not "useful." From a practical standpoint, this means that to be patentable an algorithm must be applied in a "useful" way. In *Alappat*, we held that data, transformed by a machine through a series of mathematical calculations to produce a smooth waveform display on a rasterizer monitor, constituted a practical application of an abstract idea (a mathematical algorithm, formula, or calculation), because it produced "a useful, concrete and tangible result" — the smooth waveform.

Similarly, in Arrhythmia Research Technology Inc. v. Corazonix Corp., 958 F.2d 1053, 22 USPQ2d 1033 (Fed.Cir. 1992), we held that the transformation of electrocardiograph signals from a patient's heartbeat by a machine through a series of mathematical calculations constituted a practical application of an abstract idea (a mathematical algorithm, formula, or calculation), because it corresponded to a useful, concrete or tangible thing — the condition of a patient's heart.

Today, we hold that the transformation of data, representing discrete dollar amounts, by a machine through a series of mathematical calculations into a final share price, constitutes a practical application of a mathematical algorithm, formula, or calculation, because it produces "a useful, concrete and tangible result" — a final share price momentarily fixed for recording and reporting purposes and even accepted and relied upon by regulatory authorities and in subsequent trades.

The district court erred by applying the *Freeman-Walter-Abele* test to determine whether the claimed subject matter was an unpatentable abstract idea. . . .

After *Diehr* and *Chakrabarty*, the *Freeman-Walter-Abele* test has little, if any, applicability to determining the presence of statutory subject matter. As we pointed out in *Alappat*, 33 F.3d at 1543, 31 USPQ2d at 1557, application of the test could be misleading, because a process, machine, manufacture, or composition of matter employing a law of nature, natural phenomenon, or abstract idea is patentable subject matter even though a law of nature, natural phenomenon, or abstract idea would not, by itself, be entitled to such protection. The test determines the presence of, for example, an algorithm. Under *Benson*, this may have been a sufficient indicium of nonstatutory subject matter. However, after *Diehr* and *Alappat*, the mere fact that a claimed invention involves inputting numbers, calculating numbers, outputting numbers, and storing numbers, in and of itself, would not render it nonstatutory subject matter, unless, of course, its operation does not produce a "useful, concrete and tangible result." *Alappat*, 33 F.3d at 1544, 31 USPQ2d at 1557. . . .

The question of whether a claim encompasses statutory subject matter should not focus on which of the four categories of subject matter a claim is directed to — process, machine, manufacture, or composition of matter — but rather on the essential characteristics of the subject matter, in particular, its practical utility. Section 101 specifies that statutory subject matter must also satisfy the other "conditions and requirements" of Title 35, including novelty, nonobviousness, and adequacy of disclosure and notice. See In re

Warmerdam, 33 F.3d 1354, 1359, 31 USPQ2d 1754, 1757-58 (Fed.Cir. 1994). For purpose of our analysis, as noted above, claim 1 is directed to a machine programmed with the Hub and Spoke software and admittedly produces a "useful, concrete, and tangible result." *Alappat*, 33 F.3d at 1544, 31 USPQ2d at 1557. This renders it statutory subject matter, even if the useful result is expressed in numbers, such as price, profit, percentage, cost, or loss.

COMMENTS & QUESTIONS

1. Any thought that the new test announced in *State Street* was an aberration, or that it could easily be reconciled with prior precedent, was put to rest in AT&T v. Excel Communications, 172 F.3d 1352 (Fed. Cir. 1999). In that case the court held that method claims to a method for "generating a message record for an interexchange call" and recording who the call should be billed to, are patentable subject matter. The court applied *State Street*'s "useful, concrete and tangible result" test, concluding that the generation of billing records was clearly useful. The court specifically rejected the argument that a patentable software claim must have physical structure associated with it, noting that physical transformation is only one of several possible ways to bring about a useful result. In sustaining the patent, the court laid to rest much of the history of the mathematical algorithm exception:

> A mathematical formula alone, sometimes referred to as a mathematical algorithm, viewed in the abstract, is considered unpatentable subject matter. Courts have used the terms "mathematical algorithm," "mathematical formula," and "mathematical equation," to describe types of nonstatutory mathematical subject matter without explaining whether the terms are interchangeable or different. Even assuming the words connote the same concept, there is considerable question as to exactly what the concept encompasses. See, e.g., *Diehr*, 450 U.S. at 186 n. 9, 101 S.Ct. 1048 ("The term 'algorithm' is subject to a variety of definitions . . . [Petitioner's] definition is significantly broader than the definition this Court employed in *Benson* and *Flook*."); accord In re Schrader, 22 F.3d 290, 293 n. 5, 30 USPQ2d 1455, 1457 n. 5 (Fed.Cir. 1994).
>
> This court recently pointed out that any step-by-step process, be it electronic, chemical, or mechanical, involves an "algorithm" in the broad sense of the term. Because §101 includes processes as a category of patentable subject matter, the judicially-defined proscription against patenting of a "mathematical algorithm," to the extent such a proscription still exists, is narrowly limited to mathematical algorithms in the abstract. See also *Benson*, 409 U.S. at 65, 93 S.Ct. 253 (describing a mathematical algorithm as a "procedure for solving a given type of mathematical problem").
>
> Since the process of manipulation of numbers is a fundamental part of computer technology, we have had to reexamine the rules that govern the patentability of such technology. The sea-changes in both law and technology stand

as a testament to the ability of law to adapt to new and innovative concepts, while remaining true to basic principles. In an earlier era, the PTO published guidelines essentially rejecting the notion that computer programs were patentable. As the technology progressed, our predecessor court disagreed, and, overturning some of the earlier limiting principles regarding §101, announced more expansive principles formulated with computer technology in mind. In our recent decision in State Street, this court discarded the so-called "business method" exception and reassessed the "mathematical algorithm" exception, see 149 F.3d at 1373-77, 47 USPQ2d at 1600-04, both judicially-created "exceptions" to the statutory categories of §101. As this brief review suggests, this court (and its predecessor) has struggled to make our understanding of the scope of §101 responsive to the needs of the modern world. . . .

The State Street formulation, that a mathematical algorithm may be an integral part of patentable subject matter such as a machine or process if the claimed invention as a whole is applied in a "useful" manner, follows the approach taken by this court en banc in In re Alappat, 33 F.3d 1526, 31 USPQ2d 1545 (Fed.Cir. 1994). . . . [T]he *Alappat* inquiry simply requires an examination of the contested claims to see if the claimed subject matter as a whole is a disembodied mathematical concept representing nothing more than a "law of nature" or an "abstract idea," or if the mathematical concept has been reduced to some practical application rendering it "useful." Id. at 1544, 31 USPQ2d at 1557. In *Alappat*, we held that more than an abstract idea was claimed because the claimed invention as a whole was directed toward forming a specific machine that produced the useful, concrete, and tangible result of a smooth waveform display. See id. at 1544, 31 USPQ2d at 1557.

In both *Alappat* and *State Street*, the claim was for a machine that achieved certain results. In the case before us, because Excel does not own or operate the facilities over which its calls are placed, AT&T did not charge Excel with infringement of its apparatus claims, but limited its infringement charge to the specified method or process claims. Whether stated implicitly or explicitly, we consider the scope of §101 to be the same regardless of the form — machine or process — in which a particular claim is drafted.

Id at 23–24.

2. Related to the record-generation claim in *AT&T,* but potentially even more far-reaching, is a new type of patent claim to "propagated signals." These claims are directed to "a manufactured transient phenomenon, such as an electrical, optical, or acoustical signal." Jeffrey R. Kuester et al., A New Frontier in Patents: Patent Claims to Propagated Signals, 17 J. Marshall J. Comp. & Info. L. 75 (1998). The claim would be directly infringed by the sending of data covered by the claim over computer wires, telephone lines, or the airwaves.

After *State Street*, is there any reason such a claim would not be patentable? Does it produce a useful, concrete, and tangible result?

3. The PTO revised its Examination Guidelines in late 1998 in response to the decision in *State Street*, adding several "training examples" dealing with business, artificial intelligence, and mathematical processing claims.

The examples show a reluctance on the part of the PTO to abandon the physical transformation approach that characterized the caselaw through *Alappat.*

4. *State Street* also opened the door to the patenting of business methods by rejecting the preexisting "business method" exception to patentability. See supra Chapter III. For a review of some recently issued patents on Internet-related business techniques, see Robert P. Merges, As Many as Six Impossible Patents Before Breakfast: Property Rights for Business Concepts and Patent System Reform, 14 Berkeley Tech. L.J. 577 (1999) (describing Walker Digital, a patent-driven Internet commerce "idea lab" company responsible for the patented "Priceline.com" reverse auction airline ticket Internet service).

PROBLEMS

Problem 7-9. Yoshimoto, a company that makes software for video games, seeks a patent on a new and nonobvious computer program. Yoshimoto claims the program "implemented in a read-only memory (ROM) readable by a computer." Is this claim patentable subject matter? How would this question be answered under (1) *Benson?* (2) *Diehr?* (3) *Alappat?* (4) the PTO Guidelines?

Problem 7-10. Linus, a programmer who specializes in computer-assisted design tools, develops a new and more efficient way of cataloguing data in a computer representative of "virtual objects." The virtual objects are in turn computer graphics used to model real-world objects. She claims "a data structure resident in a computer-readable memory." The data structure is further described and defined in the claims, but the memory is not. Is Linus's claim patentable under the PTO guidelines? Is there any form of software such a claim would not cover?

What other types of legal protection might Linus be able to obtain?

Problem 7-11. Safe Flight is an aircraft instrument company. It seeks to obtain a patent on avionic windshear equipment designed to give warning to a plane's pilot on takeoff or landing when the plane encounters violent shifts in wind direction ("windshear"). This can result in disaster if the pilot cannot compensate quickly and properly for the changing conditions. Windshear is measured in Safe Flight's invention by means of an algorithm that compares the rate of change of the instantaneous airspeed (the speed of an aircraft measured against the surrounding air) with the change in groundspeed (the speed of an aircraft measured against the ground, which the patentee calls "horizontal inertial acceleration"). Safe Flight claims a "windshear measurement and control system" in an aircraft cockpit that employs the algorithm to constantly recalculate windshear during landing, and automatically generates a warning buzzer if windshear exceeds a certain threshold level.

Is Safe Flight's claim patentable subject matter under section 101? Should it be? Would it be patentable if it were not implemented in a computer?

2. Examination and Validity of Software Patents

Deciding whether (and under what circumstances) computer programs qualify as patentable subject matter is only the first hurdle in obtaining a software patent. A valid patent must pass four other tests in examination in the Patent Office: it must be useful (35 U.S.C. §101), novel (35 U.S.C. §102), nonobvious (35 U.S.C. §103), and clearly described in the patent application (35 U.S.C. §112 ¶1). Utility is not a particularly strict test, and most computer programs that work will meet it easily. But each of the other steps in the patent examination process presents unique issues in the software context.

a. Novelty and Nonobviousness

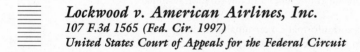

Lockwood v. American Airlines, Inc.
107 F.3d 1565 (Fed. Cir. 1997)
United States Court of Appeals for the Federal Circuit

LOURIE, Circuit Judge.

Lawrence B. Lockwood appeals from the final judgment of the United States District Court for the Southern District of California granting summary judgment in favor of American Airlines, Inc. In that summary judgment, the court held that (1) U.S. Patent Re. 32,115, U.S. Patent 4,567,359, and U.S. Patent 5,309,355 were not infringed by American's SABREvision reservation system, and that (2) the '355 patent and the asserted claims of the '359 patent were invalid under 35 U.S.C. §102 and 35 U.S.C. §103, respectively. Lockwood v. American Airlines, Inc., 834 F.Supp. 1246, 28 USPQ2d 1114 (S.D.Cal. 1993), *req. for reconsideration denied,* 847 F.Supp. 777 (S.D.Cal. 1994) (holding the '115 and '359 patents not infringed); Lockwood v. American Airlines, Inc., 877 F.Supp. 500, 34 USPQ2d 1290 (S.D.Cal. 1994) (holding the asserted claims of the '355 patent invalid and not infringed);

Lockwood v. American Airlines, Inc., 37 USPQ2d 1534, 1995 WL 822659 (S.D.Cal. 1995) (holding the '359 patent invalid). Because the district court correctly determined that there were no genuine issues of material fact in dispute and that American was entitled to judgment as a matter of law, we affirm.

Background

The pertinent facts are not in dispute. Lockwood owns the '115, '355, and '359 patents, all of which relate to automated interactive sales terminals that provide sales presentations to customers and allow the customers to order goods and services. Lockwood sued American asserting that American's SABREvision airline reservation system infringed all three patents. SABRE-vision is used by travel agents to access schedule and fare information, to book itineraries, and to retrieve photographs of places of interest, including hotels, restaurants, and cruises, for display to consumers. It improves upon American's SABRE reservation system, which originated in the 1960s and which cannot display photographs. . . .

The '359 patent discloses a system of multiple interactive self-service terminals that provide audio-visual sales presentations and dispense goods and services from multiple institutions. Claim 1, the only independent claim, reads in pertinent part:

> A system for automatically dispensing information, goods, and services for a plurality of institutions in a particular industry, comprising:
> . . .
>   at least one customer sales and information terminal . . .
> . . .
>   said sales and information terminal including:
>   audio-visual means for interaction with a customer, comprising:
>   means for storing a sequence of audio and video information to be selectively transmitted to a customer;
>   means for transmitting a selected sequence of said stored information to the customer;
>   customer operated input means for gathering information from a customer. . . .

The district court held that SABREvision did not infringe the '359 patent because it lacked the "audio-visual means" and "customer operated input means." The court also held the '359 patent invalid because it would have been obvious in light of the original SABRE system in combination with the self-service terminal disclosed in U.S. Patent 4,359,631, which issued in 1982 and was subsequently reissued as the '115 patent. . . .

Discussion . . .

A. *Validity*

The district court held that the asserted claims of the '359 patent would have been obvious in light of the prior art '631 patent and the original SABRE system. A determination of obviousness under 35 U.S.C. §103 is a legal conclusion involving factual inquiries. Uniroyal, Inc. v. Rudkin-Wiley Corp., 837 F.2d 1044, 1050, 5 USPQ2d 1434, 1438 (Fed.Cir. 1988). Lockwood argues that the subject matter of the '359 claims would not have been obvious and that the district court impermissibly drew adverse factual inferences in concluding that the patent was invalid. Lockwood first argues that the district court erred in concluding that the SABRE system qualified as prior art.

American submitted an affidavit averring that the SABRE system was introduced to the public in 1962, had over one thousand connected sales desks by 1965, and was connected to the reservation systems for most of the other airlines by 1970. Lockwood does not dispute these facts, but argues that because "critical aspects" of the SABRE system were not accessible to the public, it could not have been prior art. American's expert conceded that the essential algorithms of the SABRE software were proprietary and confidential and that those aspects of the system that were readily apparent to the public would not have been sufficient to enable one skilled in the art to duplicate the system. However, American responds that the public need not have access to the "inner workings" of a device for it to be considered "in public use" or "used by others" within the meaning of the statute.

We agree with American that those aspects of the original SABRE system relied on by the district court are prior art to the '359 patent. The district court held that SABRE, which made and confirmed reservations with multiple institutions (e.g., airlines, hotels, and car rental agencies), combined with the terminal of the '631 patent rendered the asserted claims of the '359 patent obvious. The terminal of the '631 patent admittedly lacked this "multiple institution" feature. It is undisputed, however, that the public was aware that SABRE possessed this capability and that the public had been using SABRE to make travel reservations from independent travel agencies prior to Lockwood's date of invention.

If a device was "known or used by others" in this country before the date of invention or if it was "in public use" in this country more than one year before the date of application, it qualifies as prior art. See 35 U.S.C. §102(a) and (b) (1994). Lockwood attempts to preclude summary judgment by pointing to record testimony that one skilled in the art would not be able to build and practice the claimed invention without access to the secret aspects of SABRE. However, it is the claims that define a patented invention. See Constant v. Advanced Micro-Devices, Inc., 848 F.2d 1560, 1571, 7

USPQ2d 1057, 1064 (Fed.Cir. 1988). As we have concluded earlier in this opinion, American's public use of the high-level aspects of the SABRE system was enough to place the claimed features of the '359 patent in the public's possession. See In re Epstein, 32 F.3d 1559, 1567-68, 31 USPQ2d 1817, 1823 (Fed.Cir. 1994) ("Beyond this 'in public use or on sale' finding, there is no requirement for an enablement-type inquiry."). Lockwood cannot negate this by evidence showing that other, unclaimed aspects of the SABRE system were not publicly available. Moreover, the '359 patent itself does not disclose the level of detail that Lockwood would have us require of the prior art. For these reasons, Lockwood fails to show a genuine issue of material fact precluding summary judgment.

Lockwood further argues that even if the SABRE system is effective prior art, the combination of that system and the '631 patent would not have yielded the invention of the '359 patent. The terminal in the claims of the '359 patent includes a number of means-plus-function limitations, subject to 35 U.S.C. §112, ¶6, including "means for gathering information from a customer" and "means for storing a sequence of audio and video information to be selectively transmitted to a customer." Means-plus-function clauses are construed "as limited to the corresponding structure[s] disclosed in the specification and equivalents thereof." In re Donaldson Co., 16 F.3d 1189, 1195, 29 USPQ2d 1845, 1850 (Fed.Cir. 1994) (en banc); see 35 U.S.C. §112, ¶6 (1994). Lockwood argues that the structures disclosed in the '359 patent differ substantially from the terminal disclosed in the '631 patent and that, at the very least, his expert's declaration raised genuine issues of material fact sufficient to preclude summary judgment.

We do not agree. We believe that American has met its burden, even in light of the presumption of patent validity, to show that the means limitations relating to the terminal in the claims of the '359 patent appear in the '631 specification. Lockwood has failed to respond by setting forth specific facts that would raise a genuine issue for trial. Specifically, Lockwood has not alleged that the '631 disclosure lacks the structures disclosed in the '359 patent specification or their equivalents. As the district court noted, Lockwood's expert, Dr. Tuthill, relied on structures that are not mentioned in either the '631 or the '359 patents. For example, Tuthill states that the claimed invention differs from the '631 patent because the terminal described in the '631 patent uses a "backward-chaining" system to solve problems while the '359 patent uses a "forward-chaining" system. Yet neither the '359 nor the '631 patents mentions backward- or forward-chaining. Nor does the '359 specification describe any hardware or software structure as being limited to any particular problem-solving technique. In addition, Lockwood argues that the hardware and software disclosed in the two patents are not equivalent to each other. However, the '359 patent claims the hardware and software in broad terms, and the patents both describe similar computer controlled self-service terminals employing video disk players that store and retrieve audio-visual information. For example, with regard to the "means for controlling said storage and transmitting means," Lockwood's expert avers that the "structure described in the '359 patent which corresponds to this means is the processor unit and the application program which the processor executes." Yet, the only software descriptions in the '359 patent consist of high level exemplary functional flowcharts. Lockwood's arguments and his expert's statements are thus conclusory. They fail to identify which structures in the '359 patent are thought to be missing from the '631 patent disclosure. Accordingly, we agree with the district court that Lockwood's and his expert's declarations have not adequately responded to American's motion by raising genuine issues of material fact, and we therefore conclude that the district court properly held the asserted claims of the '359 patent to have been obvious as a matter of law. . . .

AFFIRMED.

COMMENTS AND QUESTIONS

1. The court holds that the SABRE system was "known or used by others" before the critical date of the '359 patent. Is affirmatively secret use sufficient to constitute prior art? Has the public really "known or used" the invention if they are prevented from seeing how it works or duplicating it? Does it matter whether American protected the SABRE system as a trade secret during the 1960s?

2. The court notes that Lockwood has not disclosed much more structural detail in his patent specification than the prior art '631 patent and therefore cannot claim to have differentiated his invention from that prior patent. Does this holding improperly conflate the enablement requirement and the test for obviousness?

3. Is a "design alternative" that accomplishes the same goal as a prior program but in a different way necessarily obvious? Or should programmers be entitled to patent their new designs that accomplish well-known goals?

4. A study by Greg Ahronian confirms that examiners cite less prior art in allowing software patents than in allowing any other form of patent. Why aren't patent examiners citing more prior art? Knowledgeable observers of the patent system offer at least three explanations. First, because software was not thought patentable until recently, the Patent Office has only recently taken steps to hire patent examiners qualified in computer software or related fields. During the 1980s and the early 1990s, the flood of software patent applications was handled largely by people operating outside their area of expertise. While the Meyer excerpt suggests that this may be a problem for patent applicants, it is at least as big a problem for the public. Abundant evidence indicates that the Patent Office has issued software patents on a number of applications that did not meet the standard tests of novelty and nonobviousness. For anecdotes discussing some of the more extreme examples, see Simson L. Garfinkel, Patently Absurd, Wired, July 1994, at 104.

Second, for similar reasons, the PTO's classification system is not equipped to handle software patents. As a result, software patents tend to be classified according to the field in which the software will ultimately be used, rather than according to the nature of the software invention. This makes it much harder for examiners to find what prior art does exist. This problem too could be solved, either by reclassification or by an increased emphasis on computer search systems.

The final problem with prior art in the software field is more enduring: prior art in this particular industry may be difficult or, in some cases, impossible to find. As Julie Cohen explains:

> [I]n the field of computers and computer programs, much that qualifies as prior art lies outside the areas in which the PTO has traditionally looked — previously issued patents and previous scholarly publications. Many new developments in computer programming are not documented in scholarly publications at all. Some are simply incorporated into products and placed on the market; others are discussed only in textbooks or user manuals that are not available to examiners on line. In an area that relies so heavily on published, "official" prior art, a rejection based on "common industry knowledge" that does not appear in the scholarly literature is unlikely. Particularly where the examiner lacks a computer science background, highly relevant prior art may simply be missed. In the case of the multimedia data retrieval patent granted to Compton's New Media,[46] industry criticism prompted the PTO to reexamine the patent and ultimately to reject it because it did not represent a novel and nonobvious advance over existing technology. However, it would be inefficient, and probably impracticable, to reexamine every computer program-related patent, and the PTO is unlikely to do so.

Julie E. Cohen, Reverse Engineering and the Rise of Electronic Vigilantism: Intellectual Property Implications of "Lock-Out" Technologies, 68 S. Cal. L. Rev. 1091, 1178 (1995). The current PTO Examination Guidelines for Computer-Related Inventions direct that examiners must "conduct a thorough search of the prior art," but they give no indication of how to accomplish that task.

One initiative that has received at least verbal support from a cross-section of the software industry is the work of the Software Patent Institute (SPI). Funded by a number of software publishers, the SPI seeks to create a "prior art" database of software techniques with an emphasis on those that have not been patented and are not otherwise available. They also hope to offer educational services directed to patent examiners, to address complaints about the quality of patent examinations in this field. Such online collections of software prior art may help to alleviate some of the problems Professor

---

46. Compton's received a patent on an application filed in the late 1980s which purported to cover basic use of hypertext. When Compton's announced the existence of the patent at a computer trade show, and offered everyone in the multimedia industry a chance to license it, public outrage prompted the PTO to initiate a reexamination proceeding sua sponte. The PTO rejected the application after considering a number of prior art references submitted by third parties.

Cohen describes. But they are unlikely to address other problems, such as the large body of "prior art" published only in object code form in certain commercial or semi-commercial programs. This "secret" prior art in the software industry is discussed in more detail in the next section.

Do the problems discovering software prior art suggest that the Patent Office should be more cautious about granting inventions in this industry than in other technologies? If not, what should be done about this problem?

5. For a collection of sources of prior art in the software industry, see Gregory A. Stobbs, Software Patents 109-147 (1995).

Note on Obviousness and Computer-Implemented Inventions

It is not enough that an applicant for a patent simply have invented something new. To qualify for patent protection, the new invention must also show an "inventive leap" over the prior art — it must not be merely obvious to one skilled in the art. 35 U.S.C. §103. Application of the non-obviousness standard is never easy, since it requires examiners and courts to make a judgment call regarding what would have been considered obvious to scientists in the field at the time the invention was made. But as Cohen points out, determining obviousness is particularly problematic in the case of inventions that use a computer to solve problems outside the field of programming itself:

> Intuitively, the most troubling aspect of many computer program-related patents is that they appear to reward the inventor for recognizing the obvious — that a given function may be performed more efficiently or more accurately if computerized — and using general purpose computer equipment and standard programming techniques to computerize it. Other computer program-related patents simply reward the programmer for developing otherwise unpatentable mathematical formulas. . . .
>
> In response to *Iwahashi*, Richard Stern, former chief of the Department of Justice's Intellectual Property Section, proposed reconceiving the standard for nonobviousness for computer program-related inventions. His solution, which may be termed the "innovative programmer" standard, adds a third step to the *Freeman-Walter-Abele* test. If a claimed invention recites a mathematical algorithm, but appears to be statutory subject matter when taken as a whole, the examiner must ask whether the claimed invention would have been obvious to "a person of ordinary skill . . . who: (a) knew the particular algorithm; (b) desired to accomplish the function or task to be performed; and (c) desired to do so with the aid of a computer. . . ." If not, it is nonobvious, and so patentable. By taking general purpose computer equipment and the mathematical algorithm as part of the prior art for purposes of assessing nonobviousness, the innovative programmer standard is intended to avoid [granting patents to the first person to implement a known physical process on a computer, regardless of how straightforward that implementation might be].

Cohen, supra, at 1170. Does the Stern-Cohen proposal make sense? Is it consistent with patent law outside the computer context, or would it require the creation of a new, higher standard of examination specific to computer-implemented inventions?

The first case involving the obviousness of a software-implemented invention is, perhaps surprisingly, a Supreme Court case from the 1970s. In Dann v. Johnston, 425 U.S. 219 (1976), the Court held a patent on a "machine system for automatic record-keeping of bank checks and deposits" invalid for obviousness. The Court took a rather broad view of obviousness in the computer industry, focusing on whether analogous systems to the patentee's had been implemented in computer before, rather than analyzing the precise differences between the patentee's program and the prior art programs. The clear implication of the opinion is that if a reasonably skilled programmer could produce a program analogous to the patented one, and if there was motivation in the prior art to do so, the patented program is obvious.

On the other hand the Federal Circuit in In re Zurko, 111 F.3d 887 (Fed. Cir. 1997) held that a patented invention was nonobvious even though each of the elements of the invention could be found in the prior art, where the prior art did not identify the problem to be solved. This offers another outlet for patentees: they can focus their invention on solving a previously unknown problem rather than on the particular elements of their solution.

b. Enablement of Software Inventions

35 U.S.C. §112, ¶1

The specification shall contain a written description of the invention, and of the manner and process of making and using it, in such full, clear, concise and exact terms as to enable any person skilled in the art to which it pertains, or with which it is most nearly connected, to make and use the same, and shall set forth the best mode contemplated by the inventor of carrying out his invention.

The primary issue for software patents under section 112, the disclosure provision of the patent code, has been whether a patent applicant must deposit source code to meet the enablement (i.e., "how to make and use") requirement of section 112, ¶1. Two cases in particular have defined the terms of this debate.

In White Consolidated Industries, Inc. v. Vega Servo-Control, Inc., 713 F.2d 788 (Fed. Cir. 1983), the Federal Circuit invalidated a patent for a machine tool control system that was run by a computer program. Part of the invention was a programming language translator designed to convert an input program into machine language so that the system could then execute the program. The patent specification identified an example of a translator program, the so-called SPLIT program, which was a trade secret of Sundstrand, a company that later became the plaintiff White Consolidated. When the application was filed, the SPLIT program was available exclusively from Sundstrand.

The defendant, Vega, asserted patent invalidity as a defense to the infringement claim. Specifically, Vega argued that the SPLIT program was the only suitable translator program, and merely identifying it was not sufficient to meet the standards of section 112. White claimed that widely available equivalent translators could also be used and that in any event the specification described the characteristics of the necessary translator program in terms that enabled a programmer of ordinary skill to create it from scratch.

The court held that the program translator was an integral part of the invention, and that mere identification of it was not sufficient to discharge the applicant's duty under section 112. The court seemed concerned that maintaining the translator program as a trade secret would allow White to extend the patent beyond the statutory term.

In Northern Telecom, Inc. v. Datapoint Corp., 908 F.2d 931 (Fed. Cir.), *cert. denied*, 498 U.S. 920 (1990), the Federal Circuit once again confronted the enablement requirement in the context of an invention containing a computer program. In *Northern Telecom*, the court reversed the district court's holding of invalidity on enablement grounds. Here the patent claimed an improved method of entering, verifying, and storing (or "batching") data with a special data entry terminal. The district court invalidated certain claims of the patent on the grounds that they were inadequately disclosed under section 112.

In a decision reversing this aspect of the district court's ruling, the Federal Circuit held that when claims pertain to a computer program which implements a claimed device or method, the enablement requirement varies according to the nature of the claimed invention as well as the role and complexity of the computer program needed to implement it. Under the facts in this case, the core of the claimed invention was the combination of components or steps rather than the details of the program the applicant actually used. The court noted expert testimony that various programs could be used to implement the invention, and that it would be "relatively straightforward [in light of the specification] for a skilled computer programmer to design a program to carry out the claimed invention." Id. at 941-42. The court continued:

> The computer language is not a conjuration of some black art, it is simply a highly structured language. . . . [T]he conversion of a complete thought (as expressed in English and mathematics, i.e. the known input, the desired output, the mathematical expressions needed and the methods of using those expressions) into a language a machine understands is necessarily a mere clerical function to a skilled programmer.

Id. The court refused to state categorically that a program listing would never have to be disclosed, however, especially in cases such as *White*, where it was estimated that it would take two years for a skilled programmer to produce a working program of the type called for by the specification.

COMMENTS AND QUESTIONS

1. Failure to disclose the specifics of a program that would take two years to re-create renders a patent invalid under section 112; what if the program would take one year to re-create? How about six months? One month?

2. For more on this issue, see Gregory A. Stobbs, Software Patents 170-223 (1995) (detailing means of describing software inventions in the specification, and giving examples); David Bender & Anthony R. Barkume, Disclosure Requirements for Software-related Patents, 8 Computer Law. 1 (1991); Michael Bondi, Upholding the Disclosure Requirements of 35 U.S.C. §112 Through the Submission of Flow Charts with Computer Software Patent Applications, 5 Software L.J. 635 (1992); D. C. Toedt III, Patents for Inventions Utilizing Computer Software: Some Practical Pointers, 9 Computer Law. 12 (1992) (suggesting disclosure of "pseudocode," i.e., cryptic, general-term code not in a particular programming language, to satisfy section 112; and discussing pros and cons of disclosing actual source code). See also Thomas P. Burke, Note, Software Patent Protection: Debugging the Current System, 69 Notre Dame L. Rev. 1115, 1158-1160 (1994):

> A software patent without source code is like a law review piece filled with case names but missing citations to case reporters. A person of ordinary skill in legal research might be able to track down the full-text of all the opinions. Marbury v. Madison would be found quicker than a state trial court opinion. But, would anyone think that such a practice was enabling or the best mode? As it is now, the disclosure requirements can be met using such devices as specifications, flow-charts, and pseudo-code. Professor Randall Davis of MIT summed it up at the National Research Council in 1990:
>
>> There is almost no way to visualize software. Sure, we have flow charts, we have data-flow diagrams, we have control flow diagrams, and everybody knows how basically useless those are. Flow charts are documentation you write afterward — because management requires them, not because they are a useful tool.
>
> A patent is most similar to a real property deed specifying the metes and bounds for a parcel of land. Both documents are not easily understood but succeed if they secure the owners' interests in the specified claims. If the goal is to inform the world of an invention, software professionals have avenues more timely and less expensive than pursuing a patent application. In fostering the trade-off between the interests of inventors and the public, the source code is the best way to explain an algorithm.
>
> Under this proposal, a computer system's complete source code would not have to be appendixed to the patent. The applicant would only have to include the source code directly relevant to enabling the claim language. In cases where

claims are broadly written (as in a means-plus-function apparatus claim that covers the automation of an entire industry), a nearly complete program listing would be required.

3. Even where disclosure is adequate under the enablement standard, it can run into problems under the "best mode" aspect of section 112. An inventor runs afoul of the best mode requirement if she intentionally withholds from the patent specification information that discloses the best means of which she is aware for practicing her invention. Does the best mode requirement compel the disclosure of source code, on the theory that the actual way in which the programmer has implemented an invention must be the "best" way known to the inventor? See In re Hayes Microcomputer Products, Inc. Patent Litigation, 982 F.2d 1527 (Fed. Cir. 1992) (maker of Hayes modems was not required to disclose secret settings used on those modems in patent application, because settings were arbitrarily chosen).

In Fonar v. General Elec. Co., 107 F.3d 1543 (Fed. Cir. 1997), the Federal Circuit not only concluded that source code need not be disclosed but decided that even flowcharts were unnecessary — a broad description of what the software did was sufficient:

> As a general rule, where software constitutes part of a best mode of carrying out an invention, description of such a best mode is satisfied by a disclosure of the functions of the software. This is because, normally, writing code for such software is within the skill of the art, not requiring undue experimentation, once its functions have been disclosed. It is well established that what is within the skill of the art need not be disclosed to satisfy the best mode requirement as long as that mode is described. Stating the functions of the best mode software satisfies that description test. We have so held previously and we so hold today. See In re Hayes Microcomputer Prods., Inc. Patent Litigation, 982 F.2d 1527, 1537-38, 25 USPQ2d 1241, 1248-49 (Fed.Cir. 1992); In re Sherwood, 613 F.2d 809, 816-17, 204 USPQ 537, 544 (CCPA 1980). Thus, flow charts or source code listings are not a requirement for adequately disclosing the functions of software. See *Sherwood*, 613 F.2d at 816-17, 204 USPQ at 544. Here, substantial evidence supports a finding that the software functions were disclosed sufficiently to satisfy the best mode requirement. See *Hayes*, 982 F.2d at 1537, 25 USPQ2d at 1248-49 (stating that there was no best mode violation where the specification failed to disclose a firmware listing or flow charts, but did disclose sufficient detail to allow one skilled in the art to develop a firmware listing for implementing the invention).

Id at ___.

Most software patent cases have taken a similar position. In two cases decided in 1997, the Federal Circuit concluded that the patentees satisfied the written description and best mode requirements for inventions partially implemented in software even though they did not use the terms "computer" or "software" anywhere in the specification! See Robotic Vision Systems, Inc. v. View Engineering, Inc., 112 F.3d 1163 (Fed. Cir. 1997) (it would be "plainly apparent" to one of ordinary skill in the art that software would be required, and it was not necessary to further disclose the nature of the code actually used); In re Dossel, 115 F.3d 942 (Fed. Cir. 1997) (means-plus-function claim which includes a "means for reconstructing data" is obviously implemented in software and so disclosure of that fact is not required; software program itself did not have to be disclosed).

4. The PTO's Examination Guidelines for Computer-Implemented Inventions clearly assume that source code need not be disclosed in many cases. In the section discussing enablement, the guidelines note that "[i]n many instances, an applicant will describe a programmed computer by outlining the significant elements of the programmed computer using a functional block diagram." Examiners facing such a disclosure are merely instructed to make sure that the block diagram adequately discloses the existence of each hardware element as well as the software elements.

3. Infringement

Pennwalt Corp. v. Durand-Wayland, Inc.
United States Court of Appeals for the Federal Circuit
833 F.2d 931 (Fed. Cir. 1987) (en banc), cert. denied, 485 U.S. 961 (1988)

BISSELL, Circuit Judge.

This appeal and cross-appeal are from a judgment of the United States District Court for the Northern District of Georgia, 225 U.S.P.Q. (BNA) 558 (N.D. Ga. 1984). The district court found that Durand-Wayland's accused devices do not infringe any claim, literally or under the doctrine of equivalents. Unable to view that finding as clearly erroneous under Fed. R. Civ. P. 52(a), we affirm the judgment of noninfringement. . . .

Background

Pennwalt sued Durand-Wayland for infringing claims 1, 2, 10 and 18 (claims-at-issue) of its U.S. Patent No. 4,106,628 (the '628 patent) on an invention of Aaron J. Warkentin and George A. Mills, entitled "Sorter for Fruit and the Like." Following a nonjury trial on the issues of patent infringement and validity, the district court, on March 22, 1984, issued an opinion concluding that (1) the claims-at-issue were not anticipated by the prior art, (2) the '628 patent had not run afoul of the "on sale" bar of 35 U.S.C. §102(b) (1982), and (3) the accused devices did not infringe any of the claims-at-issue, either literally or under the doctrine of equivalents. . . .

Opinion

The '628 patent claims a sorter. The principal object of the invention is to provide a rapid means for sorting items, such as fruit, by color, weight, or a combination of these two characteristics. The sorter recited in claims 1 and 2 conveys items along a track having an electronic-weighing device that produces an electrical signal proportional to the weight of the item, along with signal comparison means, clock means, position indicating means, and discharge means, each of which performs specified functions. The specification describes the details of a "hard-wired" network consisting of discrete electrical components which perform each step of the claims, e.g., by comparing the signals from the weighing device to reference signals and sending an appropriate signal at the proper time to discharge the item into the container corresponding to its weight. The combined sorter of claims 10 and 18 is a multifunctional apparatus whereby the item is conveyed across the weighing device and also carried past an optical scanner that produces an electrical signal proportional to the color of the item. The signals from the weighing device and color sensor are combined and an appropriate signal is sent at the proper time to discharge the item into the container corresponding to its color and weight.

Durand-Wayland manufactures and sells two different types of sorting machines. The first accused device, the "Microsizer," sorts by weight only and employs software labeled either Version 2 or Version 5. The second accused device employs software labeled Version 6 and sorts by both color and weight through the use of the "Microsizer" in conjunction with a color detection apparatus called a "Microsorter." . . .

I. Literal Infringement

Pennwalt asserts on appeal that all limitations set forth in claims 1 and 2 and some limitations set forth in claims 10 and 18 can be read literally on the accused devices. Pennwalt contends that the district court erred in interpreting the claims by going beyond the means-plus-function language of a claim limitation and comparing the structure in the accused devices with the structure disclosed in the specification. Such comparison allegedly resulted in the court's reading nonexistent structural limitations into the claims. Pennwalt relies on the statement in Graver Tank & Mfg. Co. v. Linde Air Prods. Co., 339 U.S. 605, 607, 94 L. Ed. 1097, 70 S. Ct. 854 (1950): "If accused matter falls clearly within the claim, infringement is made out and that is the end of it." In view of the literal breadth of means-plus-function language in the claims, that "test" for literal infringement would encompass any means that performed the function of a claim element. 35 U.S.C. §112 (1982). This is not the "test." . . .

[S]ection 112, paragraph 6, rules out the possibility that any and every means which performs the function specified in the claim literally satisfies that limitation. While encompassing equivalents of those disclosed in the specification, the provision, nevertheless, acts as a restriction on the literal satisfaction of a claim limitation. Data Line Corp. v. Micro Technologies, Inc., 813 F.2d 1196, 1201, 1 U.S.P.Q.2d (BNA) 2052, 2055 (Fed. Cir. 1987). If the required function is not performed exactly in the accused device, it must be borne in mind that section 112, paragraph 6, equivalency is not involved. Section 112, paragraph 6, plays no role in determining whether an equivalent function is performed by the accused device under the doctrine of equivalents.

Thus, it was not legal error (as Pennwalt asserts) for the district court to have made a comparison between Durand-Wayland's structure and the structure disclosed in the specification for performing a particular function. The statute means exactly what it says: To determine whether a claim limitation is met literally, where expressed as a means for performing a stated function, the court must compare the accused structure with the disclosed structure, and must find equivalent structure as well as identity of claimed function for that structure. . . .

We need not, and do not, determine whether the district court correctly found no equivalency in structure because the district court also found that the accused devices, in any event, did not perform the same functions specified in the claim. For example, the district court found that the accused devices had no position indicating means which tracked the location of the item being sorted. That finding negates the possibility of finding literal infringement.

II. Infringement under the Doctrine of Equivalents

Under the doctrine of equivalents, infringement may be found (but not necessarily) if an accused device performs substantially the same overall function or work, in substantially the same way, to obtain substantially the same overall result as the claimed invention. Perkin-Elmer Corp. v. Computervision Corp., 732 F.2d 888, 901-02 (Fed. Cir.), *cert. denied,* 469 U.S. 857, 83 L. Ed. 2d 120, 105 S. Ct. 187 (1984); Graver Tank, 339 U.S. at 608. That formulation, however, does not mean one can ignore claim limitations. As this court recently stated in Perkin-Elmer Corp. v. Westinghouse Elec. Corp., 822 F.2d 1528 (Fed. Cir. 1987):

> One must start with the claim, and though a "non-pioneer" invention may be entitled to some range of equivalents, a court may not, under the guise of applying the doctrine of equivalents, erase a plethora of meaningful structural and functional limitations of the claim on which the public is entitled to rely in avoiding infringement. . . . Though the doctrine of equivalents is designed to do equity, and to relieve an inventor from a semantic strait jacket when equity requires, it is not designed to permit wholesale redrafting of a claim to cover

non-equivalent devices, i.e., to permit a claim expansion that would encompass more than an insubstantial change. (Citations omitted.)

. . . In applying the doctrine of equivalents, each limitation must be viewed in the context of the entire claim. . . . "It is . . . well settled that each element of a claim is material and essential, and that in order for a court to find infringement, the plaintiff must show the presence of every element or its substantial equivalent in the accused device." Lemelson v. United States, 752 F.2d 1538, 1551, 224 U.S.P.Q. (BNA) 526, 533 (Fed. Cir. 1985) (footnote omitted). To be a "substantial equivalent," the element substituted in the accused device for the element set forth in the claim must not be such as would substantially change the way in which the function of the claimed invention is performed.

Id. at 1532-33, 3 U.S.P.Q.2d at 1324-25.

Pennwalt argues that the "accused machines simply do in a computer what the patent illustrates doing with hard-wired circuitry," and asserts that "this alone is insufficient to escape infringement," citing Decca Ltd. v. United States, 210 Ct. Cl. 546, 544 F.2d 1070, 1080-81, 191 U.S.P.Q. (BNA) 439, 447 (1976). If Pennwalt was correct that the accused devices differ only in substituting a computer for hard-wired circuitry, it might have a stronger position for arguing that the accused devices infringe the claims. The claim limitations, however, require the performance of certain specified functions. Theoretically, a microprocessor could be programmed to perform those functions. However, the district court found that the microprocessor in the accused devices was not so programmed.

After a full trial, the district court made findings that certain functions of the claimed inventions were "missing" from the accused devices and those which were performed were "substantially different." Pennwalt, 225 U.S.P.Q. at 572. The district court observed that "because the 'Microsizer' uses different elements and different operations (on the elements it does use) than the elements and operations disclosed in the patent-in-suit to achieve the desired results, infringement can only be found if the different elements and operations are the legal equivalents of those disclosed in the patent-in-suit." Id. It is clear from this that the district court correctly relied on an element-by-element comparison to conclude that there was no infringement under the doctrine of equivalents, because the accused devices did not perform substantially the same functions as the Pennwalt invention. For example, the district court found in part:

The machine described in the patent-in-suit uses shift registers that respond to "clock pulses" in order to indicate the various positions of the items to be sorted before each item is discharged. The "Microsizer" does not have any "indicating means" to determine positions of the items to be sorted since the microprocessor stores weight and color data, not the positions of the items to be sorted. After a piece of fruit has been analyzed by the "Microsorter" and while it is in transit from the optical detection means to the weight scale, the color value determined by the "Microsorter" is sorted in a color value queue. A color value queue pointer (which changes in value) points to the location of the data corresponding to the next piece of fruit to reach the weight scale. A weight value queue pointer (which also changes in value) is used to correspond to the number of cups between the weight scale and the drop location. The microprocessor software utilizes a random access memory that stores the digital numbers which resulted from the conversion of the analog signals generated by the "Microsorter" and the weight scale, and the queue pointers (under clock control) point to the memory location that has the data about a piece of fruit. The data is never "shifted" around, but rather is just stored in memory until the software routines call for data to be utilized in subsequent portions of the program(s). Thus, the "Microsizer" has neither a "first position indicating means" nor a "second position indicating means." The machine described in the patent-in-suit produces signals that indicate where the fruit is, i.e. track the progression of each cup. The "Microsizer" does not.

Id. at 569. . . .

The trial court found that the accused devices do not have any position indicating means to determine positions of the items to be sorted. Specifically with respect to claims 10 and 18, the court correctly held that "the microprocessor stores weight and color data, not the positions of the items to be sorted." Pennwalt, 225 U.S.P.Q. at 569. Since each of the claims-at-issue requires a position indicating means and the same analysis applies to each, we set forth only the relevant language of claim 10:

> first position indicating means responsive to a signal from said clock means and said signal from said second comparison means for continuously indicating the position of an item to be sorted while the item is in transit between said optical detection means and said electronic weighing means,
>
> second position indicating means responsive to the signal from said clock means, the signal from said first comparison means and said first position indicating means for generating a signal continuously indicative of the position of an item to be sorted after said item has been weighed. . . .

Second, the district court correctly rejected Pennwalt's assertion that the memory component of the Durand-Wayland sorter which stores information as to weight and color of an item performed substantially the same functions as claimed for the position indicating means. The district court found that a memory function is not the same or substantially the same as the function of "continuously indicating" where an item is physically located in a sorter. On this point the record is indisputable that before the words "continuously indicating" were added as an additional limitation, the claim was unpatentable in view of prior art which, like the accused machines, stores the information with respect to sorting criteria in memories, but did not "continuously" track the location. See, e.g., U.S. Patent No. 3,289,832 issued to Ramsey.

Thus, the facts here do not involve later-developed computer technology which should be deemed within the scope of the claims to avoid the pirating of an invention. On the contrary, the inventors could not obtain a patent with claims in which the functions were described more broadly. Having secured claims only by including very specific functional limitations, Pennwalt now seeks to avoid those very limitations under the doctrine of equivalents. This it cannot do. Graham v. John Deere Co., 383 U.S. 1, 33-34, 15 L. Ed. 2d 545, 86 S. Ct. 684 (1965); Chemical Eng'g Corp., 795 F.2d at 1572 n.8, 230 U.S.P.Q. at 391 n.8; see also Exhibit Supply Co. v. Ace Patents Corp., 315 U.S. 126, 136, 86 L. Ed. 736, 62 S. Ct. 513 (1942); Coleco Indus., Inc. v. United States Int'l Trade Comm'n, 65 C.C.P.A. 105, 573 F.2d 1247, 1257-58, 197 U.S.P.Q. (BNA) 472, 479-80 (1978). Simply put, the memory components of the Durand-Wayland sorter were not programmed to perform the same or an equivalent function of physically tracking the items to be sorted from the scanner to the scale or from the scale to its appropriate discharge point as required by the claims.

[Affirmed. The dissent's discussion of the proper legal test for applying the doctrine of equivalents is omitted. See supra Chapter 3.]

COMMENTS AND QUESTIONS

1. To similar effect as *Pennwalt* is Wiener v. NEC Electronics, Inc., 102 F.3d 534 (Fed. Cir. 1996). In that case the Federal Circuit upheld the district court's finding of noninfringement under the doctrine of equivalents because there were substantial differences between the patent's requirement that a computer program "call on" columns of data one byte at a time and the defendant's product, where the columns alleged to be equivalent were not in the data matrix and therefore were not called upon to read data). The court rejected the "conclusory" declaration of plaintiff's expert that the two processes were identical. In General Electric v. Nintendo of America, 179 F.3d 1350 (Fed. Cir. 1999), the court held that Nintendo's video game systems did not infringe GE's television switch patents because the patents, written in means-plus-function format, did not disclose a function for the switches identical to Nintendo's function. In Digital Biometrics, Inc. v. Identix, Inc., 149 F.3d 1335 (Fed. Cir. 1998) the court narrowly construed a patent claim to "image arrays" storing a two-dimensional slice of video data that were merged into a "composite array" storing a fingerprint image. The court held that the defendant's systems, which constructed the composite array directly rather than by using two-dimensional slices, did not create "image arrays" within the meaning of the claims.

The trend in software infringement cases seems to be towards interpreting the claims narrowly. Does this make sense in light of what we know about the validity of software patents? Does it render software patents ineffective because of the fast-changing nature of the technology?

One possible explanation for the narrow construction, given software patents in infringement cases to date, is that most of those patents have claim elements written in means-plus-function format. A growing number of software patent claims are drafted in means-plus-function language, largely to meet the "structure" requirements the courts have imposed on patentees under section 101. How should the "corresponding structure or equivalents thereof" in the specification be interpreted for infringement purposes? In particular, suppose that a patentee, to come within the dictates of section 101, claims a "means for processing data" of a particular type. In the specification, the patentee discloses the use of an IBM-compatible personal computer with a Pentium microprocessor. Is this claim infringed by a defendant who runs the same software on a 486 microprocessor? An Apple computer using a Motorola microprocessor? A Sun Microsystems SPARC workstation? A Cray supercomputer? A cash register containing a microprocessor?

2. Is a computer-implemented process "equivalent" to an older process, whether physical or analog? The Federal Circuit has been struggling with this issue for some time. Compare Texas Instruments, Inc. v. International Trade Commn., 805 F.2d 1558 (Fed. Cir. 1986) (modern calculators did not infringe TI's original patent on the integrated circuit) with Hughes Aircraft Co. v. United States, 717 F.2d 1351 (Fed. Cir. 1983) (patent on means for controlling satellites from the ground via telemetry was infringed by use of on-board microprocessors to control satellites). For a case applying the doctrine of equivalents in a software context, see Safe Flight Instrument Corp. v. Sundstrand Data Control, Inc., 706 F. Supp. 1146 (D. Del. 1989), *aff'd*, 899 F.2d 1228 (Fed. Cir. 1990), *cert. denied*, 498 U.S. 919 (1990).

Significant light was shed on this important question in Alpex Corp. v. Nintendo of America, 102 F.3d 1214 (Fed. Cir. 1996). There the court reversed a jury verdict of infringement in a case involving an old video-game patent asserted against modern technology. Differences in the nature of the devices doomed the infringement claim.

In addition to the doctrine of equivalents, patent infringement analysis also requires application of the "reverse doctrine of equivalents." As the name suggests, the reverse doctrine of equivalents operates to protect patent defendants whose products fall within the literal scope of patent claims, but which are in fact substantially different from the patented invention. This doctrine may be of particular importance in the software industry, because the application of computer technology has resulted in major changes in the way in which certain physical operations are performed.

3. In 1994, a jury ruled that Microsoft had infringed data compression software patents held by Stac Electronics, and awarded damages of $120 million to Stac. See J. Burgess, Microsoft Found Guilty of Patent Infringement: Software Giant Ordered to Pay $120 Million, Washington Post, Feb. 24, 1994, at D11. Although two software patents were asserted, the jury failed to specify whether one or both of the patents were infringed. Interestingly, the jury also found that Stac had misappropriated Microsoft's trade secrets relating to data compression, and awarded Microsoft $13.6 million in damages. According to reports in the press, Microsoft tried to license the patents from Stac, but the parties failed to reach agreement. After the verdict, Microsoft temporarily excised the infringing code from its MS-DOS operating system software, but later settled the case while the appeal was pending.

The case clearly shows the potential impact of software patents — Stac was able to force Microsoft to temporarily pull a multimillion dollar product from the market. To some, this case serves as an example of why patents for software are so problematic. See, e.g., L. M. Fisher, The Executive Computer: Will Users Be the Big Losers in Software Patent Battles?, N.Y. Times, May 6, 1994, at §3, p. 7.

Judging from the volume of issued software patents, the flood of software patent litigation is just beginning. For example, E-Data has sued a number of companies (and notified thousands of others of potential infringement) on its U.S. Patent 4,528,643, a "System for reproducing information in material objects at a point of sale location."

E. TRADEMARK AND TRADE DRESS

As a result of the network externalities generated by computer technology, trademark and trade dress protection takes on added importance in achieving success in the marketplace. Consumers place significant value on obtaining application programs that are compatible with popular operating systems and peripheral equipment that connects with their base computer systems. Equipment and software manufacturers, therefore, use trademark as an important part of their overall strategy to protect and market their products. In this way, trademark protection can significantly enhance the protection of new technology. On the other hand, trademark protection can also potentially inhibit competition by limiting the ability of new entrants to inform consumers of the compatible attributes of their products.

In the United States, protection for trademarks and trade dress is provided by the Lanham Act, 15 U.S.C. §1051 et seq. and by state common law. Trademark law generally applies to computers and computer companies in the same way that it does to other industries. For instance, Apple Computer is entitled to prevent competitors from selling computers with names confusingly similar to Apple (such as Pineapple). Apple Computer v. Formula Intl., 562 F. Supp. 775 (C.D. Cal. 1983), *aff'd*, 725 F.2d 521 (9th Cir. 1984). In this chapter, though, we are not concerned with "ordinary" trademark issues that happen to come up in the computer context. Rather, we focus on several circumstances in which the computer industry presents unique trademark issues. As you work through these cases and problems, assess the degree to which trademark law promotes (or retards) innovation in the computer industry.

1. Protecting Programs Through Trademark

Note on Protection of Product Configurations

Trademarks are traditionally thought of as words or other written symbols or logos describing a product, service, or company. However, the Lanham Act also protects "trade dress" — a term that encompasses the total design of product packaging, including its shape and color. Trade dress has recently been expanded to protect the configuration of the marked product itself, on the same terms as other forms of trade dress. See Two Pesos, Inc. v. Taco Cabana, 505 U.S. 763 (1992) (total design of the interior of a fast-food Mexican restaurant was entitled to Lanham Act protection).

There remains some disagreement among the courts concerning precisely how to treat product configurations, however. In Duraco Prods. v. Joy Plastic Enterprises, 40 F.3d 1431 (3d Cir. 1994), the Third Circuit rejected the proposition (arguably mandated by *Two Pesos*) that traditional trade dress doctrine applies to product configurations. Instead, the Third Circuit developed a new, stricter test for protecting product configurations. Under the *Duraco* test, a product configuration qualifies as inherently distinctive (and therefore protectable without proof of secondary meaning) only if it is (1) unusual and memorable, (2) capable of being conceptually separated from the product itself, and (3) likely to serve primarily a trademark rather than a utilitarian function. By contrast, the Eighth Circuit has specifically rejected the *Duraco* test, holding that product configurations need not be "memorable" or "striking in appearance" to qualify as inherently distinctive. Stuart Hall Co. v. Ampad Corp., 51 F.3d 780 (8th Cir. 1995).

This dispute is relevant to the computer industry because some — and perhaps all — of the products capable of trademark protection in that industry qualify as product configurations. Certainly the shape of the computer at issue in Digital Equip. Corp. v. C. Itoh & Co., 229 U.S.P.Q. 598 (D.N.J. 1985), is a product configuration. Arguably, so is the layout of icons and menus on a computer screen. Cf. Ex parte Donaldson, 26 U.S.P.Q.2d 1250 (Bd. Pat. App. & Int. 1992) (design patents case). If the Third Circuit standard prevails, trade dress protection may be harder to obtain for GUIs and other software configurations.

COMMENTS AND QUESTIONS

Protecting software product configurations obviously overlaps not only with copyright law but also with design patent law (which protects novel computer icons). Should one mode of protection preclude other forms, or is overlapping protection desirable? Note that some courts are not willing to permit product configurations that are the subject of a utility patent to be protected under trade dress law as well, perhaps out of a concern that the owner of the design would be able to extend protection beyond the term of the patent. See Vornado Air Circulation Systems v. Duracraft Corp., 58 F.3d 1498 (10th Cir. 1995), *cert. denied,* 116 S. Ct. 753 (1996).

In previous sections, we have discussed various means of protecting computer software and hardware against imitation by others. In a recent article, Lauren Fisher Kellner suggests that trademark could also be used to protect the "look and feel" of computer software against imitation.

Lauren Fisher Kellner, Trade Dress Protection for Computer User Interface "Look and Feel"
61 U. Chi. L. Rev. 1011 (1994)

An employee on the way back to his office glances at the illuminated screen of a nearby computer. He sees the small trash can in the lower right corner; graphics of labeled folders, documents, and applications in the center; and a strip of words across the top of the screen. He immediately assumes that the computer is an Apple Macintosh. A spreadsheet programmer determines that an application program requires her to use a backslash to access multi-layered command menus and to strike the first letter of layered menu items in sequence in order to execute any series of commands. She knows the software must be Lotus 1-2-3.

Apple and Lotus realize that computer users rely on these elements to identify computer programs. Accordingly, both companies have invested time and money to build consumer recognition of their programs and have sought legal protection for their user interfaces. . . .

Both the academic literature and litigants' briefs have focused on copyright as a way to protect a developer's intellectual property rights in a software product's "look and feel." Trademark law, however, would be a better source of protection. Trademark law recognizes consumers' interests in identifying a product's source, and trade dress law, a subset of trademark law, protects a product's overall appearance to the purchasing public. While trade dress infringement claims have become more common in all product areas over the past few years, only a few attorneys have aggressively sought trade dress protection for the "look and feel" of user interfaces. . . .

The first requirement for trademark protection of any product feature or set of features is "distinctiveness" — the feature must be able to indicate the source of the product. . . .

Some user interfaces create an inherently distinctive visual impression. For example, Berkeley Systems's After Dark screen saver program features a stylized underwater scene: colorful fish, gurgling noises, and winged toasters. That visual display is unique and unusual enough to qualify as inherently distinctive. In general, then, program developers that create unique visual displays should be able to meet trademark law's distinctiveness requirement without showing secondary meaning. . . .

The trend toward standardization of user interfaces presents a potential obstacle to manufacturers' attempts to establish distinctiveness. Most user interfaces are unique when introduced. But microcomputer markets are moving toward "standard user interfaces" among programs in an attempt to reduce the time it takes to learn a new application. This "family look" promotes sales *within* product lines (among all programs designed to run on one type of computer) and *across* product lines (between programs designed for different computer systems) by increasing consumer confidence in and familiarity with standard techniques. Standardization may make it difficult, if not impossible, for a plaintiff to establish distinctiveness because "in markets with standard user interfaces, programs are *expected* to look and feel the same." [Gregory Wrenn, Federal Intellectual Property Protection for Computer Software Audiovisual Look and Feel: The Lanham, Copyright, and Patent Acts, 4 High Tech. L.J. 279, 284 (1989).]

[Kellner acknowledges that "the likelihood of point-of-sale consumer confusion seems low because computers and software are typically sold in clearly marked packages. It is thus unlikely (though conceivable) that a purchaser intending to buy Lotus 1-2-3 would end up buying, say, VP-Planner instead."]

In contrast, potential consumers of computer products may see a computer's user interface on co-workers' desks or in store windows and assume that they are viewing the Apple Macintosh or Lotus 1-2-3 interface. By the time they eventually discover that they were really viewing Microsoft Windows or VP-Planner, they may already have invested time and energy in researching and shopping for the wrong product. Microsoft Windows or VP-Planner retailers could then convince these potential consumers that the look-alike products are indistinguishable from the Macintosh or Lotus 1-2-3; they will have used Apple's or Lotus's reputation and customer recognition to lure customers into investigating and ultimately purchasing look-alike products. Trade dress law attempts to prevent precisely this kind of misappropriation of another product's good reputation.

COMMENTS AND QUESTIONS

1. "Distinctiveness" is a term of art which encompasses trademarks that are arbitrary, suggestive, or fanciful, or that are descriptive of the products they identify but that have acquired "secondary meaning" in the minds of consumers as identifying a product from a particular source. Trademark protection does not extend to trademarks or trade dress that are merely descriptive of the things being identified. Further, in no case can a "generic" term or dress be protected. The preceding article suggests that many user interfaces are inherently distinctive because the way the interface works is unrelated to the nature of the functions performed. In practice, however, individual elements of the user interface may merely be descriptive of the functions performed. How would a court classify the "trash can" icon in the Macintosh

GUI (graphical user interface), which serves to delete files? The Lotus 1-2-3 program's choice of the "D" key for delete, "F" for file, or "Q" for quit? See Lotus v. Borland, discussed supra. Must these individual elements be distinctive in order to be protected, or is it sufficient that the interface *as a whole* is not "descriptive" of a computer operating system?

Alternatively, one might treat items such as computer graphics as a "product configuration." In some circuits, product configurations are subject to different rules regarding distinctiveness. See the comment following this section.

User interfaces may also run afoul of the "genericide" doctrine. If a particular interface has become the de facto standard in the industry, so that people naturally use that interface for a certain type of program, the interface is likely to be generic and therefore unprotectable. A good example is the QWERTY keyboard, an "interface" that may have been arbitrary as a means of organizing letters when it was introduced,[15] but has since become a universal keyboard standard. Because consumers do not associate the QWERTY keyboard with a particular maker but rather with the general idea of a typewriter, that keyboard is almost certainly generic and therefore incapable of trademark protection. The same fate may befall certain elements of graphical user interfaces, such as pull-down menus. Even if the use of pull-down menus once identified the computer as an Apple, their prevalence in numerous programs today suggests that they may be unprotectable.

Does it make economic sense that the very success of a GUI should be its undoing from the standpoint of trademark law? Quite possibly. One effect of the genericness rule is to prevent a particular company from using trademark law to corner the market on an industry standard, and thereby foreclose competition in the underlying products. We discuss this issue in more detail in the next section.

2. Kellner acknowledges that consumers are unlikely to be confused between different GUIs at the point of sale. There is no question that Apple and Microsoft, for example, package their products very differently. Neither is attempting to trade on the business goodwill of the other. But she suggests that post-sale confusion is still possible. Consumers will see one operating system in action, think it is the other, and proceed to buy the wrong computer. The last step in this chain is critical, since the Lanham Act is concerned with the likelihood of confusion among *consumers* of the products at issue.

Whether you think such post-sale confusion is likely depends in part on your assessment of the sophistication of the purchasing public. Several courts have found confusion unlikely in the computer context because the buyers were relatively sophisticated and they were purchasing "big-ticket" items. See, e.g., Intel v. Advanced Micro Devices, 756 F. Supp. 1292

---

15. In fact, rumor has it that the QWERTY keyboard was designed in part to reduce the jamming of keys in early manual typewriters, both by slowing down the typists and by physically separating letters often used together. Do these facts shed any light on the arbitrary nature of the keyboard?

(N.D. Cal. 1991) (purchasers of Intel microprocessors were computer hardware manufacturers, not end users; hardware manufacturers were sophisticated enough to distinguish between Intel 386 and AMD 386 chips); Engineering Dynamics, Inc. v. Structural Software, Inc., 26 F.3d 1335, 1350 (5th Cir. 1994) (competing software programs did not create a likelihood of confusion among sophisticated users in a limited market; consumers pay little attention to trade dress in determining which product to purchase).

3. Even if consumers are not confused, might programmers have a cause of action for trademark dilution? Can the Macintosh GUI be said to be "famous"? If so, does the use of pull-down menus by other companies reduce the identification value of the Apple mark?

4. How does protecting GUIs under trademark law comport with the purposes of the Lanham Act? Does it overreach — destroying useful competition in an attempt to prevent relatively minor instances of consumer confusion?

5. The most significant limitation on trade dress protection for computer programs is functionality. Trademark protection does not extend to any part of the dress that is "functional." Because (as discussed above) most of the effort in software design goes into making programs efficient and easy to use, the functionality rule serves to prevent companies from "capturing" new innovation indefinitely through trademark law.

Recall that much of the debate over the appropriate scope of protection for GUIs in the copyright context had to do with whether elements of the interface were dictated by functional considerations. See Apple Computer v. Microsoft Corp., 35 F.3d 1435 (9th Cir. 1994). Isn't that debate likely to be replicated in the trademark context, with courts disagreeing over whether particular portions of the GUI are economically or aesthetically functional? If so, is there any advantage to seeking trademark rather than copyright protection?

One possible difference between trademark and copyright law may be the way in which functionality is tested. Courts and commentators often debate whether the functionality of trade dress is to be tested by looking at the dress as a whole, or by evaluating the purpose of individual elements. On this point, see Stormy Clime Ltd. v. ProGroup, Inc., 809 F.2d 971 (2d Cir. 1987) (combination and organization of functional elements was itself functional); LeSportsac, Inc. v. K Mart Corp., 754 F.2d 71 (2d Cir. 1985) (arrangement of features on nylon luggage was ornamental rather than functional). This debate, too, has its parallels in the copyright context, though it is all but resolved that copyright focuses on the individual elements rather than on the program as a whole in determining functionality. Compare Computer Associates v. Altai, 982 F.2d 693 (2d Cir. 1992) with Whelan Associates v. Jaslow Dental Laboratory, 797 F.2d 1222 (3d Cir. 1986).

An "element by element" approach may be particularly appropriate in the software context, where alleged infringers take only a few key pieces of the plaintiff's trade dress (such as overlapping folders and pull down menus from the Macintosh interface). If the elements taken are all functional, presumably no court would find trademark infringement even if the interface as a whole were protectable.

Is there reason for trademark owners to prefer one of these approaches over the other? It is reasonable to assume that testing functionality "as a whole" will allow greater leeway for protecting partially functional trade dress, but the reverse may be true: that if any part of the interface is functional, none of it is protectable.

PROBLEMS

Problem 7-12. Apple's Macintosh graphical user interface has three main components which are copied by Kiwi Computer in its competing operating system. The three components are "pull-down menus," in which the major classes of user commands are written in a bar at the top of the screen that can be "opened" by clicking on the relevant class of commands with a mouse and dragging the mouse downwards; a series of icons on the right-hand side of the screen to perform other common commands or to open particular folders; and "overlapping windows," which show the most recently used document or program "on top of" other open applications.

Apple contends that none of these three components is functional and points to other ways in which Kiwi could have implemented a graphical user interface. For example, Apple argues that Kiwi could have used "pull-up" or "pull-across" menus, or an expandable list of commands controlled by the keyboard or by voice rather than by the mouse. Kiwi responds that it is easier and more logical for most computer users to drag the mouse downward than to push it up or across. Apple points to alternatives to its overlapping windows such as "tiled windows" that appear side by side; Kiwi responds that tiled windows are inefficient because they limit both the number of windows that can be open and the size of each open window.

Are Apple's design features functional? Does your answer depend on the legal standard you apply?

Problem 7-13. Intel manufactures generations of microprocessor chips with identifying number sequences. For reference, these number sequences are shortened to the last three digits and are commonly referred to by computer users in this way (e.g., the 286, 287, 386, 387, 486). Advanced Micro Devices, a competing microprocessor manufacturer, refers to its comparable chips by the same number sequences. Has AMD infringed an Intel trademark?

> Intel refers to its latest generation of chip as the "Pentium." From a trademark perspective, what are the advantages and disadvantages of this change in naming convention?
>
> *Problem 7-14.* Sun Microsystems is a leading manufacturer of computer workstation systems. A recent entrant in the computer workstation software market is thinking about using the name SunRiver. Would Sun Microsystems be able to enjoin the use of this name? What information would you need to know to answer this question properly?

2. Compatibility and Standardization

A second way in which trademark law may operate to influence the market for computer products directly is in the areas of standardization and compatibility.[16] Because computer hardware and software are sold in complex markets with high-network externalities both among and between products, there is tremendous value to be gained not only in identifying your own product but in describing for consumers how it interacts with other products on the market. For example, the sellers of applications programs want to be able to tell their customers whether or not their program will run with a particular operating system. Because identifying the operating system often involves the use of a competitor's trademark, however, trademark law stands as a potential barrier to this form of compatibility identification.

≡≡≡ *Creative Labs, Inc v. Cyrix Corporation*
United States District Court for the Northern District of California.
43 U.S.P.Q.2d 1778 (N.D. Cal. 1997)

WILKEN, District Judge.

Plaintiffs Creative Labs and Creative Technology (collectively "Creative") move for a preliminary injunction against Defendants Cyrix Corporation and Tiger Direct, Inc. ("Tiger") for copyright infringement and false advertising. Defendants oppose the motion. The matter was heard on May 2, 1997. On May 2, 1997, Creative also filed an ex parte application for an order to show cause re contempt. Having considered all of the papers filed by the parties and oral argument on the motion, the Court GRANTS the motion for a preliminary injunction and DENIES the ex parte application for an order to show cause.

---

16. We have already encountered standardization in the previous section, in the context of genericness.

Background

Plaintiff Creative is the maker of the Sound Blaster line of sound cards, devices that interact with computer hardware and software to create sound effects. Sound Blaster is currently the dominant sound card on the market. Defendant Cyrix has recently introduced a microprocessor, the Media GX, which is capable of producing audio effects without the assistance of a sound card. The audio component of the Media GX is referred to as Xpress AUDIO.

Cyrix has advertised XpressAUDIO as "compatible with Sound Blaster" and as "fully compatible with Sound Blaster." Tiger, a computer manufacturer which plans to introduce a line of computers that use the Media GX, claims that its products feature "integrated SoundBlaster 16/Pro compatible audio." Compaq uses the Media GX in one of its computers, the Presario 2100, but does not advertise that the Presario 2100 is Sound Blaster compatible.

Creative tested the Media GX on a Presario 2100 computer to determine whether Cyrix's claims concerning XpressAUDIO's compatibility with Sound Blaster were accurate. In a study testing 200 computer games, Creative discovered that sixteen games, or 8% of the total tested, did not run properly on the Presario 2100. Creative also found that the Presario 2100 did not support two functions supported by Sound Blaster: Adaptive Delta Pulse Code Modulation ("ADPCM") and Musical Instrument Digital Interface ("MIDI").

Creative also learned that Cyrix was making some Creative Labs software programs available for copying by the public on Cyrix's website.

Creative filed suit against Cyrix for copyright infringement, 17 U.S.C. §510 et seq.; against Cyrix and Tiger for false advertising and trademark dilution, 15 U.S.C. §1125(a), (c); and against Cyrix, Tiger, and Compaq for trademark infringement, 15 U.S.C. §§1114(a), 1125(a), and unfair competition, Cal.Bus. & Prof.Code §17200 et seq, The Court granted Creative's application for a temporary restraining order against Cyrix and Tiger. Creative now seeks a preliminary injunction enjoining all Defendants from trademark infringement, Cyrix and Tiger from false advertising, and Cyrix from copyright violations. In this order, the Court considers only the claims concerning false advertising and copyright infringement.

Discussion . . .

B. *False Advertising*

Creative alleges that Cyrix and Tiger advertisements which assert that the XpressAUDIO system is "compatible" with Sound Blaster constitute false advertising in violation of Lanham Act §43(a), 15 U.S.C. §1125(a). The elements of a Lanham Act false advertising claim are:

> (1) a false statement of fact by the defendant in a commercial advertisement about its own or another's product; (2) the statement actually deceived or has the tendency to deceive a substantial segment of its audience; (3) the deception is material, in that it is likely to influence the purchasing decision; (4) the defendant caused its false statement to enter interstate commerce; and (5) the plaintiff has been or is likely to be injured as a result of the false statement, either by direct diversion of sales from itself to defendant or by a lessening of the goodwill associated with its products.

Southland Sod Farms v. Stover Seed Co., 108 F.3d 1134, 1139 (9th Cir. 1997). Plaintiffs do not need to prove injury to be entitled to injunctive relief. Id. at 1145-46.

Plaintiffs seek to enjoin Cyrix and Tiger from advertising that their systems are Sound Blaster compatible. Cyrix and Tiger respond that XpressAUDIO is compatible with Sound Blaster. The dispute thus turns on the definition of the term "compatible." Creative maintains that competing computer products are compatible only if "the first product can be used in place of the second product without producing any difference in performance and that the first product has the same capabilities and functions as the second product." Creative supplied the declaration of an individual who works in the computer games industry asserting that the phrase "Sound Blaster compatible" indicates that the hardware "will properly play any software title that plays properly on a genuine Creative Labs Sound Blaster." Creative also refers to three dictionary definitions to support its interpretation. One provides that "[c]ompatibility means that the hardware ideally operates in all respects like the standard on which it is based." The second dictionary defines compatibility as, "[t]he capability of a peripheral [or] program . . . to function with or substitute for a given make and model of computer. . . . To be truly compatible, a program or device should operate on a given system without modification; all features should operate as intended, and a computer claiming to be compatible with another should run all the other computer's software without modification." By referring to a device for measuring how IBM-compatible personal computers are, however, the definition suggests that compatibility is not necessarily a matter of all or nothing. The third dictionary provides that compatibility is "[t]he extent to which a given piece of hardware or software conforms to an accepted standard. . . . This implies that the device will perform in every way just like the standard device." Cyrix cites a dictionary which defines the term "compatible" as describing a product which meets some, but not all, parts of a specification.

Princeton Graphics Operating, L.P. v. NEC Home Electronics (U.S.A.), Inc., 732 F.Supp. 1258 (S.D.N.Y. 1990), supports Creative's definition of compatibility. In *Princeton Graphics*, the court chose the more restrictive definition because of the importance of precise definitions in the computer industry. Id. at 1261. It also believed that it is appropriate to apply a more precise definition of compatibility when a well-known industry standard is being used. Id. at 1262 n. 9. The Court therefore finds that a product advertising itself as Sound Blaster compatible must support the same functions as Sound Blaster.

Creative argues that XpressAUDIO is not compatible with Sound Blaster because it does not support all games that can function with Sound Blaster. Creative relies on a study in which sixteen computer games, 8% of all games tested, did not function properly on a Presario 2100 computer. The study, however, does not establish that those failures were due to incompatibilities between XpressAUDIO and Sound Blaster. Cyrix counters that it did not encounter problems when running six of these games on a properly configured computer. Id. Cyrix's own study indicates a failure rate of approximately 2%. Even if the failure rate of games played on computers with XpressAUDIO is closer to 2% than 8%, the evidence indicates that some games that function with Sound Blaster do not function with XpressAUDIO.

Creative also argues that XpressAUDIO is not Sound Blaster compatible because it does not support two specific functions supported by Sound Blaster: ADPCM and MIDI. Cyrix concedes that XpressAUDIO does not support ADPCM, but it maintains that few games employ ADPCM, that those games are not currently being sold, and that ADPCM does not meet consumer expectations of sound quality. The lack of ADPCM support, while perhaps insufficient to establish lack of compatibility alone, supports such a finding.

Creative also maintains that XpressAUDIO does not support Sound Blaster's MIDI function. XpressAUDIO, however, does support MIDI. The MIDI function can be turned off by the computer manufacturer. Compaq has turned off the MIDI feature on the XpressAUDIO systems used in Presario 2100 computers. Cyrix has not advertised that Presario 2100s are Sound Blaster compatible.

Because some computer games that function with Sound Blaster do not function with XpressAUDIO, the Court finds that XpressAUDIO is probably not compatible with Sound Blaster. Cyrix and Tiger's claims that systems using XpressAUDIO are Sound Blaster compatible will probably mislead consumers who would interpret the claim of Sound Blaster compatibility to mean that any product that functions with Sound Blaster would also function with XpressAUDIO.

Because Creative has established the likelihood of consumer confusion, it has also established the possibility of irreparable harm. The Court also finds that injunctive relief would further the public's interest in being protected from false trade descriptions. See U-Haul Intl., Inc. v. Jartran, Inc., 681 F.2d 1159, 1162 (9th Cir. 1982).

The Court therefore grants Creative's motion for a preliminary injunction enjoining Cyrix and Tiger from claiming that XpressAUDIO and computer systems using XpressAUDIO are Sound Blaster compatible. . . .

COMMENTS & QUESTIONS

1. In *Princeton Graphics* (cited in *Creative Labs*), Princeton, the maker of a monitor for IBM computers, sued NEC, a competing maker of monitors because NEC had claimed its monitor was "compatible" with the new IBM video standard (a claim Princeton could not make for its own monitors). The court determined that the term "compatible" had a precise meaning specific to the computer industry and that it required a high degree of inter-operability:

> We find that within the "retail channel," the term "compatible" does not have the broad and flexible meaning as suggested by defendant when, as here, there is a possibility that a more precise definition may be applied.[21] Indeed, if there was one over-arching impression left on this court after the testimony given in this case it was that the computer industry is concerned with and depends upon accuracy. Thus, the testimony confirms our view that in an industry which depends upon accuracy, a lack of precision in the use of common terms, particularly in circumstances where those terms have the potential to be specific, would be an anomaly.
>
>   . . . We therefore find that in light of the evidence the definition of "compatible" as understood in the "retail channel" of sophisticated users has a clear and definite meaning — i.e., when a clearly defined standard, like IBM's VGA standard, exists and is widely accepted within the industry, a "compatible" product must meet that standard or at least perform in a manner equivalent to the standard's requirements.[22]

Princeton Graphics Operating v. NEC Home Electronics USA, 732 F. Supp. 1258 (S.D.N.Y. 1990). The court went on to conclude that falsely advertising compatibility could affect market purchasing decisions, and therefore cause actionable injury:

21. We make no finding as to the understanding of the term "compatible" in the broader consumer market.

22. We find that the dictionary and glossary definitions offered by defendant were of little probative value. Either the exhibit indicated only what the understanding might be within the general population and not the "retail channel," or the definition was worded in such a way to lend support to either party's position. The definitions set forth in the exhibits did little to indicate the common construction and understanding of the term within the "retail channel" when applied to concrete products.

Further, defendant offered articles published in trade magazines as evidence that "compatible" had a broad and flexible meaning within the "retail channel." However, the context of the use of "compatibility" varied in those articles. In certain articles "compatibility" was used in relation to "Super VGA" capability or in relation to a monochrome gas plasma video display's VGA capability. Thus, in those articles "compatibility" was being discussed in a different context than the instant situation. For example, testimony at trial indicated that "Super VGA" (800 x 600 resolution) is not an IBM standard. Accordingly, defendant's articles are not of great persuasive weight since we believe that a broader understanding of compatible is favored by the "retail channel" in the absence of a clear industry standard.

Moreover, although the articles were not offered for the truth of their contents, we note only one short article offered by defendant explicitly stated that the MultiSync was "fully compatible" with VGA, and that explicit representation was only contained in a caption to the article's accompanying illustration.

We believe that plaintiff has a presented a reasonable basis for its belief it was likely that defendant's advertising caused it damage. First, it is logical to conclude that had consumers been faced with two monitor products in the spring and summer of 1987, neither of which claimed PS/2 compatibility, some might have chosen plaintiff's. It is also reasonable to assume that had there been no compatibility claim on the part of defendant in the spring of 1987, some consumers would have waited until the summer of 1987 to buy plaintiff's PS/2 compatible UltraSync. Accordingly, we conclude that plaintiff has standing to bring this action.[23]

Id at 27.

Does holding NEC to such a high threshold for compatibility help or hurt consumers? On the one hand, preventing NEC from falsely claiming that it is fully compatible with IBM's VGA standard prevents consumers from buying the monitor under false pretenses. In this sense, strictly defining compatibility ensures that consumers will know what they are getting. On the other hand, this strict definition means that at that time, no one except IBM could claim they made a VGA-compatible monitor. If consumers are unwilling to buy products unless they are advertised as compatible, the effect is to give IBM a monopoly on monitor sales associated with IBM computers. From a consumer standpoint, this may actually be less desirable. If a consumer is willing to settle for a cheaper, 95 percent compatible monitor, shouldn't they be allowed to do so?

Should NEC or Cyrix be allowed to advertise their "near-compatibility"? Under these decisions, how could they do so?

2. Both *Creative Labs* and *Princeton Graphics* arose under the false advertising provisions of section 43(a) of the Lanham Act. The Lanham Act's prohibition on false advertising operates in parallel to its protection of trademarks. (Because Princeton Graphics did not own the IBM or VGA trademarks, it had no right to bring a claim for trademark infringement.) Presumably IBM could have brought a similar case for false advertising. See Intel Corp. v. Advanced Micro Devices, Comp. Indus. Lit. Rptr. 11730 (N.D. Cal. Aug. 27, 1990) (enjoining AMD from claiming that its products were compatible with Intel's 287 microprocessor).

3. In *Princeton Graphics* the issue was whether NEC's monitor was, in fact, compatible with the VGA standard. Presumably, Princeton would not have had a cause of action against NEC if its statements regarding compatibility were accurate. Would IBM have a cause of action for trademark infringement? After all, NEC is in some sense trading on the value of the IBM standard by associating its product with IBM in the minds of consumers.

---

23. Defendant asserts that since many customers in the "retail channel" research and test a monitor's capabilities before buying, even if defendant's claims were false it is unlikely that many in the "retail channel" were misled. Since the action has been bifurcated between liability and relief, during this liability phase plaintiff was not required to show proof of actual loss and to what extent any within the "retail channel" were actually influenced by defendant's claims to purchase MultiSync monitors. Thus, our conclusion that plaintiff has standing to bring this Lanham Act action does not imply that this court has reached any conclusion as to damages.

The answer to this question turns out to be complex. The doctrine of "non-trademark use" provides that it is acceptable to use a party's trademark *to refer to that party's product*. The Lanham Act is concerned only with the appropriation of that trademark to refer to another product. See New Kids on the Block v. News America Publishing, 971 F.2d 302 (9th Cir. 1992). Just as it is allowable to identify a competitor's product accurately in a comparative advertisement, shouldn't it be allowable to make an accurate statement that your product works well with another product?

In a parallel case to *Creative Labs*, Creative claimed that Cyrix's advertisements constituted trademark infringement because they referred to Creative too often and might therefore create confusion as to sponsorship or affiliation in the minds of consumers. How should this argument be evaluated? If a consumer sees an NEC advertisement and assumes either (1) that IBM is selling the NEC monitor or (2) that IBM has endorsed or somehow approved the NEC monitor, the consumer's purchasing decision may be influenced. Confusion as to source and sponsorship are actionable under the Lanham Act. Companies such as IBM and Microsoft may try to prevent the use of their trademarks as compatibility identifiers either because they are afraid of being associated with an inferior product or because restricting the use of the marks for compatibility purposes will boost their own sales of competing products.

Should trademark law prevent true statements concerning compatibility? The answer depends on the relative weight you place on competing policies. If you believe that a trademark confers strong property rights, and that trademark owners are right to be concerned about free riders selling inferior goods on the basis of their compatibility, you might favor trademark protection in this context. On the other hand, if you believe that compatibility ought to be encouraged, you are likely to think consumers will benefit from access to more accurate information about product alternatives.

4. The setting of industry standards, particularly where it is done by a group of industry representatives, may be protected as a collective or certification mark.[17] A certification mark is a mark used to designate goods that meet certain quality or accuracy standards. Rather than belonging to a particular company, use of the certification mark is open to anyone who meets the standards set out by the owner of the mark. If VGA was protected by a certification mark, IBM could sue anyone who falsely claimed to meet the VGA compatibility standard but would have to permit use of its mark to identify goods which were in fact in compliance with the standard.

A number of companies have already established such certification standards. Examples include "SPARC" for computer equipment that meets standards set by Sun Microsystems and "Windows — Ready to Run" for third-party programs that run on Microsoft's Windows operating system. See Mark F. Radcliffe, Trademarks in the Computer Industry: A New Role, 18:

---

17. 15 U.S.C. §1054 provides that certification marks may be registered under the same standards as a normal trademark.

2 New Matter 1 (1993). Standards may also be set by third-party organizations such as ISO or ANSI. Companies who submit their product specifications as a standard to such groups generally give up control over the trademark to the standard-setting organization, which will permit anyone who meets the specification to use the mark. A notable exception is Sun's Java trademark. Sun has applied to ISO to have Java declared a standard but is seeking to reserve the rights to the term "Java-compatible" to itself. At this writing, Sun was locked in litigation with Microsoft over whether Microsoft could refer to its altered version of Java as "Java-compatible." See Sun Microsystems v. Microsoft Corp., 999 F. Supp. 1301 (N.D. Cal. 1998); Mark A. Lemley & David McGowan, Could Java Change Everything? The Competitive Propriety of a Proprietary Standard, 43 Antitrust Bull. 715 (1998).

Is the use of certification marks a reasonable compromise between the competing policies outlined above? Does it solve the problem of free riding?

PROBLEM

Problem 7-15. Microsoft, one of the largest sellers of computer operating systems and application software, seeks to trademark its term "Windows" for its operating system interface. Should Microsoft be entitled to a trademark? How would you classify the mark? Against whom could Microsoft enforce the mark? How could manufacturers of application programs that make use of the "Windows" interface advertise compatibility with that interface?

---

3. Domain Names as Trademarks

Panavision International, L.P. v. Toeppen
United States Court of Appeals for the Ninth Circuit.
141 F.3d 1316 (9th Cir. 1998)

DAVID R. THOMPSON, Circuit Judge:
    This case presents two novel issues. We are asked to apply existing rules of personal jurisdiction to conduct that occurred, in part, in "cyberspace." In addition, we are asked to interpret the Federal Trademark Dilution Act as it applies to the Internet. [The discussion of personal jurisdiction is omitted].
    Panavision accuses Dennis Toeppen of being a "cyber pirate" who steals valuable trademarks and establishes domain names on the Internet using these trademarks to sell the domain names to the rightful trademark owners. . . .

We also conclude Panavision was entitled to summary judgment under the federal and state dilution statutes. Toeppen made commercial use of Panavision's trademarks and his conduct diluted those marks.

I
Background

The Internet is a worldwide network of computers that enables various individuals and organizations to share information. The Internet allows computer users to access millions of web sites and web pages. A web page is a computer data file that can include names, words, messages, pictures, sounds, and links to other information.

Every web page has its own web site, which is its address, similar to a telephone number or street address. Every web site on the Internet has an identifier called a "domain name." The domain name often consists of a person's name or a company's name or trademark. For example, Pepsi has a web page with a web site domain name consisting of the company name, Pepsi, and .com, the "top level" domain designation: Pepsi.com.

The Internet is divided into several "top level" domains: .edu for education; .org for organizations; .gov for government entities; .net for networks; and .com for "commercial" which functions as the catchall domain for Internet users.

Domain names with the .com designation must be registered on the Internet with Network Solutions, Inc. ("NSI"). NSI registers names on a first-come, first-served basis for a $100 registration fee. NSI does not make a determination about a registrant's right to use a domain name. However, NSI does require an applicant to represent and warrant as an express condition of registering a domain name that (1) the applicant's statements are true and the applicant has the right to use the requested domain name; (2) the "use or registration of the domain name . . . does not interfere with or infringe the rights of any third party in any jurisdiction with respect to trademark, service mark, trade name, company name or any other intellectual property right"; and (3) the applicant is not seeking to use the domain name for any unlawful purpose, including unfair competition.

A domain name is the simplest way of locating a web site. If a computer user does not know a domain name, she can use an Internet "search engine." To do this, the user types in a key word search, and the search will locate all of the web sites containing the key word. Such key word searches can yield hundreds of web sites. To make it easier to find their web sites, individuals and companies prefer to have a recognizable domain name.

Panavision holds registered trademarks to the names "Panavision" and "Panaflex" in connection with motion picture camera equipment. Panavision promotes its trademarks through motion picture and television credits and other media advertising.

In December 1995, Panavision attempted to register a web site on the Internet with the domain name Panavision.com. It could not do that, however, because Toeppen had already established a web site using Panavision's trademark as his domain name. Toeppen's web page for this site displayed photographs of the City of Pana, Illinois.

On December 20, 1995, Panavision's counsel sent a letter from California to Toeppen in Illinois informing him that Panavision held a trademark in the name Panavision and telling him to stop using that trademark and the domain name Panavision.com. Toeppen responded by mail to Panavision in California, stating he had the right to use the name Panavision.com on the Internet as his domain name. Toeppen stated:

> If your attorney has advised you otherwise, he is trying to screw you. He wants to blaze new trails in the legal frontier at your expense. Why do you want to fund your attorney's purchase of a new boat (or whatever) when you can facilitate the acquisition of 'PanaVision.com' cheaply and simply instead?

Toeppen then offered to "settle the matter" if Panavision would pay him $13,000 in exchange for the domain name. Additionally, Toeppen stated that if Panavision agreed to his offer, he would not "acquire any other Internet addresses which are alleged by Panavision Corporation to be its property."

After Panavision refused Toeppen's demand, he registered Panavision's other trademark with NSI as the domain name Panaflex.com. Toeppen's web page for Panaflex.com simply displays the word "Hello."

Toeppen has registered domain names for various other companies including Delta Airlines, Neiman Marcus, Eddie Bauer, Lufthansa, and over 100 other marks. Toeppen has attempted to "sell" domain names for other trademarks such as intermatic.com to Intermatic, Inc. for $10,000 and americanstandard.com to American Standard, Inc. for $15,000.

Panavision filed this action against Toeppen in the District Court for the Central District of California. Panavision alleged claims for dilution of its trademark under the Federal Trademark Dilution Act of 1995, 15 U.S.C. §1125(c), and under the California Anti-dilution statute, California Business and Professions Code §14330. Panavision alleged that Toeppen was in the business of stealing trademarks, registering them as domain names on the Internet and then selling the domain names to the rightful trademark owners. The district court determined it had personal jurisdiction over Toeppen, and granted summary judgment in favor of Panavision on both its federal and state dilution claims. This appeal followed.

II
Discussion . . .

B. Trademark Dilution Claims

The Federal Trademark Dilution Act provides:

> The owner of a famous mark shall be entitled . . . to an injunction against another person's commercial use in commerce of a mark or trade name, if such use begins after the mark has become famous and causes dilution of the distinctive quality of the mark. . . .

15 U.S.C. §1125(c).

The California Anti-dilution statute is similar. See Cal. Bus. & Prof.Code §14330. It prohibits dilution of "the distinctive quality" of a mark regardless of competition or the likelihood of confusion. The protection extends only to strong and well recognized marks. Panavision's state law dilution claim is subject to the same analysis as its federal claim.

In order to prove a violation of the Federal Trademark Dilution Act, a plaintiff must show that (1) the mark is famous; (2) the defendant is making a commercial use of the mark in commerce; (3) the defendant's use began after the mark became famous; and (4) the defendant's use of the mark dilutes the quality of the mark by diminishing the capacity of the mark to identify and distinguish goods and services. 15 U.S.C. §1125(c).

Toeppen does not challenge the district court's determination that Panavision's trademark is famous, that his alleged use began after the mark became famous, or that the use was in commerce. Toeppen challenges the district court's determination that he made "commercial use" of the mark and that this use caused "dilution" in the quality of the mark.

1. Commercial Use

Toeppen argues that his use of Panavision's trademarks simply as his domain names cannot constitute a commercial use under the Act. Case law supports this argument. See Panavision International, L.P. v. Toeppen, 945 F.Supp. 1296, 1303 (C.D.Cal. 1996) ("Registration of a trade[mark] as a domain name, without more, is not a commercial use of the trademark and therefore is not within the prohibitions of the Act."); Academy of Motion Picture Arts & Sciences v. Network Solutions, Inc., 989 F.Supp. 1276, 1997 WL 810472 (C.D.Cal. Dec. 22, 1997) (the mere registration of a domain name does not constitute a commercial use); Lockheed Martin Corp. v. Network Solutions, Inc., 985 F.Supp. 949 (C.D.Cal. 1997) (NSI's acceptance of a domain name for registration is not a commercial use within the meaning of the Trademark Dilution Act).

Developing this argument, Toeppen contends that a domain name is simply an address used to locate a web page. He asserts that entering a domain name on a computer allows a user to access a web page, but a domain name is not associated with information on a web page. If a user were to type Panavision.com as a domain name, the computer screen would display Toeppen's web page with aerial views of Pana, Illinois. The screen would not provide any information about "Panavision," other than a "location window" which displays the domain name. Toeppen argues that a user who types in Panavision.com, but who sees no reference to the plaintiff Panavision on Toeppen's web page, is not likely to conclude the web page is related in any way to the plaintiff, Panavision.

Toeppen's argument misstates his use of the Panavision mark. His use is not as benign as he suggests. Toeppen's "business" is to register trademarks as domain names and then sell them to the rightful trademark owners. He "act[s] as a 'spoiler,' preventing Panavision and others from doing business on the Internet under their trademarked names unless they pay his fee." *Panavision*, 938 F.Supp. at 621. This is a commercial use. See Intermatic Inc. v. Toeppen, 947 F.Supp. 1227, 1230 (N.D.Ill. 1996) (stating that "[o]ne of Toeppen's business objectives is to profit by the resale or licensing of these domain names, presumably to the entities who conduct business under these names.").

As the district court found, Toeppen traded on the value of Panavision's marks. So long as he held the Internet registrations, he curtailed Panavision's exploitation of the value of its trademarks on the Internet, a value which Toeppen then used when he attempted to sell the Panavision.com domain name to Panavision.

In a nearly identical case involving Toeppen and Intermatic Inc., a federal district court in Illinois held that Toeppen's conduct violated the Federal Trademark Dilution Act. *Intermatic*, 947 F.Supp. at 1241. There, Intermatic sued Toeppen for registering its trademark on the Internet as Toeppen's domain name, intermatic.com. It was "conceded that one of Toeppen's intended uses for registering the Intermatic mark was to eventually sell it back to Intermatic or to some other party." Id. at 1239. The court found that "Toeppen's intention to arbitrage the 'intermatic.com' domain name constitute[d] a commercial use." Id. See also Teletech Customer Care Management, Inc. v. Tele-Tech Co., 977 F.Supp. 1407 (C.D.Cal. 1997) (granting a preliminary injunction under the Trademark Dilution Act for use of a trademark as a domain name).

Toeppen's reliance on Holiday Inns, Inc. v. 800 Reservation, Inc., 86 F.3d 619 (6th Cir. 1996), cert. denied, __ U.S. __, 117 S.Ct. 770, 136 L.Ed.2d 715 (1997) is misplaced. In *Holiday Inns*, the Sixth Circuit held that a company's use of the most commonly misdialed number for Holiday Inns' 1-800 reservation number was not trademark infringement.

Holiday Inns is distinguishable. There, the defendant did not use Holiday Inns' trademark. Rather, the defendant selected the most commonly misdialed telephone number for Holiday Inns and attempted to capitalize on consumer confusion.

A telephone number, moreover, is distinguishable from a domain name because a domain name is associated with a word or phrase. A domain name is similar to a "vanity number" that identifies its source. Using *Holiday Inns* as an example, when a customer dials the vanity number "1-800-Holiday," she expects to contact Holiday Inns because the number is associated with that company's trademark. A user would have the same expectation typing the domain name HolidayInns.com. The user would expect to retrieve Holiday Inns' web page.

Toeppen made a commercial use of Panavision's trademarks. It does not matter that he did not attach the marks to a product. Toeppen's commercial use was his attempt to sell the trademarks themselves.[5] Under the Federal Trademark Dilution Act and the California Anti-dilution statute, this was sufficient commercial use.

2. Dilution

"Dilution" is defined as "the lessening of the capacity of a famous mark to identify and distinguish goods or services, regardless of the presence or absence of (1) competition between the owner of the famous mark and other parties, or (2) likelihood of confusion, mistake or deception." 15 U.S.C. §1127.

Trademark dilution on the Internet was a matter of Congressional concern. Senator Patrick Leahy (D-Vt.) stated:

> [I]t is my hope that this anti-dilution statute can help stem the use of deceptive Internet addresses taken by those who are choosing marks that are associated with the products and reputations of others.

141 Cong. Rec. §19312-01 (daily ed. Dec. 29, 1995) (statement of Sen. Leahy). See also Teletech Customer Care Management, Inc. v. Tele-Tech Co., Inc., 977 F.Supp. 1407, 1413 (C.D.Cal. 1997).

---

5. See Boston Pro. Hockey Assoc., Inc. v. Dallas Cap & Emblem Mfg., Inc., 510 F.2d 1004 (1975), which involved the sale of National Hockey League logos. The defendant was selling the logos themselves, unattached to a product (such as a hat or sweatshirt). The court stated: "The difficulty with this case stems from the fact that a reproduction of the trademark itself is being sold, unattached to any other goods or services." Id. at 1010. The court concluded that trademark law should protect the trademark itself. "Although our decision here may slightly tilt the trademark laws from the purpose of protecting the public to the protection of the business interests of plaintiffs, we think that the two become . . . intermeshed. . . ." Id. at 1011. "Whereas traditional trademark law sought primarily to protect consumers, dilution laws place more emphasis on protecting the investment of the trademark owners." *Panavision*, 945 F.Supp. at 1301.

To find dilution, a court need not rely on the traditional definitions such as "blurring" and "tarnishment." Indeed, in concluding that Toeppen's use of Panavision's trademarks diluted the marks, the district court noted that Toeppen's conduct varied from the two standard dilution theories of blurring and tarnishment. *Panavision*, 945 F.Supp. at 1304. The court found that Toeppen's conduct diminished "the capacity of the Panavision marks to identify and distinguish Panavision's goods and services on the Internet." Id. See also *Intermatic*, 947 F.Supp. at 1240 (Toeppen's registration of the domain name, "lessens the capacity of Intermatic to identify and distinguish its goods and services by means of the Internet.").

This view is also supported by *Teletech*. There, TeleTech Customer Care Management Inc., ("TCCM"), sought a preliminary injunction against Tele-Tech Company for use of TCCM's registered service mark, "TeleTech," as an Internet domain name. *Teletech*, 977 F.Supp. at 1410. The district court issued an injunction, finding that TCCM had demonstrated a likelihood of success on the merits on its trademark dilution claim. Id. at 1412. The court found that TCCM had invested great resources in promoting its servicemark and Teletech's registration of the domain name teletech.com on the Internet would most likely dilute TCCM's mark. Id. at 1413.

Toeppen argues he is not diluting the capacity of the Panavision marks to identify goods or services. He contends that even though Panavision cannot use Panavision.com and Panaflex.com as its domain name addresses, it can still promote its goods and services on the Internet simply by using some other "address" and then creating its own web page using its trademarks.

We reject Toeppen's premise that a domain name is nothing more than an address. A significant purpose of a domain name is to identify the entity that owns the web site. "A customer who is unsure about a company's domain name will often guess that the domain name is also the company's name." Cardservice Intl. v. McGee, 950 F.Supp. 737, 741 (E.D.Va. 1997). "[A] domain name mirroring a corporate name may be a valuable corporate asset, as it facilitates communication with a customer base." MTV Networks, Inc. v. Curry, 867 F.Supp. 202, 203-204 n. 2 (S.D.N.Y. 1994).

Using a company's name or trademark as a domain name is also the easiest way to locate that company's web site. Use of a "search engine" can turn up hundreds of web sites, and there is nothing equivalent to a phone book or directory assistance for the Internet. See *Cardservice*, 950 F.Supp. at 741.

Moreover, potential customers of Panavision will be discouraged if they cannot find its web page by typing in "Panavision.com," but instead are forced to wade through hundreds of web sites. This dilutes the value of Panavision's trademark. We echo the words of Judge Lechner, quoting Judge Wood: "Prospective users of plaintiff's services who mistakenly access defendant's web site may fail to continue to search for plaintiff's own home page, due to anger, frustration or the belief that plaintiff's home page does not exist." Jews for Jesus v. Brodsky, 993 F.Supp. 282, 306-07 (D.N.J. 1998)

(Lechner, J., quoting Wood, J. in Planned Parenthood v. Bucci, 1997 WL 133313 at *4); see also *Teletech*, 977 F.Supp. at 1410 (finding that use of a search engine can generate as many as 800 to 1000 matches and it is "likely to deter web browsers from searching for Plaintiff's particular web site").

Toeppen's use of Panavision.com also puts Panavision's name and reputation at his mercy. See *Intermatic*, 947 F.Supp. at 1240 ("If Toeppen were allowed to use 'intermatic.com,' Intermatic's name and reputation would be at Toeppen's mercy and could be associated with an unimaginable amount of messages on Toeppen's web page.").

We conclude that Toeppen's registration of Panavision's trademarks as his domain names on the Internet diluted those marks within the meaning of the Federal Trademark Dilution Act, 15 U.S.C. §1125(c), and the California Anti-dilution statute, Cal.Bus. & Prof.Code §14330.

III
Conclusion

Toeppen engaged in a scheme to register Panavision's trademarks as his domain names on the Internet and then to extort money from Panavision by trading on the value of those names. Toeppen's actions were aimed at Panavision in California and the brunt of the harm was felt in California. The district court properly exercised personal jurisdiction over Toeppen.

We also affirm the district court's summary judgment in favor of Panavision under the Federal Trademark Dilution Act, 15 U.S.C. §1125(c), and the California Anti-dilution statute, Cal.Bus. & Prof.Code §14330. Toeppen made commercial use of Panavision's trademarks and his conduct diluted those marks.

AFFIRMED.

COMMENTS & QUESTIONS

1. The interaction between NSI's "first-come, first-serve" rule for allocating Internet domain names and trademark law has been explored elsewhere. See, e.g., Dan L. Burk, Trademarks Along the Infobahn: A First Look at the Emerging Law of Cybermarks, 1 Rich. J.L. & Tech. 1 (1995); Gary W. Hamilton, Trademarks on the Internet: Confusion, Collusion, or Dilution?, 4 Tex. Intell. Prop. L.J. 1 (1995); David J. Loundy, A Primer on Trademark Law and Internet Addresses, 15 John Marshall J. Comp. & Info. L. 465 (1997); James West Marcovitz, ronald@mcdonalds.com — "Owning a Bitchin' " Corporate Trademark as an Internet Address — Infringement?, 17 Cardozo L. Rev. 85 (1995); Carl Oppedahl, Remedies in Domain Name Lawsuits: How is a Domain Name Like a Cow?, 15 John Marshall J. Comp. & Info. L. 437 (1997). See also <http://www.ll.georgetown.edu/lc/in-

ternic/domain1.html> (cataloguing information about domain name trademark disputes).

Essentially, as to "cybersquatters" (those who obtain a domain name from NSI in order to deprive the trademark owner of it, in order to sell it, or in order to confuse consumers), the rule in the cases is straightforward: the trademark owner wins. Numerous cases have established liability for using another's trademark as a domain name. See. e,g., Comp Examiner Agency v. Juris, Inc., 1996 U.S. Dist. LEXIS 20259 (C.D. Cal. April 25, 1996) (injunction vs. direct competitor); Actmedia, Inc. v. Active Media Int'l, 1996 US Dist. LEXIS 20814 (N.D. Ill. July 17, 1996) (same); Cardservice Int'l v. McGee, 950 F. Supp. 737 (E.D. Va. 1997) (same); Planned Parenthood Federation v. Bucci, 42 USPQ2d 1430 (S.D.N.Y. 1997) (anti-abortion activist can't use Planned Parenthood name); Intermatic v. Toeppen, 947 F. Supp. 1227 (N.D. Ill. 1996) (same dilution analysis as *Panavision*); Hasbro, Inc. v. Internet Entertainment Group, 40 USPQ2d 1479 (W.D. Wash 1996) (adult site dilutes famous name for children's game); Toys'r'Us v. Akkaoui, 40 USPQ2d 1836 (N.D. Cal. 1996) (dilution of " 'r Us" family of marks by defendant's "adultsrus" domain name); Inset Systems v. Instruction Set, Inc., 937 F. Supp. 161 (D. Conn. 1996) (dictum).

From a policy perspective this result seems unobjectionable, but consider how well it really comports with trademark doctrine. Were consumers "confused" by Toeppen's use of the panavision.com domain? Perhaps initially they might have gone to www.panavision.com expecting to find the Web site for the corporation. Any such confusion surely would have been quickly dispelled, since Toeppen's site had nothing to do with Panavision's business.[18] The obvious alternative theory is dilution, but dilution requires the "commercial use in commerce" of a diluting mark by the defendant. 15 U.S.C. §1125(c). Has Toeppen really engaged in "commercial use" of the Panavision mark? What if he did not wish to sell it to Panavision at all but simply wanted it for his own private use, or to deprive Panavision of the ability to use the mark? Alternatively, suppose he devoted the site to a political diatribe against Panavision's employment practices. It is hard to see how this would constitute dilution in the traditional sense. See also 15 U.S.C. §1125(c), which exempts from dilution any "noncommercial use" of the mark.

These issues were placed in stark relief in Planned Parenthood Federation v. Bucci, 42 USPQ2d 1430 (S.D.N.Y. 1997), aff'd without opinion, 152 F.3d 920 (2d Cir. 1998). There an anti-abortion activist had registered "plannedparenthood.com" and used it to distribute anti-abortion literature to visitors who thought they were reaching the Planned Parenthood agency (which was actually located at "plannedparenthood.org"). The court held that Bucci's actions violated the Lanham Act and created a per se rule that

---

18. A different (and much easier) case for infringement is presented by cases in which one firm registers a competitor's domain name and uses it to put out information about the two products. This was the case, for example, when Princeton Review registered kaplan.com and devoted the site to comparing the two companies.

use on the Internet was "use in commerce." This rule, if widely adopted, would give courts strong power to prevent "cybersquatting." For a similar result, see Jews for Jesus v. Brodsky, 46 U.S.P.Q.2d 1652 (D.N.J. 1998), aff'd without opinion, 159 F.3d 1351 (3d Cir. 1998). But cf. Playboy Enter. Inc. v. Universal Tel-A-Talk, 48 U.S.P.Q.2d 1779 (E.D. Pa. 1998) (use of the Playboy term as a link to Playboy's Web site was not trademark infringement or counterfeiting).

2. Not all domain name trademark cases involve cybersquatting, however. Sometimes two legitimate trademark owners with the same name dispute who should own a domain name. These problems often arise because trademarks that are widely separated in geographic or product space are thrown together on the Internet, which is worldwide and cuts across product lines. For example, "Roadrunner" is the name of a computer company, a moving and storage company, and a Warner Brothers cartoon character. In the real world all of these uses can coexist because the products are so different that no one would be likely to confuse them. However, on the Internet, only one company can register "roadrunner.com".

Courts faced with disputes of this sort have a number of choices, but none seem especially satisfactory. They could uphold NSI's standard "first-come, first-served" approach to registration. Whoever wins the "race" to the registrar gets the name, and everyone else loses. This result is certainly easy to administer; indeed, the courts need not get involved at all. It may end up increasing consumer confusion, however, especially where the larger of two coexisting companies loses the race. Second, courts (or NSI) might try to decide which claimant to a name was the "best" or most famous. This may produce the fairest result in the final analysis but only if there is some way to make that determination. NSI's policy of placing domain names "on hold" if a registered trademark owner complains has led to some problems here, particularly when both the registrant and the complainant had legitimate claims to the name. See Giacalone v. Network Solutions, 1996 U.S. Dist. LEXIS 20807 (N.D. Cal. June 14, 1996) (preliminary injunction against NSI placing a registered domain name on hold at the request of a trademark owner). Finally, courts might force the competing applicants to coexist — either by choosing similar names ("road-runner.com" vs. "roadrunner.com"), or by sharing a welcome page. The parties in Mattel v. Hasbro (C.D. Cal. 1997) chose the latter option. Hasbro owned the trademark rights to the game "Scrabble" in North America, and Mattel owned the rights elsewhere. The parties settled a trademark dispute by agreeing to create a joint page at scrabble.com that linked to both of their respective sites.

3. A third set of cases involve suits by both sides against NSI for their role in domain name registration. Irate trademark owners sued NSI for allowing cybersquatters to register their marks and demanded that the domain name be handed over to them. When NSI started putting domain names on hold, domain name owners sued NSI for interfering with their rights. (The *Giacalone* case cited above is an example of the latter.) While either or both sides may have a claim based in property rights or tort, one court recently ruled that NSI's domain name registration policies do not themselves impli-

cate the trademark laws. Lockheed Martin Corp. v. Network Solutions Inc., 985 F. Supp. 949 (C.D. Cal. 1997). The court concluded that the mere act of registering a domain name by placing the entry on the DNS server database was not "use" necessary for a violation of the Lanham Act. The court opined that "The solution to the current difficulties faced by trademark owners on the Internet lies in . . . technical innovation, not in attempts to assert trademark rights over legitimate non-trademark uses of this important new means of communication." Id.

Is the court correct that accepting a domain name registration that infringes a trademark is a "legitimate non-trademark use"? Is this case distinguishable from the result in *Panavision*?

NSI's role in registering domain names is likely to change soon. In November 1998, the Department of Commerce entered into an agreement with the Internet Corporation for Assigned Names and Numbers (ICANN), a newly created non-profit corporation, to transition the management of certain key Internet functions (including the management of the Domain Name System) away from sole control by NSI. The U.S. government also amended its agreement with Network Solutions, once the sole provider of domain name registration services in the .com, .net and .org top-level domains. The expanded competition in domain name registration services will, in theory, lead to greater choice in services and prices. An individual wishing to register a domain name will be able to choose among many registrars. The registrar in turn passes along the information to the registry, which maintains the authoritative list of domain names and addresses. Network Solutions continues to maintain extensive control over the registry (even though five other "test" registrars were approved by ICANN in April 1999, only two had begun to operate by July 1999).

4. Finally, there are numerous trademark disputes on the Internet that do not involve domain names but rather the content of Web sites themselves. For a discussion of many of these cases, and some of the interesting issues they present, see Dan L. Burk, Trademark Doctrines for Global Electronic Commerce, 49 S. Carolina L. Rev. 695 (1998).

PROBLEMS

Problem 7-16: Web pages are written in Hypertext Markup Language (HTML). For the most part, text written in HTML actually shows up on the Web page visible to users. However, if the page designer prefaces a line with the tag "Meta", words in that line will not appear on the screen. The words in this "metatag" are a part of the Web page, however, and Internet search engines looking for keywords will find them.

Early on some Web site designers realized that they could increase the number of "hits" on their site by including popular keywords in metatags. Thus site owners might include "games" (or "warez" or "software" or "sex") in their metatag in an effort to draw Internet users into their site even if their site has nothing to do with software or sex.

Ace Widgets and Basic Widgets are the two largest companies in the fiercely competitive widget manufacturing industry. Ace puts up a Web page extolling the virtues of its widgets. If Ace includes the word "basic" in its metatag, is it liable for trademark infringement? False advertising?

Problem 7-17: New versions of Netscape's Navigator browser include a "What's Related?" feature. This feature allows someone visiting a particular Web site to click a button to get a list of sites that Netscape has determined to be "related" to the current page. The "relatedness" determination is based on an algorithm that matches keywords from different pages and the number of links the pages have in common.

Kids Adventure, a children-friendly Internet gaming site, is appalled to learn that visitors to their site who click the "What's Related?" button are given a list of sites including their major competitor's site and a site where child pornography can be found.

Does Kids Adventure have a cause of action against Netscape for falsely suggesting a connection between these sites? Would it have a cause of action against the sites themselves if it turned out they had manipulated the content of their own sites to get on Netscape's relatedness list?

F. SUI GENERIS PROTECTION OF COMPUTER TECHNOLOGY

As the previous sections suggest, computer technology does not fit neatly within any of the categories of intellectual property law. Different aspects of computers and computer programs are eligible for protection as trade secrets or by means of patents, copyright, or trademarks. Further, many important works — particularly computer programs — receive overlapping protection from several sources.

None of these sources is a perfect "fit" with computer software. Patent, copyright, trade secret, and trademark law all protect some aspects of computer programs, but what they protect is not necessarily what we ideally want to encourage. To address this problem, numerous commentators have suggested that we need a new intellectual property statute, one directed specifically at computer technology and responsive to the particular needs of the industry. In 1984, Congress responded in part to these concerns by passing the Semiconductor Chip Protection Act ("SCPA"), 17 U.S.C. §§901-914. That act provided specific protection to semiconductor "mask works." Its purpose and history were described by the Federal Circuit in Brooktree Corp. v. Advanced Micro Devices, 977 F.2d 1555 (Fed. Cir. 1992):

> The Semiconductor Chip Protection Act of 1984, Pub. L. 98-620, Title III, 98 Stat. 3347, codified at 17 U.S.C. §§901-914, arose from concerns that existing intellectual property laws did not provide adequate protection of proprietary rights in semiconductor chips that had been designed to perform a particular function. The Act, enacted after extensive congressional consideration and hearings over several years, adopted relevant aspects of existing intellectual property law, but for the most part created a new law, specifically adapted to the protection of design layouts of semiconductor chips.
>
> Chip design layouts embody the selection and configuration of electrical components and connections in order to achieve the desired electronic functions. The electrical elements are configured in three dimensions, and are built up in layers by means of a series of "masks" whereby, using photographic depositing and etching techniques, layers of metallic, insulating, and semiconductor material are deposited in the desired pattern on a wafer of silicon. This set of masks is called a "mask work," and is part of the semiconductor chip product. The statute defines a mask work as:
>
>> a series of related images, however fixed or encoded
>>   (A) having or representing the predetermined, three dimensional pattern of metallic, insulating, or semiconductor material present or removed from the layers of a semiconductor chip product; and
>>   (B) in which series the relation of the images to one another is that each image has the pattern of the surface of one form of a semiconductor chip product.
>
> 17 U.S.C. §901(a)(2). The semiconductor chip product in turn is defined as:
>
>> the final or intermediate form of any product —
>>   (A) having two or more layers of metallic, insulating, or semiconductor material, deposited or otherwise placed on, or etched away or otherwise removed from, a piece of semiconductor material in accordance with a predetermined pattern; and
>>   (B) intended to perform electronic circuitry functions.
>
> 17 U.S.C. §901(a)(1).
>
> The design of a satisfactory chip layout may require extensive effort and be extremely time consuming, particularly as new and improved electronic capabilities are sought to be created. A new semiconductor chip may incur large research and development costs, yet after the layout is imprinted in the mask work and the chip is available in commerce, it can be copied at a fraction of the cost to the originator. Thus there was concern that widespread copying of new chip layouts would have adverse effects on innovative advances in semiconductor technology, as stated in the Senate Report:
>
>> In the semiconductor industry, innovation is indispensable; research breakthroughs are essential to the life and health of the industry. But research and innovation in

the design of semiconductor chips are threatened by the inadequacies of existing legal protection against piracy and unauthorized copying. This problem, which is so critical to this essential sector of the American economy, is addressed by the Semiconductor Chip Protection Act of 1984. . . .

The Semiconductor Chip Protection Act of 1984, . . . would prohibit "chip piracy" — the unauthorized copying and distribution of semiconductor chip products copied from the original creators of such works.

S.Rep. No. 425, 98th Cong., 2d Sess., 1 (1984) (hereinafter *Senate Report*).

In the evolution of the Semiconductor Chip Protection Act it was first proposed simply to amend the Copyright Act, 17 U.S.C. §101 et seq., to include semiconductor chip products and mask works as subject of copyright. See H.R. 1028, 98th Cong., 1st Sess. (1983). However, although some courts had interpreted copyright law as applicable to computer software imbedded in a semiconductor chip, see Apple Computer, Inc. v. Franklin Computer Corp., 714 F.2d 1240, 1249, 219 U.S.P.Q. 113, 121 (3d Cir. 1983), *cert. dismissed*, 464 U.S. 1033, 104 S. Ct. 690, 79 L. Ed. 2d 158 (1984), it was uncertain whether the copyright law could protect against copying of the pattern on the chip itself, if the pattern was deemed inseparable from the utilitarian function of the chip. Indeed, the Copyright Office had refused to register patterns on printed circuit boards and semiconductor chips because no separate artistic aspects had been demonstrated. Copyright Protection for Semiconductor Chips: Hearings on H.R. 1028 Before the Subcomm. on Courts, Civil Liberties, and the Administration of Justice of the House Comm. on the Judiciary, 98th Cong., 1st Sess., 77 (1983) (hereinafter *1983 House Hearings*) (statement of Dorothy Schrader, Associate Register of Copyrights for Legal Affairs). Concern was also expressed that extension of the copyright law to accommodate the problems of mask works would distort certain settled copyright doctrines, such as fair use. 1983 House Hearings at 16-17 (statement of Jon A. Baumgarten, Copyright Counsel, Association of American Publishers, Inc.).

The patent system alone was deemed not to provide the desired scope of protection of mask works. Although electronic circuitry and electronic components are within the statutory subject matter of patentable invention, see 35 U.S.C. §101, and some original circuitry may be patentable if it also meets the requirements of the Patent Act, as is illustrated in this case, Congress sought more expeditious protection against copying of original circuit layouts, whether or not they met the criteria of patentable invention. Senate Report at 8.

The Semiconductor Chip Protection Act of 1984 was an innovative solution to this new problem of technology-based industry. While some copyright principles underlie the law, as do some attributes of patent law, the Act was uniquely adapted to semiconductor mask works, in order to achieve appropriate protection for original designs while meeting the competitive needs of the industry and serving the public interest.

The Semiconductor Chip Protection Act provides for the grant of certain exclusive rights to owners of registered mask works, including the exclusive right "to reproduce the mask work by optical, electronic, or any other means", and the exclusive right "to import or distribute a semiconductor chip product in which the mask work is embodied". 17 U.S.C. §905. Mask works that are not "original", or that consist of "designs that are staple, commonplace, or familiar in the semiconductor industry, or variations of such designs, combined in a way that, considered as a whole, is not original", are excluded from protection. 17 U.S.C. §902(b). Protection is also not extended to any "idea, procedure, process, system, method of operation, concept, principle, or discovery, regardless of the form in which it is described, explained, illustrated or embodied" in the mask work. 17 U.S.C. §902(c). The sponsors and supporters of this legislation foresaw that there would be areas of uncertainty in application of this new law to particular situations, and referred to "gray areas" wherein factual situations could arise that would not have easy answers. Those areas are emphasized by both parties in the assignments of error on this appeal. . . .

COMMENTS AND QUESTIONS

1. One of the most remarkable aspects of the SCPA has been the dearth of litigation. This has been surprising since congressional testimony prior to the SCPA's enactment indicated that piracy was rampant. A number of explanations have been offered. First, most "piracy" prior to the act may have been due to uncertainty about the scope of mask work protection under the copyright law. The SCPA clarified the law, thereby leading firms to conform their behavior to the requirements of the act. Alternatively, Ron Laurie, a leading computer law practitioner, suggests that the outright piracy alluded to during the hearings did not exist or has become technologically obsolete. Highly integrated modern mask works cannot be quickly and accurately duplicated without using a fabrication process virtually identical to that of the original manufacturer. Since such processes are highly proprietary, it is unlikely that pirates could duplicate complex, high density chip designs without reverse engineering. Ron Laurie, The First Year's Experience Under the Chip Act or "Where Are the Pirates Now That We Need Them?" 3 Computer Lawyer 11, 21 (Feb. 1986).

2. At least one new form of sui generis protection is under consideration at this writing. A proposal before the World Intellectual Property Organization would provide potentially perpetual protection for "databases," broadly defined as any "collection, assembly or compilation of works, data, information or other materials arranged in a systematic or methodical way." A similar bill was introduced in Congress in 1996, but it did not pass. If the problem with the SCPA was that it was too narrow and technology-specific, this new proposal runs the opposite risk: it is so broad that it covers potentially every source of information, including books and articles.

3. How well does the sui generis model of the SCPA work in the software context? As the principal software copyright cases have highlighted, the highly utilitarian nature of computer programming poses substantial problems for judges seeking to maintain the integrity of the copyright law. In the *Altai* case, Judge Walker commented

> Generally, we think that copyright registration — with its indiscriminating availability — is not ideally suited to deal with the highly dynamic technology

of computer science. Thus far, many of the decisions in this area reflect the courts' attempt to fit the proverbial square peg in a round hole. The district court, see Computer Assocs., 775 F. Supp. at 560, and at least one commentator have suggested that patent registration, with its exacting up-front novelty and non-obviousness requirements, might be the more appropriate rubric of protection for intellectual property of this kind. See Randell M. Whitmeyer, Comment, A Plea for Due Processes: Defining the Proper Scope of Patent Protection for Computer Software, 85 Nw. U. L. Rev. 1103, 1123-25 (1991); see also Lotus Dev. Corp. v. Borland Int'l, Inc., 788 F. Supp. 78, 91 (D. Mass. 1992) (discussing the potentially supplemental relationship between patent and copyright protection in the context of computer programs). In any event, now that more than 12 years have passed since CONTU issued its final report, the resolution of this specific issue could benefit from further legislative investigation — perhaps a CONTU II.

Judge Boudin, in his thoughtful concurrence in *Lotus*, expressly acknowledged courts' limitations in confronting the challenges posed by legal protection for computer software. He concluded by noting that "Some solutions (e.g., a very short copyright period for menus) are not options at all for courts but might be for Congress. In all events, the choices are important ones of policy, not linguistics, and they should be made with the underlying considerations in view."

This section considers alternatives and assesses the potential for the development of a sui generis regime for the protection of computer software.

≡ *Peter S. Menell, Tailoring Legal Protection for*
≡ *Computer Software*
≡   *39 Stan. L. Rev. 1329 (1987)*

. . .

B. Tailoring Legal Protection for Operating Systems

Part of the reason for copyright's inability to promote economic efficiency in the provision of computer products is that the public goods and network externality problems suggest conflicting modes of legal protection. Public goods problems are alleviated by expanding legal protection for intellectual work. External benefits from networks are promoted by facilitating access to a standard. Thus, the difficult policy question is how to promote standardization while at the same time encouraging continuing innovation (along the entire spectrum from software to hardware). By closely tailoring legal protection to reward desired innovation while permitting reasonable access to industry standards, it is possible to reach a satisfactory accommodation of these apparently conflicting objectives.

In theory, patent law is more appropriate than copyright for protecting the intellectual work contained in computer operating systems. A patent protects new and useful processes and machines. Given the interchangeability of hardware and software, it seems logical to protect computer operating systems and dedicated computers that embody a particular operating system with the same form of legal protection. Because patent law protects ideas, those who create patentable operating systems could be better assured of appropriating a substantial portion of the benefits of their efforts.

As we noted in Part III, it is difficult to obtain patent protection for computer programs. [Recall that this article was written in 1987. — EDS.] It should be pointed out, however, that although the scope of patent protection for computer software is uncertain, some of the recent cases that have upheld the patentability of computer programs involved programs that manipulate the internal operations of a computer. Moreover, the importance of network externalities flowing from widespread access to a common mini- and microcomputer operating system suggests that legal protection should be hard to come by and relatively short in duration.

To encourage innovation in operating system technology, Congress should consider creating a hybrid form of patent protection specifically tailored to accommodate the market failures endemic to the provision of computer operating systems.[190] As with traditional patent law, the standard for protection should be novelty, nonobviousness, and usefulness; dominant firms (or anyone else) should not be able to "lock up" an industry standard simply by expressing it in a unique way.

To be feasible, the modified form of patent protection for computer operating systems should be based on a timely examination of patent applications. And given the rapid pace of technological change in the computer field and the interest in promoting access to industry standards, patent protection for operating systems should be shorter in duration than traditional patent protection.

In order to promote continued innovation in widely used operating systems, the operating system patent code should, like the Semiconductor Chip Protection Act, permit some limited form of reverse engineering. And like traditional patent law, the hybrid code should allow consumers to buy a ROM chip or other device containing a patented operating system and modify it for sale to a third person.

Because traditional patent law affords absolute protection, however, it would inhibit realization of network externalities from operating systems sat-

---

190. Congress has recently followed a sui generis approach in designing legal protection for semiconductor chips. . . . Other commentators have also urged Congress to create a hybrid form of legal protection for computer software. See Davidson, supra note 63, at 673-82; Galbi, Proposal for New Legislation to Protect Computer Programming, 17 J. Copyright Soc'y 280 (1970); Karjala, supra note 63, at 61-81; Samuelson, Creating a New Kind of Intellectual Property: Applying the Lessons of the Chip Law to Computer Programs, 70 Minn. L. Rev. 471, 507, 529-31 (1985); Stern, The Case of the Purloined Object Code: Can It Be Solved? (Part 2), Byte, Oct. 1982, at 210, 222.

isfying the above subject matter requirements. In order to facilitate realization of network externalities, therefore, the hybrid patent code should contain a flexible compulsory licensing provision. Such a provision would promote access to an industry standard while assuring rewards to the creator of an innovative and socially valuable operating system. It would also limit the ability of dominant firms in the industry to engage in anticompetitive practices.

The need for compulsory licensing as a means for promoting competition and rewarding innovation is brought into focus by the decision of the Ninth Circuit Court of Appeals in Digidyne Corp. v. Data General Corp. In Data General, the defendant (Data General), a manufacturer of computers, refused to license its RDOS operating system to firms using a central processing unit other than Data General's "NOVA" system. Recognizing the anticompetitive effects of this practice in a market with network externalities,[196] the Ninth Circuit held that Data General's licensing practices were an unlawful tying arrangement that violated federal antitrust law.

In light of the strong network externalities flowing from compatibility, computer operating systems serve as "essential facilities" in computer hardware markets. Unless a firm can get onto the network, its products will be at a great disadvantage relative to those that can run the vast stock of application programs designed for the industry standard. The operating system royalty rate per use should be set so as to compensate true innovators for the cost of building a useful "highway" for the market plus a fair profit (adjusted for the risk of failure). For high volume products, these rates would probably be low. In the microcomputer market, for example, the rate would probably be less than one dollar for access to the major operating systems (assuming that the Apple and IBM operating systems merited hybrid patent protection at all.)

A patent code for operating systems based on the above outline strikes a preferable balance of the conflicting policy concerns raised by computer operating systems. By providing solid protection for truly innovative and useful operating systems, the code would reward innovation in operating systems. The limits on this regime of protection — moderate duration, reverse engineering, adaptation — and the provision for compulsory licensing would promote access to operating systems that emerge as industry standards, wide diffusion of computer products, and innovation in hardware products. The code would also avoid wasteful expenditure of resources on efforts to emulate an industry standard.

The proposed operating system code would probably entail somewhat higher administrative costs than the current system. Patent examinations,

---

196. The court accepted the plaintiffs' proof of market power on the basis of software "lock-in." Software lock-in occurs when a computer user develops or purchases application software designed to run on a particular operating system. This installed base locks the consumer into the hardware products of the owner of that operating system if competitors cannot gain access to the operating system and the costs of converting software to run on different operating systems are high. But see Helein, Software Lock-in and Antitrust Tying Arrangements: The Lessons of Data General, 5 Computer/L.J. 329, 337, 342-43 (1985) (suggesting that conversion costs might not be so high as to justify a finding of market power).

though streamlined, would be significantly more expensive than the cost of copyright registration. Moreover, compulsory licensing proceedings, as well as the cost of monitoring use of protected operating systems, would add to the expense of the system. If the royalty rates were low (as the microcomputer example indicates), however, members of the industry could be expected to cooperate in ensuring that patent owners were properly compensated.

C. Tailoring Legal Protection for Application Programs

As with legal protection for operating systems, Congress should consider creating a special form of legal protection for application programs. Given the importance of improving existing programs as a primary mode of technological innovation and the presence of some network externalities, legal protection should be significantly shorter in duration than traditional copyright protection. The relatively short commercial life of most application programs indicates that legal protection should be correspondingly short.

The regime for protecting application programs should also allow for reverse engineering. In designing legal protection for semiconductor chips, Congress recognized the importance of reverse engineering in enabling researchers to advance a field in which innovations are cumulative. A limited reverse engineering provision in the application software code would similarly promote the advancement of application software technology.

Congress should also consider the desirability of a limited form of compulsory licensing of application packages. In order to realize the benefits of network externalities and to promote creativity in the integration of software programs, it would seem worthwhile to allow limited access to application programs, particularly those that emerge as industry standards. This could be achieved without dulling primary creative incentives by delaying the availability of compulsory licensing for a limited period to allow the creator of the program to reap the rewards of commercial success.

▬▬
▬▬
▬▬
▬▬
▬▬
▬▬
▬▬
Pamela Samuelson, Randall Davis, Mitchell D. Kapor, & J. H. Reichman, A Manifesto Concerning the Legal Protection of Computer Programs
94 Colum. L. Rev. 2308 (1994)

In brief, we have concluded that while copyright law can provide appropriate protection for some aspects of computer programs, other valuable aspects of programs, such as the useful behavior generated when programs are in operation and the industrial design responsible for producing this behavior, are vulnerable to rapid imitation that, left unchecked, would undermine incentives to invest in software development. These aspects may need some legal protection against cloning that existing legal regimes cannot provide. Most of the considerable controversy about software protection, within the software industry and the legal community, has arisen when software developers have tried to use existing legal regimes to protect these kinds of program innovations. . . .

2.1 Program Innovations Are Vulnerable to Trivial Acquisitions of Equivalence That Can Cause Market Distortions

2.1.1 *Programs Reveal a Substantial Amount of Their Know-How in Products Distributed in the Market . . .*

Computer programs are an unusual kind of industrial product because the bulk of the know-how required to create them is accessible on or near the face of the product distributed in the marketplace. There are several reasons why this is so: (1) Software products need little specialized mass-production know-how; (2) they are particularly rich in design know-how; and (3) they are more susceptible in some respects to reverse engineering than traditional industrial products.

Software is easy to mass-produce because it is an information product. There is no steel to cut or bend, no wires to attach, and no parts to be cast in plastic or rubber. Mass production is as simple as loading copies onto floppy disks or tapes, a routine, inexpensive, and low-technology undertaking. Because there are no special processes required for mass production of the software products, there is no opportunity to accumulate specialized know-how about mass production. Lack of mass-production know-how matters because this is the kind of know-how that can most easily be kept out of the product, i.e., maintained as a trade secret. . . .

Software is, instead, rich in surface design know-how. This is particularly true for interactive software, i.e., programs that involve a great deal of hands-on usage. For such software, the metaphor or conception of the task is particularly important. . . .

Software is not only especially rich in design know-how; much of the value of software arises from it. . . .

The difficult task in creating interactive software, the innovations embodied in it, and the value of those innovations are typically found in conception and design: determining how to think about the task (e.g., the desktop metaphor), and then determining what capabilities the program should have (e.g., what can be done with documents, or folders). Determining the answers to these questions requires both informed insight and the time-consuming and expensive process of having real users test the software for extended periods. We can summarize by saying that interactive software is, by its nature, rich in design know-how.

Such know-how is, inescapably, present on the face of the product, and immediately evident on inspection. The hard-won insights and innovations embodied in surface design are prominently displayed by the program in operation; they are also explained in online help, and further detailed in the manual. Insights not readily explained by the tool or the manual will often become evident to a sophisticated user of the program who runs the program over a variety of examples. Since program innovation often lies in finding a way to make program behavior easier for people to use, the behavior and its insights must be evident to the user. . . .

We stress the triviality of some acquisitions of behavioral equivalence because this triviality can be the source of market-destructive effects. The particular means a firm uses to acquire behavioral equivalence is not important from a market-preservation standpoint. The crucial concern is whether it is trivially easy and quick to copy a software innovation that was very expensive to develop. If the disproportion in cost of copying and cost of innovation is substantial enough, it can destroy the innovator's opportunity to recoup its expenses, and consequently, can destroy incentives to invest in software innovation. . . .

2.2.2 Why Patent Law Is Ill-Suited to Protecting Software Innovation . . .

Apart from subject-matter concerns, one reason that patents have limited application in the protection of behavior is that patents typically issue for particular methods of achieving results, rather than for the results themselves. It is quite possible to produce functionally indistinguishable program behaviors through use of more than one method. This means that holding a patent on one method of generating certain results could not prevent the use of another method, even if those results were the program's principal source of value. Hence, patents on methods would not protect behavior and therefore would not protect the primary entity of value in software. . . .

2.2.3 Why Copyright Law Is Ill-Suited to Protecting Software Innovations . . .

Copyright law is mismatched to software, in part, because it does not focus on the principal source of value in a program (its useful behavior). As explained above, program text and behavior are largely independent, so that protecting program texts does not prevent second comers from copying valuable program behavior. The ability to copy valuable behavior legally would sharply reduce incentives for innovation, and thus thwart the policy behind legal protection. As we explain below, the right way out of this bind is not to conceive of behavior as a nonliteral element of the program's text, as some courts have done, but instead to regard it as the entity that an appropriate legal regime should protect.

[The authors propose a relatively brief period (one to five years) of automatic anticloning protection for program behavior and industrial designs of programs in order to reward software development appropriately. Such a system would augment existing intellectual property protection. The authors also propose that a registration system be developed by which innovative components of programs, such as command hierarchies and internal interfaces, would be licensed in a manner similar to the management of copyrights by copyright collectives.]

COMMENTS AND QUESTIONS

1. Does Professor Menell's proposal give too little protection for computer software? Since the mid-1980s, lawyers for IBM have argued vehemently for affording broad protection under copyright for nonliteral elements of program code (structure, sequence, and organization) and compatibility elements. See Anthony L. Clapes, Patrick Lynch, and Mark R. Steinberg, Silicon Epics and Binary Bards: Determining the Proper Scope of Copyright Protection for Computer Programs, 34 UCLA L. Rev. 1493 (1987); Anthony L. Clapes, Software, Copyright, and Competition (1989); Anthony L. Clapes and Jennifer M. Daniels, Revenge of the Luddites: A Closer Look at Computer Associates v. Altai, 9 Computer Lawyer 11 (1992). Their basic argument is that

> Computer programs are works of authorship in which the range and variety of expression are broad and deep. These works of authorship exhibit all the attributes of literary works of a kind with which the general public and the copyright laws are already quite conversant. Although written in languages that are unfamiliar to most, they are not qualitiatively different from other literary works in ways that should affect the scope of legal protection available to their detailed design elements.

34 UCLA L. Rev. at 1583. They analogize computer programming to the writing of poetry and the composing of opera. Is this a more persuasive analogy than that between computer programming and engineering and other endeavors traditionally protected by patent law? Does it justify broad copyright protection for software? Is it an adequate basis for discouraging the study of sui generis regimes?

2. Does the Manifesto by Samuelson et al. give too much reward to computer programmers? In a series of comments on the Manifesto, a number of scholars have made the point that awarding exclusive property rights comes

at a cost, both to consumers who must pay a higher price and to future generations who cannot build as easily on past works. See Paul Goldstein, Comments on A Manifesto Concerning the Legal Protection of Computer Programs, 94 Colum. L. Rev. 2573, 2574-75 (1994); Dennis S. Karjala, Misappropriation as a Third Intellectual Property Paradigm, 94 Colum. L. Rev. 2594, 2597 (1994); Peter S. Menell, The Challenges of Reforming Intellectual Property Protection for Computer Software, 94 Colum. L. Rev. 2644, 2647 (1994).

What is the appropriate level of protection for computer programs? Is it consistent with efforts to reward every inventive, creative, or time-consuming aspect of program development? Or is it something less than that? Most economists would suggest that intellectual property rights ought to extend no further than necessary to encourage optimal investment in research and development. But how much protection is "just enough"?

3. How would Menell protect screen displays (if at all)? Mathematical algorithms embodied in computer software? Computer languages? Source code? Object code? Would each of these parts of a computer program require a different statute to protect it? Or should protection be subsumed within a broader "software" framework?

It is interesting to compare Menell's suggestion that a sui generis statute should *replace* copyright and patent with the Manifesto's suggestion that further protection should be *added to* that given by existing law. Both papers evaluate software from an economic standpoint; what accounts for their radically different conclusions?

4. Does Menell's suggestion to treat application programs differently from operating systems make sense? Are the differences between the two types of programs — something he discusses in detail in the balance of his paper — sufficient to warrant different statutory treatment? To the extent that there are differences, such as the standardization problem with operating systems, are they appropriately addressed in other ways? See Chapter 8, infra (discussing application of the antitrust laws to intellectual property).

A larger issue here relates to the appropriate level of generality of a software protection statute. On the one hand, more specific laws are easier to tailor to the particular needs of an industry and afford the legislature opportunities to clarify the law. On the other hand, it is hard to anticipate all possible circumstances in drafting legislation, and judicial flexibility may both benefit litigants in the long run and guard against technological advances that render the statute irrelevant.

5. In 1991 the European Community adopted the Council Directive on the Legal Protection of Computer Programs. Council Directive 91/250, 1991 O.J. (L 122) 42. Article 1.2 provides that the Directive protects the "expression" of any form of a computer program, while "[i]deas and principles which underlie any element of a computer program, including those which underlie its interfaces, are not protected." Article 5.3 entitles users of programs "to observe, study or test the functioning of the program in order

to determine the ideas and principles of the program, if he does so while performing any of the acts of loading, displaying, running, transmitting or storing the program. . . ." Article 6 (Decompilation) provides:

1. The authorization of the rightholder shall not be required where reproduction of the code and translation of its form . . . are indispensable to obtain information necessary to achieve the interoperability of an independently created computer program with other programs, provided that the following conditions are met:

(a) these acts are performed by the licensee or by another person having a right to use a copy of a program, or on their behalf by a person authorized to do so;

(b) the information necessary to achieve interoperability has not previously been available to the persons referred to in subparagraph (a); and

(c) these acts are confined to the parts of the original program which are necessary to achieve interoperability.

2. The provisions of paragraph 1 shall not permit the information obtained through its application:

(a) to be used for goals other than to achieve the interoperability of the independently created computer program;

(b) to be given to others, except when necessary for the interoperability of the independently created computer program; or

(c) to be used for the development, production or marketing of a computer program substantially similar in its expression or for any other act which infringes copyright. . . .

Is the EC Directive more or less restrictive than U.S. law in enabling a competitor to copy nonliteral elements of computer programs? to reverse engineer computer programs? to achieve operating system compatibility? user interface compatibility? See Leo Raskind, Protecting Computer Software in the European Economic Community: The Innovative New Direction, 18 Brooklyn J. Intl. L. 729, 745-46 (1992); Thomas C. Vinje, The Development of Interoperable Products Under the EC Software Directive, 8 Computer Lawyer 1 (Nov. 1991); Jerome Huet and Jane C. Ginsburg, Computer Programs in Europe: A Comparative Analysis of the EC Software Directive, 30 Colum. J. Transnational L. 327, 362-63 (1992).

6. For a detailed discussion of sui generis proposals, including other proposals for reform, see Mary Mills, Note, New Technology and the Limits of Copyright Law: An Argument for Finding Alternatives to Copyright Legislation in an Era of Rapid Technological Change, 65 Chicago-Kent L. Rev. 307 (1989); John Phillips, Note, Sui Generis Intellectual Property Protection for Computer Software, 60 Geo. Wash. L. Rev. 997 (1992); Leo Raskind, The Uncertain Case for Special Legislative Protection for Computer Software, 47 U. Pitt. L. Rev. 1131 (1986); Pamela Samuelson, Modifying Copyrighted Software: Adjusting Copyright Doctrine to Accommodate a

Technology, 28 Jurimetrics J. 179 (1986); Richard Stern, The Bundle of Rights Suited to New Technology, 47 U. Pitt. L. Rev. 1229, 1262-67 (1986); Pamela Samuelson, Creating a New Kind of Intellectual Property: Applying the Lessons of the Chip Law to Computer Programs, 70 Minn. L. Rev. 471 (1985).

7. Several authors questioned the timing of the call for new legislation in the Samuelson et al. Manifesto. These authors argue that copyright (and, more recently, patent) law has settled down into a workable form of protection over the past few years and that drafting an entirely new statute would disturb that settling process. Further, they point out that GATT-TRIPs have just adopted copyright and patent law as the dominant modes of legal protection for software throughout the world and that changing the rules of the game now would be a difficult and time-consuming process. See Jane C. Ginsburg, Four Reasons and a Paradox: The Manifest Superiority of Copyright over Sui Generis Protection of Computer Software, 94 Colum. L. Rev. 2559 (1994). Indeed, even Menell suggests the time for reform may have passed.

Peter S. Menell, The Challenges of
Reforming Intellectual Property Protection
for Computer Software
94 Colum. L. Rev. 2644, 2651-54 (1994)

. . .

The political economy of reforming intellectual property laws to accommodate new technologies creates a particularly perplexing dilemma. From a political economy standpoint, the opportunity for comprehensive reform is most propitious before interest groups form around a new technology. Unfortunately, policymakers usually do not have sufficient understanding of the path of such technology and the implications for an appropriate intellectual property regime during this nascent stage of development. Policymakers thus are left in the unenviable position of either creating a regime before they adequately understand the problem or waiting until the contours of the problem emerge, at which point economic interests have vested and reform, if it is possible at all, is severely constrained.

The history of legal protection for computer software illustrates this dilemma well. At the time CONTU was created, there was a real political opportunity for establishing a novel comprehensive legal regime to govern computer software. Neither patent nor copyright had yet been important in the software area. The industry had developed principally through trade secret protection. Hence the range of options was rather large. On the other hand, there was relatively little appreciation of the problems that lay ahead, in particular the possibility of impeding network externalities and the potential for misapplication of copyright protection to functionality. The outcome of the CONTU process was not surprisingly a rather naive set of recommendations, closely approximating the status quo. More recently, Congress passed the Semiconductor Chip Protection Act, a modest protection regime crafted significantly by the affected interests. Thus far, sui generis regimes for software more generally . . . have generated little enthusiasm in the legislature.

In view of the substantial vesting of economic interests that has already occurred around the existing legal regime for computer software, it is difficult to imagine Congress giving serious attention to a comprehensive reform effort without a crying need. After many years of confusion, the existing legal regime appears to be muddling through. In the view of most commentators, the recent *Altai* and *Sega* decisions have correctly resolved two of the three major problems that have plagued copyright protection for software. The *Lotus* case, which presents the scope of copyright protection for user interfaces, could potentially resolve the last major difficulty with the application of copyright law to computer software.[19] Patent protection for software may prove to be problematic, although the extent of problems to date has been modest. . . .

Any attempt at *sui generis* reform would face not only the private interests vested in federal copyright law, but also the inertia flowing from the hard-won battle to establish an international copyright regime for computer software. . . . [T]he United States government devoted substantial effort over the past decade to browbeating most of the developed nations into following its path. Neither the U.S. government nor the many entities desiring uniform protection for their products across national borders are interested in starting a new fight.

In this political climate it would be more efficacious to address the individual problems most amenable to reform than to attempt a comprehensive, sui generis scheme. The two most important targets would appear to be concerns about patent protection and impediments to the nascent field of multimedia software products.

Whether patent protection will prove to be a serious problem for the software industry is a matter of controversy. Some of the concerns that have been raised, such as inconsistency in the examination process and the lack of a comprehensive repository of prior art, can and are being addressed without legislative changes. Nonetheless, other aspects of the patent system, such as the long term of protection and the lack of compulsory licensing, pose serious potential problems. Congress may be politically unable to address these problems once powerful economic interests vest in strong patent portfolios. Hence it would be opportune for Congress to take up this matter in the very near future, especially in light of the race among software firms to build patent portfolios.

---

19. This article was published prior to the First Circuit's decision in Lotus v. Borland, supra section C.1. Most intellectual property professors, including the author of this article, supported the ultimate outcome of that case. See Brief Amicus Curiae of Dennis S. Karjala and Peter S. Menell, Lotus Dev. Corp. v. Borland Intl. Inc. (U.S. 1995) (No. 94-2003), reprinted at 10 High Tech. L.J. 177 (1995); Brief Amicus Curiae of Copyright Law Professors, Lotus Dev. Corp. v. Borland Intl. Inc. (U.S. 1995) (No. 94-2003) (brief of 35 copyright law professors). — EDS.

Similarly, Congress may have a window of opportunity for addressing the legal issues posed by the emerging multi-media field. The basic problem — developing an efficient and equitable means for obtaining rights to use copyrighted works — is not new. Yet multi-media technology offers powerful methods for combining vast amounts of material in useful ways. Under existing copyright law, developers must either negotiate express licenses for each work used or make such limited use as to come within the fair use defense. New institutions that facilitate the permissions process have started to develop. Nonetheless, this is a propitious time for the legislature to consider alternative regimes — such as limited compulsory licensing — to spur this technology, particularly in the educational sector.

COMMENTS AND QUESTIONS

1. Do you agree with Menell that the time for ambitious restructuring of intellectual property protection for computer software has passed? Are his proposals — to amend the patent law to reduce the term of protection for software and to introduce compulsory licensing of program elements featuring network externalities and to amend copyright to foster multi-media development — advisable? the best that we can do? Aren't these proposals creating at least modest sui generis regimes for software?

2. Are the intellectual property laws simply the outcome of self-dealing by those who stand most to benefit from strong protection? What countervailing forces discourage overly protectionist intellectual property laws? How can intellectual property laws be made more responsive to the broader public interests in such legislation?

In this respect, the intellectual property laws themselves may be unable to do anything about one form of de facto sui generis protection: unilateral protection via contract. Mass-market software is almost invariably sold in a box with what is called a "shrinkwrap license," a piece of paper inside the box which purports to set the terms of the transaction between the parties. Shrinkwrap license terms are drafted by software vendors, and purchasers are not aware of those terms until after the purchase is complete. For this reason, most courts to consider the issue have refused to enforce shrinkwrap licenses. See Step-Saver Data Sys. v. Wyse Technology, 939 F.2d 91 (3d Cir. 1991); Vault Corp. v. Quaid Software Ltd., 847 F.2d 255 (5th Cir. 1988); Arizona Retail Systems v. The Software Link, Inc., 831 F. Supp. 759 (D. Ariz. 1993); Foresight Resource Corp. v. Pfortmiller, 719 F. Supp. 1009 (D. Kan. 1989); Mark A. Lemley, Intellectual Property and Shrinkwrap Licenses, 68 S. Cal. L. Rev. 1239 (1995). However, the Seventh Circuit recently held that shrinkwrap licenses are enforceable even when they vary from federal intellectual property rights, unless their terms are unconscionable. ProCD v. Zeidenberg, 86 F.3d 1447 (7th Cir. 1996).

Because shrinkwrap licenses are cheap to write, and because under the reasoning of cases like *ProCD* they create a form of privity between software vendors and anyone who picks up the software, they are potentially an extremely powerful form of substitute intellectual property protection. Indeed, the most troubling thing about shrinkwrap licenses from an intellectual property perspective may be that they are *too* powerful, since they give software vendors the unilateral power to rewrite the rules of intellectual property to their own liking. (For example, virtually all shrinkwrap licenses contain terms forbidding the user from reverse engineering the program.)

3. Does the experience under the SCPA, which appears to have been obsolete by the time it was enacted, suggest that Congress is too slow an institution to respond to technological change? Would it be better for Congress to provide more flexibility in the general intellectual property laws — patent, copyright, and trademark — to enable agencies (the PTO and the Copyright Office) and courts to adjust these laws to address new technologies?

4. Richard Stern has proposed that the United States enact a New Technology Law in which Congress would delegate authority to an executive agency (perhaps the PTO within the Department of Commerce) to develop a technology-specific intellectual property system. See Richard Stern, The Bundle of Rights Suited to New Technology, 47 U. Pitt. L. Rev. 1229, 1262-67 (1986). Such agency would define the subject matter of protection, determine the term of protection (not to exceed 10 years), establish the requirements for protection (e.g., registration, examination, deposit), delineate the standard for protection (degree of technical advance), define the rights accorded protected works, and determine the remedies for infringement. Would such an institution better ensure that intellectual property protection appropriately protects new forms of technology? What drawbacks — legal, practical, and political — do you see to such an approach?

5. In view of the globalization of product markets and the growth of international intellectual property law, is the United States the appropriate jurisdiction for determining intellectual property law?

8

Intellectual Property and Competition Policy

A. PRINCIPLES OF ANTITRUST LAW

1. The Scope of Antitrust Law

Antitrust law protects competition and the competitive process by preventing certain types of conduct that threaten a free market. For example, antitrust law prohibits competitors from agreeing on the price they will charge. It prohibits certain "predatory" practices designed to exclude competitors from the market, and it places certain limits on the behavior of firms with market power.

Competition is good for a variety of reasons. Basic economics teaches that firms in competition will produce more and charge lower prices than monopolists. Monopolists not only take money away from consumers by raising prices, but they impose a "deadweight loss" on society by reducing their output below the level that consumers would be willing to purchase at a competitive price. Monopoly has other problems as well. It inherently reduces consumer choice, and monopolists will usually have fewer incentives to innovate than do competitive firms.

The first United States antitrust law was passed in 1890. The Sherman Act, as the statute is called, was a reaction to populist pressure on Congress to do something about the "trusts" that had come to dominate the American business landscape. The Sherman Act granted broad powers to government to break up trusts and other conspiracies in restraint of trade. The Sherman Act is still in force today; it provides in part:

> **Section 1 [15 U.S.C. §1].** Every contract, combination in the form of trust or otherwise, or conspiracy, in restraint of trade or commerce

1101

among the several States, or with a foreign nation, is declared to be illegal. . . .

      Section 2 [15 U.S.C. §2]. Every person who shall monopolize, or attempt to monopolize, or combine or conspire with any other person or persons, to monopolize any part of the trade or commerce among the several States, or with foreign nations, shall be deemed guilty of a felony. . . .

      By their terms, these laws are broad indeed. In some sense, *every* contract necessarily restrains trade, since it forecloses options that were once open. Under the literal terms of section 1, employment contracts and sales contracts might be considered antitrust violations.

      In practice, courts quickly gave the Sherman Act a more restrictive interpretation. The Supreme Court read section 1 as prohibiting only *unreasonable* restraints of trade. Much antitrust jurisprudence in the past century has attempted to delineate reasonable and unreasonable restraints of trade.

      The courts also limited the reach of section 2, to ensure that successful businesses would not be punished for their success. Because the antitrust laws provide for felony criminal punishments, private treble-damage actions and injunctions, and plaintiffs' attorney fees, the possibility of overdeterring legitimate business conduct is a real concern. To avoid this problem, courts have distinguished between merely possessing a monopoly and actively acquiring or maintaining a monopoly through anticompetitive conduct. Section 2 prohibits only the latter.

      Congress has never amended the core sections of the Sherman Act. In 1914, however, in response to what it perceived as lax judicial enforcement of the antitrust laws, Congress passed the Clayton Act. The Clayton Act contains a number of specific provisions that served to clarify the Sherman Act. For example, section 3 of the Clayton Act enumerates and prohibits certain types of agreements that are also included within the general scope of Sherman Act section 1.

      The Clayton Act was amended in 1950 to prohibit certain types of mergers. Section 7 of the Clayton Act prohibits mergers and acquisitions (of a corporation, stock, or assets) where "the effect of such acquisition may be substantially to lessen competition or to tend to create a monopoly" in "any line of commerce . . . in any section of the country."

      Thus the antitrust laws prevent businesses from acting anticompetitively in three principal ways.

      Monopolization.   It is not illegal to have a monopoly. However, monopolists and firms gaining market power need to be very careful about their behavior. A monopolist violates section 2 if it has "market power" (defined as the power to raise prices or exclude competition in a relevant market) and engages in anticompetitive conduct designed to acquire, maintain or extend that power. Over time, courts have identified several anticompetitive practices (and the circumstances in which they are actionable); we will discuss several in the balance of this chapter. A company may also be guilty of "attempted monopolization" if it intends to monopolize a market, engages in anticompetitive conduct, and has a "dangerous probability of successful monopolization." See, e.g., Spectrum Sports v. McQuillen, 506 U.S. 447 (1993).

To find that a defendant has "monopolized" a market, the court must first decide what the relevant market is. Specifically, the court must identify a product or set of products and a geographic region within which most consumers make their purchases. Controlling such a market will allow a monopolist to raise prices without losing customers to competitors from outside the market. The definition of a relevant market and the analysis of power in that market are both extremely complex questions. A good deal of legal and economic work has gone into the attempt to define exactly what market power is. For a more detailed discussion of this issue, see United States Department of Justice and Federal Trade Commission, Revised Merger Guidelines §1 (April 2, 1992).

Agreements. Courts have identified two basic types of agreements that may be in restraint of trade — agreements among competitors (called "horizontal restraints") and agreements between buyers and sellers (called "vertical restraints").[1] Vertical restraints are generally less threatening to competition than horizontal restraints. With the exception of vertical price fixing, they are generally judged under the "rule of reason." Under the rule of reason, courts balance the anticompetitive harms of a restraint against its procompetitive benefits. Only those restraints that produce harms significantly in excess of benefits to competition are deemed unreasonable.

Horizontal restraints are more troubling because they may allow the participants to create a cartel, which can then behave anticompetitively, much as a monopolist would. At first, most agreements between competitors were deemed illegal "per se," without any necessity for a weighing of harms and benefits to competition. Today the Supreme Court has retreated from that position, recognizing that certain agreements among competitors may be efficient and procompetitive. Most horizontal restraints are now judged under the rule of reason. Only certain forms of "naked" agreements to fix prices or divide territories remain illegal per se.

Mergers. Section 7 of the Clayton Act is intended to catch monopolies "in their incipiency." Because of this, the standard for proving the existence of market power is significantly lower in the case of a merger than in a section 2 monopolization claim. If a merger would substantially lessen competition in a relevant market, the courts may prohibit it. As with most other areas of antitrust law, however, courts are usually willing to consider a merger's potential benefits to competition in deciding whether it should be prevented.

---

1. Actually, the term "vertical restraints" refers to a whole class of transactions between companies in a vertical relationship in the chain of distribution, including dealers, franchisors, distributors, resellers, etc.

Most merger enforcement takes place at the Antitrust Division of the United States Department of Justice and at the Federal Trade Commission. The Hart-Scott-Rodino Act of 1976 requires companies to file a statement with the division and the FTC and to seek approval for mergers over a certain size. If the division or the FTC feels that a merger will restrain competition, it can challenge the merger in court before the merger takes place. 15 U.S.C. §7A.

Note on Antitrust Theory

The antitrust laws clearly reflect some congressional judgment that "big is bad." But *why* it is bad has been the subject of considerable debate during the past hundred years. At least three different rationales have been advanced to explain the antitrust laws. The first view may be called the "populist" view of antitrust. Populists take the position that big is *intrinsically* bad. It may be bad for a variety of reasons: because it concentrates wealth, becauses it reduces product diversity, or because it concentrates political power, for example. But regardless of the reason, populists would use the antitrust laws as a weapon against all monopolies, and against oligopolies as well. Populists can find substantial support for their position in the legislative history of the Sherman Act and in the political climate in the late nineteenth century. For an elaboration of the populist view, see Walter Adams, The Bigness Complex (1986); Victor H. Kramer, The Supreme Court and Tying Arrangements: Antitrust as History, 69 Minn. L. Rev. 1013 (1985). However, the populist view of antitrust is inconsistent with the more recent trend towards permitting mergers and agreements that are justified on efficiency grounds.

A second group sees antitrust as designed to protect small businesses from being driven out of the market. This view is related to the populist view, since both are likely to vigorously oppose monopolies and growing concentration in an industry. But they differ in the instrumental value served by preventing monopoly. Advocates of this view tend to view antitrust as an "unfair competition" statute, and to apply the antitrust laws to a number of practices that are bad for competitors but good for consumers. For example, small business advocates are likely to be skeptical of price cuts by large national companies, which they fear will "squeeze out" local independents. The legislative history of the Clayton Act and subsequent antitrust laws lends some support to this view, but it too is inconsistent with the recent trend of antitrust decisions. Indeed, a fundamental maxim of modern antitrust law is that it "protects competition, not competitors." Brown Shoe Co. v. United States, 370 U.S. 294 (1962).

A third theory of the antitrust laws — and the theory which clearly dominates modern antitrust analysis — is the "economic" or "social welfare" model. This approach sees the purpose of the antitrust laws as promoting social welfare by ensuring that markets work freely and without interference. Antitrust law is aimed at practices that corrupt the market or usurp its function. On this view, antitrust is particularly concerned with preventing horizontal restraints of trade. While there is relatively little in the history of the antitrust laws that suggests Congress was primarily concerned with economics,[2] this approach has gained great currency with courts and scholars in the last 25 years. See, e.g., Robert H. Bork, The Antitrust Paradox: A Policy at War with Itself (1978); Richard Posner, Antitrust Law: An Economic Perspective (1976). Most antitrust experts now use economic effects as the basis for their analysis, although they continue to disagree vigorously over *how* those effects are to be analyzed.

2. Intellectual Property and Antitrust Law

Intellectual property law grants certain exclusive rights to the creators, inventors, or discoverers of certain intangible but valuable assets. In previous chapters, we have discussed the nature of these rights at length. There is substantial disagreement as to whether intellectual property rights are "property" in the ordinary sense of the term. But whatever they are, intellectual property rights give their owner the right to exclude others from using or copying his invention or creation.

Traditionally, the conventional wisdom was that the antitrust laws and the intellectual property laws are in conflict. See, e.g., Stephen Calkins, Patent Law: The Impact of the 1988 Patent Misuse Reform Act and Noerr-Pennington Doctrine on Misuse Defenses and Antitrust Counterclaims, 38 Drake L. Rev. 175, 176 n.1 (1989) (noting this conflict). Baldly stated, the conflict arises because the intellectual property laws grant "monopolies" to inventors, while the goal of the antitrust laws is to prevent or restrict monopoly.

However, scholars are increasingly taking the position that the two laws are not in conflict at all. Rather, they are complementary efforts to promote an efficient marketplace and long-run, dynamic competition through innovation. Paul Goldstein was an early proponent of the idea that the intellectual property laws were designed to promote competition. See Paul Goldstein, The Competitive Mandate: From Sears to Lear, 59 Cal. L. Rev. 971 (1971). So too was Ward Bowman, Jr.

≣ *Ward Bowman, Jr., Patent and Antitrust Law:*
≣ *A Legal and Economic Appraisal*
≣ U. Chicago Press (1973)

. . .

Antitrust law and patent law are frequently viewed as standing in diametric opposition. How can there be compatibility between anti-trust law, which promotes competition, and patent law, which promotes monopoly? In

---

2. But cf. F. M. Scherer, Efficiency, Fairness, and the Early Contributions of Economists to the Antitrust Debate, 29 Washburn L.J. 243, 250-51 (1990) (arguing that the views of mainstream economists in 1890 were consistent with the goals of the Sherman Act).

terms of the economic goals sought, the supposed opposition between these laws is lacking. Both antitrust law and patent law have a common central economic goal: *to maximize wealth by producing what consumers want at the lowest cost.* In serving this common goal, reconciliation between patent and antitrust law involves serious problems of assessing effects, but not conflicting purposes. Antitrust law does not demand competition under all circumstances. Quite properly, it permits monopoly when monopoly makes for greater output than would the alternative of an artificially fragmented (inefficient) industry. The patent monopoly fits directly into this scheme insofar as its central aim is achieved. It is designed to provide something which consumers value and which they could not have at all or have as abundantly were no patent protection afforded.

Antitrust Law

Under the Sherman Act agreements in restraint of trade (section 1) and monopolization or an attempt to monopolize (section 2) are condemned as illegal. The economic rationale for this condemnation is that monopoly makes it possible to restrict output and raise prices so that consumers pay more for and get less of the things they want most. The output restriction made possible by monopoly idles, or transfers to less urgent uses, those resources, which but for the monopolistic restriction would be more efficiently employed. Conversely, a competitive market process allocates scarce resources to those uses the community values most highly. Insofar as the antitrust laws are successful, they promote a market-oriented, profit-incentive process unimpeded by artificial roadblocks to efficiency. Such is the rationale of market competition and the antitrust laws that support it.

Patent Law

Patent law, thought by some to be an exception to a general rule in favor of competition, shares with antitrust law its central purpose — efficiently providing those things consumers value. But the means are different. Patent law pursues this goal by encouraging the invention of new and better products. Invention, like other forms of productive activity, is not costless. Those who undertake it, therefore, must be rewarded. And so elusive a commodity as an idea which qualifies as invention is peculiarly susceptible to being freely appropriated by others. A patent is a legal device to insure that there can be a property right in certain ideas. Thus the temporary right of a patentee to exclude others is a means of preventing "free riding" so that the employment of useful private resources may be remunerated. Without a patent system, prevention of "free riding" would be severely limited. Ability to keep secrets and to enforce private "know-how" contracts would, without patent law, provide inventors very limited protection from rapid and widespread copying by others. Central to the economic justification of a patent system is the presumption that without the patent right, too few resources would be devoted to invention.

The "exclusive right to make, use and vend the invention or discovery," which Congress has long granted patentees, is thus a legal monopoly exempt from the more general proscription of trade restraints and monopolization under early common law and more recent antitrust statutes.

Goal Evaluation

The goal of both antitrust law and patent law is to maximize allocative efficiency (making what consumers want) and productive efficiency (making these goods with the fewest scarce resources). In achieving this goal under either antitrust or patent law the detriment to be avoided is output restriction. This may arise from monopolization which diverts production from more urgent to less urgent use or from legal rules requiring inefficient methods of production. The evil then may be viewed as net output restriction after efficiency increases are accounted for. Both antitrust and patent law seek output expansion not output restriction. Competition deserves support insofar as it brings about this result. And so it is with patents. The temporary monopoly afforded by a patent, once a particular invention has come into being, will have all the output-restrictive disabilities of any monopoly. The argument for patents is that without this temporary monopoly there would be insufficient profit incentives to produce the invention, and that because an invention is profitable only if consumers are willing to pay what the patentee charges, the consumers are therefore better off than they would be without the invention, even if they are charged "monopoly" prices. If this is so, a trade-off (some monopoly restraint for greater output in the long run) is in the interest of socially desirable resource allocation. An appraisal of alleged conflicts between antitrust law and patent law depends upon understanding the role of profits in providing the incentive for undertaking efficient production of those things consumers value. . . .

As a public policy tool perfect competition does provide useful insights into the equilibrating forces of a competitive process unencumbered by either external or internal impediments to the flow of resources responding to price and profit signals. Monopoly and agreements in restraint of trade are such impediments, and, as has been stressed, they are condemned for this reason. This does not mean, however, and the antitrust laws have never interpreted to mean, that every departure from the perfect competitive model becomes illegal. Rather, these laws, when consistently and rationally applied, view competition as a process designed to work for the benefit of consumers. They take the world as it is and evaluate available alternatives. And the allocation of resources cannot be fully evaluated unless productive efficiencies (cost savings or better products) are assessed along with restraints. Stressing perfect competition as an ideal involves the danger of underplaying the importance

of productive efficiencies: savings arising from large size and efficiencies arising from innovation in manufacturing and from company organization and
management, as well as from dissemination of knowledge through sales effort
and advertising. Employee relations and myriad other activities also are
sources of productive efficiency. None of these commonplace examples would
exist if competition were perfect. These examples, moreover, are not output
restricting. Imperfect competition is not monopoly. Dealing with imperfections is what competition as a process rather than a static model is all
about. . . .

Efficiency and Patents

If the temporary monopoly afforded to inventors is to be justified as on
balance desirable, its existence and the uses to which it is put must face up
to the kind of restraint/efficiency assessment applied under antitrust law.
Analytically, the combination of two competing patents raises the same questions as the merger of two competing companies; both combine competing
assets. That the productive efficiencies predictable from a patent pool may
seem *de minimis* does not change the nature of the analysis. It may, however,
seem more analogous to the antitrust cartel rule than to the rule on mergers.
And, of course, jointly fixing the price to be charged for competing patents
is an even more obvious analogy.

The problem of assessing permissible output restriction under patent law
is, in one important sense, different from that under antitrust law. The legal
propriety of a basic patent monopoly has to be recognized. Consequently,
evaluating whether certain patent licensing practices should be sanctioned
will involve the proper scope of the legal monopoly. Is more being monopolized than what the patent grants, or is the practice merely maximizing the
reward attributable to the competitive advantage afforded by a patent? . . .
[M]any cases of alleged patent misuse — many, indeed, which are said to
violate antitrust law — involve vertical contract relationships between patentee and licensee in which no such extension of scope is explicable. Many
such contracts between patentees and licensees, moreover, lead to productive
efficiencies unattainable if the practices are prohibited. Given a legal patent
monopoly, it is important not only to appraise whether the monopoly has
been broadened — scope extension — but also to insure that prohibiting alleged misuses does not leave the holder of the patent a permissible alternative
which, although less profitable, is more restrictive than the alleged misuse.
The "misuse" may actually be the more efficient use of the patent from the
social as well as the private point of view. It pays monopolists as well as
competitors to be efficient, as has been stressed. . . .

Growth makes an economy richer in successive years. Achieving continuously more efficient resource allocation is not only compatible with growth
but is an essential aspect of it. But economic growth is more often associated
with a decision to produce more and newer products for tomorrow by fore-

going consumption today. This may be done by voluntary savings and investment or by a centralized decision about what should be growing. In neither case, however, is there any compelling reason to believe that tomorrow is necessarily more important than today. And this is as true for that unique commodity information, the subject matter of patent law, as it is for other unpatentable commodities. Thus, even though the patent legislation looks to the promotion of the progress of science and useful arts, and the antitrust law prohibits artificial barriers to the entry of new products or new production methods, neither need favor or disfavor investment over current consumption, and neither need favor particular forms of investment. The allocative function which antitrust law should support and patent law should supplement is market motivation, making it profitable to produce the wide variety of things that people choose. Free exercise of choice is not limited to choices among goods, or choices between goods and services, or choices between either of these and leisure. Timing — the choice between having now and having later — is also involved.

The patent law is often described as a special inducement to growth. But, as has been indicated, the patent law, like antitrust law, does not foster growth for growth's sake. Rather, it is designed to make possible those new products and new production methods which customers deem worth paying for. Promoting science and the useful arts . . . is not to be viewed as a reasonably exact facsimile of a crop-support program for the over-production of invention. If the patent system works as it is designed to work, growth can be expected to be fostered. Scarce resources are diverted from current consumption to better products for tomorrow. Growth under a patent system is thus not a separate goal competing with consumer choice. It is the result of it. . . .

COMMENTS AND QUESTIONS

1. In the first half of this century, Joseph Schumpeter advanced the classic incentive theory that underlies much of intellectual property law. In his view, competition will lead to a focus on short-run marginal cost, to the exclusion of (long-run efficient) capital investments in research and development. In the absence of intellectual property protection, the result of competition is insufficient innovation. If that is true (and we have certainly assumed that it is for much of this book), what does it say about the wisdom of applying the antitrust laws to intellectual property? Is it a good idea to enforce competition in high-technology industries? Or should the goal of antitrust law be modified somehow in this context?

2. Recent advances in economic thought have caused many economists to reconsider the arguments Bowman makes in the last paragraph of the excerpt. In particular, a significant paper by Richard Levin et al. forcefully demonstrates that different industries depend on different types of appropriability mechanisms. Richard Levin, Alvin Klevorick, Richard Nelson & Sidney

Winter, Appropriating the Returns from Industrial Research and Development, 3 Brookings Papers on Economic Activity 783 (1987). The fact that appropriability varies from industry to industry significantly complicates the simple economic model of patents that Bowman sets forth. Further, work by David Teece and others suggests that firms within an industry will adapt strategically to the appropriability environment of that industry by taking advantage of what Teece calls "co-specific assets." David J. Teece, Profiting from Technological Innovation: Implications for Integration, Collaboration, Licensing and Public Policy, 15 Res. Pol. 285 (1986). The upshot of this recent work is that the economic model of patent incentives is substantially more complicated than it appeared to be a few years ago.

3. Consider carefully what Bowman is arguing. He asserts that intellectual property and antitrust are not ends in themselves but ways toward a larger goal of wealth maximization. As a matter of policy, he is probably correct.

Bowman's conclusion that the patent laws and the antitrust laws do not have conflicting purposes can be misleading, however. Bowman is best understood as saying that both patent law and antitrust law are tools to be used in promoting wealth maximization. They are designed to be used together to achieve a certain result. However, they strive toward that result in ways that are often in tension. Antitrust law seeks to maximize wealth by preventing monopoly, while patent law may in some cases encourage monopoly.

It is too facile to say that the patent laws give a monopoly to the patentee. Most patents cover a product or products which are only a small subset of a *product market*. It is therefore possible to have a perfectly competitive industry composed of patent holders, if each holds a patent on one of several possible products in a market. Only where an innovation creates an entirely new market, or represents a quantum advance in an old one, is the patent likely to confer an economic monopoly directly.

There is room, therefore, for the patent and antitrust laws to coexist in the service of long-run, dynamic efficiencies. In this scheme, patents constitute an exception to the reach of the antitrust laws, and antitrust constrains what a patentee can do with its patent. Efficient wealth maximization requires that a line be drawn between conduct which is permissible and that which is impermissible. Drawing that line is the subject of this chapter.

B. MONOPOLIZATION

1. Defining the Market

The first step in any antitrust inquiry is to define the relevant product and geographic markets. Monopolization does not occur in a vacuum. To be liable, an antitrust defendant must have monopolized *something, somewhere*.

Market definition and market power analysis are well-developed in antitrust law; we will not repeat the general issues here.

There are, however, some market definition issues that are specific to intellectual property law. The first of these is the long-standing presumption that intellectual property rights confer market power. The Supreme Court has reasoned that, since a patent confers the right to exclude competitors *from practicing the patented invention*, it must confer monopoly power over that invention. Thus the Court has on several occasions presumed market power from the existence of patents or copyrights. E.g., Jefferson Parish Hosp. Dist. v. Hyde, 466 U.S. 2 (1984); United States v. Loew's, Inc., 371 U.S. 38 (1962); see also Digidyne v. Data General Corp., 734 F.2d 1336 (9th Cir. 1984), *cert. denied*, 473 U.S. 908 (1985).

This presumption that intellectual property rights confer market power has little basis in fact. Patents grant the right to exclude in a tightly defined technological domain. In most cases, this does not translate into what an economist would call a "monopoly," since the technological domain is rarely coextensive with an economic product market. For example, consider a typical mechanical patent, say on a new type of muffler for lawn mowers. This typical patent does not confer the right to exclude others from making lawn mowers, of course; nor even from making most types of mufflers for lawn mowers. Only the new muffler is covered. Whether one considers the market for lawn mowers, for lawn mower components, or even lawn mower mufflers, this patent does not confer a monopoly. *Only* in the very limited market defined by this new type of muffler and its close technological substitutes does the patent holder have exclusionary power, i.e., a "monopoly."[3] Most patents are like that. Occasionally, however, a patent is granted on a "pioneering" or "basic" invention — e.g., the lightbulb, or the laser, or the communication satellite, or recombinant DNA techniques. In these rare instances, the patent may confer exclusionary power over an entire market and hence qualify as a true economic monopoly.

Commentators have been virtually unanimous in their criticism of the market power presumption. See, e.g., Philip Areeda & Louis Kaplow, Antitrust Analysis 441 (1987); Herbert Hovenkamp, Economics and Federal Antitrust Law §8.3, at 219 (1985); Mark A. Lemley, Comment, The Economic Irrationality of the Patent Misuse Doctrine, 78 Cal. L. Rev. 1599, 1626-28 (1990); William Montgomery, Note, The Presumption of Economic Power for Patented and Copyrighted Products in Tying Arrangements, 85 Colum.

---

3. It is sometimes asserted that a patent is merely a property right, as opposed to a monopoly, and that the exclusionary power of a patent should not be deemed any more forceful than property rights in one's house. This is misleading, since unlike a house, or any piece of tangible (real or personal) property, a patent grants its owner the right to exclude others from doing certain things with their own physical property. For example, my ownership of a house does not ordinarily limit what you may do with your own house, but my "ownership" of a patented invention will prevent you from making or using the same invention, even if you developed it independently of me.

L. Rev. 1140, 1156 (1985). Perhaps in response to this criticism, some courts have backed away from the market power presumption in recent years. See Abbott Labs. v. Brennan, 952 F.2d 1346, 1354-55 (Fed. Cir. 1991), *cert. denied,* 505 U.S. 1205 (1992); A. I. Root Co. v. Computer/Dynamics, Inc., 806 F.2d 673 (6th Cir. 1986); Mitsubishi Elec. Corp. v. IMS Tech. Inc., 44 U.S.P.Q.2d 1904 (N.D. Ill. 1997); 3 P.M., Inc. v. Basic Four Corp., 591 F. Supp. 1350 (E.D. Mich. 1984). Moreover, the new United States Antitrust Guidelines for the Licensing and Acquisition of Intellectual Property ("Intellectual Property Guidelines"), issued jointly by the Federal Trade Commission and the Antitrust Division of the Department of Justice, specifically reject the market power presumption:

> Market power is the ability profitably to maintain prices above, or output below, competitive levels for a significant period of time. The Agencies will not presume that a patent, copyright or trade secret necessarily confers market power upon its owner. Although the intellectual property right confers the power to exclude with respect to the *specific* product, process, or work in question, there will often be sufficient actual or potential close substitutes for such product, process or work to prevent the exercise of market power.

Intellectual Property Guidelines §2.2 (1995). Finally, legislation is periodically introduced in Congress to abolish the market power presumption. See, e.g., H.R. 2674, 104th Cong., 1st Sess. (1995).

The economic literature on the market power issue is complex. See, for example, David J. Teece, Raymond Hartman & Will Mitchell, Assessing Market Power in Regimes of Rapid Technological Change, 2 Indus. & Corp. Change 3, 7 (1993) ("The more innovative the new product or process, the greater the conventional market power will appear, because price changes will have little or no influence on demand for a truly innovative product."); id., at 29 (proposing to measure market power by performance attributes, not price changes). See also Kenneth Burchfiel, Patent Misuse and Antitrust Reform: "Blessed Be the Tie," 4 Harv J.L. & Tech. 1 (1991) (suggesting simpler tests for determining market power, including: (1) whether a royalty was paid by licensees; (2) whether accused infringers chose to keep infringing during litigation; and (3) the profit derived from the patented product); Jesse Markham, Concentration: A Stimulus or Retardant to Innovation?, in Industrial Concentration: The New Learning 247 (Harvey Goldschmid, H. Michael Mann & J. Fred Watson eds. 1974) (summarizing empirical tests of the Schumpeterian hypothesis).

The Antitrust Division takes the position that markets may need to be defined differently in the intellectual property context. In addition to evaluating the effect of intellectual property rights on markets for goods, the division has defined two additional types of markets:

3.2.2 Technology Markets

Technology markets consist of intellectual property that is licensed (the "licensed technology") and its close substitutes — that is, the technologies or goods that are close enough substitutes significantly to constrain the exercise of market power with respect to the intellectual property that is licensed.[18] When rights to intellectual property are marketed separately from the products in which they are used, the Agencies may rely on technology markets to analyze the competitive effects of a licensing arrangement.

To identify a technology's close substitutes and thus to delineate the relevant technology market, the Agencies will, if the data permit, identify the smallest group of technologies and goods over which a hypothetical monopolist of those technologies and goods likely would exercise market power — for example, by imposing a small but significant and nontransitory price increase. The Agencies recognize that technology often is licensed in ways that are not readily quantifiable in monetary terms. In such circumstances, the Agencies will delineate the relevant market by identifying other technologies and goods which buyers would substitute at a cost comparable to that of using the licensed technology.

In assessing the competitive significance of current and likely potential participants in a technology market, the Agencies will take into account all relevant evidence. When market share data are available and accurately reflect the competitive significance of market participants, the Agencies will include market share data in this assessment. The Agencies also will seek evidence of buyers' and market participants' assessments of the competitive significance of technology market participants. . . . When market share data or other indicia of market power are not available, and it appears that competing technologies are comparably efficient, the Agencies will assign each technology the same market share. For new technologies, the Agencies generally will use the best available information to estimate market acceptance over a two-year period, beginning with commercial introduction.

3.2.3 Research and Development: Innovation Markets

. . . A licensing arrangement may have competitive effects on innovation that cannot be adequately addressed through the analysis of goods or technology markets. For example, the arrangement may affect the development of goods that do not yet exist. Alternatively, the arrangement may affect the development of new or improved goods or processes in geographic

---

18. For example, the owner of a process for producing a particular good may be constrained in its conduct with respect to that process not only by other processes for making that good, but also by other goods that compete with the downstream good and by the processes used to produce those other goods.

markets where there is not actual or likely potential competition in the relevant goods.

An innovation market consists of the research and development directed to particular new or improved goods or processes, and the close substitutes for that research and development. The close substitutes are research and development efforts, technologies, and goods that significantly constrain the exercise of market power with respect to the relevant research and development, for example by limiting the ability and incentive of a hypothetical monopolist to retard the pace of research and development. The Agencies will delineate an innovation market only when the capabilities to engage in the relevant research and development can be associated with specialized assets or characteristics of specific firms.

In assessing the competitive significance of current and likely potential participants in an innovation market, the Agencies will take into account all relevant evidence. When market share data are available and accurately reflect the competitive significance of market participants, the Agencies will include market share data in this assessment. The Agencies also will seek evidence of buyers' and market participants' assessments of the competitive significance of innovation market participants. . . . The Agencies may base the market shares of participants in an innovation market on their shares of identifiable assets or characteristics upon which innovation depends, on shares of research and development expenditures, or on shares of a related product. When entities have comparable capabilities and incentives to pursue research and development that is a close substitute for the research and development activities of the parties to a licensing arrangement, the Agencies may assign equal market shares to such entities.

Intellectual Property Guidelines §§3.2.2, 3.2.3 (citations omitted).

COMMENTS AND QUESTIONS

1. Does defining separate technology and innovation markets make sense? Another way of approaching this question is to ask whether these market definitions capture any anticompetitive conduct that could not be challenged under a traditional product market definition. One possible answer is that alleging a market for innovation allows the Antitrust Division to challenge mergers or anticompetitive conduct earlier than it could in product markets, since it need not wait until an actual product market is affected. Whether this is good or bad depends on your confidence in the ability of courts to define innovation markets and assess anticompetitive effects accurately.

But it is also possible that technology and innovation markets allow the antitrust laws to reach conduct that could not be attacked under traditional product market definition. For example, a dominant company in a product market may be able to merge with or drive out a competitor whose current market share is insignificant, but who has the best technology and is therefore well placed to challenge the dominant firm. If nothing else, technology markets may make market power analysis more forward-looking, by requiring the parties to focus on likely future market power rather than just current market shares. At the same time, technology and innovation markets may be used by the government to attack mergers that do not present product market overlaps at all.

2. How can we define a market for "research and development"? The Intellectual Property Guidelines suggest several possible ways of determining market shares, including arbitrarily assigning each participant an equal share. Is this reasonable? Is R&D expenditure a fair measure? One problem with the guidelines' assumption that innovation monopolists can profitably retard or restrict innovation is that it depends on a fairly static model of monopoly. Monopolists who cut back on research and development may reap the rewards in the short term but are likely to be left behind as new companies "enter the market" for research and development. Can an argument be made the other way — that the real danger is that a dominant firm will extend its product monopoly by spending excessively on research and development?

PROBLEM[4]

Problem 8-1. Two companies that specialize in advanced metallurgy agree to cross-license future patents relating to the development of a new component for aircraft jet turbines. Innovation in the development of the component requires the capability to work with very high tensile strength materials for jet engines. Aspects of the licensing arrangement raise the possibility that competition in research and development of this and related components will be lessened. The Antitrust Division is considering whether to define an innovation market in which to evaluate the competitive effects of the arrangement. What factors should it take into account in making this decision?

---

2. Intellectual Property and Anticompetitive Conduct

It is not enough to define a market and determine that an intellectual property owner is a monopolist. The reason is that monopoly alone is not illegal. Rather, it must be combined with some form of anticompetitive conduct. Assuming that the intellectual property owner has market power in a particular case, what conduct will suffice? It is in this context that the "con-

---

4. This problem is presented as Example 3 in the Intellectual Property Guidelines.

flict" between intellectual property and the antitrust laws arises most starkly. Certainly, antitrust law must make some accommodation for monopolies that result from intellectual property, if those intellectual property rights are to have any meaning. In particular, it is evident that merely possessing and enforcing a patent, copyright, trademark or trade secret against competitors ought not to violate the antitrust laws:

> Although the conflict between the patent and the antitrust laws has long been thought troublesome, it is in fact even more deep-seated than is generally perceived. Consider a patentee that intends to employ a particular restriction, practice, or strategy in exploiting its patent. Limiting the analysis to the antitrust issues, which is the intended scope of this Article, one might initially conclude that the practice should be held permissible only if it does not violate the antitrust laws.
>
> Not long after the passage of the Sherman Act, however, courts realized that this approach is too facile. A practice is typically deemed to violate the antitrust laws because it is anticompetitive. But the very purpose of a patent grant is to reward the patentee by limiting competition, in full recognition that monopolistic evils are the price society will pay. Generalizing from this principle, one could reverse the initial conclusion, arguing that any action by a patentee in violation of the antitrust laws is privileged under the patent statute.
>
> Courts subsequently recognized that this conclusion was also too simplistic, because the patent statute was plainly not intended to bestow upon each patentee carte blanche in all its endeavors. For example, a patentee who negotiates a favorable royalty by holding a prospective licensee at gunpoint clearly will not be relieved from the proscriptions of either criminal or contract law. The question is whether one should view antitrust law any differently. At a minimum, it seems clear that a firm having one otherwise insignificant patent may not freely engage in price-fixing, mergers, predatory pricing, or anything else it wishes solely on that account. The statutory limitation of the patent grant to seventeen years illustrates the position, now generally accepted by commentators, that the reward for inventive activity should not be unbounded. This realization, however, does not indicate whether any or all practices in violation of the antitrust laws are out of bounds.

Louis Kaplow, The Patent-Antitrust Intersection: A Reappraisal, 97 Harv. L. Rev. 1815 (1984).

If we start from the proposition that it cannot be illegal to possess a patent (or other intellectual property right), what role remains for antitrust law? Two types of factual situations may still present antitrust problems. First, the patentee may have acquired his patent illegally (for example, through fraud). Second, a patentee may use his patent to acquire or maintain power beyond that granted by the intellectual property right itself. In both situations, antitrust courts are properly concerned with the effects that anticompetitive use of the patent may have. See generally Beal Corp. Liquidating Trust v. Valleylab, Inc., 927 F. Supp. 1350 (D. Colo. 1996) (patent law does not preclude antitrust action regarding patent license, where effect of license is to expand the scope of the patent).

a. Enforcing Intellectual Property Rights

Walker Process Equipment, Inc. v. Food Machinery & Chemical Corp.
Supreme Court of the United States
382 U.S. 172 (1965)

Mr. J. CLARK delivered the opinion of the Court.

The question before us is whether the maintenance and enforcement of a patent obtained by fraud on the Patent Office may be the basis of an action under §2 of the Sherman Act. . . . [Food Machinery sued Walker for patent infringement, and Walker counterclaimed for violations of the antitrust laws.] Walker then amended its counterclaim to charge that Food Machinery had "illegally monopolized interstate and foreign commerce by fraudulently and in bad faith obtaining and maintaining . . . its patent . . . well knowing that it had no basis for . . . a patent." It alleged fraud on the basis that Food Machinery had sworn before the Patent Office that it neither knew nor believed that its invention had been in public use in the United States for more than one year prior to filing its patent application when, in fact, Food Machinery was a party to prior use within such time. [The District Court dismissed Walker's counterclaim, and the Court of Appeals affirmed.]

. . . The gist of Walker's claim is that since Food Machinery obtained its patent by fraud it cannot enjoy the limited exception to the prohibitions of §2 of the Sherman Act, but must answer under that section and §4 of the Clayton Act in treble damages to those injured by any monopolistic action taken under the fraudulent patent claim. Nor can the interest in protecting patentees from "innumerable vexatious suits" be used to frustrate the assertion of rights conferred by the antitrust laws. It must be remembered that we deal only with a special class of patents, i.e. those procured by intentional fraud.

Under the decisions of this Court a person sued for infringement may challenge the validity of the patent on various grounds, including fraudulent procurement. In fact, one need not await the filing of a threatened suit by the patentee; the validity of the patent may be tested under the Declaratory Judgment Act. . . . At the same time, we have recognized that an injured party may attack the misuse of patent rights. To permit recovery of treble damages for the fraudulent procurement of the patent coupled with violations of §2 accords with these long-recognized procedures. It would also promote the purposes so well expressed in Precision Instrument [Mfg. Co. v. Automotive Maintenance Machinery, 324 U.S. 806 (1945)] at 816:

> A patent by its very nature is affected with a public interest. . . . [It] is an exception to the general rule against monopolies and to the right to access to a free and open market. The far-reaching social and economic consequences of a patent, therefore, give the public a paramount interest in seeing that patent

monopolies spring from backgrounds free from fraud or other inequitable conduct and that such monopolies are kept within their legitimate scope.

III

Walker's counterclaim alleged that Food Machinery obtained the patent by knowingly and willfully misrepresenting facts to the Patent Office. Proof of this assertion would be sufficient to strip Food Machinery of its exemption from the antitrust laws. By the same token, Food Machinery's good faith would furnish a complete defense. This includes an honest mistake as to the effect of prior installation upon patentability — so-called "technical fraud."

To establish monopolization or attempt to monopolize a part of trade or commerce under §2 of the Sherman Act, it would then be necessary to appraise the exclusionary power of the illegal patent claim in terms of the relevant market for the product involved. Without a definition of that market there is no way to measure Food Machinery's ability to lessen or destroy competition. It may be that the device — knee-action swing diffusers — used in sewage treatment systems does not comprise a relevant market. There may be effective substitutes for the device which do not infringe the patent. This is a matter of proof, as is the amount of damages suffered by Walker.

[The Court remanded the case with instructions to consider the antitrust claim on its merits.]

Brunswick Corporation v. Riegel Textile Corporation
United States Court of Appeals for the Seventh Circuit
752 F.2d 261 (7th Cir. 1984), cert. denied, 472 U.S. 1018 (1985)

POSNER, Circuit Judge.

Brunswick Corporation appeals from the dismissal, on the pleadings, of its antitrust suit against Riegel Textile Corporation. 578 F. Supp. 893 (N.D. Ill. 1983). The appeal requires us to consider aspects of the relationship between patent and antitrust law.

The complaint alleges that in 1967 Brunswick invented a new process for making "antistatic yarn," which is used to make garments worn in hospital operating rooms and other areas where there are volatile gases that could be ignited by static electricity. Brunswick, which is not itself a textile manufacturer, disclosed its invention to Riegel, which is. Riegel promised to keep the invention secret. In April 1970 Brunswick applied for a patent on the new process and in August Riegel did likewise — in breach of its agreement with Brunswick. (Riegel denies that this was a breach, but as Brunswick has been given no chance to substantiate the allegations of its complaint we must treat them as true for purposes of this appeal.) Without considering Brunswick's application the Patent Office issued a patent to Riegel in 1972. The Patent

Office discovered the Brunswick application in 1973, and in 1975 instituted a patent-interference proceeding to determine priority of invention between Riegel and Brunswick. See 35 U.S.C. §135; 1 Rosenberg, Patent Law Fundamentals §10.02 (2d ed. 1984). That proceeding was still pending before the Patent Office when Brunswick brought this lawsuit in 1982, but since the Patent Office has held that although Brunswick indeed invented the process first, its patent application was invalid. Brunswick has challenged this ruling in another lawsuit in the Northern District of Illinois, and it has also sued Riegel in an Illinois state court for unfair competition.

Brunswick's complaint in this case is that by procuring a patent by fraud and then defending the patent's validity groundlessly in the patent-interference proceeding, Riegel monopolized in the production of antistatic yarn in violation of section 2 of the Sherman Act, 15 U.S.C. §2. The complaint also describes Riegel's misconduct as an attempt to monopolize and as a conspiracy (with Riegel's agents and employees) to monopolize, which are also forbidden by section 2, except that a conspiracy between a corporation and its employees is not actionable under antitrust law. University Life Ins. Co. v. Unimarc Ltd., 699 F.2d 846, 852 (7th Cir. 1983). The district court dismissed the suit on alternative grounds: the complaint fails to state an antitrust cause of action; the suit is barred by the antitrust statute of limitations.

Getting a patent by means of fraud on the Patent Office can, but does not always, violate section 2 of the Sherman Act. See, e.g., Walker Process Equipment, Inc. v. Food Machinery & Chem. Corp., 382 U.S. 172, 15 L. Ed. 2d 247, 86 S. Ct. 347 (1965); United States v. Singer Mfg. Co., 374 U.S. 174, 196-97, 10 L. Ed. 2d 823, 83 S. Ct. 1773 (1963); American Cyanamid Co. v. FTC, 363 F.2d 757, 770-71 (6th Cir. 1966); see generally 3 Areeda & Turner, Antitrust Law ¶¶707a-b, d, f (1978). A patent entitles the patentee to prevent others from making or selling the patented product or, as here, using the patented production process, and he may be able to use this legal right to restrict competition. If antistatic yarn cannot be produced efficiently other than by using Riegel's patented process, Riegel may be able to exclude competition in the sale of such yarn, in which event it may have "monopoly power" — the "power to control prices or exclude competition." United States v. E. I. du Pont de Nemours & Co., 351 U.S. 377, 391-92, 100 L. Ed. 1264, 76 S. Ct. 994 (1956) (plurality opinion). And to create (or attempt to create, or conspire to create) monopoly power by improper means is to monopolize (or attempt to monopolize, or conspire to monopolize) within the meaning of section 2 of the Sherman Act. See, e.g., United States v. Grinnell Corp., 384 U.S. 563, 570-71, 16 L. Ed. 2d 778, 86 S. Ct. 1698 (1966).

But "may" is not "does"; and for a patent fraud actually to create or threaten to create monopoly power, and hence violate section 2, three conditions must be satisfied besides proof that the defendant obtained a patent by fraud:

1. The patent must dominate a real market. See Walker Process Equipment, Inc. v. Food Machinery & Chem. Corp., supra, 382 U.S. at 177-78;

American Hoist & Derrick Co. v. Sowa & Sons, Inc., 725 F.2d 1350, 1366-67, 220 U.S.P.Q. (BNA) 763 (Fed. Cir. 1984); Handgards, Inc. v. Ethicon, Inc., 601 F.2d 986, 993 n.13 (9th Cir. 1979). Although the Patent Office will not issue a patent on an invention that has no apparent utility, the invention need not have any commercial value at all (other products or processes may be superior substitutes), and it certainly need not have enough value to enable the patentee to drive all or most substitutes from the market. If a patent has no significant impact in the marketplace, the circumstances of its issuance cannot have any antitrust significance.

2. The invention sought to be patented must not be patentable. If the invention is patentable, it does not matter from an antitrust standpoint what skullduggery the defendant may have used to get the patent issued or transferred to him. The power over price that patent rights confer is lawful, and is no greater than it otherwise would be just because the person exercising the rights is not the one entitled by law to do so. The distinction between a fraud that leads the Patent Office to issue a patent on an unpatentable invention (as in a case where the patent applicant concealed from the Patent Office the fact that the invention already was in the public domain) and one that merely operates to take the patent opportunity away from the real inventor (who but for the fraud would have gotten a valid patent that would have yielded him a royalty measured by the monopoly power that the patent conferred) is supported by analogy to cases holding that fraud on the Patent Office, to be actionable as patent fraud, must be material in the sense that the patent would not have been issued but for the misconduct. See, e.g., E. I. du Pont de Nemours & Co. v. Berkley & Co., 620 F.2d 1247, 1274 (8th Cir. 1980); Norton v. Curtiss, 57 C.C.P.A. 1384, 433 F.2d 779, 794, 167 U.S.P.Q. (BNA) 532 (C.C.P.A. 1970). Equally, for a fraud to be material in an antitrust sense the plaintiff must show that but for the fraud no patent would have been issued to anyone. If a patent would have been issued to someone, the fraud could but have diverted market power from the one who had the right to possess and exploit it to someone else.

3. The patent must have some colorable validity, conferred for example by the patentee's efforts to enforce it by bringing patent-infringement suits. Indeed, some formulations of the antitrust offense of patent fraud make it seem that the offense is not the fraudulent procuring of a patent in circumstances that create monopoly power but the bringing of groundless suits for patent infringement. See, e.g., Handgards, Inc. v. Ethicon, Inc., supra, 601 F.2d at 993 and n.13. This metamorphosis is natural because most patent-antitrust claims are asserted as counter-claims to patent-infringement suits and because the abusive prosecution of such suits could violate the antitrust laws even if the patent had not been obtained by fraud. See, e.g., id. at 994. But enforcement actions are not a sine qua non of monopolizing by patent fraud. Since a patent known to the trade to be invalid will not discourage competitors from making the patented product or using the patented process, and so will not confer monopoly power, suing an infringer is some evidence that the patent has (or at least the patentee is seeking to clothe it with) some

colorable validity that might deter competitors. But it is not indispensable evidence; the concern of section 2 is with the exclusion of competition, not with the particular means of exclusion. Indeed, one might argue that just by virtue of being issued, a patent would have some apparent validity and that no more should be necessary. But this would go too far the other way. Since patents are issued in ex parte proceedings, and since by hypothesis the patent applicant had to use fraud to persuade the Patent Office to issue the patent, the patent might not fool anybody in the defendant's market.

Let us see whether these three conditions are satisfied by the complaint in this case. Right off the bat there is a problem with condition (1), as Brunswick's complaint does not allege that the business of making and selling antistatic yarn is an economically meaningful market. However, facts are pleaded that allow inference that antistatic yarn probably does not have good substitutes, in which event its sole producer could maintain its price significantly above the cost of production and sale.

There is a serious problem, however, with condition (2). Far from alleging that the process for making antistatic yarn that Riegel patented is not patentable, the complaint alleges that it is. Brunswick's only objection is to the patentee's identity; it thinks that it rather than Riegel should be the patentee. But as we have already suggested, to say that a patent should have been issued because the invention covered by it is patentable, but should have been issued to a different person and would have been but for fraud (the breach of the promise to Brunswick not to disclose its invention), is to say in effect that the patentee stole the patent from its rightful owner; and stealing a valid patent is not at all the same thing, from an antitrust standpoint, as obtaining an invalid patent. Until unmasked in an infringement or cancellation[5] or other proceeding, a patent on an unpatentable invention may create a monopoly by discouraging (through litigation or other means) others from making the patented product, just as a valid patent may, but the monopoly that such a patent creates is illegal, and hence actionable under antitrust law. The theft of a perfectly valid patent, in contrast, creates no monopoly power; it merely shifts a lawful monopoly into different hands. This has no antitrust significance, although it hurts the lawful owner of the monopoly power.

The purpose of the antitrust laws as it is understood in the modern cases is to preserve the health of the competitive process — which means, so far as a case such as this is concerned, to discourage practices that make it hard for consumers to buy at competitive prices — rather than to promote the welfare of particular competitors. This point was implicit in the famous dictum of Brown Shoe Co. v. United States, 370 U.S. 294, 320, 8 L. Ed. 2d 510, 82 S. Ct. 1502 (1962), that antitrust law (the Court was speaking of section 7 of the Clayton Act, but the point has been understood to be general) is concerned "with the protection of *competition*, not *competitiors*" (emphasis in original), and has been repeated with growing emphasis in recent years by

---

5. There is no procedure for "cancellation" of a patent. Judge Posner was apparently referring to "reexamination" of the patent by the PTO at the request of a third party. — EDS.

this and other courts. See, e.g., Sutliff, Inc. v. Donovan Cos., 727 F.2d 648, 655 (7th Cir. 1984), and cases cited here. True, competitors as well as consumers still have standing to complain about antitrust violations, but that is because competitors are thought to be effective (maybe indispensable) surrogates for the many consumers who do not realize they are the victims of monopolistic practices, or if they do may lack incentives to bring suit because the harm to an individual consumer may be tiny even though the aggregate harm is immense. See Landes, Optimal Sanctions for Antitrust Violations, 50 U. Chi. L. Rev. 652, 671-72 (1983); cf. In re Industrial Gas Antitrust Litigations, 681 F.2d 514, 520 (7th Cir. 1982). If no consumer interest can be discerned even remotely in a suit brought by a competitor — if, as here, a victory for the competitor can confer no benefit, certain or probable, present or future, on consumers — a court is entitled to question whether a violation of antitrust law is being charged. See generally Easterbrook, The Limits of Antitrust, 63 Tex. L. Rev. 1, 33-39 (1984). If injury to a competitior, caused by wrongful conduct, were enough to bring the antitrust laws into play, the whole state tort law of unfair competition would be absorbed into federal antitrust law; it has not been: "unfair competition, as such, does not violate the antitrust laws." Sutliff, Inc. v. Donovan Cos., supra, 727 F.2d at 655; see also Car Carriers, Inc. v. Ford Motor Co., 745 F.2d 1101, 1107-08 (7th Cir. 1984); Havoco of America, Ltd. v. Shell Oil Co., 626 F.2d 549, 544-59 (7th Cir. 1980).

We cannot find the consumer interest in this case. Brunswick is complaining not because Riegel is gouging the consumer by charging a monopoly price for antistatic yarn, but because Riegel took away a monopoly that rightfully belonged to Brunswick as the real inventor. It is true that when Riegel got its patent Brunswick's patent application was still pending. But there is no suggestion that any other competitors were in the picture. If Riegel had not committed the alleged fraud, Brunswick would have had the whole field to itself — would have had the monopoly of antistatic yarn that it accuses Riegel of having stolen. This would be true even if Brunswick had not tried to get a patent on its process for making antistatic yarn. It still could, and as a rational profit-maximizer presumably would, have tried to license the invention as a trade secret; and a trade secret known only to, and licensed by, one firm may create as much monopoly power as a patent (more, even, if the secret can be kept for more than 17 years).

The nature of the remedy sought in an antitrust case is often, as here, an important clue to the soundness of the antitrust claim. Brunswick is asking, as a main part of the remedy, for an order transferring ownership of the patent to Riegel to itself. There is no contention that in asking for this Brunswick is motivated by altruism. It wants to make as much money as it can from the patent — as much as Riegel made, or, if possible, even more. There is nothing discreditable in this ambition but we do not see how consumers can benefit from its achievement. The nature of the remedy sought shows that Brunswick, far from contesting the propriety of a patent monopoly of antistatic yarn, makes that propriety the very foundation for the judicial relief that it seeks.

It makes no difference that Brunswick, which as we said is not a textile manufacturer, says that if it owned the patent it would license production to several manufacturers. There would then be more manufacturers of antistatic yarn than there are today, but there would not be more competition if the "competitors" were constrained by the terms of the patent license to charge the monopoly price. And they would be. As rational profit-maximizer Brunswick would charge its licensees a royalty designed to extract from them all the monopoly profits that the patent made possible; and the licensees would raise their prices to consumers to cover the royalty expense. The price to the consumer would be the same as it is, today, with Riegel the only seller in the market.

It is not a purpose of antitrust law to confer patents or to resolve disputes between rival applicants for a patent. From the standpoint of antitrust law, concerned as it is with consumer welfare, it is a matter of indifference whether Riegel or Brunswick exploits a monopoly of antistatic yarn. Cf. Products Liability Ins. Agency, Inc. v. Crum & Forster Ins. Cos., 682 F.2d 660, 665 (7th Cir. 1982). Indeed, if anything, competitive pricing is more likely if Brunswick loses this suit that if it wins it. If Brunswick is confident that Riegel's patent is invalid, it can go into the antistatic-yarn business itself, with little fear of being held liable for patent infringement; and by entering, it will inject some competition into that market for the first time. Brunswick argues that it could not induce textile manufacturers to produce antistatic yarn under license from it since they would fear that Riegel would sue them, however baselessly, for patent infringement. But when a patentee (or, as in this case, a patent applicant) licenses his patent to other firms, he typically agrees to indemnify them for any costs incurred in patent-infringement suits brought against them. Brunswick, a large corporation, can afford to indemnify its licensees and would promise to do so if it really believed that it and not Riegel was the lawful owner of the patent. . . .

The third condition for a patent fraud to violate section 2 of the Sherman Act — that the defendant has made efforts to give the color of validity to his patent on an unpatentable invention — cannot be met in a case where the plaintiff himself asserts that the underlying invention is patentable. This reinforces our conclusion that the complaint states no antitrust cause of action. . . .

COMMENTS AND QUESTIONS

1. The Supreme Court's decision in *Walker Process* has given rise to a whole series of antitrust claims (such as *Brunswick*) that are based on fraudulent procurement of a patent. Indeed, these antitrust suits have come to be referred to as "*Walker Process* claims."

However, the idea that fraudulent procurement of a patent itself violates the antitrust laws is highly misleading. As the last section of the Court's opinion makes clear, the only effect of the defendant's fraud was that it lost the protection of the patent laws. The Court still required that Walker prove all the substantive elements of a section 2 violation (market power, anticompetitive conduct, and, for attempt claims, intent to monopolize). Indeed, Justice Harlan concurred precisely in order to emphasize that point:

> We hold today that a treble-damage action for monopolization . . . may be maintained under §4 of the Clayton Act if two conditions are satisfied: (1) the relevant patent is shown to have been produced by knowing and willful fraud . . . ; and (2) all the elements otherwise necesary to establish a §2 monopolization charge are proved. Conversely, such a private cause of action would *not* be made out if the plaintiff . . . failed to prove the elements of a §2 charge even though he has established actual fraud in the procurement of the patent and the defendant's knowledge of that fraud.

382 U.S. at 179 (Harlan, J., concurring). Thus *Walker Process* is limited to removing the cloak of protection afforded by patent law to patents themselves, where the patent in question was obtained by fraud.

2. Does *Brunswick*'s requirement that a plaintiff claiming patent fraud show that the innovation was not patentable at all make sense? If the plaintiff merely intends to take over the patent monopoly from its current owner, it is hard to see how competition will benefit from the antitrust claim. Upon reflection, though, there may be good reasons for allowing would-be patentees to bring patent fraud claims. For even though competition may not benefit in the short run from such enforcement, enforcement will prevent companies from "free-riding" on the patent laws by making sure that those laws reward only those who submit patentable innovations. And if the rightful owner of the patent is not able to enforce its rights, who will?

The real question here seems to be whether antitrust is the correct cause of action. After all, each of the claims that constitute "fraud" can be raised in other contexts — as defenses to a patent infringement suit, or in a declaratory judgment action to invalidate the patent, or in a patent interference. Should an antitrust plaintiff be entitled to treble damages and attorney fees in addition to such remedies, where the only error was that the patent was awarded to the wrong person?

3. The antitrust status of patents obtained by fraud is, of course, closely related to the definition of such fraud. We discussed this problem at some length in Chapter 3, when we considered the patent defense of inequitable conduct. Does it make sense to punish fraudulent patents both in the patent law and through antitrust law?

Note on Antitrust Petitioning Immunity

Even if a patent is lawfully acquired, using it as an anticompetitive weapon might violate the antitrust laws. After all, baseless infringement suits on a valid patent are just as troubling as infringement suits on a baseless patent. Several courts have held that "anticompetitive litigation," standing alone, can suffice to violate the antitrust laws. See, e.g., Handgards, Inc. v. Ethicon, Inc., 601 F.2d 986, 996 (9th Cir. 1979), *cert. denied*, 444 U.S. 1025 (1980); CVD, Inc. v. Raytheon Co., 769 F.2d 842, 851 (1st Cir. 1985), *cert. denied*, 475 U.S. 1016 (1986). Still other courts have held that litigation may constitute anticompetitive conduct sufficient to satisfy one element of the monopolization test. See Daralyn Durie & Mark A. Lemley, The Antitrust Liability of Labor Unions for Anticompetitive Litigation, 80 Cal. L. Rev. 757, 778 n.128 (1992) (collecting cases).

There is a significant constitutional limitation on antitrust claims or counterclaims that are based on the filing of an infringement suit. Before considering the merits of an antitrust counterclaim, courts must first decide whether the patentee is immune from antitrust liability under the "*Noerr-Pennington* doctrine." That doctrine protects antitrust defendants from liability for "petitioning the government." The Supreme Court has consistently held that filing a lawsuit or an action before an administrative agency is "petitioning" and therefore is presumptively entitled to immunity from antitrust suit.

There is an exception to antitrust immunity for "sham litigation," however. If a lawsuit is a sham, rather than a "genuine effort . . . to influence" the decisionmaker, it is not entitled to antitrust immunity, and the counterclaim will be evaluated on the merits. Allied Tube & Conduit Corp. v. Indian Head, Inc., 486 U.S. 492 (1988). Not surprisingly, there has been heated debate over the precise scope of the "sham exception." The Supreme Court visited this issue in the 1993 case of Professional Real Estate Investors v. Columbia Pictures, Inc., 508 U.S. 49 (1993). In that case, Columbia Pictures had brought a copyright infringement suit against PREI based on PREI's performance of copyrighted movies in guests' hotel rooms, and PREI counterclaimed on the grounds that Columbia had conspired to monopolize the market and restrain trade. Columbia lost its copyright case on summary judgment. But the district court held that PREI was not entitled to pursue its antitrust claim because Columbia's copyright suit, though unsuccessful, was not a sham.

The Supreme Court affirmed the application of antitrust immunity to PREI's counterclaim. In doing so, it set out a new, two-part test to determine whether a lawsuit is a sham. "First, the lawsuit must be objectively baseless in the sense that no reasonable litigant could realistically expect success on the merits." *PREI*, supra. Second, "the court should focus on whether the baseless lawsuit conceals an attempt to interfere directly with the business relationships of a competitor through the use of governmental *process* — as opposed to the *outcome* of that process — as an anticompetitive weapon." Id. (emphasis in original). These are commonly referred to as the "objectively baseless" and "subjectively baseless" tests, respectively. Only if the suit is a sham under this definition will the court proceed to consider the substantive elements of an antitrust violation. The effect of this new test is to make it extremely difficult to prevail in an antitrust counterclaim based on wrongful enforcement of an intellectual property right.

Interestingly, the Federal Circuit has held that *Noerr-Pennington* immunity does not apply to allegations of fraudulent patent procurement under *Walker Process*, but only to allegations of sham litigation. Nobelpharma AB v. Implant Innovations, Inc., 141 F.3d 1059 (Fed. Cir. 1998).

b. Other Anticompetitive Conduct

Data General Corp. v. Grumman Systems Support Corp.
United States Court of Appeals for the First Circuit
36 F.3d 1147 (1st Cir. 1994)

STAHL, Circuit Judge.

While this case raises numerous issues touching on copyright law, Grumman's most intriguing argument — presented below as both a defense and a counterclaim — is that DG illegally maintained its monopoly in the market for service of DG computers by unilaterally refusing to license ADEX to Grumman and other competitors. The antitrust claims are intriguing because they present a curious conflict, namely, whether (and to what extent) the antitrust laws, in the absence of any statutory exemption, must tolerate short-term harm to the competitive process when such harm is caused by the otherwise lawful exercise of an economically potent "monopoly" in a copyrighted work.

After a careful analysis, we affirm [a jury verdict for Data General on the question of copyright infringement] on all but one relatively minor issue concerning the calculation of damages.

I. Background

DG and Grumman are competitors in the market for service of computers manufactured by DG, and the present litigation stems from the evolving nature of their competitive relationship. DG not only designs and manufactures computers, but also offers a line of products and services for the maintenance and repair of DG computers. Although DG has no more than a 5% share of the highly competitive "primary market" for mini-computers, DG occupies approximately 90% of the "aftermarket" for service of DG computers. As a group, various "third party maintainers" ("TPMs") earn roughly 7% of the service revenues; Grumman is the leading TPM with approximately 3% of the available service business. The remaining equipment owners (typically large companies in the high technology industry) generally maintain their own computers and peripherals, although they occasionally need outside service on a "time and materials" basis. . . .

From 1976 until some point in the mid-1980s, DG affirmatively encouraged the growth of TPMs with relatively liberal policies concerning TPM access to service tools. DG sold or licensed diagnostics directly to TPMs, and allowed TPMs to use diagnostics sold or licensed to DG equipment owners. DG did not restrict access by TPMs to spare parts manufactured by DG or other manufacturers. DG allowed (or at least tolerated) requests by TPMs for DG's repair depot to fix malfunctioning circuit boards, the heart of a computer's central processing unit ("CPU"). DG sold at least some schematics and other documentation to TPMs. DG also sold TPMs engineering change order kits. And finally, DG training classes were open to TPM field engineers. Grumman suggests that DG's liberal policies were beneficial to DG because increased capacity (and perhaps competition) in the service aftermarket would be a selling point for DG equipment.

3. Increased Restrictions

In the mid-1980s, DG altered its strategy. With the goal of maximizing revenues from its service business, DG began to refuse to provide many service tools directly to TPMs. DG would not allow TPMs to use the DG repair depot, nor would it permit TPMs to purchase schematics, documentation, "change order" kits, or certain spare parts. DG no longer allowed TPM technicians to attend DG training classes. Finally, DG developed and severely restricted the licensing of ADEX, a new software diagnostic for its MV computers. The MV series was at once DG's most advanced computer hardware and an increasingly important source of sales and service revenue for DG.

A number of items unavailable to TPMs directly from DG were either available to all equipment owners (even customers of TPMs) from DG, or were available to TPMs from sources other than DG. For example, DG depot service, change order kits, and at least some documentation were available to all equipment owners. There is also evidence that Grumman had its own repair depot and that Grumman could make use of repair depots run by other service organizations (sometimes called "fourth party maintainers"). Likewise, there is evidence that TPMs could purchase at least some spare parts from sources other than DG.

The situation was different with respect to ADEX. DG service technicians would use ADEX [a computer diagnostic program developed by Data General] in performing service for DG equipment owners. DG would also license ADEX for the exclusive use of the in-house technicians of equipment owners who perform most of their own service. However, DG would not license ADEX to its own service customers or to the customers of TPMs. Nor was ADEX available to TPMs from sources other than DG. At least two other diagnostics designed to service DG's MV computers may have become available as early as 1989, but no fully functional substitute was available when this case was tried in 1992.

Grumman found various ways to skirt DG's ADEX restrictions. Some former DG employees, in violation of their employment agreements, brought copies of ADEX when they joined Grumman. In addition, DG field engineers often stored copies of ADEX at the work sites of their service customers, who were bound to preserve the confidentiality of any DG proprietary information in their possession. Although DG service customers had an obligation to return copies of ADEX to DG should they cancel their service agreement and switch to a TPM, few customers did so. It is essentially undisputed that Grumman technicians used and duplicated copies of ADEX left behind by DG field engineers. There is also uncontroverted evidence that Grumman actually acquired copies of ADEX in this manner in order to maintain libraries of diagnostics so that Grumman technicians could freely duplicate and use any copy of ADEX to service any of Grumman's customers with DG's MV computers.

C. The Present Litigation

In 1988, DG filed suit against Grumman in the United States District Court for the District of Massachusetts. . . . In one count, DG alleged that Grumman's use and duplication of ADEX infringed DG's ADEX copyrights, and requested injunctive relief, 17 U.S.C. §502 (1988), as well as actual damages and profits, 17 U.S.C. §504(b) (1988). In another count, DG alleged that Grumman had violated Massachusetts trade secrets law by misappropriating copies of ADEX in violation of confidentiality agreements binding on former DG employees and DG service customers. . . .

b. Antitrust Defenses

Grumman claimed that DG could not maintain its infringement action because DG had used its ADEX copyrights to violate Sections 1 and 2 of the Sherman Antitrust Act, 15 U.S.C. §§1 and 2 (1988 & Supp. IV 1992). Specifically, Grumman charged that DG misused its copyrights by (1) tying the availability of ADEX to a consumer's agreement either to purchase DG support services (a "positive tie") or not to purchase support services from TPMs (a "negative tie"), and (2) willfully maintaining its monopoly in the support services aftermarket by imposing the alleged tie-in and refusing to deal with TPMs. . . .

In rejecting Grumman's motion for reconsideration of the grant of summary judgment on the monopolization claim, the district court also directly addressed Grumman's contention that DG's refusal to license ADEX to TPMs constitutes exclusionary conduct. The court stated that DG's refusal to license ADEX to TPMs was not exclusionary because "DG offers to the public a license to use MV/ADEX on any computer owned by the customer," and therefore DG " 'did not withhold from one member of the public a service offered to the rest[.]' " *Grumman III,* slip op. at 5 (citing Olympia Equip. Leasing Co. v. Western Union Tel. Co., 797 F.2d 370, 377 (7th Cir. 1986), *cert. denied,* 480 U.S. 934, 94 L. Ed. 2d 765, 107 S. Ct. 1574 (1987)). . . .

B.   *Grumman's Antitrust Counterclaims . . .*

2.   Monopolization . . .

a.   Unilateral Refusals to Deal

Because a monopolization claim does not require proof of concerted activity, even the unilateral actions of a monopolist can constitute exclusionary conduct. See 15 U.S.C. §2 (referring to "every person who shall monopolize . . . or combine or conspire with any other person . . . to monopolize"); Moore v. Jas. H. Matthews & Co., 473 F.2d 328, 332 (9th Cir. 1973) (observing that "section 2 is not limited to concerted activity"). Thus, a monopolist's unilateral refusal to deal with its competitors (as long as the refusal harms the competitive process) may constitute prima facie evidence of exclusionary conduct in the context of a Section 2 claim. See Kodak, 112 S. Ct. at 2091 n.32 (citing Aspen Skiing Co. v. Aspen Highlands Skiing Corp., 472 U.S. 585, 602-05, 86 L. Ed. 2d 467, 105 S. Ct. 2847 (1985)). A monopolist may nevertheless rebut such evidence by establishing a valid business justification for its conduct. See Kodak, 112 S. Ct. at 2091 n.32 (suggesting that monopolist may rebut an inference of exclusionary conduct by establishing "legitimate competitive reasons for the refusal"); Aspen Skiing, 472 U.S. at 608 (suggesting that sufficient evidence of harm to consumers and competitors triggers further inquiry as to whether the monopolist has "persuaded the jury that its [harmful] conduct was justified by [a] normal business purpose"). In general, a business justification is valid if it relates directly or indirectly to the enhancement of consumer welfare. Thus, pursuit of efficiency and quality control might be legitimate competitive reasons for an otherwise exclusionary refusal to deal, while the desire to maintain a monopoly market share or thwart the entry of competitors would not. See Kodak, 112 S. Ct. at 2091 (discussing the validity and sufficiency of various business justifications); Aspen Skiing, 472 U.S. at 608-11 (same); see generally 7 Areeda & Turner ¶ 1504, at 377-83; 9 Areeda & Turner ¶¶ 1713, 1716-17, at 148-61, 185-239. In essence, a unilateral refusal to deal is prima facie exclusionary if there is evidence of harm to the competitive process; a valid business justification requires proof of countervailing benefits to the competitive process.

Despite the theoretical possibility, there have been relatively few cases in which a unilateral refusal to deal has formed the basis of a successful Section 2 claim. Several of the cases commonly cited for a supposed duty to deal were actually cases of joint conduct in which some competitors joined to frustrate others. See Associated Press v. United States, 326 U.S. 1, 89 L. Ed. 2013, 65 S. Ct. 1416 (1945); United States v. Terminal R.R. Ass'n, 224 U.S. 383, 56 L. Ed. 810, 32 S. Ct. 507 (1912). Prior to Aspen Skiing, the case that probably came closest to condemning a true unilateral refusal to deal was Otter Tail Power Co. v. United States, 410 U.S. 366, 35 L. Ed. 2d 359, 93 S. Ct. 1022 (1973), which condemned the refusal of a wholesale power supplier either to sell wholesale power to municipal systems or to "wheel power" when Otter Tail's retail franchises expired and local municipalities sought to supplant Otter Tail's local distributors. The case not only involved a capital-intensive public utility facility — which could not effectively be duplicated and occupied a distinct separate market — but the Supreme Court laid considerable emphasis on "supported" findings in the district court "that Otter Tail's refusals to sell at wholesale or to wheel were solely to prevent municipal power systems from eroding its monopolistic position." 410 U.S. at 378.

In Aspen Skiing, the Court criticized a monopolist's unilateral refusal to deal in a very different situation, casting serious doubt on the proposition that the Court has adopted any single rule or formula for determining when a unilateral refusal to deal is unlawful. In that case, an "all-Aspen" ski ticket — valid at any mountain in Aspen — had been developed and jointly marketed when the three (later four) ski areas in Aspen were owned by independent entities. 472 U.S. at 589. Some time after Aspen Skiing Company ("Ski Co.") came into control of three of the four ski areas, Ski Co. refused to continue a joint agreement with Aspen Highlands Skiing Corp. ("Highlands"), the owner of the fourth area. Id. at 592-93. Although there was no "essential facility" involved, the Court found that it was exclusionary for Ski Co., as a monopolist, to refuse to continue a presumably efficient "pattern of distribution that had originated in a competitive market and had persisted for several years." Id. at 603.

It is not entirely clear whether the Court in Aspen Skiing merely intended to create a category of refusal-to-deal cases different from the essential facilities category or whether the Court was inviting the application of more general principles of antitrust analysis to unilateral refusals to deal. We follow the parties' lead in assuming that Grumman need not tailor its argument to a preexisting "category" of unilateral refusals to deal.

b. Unilateral Refusals to License

DG attempts to undermine Grumman's monopolization claim by proposing a powerful irrebuttable presumption: a unilateral refusal to license a copyright can never constitute exclusionary conduct. We agree that some type of presumption is in order, but reach that conclusion only after an exhaustive inquiry touching on the general character of presumptions, the role of market analysis in the copyright context, existing responses to the tension between the antitrust and patent laws, the nature of the rights extended by the copyright laws, and our duty to harmonize two conflicting statutes.

(1)   The Propriety of a Presumption

We begin our analysis with two observations. First, DG's rule of law could be characterized as either an empirical assumption or a policy preference. For example, if we were convinced that refusals to license a copyright always have a net positive effect on the competitive process, we might adopt a presumption to this effect in order to preclude wasteful litigation about a known fact. On the other hand, if we were convinced that the rights enumerated in the Copyright Act should take precedence over the responsibilities set forth in the Sherman Act, regardless of the realities of the market, we might adopt a blanket rule of preference. DG's argument contains elements of both archetypal categories of presumptions.

Second, we note that the phrase "competitive process" may need some refinement in order to evaluate either an empirical assumption or a policy presumption concerning the desirability of unilateral refusals to license a copyright. Antitrust law generally seeks to punish and prevent harm to consumers in particular markets, with a focus on relatively specific time periods. See, e.g., Jefferson Parish, 466 U.S. at 18 (holding that "any inquiry into the validity of a tying arrangement must focus on the market or markets in which the two products are sold, for that is where the anticompetitive forcing has its impact"). Thus, in determining whether conduct is exclusionary in the context of a monopolization claim, we ordinarily focus on harm to the competitive process in the relevant market and time period. See generally 3 Areeda & Turner ¶517-28, at 346-88, ¶¶ 533-36, at 406-431. Confining the competitive process in this way assists courts in deciding particular disputes based primarily on case-specific adjudicative facts rather than generally-applicable "legislative" facts or assumptions. The use and protection of copyrights also affects the "competitive process," but it may not be appropriate to judge the effect of the use of a copyright by looking only at one market or one time period.

We now consider what appears to be an empirical proclamation from DG: "The refusal to make one's innovation available to rivals . . . is procompetitive conduct." As support, DG cites Grinnell, 384 U.S. at 570-71, in which the Court held that willful maintenance of monopoly does not include "growth or development as a consequence of a superior product." It is not the superiority of a work that allows the author to exclude others, however, but rather the limited monopoly granted by copyright law. Moreover, one reason why the Copyright Act fosters investment and innovation is that it may allow the author to earn monopoly profits by licensing the copyright to others or reserving the copyright for the author's exclusive use. See Sony Corp. of Am. v. Universal City Studios, Inc., 464 U.S. 417, 429, 78 L. Ed. 2d 574, 104 S. Ct. 774 (1984) (explaining that the limited copyright monopoly "is intended to motivate the creative activity of authors and inventors by the provision of a special reward"). Thus, at least in a particular market and for a particular period of time, the Copyright Act tolerates behavior that may harm both consumers and competitors. Cf. SCM Corp. v. Xerox Corp., 645 F.2d 1195, 1203 (2d Cir. 1981) ("The primary purpose of the antitrust laws — to preserve competition — can be frustrated, albeit temporarily, by a holder's exercise of the patent's inherent exclusionary power during its term."), *cert. denied,* 455 U.S. 1016, 72 L. Ed. 2d 132, 102 S. Ct. 1708 (1982).

DG does not in fact argue that consumers are better off in the short term because of the inability of TPMs to license ADEX. Instead, DG suggests that allowing copyright owners to exclude others from the use of their works creates incentives which ultimately work to the benefit of consumers in the DG service aftermarket as well as to the benefit of consumers generally. In other words, DG seeks to justify any immediate harm to consumers by pointing to countervailing long-term benefits. Certainly, a monopolist's refusal to license others to use a commercially successful patented idea is likely to have more profound anti-competitive consequences than a refusal to allow others to duplicate the copyrighted expression of an unpatented idea (although such differences may become less pronounced if copyright law becomes increasingly protective of intellectual property such as computer software). But by no means is a monopolist's refusal to license a copyright entirely "procompetitive" within the ordinary economic framework of the Sherman Act. Accordingly, it may be inappropriate to adopt an empirical assumption that simply ignores harm to the competitive process caused by a monopolist's unilateral refusal to license a copyright. Even if it is clear that exclusive use of a copyright can have anti-competitive consequences, some type of presumption may nevertheless be appropriate as a matter of either antitrust law or copyright law.

(2) Antitrust Law and the Accommodation of Patent Rights

Antitrust law is somewhat instructive. Although creation and protection of original works of authorship may be a national pastime, the Sherman Act does not explicitly exempt such activity from antitrust scrutiny and courts should be wary of creating implied exemptions. See Square D Co. v. Niagara Frontier Tariff Bureau, Inc., 476 U.S. 409, 421, 90 L. Ed. 2d 413, 106 S. Ct. 1922 (1986) ("Exemptions from the antitrust laws are strictly construed and strongly disfavored."); cf. Flood v. Kuhn, 407 U.S. 258, 32 L. Ed. 2d 728, 92 S. Ct. 2099 (1972) (holding that the longstanding judicially created exemption of professional baseball from the Sherman Act is an established "aberration" in which Congress has acquiesced). The Supreme Court has suggested that an otherwise reasonable yet anti-competitive use of a copyright should not "be deemed a per se violation of the Sherman Act," Broadcast Music, Inc. v. CBS, Inc., 441 U.S. 1, 19, 60 L. Ed. 2d 1, 99 S. Ct. 1551 (1979), but a monopolistic refusal to license might still violate the rule of reason, see Rural Tel. Serv. Co. v. Feist Publications, Inc., 957 F.2d 765, 767-69 (10th Cir.) (analyzing reasonableness of monopolist's unilateral refusal to license copyrighted telephone listings to a competing distributor of telephone directories), *cert. denied,* 121 L. Ed. 2d 429, 113 S. Ct. 490 (1992).[63] Should an antitrust plaintiff be allowed to demonstrate the anti-competitive effects of a monopolist's unilateral refusal to grant a copyright license? Would the monopolist then have to justify its refusal to license by introducing evidence that the protection of the copyright laws enabled the author to create a work which advances consumer welfare?

The courts appear to have partly settled an analogous conflict between the patent laws and the antitrust laws, treating the former as creating an implied limited exception to the latter. In Simpson v. Union Oil Co., 377 U.S. 13, 24, 12 L. Ed. 2d 98, 84 S. Ct. 1051 (1964), the Supreme Court stated that "the patent laws which give a 17-year monopoly on 'making, using, or selling the invention' are in pari materia with the antitrust laws and modify them pro tanto." Similarly, we have suggested that the exercise of patent rights is a "legitimate means" by which a firm may maintain its monopoly power. Barry Wright, 724 F.2d at 230. Other courts have specifically held that a monopolist's unilateral refusal to license a patent is ordinarily not properly viewed as exclusionary conduct. See Miller Insituform, 830 F.2d at 609 ("A patent holder who lawfully acquires a patent cannot be held liable under Section 2 of the Sherman Act for maintaining the monopoly power he lawfully acquired by refusing to license the patent to others."); Westinghouse, 648 F.2d at 647 (finding no antitrust violation because "Westinghouse has done no more than to license some of its patents and refuse to license others"); SCM Corp., 645 F.2d at 1206 (holding that "where a patent has been lawfully acquired, subsequent conduct permissible under the patent laws cannot trigger any liability under the antitrust laws"); see also 3 Areeda & Turner ¶ 704, at 114 ("The patent is itself a government grant of monopoly and is therefore an exception to usual antitrust rules."). This exception is inoperable if the patent was unlawfully "acquired." SCM Corp., 645 F.2d at 1208-09 (analyzing legality of Xerox's acquisition of plain-paper copier patent); see generally 3 Areeda & Turner ¶705-707, at 117-45 (discussing effect of patent

---

63. It is in any event well settled that concerted and contractual behavior that threatens competition is not immune from antitrust inquiry simply because it involves the exercise of copyright privileges. See, e.g., Kodak, 112 S. Ct. at 2089 n.29 ("The Court has held many times that power gained through some natural and legal advantage such as a patent, copyright, or business acumen can give rise to liability if 'a seller exploits his dominant position in one market to expand his empire into the next.' ") (quoting Times-Picayune Publishing Co. v. United States, 345 U.S. 594, 611, 97 L. Ed. 1277, 73 S. Ct. 872 (1953) (tying case)); United States v. Paramount Pictures, Inc., 334 U.S. 131, 143, 92 L. Ed. 1260, 68 S. Ct. 915 (1948) (holding that horizontal conspiracy to engage in price-fixing in copyright licenses is illegal per se); id. at 159 (holding that block-booking of motion pictures — "a refusal to license one or more copyrights unless another copyright is accepted" — is an illegal tying arrangement); Straus v. American Publishers' Ass'n, 231 U.S. 222, 234, 58 L. Ed. 192, 34 S. Ct. 84 (1913) ("No more than the patent statute was the copyright act intended to authorize agreements in unlawful restraint of trade . . . ."); Digidyne Corp. v. Data General Corp., 734 F.2d 1336 (9th Cir. 1984) (affirming finding of illegal tie between copyrighted software and computer hardware), *cert. denied,* 473 U.S. 908, 87 L. Ed. 2d 657, 105 S. Ct. 3534 (1985); cf. Miller Insituform, Inc. v. Insituform of N. Am., Inc., 830 F.2d 606, 608-09 & n.4 (6th Cir. 1987) (describing ways in which patent holder may violate the antitrust laws), *cert. denied,* 484 U.S. 1064, 98 L. Ed. 2d 988, 108 S. Ct. 1023 (1988); United States v. Westinghouse Elec. Corp., 648 F.2d 642, 646-47 (9th Cir. 1981) (same).

acquisition, internal development of patents, and improprieties in patent procurement on applicability of antitrust laws).

The "patent exception" is largely a means of resolving conflicting rights and responsibilities, i.e., a policy presumption. See, e.g., Miller Insituform, 830 F.2d at 609 (declaring summarily that "there is no adverse effect on competition since, as a patent monopolist, [the patent holder] had [the] exclusive right to manufacture, use, and sell his invention"). At the same time, the exception is grounded in an empirical assumption that exposing patent activity to wider antitrust scrutiny would weaken the incentives underlying the patent system, thereby depriving consumers of beneficial products. See, e.g., SCM Corp., 645 F.2d at 1209 (holding that imposition of antitrust liability for an arguably unreasonable refusal to license a lawfully acquired patent "would severely trample upon the incentives provided by our patent laws and thus undermine the entire patent system").

(3) Copyright Law

Copyright law provides further guidance. The Copyright Act expressly grants to a copyright owner the exclusive right to distribute the protected work by "transfer of ownership, or by rental, lease, or lending." 17 U.S.C. §106. Consequently, "the owner of the copyright, if [it] pleases, may refrain from vending or licensing and content [itself] with simply exercising the right to exclude others from using [its] property." Fox Film Corp. v. Doyal, 286 U.S. 123, 127, 76 L. Ed. 1010, 52 S. Ct. 546 (1932). See also Stewart v. Abend, 495 U.S. 207, 229, 109 L. Ed. 2d 184, 110 S. Ct. 1750 (1990). We may also venture to infer that, in passing the Copyright Act, Congress itself made an empirical assumption that allowing copyright holders to collect license fees and exclude others from using their works creates a system of incentives that promotes consumer welfare in the long term by encouraging investment in the creation of desirable artistic and functional works of expression. See Feist Publications, Inc. v. Rural Tel. Serv. Co., 499 U.S. 340, 111 S. Ct. 1282, 1290, 113 L. Ed. 2d 358 (1991) ("The primary objective of a copyright is not to reward the labor of authors, but 'to promote the Progress of Science and useful Arts.' ") (quoting U.S. Const. art. I. §8, cl. 8); Sony Corp., 464 U.S. at 429 (discussing goals and incentives of copyright protection); Twentieth Century Music Corp. v. Aiken, 422 U.S. 151, 156, 45 L. Ed. 2d 84, 95 S. Ct. 2040 (1975) ("The immediate effect of our copyright law is to secure a fair return for an 'author's' creative labor. But the ultimate aim is, by this incentive, to stimulate artistic creativity for the general public good."). We cannot require antitrust defendants to prove and reprove the merits of this legislative assumption in every case where a refusal to license a copyrighted work comes under attack. Nevertheless, although "nothing in the copyright statutes would prevent an author from hoarding all of his works during the term of the copyright," Stewart, 495 U.S. at 228-29 (emphasis added), the Copyright Act does not explicitly purport to limit the scope of the Sherman Act. And, if the Copyright Act is silent on the subject generally, the silence is particularly acute in cases where a monopolist harms consumers in the monopolized market by refusing to license a copyrighted work to competitors.

We acknowledge that Congress has not been entirely silent on the relationship between antitrust and intellectual property laws. Congress amended the patent laws in 1988 to provide that "no patent owner otherwise entitled to relief for infringement . . . of a patent shall be denied relief or deemed guilty of misuse or illegal extension of the patent right by reason of [the patent owner's] refusal to license or use any rights to the patent." 35 U.S.C. §271(d) (1988). Section 271(d) clearly prevents an infringer from using a patent misuse defense when the patent owner has unilaterally refused a license, and may even herald the prohibition of all antitrust claims and counterclaims premised on a refusal to license a patent. See Richard Calkins, Patent Law: The Impact of the 1988 Patent Misuse Reform Act and Noerr-Pennington Doctrine on Misuse Defenses and Antitrust Counterclaims, 38 Drake L. Rev. 192-97 (1988-89). Nevertheless, while Section 271(d) is indicative of congressional "policy" on the need for antitrust law to accommodate intellectual property law, Congress did not similarly amend the Copyright Act.

(4) Harmonizing the Sherman Act and the Copyright Act

Since neither the Sherman Act nor the Copyright Act works a partial repeal of the other, and since implied repeals are disfavored, e.g., Watt v. Alaska, 451 U.S. 259, 267, 68 L. Ed. 2d 80, 101 S. Ct. 1673 (1981), we must harmonize the two as best we can, id., mindful of the legislative and judicial approaches to similar conflicts created by the patent laws. We must not lose sight of the need to preserve the economic incentives fueled by the Copyright Act, but neither may we ignore the tension between the two very different policies embodied in the Copyright Act and the Sherman Act, both designed ultimately to improve the welfare of consumers in our free market system. Drawing on our discussion above, we hold that while exclusionary conduct can include a monopolist's unilateral refusal to license a copyright, an author's desire to exclude others from use of its copyrighted work is a presumptively valid business justification for any immediate harm to consumers.[64]

c. DG's Refusal to License ADEX to non-CMOs

Having arrived at the applicable legal standards, we may resolve Grumman's principal allegation of exclusionary conduct. Although there may

---

64. Wary of undermining the Sherman Act, however, we do not hold that an antitrust plaintiff can never rebut this presumption, for there may be rare cases in which imposing antitrust liability is unlikely to frustrate the objectives of the Copyright Act.

be a genuine factual dispute about the effect on DG equipment owners of DG's refusal to license ADEX to TPMs, DG's desire to exercise its rights under the Copyright Act is a presumptively valid business justification. . . .

Grumman attempts to analogize this case to Aspen Skiing by focusing on the fact that DG once encouraged firms to enter the DG service aftermarket by allowing liberal access to service tools, but no longer does so. The analytical framework of Aspen Skiing cannot function in these circumstances, however, because we are unable to view DG's market practices in both competitive and noncompetitive conditions. While TPMs have made inroads in the market for service of DG computers, DG has always been a monopolist in that market, and competitive conditions have never prevailed. Therefore, it would not be "appropriate to infer" from DG's change of heart that its former policies "satisfy consumer demand in free competitive markets." Aspen Skiing, 472 U.S. at 603.

Nor does it appear that Grumman would be able at trial to overcome the presumption on any other theory. There is no evidence that DG acquired its ADEX copyrights in any unlawful manner; indeed, the record suggests that DG developed all its software internally. Cf. 3 Areeda & Turner ¶706, at 127-28 (arguing that although an internally developed patent may be as exclusionary as one acquired from outside a firm, labelling the former as exclusionary would "discourage progressiveness by monopolists"). And, while there is evidence that DG knew that developing a "proprietary position" in the area of diagnostic software would help to maintain its monopoly in the aftermarket for service of DG computers, there is also evidence that DG set out to create a state-of-the-art diagnostic that would help to improve the quality of DG service. Cf. id. ¶706, at 128-29 (suggesting that "nearly all commercial research rests on a mixture of motivations" and that a search for an overriding "antisocial" motivation would be unilluminating). In fact, there is clearly some evidence that ADEX is a significant benefit to owners of DG's MV computers. ADEX is a better product than any other diagnostic for MV computers. The use of ADEX appears to have increased the efficiency and reduced the cost of service because technicians can locate problems more quickly and, through the use of the software's "remote assistance" capability, can arrive at customer sites having determined ahead of time what replacement parts are necessary. In addition to the possibility of lower prices occasioned by such gains in efficiency, ADEX also promises to lower prices through gains in effectiveness. For example, customers may save the cost of replacing expensive hardware components because the use of advanced diagnostics increases the possibility that technicians can locate a problem and repair the component.

COMMENTS AND QUESTIONS

1. Consider the court's conclusion that a refusal to license a copyrighted work is a presumptively valid business justification. How does this square with the rationales behind the copyright law? Is it inconsistent with the ultimate

goal of copyright — to promote the *dissemination* of works of authorship to the general public?

2. Does it make a difference that this case arises under copyright and not under patent law? The *Data General* court notes that Congress recently amended the Patent Act to provide that refusing to license a patent does not constitute patent misuse (though the statute is silent on whether refusal to license can violate the antitrust laws). Are there different policies in the patent and copyright laws that justify treating the two differently? In particular, consider whether the fact that the information contained in a patent is disclosed to the world makes a patentee's refusal to license less troubling than the same action by a copyright owner.

3. Some recent cases have cast doubt on the strength of the presumption that an intellectual property owner is always justified in unilaterally refusing to license its patents. In Kodak v. Image Tech. Servs., 125 F.3d 1195 (9th Cir. 1997), the court faced an appeal from a jury verdict that Kodak had tied the purchase of replacement parts for its copiers to the purchase of service from Kodak, in an effort to drive independent service organizations out of business. (A prior Supreme Court ruling had accepted the legal theory of monopolization by tying in this case.) One of Kodak's arguments on appeal was that because some of its spare parts were patented, it was free to refuse to license those patents to anyone it chose. The Ninth Circuit rejected that argument, reasoning that the "refusal to license" asserted in that case was really a part of an unlawful tying arrangement. This result is perfectly consistent with the existing case law on patents and tying arrangements. Indeed the court applied the *Data General* presumption; it simply concluded that the presumption was rebutted in this case. Strong language in the opinion suggests, however, that the Ninth Circuit is skeptical about the ability of patent owners to refuse to license under all conditions. At least one court has expressly rejected the Ninth Circuit's holding, concluding that a refusal to license can *never* violate the antitrust laws. In re Independent Service Organizations Antitrust Litigation, 989 F. Supp. 1131 (D. Kan. 1997).

An even stronger challenge to the ability of patent owners to control licensing was presented in Intergraph Corp. v. Intel Corp., 3 F. Supp. 2d 1255 (N.D. Ala. 1998). In that case, Intergraph obtained an order from the district court requiring Intel to continue a long-standing business relationship with Intergraph by providing Intergraph with chips, technical information, and assistance regarding its new Pentium II chip architecture in advance of the chip's release. The district court acknowledged that Intel owned patents, copyrights, and mask-work registrations governing its chip architecture, but concluded that Intel's intellectual property rights did not protect it from having to continue dealing with Intergraph. The Federal Circuit reversed the district court, concluding that Intel's decision to stop supplying Intergraph with chips and technical information did not monopolize a market and in any event was a legitimate exercise of its intellectual property rights. Intergraph Corp. v. Intel Corp., 52 U.S.P.Q.2d 1641 (Fed. Cir. 1999).

An interesting development in the patent context is C.R. Bard v. M3 Systems, 157 F.3d 1340 (Fed. Cir. 1998), where the Federal Circuit (over a vigorous dissent by Judge Newman) held that deliberate acts to create incompatibility between a patented system and a competitor's product could be the sort of predatory conduct that was actionable under the antitrust laws.

Note on Trade Secret and the Antitrust Laws

State trade secret statutes provide inventors with quasi-exclusive rights to their inventions. They are entitled to prevent others from taking the idea from them, but only for so long as the inventor maintains the idea as a secret. Further, trade secret owners cannot prevent others from developing the same technology independently, or from learning the secret through "reverse engineering." Nonetheless, within those constraints the trade secret laws do confer a right to exclude potential competitors from the use of a new technology.

There has been relatively little litigation concerning the interaction between the federal antitrust laws and state trade secret laws. In Perfumer's Workshop v. Roure Bertrand du Pont, 737 F. Supp. 785, 790-91 (S.D.N.Y. 1990), the court established a basic proposition analogous to the rule in patent and copyright cases: monopoly power conferred by the existence of a trade secret does not violate the antitrust laws. The rationale behind this basic conclusion is more complex. The rationale used in the patent and copyright cases — that we must interpret the two federal statutes in light of each other — fails here, because trade secrets are protected only by state law. It might be thought, then, that federal antitrust law should preempt state trade secret law whenever the two come into conflict.

A better approach may be that there is no general conflict between the two laws at all, because on balance trade secret laws are pro- rather than anticompetitive. As one commentator has argued:

> The law of trade secrets seems to have mixed effects on competition. The free exchange of ideas may be impeded by the existence of a strong trade secret law and this could stifle innovation with the consequence that economic expansion would be less than it otherwise would be. Disclosing a secret in confidence to an employee, contractor, supplier or customer can legally deprive such persons of the opportunity to compete with the discloser by using that secret, and in this sense the discloser has a power of exclusion and a power to impede employee mobility. However, inevitable leaks, accidental disclosures, and efficient reverse engineering limit the effectiveness of any program of secrecy.
>
> On the other hand, trade secret law serves a positive function in the promotion of competition by stimulating innovation by providing a needed lead time within which development costs can be at least partially recovered. On balance, because of the relatively small and speculative harm to competition and because of the probable benefits to competition through the basic incentive of lead time, trade secret law does not seem inimical to free competition.

Gordon L. Doerfer, The Limits on Trade Secret Law Imposed by Federal Patent and Antitrust Supremacy, 80 Harv. L. Rev. 1432, 1461-62 (1967). In addition to the benefits of lead time, a powerful argument can be made that trade secret laws encourage licensing, and therefore encourage wider dissemination of information than would take place absent any such legal protection.

Do these rationales for coexistence between the antitrust and trade secret laws suggest that trade secrets should be treated differently from patents and copyrights when it comes to antitrust analysis? If so, how? Trade secret-antitrust cases have generally been of two types: (1) cases where the antitrust defendant allegedly used trade secrecy to extend the term of an expired patent, see United States v. Pilkington plc, Civ. No. 94-345-TUC-WDB (D. Ariz. filed May 25, 1994); and (2) cases where the trade secret license is used as a "subterfuge[ ] enabling the participants to divide markets and fix prices," A & E Plastik Pak Co. v. Monsanto Co., 396 F.2d 710, 715 (9th Cir. 1968). Arguably, both sets of cases can fit comfortably within the normal range of patent-antitrust analysis.

Note on a Monopolist's Duty to Disclose

Antitrust law sometimes imposes burdens on monopolists that it does not impose on ordinary competitors. One of the most contentious issues in antitrust law is the duty of monopolists to give competitors access to their facilities. In limited circumstances, courts have been willing to require monopolists to provide such access on nondiscriminatory terms. For example, the Supreme Court has held that a group of railroads which together owned the central railroad switching yard in St. Louis, Missouri could not deny access to that yard to nonmembers. United States v. Terminal Ry., 212 U.S. 1 (1912). That case reasoned that because it was the only rail yard serving St. Louis, and because of the importance of the city as a hub, the rail yard was an "essential facility" from which its owners could not exclude others to the detriment of competition. A more recent case held that the owner of ski resorts with local market power could not discontinue a multi-area ski lift ticket it had shared with a smaller competitor in the absence of a legitimate business justification, even though the Court was unwilling to describe the lift ticket as an "essential facility." Aspen Skiing Co. v. Aspen Highlands Skiing Corp., 472 U.S. 585 (1985).

With these limited exceptions, however, the general rule is that businesses — even monopolists — do not have to give their competitors access to their facilities. In Berkey Photo, Inc. v. Eastman Kodak Co., 603 F.2d 263 (2d Cir. 1979), *cert. denied*, 444 U.S. 1093 (1980), the court rejected Berkey's argument that Kodak owed it a duty to pre-disclose its forthcoming products, so that Berkey could prepare compatible products. The court recognized that Kodak's size and advance knowledge of its own plans gave it an advantage over Berkey. But, the court said,

> a large firm does not violate section 2 simply by reaping the competitive rewards attributable to its efficient size, nor does an integrated business offend the Sherman Act whenever one of its departments benefits from association with a division possessing a monopoly in its own market. So long as we allow a firm to

compete in several fields, we must expect it to seek the competitive advantages of its broad-based activity. . . .

Berkey postulates that Kodak had a duty to disclose limited types of information to certain competitors under specific circumstances. But it is difficult to comprehend how a major corporation, accustomed though it is to making business decisions with antitrust considerations in mind, could possess the omniscience to anticipate all the instances in which a jury might one day in the future retrospectively conclude that predisclosure was warranted. And it is equally difficult to discern workable guidelines that a court might set forth to aid the firm's decision. For example, how detailed must the information conveyed be? And how far must research have progressed before it is "ripe" for disclosure? These inherent uncertainties would have an inevitable chilling effect on innovation. They go far, we believe, towards explaining why no court has ever imposed the duty Berkey seeks to create here.

Id.

The already difficult line between these two strands of cases is further complicated when the subject matter at issue is intellectual property. An intellectual property right is an exclusive right. It would be stripped of much of its value if antitrust law were to compel intellectual property owners to share it with their competitors. But there may be circumstances in which access to a patent or copyright has become an "essential facility" for competitors in the same or in a downstream market. In those circumstances, courts must face a direct conflict between the rights granted by intellectual property law and the prohibitions of antitrust law. Cf. David Scheffman, The Application of Raising Rivals' Costs Theory to Antitrust, 37 Antitrust Bull. 187 (1992) (suggesting that raising the cost of a key input to a rival creates a problem similar to the problem of denying access to an essential facility).

In Intergraph Corp. v. Intel Corp., 3 F. Supp. 2d 1255 (N.D. Ala. 1998), discussed supra, the district court concluded that Intel's as-yet-unreleased Pentium II chip architecture was an "essential facility" because hardware and peripheral manufactures had to be able to make products compatible with the architecture sold by Intel, the dominant firm in the microprocessor industry. It therefore ordered Intel to provide advanced product information and technical support (as well as a guaranteed supply of chips) to Intergraph. This is precisely the sort of relief Berkey Photo sought. What, if anything, distinguishes the cases? If Intel's chip architecture is really an "essential facility," does Intel have a similar obligation to provide this information to its competitors in the microprocessor business? What does this case suggest about Microsoft's power to deny competitors access to the APIs of its Windows operating system? Note that the Federal Circuit reversed the "essential facility" finding on appeal. Intergraph Corp. v. Intel Corp., __ F.3d __ (Fed. Cir. 1999). Cf. Aldridge v. Microsoft Corp., 995 F. Supp. 728 (S.D. Tex. 1998) (holding that Microsoft's Windows 95 operating system was potentially subject to an essential facilities claim but that the portions at issue in this case were not, in fact, essential facilities).

This problem has arisen on several different occasions in the computer industry. In each case, the issue is one of standardization. One company's system (protected by intellectual property) has become the de facto standard for an entire industry. Is the company entitled therefore to control the entire industry? To date, courts have not answered this question. Instead, they have dealt with these problems as problems of copyright law, not considering antitrust issues at all. See, e.g., Apple Computer v. Franklin Computer, 714 F.2d 1240 (3d Cir. 1983), *cert. dismissed*, 462 U.S. 1033 (1984) (Franklin had no right to create programs compatible with the Apple II operating system that would justify access to Apple's operating system code); Sega Ents. Ltd. v. Accolade, Inc., 977 F.2d 1510 (9th Cir. 1992) (Accolade had right under copyright fair use doctrine to reverse engineer Sega's copyrighted code in order to create games that would work on Sega's video game system). Issues of standardization and the role of monopolists with intellectual property will recur, in the computer industry and elsewhere.

As we shall see in the next section, antitrust law normally is intensely skeptical of agreements and exchanges of information between competitors. The plaintiff in *Berkey Photo* seems to be asking that the court *require* that competitors share information about their new product specifications. Shouldn't the court be concerned about helping facilitate a cartel between the parties in an "essential facilities" case?

C. AGREEMENTS TO RESTRAIN TRADE

1. Vertical Restraints

a. *Tying Arrangements*

In a tying arrangement, the seller forces the buyer to buy a product he does not want (the tied product) in order to get a product he does want (the tying product). According to antitrust cases, sellers with market power in the tying product can leverage that monopoly into the tied product by forcing captive buyers to buy the tied product exclusively from them. Whether this strategy is economically viable has been the subject of heated debate, as we discuss below. Tying arrangements, like joint ventures, have proven difficult to pigeonhole for antitrust purposes. They have been attacked under section 1 of the Sherman Act as involving a conspiracy,[6] under section 2 of the Sherman Act as involving monopolization of the tied product market, and under section 3 of the Clayton Act as requiring an exclusive dealing arrangement.[7]

---

6. The presumed "conspiracy" here is between the coercing seller and the coerced buyer. This is hardly the sort of collusion one normally thinks of as subject to section 1. But see Perma-Life Mufflers, Inc. v. International Parts Co., 392 U.S. 134 (1968) (defendant may be guilty of conspiracy with plaintiff he coerces under section 1).

7. Only some tying arrangements fall within the scope of this last provision. A tying arrangement can be either exclusive or nonexclusive, depending on whether it requires buyers to buy *all* of their supply of the tied product from the seller or only a specified amount of it.

Tying is one of the theories most commonly asserted against intellectual property owners. A common antitrust complaint is that a patentee, who has some power in the (patented) tying product by virtue of her intellectual property right, is requiring buyers to purchase unpatented tied products as well.

International Salt Co., Inc. v. United States
Supreme Court of the United States
332 U.S. 392 (1947)

Mr. Justice JACKSON delivered the opinion of the Court.

The Government brought this civil action to enjoin the International Salt Company, appellant here, from carrying out provisions of the leases of its patented machines to the effect that lessees would use therein only International's salt products. The restriction is alleged to violate §1 of the Sherman Act, and §3 of the Clayton Act. Upon appellant's answer and admissions of fact, the Government moved for summary judgment under Rule 56 of the Rules of Civil Procedure, upon the ground that no issue as to a material fact was presented and that, on the admissions, judgment followed as a matter of law. Neither party submitted affidavits. Judgment was granted and appeal was taken directly to this Court.

It was established by pleadings or admissions that the International Salt Company is engaged in interstate commerce in salt, of which it is the country's largest producer for industrial uses. It also owns patents on two machines for utilization of salt products. One, the "Lixator," dissolves rock salt into a brine used in various industrial processes. The other, the "Saltomat," injects salt, in tablet form, into canned products during the canning process. The principal distribution of each of these machines is under leases which, among other things, require the lessees to purchase from appellant all unpatented salt and salt tablets consumed in the leased machines.

Appellant had outstanding 790 leases of an equal number of "Lixators," all of which leases were on appellant's standard form containing the tying clause and other standard provisions; of 50 other leases which somewhat varied the terms, all but 4 contained the tying clause. It also had in effect 73 leases of 96 "Saltomats," all containing the restrictive clause. In 1944, appellant sold approximately 119,000 tons of salt, for about $500,000, for use in these machines.

The appellant's patents confer a limited monopoly of the invention they reward. From them appellant derives a right to restrain others from making, vending or using the patented machines. But the patents confer no right to restrain use of, or trade in, unpatented salt. By contracting to close this market for salt against competition, International has engaged in a restraint of trade for which its patents afford no immunity from the antitrust laws. Morton Salt Co. v. G. S. Suppiger Co., 314 U.S. 488; Mercoid Corp. v. Mid-Continent Investment Co., 320 U.S. 661; Mercoid Corp. v. Minneapolis-Honeywell Co., 320 U.S. 680.

Appellant contends, however, that summary judgment was unauthorized because it precluded trial of alleged issues of fact as to whether the restraint was unreasonable within the Sherman Act or substantially lessened competition or tended to create a monopoly in salt within the Clayton Act. We think the admitted facts left no genuine issue. Not only is price-fixing unreasonable, per se, United States v. Socony-Vacuum Oil Co., 310 U.S. 150; United States v. Trenton Potteries Co., 273 U.S. 392, but also it is unreasonable, per se, to foreclose competitors from any substantial market. Fashion Originators Guild v. Federal Trade Commission, 114 F.2d 80, *affirmed,* 312 U.S. 457. The volume of business affected by these contracts cannot be said to be insignificant or insubstantial and the tendency of the arrangement to accomplishment of monopoly seems obvious. Under the law, agreements are forbidden which "tend to create a monopoly," and it is immaterial that the tendency is a creeping one rather than one that proceeds at full gallop; nor does the law await arrival at the goal before condemning the direction of the movement.

Appellant contends, however, that the "Lixator" contracts are saved from unreasonableness and from the tendency to monopoly because they provided that if any competitor offered salt of equal grade at a lower price, the lessee should be free to buy in the open market, unless appellant would furnish the salt at an equal price; and the "Saltomat" agreements provided that the lessee was entitled to the benefit of any general price reduction in lessor's salt tablets. The "Lixator" provision does, of course, afford a measure of protection to the lessee, but it does not avoid the stifling effect of the agreement on competition. The appellant had at all times a priority on the business at equal prices. A competitor would have to undercut appellant's price to have any hope of capturing the market, while appellant could hold that market by merely meeting competition. We do not think this concession relieves the contract of being a restraint of trade, albeit a less harsh one than would result in the absence of such a provision. The "Saltomat" provision obviously has no effect of legal significance since it gives the lessee nothing more than a right to buy appellant's salt tablets at appellant's going price. All purchases must in any event be of appellant's product. . . .

United States v. Microsoft Corp., 147 F.3d 935 (D.C. Cir. 1998). In a much-reported decision, the D.C. Circuit considered the legality of Microsoft's bundling its Internet Explorer browser into its Windows 95 operating system.[8] The fundamental facts of the case are that Microsoft had originally sold Windows 95 and Internet Explorer separately but proceeded through various upgrades to link the two more closely together, selling Windows 95 only in conjunction with Internet Explorer (though not vice versa). The Justice Department charged that this violated a 1994 consent decree. For purposes of interpreting that consent decree, the key question

---

8. The inclusion of the browser in the later Windows 98 operating system is being litigated in a separately filed case, still pending at this writing. A decision by the district court in that case was issued in November 1999.

was whether the two bundled programs constituted a permissible "integrated product." The district court issued an injunction precluding Microsoft from offering Windows 95 only on the condition that the user also take Internet Explorer.

The D.C. Circuit reversed. The majority held that Microsoft's actions did not violate the consent decree. Two of the judges went beyond that question, however, and addressed the propriety of a tying theory in far-reaching dictum:

> The point of the test is twofold and may be illustrated by its application to the paradigm case of the Novell complaint and the subsequent release of Windows 95. First, "integration" suggests a degree of unity, something beyond merely placing disks in the same box. If an OEM or end user (referred to generally as "the purchaser") could buy separate products and combine them himself to produce the "integrated product," then the integration looks like a sham. If Microsoft had simply placed the disks for Windows 3.11 and MS-DOS in one package and covered it with a single license agreement, it would have offered purchasers nothing they could not get by buying the separate products and combining them on their own.[11]
>
> Windows 95, by contrast, unites the two functionalities in a way that purchasers could not; it is not simply a graphical user interface running on top of MS-DOS. Windows 95 is integrated in the sense that the two functionalities — DOS and graphical interface — do not exist separately: the code that is required to produce one also produces the other. Of course one can imagine that code being sold on two different disks, one containing all the code necessary for an operating system, the other with all the code necessary for a graphical interface. But as the code in the two would largely overlap, it would be odd to speak of either containing a discrete functionality. Rather, each would represent a disabled version of Windows 95. The customer could then "repair" each by installing them both on a single computer, but in such a case it would not be meaningful to speak of the customer "combining" two products. Windows 95 is an example of what Professor Areeda calls "physical or technological interlinkage that the customer cannot perform." X Areeda, Antitrust Law §1746b at 227, 228 (1996).
>
> So the combination offered by the manufacturer must be different from what the purchaser could create from the separate products on his own. The second point is that it must also be better in some respect; there should be some technological value to integration. Manufacturers can stick products together in

---

11. The same analysis would apply to peripherals. If, for example, Microsoft tried to bundle its mouse with the operating system, it would have to show that the mouse/operating system package worked better if   combined by Microsoft than it would if combined by OEMs. This is quite different from showing that the mouse works better with the operating system than other mice do. See X Areeda, Elhauge & Hovenkamp, Antitrust Law ¶1746b. Problems seem unlikely to arise with peripherals, because their physical existence makes it easier to identify the act of combination. It seems unlikely that a plausible claim could be made that a mouse and an operating system were integrated in the sense that neither could be said to exist separately. An operating system used with a different mouse does not seem like a different product. But Windows 95 without IE's code will not boot, and adding a rival browser will not fix this. If the add/remove utility is run to hide the IE 4 technologies, Windows 95 reverts to an earlier version, OEM service release ("OSR") 2.0.

ways that purchasers cannot without the link serving any purpose but an anticompetitive one. The concept of integration should exclude a case where the manufacturer has done nothing more than to metaphorically "bolt" two products together, as would be true if Windows 95 were artificially rigged to crash if IEXPLORE.EXE were deleted. Cf. ILC Peripherals Leasing Corp. v. International Business Machines Corp., 448 F.Supp. 228, 233 (N.D.Cal. 1978) ("If IBM had simply bolted a disk pack or data module into a drive and sold the two items as a unit for a single price, the 'aggregation' would clearly have been an illegal tying arrangement.") aff'd per curiam sub nom. Memorex Corp. v. International Business Machines Corp., 636 F.2d 1188 (9th Cir. 1980); X Areeda, Elhauge & Hovenkamp, Antitrust Law ¶ 1746 at 227 (discussing literal bolting). Thus if there is no suggestion that the product is superior to the purchaser's combination in some respect, it cannot be deemed integrated.[12]

It might seem difficult to put the two elements discussed above together. If purchasers cannot combine the two functionalities to make Windows 95, it might seem that there is nothing to test Windows 95 against in search of the required superiority. But purchasers can combine the functionalities in their stand-alone incarnations. They can install MS-DOS and Windows 3.11. The test for the integration of Windows 95 then comes down to the question of whether its integrated design offers benefits when compared to a purchaser's combination of corresponding stand-alone functionalities. The decree's evident embrace of Windows 95 as a permissible single product can be taken as manifesting the parties' agreement that it met this test.

The short answer is thus that integration may be considered genuine if it is beneficial when compared to a purchaser combination. But we do not propose that in making this inquiry the court should embark on product design assessment. In antitrust law, from which this whole proceeding springs, the courts have recognized the limits of their institutional competence and have on that ground rejected theories of "technological tying." A court's evaluation of a claim of integration must be narrow and deferential. As the Fifth Circuit put it, "[S]uch a violation must be limited to those instances where the technological factor tying the hardware to the software has been designed for the purpose of tying the products, rather than to achieve some technologically beneficial result. Any other conclusion would enmesh the courts in a technical inquiry into the justifiability of product innovations." Response of Carolina, Inc. v. Leasco Response, Inc., 537 F.2d 1307, 1330 (5th Cir. 1976). . . .

We emphasize that this analysis does not require a court to find that an integrated product is superior to its stand-alone rivals. See ILC Peripherals Leasing Corp. v. International Business Machines Corp., 458 F.Supp. 423, 439 (N.D.Cal. 1978) ("Where there is a difference of opinion as to the advantages of two alternatives which can both be defended from an engineering standpoint, the court will not allow itself to be enmeshed 'in a technical inquiry into the justifiability of product innovations.' ") (quoting *Leasco*, 537 F.2d at 1330),

---

12. Thus of course we agree with the separate opinion that "commingling of code . . . alone is not sufficient evidence of true integration." Commingling for an anticompetitive purpose (or for no purpose at all) is what we refer to as "bolting."

aff'd per curiam sub nom. Memorex Corp. v. IBM Corp., 636 F.2d 1188 (9th Cir. 1980). We do not read §IV(E)(i) to "put[] judges and juries in the unwelcome position of designing computers." IX Areeda, Antitrust Law ¶1700j at 15 (1991). The question is not whether the integration is a net plus but merely whether there is a plausible claim that it brings some advantage. Whether or not this is the appropriate test for antitrust law generally, we believe it is the only sensible reading of §IV(E)(i).

Judge Wald dissented from this opinion. She suggested a somewhat less deferential interpretation of the consent decree — and of tying law generally: "I think the prohibition [in the consent decree on tying] and the proviso [allowing integrated products] could reasonably be construed to state that Microsoft may offer an 'integrated' product to OEMs under one license only if the integrated product achieves synergies great enough to justify Microsoft's extension of its monopoly to an otherwise distinct market." Id. at __. Judge Wald would balance the productive efficiencies against the harm to competition in determining whether an antitrust violation had occurred.

COMMENTS AND QUESTIONS

1. Is the deference shown by the majority in *Microsoft* appropriate? Under Judge Williams' opinion, how likely is it that any "integrated" product will ever be held to be a tie? Does the opinion have implications beyond the integration of software products, to cover any sort of synergistic combination of products?

Surely there is some value to embedding Internet Explorer in Windows 95; it is easier for novice computer users to access the Internet without having to obtain separate software to do it. Is this value enough to outweigh the elimination of competition in the browser market (which will almost certainly result from Microsoft's bundling)?

The pros and cons of browser bundling are discussed in Mark A. Lemley & David McGowan, Could Java Change Everything? The Competitive Propriety of a Proprietary Standard, 43 Antitrust Bull. 715 (1998).

2. In *International Salt*, the patentee allegedly tied the sale of its patented machine to the sale of a basic staple commodity, salt. What is wrong with that? The theory of leveraging is that the defendant will use its power in the tying market to obtain power in the tied product market (in this case, the market for salt). This power is considered to be beyond the scope of the patent grant, since it restrains trade in a separate market. But was there really any danger that International Salt would gain market power over such a basic commodity as salt? If not, is there any reason to be troubled by such a tying arrangement? See Joel Dirlam & Alfred Kahn, Fair Competition: The Law and Economics of Antitrust Policy 97 (1954) (criticizing *International Salt* for failing to inquire into market power).

On this point, the Supreme Court is sharply divided. In Jefferson Parish Hosp. Dist. v. Hyde, 466 U.S. 2 (1984), the Court split 5–4 on the standard to be used to evaluate tying cases. The majority concluded that tying was illegal "per se," but only in circumstances in which the defendant had market power in the tying product. This creates a "hybrid" rule somewhere between the "per se" rule and the rule of reason. The four dissenters would have required proof of three separate elements in tying cases: (a) a demonstration that there were really two separate products at stake; (b) proof of market power in the tying product; and (c) a dangerous probability of acquiring power in the tied product. Even then, the dissent would treat tying arrangements under the rule of reason. Decisions subsequent to *Jefferson Parish* have left some question as to whether the majority rule in that case still stands. See Eastman Kodak Co. v. Image Technical Services, 504 U.S. 451 (1992) (appearing to apply the rule of reason in a Section 2 tying case).

3. Consider the last paragraph of the court's opinion. Is there really a tie here? Can't the "bound" parties get out of the contract if they find a better deal elsewhere? The court suggests that this "right of first refusal" is anticompetitive, but might it not serve the laudable (or at least inoffensive) purpose of keeping International Salt informed about the competitive market price? On this point, see Mark Grady & Jay Alexander, Patent Law and Rent Dissipation, 78 Va. L. Rev. 305 (1992).

4. Intellectual property tying arrangements are not always between patented machines and staple products. In Automatic Radio Mfg. Co. v. Hazeltine Research, 339 U.S. 827 (1950), Hazeltine Research, a patent holding company, had licensed a group of 570 patents and 200 patent applications in a single block. Automatic Radio accused Hazeltine of "tying" desirable patents to undesirable patents. The Court disagreed:

> [P]etitioner urges that this case "is identical in principle" with the "Tie-in" cases. It is contended that the licensing provision requiring royalty payments of a percentage of the sales of the licensee's products constitutes a misuse of patents because it ties in a payment on unpatented goods. Particular reliance is placed on language from United States v. U.S. Gypsum Co., 333 U.S. 364, 389, 400. That case was a prosecution under the Sherman Act for an alleged conspiracy of Gypsum and its licensees to extend the monopoly of certain patents and to eliminate competition by fixing prices on patented and unpatented gypsum board. The license provisions based royalties on all sales of gypsum board, both patented and unpatented. It was held that the license provisions, together with evidence of an understanding that only patented board would be sold, showed a conspiracy to restrict the production of unpatented products which was an invalid extension of the area of the patent monopoly. 333 U.S. at 397. There is no indication here of conspiracy to restrict production of unpatented or any goods to effectuate a monopoly, and thus the Gypsum case does not aid petitioner. That which is condemned as against public policy by the "Tie-in" cases is the extension of the monopoly of the patent to create another monopoly or restraint of competition — a restraint not countenanced by the patent grant. See, e. g., Mercoid Corp. v. Mid-Continent Investment Co., 320 U.S. 661, 665-666; Morton Salt Co. v. Suppiger Co., 314 U.S. 488; Ethyl Gasoline Corp. v. United States, 309 U.S. 436, 456. The principle of those cases cannot be contorted to circumscribe the instant situation. This royalty provision does not

create another monopoly; it creates no restraint of competition beyond the legitimate grant of the patent. The right to a patent includes the right to market the use of the patent at a reasonable return. See 46 Stat. 376, 35 U. S. C. §40; Hartford-Empire Co. v. United States, 323 U.S. 386, 417, 324 U.S. 570, 574.

The licensing agreement in issue was characterized by the District Court as essentially a grant by Hazeltine to petitioner of a privilege to use any patent or future development of Hazeltine in consideration of the payment of royalties. Payment for the privilege is required regardless of use of the patents. The royalty provision of the licensing agreement was sustained by the District Court and the Court of Appeals on the theory that it was a convenient mode of operation designed by the parties to avoid the necessity of determining whether each type of petitioner's product embodies any of the numerous Hazeltine patents. 77 F. Supp. at 496. The Court of Appeals reasoned that since it would not be unlawful to agree to pay a fixed sum for the privilege to use patents, it was not unlawful to provide a variable consideration measured by a percentage of the licensee's sales for the same privilege. 176 F.2d at 804. Numerous District Courts which have had occasion to pass on the question have reached the same result on similar grounds, and we are of like opinion.

Id. Can the bundling of patent licenses ever constitute actionable tying of products under the antitrust laws? Does it matter whether the patents confer actual market power? Whether the licenses are exclusive or nonexclusive?[9]

Even if package licensing is not strictly "tying" of multiple products, is it nonetheless objectionable? See George Stigler, United States v. Loew's, Inc.: A Note on Block Booking, 1963 Sup. Ct. Rev. 152 (discussing package licensing as a means of price discrimination and/or cross-subsidization).

5. Several companies have alleged that after they asserted their intellectual property rights against Intel (either by filing suit or by entering into license negotiations), Intel threatened to "cut them off" from the supply of its crucial technology unless they dropped their suits. In effect the allegation is that Intel is using its power in the market for microprocessors to obtain royalty-free licenses to everyone else's intellectual property. Is this a reciprocal tying arrangement of a sort, between Intel's technology and the intellectual property asserted against it? If so, does it violate the antitrust laws? See Intergraph Corp. v. Intel Corp., 3 F. Supp. 2d 1255 (N.D. Ala. 1998) 52 U.S.P.Q. 1641 (Fed. Cir. 1999); In re Intel Corp., FTC Dock. 9288 (complaint filed June 8, 1998).

Intel and the FTC entered into a consent decree in 1999. Under that decree Intel agreed not to terminate its supply of chips or information to a buyer merely because that buyer sued it for intellectual property infringement. Intel retained the right to terminate supply for a number of legitimate business reasons, however. Further, the consent decree provided that Intel could

---

9. *Automatic Radio* was significantly limited by the Supreme Court in Zenith Radio Corp. v. Hazeltine Research, Inc., 395 U.S. 100 (1969). *Zenith* concludes that charging a single rate for a group of licenses is allowable because the patentee can charge anything it wants for licenses, but that compelling the purchase of other patented or unpatented products is unlawful. In other words, Hazeltine can charge Zenith $1000 to use patent *A*, but Hazeltine cannot force Zenith to use patents *A* and *B* together, even if it only charges $1000 for the set.

Is this distinction logical?

cut off any purchaser who sued it for infringement if that purchaser sought to enjoin the sale of Intel's core products. In essence, therefore, Intel retained the power to demand a compulsory license of intellectual property rights from the companies with which it does business.

Is the consent decree an appropriate compromise between the positions of the parties? Is it likely to have procompetitive effects?

6. Another common "tie" involving intellectual property is between patented machines and "nonstaple" products. (A nonstaple product is a product that is useful only in connection with the machine to which it is tied. Thus certain specially made replacement parts are nonstaple products.) Does a patent on a machine confer the right to a monopoly over replacement parts usable only on that machine? Congress resolved this issue in the patent misuse context in 1952. After a number of Supreme Court decisions had found patentees guilty of patent misuse for tying their patent machines to nonstaple products, Congress explicitly provided in 35 U.S.C. §271(d) that tying of nonstaple products did not constitute misuse. See Dawson Chemical v. Rohm & Haas Co., 448 U.S. 176, 209 (1980) (reviewing cases and legislative history); Mark A. Lemley, Comment, The Economic Irrationality of the Patent Misuse Doctrine, 78 Cal. L. Rev. 1599, 1609-10 (1990).

In the antitrust context, the tying of nonstaple goods and services remains a major issue. A 1992 Supreme Court case (not involving intellectual property) held that an "aftermarket" for parts or service for a single brand of machines could be a relevant antitrust market. Eastman Kodak Co. v. Image Technical Servs. Inc., 504 U.S. 451 (1992).

On the other hand, consider Advanced Computer Services v. MAI Systems, 854 F. Supp. 356 (E.D. Va. 1994). MAI Systems has attempted for several years to prevent independent computer repair services from servicing its computers. In 1993, MAI convinced the Ninth Circuit that independent repairers were engaged in copyright infringement, since they necessarily made a "copy" of the operating system when they started the computer in an attempt to repair it. MAI Systems v. Peak Computing, 991 F.2d 511 (9th Cir. 1993), *cert. dismissed*, 510 U.S. 1033 (1994). After this decision, a number of computer repairers filed an antitrust claim against MAI, alleging that MAI was tying the sale of its operating system to the sale of repair service. The District Court granted summary judgment for MAI on the antitrust claim, reasoning that there was no separate "market" for MAI's operating system, and MAI lacked market power in the broader market for minicomputer operating systems. ACS, 845 F. Supp. at 356.

In the wake of *Kodak*, a number of similar cases were reopened in the computer industry. See, e.g., Virtual Maintenance, Inc. v. Prime Computer, Inc., 957 F.2d 1318 (6th Cir. 1992), *vacated and remanded*, 506 U.S. 910 (1992), *on remand*, 995 F.2d 1324 (6th Cir. 1993); Datagate, Inc. v. Hewlett-Packard Co., 941 F.2d 864 (9th Cir. 1991), *cert. denied*, 503 U.S. 984 (1992). A large number of post-*Kodak* cases turn on whether the plaintiff can characterize the tied product as a genuinely separate product market. See, e.g., Tricom Inc. v. Electronic Data Sys., 902 F. Supp. 741 (E.D. Mich.

1995) (conditioning of a software lease agreement on the purchase of CPU time from the software dealer was actionable tying arrangement).

7. Is there any reason to treat intellectual property owners differently from companies whose power in aftermarkets is "inherent" in their product? The answer may depend both on whether you believe that intellectual property law justifies certain restraints on trade in order to encourage innovation, and on whether you think there is a real danger that power in an equipment market could profitably be leveraged into aftermarkets. Economists have debated the validity of "leveraging" theory for some time. Both Robert Bork and Richard Posner have advanced the idea that leveraging is economically irrational. See Robert Bork, The Antitrust Paradox: A Policy at War with Itself (1978):

> The fallacy of the cases on tying arrangements may be shown through a hypothetical example based on the facts of *International Salt*. Suppose that a food canner is just willing to pay $100 for a one-year lease of a salt-dispensing machine. How is it possible for the lessor of the machine to make him pay $100 and, in addition, require him to take all his salt from the lessor? If the requirement is necessary, the lessee is giving up something he values. If the lessor has charged the full value of the machine, he cannot then charge still more in the form of coercion to take what amounts to a requirements contract for salt. That is double counting of monopoly power. The tying arrangement, whatever else it may accomplish, is obviously not a means of gaining two monopoly profits from a single monopoly.
>
> The argument is identical with respect to reciprocity. In *Consolidated Foods* the Supreme Court struck down Consolidated's acquisition of Gentry, a small manufacturer of dehydrated onion and garlic, on the theory that Consolidated, a large food processor, might condition its purchases from food suppliers upon their willingness to purchase from Gentry. There is, however, no way in which that practice could be anti-competitive. Suppose that both Consolidated and Gentry buy and sell in fully competitive markets. Consolidated will then have no market power with which to force its suppliers to purchase from Gentry. The suppliers can turn to other customers. Suppose, however, that all markets involved display large elements of market power. Consolidated, before its acquisition of Gentry, may be presumed to have negotiated the best price it could from its suppliers. The acquisition of Gentry does not alter Consolidated's purchasing power, but the attempt after the acquisition to force Gentry's onions and garlic on suppliers as a condition of continued purchases by Consolidated is merely a way of demanding a still lower price from the suppliers. If the tactic works, Consolidated merely learns that it was paying too high a price to begin with: it would be better off renegotiating prices than cramming unwanted onions and garlic down its suppliers' throats. Nor is the theory improved by assuming that Consolidated understands all this and agrees to pay suppliers more in return for purchases from Gentry as a tactic of monopolizing the onion market. Rival sellers in Gentry's market can respond in a variety of ways, the most obvious being a price cut that just matches the price increase laid out by Consolidated. In that kind of price war, Consolidated has no advantages.
>
> But perhaps the most concise devastation of the law's tie-in theory was provoked by the Supreme Court's 1962 *Loew's* decision. Distributors of pre-

1948 copyrighted motion-picture feature films for television exhibition engaged in the practice of block booking their films to television stations. The stations had to take entire groups of films and could not pick and choose particular films from the proffered packages. The Court majority held, of course, that the practice violated Section 1 of the Sherman Act. The opinion's economic analysis of block booking was confined to a recitation of Justice Frankfurter's by then clearly indefensible dictum on the purpose of tying arrangements.

The inadequacy of the Supreme Court's theory of tying was well stated by George Stigler in a critique of the *Loew's* opinion:

> Consider the following simple example. One film, Justice Goldberg cited 'Gone With the Wind', is worth $10,000 to the buyer, while a second film, the Justice cited 'Getting Gertie's Garter', is worthless to him. The seller could sell the one for $10,000 and throw away the second, for no matter what its cost, bygones are forever bygones. Instead the seller compels the buyer to take both. But surely he can obtain no more than $10,000, since by hypothesis this is the value of both films to the buyer. Why not, in short, use his monopoly power directly on the desirable film? It seems no more sensible, on this logic, to blockbook the two films than it would be to compel the exhibitor to buy 'Gone With the Wind' and seven Ouija boards, again for $10,000.

To these queries the law has no answers. Indeed, it has so far successfully managed to ignore the existence of the queries.

Id. at 373-74; cf. Melissa Hamilton, Software Tying Arrangements Under the Antitrust Laws: A More Flexible Approach, 71 Denv. L. Rev. 607 (1994) (making the rather different argument that tying is efficient in the software context because it allows programmers to continue to upgrade their products). The logic of Judge Bork's argument is appealing. But is it correct? Louis Kaplow has offered a powerful critique of the Bork-Posner theory. Louis Kaplow, Extension of Monopoly Power Through Leverage, 85 Colum. L. Rev. 515 (1985). There, he argues that

> There are a number of deficiencies in the analysis of recent commentators who have attempted to proclaim the death of leverage theory. The basic mistake in their central thesis is that antitrust law should be indifferent to the exploitation of monopoly power because extant power is a fixed sum and thus will result in the same damage regardless of how it is deployed. Although of some superficial appeal, it can readily be demonstrated that their analysis is strongly counterintuitive. Consider the case of a terrorist on the loose with one stick of dynamite. The fixed sum thesis posits that since the power is fixed — that is, the terrorist has one and only one dynamite stick — we should be indifferent to where the dynamite is placed. It is all too obvious, however, that the potential damage resulting from power in this context, as well as in virtually any other we can imagine, is overwhelmingly dependent upon how it may be used. . . .
>
> The position of these commentators [Bork, Posner, others] can be given more meaning only by developing it more extensively. The most reasonable interpretation, I believe, is that their two categories represent an implicit attempt to distinguish between short-run and long-run phenomena — or, to use more technical language, between static and dynamic models. Profit-maximizing practices are meant to refer roughly to those actions that can have fairly direct and

immediate effects, while monopoly extension refers to behavior designed to have implications on the magnitude of profits and welfare loss in the future. The prototypical example of a profit maximization device is a pricing decision by a firm with market power, a decision which can be implemented rather quickly. By contrast, practices designed to affect the market share and elasticity of market demand might be labelled monopoly extension devices. These practices do not increase short-run profits, and might even decrease them. The firm's motivation is to change the structural conditions it faces in the future in order that it may receive greater profits in the future.

Id. at 516, 523-24. Professor Kaplow goes on to point out a number of reasons why leveraging may be possible in the real world. His reasons include the dynamic nature of markets, in which firms may give up short-run profits to maximize long-run gains, and inherent market imperfections that may allow monopolists to gain limited power in new markets through leveraging. See also George Stigler, United States v. Loew's, Inc.: A Note on Block Booking, 1963 Sup. Ct. Rev. 152 (explaining price discrimination potential inherent in tying).

Despite scholarly criticism, and to the surprise of many, the Eleventh Circuit held in 1999 that block-booking of television shows was illegal per se as a tying arrangement. MCA Television Ltd. v. Public Interest Corp., 171 F.3d 1265 (11th Cir. 1999).

8. Some forms of intellectual property may confer or promote power over price by their very nature. Trademarks, for example, are designed to differentiate products in the minds of consumers. The result of product differentiation is that producers have a form of constrained monopoly power, even in otherwise competitive industries. (For example, users of Colgate toothpaste may put up with a price increase rather than buy a competitor's product, because of "brand loyalty" generated by trademarks.) In the 1970s, the FTC investigated leaders in the packaged breakfast cereal industry, operating on the theory that brand names give each cereal a certain degree of market power.

PROBLEMS

Problem 8-2. Advanced Micro Devices (AMD) and Intel both make microprocessor chips. Chip designs tend to travel in waves, or "generations," which reflect newer technology and which successively become the industry standard. The old generation of chips were known as 286 chips. They were replaced by the next generation, the 386. Intel has since developed three new generations of chips, the 486, the Pentium, and the Pentium Pro. In 1991, AMD filed suit against Intel, alleging that Intel was intimidating chip consumers into buying Intel's rather than AMD's 386 chips by threatening to withhold 486 chips from any firm that defects to AMD. The suit claims that this is an illegal tying arrangement.

How should such a "temporal tying arrangement" be treated under the antitrust laws? Is leveraging a problem here? Does it matter if Intel obtains a patent on the 486?

Problem 8-3. Tastee Turkee (TT) is a national fast-food franchise that specializes in roast turkey meals. TT does not operate any of its own restaurants. However, it has a series of requirements that franchisees must meet in order to use the Tastee Turkee name. In addition to requirements regarding color scheme, logo, building, and products, TT requires that its franchisees purchase all of their packaging materials, food, and kitchen equipment from TT itself or from approved TT vendors.

Cantor is a franchisee who believes that he can obtain food and packaging material more cheaply from his own sources. When TT refuses to allow him to do so, he brings suit under the antitrust laws. Cantor's theory is that TT has tied the "Tastee Turkee" trademark and image to the purchase of various staple goods.

Is this a "tying arrangement"? If so, does it violate the antitrust laws?

b. Licensing Restrictions

One of the most difficult antitrust issues surrounding intellectual property rights concerns the limitations placed on the licensing of intellectual property. Intellectual property owners may wish to license their rights to others for a number of reasons. The owners may be ill-equipped to make the protected product themselves; they may want a revenue stream without having to invest in producing and selling the product; they may wish to reserve one geographic or product market to themselves, while allowing others to exploit the intellectual property right elsewhere; or they may simply feel that broad dissemination of their product will redound to their benefit (for example, because there is value in having their product become an industry standard). Economic theory encourages licensing, because it allows the market to transfer use of the intellectual property right to the most productive user of that right.

There is no antitrust problem with licensing per se. Indeed, the Antitrust Division's Intellectual Property Guidelines take the position that licensing is essentially procompetitive:

Intellectual property typically is one component among many in a production process and derives value from its combination with complementary factors. Complementary factors of production include manufacturing and distribution facilities, workforces, and other items of intellectual property. The owner of intellectual property has to arrange for its combination with other necessary factors to realize its commercial value. Often, the owner finds it most efficient to contract with others for these factors, to sell rights to the intellectual property, or to enter into a joint venture arrangement for its development, rather than supplying these complementary factors itself.

Licensing, cross-licensing, or otherwise transferring intellectual property (hereinafter "licensing") can facilitate integration of the licensed property with complementary factors of production. This integration can lead to more efficient exploitation of the intellectual property, benefiting consumers through the reduction of costs and the introduction of new products. Such arrangements increase the value of intellectual property to consumers and to the developers of the technology. By potentially increasing the expected returns from intellectual property, licensing also can increase the incentive for its creation and thus promote greater investment in research and development. . . .

Field-of-use, territorial and other limitations on intellectual property licenses may serve procompetitive ends by allowing the licensor to exploit its property as efficiently and effectively as possible. These various forms of exclusivity can be used to give a licensee an incentive to invest in the commercialization and distribution of products embodying the licensed intellectual property and to develop additional applications for the licensed property. The restrictions may do so, for example, by protecting the licensee against free-riding on the licensee's investments by other licensees or by the licensor. They may also promote the licensor's incentive to license, for example, by protecting the licensor from competition in the licensor's own technology in a market niche that it prefers to keep to itself. These benefits of licensing restrictions apply to patent, copyright, and trade secret licenses, and to know-how agreements.

Intellectual Property Guidelines §2.3. For many of these reasons, the division's prior set of guidelines governing intellectual property licensing (promulgated by the Reagan Administration in 1988, and contained in the International Guidelines) took a very "hands-off" approach to intellectual property licensing transactions. That has changed with the current guidelines, however. While the new guidelines recognize the procompetitive benefits of licensing, they also identify some licensing arrangements that are cause for antitrust concern:

> [A]ntitrust concerns may arise when a licensing arrangement harms competition among entities that would have been actual or likely potential competitors in a relevant market in the absence of a license (entities in a "horizontal relationship"). A restraint in a licensing arrangement may harm such competition, for example, if it facilitates market division or price fixing. In addition, license restrictions with respect to one market may harm such competition in another market by anticompetitively foreclosing access to, or significantly raising the price of, an important input, or by facilitating coordination to increase price or reduce output.

Id., §3.1.

COMMENTS AND QUESTIONS

1. Consider the Antitrust Division's new policy statement — that it is concerned about agreements that reduce competition which would have existed but for the license. On the one hand, this policy can be (and has been) criticized as being too aggressive. By declaring that intellectual property is like any other form of property, the guidelines narrowly circumscribe the terms of licensing agreements. The effect may be to prevent some efficient licensing agreements from being signed, even though the procompetitive effects of the agreement would outweigh its negative effects on preexisting competition.

On the other hand, perhaps the policy can be criticized from the other side for not being agressive enough. Why should the antitrust laws be concerned only with protecting competition that would have existed absent the license? If licensing is favored because it is procompetitive, doesn't antitrust law have some obligation to see that those procompetitive benefits which result from the license itself are not lost because of restrictive license provisions?

2. One way of resolving this debate is to distinguish between the effects inherent to the license, and the effects that result from restrictive provisions ancillary to the license. For example, an agreement to license a patent may promote competition in the market for the patented product, but the agreement may also contain a term dividing geographic territories in the sale of unpatented products or setting resale prices. To the extent that these restrictive terms can be severed from the balance of the license, antitrust law may be able to challenge them without threatening the license itself.

3. The guidelines also establish that most licensing agreements will be treated under the rule of reason, rather than the per se rule. Intellectual Property Guidelines §3.4. Does this provision help to alleviate any concern that the guidelines are too aggressive in their treatment of intellectual property licensing?

Note on Intellectual Property and Exclusive Dealing

Section 3 of the Clayton Act prohibits some "exclusive dealing arrangements" under the rule of reason.[10] Exclusive dealing refers to agreements that prevent the agreeing party from buying or selling the goods of a competitor. For example, a company with a significant share of the market may attempt to drive its competitors out of business by requiring that all its customers buy

---

10. Strictly speaking, section 3 of the Clayton Act applies only to transactions for the sale of goods. However, courts have interpreted sections 1 and 2 of the Sherman Act (which do extend to service transactions) to cover most of the same exclusive dealing arrangements as Clayton Act §3.

exclusively from it. If the customers consider the exclusive dealer to be their most important source of supply, they will probably agree to purchase all their goods from that dealer. The result may be that the exclusive dealer expands its market power into an actual monopoly.

On the other hand, exclusive dealing arrangements may have procompetitive justifications. Agreements to buy everything you need from (or sell everything you produce to) one source may significantly reduce transactions costs and allow smaller companies to take advantage of discounts resulting from economies of scale in purchasing. The problem for antitrust law is to distinguish exclusive dealing arrangements that restrict competition from those that merely serve to make business transactions easier. To draw this line, antitrust law requires some showing of market power to make out a section 3 violation.

Intellectual property licenses often contain exclusivity provisions. Such provisions can bind either the licensor or the licensee (or conceivably both). The licensor might agree to grant an "exclusive license" for his patent. An exclusive license gives only the licensee the right to work the patent. The licensor cannot license anyone else to work the patent and in some circumstances gives up the right to work it himself. On the other end, a license may bind the licensee not to use or sell products other than the patentee's product. In either case, the contract expressly requires that one party deal exclusively with the other. The Antitrust Division's Intellectual Property Guidelines take the position that exclusive licenses do not raise antitrust concerns unless the parties to the agreement (or other potential licensees) would have been actual or potential competitors absent the exclusivity provision. See Intellectual Property Guidelines §4.1.2.

Exclusive dealing practices need not be express. Certain types of licensing provisions have the practical effect of coercing the licensee into using only the licensor's product. One example is the nonmetered royalty. In a recent complaint, the Antitrust Division charged that Microsoft had dominated the market for installed computer operating systems by requiring that computer manufacturers pay royalties to Microsoft on each computer shipped, regardless of whether the computer had a Microsoft operating system installed. Because the manufacturers had to pay for Microsoft's operating system anyway, they were extremely unlikely to install a competitor's operating system, since if they did they would have had to pay twice for an operating system. United States v. Microsoft, Inc., 159 F.R.D. 318 (D.D.C.), *rev'd on other grounds,* 56 F.3d 1448 (D.C. Cir. 1995). When Microsoft settled the division's complaint, it agreed to stop the use of nonmetered royalties.

Yet another Microsoft case offers some interesting thoughts on the role of intellectual property law in justifying licensing restrictions. In United States v. Microsoft Corp., 1998-2 Trade Cas. ¶72,261 (D.D.C. 1998), the case brought against Windows 98, Microsoft defended its allegedly anticompetitive licensing restrictions in part on the ground that its license terms "merely highlight and expressly state the rights that Microsoft already enjoys under federal copyright law" and therefore could not violate the antitrust laws. The court rejected this argument, noting that the copyright laws did not give Microsoft "unlimited" power over its software. It wrote:

> [C]opyright law does not give Microsoft blanket authority to license (or refuse to license) its intellectual property as it sees fit. A copyright does not give its holder immunity from laws of general applicability, including the antitrust laws. Copyright holders are restricted in their ability to extend their control to other markets. They may not prevent the development and use of interoperable programs by competitors. Antitrust liability may also attach to other anticompetitive licensing restrictions involving copyrighted works.

Id. Is this approach consistent with the general principles of antitrust and intellectual property law we have discussed so far?

PROBLEM

Problem 8-4. Microsoft Corporation succeeds in registering a trademark in the name "Windows" to describe a graphical user interface on a computer operating system. Microsoft agrees to license the "Windows" trademark to certain companies who wish to indicate that their application programs are "Windows-compatible." However, Microsoft requires as a condition of the license that the companies sell only application programs that run exclusively on Microsoft Windows operating systems. Has Microsoft violated the antitrust laws? What role should market power play in the analysis? Does the Windows trademark itself confer market power?

Note on Other Common Licensing Restrictions

Besides exclusivity, a number of other common features of intellectual property licenses raise potential antitrust issues. We consider several here.

Grantback Clauses. Intellectual property owners sometimes require as a condition of the license that the licensee "grant back" to the licensor the rights to use any improvements the licensee makes in the technology. Some grantback clauses simply require that the licensee agree not to sue the licensor for infringing on any improvement patents that result from the licensee's work. Others are more strict, however, requiring the assignment or exclusive license of the rights to any improvement.

One reason licensors use grantback clauses is obvious — they don't want to be held "hostage" by their licensees or lose the right to practice their invention. By retaining the right to use any improvements the licensee makes, the original intellectual property owner can ensure that she stays current in her technology. Grantback clauses may also help to avoid the "blocking patents" problem, by ensuring that the senior intellectual property owner always has the dominant right.

The precise nature of the grantback clause may differ from case to case. Some patentees require that their licensee *assign* any improvement patents back to the licensee. In this instance, the grantback clause avoids the blocking patents problem entirely, since the original patentee ends up owning both the original and the improvement patents. By contrast, if the original patentee requires only that the improver "license back" the right to use the improvement, the grantback clause looks more like an ex ante cross-licensing solution to a prospective blocking patents problem.

On the other hand, grantback clauses may be used to stifle competition. Grantbacks allow a dominant industry player to license her intellectual property freely, without worrying that her licensees may develop an improved process for manufacturing her product and therefore compete with her. Indeed, such licenses may be an effective way of *preventing* serious competition, since they allow the dominant player to "capture" potential competitors and take advantage of their ideas. Further, because licensees no longer have the exclusive (or perhaps any) right to their improvements, they may be less willing to invest in research and development of such improvements.

Traditionally, grantback clauses have not been considered to violate the antitrust laws.[11] The Antitrust Division's current Intellectual Property Guidelines, however, conclude that "[g]rantbacks may adversely affect competition . . . if they substantially reduce the licensee's incentives to engage in research and development and thereby limit rivalry in innovation markets." In such circumstances, federal agencies will weigh any harm to competition against "offsetting procompetitive effects, such as (1) promoting the dissemination of licensees' improvements to the licensed technology, (2) increasing the licensors' incentives to disseminate the licensed technology, or (3) otherwise increasing competition and output in a relevant technology or innovation market." Intellectual Property Guidelines, §5.6. The guidelines encourage the use of nonexclusive grantback clauses.

Is there reason to treat patent grantbacks differently from copyright grantbacks, because of the difference between the "blocking patents" rule and the derivative works rule in copyright?

Field-of-use Restrictions. Alluded to earlier, field-of-use restrictions in intellectual property licenses grant licensees the right to work a patent in only a limited area. The area may be limited either geographically or by product market. Such restrictions may enable a patentee to exploit his technology efficiently in several markets while maintaining the exclusivity granted him by the patent laws. But field-of-use restrictions may also be an efficient means

---

11. A notable exception is United Shoe Machinery Co. v. United States, 258 U.S. 451 (1922).

of enforcing a territorial market division agreement between horizontal competitors. Horizontal market division can have the effect of creating mini-monopolies in what would otherwise be a competitive market. If the patent licensed is a sham, or is only a minor part of the business of the licensor and licensee, it is possible that the patent license is really a "front" for such a market division scheme. Territorial market division outside the scope of a patent is illegal per se under the rule of United States v. Topco Associates, Inc., 405 U.S. 596 (1972); 35 U.S.C. §261.

Are there legitimate reasons for some field-of-use restrictions that inhere in the nature of the intellectual property laws? Consider, for example, a geographic market division scheme that grants each licensee an exclusive territory in a different country. Because of the national nature of patents, this means that each licensee receives the exclusive right to work the patent in one country (although not the exclusive worldwide right to sell the technology). Is there anything wrong with such a scheme? See United States v. Westinghouse Elec. Corp., 648 F.2d 642 (9th Cir. 1981) (approving such an arrangement).

Extensions of Patent Term. This is primarily an issue in patent law, because patents have a shorter term than copyrights. A license may seek to extend the exclusive rights of the patent laws beyond the old 17-year (or the new 20-year) patent term, either by compelling the continued payment of royalties or by imposing collateral obligations (such as exclusive dealing arrangements or grantback clauses) that do not terminate when the patent expires.

Whether extensions of the patent term by contract should be of concern to anyone but the contracting parties themselves has been a matter of some debate. Compare Robert Merges, Reflections on Current Legislation Affecting Patent Misuse, 70 J. Pat. & Trademk. Off. Society 793, 801-02 (1988) (patent term extensions are anticompetitive) with Mark A. Lemley, Comment, The Economic Irrationality of the Patent Misuse Doctrine, 78 Cal. L. Rev. 1599, 1630 (1990) (agreements to extend patent term are not anticompetitive). Is the issue here the same that we have been considering repeatedly throughout this book — the right of private parties to change the scope of the intellectual property laws by contract? Or do the antitrust laws put a different spin on the debate?

Finally, note that while patents and copyrights have fixed dates on which they expire, trademarks and trade secrets do not. Licensees of multiple intellectual property rights (i.e., both patents and trade secrets) sometimes allege that the intellectual property owner is illegally extending the patent term by continuing to enforce her trade secret rights after her patents have expired. The Justice Department made such a claim in United States v. Pilkington plc, Civ. No. 94-345-TUC-WDB (D. Ariz. filed May 25, 1994). Assuming that the licensor still has valid trade secrets, is there anything wrong with continuing such a "group license"? See Christianson v. Colt Indus. Operating Corp., 870 F.2d 1292, 1303 (7th Cir. 1989), *cert. denied*, 493 U.S. 822 (1989) (trade secrets may extend beyond the expiration of a patent on the same technology, provided that section 112 of the Patent Act did not require the patentee to disclose the secrets in the patent itself).

2. Horizontal Restraints

a. Pooling and Cross-Licensing

Cartels — loosely defined as agreements among competitors to restrict output and raise prices — are illegal under section 1 of the Sherman Act, 15 U.S.C. §1. When competitors have agreed to set prices or output levels or divide markets, the Sherman Act may be violated without regard to proof of market power or effect. This is the "per se" rule. However, an early Supreme Court case concluded that it may not be appropriate to subject price agreements involving patented products to the per se rule.

United States v. General Electric Company et al.
Supreme Court of the United States
272 U.S. 476 (1926)

. . .

Mr. Chief Justice TAFT delivered the opinion of the Court.

This is a bill in equity brought by the United States in the District Court for the Northern District of Ohio to enjoin the General Electric Company, the Westinghouse Electric and Manufacturing Company, and the Westinghouse Lamp Company from further violation of the Anti-Trust Act of July 2, 1890. 26 Stat. 209, c. 647. The bill made two charges, one that the General Electric Company in its business of making and selling incandescent electric lights had devised and was carrying out a plan for their distribution throughout the United States by a number of so-called agents, exceeding 21,000, to restrain interstate trade in such lamps and to exercise a monopoly of the sale thereof; and, second, that it was achieving the same illegal purpose through a contract of license with the defendants, the Westinghouse Electric and Manufacturing Company and the Westinghouse Lamp Company. . . .

The Government alleged that the system of distribution adopted was merely a device to enable the Electric Company to fix the resale prices of lamps in the hands of purchasers, that the so-called agents were in fact wholesale and retail merchants, and the lamps passed through the ordinary channels of commerce in the ordinary way, and that the restraint was the same and just as unlawful as if the so-called agents were avowed purchasers handling the lamps under resale price agreements. The Electric Company answered that its distributors were bona fide agents, that it had the legal right to market its lamps and pass them directly to the consumer by such agents, and at prices and by a system prescribed by it and agreed upon between it and its agents, there being no limitation sought as to resale prices upon those who purchased from such agents. . . .

But it is said that the system of distribution is so complicated and involves such a very large number of agents distributed throughout the entire country, that the very size and comprehensiveness of the scheme brings it within the Anti-Trust law. We do not question that in a suit under the Anti-Trust Act the circumstance that the combination effected secures domination of so large a part of the business affected as to control prices is usually most important in proof of a monopoly violating the Act. But under the patent law the patentee is given by statute a monopoly of making, using and selling the patented article. The extent of his monopoly in the articles sold and in the territory of the United States where sold is not limited in the grant of his patent, and the comprehensiveness of his control of the business in the sale of the patented article is not necessarily an indication of illegality of his method. As long as he makes no effort to fasten upon ownership of the articles he sells control of the prices at which his purchaser shall sell, it makes no difference how widespread his monopoly. It is only when he adopts a combination with others, by which he steps out of the scope of his patent rights and seeks to control and restrain those to whom he has sold his patented articles in their subsequent disposition of what is theirs, that he comes within the operation of the Anti-Trust Act. The validity of the Electric Company's scheme of distribution of its electric lamps turns, therefore, on the question whether the sales are by the company through its agents to the consumer, or are in fact by the company to the so-called agents at the time of consignment. The distinction in law and fact between an agency and a sale is clear. For the reasons already stated, we find no ground for inference that the contracts made between the company and its agents are, or were intended to be, other than what their language makes them.

The Government relies in its contention for a different conclusion on the case of Dr. Miles Medical Company v. John D. Park & Sons Company; 220 U.S. 373. That case was a bill in equity brought by the Miles Medical Company to enjoin Park & Sons Company from continuing an alleged conspiracy with a number of wholesale and retail dealers in proprietary medicines, to induce the persons who had entered into certain agency contracts, to the number of 21,000 through the country, to break their contracts of agency with the Medical Company, to the great injury of that company. The agency concerned the sale of proprietary medicines prepared by secret methods and formulas and identified by distinctive packages and trade-marks. The company had an extensive trade throughout the United States and certain foreign countries. It had been its practice to sell its medicines to jobbers and wholesale druggists, who in turn sold to retail druggists for sale to the customer. It had fixed not only the price of its own sales to jobbers and wholesale dealers but also the prices of jobbers and small dealers. The defendants had inaugurated a cut-rate or cut-price system which had caused great damage to the complainant's business, injuriously affected its reputation and depleted the sales of its remedies. The bill was demurred to, on the ground that the methods set forth in the bill, by which attempt was made to control the sales or prices to consumers, was illegal both at common law and under the Anti-Trust Act, and deprived the bill of any equity. This was the issue considered by the Court.

The plan of distribution of the Miles Medical Company resembled in many details the plan of distribution in the present case, except that the subject matter there was medicine by a secret formula, and not a patented article. But there were certain vital differences. These led the Circuit Court of Appeals (164 Fed. 803) to declare that the language of the so-called contracts of agency was false in its purport, and merely used to conceal what were really sales to the so-called agents. This conclusion was sustained by certain allegations in the bill inconsistent with the contracts of agency, to the effect that the Medical Company did sell to these so-called agents the medical packages consigned. This Court, however, without reference to these telltale allegations of the bill, found in the contracts themselves and their operation plain provision for purchases by the so-called agents, which necessarily made the contracts, as to an indefinite amount of the consignments to them, contracts of sale rather than of agency. The Court therefore held that the showing made was of an attempt by the Miles Medical Company, through its plan of distribution, to hold its purchasers, after the purchase at full price, to an obligation to maintain prices on a resale by them. This is the whole effect of the *Miles Medical* case. . . .

We are of opinion, therefore, that there is nothing as a matter of principle, or in the authorities, which requires us to hold that genuine contracts of agency like those before us, however comprehensive as a mass or whole in their effect, are violations of the Anti-Trust Act. The owner of an article, patented or otherwise, is not violating the common law, or the Anti-Trust law, by seeking to dispose of his article directly to the consumer and fixing the price by which his agents transfer the title from him directly to such consumer. The first charge in the bill can not be sustained.

Second. Had the Electric Company, as the owner of the patents entirely controlling the manufacture, use and sale of the tungsten incandescent lamps, in its license to the Westinghouse Company, the right to impose the condition that its sales should be at prices fixed by the licensor and subject to change according to its discretion? The contention is also made that the license required the Westinghouse Company not only to conform in the matter of the prices at which it might vend the patented articles, but also to follow the same plan as that which we have already explained the Electric Company adopted in its distribution. It does not appear that this provision was express in the license, because no such plan was set out therein; but even if the construction urged by the Government is correct, we think the result must be the same.

The owner of a patent may assign it to another and convey, (1) the exclusive right to make, use and vend the invention throughout the United States, or, (2) an undivided part or share of that exclusive right, or (3) the exclusive right under the patent within and through a specific part of the

United States. But any assignment or transfer short of one of these is a license, giving the licensee no title in the patent and no right to sue at law in his own name for an infringement. Waterman v. Mackenzie, 138 U.S. 252, 255; Gayler v. Wilder, 10 How. 477, 494, 495; Moore v. Marsh, 7 Wall. 515, and Crown Company v. Nye Tool Works, 261 U.S. 24, 30. Conveying less than title to the patent, or part of it, the patentee may grant a license to make, use and vend articles under the specifications of his patent for any royalty or upon any condition the performance of which is reasonably within the reward which the patentee by the grant of the patent is entitled to secure. It is well settled, as already said, that where a patentee makes the patented article and sells it, he can exercise no future control over what the purchaser may wish to do with the article after his purchase. It has passed beyond the scope of the patentee's rights. Adams v. Burke, 17 Wall. 453; Bloomer v. McQuewan, 14 How. 539; Mitchell v. Hawley, 16 Wall. 544; Hobbie v. Jennison, 149 U.S. 355; Keeler v. Standard Folding Bed Co., 157 U.S. 659. But the question is a different one which arises when we consider what a patentee who grants a license to one to make and vend the patented article may do in limiting the licensee in the exercise of the right to sell. The patentee may make and grant a license to another to make and use the patented articles, but withhold his right to sell them. The licensee in such a case acquires an interest in the articles made. He owns the material of them and may use them. But if he sells them, he infringes the right of the patentee, and may be held for damages and enjoined. If the patentee goes further, and licenses the selling of the articles, may he limit the selling by limiting the method of sale and the price? We think he may do so, provided the conditions of sale are normally and reasonably adapted to secure pecuniary reward for the patentee's monopoly. One of the valuable elements of the exclusive right of a patentee is to acquire profit by the price at which the article is sold. The higher the price, the greater the profit, unless it is prohibitory. When the patentee licenses another to make and vend, and retains the right to continue to make and vend on his own account, the price at which his licensee will sell will necessarily affect the price at which he can sell his own patented goods. It would seem entirely reasonable that he should say to the licensee, "Yes, you may make and sell articles under my patent, but not so as to destroy the profit that I wish to obtain by making them and selling them myself." He does not thereby sell out-right to the licensee the articles the latter may make and sell, or vest absolute ownership in them. He restricts the property and interest the licensee has in the goods he makes and proposes to sell.

This question was considered by this Court in the case of Bement v. National Harrow Company, 186 U.S. 70. A combination of manufacturers owning a patent to make float spring tool harrows, licensed others to make and sell the products under the patent, on condition that they would not during the continuance of the license sell the products at a less price, or on more favorable terms of payment and delivery to purchasers, than were set forth in a schedule made part of the license. That was held to be a valid use of the patent rights of the owners of the patent. It was objected that this

made for a monopoly. The Court, speaking by Mr. Justice Peckham, said
(p. 91):

> The very object of these laws is monopoly, and the rule is, with few excep-
> tions, that any conditions which are not in their very nature illegal with regard
> to this kind of property, imposed by the patentee and agreed to by the licensee
> for the right to manufacture or use or sell the article, will be upheld by the
> courts. The fact that the conditions in the contracts keep up the monopoly or
> fix prices does not render them illegal.

. . . The authority of Bement v. National Harrow Company has not
been shaken by the cases we have reviewed.

For the reasons given, we sustain the validity of the license granted by
the Electric Company to the Westinghouse Company. The decree of the
District Court dismissing the bill is

Affirmed.

COMMENTS AND QUESTIONS

1. *General Electric* stands for two distinct propositions. The first has to
do with when restrictions on resale run afoul of the rule against "resale price
maintenance." Resale price maintenance refers to attempts by a product's
original seller to control the price at which the buyer may resell the product.
The general rule in antitrust law has long been that resale price maintenance
agreements (which are really vertical, not horizontal, agreements) are illegal
per se. *General Electric* applies the prohibition on resale price maintenance
with full force to the sale of patented products. Patentees lose control over
their patent products once they have sold them. See United States v. Univis
Lens Co., 316 U.S. 241 (1942) (patents); Bobbs-Merrill Co. v. Straus, 210
U.S. 339 (1908) (same rule applies to copyrights). The presence of intellectual property rights has only a marginal effect on the Court's resolution of
that issue.

2. The second rule of *General Electric* allows a patent holder to control
the price at which a *licensee* who is not a reseller sells the patented product.
This rule has been roundly criticized. See, e.g., P. Areeda & L. Kaplow,
Antitrust Analysis 447-49 (1988). The economic problem with *General Electric* is straightforward. A cartel is a group of competitors who agree to reduce
their output in order to raise prices and capture more profits for themselves.
Because they price above the competitive level and thus produce excess profits, each cartel member has an incentive to "cheat" by expanding its output
to capture a larger percentage of the profits. If all members cheat, though,
output will increase back to the competitive level, and prices will fall. Cartels,
therefore, are inherently unstable.

The only way to run a cartel effectively over the long run is to employ
some sort of "policing" mechanism to prevent cheating. The difficulty with

most such mechanisms is that they must be secret, since cartels are illegal (at least in the United States). The *General Electric* decision has the effect of providing the perfect policing mechanism for cartel enforcers. By patenting a product and licensing it to its competitors, a company can control the prices at which its competitors sell under *General Electric*. What is more, since such control is legal, the patent holder can enforce the cartel in court, bringing breach of contract actions against any competitor who tries to cut prices! As a result, *General Electric* would appear to allow a patent holder in a competitive or oligopolistic industry to convert its control over one product into effective control over an entire market.

Of course, we do not want to discourage all such licensing. Rather, the trick is to determine whether the license really does enable competitors to take advantage of a new technology, or whether the license is merely a sham designed to facilitate a cartel. How might we distinguish these two situations?

3. Is there a good rationale for giving a patentee control over licensee pricing? The Court suggests that the patentee may wish to avoid losing profits to its licensee. This rationale makes sense where the patentee and the licensee are likely to compete directly, but not otherwise. Does it suggest that the courts should treat exclusive licenses differently from nonexclusive licenses? Is there an economic reason to be less concerned about one or the other?

4. Subsequent courts have not been kind to *General Electric*. While it has never been expressly overruled, it has been so limited that it has little force today.

United States v. Line Material Co. et al.
Supreme Court of the United States
333 U.S. 287 (1948)

. . .

Mr. Justice REED delivered the opinion of the Court.

The United States sought an injunction under §§1 and 4 of the Sherman Act in the District Court against continuance of violations of that Act by an allegedly unlawful combination or conspiracy between appellees, through contracts, to restrain interstate trade in certain patented electrical devices. The restraint alleged arose from a cross-license arrangement between the patent owners, Line Material Company and Southern States Equipment Corporation, to fix the sale price of the devices, to which arrangement the other appellees, licensees to make and vend, adhered by supplemental contracts. . . .

On consideration of the agreements and the circumstances surrounding their negotiation and execution, the District Court found that the arrangements, as a whole, were made in good faith, to make possible the manufacture by all appellees of the patented devices, to gain a legitimate return to the patentees on the inventions; and that, apart from the written agreements, there was no undertaking between the appellees or any of them to fix prices.

Being convinced, as we indicated at the first of this opinion, that the *General Electric* case controlled and permitted such price arrangements as are disclosed in the contracts, the District Court dismissed the complaint. The Government attacks the rationale of the *General Electric* case and urges that it be overruled, limited and explained or differentiated. . . .

General Electric is a case that has provoked criticism and approval. It had only bare recognition in Ethyl Gasoline Corp. v. United States, 309 U.S. 436, 456. That case emphasized the rule against the extension of the patent monopoly, p. 456, to resale prices or to avoid competition among buyers. Pages 457-58. We found it unnecessary to reconsider the rule in United States v. Masonite Corp., 316 U.S. 265, 277, although the arrangement there was for sale of patented articles at fixed prices by dealers whom the patentee claimed were del credere agents. As we concluded the patent privilege was exhausted by a transfer of the articles to certain agents who were part of the sales organization of competitors, discussion of the price-fixing limitation was not required. In Katzinger Co. v. Chicago Mfg. Co., 329 U.S. 394, 398, where a suit was brought to recover royalties on a license with price limitations, this Court refused to examine the *General Electric* rule because of the claimed illegality of the Katzinger patent. If the patent were invalid, the price-fixing agreement would be unlawful. We affirmed the action of the Circuit Court of Appeals in remanding the case to the District Court to determine the validity of the patent. The *General Electric* case was cited with approval in Carbice Corp. v. American Patents Development Corp., 283 U.S. 27, 31. Other courts have explained or distinguished the *General Electric* rule. As a reason for asking this Court to reexamine the rule of the *General Electric* case, the Government states that price maintenance under patents through various types of agreements is involved in certain pending cases. Furthermore, the point is made that there is such a "host of difficult and unsettled questions" arising from the *General Electric* holding that the simplest solution is to overrule the precedent on the power of a patentee to establish sale prices of a licensee to make and vend a patented article.

Such a liquidation of the doctrine of a patentee's power to determine a licensee's sale price of a patented article would solve problems arising from its adoption. Since 1902, however, when Bement v. National Harrow Co., 186 U.S. 70, was decided, a patentee has been able to control his licensee's sale price within the limits of the patent monopoly. Litigation that the rule has engendered proves that business arrangements have been repeatedly, even though hesitatingly, made in reliance upon the contractors' interpretation of its meaning. Appellees urge that Congress has taken no steps to modify the rule. Such legislative attitude is to be weighed with the counterbalancing fact that the rule of the *General Electric* case grew out of a judicial determination. The writer accepts the rule of the *General Electric* case as interpreted by the third subdivision of this opinion. As a majority of the Court does not agree with that position, the case cannot be reaffirmed on that basis. Neither is there a majority to overrule *General Electric*. In these circumstances, we must proceed to determine the issues on the assumption that *General Electric* continues as a precedent. Furthermore, we do not think it wise to undertake to explain, further than the facts of this case require, our views as to the applicability of patent price limitation in the various situations listed by the Government. . . .

III. The Determination of the Issue

Under the above-mentioned assumption as to *General Electric,* the ultimate question for our decision on this appeal may be stated, succinctly and abstractly, to be as to whether in the light of the prohibition of §1 of the Sherman Act, note 1, supra, two or more patentees in the same patent field may legally combine their valid patent monopolies to secure mutual benefits for themselves through contractual agreements, between themselves and other licensees, for control of the sale price of the patented devices. . . .

It is equally well settled that the possession of a valid patent or patents does not give the patentee any exemption from the provisions of the Sherman Act beyond the limits of the patent monopoly. By aggregating patents in one control, the holder of the patents cannot escape the prohibitions of the Sherman Act. See Standard Sanitary Mfg. Co. v. United States, 226 U.S. 20; United States v. United States Gypsum Co., post, p. 364. During its term, a valid patent excludes all except its owner from the use of the protected process or product. United States v. United Shoe Machinery Co., 247 U.S. 32, 58; Special Equipment Co. v. Coe, 324 U.S. 370, 378. This monopoly may be enjoyed exclusively by the patentee or he may assign the patent "or any interest therein" to others. Rev. Stat. §4898, as amended 55 Stat. 634. As we have pointed out, a patentee may license others to make and vend his invention and collect a royalty therefor. Thus we have a statutory monopoly by the patent and by the Sherman Act a prohibition, not only of monopoly or attempt to monopolize, but of every agreement in restraint of trade. . . . It is not the monopoly of the patent that is invalid. It is the improper use of that monopoly. . . .

We are thus called upon to make an adjustment between the lawful restraint on trade of the patent monopoly and the illegal restraint prohibited broadly by the Sherman Act. That adjustment has already reached the point, as the precedents now stand, that a patentee may validly license a competitor to make and vend with a price limitation under the *General Electric* case and that the grant of patent rights is the limit of freedom from competition. . . .

We turn now to the situation here presented of an agreement where one of the patentees is authorized to fix prices under the patents. The argument of respondents is that if a patentee may contract with his licensee to fix prices, it is logical to permit any number of patentees to combine their patents and authorize one patentee to fix prices for any number of licensees. In this present agreement Southern and Line have entered into an arrangement by which Line is authorized to and has fixed prices for devices produced under the Lemmon and Schultz patents. It seems to us, however, that such argument fails to take into account the cumulative effect of such multiple agreements in establishing an intention to restrain. The obvious purpose and effect of the agreement was to enable Line to fix prices for the patented devices. Even where the agreements to fix prices are limited to a small number of patentees, we are of the opinion that it crosses the barrier erected by the Sherman Act against restraint of trade though the restraint is by patentees and their licensees. . . .

We think that this general rule against price limitation clearly applies in the circumstances of this case. Even if a patentee has a right in the absence of a purpose to restrain or monopolize trade, to fix prices on a licensee's sale of the patented product in order to exploit properly his invention or inventions, when patentees join in an agreement as here to maintain prices on their several products, that agreement, however advantageous it may be to stimulate the broader use of patents, is unlawful per se under the Sherman Act. It is more than an exploitation of patents. There is the vice that patentees have combined to fix prices on patented products. It is not the cross-licensing to promote efficient production which is unlawful. There is nothing unlawful in the requirement that a licensee should pay a royalty to compensate the patentee for the invention and the use of the patent. The unlawful element is the use of the control that such cross-licensing gives to fix prices. The mere fact that a patentee uses his patent as whole or part consideration in a contract by which he and another or other patentees in the same patent field arrange for the practice of any patent involved in such a way that royalties or other earnings or benefits from the patent or patents are shared among the patentees, parties to the agreement, subjects that contract to the prohibitions of the Sherman Act whenever the selling price, for things produced under a patent involved, is fixed by the contract or a license authorized by the contract. Licensees under the contract who as here enter into license arrangements, with price-fixing provisions, with knowledge of the contract, are equally subject to the prohibitions.

The decree of the District Court is reversed and the case is remanded for the entry of an appropriate decree in accordance with this opinion.

Mr. Justice DOUGLAS, with whom Mr. Justice BLACK, Mr. Justice MURPHY and Mr. Justice RUTLEDGE join, concurring.

While I have joined in the opinion of the Court, its discussion of the problem is for me not adequate for a full understanding of the basic issue presented. My view comes to this — it is a part of practical wisdom and good law not to permit United States v. General Electric Co., 272 U.S. 476, to govern this situation, though if its premise be accepted, logic might make its application to this case wholly defensible. But I would be rid of United States v. General Electric Co. . . .

COMMENTS AND QUESTIONS

1. Four Justices were explicit in their desire to overrule *General Electric*. Justice Reed said he did not wish to overrule it. What does he advocate instead? Consider Part III of the opinion. Justice Reed concludes that where "patent pooling" is involved (that is, where two or more different patentees in an industry share patents), *General Electric* does not apply and the cooperative activity is illegal. Does this distinction make sense? Why should cooperation between cartel members be illegal if two of them happen to hold patents, but legal if only one does? Or is the Court's decision in *Line Material* merely a way of distinguishing between vertical licensing of a single patent (which we want to encourage) and horizontal cross-licensing between competitors, which may facilitate a cartel?

One problem with this latter theory is that not all patent pooling is anticompetitive. Indeed, an argument can be made that patent pooling is more desirable than controls on the licensing of a single patent. The argument focuses on the problem of "blocking patents" discussed in Chapter 3. Where one company patents an original invention and a second company (possibly but not necessarily a licensee) patents an improvement to that invention, the law leaves the two companies in a rather peculiar situation. In order to make the most efficient use of the invention, each company wants to use both the original invention and its improvement. But alone, neither company can do so without infringing the other's patent. This is the blocking patents problem. One obvious solution is for the companies to cross-license their patents. This allows both companies to use the invention efficiently, benefiting society and preserving some measure of competition. Why should such an arrangement be subject to greater antitrust scrutiny than the sort of cartel facilitation apparent in *General Electric*?

The problem of "blocking patents" is a common one in the patent-antitrust literature. In addition to the examples cited by the Court, see Eleanor M. Fox & Lawrence A. Sullivan, Antitrust 353-54 (1989); Robert P. Merges, Intellectual Property Rights and Bargaining Breakdown: The Case of Blocking Patents, 62 Tenn. L. Rev. 75 (1994). Blocking patents can take one of two forms. First, two or more companies may have valid patents on different steps of a process in which each step is necessary to make the end product. This can be thought of as "pure" technical necessity. Without cross-licensing, neither company would be able to produce the product at all. In this circumstance, the economic case against applying antitrust law is fairly clear: preventing patent pooling will reduce rather than increase market participation. In this circumstance, at least some courts have been willing to disregard *Line Material* and permit pools of blocking patents. For example, in Carpet Seaming Tape Licensing Corp. v. Best Seam, Inc., 616 F.2d 1133 (9th Cir. 1980), *cert. denied*, 464 U.S. 818 (1983), the court held:

> A well-recognized legitimate purpose for a pooling agreement is exchange of blocking patents. The trial court . . . demonstrates a misunderstanding as to

the nature of the relationship which can give rise to a blocking situation. The trial judge found that the Winkler and Clymin patents did not block the Burgess patents, but ignored the possibility that the Burgess patents may well have blocked the Winkler and Clymin patents despite well-established law that patents on basic processes and products may block patents on improvements to those products and processes. As the Supreme Court stated in Standard Oil Co. v. United States, 283 U.S. at 172, n.5:

> "This is often the case where patents covering improvements of a basic process, owned by one manufacturer, are granted to another. A patent may be rendered quite useless, or 'blocked,' by another unexpired patent which covers a vitally related feature of the manufacturing process. Unless some agreement can be reached, the parties are hampered and exposed to litigation. And, frequently, the cost of litigation to a patentee is greater than the value of a patent for a minor improvement."
>
> . . .

The economics underlying this attitude toward the accumulation of improvement patents is ably and succinctly stated by Professor Ward S. Bowman:

> "If . . . one patent was subservient to the other, and improvement patent unusable without infringing the basic patent, then combining or pooling them eliminates no user alternative. In terms of possible trade restraint, this case is indistinguishable from a vertical merger. The two patents combined . . . could not restrict output or raise price any more than if the two were exploited separately."

W. Bowman, Patent and Antitrust Law at 201 (1973).

See also Clorox Co. v. Sterling Winthrop, 932 F. Supp. 469 (E.D.N.Y. 1996) (agreement settling trademark dispute over the terms Lysol and Pine-Sol by limiting the product markets each could sell to was not anticompetitive, since owner of Pine-Sol can compete with Lysol using other marks).

A more common problem involves a number of patents whose validity is open to question. This may be thought of as "practical technical necessity." Rather than entering into prohibitively expensive and protracted litigation over the validity of their respective patents, can firms in this situation settle their differences by pooling their patents (at least one of which is presumably valid)? Standard Oil Co. v. United States, 283 U.S. 163 (1931), seems to suggest that such an agreement would be treated under the rule of reason. Of course, that does not mean that the agreement will always be upheld. If the companies with blocking patents together would comprise only a small part of the industry, their agreement would presumably survive rule of reason scrutiny. But if the new technology encompassed by the patents would dominate a market, even *Standard Oil* presents an antitrust problem. Is this a desirable outcome? On the one hand, allowing the agreement would get the product to market more quickly and inexpensively. Further, because each firm in our hypothetical situation holds an arguably valid patent, one firm is likely to end up with a monopoly in the product anyway (the firm whose patent is upheld). On the other hand, allowing competitors to resolve their differences by cross-licensing their patents could foreclose any incentive for those firms to compete in the future. Instead of a monopoly patent holder being faced with several competitors trying to unseat it, the patent cartel would face no foreseeable competition. Whether you believe the rule of reason or outright legality should govern this situation depends on your assessment of the relative costs of each approach.

2. Does it matter for antitrust analysis whether the parties to the pool are in the same market? Is the existence of the pool presumptive evidence that they are? Market definition problems have plagued patent pooling cases. See Roger B. Andewelt, Analysis of Patent Pools Under the Antitrust Laws, 53 Antitrust L.J. 611 (1984).

3. Can an argument be made in favor of patent pooling on the grounds that it reduces transactions costs? Consider the following excerpt:

> Patent pools arise because it makes sense for firms in industries characterized by crossing and conflicting IPRs [intellectual property rights] to "institutionalize" technology exchange. In many cases, pools are creatures of necessity. For example, where different firms hold patents on the basic building blocks of the industry's products, they will have to cross-license to produce at all. This was the case, for example, with the aircraft industry in the early days of the twentieth century, and with sewing machines. Even where no single patent or set of patents is essential, however, firms in an industry often find that they engage in such frequent negotiations that a regularized institution with formal rules, or even general guidelines, is helpful in reducing transaction costs. An example of a pool such as this is the one formed by the early shoe machinery industry. The economic literature on institutions explains this quite well; to use one popular metaphor, the "repeat-play" nature of an institution makes it easier to reach agreement on any particular issue, because disparities tend to balance out over many transactions. . . .
>
> All patent pools share one fundamental characteristic: they provide a regularized transactional mechanism in place of the statutory property rule baseline which requires an individual bargain for each transaction. But my review of particular pools shows that this general structure has encompassed a diversity of organizational forms. . . .
>
> Many patent pools are in essence contracts. Firms agree to consolidate patents and license them collectively. The royalties from licensing the patents, and sometimes from sales of products made by one or more parties to the pooling agreement, are divided according to a contractual formula. . . . In this simple example, the contract integrated numerous transactions that would otherwise have been negotiated separately. And, most importantly for present purposes, it translated the contribution of each major patent holder into a precise percentage of the royalty stream.

Robert P. Merges, Contracting into Liability Rules: Intellectual Property Rights and Collective Rights Organizations, 84 Cal. L. Rev. 1293 (1996).

Does a similar argument justify copyright pools? Why or why not?

PROBLEM

Problem 8-5. Discworld sues the two largest makers of compact discs, alleging that their patented processes infringe its pioneer patent on compact disc technology. After years of discovery, the parties settle the suit by agreeing that each company will license all its patents to the other two, forming a "pool." The settlement agreement contains a "grantback" clause, under which each party agrees to license to pool members any improvements in its technology. The parties to the pool also agree that they will license the collective patents to outside companies, but only if those outside companies agree to use only the technology developed and patented by members of the pool.

NextDisc, Inc., a new entrant into the compact disc market, is approached by the pool and asked to take a license. NextDisc considers the requested royalty to be outrageously high. Rather than taking a license, NextDisc sues the pool members for violating the antitrust laws. What result?

---

b. Industry Standardization

Broadcast Music, Inc., et al. v. Columbia Broadcasting System, Inc.
Supreme Court of the United States
441 U.S. 1 (1979)

Mr. Justice WHITE delivered the opinion of the Court.

This case involves an action under the antitrust and copyright laws brought by respondent Columbia Broadcasting System, Inc. (CBS), against petitioners, American Society of Composers, Authors and Publishers (ASCAP) and Broadcast Music, Inc. (BMI), and their members and affiliates. The basic question presented is whether the issuance by ASCAP and BMI to CBS of blanket licenses to copyrighted musical compositions at fees negotiated by them is price fixing per se unlawful under the antitrust laws. . . .

BMI, a nonprofit corporation owned by members of the broadcasting industry, was organized in 1939, is affiliated with or represents some 10,000 publishing companies and 20,000 authors and composers, and operates in much the same manner as ASCAP. Almost every domestic copyrighted composition is in the repertory either of ASCAP, with a total of three million compositions, or of BMI, with one million.

Both organizations operate primarily through blanket licenses, which give the licensees the right to perform any and all of the compositions owned by the members or affiliates as often as the licensees desire for a stated term. Fees for blanket licenses are ordinarily a percentage of total revenues or a flat dollar amount, and do not directly depend on the amount or type of music used. Radio and television broadcasters are the largest users of music, and almost all of them hold blanket licenses from both ASCAP and BMI. Until this litigation, CBS held blanket licenses from both organizations for its television network on a continuous basis since the late 1940's and had never attempted to secure any other form of license from either ASCAP or any of its members. Id., at 752-754.

The complaint filed by CBS charged various violations of the Sherman Act and the copyright laws. CBS argued that ASCAP and BMI are unlawful monopolies and that the blanket license is illegal price fixing, an unlawful tying arrangement, a concerted refusal to deal, and a misuse of copyrights. The District Court, though denying summary judgment to certain defendants, ruled that the practice did not fall within the per se rule. 337 F. Supp. 394, 398 (SDNY 1972). After an 8-week trial, limited to the issue of liability, the court dismissed the complaint, rejecting again the claim that the blanket license was price fixing and a per se violation of §1 of the Sherman Act, and holding that since direct negotiation with individual copyright owners is available and feasible there is no undue restraint of trade, illegal tying, misuse of copyrights, or monopolization. 400 F. Supp., at 781-783. . . .

[The Court of Appeals reversed.]

A

As a preliminary matter, we are mindful that the Court of Appeals' holding would appear to be quite difficult to contain. If, as the court held, there is a *per se* antitrust violation whenever ASCAP issues a blanket license to a television network for a single fee, why would it not also be automatically illegal for ASCAP to negotiate and issue blanket licenses to individual radio or television stations or to other users who perform copyrighted music for profit? Likewise, if the present network licenses issued through ASCAP on behalf of its members are *per se* violations, why would it not be equally illegal for the members to authorize ASCAP to issue licenses establishing various categories of uses that a network might have for copyrighted music and setting a standard fee for each described use?

Although the Court of Appeals apparently thought the blanket license could be saved in some or even many applications, it seems to us that the *per se* rule does not accommodate itself to such flexibility and that the observations of the Court of Appeals with respect to remedy tend to impeach the *per se* basis for the holding of liability.

CBS would prefer that ASCAP be authorized, indeed directed, to make all its compositions available at standard per-use rates within negotiated categories of use. 400 F. Supp., at 747 n. 7.[28] But if this in itself or in conjunction

---

28. Surely, if ASCAP abandoned the issuance of all licenses and confined its activities to policing the market and suing infringers, it could hardly be said that member copyright owners

with blanket licensing constitutes illegal price fixing by copyright owners, CBS urges that an injunction issue forbidding ASCAP to issue any blanket license or to negotiate any fee except on behalf of an individual member for the use of his own copyrighted work or works.[29] Thus, we are called upon to determine that blanket licensing is unlawful across the board. We are quite sure, however, that the *per se* rule does not require any such holding.

B

In the first place, the line of commerce allegedly being restrained, the performing rights to copyrighted music, exists at all only because of the copyright laws. Those who would use copyrighted music in public performances must secure consent from the copyright owner or be liable at least for the statutory damages for each infringement and, if the conduct is willful and for the purpose of financial gain, to criminal penalties. Furthermore, nothing in the Copyright Act of 1976 indicates in the slightest that Congress intended to weaken the rights of copyright owners to control the public performance of musical compositions. Quite the contrary is true. Although the copyright laws confer no rights on copyright owners to fix prices among themselves or otherwise to violate the antitrust laws, we would not expect that any market arrangements reasonably necessary to effectuate the rights that are granted would be deemed a per se violation of the Sherman Act. Otherwise, the commerce anticipated by the Copyright Act and protected against restraint by the Sherman Act would not exist at all or would exist only as a pale reminder of what Congress envisioned.

C

More generally, in characterizing this conduct under the *per se* rule, our inquiry must focus on whether the effect and, here because it tends to show effect, see United States v. United States Gypsum Co., 438 U.S. 422, 436 n. 13 (1978), the purpose of the practice are to threaten the proper operation of our predominantly free-market economy — that is, whether the practice facially appears to be one that would always or almost always tend to restrict competition and decrease output, and in what portion of the market, or in-

---

would be in violation of the antitrust laws by not having a common agent issue per-use licenses. Under the copyright laws, those who publicly perform copyrighted music have the burden of obtaining prior consent. Cf. Zenith Radio Corp. v. Hazeltine Research, Inc., 395 U.S., at 139-140.

29. In its complaint, CBS alleged that it would be "wholly impracticable" for it to obtain individual licenses directly from the composers and publishing houses, but it now says that it would be willing to do exactly that if ASCAP were enjoined from granting blanket licenses to CBS or its competitors in the network television business.

stead one designed to "increase economic efficiency and render markets more, rather than less, competitive." Id., at 441 n.16; see National Society of Professional Engineers v. United States, 435 U.S., at 688; Continental T.V., Inc. v. GTE Sylvania Inc., 433 U.S., at 50 n.16; Northern Pac. R. Co. v. United States, 356 U.S., at 4.

The blanket license, as we see it, is not a "naked restrain[t] of trade with no purpose except stifling of competition," White Motor Co. v. United States, 372 U.S. 253, 263 (1963), but rather accompanies the integration of sales, monitoring, and enforcement against unauthorized copyright use. See L. Sullivan, Handbook of the Law of Antitrust §59, p. 154 (1977). As we have already indicated, ASCAP and the blanket license developed together out of the practical situation in the marketplace: thousands of users, thousands of copyright owners, and millions of compositions. Most users want unplanned, rapid, and indemnified access to any and all of the repertory of compositions, and the owners want a reliable method of collecting for the use of their copyrights. Individual sales transactions in this industry are quite expensive, as would be individual monitoring and enforcement, especially in light of the resources of single composers. Indeed, as both the Court of Appeals and CBS recognize, the costs are prohibitive for licenses with individual radio stations, nightclubs, and restaurants, 562 F.2d, at 140 n.26, and it was in that milieu that the blanket license arose.

A middleman with a blanket license was an obvious necessity if the thousands of individual negotiations, a virtual impossibility, were to be avoided. Also, individual fees for the use of individual compositions would presuppose an intricate schedule of fees and uses, as well as a difficult and expensive reporting problem for the user and policing task for the copyright owner. Historically, the market for public-performance rights organized itself largely around the single-fee blanket license, which gave unlimited access to the repertory and reliable protection against infringement. When ASCAP's major and user-created competitor, BMI, came on the scene, it also turned to the blanket license.

With the advent of radio and television networks, market conditions changed, and the necessity for and advantages of a blanket license for those users may be far less obvious than is the case when the potential users are individual television or radio stations, or the thousands of other individuals and organizations performing copyrighted compositions in public. But even for television network licenses, ASCAP reduces costs absolutely by creating a blanket license that is sold only a few, instead of thousands, of times, and that obviates the need for closely monitoring the networks to see that they do not use more than they pay for. ASCAP also provides the necessary resources for blanket sales and enforcement, resources unavailable to the vast majority of composers and publishing houses. Moreover, a bulk license of some type is a necessary consequence of the integration necessary to achieve these efficiencies, and a necessary consequence of an aggregate license is that its price must be established.

Mr. Justice STEVENS, dissenting.

The Court holds that ASCAP's blanket license is not a species of price fixing categorically forbidden by the Sherman Act. I agree with that holding. The Court remands the cases to the Court of Appeals, leaving open the question whether the blanket license as employed by ASCAP and BMI is unlawful under a rule-of-reason inquiry. I think that question is properly before us now and should be answered affirmatively. . . .

COMMENTS AND QUESTIONS

1. There is no "technical necessity" for the copyright collective in *Broadcast Music* in the same sense that there is for a cross-license of blocking patents. It is certainly possible in theory to require each copyright holder to sell licenses independently, and to punish cooperation in the market for copyright licenses in the same way that section 1 punishes other cooperation among competitors. But the result would be that most licenses currently sold would not be sold, because of prohibitive transaction costs. "Economic necessity" therefore requires that we exempt such sales from the per se rule. The Court concludes that the rule of reason applies, and remands the case.

The transactions cost rationale seems to require that copyright collectives have a certain degree of market power. After all, if the industry were composed of thousands of small "collectives," transactions costs would still be a significant barrier to copyright licenses. In fact, the "efficient scale" of a copyright collective would seem to be the entire industry. Given this fact, should antitrust courts be less concerned about the market power of copyright collectives? Or, on the contrary, should it subject the industry to pervasive regulation of the sort found in other "natural monopoly" contexts?

2. The Court reasons that since the "market" in question was created by the copyright collectives in the first place, they ought to be protected from per se antitrust scrutiny. But is it fair to say that ASCAP and BMI "created" the market? Certainly, they facilitate the granting of copyright permissions, although they do so at the expense of competition among copyright owners in the sale of permissions. It may be that, absent large organizations such as BMI with significant bargaining power, most musicians would trade away any performance right royalty in exchange for the public exposure that comes from having their songs played on the radio. If this is true, BMI skews the ordinary market distribution in favor of copyright owners, with the result that radio stations pay a higher price.

There are two possible responses to this argument. First, the transactions cost problem discussed above may be so great that absent a copyright collective, artists and broadcasters might never connect at all. As a result, broadcasters would be unable to play songs by artists who would like to have their songs played. Second, BMI may actually restore equity to the bargaining situation by providing countervailing power to match the power of the broadcasters. Individual copyright owners might feel they have no choice but to give up their royalties; they are likely to be in a much better bargaining position as part of a large collective organization.

3. The music industry is not the only one to have developed collectives of copyright owners. In the publishing industry, an organization called the Copyright Clearance Center (CCC) has taken a leading role in granting permissions and collecting fees from people who photocopy pages out of books or periodicals. The CCC's standard fee for photocopying permissions is fairly high — 25 cents per page. (Thus the royalty for copying a 200-page book is $50. This price does not include any reproduction cost.) While the CCC is in much the same situation as BMI or ASCAP, the people with whom it is bargaining (individuals, corporations, and universities) are generally smaller than broadcasters and have less bargaining power. Does this fact make CCC's existence as the dominant publishing collective troublesome from an antitrust perspective?

Note on Intellectual Property and Standard-Setting Organizations

In industries characterized by large standardization externalities — that is, where the market naturally tends towards a single product or type of product — group standard-setting organizations can serve valuable procompetitive purposes. They may allow a number of companies to make and sell competing products that are compatible with each other, thus encouraging more flexible use of the products by consumers and preventing one firm from dominating the market with a de facto standard. See Mark A. Lemley, Antitrust and the Internet Standardization Problem, 28 Conn. L. Rev. 1041 (1996). Unfortunately, standard-setting organizations are sometimes subject to "capture" by a particular participant. In the context of high-technology markets, the most likely means of capturing a standard is by the strategic use of intellectual property rights.[12]

For example, in 1992, the Video Electronics Standards Association (VESA) adopted a computer hardware standard called the VL-Bus standard, which governs the transmission of information between a computer's CPU and its peripheral devices. Each of the members voting to adopt the standard, including Dell Computer Corporation, was required by VESA rules to affirm that it did not own any patent rights that covered the VL-Bus stan-

---

12. In more traditional products markets, capture of standard-setting organizations sometimes takes the form of controlling the organization and using it to vote down a competitor's standard. This is what apparently happened in Allied Tube & Conduit Corp. v. Indian Head, Inc., 486 U.S. 492 (1988). That approach will not have the same effect in industries such as the software market unless the capturing party has intellectual property rights in the dominant standard. Absent such rights, competitors will alter their products to comply with the winning standard. This may require some time and expense, but the expense will be well worth it to competitors if the lock-in effects of the standard are significant.

dard.[13] Dell's representative did in fact make such a statement. Nonetheless, Dell asserted a patent against other VESA members for using the VL-Bus standard eight months later, after the VL-Bus standard had been widely adopted. By working to adopt as a group standard a technology Dell allegedly knew was proprietary,[14] Dell could obtain the help of its competitors in establishing a standard that it would ultimately be able to control.

The competitive harms of this form of capture are relatively clear. Not only does the capturing party end up with exclusive control over the market standard, converting a group standard-setting process into a de facto one, but the capturing party can use the group standard to achieve a dominant position it could not have attained in an open standards competition. Had Dell announced up front that the standards it was backing were proprietary, it is unlikely that the affected industry would have chosen those standards. At the very least, those standards would have faced stiffer competition than they did.

The most likely avenue of antitrust attack against such capture does not involve section 1 at all but rather is an attempted monopolization claim under section 2. Attempted monopolization has three elements — intent to monopolize, anticompetitive conduct in furtherance of that intent, and a dangerous probability of success. Spectrum Sports v. McQuillen, 506 U.S. 970 (1993). Assuming the failure to disclose relevant intellectual property rights was intentional and not an oversight, the first element should be easy to satisfy. Efforts to capture an industry standard in any given case would constitute anticompetitive conduct precisely in the situation where those efforts are likely to lead to monopolization — that is, where the standard being set is one that will likely dominate the industry. Market power may be the necessary result of patent enforcement in some cases, while in others the patent owner's control over the market stems from a failure of information in the market, a failure that the patent owner herself has induced. While the fact that the antitrust defendant does own intellectual property rights governing the technology suggests some caution in applying the antitrust laws, the mere possession of an intellectual property right will not protect its owner from a charge of dominating a market by extending the scope of that right. In the Dell case, the FTC entered into a consent decree in which Dell agreed not to assert its intellectual property rights in the VL-Bus. See In re Dell Computer Corp., 121 FTC 616 (FTC 1995).

Of course, not all patents covering standards will necessarily be anticompetitive. While one approach to standards is to require them to be intellectual property-free (the approach of the Internet Engineering Task Force (IETF),

---

13. Many standard-setting organizations, including the American National Standards Institute (ANSI) and Semiconductor Equipment and Materials International (SEMI) have similar rules.

14. Whether Dell in fact knew this is a matter of some dispute. In her dissent to the Federal Trade Commission's proposed consent decree involving Dell, Commissioner Mary Azcuenaga claimed that there was "no evidence to support such a finding of intentional conduct."

at least until recently), intellectual property can coexist with procompetitive standard-setting. For example, the American National Standards Institute and other groups do not require that an intellectual property owner give up any claim to a standard, but merely that they license their intellectual property rights on a reasonable basis. Other examples of reasonable and even procompetitive uses of intellectual property in the standard-setting context are possible. It is only in that subset of cases where the patent is used as a competitive weapon that concerns about market control are implicated.[15]

c. *Joint Ventures*

A form of cooperative behavior among competitors that has grown especially popular in high technology industries is the joint venture, or strategic alliance. A joint venture occupies a position somewhere between a cartel and a merger. Whereas members of a cartel research, develop, produce, and market goods independently of each other, merely coordinating their output and pricing strategies, joint venturers are competitors who actively cooperate in research and development, production, or marketing. At the same time, they may compete vigorously with respect to sales, output, and pricing decisions.

Because of this unusual structure, joint ventures defy traditional antitrust analysis. See Thomas M. Jorde & David Teece, Innovation, Cooperation, and Antitrust: Balancing Cooperation and Competition, 4 High Tech. L.J. 1 (1989); Joseph Brodley, Joint Ventures and Antitrust Policy, 95 Harv. L. Rev. 1521 (1982). Courts have treated them under either section 1 of the Sherman Act or under section 7 of the Clayton Act, depending on whether the court analogized the joint venture to a cartel or to a merger. At the same time, a growing number of scholars have argued that joint ventures deserve to be treated differently — and generally more favorably — than other horizontal agreements under section 1. These economists point to the competitive efficiencies that can result from the integration of joint venture assets, and to the fact that joint ventures are better for competition than the alternative of merger between the same companies. Research and development joint ventures are particularly important in innovative industries, according to this argument, because of the high research costs in those industries and the speculative nature of new research projects. See Gene Grossman & Carl Shapiro, Research Joint Ventures: An Antitrust Analysis, 2 J.L. Econ. & Org. 315 (1986).

---

15. In rare cases, a rule *precluding* patents on standards might be found to be anticompetitive. In In re Am. Society of Sanitary Eng., 106 FTC 324, 328-29 (1985), the FTC alleged that a standard-setting organization could not refuse to consider revising its standards to include a new product solely on the grounds that that product was patented. The case was settled by consent decree. It is significant that the standard in question was inclusive rather than exclusive, and so allowing the complainant's product to be included would not have restricted the rights of other members to make use of other technology covered by the standard. Nonetheless, the case should serve as a caution for rules such as IETF's requiring that participants relinquish their intellectual property rights.

Because of their unique nature, courts have generally subjected joint ventures only to rule of reason treatment under section 1. See Northrop Corp. v. McDonnell Douglas Corp., 705 F.2d 1030 (9th Cir. 1983) (reasoning that per se condemnation was inappropriate because of the novelty of joint ventures and their potential procompetitive effects).

Courts have also treated joint ventures under section 7 of the Clayton Act. United States v. Penn-Olin Corp., 378 U.S. 158 (1964) involved a joint venture between Pennsalt Chemicals and Olin Mathieson Chemical to produce and sell sodium chlorate. The Court held that section 7 applied to third companies created under joint ownership of two parents. In applying section 7, it wrote:

> The District Court found that "Pennsalt and Olin each possessed the resources and general capability needed to build its own plant in the southeast and to compete with Hooker and [American Potash] in that market. Each could have done so if it had wished." 217 F. Supp. 110, 129.[5] In addition, the District Court found that, contrary to the position of the management of Olin and Pennsalt, "the forecasts of each company indicated that a plant could be operated with profit." Ibid.
>
> The District Court held, however, that these considerations had no controlling significance, except "as a factor in determining whether as a matter of probability *both* companies would have entered the market as individual competitors if Penn-Olin had not been formed. Only in this event would potential competition between the two companies have been foreclosed by the joint venture." Id., at 130. In this regard the court found it "impossible to conclude that as a matter of reasonable probability *both* Pennsalt and Olin would have built plants in the southeast if Penn-Olin had not been created." Ibid. The court made no decision concerning the probability that one would have built "while the other continued to ponder." It found that this "hypothesized situation affords no basis for concluding that Penn-Olin had the effect of substantially lessening competition." Ibid. That would depend, the court said, "upon the competitive impact which Penn-Olin will have as against that which might have resulted if Pennsalt or Olin had been an individual market entrant." Ibid. The court found that this impact could not be determined from the record in this case. "Solely as a matter of theory," it said, ". . . no reason exists to suppose that Penn-Olin will be a less effective competitor than Pennsalt or Olin would have been. The contrary conclusion is the more reasonable." Id., at 131.

---

5. The court explained further: "At the time when the joint venture was agreed upon Pennsalt and Olin each had an extensive background in sodium chlorate. Pennsalt had years of experience in manufacturing and selling it. Although Olin had never been a commercial manufacturer, it possessed a substantially developed manufacturing technique of its own, and also had available to it a process developed by Vickers-Krebs with whom it had been negotiating to construct a plant. Olin had contacts among the southeastern pulp and paper mills which Pennsalt lacked, but Pennsalt's own estimates indicate that in a reasonable time it would develop adequate business to support a plant if it decided to build. A suitable location for a plant was available to each company — Calvert City, Kentucky for Pennsalt, and the TVA area around Chattanooga, Tennessee for Olin. The financing required would not have been a problem for either company." Ibid.

We believe that the court erred in this regard. Certainly the sole test would not be the probability that *both* companies would have entered the market. Nor would the consideration be limited to the probability that one entered alone. There still remained for consideration the fact that Penn-Olin eliminated the potential competition of the corporation that might have remained at the edge of the market, continually threatening to enter. Just as a merger eliminates actual competition, this joint venture may well foreclose any prospect of competition between Olin and Pennsalt in the relevant sodium chlorate market. The difference, of course, is that the merger's foreclosure is present while the joint venture's is prospective. Nevertheless, "[potential] competition . . . as a substitute for . . . [actual competition] may restrain producers from overcharging those to whom they sell or underpaying those from whom they buy. . . . Potential competition, insofar as the threat survives [as it would have here in the absence of Penn-Olin], may compensate in part for the imperfection characteristic of actual competition in the great majority of competitive markets." Wilcox, Competition and Monopoly in American Industry, TNEC Monograph No. 21 (1940) 7-8. Potential competition cannot be put to a subjective test. It is not "susceptible of a ready and precise answer." As we found in United States v. El Paso Natural Gas Co., supra, at 660, the "effect on competition . . . is determined by the nature or extent of that market and by the nearness of the absorbed company to it, that company's eagerness to enter that market, its resourcefulness, and so on." The position of a company "as a competitive factor . . . was not disproved by the fact that it had never sold . . . there. . . . [I]t is irrelevant in a market . . . where incremental needs are booming." The existence of an aggressive, well equipped and well financed corporation engaged in the same or related lines of commerce waiting anxiously to enter an oligopolistic market would be a substantial incentive to competition which cannot be underestimated. . . .

Does it make sense to treat a new joint venture as a merger between the two parents? The Court says elsewhere that "[r]ealistically, the parents would not compete with their progeny." Is that a fair assessment of all joint ventures? Or does it apply only where the joint venture itself will be the exclusive outlet for the products designed?

The issue before the Ninth Circuit in *Northrop* was whether the rule of reason or per se illegality was the appropriate standard to apply under section 1. *Northrop* does not preclude application of section 1 to joint ventures in all cases. It does, however, cite language from the Supreme Court in *Maricopa County* to the effect that joint ventures in which all capital, profits, and losses are shared are properly treated as a single firm rather than two cooperating entities. At some point, therefore, section 1 must give way to merger analysis under section 7.[16] As the facts of *Northrop* help demonstrate, it is far from clear when that point is reached. The court suggests that cooperation on a single project is sufficient, even though the firms have other enterprises that are wholly separate, but how much cooperation is required? Does it

---

16. "Conspiracies" within the same firm are not actionable under section 1 of the Sherman Act. See Copperweld v. Independence Tube, 467 U.S. 752 (1984).

matter whether the joint venture is solely for research or whether it extends into later aspects of the production process?

Joint venture agreements often contain some means of dividing the product or territorial markets. In *Northrop*, this agreement was that McDonnell would build only carrier-based aircraft and Northrop would build only land-based aircraft. What sort of scrutiny should these agreements receive? The *Northrop* court was quite deferential to the agreement in that case, for two reasons. First, the court read the agreement as not foreclosing head-to-head competition between the companies, citing evidence that such competition had in fact occurred. Second, the court reasoned that the joint venture was on balance procompetitive, since without the joint venture neither company would have entered the market. Since the joint venture produced one additional competitor (the joint venture), it promoted competition.

The court's first reason for allowing market-splitting agreements is suspect. In *Northrop*, there was head-to-head competition only because McDonnell began to sell in Northrop's "product territory" (land-based fighters). Had the parties enforced their agreement to divide products, they would not have competed with each other. Territorial market division agreements are common in joint ventures, but they may not be a necessary part of the joint venture. In the absence of market power, the contractual restraints in the joint venture agreement should be of little concern. If there is market power among the joint venture participants, it is important to know whether the joint venture would have succeeded absent the market-division mechanism. If so, then the existence of that mechanism reduces competition with no offsetting benefit to society, since the choice is between a competitive joint venture and a collusive one. It seems reasonable to evaluate such agreements separately from the rest of the joint venture, and to strike down such agreements without holding the entire joint venture unlawful.

The court's second reason, that the joint venture was on balance pro-competitive, has much economic force in principle. The force it has in practice, though, is highly dependent on the facts of individual cases. It is important to know whether either or both of the joint venturers would have entered the market absent the joint venture. Certainly, if both would have entered and could have done so successfully,[17] the effect of the joint venture would be to reduce competition. This is the point of the *Penn-Olin* case.

Penn-Olin goes farther than this, of course. The Court concludes that foreclosing "potential competition" by the joint venturers may still make the joint venture anticompetitive. While this may be true in certain limited circumstances, the Court may place too much emphasis on the effect on potential competition. The doctrine of "contestable markets," or potential

---

17. As Jorde and Teece note below, the mere fact that the joint venturers would have entered the market does not alone compel the conclusion that the joint venture reduced competition. For example, if each firm could have entered only at the fringes of a market dominated by an existing firm, a single joint venture which can compete on an equal basis with that dominant firm may be more desirable than two fringe competitors.

competition, ironically advanced in the 1980s as a reason to *weaken* antitrust scrutiny, see W. Baumol, J. Panzer, and R. Willig, Contestable Markets and the Theory of Industry Structure (1982), has come under empirical attack. See, e.g., Randal L. Reed and Don E. Waldman, Mergers and Air Fares: "Contestable Markets" in the Airline Industry, 20:3 Antitrust L. & Econ. Rev. 15 (1988); Steven Morrison & Clifford Winston, Empirical Implications and Tests of the Contestability Hypothesis, 30 J.L. & Econ. 53 (1987).

The existence of these difficult factual issues in antitrust scrutiny of joint ventures counsels against the application of a per se rule to joint ventures. But joint venturers should be careful not to mistake the absence of a per se rule for the absence of any antitrust scrutiny at all. Because joint ventures receive special treatment, courts must also be careful not to let ordinary cartels be characterized as joint ventures. See Timken Roller Bearing Co. v. United States, 341 U.S. 592 (1951).

Note on the National Cooperative Research Act

Responding to many of the criticisms described above, and to encourage the formation of research and development joint ventures, Congress in 1984 passed the National Cooperative Research Act (NCRA), 15 U.S.C. §4301 et seq. NCRA provides that all research and development joint ventures must be judged under the rule of reason, and that prevailing defendants in an antitrust suit against a joint venture are entitled to collect attorney fees. In addition, it allowed joint venturers to register the venture with the Department of Justice. Registered joint ventures were subject only to single, rather than treble damages.

A number of joint ventures were registered with the Department of Justice after passage of the 1984 Act. Most of these were efforts by small companies, however, and NCRA has received very little in the way of judicial elaboration. In an influential article, Thomas Jorde and David Teece argued that the problem was twofold: NCRA didn't extend to joint ventures for development, production, or marketing, and it didn't offer enough protection to joint venturers. Thomas Jorde & David Teece, Innovation, Cooperation, and Antitrust: Balancing Competition and Cooperation, 4 High Tech. L.J. 1 (1989). They argued:

> While Congress should be applauded for its initial effort in this area, it is now clear that the Act fails to address fully the needs of innovation, and hence has not generated the hoped for response from private business. . . .
>
> Perhaps because it was Congress' first major effort in this area, it is not surprising that the NCRA failed to take important additional steps that we believe would provide greater incentives for cooperative innovation and commercialization arrangements without threatening competition.

> a. The Importance of Commercialization to Successful
> Innovation Is Not Recognized

The substantive protections provided by the NCRA — guaranteed rule of reason treatment and reduction of damages — extend only to research and narrowly confined to marketing intellectual property developed through a joint R&D program. Treatment of commercialization agreements is thus left uncertain, to be determined only by interpretation of the "reasonably required" standard.

In our view, the NCRA is not sufficiently generous toward cooperative commercialization efforts in the context of cooperative innovation. By limiting protection to joint marketing of intellectual property only, for example, the NCRA unwisely precludes joint manufacturing and production of innovative products and processes, which in turn might provide the cooperating ventures with significant feedback information to aid in further innovation and product development. To be sure, it is possible that one might interpret the language "reasonably required" to include all commercialization efforts that aid continued and next generation research of the kind that we have described as critical to cooperative innovation. Whatever the proper reading, we think it would be better to recognize forthrightly that commercialization may be integral to successful, ongoing innovation efforts. To examine commercialization only in terms of whether it is "ancillary" to or "reasonably required" for joint research is likely only to perpetuate analytical confusion and unpredictability when the rule of reason is applied to cooperative innovation arrangments.

b.   Rule of Reason Analysis Is Not Adequately Defined

The NCRA gives little guidance concerning the substantive content of its rule of reason approach. To its credit, it does, in requiring that markets be defined in the context of research, implicitly acknowledge the need to define market power in terms of identifiable markets that are tailored to the special characteristics of research and innovation. However, the NCRA fails either to create a market-power-based safe harbor or to specify factors to be considered within rule of reason analysis. It simply requires consideration of "all relevant factors affecting competition," paying no special attention to the special characteristics of the innovation process in a quickly changing industry. In this sense, it does not adequately reflect Congress' appreciation of the importance of innovation to American competitiveness.

c.   The Registration Procedure Fails to Provide
     Effective Protection from Antitrust Exposure

While the NCRA's elimination of treble damages for registered ventures is an important step forward, cooperating firms are still not protected from antitrust litigation. The cost of defending antitrust suits is not materially reduced by the exceedingly narrow circumstances in which the Act permits an award of attorneys' fees to prevailing defendants. Moreover, single damages are still available. We believe that if an approval procedure existed under which procompetitive arrangements could obtain exemptions from further antitrust exposure to private damage actions, then many more competitively beneficial ventures would utilize the NCRA.

3. Registrations Under the NCRA Are Disappointing but
   Not Surprisingly Low

Businesses seem to have recognized the limited nature of the steps taken by the NCRA. Only 111 separate cooperative ventures registered under the NCRA between 1984 and June 1988. We have attempted to classify these according to the field of endeavor of the contemplated cooperation, aggregating filings into the categories identified in [Table 8-1].

Several observations should be made. First, the number of registrations is small. Second, we have the impression that most of these are very modest endeavors. At least one pre-existing and substantial entity, Bellcore, has registered under the NCRA, but with the exception of MCC and the SRC, we believe that most of the endeavors operating under the Act have budgets of less than $5 million per year. Third, we are struck by the total number of endeavors aimed at solving environmental and health problems — about 30 percent of the total non-redundant registrations. The Motor Vehicle Manufacturers Association, for instance, has 15 endeavors dealing with research on diesel emissions, benzene emissions, acid rain, long-range transport of air pollutants, and vehicle side impact test procedures. In short, the NCRA appears to have been employed largely for endeavors aimed at solving industry problems that are not of great competitive moment.

Id. To solve this problem, Jorde and Teece proposed that Congress do two things. First, they suggested that the definition of a "joint venture" be expanded to include production, testing, and marketing joint ventures. Second, they suggested creating a "safe harbor" protecting joint ventures with less than a 20-25% market share from any private antitrust litigation. Jorde and Teece drafted proposed legislation to this end. Congress enacted a similar bill in 1993 to revise NCRA. The new act includes production and testing (but not marketing) within the definition of a joint venture. It still does not contain a safe harbor provision, however, and it is limited to domestic companies.

TABLE 8-1
Registration of Cooperative Endeavors Under the National Cooperative Research Act, January 1985-June 1988

Categories	Numbers	Percent of total
Advanced Materials	4	3.6
Biotechnology	3	2.7
Energy	10	9.0
Environmental, Health, and Medical	33	29.7
Information	33	29.7
Manufacturing	28	25.2
Total	111	100

COMMENTS AND QUESTIONS

1. Is there any reason not to enact the Jorde-Teece safe harbor proposal? The safe harbor will not prevent all antitrust litigation against small joint ventures — courts will still be called upon to define the market in order to determine market share, for example. But it would offer some protection to small joint ventures that do not seem an immediate threat to the competitive landscape.

In this regard, it is worth noting that the Antitrust Division's Intellectual Property Guidelines do adopt a "safety zone" in licensing cases, along the lines that Jorde and Teece suggested. Under this safety zone, the Division will not challenge restraints that are not illegal per se unless the licensor and licensees collectively account for more than 20 percent of the market. Intellectual Property Guidelines §4.3. The division's policy should prevent Justice Department prosecution of small joint ventures, but it does not bar private suits.

For a potential argument against the proposal, see Joseph Brodley, Joint Ventures and Antitrust Policy, 95 Harv. L. Rev. 1521 (1982). Professor Brodley suggests that applying the rule of reason to joint ventures "leads to unfocused, protracted litigation that places the party with the burden of proof at a severe disadvantage. Because the burden of proof normally falls on the government, the result is likely to be significant underenforcement of the antitrust laws." He suggests that courts allocate the burden of proof to the defendant joint venturers in cases where certain factors allow the court to presume a threshold risk of anticompetitive harm.

PROBLEM

Problem 8-6. In the summer of 1991, personal computer giants Apple and IBM announced plans to enter into a joint venture to produce a new type of disk operating system (OS) called Pink. Pink would be used in connection with a new generation of high-speed microprocessors developed by Motorola that use reduced-instruction-set computing, or RISC. Pink has the added virtue of being able to run on several different types of microprocessors. Apple and IBM would each produce (compatible) computers using the new OS and microprocessors.

At the time of the agreement, both the OS market and the microprocessor market were "slaves" to the personal computer market, since every personal computer had to incorporate an OS and a microprocessor compatible not only with the computer but with each other. IBM-compatible machines (though not IBM's own machines) dominated the personal computer market, with Apple the most significant maker of noncompatible computers. Microsoft (which makes MS-DOS for use with IBM computers) dominated the OS market, and Intel dominated the microprocessor market. Apple also participates in the OS market, making Apple-compatible OS, and Motorola does the same in the microprocessor market. However, because the computer "sets" must be integrated, there was at that time little or no direct competition between Apple and Microsoft in the OS market, or between Intel and Motorola in the microprocessor market. The Apple-IBM joint venture prepared to enter the OS market in force, and analysts speculated that it might well dominate that market. The venture would boost Motorola's position in the microprocessor market, but Pink's ability to run with different microprocessors would also allow genuine competition in that market. Finally, the joint venture would have a significant presence in the personal computer market, where it would greatly enhance compatibility among computers, which is universally regarded as desirable from an information-flow standpoint.

How should such a scheme be evaluated under the antitrust laws? Should it survive antitrust scrutiny? Do subsequent events in the computer industry shed any light on how to approach this issue?

D. MERGERS AND ACQUISITIONS

≣ *SCM Corporation v. Xerox Corporation*
United States Court of Appeals for the Second Circuit
645 F.2d 1195 (2d Cir. 1981), cert. denied, 455 U.S. 1016 (1982)

MESKILL, Circuit Judge:

The plaintiff, SCM Corporation (SCM), appeals from an order entered in the United States District Court for the District of Connecticut, Jon O. Newman, Judge, dismissing its claim for monetary damages asserted in this private antitrust action for injuries sustained as a result of alleged exclusionary acts committed by the defendant, Xerox Corporation (Xerox), in violation of §§1 and 2 of the Sherman Act, 15 U.S.C. §§1, 2 (1976), and §7 of the Clayton Act, 15 U.S.C. §18 (1976). The trebled amount of damages calculated by the jury on this claim totalled $111.3 million. The principal anticompetitive acts alleged by SCM concerned patent acquisitions made by Xerox. SCM averred that Xerox's acquisition of certain patents and subsequent refusal to license those patents excluded SCM from competing effectively in a relevant product market and submarket dominated by Xerox products that embraced the patented art. Judge Newman ruled below that a need to accommodate the antitrust and patent laws precluded damage liability predicated upon Xerox's refusal to license its patents; however, he left open the possibility of granting the plaintiff equitable relief. Judge Newman certified his order for interlocutory review pursuant to 28 U.S.C. §1292(b) (1976) and, as developed below, we exercised our discretion under that section to accept this appeal. Without commenting upon Judge Newman's remedial theory, we affirm the denial of monetary damages in connection with SCM's exclusionary claim based upon our determination that none of Xerox's patent-related conduct, the only conduct alleged by SCM to have caused it any harm, contributed to any antitrust violation. . . .

SCM has argued that Xerox's acquisition of its patents and subsequent exercise of the exclusionary power in them violated the antitrust laws and injured SCM. Xerox contends that its acquisition of the patents was lawful and its decision not to license its patents for plain-paper copying constituted a lawful exercise of patent power. Xerox does not dispute that it achieved monopoly power in the relevant market and submarket by 1969, but contends that an examination of the circumstances under which this feat was accomplished reveals the lawfulness of the monopoly it attained. Our analysis commences with a review of the relationship between the patent and antitrust laws. . . .

II

The law is unsettled concerning the effect under the antitrust laws, if any, that the evolution of a patent monopoly into an economic monopoly might have upon a patent holder's right to exercise the exclusionary power ordinarily inherent in a patent. Indeed, implicit in Judge Newman's decision below is a deep concern over the uncertain antitrust law implications just such an event might have had in this case. His thoughtful analysis of the relationship between the patent and antitrust laws led him to conclude that "the need to accommodate the patent laws with the antitrust laws precludes the imposition of damage liability . . . for a unilateral refusal to license valid patents." 463 F. Supp. at 1012-13. Judge Newman opined that whether or not Xerox's refusal to license the patents it acquired under the 1956 agreement transgressed any provisions of the antitrust laws, monetary damage liability could not be imposed upon Xerox without seriously undermining the patent system. The district court's thesis rests on the assumption that despite the lawfulness of a patent's acquisition, "[i]n some circumstances, [a] refusal to license may be considered a §2 violation." 463 F. Supp. at 1012.

SCM has contended that a unilateral refusal to license a patent should be treated like any other refusal to deal by a monopolist, see generally Otter Tail Power Co. v. United States, 410 U.S. 366 (1973); Lorain Journal Co. v. United States, 342 U.S. 143 (1951); Eastman Kodak Co. v. Southern Photo Materials Co., 273 U.S. 359 (1927), where the patent has afforded its holder monopoly power over an economic market. While, as SCM suggests, a concerted refusal to license patents is no less unlawful than other concerted refusals to deal, in such cases the patent holder abuses his patent by attempting to enlarge his monopoly beyond the scope of the patent granted him. See, e.g., Zenith Radio Corp. v. Hazeltine Research, Inc., supra, 305 U.S. at 118-19; United States v. Singer Manufacturing Co., 374 U.S. 174, 192-97 (1963); United States v. Line Mutual Co., 333 U.S. 287, 314-15 (1948); Hartford-Empire Co. v. United States, 323 U.S. 386, 406-07 (1945); United States v. Masonite Corp., 316 U.S. 265, 277 (1942). Where a patent holder, however, merely exercises his "right to exclude others from making, using, or selling the invention," 35 U.S.C. §154 (1976), by refusing unilaterally to license his patent for its seventeen-year term, see, e.g., Bement v. National Harrow Co., 186 U.S. 70, 88-90 (1902), such conduct is expressly permitted by the patent laws. "The heart of [the patentee's] legal monopoly is the right to invoke the State's power to prevent others from utilizing his discovery without his consent." Zenith Radio Corp. v. Hazeltine Research, Inc., supra, 395 U.S. at 135 (citing Crown Die & Tool Co. v. Nye Tool & Machine Works, 261 U.S. 24 (1923); Continental Paper Bag Co. v. Eastern Paper Bag Co., 210 U.S. 405 (1908)). Simply stated, a patent holder is permitted to maintain his patent monopoly through conduct permissible under the patent laws.

No court has ever held that the antitrust laws require a patent holder to forfeit the exclusionary power inherent in his patent the instant his patent monopoly affords him monopoly power over a relevant product market. In *Alcoa* this Court never questioned the legality of the economic monopoly Alcoa maintained by virtue of the two successive patents it had acquired. United States v. Aluminum Co. of America, supra, 148 F.2d at 422, 430. Indeed, Judge Learned Hand termed Alcoa's economic monopoly during the terms of those patents "lawful." 148 F.2d at 430. We do not interpret Judge Wyzanski's decision in United States v. United Shoe Machinery Corp., 110 F. Supp. 295 (D. Mass. 1953), *aff'd per curiam,* 347 U.S. 521 (1954), as supporting SCM's argument to the contrary. In *United Shoe,* the primary vehicle found to have been employed by United Shoe in achieving and maintaining its monopoly was its lease-only system of distributing its machines. 110 F. Supp. at 344. The patent acquisitions scrutinized by Judge Wyzanski occurred after United Shoe possessed substantial market power and were not "one of the principal factors . . . enabling [United Shoe] to achieve and hold its share of the market." 110 F. Supp. at 312. Thus, contrary to appellant's contention, the *United Shoe* case stands in stark contrast to the one at bar where the patents were acquired prior to the appearance of the relevant product market and where the patents themselves afforded Xerox the power to achieve eventual market dominance.

In *Alcoa* Judge Learned Hand stated that the "successful competitor, having been urged to compete, must not be turned upon when he wins." 148 F.2d at 430. And while that statement was made in regard to a hypothetical situation where only one of a group of competitors ultimately survives, it at least indicates a concern Judge Hand had for preserving those economic incentives that provide the primary impetus for competition. Subsequently, the Supreme Court in United States v. Grinnell Corp., 384 U.S.

563 (1966), amplified this consideration when it set forth the elements of a §2 violation as follows:

> The offense of monopoly under §2 of the Sherman Act has two elements: (1) the possession of monopoly power in the relevant market and (2) the willful acquisition or maintenance of that power *as distinguished from growth or development as a consequence of a superior product, business acumen, or historic accident.*

Id. at 570-71 (emphasis added).

Thus, in Berkey Photo, Inc. v. Eastman Kodak Co., 603 F.2d 263, 275 (2d Cir. 1979), *cert. denied,* 444 U.S. 1093 (1980), this Court stated that "[w]e tolerate the existence of monopoly power . . . only insofar as necessary to preserve competitive incentives and to be fair to the firm that has attained its position innocently." In United States v. Griffith, 334 U.S. 100 (1948), the Supreme Court declared, however, that "the use of monopoly power, however lawfully acquired, to foreclose competition, to gain a competitive advantage, or to destroy a competitor, is unlawful." Id. at 107. Echoing the same consideration in *Berkey,* Judge Kaufman stated that while "[t]he mere possession of monopoly power does not *ipso facto* condemn a market participant . . . , the firm must refrain at all times from conduct directed at smothering competition." Berkey Photo, Inc. v. Eastman Kodak Co., supra, 603 F.2d at 275.

The tension between the objectives of preserving economic incentives to enhance competition while at the same time trying to contain the power a successful competitor acquires is heightened tremendously when the patent laws come into play. As the facts of this case demonstrate, the acquisition of a patent can create the potential for tremendous market power.

III

Patent *acquisitions* are not immune from the antitrust laws. Surely, a §2 violation will have occurred where, for example, the dominant competitor in a market acquires a patent covering a substantial share of the same market that he knows when added to his existing share will afford him monopoly power. See generally Kobe, Inc. v. Dempsey Pump Co., 198 F.2d 416 (10th Cir.), *cert. denied,* 344 U.S. 837 (1952); United States v. Besser Manufacturing Co., 96 F. Supp. 304, 310-11 (E.D. Mich. 1951), *aff'd,* 343 U.S. 444 (1952). That the asset acquired is a patent is irrelevant; in such a case the patented invention already has been commercialized successfully, and the magnitude of the transgression of the antitrust laws' proscription against willful aggregations of market power outweighs substantially the negative effect that the elimination of that class of purchasers for commercialized patents places upon the patent system.

The patent system would be seriously undermined, however, were the threat of potential antitrust liability to attach upon the acquisition of a patent at a time prior to the existence of the relevant market and, even more disconcerting, at a time prior to the commercialization of the patented art. As SCM itself admits, the procurement of a patent by the inventor will not violate §2 even where it is likely that the patent monopoly will evolve into an economic monopoly; yet SCM would deny the same reward to anyone but the patentee.[9]

If the antitrust laws were interpreted to proscribe the natural evolution of a patent monopoly into an economic monopoly, then Judge Newman's concern would be well founded. If the threat of treble damage liability for refusing to license were imbedded in the minds of potential patent holders as a likely prospect incident to every successful commercial exploitation of a patented invention, the efficacy of the economic incentives afforded by our patent system might be severely diminished.

Nevertheless, it is especially clear that the economic incentives provided by the patent laws were intended to benefit only those persons who lawfully acquire the rights granted under our patent system. Cf. Walker Process Equipment, Inc. v. Food Machinery & Chemical Corp., 382 U.S. 172 (1965) (patent obtained by fraud on Patent Office). Where a patent in the first instance has been lawfully acquired, a patent holder ordinarily should be allowed to exercise his patent's exclusionary power even after achieving commercial success; to allow the imposition of treble damages based on what a reviewing court might later consider, with the benefit of hindsight, to be too much success would seriously threaten the integrity of the patent system. Where, however, the acquisition itself is unlawful, the subsequent exercise of the ordinarily lawful exclusionary power inherent in the patent would be a continuing wrong, a continuing unlawful exclusion of potential competitors.

---

9. Notwithstanding that "[t]he law . . . recognizes that [a patentee] may assign to another his patent, in whole or in part, and may license others to practice his invention." Zenith Radio Corp. v. Hazeltine Research, Inc., supra, 395 U.S. at 135, SCM argues that a distinction should be made between the exploitation of a patent by an inventor and an investor. We assume, therefore, that had Chester Carlson possessed the resources to commercialize xerography to the same extent as did Xerox, SCM would not have challenged a refusal by the inventor to license his patents as violative of the antitrust laws. Investors, however, play a key role, if not an indispensable one today, in both the inventive process and commercialization of inventions. And it is fair to say, we think, that the contribution of the inventor in both the funding of research that leads to inventions and the promotion that necessarily must follow to achieve successful commercialization is of comparable value. See generally Picard v. United Aircraft Corp., 128 F.2d 632, 642 (2d Cir.), *cert. denied*, 317 U.S. 651 (1942) (Frank, J., concurring). In either case, the ultimate intended beneficiary of the patent laws — the public — is equally benefitted. See generally Mannington Mills, Inc. v. Congoleum Industries, Inc., 610 F.2d 1059, 1070-71 (3d Cir. 1979); United States v. Parker-Rust-Proof Co., 61 F. Supp. 805, 808 (E.D. Mich. 1945); In re Anthony, 414 F.2d 1383, 1398 (C.C.P.A. 1969); In re Herr, 377 F.2d 610, 619 (C.C.P.A. 1967). Since Xerox participated financially in both the inventive process by funding research at Battelle and the subsequent commercialization of xerography by bringing the first plain-paper copier to the market, we see little reason to deny Xerox the full benefit of the patents it acquired on this basis alone.

Without passing upon the validity of Judge Newman's theory to preclude antitrust damage liability in all cases where the injury is predicated upon a patent holder's refusal to license, we hold that where a patent has been lawfully acquired, subsequent conduct permissible under the patent laws cannot trigger any liability under the antitrust laws. This holding, we believe, strikes an adequate balance between the patent and antitrust laws. Therefore, to determine whether Xerox incurred any antitrust damage liability to SCM in 1969, our inquiry must now shift to determining whether the acquisition of the Carlson and Battelle patents pursuant to the 1956 agreement violated either the Sherman Act or the Clayton Act. Because the essence of a patent is the monopoly or exclusionary power it confers upon the holder, analyzing the lawfulness of the acquisition of a patent necessitates that we primarily focus upon the circumstances of the acquiring party and the status of the relevant product and geographic markets at the time of acquisition.

IV

Section 2 of the Sherman Act

Turning to the facts of this case, the patents about which we are concerned were acquired in 1956, four years prior to the production of the 914, Xerox's first automatic plain-paper copier, and at least eight years prior to the appearance of the relevant market and submarket. In 1956 Xerox had achieved success in the commercialization of xerography but not in the field of automatic plain-paper copying. There is evidence in the record, however, that Xerox in 1956 valued a non-exclusive license in xerography at $70 million and that key personnel at Xerox believed that they already possessed the necessary technology in 1956 to produce a plain-paper copier. Notwithstanding their optimistic forecasts, however, the confidence of the Xerox organization was still tempered by the risks of producing a new, technologically-sophisticated product line. Thus, in 1958 Xerox considered the possibility of having IBM manufacture and market the 914. But before Xerox's management made a final decision on the matter, IBM informed them that it was not interested in manufacturing or marketing the 914 or another model, the 813, having concluded that both were a bad business risk. While SCM argues that IBM's turndown was directed exclusively at the models 914 and 813 and did not amount to a rejection of plain-paper copying entirely, at the very least the episode demonstrates that as of 1958 the achievement of commercial success in plain-paper copying was not a foregone conclusion.

It was also in 1956 that Xerox acquired non-exclusive licenses from Horizons Corporation, another research organization, covering a small number of xerographic patents.[11] Additionally, Xerox began to cultivate relationships

---

11. In 1960 Horizons Corporation agreed to assign these patents to Xerox.

with its international family of companies in 1956. But SCM has not argued that either the Horizons patents or the international agreements caused it any injury. Rather, SCM contends that these facts constitute proof of Xerox's willful acquisition of monopoly power over the relevant market and submarket that came into being, according to the jury, between eight and thirteen years later. Likewise, other aspects of the 1956 agreement with Battelle, such as the promise by Battelle to transfer to Xerox all know-how it developed and patents it obtained in the future were claimed to be additional evidence of Xerox's willful acquisition of its market dominance. But the promise to transfer all xerographic know-how developed and patents obtained in the future was conditioned on Xerox's promise to contribute at least $25,000 a year for research that would help develop the know-how and patents. There appears to be little distinction, if any, between patents obtained under a contract with a research organization and patents generated internally by a company, see P. Areeda & D. Turner, Antitrust Law: An Analysis of Antitrust Principles and Their Application ¶704e (1978), and ordinarily there is no limitation on a company's freedom to generate its own patents. See generally Automatic Radio Manufacturing Co. v. Hazeltine Research, Inc., 339 U.S. 827, 834 (1950). The jury's specific finding that by 1969 Xerox had not obtained any patents primarily for the purpose of blocking the development and marketing of competitive products laid to rest any suspicion that either Xerox's internal R & D program or its R & D work subcontracted to Battelle was driven principally by anticompetitive animus.[12] In any event, none of Xerox's conduct other than the acquisition of the Carlson and Battelle Patents under the 1956 agreement caused SCM any harm. But even more important, all of the events described occurred between eight and thirteen years prior to the appearance of the relevant product market and submarket defined by SCM.

In scrutinizing acquisitions of patents under §2 of the Sherman Act, the focus should be upon the market power that will be conferred by the patent *in relation* to the market position then occupied by the acquiring party. We agree with Professors Areeda and Turner that whether limitations should be imposed on the patent rights of an acquiring party should be dictated by the extent of the power already possessed by that party in the relevant market into which the products embodying the patented art enter. See Areeda & Turner, supra, at ¶819. Therefore, that Xerox acquired the patents in this case four years prior to the production of the first plain-paper copier and at least eight years prior to the appearance of the relevant product market and submarket over which those patents eventually afforded it monopoly power would seem to dispose entirely of SCM's 1969 exclusion claim under §2.

SCM argues, however, that

> [t]o uphold the jury's verdicts in this case, this Court need hold only that an agreement to purchase patents that eliminate an existing potential for compe-

---

12. SCM also argues that the two-year covenants not to compete imposed by Xerox on its employees up until 1972 also evidence Xerox's willful acquisition of monopoly power. We find this argument wholly without merit.

tition in a reasonably foreseeable economic market can be found to be unreasonable if it (a) results in the acquisition of persistent, substantial, real-world economic monopoly power and (b) imposes a restraint on competition that is greater than reasonably necessary to induce the purchaser to develop and market the product involved.

SCM's proposition is that even prior to the commercialization of the patented invention and prior to the appearance of the relevant market over which Xerox eventually achieved monopoly power a §2 violation occurred. SCM suggests that the antitrust laws impose a limitation on the extent of the rights in a patent a purchaser may acquire, and that in some instances a patent with its inherent exclusionary power may not be transferred *in toto*. The limitation that SCM would impose, however, turns not upon the market position of the acquiring party, but rather, upon the potential for commercial success a particular patent may hold. Thus, SCM argues that a purchaser of a patent is entitled only to the rights in a patent reasonably necessary to induce his investment to commercialize the patent. Presumably, under SCM's proposed rule, where the commercial success of a patented invention virtually is guaranteed, no person other than the inventor can hold exclusive rights in the patent, at least where it is foreseeable that the products generated under the patent will create their own relevant product market.

SCM contends that the test it proposes represents the appropriate rule of reason analysis to be employed in patent acquisition cases. By introducing the concept of foreseeability, SCM seeks to escape an unfavorable disposition of its case that it apparently feared might be based upon the absence of the relevant product market and submarket at the time of the patent acquisitions in 1956. Implicit in the jury's findings was that it was reasonably foreseeable in 1956 that the agreement with Battelle would permit Xerox to obtain monopoly power in a relevant product market.[13] While sufficient evidence was presented by Xerox to support a contrary finding, we are unable to hold, as a matter of law, that no rational jury could find that a reasonably foreseeable effect of the 1956 agreement was the eventual acquisition by Xerox of monopoly power in a relevant market. But notwithstanding the jury's implicit finding, we conclude that, under the facts presented here, the policies of the patent laws preclude the imposition of antitrust liability.

It is undisputed that the first automatic plain-paper copier was not produced by Xerox until four years after the 1956 agreement was executed. Additionally, while Xerox concedes that its plain-paper copiers eventually formed an independent relevant product market, SCM has not challenged on appeal the jury's finding that this event did not occur until some time after

---

13. This finding was implicit in the jury's affirmative answer to question 20: "Was the probable effect of Xerox's acquisition of patents pursuant to the 1956 Xerox-Battelle agreement, when the agreement was made, substantially to lessen competition or to tend to create a monopoly in any relevant market or sub-market that you have found to exist?" In explaining this question, Judge Newman instructed the jurors to determine "whether the acquisition at the time it was made was reasonably probable to have the proscribed effect in the reasonably foreseeable future."

1964, eight years after the agreement. Furthermore, Xerox contributed in a very substantial way to the development of an automatic plain-paper copier by investing in research and development not only after 1956 but also for almost a decade before the agreement. Moreover, the party from whom Xerox purchased the patent under the 1956 agreement was not a potential competitor. We believe that, under the circumstances presented here, to impose antitrust liability upon Xerox would severely trample upon the incentives provided by our patent laws and thus undermine the entire patent system. Therefore, irrespective of the jury's implicit finding that Xerox's commercial success was reasonably foreseeable in 1956, Xerox was lawfully entitled to purchase the patents it did pursuant to the agreement it made with Battelle that year.

With respect to Xerox's subsequent unilateral refusal to license the Carlson and Battelle patents, which we have held were lawfully acquired, that conduct was permissible under the patent laws and, therefore, did not give rise to any liability under §2. . . .

Section 7 of the Clayton Act

Section 7 of the Clayton Act proscribes a corporation from acquiring the whole or any part of the assets of another corporation where "the effect of such acquisition may be substantially to lessen competition, or to tend to create a monopoly [in any line of commerce]," 15 U.S.C. §18 (1976). Since a patent is a form of property, see generally Transparent-Wrap Machine Corp. v. Stokes & Smith Co., 329 U.S. 637, 643 (1947), and thus an asset, there seems little reason to exempt patent acquisitions from scrutiny under this provision. See United States v. Lever Brothers Co., 216 F. Supp. 887, 889 (S.D.N.Y. 1963); see generally L. Sullivan, Handbook of the Law of Antitrust, §180 (1977); 16a J. von Kalinowski, Business Organizations: Antitrust Laws and Trade Regulation §16.05 (1980); Kessler & Stern, Competition, Contract, and Vertical Integration, 69 Yale L.J. 1, 75-78 (1959).

Section 7 principally was designed to curtail the anti-competitive consequences of corporate acquisitions in their "incipiency." Brown Shoe Co. v. United States, 370 U.S. 294, 317 (1962). Thus, the analysis ordinarily employed in determining the lawfulness of a corporate acquisition under §7 is prospective in nature. The jury found that the probable effect of the 1956 Xerox-Battelle agreement was substantially to lessen competition or to tend to create a monopoly in the relevant product market and submarket that appeared between eight and thirteen years later. The jury additionally found that, as of 1964 and 1969, the probable effect of Xerox's continued holding of patents acquired pursuant to the 1956 Xerox-Battelle agreement was substantially to lessen competition or to tend to create a monopoly. We conclude that neither of these findings can stand as a matter of law.

While the Supreme Court in Brown Shoe Co. v. United States, supra, 370 U.S. at 323, stated that the language contained in §7 is indicative that

Congress' "concern was with probabilities, not certainties," the speculative aspect of this antitrust law was intended to allow courts to appreciate immediately the potential consequences that a particular acquisition might have upon an *existing* line of commerce. Thus in *Brown Shoe*, the Supreme Court stated:

> Because §7 of the Clayton Act prohibits any merger which may substantially lessen competition "in *any* line of commerce," (emphasis supplied) it is necessary to examine the effects of a merger in each such economically significant submarket to determine if there is a reasonable probability that the merger will substantially lessen competition.

370 U.S. at 325. The existing market provides the framework in which the probability and extent of an adverse impact upon competition may be measured. In the case at bar it would have been impossible to examine the effects of the Xerox-Battelle agreement upon the relevant product market and submarket in 1956 because those markets did not come into being until sometime between 1964 and 1969. The jury was instructed that it should include in its considerations whether the appearance of the relevant market and submarket and Xerox's domination of those markets was reasonably foreseeable in 1956.

Judge Newman concisely stated below that "[s]ection 7 is concerned with undue concentrations of power and the anti-competitive effects of permitting one entity with market power to strengthen its position by acquisition." 463 F. Supp. at 1001-02. Where, as here, it is conceded that the relevant product market and submarket did not exist until eight years following the patent acquisitions and that Xerox possessed no power whatsoever in even the inchoate market and submarket until 1960 when it introduced its 914 copier, as a matter of law the 1956 agreement did not violate §7 at the time it was made. Finally, our decision regarding SCM's foreseeability argument under §§1 and 2 of the Sherman Act disposes of its argument propounded under §7 along those lines.

SCM argues alternatively that the Supreme Court's decisions in *du Pont-GM,* supra, and United States v. ITT Continental Baking Co., 420 U.S. 223 (1975), require that we affirm the jury's second finding under §7 that Xerox's continued "holding" of the patents it obtained under the Xerox-Battelle agreement in 1956 violated §7 as of 1969. In *du Pont-GM,* the government commenced its action approximately thirty years after du Pont had purchased a twenty-three percent stock interest in General Motors Corporation, and alleged that du Pont had used its stock ownership to attain a "commanding position as General Motors' supplier of automotive finishes and fabrics," 353 U.S. at 588-89. There, the Court held that

> any acquisition by one corporation of all or any part of the stock of another corporation, competitor or not, is within the reach of the section *whenever the reasonable likelihood* appears that the acquisition will result in a restraint of commerce or in the creation of a monopoly in any line of commerce.

353 U.S. at 592 (emphasis added). Subsequently, in *ITT Continental Baking*, the Court reaffirmed, albeit in dictum, its holding in *du Pont-GM* that the term acquisition as it is employed in §7 comprehends both the initial "acquiring" and subsequent "retaining" of the stock of another corporation. 420 U.S. at 241-42. Relying on these cases, SCM would have us hold that Xerox's acquisition of the Carlson and Battelle patents in 1956 became actionable under §7 as soon as the monopoly afforded by the patents burgeoned into an economic monopoly. Whatever the meaning that may be ascribed to §7 in other contexts, the patent laws circumscribe the scope of the provision here. Where a corporation's acquisition of a patent is not violative of §7, as was the case here, its subsequent holding of the patent cannot later be deemed violative of this section. Where a company has acquired patents lawfully, it must be entitled to hold them free from the threat of antitrust liability for the seventeen years that the patent laws provide. To hold otherwise would unduly trespass upon the policies that underlie the patent law system. The restraint placed upon competition is temporally limited by the term of the patents, and must, in deference to the patent system, be tolerated throughout the duration of the patent grants. . , .

COMMENTS AND QUESTIONS

1. Automatic Radio Mfg. v. Hazeltine Research, Inc., 339 U.S. 827, 834 (1950), established the proposition that "[t]he mere accumulation of patents, no matter how many, is not in and of itself illegal." However, the strength of that case was diluted by the amendment of the Clayton Act the same year. Section 7 of the Clayton Act prohibits the acquisition of any company, stock, *or asset* where such acquisition is likely to restrain competition. Courts have viewed patents and copyrights as "assets" subject to the provisions of that section. *SCM* does so; see also United States v. Lever Bros., 216 F. Supp. 887, 889 (S.D.N.Y. 1963); United States v. Columbia Pictures Corp., 189 F. Supp. 153, 182-83 (S.D.N.Y. 1960). SCM recognizes that in some circumstances section 7 would prohibit patent acquisitions (or mergers between companies holding patents in the same market), even though in-house development of the same intellectual property would be perfectly legal.

Most patents are not commercialized by the original inventor.[18] Companies that purchase a patent outright are subject to the Clayton Act; companies that develop a number of patents "in-house" are not. Is there more reason to be concerned about acquisitions than about internal growth? If not,

---

18. In some fields such as biotechnology, however, "bundled" patenting of groups of similar inventions are becoming increasingly common. For example, the designer of a synthetic analogue to a naturally-occurring chemical may patent a whole series of similar analogues to prevent imitation by competitors. Large companies and universities (which have major research departments) are more likely to engage in bundled patenting than are small inventors. To the extent that these patents are commercialized (and they often are not), they will normally be commercialized by the same firm.

why the differential treatment? One reason may have to do with the nature of the antitrust remedy. It is much easier to enjoin a proposed purchase or merger, or to order divestiture of a recently acquired asset, than it is to fashion a remedy against a company that has developed and patented its own research.

More common than outright purchase of patent rights is the long-term licensing of those rights by inventors to companies that will commercialize the invention. Do such licenses constitute the "acquisition" of an asset within the meaning of section 7? The law is not clear. Should licenses fall within the scope of the Clayton Act? Does it matter whether the license is exclusive or nonexclusive? The Antitrust Division's Intellectual Property Guidelines provide that exclusive licenses will be treated like acquisitions under section 7. Intellectual Property Guidelines §5.7.

PROBLEMS

Problem 8-7. In the summer of 1991, Borland International announced its plan to buy Ashton-Tate Corp. Both companies were major players in the database software market for personal computers. Together, they would presumably become the dominant player in that market, reducing competition. However, the merged company also planned to enter the database software market for minicomputers, a market dominated at that time by Oracle Corp. Should antitrust law permit the expected increase in competition in that market to offset the damage to competition in the PC market?

The Antitrust Division expressed concern that the combined intellectual property libraries of Borland and Ashton-Tate would enable the combined company to prevent further entry into the database market. In particular, the division was concerned that large potential entrants like Lotus Corp. will be deterred by the threat of intellectual property litigation. How would you evaluate a proposal by the division, as a condition of permitting the merger, to require the new company to give up its right to sue new entrants for copyright infringement?

Problem 8-8. Megaplex, Inc., is a major petrochemical manufacturer. Concerned that Newco is "gaining" on it in the market for new gasoline additives through greater research prowess, Megaplex approaches several solo inventors who hold patents in the field and "buys" the rights to the patents from the inventors. Armed with a patent portfolio that Newco will have difficulty "inventing around," Megaplex manages to maintain its market share in spite of Newco's research efforts. Has Megaplex violated the antitrust laws?

Table of Cases

Table of Statutes

Index